NANOSCIENCE
AND TECHNOLOGY

**A Collection of Reviews
from Nature Journals**

NANOSCIENCE
AND TECHNOLOGY

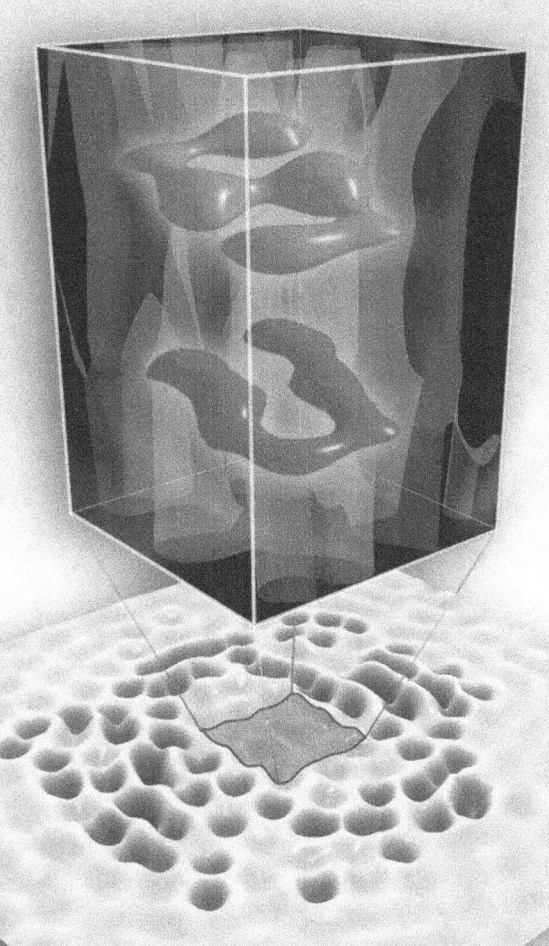

A Collection of
Reviews from
Nature Journals

Edited by
Peter Rodgers
With an Introduction by James Heath

World Scientific

nature publishing group npg

Published by

Macmillan Publishers Ltd
(trading as Nature Publishing Group)
4–6 Crinan Street
London N1 9XW
United Kingdom

and

World Scientific Publishing Co. Pte. Ltd.
5 Toh Tuck Link, Singapore 596224
USA office: 27 Warren Street, Suite 401-402, Hackensack, NJ 07601
UK office: 57 Shelton Street, Covent Garden, London WC2H 9HE

Library of Congress Cataloging-in-Publication Data
Nanoscience and technology : a collection of reviews from Nature journals / edited by Peter Rodgers;
   with an introduction by James Heath.
     p. cm.
   Includes bibliographical references.
   ISBN-13 978-981-4282-68-0
   ISBN-10 981-4282-68-5
   ISBN-13 978-981-4282-69-7 (pbk)
   ISBN-10 981-4282-69-3 (pbk)
  1. Nanotechnology. 2. Nanoscience. I. Title. II. Rodgers, Peter.
  T174.7 .N3583 2010
  620.5--dc22

                            2010294131

British Library Cataloguing-in-Publication Data
A catalogue record for this book is available from the British Library.

Cover image: "Writing with electrons", copyright Hari Manoharan/Stanford University.

CONTENTS

EDITOR'S INTRODUCTION

Review articles have an important role in science. New results are published in research journals, and the most significant results are eventually included in monographs, and some even make it as far as the textbooks read by undergraduates. Review articles occupy the space between journals and monographs, surveying recent progress in a well-defined area of research for the benefit of those already in the field, while also acting as an introduction for those researchers who might be thinking of moving into a new field, and for those who simply want to learn more about the subject.

Review articles are particularly important in areas that are new and/or fast-moving, which includes many areas within nanoscience and technology. That is why *Nature Nanotechnology*, *Nature Materials* and many other journals from Nature Publishing Group publish review articles and progress articles (which are essentially short review articles). This book brings together 35 review and progress articles from eight different *Nature* journals to describe the state-of-the-art in many areas of nanoscience and technology. The six chapter titles give an indication of the broad range of topics covered: nanomaterials and nanostructures; molecular machines and devices; nanoelectronics; nanophotonics; nanobiotechnology and nanomedicine; and selected applications.

The research reported in these articles ranges from the most basic physics and chemistry through to applications as diverse as plasmonic nanosensors and water purification. This spectrum of pure-to-applied research can often be seen in the same article: carbon nanotubes, for example, are fascinating test beds for exploring a variety of fundamental physical phenomena in one dimension, while also displaying considerable potential as a structural, photonic and electronic material (see p. 174). And despite the fact that it is almost 20 years since Iijima's breakthrough paper on nanotubes, researchers are still developing better ways to produce and process them (see p. 3). The other areas of nanoscience and technology covered in this collection are equally rich and complex.

The articles are prefaced by a new essay by James Heath of the California Institute of Technology that looks at the challenges and opportunities facing researchers as they work to convert the promise and potential of nanoscience and technology into useful products and processes. Focusing on applications in energy and health, Heath outlines the fundamental challenges associated with each application and discusses how nanotechnology can be harnessed to tackle these challenges.

One important area that is underrepresented in this collection is the impact of engineered nanomaterials on the environment and health. This does not reflect a lack of interest in this area: *Nature Nanotechnology*, for instance, has published research on the effects of exposing mice to carbon nanotubes,[1,2] but it has not yet published any review articles on nanotoxicology, or the effects of nanoparticles on the environment, because these subjects have already been reviewed elsewhere (see, for example, Refs 3 and 4). However, the chapter on nanobiotechnology and nanomedicine covers a wide range of topics including nanoparticle therapeutics (p. 239) and the use of nanopores to sequence DNA (p. 261).

And there has not, yet, been enough research into public attitudes to the risks and benefits associated with nanotechnology to warrant a full-scale review article, although clear messages have emerged from a number of opinion surveys: the general public know very little about nanotechnology; the public response to nanotechnology is not homogeneous and depends on a number of social factors (such as gender and religious beliefs); and simply telling the public more about nanotechnology will not necessarily make them more likely to accept it (see Ref. 5 for a summary).

Of course, a review article is almost out of date as soon as it is published, and this problem is compounded in a collection like this, but the basics of any field of science that is mature enough for a review article do not change, and I am confident that the articles in this collection will stand the test of time.

Acknowledgements

Thanks to Hollie Cayzer, Cecilia Högström, Angelina Storti and Theresa Tuson for obtaining permission to reproduce figures, to Jane Morris for copy-editing the introductory pieces, to Kelly Hopkins for collecting all the articles together, to Jason Wilde for project management, to colleagues on various NPG journals for commissioning these great articles and, finally, to the authors for writing them all.

When citing any of these reviews, please cite the original version. Full citation details are available on the contents pages of each chapter.

References

1. Schipper, M.L. *et al.* A pilot toxicology study of single-walled carbon nanotubes in a small sample of mice. *Nature Nanotech.* **3**, 216–221 (2008).
2. Poland, C.A. *et al.* Carbon nanotubes introduced into the abdominal cavity of mice show asbestos-like pathogenicity in a pilot study. *Nature Nanotech.* **3**, 423–428 (2008).
3. Lewinski, N., Colvin, V. & Drezek, R. Cytotoxicity of nanoparticles. *Small* **4**, 26–49 (2008).
4. Handy, R.D., Owen, R. & Valsami-Jones, E. The ecotoxicology of nanoparticles and nanomaterials: current status, knowledge gaps, challenges and future needs. *Ecotoxicology* **17**, 315–325 (2008).
5. Currall, S.C. New insights into public perceptions. *Nature Nanotech.* **4**, 79–80 (2009).

Peter Rodgers
Chief Editor
Nature Nanotechnology

CHALLENGES AND OPPORTUNITIES FOR NANOSCIENCE AND TECHNOLOGY

James R. Heath*

Last year I testified before the science subcommittee of the US Senate on the subject of the reauthorization of the National Nanotechnology Initiative (NNI).[1] When this initiative was launched by President Bill Clinton in 2001, the number of researchers working on nanotechnology was relatively small and the field was not even fully defined. That soon changed as the NNI started to invest more than $1billion per year in nanoscience and technology at universities, national laboratories and elsewhere. Other governments soon followed suit, and nanotechnology institutes began to spring up around the globe. Although the field was extremely long on promise, it fell equally short on delivery. Almost any idea that could even be partially defended could garner some sort of grant funding.

Around that time, which coincided with the dot.com boom, I was at a party near the heart of Silicon Valley, and venture capitalists were offering to fund nanotechnology start-up companies that had, as their basis, perhaps one refereed journal article and some notes on a cocktail napkin. The wild landscape that surrounded nanotechnology at that time is now much tamer, and the number of researchers (and papers and journals) in the field has grown dramatically. Moreover, funding agencies, companies, venture capitalists and so forth now demand that any proposals in nanoscience and technology they are going to fund must have a significant scientific foundation. Perhaps the biggest sign of how the field has matured, however, is that lawyers and legislators have now become interested in regulating nanotechnology, as I witnessed during my visit to the senate. The need to regulate nanotechnology was the main theme of the hearing I took part in. It felt a bit like the taming of the American west.

In the early 1990s when nanoscience was first coming into its own, there were two basic drivers for the field. The first was the goal of developing a bottom-up, or biologically inspired, manufacturing approach. Such an approach, in principle, would integrate organic, inorganic and biomaterials to yield novel nanosystems with properties that arose from the collective interactions of the components. The second goal was to develop advanced-performance electronic and photonic devices through the integration of extreme scaling (that is, making everything smaller) with non-traditional, high-performance electronic components. The value proposition for each of these two goals was that the pathway towards achieving either one would result in capabilities that would enable many applications. Work over the past several years has begun to establish the scientific fundamentals of nanotechnology more firmly, and the nanotechnology application space is becoming increasingly better defined. The reviews contained within this book provide a marvellous view of that progress, and give evidence that the promise of the field is now beginning to be balanced with a significant amount of delivery.

*Division of Chemistry and Chemical Engineering and the Kavli Nanoscience Center, California Institute of Technology, MC 127-72, Pasadena, California 91125, USA.

For those of us who have worked in the field for a long time, nanotechnology still seems young and filled with remarkable potential. In this article I attempt to project where nanotechnology is headed over the next decade or two. New technologies evolve exponentially (an example is the rapidly dropping cost of DNA sequencing), with the result that enabled applications can emerge in unpredictable and often disruptive ways (consider how extremely inexpensive transistors enabled the internet, iPods and cellular phones). Thus, application-specific predictions regarding the future of emerging technologies almost always miss the mark. Nevertheless, there are some major scientific and technological challenges that will certainly drive a large fraction of nanotechnology research for the next several years, and those problems provide a more reliable framework for a future-of-nanotechnology discussion.

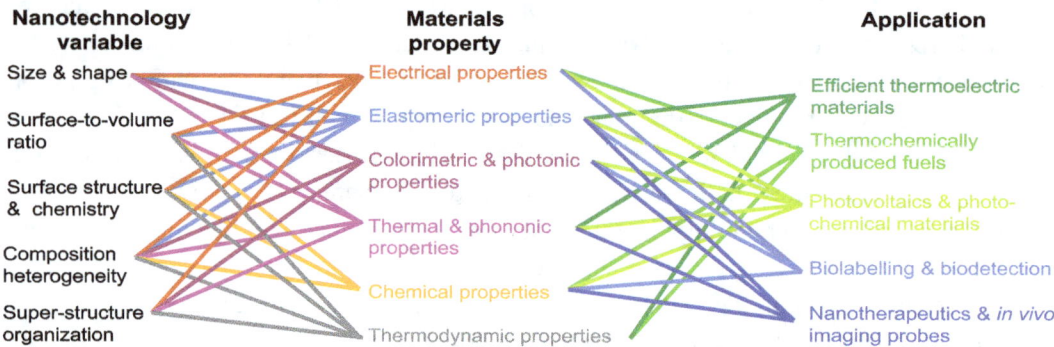

Figure 1. The left column of this figure shows what I call the nanotechnology variables (parameters that can be controlled in experiments). The middle column shows the various materials properties that can be controlled by some or all of the nanotechnology variables. The right column lists five selected applications in the fields of energy and health that depend on some or all of the materials properties.

I will begin by summarizing the value of nanotechnology, which can be framed within a discussion of what I call nanotechnology variables — that is, experimentally controllable parameters (such as size and surface-area-to-volume ratio; Fig. 1) that serve to distinguish nanotechnologies from more traditional technologies. I will then turn to the current state-of-the-art in nanotechnology within the context of the original goals of the field. The bulk of this article will then be devoted to covering two of the newer challenges. The more important of these involves the development of inexpensive, clean and renewable energy. A second major challenge is to provide broadly available and cost-effective healthcare in the developed and developing worlds. Finding solutions to these problems will require coordinated international efforts involving scientists, politicians, an informed public and the corporate world. However, if the science that provides the foundation for solving these problems is not correct, then not much else matters. For these problems, nanotechnologies will have important supporting roles in improving the performance of existing products and processes. First, however, let's turn to some of the reasons that make nanotechnologies attractive for a broad range of problems.

The value proposition of nanotechnology

For bulk materials, certain fundamental properties — such as electrical and thermal conductivity, colour, melting point, elasticity, electronic structure and so on — are intrinsic to the material structure and composition. A unique hallmark of nanomaterials is that each of these fundamental properties has one or more associated length scales and, as a consequence, the properties may themselves be controlled by experimentally tuning one or more of the nanotechnology variables (Fig. 1). This adds a tremendous flexibility to nanomaterials, in ways that are only partially understood. It can also make the optimization of a given physical property challenging, as so many variables can influence that property. Whatever the challenges, these variables set nanotechnologies apart, and give them significant and unique value for a host of applications. Some potential and existing applications are presented in the right-hand column of Fig. 1, and these are representative of nanotechnologies only in that they span the range from existing commercial products to avenues that are still evolving at the basic science level. They also relate to the two problems at hand — energy and healthcare — and at least some important proof-of-concept demonstrations exist for all the applications listed.

The early goal of building advanced electronic and photonic technologies using a combination of novel materials integrated into near molecular- or macromolecular-scale devices has resulted in some impressive scientific demonstrations. A partial list includes molecular electronic memories,[2] circuits based on nanowires[3,4] and carbon nanotubes,[5] novel circuit architectures, new photonic materials,[6] and new patterning approaches.[7] Photonics is a younger technology, and so nanotechnology advances in that field are likely to have a shorter-term impact. For information technology, there will certainly be some niche applications that directly arise from this work, perhaps through patterning methods and new materials. In addition, the nanotechnology-based metrology tools already have important roles in commercial manufacturing. It is even possible that a disruptive technology in the form of a new switching device that can replace the silicon transistor will yet emerge. Nevertheless, the semiconductor industry has successfully continued its focus on pushing silicon-based (top-down) manufacturing towards higher performance and lower-cost devices. This, in itself, is increasingly a form of nanotechnology, but nanowires, nanotubes and molecular electronic devices do not have major roles in this push at present.

Instead, the real value of the nanoscience effort has been threefold. First, a whole new set of nanomaterials (quantum dots, molecular machines, nanowires and so on) is now broadly available. Second, manufacturing approaches that include the programmed formation and assembly of nanostructures, the integration of disparate materials types, and novel patterning and pattern-replication approaches, are now established. Third, an accompanying knowledge is emerging of how the nanotechnology variables of Fig. 1 may be controlled to influence a given set of physical properties for a given nanosystem, and this is having a profound influence on a whole new set of challenges, some of which are discussed below.

A full discussion of the graph in Fig. 1 is beyond the scope of this article, and so I will limit the discussion to the few applications listed, connecting those opportunities back to the nanotechnology variables, and I will attempt to elucidate the fundamental limits of the applications, and how nanotechnology can be harnessed to help approach those limits. The top three applications in Fig. 1 (green font) are all related to the problem of clean energy, whereas the last two applications are related to disease diagnostics and therapeutics.

Selected energy applications

Compelling arguments can be made that the Sun will provide our eventual, primary source of clean and renewable energy,[8] and so the scientific challenges are to find cost-effective ways to harvest, store, convert and transport this energy. The most important of these challenges are energy harvesting and conversion. If solar energy can be efficiently captured and converted into transportable fuels, then existing infrastructure can go a long way towards providing a solution for the rest of the energy problem. The three energy applications listed in Fig. 1 are for energy capture and conversion — the first two (thermoelectric materials and thermochemically produced fuels) treat the Sun as a source of heat, whereas the third (photovoltaic devices and photoelectrochemical cells) treats it as a source of photons. Between them these three applications illustrate emerging new science and also serve as examples of how the nanotechnology variables can be harnessed for specific applications.

Thermoelectric materials convert thermal gradients into electric fields for power generation (or convert electricity into thermal gradients for refrigeration). Thermoelectric nanomaterials are promising because the thermoelectric efficiency of a material depends on its electronic, thermal, phononic and thermodynamic properties, which, in turn, are connected to virtually every nanotechnology variable listed in Fig. 1.

A thermoelectric device is effectively a heat engine with no moving parts, with an efficiency that is somewhat analogous to a Carnot engine, so that maximizing the temperature difference between the hot and cold sides, and minimizing entropic losses, will lead to higher efficiencies.[9] Formally, the efficiency of a thermoelectric material is defined by a dimensionless parameter $ZT = \sigma S^2 T / \kappa$, where T is temperature, σ is electrical conductivity, S is the thermopower (which correlates with entropy), and κ is the thermal conductivity. For bulk materials, maximizing ZT is challenging because optimizing one physical parameter often adversely affects another. However, at the nanoscale, some of these co-dependencies can be separated. For example, in a semimetal or highly doped semiconductor, the thermopower is proportional to the energy derivative of the density of electronic states. In a nanostructured semiconductor or semimetal, the density of electronic states can have sharp peaks, and this can, in theory, produce a high thermopower.[10] Harnessing this effect to produce high-ZT materials has had only limited success, possibly because lattice distortions can provide energetically favourable routes for lifting electronic-state degeneracies.

A promising avenue for developing better thermoelectric materials is to exploit size-dependent thermal and phononic affects. The thermal conductivity in nanoscale semiconductor wires (and particle composites[11]) may be dramatically lowered without significantly influencing the electrical conductivity or the thermopower.[12,13] Much (but not all) of this effect can be attributed to large surface-to-volume ratios, which increase phonon scattering rates.[14,15] Also, long-wavelength acoustic phonons in nanostructures can sometimes be partially defined by the overall size and shape of the structure. In a nanowire, this means that the length rather than the cross-section of the nanowire limits the phonon mean free path. A charge carrier injected into a nanowire can relax by giving off heat to the lattice (this is the origin of electrical resistance), and the first step in this relaxation process is a collision between the electron and a long-wavelength phonon that conserves momentum (that is, is non-dissipative). The long-wavelength phonon can carry the heat to the other end of the nanowire, thus increasing the thermopower. As the thermal conductivity, κ, is averaged over all phonon modes, it is possible to dramatically reduce the thermal conductivity and increase the thermopower at the same time.[12]

Heat transport in nanostructures has recently emerged as an exciting new area for study. New physics is being reported,[16,17] but quantitative models that capture this physics and provide guidance for materials design are still lacking. One challenge is that any real thermocooling or thermopower application requires both hole and electron (p- and n-type) thermoelectric materials, as well as high values of ZT, which may be difficult to achieve.

Thermochemically produced fuels treat the Sun as a source of heat, with solar concentrators providing the energy necessary to drive the endothermic part of a thermochemical cycle. The cycle converts the kinetic energy of heat into the potential energy of chemical bonds. Such a cycle can, for example, use CO_2 or H_2O as feedstocks and generate H_2, CO or hydrocarbon fuels.[18] An example of a thermochemical cycle designed to split water via the reduction and subsequent oxidation of a metal oxide is:

$$T_{HIGH} \qquad MO_x \rightarrow MO_{x-\delta} + \frac{\delta}{2}O_2$$

$$T_{LOW} \qquad \delta H_2O + MO_{x-\delta} \rightarrow \delta H_2 + MO_x$$

$$Net \qquad H_2O \rightarrow \frac{1}{2}O_2 + H_2$$

If the efficiency of this process is defined as the power stored in the free energy of the products divided by the solar power from the concentrator, then the Carnot efficiency limit applies, so that a bigger temperature difference $(T_{HIGH} - T_{LOW})$ translates into a higher efficiency. This is distinct from approaches that treat the Sun as a light source, which are designed for lower-temperature operation.

The solar thermal harvester itself is a decidedly macroscopic system — large mirrors concentrate solar heat onto collectors, which then often drive steam or Stirling (compressed gas) engines — and the electricity produced with this approach can cost as little as 10–15 cents per kilowatt hour. Although there are likely to be a number of opportunities for nanomaterials for the more efficient capture and storage of solar heat, the clearest opportunities lie within the relatively new use of solar–thermal technology to produce transportable fuels. As would be expected, most mature thermochemical processes do not rely on nanomaterials.[19] Nevertheless, nanostructured materials hold significant promise as the base materials (the MO_x in the equations above) for promoting thermochemical cycles. The nanotechnology variable of surface-to-volume ratio can be harnessed to increase the rate of a heterogeneous chemical process, such as the reduction or oxidation of the metal oxide at the gas/solid interface. Moreover, high surface-to-volume ratio materials can exhibit relatively high rates of oxygen diffusion — thus potentially accelerating the thermochemical cycle. In fact, similar arguments can be made for nanostructured lightweight battery materials.[20] Compositional heterogeneity at the nanoscale (for example, doping or catalyst addition) can be harnessed to increase the melting point of the oxide, thus permitting higher-temperature operation, or potentially generating different product fuels. However, trade-offs can include reduced melting points that are characteristic of finite-sized materials, or the appearance of new (and perhaps undesired) catalytic processes that occur at surface structures that are more common at the nanoscale than for bulk surfaces.

Developing a quantitative understanding of the surface science of nanomaterials will be a central theme of fundamental nanoscience research for years to come. It will require the development of new analytical tools and chemical approaches, and it will probably have the same profound influence on nanotechnologies that traditional surface (and interface) science has had on the semiconductor industry. Surface science is central to all of the applications discussed in this article.

Photovoltaic devices convert light into electrical power, and **photoelectrochemical cells** add an additional energy-conversion step by electrochemically generating transportable fuels, such as H_2, by the electrochemical splitting of H_2O.

For photovoltaics, the major challenge is to efficiently capture the incident solar photon flux and convert it into electrical power while minimizing resistive losses. Economic considerations are very important: about $100 m^{-2} is a viable target for an array of 10%-efficient solar cells — this cost per unit area is more expensive than paint[8] but less than a good carpet. Although the thermodynamic limit of the efficiency of a solar cell can exceed 90% (Ref. 21), the Shockley limit is more realistic, and is around 30% for a conventional p–n junction silicon solar cell.[22,23] The Shockley limit assumes a p–n homojunction solar cell, with an energy bandgap. In such a junction, a photo-excited electron–hole pair is separated at the junction, with the hole and the electron travelling through the p-doped and n-doped regions of the junction, respectively, to be collected as current at the contacts to the semiconductor. The solar spectrum spans many wavelengths, but photons with energy less than the bandgap are not used, and photo-excited carriers with energies greater than the bandgap quickly relax to the band edges, giving off their excess energy as heat to the lattice. This relaxation mechanism alone limits the achievable sunlight-conversion efficiency within single-crystal silicon photocells to 44%.

Finding ways to match, or even beat, the Shockley limit leads naturally to an exploration of nanotechnologies. For example, minimizing the time between when a photon is absorbed, and the electron–hole pair is separated, implies the need for very small junctions in which the area of the interface between the p- and n-type regions is large — such as might be achieved by making interpenetrating super-structures comprising both p- and n-type conductors. The discussion on phonon physics in the section on thermoelectric materials above is also relevant, as designing systems in which the relaxation rate of photo-excited carriers is reduced is another promising avenue of study, but one that has received only slight attention so far.[24,25] The surface structure of nanoscale photoconductors is of key importance, as a roughened or charged surface can have an adverse effect on performance by scattering charge carriers, reducing carrier mobilities and enhancing non-radiative recombination rates. Certain chemical passivation strategies have provided encouraging results,[26] but no clear strategies have yet emerged. One approach that has proved promising, but is currently too expensive for most applications, is to build a thin-film superlattice with layers characterized by graded bandgaps, so that the lowest-energy bandgap material is at the bottom, and the largest is at the top.[27] This allows, within the restrictions of the Shockley limit, a more efficient absorption and use of solar light, as each layer within the stack absorbs light near the bandgap of the material in that layer, thus limiting losses caused by non-radiative processes.

For photoelectrochemically generated fuels, the photogenerated and separated electron and hole pairs are used to drive multi-electron electrochemical processes, such as the splitting of water or reduction of CO_2 (Ref. 8). Although photoelectrochemical cells operate at moderate temperatures, many of the challenges associated with thermochemical fuel production remain, such as the integration of catalysts, and the use of high surface-to-volume materials to efficiently drive heterogeneous chemical processes. The development of high-efficiency nanotechnology-based photovoltaics and photoelectrochemical fuel generators is significantly more complex than what is required for photothermal fuels production, largely because the demands on the quality of the surfaces, interfaces and superstructure are more extreme, and these are coupled with the need to fabricate that platform at very low cost. Nevertheless, this is an area where even incremental advances can translate into huge economic benefits, and that incentive will provide a pathway for the incorporation of nanotechnologies over the next decade or so.

Selected health applications

Over the next 10–20 (or more) years, healthcare will evolve from reactive medicine (disease is detected and treated in late stages) to a personalized, proactive and preventative medicine (presymptomatic disease is detected and treated on an individual basis). At the heart of this transformation will be a host of new, miniaturized technologies that permit biological information to be acquired and analysed quickly and cheaply.

Consider how cancer is detected and treated today. An initial diagnosis typically occurs through imaging (for example, mammography), physical inspection or an endoscopic procedure that is carried out on a patient who presents with certain complaints. If a potential cancer is found, it is biopsied and analysed. If cancerous, a major surgery is often performed, followed by radiation and chemotherapies — both of which have serious, detrimental side effects. For many cancers, multiple treatment options are possible, and so a patient is first placed on one treatment, followed by a watch-and-wait period. If the patient does not respond within a couple of months, then a second treatment is prescribed, and so on.

As genomic sequencing becomes more affordable, patients will have at least some information relevant to their probabilistic health future, which can be used to schedule physician visits before the development of symptoms. Given advances in the knowledge of disease biomarkers such as proteins, circulating tumour cells[28] and microRNAs,[29] it is likely that, for a given disease, a panel of blood-based biomarkers will be assessed to provide presymptomatic information relevant to the onset and progression of the disease. Therapies that are targeted to attack just the diseased cells will be administered, and a second set of biomarker measurements will be carried out to identify potential positive and adverse responses, or to rapidly adjust or alter the therapy. Following therapy, the patient will continue to be monitored using a panel of biomarkers that are indicative of disease recurrence.

Nanotechnology will have several roles in this evolution of healthcare.

Biolabelling and biodetection are major components of *in vitro* disease diagnostics, and nanotechnology already has a minor role in these areas. In fact, quantum dot biolabels are the prototypes that illustrate how the optimization of multiple nanotechnology variables (size and shape, surface structure and chemistry, composition heterogeneity) can lead to a robust commercial product.[30] Given the richness of the human blood proteome,[31] it is likely blood-based proteins (and potentially microRNAs) will eventually provide the major window for monitoring a patient's evolving health and disease, so I will focus on this area. Genome sequencing, which is rapidly becoming cost-effective, will provide the window into predictive medicine. For the case of proteins, the appropriate question is: "what additional technology advances are needed to drop protein biomarker measurements down to a penny (or less) per protein?". Answering this question can provide a guide for determining which technologies might and might not have roles in future healthcare.

For *in vitro* diagnostics, lab-on-a-chip microtechnologies[32] will have a more important role than nanotechnologies. Compared with existing assay techniques, microtechnologies can allow similar measurements, but from greatly reduced sample sizes. Small sample sizes have benefits that range from rapid assays to assays that can resolve information at the single-cell level — both of which will be important in future healthcare. A less mature, but extremely valuable advantage, is that microtechnologies can potentially integrate the steps of biospecimen handling, purification and so on, with the actual assay.[33] Looking slightly further ahead, the development of miniaturized platforms designed for the measurement of large panels of biomarkers, from sample sizes as small as a single cell (for example, circulating tumour cells[34] or stem cells[35]) will become increasingly important.

Nanotechnologies that find use in amplifying signals from existing protein assay technologies will probably have roles in increasing the sensitivity of *in vitro* diagnostics measurements.[36] However, many biomolecular detection nanotechnologies, such as nanowire[37] or nanocantilever[38] biosensors, are simply too expensive for wide-scale use.

Such nanosensors, when entrained within microfluidic channels, can rapidly detect specific biomolecules from small quantities of serum or other tissues. However, conventional assays can do the same,[39] and for less cost. For a highly multiplexed conventional (antibody-based) protein assay, the major costs are the antibodies that are used to capture and detect specific proteins. Alternative protein capture agents[40,41] that show the selectivity and sensitivity of antibodies, but also exhibit chemical, biochemical and physical stability, would provide a transformative technology that would influence all assay platforms, from conventional to nanotechnology-based.

Nanotherapeutics and *in vivo* imaging probes are nanoparticles that represent a significant step beyond quantum dot biolabels, and will be a very high-impact nanotechnology. I will focus my discussion on nanotherapeutics because nanotechnology-based *in vivo* imaging probes, although important, possess only a subset of the requirements that are needed for an effective nanotherapeutic.

Although some nanotherapeutics are plasmonic particles that act on tumours via physical effects (infrared heating to induce cell death),[42] most nanotherapeutics are organic nanoparticles that contain, within their core, $\sim 10^3$ copies of a known drug, which can range from a traditional small-molecule drug to a siRNA oligomer.[43] The shell of the particle is engineered for a host of possible functions that are independent of the payload. These functions can include controlled circulation times (up to days), evasion of the immune system, controlled targeting and delivery, controlled release, and controlled decomposition and clearance. To achieve these functions, virtually every physical and chemical property of the particle — including its size and shape, elastic modulus, surface charge, composition, the way in which the drug is released and so on — must be carefully engineered through control of the nanotechnology variables. However, the demonstrated and potential benefits are tremendous compared with standard therapeutics.[44]

To appreciate the potential value of a nanotherapeutic, first consider the limitations of a small-molecule chemotherapy. Chemotherapeutics have very short circulation times, which means that high-frequency dosing is often required. Small molecules can also leak out of the circulatory system and damage healthy tissues. In addition, resistance to chemotherapies can develop when certain cell surface proteins become over-expressed so that they pump the drugs out of diseased cells as soon as they are internalized. Although antibody therapeutics and antibody–drug conjugates can avoid some of these limitations, only nanoparticles have the potential to avoid all of them. For some emerging drug concepts, such as siRNA therapeutics, delivery is a huge challenge, and nanoparticles provide one of the most promising avenues for solving that problem.

This field is evolving so rapidly that almost any review is out of date by the time it is published. Nevertheless, I'll make some predictions. First, it is likely that, over the next decade, nanoparticle-encapsulated chemotherapies will go a long way towards removing the toxic side effects, and perhaps some of the acquired drug resistances, that are associated with traditional chemotherapies. In addition, nanoparticles that adopt new functions that include accurate targeting of diseased tissues, the ability to penetrate deep into tissues, avoidance of the liver (a common issue with nanoparticles) or enzymatically triggered drug release (to increase release specificity) are also likely to emerge. In the more distant future we will likely see nanotherapeutics with release mechanisms that are customized for precisely controlled release rates, dependent upon the mechanism of the therapeutic to be delivered. It is also likely that nanotherapeutics will appear that have their size, shape[45] and elastic modulus optimized to permit passage across the blood/brain barrier.

Conclusions

Over the past 10–15 years, nanotechnology has advanced beyond the stage of an infant scientific field to now provide a toolkit that is poised to help solve a number of pressing problems. For even the small number of applications that I have described here, the relevant nanotechnologies span the range from inorganic to organic to biologically derived nanomaterials. A common hallmark that sets these various nanomaterials apart from more traditional bulk or molecular materials are the experimental handles, or the nanotechnology variables, that give scientists and engineers a whole new design space for optimization of properties.

References

1. Heath, J.R. Testimony before the 110th Congress, Senate Subcommittee on Science, Technology and Innovation, National Nanotechnology Initiative: Charting the Course for Reauthorization. (April 24, 2008); available at <http://commerce.senate.gov/public/index.cfm>.
2. Green, J.E. *et al.* A 160,000 bit molecular electronic memory circuit patterned at 10^{11} bits per square centimetre. *Nature* **445**, 414–417 (2007).
3. Javey, A., Nam, S., Friedman, R.S., Yan, H. & Lieber, C.M. Layer-by-layer assembly of nanowires for three-dimensional, multifunctional electronics. *Nano Lett.* **7**, 773–777 (2007).
4. Sheriff, B.A., Wang, D., Heath, J.R. & Kurtin, J.N. Complementary symmetry nanowire logic circuits: experimental demonstrations and *in silico* optimizations. *ACS Nano* **2**, 1789–1798 (2008).
5. Avouris, P. Carbon nanotube electronics and photonics. *Phys. Today* **62**, 34–40 (January 2009).
6. Atwater, H.A. The promise of plasmonics. *Sci. Am.* **296**, 38–45 (April 2007).
7. Rothemund, P.W.K. Scaffolded DNA origami for nanoscale shapes and patterns. *Nature* **440**, 297–302 (2006).
8. Lewis, N.S. & Nocera, D.G. Powering the planet: chemical challenges in solar energy utilization. *Proc. Natl. Acad. Sci. USA* **103**, 15729–15735 (2006).
9. Humphrey, T.E. & Linke, H. Reversible thermoelectric materials. *Phys. Rev. Lett.* **94**, 096601 (2005).
10. Hicks, L.D. & Dresselhaus, M.S. Thermoelectric figure of merit of a one dimensional conductor. *Phys. Rev. B* **47**, 16631–16634 (1993).
11. Poudel, B. *et al.* High-thermoelectric performance of nanostructured bismuth antimony telluride bulk alloys. *Science* **320**, 634–638 (2008).
12. Boukai, A. *et al.* Silicon nanowires as highly efficient thermoelectric materials. *Nature* **451**, 168–171 (2008).
13. Hochbaum A.L. *et al.* Enhanced thermoelectric performance of rough silicon nanowires. *Nature* **451**, 163–165 (2008).
14. Cahill, D.G., Watson, S.K. & Pohl, R.O. Lower limit to the thermal conductivity of disordered crystals. *Phys. Rev. B* **46**, 6131–6140 (1992).
15. Martin, P., Aksamija, Z., Pop, E. & Ravaioli, U. Impact of phonon-surface roughness scattering on thermal conductivity of thin Si nanowires. *Phys. Rev. Lett.* **102**, 125503 (2009).
16. Chang, C.W., Okawa, D., Majumdar, A. & Zettl, A. Solid-state thermal rectifier. *Science* **314**, 1121–1124 (2006).
17. Chang, C.W., Okawa, D., Garcia, H., Majumdar, A. & Zettl, A. Breakdown of Fourier's law in nanotube thermal conductors. *Phys. Rev. Lett.* **101**, 075903 (2008).
18. Kodama, T. & Gokon, N. Thermochemical cycles for high-temperature solar hydrogen production. *Chem. Rev.* **107**, 4048–4077 (2007).
19. Perret, R. *et al.* *II.I.1 Development of Solar-Powered Thermochemical Production of Hydrogen from Water* (DOE Hydrogen Program, Annual Progress Report, US Department of Energy, 2006); available at <http://www.hydrogen.energy.gov>.
20. Chan, C.K. *et al.* High-performance lithium battery anodes using silicon nanowires. *Nature Nanotech.* **3**, 31–35 (2008).

21. Luque, A. & Marti, A. Entropy production in photovoltaic conversion. *Phys. Rev. B* **55**, 6994–6999 (1997).

22. Shockley, W. & Queisser, H.J. Detailed balance limits of efficiency of p–n junction solar cells. *J. Appl. Phys.* **32**, 510–519 (1961).

23. Tiedje, T., Yablonovitch, E., Cody, G.D. & Brooks, B.G. Limiting efficiency of silicon solar cells. *IEEE Trans. Electron. Dev.* **31**, 711–716 (1984).

24. Sykora, M. *et al.* Size-dependent intrinsic radiative decay rates of silicon nanocrystals at large confinement energies. *Phys. Rev. Lett.* **100**, 067401 (2008).

25. Pandey, A. & Guyot-Sionnest, P. Slow electron cooling in colloidal quantum dots. *Science* **322**, 929–932 (2008).

26. Green, J.E., Wong, S.J. & Heath, J.R. Hall mobility measurements and chemical stability of ultra thin, methylated Si(111)-on-insulator films. *J. Phys. Chem. C* **112**, 5185–5189 (2008).

27. Barnham, K.W.J. & Duggan, G. A new approach to high-efficiency multi-band-gap solar cells. *J. Appl. Phys.* **67**, 3490–3493 (1990).

28. Nagrath, S. *et al.* Isolation of rare circulating tumour cells in cancer patients by microchip technology. *Nature* **450**, 1235–1239 (2007).

29. Mitchell, P.S. *et al.* Circulating microRNAs as stable blood-based marker for cancer detection. *Proc. Natl Acad. Sci. USA* **105**, 10513–10518 (2008).

30. Alivisatos, A.P., Gu, W. & Larabell, C. Quantum dots as cellular probes. *Annu. Rev. Biomed. Eng.* **7**, 55–76 (2005).

31. Lathrop, J.T., Anderson, N.L., Anderson, N.G. & Hammond, D.J. Therapeutic potential of the plasma proteome. *Curr. Opin. Mol. Ther.* **5**, 250–257 (2003).

32. Melin, J. & Quake, S.R. Microfluidic large-scale integration: The evolution of design rules for biological automation. *Annu. Rev. Biophys. Biomol. Struct.* **36**, 213–321 (2007).

33. Fan, R. *et al.* Integrated barcode chips for rapid, multiplexed analysis of proteins in microliter quantities of blood. *Nature Biotechnol.* **26**, 1373–1378 (2008).

34. Cristofanilli, M. *et al.* Circulating tumor cells, disease progression, and survival in metastatic breast cancer. *N. Engl. J. Med.* **351**, 781–791 (2004).

35. Zhong, J.F. *et al.* A microfluidic processor for gene expression profiling of single human embryonic stem cells. *Lab on a Chip* **8**, 68–74 (2008).

36. Nam, J-M, Thaxton, C.S. & Mirkin, C.A. Nanoparticle-based bio-bar codes for the ultrasensitive detection of proteins. *Science* **301**, 1884–1886 (2003).

37. Cui, Y. *et al.* Nanowire nanosensors for highly sensitive and selective detection of biological and chemical species. *Science* **293**, 1289–1292 (2001).

38. Braun, T. *et al.* Quantitative time-resolved measurement of membrane protein-ligand interactions using microcantilever array sensors. *Nature Nanotech.* **4**, 179–185 (2009).

39. Zimmerman, M. *et al.* Modeling and optimization of high sensitivity, low-volume microfluidic-based surface immunoassays. *Biomed. Microdev.* **7**, 99–110 (2005).

40. Thiel, K. Oligo oligarchy – the surprisingly small world of aptamers. *Nature Biotechnol.* **22**, 649–651 (2004).

41. Agnew, H.D. *et al.* Iterative in situ click chemistry creates antibody-like protein capture agents. *Angew. Chem Int. Ed. Engl.* **48**, 1–5 (2009).

42. Lal, S., Clare, S.E. & Halas, N.J. Nanoshell-enabled photothermal cancer therapy: Impending clinical impact. *Acc. Chem. Res.* **41**, 1842–1851 (2008).

43. Bartlett, D.W., Su, H., Hildebrandt, I.J., Weber, W.A. & Davis, M.E. Impact of tumor-specific targeting on the biodistribution and efficacy of siRNA nanoparticles measured by multimodality in vivo imaging. *Proc. Natl. Acad. Sci. USA* **104**, 15549–15554 (2007).

44. Heath, J.R. & Davis. M.E. Nanotechnology and cancer. *Ann. Rev. Med.* **59**, 251–265 (2008).

45. Geng, Y. *et al.* Shape effects of filaments versus spherical particles in flow and drug delivery. *Nature Nanotech.* **2**, 249–255 (2007).

NANOMATERIALS AND NANOSTRUCTURES

Progress towards monodisperse single-walled carbon nanotubes

The defining characteristic of a nanomaterial is that its properties vary as a function of its size. This size dependence can be clearly observed in single-walled carbon nanotubes, where changes in structure at the atomic scale can modify the electronic and optical properties of these materials in a discontinuous manner (for example, changing metallic nanotubes to semiconducting nanotubes and vice versa). However, as most practical technologies require predictable and uniform performance, researchers have been aggressively seeking strategies for preparing samples of single-walled carbon nanotubes with well-defined diameters, lengths, chiralities and electronic properties (that is, uniformly metallic or uniformly semiconducting). This review highlights post-synthetic approaches for sorting single-walled carbon nanotubes — including selective chemistry, electrical breakdown, dielectrophoresis, chromatography and ultracentrifugation — and progress towards selective growth of monodisperse samples.

MARK C. HERSAM

Department of Materials Science and Engineering and Department of Chemistry, Northwestern University, Evanston, Illinois 60208-3108, USA
e-mail: m-hersam@northwestern.edu

Single-walled carbon nanotubes (SWNTs) are hollow cylinders of carbon with diameters on the order of one nanometre, lengths ranging from tens of nanometres to centimetres, and walls that are one atomic-layer thick (Fig. 1a). Like other nanomaterials, the properties of SWNTs depend on their size and atomic structure. The resulting diverse and exemplary properties of SWNTs have inspired a vast range of proposed applications including transistors, logic gates, interconnects, conductive films, field emission sources, infrared emitters, sensors, scanning probes, nanomechanical devices, mechanical reinforcements, hydrogen storage elements and catalytic supports[1-6]. Over the past 15 years, researchers have demonstrated most of these applications in research laboratories by undergoing painstaking selection of SWNTs with appropriate properties.

SWNTs are thermodynamically stable forms of carbon that are produced when a carbonaceous feedstock is exposed to a metal catalyst at high temperatures. The most common means of synthesizing SWNTs include: (1) arc discharge[7], where a plasma is struck between graphite rods; (2) laser ablation[8], where a high-intensity laser beam is focused on a graphite rod; (3) chemical vapour deposition[9], where a carbon-bearing gas is heated in a furnace. Although some progress has been made in controlling the properties of SWNTs during growth, none of these synthetic techniques produce identical populations of SWNTs. The resulting lack of uniformity in the properties of SWNT samples is one of the primary reasons why SWNTs are rarely used in commercial applications today. In an effort to overcome this limitation, researchers have invested substantial effort towards the development of post-synthetic sorting schemes and selective growth methods for the production of bulk samples of monodisperse SWNTs[10-12]. This review highlights progress towards this goal with an emphasis on recent developments.

SWNT STRUCTURE–PROPERTY RELATIONSHIPS

The structure of a carbon nanotube is typically discussed in terms of graphene, which is a single layer of sp^2-bonded carbon atoms in a honeycomb lattice. In particular, a SWNT can be made by rolling up graphene to form a seamless cylinder. In many circumstances, SWNTs will bond to each other through van der Waals forces to form bundles, whereas some synthetic methods will also form concentric cylinders of carbon that are called multiwalled carbon nanotubes (MWNTs; see Fig. 1a).

The diameter of a SWNT typically ranges between 0.5 nm and 2 nm, whereas its length can vary over several orders of magnitude (from 10 nm to 1 cm). The length can be important for some applications, but the electronic properties of a SWNT depend on the direction in which the graphene was rolled up to form the nanotube. This direction is called the chiral vector. Each carbon atom in graphene can be identified with a pair of integers (n,m) and a pair of lattice vectors $(\mathbf{a}_1, \mathbf{a}_2)$, leading to the definition of the chiral vector, $C_h = n\mathbf{a}_1 + m\mathbf{a}_2$ (Fig. 1b). The magnitude of the chiral vector determines the diameter, $d = C_h/\pi$. Approximately 67% of SWNTs are semiconducting and 33% of SWNTs are metallic at room temperature.

For semiconducting SWNTs, the bandgap varies inversely with diameter. Furthermore, for both metallic and semiconducting SWNTs, the dominant optical transitions vary with diameter and chiral vector. Consequently, to achieve uniform electrical and optical properties, SWNTs need to be monodisperse in both their diameter

When citing this review, please cite the original version as shown on the contents page of this chapter.

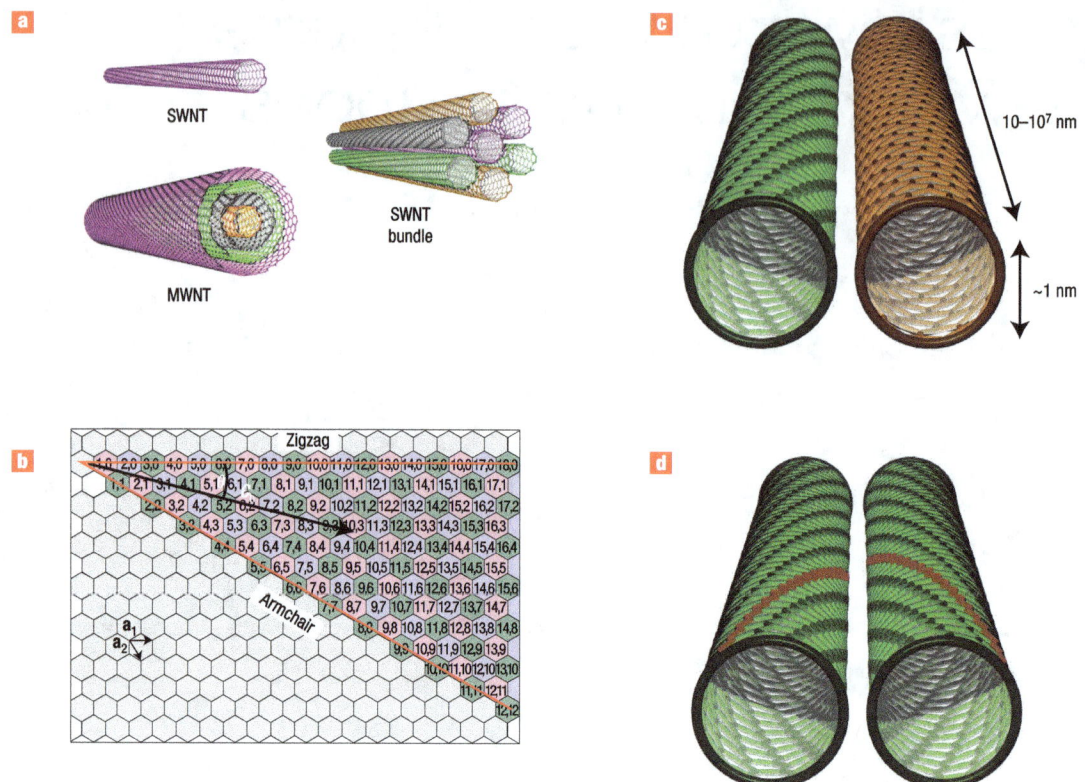

Figure 1 Typical carbon nanotube structures. **a**, Schematic representations of a single-walled carbon nanotube (SWNT), a multiwalled carbon nanotube (MWNT) and a bundle of SWNTs. **b**, Conceptually, a SWNT is formed by rolling up graphene along a chiral vector to form a cylinder. The circumference of the SWNT is determined by its chiral vector $C_h = n\mathbf{a}_1 + m\mathbf{a}_2$, where (n,m) are integers known as the chiral indices and $\mathbf{a}_1$ and $\mathbf{a}_2$ are the unit vectors of the graphene lattice. Nanotubes with $n = m$ (known as armchair nanotubes) and those with $n - m = 3j$, where $j = 0,1,2,3,\ldots$, are metallic at room temperature (labelled green). Carbon nanotubes with $n - m = 3j + 1$ (labelled pink) and $n - m = 3j + 2$ (labelled purple) are semiconductors with a bandgap that varies inversely with diameter. Nanotubes with $m = 0$ are known as zigzag nanotubes and can be either metallic or semiconducting. θ is the chiral angle. **c**, Schematic representation of two SWNTs with nearly identical diameters but different chiral indices. **d**, Schematic representation of two SWNTs with identical chiral vectors but different chiral handedness. Reproduced from ref. 14. Courtesy of Mike Arnold.

and electronic type because SWNTs with nearly identical diameters can have different chiral vectors and thus different electronic properties (Fig. 1c). In addition, SWNTs with identical chiral vectors can possess different chiral handedness (Fig. 1d), which will influence the interaction of SWNTs with circularly polarized light[13].

To achieve optimal performance in all conceivable applications, populations of SWNTs should be monodisperse with respect to electronic type, diameter, length and chiral handedness. Experimental methods for determining the degree of monodispersity include microscopy (for example, atomic force microscopy, scanning tunnelling microscopy and electron microscopy), optical spectroscopy (for example, optical absorbance, photoluminescence and Raman spectroscopy), and charge transport measurements. However, beyond achieving the highest possible purity, an effective SWNT sorting strategy should also be: (1) scalable (that is, it should be possible to ramp up the process to produce enough material to keep pace with anticipated increases in demand); (2) compatible with the wide range of SWNT lengths and diameters that are present in as-synthesized SWNT material; (3) nondestructive (that is, the outstanding properties of SWNTs should not be degraded during sorting); (4) iteratively repeatable (that is, it should be possible to repeat the sorting process to achieve improved purity levels); (5) affordable, thus enabling economical incorporation into target applications. Ultimately, a

SWNT sorting strategy can be deemed to have successfully met these additional criteria once its resulting material has found widespread use in commercial applications.

SELECTIVE CHEMISTRY

In an effort to amplify the differences between various SWNT species in a polydisperse mixture, many researchers have explored chemistries with reactivities that vary as a function of SWNT electronic type, diameter and/or chiral handedness. Following selective chemical functionalization, the SWNTs are then sorted using a variety of techniques including ultracentrifugation, electrophoretic separation and/or chromatographic methods. As the quality of the separation is often limited by the selectivity of the initial chemical reaction, selective chemistries will first be discussed before moving on to specific sorting techniques.

For an overview of carbon nanotube chemistry, the reader is referred to a number of recent reviews of the subject[15–17]. As summarized in Fig. 2, chemistry on SWNTs can be classified into five categories: (1) covalent sidewall chemistry; (2) covalent chemistry at defect sites or open ends; (3) non-covalent surfactant encapsulation; (4) non-covalent polymer wrapping; (5) molecular insertion into the SWNT interior. Among these chemical routes, chemistry on the exterior sidewalls has proven to be the most selective to date.

Covalent sidewall chemistries have been developed that discriminate as a function of electronic type and diameter[18]. For example, selective covalent sidewall functionalization of metallic SWNTs has been achieved with diazonium salts in aqueous solution[19,20]. Selectivity is attributed to the availability of electrons near the Fermi level in metallic SWNTs that stabilize a charge-transfer state preceding bond formation. Following covalent functionalization with p-hydroxybenzene diazonium salt, a negative charge can be induced on the metallic SWNTs through deprotonation in alkaline solutions, thus enabling subsequent separation by electronic type using free solution electrophoresis[21]. Alternatively, diazonium compounds with a long alkyl tail render the metallic SWNTs selectively soluble in tetrahydrofuran (THF)[22]. Because diazonium covalent functionalization significantly perturbs the electronic and optical properties of metallic SWNTs, this chemistry has also been used to render metallic SWNTs inactive in thin-film electronic devices[23,24].

A similar effect has been observed using dichlorocarbene covalent chemistry. In particular, dichlorocarbene opens an energy gap at the Fermi level, thus converting metallic SWNTs to semiconductors[25,26]. Even more perturbatively, nitronium ions selectively attack small-diameter metallic SWNTs, reducing them to amorphous carbon[27–29]. Binary sulphonic acid mixtures have also been used to dissolve SWNTs through direct protonation[30]. As the protonation is sensitive to geometric strain, this approach provides diameter selectivity. Solution-phase ozonolysis similarly shows a dependence on diameter due to diameter-dependent π-orbital misalignment and/or pyramidalization[31].

Selective covalent chemistry has also been driven optically. Examples include selective osmylation of metallic SWNTs driven by ultraviolet light[32] and hydrogen peroxide oxidative degradation of semiconducting SWNTs[33]. The latter approach also provides diameter selectivity because the wavelength of light can be tuned to excite specific diameter-dependent SWNT optical transitions. Selective oxidation of semiconducting SWNTs has been attributed to hole-doping by hydrogen peroxide[34,35]. Finally, selective covalent sidewall functionalization of semiconducting SWNTs has been achieved by cycloaddition of azomethine ylides derived from trialkylamine-N-oxides labelled with polycyclic aromatics[36].

Because covalent sidewall chemistry often induces substantial and/or irreversible changes to the properties of SWNTs, selective non-covalent chemistries have concurrently been pursued in an effort to minimize perturbation to the SWNT following functionalization. Selectivity by electronic type has been studied extensively for non-covalent chemistries based on amines. In particular, selective functionalization of metallic SWNTs has been achieved with propylamine and isopropylamine in THF[37,38]. Conversely, if the SWNTs are first oxidized in acidic solution, octadecylamine selectively interacts with semiconducting SWNTs[39–41]. In addition, charge-transfer complex formation with metallic SWNTs has been demonstrated for bromine[42,43] and charge-transfer aromatics[44]. Diameter-dependent binding has been observed for common surfactants (for example, sodium dodecyl sulphate, sodium dodecylbenzene sulphonate, and sodium cholate) in aqueous solution[45]. The diameter selectivity of these surfactants has subsequently been amplified by hydrogen peroxide degradation of those SWNTs that are not surfactant-encapsulated[46].

Selective non-covalent functionalization of semiconducting SWNTs has been realized with porphyrin chemistry[47]. More recently, using chiral diporphyrin molecules, selective chemistry has been observed as a function of SWNT chiral handedness[48]. The degree of chiral selectivity can be optimized by controlling the dihedral angle between the porphyrins[49]. Non-covalent polymer wrapping has also shown selectivity as a function of diameter and electronic type for

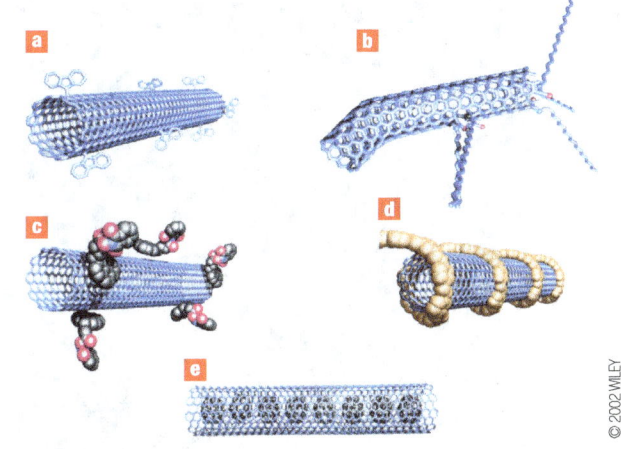

Figure 2 Strategies for chemically functionalizing SWNTs. **a**, Covalent sidewall chemistry. **b**, Covalent chemistry at defects or open ends. **c**, Non-covalent surfactant encapsulation. **d**, Non-covalent polymer wrapping. **e**, Molecular insertion into the SWNT interior. Reproduced from ref. 15.

small-diameter SWNTs. In particular, fluorene-based polymers provide diameter selectivity[50–52], whereas DNA possesses an affinity for semiconducting SWNTs[53].

The interaction of solvents with SWNTs as a function of electronic type has also been studied. In particular, solvents with aromatic backbones and electron-withdrawing groups tend to withdraw electrons selectively from metallic SWNTs[54]. It is important to note that the vast majority of work on selective non-covalent functionalization has focused on relatively small-diameter SWNTs (diameter <1.2 nm). A noteworthy exception is the use of oligo-acene adducts as a selective non-covalent anchor to large-diameter SWNTs (diameter ~1.6 nm)[55].

SELECTIVE DESTRUCTION

For many applications, the presence of unwanted species of SWNTs compromises optimal performance. For example, in a field-effect transistor (FET), metallic SWNTs will hinder effective electronic switching. In these cases, selective removal of the unwanted SWNT species is sufficient. As described above, many covalent functionalization strategies can achieve this objective using solution-phase chemistry. In addition, a variety of conditions have been identified for gas-phase selective etching of targeted SWNT species. Specific examples include selective oxidation of SWNTs with small diameters and high chiral angles following annealing in an oxygen-rich atmosphere[56,57]. Gas-phase reaction with fluorine gas followed by annealing has been shown to selectively etch metallic SWNTs with diameters less than 1.1 nm (ref. 58) For larger diameter SWNTs (diameter ~1.4–2 nm), gas-phase plasma hydrocarbonation reactions selectively etch metallic SWNTs in thin films[59]. This process also narrows the SWNT diameter distribution to 1.3–1.6 nm, which is the ideal range for SWNT integrated circuit technology. It should also be noted that the plasma hydrocarbonation approach is compatible with conventional semiconductor processing, thus providing opportunities for the fabrication of integrated circuits that combine silicon and SWNT devices.

Additional selective destruction approaches exploit controlled SWNT breakdown by current flow or optical irradiation. The seminal paper[60] on selective electrical breakdown of metallic SWNTs is

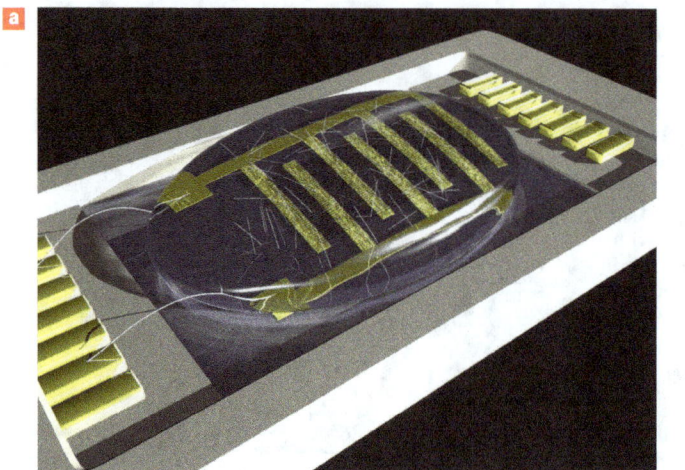
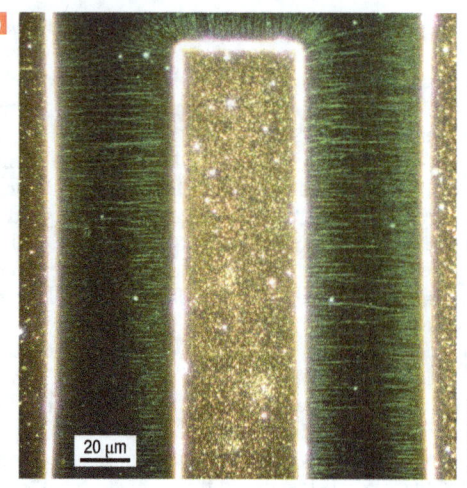

Figure 3 Dielectrophoresis of SWNTs. **a**, Schematic showing the typical experimental setup for dielectrophoresis of a SWNT solution using a microelectrode array. In this case, the electrode spacing is 50 micrometres. **b**, Dark-field microscope image following dielectrophoresis. The Rayleigh scattered light suggests that the SWNTs have aligned with the electric field. Reproduced from ref. 68.

popularly referred to as 'constructive destruction'. In this approach, a gate bias is applied to deplete semiconducting SWNTs in a thin-film FET geometry. Large currents are then passed through the conductive metallic SWNTs until they are electrically destroyed. Incidentally, a variation of this approach also allows individual shells of MWNTs to be removed electrically[61]. Optically driven approaches have focused on selective oxidation in hydrogen peroxide or air. The hydrogen peroxide approach selectively oxidizes semiconducting SWNTs with additional diameter selectivity provided by the wavelength of the incident light[33]. Laser irradiation in air, on the other hand, provides selective photolysis-assisted oxidation of metallic SWNTs[62]. Similarly, xenon-lamp irradiation leads to preferential destruction of metallic SWNTs[63]. In all selective destruction approaches, care has to be taken to prevent collateral damage to adjacent SWNTs during the destructive process. In addition, as selective destruction is typically an irreversible process, the eliminated SWNTs cannot be harvested for other applications.

ELECTROPHORETIC SEPARATION

As SWNTs possess dimensions similar to many biomolecules, there have been a number of efforts to use techniques borrowed from the life sciences — such as electrophoretic separation — to sort SWNTs by their physical and electronic properties. Conventional electrophoresis sorts SWNTs according to their relative mobility through a gel, capillary or solution in response to a direct current (d.c.) electric field. In the case of gel electrophoresis, the SWNTs with the smallest molecular weight travel most quickly, thus leading to sorting by SWNT length[64,65]. However, owing to a diameter-dependence of SWNT cutting during the ultrasonication process that disperses SWNTs in solution, the length sorting of gel electrophoresis also implies some diameter sorting. Similarly, capillary electrophoresis leads to length sorting[66], although this technique can additionally be used to separate individual SWNTs from bundled SWNTs[67]. Free solution electrophoresis has also been used for separation by SWNT electronic type (see section on Selective chemistry[21]).

Another approach for SWNT electronic type sorting is alternating current (a.c.) dielectrophoresis. Dielectrophoresis sorts SWNTs according to their dielectric constant in the presence of an a.c. electric field. Owing to differences in the dielectric constants

of metallic and semiconducting SWNTs, effective electronic type sorting can be achieved. Fig. 3a depicts the original geometry used for dielectrophoretic sorting of SWNTs[68]. A drop of SWNTs suspended in solution is placed on a microelectrode array. Upon application of an a.c. electric field, the more polarizable species (that is, the metallic SWNTs) are selectively deposited onto the substrate between the microelectrodes. The electric field also induces SWNT alignment as seen in Fig. 3b. Subsequent work has demonstrated that sorting by electronic type improves with increasing frequency of the electric field[69,70]. In addition, the concurrent use of cationic and anionic surfactants improves electronic type separation by neutralizing surface charges on the semiconducting SWNTs[71]. The obvious disadvantage of dielectrophoretic sorting with microelectrode arrays is limited throughput (for example, the original experiments isolated 100 picograms of metallic SWNTs)[68]. Consequently, recent work has focused on attempts to scale-up dielectrophoresis using larger electrodes[72] or dielectrophoretic field-flow fractionation (FFF)[73]. In addition to electronic type separation, the latter approach also has shown evidence for diameter sorting among semiconducting SWNTs. Dielectrophoretic FFF also naturally follows from previous work to sort SWNTs by length using conventional FFF.[74,75]

CHROMATOGRAPHY

Again drawing inspiration from biochemistry, several chromatographic methods have been explored for sorting SWNTs. For separation by SWNT diameter or electronic type, ion-exchange chromatography (IEX) of DNA-encapsulated SWNTs has shown the most promise. In this approach, small diameter (<1.2 nm) SWNTs are first individually encapsulated with single-stranded DNA in aqueous solution. Following IEX, separation by electronic type and diameter has been observed from optical absorbance measurements. The optical property differences in the resulting IEX fractions are so striking that they can be observed as distinguishable colours with the naked eye as shown in Fig. 4a (ref. 76). Fluorescence and Raman spectroscopy also confirm successful SWNT sorting[77]. Subsequent research established that the sorting quality depends on DNA sequence, with a sequence of $(GT)_n$ where n is 10 to 45 leading to the best results[78]. Theoretical work has concluded that the effective charge density of the DNA–SWNT hybrid, primarily governed by the DNA helical pitch, is the dominant factor governing the IEX sorting process[79].

Although the initial results of IEX were encouraging, the authors acknowledged that electronic type and diameter sorting are convoluted by SWNT length variations[76]. Therefore, the sorting results of IEX could presumably be improved if the SWNTs were first sorted by length. Several groups have demonstrated effective length sorting of SWNTs by chromatographic methods. In particular, size exclusion chromatography (SEC) has been repeatedly demonstrated for SWNT length separation[80–84]. In addition, gel permeation chromatography (GPC) has enabled length sorting for zwitterion-functionalized SWNTs[85]. By sequentially applying SEC and IEX, SWNTs with narrow diameter, length and chiral angle distributions have been achieved, as shown in Fig. 4b (ref. 86). In particular, the combined SEC/IEX approach has enabled sorting between (9,1) and (6,5) SWNTs — two species of SWNTs with the same diameter but different chiral angles. Despite the successes of chromatographic sorting, it should be acknowledged that the vast majority of work has been limited to small-diameter (<1.2 nm) SWNTs functionalized with DNA. Recent attempts to use chromatographically sorted larger-diameter SWNTs (~1.5 nm) for nanoelectronics were unsuccessful, thus prompting the authors to call for an alternative separation method in this diameter range[87].

ULTRACENTRIFUGATION

The final post-synthetic sorting approach to be discussed is ultracentrifugation. Conventional ultracentrifugation occurs in a constant density medium and sorts SWNTs according to their sedimentation coefficient. As the sedimentation coefficient depends on several parameters including buoyant density and molecular weight, SWNT sorting by conventional ultracentrifugation typically suffers from convolution among multiple structural parameters (for example, diameter and length). A noteworthy exception is the separation of individual surfactant-encapsulated SWNTs from SWNT bundles in aqueous solution[88]. In addition, when coupled with selective chemistry that induces bundling or selective dispersion of SWNTs of a particular electronic type, conventional ultracentrifugation can be used to sort metal and semiconducting SWNTs[37,42,89].

Alternatively, density gradient ultracentrifugation (DGU) allows SWNTs to be sorted exclusively by their buoyant density[90]. In DGU, a density gradient is intentionally formed in the centrifuge tube, after which the SWNTs are loaded. During ultracentrifugation, the SWNTs then sediment through the gradient until they reach their respective isopycnic points (that is, the point where the SWNT density matches the density of the gradient). The net result is that the SWNTs will form layers in the centrifuge tube according to their buoyant density. Assuming that the encapsulating surfactant layer is uniform for all SWNTs in solution, the SWNT buoyant density will depend on the SWNT diameter. A hydrodynamic model that describes the motion of SWNTs during DGU has recently been reported[91].

Experimentally, the DGU approach was first used to sort DNA–SWNT hybrids by diameter[92]. However, the DNA encapsulation chemistry possessed several disadvantages including irreversible wrapping, high cost and inability to disperse large-diameter (>1.2 nm) SWNTs. Consequently, subsequent work focused on surfactant-encapsulated SWNTs[93]. Sodium cholate encapsulation led to clear diameter sorting by DGU as evidenced by the visibly distinguishable coloured bands shown in Fig. 5a,b. When using co-surfactant mixtures of sodium cholate and sodium dodecyl sulphate, electronic-type sorting by DGU was achieved as seen in Fig. 5c. The latter sorting was attributed to inequivalent binding of the two surfactants as a function of the SWNT polarizability, which produced differences in the density of the SWNT–surfactant hybrid depending on SWNT electronic type. Following iterative application of the DGU technique, purities

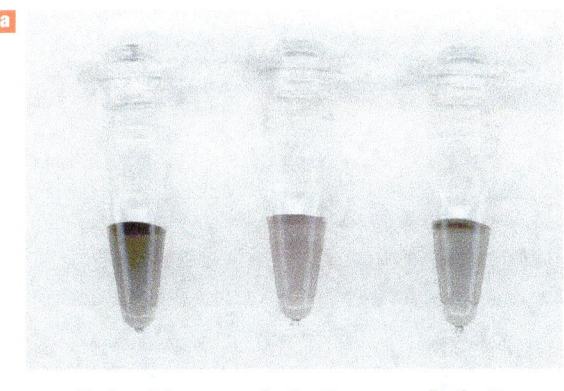

Starting solution   Fraction 47   Fraction 49

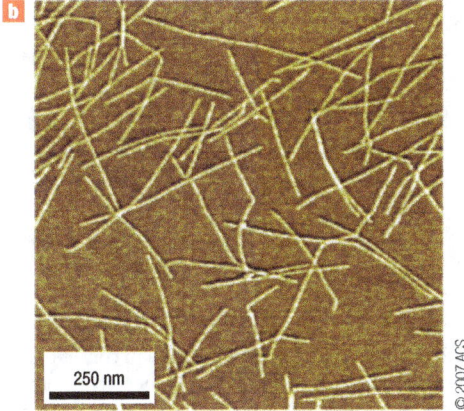

250 nm

© 2007 ACS

Figure 4 Chromatographic sorting of SWNTs. **a**, Ion-exchange chromatography enables sorting of DNA-encapsulated SWNTs by electronic type and diameter. The resulting purity is sufficiently high to see visible colour differences in the collected fractions. In particular, the pink middle fraction is enriched in metallic SWNTs compared with the starting solution or the later fraction. Reproduced from ref. 76. **b**, Sequential sorting by size exclusion chromatography and ion-exchange chromatography results in SWNTs of uniform diameter and length as observed in this atomic force microscopy image. Reproduced from ref. 86.

approaching 99% have been achieved. In addition, strong evidence for electronic-type sorting has been provided in the form of optical absorbance spectroscopy (Fig. 5d) and thin-film FET charge transport measurements. The optical purity of SWNTs sorted by DGU has enabled fundamental studies of SWNT exciton dynamics[94], exciton energy transfer[95] and photoluminescence quantum yields[96]. In addition, DGU measurements have provided insight into the diameter-dependence of endohedral water in SWNTs[97], whereas DGU-sorted metallic SWNTs have been used as coloured semitransparent conductive coatings[98–100].

SELECTIVE GROWTH

One of the long-standing goals of the SWNT field is the development of synthetic strategies that provide narrow distributions of SWNT structure and/or electronic type. In this manner, the many advantages of SWNT growth (for example, patterned growth[101], horizontal alignment[102], vertical alignment[103], high throughput[104] and millimetre length[105]) could be directly combined with monodisperse SWNTs.

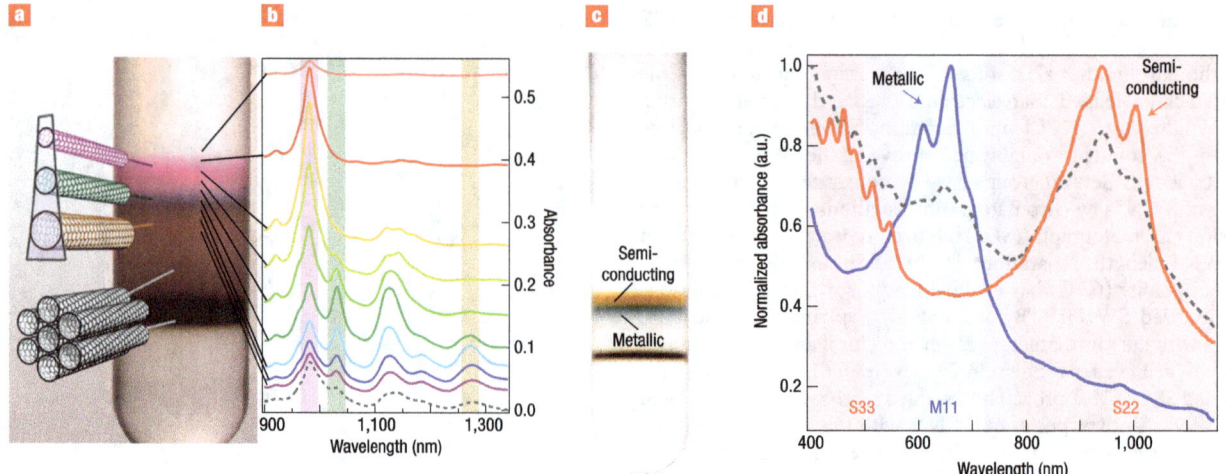

Figure 5 Sorting SWNTs using density gradient ultracentrifugation (DGU). **a**, Small-diameter (0.7–1.1 nm) SWNTs encapsulated with sodium cholate are sorted by diameter following DGU. Under these conditions, smaller-diameter SWNTs are more buoyant and thus settle at higher points in the centrifuge tube. Larger diameter and bundled SWNTs are less buoyant and thus settle at progressively lower points in the centrifuge tube. The observation of visible colours in the topmost bands verifies sorting by SWNT diameter. **b**, Optical absorbance spectra for different fractions, further confirm sorting by diameter. **c**, Large-diameter (1.1–1.6 nm) SWNTs encapsulated with a co-surfactant mixture of sodium cholate and sodium dodecyl sulphate are sorted according to electronic type following DGU. **d**, Optical absorbance spectra confirm sorting by electronic type as metallic fractions are peaked in the M11 region of the absorbance spectrum whereas semiconducting fractions are peaked in the S22 and S33 regions of the absorbance spectrum. Note: M11 = first-order optical transitions for metallic SWNTs; S22 and S33 = second-order and third-order optical transitions for semiconducting SWNTs respectively. Reproduced from ref. 93.

Through exhaustive searches of possible growth conditions, researchers have achieved varying degrees of SWNT diameter control by controlling temperature[106], pressure[107–109], carbon feedstock[110,111], catalyst particle size[112] and catalyst material[113–115]. Among these parameters, the development of new catalysts has shown the most promise. In particular, bimetallic catalysts (for example, CoMo[109,111,113,114] and FeRu[115]) have enabled narrow SWNT diameter distributions in the small-diameter limit (< 1.2 nm). As fewer pairs of chiral indices are available in the small-diameter limit, these catalysts have also enabled narrow distributions of (n,m). Isolated examples of enriched semiconductor SWNT growth via plasma enhanced chemical vapour deposition[116] and enriched metallic SWNT growth via monohydroxy alcohol homologue carbon feedstocks[117] have also been reported.

However, it has been speculated that specific (n,m) growth (especially in the large diameter limit) may not be possible because diameter variations that span many chiral indices are inevitable owing to thermal vibrations in the catalyst particles at elevated growth temperatures[42]. A possible strategy for circumventing these stochastic variations is to grow epitaxially from seed SWNTs[118]. In this manner, amplification of an isolated SWNT species could be achieved. This 'cloning' approach has recently been used to lengthen pre-existing SWNTs as confirmed by atomic force microscopy[119]. In addition, conditions that promote continued growth of SWNTs without nucleating new SWNT species have been identified[120,121].

CONCLUSIONS AND FUTURE OUTLOOK

Recent years have seen tangible progress toward the ultimate goal of large-scale, economical production of monodisperse SWNTs. Among the many post-synthetic techniques that have been developed, no definitive winner has yet emerged. Furthermore, selective growth strategies have not matched or exceeded the degree of control demonstrated by post-synthetic sorting approaches. In the absence of a revolutionary breakthrough, it appears that the optimal solution will include clever refinements or combinations of pre-existing sorting and/or growth techniques.

Even if the polydispersity problem is unambiguously solved, other challenges remain for SWNTs in real-world technology. For example, many applications will require highly precise means of assembling monodisperse SWNTs into predetermined geometries with nanometre precision[122]. In cases where SWNTs will serve as a replacement material in an established technology, integration and process compatibility issues will need to be addressed. Relevant economic and performance metrics also must be convincingly met. For the overwhelming majority of examples in this review, researchers have been content to demonstrate improvements in SWNT monodispersity using analytical techniques (for example, optical absorbance, photoluminescence, Raman spectroscopy, scanning probe microscopy, and so on). However, widely accepted standards for establishing SWNT purity levels from these techniques remain elusive. Developing such standards should thus be a high priority for SWNT researchers. In the meantime, in an effort to convince the private sector to make meaningful investments in SWNT technology, novel or substantially improved device applications enabled by monodisperse SWNTs will likely become the focus of future work in this field.

doi:10.1038/nnano.2008.135

Published online: 30 May 2008.

References
1. Thostenson, E. T., Ren, Z. F. & Chou, T. W. Advances in the science and technology of carbon nanotubes and their composites: A review. *Compos. Sci. Technol.* **61**, 1899–1912 (2001).
2. Mahar, B., Laslau, C., Yip, R. & Sun, Y. Development of carbon nanotube-based sensors - A review. *IEEE Sens. J.* **7**, 266–284 (2007).
3. Kaushik, B. K., Goel, S. & Rauthan, G. Future VLSI interconnects: optical fiber or carbon nanotube - a review. *Microelectron. Int.* **24**, 53–63 (2007).
4. Avouris, P., Chen, Z. H. & Perebeinos, V. Carbon-based electronics. *Nature Nanotech.* **2**, 605–615 (2007).
5. Charlier, J. C., Blase, X. & Roche, S. Electronic and transport properties of nanotubes. *Rev. Mod. Phys.* **79**, 677–732 (2007).

6. Anantram, M. P. & Leonard, F. Physics of carbon nanotube electronic devices. *Rep. Prog. Phys.* **69,** 507–561 (2006).
7. Ebbesen, T. W. & Ajayan, P. M. Large-scale synthesis of carbon nanotubes. *Nature* **358,** 220–222 (1992).
8. Guo, T., Nikolaev, P., Thess, A., Colbert, D. T. & Smalley, R. E. Catalytic growth of single-walled nanotubes by laser vaporization. *Chem. Phys. Lett.* **243,** 49–54 (1995).
9. Endo, M. *et al.* The production and structure of pyrolytic carbon nanotubes. *J. Phys. Chem. Solids* **54,** 1841–1848 (1993).
10. Krupke, R. & Hennrich, F. Separation techniques for carbon nanotubes. *Adv. Engineer. Mat.* **7,** 111–116 (2005).
11. Banerjee, S., Hemraj-Benny, T. & Wong, S. S. Routes towards separating metallic and semiconducting nanotubes. *J. Nanosci. Nanotechnol.* **5,** 841–855 (2005).
12. Haddon, R. C., Sippel, J., Rinzler, A. G. & Papadimitrakopoulos, F. Purification and separation of carbon nanotubes. *Mater. Res. Bull.* **29,** 252–259 (2004).
13. Ivchenko, E. L. & Spivak, B. Chirality effects in carbon nanotubes. *Phys. Rev. B* **66,** 155404 (2002).
14. Arnold, M. S. *Carbon Nanotubes: Photophysics, Biofunctionalization, and Sorting via Density Differentiation.* PhD Thesis, Northwestern Univ. (2006).
15. Hirsch, A. Functionalization of single-walled carbon nanotubes. *Angew. Chem. Int. Ed.* **41,** 1853–1859 (2002).
16. Banerjee, S., Hemraj-Benny, T. & Wong, S. S. Covalent surface chemistry of single-walled carbon nanotubes. *Adv. Mat.* **17,** 17–29 (2005).
17. Tasis, D., Tagmatarchis, N., Bianco, A. & Prato, M. Chemistry of carbon nanotubes. *Chem. Rev.* **106,** 1105–1136 (2006).
18. Campidelli, S., Meneghetti, M. & Prato, M. Separation of metallic and semiconducting single-walled carbon nanotubes via covalent functionalization. *Small* **3,** 1672–1676 (2007).
19. Strano, M. S. *et al.* Electronic structure control of single-walled carbon nanotube functionalization. *Science* **301,** 1519–1522 (2003).
20. Strano, M. S. Probing chiral selective reactions using a revised Kataura plot for the interpretation of single-walled carbon nanotube spectroscopy. *J. Am. Chem. Soc.* **125,** 16148–16153 (2003).
21. Kim, W. J., Usrey, M. L. & Strano, M. S. Selective functionalization and free solution electrophoresis of single-walled carbon nanotubes: Separate enrichment of metallic and semiconducting SWNT. *Chem. Mater.* **19,** 1571–1576 (2007).
22. Toyoda, S. *et al.* Separation of semiconducting single-walled carbon nanotubes by using a long-alkyl-chain benzenediazonium compound. *Chem. – Asian J.* **2,** 145–149 (2007).
23. An, L., Fu, Q. A., Lu, C. G. & Liu, J. A simple chemical route to selectively eliminate metallic carbon nanotubes in nanotube network devices. *J. Am. Chem. Soc.* **126,** 10520–10521 (2004).
24. Balasubramanian, K., Sordan, R., Burghard, M. & Kern, K. A selective electrochemical approach to carbon nanotube field-effect transistors. *Nano Lett.* **4,** 827–830 (2004).
25. Kamaras, K., Itkis, M. E., Hu, H., Zhao, B. & Haddon, R. C. Covalent bond formation to a carbon nanotube metal. *Science* **301,** 1501–1501 (2003).
26. Hu, H. *et al.* Sidewall functionalization of single-walled carbon nanotubes by addition of dichlorocarbene. *J. Am. Chem. Soc.* **125,** 14893–14900 (2003).
27. An, K. H. *et al.* A diameter-selective attack of metallic carbon nanotubes by nitronium ions. *J. Am. Chem. Soc.* **127,** 5196–5203 (2005).
28. Yang, C. M. *et al.* Selective removal of metallic single-walled carbon nanotubes with small diameters by using nitric and sulfuric acids. *J. Phys. Chem. B* **109,** 19242–19248 (2005).
29. An, K. H. *et al.* A diameter-selective chiral separation of single-wall carbon nanotubes using nitronium ions. *J. Electron. Mater.* **35,** 235–242 (2006).
30. Ramesh, S. *et al.* Diameter selection of single-walled carbon nanotubes through programmable solvation in binary sulfonic acid mixtures. *J. Phys. Chem. C* **111,** 17827–17834 (2007).
31. Banerjee, S. & Wong, S. S. Demonstration of diameter-selective reactivity in the sidewall ozonation of SWNTs by resonance Raman spectroscopy. *Nano Lett.* **4,** 1445–1450 (2004).
32. Banerjee, S. & Wong, S. S. Selective metallic tube reactivity in the solution-phase osmylation of single-walled carbon nanotubes. *J. Am. Chem. Soc.* **126,** 2073–2081 (2004).
33. Yudasaka, M., Zhang, M. & Iijima, S. Diameter-selective removal of single-wall carbon nanotubes through light-assisted oxidation. *Chem. Phys. Lett.* **374,** 132–136 (2003).
34. Miyata, Y., Maniwa, Y. & Kataura, H. Selective oxidation of semiconducting single-wall carbon nanotubes by hydrogen peroxide. *J. Phys. Chem. B* **110,** 25–29 (2006).
35. Lu, J. *et al.* Why semiconducting single-walled carbon nanotubes are separated from their metallic counterparts. *Small* **3,** 1566–1576 (2007).
36. Menard-Moyon, C., Izard, N., Doris, E. & Mioskowski, C. Separation of semiconducting from metallic carbon nanotubes by selective functionalization with azomethine ylides. *J. Am. Chem. Soc.* **128,** 6552–6553 (2006).
37. Maeda, Y. *et al.* Large-scale separation of metallic and semiconducting single-walled carbon nanotubes. *J. Am. Chem. Soc.* **127,** 10287–10290 (2005).
38. Maeda, Y. *et al.* Dispersion and separation of small-diameter single-walled carbon nanotubes. *J. Am. Chem. Soc.* **128,** 12239–12242 (2006).
39. Chattopadhyay, D., Galeska, L. & Papadimitrakopoulos, F. A route for bulk separation of semiconducting from metallic single-wall carbon nanotubes. *J. Am. Chem. Soc.* **125,** 3370–3375 (2003).
40. Samsonidze, G. G. *et al.* Quantitative evaluation of the octadecylamine-assisted bulk separation of semiconducting and metallic single-wall carbon nanotubes by resonance Raman spectroscopy. *Appl. Phys. Lett.* **85,** 1006–1008 (2004).
41. Kim, S. N., Luo, Z. T. & Papadimitrakopoulos, F. Diameter and metallicity dependent redox influences on the separation of single-wall carbon nanotubes. *Nano Lett.* **5,** 2500–2504 (2005).
42. Chen, Z. H. *et al.* Bulk separative enrichment in metallic or semiconducting single-walled carbon nanotubes. *Nano Lett.* **3,** 1245–1249 (2003).
43. Park, N. *et al.* Band gap sensitivity of bromine adsorption at carbon nanotubes. *Chem. Phys. Lett.* **403,** 135–139 (2005).
44. Lu, J. *et al.* Selective interaction of large or charge-transfer aromatic molecules with metallic single-wall carbon nanotubes: Critical role of the molecular size and orientation. *J. Am. Chem. Soc.* **128,** 5114–5118 (2006).
45. McDonald, T. J., Engtrakul, C., Jones, M., Rumbles, G. & Heben, M. J. Kinetics of PL quenching during single-walled carbon nanotube rebundling and diameter-dependent surfactant interactions. *J. Phys. Chem. B* **110,** 25339–25346 (2006).
46. McDonald, T. J., Blackburn, J. L., Metzger, W. K., Rumbles, G. & Heben, M. J. Chiral-selective protection of single-walled carbon nanotube photoluminescence by surfactant selection. *J. Phys. Chem. C* **111,** 17894–17900 (2007).
47. Li, H. P. *et al.* Selective interactions of porphyrins with semiconducting single-walled carbon nanotubes. *J. Am. Chem. Soc.* **126,** 1014–1015 (2004).
48. Peng, X. *et al.* Optically active single-walled carbon nanotubes. *Nature Nanotech.* **2,** 361–365 (2007).
49. Peng, X., Komatsu, N., Kimura, T. & Osuka, A. Improved optical enrichment of SWNTs through extraction with chiral nanotweezers of 2,6-pyridylene-bridged diporphyrins. *J. Am. Chem. Soc.* **129,** 15947–15953 (2007).
50. Chen, F. M., Wang, B., Chen, Y. & Li, L. J. Toward the extraction of single species of single-walled carbon nanotubes using fluorene-based polymers. *Nano Lett.* **7,** 3013–3017 (2007).
51. Nish, A., Hwang, J. Y., Doig, J. & Nicholas, R. J. Highly selective dispersion of single walled carbon nanotubes using aromatic polymers. *Nature Nanotech.* **2,** 640–646 (2007).
52. Hwang, J. Y. *et al.* Polymer structure and solvent effects on the selective dispersion of single-walled carbon nanotubes. *J. Am. Chem. Soc.* **130,** 3543–3553 (2008).
53. Fantini, C., Jorio, A., Santos, A. P., Peressinotto, V. S. T. & Pimenta, M. A. Characterization of DNA-wrapped carbon nanotubes by resonance Raman and optical absorption spectroscopies. *Chem. Phys. Lett.* **439,** 138–142 (2007).
54. Shin, H.-J. *et al.* Tailoring electronic structures of carbon nanotubes by solvent with electron-donating and -withdrawing groups. *J. Am. Chem. Soc.* **130,** 2062–2066 (2008).
55. Tromp, R. M., Afzali, A., Freitag, M., Mitzi, D. B. & Chen, Z. Novel strategy for diameter-selective separation and functionalization of single-wall carbon nanotubes. *Nano Lett.* **8,** 469–472 (2008).
56. Nagasawa, S., Yudasaka, M., Hirahara, K., Ichihashi, T. & Iijima, S. Effect of oxidation on single-wall carbon nanotubes. *Chem. Phys. Lett.* **328,** 374–380 (2000).
57. Miyata, Y. *et al.* Chirality-dependent combustion of single-walled carbon nanotubes. *J. Phys. Chem. C* **111,** 9671–9677 (2007).
58. Yang, C. M. *et al.* Preferential etching of metallic single-walled carbon nanotubes with small diameter by fluorine gas. *Phys. Rev. B* **73,** 075419 (2006).
59. Zhang, G. Y. *et al.* Selective etching of metallic carbon nanotubes by gas-phase reaction. *Science* **314,** 974–977 (2006).
60. Collins, P. C., Arnold, M. S. & Avouris, P. Engineering carbon nanotubes and nanotube circuits using electrical breakdown. *Science* **292,** 706–709 (2001).
61. Collins, P. G., Hersam, M., Arnold, M., Martel, R. & Avouris, P. Current saturation and electrical breakdown in multiwalled carbon nanotubes. *Phys. Rev. Lett.* **86,** 3128–3131 (2001).
62. Huang, H. J., Maruyama, R., Noda, K., Kajiura, H. & Kadono, K. Preferential destruction of metallic single-walled carbon nanotubes by laser irradiation. *J. Phys. Chem. B* **110,** 7316–7320 (2006).
63. Zhang, Y., Zhang, Y., Xian, X., Zhang, J. & Liu, Z. Sorting out semiconducting single-walled carbon nanotube arrays by preferential destruction of metallic tubes using xenon-lamp irradiation. *J. Phys. Chem. C* **112,** 3849–3856 (2008).
64. Heller, D. A. *et al.* Concomitant length and diameter separation of single-walled carbon nanotubes. *J. Am. Chem. Soc.* **126,** 14567–14573 (2004).
65. Vetcher, A. A. *et al.* Fractionation of SWNT/nucleic acid complexes by agarose gel electrophoresis. *Nanotechnology* **17,** 4263–4269 (2006).
66. Doorn, S. K. *et al.* High resolution capillary electrophoresis of carbon nanotubes. *J. Am. Chem. Soc.* **124,** 3169–3174 (2002).
67. Doorn, S. K. *et al.* Capillary electrophoresis separations of bundled and individual carbon nanotubes. *J. Phys. Chem. B* **107,** 6063–6069 (2003).
68. Krupke, R., Hennrich, F., von Lohneysen, H. & Kappes, M. M. Separation of metallic from semiconducting single-walled carbon nanotubes. *Science* **301,** 344–347 (2003).
69. Krupke, R., Hennrich, F., Kappes, M. M. & Lohneysen, H. V. Surface conductance induced dielectrophoresis of semiconducting single-walled carbon nanotubes. *Nano Lett.* **4,** 1395–1399 (2004).
70. Krupke, R., Linden, S., Rapp, M. & Hennrich, F. Thin films of metallic carbon nanotubes prepared by dielectrophoresis. *Adv. Mater.* **18,** 1468–1470 (2006).
71. Hong, S., Jung, S., Choi, J., Kim, Y. & Baik, S. Electrical transport characteristics of surface-conductance-controlled, dielectrophoretically separated single-walled carbon nanotubes. *Langmuir* **23,** 4749–4752 (2007).
72. Lutz, T. & Donovan, K. J. Macroscopic scale separation of metallic and semiconducting nanotubes by dielectrophoresis. *Carbon* **43,** 2508–2513 (2005).
73. Peng, H. Q., Alvarez, N. T., Kittrell, C., Hauge, R. H. & Schmidt, H. K. Dielectrophoresis field flow fractionation of single-walled carbon nanotubes. *J. Am. Chem. Soc.* **128,** 8396–8397 (2006).
74. Liu, J. *et al.* Fullerene pipes. *Science* **280,** 1253–1256 (1998).
75. Chen, B. L. & Selegue, J. P. Separation and characterization of single-walled and multiwalled carbon nanotubes by using flow field-flow fractionation. *Anal. Chem.* **74,** 4774–4780 (2002).
76. Zheng, M. *et al.* DNA-assisted dispersion and separation of carbon nanotubes. *Nature Mater.* **2,** 338–342 (2003).
77. Strano, M. S. *et al.* Understanding the nature of the DNA-assisted separation of single-walled carbon nanotubes using fluorescence and Raman spectroscopy. *Nano Lett.* **4,** 543–550 (2004).
78. Zheng, M. *et al.* Structure-based carbon nanotube sorting by sequence-dependent DNA assembly. *Science* **302,** 1545–1548 (2003).
79. Lustig, S. R., Jagota, A., Khripin, C. & Zheng, M. Theory of structure-based carbon nanotube separations by ion-exchange chromatography of DNA/CNT hybrids. *J. Phys. Chem. B* **109,** 2559–2566 (2005).
80. Duesberg, G. S., Muster, J., Krstic, V., Burghard, M. & Roth, S. Chromatographic size separation of single-wall carbon nanotubes. *Appl. Phys. A* **67,** 117–119 (1998).
81. Farkas, E., Anderson, M. E., Chen, Z. H. & Rinzler, A. G. Length sorting cut single wall carbon nanotubes by high performance liquid chromatography. *Chem. Phys. Lett.* **363,** 111–116 (2002).
82. Huang, X. Y., McLean, R. S. & Zheng, M. High-resolution length sorting and purification of DNA-wrapped carbon nanotubes by size-exclusion chromatography. *Anal. Chem.* **77,** 6225–6228 (2005).
83. Arnold, K., Hennrich, F., Krupke, R., Lebedkin, S. & Kappes, M. M. Length separation studies of single walled carbon nanotube dispersions. *Phys. Status Solidi B* **243,** 3073–3076 (2006).

84. Bauer, B. J., Fagan, J. A., Hobbie, E. K., Chun, J. & Bajpai, V. Chromatographic fractionation of SWNT/DNA dispersions with on-line multi-angle light scattering. *J. Phys. Chem. C* **112**, 1842–1850 (2008).

85. Chattopadhyay, D., Lastella, S., Kim, S. & Papadimitrakopoulos, F. Length separation of Zwitterion-functionalized single wall carbon nanotubes by GPC. *J. Am. Chem. Soc.* **124**, 728–729 (2002).

86. Zheng, M. & Semke, E. D. Enrichment of single chirality carbon nanotubes. *J. Am. Chem. Soc.* **129**, 6084–6085 (2007).

87. Zhang, L. *et al.* Assessment of chemically separated carbon nanotubes for nanoelectronics. *J. Am. Chem. Soc.* **130**, 2686–2691 (2008).

88. O'Connell, M. J. *et al.* Band gap fluorescence from individual single-walled carbon nanotubes. *Science* **297**, 593–596 (2002).

89. Wei, L. *et al.* Selective enrichment of (6,5) and (8,3) single-walled carbon nanotubes via cosurfactant extraction from narrow (n,m) distribution samples. *J. Phys. Chem. B* **112**, 2771–2774 (2008).

90. Green, A. A. & Hersam, M. C. Ultracentrifugation of single-walled nanotubes. *Mater. Today* **10**, 59–60 (December 2007).

91. Nair, N., Kim, W.-J., Braatz, R. D. & Strano, M. S. Dynamics of surfactant-suspended single-walled carbon nanotubes in a centrifugal field. *Langmuir* **24**, 1790–1795 (2008).

92. Arnold, M. S., Stupp, S. I. & Hersam, M. C. Enrichment of single-walled carbon nanotubes by diameter in density gradients. *Nano Lett.* **5**, 713–718 (2005).

93. Arnold, M. S., Green, A. A., Hulvat, J. F., Stupp, S. I. & Hersam, M. C. Sorting carbon nanotubes by electronic structure using density differentiation. *Nature Nanotech.* **1**, 60–65 (2006).

94. Zhu, Z. P. *et al.* Pump-probe spectroscopy of exciton dynamics in (6,5) carbon nanotubes. *J. Phys. Chem. C* **111**, 3831–3835 (2007).

95. Qian, H. *et al.* Exciton energy transfer in pairs of single-walled carbon nanotubes. *Nano Lett.* **8**, 1363–1367 (2008).

96. Crochet, J., Clemens, M. & Hertel, T. Quantum yield heterogeneities of aqueous single-wall carbon nanotube suspensions. *J. Am. Chem. Soc.* **129**, 8058–8059 (2007).

97. Hennrich, F. *et al.* Diameter sorting of carbon nanotubes by gradient centrifugation: Role of endohedral water. *Phys. Status Solidi B* **244**, 3896–3900 (2007).

98. Green, A. A. & Hersam, M. C. Colored semitransparent conductive coatings consisting of monodisperse metallic single-walled carbon nanotubes. *Nano Lett.* **8**, 1417–1422 (2008).

99. Yanagi, K., Miyata, Y. & Kataura, H. Optical and conductive characteristics of metallic single-wall carbon nanotubes with three basic colors: Cyan, magenta, and yellow. *Appl. Phys. Express* **1**, 034003 (2008).

100. Miyata, Y., Yanagi, K., Maniwa, Y. & Kataura, H. Highly stabilized conductivity of metallic single wall carbon nanotube thin films. *J. Phys. Chem. C* **112**, 3591–3596 (2008).

101. Kong, J., Soh, H. T., Cassell, A. M., Quate, C. F. & Dai, H. J. Synthesis of individual single-walled carbon nanotubes on patterned silicon wafers. *Nature* **395**, 878–881 (1998).

102. Kang, S. J. *et al.* High-performance electronics using dense, perfectly aligned arrays of single-walled carbon nanotubes. *Nature Nanotech.* **2**, 230–236 (2007).

103. Hata, K. *et al.* Water-assisted highly efficient synthesis of impurity-free single-walled carbon nanotubes. *Science* **306**, 1362–1364 (2004).

104. Cassell, A. M., Raymakers, J. A., Kong, J. & Dai, H. J. Large scale CVD synthesis of single-walled carbon nanotubes. *J. Phys. Chem. B* **103**, 6484–6492 (1999).

105. Huang, S. M., Cai, X. Y. & Liu, J. Growth of millimeter-long and horizontally aligned single-walled carbon nanotubes on flat substrates. *J. Am. Chem. Soc.* **125**, 5636–5637 (2003).

106. Bandow, S. *et al.* Effect of the growth temperature on the diameter distribution and chirality of single-wall carbon nanotubes. *Phys. Rev. Lett.* **80**, 3779–3782 (1998).

107. Nikolaev, P. *et al.* Gas-phase catalytic growth of single-walled carbon nanotubes from carbon monoxide. *Chem. Phys. Lett.* **313**, 91–97 (1999).

108. Hiraoka, T., Bandow, S., Shinohara, H. & Iijima, S. Control on the diameter of single-walled carbon nanotubes by changing the pressure in floating catalyst CVD. *Carbon* **44**, 1853–1859 (2006).

109. Wang, B. *et al.* Pressure-induced single-walled carbon nanotube (n,m) selectivity on Co-Mo catalysts. *J. Phys. Chem. C* **111**, 14612–14616 (2007).

110. Maruyama, S., Kojima, R., Miyauchi, Y., Chiashi, S. & Kohno, M. Low-temperature synthesis of high-purity single-walled carbon nanotubes from alcohol. *Chem. Phys. Lett.* **360**, 229–234 (2002).

111. Wang, B. *et al.* (n,m) selectivity of single-walled carbon nanotubes by different carbon precursors on Co-Mo catalysts. *J. Am. Chem. Soc.* **129**, 9014–9019 (2007).

112. Sinnott, S. B. *et al.* Model of carbon nanotube growth through chemical vapor deposition. *Chem. Phys. Lett.* **315**, 25–30 (1999).

113. Kitiyanan, B., Alvarez, W. E., Harwell, J. H. & Resasco, D. E. Controlled production of single-wall carbon nanotubes by catalytic decomposition of CO on bimetallic Co-Mo catalysts. *Chem. Phys. Lett.* **317**, 497–503 (2000).

114. Bachilo, S. M. *et al.* Narrow (n,m)-distribution of single-walled carbon nanotubes grown using a solid supported catalyst. *J. Am. Chem. Soc.* **125**, 11186–11187 (2003).

115. Li, X. L. *et al.* Selective synthesis combined with chemical separation of single-walled carbon nanotubes for chirality selection. *J. Am. Chem. Soc.* **129**, 15770–15771 (2007).

116. Li, Y. M. *et al.* Preferential growth of semiconducting single-walled carbon nanotubes by a plasma enhanced CVD method. *Nano Lett.* **4**, 317–321 (2004).

117. Wang, Y. *et al.* Direct enrichment of metallic single-waited carbon nanotubes induced by the different molecular composition of monohydroxy alcohol homologues. *Small* **3**, 1486–1490 (2007).

118. Wang, Y. H. *et al.* Continued growth of single-walled carbon nanotubes. *Nano Lett.* **5**, 997–1002 (2005).

119. Smalley, R. E. *et al.* Single wall carbon nanotube amplification: En route to a type-specific growth mechanism. *J. Am. Chem. Soc.* **128**, 15824–15829 (2006).

120. Ogrin, D. *et al.* Amplification of single-walled carbon nanotubes from designed seeds: Separation of nucleation and growth. *J. Phys. Chem. C* **111**, 17804–17806 (2007).

121. Iwasaki, T., Robertson, J. & Kawarada, H. Mechanism analysis of interrupted growth of single-walled carbon nanotube arrays. *Nano Lett.* **8**, 886–890 (2008).

122. Yan, Y. H., Chan-Park, M. B. & Zhang, Q. Advances in carbon-nanotube assembly. *Small* **3**, 24–42 (2007).

Acknowledgements

This work was supported by the US Army Telemedicine and Advanced Technology Research Center (DAMD17-05-1-0381), the National Science Foundation (DMR-0520513, EEC-0647560, and DMR-0706067), and the Department of Energy (DE-FG02-03ER15457). An Alfred P. Sloan Research Fellowship is also acknowledged. M.C.H. also thanks M. S. Arnold and A. A. Green for helpful discussions.

The rise of graphene

Graphene is a rapidly rising star on the horizon of materials science and condensed-matter physics. This strictly two-dimensional material exhibits exceptionally high crystal and electronic quality, and, despite its short history, has already revealed a cornucopia of new physics and potential applications, which are briefly discussed here. Whereas one can be certain of the realness of applications only when commercial products appear, graphene no longer requires any further proof of its importance in terms of fundamental physics. Owing to its unusual electronic spectrum, graphene has led to the emergence of a new paradigm of 'relativistic' condensed-matter physics, where quantum relativistic phenomena, some of which are unobservable in high-energy physics, can now be mimicked and tested in table-top experiments. More generally, graphene represents a conceptually new class of materials that are only one atom thick, and, on this basis, offers new inroads into low-dimensional physics that has never ceased to surprise and continues to provide a fertile ground for applications.

A. K. GEIM AND K. S. NOVOSELOV

Manchester Centre for Mesoscience and Nanotechnology, University of Manchester, Oxford Road, Manchester M13 9PL, UK

*e-mail: geim@man.ac.uk; kostya@graphene.org

Graphene is the name given to a flat monolayer of carbon atoms tightly packed into a two-dimensional (2D) honeycomb lattice, and is a basic building block for graphitic materials of all other dimensionalities (Fig. 1). It can be wrapped up into 0D fullerenes, rolled into 1D nanotubes or stacked into 3D graphite. Theoretically, graphene (or '2D graphite') has been studied for sixty years[1–3], and is widely used for describing properties of various carbon-based materials. Forty years later, it was realized that graphene also provides an excellent condensed-matter analogue of (2+1)-dimensional quantum electrodynamics[4–6], which propelled graphene into a thriving theoretical toy model. On the other hand, although known as an integral part of 3D materials, graphene was presumed not to exist in the free state, being described as an 'academic' material[5] and was believed to be unstable with respect to the formation of curved structures such as soot, fullerenes and nanotubes. Suddenly, the vintage model turned into reality, when free-standing graphene was unexpectedly found three years ago[7,8] — and especially when the follow-up experiments[9,10] confirmed that its charge carriers were indeed massless Dirac fermions. So, the graphene 'gold rush' has begun.

MATERIALS THAT SHOULD NOT EXIST

More than 70 years ago, Landau and Peierls argued that strictly 2D crystals were thermodynamically unstable and could not exist[11,12]. Their theory pointed out that a divergent contribution of thermal fluctuations in low-dimensional crystal lattices should lead to such displacements of atoms that they become comparable to interatomic distances at any finite temperature[13]. The argument was later extended by Mermin[14] and is strongly supported by an omnibus of experimental observations. Indeed, the melting temperature of thin films rapidly decreases with decreasing thickness, and the films become unstable (segregate into islands or decompose) at a thickness of, typically, dozens of atomic layers[15,16]. For this reason, atomic monolayers have so far been known only as an integral part of larger 3D structures, usually grown epitaxially on top of monocrystals with matching crystal lattices[15,16]. Without such a 3D base, 2D materials were presumed not to exist, until 2004, when the common wisdom was flaunted by the experimental discovery of graphene[7] and other free-standing 2D atomic crystals (for example, single-layer boron nitride and half-layer BSCCO)[8]. These crystals could be obtained on top of non-crystalline substrates[8–10], in liquid suspension[7,17] and as suspended membranes[18].

Importantly, the 2D crystals were found not only to be continuous but to exhibit high crystal quality[7–10,17,18]. The latter is most obvious for the case of graphene, in which charge carriers can travel thousands of interatomic distances without scattering[7–10]. With the benefit of hindsight, the existence of such one-atom-thick crystals can be reconciled with theory. Indeed, it can be argued that the obtained 2D crystallites are quenched in a metastable state because they are extracted from 3D materials, whereas their small size (<<1 mm) and strong interatomic bonds ensure that thermal fluctuations cannot lead to the generation of dislocations or other crystal defects even at elevated temperature[13,14]. A complementary viewpoint is that the extracted 2D crystals become intrinsically stable by gentle crumpling in the third dimension[18,19] (for an artist's impression of the crumpling, see the cover of this issue). Such 3D warping (observed on a lateral scale of ≈10 nm)[18] leads to a gain in elastic energy but suppresses thermal vibrations (anomalously large in 2D), which above a certain temperature can minimize the total free energy[19].

BRIEF HISTORY OF GRAPHENE

Before reviewing the earlier work on graphene, it is useful to define what 2D crystals are. Obviously, a single atomic plane is a 2D

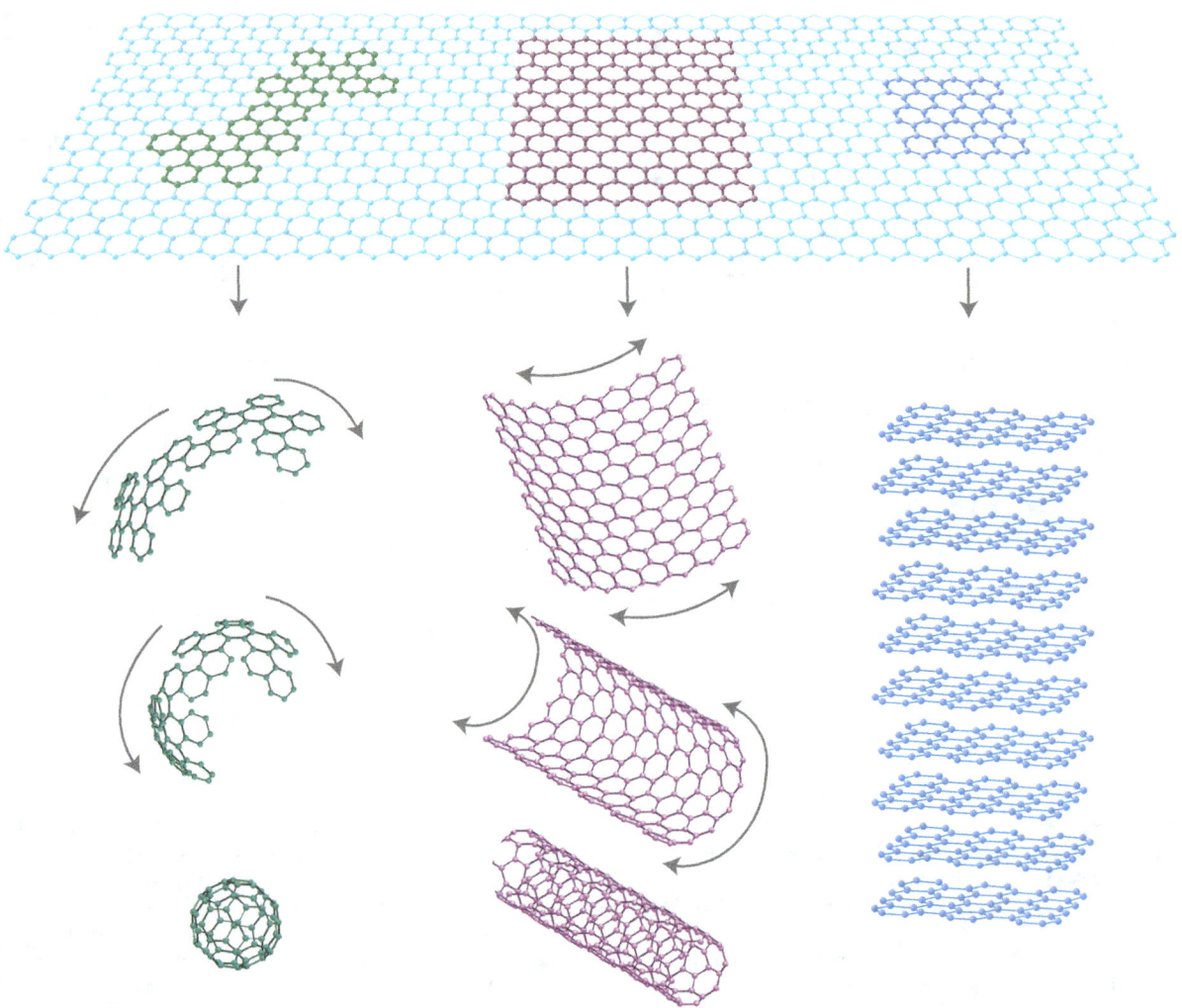

Figure 1 Mother of all graphitic forms. Graphene is a 2D building material for carbon materials of all other dimensionalities. It can be wrapped up into 0D buckyballs, rolled into 1D nanotubes or stacked into 3D graphite.

crystal, whereas 100 layers should be considered as a thin film of a 3D material. But how many layers are needed before the structure is regarded as 3D? For the case of graphene, the situation has recently become reasonably clear. It was shown that the electronic structure rapidly evolves with the number of layers, approaching the 3D limit of graphite at 10 layers[20]. Moreover, only graphene and, to a good approximation, its bilayer has simple electronic spectra: they are both zero-gap semiconductors (they can also be referred to as zero-overlap semimetals) with one type of electron and one type of hole. For three or more layers, the spectra become increasingly complicated: Several charge carriers appear[7,21], and the conduction and valence bands start notably overlapping[7,20]. This allows single-, double- and few- (3 to <10) layer graphene to be distinguished as three different types of 2D crystals ('graphenes'). Thicker structures should be considered, to all intents and purposes, as thin films of graphite. From the experimental point of view, such a definition is also sensible. The screening length in graphite is only ≈5 Å (that is, less than two layers in thickness)[21] and, hence, one must differentiate between the surface and the bulk even for films as thin as five layers[21,22].

Earlier attempts to isolate graphene concentrated on chemical exfoliation. To this end, bulk graphite was first intercalated[23] so that

graphene planes became separated by layers of intervening atoms or molecules. This usually resulted in new 3D materials[23]. However, in certain cases, large molecules could be inserted between atomic planes, providing greater separation such that the resulting compounds could be considered as isolated graphene layers embedded in a 3D matrix. Furthermore, one can often get rid of intercalating molecules in a chemical reaction to obtain a sludge consisting of restacked and scrolled graphene sheets[24-26]. Because of its uncontrollable character, graphitic sludge has so far attracted only limited interest.

There have also been a small number of attempts to grow graphene. The same approach as generally used for the growth of carbon nanotubes so far only produced graphite films thicker than ≈100 layers[27]. On the other hand, single- and few-layer graphene have been grown epitaxially by chemical vapour deposition of hydrocarbons on metal substrates[28,29] and by thermal decomposition of SiC (refs 30–34). Such films were studied by surface science techniques, and their quality and continuity remained unknown. Only lately, few-layer graphene obtained on SiC was characterized with respect to its electronic properties, revealing high-mobility charge carriers[32,33]. Epitaxial growth of graphene offers probably the only viable route towards electronic applications and, with so much at stake, rapid progress in this direction is expected. The approach that seems promising but has not been attempted yet is the use of the previously demonstrated epitaxy on catalytic surfaces[28,29] (such as Ni or Pt) followed by the deposition of an insulating support on top of graphene and chemical removal of the primary metallic substrate.

THE ART OF GRAPHITE DRAWING

In the absence of quality graphene wafers, most experimental groups are currently using samples obtained by micromechanical cleavage of bulk graphite, the same technique that allowed the isolation of graphene for the first time[7,8]. After fine-tuning, the technique[8] now provides high-quality graphene crystallites up to 100 μm in size, which is sufficient for most research purposes (see Fig. 2). Superficially, the technique looks no more sophisticated than drawing with a piece of graphite[8] or its repeated peeling with adhesive tape[7] until the thinnest flakes are found. A similar approach was tried by other groups (earlier[35] and somewhat later but independently[22,36]) but only graphite flakes 20 to 100 layers thick were found. The problem is that graphene crystallites left on a substrate are extremely rare and hidden in a 'haystack' of thousands of thick (graphite) flakes. So, even if one were deliberately searching for graphene by using modern techniques for studying atomically thin materials, it would be impossible to find those several micrometre-size crystallites dispersed over, typically, a 1-cm² area. For example, scanning-probe microscopy has too low throughput to search for graphene, whereas scanning electron microscopy is unsuitable because of the absence of clear signatures for the number of atomic layers.

The critical ingredient for success was the observation that graphene becomes visible in an optical microscope if placed on top of a Si wafer with a carefully chosen thickness of SiO₂, owing to a feeble interference-like contrast with respect to an empty wafer. If not for this simple yet effective way to scan substrates in search of graphene crystallites, they would probably remain undiscovered today. Indeed, even knowing the exact recipe[8], it requires special care and perseverance to find graphene. For example, only a 5% difference in SiO₂ thickness (315 nm instead of the current standard of 300 nm) can make single-layer graphene completely invisible. Careful selection of the initial graphite material (so that it has largest possible grains) and the use of freshly cleaved and cleaned surfaces of graphite and SiO₂ can also make all the difference. Note that graphene was recently[37,38] found to have a clear signature in Raman microscopy, which makes this technique useful for quick inspection of thickness, even though potential crystallites still have to be first hunted for in an optical microscope.

Similar stories could be told about other 2D crystals (particularly, dichalcogenide monolayers) where many attempts were made to split these strongly layered materials into individual planes[39,40]. However, the crucial step of isolating monolayers to assess their properties individually was never achieved. Now, by using the same approach as demonstrated for graphene, it is possible to investigate potentially hundreds of different 2D crystals[8] in search of new phenomena and applications.

FERMIONS GO BALLISTIC

Although there is a whole new class of 2D materials, all experimental and theoretical efforts have so far focused on graphene, somehow ignoring the existence of other 2D crystals. It remains to be seen whether this bias is justified, but the primary reason for it is clear: the exceptional electronic quality exhibited by the isolated graphene crystallites[7–10]. From experience, people know that high-quality samples always yield new physics, and this understanding has played a major role in focusing attention on graphene.

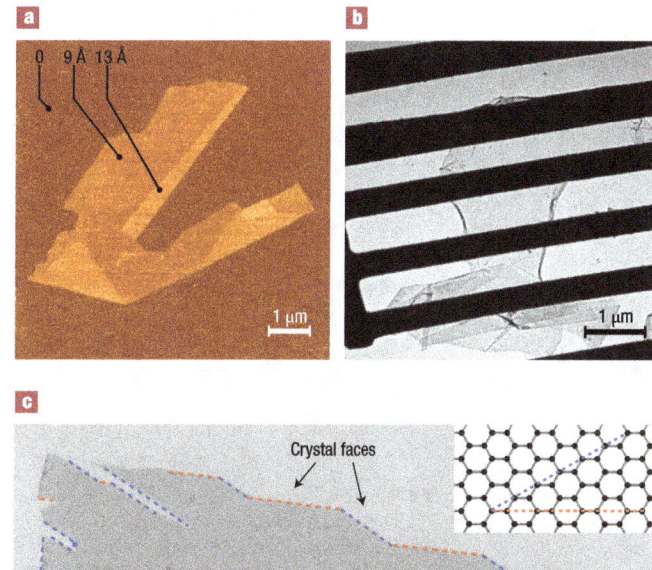

Figure 2 One-atom-thick single crystals: the thinnest material you will ever see. **a**, Graphene visualized by atomic force microscopy (adapted from ref. 8). The folded region exhibiting a relative height of ≈4 Å clearly indicates that it is a single layer. (Copyright National Academy of Sciences, USA.) **b**, A graphene sheet freely suspended on a micrometre-size metallic scaffold. The transmission electron microscopy image is adapted from ref. 18. **c**, Scanning electron micrograph of a relatively large graphene crystal, which shows that most of the crystal's faces are zigzag and armchair edges as indicated by blue and red lines and illustrated in the inset (T.J. Booth, K.S.N, P. Blake and A.K.G. unpublished work). 1D transport along zigzag edges and edge-related magnetism are expected to attract significant attention.

Graphene's quality clearly reveals itself in a pronounced ambipolar electric field effect (Fig. 3) such that charge carriers can be tuned continuously between electrons and holes in concentrations n as high as 10^{13} cm^{-2} and their mobilities μ can exceed 15,000 cm² V^{-1} s^{-1} even under ambient conditions[7–10]. Moreover, the observed mobilities weakly depend on temperature T, which means that μ at 300 K is still limited by impurity scattering, and therefore can be improved significantly, perhaps, even up to ≈100,000 cm² V^{-1} s^{-1}. Although some semiconductors exhibit room-temperature μ as high as ≈77,000 cm² V^{-1} s^{-1} (namely, InSb), those values are quoted for undoped bulk semiconductors. In graphene, μ remains high even at high n (>10^{12} cm^{-2}) in both electrically and chemically doped devices[41], which translates into ballistic transport on the submicrometre scale (currently up to ≈0.3 μm at 300 K). A further indication of the system's extreme electronic quality is the quantum Hall effect (QHE) that can be observed in graphene even at room temperature, extending the previous temperature range for the QHE by a factor of 10 (ref. 42).

An equally important reason for the interest in graphene is a particular unique nature of its charge carriers. In condensed-matter physics, the Schrödinger equation rules the world, usually being quite sufficient to describe electronic properties of materials. Graphene is an exception — its charge carriers mimic relativistic particles and are more easily and naturally described starting with the Dirac equation rather than the Schrödinger equation[4–6,43–48].

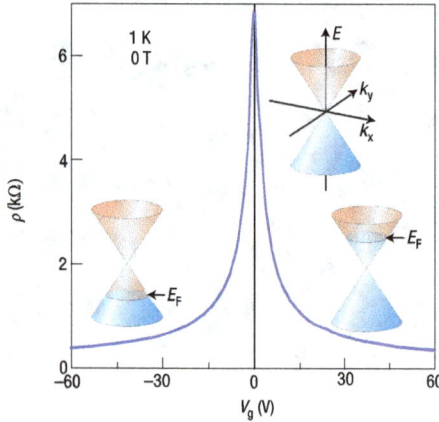

Figure 3 Ambipolar electric field effect in single-layer graphene. The insets show its conical low-energy spectrum $E(k)$, indicating changes in the position of the Fermi energy E_F with changing gate voltage V_g. Positive (negative) V_g induce electrons (holes) in concentrations $n = \alpha V_g$ where the coefficient $\alpha \approx 7.2 \times 10^{10}$ cm^{-2} V^{-1} for field-effect devices with a 300 nm SiO$_2$ layer used as a dielectric[7-9]. The rapid decrease in resistivity ρ on adding charge carriers indicates their high mobility (in this case, $\mu \approx 5{,}000$ cm^2 V^{-1} s^{-1} and does not noticeably change with increasing temperature to 300 K).

Although there is nothing particularly relativistic about electrons moving around carbon atoms, their interaction with the periodic potential of graphene's honeycomb lattice gives rise to new quasiparticles that at low energies E are accurately described by the (2+1)-dimensional Dirac equation with an effective speed of light $v_F \approx 10^6$ m^{-1}s^{-1}. These quasiparticles, called massless Dirac fermions, can be seen as electrons that have lost their rest mass m_0 or as neutrinos that acquired the electron charge e. The relativistic-like description of electron waves on honeycomb lattices has been known theoretically for many years, never failing to attract attention, and the experimental discovery of graphene now provides a way to probe quantum electrodynamics (QED) phenomena by measuring graphene's electronic properties.

QED IN A PENCIL TRACE

From the point of view of its electronic properties, graphene is a zero-gap semiconductor, in which low-E quasiparticles within each valley can formally be described by the Dirac-like hamiltonian

$$\hat{H} = \hbar v_F \begin{pmatrix} 0 & k_x - ik_y \\ k_x + ik_y & 0 \end{pmatrix} = \hbar v_F \boldsymbol{\sigma} \cdot \mathbf{k} , \qquad (1)$$

where $\mathbf{k}$ is the quasiparticle momentum, $\boldsymbol{\sigma}$ the 2D Pauli matrix and the k-independent Fermi velocity v_F plays the role of the speed of light. The Dirac equation is a direct consequence of graphene's crystal symmetry. Its honeycomb lattice is made up of two equivalent carbon sublattices A and B, and cosine-like energy bands associated with the sublattices intersect at zero E near the edges of the Brillouin zone, giving rise to conical sections of the energy spectrum for $|E| < 1$ eV (Fig. 3).

We emphasize that the linear spectrum $E = \hbar v_F k$ is not the only essential feature of the band structure. Indeed, electronic states near zero E (where the bands intersect) are composed of states belonging to the different sublattices, and their relative contributions in the make-up of quasiparticles have to be taken into account by, for

example, using two-component wavefunctions (spinors). This requires an index to indicate sublattices A and B, which is similar to the spin index (up and down) in QED and, therefore, is referred to as pseudospin. Accordingly, in the formal description of graphene's quasiparticles by the Dirac-like hamiltonian above, $\boldsymbol{\sigma}$ refers to pseudospin rather than the real spin of electrons (the latter must be described by additional terms in the hamiltonian). Importantly, QED-specific phenomena are often inversely proportional to the speed of light c, and therefore enhanced in graphene by a factor $c/v_F \approx 300$. In particular, this means that pseudospin-related effects should generally dominate those due to the real spin.

By analogy with QED, one can also introduce a quantity called chirality[6] that is formally a projection of $\boldsymbol{\sigma}$ on the direction of motion $\mathbf{k}$ and is positive (negative) for electrons (holes). In essence, chirality in graphene signifies the fact that k electron and $-k$ hole states are intricately connected because they originate from the same carbon sublattices. The concepts of chirality and pseudospin are important because many electronic processes in graphene can be understood as due to conservation of these quantities[6,43-48].

It is interesting to note that in some narrow-gap 3D semiconductors, the gap can be closed by compositional changes or by applying high pressure. Generally, zero gap does not necessitate Dirac fermions (that imply conjugated electron and hole states), but in some cases they might appear[5]. The difficulties of tuning the gap to zero, while keeping carrier mobilities high, the lack of possibility to control electronic properties of 3D materials by the electric field effect and, generally, less pronounced quantum effects in 3D limited studies of such semiconductors mostly to measuring the concentration dependence of their effective masses m (for a review, see ref. 49). It is tempting to have a fresh look at zero-gap bulk semiconductors, especially because Dirac fermions have recently been reported even in such a well-studied (small-overlap) 3D material as graphite[50,51].

CHIRAL QUANTUM HALL EFFECTS

At this early stage, the main experimental efforts have been focused on the electronic properties of graphene, trying to understand the consequences of its QED-like spectrum. Among the most spectacular phenomena reported so far, there are two new ('chiral') quantum Hall effects (QHEs), minimum quantum conductivity in the limit of vanishing concentrations of charge carriers and strong suppression of quantum interference effects.

Figure 4 shows three types of QHE behaviour observed in graphene. The first one is a relativistic analogue of the integer QHE and characteristic of single-layer graphene[9,10]. It shows up as an uninterrupted ladder of equidistant steps in the Hall conductivity σ_{xy} which persists through the neutrality (Dirac) point, where charge carriers change from electrons to holes (Fig. 4a). The sequence is shifted with respect to the standard QHE sequence by ½, so that $\sigma_{xy} = \pm 4e^2/h\ (N + \frac{1}{2})$ where N is the Landau level (LL) index and factor 4 appears due to double valley and double spin degeneracy. This QHE has been dubbed 'half-integer' to reflect both the shift and the fact that, although it is not a new fractional QHE, it is not the standard integer QHE either. The unusual sequence is now well understood as arising from the QED-like quantization of graphene's electronic spectrum in magnetic field B, which is described[45,52-54] by $E_N = \pm v_F\sqrt{2e\hbar BN}$ where $\pm$ refers to electrons and holes. The existence of a quantized level at zero E, which is shared by electrons and holes (Fig. 4c), is essentially everything one needs to know to explain the anomalous QHE sequence[52-56]. An alternative explanation for the half-integer QHE is to invoke the coupling between pseudospin and orbital motion, which gives rise to a geometrical phase of π accumulated along cyclotron trajectories, which is often referred to as Berry's

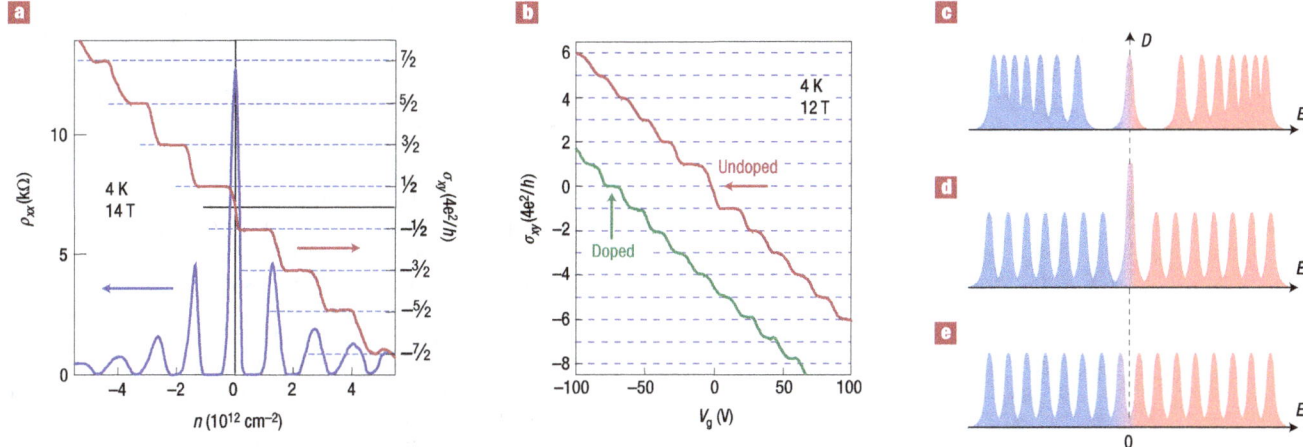

Figure 4 Chiral quantum Hall effects. **a**, The hallmark of massless Dirac fermions is QHE plateaux in σ_{xy} at half integers of $4e^2/h$ (adapted from ref. 9). **b**, Anomalous QHE for massive Dirac fermions in bilayer graphene is more subtle (red curve[56]): σ_{xy} exhibits the standard QHE sequence with plateaux at all integer N of $4e^2/h$ except for $N=0$. The missing plateau is indicated by the red arrow. The zero-N plateau can be recovered after chemical doping, which shifts the neutrality point to high V_g so that an asymmetry gap ($\approx 0.1\,$eV in this case) is opened by the electric field effect (green curve[60]). **c–e**, Different types of Landau quantization in graphene. The sequence of Landau levels in the density of states D is described by $E_N \propto \sqrt{N}$ for massless Dirac fermions in single-layer graphene (**c**) and by $E_N \propto \sqrt{N(N-1)}$ for massive Dirac fermions in bilayer graphene (**d**). The standard LL sequence $E_N \propto N+\frac{1}{2}$ is expected to recover if an electronic gap is opened in the bilayer (**e**).

phase[9,10,57]. The additional phase leads to a π-shift in the phase of quantum oscillations and, in the QHE limit, to a half-step shift.

Bilayer graphene exhibits an equally anomalous QHE (Fig 4b)[56]. Experimentally, it shows up less spectacularly. The standard sequence of Hall plateaux $\sigma_{xy} = \pm N4e^2/h$ is measured, but the very first plateau at $N = 0$ is missing, which also implies that bilayer graphene remains metallic at the neutrality point[56]. The origin of this anomaly lies in the rather bizarre nature of quasiparticles in bilayer graphene, which are described[58] by

$$\hat{H} = -\frac{\hbar^2}{2m}\begin{pmatrix} 0 & (k_x - ik_y)^2 \\ (k_x + ik_y)^2 & 0 \end{pmatrix}. \qquad (2)$$

This hamiltonian combines the off-diagonal structure, similar to the Dirac equation, with Schrödinger-like terms $\hat{p}^2/2m$. The resulting quasiparticles are chiral, similar to massless Dirac fermions, but have a finite mass $m \approx 0.05m_0$. Such massive chiral particles would be an oxymoron in relativistic quantum theory. The Landau quantization of 'massive Dirac fermions' is given[58] by $E_N = \pm \hbar\omega\sqrt{N(N-1)}$ with two degenerate levels $N = 0$ and 1 at zero E (ω is the cyclotron frequency). This additional degeneracy leads to the missing zero-E plateau and the double-height step in Fig. 4b. There is also a pseudospin associated with massive Dirac fermions, and its orbital rotation leads to a geometrical phase of 2π. This phase is indistinguishable from zero in the quasiclassical limit ($N \gg 1$) but reveals itself in the double degeneracy of the zero-E LL (Fig. 4d)[56].

It is interesting that the 'standard' QHE with all the plateaux present can be recovered in bilayer graphene by the electric field effect (Fig. 4b). Indeed, gate voltage not only changes n but simultaneously induces an asymmetry between the two graphene layers, which results in a semiconducting gap[59,60]. The electric-field-induced gap eliminates the additional degeneracy of the zero-E LL and leads to the uninterrupted QHE sequence by splitting the double step into two (Fig. 4e)[59,60]. However, to observe this splitting in the QHE measurements, the neutrality region needs to be probed at finite gate voltages, which can be achieved by additional chemical doping[60]. Note that bilayer graphene is the only known material in

which the electronic band structure changes significantly via the electric field effect, and the semiconducting gap ΔE can be tuned continuously from zero to ≈ 0.3 eV if SiO_2 is used as a dielectric.

CONDUCTIVITY 'WITHOUT' CHARGE CARRIERS

Another important observation is that graphene's zero-field conductivity σ does not disappear in the limit of vanishing n but instead exhibits values close to the conductivity quantum e^2/h per carrier type[9]. Figure 5 shows the lowest conductivity $\sigma_{\min}$ measured near the neutrality point for nearly 50 single-layer devices. For all other known materials, such a low conductivity unavoidably leads to a metal–insulator transition at low T but no sign of the transition has been observed in graphene down to liquid-helium T. Moreover, although it is the persistence of the metallic state with σ of the order of e^2/h that is most exceptional and counterintuitive, a relatively small spread of the observed conductivity values (see Fig. 5) also allows speculation about the quantization of $\sigma_{\min}$. We emphasize that it is the resistivity (conductivity) that is quantized in graphene, in contrast to the resistance (conductance) quantization known in many other transport phenomena.

Minimum quantum conductivity has been predicted for Dirac fermions by a number of theories[5,45,46,48,61–65]. Some of them rely on a vanishing density of states at zero E for the linear 2D spectrum. However, comparison between the experimental behaviour of massless and massive Dirac fermions in graphene and its bilayer allows chirality- and masslessness-related effects to be distinguished. To this end, bilayer graphene also exhibits a minimum conductivity of the order of e^2/h per carrier type[56,66], which indicates that it is chirality, rather than the linear spectrum, that is more important. Most theories suggest $\sigma_{\min} = 4e^2/h\pi$, which is about π times smaller than the typical values observed experimentally. It can be seen in Fig. 5 that the experimental data do not approach this theoretical value and mostly cluster around $\sigma_{\min} = 4e^2/h$ (except for one low-μ sample that is rather unusual by also exhibiting 100%-normal weak localization behaviour at high n; see below). This disagreement has become known as 'the mystery of a missing pie', and it remains unclear whether it is due

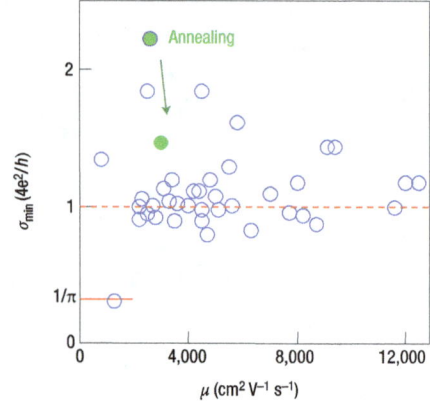

Figure 5 Minimum conductivity of graphene. Independent of their carrier mobility μ, different graphene devices exhibit approximately the same conductivity at the neutrality point (open circles) with most data clustering around $\approx 4e^2/h$ indicated for clarity by the dashed line (A.K.G. and K.S.N., unpublished work; includes the published data from ref. 9). The high-conductivity tail is attributed to macroscopic inhomogeneity. By improving the homogeneity of the samples, σ_{min} generally decreases, moving closer to $\approx 4e^2/h$. The green arrow and symbols show one of the devices that initially exhibited an anomalously large value of σ_{min} but after thermal annealing at ≈ 400 K its σ_{min} moved closer to the rest of the statistical ensemble. Most of the data are taken in the bend resistance geometry where the macroscopic inhomogeneity plays the least role.

to theoretical approximations about electron scattering in graphene, or because the experiments probed only a limited range of possible sample parameters (for example, length-to-width ratios[48]). To this end, note that close to the neutrality point ($n \leq 10^{11}$ cm^{-2}) graphene conducts as a random network of electron and hole puddles (A.K.G. and K.S.N., unpublished work). Such microscopic inhomogeneity is probably inherent to graphene (because of graphene sheet's warping/rippling)[18,67] but so far has not been taken into account by theory. Furthermore, macroscopic inhomogeneity (on the scale larger than the mean free path l) is also important in measurements of σ_{min}. The latter inhomogeneity can explain a high-σ tail in the data scatter in Fig. 5 by the fact that σ reached its lowest values at slightly different gate voltage (V_g) in different parts of a sample, which yields effectively higher values of experimentally measured σ_{min}.

WEAK LOCALIZATION IN SHORT SUPPLY

At low temperatures, all metallic systems with high resistivity should inevitably exhibit large quantum-interference (localization) magnetoresistance, eventually leading to the metal–insulator transition at $\sigma \approx e^2/h$. Such behaviour was thought to be universal, but it was found missing in graphene. Even near the neutrality point where resistivity is highest, no significant low-field ($B < 1$ T) magnetoresistance has been observed down to liquid-helium temperatures[67], and although sub-100 nm Hall crosses did exhibit giant resistance fluctuations (K.S.N. *et al.* unpublished work), those could be attributed to changes in the percolation through electron and hole puddles and size quantization. It remains to be seen whether localization effects at the Dirac point recover at lower T, as the phase-breaking length becomes increasingly longer[68], or the observed behaviour indicates a "marginal Fermi liquid"[44,69], in which the phase-breaking length goes to zero with decreasing E. Further experimental studies are much needed in this regime, but it is difficult to probe because of microscopic inhomogeneity.

Away from the Dirac point (where graphene becomes a good metal), the situation has recently become reasonably clear. Universal conductance fluctuations were reported to be qualitatively normal in this regime, whereas weak localization magnetoresistance was found to be somewhat random, varying for different samples from being virtually absent to showing the standard behaviour[67]. On the other hand, early theories had also predicted every possible type of weak-localization magnetoresistance in graphene, from positive to negative to zero. Now it is understood that, for large n and in the absence of inter-valley scattering, there should be no magnetoresistance, because the triangular warping of graphene's Fermi surface destroys time-reversal symmetry within each valley[70]. With increasing inter-valley scattering, the normal (negative) weak localization should recover. Changes in inter-valley scattering rates by, for example, varying microfabrication procedures can explain the observed sample-dependent behaviour. A complementary explanation is that a sufficient inter-valley scattering is already present in the studied samples but the time-reversal symmetry is destroyed by elastic strain due to microscopic warping of a graphene sheet[67,71]. The strain in graphene has turned out to be somewhat similar to a random magnetic field, which also destroys time-reversal symmetry and suppresses weak localization. Whatever the mechanism, theory expects (approximately[72]) normal universal conductance fluctuations at high n, in agreement with the experiment[67].

PENCILLED-IN BIG PHYSICS

Owing to space limitations, we do not attempt to overview a wide range of other interesting phenomena predicted for graphene theoretically but as yet not observed experimentally. Nevertheless, let us mention two focal points for current theories. One of them is many-body physics near the Dirac point, where interaction effects should be strongly enhanced due to weak screening, the vanishing density of states and graphene's large coupling constant $e^2/\hbar v_F \approx 1$ ("effective fine structure constant"[69,73]). The predictions include various options for the fractional QHE, quantum Hall ferromagnetism, excitonic gaps, and so on. (for example, see refs 45,73–80). The first relevant experiment in ultra-high B has reported the lifting of spin and valley degeneracy[81].

Second, graphene is discussed in the context of testing various QED effects, among which the *gedanken* Klein paradox and *zitterbewegung* stand out because these effects are unobservable in particle physics. The notion of Klein paradox refers to a counterintuitive process of perfect tunnelling of relativistic electrons through arbitrarily high and wide barriers. The experiment is conceptually easy to implement in graphene[47]. *Zitterbewegung* is a term describing jittery movements of a relativistic electron due to interference between parts of its wavepacket belonging to positive (electron) and negative (positron) energy states. These quasi-random movements can be responsible for the finite conductivity $\approx e^2/h$ of ballistic devices[46,48], are hypothesized to result in excess shot noise[48] and might even be visualized by direct imaging[82,83] of Dirac trajectories. In the latter respect, graphene offers truly unique opportunities because, unlike in most semiconductor systems, its 2D electronic states are not buried deep under the surface, and can be accessed directly by tunnelling and other local probes. Many interesting results can be expected to arise from scanning-probe experiments in graphene. Another tantalizing possibility is to study QED in a curved space (by controllable bending of a graphene sheet), which allows certain cosmological problems to be addressed[84].

2D OR NOT 2D

In addition to QED physics, there are many other reasons that should perpetuate active interest in graphene. For the sake of brevity, they can be summarized by referring to analogies with carbon nanotubes and 2D electron gases in semiconductors. Indeed, much of the fame and glory of nanotubes can probably be credited to graphene, the very material they are made of. By projecting the accumulated knowledge about carbon nanotubes onto their flat counterpart and bearing in mind the rich physics brought about by semiconductor 2D systems, a reasonably good sketch of emerging opportunities can probably be drawn.

The relationship between 2D graphene and 1D carbon nanotubes requires a special mention. The current rapid progress on graphene has certainly benefited from the relatively mature research on nanotubes that continue to provide a near-term guide in searching for graphene applications. However, there exists a popular opinion that graphene should be considered simply as unfolded carbon nanotubes and, therefore, can compete with them in the myriad of applications already suggested. Partisans of this view often claim that graphene will make nanotubes obsolete, allowing all the promised applications to reach an industrial stage because, unlike nanotubes, graphene can (probably) be produced in large quantities with fully reproducible properties. This view is both unfair and inaccurate. Dimensionality is one of the most defining material parameters, and as carbon nanotubes exhibit properties drastically different from those of 3D graphite and 0D fullerenes, 2D graphene is also quite different from its forms in the other dimensions. Depending on the particular problem in hand, graphene's prospects can be sometimes superior, sometimes inferior, and most often completely different from those of carbon nanotubes or, for the sake of argument, of graphite.

GRAPHENIUM INSIDE

As concerns applications, graphene-based electronics should be mentioned first. This is because most efforts have so far been focused in this direction, and such companies as Intel and IBM fund this research to keep an eye on possible developments. It is not surprising because, at the time when the Si-based technology is approaching its fundamental limits, any new candidate material to take over from Si is welcome, and graphene seems to offer an exceptional choice.

Graphene's potential for electronics is usually justified by citing high mobility of its charge carriers. However, as mentioned above, the truly exceptional feature of graphene is that μ remains high even at highest electric-field-induced concentrations, and seems to be little affected by chemical doping[41]. This translates into ballistic transport on a submicrometre scale at 300 K. A room-temperature ballistic transistor has long been a tantalizing but elusive aim of electronic engineers, and graphene can make it happen. The large value of v_F and low-resistance contacts without a Schottky barrier[7] should help further reduce the switching time. Relatively low on–off ratios (reaching only ≈ 100 because of graphene's minimum conductivity) do not seem to present a fundamental problem for high-frequency applications[7], and the demonstration of transistors operational at THz frequencies would be an important milestone for graphene-based electronics.

For mainstream logic applications, the fact that graphene remains metallic even at the neutrality point is a major problem. However, significant semiconductor gaps ΔE can still be engineered in graphene. As mentioned above, ΔE up to 0.3 eV can be induced in bilayer graphene but this is perhaps more interesting in terms of tuneable infrared lasers and detectors. For single-layer graphene, ΔE can be induced by spatial confinement or lateral-superlattice potential. The latter seems to be a relatively straightforward solution because sizeable gaps should naturally occur in graphene epitaxially grown on top of crystals with matching lattices such as boron nitride or the same SiC (refs 30–34), in which superlattice effects are undoubtedly expected.

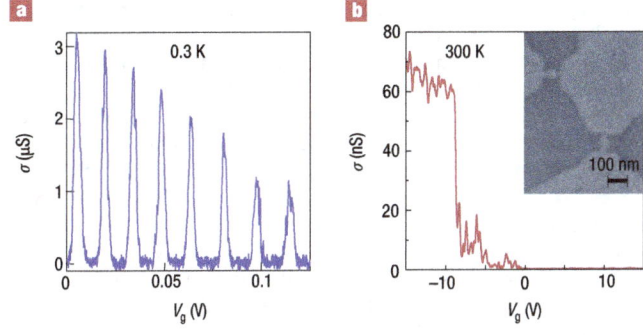

Figure 6 Towards graphene-based electronics. To achieve transistor action, nanometre ribbons and quantum dots can be carved in graphene (L. A. Ponomarenko, F. Schedin, K. S. N. and A. K. G., in preparation). **a**, Coulomb blockade in relatively large quantum dots (diameter ≈ 0.25 μm) at low temperature. Conductance σ of such devices can be controlled by either the back gate or a side electrode also made from graphene. Narrow constrictions in graphene with low-temperature resistance much larger than 100 kΩ serve as quantum barriers. **b**, 10-nm-scale graphene structures remain remarkably stable under ambient conditions and survive thermal cycling to liquid-helium temperature. Such devices can show a high-quality transistor action even at room temperature so that their conductance can be pinched-off completely over a large range of gate voltages near the neutrality point. The inset shows a scanning electron micrograph of two graphene dots of ≈ 40 nm in diameter with narrower (<10 nm) constrictions. The challenge is to make such room-temperature quantum dots with sufficient precision to obtain reproducible characteristics for different devices, which is hard to achieve by standard electron-beam lithography and isotropic dry etching.

Owing to graphene's linear spectrum and large v_B the confinement gap is also rather large[85–87] ΔE (eV) $\approx \alpha \hbar v_F/d \approx 1/d$ (nm), compared with other semiconductors, and it requires ribbons with width d of about 10 nm for room-temperature operation (coefficient α is $\approx$½ for Dirac fermions)[87]. With the Si-based technology rapidly advancing into this scale, the required size is no longer seen as a significant hurdle, and much research is expected along this direction. However, unless a technique for anisotropic etching of graphene is found to make devices with crystallographically defined faces (for example, zigzag or armchair), one has to deal with conductive channels having irregular edges. In short channels, electronic states associated with such edges can induce a significant sample-dependent conductance[85–87]. In long channels, random edges may lead to additional scattering, which can be detrimental for the speed and energy consumption of transistors, and in effect, cancel all the advantages offered by graphene's ballistic transport. Fortunately, high-anisotropy dry etching is probably achievable in graphene, owing to quite different chemical reactivity of zigzag and armchair edges.

An alternative route to graphene-based electronics is to consider graphene not as a new channel material for field-effect transistors (FET) but as a conductive sheet, in which various nanometre-size structures can be carved to make a single-electron-transistor (SET) circuitry. The idea is to exploit the fact that, unlike other materials, graphene nanostructures are stable down to true nanometre sizes, and possibly even down to a single benzene ring. This allows the exploration of a region somewhere in between SET and molecular electronics (but by using the top-down approach). The advantage is that everything including conducting channels, quantum dots, barriers and interconnects can be cut out from a graphene sheet, whereas other material characteristics are much less important for the SET architecture[88,89] than for traditional FET circuits. This approach is illustrated in Fig. 6, which shows a SET made entirely from graphene by using electron-beam lithography and dry etching (Fig. 6b, inset). For a minimum feature size of $\approx$10 nm the combined Coulomb and confinement gap reaches >$3kT$, which should allow a SET-like circuitry operational at room temperature (Fig. 6b), whereas resistive (rather than traditional tunnel) barriers can be used to induce Coulomb blockade. The SET architecture is relatively well developed[88,89], and one of the main reasons it has failed to impress so far is difficulties with the extension of its operation to room temperature. The fundamental cause for the latter is a poor stability of materials for true-nanometre sizes, at which the Si-based technology is also likely to encounter fundamental limitations, according to the semiconductor industry roadmap. This is where graphene can come into play.

It is most certain that we will see many efforts to develop various approaches to graphene electronics. Whichever approach prevails, there are two immediate challenges. First, despite the recent progress in epitaxial growth of graphene[33,34], high-quality wafers suitable for industrial applications still remain to be demonstrated. Second, individual features in graphene devices need to be controlled accurately enough to provide sufficient reproducibility in their properties. The latter is exactly the same challenge that the Si technology has been dealing with successfully. For the time being, to make proof-of-principle nanometre-size devices, one can use electrochemical etching of graphene by scanning-probe nanolithography[90].

GRAPHENE DREAMS

Despite the reigning optimism about graphene-based electronics, 'graphenium' microprocessors are unlikely to appear for the next 20 years. In the meantime, many other graphene-based applications are likely to come of age. In this respect, clear parallels with nanotubes allow a highly educated guess of what to expect soon.

The most immediate application for graphene is probably its use in composite materials. Indeed, it has been demonstrated that a graphene powder of uncoagulated micrometre-size crystallites can be produced in a way scaleable to mass production[17]. This allows conductive plastics at less than one volume percent filling[17], which in combination with low production costs makes graphene-based composite materials attractive for a variety of uses. However, it seems doubtful that such composites can match the mechanical strength of their nanotube counterparts because of much stronger entanglement in the latter case.

Another enticing possibility is the use of graphene powder in electric batteries that are already one of the main markets for graphite. An ultimately large surface-to-volume ratio and high conductivity provided by graphene powder can lead to improvements in the efficiency of batteries, taking over from the carbon nanofibres used in modern batteries. Carbon nanotubes have also been considered for this application but graphene powder has an important advantage of being cheap to produce[17].

One of the most promising applications for nanotubes is field emitters, and although there have been no reports yet about such use of graphene, thin graphite flakes were used in plasma displays (commercial prototypes) long before graphene was isolated, and many patents were filed on this subject. It is likely that graphene powder can offer even more superior emitting properties.

Carbon nanotubes have been reported to be an excellent material for solid-state gas sensors but graphene offers clear advantages in this particular direction[41]. Spin-valve and superconducting field-effect transistors are also obvious research targets, and recent reports describing a hysteretic magnetoresistance[91] and substantial bipolar supercurrents[92] prove graphene's major potential for these applications. An extremely weak spin-orbit coupling and the absence of hyperfine interaction in ^{12}C-graphene make it an excellent if not

ideal material for making spin qubits. This guarantees graphene-based quantum computation to become an active research area. Finally, we cannot omit mentioning hydrogen storage, which has been an active but controversial subject for nanotubes. It has already been suggested that graphene is capable of absorbing a large amount of hydrogen[93], and experimental efforts in this direction are duly expected.

AFTER THE GOLD RUSH

It has been just over two years since graphene was first reported, and despite remarkably rapid progress, only the very tip of the iceberg has been uncovered so far. Because of the short timescale, most experimental groups working now on graphene have not published even a single paper on the subject, which has been a truly frustrating experience for theorists. This is to say that, at this time, no review can possibly be complete. Nevertheless, the research directions explained here should persuade even die-hard sceptics that graphene is not a fleeting fashion but is here to stay, bringing up both more exciting physics, and perhaps even wide-ranging applications.

doi:10.1038/nmat1849

References

1. Wallace, P. R. The band theory of graphite. *Phys. Rev.* **71**, 622–634 (1947).
2. McClure, J. W. Diamagnetism of graphite. *Phys. Rev.* **104**, 666–671 (1956).
3. Slonczewski, J. C. & Weiss, P. R. Band structure of graphite. *Phys. Rev.* **109**, 272–279 (1958).
4. Semenoff, G. W. Condensed-matter simulation of a three-dimensional anomaly. *Phys. Rev. Lett.* **53**, 2449–2452 (1984).
5. Fradkin, E. Critical behavior of disordered degenerate semiconductors. *Phys. Rev. B* **33**, 3263–3268 (1986).
6. Haldane, F. D. M. Model for a quantum Hall effect without Landau levels: Condensed-matter realization of the 'parity anomaly'. *Phys. Rev. Lett.* **61**, 2015–2018 (1988).
7. Novoselov, K. S. *et al.* Electric field effect in atomically thin carbon films. *Science* **306**, 666–669 (2004).
8. Novoselov, K. S. *et al.* Two-dimensional atomic crystals. *Proc. Natl Acad. Sci. USA* **102**, 10451–10453 (2005).
9. Novoselov, K. S. *et al.* Two-dimensional gas of massless Dirac fermions in graphene. *Nature* **438**, 197–200 (2005).
10. Zhang, Y., Tan, J. W., Stormer, H. L. & Kim, P. Experimental observation of the quantum Hall effect and Berry's phase in graphene. *Nature* **438**, 201–204 (2005).
11. Peierls, R. E. Quelques proprietes typiques des corpses solides. *Ann. I. H. Poincare* **5**, 177–222 (1935).
12. Landau, L. D. Zur Theorie der phasenumwandlungen II. *Phys. Z. Sowjetunion* **11**, 26–35 (1937).
13. Landau, L. D. & Lifshitz, E. M. *Statistical Physics, Part I* (Pergamon, Oxford, 1980).
14. Mermin, N. D. Crystalline order in two dimensions. *Phys. Rev.* **176**, 250–254 (1968).
15. Venables, J. A., Spiller, G. D. T. & Hanbucken, M. Nucleation and growth of thin films. *Rep. Prog. Phys.* **47**, 399–459 (1984).
16. Evans, J. W., Thiel, P. A. & Bartelt, M. C. Morphological evolution during epitaxial thin film growth: Formation of 2D islands and 3D mounds. *Sur. Sci. Rep.* **61**, 1–128 (2006).
17. Stankovich, S. *et al.* Graphene-based composite materials. *Nature* **442**, 282–286 (2006).
18. Meyer, J. C. *et al.* The structure of suspended graphene sheets. *Nature* (in the press); doi:10.1038/nature05545.
19. Nelson, D. R., Piran, T. & Weinberg, S. *Statistical Mechanics of Membranes and Surfaces* (World Scientific, Singapore, 2004).
20. Partoens, B. & Peeters, F. M. From graphene to graphite: Electronic structure around the K point. *Phys. Rev. B* **74**, 075404 (2006).
21. Morozov, S. V. *et al.* Two-dimensional electron and hole gases at the surface of graphite. *Phys. Rev. B* **72**, 201401 (2005).
22. Zhang, Y., Small, J. P., Amori, M. E. S. & Kim, P. Electric field modulation of galvanomagnetic properties of mesoscopic graphite. *Phys. Rev. Lett.* **94**, 176803 (2005).
23. Dresselhaus, M. S. & Dresselhaus, G. Intercalation compounds of graphite. *Adv. Phys.* **51**, 1–186 (2002).
24. Shioyama, H. Cleavage of graphite to graphene. *J. Mater. Sci. Lett.* **20**, 499–500 (2001).
25. Viculis, L. M., Mack, J. J., & Kaner, R. B. A chemical route to carbon nanoscrolls. *Science* **299**, 1361 (2003).
26. Horiuchi, S. *et al.* Single graphene sheet detected in a carbon nanofilm. *Appl. Phys. Lett.* **84**, 2403–2405 (2004).
27. Krishnan, A. *et al.* Graphitic cones and the nucleation of curved carbon surfaces. *Nature* **388**, 451–454 (1997).
28. Land, T. A., Michely, T., Behm, R. J., Hemminger, J. C. & Comsa, G. STM investigation of single layer graphite structures produced on Pt(111) by hydrocarbon decomposition. *Surf. Sci.* **264**, 261–270 (1992).
29. Nagashima, A. *et al.* Electronic states of monolayer graphite formed on TiC(111) surface. *Surf. Sci.* **291**, 93–98 (1993).
30. van Bommel, A. J., Crombeen, J. E. & van Tooren, A. LEED and Auger electron observations of the SiC(0001) surface. *Surf. Sci.* **48**, 463–472 (1975).
31. Forbeaux, I., Themlin, J.-M. & Debever, J. M. Heteroepitaxial graphite on 6H-SiC(0001): Interface formation through conduction-band electronic structure. *Phys. Rev. B* **58**, 16396–16406 (1998).
32. Berger, C. *et al.* Ultrathin epitaxial graphite: 2D electron gas properties and a route toward graphene-based nanoelectronics. *J. Phys. Chem. B* **108**, 19912–19916 (2004).

33. Berger, C. *et al.* Electronic confinement and coherence in patterned epitaxial graphene. *Science* **312,** 1191–1196 (2006).

34. Ohta, T., Bostwick, A., Seyller, T., Horn, K. & Rotenberg, E. Controlling the electronic structure of bilayer graphene. *Science* **313,** 951–954 (2006).

35. Ohashi, Y., Koizumi, T., Yoshikawa, T., Hironaka, T. & Shiiki, K. Size effect in the in-plane electrical resistivity of very thin graphite crystals. *TANSO* 235–238 (1997).

36. Bunch, J. S., Yaish, Y., Brink, M., Bolotin, K. & McEuen, P. L. Coulomb oscillations and Hall effect in quasi-2D graphite quantum dots. *Nano Lett.* **5,** 287–290 (2005).

37. Ferrari, A. C. *et al.* Raman spectrum of graphene and graphene layers. *Phys. Rev. Lett.* **97,** 187401 (2006).

38. Gupta, A., Chen, G., Joshi, P., Tadigadapa, S. & Eklund, P. C. Raman scattering from high-frequency phonons in supported n-graphene layer films. *Nano Lett.* **6,** 2667–2673 (2006).

39. Divigalpitiya, W. M. R., Frindt, R. F. & Morrison, S. R. Inclusion systems of organic molecules in restacked single-layer molybdenum disulfide. *Science* **246,** 369–371 (1989).

40. Klein, A., Tiefenbacher, S., Eyert, V., Pettenkofer, C. & Jaegermann, W. Electronic band structure of single-crystal and single-layer WS₂: Influence of interlayer van der Waals interactions. *Phys. Rev. B* **64,** 205416 (2001).

41. Schedin, F. *et al.* Detection of individual gas molecules by graphene sensors. Preprint at http://arxiv.org/abs/cond-mat/0610809 (2006).

42. Novoselev, K. S. *et al.* Room-temperature quantum Hall effect in graphene. *Science* (in the press); doi:10.1126/science.1137201.

43. Schakel, A. M. J. Relativistic quantum Hall effect. *Phys. Rev. D* **43,** 1428–1431 (1991).

44. González, J., Guinea, F. & Vozmediano, M. A. H. Unconventional quasiparticle lifetime in graphite. *Phys. Rev. Lett.* **77,** 3589–3592 (1996).

45. Gorbar, E. V., Gusynin, V. P., Miransky, V. A. & Shovkovy, I. A. Magnetic field driven metal-insulator phase transition in planar systems. *Phys. Rev. B* **66,** 045108 (2002).

46. Katsnelson, M. I. Zitterbewegung, chirality, and minimal conductivity in graphene. *Eur. Phys. J. B* **51,** 157–160 (2006).

47. Katsnelson, M. I, Novoselov, K. S. & Geim, A. K. Chiral tunnelling and the Klein paradox in graphene. *Nature Phys* **2,** 620–625 (2006).

48. Tworzydlo, J., Trauzettel, B., Titov, M., Rycerz, A. & Beenakker, C. W. J. Quantum-limited shot noise in graphene. *Phys. Rev. Lett.* **96,** 246802 (2006).

49. Zawadzki, W. Electron transport phenomena in small-gap semiconductors. *Adv. Phys.* **23,** 435–522 (1974).

50. Luk'yanchuk, I. A. & Kopelevich, Y. Dirac and normal fermions in graphite and graphene: Implications of the quantum Hall effect. *Phys. Rev. Lett.* **97,** 256801 (2006).

51. Zhou, S. Y. *et al.* First direct observation of Dirac fermions in graphite. *Nature Phys.* **2,** 595–599 (2006).

52. Zheng, Y. & Ando, T. Hall conductivity of a two-dimensional graphite system. *Phys. Rev. B* **65,** 245420 (2002).

53. Gusynin, V. P. & Sharapov, S. G. Unconventional integer quantum Hall effect in graphene *Phys. Rev. Lett.* **95,** 146801 (2005).

54. Peres, N. M. R., Guinea, F. & Castro Neto, A. H. Electronic properties of disordered two-dimensional carbon. *Phys. Rev. B* **73,** 125411 (2006).

55. MacDonald, A. H. Quantized Hall conductance in a relativistic two-dimensional electron gas. *Phys. Rev. B* **28,** 2235–2236 (1983).

56. Novoselov, K. S. *et al.* Unconventional quantum Hall effect and Berry's phase of 2π in bilayer graphene. *Nature Phys.* **2,** 177–180 (2006).

57. Mikitik, G. P. & Sharlai, Yu.V. Manifestation of Berry's phase in metal physics. *Phys. Rev. Lett.* **82,** 2147–2150 (1999).

58. McCann, E. & Fal'ko, V. I. Landau-level degeneracy and quantum Hall effect in a graphite bilayer. *Phys. Rev. Lett.* **96,** 086805 (2006).

59. McCann, E. Asymmetry gap in the electronic band structure of bilayer graphene. *Phys. Rev. B* **74,** 161403 (2006).

60. Castro, E. V. *et al.* Biased bilayer graphene: semiconductor with a gap tunable by electric field effect. Preprint at http://arxiv.org/abs/cond-mat/0611342 (2006).

61. Lee, P. A. Localized states in a *d*-wave superconductor. *Phys. Rev. Lett.* **71,** 1887–1890 (1993).

62. Ludwig, A. W. W., Fisher, M. P. A., Shankar, R. & Grinstein, G. Integer quantum Hall transition: An alternative approach and exact results. *Phys. Rev. B* **50,** 7526–7552 (1994).

63. Ziegler, K. Delocalization of 2D Dirac fermions: the role of a broken symmetry. *Phys. Rev. Lett.* **80,** 3113–3116 (1998).

64. Ostrovsky, P. M., Gornyi, I. V. & Mirlin, A. D. Electron transport in disordered graphene. *Phys. Rev. B* **74,** 235443 (2006).

65. Nomura, K. & MacDonald, A. H. Quantum transport of massless Dirac fermions in graphene. Preprint at http://arxiv.org/abs/cond-mat/0606589 (2006).

66. Nilsson, J., Castro Neto, A. H., Guinea, F. & Peres, N. M. R. Electronic properties of graphene multilayers. Preprint at http://arxiv.org/abs/cond-mat/0604106 (2006).

67. Morozov, S. V. *et al.* Strong suppression of weak localization in graphene. *Phys. Rev. Lett.* **97,** 016801 (2006).

68. Aleiner, I. L. & Efetov, K. B. Effect of disorder on transport in graphene. *Phys. Rev. Lett.* **97,** 236801 (2006).

69. Das Sarma, S., Hwang, E. H., Tse, W. K. Is graphene a Fermi liquid? Preprint at http://arxiv.org/abs/cond-mat/0610581 (2006).

70. McCann, E. *et al.* Weak localisation magnetoresistance and valley symmetry in graphene. *Phys. Rev. Lett.* **97,** 146805 (2006).

71. Morpurgo, A. F. & Guinea, F. Intervalley scattering, long-range disorder, and effective time reversal symmetry breaking in graphene. *Phys. Rev. Lett.* **97,** 196804 (2006).

72. Rycerz, A., Tworzydlo, J. & Beenakker, C. W. J. Anomalously large conductance fluctuations in weakly disordered graphene. Preprint at http://arxiv.org/abs/cond-mat/0612446 (2006).

73. Nomura, K. & MacDonald, A. H. Quantum Hall ferromagnetism in graphene. *Phys. Rev. Lett.* **96,** 256602 (2006).

74. Yang, K., Das Sarma, S. & MacDonald, A. H. Collective modes and skyrmion excitations in graphene SU(4) quantum Hall ferromagnets. *Phys. Rev. B* **74,** 075423 (2006).

75. Apalkov, V. M. & Chakraborty, T. The fractional quantum Hall states of Dirac electrons in graphene. *Phys. Rev. Lett.* **97,** 126801 (2006).

76. Khveshchenko, D. V. Composite Dirac fermions in graphene. Preprint at http://arxiv.org/abs/cond-mat/0607174 (2006).

77. Alicea, J. & Fisher, M. P. A. Graphene integer quantum Hall effect in the ferromagnetic and paramagnetic regimes, *Phys. Rev. B* **74,** 075422 (2006).

78. Khveshchenko, D. V. Ghost excitonic insulator transition in layered graphite. *Phys. Rev. Lett.* **87,** 246802 (2001).

79. Abanin, D. A., Lee, P. A. & Levitov, L. S. Spin-filtered edge states and quantum Hall effect in graphene. *Phys. Rev. Lett.* **96,** 176803 (2006).

80. Toke, C., Lammert, P. E., Crespi, V. H. & Jain, J. K. Fractional quantum Hall effect in graphene. *Phys. Rev. B* **74,** 235417 (2006).

81. Zhang, Y. *et al.* Landau-level splitting in graphene in high magnetic fields. *Phys. Rev. Lett.* **96,** 136806 (2006).

82. Schliemann, J., Loss, D. & Westervelt, R. M. Zitterbewegung of electronic wave packets in III-V zinc-blende semiconductor quantum wells. *Phys. Rev. Lett.* **94,** 206801 (2005).

83. Topinka, M. A., Westervelt, R. M. & Heller, E. J. Imaging electron flow. *Phys. Today* **56,** 47–53 (2003).

84. Cortijo, A. & Vozmediano, M. A. H. Effects of topological defects and local curvature on the electronic properties of planar graphene. *Nucl. Phys. B* **763,** 293–308 (2007).

85. Nakada, K., Fujita, M., Dresselhaus, G. & Dresselhaus, M. S. Edge state in graphene ribbons: Nanometer size effect and edge shape dependence. *Phys. Rev. B* **54,** 17954–17961 (1996).

86. Brey, L. & Fertig, H. A. Electronic states of graphene nanoribbons. *Phys. Rev. B* **73,** 235411 (2006).

87. Son, Y.W, Cohen, M. L. & Louie, S. G. Energy gaps in graphene nanoribbons *Phys. Rev. Lett.* **97,** 216803 (2006).

88. Tilke, A. T., Simmel, F. C., Blick, R.H, Lorenz, H. & Kotthaus, J. P. Coulomb blockade in silicon nanostructures. *Prog. Quantum Electron.* **25,** 97–138 (2001).

89. Takahashi, Y., Ono, Y., Fujiwara, A. & Inokawa, H. Silicon single-electron devices. *J. Phys. Condens. Matter* **14,** R995–R1033 (2002).

90. Tseng, A. A., Notargiacomo A. & Chen T. P. Nanofabrication by scanning probe microscopy lithography: A review. *J. Vac. Sci. Tech. B* **23,** 877–894 (2005).

91. Hill, E. W., Geim, A. K., Novoselov, K., Schedin, F. & Blake, P. Graphene spin valve devices. *IEEE Trans. Magn.* **42,** 2694–2696 (2006).

92. Heersche, H. B., Jarillo-Herrero, P., Oostinga, J. B., Vandersypen, L. M. K. & Morpurgo, A. F. Bipolar supercurrent in graphene. *Nature* (in the press); doi:10.1038/nature05555.

93. Sofo, O., Chaudhari, A. & Barber, G. D. Graphane: a two-dimensional hydrocarbon. Preprint at http://arxiv.org/abs/cond-mat/0606704 (2006).

Acknowledgements

We are most grateful to Irina Grigorieva, Alberto Morpurgo, Uli Zeitler, Antonio Castro Neto and Allan MacDonald for many useful comments that helped to improve this review. The image of crumpled graphene on the cover of this issue was kindly provided by Jannik Meyer. The work was supported by EPSRC (UK), the Royal Society and the Leverhulme trust.

Multiferroics: progress and prospects in thin films

Multiferroic materials, which show simultaneous ferroelectric and magnetic ordering, exhibit unusual physical properties — and in turn promise new device applications — as a result of the coupling between their dual order parameters. We review recent progress in the growth, characterization and understanding of thin-film multiferroics. The availability of high-quality thin-film multiferroics makes it easier to tailor their properties through epitaxial strain, atomic-level engineering of chemistry and interfacial coupling, and is a prerequisite for their incorporation into practical devices. We discuss novel device paradigms based on magnetoelectric coupling, and outline the key scientific challenges in the field.

R. RAMESH[1] AND NICOLA A. SPALDIN[2]

[1]Department of Materials Science and Engineering, and Department of Physics, University of California, Berkeley, California 94720, USA.
[2]Materials Department, University of California, Santa Barbara, California 93106, USA.

e-mail: rramesh@berkeley.edu; nicola@mrl.ucsb.edu

The past few years have seen a tremendous flurry of research interest in materials that show simultaneous ferromagnetic order (or any other kind of magnetic) and ferroelectric ordering (Fig. 1). Such 'magnetoelectric multiferroics' were studied to some degree in the 1960s and 1970s (ref. 1) but then languished, in large part because single-phase materials with both properties could not be widely produced. The renaissance of magnetoelectric multiferroics[2–4] has been fuelled by developments in, and collaborations between, many areas of theory and experiment. First, the production of high-quality single-crystalline samples, in some cases through high-pressure routes, has led to the identification of new types of multiferroics[5,6] with fundamentally new mechanisms for ferroelectricity. Second, improved first-principles computational techniques have aided in the design of new multiferroics, and provided understanding of the factors that promote coupling between magnetic and ferrolectric order parameters[7,8]. Finally, advances in thin-film growth techniques have provided routes to structures and phases that are inaccessible by traditional chemical means, and have allowed the properties of existing materials to be modified by strain engineering[9]. This availability of high-quality thin-film samples, in conjunction with a broad spectrum of analytical tools, has improved our ability to accurately characterize multiferroic behaviour, and has opened the door to the design of practical devices based on magnetoelectric coupling. In our opinion, these developments will prevent a recurrence of the downturn in interest that occurred in the 1970s, as thin-film multiferroics begin to reveal a range of fascinating phenomena as well as stimulate exploration of new device heterostructures.

Here we review recent progress in, and future prospects for, thin-film, heterostructure and nanostructure multiferroics; the accompanying review by Cheong and Mostovoy (page 13 of this issue) focuses on the corresponding bulk materials[10]. We discuss three types

of thin-film architectures (Fig. 2): single-phase thin films, in which heteroepitaxial strain and crystal chemistry are the key variables in controlling and improving magnetoelectric coupling; horizontal heterostructures, in which the principles of heteroepitaxy at the interfaces can be used to control and initiate magnetoelectric coupling at the atomic scale; and nanoscale 'vertical heterostructures', in which coupling occurs through vertical heteroepitaxy. The remainder of this review is organized as follows. First, we review the basic chemistry underlying the scarcity of magnetic ferroelectrics and the routes to combining magnetism and ferroelectricity that have overcome their contra-indication. Then we focus on multiferroics that are based on the three types of model thin-film architectures shown in Fig. 2. Finally, we discuss device architectures based on magnetoelectric coupling in thin films, before summarizing the prospects for research and technology in thin-film multiferroics.

CONTRA-INDICATION BETWEEN MAGNETISM AND FERROELECTRICITY

The scarcity of ferromagnetic ferroelectrics[7] is now well understood to result from the contra-indication between the conventional mechanism for cation off-centring in ferroelectrics (which requires formally empty d orbitals), and the formation of magnetic moments (which usually results from partially filled d orbitals)[3]. For ferroelectricity and magnetism to coexist in a single phase, therefore, the atoms that move off centre to form the electric dipole moment should be different from those that carry the magnetic moment. In principle, this could be achieved through either an alternative (non-d-electron) mechanism for magnetism, or through an alternative mechanism for ferroelectricity; in practice only the latter route has been pursued, and the exploration of multiferroics with different forms of magnetism is an open area for future research. In the magnetic perovskite-structure oxides and related materials, multiferroism is most commonly achieved by making use of the stereochemical activity of the lone pair on the large (A-site) cation to provide the ferroelectricity, while keeping the small (B-site) cation magnetic. This is the mechanism for ferroelectricity in the Bi-based magnetic ferroelectrics, the most widely studied of which is bismuth ferrite, $BiFeO_3$ (ref. 11). (Note that $BiFeO_3$ adopts the perovskite, not the 'ferrite' structure, in spite

When citing this review, please cite the original version as shown on the contents page of this chapter.

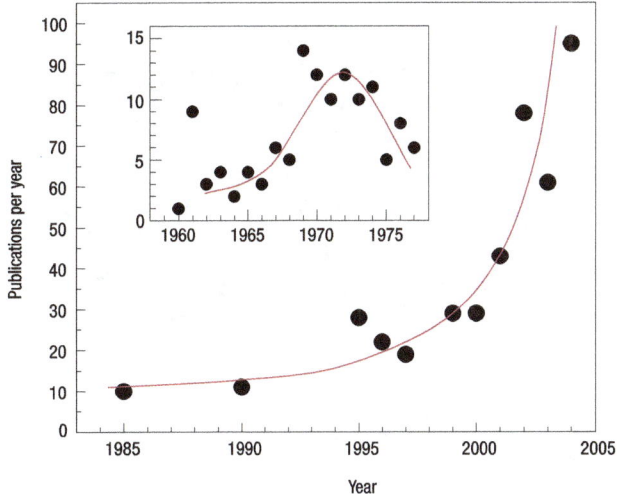

Figure 1 Publications per year with 'magnetoelectric' as a keyword. Reprinted with permission from ref. 2. Copyright (2005) IOP.

of its nomenclature.) A second route to multiferroism is provided by 'geometrically driven' ferroelectricity, which is compatible with the coexistence of magnetism; the antiferromagnetic ferroelectrics YMnO$_3$ (refs 12,13) and BaNiF$_4$ (ref. 14) fall into this class. A particularly appealing, recently identified mechanism occurs in TbMnO$_3$, in which ferroelectricity is induced by the formation of a symmetry-lowering magnetic ground state that lacks inversion symmetry[5]. The resulting polarization is small, but because it is caused directly by the magnetic ordering, strong and possibly new magnetoelectric interactions should be expected. A search for the opposite effect — weak ferromagnetism that is induced by a symmetry-lowering ferroelectric distortion — is under way[15,16]. Finally, certain non-centrosymmetric charge-ordering arrangements can cause ferroelectricity in magnetic materials; here LuFe$_2$O$_4$ has generated recent attention[6,17]. We also point out that, to be ferroelectric, a material must be insulating (otherwise the mobile charges would screen out the electric polarization). This is an additional constraint, because many ferromagnets tend to be metallic, with most magnetic insulators having antiferromagnetic ordering. The need for insulating behaviour can also cause problems if samples are leaky, as this can suppress ferroelectric behaviour even if the structure is non-centrosymmetric. This is a common problem in the case of magnetic ferroelectrics, because magnetic transition metal ions are often able to accommodate a wider range of valence states than their diamagnetic counterparts, leading in turn to non-stoichiometry and hopping conductivity.

SINGLE-COMPONENT THIN-FILM MULTIFERROICS

Although the four routes to multiferroism described in the previous section have led to the identification of an array of new multiferroics, most of these have only been prepared in single crystal, ceramic or powder form. To our knowledge only two classes of single-phase multiferroics — the hexagonal manganites and the Bi-based perovskites — have been prepared as single-phase films. We summarize the literature here, highlighting where the properties of the thin films are fundamentally different from those of the corresponding bulk samples.

Perhaps the first multiferroic to be investigated in thin-film form[18] was the hexagonal manganite YMnO$_3$, which is appealing

because its geometric ferroelectricity leads to a uniaxial polarization perpendicular to the plane of the film. The first study[18] used radiofrequency magnetron sputtering to obtain epitaxial (0001) films on (111)MgO and (0001)ZnO/(0001)sapphire, and polycrystalline films on (111)Pt/(111)MgO. Soon after, it was shown[19] that the metastable, non-ferroelectric cubic perovskite structure could be stabilized in thin films using appropriate conditions and substrates. Subsequently, YMnO$_3$ films have been grown on a range of substrates (Si(100), Pt/TiO$_2$/SiO$_2$/Si, Pt(111)/Ti/SiO$_2$/Si, (111)Pt/ (0001)sapphire, SiO$_2$-buffered silicon, GaN-on-sapphire) using a range of techniques (sputtering, spin coating, sol–gel processes, pulsed laser deposition, metal-organic chemical vapour deposition (MOCVD) and molecular beam epitaxy (MBE))[20–27]. Although the thin-film and bulk properties are qualitatively similar, the thin-film samples show the usual reduction in ferroelectric polarizations and dielectric response compared with the corresponding single-crystal values. Preparation and characterization of other rare-earth hexagonal manganites in thin-film form, including those that form the competing perovskite phase in the bulk[28] (S. B. Ogale, personal communication), are under way.

Perovskite-structure bismuth ferrite is currently the most studied single-component multiferroic, in part because its large polarization and high Curie temperature (~820 °C) make it appealing for applications in ferroelectric non-volatile memories and high-temperature electronics. An early bulk sample[29] yielded a small polarization value; measured polarizations for thin films grown by a variety of techniques[11,30–33] initially showed a large spread of values (for a summary table see ref. 34). However, the measured values are now converging to ~90 μC cm^{-2} along the [111] direction of the pseudo-cubic perovskite unit cell, consistent with first-principles calculations[34]. The large difference between the thin-film and bulk values, initially attributed to epitaxial strain[11], could also result from leakage effects in crystals caused by defect chemistry or the existence of second phases, or mechanical constraints in granular bulk ceramics. There are also intriguing experimental and theoretical reports of a tetragonal phase, with $c/a = 1.26$ and an even larger ferroelectric polarization of ~150 μC cm^{-2} (refs 35,36); this is yet to be confirmed by independent reports.

The magnetic properties of BiFeO$_3$ thin films can also be markedly different from those of the bulk. BiFeO$_3$ has long been known in its bulk form to be an antiferromagnet with Néel temperature $T_N \approx 643$ K (ref. 37). The Fe magnetic moments are coupled ferromagnetically within the pseudo-cubic (111) planes and antiferromagnetically between adjacent planes. If the magnetic moments are oriented perpendicular to the [111] direction, as predicted by first-principles calculations[15], the symmetry also permits a canting of the antiferromagnetic sublattices resulting in a macroscopic magnetization — 'weak ferromagnetism'[38,39]. However, superimposed on the antiferromagnetic ordering, there is a spiral spin structure in which the antiferromagnetic axis rotates through the crystal with an incommensurate long-wavelength period of ~620 Å (ref. 40). This spiral spin structure leads to a cancellation of any macroscopic magnetization and also inhibits the observation of the linear magnetoelectric effect[41]. However, a significant magnetization (~0.5μ_B per unit cell) and a strong magnetoelectric coupling have been observed in epitaxial thin films[11], suggesting that the spiral spin structure could be suppressed[42]. More work is needed to characterize the magnetic behaviour of BiFeO$_3$ thin films fully — direct experimental confirmation of the spin structure would be highly desirable — and the origin of the large magnetization in ultrathin, highly strained BiFeO$_3$ (refs 11,43) is still unclear.

Finally, we mention that there has been some progress in integrating BiFeO$_3$ with silicon, desirable for Si-CMOS (complementary metal oxide semiconductor) electronics applications, and with GaN, for high-temperature electronics

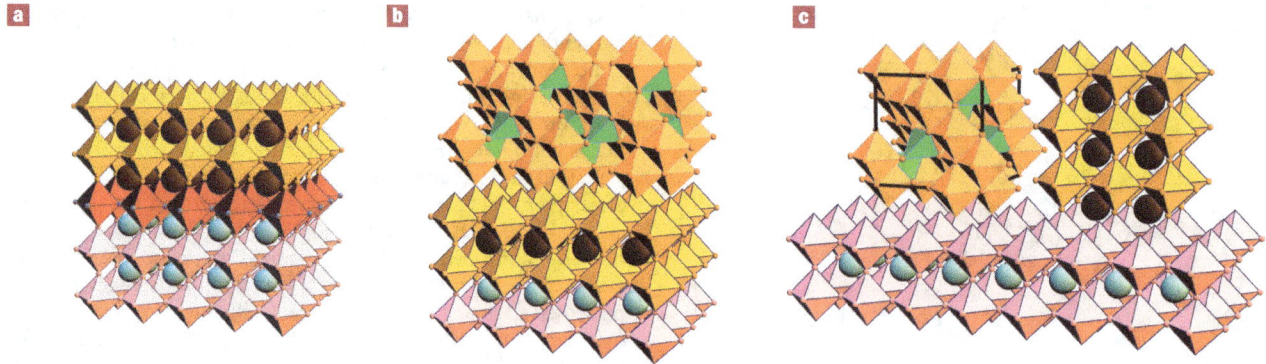

Figure 2 Schematics of the three types of model thin-film architectures reviewed here. **a**, Single-phase, epitaxial films grown on single-crystal substrates; **b**, horizontal heterostructures in which a magnetic phase is epitaxially interleaved with a ferroelectric (piezoelectric) phase to create an engineered magnetoelectric; **c**, the vertical analogue of **b**, in which nanopillars (or nanodots) of one phase are embedded epitaxially in a matrix of the other phase.

applications. Epitaxial growth on silicon has been helped by the use of $SrTiO_3$ as a template[44] and a similar approach has been used to grow epitaxial $BiFeO_3$ on GaN. However, detailed characterization of the domain structure and dynamics, and reduction of the leakage current density and the coercive field, are critical challenges that need to be addressed before $BiFeO_3$-based films will be candidates for integrated microelectronic devices such as storage elements in non-volatile ferroelectric memories.

$BiFeO_3$ is unusual among the Bi-based perovskites in that it can be made under ambient conditions; most others have traditionally needed high-pressure synthesis[45]. Recently, the stabilization of perovskite $BiMnO_3$ through thin-film growth[46–48] has shown the usefulness of thin-film techniques in accessing Bi-based perovskites without high-pressure apparatus. Unlike $BiFeO_3$, the unusual orbital ordering in $BiMnO_3$ (refs 49,50) leads to ferromagnetism below ~105 K; thin-film $BiMnO_3$ is therefore being studied as a potential barrier material for magnetically and electrically controllable tunnel junctions[51]. Optical second-harmonic measurements in the presence of an electric field[47] and the writing of polarization bits using a Kelvin force microscope[52] clearly identify the existence of polarization in $BiMnO_3$ films. But the high levels of leakage have prevented a direct transport measurement of ferroelectric response. Interestingly, the reported ferroelectricity in the bulk samples[53] has been called into question by a reanalysis of the diffraction data[54] and by density functional calculations (P. Baettig and N. A. Spaldin, unpublished results). First-principles calculations for the experimentally reported structures[55] indicate a small polar canting of an otherwise antiferroelectric structure (a 'weak ferroelectricity') whereas full structural optimizations find the nonpolar antiferroelectric structure to be lower in energy. Further work on $BiMnO_3$ to resolve this issue, and to control and mitigate leakage effects in films, is certainly to be encouraged, as this is one of the most promising materials for stabilizing a robust ferromagnetic state coupled with ferroelectricity.

Finally, we note that a range of other multiferroics with lone-pair-active A-sites and magnetic transition metal B-sites are now being explored in thin-film form. Thin-film $BiCrO_3$ was recently reported on a range of substrates[56], and the newly discovered 'super-tetragonal' ferroelectric $PbVO_3$ (ref. 57) has been stabilized in thin-film form (L. W. Martin *et al.*, manuscript in preparation). Clearly many potential combinations of lone pairs and transition metals remain to be explored, but in all cases, the stoichiometry must be carefully controlled so that the transition metal *d* electrons provide magnetism without also contributing to electronic transport.

In addition to modifying properties through size quantization (such as the suppression of the spin spiral in $BiFeO_3$) and stabilizing

phases that are inaccessible through conventional bulk techniques, thin-film growth techniques provide the ability to vary the lattice mismatch between the film and the substrate, so as to introduce epitaxial strain in the thin-film material. In conventional ferroelectrics, strain effects can lead to a substantial increase of the spontaneous polarization and Curie temperature[58], and can even drive paraelectric materials such as $SrTiO_3$ into the ferroelectric phase[59]. Because the mechanism for ferroelectricity in multiferroic materials can differ from that in conventional perovskite ferroelectrics, it is unclear whether similar strain effects should be expected in thin-film multiferroics. Indeed, first-principles density functional calculations for multiferroic $BiFeO_3$ showed[36,60] that the strain dependence of the ferroelectric polarization in $BiFeO_3$ is rather weak compared with conventional ferroelectric materials, and that this is due to the high stability of the ferroelectric state in $BiFeO_3$, indicated by the large ionic displacements from the centrosymmetric reference structure and the high ferroelectric Curie temperature. As a result, the changes in relative ionic positions when the lattice is strained are small. Although not directly related to the multiferroicity, a similar weak strain dependence should be expected in other Bi-based multiferroics, in which the stereochemically active Bi^{3+} lone pairs cause high critical temperatures and large polarizations. This indicates that for strain effects to be used in engineering the polarization of new multiferroics, softer ferroelectrics that are closer to the ferroelectric instability would have to be used.

Larger qualitative effects can be expected when heteroepitaxial constraints lead to changes in crystal symmetry, as the symmetry determines the easy axes of magnetization, and the presence or absence of spin canting. For example, $BiFeO_3$ undergoes a reduction from rhombohedral to monoclinic symmetry when grown on a [100] $SrTiO_3$ substrate. This should lead to profound changes in the magnetoelectric switching behaviour[61] as follows. In rhombohedral $BiFeO_3$, the spontaneous electric polarization is directed along one of the eight [111] axes of the perovskite structure and is accompanied by a ferroelastic strain resulting from the distortion of the lattice. The preferred orientation of the antiferromagnetically aligned spins (in the absence of the long-wavelength spiral mentioned above) is in the (111) plane perpendicular to the rhombohedral axis, with six equivalent easy axes within the plane[15]. The orientation of the antiferromagnetic sublattice magnetization is therefore coupled to the ferroelectric polarization through the ferroelastic strain state of the system. Switching of the polarization by either 109° or 71° changes the rhombohedral axis of the system and is therefore accompanied by a switching of the ferroelastic domain state, which in turn changes the orientation of the easy magnetization plane[61]. In monoclinic BiFeO$_3$, however, first-principles calculations show that the six-fold degeneracy of the easy plane is lifted, and an orientation of the antiferromagnetic axis parallel to the [1$\bar{1}$0] direction is preferred, that is, perpendicular to the rhombohedral axis but simultaneously parallel to the (001) plane. Because the [1$\bar{1}$0] direction is perpendicular to both the [111] and [11$\bar{1}$] directions but not perpendicular to [$\bar{1}$1$\bar{1}$], in this case we expect the antiferromagnetic axis to change during the 109° ferroelectric switching but not during the 71° switching. Experimental studies are in progress to clarify the role of such strains in magnetoelectric coupling. In epitaxial thin films and heterostructures, we believe this will be a fruitful direction for research as we endeavour to understand the interactions and coupling between the order parameters.

The question of a fundamental size limit for ferroelectricity is currently of considerable interest[62], both in terms of understanding the basics of cooperative ferroelectric behaviour, and in response to the continued drive for miniaturization of electronic devices. Indeed it was long believed, from empirical evidence and phenomenological arguments, that there should be a critical size of the order of hundreds of ångströms below which a spontaneous electric polarization could not be sustained in a material (for a review of the early literature, see ref. 63). But more recent experimental work[64,65], corroborated by first-principles calculations[66,67], has indicated that the critical size in conventional ferroelectrics is orders of magnitude smaller than previously thought, and that ferroelectricity can persist down to vanishingly small sizes. We now need similar studies on multiferroics, in which the mechanism for ferroelectricity is unconventional, and in which coupling to the magnetism could affect the size dependence. Likewise, the size dependence of the magnetic ordering in multiferroics (be it ferro-, antiferro- or a more complicated state of order), and the behaviour of the magnetoelectric coupling with film thickness, remain to be explored.

HORIZONTAL MULTILAYER HETEROSTRUCTURES

The ability to grow high-quality films with precisely controlled composition, atomic arrangements and interfaces has added to the toolbox for the creation of new functional materials, and holds particular promise for the rational design of new multiferroics. A number of routes to new multiferroics have been pursued; here we review progress in the use of atomic-level layering to engineer specific magnetic ordering through superexchange interactions, and the use of compositional ordering that breaks inversion symmetry to increase polarizations. We also outline new ideas regarding interfacial multiferroism, and describe some early demonstrations of electric field effects using multiferroics. Horizontal heterostructures consisting of alternating layers of conventional ferromagnets and ferroelectrics[68,69] have been reviewed elsewhere[70].

Motivated by the report[71] of the possibility of single-atomic-layer superlattice ordering in double perovskite La$_2$FeCrO$_6$, Baettig and Spaldin[72] proposed the analogous Bi-based compound as a potential multiferroic material. Their density functional calculations for the (111) ordered structure indicated an $R3c$ ground state with a polarization of ~80 μC cm^2, and ferrimagnetic ordering with a magnetization of ~160 e.m.u. cm^3 ($2\mu_B$ per formula unit). The expansion of the set of multiferroics to include such ferrimagnets is appealing from the perspective of obtaining the robust insulating behaviour required to sustain a ferroelectric polarization, because many ferromagnetic materials are metallic, whereas magnetic insulators tend to show antiferromagnetic coupling between the magnetic ions. Although there has been considerable progress on the growth of a range of layered double perovskites (see for example refs 73 and 71), work on Bi-based double layered perovskites is in its infancy. A number of attempts are being made to synthesize Bi$_2$FeCrO$_6$, although achieving the required (111) layering of the

Fe^{3+} and Cr^{3+} ions is challenging because of their similar size and charge. More promising are Bi-based double perovskites with B-site cations of different charge, such as Bi$_2$MnNiO$_6$ which has so far been synthesized only at high pressure[74].

The advent of MBE and related growth tools was an important step in our ability to create atomically sharp heterostructures with interfacial behaviour that is entirely different from that of the constituent materials. One example is the recent demonstration[75] of new electronic properties at the interface between a band insulator and a Mott insulator in heteroepitaxial perovskites.

Importantly for the study of ferroelectrics, the use of layer-by-layer growth to induce compositional ordering that breaks inversion symmetry has produced new heterostructures with higher polarization and large nonlinear optical response. The prototype, first predicted using density functional theory computations[76], is the layering of CaTiO$_3$, BaTiO$_3$ and SrTiO$_3$ in A–B–C–A–B–C... arrangements to lift the inversion centre[77,78]. This approach has been suggested as a new route to multiferroics, as it circumvents the contra-indication between magnetism and ferroelectricity by constraining the magnetic ions in a polar arrangement in spite of their natural tendency to remain centrosymmetric (A. J. Hatt and N. A. Spaldin, manuscript in preparation).

Several interesting interfacial magnetoelectric effects have been proposed theoretically and are particularly exciting areas for future experimental study. Duan et al.[79] have used first-principles density functional theory to demonstrate a magnetoelectric effect in a ferromagnetic/ferroelectric heterostructure that arises from a purely electronic mechanism, not mediated by strain. Using a model Fe/BaTiO$_3$ horizontal superlattice, they showed that the displacement of atoms at the interface due to the ferroelectric instability changes the overlap between atomic orbitals at the interface, which in turn influences the magnetization of the magnetic layer resulting in a magnetoelectric effect. Such effects, they argue, should be possible in both vertical and horizontal heterostructures. In addition, Stengel et al.[80] used first-principles calculations of metal–insulator heterostructures in the presence of a finite electric field to show that a polarized dielectric induces an electric polarization in the adjacent metal; if the metal is ferromagnetic this should result in a region of coexisting electrical polarization and magnetization at the interface.

Early experimental studies on heterostructures comprising a ferro- or piezoelectric and a carrier-mediated magnet (such as a diluted magnetic semiconductor[82] or a double-exchange manganite[82,83]) suggest the possibility of artificially engineered multiferroics in which the coupling is mediated through an electric field effect at the interface. The prototypical examples of such 'field-effect multiferroic heterostructures' are Co-doped TiO$_2$/PZT (ref. 82), and lanthanum calcium manganite/PZT (ref. 83). In both cases, the piezoelectric PZT influences the magnetic behaviour by modulating the carrier density in the magnetic layer through a conventional electric field effect. Much more detailed work needs to be done in this promising area. In particular, the origin of magnetism in Co-doped TiO$_2$ is a controversial topic, with a considerable spread of experimental data in the published literature.

Finally, we mention that the field of horizontal heterostructure multiferroics is starting to benefit immensely from the use of a variety of surface-sensitive electronic probes such as angle-resolved photoemission (ARPES). An emerging area of research involves the probing of surface and near-surface interface electronic structure in situ. For this, the deposition process must be connected in situ to the probe; although this is common in semiconductor heteroepitaxy, it is only just evolving in complex oxide heteroepitaxy. Such systems, shown schematically in Fig. 3, are currently being designed in several laboratories around the world, and promising preliminary results are emerging[84].

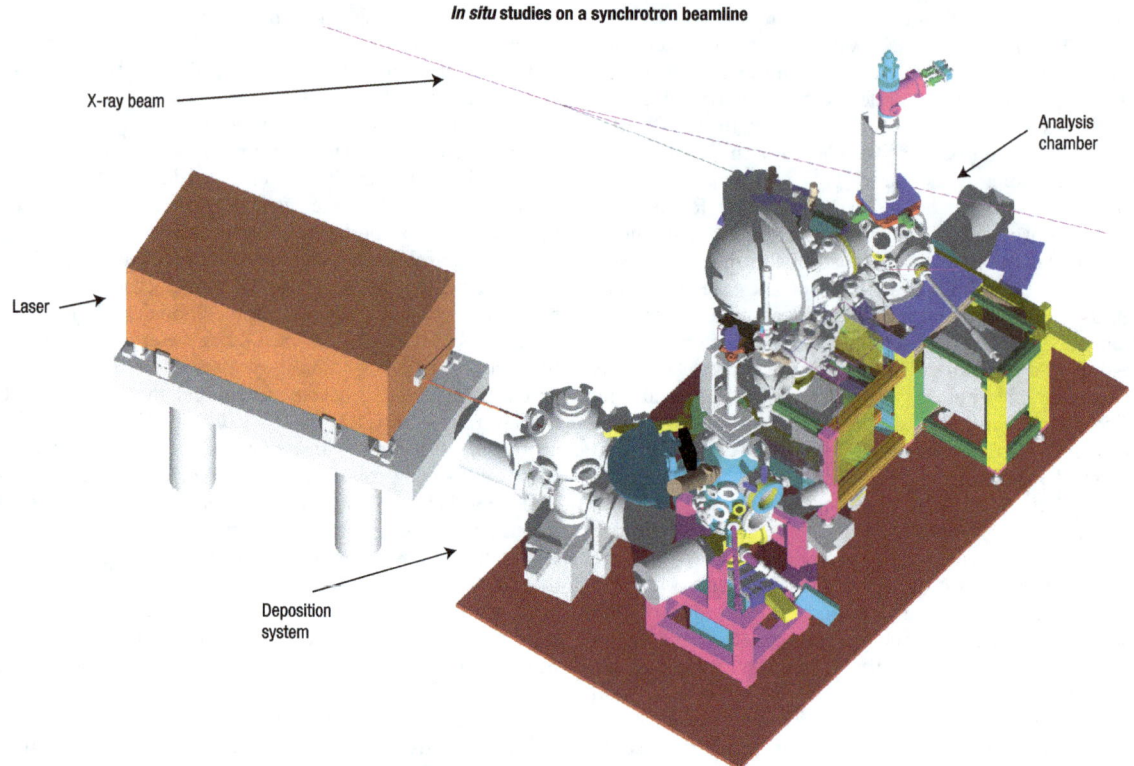

In situ studies on a synchrotron beamline

X-ray beam

Laser

Analysis chamber

Deposition system

Figure 3 A schematic illustration of a thin-film deposition system (in this case a laser MBE system) that is attached to a synchrotron beamline. Such a system makes it easier to probe the electronic structure of surfaces and interfaces as the heterostructure is being grown. In the field of complex oxides, such *in situ* facilities are just emerging.

VERTICAL HETEROSTRUCTURES

Vertical heterostructures such as the nanopillar geometry shown in Fig. 2c offer numerous advantages over the conventional 'horizontal' heterostructures discussed above. First, they have a larger interfacial surface area and are intrinsically heteroepitaxial in three dimensions; this should allow for stronger coupling between ferroelectric and magnetic components. In addition, substrate-imposed mechanical clamping, which is known to suppress both the piezoelectric response and the magnetoelectric coupling mediated by lattice deformation in thin-film-on-substrate geometries, should be reduced in the vertical architecture[85,86]. The prototypical multiferroic vertical nanostructure consists of a magnetic spinel phase which is epitaxially embedded into the ferroelectric matrix. The first example[87] contained $CoFe_2O_4$ pillars in a $BaTiO_3$ matrix, but many different combinations of perovskites ($BaTiO_3$, $PbTiO_3$, $BiFeO_3$ and $SrTiO_3$) and spinels ($CoFe_2O_4$, $NiFe_2O_4$ and Fe_3O_4) have since been grown. However, the design and control of such heterostructures remains a challenge.

Recent work has demonstrated very strong magnetoelectric coupling in such nanostructures, through switching of the magnetization on reversal of the ferroelectric polarization[88]. Detailed studies in progress suggest that the switching is mediated by strong mechanical coupling between the two lattices, which leads to a time-dependent modulation of the magnetic anisotropy in the nanopillar. The strength of the coupling suggests a key role for heteroepitaxy, but does not lead to controllable switching of the magnetization; control can only be achieved if an additional weak magnetic field is superimposed to lift the time-reversal symmetry.

This approach of using vertically heteroepitaxial nanostructures is relatively new, and much research is still needed to understand the nature of the heteroepitaxy as well as the coupling mechanisms. In terms of nanostructure assembly, controlling the degree of order among the nanopillars is possibly the most interesting, albeit difficult problem. Indeed, this has been an ongoing quest in the field of semiconductor quantum dot nanostructures, with recent progress also driven by directed assembly processes[89]. The formation of self-assembled, vertical nanostructures with long-range ordering will undoubtedly have great impact, not only in the field of multiferroics, but in a broad range of photonic applications. In addition, many open questions regarding the magnetoelectric coupling remain: what is the timescale of the coupling process? How does the time dependence of the magnetism relate to the ferroelectric switching? What is the smallest value of magnetic field that can lead to full switching of the magnetization? How does the behaviour depend on the chemistry of the nanopillars? Is there a critical dimension below which the coupling will disappear or change in nature? These areas are ripe for future research.

COUPLING PHENOMENA AND DEVICE STRUCTURES

Given the successful incorporation of magnetic and ferroelectric materials into a variety of technologies, the question arises: what are the applications that are uniquely made possible through multiferroic materials and/or magnetoelectricity? Earlier reviews[90,91] outlined many possible applications of bulk multiferroics, with a strong emphasis on high-frequency devices such as filters and oscillators that could be tuned by magnetic fields. Indeed, recent work on bulk multilayers[92,93] indicates their promise in electrically tuneable microwave applications such as filters, oscillators and phase shifters (in which the ferri-, ferro- or antiferro-magnetic

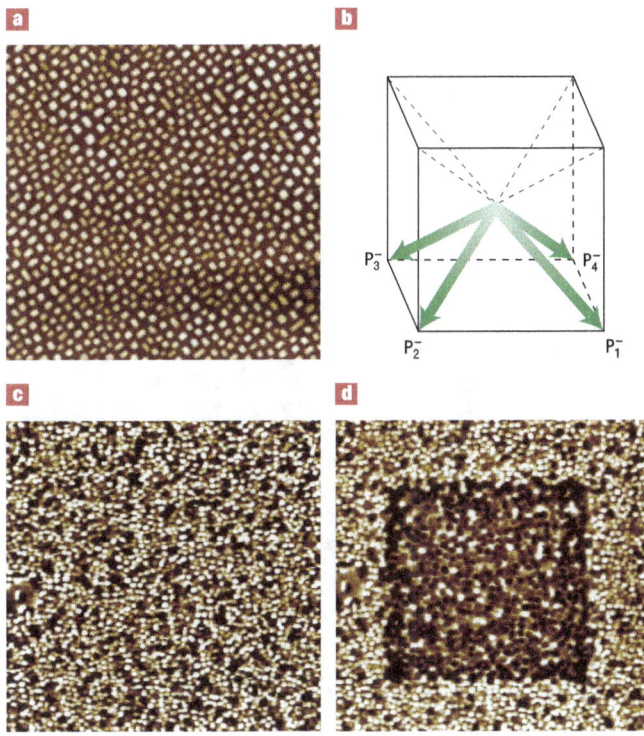

Figure 4 Thin-film multiferroic nanostructures. **a**, AFM image of the surface of self-assembled $BiFeO_3$–$CoFe_2O_4$ nanostructures in which the $CoFe_2O_4$ nanopillars appear in bright contrast; **b**, schematic illustrating the four possible polarization directions in the matrix $BiFeO_3$ phase; **c**, magnetic force microscopy image of the nanostructure that has been pre-magnetized in a magnetic field of 2 T. Note the bright contrast in all the nanopillars, indicating that they are all magnetized in the same direction; the magnetic coercive field of the nanopillars is ~3 kOe (not shown); **d**, MFM image of the same region as in c, after the application of a 16 V d.c. field to the central boxed area along with a small magnetic field of about 700 Oe. Note that all nanopillars have switched their magnetization direction to the black state. Reprinted from ref. 88. Copyright (2005) American Chemical Society.

resonance is tuned electrically instead of magnetically), and as low-frequency, high-resolution magnetic field sensors[94]. Coupling in this case is believed to result from the macroscopic mechanical coupling of the two materials at the bonded interface. In such bilayers, coupling coefficients (which are proportional to the magnetoelectric susceptibilities) as large as 1–5 Oe cm kV^{-1} at microwave frequencies have been reported; these are an order of magnitude higher than in polycrystalline composites, pointing to the key role of crystallographic orientation control in optimizing the magnetoelectric response.

In this section we discuss recent progress in the development of thin-film multiferroic devices, and illustrate the potential of multiferroic/magnetoelectric thin films and heterostructures with a few key examples. Although less advanced than the bulk multilayers mentioned above, large coupling effects have been observed in thin films; for example in nanopillar heterostructures, switching of the ferroelectric polarization by an applied electric field leads to a reversal of the magnetization direction (Fig. 4)[88]. We focus specifically on approaches by which magnetic responses can be controlled through the application of an electric field in thin-film heterostructures and nanostructures. Although the number of studies has been small, we expect this to change in the immediate future, as the

integration of thin-film heterostructure devices with conventional silicon electronics provides impetus. Of course, the converse effect, tuning electrical behaviour with a magnetic field, is also an enticing possibility, although the need to generate and apply large magnetic fields (of several tesla) is an issue that must be resolved before this approach is practical.

First we review experimental methods for measuring magnetoelectric coupling in thin films. The classical method for probing magnetoelectric coupling is to measure the magnetoelectric response, $\partial P/\partial H$ or $\partial M/\partial E$ (where P is electric polarization, H is magnetic field, M is magnetization and E is electric field), directly as a function of temperature; this is analogous to the measurement of dielectric or magnetic susceptibilities. In thin films, however, such measurements are often complicated by leakage in the dielectric and other parasitic effects. Furthermore, the magnetoelectric coefficients are often small, leading to response signals that are only a few nanovolts and therefore require lock-in detection. Because of these difficulties, a variety of approaches has been developed to probe magnetoelectric coupling, many of which offer high spatial resolution; these approaches take advantage of the diversity of interactions between magnetoelectric materials and a broad spectrum of electromagnetic radiation (X-rays, visible optics, infrared optics, microwave and millimetre waves). A particularly important development is the evolution of chemical and magnetic state-specific imaging using photo-excited electrons in photoemission electron microscopy (PEEM). State-of-the-art PEEM systems have spatial resolution of the order of 20–50 nm, and with the PEEM 3 microscope at the Advanced Light Source at Lawrence Berkeley Laboratory, the possibility of 10-nm resolution is imminent. This technique has been used successfully by Scholl and co-workers[95,96] to probe the exchange bias coupling between ferromagnetic cobalt and antiferromagnetic $LaFeO_3$. Using a combination of X-ray magnetic circular dichroism PEEM to probe the domain structure of the ferromagnet, and linear dichroism PEEM to probe the antiferromagnet, they have demonstrated the existence of exchange bias coupling in this system. A similar approach in our laboratory will explore the coupling between a range of ferro- and ferrimagnets (Co, $CoFe_2O_4$, $NiFe_2O_4$) and ferroelectric $BiFeO_3$. Magneto-optic measurements in the visible range using the Kerr effect (MOKE) provide another route to exploring magnetoelectric coupling in thin films. These studies are conceptually similar to conventional MOKE measurements, but the modulation of the MOKE response is through an electric field. In addition, ferro- or antiferromagnetic resonance spectroscopy is a powerful technique for probing coupling effects at microwave or millimetre wave frequencies, and, as mentioned above, presents the most obvious route to practical applications.

Next we discuss two routes to controlling the magnetic behaviour using electric fields: through modifying the amount or direction of spin canting, and through influencing the exchange bias coupling. A common feature of antiferromagnetic oxides, including the multiferroic perovskites, is the presence of canted magnetism on the transition metal sublattice. Such a canting arises primarily from the Dzyaloshinskii–Moriya interaction[38,39], and its existence and magnitude are determined by the symmetry of the crystal and the strength of the spin–orbit interaction respectively. Typical canting angles are ~0.5°, which can lead to weak ferromagnetism with modest remanent magnetization of around 1 to 10 e.m.u. cm^{-3}. Because the direction of the canting is determined by the symmetry of the crystal, it has been suggested[15] that it should be possible to find multiferroics in which reversal of the ferroelectric polarization using an electric field causes a simultaneous reversal of the canting and hence of the magnetization. First-principles calculations[15] on $BiFeO_3$ showed that it is not a suitable candidate; here the spin canting couples to the antiferrodistortive rotations of the FeO_6 octahedra rather than to the ferroelectric distortion of the lattice, and therefore 180° switching of the ferroelectric polarization should not affect the magnetic state.

nature materials | VOL 6 | JANUARY 2007 | www.nature.com/naturematerials

In addition, calculations showed[15] that epitaxial strain does not significantly affect the canting angle in $BiFeO_3$, although a change in crystal symmetry induced by heteroepitaxy would of course change the allowed canting directions. More interestingly, the effects of non-180° polarization reorientation in $BiFeO_3$ thin films have recently been examined[61], and the orientation of the antiferromagnetic vector has been manipulated using an electric field; this could be a promising route to orientation control of weak ferromagnetism. And electric-field switching of weak antiferromagnetism has been demonstrated computationally[14] in $BaNiF_4$; this could lead to electric field control of magnetization through the exchange bias routes discussed below. The most important open question here is whether the canting angle can be increased, because an order of magnitude increase in canting angle would yield magnetizations that are comparable to those in existing magnetic devices. The dependence of canting angle on epitaxial strain, on symmetry modification through heteroepitaxy, on the chemistry of the non-magnetic ions, and on the crystal structure are ripe areas for further study.

The concept of exchange bias coupling in magnetic materials is well known, and is extensively used in magnetic sensing and storage applications[97]. Exchange bias occurs when the strong magnetic exchange interactions at the interface between an antiferromagnet (typically a metal such as Mn or a semiconductor such as MnO) and a ferromagnet pin the orientation of the spins in the ferromagnetic layer (Fig. 5). The availability of ferroelectric antiferromagnets, such as the multiferroic perovskites, presents an exciting opportunity for modulating or controlling the magnetic structure by the application of an electric field[98,99] as follows. Coupling between ferroelectricity and antiferromagnetism provides permanent, non-volatile control of the orientation of the antiferromagnetic axis[61], then coupling between the antiferromagnetic component and an adjacent ferromagnet should allow switching of the ferromagnetism through the application of an electric field. This promises to be an exciting area of research and suggests the possibility of new device technologies, particularly in the spintronic applications discussed next[100]. Many laboratories worldwide have taken up the challenge, and although these efforts are in their infancy, it is likely that the topic will be extensively researched in the immediate future. Of course, such effects do not necessarily require a multiferroic; a large magnetoelectric coupling would also be sufficient[101].

The use of magnetoelectric coupling and multiferroics in spintronics is a rapidly emerging area of research, and a number of possible device architectures have been proposed[102]. This theme, related to the one above, focuses on controlling and manipulating the spins of carriers in a ferromagnet using an electric field. One specific manifestation is a multiferroic tunnel junction which is used as a spin filter device with the potential to control both electrically and magnetically[51,103]. Experimentally, this seems to be difficult to accomplish because so many variables affect behaviour at such length scales. First, the multiferroic tunnel barrier must be thin; of the order of 1–2 nm thick. The growth of epitaxial thin-film multiferroics with high structural quality containing only a few unit cells is challenging and will undoubtedly be a focal point in the coming years. In addition, the stability of both magnetic and ferroelectric order parameters must be maintained down to these small thicknesses. As we discussed earlier, the robustness of ferroelectricity in thin films is of tremendous current interest, with state-of-the-art experiments and first-principles computations indicating a critical thickness of a few unit cells, strongly dependent on the screening properties of the electrodes[64,67]. How the critical behaviour is affected by the simultaneous presence of magnetic ordering in multiferroics is an open question.

Finally, we discuss the potential applications of multiferroics in electrically controlled spin wave devices. The excitation of spin waves by an alternating electric field, and of ferroelectric oscillations by a

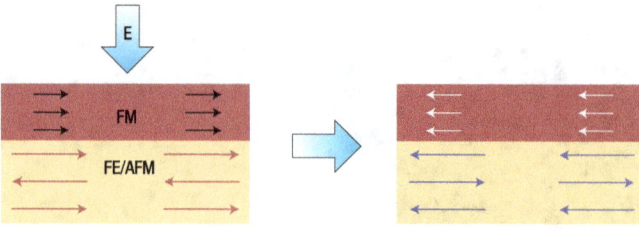

Figure 5 Essence of an electrically tuneable magnetic device. This schematic shows the essential features of such a device consisting of a multiferroic (ferroelectric antiferromagnet) in contact with a ferromagnet. Application of an electric field to the multiferroic causes a switching of the ferroelectric polarization: this in turn, through the internal coupling between ferroelectricity and antiferromagnetism, changes the AFM order which subsequently changes the state of magnetism in the ferromagnet owing to its coupling to the antiferromagnet through exchange bias coupling. This is a conceptual device, and there are several current attempts worldwide to create this or similar devices.

magnetic field, was discussed theoretically in a landmark 1982 review by Smolenskii and Chupis[91]. They proposed that such excitations, which should be strongest in ferroelectric ferrimagnets, could be used to produce magnetoelectric generators or spin wave amplifiers driven by electric field or current. However, there has been very limited subsequent work on the influence of electric fields on spin waves. Experimental results are just beginning to emerge, with the first demonstration of hybrid excitations in multiferroic manganites recently reported[104]. We believe that this should be an area of strong focus in future.

We hope that this review has captured the sense of excitement and anticipation that is prevalent in the field of magnetoelectric multiferroics, and indeed in the entire area of complex oxide thin-film heterostructures. Discoveries are being made and reported almost weekly; we have attempted to put these developments into perspective, with a particular focus on their prospects for thin-film-based devices. Thin-film multiferroic systems provide an outstanding opportunity for making use of a rich spectrum of physical responses. Among these, the potential to electrically control magnetism in its many manifestations to create new devices within the framework of spintronics, information storage and communication, forms the spur that is energizing worldwide research activity.

As in any rapidly evolving field, it is impossible to do justice to all that is going on. We apologize in advance for any omissions, and hope that they will provide the motivation for additional articles and commentaries.

doi:10.1038/nmat1805

References

1. Freeman, A. J. & Schmid, H. (eds.) *Magnetoelectric Interaction Phenomena in Crystals* (Gordon and Breach, London, 1995).
2. Fiebig, M. Revival of the magnetoelectric effect. *J. Phys. D* **38**, R1–R30 (2005).
3. Spaldin, N. A. & Fiebig, M. The renaissance of magnetoelectric multiferroics. *Science* **309**, 391–392 (2005).
4. Eerenstein, W., Mathur, N. D. & Scott, J. Multiferroic and magnetoelectric materials. *Nature* **44**, 759–765 (2006).
5. Kimura, T. *et al.* Magnetic control of ferroelectric polarization. *Nature* **426**, 55–58 (2003).
6. Ikeda, N. *et al.* Ferroelectricity from iron valence ordering in the charge-frustrated system $LuFe_2O_4$. *Nature* **436**, 1136–1138 (2005).
7. Hill, N. A. Why are there so few magnetic ferroelectrics? *J. Phys. Chem. B* **104**, 6694–6709 (2000).
8. Ederer, C. & Spaldin, N. A. Recent progress in first-principles studies of magnetoelectric multiferroics. *Curr. Opin. Solid State Mater. Sci.* **9**, 128–139 (2005).
9. Schlom, D. G. *et al.* Oxide nano-engineering using MBE. *Mater. Sci. Eng. B* **87**, 282–291 (2001).
10. Cheong, S.-W. & Mostovoy, M. Multiferroics: A magnetic twist for ferroelectricity. *Nature Mater.* **6**, 13–20 (2006).
11. Wang, J. *et al.* Epitaxial $BiFeO_3$ multiferroic thin film heterostructures. *Science* **299**, 1719 (2003).

12. van Aken, B. B., Palstra, T. T. M., Filippetti, A. & Spaldin, N. A. The origin of ferroelectricity in magnetoelectric $YMnO_3$. *Nature Mater.* **3,** 164–170 (2004).
13. Fennie, C. J. & Rabe, K. M. Ferroelectric transition in $YMnO_3$ from first principles. *Phys. Rev. B* **72,** 100103(R) (2005).
14. Ederer, C. & Spaldin, N. A. $BaNiF_4$: an electric field-switchable weak antiferromagnet. *Phys. Rev. B* **74,** 1 (2006).
15. Ederer, C. & Spaldin, N. A. Weak ferromagnetism and magnetoelectric coupling in bismuth ferrite. *Phys. Rev. B* **71,** 060401(R) (2005).
16. Ederer, C. & Spaldin, N. A. Origin of ferroelectricity in the multiferroic barium fluorides $BaMF_4$. *Phys. Rev. B* **74,** 024102 (2006).
17. Subramanian, M. A. *et al.* Giant room-temperature magnetodielectric response in the electronic ferroelectric $LuFe_2O_4$. *Adv. Mater.* **18,** 1737–1739 (2006).
18. Fujimura, N., Ishida, T., Yoshimura, T. & Ito, T. Epitaxially grown $YMnO_3$ film: new candidate for nonvolatile memory devices. *Appl. Phys. Lett.* **69,** 1011–1013 (1996).
19. Salvador, P. A., Doan, T.-D., Mercey, B. & Raveau, B. Stabilization of $YMnO_3$ in a perovskite structure as a thin film. *Chem. Mater.* **10,** 2592–5 (1998).
20. Yoo, D. C., Lee, J. Y., Kim, I. S. & Kim Y. T. Microstructure control of $YMnO_3$ thin films on Si (100) substrates. *Thin Solid Films* **416,** 62–65 (2002).
21. Suzuki, K., Fu, D. S., Nishizawa, K., Miki, T. & Kato, K. Ferroelectric property of alkoxy-derived $YMnO_3$ films crystallized in argon. *Jpn J. Appl. Phys.* **42,** 5692–5695 (2003).
22. Zhou, L., Wang, Y. P., Liu, Z. G., Zou, W. Q. & Du, Y. W. Structure and ferroelectric properties of ferroelectromagnetic $YMnO_3$ thin films prepared by pulsed laser deposition. *Phys. Status Solidi A* **201,** 497–501 (2004).
23. Kim, K. T. & Kim, C. L. The effects of drying temperature on the crystallization of $YMnO_3$ thin films prepared by sol-gel method using alkoxides. *J. Eur. Ceram. Soc.* **24,** 2613–2617 (2004).
24. Shigemitsu, N. *et al.* Pulsed-laser-deposited $YMnO_3$ epitaxial films with square polarization-electric field hysteresis loop and low-temperature growth. *Jpn J. Appl. Phys.* **43,** 6613–6616 (2004).
25. Kim, D. *et al.* C-axis oriented MOCVD $YMnO_3$ thin film and its electrical characteristics in MFIS FeTRAM. *Integr. Ferroelectr.* **68,** 75 (2004).
26. Chye, Y. *et al.* Molecular beam epitaxy of $YMnO_3$ on *c*-plane GaN. *Appl. Phys. Lett.* **88,** 132903 (2006).
27. Posadas, A. *et al.* Epitaxial growth of multiferroic $YMnO_3$ on GaN. *Appl. Phys. Lett.* **87,** 17195 (2005).
28. Balasubramanian, K. R. *Phase Competition and Thin Film Growth of Layered Ferroelectrics and Related Perovskite Phases* Thesis, Carnegie Mellon Univ. (2006).
29. Teague, J. R., Gerson, R. & James, W. J. Dielectric hysteresis in single crystal $BiFeO_3$. *Solid State Commun.* **8,** 1073–1074 (1970).
30. Li, J. *et al.* Dramatically enhanced polarization in (001), (101), and (111) $BiFeO_3$ thin films due to epitaxial-induced transitions. *Appl. Phys. Lett.* **84,** 5261–5263 (2004).
31. Qi, X. *et al.* Epitaxial growth of $BiFeO_3$ thin films by LPE and sol-gel methods. *J. Magn. Magn. Mater.* **283,** 415–421 (2004).
32. Qi, X., Dho, J., Tomov, R., Blamire, M. G. & MacManus-Dricoll, J. L. Greatly reduced leakage current and conduction mechanism in aliovalent-ion-doped $BiFeO_3$. *Appl. Phys. Lett.* **86,** 062903 (2005).
33. Qi, X. *et al.* High resolution X-ray diffraction and transmission electron microscopy of multiferroic $BiFeO_3$. *Appl. Phys. Lett.* **86,** 071913 (2005).
34. Neaton, J. B., Ederer, C., Waghmare, U. V., Spaldin, N. A. & Rabe, K. M. First-principles study of spontaneous polarization in multiferroic $BiFeO_3$. *Phys. Rev. B* **71,** 014113 (2005).
35. Yun, K. Y., Ricinschi, D., Kanashima, T., Noda, M. & Okuyama, M. Giant ferroelectric polarization beyond 150 $\mu C/cm^2$ in $BiFeO_3$ thin film. *Jpn J. Appl. Phys.* **43,** L647–L648 (2004).
36. Ederer, C. & Spaldin, N. A. Effect of epitaxial strain on the spontaneous polarization of thin film ferroelectrics. *Phys. Rev. Lett.* **95,** 257601 (2005).
37. Kiselev, S. V., Ozerov, R. P. & Zhdanov, G. S. Detection of magnetic order in ferroelectric $BiFeO_3$ by neutron diffraction. *Sov. Phys. Dokl.* **7,** 742–744 (1963).
38. Dzyaloshinskii, I. E. Thermodynamic theory of "weak" ferromagnetism in antiferromagnetic substances. *Sov. Phys. JETP* **5,** 1259–1272 (1957).
39. Moriya, T. Anisotropic superexchange interaction and weak ferromagnetism. *Phys. Rev.* **120,** 91–98 (1960).
40. Sosnowska, I., Peterlin-Neumaier, T. & Streichele, E. Spiral magnetic ordering in bismuth ferrite. *J. Phys. C* **15,** 4835–4846 (1982).
41. Popov, Y. F. *et al.* Linear magnetoelectric effect and phase transitions in bismuth ferrite, $BiFeO_3$. *JETP Lett.* **57,** 69–73 (1993).
42. Bai, F. *et al.* Destruction of spin cycloid in $(111)_c$-oriented $BiFeO_3$ thin films by epitaxial constraint: enhanced polarization and release of latent magnetization. *Appl. Phys. Lett.* **86,** 032511 (2005).
43. Bea, H. *et al.* Unravelling the origin of the controversial magnetic properties of $BiFeO_3$ thin films. Preprint at <http://arxiv.org/cond-mat/0606441>(2006).
44. Wang, J. *et al.* Epitaxial $BiFeO_3$ thin films on Si. *Appl. Phys. Lett.* **85,** 2574–2576 (2004).
45. Azuma, M. *et al.* Magnetic ferroelectrics Bi, Pb-3d transition metal perovskites. *Trans. Mater. Res. Soc. Jpn* **31,** 41–46 (2006).
46. dos Santos, A. M. *et al.* Epitaxial growth and properties of metastable $BiMnO_3$ thin films. *Appl. Phys. Lett.* **84,** 91–93 (2004).
47. Sharan, A. *et al.* Bismuth manganite: a multiferroic with a large nonlinear optical response. *Phys. Rev. B* **69,** 214109 (2004).
48. Eerenstein, W., Morrison, F. D., Scott, J. F. & Mathur, N. Growth of highly resistive $BiMnO_3$ films. *Appl. Phys. Lett.* **87,** 101906 (2005).
49. Atou, T., Chiba, H., Ohoyama, K., Yamaguichi, Y. & Syono, Y. Structure determination of ferromagnetic perovskite $BiMnO_3$. *J. Solid State Chem.* **145,** 639–642 (1999).
50. dos Santos, A. M. *et al.* Orbital ordering as the determinant for ferromagnetism in biferroic $BiMnO_3$. *Phys. Rev. B* **66,** 064425 (2002).
51. Gajek, M. *et al.* Spin filtering through ferromagnetic $BiMnO_3$ tunnel barriers. *Phys. Rev. B* **72,** 020406 (2005).
52. Son, J. Y., Kim, B. G., Kim, C. H. & Cho, J. H. Writing polarization bits on the multiferroic $BiMnO_3$ thin film using Kelvin probe force microscope. *Appl. Phys. Lett.* **84,** 4971–4973 (2004).
53. dos Santos, A. M. *et al.* Evidence for the likely occurence of magnetoferroelectricity in the simple perovskite $BiMnO_3$. *Solid State Commun.* **122,** 49–52 (2002).
54. Belik, A. A. *et al.* $BiScO_3$: Centrosymmetric $BiMnO_3$-type oxide. *J. Am. Chem. Soc.* **128,** 706–707 (2006).
55. Shishidou, T., Mikamo, N., Uratani, Y., F.Ishii & Oguchi, T. First-principles study on the electronic structure of bismuth transition metal oxides. *J. Phys. Cond. Mat.* **16,** S5677–S5683 (2004).
56. Murakami, M. *et al.* Fabrication of multiferroic epitaxial $BiCrO_3$ thin films. *Appl. Phys. Lett.* **88,** 152902 (2006).
57. Shpanchenko, R. V. *et al.* Synthesis, structure, and properties of new perovskite $PbVO_3$. *Chem. Mater.* **16,** 3267–3273 (2004).
58. Choi, K. J. *et al.* Enhancement of ferroelectricity in strained $BaTiO_3$ thin films. *Science* **306,** 1005–1009 (2004).
59. Haeni, J. H. *et al.* Room-temperature ferroelectricity in strained $SrTiO_3$. *Nature* **430,** 758–761 (2004).
60. Ederer, C. & Spaldin, N. A. Influence of strain and oxygen vacancies on the magnetoelectric properties of multiferroic bismuth ferrite. *Phys. Rev. B* **71,** 224103 (2005).
61. Zhao, T. *et al.* Electrically controllable antiferromagnets: Nanoscale observation of coupling between antiferromagnetism and ferroelectricity in multiferroic $BiFeO_3$. *Nature Mater.* **5,** 823–829 (2006).
62. Spaldin, N. A. Fundamental size limits in ferroelectricity. *Science* **304,** 1606 (2004).
63. Shaw, T. M., Trolier-McKinstry, S. & McIntyre, P. C. The properties of ferroelectric films at small dimensions. *Annu. Rev. Mater. Sci.* **30,** 263–298 (2000).
64. Ahn, C. H., Rabe, K. M. & Triscone, J.-M. Ferroelectricity at the nanoscale: Local polarization in oxide thin films and heterostructures. *Science* **303,** 488–491 (2004).
65. Fong, D. D. *et al.* Ferroelectricity in ultrathin perovskite films. *Science* **304,** 1650–1653 (2004).
66. Ghosez, P. & Rabe, K. M. Microscopic model of ferroelectricity in stress-free $PbTiO_3$ thin films. *Appl. Phys. Lett.* **76,** 2767–2769 (2000).
67. Junquera, J. & Ghosez, P. Critical thickness for ferroelectricity in perovskite ultrathin films. *Nature* **422,** 506–509 (2003).
68. Murugavel, P., Saurel, D., Prellier, W., Simon, C. & Raveau, B. Tailoring of ferromagnetic $Pr_{0.85}Ca_{0.15}MnO_3$/ferroelectric $Ba_{0.6}Sr_{0.4}TiO_3$ superlattices for multiferroic properties. *Appl. Phys. Lett.* **84,** 4424–4426 (2004).
69. Singh, M. P., Prellier, W., Simon, C. & Raveau, B. Magnetocapacitance effect in perovskite-superlattice based multiferroics. *Appl. Phys. Lett.* **87,** 022505 (2005).
70. Prellier, W., Singh, M. P. & Murugavel, P. The single-phase multiferroic oxides: from bulk to thin film. *J. Phys. Cond. Mat.* **17,** 7753 (2005).
71. Ueda, K., Tabata, H. & Kawai, T. Ferromagnetism in $LaFeO_3$-$LaCrO_3$ superlattices. *Science* **280,** 1064–1066 (1998).
72. Baettig, P. & Spaldin, N. A. Ab initio prediction of a multiferroic with large polarization and magnetization. *Appl. Phys. Lett.* **86,** 012505 (2005).
73. Asano, H. *et al.* Pulsed-laser-deposited epitaxial $Sr_2FeMoO_{6-\gamma}$ thin films: Positive and negative magnetoresistance regimes. *Appl. Phys. Lett.* **74,** 3696–3698 (1999).
74. Azuma, M. *et al.* Designed ferromagnetic, ferroelectric Bi_2NiMnO_6. *J. Am. Chem. Soc.* **127,** 8889–8892 (2005).
75. Ohtomo, A. & Hwang, H. Y. A high-mobility electron gas at the $LaAlO_3/SrTiO_3$ heterointerface. *Nature* **427,** 423–426 (2004).
76. Sai, N., Meyer, B. & Vanderbilt, D. Compositional inversion symmetry breaking in ferroelectric perovskites. *Phys. Rev. Lett.* **84,** 5636–5639 (2000).
77. Lee, H. N., Christen, H. M., Chisholm, M. F., Rouleau, C. M. & Lowndes, D. H. Strong polarization enhancement in asymmetric three-component ferroelectric superlattices. *Nature* **433,** 395–399 (2005).
78. Warusawithana, M. P., Colla, E. V., Eckstein, J. N. & Weissman, M. B. Artificial dielectric superlattices with broken inversion symmetry. *Phys. Rev. Lett.* **90,** 036802 (2003).
79. Duan, C.-G., Jaswal, S. S. & Tsymbal, E. Y. Towards ferroelectrically-controlled magnetism: Magnetoelectric effect in $Fe/BaTiO_3$ multilayers. Preprint at <http://arxiv.org/ond-mat/0604560> (2006).
80. Stengel, M. & Spaldin, N. A. Origin of the dielectric dead layer in nanoscale capacitors. *Nature* **443,** 679–682 (2006).
81. Zhao, T. *et al.* Electric field effect in diluted magnetic insulator anatase $Co:TiO_2$. *Phys. Rev. Lett.* **94,** 126601 (2005).
82. Tanaka, H., Zhang, J. & Kawai, T. Giant electric field modulation of double exchange ferromagnetism at room temperature in the perovskite manganite/titanate p–n junction. *Phys. Rev. Lett.* **88,** 027204 (2002).
83. Wu, T. *et al.* Electroresistance and electronic phase separation in mixed-valent manganites. *Phys. Rev. Lett.* **86,** 5998–6001 (2001).
84. Toyota, D. *et al.* Thickness-dependent electronic structure of ultrathin $SrRuO_3$ films studied by *in situ* photoemission spectroscopy. *Appl. Phys. Lett.* **87,** 162508 (2005).
85. Nan, C.-W., Liu, G., Lin, Y. & Chen, H. Magnetic-field-induced electric polarization in multiferroic nanostructures. *Phys. Rev. Lett.* **94,** 197203 (2005).
86. Nagarajan, V. *et al.* Size effects in ultrathin epitaxial ferroelectric heterostructures. *Appl. Phys. Lett.* **84,** 5225–5227 (2004).
87. Zheng, H. *et al.* Multiferroic $BaTiO_3$-$CoFe_2O_4$ nanostructures. *Science* **303,** 661–663 (2004).
88. Zavaliche, F. *et al.* Electric field-induced magnetization switching in epitaxial columnar nanostructures. *Nano Lett.* **5,** 1793–1796 (2005).
89. Robinson, J. T. *et al.* Metal-induced assembly of a semiconductor island lattice: Ge truncated pyramids on Au-patterned Si. *Nano Lett.* **5,** 2070–2073 (2005).
90. Wood, V. E. & Austin, A. E. in *Magnetoelectric Interaction Phenomena in Crystals* (eds Freeman, A. J. & Schmid, H.) 181–194 (Gordon and Breach, London, 1975).
91. Smolenskii, G. A. & Chupis, I. E. Ferroelectromagnets. *Sov. Phys. Usp.* **25,** 475–493 (1982).
92. Srinivasan, G. *et al.* Magnetoelectric bilayer and multilayer structures of magnetostrictive and piezoelectric oxides. *Phys. Rev. B* **64,** 214408 (2001).

93. Srinivasan, G., Rasmussen, E. T., Levin, B. J. & Hayes, R. Magnetoelectric effects in bilayers and multilayers of magnetostrictive and piezoelectric perovskite oxides. *Phys. Rev. B* **65**, 134402 (2002).

94. Dong, S., Cheng, J., Li, J. F. & D.Viehland. Enhanced magnetoelectric effects in laminate composites of terfenol-D/Pb(Zr,Ti)O₃ under resonant drive. *Appl. Phys. Lett.* **83**, 4812–4814 (2003).

95. Scholl, A. *et al.* Observation of antiferromagnetic domains in epitaxial thin films. *Science* **287**, 10 14–1016 (2000).

96. Scholl, A., Liberati, M., Arenholz, E., Ohldag, H. & Stöhr, J. Creation of an antiferromagnetic exchange spring. *Phys. Rev. Lett.* **92**, 247201 (2004).

97. Nogués, J. & Schuller, I. Exchange bias. *J. Magn. Magn. Mater.* **192**, 203 (1999).

98. Martí, X. *et al.* Exchange bias and electric polarization with YMnO₃. *Appl. Phys. Lett.* **89**, 032510 (2006).

99. Bea, H. *et al.* Tunnel magnetoresistance and exchange bias with multiferroic BiFeO₃ epitaxial thin films. Preprint at <http://arxiv.org/ond-mat/0607563> (2006).

100. Zutíc, I., Fabian, J. & Sarma, S. D. Spintronics: Fundamentals and applications. *Rev. Mod. Phys.* **76**, 323 (2005).

101. Borisov, P., Hochstrat, A., Chen, X., Kleemann, W. & Binek, C. Magnetoelectric switching of exchange bias. *Phys. Rev. Lett.* **94**, 117203 (2005).

102. Binek, C. & Doudin, B. Magnetoelectronics with magnetoelectrics. *J. Phys. Cond. Mat.* **17**, L39-L44 (2005).

103. Tsymbal, E. Y. & Kohlstedt, H. Tunneling across a ferroelectric. *Science* **313**, 181–183 (2006).

104. Pimenov, A. *et al.* Possible evidence for electromagnons in multiferroic manganites. *Nature Phys.* **2**, 97–100 (2006).

Acknowledgements

Clearly, this work represents the cumulative efforts of many researchers around the world. Specifically, we acknowledge several key collaborators (in alphabetical order): L. Q. Chen (Pennsylvania State Univ. (PSU)), L. E. Cross (PSU), C. Ederer (Columbia Univ.), C. B. Eom (Univ. Wisconsin), M. Fiebig (Univ. Bonn), V. Gopalan (PSU), J. Kreisel (Univ. Grenoble), S. Ogale (Univ. Maryland), K. M. Rabe (Rutgers Univ.), D. Schlom (PSU), H. Schmid (Univ. Geneva), A. Scholl (ALS-LBL), J. F. Scott (Univ. Cambridge), D. Viehland (Virginia Tech.), M. Wuttig (Univ. Maryland), F. Zavaliche (Seagate Technologies) and T. Zhao (Seagate Technologies). We also thank past and present members of our research groups at Berkeley and Santa Barbara. We thank the following funding agencies: ONR, DOE and NSF (R.R.) and NSF (N.A.S.).

Correspondence should be addressed to R.R. or N.A.S.

Inorganic nanotubes and fullerene-like nanoparticles

Although graphite, with its anisotropic two-dimensional lattice, is the stable form of carbon under ambient conditions, on nanometre length scales it forms zero- and one-dimensional structures, namely fullerenes and nanotubes, respectively. This virtue is not limited to carbon and, in recent years, fullerene-like structures and nanotubes have been made from numerous compounds with layered two-dimensional structures. Furthermore, crystalline and polycrystalline nanotubes of pure elements and compounds with quasi-isotropic (three-dimensional) unit cells have also been synthesized, usually by making use of solid templates. These findings open up vast opportunities for the synthesis and study of new kinds of nanostructures with properties that may differ significantly from the corresponding bulk materials. Various potential applications have been proposed for the inorganic nanotubes and the fullerene-like phases. Fullerene-like nanoparticles have been shown to exhibit excellent solid lubrication behaviour, suggesting many applications in, for example, the automotive and aerospace industries, home appliances, and recently for medical technology. Various other potential applications, in catalysis, rechargeable batteries, drug delivery, solar cells and electronics have also been proposed.

R. TENNE

Head of Department of Materials and Interfaces; Director of the Helen and Martin Kimmel Center for Nanoscale Science; holder of the Drake Family Professorial Chair in Nanotechnology, Weizmann Institute of Science, Rehovot 76100, Israel.

e-mail: reshef.tenne@weizmann.ac.il

Chemistry is, in general, not favourable of materials with empty space because the chemical bond is not stable beyond a distance of a few angstroms. Therefore, in order to maximize the interaction between the electron clouds of nearest-neighbour atoms and thereby stabilize the chemical bond between them, atoms in materials arrange themselves in close proximity. This paradigm is reflected by 'space-filling models', which visualize the overlapping electron clouds of neighbouring atoms in a given compound or a solid. Consequently, until recently, the science and technology of hollow-closed structures was in fact an almost unexplored terrain. Pauling was among the first to investigate closed polyhedral and tubular forms of asbestos minerals, such as kaolinite (alumino-silicate)[1]. Noted also for their tendency to form polyhedral structures are the boron–carbon–hydrogen compounds (carbohydrenes), which were investigated in great detail by Lipscomb[2,3].

The discovery of carbon fullerenes by Kroto, Smalley and Curl[4] in 1985 and later on carbon nanotubes by Iijima[5] served as a turning point in the exploration of this unknown territory. Perhaps equally important was the discovery of inorganic nanotubes and fullerene-like structures[6] in 1992, establishing a new paradigm in the chemistry of nanomaterials and leading to the birth of a new field of inorganic chemistry, that is, one dealing with polyhedral (closed hollow) nanostructures.

It was in fact quite natural to assume that the virtue of fullerenes and nanotubes is not limited to carbon, and could occur in other two-dimensional layered compounds, such as WS_2 (ref. 6) or its structural analogue MoS_2 (refs 7,8). In contrast to graphite, which is made of mono-atomic carbon sheets arranged in a hexagonal honeycomb and held together by van der Waals forces, these inorganic two-dimensional solids are made of stacked molecular sheets. Each molecular sheet consists of a sixfold-bonded molybdenum layer sandwiched between two threefold-bonded sulphur layers. Here too, the sheets (walls) are held together by van der Waals forces. In analogy with graphite, the unit cell of MoS_2 is made of two layers in hexagonal arrangement (2H). However, rim Mo and S atoms, which are abundant in the nano-regime, are only four- and twofold-bonded, respectively, making these nanostructures unstable in the planar form. Therefore, by folding the molecular sheet and stitching the rim atoms together, seamless stable nanotubular (one-dimensional) and spherical (zero-dimensional) structures with all Mo and S atoms being six- and threefold-bonded, respectively, are obtained. Following this early observation, nanostructures of these kinds were shown to be ubiquitous among layered compounds and were termed inorganic fullerene-like structures and inorganic nanotubes. Much like their carbon predecessors, they come in

When citing this review, please cite the original version as shown on the contents page of this chapter.

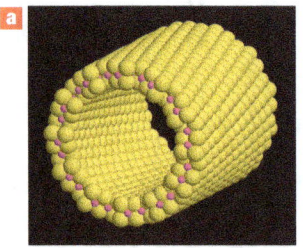
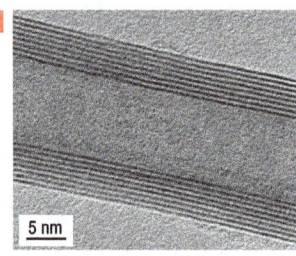

Figure 1 Presentation of WS$_2$ nanotubes, which are synthesized in large amounts by sulphidizing tungsten oxide nanoparticles in a fluidized bed reactor. **a**, Schematic presentation of a single-wall WS$_2$ nanotube with W and S atoms in pink and yellow respectively. Reproduced from ref. 76 by permission of the PCCP Owner Societies. **b**, A TEM image of a portion of a multiwall WS$_2$ nanotube with eight cylindrical and concentric layers. The distance between each layer is 0.61 nm. The c axis [001] is always normal to the surface of the nanotube.

various forms and shapes, such as multiwall and single-wall nanotubes and nested (multiwall) fullerene-like structures. Unlike carbon fullerenes and nanotubes, which can easily be synthesized in single-wall form, here, concentric (nested) multilayer structures are usually obtained. Figure 1a shows a schematic presentation of a WS$_2$ nanotube, and Fig. 1b shows a transmission electron microscope (TEM) image of a multiwall WS$_2$ nanotube.

Practically, the synthesis of pure phases of this kind proved to be relatively straightforward in several cases (MoS$_2$, WS$_2$, VO$_x$-alkylamine and titanate nanotubes), but exceedingly difficult if not impossible in other cases, such as that of NiCl$_2$ (ref. 9) and Cs$_2$O (refs 10,11). More recently, the synthesis of crystalline nanotubes from quasi-isotropic (three-dimensional) compounds, for example GaN (ref. 12) and In$_2$O$_3$ (ref. 13) was accomplished. In fact, these kind of nanotubes are not formed by the winding of a molecular sheet, but instead can be visualized as crystalline hollow nanofibres. They are usually obtained by templating and in most cases are highly faceted. Furthermore, the inner and outer surfaces of such nanotubes are generally very reactive, as can be seen, for example, by the formation of an amorphous indium oxide film on InN nanotubes[14].

Recent studies have alluded to the numerous potential applications of inorganic nanotubes and fullerene-like nanoparticles. These applications are highlighted in a recent report[15] and will be discussed in some detail below. Most importantly, fullerene-like WS$_2$ and MoS$_2$ nanoparticles were shown to exhibit superior tribological behaviour, particularly under high loads. This behaviour offers a plethora of potential applications. Various other potential applications in areas such as catalysis, energy conversion and storage, drug delivery and sensors will be discussed as well.

THE SYNTHESIS OF INORGANIC NANOTUBES AND FULLERENES

The synthesis of such nanoparticles has witnessed a substantial growth in recent years[16–20]. Therefore, this review will focus on the most recent developments, particularly those of the last three years.

The synthesis of fullerene-like structures and nanotubes of MX$_2$ (M=transition metal; X=S,Se,Te) from the respective oxide nanoparticles, is a slow diffusion-controlled process, which is limited to reactive oxides such as WO$_3$ and therefore is not possible for less reactive oxides, such as those of titanium and niobium. To try and alleviate these deficiencies, a new strategy was developed whereby metal halides or carbonyls are used as precursors. These compounds have high vapour pressure at temperatures below 300 °C and are very

unstable, reacting with H$_2$X (X=S,Se,Te) vapour instantaneously. This reaction leads to fast nucleation of incipient MX$_2$ nuclei, which grow outwards in a layer-by-layer fashion. Fullerene-like NbS$_2$ nanoparticles[21] were the first to be synthesized according to this strategy. Subsequently, TiS$_2$ nanotubes (ref. 22) and fullerene-like MoS$_2$ (MoSe$_2$) (ref. 23), WS$_2$ (ref. 24), and TiS$_2$ (ref. 25) nanoparticles were synthesized by the reaction of the respective metal halide with a sulphur reagent.

The synthesis of very uniform single-wall nanotubes of the multinary compound SbPS$_{4-x}$Se$_x$ was recently reported[26]. The sulphur-to-selenium exchange led to a red shift in the absorption coefficient of the nanotubes. Various metal oxide nanotubes were synthesized through soft chemistry ('chemie douce') processes, such as hydrothermal, sol-gel, intercalation–exfoliation, sonochemical reactions, and so on. V$_2$O$_{5-x}$-alkylamine nanotubes, which have been studied extensively in the past, were recently used as a template for the synthesis of VS$_2$ nanotubes[27]. This work is remarkable in so far as the bulk lamellar phase of VS$_2$ does not exist, which is another manifestation of the inherent stability of these nanotubular phases. Reversible electrochemical copper-ion intercalation in the VS$_2$-alkylamine nanotubes was also demonstrated in this work.

Recently, the first oxihalide nanotubes of the layered compound WO$_2$Cl$_2$ were prepared using a chemie-douce process[28]. Layered structures are also prevalent for hexavalent uranium oxo- compounds owing to the strong tendency of the U^{6+} ion to form linear uranyl ions, UO$_2^{2+}$. The synthesis of an ordered array of [(UO$_2$)$_3$(SeO$_4$)5]$^{4-}$ nanotubes in a room-temperature reaction in aqueous solution was recently reported[29].

A remarkable manifestation of the kinetic stabilization of closed fullerene-like nanostructures was demonstrated recently in the synthesis of fullerene-like nanoparticles from the layered compound Cs$_2$O (ref. 10). Films with a Cs/O ratio of approximately 2:1 are known to reduce the electron work function, increasing the sensitivity of optoelectronic detectors, used, for example, in medical imaging. Unfortunately, these films are highly reactive and are damaged by even a short exposure to low vacuum conditions. Nested closed nanoparticles

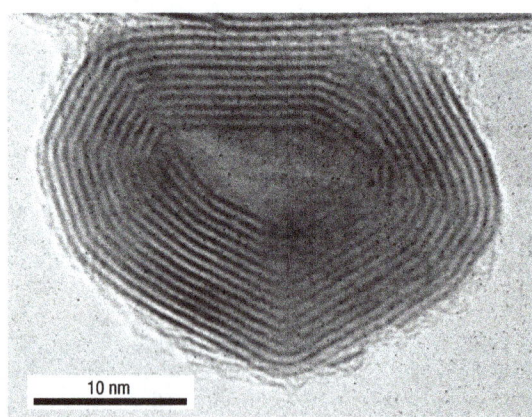

Figure 2 TEM image of a closed-cage (fullerene-like) nested Cs$_2$O nanostructure obtained by intense solar irradiation of pure crystalline Cs$_2$O powder in vacuum. The measured layer-to-layer distance is 0.635–0.645 nm (refs 10,11). Although the structure of the closed nanoparticle is not free of defects, its overall closed structure provides remarkable kinetic stabilization against reaction with oxygen, CO$_2$ and water in the ambient atmosphere. This behaviour is not unique to this particular kind of nanoparticle. It is also common to the fullerene-like nanoparticles and nanotubes of metal dihalides, such as NiCl$_2$ (ref. 9). Reproduced with permission from ref. 11. Copyright (2006) Wiley-VCH.

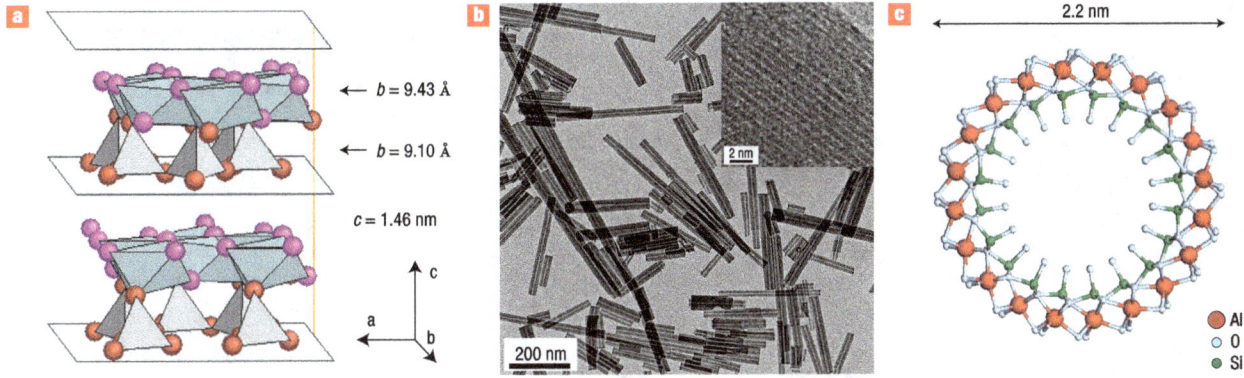

Figure 3 Naturally occurring nanotubes are made of lattices with built-in asymmetry and have been known for many years[1]. **a**, A schematic rendering of the lattice of chrysotile (magnesium hydroxide silicate). The length of each repeating unit along the *b* axis of the magnesium hydroxide octahedra (green) is longer than that of the silica tetrahedra (grey). Reprinted with permission from ref. 30. Copyright (2004) Wiley-VCH. **b**, A TEM micrograph of synthetic multiwall asbestos nanotubes (chrysotile). The inset shows an enlarged TEM image of the wall of an individual chrysotile nanotube with well-resolved magnesium hydroxide silicate molecular sheets 0.74 nm apart. (Nanotubes synthesized by N. Roveri.) **c**, A cross section of alumino-silicate (imogolite) nanotubes. Reprinted with permission from ref. 32. Copyright (2005) ACS. The structure of such nanotubes was first studied by Pauling[1] who suggested that the driving force for the formation of the nanotubes lies in the built-in asymmetry of the unit cell along the *b*-axis. Thus the larger octahedra (magnesium hydroxide in **a**, and alumina in **c**) are present on the outer perimeter of each layer in the nanotube and the smaller silica tetrahedra occupy the inner perimeter. In contrast to naturally occurring chrysotile nanotubes and fibres, the synthetic nanotubes were found to be non-toxic offering them numerous applications in future construction materials.

of Cs_2O were observed in laser-ablated powder, initially in tiny amounts and subsequently, using solar ablation, in larger quantities (Fig. 2). Remarkably though, these fullerene-like nanoparticles suffered only slow degradation in the ambient atmosphere.

Naturally occurring nanotubes of asbestos-related minerals, like chrysotile (magnesium hydroxide silicate) and kaolinite (alumino-silicate) have been known for many years and investigated in the past[1] (see Fig. 3a for a schematic drawing of the chrysotile lattice). The lattice mismatch between the magnesium hydroxide or alumina octahedra and the smaller silica tetrahedra leads to a strong built-in strain, and consequently, a bending force arises with the convex magnesium hydroxide or alumina octahedra appearing in the outer perimeter and the concave silica tetrahedra in the inner perimeter (Fig. 3c), thereby forming nanotubes (Fig. 3b). Incidentally, Pauling concluded that layered materials, such as MoS_2 and $CdCl_2$, having no asymmetric structure and therefore no built-in strain, are not likely to form closed structures like nanotubes, which is obviously not the case today. Stoichiometric nanotubes of chrysotile were obtained in large yields only quite recently[30,31]. Most astonishingly, whereas the biotoxicity of naturally occurring asbestos is well documented, synthetic chrysotile nanotubes were found to be completely benign[31], suggesting that in the future they could replace the banned natural asbestos fibres in construction materials. Also recently, single-wall alumino-germanate and silicate nanotubes with imogolite structure (see Fig. 3c) were obtained by a soft-chemistry process[32].

Nanotubes of the graphite-like boron nitride compound were among the first to be studied. Using metal oxides as catalysts, the large-scale synthesis of multiwall BN nanotubes was recently reported[33]. Further purification of the nanotubes was obtained through polymer wrapping of the nanotube backbone and secretion of obstinate boron nitride residues[34].

Crystalline nanotubes (or hollow nanofibres) of quasi-isotropic three-dimensional compounds were recently obtained, for example, by depositing a precursor on a one-dimensional template such as porous anodized aluminium oxide (AAO), and soft templates, such as elongated micelles. Similarly, faceted GaN nanotubes with hexagonal cross-sections were produced by the reaction of trimethylgallium

and ammonia on a template of an ordered array of ZnO nanowires, which was subsequently removed[12]. Carbothermal synthesis was used ingeniously to synthesize nanotubes of various three-dimensional oxides, nitrides and sulphides[13,14]. In one such example the heated graphite crucible also served to reduce the solid precursor M_2O_3 (M=Al,Ga,In) producing a volatile intermediate, M_2O, which subsequently reacted with nitrogen (or ammonia) producing MN nanotubes with faceted cross-sections. Recently, a new synthetic strategy was promoted, whereby nanorod precursors are first synthesized. These nanorod templates are converted into nanotubes through self-sacrificial chemistry. Two examples of this approach are the synthesis of spinel ($ZnAl_2O_4$) (ref. 35), and Zn_3P_2 and Cd_3P_2 nanotubes (ref. 36).

Numerous reports on the use of AAO membrane as a template for the growth of various nanotubes and nanowires have appeared in recent years. Nanotubes with walls made of gold nanoparticles instead of gold atoms only were obtained by chemical deposition of the gold nanoparticles into a silanized AAO membrane[37]. Nanotubes made of self-assembled SnO_2 nanoparticles were prepared using a similar method[38], and were used as the host material for Li intercalation batteries. Crystalline Fe_3O_4 nanotubes, which exhibited small magnetoresistance behaviour at 77 K, have also been reported[39].

THEORETICAL CALCULATIONS

With the advent of first-principle computational methods, analysis of inorganic nanotubes and fullerene-like structures became possible, bringing new insight and predicting new properties, some of them yet to be verified by experiment. The structure, energetics and the electronic structure of double-wall BN nanotubes were recently discussed[40]. The most stable double-wall tubes are those with minimum overlap between adjacent boron and nitrogen atoms of the two concentric tubes, which is fulfilled by zig-zag nanotubes. Some overlap is inevitable in armchair nanotubes, making them less stable than zig-zag nanotubes.

Numerous nanotubes with graphite-like ('sp^2'-like or sheet-like) lattice and cylindrical shape were studied using density functional

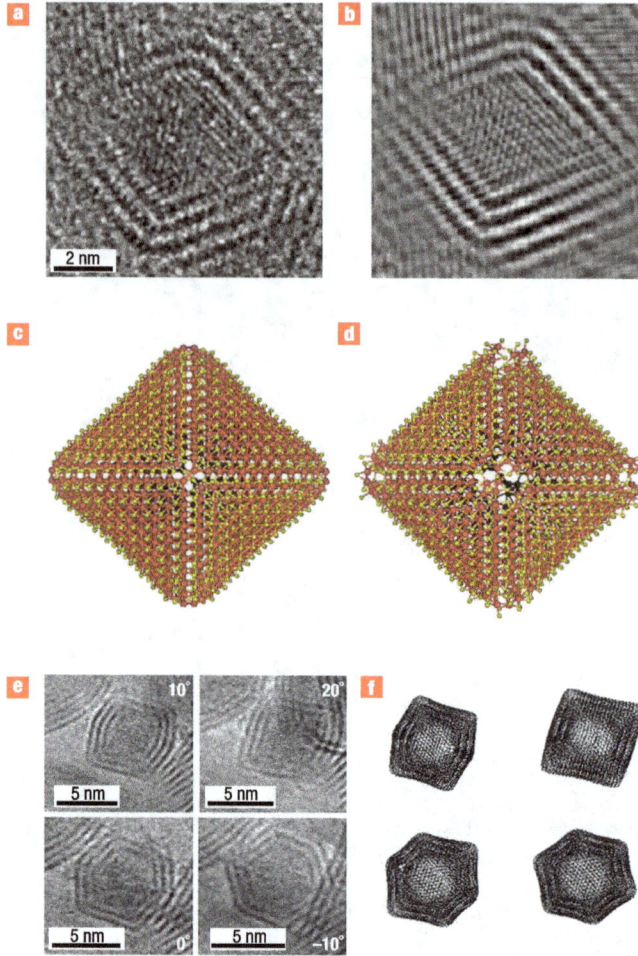

Figure 4 Hollow nano-octahedra, 3–6 nm in size and 2–5 layers thick were found in laser-ablated samples of different layered compounds, such as MoS_2, $NiCl_2$, SnS_2. They were generally found to have a chemical composition almost identical to the bulk material, somewhat analogous to C_{60} in the carbon fullerene series. They are believed to have a rhombus in each of their six corners. A typical high-resolution TEM micrograph of an octahedral MoS_2 nanoparticle (**a**) and its filtered Fourier transform image (**b**). The distance between each MoS_2 layer is 0.61 nm. A $Mo_{576}S_{1140}$ octahedron calculated from first principles using a density-functional tight-binding (DFTB) algorithm at 0 K (**c**) and after molecular dynamic (MD) simulation at 300 K (**d**). The starting model was $(MoS_2)_{576}$ (corner to corner distance of 3.8 nm), which relaxed by losing two sulphur atoms per corner (12 in total). **e**, A TEM image of a three-layered nano-octahedron at various tilt angles. **f**, The corresponding images of the semi-empirically calculated atomic model for a $(MoS_2)_{784}@(MoS_2)_{1296}@(MoS_2)_{1936}$ octahedral fullerene. Note the visual similarity between the experimentally observed nano-octahedron (**e**) and the calculated structure (**f**). According to the DFTB–MD calculations, the MoS_2 nano-octahedra were found to be metallic, but this remains to be confirmed by experiment. Reprinted with permission from ref. 43. Copyright (2006) Wiley-VCH.

theory methodology[41]. Many of the studied nanotubes are known to possess a zincblende or wurtzite ('sp^3'-like or diamond-like) lattice with tetrahedral arrangement of the atoms in the bulk. Their graphitic phase is either unknown or unstable. For example, the graphitic AlN nanotubes are higher in energy by as much as 0.68 eV per atom compared with bulk zincblende (diamond-like) AlN. Experimentally

synthesized nanotubes from group III and group V elements, such as AlN or GaP, have a zincblende (or wurzite) structure and are highly faceted. Therefore, the relevance of the computational effort to the experimental investigation of such nanotubes is not clear at this point.

$MgCl_2$ is a compound with layered $CdCl_2$ structure, in which the Mg atoms form a sheet sandwiched between two atomic sheets of chlorine. In analogy to the MoS_2 structure, the coordination numbers of the Mg and Cl are six and three, respectively. The structure of $MgCl_2$ nanotubes was elucidated recently by quantum chemical calculations[42]. Although nanotubes with different coordination numbers were also attempted, those with the layered $CdCl_2$ structure were found to be the most stable for diameters of 2.5 nm and above. Experimentally, nanotubes and fullerene-like structures of various metal halides have been synthesized in the past[9,19].

Early on it was recognized that rhombi rather than pentagonal rings, which occur in carbon fullerenes and nanotubes, are the favourable elements in polyhedra of layered compounds[7]. Fig. 4a, b shows a MoS_2 nano-octahedron consisting of four layers, each with rhombi in their six corners. Multiwall nano-octahedra, with the inner wall being made up of about 600–800 molybdenum and 1,200–1,600 sulphur atoms, respectively, are routinely found in laser-ablated samples of various compounds with layered structure[43,44], nonetheless in minute quantities.

Very recently, detailed density-functional tight-binding coupled with molecular dynamics (DFTB–MD) theoretical analysis for MoS_2 nano-octahedra was undertaken, permitting a direct comparison with experimental data[43]. Figure 4c shows a stable $Mo_{576}S_{1140}$ octahedron obtained by DFTB calculations at 0 K (left). After MD simulation to 300 K (right) the edges and corners become distorted, but the overall structure of the nano-octahedron is preserved. Furthermore, using a few other simplifications, the calculations could be extended to any size, and the phase stability could be compared with that of MoS_2 nanoparticles with different structures. Figure 4e, f compares TEM and computer generated images of MoS_2 nano-octahedra under various tilt angles. The great similarity between the experimental and calculated images is compelling and gives further credibility to that analysis. Figure 5 shows the energy diagram of the various species as a function of their size. A few important conclusions can be drawn from this combined theoretical–experimental study: (1) the nano-octahedra are stable over a limited range of about 5,000–100,000 atoms (3–6 nm). Below this size, MoS_2 nanoplatelets are more stable, whereas quasi-spherical closed-cage (fullerene-like) nanoparticles become the most stable species above the higher limit. At even larger sizes (approximately 10^7 atoms) macroscopic MoS_2 platelets become the most stable structure. (2) Among the possible stable nano-octahedra structures those with the formula Mo_nS_{2n-12} seem to be the most stable. (3) Whereas MoS_2 platelets and the quasi-spherical nested structures are semiconductors, the nano-octahedra are metallic. Despite there being only small overall chemical differences between structures, the nano-octahedra seem to be the most stable in the size range of a few nm. The structures do however have completely different physical properties.

PHYSICAL PROPERTIES

The physical characteristics of macroscopic structures, be it mechanical, optical or electronic properties, are influenced by the intrinsic properties of the material as well as by various extrinsic factors, such as impurities, defects, deviations from stoichiometry and so on. In contrast, nanostructured moieties can be made (almost) perfectly crystalline, and therefore their behaviour is exclusively determined by the intrinsic structure and the chemical bonding, rendering quantitative comparison with first-principle calculations possible. The physical and chemical properties of inorganic nanotubes are variable and largely depend on the property of the bulk compound from which they are derived. Thus, BN (ref. 40), MoS_2 (ref. 43) and $NbSe_2$ (ref. 45) nanotubes are, insulators, semiconductors and superconductors, respectively.

ELECTRONIC STRUCTURE

Whereas quantum dots exhibit a blue shift with decreasing size of the nanoparticle, both experiment and theory has shown in the past that the bandgap of semiconducting nanotubes, such as GaSe and MoS_2, shrinks with decreasing nanotube diameter. This effect can be ascribed to two factors: (1) small quantum size effect of the closed nanostructures, and (2) the deformation of the chemical bonds in the curved sheet leading to a less than perfect hybridization of the atomic orbitals. Furthermore, whereas armchair nanotubes were often found to have an indirect (Δ–Γ) electronic transition, zig-zag nanotubes possess a direct (Γ–Γ) transition suggesting that they could emit strong luminescence on optical or electrical excitation[9]. Numerous publications have been dedicated to the calculations of the band structure of nanotubes, such as AlN (ref. 41) and SiC (ref. 46) with an 'sp^2'-like (graphitic) lattice, which differs from the bulk material characterized by an 'sp^3'-diamond-like (zincblende or wurtzite) structure. Generally, their calculated properties resemble those of nanotubes from genuine layered compounds. BN nanotubes were found to have a high bandgap irrespective of their chirality. Both experiment and theory show that the bandgap of such nanotubes shrinks to the point where they become metallic (giant Stark effect) with the application of a large transversal electric field[47,48].

MECHANICAL PROPERTIES

Tensile and buckling tests were performed on individual WS_2 nanotubes with a scanning electron microscope (SEM)[49]. Figure 6 displays a WS_2 nanotube compressed against the surface of a Si wafer. The Young's modulus of the nanotube was found to be 150 GPa, that is, about one-seventh of the value of carbon nanotubes. Independent tensile tests showed that the nanotubes behave elastically almost up to their failure. The Young's modulus, yield strength and elongation to failure are 150 GPa, 16 GPa and 12%, respectively[49]. *Ab initio* calculations were on the high side but nevertheless in good agreement with the experimental data. The experimentally determined yield strength/Young's modulus ratio of 0.11 is an exceedingly high value in comparison with other high-strength materials. These observations indicate that the failure of the nanotubes is initiated by the rupture of an individual chemical bond, and that the role of macroscopic mechanisms, such as presence of dislocations, dislocations-migration and propagation of cracks along grain boundaries, are completely irrelevant here. It is also notable that the measured values of the yield strength of carbon nanotubes (20–250 GPa) (ref. 50) suffer from large scattering, probably owing to structural imperfections in their structure.

APPLICATIONS

MECHANICAL, INCLUDING TRIBOLOGICAL APPLICATIONS

The quasi-spherical shape of the fullerene-like WS_2 and MoS_2 nanoparticles and their inert sulphur-terminated surface suggested that they could serve as superior solid lubricants in the form of additives to lubrication fluids, greases and for self-lubricating coatings[15]. These advances led to a surge of interest in this technology from a very large variety of industries, having virtually no common denominator other

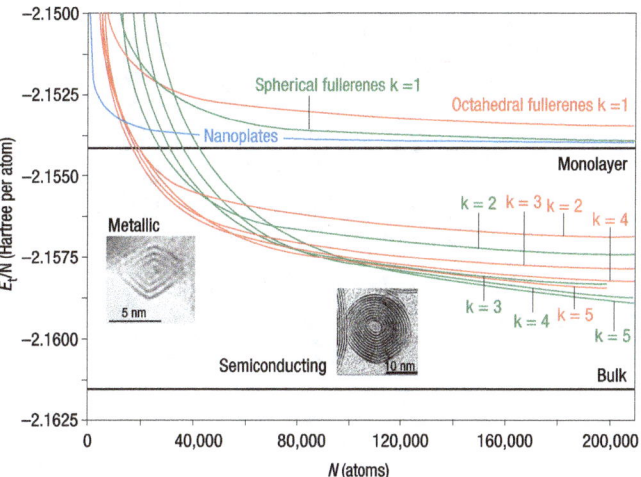

Figure 5 The energies per atom E_t/N of various MoS_2 nanostructures is presented as a function of its size (N is the total number of atoms). The calculations were carried out by a semi-empirical model[43]. The different calculated nanostructures are: multilayered octahedral fullerenes (red) and fullerene-like spherical nanoparticles (green), with different numbers of MoS_2 walls, k; triangular nanoplatelets (blue) are also calculated. Note that, according to calculations, the multiwall nano-octahedra are the most stable moieties from about 5,000–100,000 atoms (3–6 nm). This range extends to higher values than those obtained experimentally (the largest nano-octahedra observed in the experiments contained 25,000 Mo and S atoms). This discrepancy can be attributed to the approximations made in the model. Note also that the nano-octahedra are metallic according to the DFTB–MD calculations, whereas all the other forms of MoS_2 are semiconductors. This important fact is yet to be confirmed experimentally. Reprinted with permission from ref. 43. Copyright (2006) Wiley-VCH.

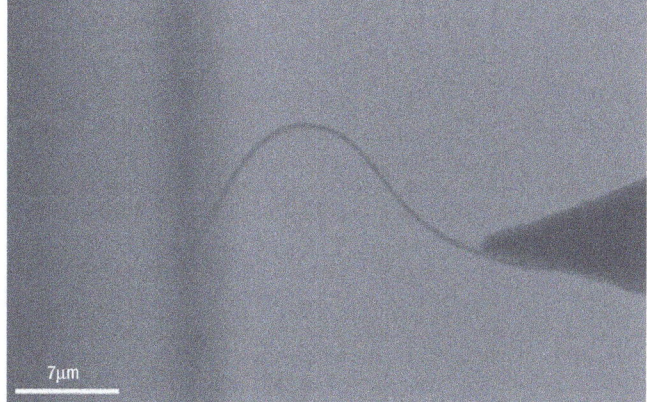

Figure 6 Tensile strength, bending and compression experiments on individual WS_2 nanotubes were carried out, providing quantitative data on the mechanical properties of these nanostructures. Shown here is an SEM image in which a WS_2 nanotube is compressed against the surface of a silicon wafer. The nanotube was glued to a cantilever by electron-beam irradiation in the presence of carbon vapours, which can easily be afforded in the SEM. It was pushed against a mirror flat silicon wafer surface using a micromanipulator. Note the flexibility of the elastic nanotube. Using known elastic models, the Young's modulus of the nanotubes was determined to be 150 GPa, that is, about one-seventh of the value of a carbon nanotube. Reprinted with permission from ref. 48. Copyright (2005) National Academy of Sciences, USA.

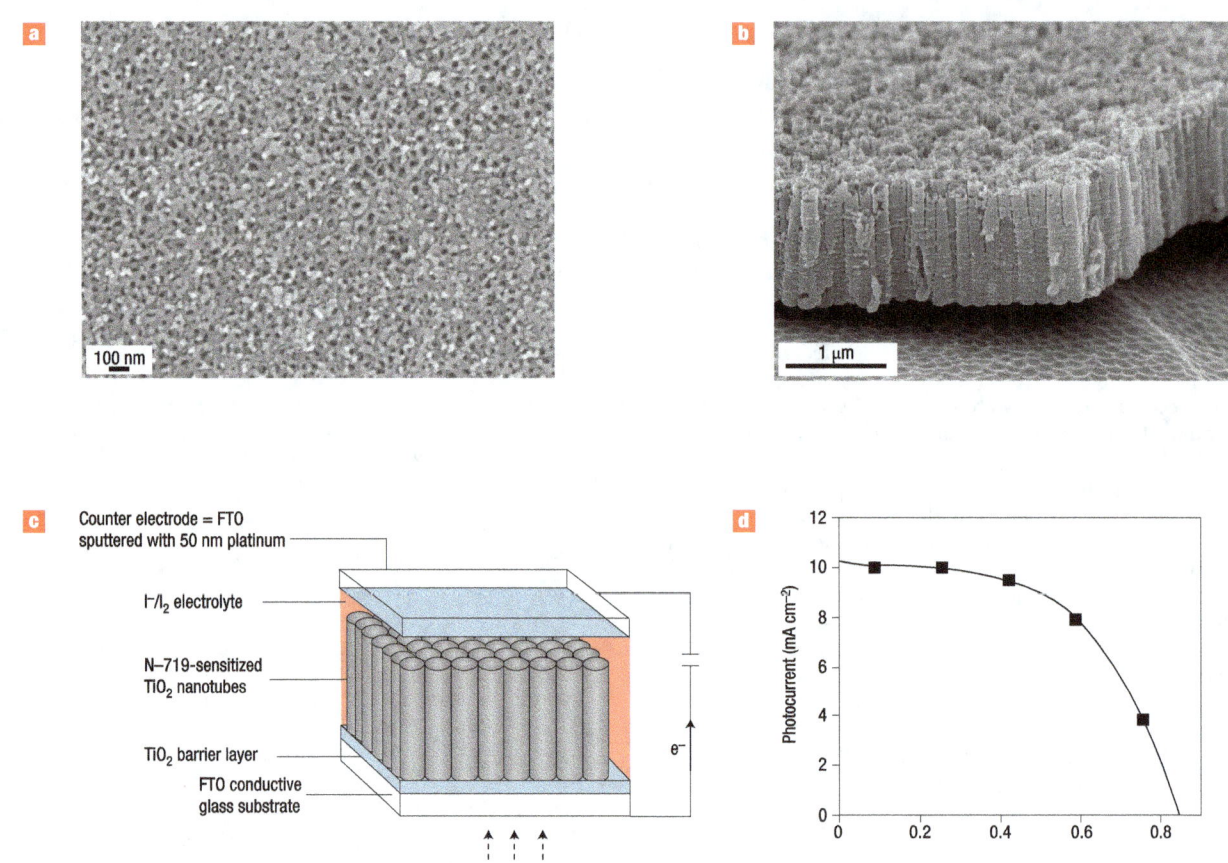

Figure 7 Anodization of titanium film, deposited on a conductive glass in acidic solutions containing HF, leads to the formation of a dense array of interconnected polycrystalline titania (TiO$_2$) nanotubes. This film can then be integrated into a dye-sensitized solar cell. **a**, An edge-on SEM image of a film made of an array of titanium nanotubes (46 nm in diameter and 17 nm wall thickness). Titanium film (4 μm) was deposited on fluorine-doped tin oxide film on glass and anodized at 12 V in 0.5% HF–acetic acid electrolyte. The film was immersed in 0.2 M TiCl$_4$ solution for 1 h and subsequently annealed for 3 h in air at 450 °C. **b**, The same film viewed by SEM, side-on. **c**, Schematics of a front-side-illuminated dye-sensitized solar cell. The film was soaked in Ru-bipyridine-based dye (N-719, Solaronix). The electrolyte contained 0.5 M LiI, 0.05 M I$_2$, 0.6 M *N*-methylbenzimidazole, 0.10 M guanidinium thiocyanate and 0.5 M tert-butylpyridine in methoxypropionitrile. A fluorine-doped tin oxide (FTO) conductive glass slide sputter-coated with 25 nm of Pt was used at the counter electrode. **d**, A current–voltage (*I–V*) curve under AM1.5 solar simulator illumination (100 mW cm^{-2}). The power output of this cell is given by *I* x *V* (8 x 0.59 = 4.7 mW) at the point of maximum power. The efficiency is power output/power input x 100 = 4.7%. The high photocurrent can be attributed to the large surface area of the nanotubes, the vectorial charge transport along the walls of the nanotube and the reduced electron–hole recombination of the anodized surface. This architecture minimizes losses occurring in nanoparticle-based photoelectrodes owing to the hopping of charge carriers across nanoparticle boundaries. Reprinted with permission from ref. 65. Copyright (2006) IOP.

than the need to reduce friction and wear. Numerous development programmes are underway, some of them at a very advanced stage or even having made products. The first medical application — alleviating friction in orthodontic wires — was recently demonstrated[51], with numerous others to follow. Toxicological tests on the fullerene-like WS$_2$ nanoparticles concluded that they are safe[52]. Mechanistic studies indicate that the spherical MS$_2$ nanoparticles exhibit rolling (sliding) friction under mild loads (<0.5 GPa) (hydrodynamic lubrication). Under heavier loads, where the spacing between the mating surfaces is small (mixed and boundary lubrication conditions), the nanoparticles slowly deform and exfoliate giving up WS$_2$ layers that stick to the underlying surfaces and provide easy shear[15,53].

In another series of experiments, fullerene-like WS$_2$ and MoS$_2$ nanoparticles were shown to withstand shockwaves of up to 25 and 30 GPa, respectively, with concurrent temperatures rising up to 1,000 °C (ref. 54). No evidence for significant structural degradation or phase changes was observed, making these materials probably the strongest cage molecules known today and offering them a plethora of applications for shielding; in cars and, in the future, perhaps also as additives to strengthen construction materials. Furthermore, Raman measurements of these nanoparticles indicated no degradation under hydrostatic pressures of 20 GPa, indicating again their high-pressure resilience[55]. It is noticeable that carbon nanotubes and fullerenes are unlikely to be stable under high pressure owing to the 1.7 GPa graphite-to-diamond transformation[56].

SENSING APPLICATIONS, ELECTRONICS AND QUANTUM HARVESTING

Room-temperature ferromagnetic VO$_x$-alkylamine nanotubes, which could be useful for spintronics, were obtained by lithium and iodine intercalation[57]. Moreover, the doping has the effect of shifting the Fermi level from the middle of the energy gap to either the valence (iodine) or the conduction band (lithium), turning this Mott-Hubbard insulator into a good conductor.

A number of interesting observations were recently made with titanate nanotubes of the formula $H_2Ti_3O_7$. By adding these nanotubes to organic light-emitting diodes, the luminosity was improved and the turn-on voltage decreased. These effects were attributed to the lower barrier for hole injection from the electrode to the organic semiconductor and to the improved hole transport in the organic film[58]. In another recent study, enhanced electrochromism was observed in films made of titanate nanotubes[59]. The titanate nanotube films were shown to exhibit faster proton diffusion and higher capacity than those made of TiO_2 (anatase) nanoparticles. Intercalating iron atoms in titanate nanotubes was shown to lead to a significant red shift of the absorption edge, and to interesting magnetic characteristics[60].

A biosensor for dopamine, which is an important marker for neurodegenerative diseases such as Parkinson's disease, based on titanate nanotube films deposited on glassy carbon electrodes, was recently reported[61]. Electrochemical measurements showed that the reduction peak of the dopamine was well separated from that of ascorbic and uric acids, which is not easily achievable in physiological solutions. This observation was attributed to the negative charge of the nanotube, attracting the positively charged dopamine moieties and excluding the negatively charged ascorbic and uric acid ions. In another series of remarkable studies, arrays of polycrystalline titania nanotubes were prepared by electrochemical anodization of titanium foil[62]. The nanotube array was coated with a 10 nm palladium film and its resistivity as a function of H_2 concentration in the gas atmosphere was monitored, showing perhaps the highest hydrogen sensitivity for any known hydrogen sensor, suggesting various possible medical and environmental applications. Optimizing the Ti anodization process, the authors have developed a water photocleavage system, which showed 7.9% photoconversion efficiency in the 320–400 nm range[63]. A dye-sensitized photoelectrochemical cell based on the titania nanotube films with overall 4.7% conversion efficiency was also reported (Fig. 7)[64,65]. The large surface area of the nanotube array, coupled with its low electron-hole recombination and efficient vectorial charge transport along the nanotube axis were among the reasons for the reported high efficiencies of the cell.

Nanofluidic devices based on polycrystalline metallic and metal oxide nanotubules have been used in the past for biomolecule diagnostics[66]. More recently, perfectly crystalline and very narrow (<20 nm) nanotubes were used as a platform for nanofluidics and unipolar charge transport[67]. The synthesized silica nanotubes were integrated in a metal oxide (KCl) solution field-effect transistor (MOSolFET) (Fig. 8). The conductance of the potassium ion (majority carrier) within the p-type nanotube channel was modulated by controlling the gate voltage. Furthermore, the translocation of individual DNA molecules in sub-femtolitre concentrations through the nanotube channel could be detected as a temporary current drop or rise.

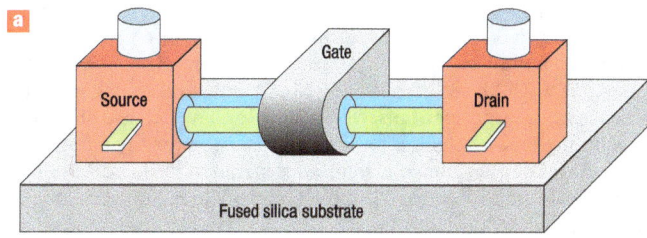

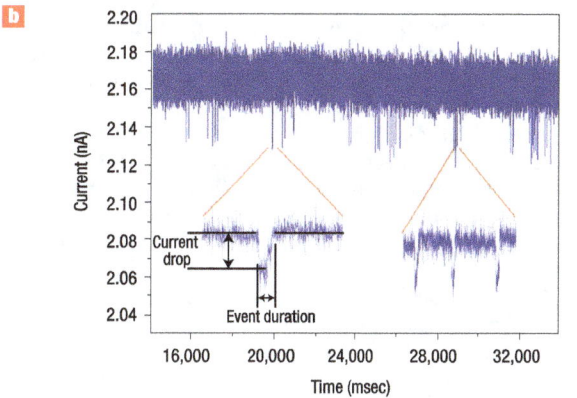

Figure 8 A single nanotube metal-oxide solution field effect transistor (MOSolFET) could be considered as the ionic analogue of the electronic field-effect transistor. The diameter of the nanotube (nanochannel) where the fluid flows is smaller than twice the thickness of the electrical double layer (Debye length) of the electrolyte solution. Therefore, there is no charge balance in the centre of the nanotube and if the surface of the nanotube is made negative by the gate electrode, the flowing solution becomes more concentrated with respect to the cations inside the nanotube channel. **a**, Schematics of the MOSolFET device with a silica nanotube (blue), which can serve as a detector for a single DNA molecule. **b**, The temporary drops in the current–time trace are markers for a single DNA molecule passage through the nanotube. Measurements were carried out in 2 M KCl solution. Reprinted with permission from ref. 67. Copyright (2006) ACS.

CATALYSIS AND BIOTECHNOLOGY

In a very elegant study, hollow nanoparticles of MoS_2, albeit with non-perfect crystallinity, were prepared by sonochemical reactions of $Mo(CO)_6$ in the presence of silica nanoparticles, which served as a template and were subsequently removed by HF etching and annealing[68]. These nanoparticles revealed high reactivity and selectivity towards hydrodesulphurization of thiophene. In a more recent study, Ni nanoparticles were deposited onto a MoS_2-nanotube support. A very reactive and selective catalyst for hydrodesulphurization of thiophene and a few of its derivatives was obtained[69]. These kinds of catalysts could play a major role in mitigating the environmental impact of sulphur-rich gasoline. Also recently, $(BiO)_2CO_3$ nanotubes were shown to exhibit a very strong antibacterial reactivity for *Helicobacter pylori*, which is implicated in peptic ulcers and gastritis[70]. Immobilization of various proteins and biomolecules, such as ferritine, cytochrome-c, strepavidin and glucose oxide onto multiwall BN nanotubes was demonstrated[71], suggesting numerous biotechnological and medical applications for such nanostructures. Various nanotubes, such as those of titanate or VO_x-alkylamine, which are produced by chemie-douce processes, are in fact nanoscrolls, which can be unwrapped, by varying the pH, for example. This attribute offers the possibility to use them as drug carriers for targeting specific tissues or tumours.

ELECTROCHEMICAL APPLICATIONS

A solution-based lithium intercalation of TiS_2 nanotubes with the formal charge of one lithium atom/TiS_2 molecule makes up one of many studies on nanotube-based intercalation batteries[22]. The same group also showed that high-purity TiS_2 nanotubes with open-ended tips can efficiently store 2.5 wt% of hydrogen at 298 K under pressure of 4 MPa (ref. 72). Detailed analysis showed that about 1 wt% of the hydrogen is chemisorbed and the rest of the hydrogen atoms are physisorbed to the nanotubes. Relatively large amounts of lithium could be inserted into $MoS_2I_{0.3}$ nanotube bundles (up to 3 mol Li per mol MoS_2), which were prepared by chemical vapour transport[73]. Comparing their electrochemical behaviour with those of bulk $2H\text{-}MoS_2$, one finds a significant increase in the amount of inserted lithium and a decrease of about 0.7 V in the onset potential for lithium insertion for the nanotube electrodes. The electron spin resonance signal of lithiated $MoS_2I_{0.3}$ nanotube bundles decreased upon air exposure, but at a much slower rate than lithiated bulk $2H\text{-}MoS_2$ crystals, indicating the greater stability of the lithiated nanotubes compared with lithiated bulk MoS_2.

A number of Li intercalation studies of VO_x-alkylamine nanotubes were reported in the past. Li intercalation was also studied in titanate[74] and SnO_2 (ref. 38) nanotubes. The energy density of the titanate nanotube-based electrode was 210 mA h g^{-1} at a current density of 0.24 A g^{-1}. The capacity remained high even at relatively high discharge rates (144 mA h g^{-1} at a current density of 3.6 A g^{-1}, respectively). Furthermore, the electrode was found to be stable during many charge/discharge cycles. The capacity of the SnO_2 nanotube-based electrode was reported to be 525 mA h g^{-1} after 80 cycles[38]. Polycrystalline $Ni(OH)_2$ tubes coated with 5 wt% $Ca(OH)_2$, $Co(OH)_2$, or $Y(OH)_3$ were prepared using an AAO template[75]. The electrochemical measurements showed that both pure and coated $Ni(OH)_2$ tube electrodes exhibited electrochemical performance superior to that of a conventional electrode made of spherical $Ni(OH)_2$ particles. The pure $Ni(OH)_2$ tube electrode displayed the highest discharge capacity of 315 mA h g^{-1}, but the $Co(OH)_2$-coated electrode exhibited higher recycling stability.

CONCLUSIONS

Inorganic nanotubes and fullerene-like nanostructures are a generic structure of inorganic layered two-dimensional compounds. In some cases, such as WS_2, MoS_2, BN and V_2O_5, fullerene-like nanoparticles and nanotubes are produced in gross amounts. However, size and shape control is still in its infancy. More recently, nanotubes of numerous inorganic compounds with non-layered structure have been prepared using various templates. Study of these novel nanostructures has led to the observation of a number of interesting properties offering numerous potential applications in tribology, high-energy-density batteries, sensors, photoconversion of solar energy and nanoelectronics.

doi:10.1038/nnano.2006.62

References

1. Pauling, L. The structure of the chlorites. *Proc. Nat. Acad. Sci. USA* **16**, 578–582 (1930).
2. Bicerano, J. Maryanick, D. S. & Lipscomb, W. N. Molecular orbital studies on large closo boron hydrides. *Inorg. Chem.* **17**, 3443–3453 (1978).
3. Lipscomb, W. N. and Massa, L. Examples of large closo boron hydride analogs of carbon fullerenes. *Inorg. Chem.* **31**, 2297–2299 (1992).
4. Kroto, H. W., Heath, J. R., O'Brien, S. C., Curl, R. F. & Smalley, R. E. C_{60}: Buckminsterfullerene. *Nature* **318**, 162–163 (1985).
5. Iijima, S. Helical microtubules of graphitic carbon. *Nature* **354**, 56–58 (1991).
6. Tenne, R., Margulis, L., Genut M. & Hodes, G. Polyhedral and cylindrical structures of tungsten disulphide. *Nature* **360**, 444–445 (1992).
7. Margulis, L., Salitra, G., Tenne, R. & Talianker, M. Nested fullerene-like structures. *Nature* **365**, 113–114 (1993).
8. Feldman, Y., Wasserman, E., Srolovitz D. J. & Tenne R. High rate gas phase growth of MoS_2 nested inorganic fullerene-like and nanotubes. *Science* **267**, 222–225 (1995).
9. Rosenfeld Hacohen Y. *et al.* Synthesis of $NiCl_2$ nanotubes and fullerene-like structures by laser ablation. *Phys. Chem. Chem. Phys.* **5**, 1644–1651 (2003).
10. Albu-Yaron, A. *et al.* Preparation and structural characterization of stable Cs_2O closed-cage structures. *Angew. Chem. Int. Edn* **44**, 4169–4172 (2005).
11. Albu-Yaron, A. *et al.* Synthesis of fullerene-like Cs_2O nanoparticles by concentrated sunlight, *Adv. Mater.* (in the press).
12. Goldberger, J. *et al.* Single-crystal gallium nitride nanotubes. *Nature* **422**, 599–602 (2003).
13. Li, Y., Bando, Y. & Golberg, D. Single-crystalline In_2O_3 nanotubes filled with In. *Adv. Mater.* **15**, 581–585 (2003).
14. Yin, L. W., Bando, Y., Golberg, D. & Li, M. Growth of single-crystal indium nitride nanotubes and nanowires by controlled carbon-nitridation reaction. *Adv. Mater.* **16**, 1833–1838 (2004).
15. Rapoport, L., Fleischer. N. & Tenne, R. Applications of WS_2 (MoS_2) inorganic nanotubes and fullerene-like nanoparticles for solid lubrication and for structural nanocomposites. *J. Mater. Chem.* **15**, 1782–1788 (2005).
16. Tenne, R. & Rao, C. N. R. Inorganic nanotubes. *Phil. Trans. R. Soc. Lond. A* **362**, 2099–2125 (2004).
17. Remskar, M. Inorganic nanotubes. *Adv. Mater.* **16**, 1497–1504 (2004).
18. Patzke, G. R., Krumeich, F. & Nesper, R. Oxidic nanotubes and nanorods–anisotropic modules for a future nanotechnology. *Angew. Chem. Int. Edn* **41**, 2446–2461 (2002).
19. Tenne, R. Advances in the synthesis of inorganic nanotubes and fullerene-like nanoparticles. *Angew. Chem. Int. Edn* **42**, 5124–5132 (2003).
20. Rao, C. N. R. & Nath, M. Inorganic nanotubes. *Dalton Trans.* 1–25 (2003).
21. Schuffenhauer, C., Popovitz-Biro, R. & Tenne, R., Synthesis of NbS_2 nanoparticles with (nested) fullerene-like structure (IF). *J. Mater. Chem.* **12**, 1587–1591 (2002).
22. Chen, J., Tao, Z. L. & Li, S. L. Lithium intercalation in open-ended TiS_2 nanotubes. *Angew. Chem. Int. Edn* **42**, 2147–2151 (2003).
23. Etzkorn, J. *et al.* Metal-organic chemical vapor deposition synthesis of hollow inorganic-fullerene-type MoS_2 and $MoSe_2$ nanoparticles. *Adv. Mater.* **17**, 2372–2375 (2005).
24. Li, X. P., Ge, J. P. & Li, Y. D. Atmospheric pressure chemical vapor deposition: An alternative route to large-scale MoS_2 and WS_2 inorganic fullerene-like nanostructures and nanoflowers. *Chem. Eur. J.* **10**, 6163–6171 (2004).
25. Margolin, A., Popovitz-Biro, R., Albu-Yaron, A., Rapoport, L. & Tenne, R. Inorganic fullerene-like nanoparticles of TiS_2. *Chem. Phys. Lett.* **411**, 162–166 (2005).
26. Malliakas, C. D. & Kanatzidis, M. G. Inorganic single wall nanotubes of $SbPS_{4-x}Se_x$ (0 <x<3) with tunable band gap *J. Am. Chem. Soc.* **128**, 6538–6539 (2006).
27. Therese, H. A. *et al.* VS_2 nanotubes containing organic-amine templates from the NT-VOx precursors and reversible copper intercalation in NT-VS_2. *Angew. Chem. Int. Edn* **44**, 262–265 (2005).
28. Armstrong, A. R., Canales, J. & Bruce, P. G. WO_2Cl_2 Nanotubes and nanowires. *Angew. Chem. Int. Edn* **43**, 4899–4902 (2004).
29. Krivovichev, S. V. *et al.* Nanoscale tubules in uranyl selenates. *Angew. Chem. Int. Edn* **44**, 1134–1136 (2005).
30. Falini, G. *et al.* Tubular-shaped stoichiometric chrysotile nanocrystals. *Chem. Eur. J.* **10**, 3043–3049 (2004).
31. Roveri, N. Geoinspired Synthetic Chrysotile Nanotubes. *J. Mater. Res.* (in the press).
32. Mukherjee, S., Bartlow, V. M. & Nair, S. Phenomenology of the growth of single-walled aluminosilicate and aluminogermanate nanotubes of precise dimensions. *Chem. Mater.* **17**, 4900–4909 (2005).
33. Zhi, C.Y., Bando Y., Tan C.C., Goldberg, D. Effective precursor for high yield synthesis of pure BN nanotubes. *Solid State Commun.* **135**, 67–70 (2005).
34. Zhi, C.Y. *et al.* Purification of boron nitride nanotubes through polymer wrapping. *J. Phys. Chem. B* **110**, 1525–1528 (2006).
35. Fan, H. J. *et al.* Monocrystalline spinel nanotube fabrication based on the Kirkendall effect. *Nature Mater.* **5**, 627–631 (2006).
36. Shen, G. *et al.* Single-crystalline nanotubes of II_3-V_2 semiconductors (submitted).
37. Sehayek, T., Lahav, M., Popovitz-Biro, R., Vaskevich, A. & Rubinstein, I. Template synthesis of nanotubes by room-temperature coalescence of metal nanoparticles. *Chem. Mater.* **17**, 3743–3748 (2005).
38. Wang, Y., Lee, J. Y. & Zeng, H. C. Polycrystalline SnO_2 nanotubes prepared via infiltration casting of nanocrystallites and their electrochemical application. *Chem. Mater.* **17**, 3899–3903 (2005).
39. Liu, Z. *et al.* Single Crystalline Magnetite Nanotubes. *J. Am. Chem. Soc.* **127**, 6–7 (2005).
40. Jhi, S.-H., Roundya, D. J., Louie, S. G. & Cohen, M. L. Formation and electronic properties of double-walled boron nitride nanotubes. *Solid State Comm.* **134**, 397–402 (2005).
41. Zhao, M., Xia, Y., Zhang, D. & Mei, L. Stability and electronic structure of AlN nanotubes. *Phys. Rev. B* **68**, 235415 (2003).
42. Linnolahti, M. & Pakkanen, T. A. Quantum chemical treatment of large nanotubes via use of line group symmetry: Structural preferences of magnesium dichloride nanotubes. *J. Phys. Chem. B* **110**, 4675–4678 (2006).
43. Enyashin, A. N. *et al.* Structure and stability of molybdenum sulfide fullerenes. *Angew. Chem. Int. Edn* (in the press).
44. Parilla, P. A. *et al.* Formation of nanooctahedra in molybdenum disulfide and molybdenum diselenide using pulsed laser vaporization. *J. Phys. Chem. B* **108**, 6197–6207 (2004).
45. Nath, M., Kar S., Raychaudhuri, A. K. & Rao, C. N. R. Superconducting $NbSe_2$ nanostructures. *Chem. Phys. Lett.* **368**, 690–695 (2003).
46. Zhao, M., Xia, Y., Li, F., Zhang, R. Q. & Lee, S. T. Strain energy and electronic structures of silicon carbide nanotubes: Density functional calculations. *Phys. Rev. B* **71**, 085312 (2005).
47. Khoo, K. H., Mazzoni, M. S. C. & Louie, S. G. Tuning the electronic properties of boron nitride nanotubes with transverse electric fields: A giant dc Stark effect. *Phys. Rev. B* **69**, 201401(R) (2004).
48. Ishigami, M., Sau, J. D., Aloni, S., Cohen, M. L. & Zettl A. Observation of the giant stark effect in boron-nitride nanotubes. *Phys. Rev. Lett.* **94**, 056804 (2005).
49. Kaplan-Ashiri, I. *et al.* On the mechanical behaviour of WS_2 nanotubes under axial tension and compression. *Proc. Natl. Acad. Sci. USA* **103**, 523–528 (2006).
50. Barber, A. H., Kaplan-Ashiri, I., Cohen, S. R., Tenne, R. & Wagner, H. D. Stochastic strength of nanotubes: An appraisal of available data. *Comp. Sci. Technol.* **65**, 2380–2384 (2005).
51. Katz, A., Redlich, M., Rapoport, L., Wagner, H. D. & Tenne, R. Improved orthodontic wires using fullerene-like WS_2 self-lubricating coatings. *Tribol. Lett.* **21**, 135–139 (2006).
52. Moore, G. E. Acute inhalation toxicity study in rats – limit test. Product safety laboratories, study number 18503. Dayton, New Jersey, USA 2005, Jan 18.
53. Joly-Pottuz, L. *et al.* Ultralow-friction and wear properties of IF-WS_2 under boundary lubrication. *Tribol. Lett.* **18**, 477–485 (2005).
54. Zhu, Y. Q. *et al.* Shock-absorbing and failure mechanism of WS_2 and MoS_2 nanoparticles with fullerene-like structure under shockwave pressures. *J. Am. Chem. Soc.* **127**, 16263–16272 (2005).
55. Joly-Pottuz, L. *et al.* Pressure-induced exfoliation of inorganic fullerene-like WS_2 particles in a Hertzian contact. *J. Appl. Phys.* **99**, 023524 (2006).

56. Bundy, F. P. The P, T phase and reaction diagram for elemental carbon. *J. Geophys. Res.* **85**, 6930–6936 (1979).

57. Krusin-Elbaum, L. *et al.* Room-temperature ferromagnetic nanotubes controlled by electron or hole doping. *Nature* **431**, 672–676 (2004).

58. Qian, L. *et al.* Improved optoelectronic characteristics of light-emitting diodes by using a dehydrated nanotube titanic acid (DNTA)-polymer nanocomposites. *J. Phys. Chem. B* **108**, 13928–13931 (2004).

59. Tokudome, H. & Miyauchi, M. Electrochromism of titanate-based nanotubes. *Angew. Chem. Int. Edn* **44**, 1974–1977 (2005).

60. Xu, X. G., Ding, X., Chen, Q. & Peng, L.-M. Electronic, optical, and magnetic properties of Fe-intercalated $H_2Ti_3O_7$ nanotubes: First-principles calculations and experiments. *Phys. Rev. B* **73**, 165403 (2006).

61. Liu, A., Wei, M., Honma, I. & Zhou, H. Biosensing properties of titanate-nanotube films: selective detection of dopamine in the presence of ascorbate and uric acid. *Adv. Funct. Mater.* **16**, 371–376 (2006).

62. Mor, G. K., Carvalho, M. A., Varghese, O. K., Pishko, M. V. & Grimes, C. A. A room-temperature TiO_2-nanotube hydrogen sensor able to self-clean photoactively from environmental contamination. *J. Mater. Res.* **19**, 628–634 (2004).

63. Ruan, C. M., Paulose, M., Vargese, O. K. & Grimes, C. A. Enhanced photoelectrochemical-response in highly ordered TiO_2 nanotube-arrays anodized in boric acid containing electrolyte. *Sol. Energ. Mater. Sol. Cell* **90**, 1283–1295 (2006).

64. Paulose, M.*et al.* Backside illuminated dye-sensitized solar cells based on titania nanotube array electrodes. *Nanotechnology* **17**, 1446–1448 (2006).

65. Paulose, M., Shankar, K., Varghese, O. K., Mor, G. K. & Grimes, C. A. Application of highly-ordered TiO_2 nanotube-arrays in heterojunction dye-sensitized solar cells. *J. Phys. D* **39**, 2498–2503 (2006).

66. Mitchell, D. T. *et al.* Smart nanotubes for bioseparations and biocatalysis. *J. Am. Chem. Soc.* **124**, 11864–11865 (2002).

67. Goldberger, J., Fan, R. & Yang, P. Inorganic nanotubes: A novel platform for nanofluidics. *Acc. Chem. Res.* **39**, 239–248 (2006).

68. Dhas, N. A. & Suslick, K. S. Sonochemical preparation of hollow nanospheres and nollow nanocrystals. *J. Am. Chem. Soc.* **127**, 2368–2369 (2005).

69. Cheng, F., Gou, X., Chen, J. & Xu, Q. Ni/MoS$_2$ nanocomposites as the catalysts for hydrodesulphurization of thiophene and thiophene derivatives. *Adv. Mater.* **18**, 2561–2564 (2006).

70. Chen, R., So, M. H., Yang, J., Deng, F., Chea, C. M. & Sun, H. Fabrication of bismuth subcarbonate nanotube arrays from bismuth citrate. *Chem. Commun.* 2265–2267 (2006).

71. Zhi, C., Bando, Y., Tang, C. & Golberg, D. Immobilization of proteins on BN nanotubes. *J. Am. Chem. Soc.* **127**, 17144–17145 (2005).

72. Chen, J., Li, S., Tao, Z., Shen, Y. & Cui, C. *J. Am. Chem. Soc.* **125**, 5284–5285 (2003).

73. Dominko, R. *et al. Electrochimica Acta* **48**, 3079–3084 (2003).

74. Li, J., Tang, Z. & Zhang, Z. H-titanate nanotube: a novel lithium intercalation host with large capacity and high rate capability. *Electrochem. Commun.* **7**, 62–67 (2005).

75. Li, W., Zhang, S. & Chen, J. Synthesis, characterization, and electrochemical application of $Ca(OH)_2$-, $Co(OH)_2$-, and $Y(OH)_3$-coated $Ni(OH)_2$ tubes. *J. Phys. Chem. B* **109**, 14025–14032 (2005).

76. Tenne, R. Issue 15 cover image. *Phys. Chem. Chem. Phys.* **6** (2004).

Acknowledgements

I am grateful to M. Bar-Sadan, I. Kaplan-Ashiri, R. Rosentsveig, A. Margolin, A, Albu-Yaron, R. Popovitz-Biro, S. R. Cohen, G. Seifert, M. Jansen, H. D. Wagner and J.M. Gordon for their help. The support of the Israeli Ministry of Science and Technology and the Israel Science Foundation is acknowledged.

The role of interparticle and external forces in nanoparticle assembly

The past 20 years have witnessed simultaneous multidisciplinary explosions in experimental techniques for synthesizing new materials, measuring and manipulating nanoscale structures, understanding biological processes at the nanoscale, and carrying out large-scale computations of many-atom and complex macromolecular systems. These advances have led to the new disciplines of nanoscience and nanoengineering. For reasons that are discussed here, most nanoparticles do not 'self-assemble' into their thermodynamically lowest energy state, and require an input of energy or external forces to 'direct' them into particular structures or assemblies. We discuss why and how a combination of self- and directed-assembly processes, involving interparticle and externally applied forces, can be applied to produce desired nanostructured materials.

YOUNJIN MIN[1], MUSTAFA AKBULUT[2], KAI KRISTIANSEN[1], YUVAL GOLAN[3] AND JACOB ISRAELACHVILI[1]*

[1]Department of Chemical Engineering, University of California Santa Barbara, Santa Barbara, California 93106, USA
[2]Department of Chemical Engineering, Princeton University, A-306 Engineering Quadrangle, Princeton, New Jersey 08544-5263, USA
[3]Department of Materials Engineering, and Ilse Katz Institute of Nanotechnology, Ben-Gurion University of the Negev, Beer-Sheva 84105, Israel
*email: jacob@engineering.ucsb.edu

In the olden days — a mere 20 years ago — people talked about hard colloidal particles and soft vesicles or micelles, which covered all sizes from the micrometre to the nanometre regime. Now we talk about nanoparticles, nanoscience and nanotechnology. Is this change in terminology scientific or faddish? Certainly, there are electronic and optical properties of materials that seem to be unique to nanoscale particles, such as quantum dots, which exhibit large changes, or differences, in properties in the size range from 1–10 nm (refs 1–3).

In this review we have attempted to make a critical analytic assessment of this question as it applies to the classical material and mechanical properties of nanoparticles and nanoparticle assemblies, such as their organization, interparticle adhesion forces, elastic properties, phase behaviour and dynamic (for example, relaxation) processes. We review how and why these properties change from the smallest particles with sizes in the 1 nm regime (for example, anisotropic liquid-crystal molecules and C_{60}), through the nanoscale regime, with sizes in the range 1–100 nm, and into the micro- and macroscale regime (colloidal particles, powders, granular and microstructured materials), with sizes greater than 200 nm.

But is size the best way to describe or define a nanoparticle scientifically? Some structures are nanosized in all three orthogonal directions, such as the nanospheres, spherical micelles and surfactant-coated core–shell particles shown in Fig. 1. Others are nanoscaled in only two directions (for example, nanorods and wires, actin and microtubules), with still others in only one direction (for example, sheets such as surfactant and lipid bilayers and biological membranes).

Some materials exhibit no unusual properties in the nanoparticle regime. The intrinsic properties of molecular clusters of simple nonpolar hydrocarbon molecules such as alkanes are 'additive', that is, independent of the size of the cluster from the single molecule to the bulk phase. In other systems, large changes do occur, but continuously, as the clusters grow from the atomic/molecular to the macroscopic/bulk. Here, individual nanoparticles, that is, those in the nanoparticle state or size regime, cannot be said to exhibit special properties, just transitional values. However, as discussed below, the properties of assemblies of such particles may exhibit peak values when the particles are in the nano-regime. Other materials can exhibit truly unique properties in the nano-regime, even for isolated particles.

Size is not the only criterion that imparts qualitative or quantitative differences to nanoparticles: geometry and shape also have a role, although most effects usually manifest themselves when at least one of the three nanoparticle dimensions falls in the nanoscale regime, from 1 to 100 nm (Fig. 1), which for spheres spans six decades in volume or number of constituent atoms.

As particles grow in size a number of intrinsic properties change, some qualitatively, others quantitatively; some affect the equilibrium (thermodynamic) properties, others the non-equilibrium (dynamic) properties such as relaxation times. Some affect individual nanoparticles, others affect assemblies of nanoparticles. Nanoparticle assemblies can be ordered or disordered, and the particles can be held together by weak non-covalent or strong covalent or metallic bonds. Others can be sintered to provide robust, porous high-surface-to-volume nanostructured materials, and in this regard they are similar to zeolites (which may be thought of as an assembly of interconnected nanoholes). All of these features and their implications will now be considered.

When citing this review, please cite the original version as shown on the contents page of this chapter.

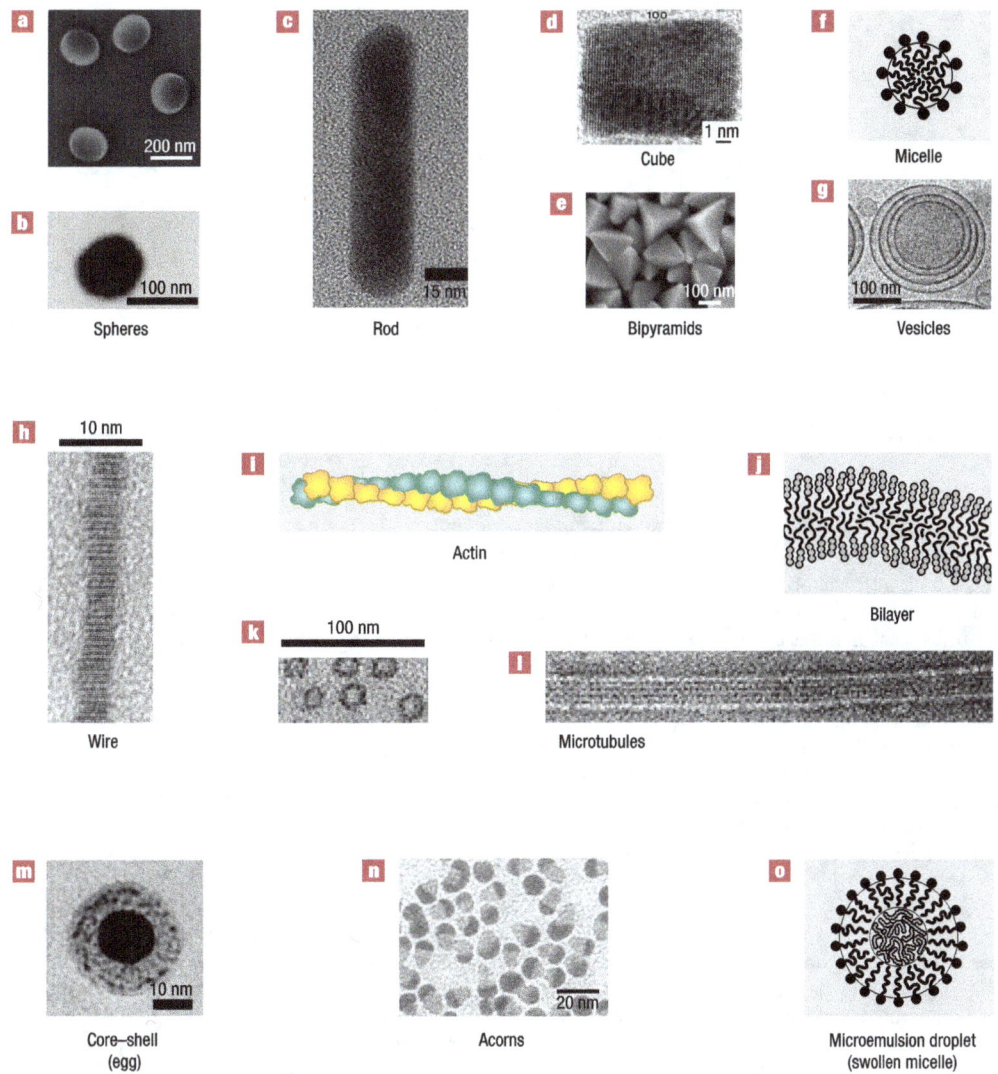

Figure 1 Different types of nanoparticles. A nanoparticle can be defined as one having at least one dimension in the nanometre regime (arbitrarily set at 1–100 nm), and can be further classified according to its chemical composition, cohesive energy, hardness, opto-electronic properties, and so on (see Table 1). **a–g**, One-component nanoparticles, simple shapes, hard or soft. SEM image of main-chain polyether nanoparticles[93] (**a**); TEM image of 60-nm gold nanosphere[94] (**b**); TEM image of Au nanorod[95] (**c**); HRTEM image of cubic Pt nanocrystal oriented along [001][96] (**d**); SEM image of Ag bipyramids[97] approximately 150 nm in edge length (**e**); a schematic micelle (**f**) and a cryo-TEM image of evolving vesicles (**g**). **h–l**, One-component nanoparticles with at least one dimension in the nanoregime. HRTEM image of PbSe nanowire[14] formed in the presence of oleic acid and *n*-tetradecylphosphonic acid (**h**); schematics of actin (**i**) and surfactant or lipid bilayer (**j**); TEM images of microtubules[98] (**k,l**; scale bar refers to both). **m–o**, Two or more components (core–shell structures and so on). TEM image of Au colloid[99] stabilized by sodium citrate and 19 layers of polyelectrolyte (**m**); HRTEM image of CoPd 'acorns'[100] (**n**); microemulsion droplet (**o**). Figures reprinted with permission from: **a** ref.93 ©2005 NPG; **b** ref. 94 © 2002 APS; **c** ref. 95 © 2006 APS; **d** ref. 96 © 2000 ACS; **e** ref. 97 © 2006 ACS; **f,g**, courtesy of R. Reeder and Janet Burns, Proctor and Gamble; **h** ref. 14 © 2005 ACS; **k,l** ref. 98 © 2005 PNSA; **m** ref. 99 © 2004 ACS; **n** ref. 100 © 2004 ACS.

HOW AND WHY ARE NANOPARTICLES DIFFERENT FROM OTHERS?

SKIN DEPTH OR SURFACE-AREA-TO-VOLUME SCALING EFFECTS

The intrinsic properties of some nanosized aggregates can be 'additive', given by summing the contributions from the individual atoms or molecules, or 'non-additive', depending on cooperative effects. For example, the optical properties such as the dielectric constant and refractive index of clusters of noble gas atoms or alkane molecules are additive, and can be calculated by summing the contributions from the isolated molecules. In contrast, many properties, especially of metals and semiconducting materials, are not additive: their a.c. conductivity, strength, melting temperatures, chemical activity,

and opto-electronic properties are often determined within a surface layer, attaining their maximum value only after a certain critical 'skin-depth' or 'proximity length', λ, from the surface has been reached. The most common expression for such a property is[4]:

$$I(x) = I_\infty - (I_\infty - I_r)e^{-(x-r)/\lambda} \qquad (1)$$

where x is the distance from the surface, $I(x\rightarrow\infty) = I_\infty$ is the bulk value per unit area of the surface in the limit of $x\rightarrow\infty$, r is the atomic or molecular size, $I(x\rightarrow r) = I_r$ is the molecular value in the limit of $x\rightarrow r$, and where λ typically lies between 1 and 100 nm (3 to 300 atomic radii, r)[5-7]. Table 1 gives the values of λ for the more important

Table 1 Skin depths and proximity lengths, λ, of some common material properties.

Property	λ	Equations for particles (or thin films) of radius (film thickness) R
Cohesive energy, latent heat*	0.3 – 2 nm	$E_r = E_{bulk}(1 - \lambda/R)$
Surface tension*	0.3 – 5 nm	Tolman equation: $\gamma_r = r\gamma_{bulk}/(R+\lambda)$
Melting point depression of metals*	1 – 3 nm	$T_r = T_{bulk}(1 - \lambda/R)$
Vapour pressure	0.3 – 10 nm	Kelvin equation: λ depends on the relative humidity or vapour pressure.
Chemical reactivity		
Covalent bonds	0.5 nm	-
Electron transfer	0.7 – 1.0 nm	'Harpooning effect'
Opto-electronic properties†		
a.c. conductivity (metals)	~1 μm at 1 GHz	$\lambda \propto 1/\sqrt{frequency}$
d.c. (conducting polymers)	~Two molecules	-
Bandgap energy	2 – 30 nm	-

*The first three properties are intimately related and interdependent.
†Detailed discussion of electronic and optical properties is outside the scope of this review.

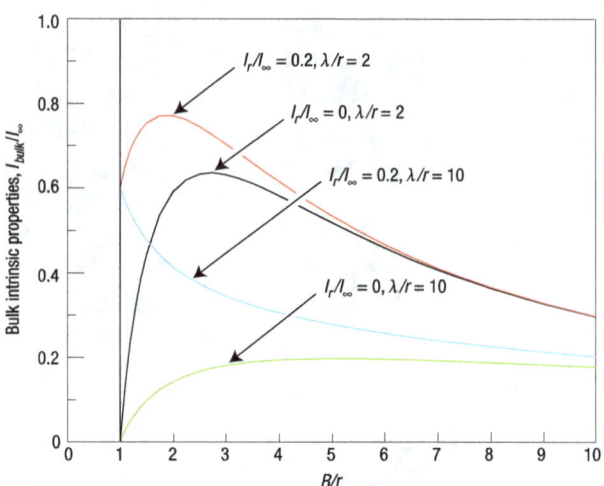

Figure 2 Illustration of how certain bulk properties of nanostructured materials can peak at a particular radius of their constituent nanoparticles. The figure shows four plots of the bulk intrinsic property of a material composed of spheres of radius R where the property of interest, I, is proportional to the total volume within a surface layer of characteristic 'skin depth' or 'proximity length' λ, as given by equation (1). Putting x = R in equation (1), the bulk intrinsic property of the material is therefore $I_{bulk} = 4\pi R^2 I(R) \times$ (particle density) $\approx 4\pi R^2 [I_\infty - (I_\infty - I_r)e^{-(R-r)/\lambda}]/4/3\pi R^3$. Normalizing R by the molecular radius, r, the bulk intrinsic property may be written as $I_{bulk} = 3(r/R)[I_\infty - (I_\infty - I_r)e^{-[(R/r)-1]r/\lambda}]$. Plots of I_{bulk} are shown for various values of I_r/I_∞ and the normalized skin depth, λ/r. Table 1 lists values of λ for a range of chemical and physical properties.

physical and chemical properties of thin films, small particles and highly curved surfaces. Figure 2 shows how proximity effects can give rise to a peak in the bulk intrinsic property of a nanostructured material I_{bulk} at some nanoscale radius of the particles, R, even though the properties of the individual nanoparticles do not peak at any particular size. For $I_\infty \gg I_r$ the peak values in I_{bulk} occur at R/r values close to $\sqrt{2\lambda/r}$, that is, determined by a combination of molecular and skin-depth dimensions. Equation (1) and the plots in Fig. 2 also show that not all combinations of values for I_r/I_∞ and λ/r exhibit a maximum at particle radii R greater than the atomic or molecular size r, that is, in the physically meaningful range of R/r>1. It can be shown that, to a good approximation, for λ in the range 1<λ/r<20, all points that fall below the curve $I_r/I_\infty = 0.5(\lambda/r)^{-3/4}$ exhibit a maximum value at R/r≥1.

EQUILIBRIUM EFFECTS

As particle sizes exceed the range of the interparticle forces between them, whether in solution or air (Fig. 3a), the phase diagrams change qualitatively: the liquid–vapour coexistence and critical point vanish, that is, there is no liquid phase. The system has a disordered and an ordered state, but nothing equivalent to a vapour–liquid coexistence regime. This is the reason why C_{60} has no liquid phase[8,9] (Fig. 3b), and most colloidal particles aggregate straight into crystals from the gas phase if the attractive forces are short-ranged (Fig. 3c). But if the forces are long-ranged (compared with the particles' size), for example, repulsive electrostatic 'double-layer' forces, the particles can have a vapour-like/liquid-like coexistence regime. This phenomenon therefore depends on the type of force acting between the particles, and can therefore 'kick in' for particles as small as C_{60} or as large as micrometre-sized colloidal particles.

Another thermodynamic scaling effect is to do with the natural polydispersity of soft self-assembling particles such as micelles (Fig. 1). With increasing size, the polydispersity in size or radius R or aggregation number N increases in absolute terms, but decreases relative to R or N, that is, the normalized distribution becomes narrower with increasing size (although the 'dimensionality' of the aggregates also have an important role in determining their polydispersity[10]).

DYNAMIC AND NON-EQUILIBRIUM EFFECTS

Some important dynamic effects are strongly influenced by particle size. First, as particles increase in size, say radius R, their van der Waals adhesion or binding energies and forces, E_{ad} and F_{ad}, increase linearly with R, that is, $E_{ad} \propto R$ (ref. 11). For atomically sized particles, E_{ad} is of order kT, but for particles larger than 1–2 nm E_{ad} becomes much larger than kT. As the adhesion (or binding) energy appears as $e^{-E_{ad}/kT}$

in equations to do with bonding lifetimes, dynamics, relaxations and equilibration times, nanoparticle systems become increasingly non-equilibrium systems, that is, kinetically trapped in whatever state they were prepared. Such systems can show 'ageing effects' such as deformational creep. Long relaxation times have serious practical implications for the preparation and processing of nanoparticle assemblies (see later), especially those that are presumed to rely on self-assembly — the spontaneous, naturally occurring ordering process, dictated by thermodynamics. When thermodynamics no longer dictates the self-assembly pathway (or does so only after a very long time), the particles become ordered by directed or engineered assembly, which is driven by the externally applied forces on the system such as compressive or shear stresses, gravity, magnetic or electric fields, as discussed later. Directed assembly depends on the method of preparation and the previous history of the sample, which must therefore be carefully chosen and controlled at all stages.

The second dynamic effect is to do with surface diffusion. In comparison with large particles, the diffusion of molecules or atoms, for example, gold, over the surface of a nanoparticle is rapid, with important consequences. For example, because the mean-square distance diffused is proportional to the time t, as described by the diffusion equation, $<x>^2 \propto Dt$, if a surface molecule takes one year to diffuse to the other side of a 10 μm colloidal particle, it will take ~30 s to diffuse to the other side of a 10 nm nanoparticle. This has profound implications for particle sintering and coalescence, diffusion-determined deformations, chemical reactions, and other changes in nanoparticle assemblies when studied over laboratory or 'engineering' timescales[12]. Similar considerations apply to the relaxation times of diffusing nanoparticles in liquids: for example, the characteristic rotational relaxation time for a sphere of volume V in a liquid of viscosity η is $\tau_{rot} = 3V\eta/kT$ (ref. 13). For a 10 nm sphere in water at room temperature we obtain $\tau_{rot} \approx 1$ μs, but 10^9 times longer

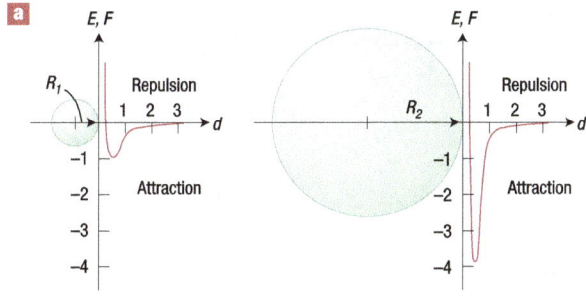

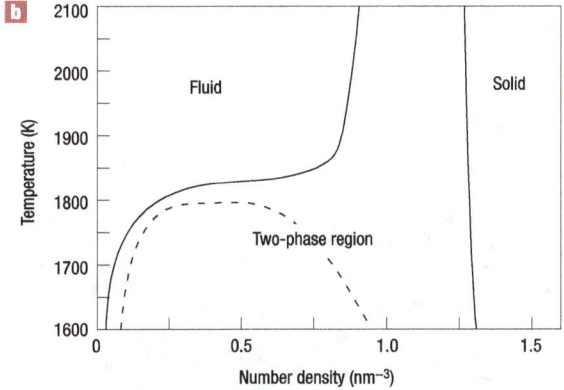

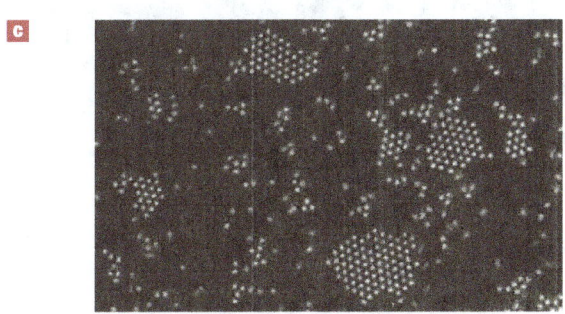

Figure 3 Effects of size on phase states of nanoparticles and small colloidal particles. **a**, Schematic force–distance functions for particles of different size but interacting through the same type of force, for example, the attractive van der Waals force. **b**, Phase diagram for nanoparticles, shown here for C_{60}, and small colloidal particles[8]. The solid lines denote the boundaries of the solid–liquid coexistence. The dashed-line denotes the (metastable) liquid–vapour coexistence line. **c**, Image of small (<1 μm) colloidal particles showing gas–solid coexistence[101]. Figures reprinted with permission from: **b** ref. 8 © 1993 NPG; **c** ref. 101 © 1980 Springer.

for a 10 μm sphere. Larger particles take much longer to become ordered and, conversely, to become disordered.

Strictly, equilibration times do not depend on the size *per se* but on the strength of the intermolecular forces and the number of atoms or molecules in the clusters (which is size- but also shape-related). In the case of nanoparticles with more than one component, such as the core–shell and acorn particles in Fig. 1, there is more than one equilibration or relaxation time. Likewise, particles that have some internal structure, such as the multiwalled carbon nanotubes (CNTs) and vesicles of Fig. 1, can have high internal energy barriers that must be overcome during their assembly, resulting in long formation times even though the strength of the 'bonds' may not be strong. Such

structures often exhibit long-lived transients that slowly evolve (age) with time.

For these reasons, most nanoparticle systems are non-equilibrium ones; to understand and control them one needs to first appreciate the strength and range of the diverse interaction forces and energies between their constituent atoms as well as between the particles themselves.

INTERNANOPARTICLE FORCES

VAN DER WAALS FORCES

Van der Waals forces arise between all molecules and particles, whether in air (vacuum) or in liquid. Between identical materials they are attractive, but can be repulsive between dissimilar materials in a third (usually liquid) medium. Bare, that is, uncoated, nanoparticles of metals, metal oxides, ceramic materials and chalcogenides have strong van der Waals forces and tend to aggregate when in inert nonpolar, such as hydrocarbon, liquids. Aggregation can lead to fusion or coalescence, for example, converting nanorods into nanowires[14,15]. Aggregation can be prevented by protecting nanoparticles via 'steric' or 'electrostatic' stabilization. Steric stabilization (see below) usually involves coating the nanoparticles with a tightly bound polymer or surfactant monolayer. Electrostatic stabilization can be used in aqueous solutions where the particles become charged and/or surrounded by a hydration layer, which again prevents their coalescence. In both of the above, strong contact adhesion is replaced by a weak, non-covalent adhesion that allows the organization of the nanoparticles to be 'manipulated', as described later.

Protocols for calculating the van der Waals interaction forces or energies between small molecules or large macroscopic bodies are well established[11]. However, those between nanoparticles are complicated because their size and separation are comparable, precluding the use of asymptotic equations; the particles are often highly anisotropic, and their non-bulk-like dielectric properties are often not known.

INTERNAL AND EXTERNAL MAGNETIC AND ELECTROSTATIC FORCES

Some interparticle forces arise both internally, between neighbouring particles, as well as from external force fields. For example, magnetic particles mutually orient their dipoles so as to attract each other via the magnetic dipole–dipole interaction, whereas an external magnetic field forces the dipoles to orient along the field, resulting in end-to-end attraction but side-to-side repulsion. In either case, the particles can develop both short- and long-range order. Figure 4a shows magnetic nanoparticles that have 'self-assembled' into rings in the absence of any external field but into strings with long-range order in the presence of a field (Fig. 4b) — an example of 'directed assembly'.

Unlike the attractive magnetic force between magnetic particles, the electrostatic force between similarly charged particles (in liquids, usually water) is repulsive. However, externally applied electric fields can give rise to similar dipole-orienting effects to those from magnetic fields[16,17], with the added feature that some nanoparticles can become mobile as they are also forced to carry charges with them, that is, a current now passes through the nanoparticle suspension[16]. Figure 4c,d show examples of well-ordered nanostructures produced by electric-fields in air[16] (c) and in liquid[17] (d). CNTs have also been found to better align during their synthesis in the presence of an electric field[18].

Naturally charged particles repel each other in aqueous electrolyte solutions via the so-called electric double-layer force. Those that are not naturally charged can be functionalized with ionic (acidic or basic) groups or highly hydrated groups to provide sufficient repulsion against the attractive van der Waals force to prevent their aggregation. But there is always a weak 'secondary' energy minimum at some finite separation of a few nanometres, which is of the scale

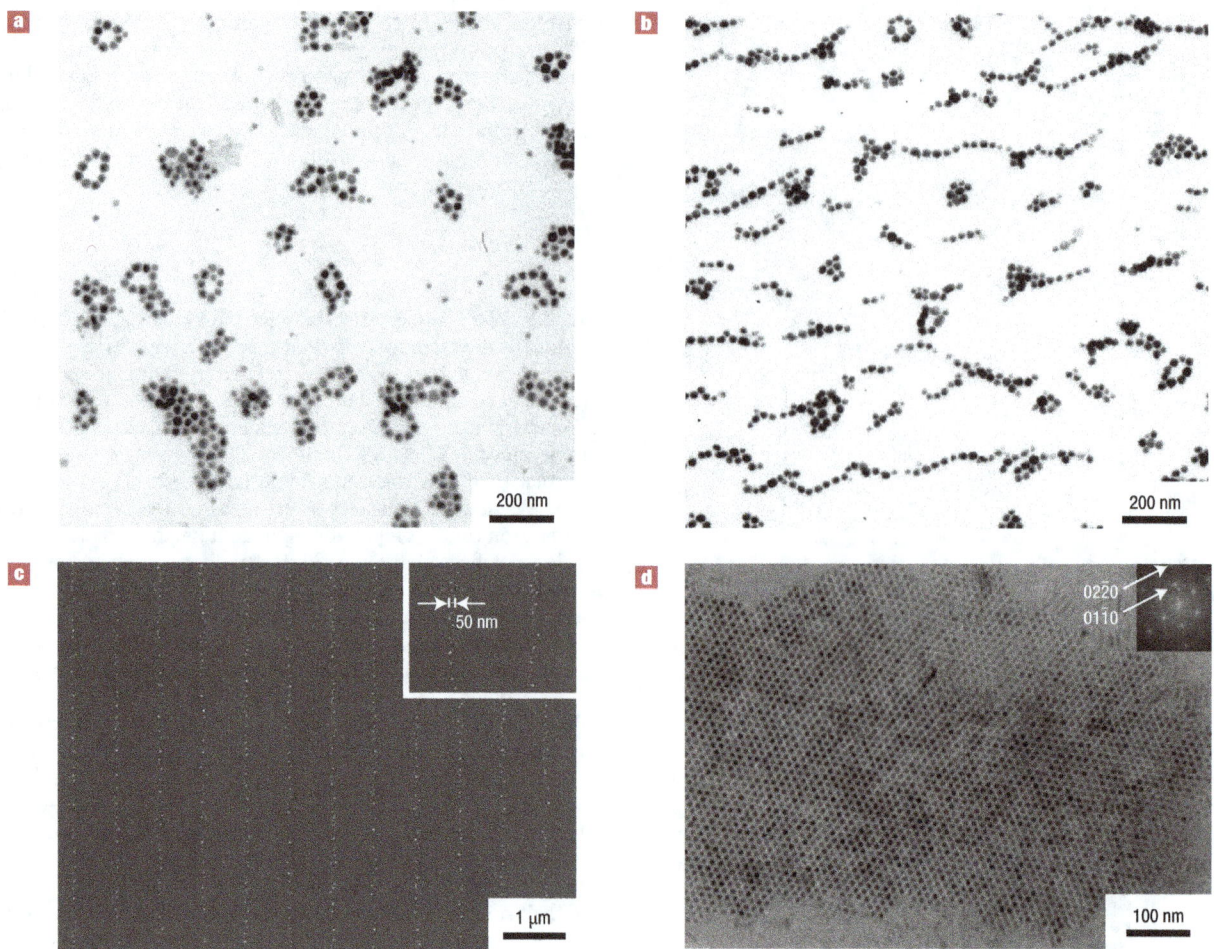

Figure 4 Nanoparticle orderings due to internal and external magnetic and electrostatic forces. **a,b**, TEM images (Philips EM-400, 80 kV) of 'self-assembled' Co nanoparticle rings — deposited on a carbon TEM grid after being magnetized for 70 min then aged for one week in zero field[23] (**a**), and of Co nanoparticle chains deposited by 'directed assembly' under the influence of a magnetic field of $B = 225$ G (**b**; ref. 23). **c**, SEM image of Ag nanoparticles on a pre-patterned silicon substrate using an ion-injection method[16]. **d**, TEM images of perpendicularly aligned Cd nanorods that have ordered into a superlattice by a combination of a d.c. electric field and slow evaporation of the solvent[17]. Inset: Fourier transform of the image showing the electron diffraction from the hexagonally ordered 2D array. Figures reprinted with permission: **a,b** ref. 23 © 2002 ACS; **c** ref. 16 © 2006 NPG; **d** ref. 17 © 2006 ACS.

of the size of nanoparticles. As mentioned earlier, such a confluence of length scales allows for delicate manipulation of the phase state of particles in solution. As the electrostatic interaction is easily tuned by changing the pH, ionic strength or type of electrolyte[11,19,20], control of nanoparticle ordering by adjusting or changing the solution conditions can provide a powerful way of controlling nanoparticle assemblies — a technique that has not been as fully exploited with nanoparticles as it has with larger colloidal particles (referred to as colloidal processing)[21,22].

Carbon nanotubes are a particularly promising class of nanoparticles because they can be made with different internal structures and exposed surfaces (hydrophobic, conducting or semiconducting). But there are serious challenges to solubilizing CNTs in liquids because of their strong van der Waals forces and hydrophobicity, which makes them aggregate in common organic and aqueous media unless coated with some stabilizing agent such as a surfactant or polyelectrolyte layer. Excess surfactant leads to an additional depletion attraction (see below), which further complicates but enriches their processing.

The combination or competition among the internal interparticle and externally applied force fields, including magnetic[23,24], electric[25,26], convective and flow-induced viscous and lubrication forces[27,28] (see later), are providing new possibilities for ordering various types of nanoparticles.

REPULSIVE STERIC, CONFINING OR JAMMING FORCES

Rough surfaces, and those covered by confining nanoparticles, repel each other with a 'steric' force, where steric here means that the opposing asperities or confined nanoparticles physically touch each other and oppose being elastically or plastically deformed, or rearranged. Such forces have been found to be linear or exponentially repulsive, the former arising in the case of adhering asperities or particles, as shown in Fig. 5a. Exponential repulsions seem to arise between surfaces confining soft, non-adhering or weakly adhering nanoparticles such as those shown in Figs 5b and 7b. They also arise between rough polymer surfaces that are also soft and weakly adhering (Fig. 5c). Such exponentially repulsive steric forces have been reproduced by computer simulations[29], but have so far defied any

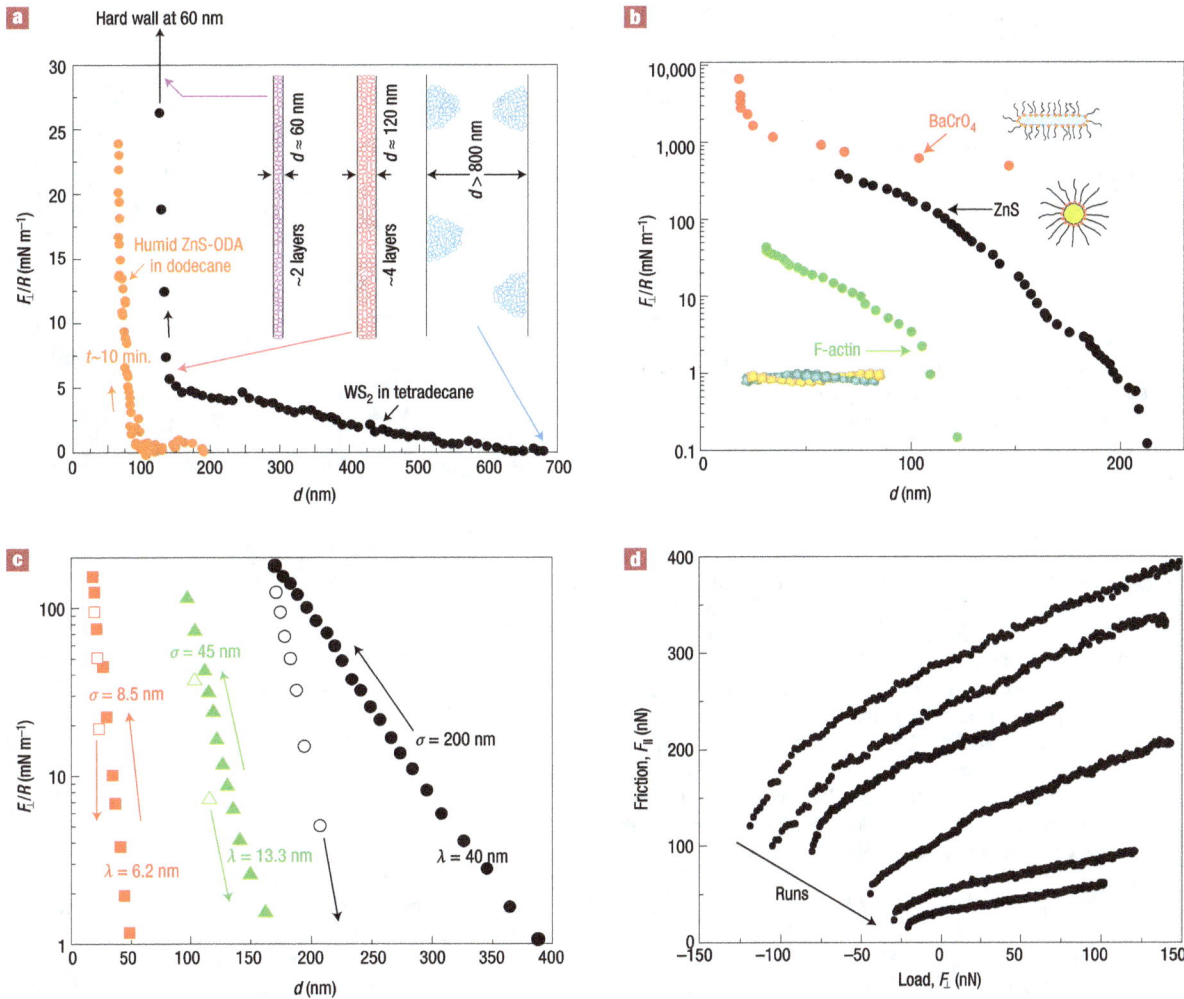

Figure 5 Typical normal ($F_\perp$) and lateral ($F_\parallel$) friction forces between surfaces confining nanoparticles or between nanostructured (rough, patterned) surfaces. **a**, Force profiles (normalized by the radius of the surfaces R) measured by SFA between two mica surfaces with adsorbed WS_2 nanoparticles in dry tetradecane[43], and in dodecane containing ZnS–ODA nanoparticles exposed to humid air[33]. Both forces are linearly repulsive. The schematics show the deduced changing nanoparticle-layer morphology with decreasing gap distance d. **b**, Forces between surfaces during approach in solutions containing F-actin (see Fig. 1), surfactant-coated ZnS spheres[33] and surfactant-coated $BaCrO_4$ rods[81]. The forces are roughly exponential in all three cases. **c**, Measured repulsive force–distance curves on approach of randomly rough surfaces of poly(vinylidene fluoridetrifluoroethylene) with different RMS roughness[27]. For surfaces with roughness, σ, varying from 3 nm (square) to 220 nm (circle), all normal forces measured on approach after the initial (slightly adhesive) contact exhibited an almost perfect exponential repulsion with decay lengths, λ, from 2.0 to 40.0 nm. Filled symbols signify 'run in' and open symbols 'run out'. **d**, AFM-measured friction-load curves[40] for a platinum-coated tip on a mica surface showing a monotonically decreasing shear strength and adhesion with repeated sliding and loading.

simple explanation. Irreversible (plastic) deformations at nanoscale junctions have been measured using the atomic force microscope (AFM) technique, as shown in Fig. 5d.

SOLVATION, STRUCTURAL AND DEPLETION FORCES

Solvation, structural or (in water) hydration forces arise between two particles or surfaces if the solvent or water molecules become ordered by the surfaces. When such ordering occurs, continuum theories of van der Waals and electrostatic double-layer forces break down, and the monotonic forces they predict become replaced or accompanied by an exponentially decaying oscillatory force with a periodicity equal to the size of the confined liquid molecules, micelles or nanoparticles[30–32]. The final, deep energy minimum at contact is often referred to as the 'depletion force' because at this point all the intervening molecules or micelles have been depleted from between

the two surfaces. Oscillatory forces mediated by spherical and especially rod-like nanoparticles can lead to quasi-ordered layers on single surfaces or between two confining surfaces, providing a means for ordering them that does not arise in bulk solution[33,34].

CAPILLARY FORCES

Attractive capillary forces arise from the Laplace pressure within the curved menisci formed by condensation of liquid or vapour bridges around two adhering particles. Such forces have been used to obtain ordered 2D and 3D colloidal crystals on substrates[35–37]. For example, Nikoobakht et al.[36] showed that capillary forces induce gold nanorods to align parallel to each other and — depending on the nanoparticle and surfactant concentrations, solvent evaporation rate, ionic strength, and other solution conditions — to assemble into well-defined one-, two- or three-dimensional structures.

When a surfactant monolayer coats a particle where the headgroup–substrate binding is ionic or dipolar (not covalent), water can penetrate into the headgroup–substrate interface and lift the monolayer off the particle surface[38] (see Fig. 7d). The hydrated nanoparticle cores end up coming into contact while separated by a water bridge, which gives rise to an attractive capillary force between them, which in turn gives rise to rigid aggregates. More significantly, the contacting cores and presence of water can catalyse the chemical sintering of the cores.

CONVECTIVE FORCES

As in the case of externally applied magnetic and electrostatic forces, externally applied convective forces, controlled by directed flow patterns, are also being used to align and order particles, both in the bulk and at surfaces[35]. Prevo et al.[39] recently described a process in which confined particles in thin films can be manipulated to order into two-dimensional crystals by the combined action of hydrodynamic and capillary forces.

FRICTION AND LUBRICATION FORCES

Friction and lubrication or shear forces are intimately coupled to the normal and convective forces between nanoparticles or between nanoparticles and surfaces. Figure 5d shows the friction forces measured by AFM between a nanosized tip of radius ~140 nm and a surface where the measured forces were well-described by adhesion–friction theories developed for macroscopic surfaces. Friction and lubrication forces have a central role in the ordering and assembly of nanoparticles in air or liquids[40–42]. They are particularly important in concentrated dispersions where particles have little room to manoeuvre except by sliding past each other. The application of shear forces to order nanoparticles has not yet been studied in detail, but shows promise as yet another method of engineered assembly[43,44].

SYNTHESIS AND PROCESSING OF NANOPARTICLES

There is much current work on the synthesis, processing and characterization of nanostructured materials, that is, assemblies of nanoparticles, including dots, rods, wires, belts and tubes. Almost any common solid element or compound can or soon will be synthesizable as a zero and/or one-dimensional material. With thousands of papers published every year on this topic, a comprehensive literature survey is impossible, and only selected representative publications are cited in this section, particularly as they concern fundamental scientific insights into nanoparticle interactions and systems.

Synthesis of nanostructures was realized many decades ago from aqueous solutions that generally contained a surfactant and oil. The resulting 'microemulsions' (Figs 1o and 6a) are clear, thermodynamically stable nanostructures with typical dimensions between 5 and 100 nm, and well-defined hierarchical morphologies or 'mesophase' structures[45,46]. A visualization of a molecularly ordered nanostructure of bis(2-ethylhexyl)sulphosuccinate sodium salt (AOT) molecules surrounding a water core is shown in Fig. 6a (ref. 47). An interesting new class of nanoparticles is emerging, where the nanoparticles themselve behave like surfactants — stabilizing microemulsion droplets by adsorbing at their interfaces (see Fig. 1o); the resulting structures are known as Pickering emulsions.

As discussed earlier, most nanoparticles exist in a stable state, but not necessarily in their true equilibrium state. For example, thermal treatment of layered materials such as graphite — the thermodynamically stable form of carbon — led to the discovery of carbon fullerenes in 1985[48] and later to CNTs[49] in which the graphite sheets are folded into spherical and tubular geometries. Analogous nested cage-like nanostructures of materials that were initially compounds such as tungsten disulphide and molybdenum disulphide were obtained in the early 1990s by thermal annealing of the

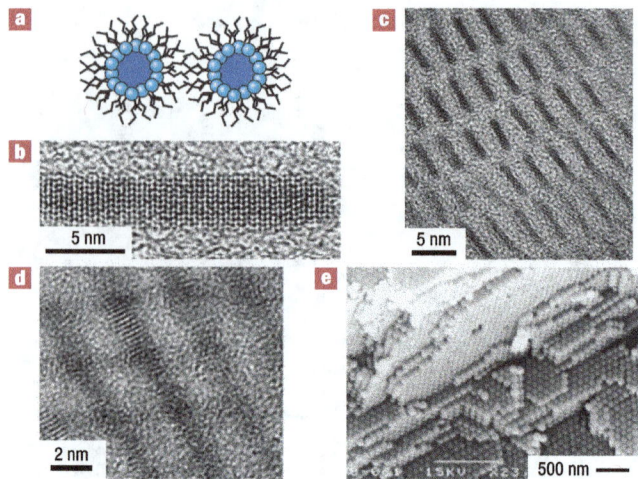

Figure 6 Different types of nanoparticle assemblies and nanostructured materials. **a**, Microemulsion or swollen micelle — nanostructures in thermodynamic equilibrium[47]. **b**, A single surfactant-coated CdSe nanorod[102]. **c**, A 2D array of ZnS nanorods synthesized in molten ODA. **d**, Two crossed arrays of ODA-coated ZnSe nanowires assembled at the air–water interface and transferred sequentially onto a solid support using Langmuir–Blodgett deposition[60]. **e**, Polystyrene spheres, 100 nm in diameter, assembled into a 3D ordered structure using a 12 mm confinement cell[72]. Figures reprinted with permission from: **a** ref. 47 © 1990 ACS; **b** ref. 102 © 2000 ACS; **c** N. Belman et al. unpublished; **d**, ref. 60 © 2006 Wiley; **e** ref. 72 © 1999 IEEE.

corresponding oxide materials[50]. This methodology has been expanded since then to synthesize 'inorganic fullerenes' and inorganic nanotubes from a large variety of materials[51].

The most successful technique for large-volume synthesis of uniform nanostructures has been the 'wet' chemical–colloid approach, which typically makes use of surface-active organic molecules such as octadecylamine (ODA) for kinetically controlling and arresting the growth. This approach has been successfully used for metals[52], oxides[53], phosphide[54,55], arsenides[56], selenides[57,58] and a continually increasing variety of other semiconducting materials[58,59]. An example of a surfactant-coated CdSe nanorod is shown in Fig. 6b.

Although successful and effective in terms of yield, uniformity and stability, the molecular mechanisms underlying the synthesis and assembly of 0D and 1D nanostructures using organic surfactants are still far from understood. It is clear that the synergistic or competitive interactions between the surfactant chains, their headgroups, and the inorganic substrate cores, involving both geometric packing effects and composition of the mixture (availability of surfactant), all have a role in determining the size and shape of the resulting nanoparticles and their higher-order assembly into hierarchical structures ('super-structures' or 'supramolecular assemblies')[60–62]. Starving the reaction mixture of one or more reactants, and cooling, are two common ways of controlling or arresting the growth processes.

Synthesis and processing were traditionally done in two distinct steps. The use of surfactants now allows for synthesis of uniform nanoparticles and their assembly into ordered two-dimensional and three-dimensional assemblies in one step, where size and shape uniformity (monodispersity) are essential for achieving well-ordered hierarchical structures. In 2003, Pradhan and Efrima[63] described a single-pot synthetic route for preparing a variety of highly uniform semiconductor nanodots, rods and wires. The synthesis used molten primary alkylamine surfactants, $CH_3-(CH_2)_n-NH_2$, which also served as the reaction medium for obtaining metal–sulphide (MS) semiconductor nanoparticles or cores, where M = Zn, Cd, Pb, Hg and others. The method has two major advantages: the xanthate precursor decomposes to give a perfectly stoichiometric ratio of metal-to-sulphide, and molten ODA solvent provides an organic (surfactant) template that co-crystallizes with the inorganic mineral cores, resulting in a uniform monolayer coating of the cores on cooling. The method was later modified for controlled synthesis of ZnSe (refs 64,65) and CdS (ref. 66) nanostructures.

These highly monodisperse nanoparticles spontaneously assemble into two-dimensional superlattices on solid supports such as carbon-coated TEM grids[67], as shown in Fig. 6c for ODA-coated ZnS nanorods. Interestingly, the side-by-side spacing between the rods can be tuned by using alkylamine molecules of different chain length (for example, tetradecylamine) as verified using TEM and synchrotron grazing-incidence X-ray diffraction. For obtaining anisotropic nanostructures an external electric field can be applied, which, thanks to the inherent electric dipole moment along the long axis of the nanorods, results in well-oriented microstrings[65].

Another method for ordering nanoparticles is based on confinement. For example, placement of nanowires at the air–water interface of a Langmuir trough confines the particles to the two-dimensional interface. On compression, the nanowires order into regular patterns[62]. Acharya et al.[60] demonstrated that such ordered two-dimensional nanowire assemblies of ZnSe can be transferred onto a variety of solid substrates. Transfer of a second two-dimensional nanowire assembly at a given angle to the first leads to the formation of a crossed nanowire structure with ultrahigh junction densities of the order of $6 \times 10^4 \ \mu m^{-2}$, as shown in Fig. 6d — another classic example of directed assembly.

Nanoparticle coalescence reactions in far-from-equilibrium assemblies can give rise to stable engineered structures that would not form by natural self-assembly. For example, controlled, irreversible, coalescence of ZnS nanorods into nanowires at the air–water interface can be achieved by heating or applying a surface pressure[67,68], and ultra-narrow PbS nanowires can be similarly coalesced by pressure to form wider nanowires[15]. Chemical, rather than pressure-induced, removal of the organic surfactant layer can also result in coalescence, as occurs when CdTe nanoparticles coalesce into luminescent nanowires[69,70] or nanosheets[70,71] depending on the applied pressure.

Three-dimensional confinement has also been used for assembling colloidal nanoparticles into ordered three-dimensional structures, as shown in Fig. 6e, where a 12 μm confinement cell was used to assemble 100-nm polystyrene spheres into a new class of solid materials termed 'photonic crystals'[72]. By coupling confinement and flow, Kumacheva et al.[73] achieved two-dimensional colloidal lattices of polystyrene beads. A template approach has also been used[74] to produce ordered three-dimensional nanomaterials: an ordered template made out of various polymers or glass beads was prepared in the desired geometry into which the colloidal semiconductors were impregnated; dissolution of the template materials led to a photonic crystal of the desired semiconductor material. Ahmed and Ryan[37] demonstrated how to avoid non-equilibrium organization: they confined a CdS nanoparticle solution between two surfaces and allowed the solvent (toluene) to evaporate very slowly, thereby giving time for the particles to organize themselves into their (presumably) equilibrium thin-film configuration.

EFFECTS OF NANOPARTICLE PROPERTIES AND FORCES ON ASSEMBLIES

The structure of self- or engineered assemblies of nanoparticles is dependent not only on the interparticle forces (see earlier) but also on many other nanoparticle properties including their shape[75], processing routes, environmental factors, and so on. For example, Jana[75] showed that cetyl-trimethylammonium-bromide-functionalized gold or silver spheroids assemble into honeycombed,

cylindrical, platelet (smectic) or ribbon structures depending on the aspect ratio. In this section we first review recent experiments, using the surface forces apparatus (SFA), on the forces between surfaces across various nanoparticle dispersions in both aqueous and organic solvents, where the effects of size, shape, solvent (including air), concentration, humidity, hardness of coating, and time were studied.

EFFECT OF NANOPARTICLE GEOMETRY

Figure 7a shows the reversible force profiles measured between inert mica surfaces across dispersions of nanospheres, nanorods, and nanowires of surfactant-coated ZnS in dodecane. All the nanoparticles have ZnS core diameters of ~1 nm and an outer diameter of ~4 nm (including the ~1.5-nm-thick hexa- or octa-decylamine layers, HDA or ODA), and lengths of 4, 8 and ~200 nm for the spheres, rods and wires, respectively. One can readily see that attractive van der Waals forces between the surfaces are negligible compared with the repulsive steric forces of confining (compressing and aligning) the nanoparticles between them. The magnitude and range of the repulsive forces are of the order wires > rods > spheres. Also, the repulsions are roughly exponential especially at small separations (high loads or pressures at $d<100$ nm) where the three curves merge. This suggests that at high compressions the ordering is determined by the diameters (~4 nm) or smallest dimension of the particles, and that all three structures form quasi layers once the gap has fallen to below ~20 particle diameters (layers). In contrast, the long-range forces are governed by the largest dimension of the particles. Simultaneous refractive index measurements showed that the volume fractions of the nanoparticles gradually increases as d decreases, reaching values close to the close-packed densities for spheres and short rods, but not curved wires[33]. These results suggest that it should be possible to produce dense and ordered nanostructures by confinement between two smooth surfaces.

STRUCTURAL TRANSITIONS IN THIN NANOPARTICLE FILMS

Figure 7b shows the results of similar measurements to Fig. 7a but on surfactant-coated CNTs in aqueous surfactant solutions. The multiwalled, close-ended CNTs were polydisperse with diameters between 10 and 50 nm and a mean length of 2 μm, that is, they were nanowires. The stabilizing surfactant was sodium dodecyl sulphate (SDS), which has a critical micelle concentration (CMC; the concentration of surfactant at which micelles begin to form spontaneously) of about 8.7 mM. It shows the effect of concentration and time on the build-up of the forces, reflecting the slow aggregation and/or adsorption of the CNTs. The repulsion is again exponential, with a decay length that does not change but simply grows out with time.

The force–distance curves (similar to pressure–volume curves of three-dimensional systems) also reveal that the adsorbed aggregates or layers undergo a phase transition under a force, as manifested by the nearly horizontal part of the interaction in SDS solutions at $2 \times$ CMC. At the transition ($d<60$ nm), the nanotubes probably undergo a structural change before they are finally squeezed out from the gap between the surfaces. The easy reordering and squeeze-out suggests that the CNT–CNT and CNT–substrate adhesion (in water) is relatively weak compared with that of ZnS nanoparticles in organic solvents (Fig. 7a). This difference is most probably due to the strongly repulsive double-layer and hydration forces between the CNTs, which do not arise in organic liquids. Lange et al.[76] have previously modulated these short-range repulsions in water to achieve high particle densities in the 'colloidal processing' of ceramic materials. The results of Fig. 7b also show how these forces can be controlled by changing the surfactant concentration (Lange et al.[76,77] modulated their colloidal assemblies by changing the ionic strength and pH). Before day two in Fig. 7b, the surfactant concentration was increased to $6 \times$ CMC after which slow adsorption resulted in

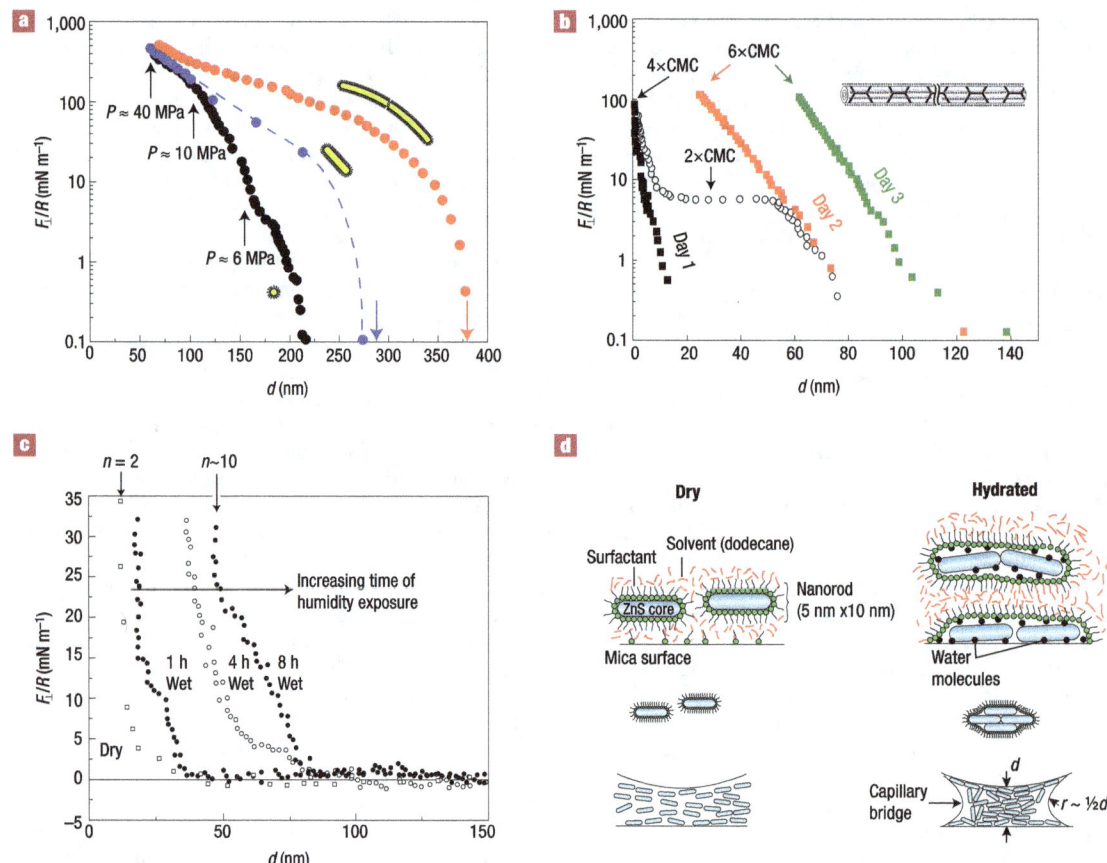

Figure 7 Effects of nanoparticle shape, size, concentration and water penetration on the forces across confined nanoparticle assemblies and films. **a**, Reversible force profiles of spheres, rods and wires (green schematics) between surfaces during approach in dry conditions. Bulk ZnS nanoparticle concentration is 2 mg ml⁻¹ in each case[33]. **b**, Three exponentially repulsive force curves of CNTs in aqueous solutions at 4×CMC and 6×CMC showing their growth with time, and one force curve at 2×CMC showing a phase transition on compression. **c**, Initially soft particles of surfactant-coated ZnS rods in dry hydrocarbon liquid give an exponential repulsion. Introduction of trace amounts of water replaces the surfactants at the hydrophilic core interfaces giving rise to capillary-force adhesion between the particles and a linear repulsion especially at very high and very low values (see Fig. 6a) because the particles are now hard[33]. n is the number of particle layers, assuming they align parallel to the surfaces. **d**, Schematics of changing nanoparticle morphology in **c** before (left) and after (right) the introduction of water into the organic liquid phase. In the dry case, the surfactant molecules are electrostatically well-attached to the ZnS core and the nanoparticles do not aggregate with each other or to the mica surface. In the wet case, the water molecules have penetrated into the interface between the surfactant headgroups and ZnS cores causing their aggregation into semi-rigid flocs, and increased adhesion to any hydrophilic, such as mica, surfaces. Depending on the amount of water, the capillary bridges can be between the nanoparticles or between the confining macroscopic surfaces. Figures reprinted with permission: **a,c,d**, ref. 33 © 2007 ACS.

an increase in the range of the repulsion, but not the characteristic decay length, with time.

STERIC INTERACTIONS OF HARD VERSUS SOFT NANOPARTICLES

Bare nanoparticles composed of metals or hard inorganic material usually have a high surface energy, strongly attractive van der Waals forces, and tend to aggregate or sinter in solution or during confinement[78]. Most nanoparticle cores are therefore synthesized with a surfactant monolayer coat[79] to reduce their adhesion and prevent them from aggregating. The surfactant coating also makes them softer and more deformable[80], allowing their assembly to be manipulated[33,81]; but they can also lose their coating in solution[38], which has important implications for their interactions and organization.

Figure 7c shows measured force curves for a system of nanoparticles that started off as 'soft', and later became 'hard' due to the stripping off of the surfactant layer. Initially, the ODA-coated ZnS nanorods interact in dry (water-free) dodecane, and the steric

repulsion is short-ranged, exponential, with little time-dependence. After water was introduced into the solvent (benignly, by increasing the relative humidity of the surrounding vapour), the repulsion grew with increasing time as the nanoparticles lost their surfactant layer and aggregated due to the increasing capillary forces between them (see earlier, and Fig. 7d). Similar effects occur with large colloidal particles in organic solvents. In addition, in the hydrated system, where the particles are now 'bare', especially at very high and very low values, the forces have become linear rather than exponential, an effect that is also seen with hard, bare WS₂ nanoparticles (Fig. 5a).

It seems that weakly adhering, soft (surfactant-coated) nanoparticles oppose being compressed with an exponential force or pressure, whereas strongly adhering, hard (and usually bare) nanoparticles do this with a linear, elastic repulsion.

EFFECTS OF NANOPARTICLE CONCENTRATION

Concentration also has a major role in internanoparticle forces. Nanoparticles in progressively more concentrated solutions

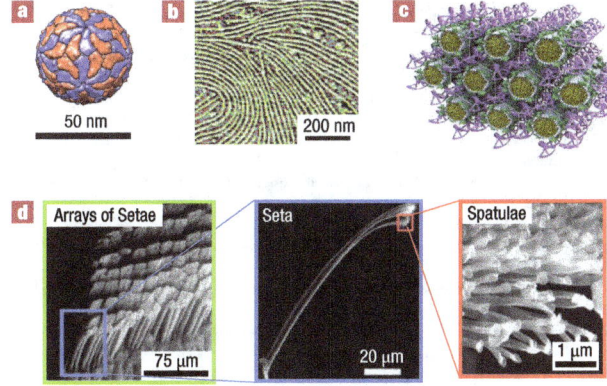

Figure 8 Different types of biological nanoparticles, and nanostructured surfaces and materials. **a**, Bluetongue virus capsid architecture[103]. **b**, Ordered monolayer state of the M13 virus on polyelectrolyte multilayers[104]. **c**, Lipid–DNA self-assembled 3D structures (courtesy of C. Safinya[105]). **d**, Gecko feet at different length scales. The foot pads of geckos are structured at all length scales from the macro- to the nanoscale, all of which act synergistically — allowing them to form strong adhesive and high friction contacts within ~10 ms, then detach with equal ease and rapidity — as the animals run on walls and ceilings[106]. Figures reprinted with permission: **a** ref. 103 © 2005 Elsevier; **b** ref. 104 © 2006 NPG; **c** ref. 105 © 2006 ACS; **d** ref. 106 © 2000 NPG.

tend to aggregate, cluster, precipitate out of solution and adsorb on surfaces. Such effects seem to influence force profiles in a similar way to the effects of time, surfactant concentration and humidity[33] (Fig. 7).

IMPLICATIONS FROM AND FOR BIOLOGICAL SYSTEMS AND MATERIALS

Biological materials and tissues are naturally structured at all length scales, that is, at the macro-, micro- and nanoscales, so that nanoparticles and nanostructures cannot be considered as some separate area of biological systems. One may, however, distinguish between isolated nanoparticles, nanostructured biosurfaces and nanostructured biological materials (as well as 'biomaterials', which are biomimetic rather than natural materials). Figure 8 shows some examples of these systems.

ISOLATED NANOPARTICLES THAT FUNCTION AS INDEPENDENT ENTITIES

Molecular biologists have been working with nanoparticles since the 1950s when viruses were first studied at the molecular level, and the term 'self-assembly' was coined[82,83]. Most virus capsids fall into the size range 10–300 nm (Fig. 8a), and their complex but highly ordered and semi-rigid nanostructures are usually the result of pure self-assembly (once all the components are present in the solution). In contrast, it is still not clear whether the equally sized but much softer (fluid-like) vesicles, liposomes, microemulsion droplets and some protein complexes (Fig. 1) are truly self-assembled structures or require directed-assembly, that is, an external source of energy, to produce and maintain them. Both single-bilayer-membrane vesicles and multiwalled liposomes are found in cells, and have been produced and used as detergents and cosmetics, and are also currently being developed as target-specific drug-delivery vehicles[84,85]. Such nanoparticles are typically drug-loaded vesicles, liposomes or polymers that are surface functionalized to recognize molecules that are unique to the target cells. The idea is to maximize delivery to the infected cells and minimize drug loss and toxicity associated with the current method of whole-body delivery, which causes damage

to healthy tissue. Using similar principles, vesicle bioreactors are also being developed to mimic cells.

Many nanoparticle systems are currently being developed for biological imaging. These include gold nanorods that are capable of generating two-photon luminescence in the near-infrared where photon adsorption by water and biological tissues is low[86], and colloidal semiconductor quantum dots, which are more responsive and often more photostable than conventional organic fluorophores[87]. One aim of this line of research is to produce tiny chemical laboratories that can circulate through the body, continuously monitoring and transmitting vital information to an outside receiver. Some proteins can be considered as nano-refineries or chemical reactors, transforming chemical 'feeds' into 'products', all at body temperature and pressure and with high precision and efficiency — something that we cannot do at present, but are learning how.

BIOLOGICAL MATERIALS AND BIOMATERIALS

The internal network of collagen in cartilage and bone is highly anisotropic, with submicrometre compartments that provide high mechanical (cushioning) strength, and that direct the flow of fluid and nutrients when strained[88,89].

NANOSTRUCTURED BIOSURFACES

Gecko feet consist of a structural hierarchy of macroscale toe pads, microscale setal arrays, and nanoscale spatula fibres (Fig. 8d) that work together to allow the animals to quickly and strongly adhere to almost any surface and then detach with equal rapidity. Inspired by these lizards, numerous attempts are being made to understand the interaction forces and articulation mechanisms for making 'smart' adhesives and robots[90].

We can both learn from biology and contribute to it. Learning from and mimicking biological assembly processes is a tremendous challenge, and promise. It is a challenge because biointeractions often involve many different types of forces acting simultaneously (both in series and in parallel) and/or sequentially in both space and time[91,92]. Some of these forces are highly specific, such as ligand–receptor (lock-and-key type) 'bonds' that actually involve a combination of different non-covalent bonds but which, unlike van der Waals or ionic bonds, cannot be described by simple general equations. It is also a promise because the possibilities seem to be endless, and include the development of new drugs and delivery methods, as mentioned above, and the fabrication of adaptable, switchable or responsive materials where the interactions between the nanoparticles can be switched or modulated reversibly.

CONCLUSIONS

Nanoparticles promise to enable us to create miniature machines and devices, chemical laboratories that can circulate in the body, and (bulk) nanostructured materials with novel properties including being reversibly adaptable. Nanoparticle science and technology is therefore bringing together the hard sciences (of physics, chemistry and biology) with engineering. The field spans the subnano to the macroscale, and covers both equilibrium and far-from-equilibrium systems with timescales that can vary from microseconds or less, to many years. Furthermore, the intermolecular and interparticle forces in nanoparticle assemblies are particularly complex. Because of this complexity the 'science' is still not well understood, and most of the advances have been empirical.

The recent advances in synthesis and characterization of nanoparticles and multifunctional molecules have therefore not always been matched by similar advances in organizing these 'building blocks' into supramolecular assemblies and materials. In this review we have tried to identify the fundamental issues and discuss practical solutions to achieving these ends.

doi: 10.1038/nmat2206

References

1. Alivisatos, A. P. Semiconductor clusters, nanocrystals, and quantum dots. *Science* **271,** 933–937 (1996).
2. Hodes, G. When small is different: Some recent advances in concepts and applications of nanoscale phenomena. *Adv. Mater.* **19,** 639–655 (2007).
3. Klabunde, K. J. S. (ed.) *Nanoscale Materials in Chemistry* (Wiley Interscience, New York, 2001).
4. Antonets, I. V., Kotov, L. N., Nekipelov, S. V. & Karpushov, E. N. Conducting and reflecting properties of thin metal films. *Tech. Phys.* **49,** 1496–1500 (2004).
5. Lenham, A. P. & Treherne, D. M. Applicability of anomalous skin-effect theory to optical constants of Cu, Ag, and Au in the Infrared. *J. Opt. Soc. Am.* **56,** 683-685 (1966).
6. Srivastava, D., Menon, M. & Cho, K. Anisotropic nanomechanics of boron nitride nanotubes: Nanostructured "skin" effect. *Phys. Rev. B* **6319,** 195413 (2001).
7. Ercolessi, F., Andreoni, W. & Tosatti, E. Melting of small gold particles: Mechanism and size effects. *Phys. Rev. Lett.***66,** 911–914 (1991).
8. Hagen, M. H. J., Meijer, E. J., Mooij, G., Frenkel, D. & Lekkerkerker, H. N. W. Does C60 have a liquid-phase. *Nature* **365,** 425–426 (1993).
9. Rey, C. & Gallego, L. J. Structures of hard-core Yukawa clusters and the tail-range dependence of the existence of a liquidlike cluster phase: Relevance to the physics of C-60. *Phys. Rev. E* **53,** 2480–2487 (1996).
10. Israelachvili, J. Self-assembly in 2 dimensions — surface micelles and domain formation in monolayers. *Langmuir* **10,** 3774–3781 (1994).
11. Israelachvili, J. N. *Intermolecular and Surface Forces* 2nd edn (Academic Press, Burlington, 1991).
12. Alcantar, N. A., Park, C., Pan, J. M. & Israelachvili, J. N. Adhesion and coalescence of ductile metal surfaces and nanoparticles. *Acta Mater.* **51,** 31–47 (2003).
13. Rosensweig, R. E. *Ferrohydrodynamics* (Cambrige Univ. Press, New York, 1985).
14. Cho, K. S., Talapin, D. V., Gaschler, W. & Murray, C. B. Designing PbSe nanowires and nanorings through oriented attachment of nanoparticles. *J. Am. Chem. Soc.* **127,** 7140–7147 (2005).
15. Patla, I. *et al.* Synthesis, two-dimensional assembly, and surface pressure-induced coalescence of ultranarrow PbS nanowires. *Nano Lett.* **7,** 1459–1462 (2007).
16. Kim, H. *et al.* Parallel patterning of nanoparticles via electrodynamic focusing of charged aerosols. *Nature Nanotech.* **1,** 117–121 (2006).
17. Ryan, K. M., Mastroianni, A., Stancil, K. A., Liu, H. T. & Alivisatos, A. P. Electric-field-assisted assembly of perpendicularly oriented nanorod superlattices. *Nano Lett.* **6,** 1479–1482 (2006).
18. Joselevich, E. & Lieber, C. M. Vectorial growth of metallic and semiconducting single-wall carbon nanotubes. *Nano Lett.* **2,** 1137–1141 (2002).
19. Sukhorukov, G. B., Antipov, A. A., Voigt, A., Donath, E. & Möhwald, H. pH-controlled macromolecule encapsulation in and release from polyelectrolyte multilayer nanocapsules. *Macromol. Rapid Comm.* **22,** 44–46 (2001).
20. Kolny, J., Kornowski, A. & Weller, H. Self-organization of cadmium sulfide and gold nanoparticles by electrostatic interaction. *Nano Lett.* **2,** 361–364 (2002).
21. Simard, J., Briggs, C., Boal, A. K. & Rotello, V. M. Formation and pH-controlled assembly of amphiphilic gold nanoparticles. *Chem. Comm.* 1943–1944 (2000).
22. Zhang, H., Zhou, Z., Yang, B. & Gao, M. Y. The influence of carboxyl groups on the photoluminescence of mercaptocarboxylic acid-stabilized CdTe nanoparticles. *J. Phys. Chem. B* **107,** 8–13 (2003).
23. Tripp, S. L., Pusztay, S. V., Ribbe, A. E. & Wei, A. Self-assembly of cobalt nanoparticle rings. *J. Am. Chem. Soc.* **124,** 7914–7915 (2002).
24. Terheiden, A., Dmitrieva, O., Acet, M. & Mayer, C. Magnetic field induced self-assembly of gas phase prepared FePt nanoparticles. *Chem. Phys. Lett.* **431,** 113–117 (2006).
25. Morkved, T. L. *et al.* Local control of microdomain orientation in diblock copolymer thin films with electric fields. *Science* **273,** 931–933 (1996).
26. Velev, O. D. & Bhatt, K. H. On-chip micromanipulation and assembly of colloidal particles by electric fields. *Soft Matter* **2,** 738–750 (2006).
27. Benz, M., Rosenberg, K. J., Kramer, E. J. & Israelachvili, J. N. The deformation and adhesion of randomly rough and patterned surfaces. *J. Phys. Chem. B* **110,** 11884–11893 (2006).
28. Huang, Y., Duan, X. F., Wei, Q. Q. & Lieber, C. M. Directed assembly of one-dimensional nanostructures into functional networks. *Science* **291,** 630–633 (2001).
29. Hyun, S., Pei, L., Molinari, J. F. & Robbins, M. O. Finite-element analysis of contact between elastic self-affine surfaces. *Phys. Rev. E* **70,** 026117 (2004).
30. Israelachvili, J. & Gourdon, D. Putting liquids under molecular-scale confinement. *Science* **292,** 867–868 (2001).
31. Gerstenberg, M. C., Pedersen, J. S. & Smith, G. S. Surface induced ordering of micelles at the solid-liquid interface. *Phys. Rev. E* **58,** 8028–8031 (1998).
32. Kocevar, K. & Musevic, I. Structural forces near phase transitions of liquid crystals. *Chem. Phys. Chem.* **4,** 1049–1056 (2003).
33. Akbulut, M. *et al.* Forces between surfaces across nanoparticle solutions: Role of size, shape, and concentration. *Langmuir* **23,** 3961–3969 (2007).
34. Min, Y. *et al.* Normal and shear forces generated during the ordering (directed assembly) of confined straight and curved nanowires. *Nano Lett.* **8,** 246 -252 (2008).
35. Lazarov, G. S., Denkov, N. D., Velev, O. D., Kralchevsky, P. A. & Nagayama, K. Formation of 2-dimensional structures from colloidal particles on fluorinated oil substrate. *J. Chem. Soc.-Faraday Trans.* **90,** 2077–2083 (1994).
36. Nikoobakht, B., Wang, Z. L. & El-Sayed, M. A. Self-assembly of gold nanorods. *J. Phys. Chem. B* **104,** 8635–8640 (2000).
37. Ahmed, S. & Ryan, K. M. Self-assembly of vertically aligned nanorod supercrystals using highly oriented pyrolytic graphite. *Nano Lett.* **7,** 2480–2485 (2007).
38. Alig, A. R. G., Akbulut, M., Golan, Y. & Israelachvili, J. Forces between surfactant-coated ZnS nanoparticles in dodecane: Effect of water. *Adv. Funct. Mater.* **16,** 2127–2134 (2006).
39. Prevo, B. G., Kuncicky, D. M. & Velev, O. D. Engineered deposition of coatings from nano- and micro-particles: A brief review of convective assembly at high volume fraction. *Colloids Surf. A* **311,** 2–10 (2007).
40. Carpick, R. W., Agrait, N., Ogletree, D. F. & Salmeron, M. Variation of the interfacial shear strength and adhesion of a nanometer-sized contact. *Langmuir* **12,** 3334–3340 (1996).
41. Ruths, M. Friction of mixed and single-component aromatic monolayers in contacts of different adhesive strength. *J. Phys. Chem. B* **110,** 2209–2218 (2006).
42. Landman, U., Luedtke, W. D., Burnham, N. A. & Colton, R. J. Atomistic mechanisms and dynamics of adhesion, nanoindentation, and fracture. *Science* **248,** 454–461 (1990).
43. Golan, Y. *et al.* Microtribology and direct force measurement of WS2 nested fullerene-like nanostructures. *Adv. Mater.* **11,** 934–937 (1999).
44. Akbulut, M., Belman, N., Golan, Y. & Israelachvili, J. Frictional properties of confined nanorods. *Adv. Mater.* **18,** 2589–2592 (2006).
45. Narayanan, R. & El-Sayed, M. A. Catalysis with transition metal nanoparticles in colloidal solution: Nanoparticle shape dependence and stability. *J. Phys. Chem. B* **109,** 12663–12676 (2005).
46. Schwuger, M. J., Stickdorn, K. & Schomacker, R. Microemulsions in technical processes. *Chem. Rev.* **95,** 849–864 (1995).
47. Smith, R. D., Fulton, J. L., Blitz, J. P. & Tingey, J. M. Reverse micelles and microemulsions in near-critical and supercritical fluids. *J. Phys. Chem.* **94,** 781–787 (1990).
48. Kroto, H. W., Heath, J. R., Obrien, S. C., Curl, R. F. & Smalley, R. E. C60: Buckminsterfullerene. *Nature* **318,** 162–163 (1985).
49. Iijima, S. Helical microtubules of graphitic carbon. *Nature* **354,** 56–58 (1991).
50. Tenne, R., Margulis, L., Genut, M. & Hodes, G. Polyhedral and cylindrical structures of tungsten disulphide. *Nature* **360,** 444–446 (1992).
51. Tenne, R. Inorganic nanotubes and fullerene-like nanoparticles. *Nature Nanotech.* **1,** 103–111 (2006).
52. Murphy, C. J. *et al.* Anisotropic metal nanoparticles: Synthesis, assembly, and optical applications. *J. Phys. Chem. B* **109,** 13857–13870 (2005).
53. Cozzoli, P. D., Kornowski, A. & Weller, H. Low-temperature synthesis of soluble and processable organic-capped anatase TiO_2 nanorods. *J. Am. Chem. Soc.* **125,** 14539–14548 (2003).
54. Kim, Y. H., Jun, Y. W., Jun, B. H., Lee, S. M. & Cheon, J. W. Sterically induced shape and crystalline phase control of GaP nanocrystals. *J. Am. Chem. Soc.* **124,** 13656–13657 2002).
55. Guzelian, A. A. *et al.* Synthesis of size-selected, surface-passivated InP nanocrystals. *J. Phys. Chem.* **100,** 7212–7219 (1996).
56. Steiner, D. *et al.* Zero-dimensional and quasi one-dimensional effects in semiconductor nanorods. *Nano Lett.* **4,** 1073–1077 (2004).
57. Peng, X. G. *et al.* Shape control of CdSe nanocrystals. *Nature* **404,** 59–61 (2000).
58. Halpert, J. E., Porter, V. J., Zimmer, J. P. & Bawendi, M. G. Synthesis of CdSe/CdTe nanobarbells. *J. Am. Chem. Soc.* **128,** 12590–12591 (2006).
59. Yu, W. W., Wang, Y. A. & Peng, X. G. Formation and stability of size-, shape-, and structure-controlled CdTe nanocrystals: Ligand effects on monomers and nanocrystals. *Chem. Mater.* **15,** 4300–4308 (2003).
60. Acharya, S., Panda, A. B., Belman, N., Efrima, S. & Golan, Y. A semiconductor-nanowire assembly of ultrahigh junction density by the Langmuir-Blodgett technique. *Adv. Mater.* **18,** 210-213 (2006).
61. Markovich, G. *et al.* Architectonic quantum dot solids. *Acc. Chem. Res.* **32,** 415–423 (1999).
62. Yang, P. D. Wires on water. *Nature* **425,** 243–244 (2003).
63. Pradhan, N. & Efrima, S. Single-precursor, one-pot versatile synthesis under near ambient conditions of tunable, single and dual band fluorescing metal sulfide nanoparticles. *J. Am. Chem. Soc.* **125,** 2050–2051 (2003).
64. Panda, A. B., Acharya, S., Efrima, S. & Golan, Y. Synthesis, assembly, and optical properties of shape- and phase-controlled ZnSe nanostructures. *Langmuir* **23,** 765–770 (2007).
65. Acharya, S., Panda, A. B., Efrima, S. & Golan, Y. Polarization properties and switchable assembly of ultranarrow ZnSe nanorods. *Adv. Mater.* **19,** 1105–1108 (2007).
66. Acharya, S., Patla, I., Kost, J., Efrima, S. & Golan, Y. Switchable assembly of ultra narrow CdS nanowires and nanorods. *J. Am. Chem. Soc.* **128,** 9294–9295 (2006).
67. Pradhan, N. & Efrima, S. Supercrystals of uniform nanorods and nanowires, and the nanorod-to-nanowire oriented transition. *J. Phys. Chem. B* **108,** 11964–11970 (2004).
68. Acharya, S. & Efrima, S. Two-dimensional pressure-driven nanorod-to-nanowire reactions in Langmuir monolayers at room temperature. *J. Am. Chem. Soc.* **127,** 3486–3490 (2005).
69. Tang, Z. Y., Kotov, N. A. & Giersig, M. Spontaneous organization of single CdTe nanoparticles into luminescent nanowires. *Science* **297,** 237–240 (2002).
70. Zhang, Z. L., Tang, Z. Y., Kotov, N. A. & Glotzer, S. C. Simulations and analysis of self-assembly of CdTe nanoparticles into wires and sheets. *Nano Lett.* **7,** 1670–1675 (2007).
71. Tang, Z. Y., Zhang, Z. L., Wang, Y., Glotzer, S. C. & Kotov, N. A. Self-assembly of CdTe nanocrystals into free-floating sheets. *Science* **314,** 274–278 (2006).
72. Xia, Y. N., Gates, B. & Park, S. H. Fabrication of three-dimensional photonic crystals for use in the spectral region from ultraviolet to near-infrared. *J. Lightwave Technol.* **17,** 1956–1962 (1999).
73. Kumacheva, E., Garstecki, P., Wu, H. K. & Whitesides, G. M. Two-dimensional colloid crystals obtained by coupling of flow and confinement. *Phys. Rev. Lett.* **91,** 128301 (2003).
74. Subramanian, G., Manoharan, V. N., Thorne, J. D. & Pine, D. J. Ordered macroporous materials by colloidal assembly: A possible route to photonic bandgap materials. *Adv. Mater.* **11,** 1261–1265 (1999).
75. Jana, N. R. Shape effect in nanoparticle self-assembly. *Angew. Chem. Int. Ed.* **43,** 1536–1540 (2004).
76. Velamakanni, B. V., Chang, J. C., Lange, F. F. & Pearson, D. S. New method for efficient colloidal particle packing via modulation of repulsive lubricating hydration forces. *Langmuir* **6,** 1323–1325 (1990).
77. Lange, F. F. Powder processing science and technology for increased reliability. *J. Am. Ceram. Soc.* **72,** 3–15 (1989).
78. Bhushan, B. *Nanotribology and Nanomechanics: An Introduction* (Springer, Berlin, 2005).
79. Xu, L. B. *et al.* Synthesis and magnetic behavior of periodic nickel sphere arrays. *Adv. Mater.* **15,** 1562–1564 (2003).
80. Cheng, J. Y., Mayes, A. M. & Ross, C. A. Nanostructure engineering by templated self-assembly of block copolymers. *Nature Mater.* **3,** 823–828 (2004).

81. Gourdon, D. *et al.* Mechanical and structural properties of BaCrO₄ nanorod films under confinement and shear. *Adv. Funct. Mater.* **14**, 238–242 (2004).

82. Fraenkelconrat, H. & Singer, B. Reconstitution of tobacco mosaic virus. 3. Improved methods and the use of mixed nucleic acids. *Biochim. Biophys Acta* **33**, 359–370 (1959).

83. Butler, P. J. G. The current picture of the structure and assembly of tobacco mosaic virus. *J. Gen. Virol.* **65**, 253–279 (1984).

84. Wijaya, A. & Flamad-Schifferli, K. High-density encapsulation of Fe₃O₄ nanoparticles in lipid vesicles. *Langmuir* **23**, 9546–9550 (2007).

85. Lian, T. & Ho, R. J. Y. Trends and developments in liposome drug delivery systems. *J. Pharm. Sci.* **90**, 667–680 (2001).

86. Wang, H. F. *et al.* In vitro and in vivo two-photon luminescence imaging of single gold nanorods. *Proc. Natl Acad. Sci. USA* **102**, 15752–15756 (2005).

87. Michalet, X. *et al.* Quantum dots for live cells, in vivo imaging, and diagnostics. *Science* **307**, 538–544 (2005).

88. Landis, W. J. The strength of a calcified tissue depends in part on the molecular-structure and organization of its constituent mineral crystals in their organic matrix. *Bone* **16**, 533–544 (1995).

89. Gwynn, I., Wade, S., Ito, K. & Richards, R. G. Novel aspects to the structure of rabbit articular cartilage. *European Cells Mater.* **4**, 18–29 (2002).

90. Lee, H., Lee, B. P. & Messersmith, P. B. A reversible wet/dry adhesive inspired by mussels and geckos. *Nature* **448**, 338–341 (2007).

91. Whitesides, G. M. & Grzybowski, B. Self-assembly at all scales. *Science* **295**, 2418–2421 (2002).

92. You, C. C., Verma, A. & Rotello, V. M. Engineering the nanoparticle-biomacromolecule interface. *Soft Matter* **2**, 190–204 (2006).

93. Yang, Z. Q., Huck, W. T. S., Clarke, S. M., Tajbakhsh, A. R. & Terentjev, E. M. Shape-memory nanoparticles from inherently non-spherical polymer colloids. *Nature Mater.* **4**, 486–490 (2005).

94. Sonnichsen, C. *et al.* Drastic reduction of plasmon damping in gold nanorods. *Phys. Rev. Lett.* **88**, 077402 (2002).

95. Pelton, M. *et al.* Ultrafast resonant optical scattering from single gold nanorods: Large nonlinearities and plasmon saturation. *Phys. Rev. B* **73**, 155419 (2006).

96. Wang, Z. L. Transmission electron microscopy of shape-controlled nanocrystals and their assemblies. *J. Phys. Chem. B* **104**, 1153–1175 (2000).

97. Wiley, B. J., Xiong, Y. J., Li, Z. Y., Yin, Y. D. & Xia, Y. A. Right bipyramids of silver: A new shape derived from single twinned seeds. *Nano Lett.* **6**, 765–768 (2006).

98. Raviv, U. *et al.* Cationic liposome-microtubule complexes: Pathways to the formation of two-state lipid-protein nanotubes with open or closed ends. *Proc. Natl Acad. Sci. USA* **102**, 11167–11172 (2005).

99. Schneider, G. & Decher, G. From functional core/shell nanoparticles prepared via layer-by-layer deposition to empty nanospheres. *Nano Lett.* **4**, 1833–1839 (2004).

100. Teranishi, T., Inoue, Y., Nakaya, M., Oumi, Y. & Sano, T. Nanoacorns: Anisotropically phase-segregated CoPd sulfide nanoparticles. *J. Am. Chem. Soc.* **126**, 9914–9915 (2004).

101. Yoshimura, S., Takano, K. & Hachisu, S. in *Polymer Colloids II* (ed. Fitch, R. M.) 139 (Plenum, New York, 1980).

102. Manna, L., Scher, E. C. & Alivisatos, A. P. Synthesis of soluble and processable rod-, arrow-, teardrop-, and tetrapod-shaped CdSe nanocrystals. *J. Am. Chem. Soc.* **122**, 12700–12706 (2000).

103. Goddard, T. D., Huang, C. C. & Ferrin, T. E. Software extensions to UCSF Chimera for interactive visualization of large molecular assemblies. *Structure* **13**, 473–482 (2005).

104. Yoo, P. J. *et al.* Spontaneous assembly of viruses on multilayered polymer surfaces. *Nature Mater.* **5**, 234–240 (2006).

105. Ewert, K. K. *et al.* A columnar phase of dendritic lipid-based cationic liposome-DNA complexes for gene delivery: Hexagonally ordered cylindrical micelles embedded in a DNA honeycomb lattice. *J. Am. Chem. Soc.* **128**, 3998–4006 (2006).

106. Autumn, K. *et al.* Adhesive force of a single gecko foot-hair. *Nature* **405**, 681–685 (2000).

Acknowledgements

Y.G. and J.I. thank the US-Israel Binational Science Foundation for Grant #2006032; MA was supported by ONR award no. N00014-05-1-0540 and by BASF Corporation-002058; the sections on friction (J.I. and Y.M.) were supported by DOE Grant DE-FG02-87ER45331, and the work on carbon nanotubes (K.K.) was supported by Norwegian Research Council grant no. 166731. This work was also partially funded by the MRSEC Program of the NSF under award number DMR05-20415. We thank Wren Greene and Noshir Pesika for their helpful comments on the ms.

Complex thermoelectric materials

Thermoelectric materials, which can generate electricity from waste heat or be used as solid-state Peltier coolers, could play an important role in a global sustainable energy solution. Such a development is contingent on identifying materials with higher thermoelectric efficiency than available at present, which is a challenge owing to the conflicting combination of material traits that are required. Nevertheless, because of modern synthesis and characterization techniques, particularly for nanoscale materials, a new era of complex thermoelectric materials is approaching. We review recent advances in the field, highlighting the strategies used to improve the thermopower and reduce the thermal conductivity.

G. JEFFREY SNYDER* AND ERIC S. TOBERER

Materials Science, California Institute of Technology, 1200 East California Boulevard, Pasadena, California 91125, USA

*e-mail: jsnyder@caltech.edu

The world's demand for energy is causing a dramatic escalation of social and political unrest. Likewise, the environmental impact of global climate change due to the combustion of fossil fuels is becoming increasingly alarming. One way to improve the sustainability of our electricity base is through the scavenging of waste heat with thermoelectric generators (Box 1). Home heating, automotive exhaust, and industrial processes all generate an enormous amount of unused waste heat that could be converted to electricity by using thermoelectrics. As thermoelectric generators are solid-state devices with no moving parts, they are silent, reliable and scalable, making them ideal for small, distributed power generation[1]. Efforts are already underway to replace the alternator in cars with a thermoelectric generator mounted on the exhaust stream, thereby improving fuel efficiency[2]. Advances in thermoelectrics could similarly enable the replacement of compression-based refrigeration with solid-state Peltier coolers[3].

Thermoelectrics have long been too inefficient to be cost-effective in most applications[4]. However, a resurgence of interest in thermoelectrics began in the mid 1990s when theoretical predictions suggested that thermoelectric efficiency could be greatly enhanced through nanostructural engineering, which led to experimental efforts to demonstrate the proof-of-principle and high-efficiency materials[5,6]. At the same time, complex bulk materials (such as skutterudites[7], clathrates[8] and Zintl phases[9]) have been explored and found that high efficiencies could indeed be obtained. Here, we review these recent advances, looking at how disorder and complexity within the unit cell as well as nanostructured materials can lead to enhanced efficiency. This survey allows us to find common traits in these materials, and distill rational design strategies for the discovery of materials with high thermoelectric efficiency. More comprehensive reviews on thermoelectric materials are well covered in several books[1,10–12] and articles[3,5,6,8,13–18].

CONFLICTING THERMOELECTRIC MATERIAL PROPERTIES

Fundamental to the field of thermoelectric materials is the need to optimize a variety of conflicting properties. To maximize the thermoelectric figure of merit (zT) of a material, a large thermopower (absolute value of the Seebeck coefficient), high electrical conductivity, and low thermal conductivity are required. As these transport characteristics depend on interrelated material properties, a number of parameters need to be optimized to maximize zT.

CARRIER CONCENTRATION

To ensure that the Seebeck coefficient is large, there should only be a single type of carrier. Mixed n-type and p-type conduction will lead to both charge carriers moving to the cold end, cancelling out the induced Seebeck voltages. Low carrier concentration insulators and even semiconductors have large Seebeck coefficients; see equation (1). However, low carrier concentration also results in low electrical conductivity; see equation (2). The interrelationship between carrier concentration and Seebeck coefficient can be seen from relatively simple models of electron transport. For metals or degenerate semiconductors (parabolic band, energy-independent scattering approximation[19]) the Seebeck coefficient is given by:

$$\alpha = \frac{8\pi^2 k_B^2}{3eh^2} m^* T \left(\frac{\pi}{3n}\right)^{2/3} , \qquad (1)$$

where n is the carrier concentration and m^* is the effective mass of the carrier.

The electrical conductivity (σ) and electrical resistivity (ρ) are related to n through the carrier mobility μ:

$$1/\rho = \sigma = ne\mu . \qquad (2)$$

Figure 1a shows the compromise between large thermopower and high electrical conductivity in thermoelectric materials that must be struck to maximize the figure of merit zT ($\alpha^2\sigma T/\kappa$), where κ is the thermal conductivity. This peak typically occurs at carrier concentrations between 10^{19} and 10^{21} carriers per cm³ (depending on the material system), which falls in between common metals and semiconductors — that is, concentrations found in heavily doped semiconductors.

EFFECTIVE MASS

The effective mass of the charge carrier provides another conflict as large effective masses produce high thermopower but low electrical conductivity. The m^* in equation (1) refers to the density-of-states

When citing this review, please cite the original version as shown on the contents page of this chapter.

effective mass, which increases with flat, narrow bands with high density of states at the Fermi surface. However, as the inertial effective mass is also related to m^*, heavy carriers will move with slower velocities, and therefore small mobilities, which in turn leads to low electrical conductivity (equation (2)). The exact relationship between effective mass and mobility is complex, and depends on electronic structure, scattering mechanisms and anisotropy. In principle, these effective mass terms can be decoupled in anisotropic crystal structures[20].

A balance must be found for the effective mass (or bandwidth) for the dominant charge carrier, forming a compromise between high effective mass and high mobility. High mobility and low effective mass is typically found in materials made from elements with small electronegativity differences, whereas high effective masses and low mobilities are found in materials with narrow bands such as ionic compounds. It is not obvious which effective mass is optimum; good thermoelectric materials can be found within a wide range of effective masses and mobilities: from low-mobility, high-effective-mass polaron conductors (oxides[14], chalcogenides[21]) to high-mobility, low-effective-mass semiconductors (SiGe, GaAs).

ELECTRONIC THERMAL CONDUCTIVITY

Additional materials design conflicts stem from the necessity for low thermal conductivity. Thermal conductivity in thermoelectrics comes from two sources: (1) electrons and holes transporting heat (κ_e) and (2) phonons travelling through the lattice (κ_l). Most of the electronic term (κ_e) is directly related to the electrical conductivity through the Wiedemann–Franz law:

$$\kappa = \kappa_e + \kappa_l \tag{3a}$$

and

$$\kappa_e = L\sigma T = ne\mu LT, \tag{3b}$$

where L is the Lorenz factor, $2.4 \times 10^{-8}\ \mathrm{J^2\,K^{-2}\,C^{-2}}$ for free electrons.

Box 1 Thermoelectric devices

The thermoelectric effects arise because charge carriers in metals and semiconductors are free to move much like gas molecules, while carrying charge as well as heat. When a temperature gradient is applied to a material, the mobile charge carriers at the hot end tend to diffuse to the cold end. The build-up of charge carriers results in a net charge (negative for electrons, e⁻, positive for holes, h⁺) at the cold end, producing an electrostatic potential (voltage). An equilibrium is thus reached between the chemical potential for diffusion and the electrostatic repulsion due to the build-up of charge. This property, known as the Seebeck effect, is the basis of thermoelectric power generation.

Thermoelectric devices contain many thermoelectric couples (Fig. B1, bottom) consisting of n-type (containing free electrons) and p-type (containing free holes) thermoelectric elements wired electrically in series and thermally in parallel (Fig. B1, top). A thermoelectric generator uses heat flow across a temperature gradient to power an electric load through the external circuit. The temperature difference provides the voltage ($V = \alpha\Delta T$) from the Seebeck effect (Seebeck coefficient α) while the heat flow drives the electrical current, which therefore determines the power output. In a Peltier cooler the external circuit is a d.c. power supply, which drives the electric current (I) and heat flow (Q), thereby cooling the top surface due to the Peltier effect ($Q = \alpha TI$). In both devices the heat rejected must be removed through a heat sink.

The maximum efficiency of a thermoelectric material for both power generation and cooling is determined by its figure of merit (zT):

$$zT = \frac{\alpha^2 T}{\rho\kappa}$$

zT depends on α, absolute temperature (T), electrical resistivity (ρ), and thermal conductivity (κ). The best thermoelectrics are semiconductors that are so heavily doped their transport properties resemble metals.

For the past 40 years, thermoelectric generators have reliably provided power in remote terrestrial and extraterrestrial locations most notably on deep space probes such as *Voyager*. Solid-state Peltier coolers provide precise thermal management for optoelectronics and passenger seat cooling in automobiles. In the future, thermoelectric systems could harness waste heat and/or provide efficient electricity through co-generation. One key advantage of thermoelectrics is their scalability — waste heat and co-generation sources can be as small as a home water heater or as large as industrial or geothermal sources.

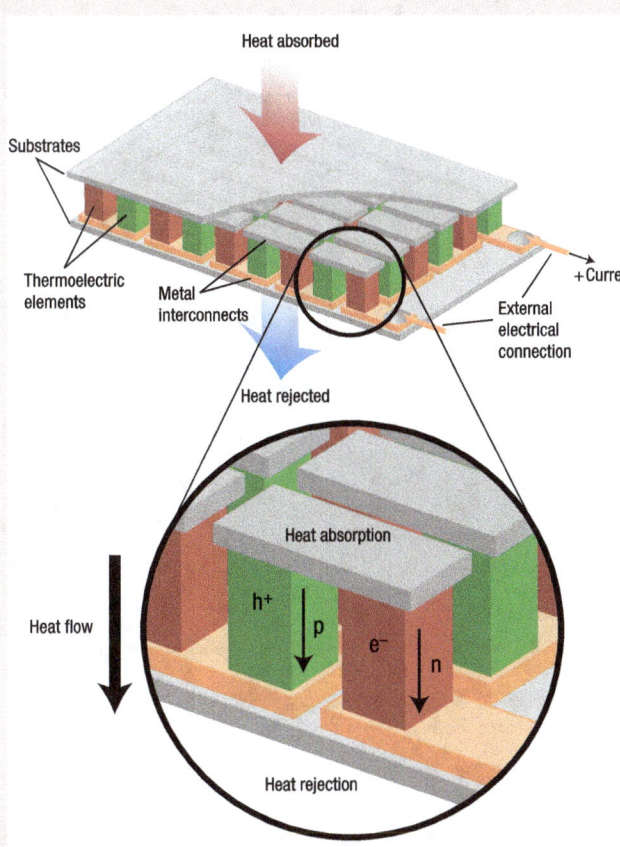

Figure B1 Thermoelectric module showing the direction of charge flow on both cooling and power generation.

The Lorenz factor can vary particularly with carrier concentration. Accurate assessment of κ_e is important, as κ_l is often computed as the difference between κ and κ_e (equation (3)) using the experimental electrical conductivity. A common source of uncertainty in κ_e occurs in low-carrier-concentration materials where the Lorenz factor can be reduced by as much as 20% from the free-electron value. Additional uncertainty in κ_e arises from mixed conduction, which introduces a bipolar term into the thermal conductivity[10]. As this term is not included in the Wiedemann–Franz law, the standard computation of κ_l erroneously includes bipolar thermal conduction. This results in a perceived increase in κ_l at high temperatures for Bi_2Te_3, PbTe and others, as shown in Fig. 2a. The onset of bipolar thermal conduction occurs at nearly the same temperature as the peak in Seebeck and electrical resistivity, which are likewise due to bipolar effects.

As high zT requires high electrical conductivity but low thermal conductivity, the Wiedemann–Franz law reveals an inherent materials conflict for achieving high thermoelectric efficiency. For materials with very high electrical conductivity (metals) or very low κ_l, the Seebeck coefficient alone primarily determines zT, as can be seen in equation (4), where $(\kappa_l/\kappa_e) \ll 1$:

$$zT = \frac{\alpha^2/L}{1 + \frac{\kappa_l}{\kappa_e}} \qquad (4)$$

LATTICE THERMAL CONDUCTIVITY

Glasses exhibit some of the lowest lattice thermal conductivities. In a glass, thermal conductivity is viewed as a random walk of energy through a lattice rather than rapid transport via phonons, and leads to the concept of a minimum thermal conductivity[22], κ_{min}. Actual glasses, however, make poor thermoelectrics because they lack the needed 'electron-crystal' properties — compared with crystalline semiconductors they have lower mobility due to increased electron scattering and lower effective masses because of broader bands. Good thermoelectrics are therefore crystalline materials that manage to scatter phonons without significantly disrupting the electrical conductivity. The heat flow is carried by a spectrum of phonons with widely varying wavelengths and mean free paths[23] (from less than 1 nm to greater than 10 μm), creating a need for phonon scattering agents at a variety of length scales.

Thermoelectrics therefore require a rather unusual material: a 'phonon-glass electron-crystal'[24]. The electron-crystal requirement stems from the fact that crystalline semiconductors have been the best at meeting the compromises required from the electronic properties (Seebeck coefficient and electrical conductivity). The phonon-glass requirement stems from the need for as low a lattice thermal conductivity as possible. Traditional thermoelectric materials have used site substitution (alloying) with isoelectronic elements to preserve a crystalline electronic structure while creating large mass contrast to disrupt the phonon path. Much of the recent excitement in the field of thermoelectrics is a result of the successful demonstration of other methods to achieve phonon-glass electron-crystal materials.

ADVANCES IN THERMOELECTRIC MATERIALS

Renewed interest in thermoelectrics is motivated by the realization that complexity at multiple length scales can lead to new mechanisms for high zT in materials. In the mid 1990s, theoretical predictions suggested that the thermoelectric efficiency could be greatly enhanced by quantum confinement of the electron charge carriers[5,25]. The electron energy bands in a quantum-confined structure are progressively narrower as the confinement increases and the dimensionality decreases. These narrow bands should produce high effective masses and therefore large Seebeck coefficients. In addition, similar sized, engineered heterostructures

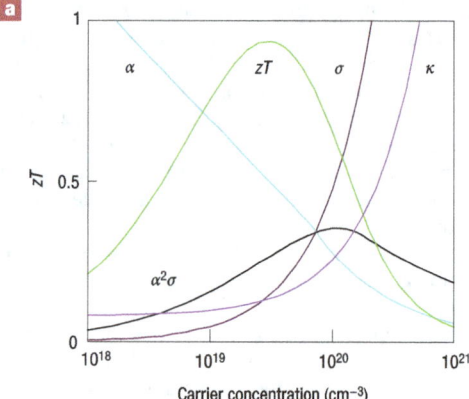

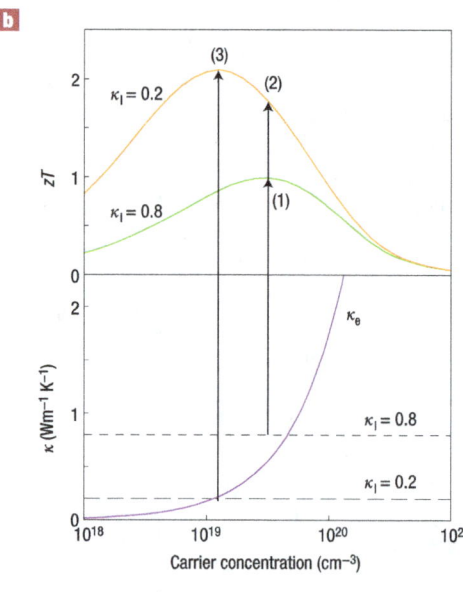

Figure 1 Optimizing zT through carrier concentration tuning. **a**, Maximizing the efficiency (zT) of a thermoelectric involves a compromise of thermal conductivity (κ; plotted on the y axis from 0 to a top value of 10 W m^{-1} K^{-1}) and Seebeck coefficient (α; 0 to 500 μV K^{-1}) with electrical conductivity (σ; 0 to 5,000 Ω^{-1}cm^{-1}). Good thermoelectric materials are typically heavily doped semiconductors with a carrier concentration between 10^{19} and 10^{21} carriers per cm^3. The thermoelectric power factor $\alpha^2\sigma$ maximizes at higher carrier concentration than zT. The difference between the peak in $\alpha^2\sigma$ and zT is greater for the newer lower-κ_l materials. Trends shown were modelled from Bi_2Te_3, based on empirical data in ref. 78. **b**, Reducing the lattice thermal conductivity leads to a two-fold benefit for the thermoelectric figure of merit. An optimized zT of 0.8 is shown at point (1) for a model system (Bi_2Te_3) with a κ_l of 0.8 Wm^{-1} K^{-1} and κ_e that is a function of the carrier concentration (purple). Reducing κ_l to 0.2 Wm^{-1} K^{-1} directly increases the zT to point (2). Additionally, lowering the thermal conductivity allows the carrier concentration to be reoptimized (reduced), leading to both a decrease in κ_e and a larger Seebeck coefficient. The reoptimized zT is shown at point (3).

may decouple the Seebeck coefficient and electrical conductivity due to electron filtering[26] that could result in high zT. Even though a high-ZT device based on these principles has yet to be demonstrated, these predictions have stimulated a new wave of interest in complex thermoelectric materials. Vital to this rebirth has been interdisciplinary collaborations: research in thermoelectrics

Box 2 State-of-the-art high-zT materials

To best assess the recent progress and prospects in thermoelectric materials, the decades of research and development of the established state-of-the-art materials should also be considered. By far the most widely used thermoelectric materials are alloys of Bi_2Te_3 and Sb_2Te_3. For near-room-temperature applications, such as refrigeration and waste heat recovery up to 200 °C, Bi_2Te_3 alloys have been proved to possess the greatest figure of merit for both n- and p-type thermoelectric systems. Bi_2Te_3 was first investigated as a material of great thermoelectric promise in the 1950s[12,16–18,84]. It was quickly realized that alloying with Sb_2Te_3 and Bi_2Se_3 allowed for the fine tuning of the carrier concentration alongside a reduction in lattice thermal conductivity. The most commonly studied p-type compositions are near $(Sb_{0.8}Bi_{0.2})_2Te_3$ whereas n-type compositions are close to $Bi_2(Te_{0.8}Se_{0.2})_3$. The electronic transport properties and detailed defect chemistry (which controls the dopant concentration) of these alloys are now well understood thanks to extensive studies of single crystal and polycrystalline material[85,86]. Peak zT values for these materials are typically in the range of 0.8 to 1.1 with p-type materials achieving the highest values (Fig. B2a,b). By adjusting the carrier concentration zT can be optimized to peak at different temperatures, enabling the tuning of the materials for specific applications such as cooling or power generation[87]. This effect is demonstrated in Fig. B2c for PbTe.

For mid-temperature power generation (500–900 K), materials based on group-IV tellurides are typically used, such as PbTe, GeTe or SnTe[12,17,18,81,88]. The peak zT in optimized n-type material is about 0.8. Again, a tuning of the carrier concentration will alter the temperature where zT peaks. Alloys, particularly with $AgSbTe_2$, have led to several reports of $zT > 1$ for both n-type and p-type materials[73,89,90]. Only the p-type alloy $(GeTe)_{0.85}(AgSbTe_2)_{0.15}$, commonly referred to as TAGS, with a maximum zT greater than 1.2 (ref. 69), has been successfully used in long-life thermoelectric generators. With the advent of modern microstructural and chemical analysis techniques, such materials are being reinvestigated with great promise (see section on nanomaterials).

Successful, high-temperature (>900 K) thermoelectric generators have typically used silicon–germanium alloys for both n- and p-type legs. The zT of these materials is fairly low, particularly for the p-type material (Fig. B2b) because of the relatively high lattice thermal conductivity of the diamond structure.

For cooling below room temperature, alloys of BiSb have been used in the n-type legs, coupled with p-type legs of $(Bi,Sb)_2(Te,Se)_3$ (refs 91,92). The poor mechanical properties of BiSb leave much room for improved low-temperature materials.

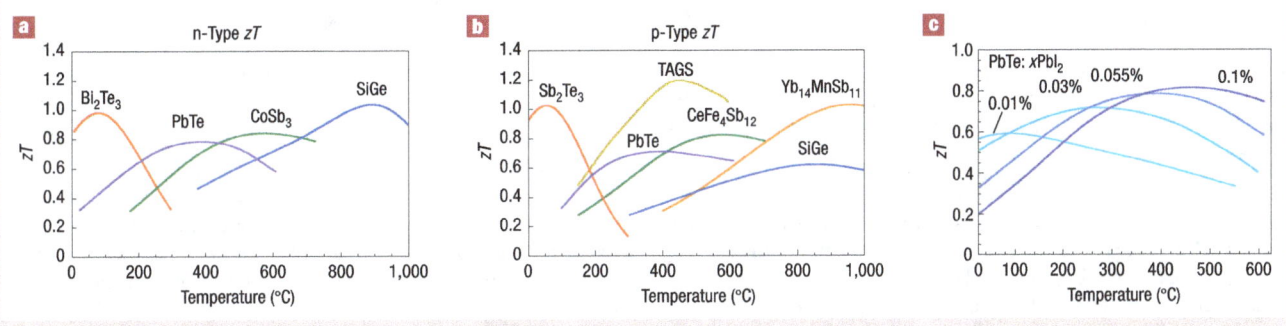

Figure B2 Figure-of-merit zT of state-of-the-art commercial materials and those used or being developed by NASA for thermoelectric power generation. **a**, p-type and **b**, n-type. Most of these materials are complex alloys with dopants; approximate compositions are shown. **c**, Altering the dopant concentration changes not only the peak zT but also the temperature where the peak occurs. As the dopant concentration in n-type PbTe increases (darker blue lines indicate higher doping) the zT peak increases in temperature. Commercial alloys of Bi_2Te_3 and Sb_2Te_3 from Marlow Industries, unpublished data; doped PbTe, ref. 88; skutterudite alloys of $CoSb_3$ and $CeFe_4Sb_{12}$ from JPL, Caltech unpublished data; TAGS, ref. 69; SiGe (doped $Si_{0.8}Ge_{0.2}$), ref. 82; and $Yb_{14}MnSb_{11}$, ref. 45.

requires an understanding of solid-state chemistry, high-temperature electronic and thermal transport measurements, and the underlying solid-state physics. These collaborations have led to a more complete understanding of the origin of good thermoelectric properties.

There are unifying characteristics in recently identified high-zT materials that can provide guidance in the successful search for new materials. One common feature of the thermoelectrics recently discovered with $zT > 1$ is that most have lattice thermal conductivities that are lower than the present commercial materials. Thus the general achievement is that we are getting closer to a 'phonon glass' while maintaining the 'electron crystal.' These reduced lattice thermal conductivities are achieved through phonon scattering across various length scales as discussed above. A reduced lattice thermal conductivity directly improves the thermoelectric efficiency, zT, (equation (4)) and additionally allows re-optimization of the carrier concentration for additional zT improvement (Fig. 1b).

There are three general strategies to reduce lattice thermal conductivity that have been successfully used. The first is to scatter phonons within the unit cell by creating rattling structures or point defects such as interstitials, vacancies or by alloying[27]. The second strategy is to use complex crystal structures to separate the electron-crystal from the phonon-glass. Here the goal is to be able to achieve a phonon glass without disrupting the crystallinity of the electron-transport region. A third strategy is to scatter phonons at interfaces, leading to the use of multiphase composites mixed on the nanometre scale[5]. These nanostructured materials can be formed as thin-film superlattices or as intimately mixed composite structures.

COMPLEXITY THROUGH DISORDER IN THE UNIT CELL

There is a long history of using atomic disorder to reduce the lattice thermal conductivity in thermoelectrics (Box 2). Early work by

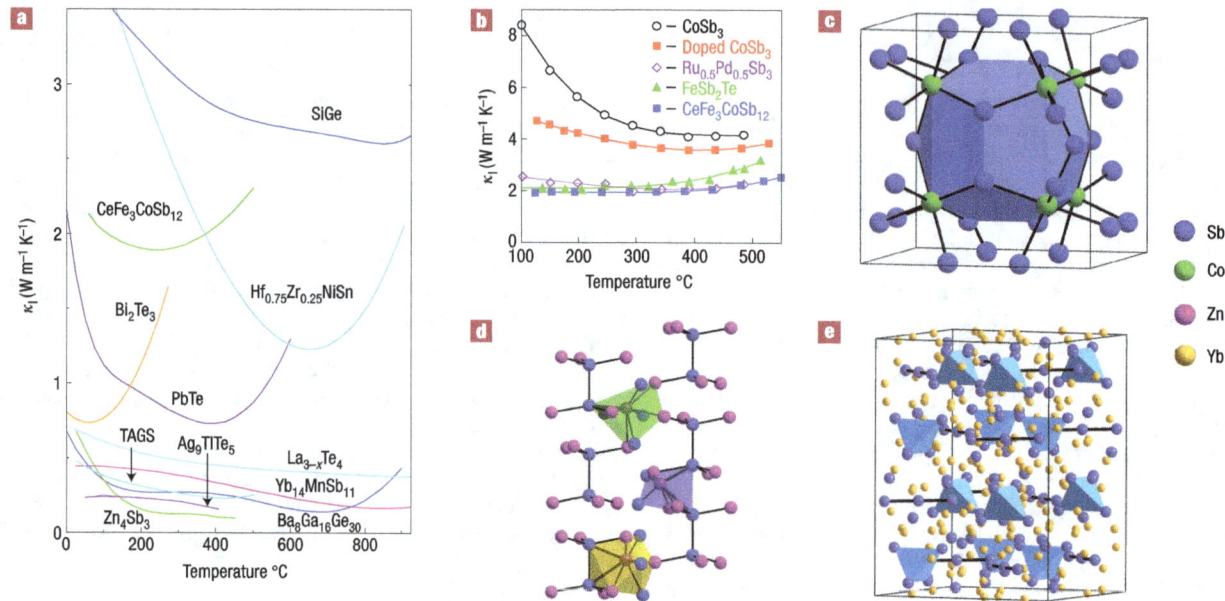

Figure 2 Complex crystal structures that yield low lattice thermal conductivity. **a,** Extremely low thermal conductivities are found in the recently identified complex material systems (such as $Yb_{14}MnSb_{11}$, ref. 45; $CeFe_3CoSb_{12}$, ref. 34; $Ba_8Ga_{16}Ge_{30}$, ref. 79; and Zn_4Sb_3, ref. 80; Ag_9TlTe_5, ref. 40; and $La_{3-x}Te_4$, Caltech unpublished data) compared with most state-of-the-art thermoelectric alloys (Bi_2Te_3, Caltech unpublished data; PbTe, ref. 81; TAGS, ref. 69; SiGe, ref. 82 or the half-Heusler alloy $Hf_{0.75}Zr_{0.25}NiSn$, ref. 83). **b,** The high thermal conductivity of $CoSb_3$ is lowered when the electrical conductivity is optimized by doping (doped $CoSb_3$). The thermal conductivity is further lowered by alloying on the Co ($Ru_{0.5}Pd_{0.5}Sb_3$) or Sb ($FeSb_2Te$) sites or by filling the void spaces ($CeFe_3CoSb_{12}$) (ref. 34). **c,** The skutterudite structure is composed of tilted octahedra of $CoSb_3$ creating large void spaces shown in blue. **d,** The room-temperature structure of Zn_4Sb_3 has a crystalline Sb sublattice (blue) and highly disordered Zn sublattice containing a variety of interstitial sites (in polyhedra) along with the primary sites (purple). **e,** The complexity of the $Yb_{14}MnSb_{11}$ unit cell is illustrated, with $[Sb_3]^{7-}$ trimers, $[MnSb_4]^{9-}$ tetrahedra, and isolated Sb anions. The Zintl formalism describes these units as covalently bound with electrons donated from the ionic Yb^{2+} sublattice (yellow).

Wright discusses how alloying Bi_2Te_3 with other isoelectronic cations and anions does not reduce the electrical conductivity but lowers the thermal conductivity[28]. Alloying the binary tellurides (Bi_2Te_3, Sb_2Te_3, PbTe and GeTe) continues to be an active area of research[29–32]. Many of the recent high-zT thermoelectric materials similarly achieve a reduced lattice thermal conductivity through disorder within the unit cell. This disorder is achieved through interstitial sites, partial occupancies, or rattling atoms in addition to the disorder inherent in the alloying used in the state-of-the-art materials. For example, rare-earth chalcogenides[18] with the Th_3P_4 structure (for example $La_{3-x}Te_4$) have a relatively low lattice thermal conductivity (Fig. 2a) presumably due to the large number of random vacancies (x in $La_{3-x}Te_4$). As phonon scattering by alloying depends on the mass ratio of the alloy constituents, it can be expected that random vacancies are ideal scattering sites.

The potential to reduce thermal conductivity through disorder within the unit cell is particularly large in structures containing void spaces. One class of such materials are clathrates[8], which contain large cages that are filled with rattling atoms. Likewise, skutterudites[7] such as $CoSb_3$, contain corner-sharing $CoSb_6$ octahedra, which can be viewed as a distorted variant of the ReO_3 structure. These tilted octahedra create void spaces that may be filled with rattling atoms, as shown in Fig. 2c with a blue polyhedron[33].

For skutterudites containing elements with low electronegativity differences such as $CoSb_3$ and $IrSb_3$, there is a high degree of covalent bonding, enabling high carrier mobilities and therefore good electron-crystal properties. However, this strong bonding and simple order leads to high lattice thermal conductivities. Thus, the challenge with skutterudites has been the reduction of the lattice thermal conductivity. Doping $CoSb_3$ to carrier concentrations that optimize

zT adds enough carriers to substantially reduce thermal conductivity[34] through electron–phonon interactions (Fig. 2b). Further reductions can be obtained by alloying either on the transition metal or the antimony site.

Filling the large void spaces with rare-earth or other heavy atoms further reduces the lattice thermal conductivity[35]. A clear correlation has been found with the size and vibrational motion of the filling atom and the thermal conductivity leading to zT values as high as 1 (refs 8,13). Partial filling establishes a random alloy mixture of filling atoms and vacancies enabling effective point-defect scattering as discussed previously. In addition, the large space for the filling atom in skutterudites and clathrates can establish soft phonon modes and local or 'rattling' modes that lower lattice thermal conductivity.

Filling these voids with ions adds additional electrons that require compensating cations elsewhere in the structure for charge balance, creating an additional source of lattice disorder. For the case of $CoSb_3$, Fe^{2+} frequently is used to substitute Co^{3+}. An additional benefit of this partial filling is that the free-carrier concentration may be tuned by moving the composition slightly off the charge-balanced composition. Similar charge-balance arguments apply to the clathrates, where filling requires replacing group 14 (Si, Ge) with group 13 (Al, Ga) atoms.

COMPLEX UNIT CELLS

Low thermal conductivity is generally associated with crystals containing large, complex unit cells. The half-Heusler alloys[8] have a simple, cubic structure with high lattice thermal conductivity ($Hf_{0.75}Zr_{0.25}NiSb$ in Fig. 2a) that limits the zT. Thus complex crystal structures are good places to look for improved materials. A good

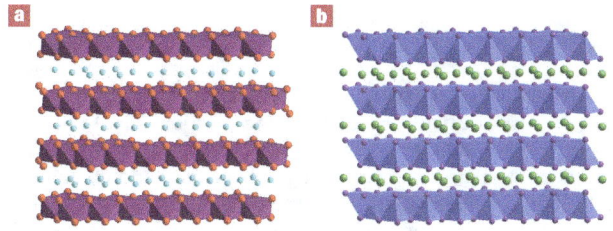

Figure 3 Substructure approach used to separate the electron-crystal and phonon-glass attributes of a thermoelectric. **a**, Na_xCoO_2 and **b**, $Ca_xYb_{1-x}Zn_2Sb_2$ structures both contain ordered layers (polyhedra) separated by disordered cation monolayers, creating electron-crystal phonon-glass structures.

example of a complex variant of Bi_2Te_3 is $CsBi_4Te_6$, which has a somewhat lower lattice thermal conductivity than Bi_2Te_3 that has been ascribed to the added complexity of the Cs layers and the few Bi–Bi bonds in $CsBi_4Te_6$ not found in Bi_2Te_3. These Bi–Bi bonds lower the bandgap compared with Bi_2Te_3, dropping the maximum zT of $CsBi_4Te_6$ below room temperature with a maximum zT of 0.8 (refs 8,36). Like Bi_2Te_3, the layering in $CsBi_4Te_6$ leads to an anisotropic effective mass that can improve the Seebeck coefficient with only minor detriment to the mobility[8]. Many ordered MTe/Bi_2Te_3-type variants (M = Ge, Sn or Pb)[37,38] are known, making up a large homologous series of compounds[39], but to date $zT < 0.6$ is found in most reports. As many of these materials have low lattice thermal conductivities but have not yet been doped to appropriate carrier concentrations, much remains to be done with complex tellurides.

Low lattice thermal conductivities are also seen in the thallium-based thermoelectric materials such as Ag_9TlTe_5 (ref. 40) and Tl_9BiTe_6 (ref. 41). Although these materials do have complex unit cells, there is clearly something unique about the thallium chemistry that leads to low thermal conductivity (0.23 W m^{-1} K^{-1} at room temperature[40]). One possible explanation is extremely soft thallium bonding, which can also be observed in the low elastic modulus these materials exhibit.

The remarkably high zT in Zn_4Sb_3 arises from the exceptionally low, glass-like thermal conductivity (Fig. 2a). In the room-temperature phase, about 20% of the Zn atoms are on three crystallographically distinct interstitial sites as shown in Fig. 2d. These interstitials are accompanied by significant local lattice distortions[42] and are highly dynamic, with Zn diffusion rates almost as high as that of superionic conductors[43]. Pair distribution function (PDF) analysis[44] of X-ray and neutron diffraction data shows that there is local ordering of the Zn interstitials into nanoscale domains. Thus, the low thermal conductivity of Zn_4Sb_3 arises from disorder at multiple length scales, from high levels of interstitials and corresponding local structural distortions and from domains of interstitial ordering. Within the unit cell, Zn interstitials create a phonon glass, whereas the more ordered Sb framework provides the electron-crystal component.

One common characteristic of nearly all good thermoelectric materials is valence balance — charge balance of the chemical valences of all atoms. Whether the bonding is ionic or covalent, valence balance enables the separation of electron energy bands needed to form a bandgap. Complex Zintl compounds have recently emerged as a new class of thermoelectrics[9] because they can form quite complex crystal structures. A Zintl compound contains a valence-balanced combination of both ionically and covalently bonded atoms. The mostly ionic cations donate electrons to the covalently bound anionic species. The covalent bonding allows higher mobility of the charge-carrier species than that found in purely ionic materials. The combination of the bonding types leads to complex structures with the possibility of multiple structural units in the same structure. One example is $Yb_{14}MnSb_{11}$ (refs 45,46), which contains $[MnSb_4]^{9-}$ tetrahedra, polyatomic $[Sb_3]^{7-}$ anions, as well as isolated Sb^{3-} anions and Yb^{2+} cations (Fig. 2e). This structural complexity, despite the crystalline order, enables extremely low lattice thermal conductivity (0.4 W m^{-1} K^{-1} at room temperature; Fig. 2a). Combined with large Seebeck coefficient and high electrical conductivity, $Yb_{14}MnSb_{11}$ results in a zT of ~1.0 at 900 °C. This zT is nearly twice that of p-type SiGe used in NASA spacecraft and has led to rapid acceptance of $Yb_{14}MnSb_{11}$ into NASA programmes for development of future thermoelectric generators. The complexity of Zintl structures also makes them ideal materials for using a substructure approach.

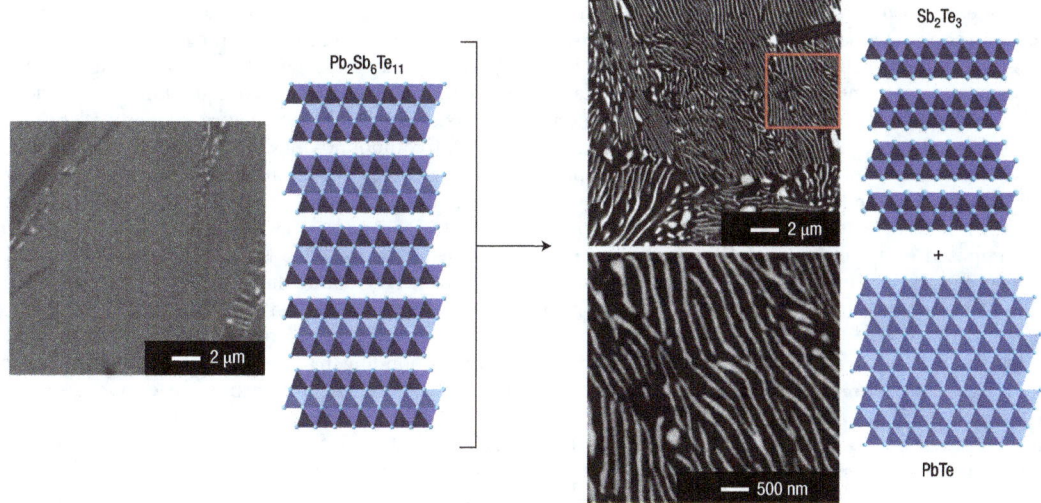

Figure 4 Nanostructured thermoelectrics may be formed by the solid-state partitioning of a precursor phase. The metastable $Pb_2Sb_6Te_{11}$ phase (left) will spontaneously assemble into lamellae of Sb_2Te_3 and PbTe (ref. 76; right). These domains are visible with backscattering scanning electron microscopy, with the dark regions corresponding to Sb_2Te_3 and the light regions to PbTe. Electron backscattering diffraction reveals that the lamellae are oriented with coherent interfaces, shown schematically (right).

Box 3 Thermoelectric measurements

Many materials have been reported with $zT > 1.5$ but few have been confirmed by others, and no devices have been assembled that show the efficiency expected from such high-zT materials. This is due to the complexity of fabricating devices, measurement uncertainty and materials complications.

The inherent difficulty in thermoelectrics is that direct efficiency measurements require nearly as much complexity as building an entire device. Thus practical assessment of the thermoelectric figure of merit typically relies on measuring the individual contributing material properties (σ, α and κ). Measurements of the thermoelectric properties are conceptually simple but results can vary considerably, particularly above room temperature where thermal gradients in the measurement system add to systematic inaccuracies. As a typical zT measurement above room temperature requires the measurement of σ, α and κ (from the density, heat capacity, C_p, and thermal diffusivity, D) each with uncertainty of 5% to 20%, the uncertainty in zT from

$$\frac{\Delta z}{z} = 2\frac{\Delta\alpha}{\alpha} + \frac{\Delta\sigma}{\sigma} + \frac{\Delta C_p}{C_p} + \frac{\Delta D}{D}$$

could easily reach 50%. Accuracy is particularly important for the Seebeck coefficient because it is squared in the calculation of zT and there are few standards with which to calibrate systems. In addition, a variety of geometric terms are required in these calculations (density, thickness and coefficient of thermal expansion[79]).

The sensitivity of the materials themselves to impurities and dopant concentrations further complicates measurements. This is because of the strong dependence of conductivity and to a lesser extent Seebeck coefficient on carrier concentration (Fig. 1a). Small inhomogeneities can result in large variations in thermoelectric properties within a sample[71], making repeatability and combining results of different measurements difficult. For example, combining Seebeck and resistivity on one sample or set of contacts and thermal conductivity on another could lead to spurious results. Likewise, reliable property values are particularly difficult to obtain when sublimation, microstructural evolution, electrochemical reactions and phase transitions are present. Thus, even the act of measuring a sample at high temperatures can alter its properties

The hallmarks of trustworthy measurements are slow, physical trends in properties. Typical materials (Box 2) show a linear or concave downward trend in Seebeck coefficient with temperature and only slow variation with chemical doping. Abrupt transitions to high-zT materials (as a function of temperature or composition) are unlikely. Thin-film samples are particularly difficult to measure. Electrical conductivity often depends critically on the perceived thickness of the conducting layer — if the substrate or quantum-well walls become conducting at any time, it can lead to an erroneously high electrical conductivity estimate of the film. Thermal conductivity and the Seebeck effect likewise depend significantly on the assumption that the substrate and insulating superlattice layers do not change with processing, atmosphere, or temperature. One should be encouraged by results of $zT > 1$ but remain wary of the uncertainties involved to avoid pathological optimism.

SUBSTRUCTURE APPROACH

One method for circumventing the inherent materials conflict of a phonon-glass with electron-crystal properties is to imagine a complex material with distinct regions providing different functions. Such a substructure approach would be analogous to the enabling features that led to high-T_c superconductivity in copper oxides. In these materials, the free charge carriers are confined to planar Cu–O sheets that are separated by insulating oxide layers. Precise tuning of the carrier concentration is essential for superconductivity. This is enabled by the insulating layers acting as a 'charge reservoir'[47] that houses dopant atoms that donate charge carriers to the Cu–O sheets. The separation of the doping regions from the conduction regions keeps the charge carriers sufficiently screened from the dopant atoms so as not to trap carriers, which would lead to a low mobility, hopping conduction mechanism rather than superconductivity.

Likewise, the ideal thermoelectric material would have regions of the structure composed of a high-mobility semiconductor that provides the electron-crystal electronic structure, interwoven with a phonon-glass. The phonon-glass region would be ideal for housing dopants and disordered structures without disrupting the carrier mobility in the electron-crystal region, much like the charge-reservoir region in high-T_c superconductors. The electron crystal regions will need to be thin, on the nanometre or ångström scale, so that phonons with a short mean free path are scattered by the phonon-glass region. Such thin, low-dimensional electron-transport regions might also be able to take advantage of quantum confinement and/or electron filtering to enhance the Seebeck coefficient. Skutterudites and clathrates represent a 0-dimensional version of the substructure approach, with isolated rattlers in an electron-crystal matrix.

The thermoelectric cobaltite oxides (Na_xCoO_2 and others such as those based on the Ca–Co–O system) may likewise be described using a substructure approach[14,48,49]. The Co–O layers form metallic layers separated by insulating, disordered layers with partial occupancies (Fig. 3a). Oxides typically have low mobilities and high lattice thermal conductivity, due to the high electronegativity of oxygen and the strong bonding of light atoms, respectively. These properties give oxides a distinct disadvantage for thermoelectric materials. The relatively large Seebeck values obtained in these systems has been attributed to spin-induced entropy[50]. The success of the cobaltite structures as thermoelectric materials may be an early example of how the substructure approach overcomes these disadvantages. The study of oxide thermoelectrics benefits greatly from the variety of structures and synthetic techniques known for oxides, as well as our understanding of oxide structure–property relationships.

Zintl compounds may be more appropriate for substructure-based thermoelectrics due to the nature of their bonding. The covalently bound anion substructures can adopt a variety of topologies — from 0-dimensional isolated single ions, dimers and polyatomic anions to extended one-, two- and three-dimensional chains, planes and nets[9]. This covalently bound substructure enables high carrier mobilities whereas the ionic cation substructure is amenable to doping and site disorder without disrupting the covalent network. The valence-balanced bonding in these materials frequently leads to bandgaps that are of a suitable size for thermoelectric applications.

The substructure approach is clearly seen in the Zintl compound $Ca_xYb_{1-x}Zn_2Sb_2$ (ref. 51), whose structure is similar to Na_xCoO_2, with sheets of disordered cations between layers of covalently bound Zn–Sb (Fig. 3b). $Ca_xYb_{1-x}Zn_2Sb_2$ demonstrates the fine-tuning ability in the ionic layer concomitant with a modest reduction in lattice thermal conductivity due to alloying of Yb and Ca. The Ca^{2+} is slightly more electropositive than Yb^{2+}, which enables a gradual changing of the carrier concentration as the Yb:Ca ratio is changed. This doping produces disorder on the cation substructure

Box 4 Thermoelectric efficiency

The efficiency of a thermoelectric device depends on factors other than the maximum zT of a material. This is primarily due to the temperature dependence of all the materials properties (α, σ, κ) that make up $zT(T)$. For example, even for state-of-the-art Bi_2Te_3, which has a peak zT value of 1.1, the effective device ZT is only about 0.7 based on the overall performance of the device as a cooler or power generator.

Here we use ZT (upper case) to distinguish the device figure of merit from lower-case $zT = \alpha^2\sigma/\kappa$, the material's figure of merit[10]. For a Peltier cooler the device ZT is most easily measured from the maximum temperature drop obtained (ΔT_{max}).

$$\Delta T_{max} = \frac{ZT_c^2}{2} .$$

For a generator, the maximum efficiency (η) is used to determine ZT:

$$\eta = \frac{\Delta T}{T_h} \cdot \frac{\sqrt{1 + ZT} - 1}{\sqrt{1 + ZT} + \frac{T_c}{T_h}} .$$

Like all heat engines, the maximum power-generation efficiency of a thermoelectric generator is thermodynamically limited by the Carnot efficiency ($\Delta T/T_h$). If temperature is assumed to be independent and n-type and p-type thermoelectric properties are matched (α, σ and κ), (an unrealistic approximation in many cases)

the maximum device efficiency is given by the above equation with $Z = z$.

To maximize efficiency across a large temperature drop, it is imperative to maximize the device ZT, and not just a peak materials zT. One method to achieve this is to tune the material to provide a large average $zT(T)$ in the temperature range of interest. For example, the peak zT in PbTe may be tuned from 300 °C to 600 °C (see Fig. B2c). For large temperature differences (needed to achieve high Carnot efficiency) segmenting with different materials that have peak zT at different temperatures (see Fig. B2a,b) will improve device ZT (ref. 93). Functionally graded materials can also be used to continuously tune zT instead of discrete segmenting[94].

For such large ΔT applications the device ZT can be significantly smaller than even the average zT due to thermoelectric incompatibility. Across a large ΔT, the electrical current required for highest-efficiency operation changes as the materials properties change with temperature or segment[95]. This imposes an additional materials requirement: the thermoelectric compatibility factors ($s = (\sqrt{(1 + zT)} \pm 1)/\alpha T$) with – for power generation[96] or + for cooling[97]) must be similar. For high efficiency, this term needs to be within about a factor of two across the different temperature ranges[95]. A compelling example of the need for compatibility matching is segmenting TAGS with SiGe. The compatibility is so poor between these materials that replacing SiGe with $Yb_{14}MnSb_{11}$ quadruples the device efficiency increase for adding the additional high-temperature segment[45].

---

but not the conducting anion substructure such that the bandgap and carrier mobility is unchanged. The disorder on the cation substructure does indeed lower the lattice thermal conductivity producing a modest increase in zT. However, the relatively simple structure of $Ca_xYb_{1-x}Zn_2Sb_2$ leads to relatively high lattice thermal conductivities (~1.5 W m^{-1} K^{-1}) — suggesting further methods to reduce the conductivity, such as nanostructuring, would lead to an improved material.

Given the broad range of phonons involved in heat transport, a substructure approach may only be one component in a high-performance thermoelectric material. Long-wavelength phonons require disorder on longer length-scales, leading to a need for hierarchical complexity. Combining a substructure approach with nanostructuring seems to be the most promising method of achieving a high-Seebeck, high-conductivity material that manages to scatter phonons at all length scales.

COMPLEX NANOSTRUCTURED MATERIALS

Much of the recent interest in thermoelectrics stems from theoretical and experimental evidence of greatly enhanced zT in nanostructured thin-films and wires due to enhanced Seebeck and reduced thermal conductivity[5]. Reduced thermal conductivity in thin-film superlattices was investigated in the 1980s[52], but has only recently been applied to enhanced thermoelectric materials. Recent efforts[53–55] on Bi_2Te_3-Sb_2Te_3 and PbTe–PbSe films and Si nanowires[56,57] have shown how phonon scattering can reduce lattice thermal conductivity to near κ_{min} values[22,58] (0.2–0.5 W m^{-1} K^{-1}). Thin films containing randomly embedded quantum dots likewise achieve exceptionally low lattice thermal conductivities[59,60]. Very high zT values (>2) have been reported in thin films but the difficulty of measurements makes them a challenge to reproduce in independent labs (Box 3). It is clear however that nanostructured thin-films and wires do exhibit lattice

thermal conductivities near (or even below[61]) κ_{min}, which results in higher material zT, but improvements of electrical and thermal contacts to these materials in a device are needed before higher device ZT (Box 4) is achieved.

The use of bulk (mm^3) nanostructured materials would avoid detrimental electrical and thermal losses and use the existing fabrication routes. The challenge for any nanostructured bulk material system is electron scattering at interfaces between randomly oriented grains leading to a concurrent reduction of both the electrical and thermal conductivities[27]. The effect of grain-boundary scattering in a silicon–germanium system has been extensively studied, as the system possesses excellent electron-crystal properties but very high thermal conductivities. In 1981, the synthesis of polycrystalline silicon germanium alloys were described, and the decrease in thermal conductivity with smaller grain size was tracked[62]. Compared with single crystals of SiGe alloys, polycrystalline materials with grains on the order of 1 µm show an enhanced zT. However, later experiments on materials with grains between 1–100 µm found that the increased phonon scattering was offset by the decrease in electrical conductivity[63]. Nevertheless, recent work suggests that truly nanostructured SiGe enhances zT (ref. 5).

The results on epitaxial thin-films suggests that the ideal nanostructured material would have thermodynamically stable, coherent, epitaxy-like, interfaces between the constituent phases to prevent grain-boundary scattering of electrons. Thus, a promising route to nanostructured bulk thermoelectric materials relies on the spontaneous partitioning of a precursor phase into thermodynamically stable phases[64]. The growth and characterization of such composite microsctructures have been studied in metals for decades because of their ability to greatly improve mechanical strength. The use of microstructure to reduce thermal conductivity in thermoelectrics, dates to the 1960s[65]. For example, crystals pulled from an InSb–Sb eutectic alloy, which forms rods of Sb as thin as 4 µm in an InSb matrix, shows a clear decrease in thermal conductivity with smaller rod diameters, in both parallel and perpendicular directions[66]. Additionally, no major decrease in electrical conductivity or Seebeck coefficient was observed. In a similar manner, several eutectics form from two thermoelectrics form such as rock salt–tetradymite (for example, $PbTe–Sb_2Te_3$) revealing a variety of layered and dendritic microstructures[64,67]. A fundamental limitation of such an approach is that rapid diffusion in the liquid phase leads to coarse microstructures[67,68].

Microstructural complexity may explain why $(AgSbTe_2)_{0.15}(GeTe)_{0.85}$ (TAGS) and $(AgSbTe_2)_x(PbTe)_{1-x}$ (LAST), first studied in the 1950s, have remained some of the materials having highest known zT. Originally believed to be a true solid solution with the rock-salt structure, recent interest has focused on the nanoscale microstructure and even phase separation that exists in these and related alloy compositions. From early on it was predicted that lattice strain in TAGS could explain the low lattice thermal conductivities of $0.3 \text{ W m}^{-1} \text{K}^{-1}$ (ref. 69). Recent work points to the presence of twin-boundary defects in TAGS as an additional source of phonon scattering[70]. Inhomogeneities on various length scales[69] have been found in LAST alloys, which may be associated with the reports of high zT; however, this also makes reproducibility a challenge. In LAST, Ag–Sb rich nanoparticles 1–10 nm in size as well as larger micrometre-sized features precipitate from the bulk[71–73]. The nanoparticles are oriented within the rock-salt crystal with coherent interfaces, therefore electronic conductivity is not significantly reduced. Conversely, the large density difference between the different regions leads to interfacial scattering of the phonons, reducing the thermal conductivity. Through these mechanisms, thermal conductivities of the order of $0.5 \text{ W m}^{-1} \text{K}^{-1}$ at 700 K have been observed. A variety of other materials have been formed that have oriented nanoparticle inclusions in a PbTe matrix[31,32,74].

As the thermoelectric properties of nanostructured materials should depend on the size and morphology of the microstructural features, the materials science of microstructural engineering should become increasingly important in the development of thermoelectric nanomaterials. Our group has focused on the partitioning of quenched, metastable phases that then transform into two phases during a controlled process[75,76]. By restricting the partitioning to solid-state diffusion at low temperatures, the resulting microstructures are quite fine. Figure 4 shows a sample of $Pb_2Sb_6Te_{11}$ after quenching and the microstructure that results on annealing at 400 °C. A lamellar spacing of 360 nm is observed, corresponding to 80-nm PbTe and 280-nm Sb_2Te_3 layer thicknesses. The lamellar spacing can be controlled from below 200 nm to several micrometres. One appealing aspect of this lamellar growth is that the low lattice mismatch between the PbTe (111) and Sb_2Te_3 (001) planes leads to coherent interfaces between the lamellae. Controlled partitioning of a precursor solid can also be done from a glass, as in the case of the alloy glass $(GeSe_2)_{70}(Sb_2Te_3)_{20}(GeTe)_{10}$, which devitrifies to fine lamellae of $GeSe_2$ and $GeSb_4Te_7$ (ref. 77). The numerous possibilities for controlling such reactions will introduce complexity at multiple length scales to thermoelectric materials engineering.

CONCLUDING REMARKS

The conflicting material properties required to produce a high-efficiency (phonon-glass electron-crystal) thermoelectric material have challenged investigators over the past 50 years. Recently, the field has undergone a renaissance with the discovery of complex high-efficiency materials that manage to decouple these properties. A diverse array of new approaches, from complexity within the unit cell to nanostructured bulk and thin-film materials, have all led to high-efficiency materials. Given the complexity of these systems, all of these approaches benefit from collaborations between chemists,

physicists and materials scientists. The global need for sustainable energy coupled with the recent advances in thermoelectrics inspires a growing excitement in this field.

doi:10.1038/nmat2090

References

1. Rowe, D. M. (ed.) *CRC Handbook of Thermoelectrics* (CRC, Boca Raton, 1995).
2. Matsubara, K. in *International Conference on Thermoelectrics* 418–423 (2002).
3. DiSalvo, F. J. Thermoelectric cooling and power generation. *Science* **285**, 703–706 (1999).
4. Rowe, D. M. (ed.) *CRC Handbook of Thermoelectrics: Macro to Nano* (CRC, Boca Raton, 2005).
5. Dresselhaus, M. S. et al. New directions for low-dimensional thermoelectric materials. *Adv. Mater.* **19**, 1043–1053 (2007).
6. Chen, G., Dresselhaus, M. S., Dresselhaus, G., Fleurial, J. P. & Caillat, T. Recent developments in thermoelectric materials. *Int. Mater. Rev.* **48**, 45–66 (2003).
7. Uher, C. in *Thermoelectric Materials Research I* (ed. Tritt, T.) 139–253 (Semiconductors and Semimetals Series 69, Elsevier, 2001).
8. Nolas, G. S., Poon, J. & Kanatzidis, M. Recent developments in bulk thermoelectric materials. *Mater. Res. Soc. Bull.* **31**, 199–205 (2006).
9. Kauzlarich, S. M., Brown, S. R. & Snyder, G. J. Zintl phases for thermoelectric devices. *Dalton Trans.* 2099–2107 (2007).
10. Goldsmid, H. J. *Applications of Thermoelectricity* (Methuen, London, 1960).
11. Tritt, T. M. (ed.) *Recent Trends in Thermoelectric Materials Research* (Academic, San Diego, 2001).
12. Heikes, R. R. & Ure, R. W. *Thermoelectricity: Science and Engineering* (Interscience, New York, 1961).
13. Sales, B. C. Electron crystals and phonon glasses: a new path to improved thermoelectric materials. *Mater. Res. Soc. Bull.* **23**, 15–21 (1998).
14. Koumoto, K., Terasaki, I. & Funahashi, R. Complex oxide materials for potential thermoelectric applications. *Mater. Res. Soc. Bull.* **31**, 206–210 (2006).
15. Mahan, G. D. in *Solid State Physics* Vol. 51 (eds Ehrenreich, H. & Spaepen, F.) 81–157 (Elsevier, 1998).
16. Rosi, F. D. Thermoelectricity and thermoelectric power generation. *Solid-State Electron.* **11**, 833–848 (1968).
17. Rosi, F. D., Hockings, E. F. & Lindenblad, N. E. Semiconducting materials for thermoelectric power generation. *RCA Rev.* **22**, 82–121 (1961).
18. Wood, C. Materials for thermoelectric energy-conversion. *Rep. Prog. Phys.* **51**, 459–539 (1988).
19. Cutler, M., Leavy, J. F. & Fitzpatrick, R. L. Electronic transport in semimetallic cerium sulfide. *Phys. Rev.* **133**, A1143–A1152 (1964).
20. Bhandari, C. M. & Rowe, D. M. in *CRC Handbook of Thermoelectrics* (ed. Rowe, D. M.) Ch. 5, 43–53 (CRC, Boca Raton, 1995).
21. Snyder, G. J., Caillat, T. & Fleurial, J.-P. Thermoelectric transport and magnetic properties of the polaron semiconductor $Fe_xCr_{3-x}Se_4$. *Phys. Rev. B* **62**, 10185 (2000).
22. Slack, G. A. (ed.) *Solid State Physics* (Academic Press, New York, 1979).
23. Dames, C. & Chen, G. in *Thermoelectrics Handbook Macro to Nano* (ed. Rowe, D. M.) Ch. 42 (CRC, Boca Raton, 2006).
24. Slack, G. A. in *CRC Handbook of Thermoelectrics* (ed. Rowe, M.) 407–440 (CRC, Boca Raton, 1995).
25. Hicks, L. D. & Dresselhaus, M. S. Effect of quantum-well structures on the thermoelectric figure of merit. *Phys. Rev. B* **47**, 12727–12731 (1993).
26. Zide, J. M. O. et al. Demonstration of electron filtering to increase the Seebeck coefficient in $In_{0.53}Ga_{0.47}As/In_{0.53}Ga_{0.28}Al_{0.19}As$ superlattices. *Phys. Rev. B* **74**, 205335 (2006).
27. Bhandari, C. M. in *CRC Handbook of Thermoelectrics* (ed. Rowe, D. M.) 55–65 (CRC, Boca Raton, 1995).
28. Wright, D. A. Thermoelectric properties of bismuth telluride and its alloys. *Nature* **181**, 834–834 (1958).
29. Kusano, D. & Hori, Y. Thermoelectric properties of p-type $(Bi_2Te_3)(0.2)$ $(Sb_2Te_3)(0.8)$ thermoelectric material doped with PbTe. *J. Jpn Inst. Met.* **66**, 1063–1065 (2002).
30. Zhu, P. W. et al. Enhanced thermoelectric properties of PbTe alloyed with Sb_2Te_3. *J. Phys. Condens. Matter* **17**, 7319–7326 (2005).
31. Poudeu, P. F. P. et al. Nanostructures versus solid solutions: Low lattice thermal conductivity and enhanced thermoelectric figure of merit in $Pb_{9.6}Sb_{0.2}Te_{10-x}Se_x$ bulk materials. *J. Am. Chem. Soc.* **128**, 14347–14355 (2006).
32. Poudeu, P. F. R. et al. High thermoelectric figure of merit and nanostructuring in bulk p-type $Na_{1-x}Pb_mSb_yTe_{m+2}$. *Angew. Chem. Int. Edn* **45**, 3835–3839 (2006).
33. Feldman, J. L., Singh, D. J., Mazin, II, Mandrus, D. & Sales, B. C. Lattice dynamics and reduced thermal conductivity of filled skutterudites. *Phys. Rev.B* **61**, R9209–R9212 (2000).
34. Fleurial, J.-P., Caillat, T. & Borshchevsky, A. in *Proc. ICT'97 16th Int. Conf. Thermoelectrics* 1–11 (IEEE Piscataway, New Jersey, 1997).
35. Sales, B. C., Mandrus, D. & Williams, R. K. Filled skutterudite antimonides: A new class of thermoelectric materials. *Science* **272**, 1325–1328 (1996).
36. Chung, D. Y. et al. A new thermoelectric material: $CsBi_4Te_6$. *J. Am. Chem. Soc.* **126**, 6414–6428 (2004).
37. Shelimova, L. E. et al. Thermoelectric properties of $PbBi_4Te_7$-based anion-substituted layered solid solutions. *Inorg. Mater.* **40**, 1146–1152 (2004).
38. Shelimova, L. E. et al. Crystal structures and thermoelectric properties of layered compounds in the $ATe-Bi_2Te_3$ (A = Ge, Sn, Pb) systems. *Inorg. Mater.* **40**, 451–460 (2004).
39. Kanatzidis, M. G. Structural evolution and phase homologies for "design" and prediction of solid-state compounds. *Acc. Chem. Res.* **38**, 359–368 (2005).
40. Kurosaki, K., Kosuga, A., Muta, H., Uno, M. & Yamanaka, S. Ag_9TlTe_5: A high-performance thermoelectric bulk material with extremely low thermal conductivity. *Appl. Phys. Lett.* **87**, 061919 (2005).
41. Wolfing, B., Kloc, C., Teubner, J. & Bucher, E. High performance thermoelectric Tl_9BiTe_6 with an extremely low thermal conductivity. *Phys. Rev. Lett.* **86**, 4350–4353 (2001).
42. Toberer, E. S., Sasaki, K. A., Chisholm, C. R. I., Haile, S. M. & Snyder, G. J. Local structure of interstitial Zn in β-Zn_4Sb_3. *Phys. Status Solidi* **1**, 253–255 (2007).

43. Chalfin, E., Lu, H. X. & Dieckmann, R. Cation tracer diffusion in the thermoelectric materials $Cu_3Mo_6Se_8$ and "beta-Zn_4Sb_3". *Solid State Ionics* **178**, 447–456 (2007).

44. Kim, H. J., Božin, E. S., Haile, S. M., Snyder, G. J. & Billinge, S. J. L. Nanoscale alpha-structural domains in the phonon-glass thermoelectric material beta-Zn_4Sb_3. *Phys. Rev. B* **75**, 134103 (2007).

45. Brown, S. R., Kauzlarich, S. M., Gascoin, F. & Snyder, G. J. $Yb_{14}MnSb_{11}$: New high efficiency thermoelectric material for power generation. *Chem. Mater.* **18**, 1873–1877 (2006).

46. Kauzlarich, S. M., Brown, S. R. & Snyder, G. J. Zintl phases for thermoelectric devices. *Dalton Trans.* 2099–2107 (2007).

47. Cava, R. J. Structural chemistry and the local charge picture of copper-oxide superconductors. *Science* **247**, 656–662 (1990).

48. Terasaki, I., Sasago, Y. & Uchinokura, K. Large thermoelectric power in $NaCo_2O_4$ single crystals. *Phys. Rev. B* **56**, 12685–12687 (1997).

49. Shin, W. & Murayama, N. Thermoelectric properties of (Bi,Pb)-Sr-Co-O oxide. *J. Mater. Res.* **15**, 382–386 (2000).

50. Wang, Y. Y., Rogado, N. S., Cava, R. J. & Ong, N. P. Spin entropy as the likely source of enhanced thermopower in $Na_xCo_2O_4$. *Nature* **423**, 425–428 (2003).

51. Gascoin, F., Ottensmann, S., Stark, D., Haile, S. M. & Snyder, G. J. Zintl phases as thermoelectric materials: Tuned transport properties of the compounds $Ca_xYb_{1-x}Zn_2Sb_2$. *Adv. Funct. Mater.* **15**, 1860–1864 (2005).

52. Yao, T. Thermal-Properties of AlAs/GaAs Superlattices. *Appl. Phys. Lett.* **51**, 1798–1800 (1987).

53. Touzelbaev, M. N., Zhou, P., Venkatasubramanian, R. & Goodson, K. E. Thermal characterization of Bi_2Te_3/Sb_2Te_3 superlattices. *J. Appl. Phys.* **90**, 763–767 (2001).

54. Caylor, J. C., Coonley, K., Stuart, J., Colpitts, T. & Venkatasubramanian, R. Enhanced thermoelectric performance in PbTe-based superlattice structures from reduction of lattice thermal conductivity. *Appl. Phys. Lett.* **87**, 23105 (2005).

55. Beyer, H. *et al.* High thermoelectric figure of merit *ZT* in PbTe and Bi_2Te_3-based superlattices by a reduction of the thermal conductivity. *Physica E* **13**, 965–968 (2002).

56. Hochbaum, A. I. *et al.* Enhanced thermoelectric performance of rough silicon nanowires. *Nature* **451**, 163–167 (2008).

57. Boukai, A. I. *et al.* Silicon nanowires as efficient thermoelectric materials. *Nature* **451**, 168–171 (2008).

58. Cahill, D. G., Watson, S. K. & Pohl, R. O. Lower limit to thermal conductivity of disordered crystals. *Phys. Rev. B* **46**, 6131–40 (1992).

59. Kim, W. *et al.* Cross-plane lattice and electronic thermal conductivities of ErAs: InGaAs/InGaAlAs superlattices. *Appl. Phys. Lett.* **88**, 242107 (2006).

60. Kim, W. *et al.* Thermal conductivity reduction and thermoelectric figure of merit increase by embedding nanoparticles in crystalline semiconductors. *Phys. Rev. Lett.* **96**, 045901 (2006).

61. Chiritescu, C. *et al.* Ultralow thermal conductivity in disordered, layered WSe_2 crystals. *Science* **315**, 351–353 (2007).

62. Rowe, D. M., Shukla, V. S. & Savvides, N. Phonon-scattering at grain-boundaries in heavily doped fine-grained silicon-germanium alloys. *Nature* **290**, 765–766 (1981).

63. Vining, C. B., Laskow, W., Hanson, J. O., Vanderbeck, R. R. & Gorsuch, P. D. Thermoelectric properties of pressure-sintered $Si_{0.8}Ge_{0.2}$ thermoelectric alloys. *J. Appl. Phys.* **69**, 4333–4340 (1991).

64. Jang, K.-W. & Lee, D.-H. in *Fourteenth International Conference on Thermoelectrics* 108 (IEEE, 1995).

65. Goldsmid, H. J. & Penn, A. W. Boundary scattering of phonons in solid solutions. *Phys. Lett. A* **27**, 523–524 (1968).

66. Liebmann, W. K. & Miller, E. A. Preparation phase-boundary energies, and thermoelectric properties of Insb-Sb eutectic alloys with ordered microstructures. *J. Appl. Phys.* **34**, 2653–2659 (1963).

67. Ikeda, T. *et al.* Solidification processing of alloys in the pseudo-binary $PbTe$-Sb_2Te_3 system. *Acta Mater.* **55**, 1227–1239 (2007).

68. Aliev, M. I., Khalilova, A. A., Arsaly, D. G., Ragimov, R. N. & Tanogly, M. Electrical and thermal properties of the GaSb-$FeGa_{1.3}$ eutectic. *Inorg. Mater.* **40**, 331–335 (2004).

69. Skrabek, E. A. & Trimmer, D. S. in *CRC Handbook of Thermoelectrics* (ed. Rowe, D. M.) 267–275 (CRC, Boca Raton, 1995).

70. Cook, B. A., Kramer, M. J., Wei, X., Harringa, J. L. & Levin, E. M. Nature of the cubic to rhombohedral structural transformation in $(AgSbTe_2)_{15}(GeTe)_{85}$ thermoelectric material. *J. Appl. Phys.* **101**, 053715 (2007).

71. Chen, N. *et al.* Macroscopic thermoelectric inhomogeneities in $(AgSbTe_2)_x(PbTe)_{1-x}$. *App. Phys. Lett.* **87**, 171903 (2005).

72. Androulakis, J. *et al.* Nanostructuring and high thermoelectric efficiency in p-type $Ag(Pb_{1-y}Sn_y)_mSbTe_{2+m}$. *Adv. Mater.* **18**, 1170–1173 (2006).

73. Hsu, K. F. *et al.* Cubic $AgPb_mSbTe_{2+m}$: Bulk thermoelectric materials with high figure of merit. *Science* **303**, 818–821 (2004).

74. Sootsman, J. R., Pcionek, R. J., Kong, H. J., Uher, C. & Kanatzidis, M. G. Strong reduction of thermal conductivity in nanostructured PbTe prepared by matrix encapsulation. *Chem. Mater.* **18**, 4993–4995 (2006).

75. Ikeda, T. *et al.* Solidification processing of alloys in the pseudo-binary $PbTe$-Sb_2Te_3 system. *Acta Mater.* **55**, 1227–1239 (2007).

76. Ikeda, T. *et al.* Self-assembled nanometer lamellae of thermoelectric PbTe and Sb_2Te_3 with epitaxy-like interfaces. *Chem. Mater.* **19**, 763–767 (2007).

77. Clavagueramora, M. T., Surinach, S., Baro, M. D. & Clavaguera, N. Thermally activated crystallization of $(GeSe_2)_{70}(Sb_2Te_3)_{20}(GeTe)_{10}$ alloy glass - morphological and calorimetric study. *J. Mater. Sci.* **18**, 1381–1388 (1983).

78. Rowe, D. M. & Min, G. Alpha-plot in sigma-plot as a thermoelectric-material performance indicator. *J. Mater. Sci. Lett.* **14**, 617–619 (1995).

79. Toberer, E. S., Christensen, M., Iversen, B. B. & Snyder, G. J. High temperature thermoelectric efficiency in $Ba_8Ga_{16}Ge_{30}$. *Phys. Rev. B* (in the press).

80. Caillat, T., Fleurial, J. P. & Borshchevsky, A. Preparation and thermoelectric properties of semiconducting Zn_4Sb_3. *J. Phys. Chem. Solids* **58**, 1119–1125 (1997).

81. Gelbstein, Y., Dashevsky, Z. & Dariel, M. P. High performance n-type PbTe-based materials for thermoelectric applications. *Physica B* **363**, 196–205 (2005).

82. Vining, C. B., Laskow, W., Hanson, J. O., Vanderbeck, R. R. & Gorsuch, P. D. Thermoelectric properties of pressure-sintered $Si_{0.8}Ge_{0.2}$ thermoelectric alloys. *J. Appl. Phys.* **69**, 4333–4340 (1991).

83. Culp, S. R., Poon, S. J., Hickman, N., Tritt, T. M. & Blumm, J. Effect of substitutions on the thermoelectric figure of merit of half-Heusler phases at 800 °C. *Appl. Phys. Lett.* **88**, 042106 (2006).

84. Goldsmid, H. J. & Douglas, R. W. The use of semiconductors in thermoelectric refrigeration. *Brit. J. Appl. Phys.* **5**, 386–390 (1954).

85. Scherrer, H. & Scherrer, S. in *Thermoelectrics Handbook Macro to Nano* (ed. rowe, D. M.) Ch. 27 (CRC, Boca Raton, 2006).

86. Kutasov, V. A., Lukyanova, L. N. & Vedernikov, M. V. in *Thermoelectrics Handbook Macro to Nano* Ch. 37 (CRC, Boca Raton, 2006).

87. Kuznetsov, V. L., Kuznetsova, L. A., Kaliazin, A. E. & Rowe, D. M. High performance functionally graded and segmented Bi_2Te_3-based materials for thermoelectric power generation. *J. Mater. Sci.* **37**, 2893–2897 (2002).

88. Fritts, R. W. in *Thermoelectric Materials and Devices* (eds. Cadoff, I. B. & Miller, E.) 143–162 (Reinhold, New York, 1960).

89. Fleischmann, H., Luy, H. & Rupprecht, J. Neuere Untersuchungen an Halbleitenden IV VI-I V VI_2 Mischkristallen. *Zeitschrift Für Naturforschung A* **18**, 646–649 (1963).

90. Fleischmann, H. Wärmeleitfähigkeit, Thermokraft und Elektrische Leitfähigkeit von Halbleitenden Mischkristallen Der Form $(A^I_{2/2} Bi^{IV}_{1-x}C^V_{x/2})D^{VI}$. *Zeitschrift Für Naturforschung A* **16**, 765–780 (1961).

91. Yim, W. M. & Amith, A. Bi-Sb alloys for magneto-thermoelectric and thermomagnetic cooling. *Solid State Electron.* **15**, 1141 (1972).

92. Sidorenko, N. A. & Ivanova, L. D. Bi-Sb solid solutions: Potential materials for high-efficiency thermoelectric cooling to below 180 K. *Inorg. Mater.* **37**, 331–335 (2001).

93. Snyder, G. J. Application of the compatibility factor to the design of segmented and cascaded thermoelectric generators. *Appl. Phys. Lett.* **84**, 2436–2438 (2004).

94. Müller, E., Drasar, C., Schilz, J. & Kaysser, W. A. Functionally graded materials for sensor and energy applications. *Mater. Sci. Eng. A* **362**, 17–39 (2003).

95. Snyder, G. J. in *Thermoelectrics Handbook Macro to Nano* (ed. Rowe, D. M.) Ch. 9 (CRC, Boca Raton, 2006).

96. Snyder, G. J. & Ursell, T. Thermoelectric efficiency and compatibility. *Phys. Rev. Lett.* **91**, 148301 (2003).

97. Ursell, T. S. & Snyder, G. J. in *Twenty-first International Conference on Thermoelectrics* ICT'02 412 (IEEE, Long Beach, California, USA, 2002).

Acknowledgements

We thank Jean-Pierre Fleurial and Thierry Caillat for discussions concerning skutterudites, Marlow Industries, Cronin Vining, Yaniv Gelbstein, Ken Kurosaki for data and discussions and JPL-NASA and the Beckman Institute at Caltech for funding.

Solid-state nanopores

The passage of individual molecules through nanosized pores in membranes is central to many processes in biology. Previously, experiments have been restricted to naturally occurring nanopores, but advances in technology now allow artificial solid-state nanopores to be fabricated in insulating membranes. By monitoring ion currents and forces as molecules pass through a solid-state nanopore, it is possible to investigate a wide range of phenomena involving DNA, RNA and proteins. The solid-state nanopore proves to be a surprisingly versatile new single-molecule tool for biophysics and biotechnology.

CEES DEKKER

Kavli Institute of Nanoscience, Delft University of Technology, Lorentzweg 1, 2628 CJ Delft, The Netherlands

e-mail: c.dekker@tudelft.nl

Have you ever struggled to pass a length of thread through the eye of a needle? If you have, you will know that it is difficult, especially if the needle's eye is small. However, advances in nanotechnology now allow single polymer molecules to be threaded through nanopores that measure just a few nanometres across. Although the passage of molecules through nanopores is commonplace in biology, it is only recently that researchers have been able to coax single DNA molecules through solid-state nanopores. This article reviews progress in this still young and rapidly expanding field, with a particular emphasis on engineered nanopores in silicon materials.

The biological cell is filled with all types of nanopores that control the trafficking of ions and molecules in and out of the cell and among the subcellular structures: examples include the ion channels that conduct ions across the cell surface; the nuclear membrane pores that control the passage of messenger RNA from the cell nucleus into the cytosol; the proteins that are secreted across pores in the membranes of cell organelles; and viruses, which dump their genomes into cells via pores that they insert into the cell membrane. Some of the transport through pores is passive, although most of it is actively controlled. Elucidating the physical processes involved in nanopore transport phenomena will remain a fertile field of research for years to come. Pioneering work on single-ion channels started decades ago with electrophysiology experiments that could measure the flow of ions through single nanopores in cell membranes[1].

In the 1990s, it was proposed that it might be possible to use nanopores as sensors for DNA[2]. If DNA could be transferred through a nanopore in a linear fashion, this might serve as a device to read off the sequence of DNA in an ultrafast way, which is of obvious interest for genomics applications. The first experimental results were reported in 1996 by Kasianowicz and co-workers at NIST on the biological pore α-haemolysin[3], which has become the protein of choice for this type of research. More recently, non-biological solid-state nanopores have opened up an even wider range of new research areas, but let us first review the wealth of results that have been obtained for α-haemolysin.

BIOLOGICAL NANOPORES

α-haemolysin is a protein secreted by *Staphylococcus aureus* bacteria as a toxin, which forms nanopores that spontaneously insert themselves into a lipid membrane. It features a transmembrane channel with a width of 1.4 nm at its narrowest point (see Fig. 1a.)[4], which allows ions to pass at a high ionic conductance of about one nS for typical conditions of one molar salt. An applied voltage of 100 mV across the membrane leads to a current of ~100 pA, which can be readily measured. A decade ago it was proposed that DNA can thread such pores, and that this would perhaps allow its sequence to be read off. The small constriction of this pore allows the passage of single-strand DNA (ssDNA), but not double-strand DNA (dsDNA) because, with a diameter of 2.2 nm, it is too wide. The idea of the basic experiment is simple: DNA is a highly charged molecule, so it can be driven through the nanopore in a linear head-to-tail fashion by an electric field (Fig. 1b). When the DNA enters the nanopore, the ionic current is reduced because part of the liquid volume that carries the ionic current is occupied by the DNA (similar to the Coulter counters used for counting and sizing particles).

Experiments on the traversal of polynucleotides through an α-haemolysin pore were first published ten years ago[3]. The data demonstrated that one could probe the traversal of individual ssDNA molecules, shown from analysis of electrical signals (the presence of DNA reduces the ion current by an order of magnitude) and from quantitative polymerase chain reaction analysis.

A host of follow-up experiments revealed that simple measurements of the passage of molecules through the nanopore (a process called translocation) could actually address a range of interesting properties of these molecules. Molecules were found to travel fast; for typical conditions, a 100 nucleotide string of C bases in ssDNA will move through in about 0.1 ms, yielding a residence time of about 1 µs per base (ref. 5). However, ssRNA molecules containing only A bases (polyA) move an order of magnitude slower than polyC, polyT and polyG molecules because polyA RNA has a secondary structure which needs to be deconstructed before it can pass through the pore[6]. As a result of the asymmetric structure of α-haemolysin, entry of a DNA molecule to the *cis* side (top in Fig. 1a) is easier than insertion at the *trans* (bottom) side[7]. This was attributed to a higher entropic barrier for entry or electrostatic repulsion of DNA from negative charge at the bottom side.

When citing this review, please cite the original version as shown on the contents page of this chapter.

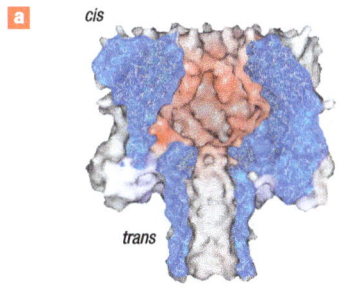

Figure 1 DNA translocation through a biological nanopore. α-haemolysin is a transmembrane protein that contains a pore that is approximately 1.4 nm wide at its narrowest point, which allows single-stranded DNA or RNA to move in and out of cells. **a**, The cross-sectional structure of α-haemolysin. The *trans* side is located towards the cytosol in the cell, whereas the *cis* side points outwards. Reproduced from ref. 11. **b**, Snapshots of a ssDNA molecule passing through such a nanopore. The DNA molecule enters the cavity at the *cis* side (left). Subsequently, the membrane voltage drives the DNA through the pore towards the *trans* side (middle, right). These snapshots are the result of a molecular dynamics simulation. Image courtesy of Aleksei Aksimentiev.

As intuitively expected, a larger applied voltage will both increase the frequency and decrease the duration of each translocation event because the molecules will move at higher velocity. The event frequency has been reported to be exponentially dependent on the voltage[7]. At constant voltage, the speed that the DNA moves through the pore is constant for all lengths — with long DNA taking an equivalently longer time to translocate — except for very short DNA (less than 12 bases), which moves at a higher velocity[5]. Translocation appears to be a thermally activated process as indicated by an exponential temperature dependence of the translocation time[8].

In experiments on DNA translocation, an α-haemolysin pore acts as an stochastic sensor which typically detects the passage of a few thousands molecules. Analyses involve plotting the induced change in current versus the translocation time in a scatter plot (an example for a solid-state nanopore is given in Fig. 3b), where each point represents the translocation of a single DNA molecule. With this type of plot even subtle differences between different populations of molecules can be discerned. For example, one can distinguish whether a ssDNA molecule enters the pore with its 3' end or its 5' end[9]. Remarkably, a 3'-threaded DNA molecule moves a factor of two slower than a 5'-threaded molecule. Molecular dynamics simulations suggested that this is due to a tilt orientation of the bases towards the 5' end at the point where the nanopore confines the DNA most stringently.

Given the motivation to develop translocation of DNA through α-haemolysin as a sequencing method, a particularly appealing early result was that the current signals were found to be different when comparing a homopolymer, made up of 100 identical bases, with another homopolymer of different type, for example comparing polyA and polyC RNA[6,8]. A difference could also be seen within an RNA molecule with a heterosequence of $A_{30}C_{70}$. The observation that one could distinguish A and C bases in the electrical signal led to optimism about the use of nanopores for sequencing. However, it was subsequently realized that the signal-to-noise ratio was insufficient to ever reach single-base resolution in translocation experiments under typical conditions. The main reason is that translocation through α-haemolysin occurs at such a fast rate: during the ~1 μs that one base spends at the narrowest constriction, the number of ions that pass the pore is in the order of 100. Statistical fluctuations in this number overwhelm the subtle differences between the four different types of bases. Sequencing by straightforward DNA translocation through α-haemolysin thus seems unrealistic. However, this is not the end of the story at all

and alternative strategies for sequencing with nanopores are being developed (see Outlook section).

A great advantage of protein nanopores is that they can be chemically engineered with advanced molecular biology techniques such as mutagenesis[10]. It is therefore possible to make well-defined local changes in the structure, for example, by exchanging one amino acid group for another, or adding a small organic molecule at a specific molecular site. Using such techniques, a variety of α-haemolysin biosensors were developed, such as ones with hydrophobic groups that bind organic molecules[11] and with a biotin-labelled chain that can probe biotin-binding proteins[12]. Similarly, the addition of a nucleotide oligomer makes it possible to probe the hybridization of complementary DNA: if the sequences match, duplex binding will occur and the DNA translocation will be halted until, after a while, the DNA is unzipped and translocates through the pore[13]. Unzipping has also been studied extensively in experiments on ssDNA hairpins[14,15] and on ssDNA with a dsDNA end[16], again achieving discrimination at single-base resolution. Recently, translocation through anthrax pores[17], which have a structure very similar to α-haemolysin, has been studied.

FABRICATION OF SOLID-STATE NANOPORES

Although biological pores have proved to be very useful for a range of interesting translocation experiments, they do exhibit a number of disadvantages such as fixed size and limited stability. Typically the pores, and also, in particular, their embedding lipid bilayer, can become unstable if changes occur in external parameters such as pH, salt concentration, temperature, mechanical stress, and so on. Fabrication of nanopores from solid-state materials presents obvious advantages over their biological counterpart such as very high stability, control of diameter and channel length, adjustable surface properties and the potential for integration into devices and arrays. Various routes have been explored to meet the challenge of fabricating pores with true nanometre dimensions.

Etching a hole into an insulating layer has been one way to create pores. Researchers have also etched holes in glass slides, albeit with larger dimensions, for electrophysiological measurements on cells[18,19]. A more delicate technique has been to guide the etching so that it occurs preferentially along one linear defect path that was created by shooting a single high-energy heavy-metal ion through a polymer layer[20]. Importantly, one can monitor the ion current across the layer during the etching, which provides feedback for when to stop etching. In this way, pores with nanometre dimensions, down to 2 nm, have been made. However these pores typically have a narrow opening angle of ~1°, which is non-ideal for translocation of polynucleotides. The surface of the etched pore can be modified, for example, by coating with a gold layer. This changes the electrostatic boundary condition of the pore and provides a template to add further organic layers through thiol self-assembly techniques.

Five years ago, Golovchenko and co-workers at Harvard reported a novel technique — ion beam sculpting — by which they could fabricate single nanopores in thin SiN membranes with true nanometre control[21]. Working with SiN and SiO_2 materials provides an advantage as these materials have been extremely well-developed and studied over the course of fifty years of microelectronics. The Harvard group used a dedicated ion beam-machine (Fig. 2a) that uses a focused ion beam to mill a tiny hole in the membrane. Feedback from ion detectors below the membrane provided signals to indicate when to stop the milling. Interestingly, they observed that, depending on the ion rate and temperature, pores could enlarge and shrink, allowing the fine-tuning of pores in the nanometre range. Pores with nanometre dimensions were made in this way and provided a starting point for DNA translocation measurements.

Our group at Delft has pursued another approach (Fig. 2b)[22]. Again, using the techniques from Si microfabrication, free-standing membranes of Si, SiN, or SiO_2 can be made. Subsequently, we can pattern holes in these using electron-beam lithography and subsequent etching. A typical diameter of such a hole is 20 nm, and this could be reduced to sub-10 nm if needed. In imaging these holes with a transmission electron microscope (TEM), we discovered that high-intensity wide-field illumination with electrons slowly modified the nanopore size. Large holes, with a diameter greater than the membrane thickness, grew in size, whereas small holes shrunk. Note that this provides a way to fine-tune the nanopores to a small size with subnanometre resolution and with direct visual feedback from a commercial TEM. This process of glassmaking at the nanoscale occurs as the SiO_2 membrane flows to decrease the surface tension and so slowly reshapes around the hole[23]. If electron-beam exposure is stopped, the structure will be entirely frozen and stable. It has also been shown that a pore can be directly drilled in a membrane by a locally focused electron beam in a TEM[22,24-26], making laborious preparatory electron-beam lithography of larger nanopores unnecessary. Again, such pores can be modified with the wide-field TEM illumination.

It is clear that the fabrication of pores with true nanometre dimensions has greatly advanced in recent years. Nanopores have now been fabricated with subnanometre control and with diameters as small as 1 nm. In addition to opening up the way to single-molecule measurements, this also leads to other applications, such as the fabrication of nanopore-based metal electrodes for electrochemistry[27]. Other groups are pursuing nanopores in multilayer membrane materials that may provide additional electronic probes for detection of DNA[28].

IONIC CONDUCTION OF SOLID-STATE NANOPORES

When these solid-state nanopores are immersed in a liquid with a monovalent salt, their ion conduction can be measured. Under good conditions, linear current–voltage characteristics can be measured up to a high voltage (1 V), and the pore conductance obtained matches that expected from the bulk conductivity of the liquid and the geometry of the nanopore. Typical measurements are taken at one molar salt. At low salt (below ~0.1 M), the effect of the negatively charged walls of SiO_2 sets in and the lowered pore conductance is dominated by the counter ions that screen the walls. Measurements of the salt dependence of the nanopores' ionic conduction shows that this picture describes the data well,

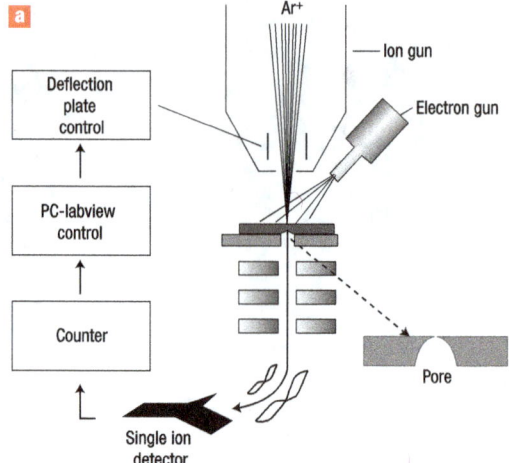

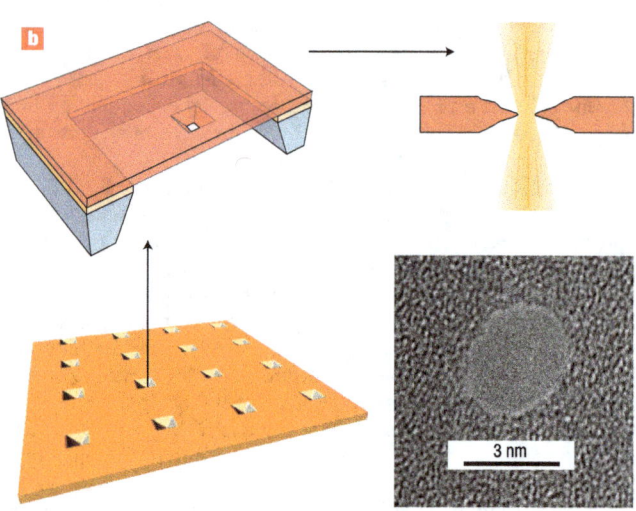

Figure 2 Fabrication of solid-state nanopores. Pores with nanometre-size diameters can be made with a variety of techniques within a thin silicon-based membrane. **a**, The focused-ion-beam technique developed by the Harvard group[21]. When the ions drilling the hole penetrate through the membrane, the beam is shut down. **b**, The TEM technique developed at Delft[22]. An electron beam drills a hole in the membrane. The result can be directly monitored in the electron microscope and the electron beam can also be used to enlarge or shrink the nanopore in a controlled way. The figure at the bottom right shows a typical TEM image of a nanopore (with a diameter of 3 nm in this example).

provided that the chemical reactivity of the silicon dioxide surface is accounted for[29].

Asymmetric current–voltage curves are not uncommon and can have multiple causes. For polymer nanopores, it has been reported that the sign and magnitude of the applied voltage can cause the pore size to vary[30]. The cause of asymmetries in Si-based nanopores, however, remains less clear. Interesting effects have been observed when the ion concentration differs on both sides of the nanopore. Resistance changes and asymmetric current–voltage curves were observed and interpreted to result from the nanopore's charged

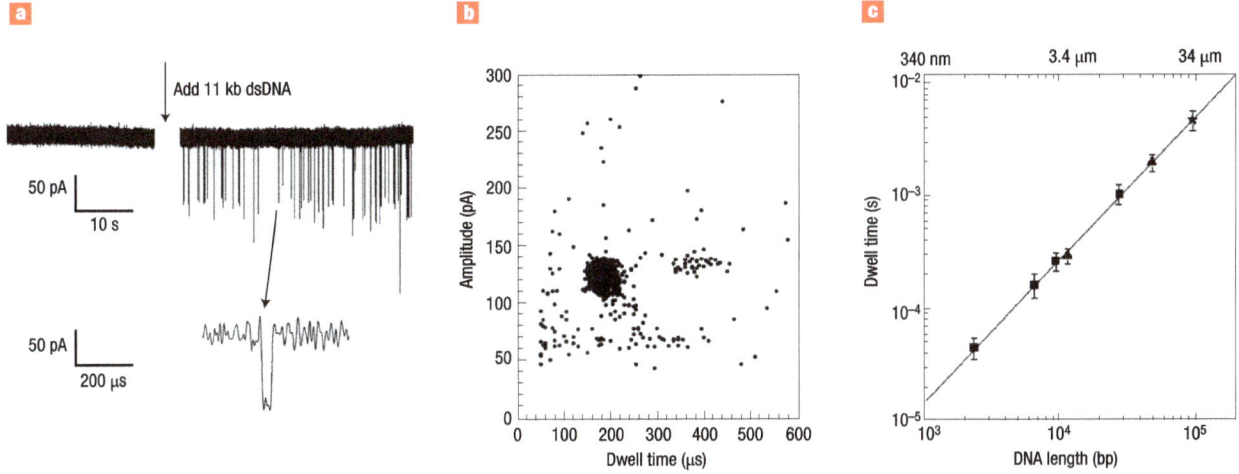

Figure 3 Translocation of dsDNA through a solid-state nanopore. **a**, An example of the ion current versus time measured for a nanopore. When 11 kbp dsDNA is added, downward spikes appear. Each spike indicates an event where a single DNA molecule traverses the 10 nm nanopore. The bottom panel zooms in on one event. **b**, A scatter plot where the amplitude of the current change is plotted versus the time of the DNA translocation. Each point is a separate translocation event. This particular example is the data for 11 kbp circular nicked dsDNA which traverses the 8 nm pore in about 200 μs without any complicated folding phenomena. On average each traversal creates a current dip of 120 pA at the applied voltage of 120 mV. **a** and **b** reproduced with permission from ref. 37. Copyright (2005) APS. **c**, Translocation time versus the contour length of the dsDNA. Surprisingly, it was found that long dsDNA molecules display a nonlinear, power-law dependence $\tau \approx L^{\alpha}$ with an exponent $\alpha = 1.27$. Reproduced with permission from ref. 39. Copyright (2005) ACS.

walls and the distribution of electric fields within the asymmetric conical-shaped pore[31].

Nanopores can exhibit undesirable phenomena such as a large variability in conductance[32] and significant noise, both as a function of time and among samples. Theoretical work on biological pores suggests that switching between different pore conformations generates 1/f noise[33], and experiments on polymer nanopores have provided evidence for this[34]. Recently, our Delft group identified a different noise source of nanopores[35], namely nanometre-sized gaseous bubbles — so-called nanobubbles. Conductance profiles show strong variations in both the magnitude of the conductance and in the noise when a single nanopore is scanned through the focus of an infrared laser beam multiple times. Differences of up to five orders of magnitude were found in the current power spectral density. A simple model of a cylindrical nanopore that contains a nanobubble explained the data well. Nanobubbles have been observed before in other experiments but the reason for their thermodynamic stability is still under debate (for a review see ref. 36).

DNA TRANSLOCATION THROUGH SOLID-STATE NANOPORES

The main aim of researchers working with solid-state nanopores has been to measure the translocation of individual DNA molecules. The first current traces that indicated dsDNA in a solid-state nanopore were reported by the Harvard group[21]. More extensive measurements were presented by Storm *et al.* from our group. Similar to the α-haemolysin experiments, molecules pass the nanopore in a linear fashion as evidenced from monitoring the conductance of a voltage-biased pore. Contrary to α-haemolysin, this is also possible with dsDNA and long molecules of up to 100 kilobases.

The basic experimental result is illustrated in Fig. 3a: when 11 kbp-long dsDNA is added to the negatively biased reservoir, downward spikes appear in the ionic current through the pore. Each spike represents a single-molecule event where an individual DNA molecule translocates the nanopore. For each event, one can

characterize the amplitude by which the current is reduced as well as the duration of the blockage. Plotting amplitude versus duration produces a scatter plot. Figure 3b shows such a scatter plot for circular 11 kbp dsDNA which displays a simple distribution around a single point. More complicated distributions are often observed and careful analysis of scatter plots has proven quite helpful to disentangle different types of translocations.

DNA with different lengths can be identified from the different durations of translocation. Indeed, it was shown that individual DNA molecules within a mixture containing different size molecules could be distinguished[37], providing a demonstration of DNA sizing that is analogous to the well-known gel electrophoresis technique but without the need to stain the DNA. Moreover, the nanopore technique requires much less material than existing techniques — as few as 1,000 molecules.

A narrow pore only allows the linear passage of DNA in a head-to-tail fashion. In wider pores, however, it has been observed that DNA can traverse in a folded manner[37,38]. Long DNA molecules form a randomly folded coil. For such a configuration, the translocation of DNA can start at a point somewhere from the side of such a coil. If this happens, two double-strand pieces of DNA enter the pore, followed by one piece when one of the two pieces has encountered its end. Indeed this was observed. More complex phenomena with up to five parallel pieces of a single dsDNA molecule were also detected.

Another interesting polymer-physics effect concerns the translocation velocity. Experiments have shown that long (say 10 kbp) DNA does not traverse a wide solid-state nanopore in a simple linear fashion at constant speed. Instead, it was found that the translocation time grows as a power of the DNA length, $\tau \approx L^{\alpha}$ with an exponent $\alpha = 1.27$ (ref. 39). This could be theoretically understood to result from the hydrodynamic drag on the section of the DNA polymer outside the pore being the dominant force counteracting the electrical driving force. This leads to a power-law scaling with an exponent of 1.22, which is in good agreement with the data[39].

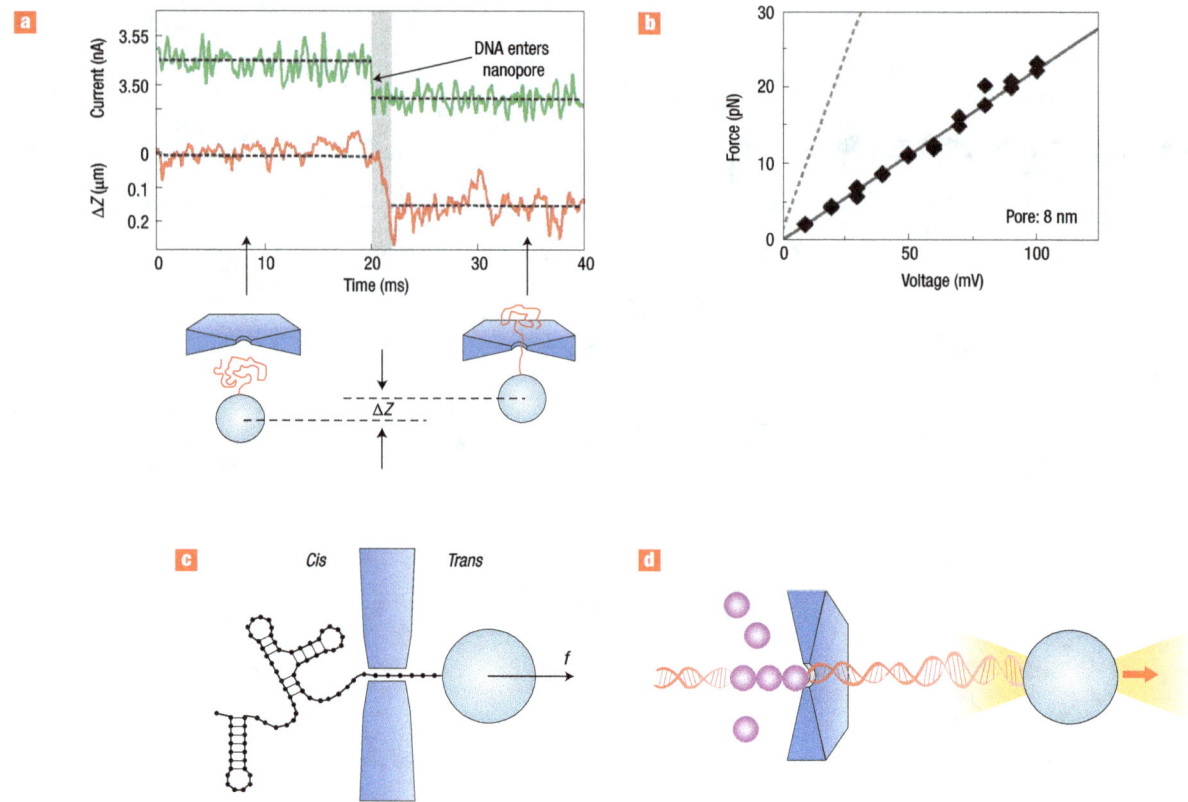

Figure 4 New developments. **a**, Force measurements can be executed on a single dsDNA molecule. DNA is attached to a bead that is controllably held in an optical tweezer. With an applied voltage, the DNA is drawn into the nanopore and as a result the bead is displaced from its equilibrium position in the trap. The displacement is a measure of the force that is acting on the DNA molecule in the nanopore. Reproduced from ref. 48. **b**, The measured force on a dsDNA molecule in an 8 nm nanopore at 20 mM KCl. The dashed line represents the force expected for DNA with a charge of 2 electrons per bp. Reproduced from ref. 48. **c**, A sketch of RNA unfolding by sequential unzipping of its hairpins, which can be monitored by pulling it through a nanopore with an applied force, *f*. It has been theoretically calculated that the signatures of individual hairpins should be discernable[55,56]. **d**, A sketch of a possible experiment where the (dis)association of DNA-binding proteins is measured. This type of experiment is interesting from the point of view of basic science, and possibly also DNA sequencing. Much thought is given to strategies where locally binding probes are a measure of the local sequence, which may be measured in nanopores.

New phenomena were also discovered for DNA translocation at low salt concentrations. Surprisingly, at 100 mM salt, the insertion of DNA does not lead to suppression of the ionic current, but to an enhancement[40,29]! In fact, the salt dependence of the magnitude of the current change showed a sign reversal when the KCl concentration was close to 0.4 M. This remarkable finding can be understood from a simple model where current decreases result from the partial blocking of the pore and current increases are attributed to motion of the counter ions that screen the charge of the DNA backbone[29].

A variety of other interesting experimental results are now appearing as the field is growing. For example, the velocity of translocating DNA can be slowed down by an order of magnitude if the salt concentration is increased, the driving voltage is decreased and the viscosity of the fluid is increased by the addition of glycerol and lowering of the temperature[41]. By melting DNA at high pH, the translocation of ssDNA can be observed[42]. For very small nanopores (~2 nm), it has been reported that ssDNA can pass, whereas dsDNA is blocked from passing because of its larger diameter[43]. Modelling of the translocation process has been initiated with extensive molecular dynamics simulations[44]. For small pores, a voltage threshold for dsDNA translocation was found, which was associated with the stretching transition that occurs in dsDNA near 60 pN (ref. 45),

suggesting that dsDNA can be drawn through very narrow pores when strongly driven.

FORCE SPECTROSCOPY WITH NANOPORES

The rich variety of experimental observations of DNA translocation events through nanopores has increasingly led to questions about the basic physical understanding of the translocation process. What are the forces acting on the DNA and how large are they? What is the role of the counter charges? How about hydrodynamic couplings? What role does entropy play on DNA entering the hole[46]? And so on.

New experimental probes — beyond the straightforward ionic conductance — are needed to address such questions. Our group has used optical tweezers for direct force measurements on DNA in nanopores[47]. Here, one end of the DNA is attached to a bead, which is trapped in the focus of an infrared laser. Subsequently, we can insert individual DNA molecules into a single nanopore and arrest the DNA during voltage-driven translocation. The force acting on the DNA then pulls the bead away from the centre of the optical trap, an effect that can be measured with high accuracy using the reflected light from the bead. These experiments directly measure the force on a DNA molecule in a solid-state nanopore (Figure 4a).

The force acting on a single dsDNA was found to rise linearly with the applied voltage, at a rate of 0.23 pN mV^{-1} (Fig.4b)[48]. Interestingly, this value is entirely independent of the KCl salt concentration as measured from 20 mM to 1 M. The force acting on a DNA molecule thus does not depend on salt, which is even more remarkable when remembering that, as mentioned above, the induced change in the ion current actually reverses sign in this range of salt concentrations. The value of 0.23 pN mV^{-1} is a factor of four lower than that expected from a naive calculation of the electrostatic force acting on DNA, which considers a DNA charge of two electrons per base pair because of the phosphate groups on the backbones. The force data have been interpreted as indicating that bound counter ions lead to an effective DNA charge of 0.50 electrons per base pair[48]. A similar value was found from the salt dependence of the current change[29]. Although this appears to nicely coincide with an early ion-condensation prediction of 0.48 electrons per base pair by Manning[49], caution is needed because a full model needs to incorporate not only the statics but also the dynamics of ion motion along DNA as well as hydrodynamic coupling.

The application of a laser to solid-state nanopores has also led to other uses. A nanopore acts as a sensitive thermometer with single nanometre dimensions because it measures the temperature-dependent conductance of the ionic liquid[50]. A laser heats up the liquid at its focal point, and its three-dimensional intensity profile was mapped out by scanning the nanopore, similar to tomography. This new technique does not rely on any optical elements and allows quantitative measurement of optical intensity or temperature distributions in aqueous environments with nanometre resolution.

New methods of force spectroscopy can also be investigated without using optical tweezers. Now that the linearity of the force versus applied voltage has been established and the force magnitude is calibrated, one can use this to simply vary the applied voltage and thus the applied force. Meller and co-workers have worked out a technique where the applied voltage is shut off as soon as the current drops due to the insertion of a DNA molecule[15]. Subsequently, the voltage can be ramped and the response of the molecule monitored.

OUTLOOK: THE S-WORD AND MORE

What more can we expect from solid-state nanopores? First and foremost, how about the big S-word, sequencing[51]? If sequencing with nanopores turns out to be feasible, the impact will be huge because it will remove the need for chemical modification, amplification and surface adsorption of the DNA. Single-molecule DNA sequencing would open up the reading of the genome of single cells and generally act as the key technology in genomics.

As mentioned above, the prospects of simply monitoring the ionic current itself during straightforward translocation are not good, mainly because the DNA moves through the pore too fast. However, optical tweezers can pull DNA through a nanopore at arbitrarily slow speeds, so this approach is worth further research.

Many alternative ideas for sequencing with nanopores exist. Theoretical calculations indicate that sequence information can be extracted from a tunnelling current through the DNA bases as measured between two transverse nanoelectrodes located right at the nanopore[52], though this claim has been disputed[53]. Another idea is to measure changes in the mutual capacitance as DNA molecules pass through a nanopore in a semiconducting multilayer[28] rather than a nanopore in a single insulating membrane. Antisense DNA oligomers may be located near the nanopores by self-assembly chemistry and may be able to probe certain sequences along a passing DNA molecule. Sequence-dependent optical probes may be added and subsequent readout at the pore may then signal the sequence. These and other ideas await experimental verification that will show whether sequencing is feasible and competitive.

The impact of solid-state nanopores is, however, by no means determined by sequencing applications. A range of both technological and scientific opportunities lie ahead. Separation, sizing and sorting can be based on molecular mass, charge, hydrophobicity, stereochemistry and so on. DNA size determination and haplotyping (deducing the constitution of an individual chromosome) are obvious first applications. Sensors can be built with solid-state nanopores — similar to that already demonstrated with α-haemolysin — by incorporating a biomolecular recognition complex (a biotin, antisense DNA or antibodies) into the pore. Further technological challenges include the integration of pores into nanofluidics[54], electronic circuitry and optoelectronic detectors, the incorporation of biological pores with solid-state pores, and the chemical functionalization of nanopores.

There are also plenty of scientific questions that can be addressed in experiments with solid-state nanopore. How are the counterions organized around the DNA, and what is their mobility and coupling to the surrounding liquid? What is the effect of the lateral confinement on the translocating molecule? Are there effects of water in the nanoscale confinement of the pore? What is the effect of the chemical composition, charge state and hydrophilicity/phobicity of the nanopore walls? What is the ultimate detection limit of local structures along the DNA: can one detect a single bound protein, a triple-strand structure, a knot, a hairpin, a mismatched base and so on? Theoretical calculations indicate that it will be feasible to unzip individual hairpins in a ssRNA molecule by slowly pulling the molecule through the pore and measuring the associated force (Fig. 4c)[55,56]. The great advantage of the nanopore technique is that — in contrast to the conventional optical tweezer techniques — one can unzip the hairpins sequentially, thus unravelling the RNA structure in a linear fashion along the molecule.

This illustrates that a nanopore can also be used as a local force actuator. This may open up new experiments where, for example, the biophysics of DNA-processing molecular motors can be studied because it is possible to follow the motor action in real time as it processes the DNA[57]. In a similar example the polymerization of DNA-binding proteins can drive DNA through a pore if its size is large enough to pass the DNA but too small to pass the proteins (Fig. 4d). Monitoring this process with a nanopore may yield parameters such as the polymerization speed and stalling forces.

It is also possible to translocate non-polynucleotide polymers, in particular to observe the passage of single protein molecules[58,59]. This raises the intriguing possibility that nanopores may provide a new route to unfolding proteins in a controlled way. Finally, the ability to fabricate pores with true nanoscale dimensions may find applications that are very different from biomolecular detection. For example, they have been used as a template for nanoelectrodes with diameters down to 2 nm (ref. 24), and have recently been used to explore new regimes for electrochemistry[60]. Larger solid-state pores can act as chip-based probes and are promising alternatives to conventional patch-clamping for electrophysiology on cells.

Given this long list of new opportunities, as well as the recent progress in the field, solid-state nanopores are set to remain an active area of research for years to come.

doi: 10.1038/nnano.2007.27

References

1. Sakmann, B. & Neher, E. (eds). *Single Channel Recording* 2nd edn (Plenum, New York, 2005).
2. Deamer, D. W. & Akeson, M. Nanopores and nucleic acids: prospects for ultrarapid sequencing. *Trends Biotechnol.* **18**, 131–180 (2000).

3. Kasianowicz, J. *et al.* Characterization of individual polynucleotide molecules using a membrane channel. *Proc. Natl Acad. Sci. USA* **93**, 13770–13773 (1996).

4. Song, L. *et al.* Structure of staphylococcal alpha-hemolysin, a heptameric transmembrane pore. *Science* **274**, 1859–66 (1996).

5. Meller, A., Nivon, L. & Branton, D. Voltage-driven DNA translocations through a nanopore. *Phys. Rev. Lett.* **86**, 3435–3438 (2001).

6. Akeson, M., Branton, D., Kasianowicz, J. J., Brandin, E. & Deamer, D. W. Microsecond time-scale discrimination among polycytidylic acid, polyadenylic acid, and polyuridylic acid as homopolymers or as segments within single RNA molecules. *Biophys. J.* **77**, 3227–3233 (1999).

7. Henrickson, S. E. *et al.* Driven DNA transport into an asymmetric nanometer-scale pore. *Phys. Rev. Lett.* **85**, 3057–3060 (2000).

8. Meller, A., Nivon, L., Brandin, E., Golovchenko, J. & Branton, D. Rapid nanopore discrimination between single oligonucleotide molecules. *Proc. Natl Acad. Sci. USA* **97**, 1079–1084 (2000).

9. Mathe, J. *et al.* Orientation discrimination of single-stranded DNA inside the -hemolysin membrane channel. *Proc. Natl Acad. Sci. USA* **102**, 12377–12382 (2005).

10. Bayley, H. *et al.* Engineered Nanopores in *NanoBiotechnology* (eds Niemeyer, C. M. & Mirkin, C. A.) (Wiley-VCH, Weinheim, 2005).

11. Gu, L. Q. *et al.* Stochastic sensing of organic analytes by a pore-forming protein containing a molecular adapter. *Nature* **398**, 686–690 (1999).

12. Movileanu, L. *et al.* Detecting protein analytes that modulate transmembrane movement of a polymer chain within a single protein pore. *Nature Biotechnol.* **18**, 1091–1095 (2000).

13. Howorka, S., Cheley, S. & Bayley, H. Sequence-specific detection of individual DNA strands using engineered nanopores. *Nature Biotechnol.* **19**, 636–639 (2001).

14. Vercoutere, W. *et al.* Rapid discrimination among individual DNA hairpin molecules at single-nucleotide resolution using an ion channel. *Nature Biotechnol.* **19**, 248–252 (2001).

15. Mathe, J. *et al.* Nanopore unzipping of individual DNA hairpin molecules. *Biophys. J.* **87**, 3205–3212 (2004).

16. Sauer-Budge, A. F. *et al.* Unzipping kinetics of double-stranded DNA in a nanopore, *Phys. Rev. Lett.* **90**, 238101 (2003).

17. Halverson, K. M. *et al.* Anthrax biosensor, protective antigen ion channel asymmetric blockade, *J. Biol. Chem.* **280**, 34056–34062 (2005).

18. Schmidt, C., Mayer, M. & Vogel, H. A chip-based biosensor for the functional analysis of single ion channels. *Angew. Chem. Int. Edn* **39**, 3137–3140 (2000).

19. Fertig, N. *et al.* Microstructured glass chip for ion-channel electrophysiology. *Phys. Rev. E* **64**, 040901 (2001).

20. Siwy, Z. & Fulinski, A. Fabrication of a synthetic nanopore ion pump. *Phys. Rev. Lett.* **89**, 198103 (2002).

21. Li, J. *et al.* Ion-beam sculpting at nanometre length scales. *Nature* **412**, 166–169 (2001).

22. Storm, A. J. *et al.* Fabrication of solid-state nanopores with single-nanometre precision. *Nature Mater.* **2**, 537–540 (2003).

23. Storm, A. J. *et al.* Electron-beam-induced deformations of SiO₂ nanostructures. *J. Appl. Phys.* **98**, 014307 (2005).

24. Heng J. B. *et al.* Sizing DNA using a nanometer-diameter pore. *Biophys. J.* **87**, 2905–2911 (2004).

25. Krapf, D. *et al.* Fabrication and characterization of nanopore-based electrodes down to 2 nm. *Nano Lett.* **6**, 105–109 (2006).

26. Zandbergen, H. W. *et al.* Sculpting nano-electrodes with a transmission electron beam for electrical and geometrical characterization of nanoparticles. *Nano Lett.* **5**, 549–553 (2005).

27. Lemay, S. G. *et al.* Lithographically fabricated nanopore-based electrodes for electrochemistry. *Anal. Chem.* **77**, 1911–1915 (2005).

28. Gracheva, M. E. *et al.* Simulation of the electric response of DNA translocation through a semiconductor nanopore-capacitor. *Nanotechnology* **17**, 622–633 (2006).

29. Smeets, R. M. M. *et al.* Salt-dependence of ion transport and DNA translocation through solid-state nanopores. *Nano Lett.* **6**, 89–95 (2006).

30. Siwy, Z. *et al.* Rectification and voltage gating of ion currents in a nanofabricated pore. *Europhys. Lett.* **60**, 349–355 (2002).

31. Siwy, Z. *et al.* Asymmetric diffusion through synthetic nanopores. *Phys. Rev. Lett.* **94**, 048102 (2005).

32. Ho, C. *et al.* Electrolytic transport through a synthetic nanometer-diameter pore. *Proc. Natl Acad. Sci. USA* **102**, 10445–10450 (2005).

33. Bezrukov, S. M. & Winterhalter, M. Examining noise sources at the single-molecule level: $1/f$ noise of an open maltoporin channel. *Phys. Rev. Lett.* **85**, 202–205 (2000).

34. Siwy, Z. & Fulinski, A. Origin of $1/f\alpha$ noise in membrane channel currents. *Phys. Rev. Lett.* **89**, 158101 (2002).

35. Smeets, R. M. M. *et al.* Nanobubbles in solid-state nanopores *Phys. Rev. Lett.*, **97**, 088101 (2006).

36. Attard, P. Nanobubbles and the hydrophobic attraction. *Adv. Colloid Interface Sci.* **104**, 75 (2003).

37. Storm, A. J. *et al.* Translocation of double-strand DNA through a silicon oxide nanopore. *Phys. Rev. E* **71**, 051903 (2005).

38. Li, J. *et al.* DNA molecules and configurations in a solid-state nanopore microscope. *Nature Mater.* **2**, 611–615 (2003).

39. Storm, A. J. *et al.* Fast DNA translocation through a solid-state nanopore. *Nano Lett.* **5**, 1193–1197 (2005).

40. Chang, H. *et al.* DNA-mediated fluctuations in ionic current through silicon oxide nanopore channels *Nano Lett.* **4**, 1551–1556 (2004).

41. Fologea, D. *et al.* Slowing DNA translocation in a solid-state nanopore. *Nano Lett.* **5**, 1734–1737 (2005).

42. Fologea, D. *et al.* Detecting single stranded DNA with a solid state nanopore, *Nano Lett.* **5**, 1905–1909 (2005).

43. Heng, J. B. *et al.* The electromechanics of DNA in a synthetic nanopore. *Biophys. J.* **90**, 1098–1106 (2006).

44. Aksimentiev, A. *et al.* Microscopic kinetics of DNA translocation through synthetic nanopores. *Biophys. J.* **87**, 2086–2097 (2004).

45. Heng, J. B. *et al.* Stretching DNA using the electric field in a synthetic nanopore. *Nano Lett.* **5**, 1883–1888 (2005).

46. Muthukumar, M. Polymer escape through a nanopore. *J. Chem. Phys.* **118**, 5174–5184 (2003).

47. Keyser, U. F., van der Does, J., Dekker, C. & Dekker, N. H. Optical tweezers for force measurements on DNA in nanopores. *Rev. Sci. Instr.* **77**, 105105 (2006).

48. Keyser, U. F. *et al.* Direct force measurements on DNA in a solid-state nanopore. *Nature Phys.* **2**, 473–477 (2006).

49. Manning, G. S. Limiting laws and counterion condensation in polyelectrolyte solutions I. Colligative properties. *J. Chem. Phys.* **51**, 924 933 (1969).

50. Keyser, U. F. *et al.* Nanopore tomography of an objective focus. *Nano Lett.* **5**, 2253 (2005).

51. Service, R. The race for $1000 genome. *Science* **311**, 1544–1546 (2006).

52. Lagerqvist, J., Zwolak, M. & Di Ventra, M. Fast DNA sequencing via transverse electronic transport. *Nano Lett.* **6**, 779–782 (2006).

53. Zhang X.-G. *et al.* First-principles transversal DNA conductance deconstructed. *Biophys. J.* **91**, L04–L06 (2006).

54. Craighead, H. G. Future lab-on-a-chip technologies for interrogating individual molecules. *Nature* **442**, 387–393 (2006).

55. Gerland, U., Bundschuh, R. & Hwa, T. Translocation of structured polynucleotides through nanopores. *Phys. Biol.* **1**, 19–26 (2004).

56. Bundschuh, R. & Gerland, U. Coupled dynamics of RNA folding and nanopore translocation. *Phys. Rev. Lett.* **95**, 208104 (2005).

57. Kafri, Y., Lubensky, D. K. & Nelson, D. R. Dynamics of molecular motors and polymer translocation with sequence heterogeneity. *Biophys. J.* **86**, 3373–3391 (2004).

58. Siwy, Z. *et al.* Protein biosensors based on biofunctionalized conical gold nanotubes. *J. Am. Chem. Soc.* **127**, 5000–5001 (2005).

59. Han, A. *et al.* Sensing protein molecules using nanofabricated pores. *Appl. Phys. Lett.* **88**, 093901 (2006).

60. Krapf, D. *et al.* Experimental observation of nonlinear ionic transport at the nanometer scale. *Nano Lett.* **6**, 2531–2535 (2006).

Acknowledgements

I have benefited from interactions with all the members from our Delft team in the past few years, Dennis M. van den Broek, Giorgia M. S. Dato, Nynke H. Dekker, Jelle van der Does, Stijn Dorp, Hendrik A. Heering, Michiel van den Hout, Ulrich F. Keyser, Bernard M. Koeleman, Diego Krapf, Serge G. Lemay, X. Sean Ling, Ralph M. M. Smeets, Derek Stein, Arnold J. Storm, Meng-Yue Wu, and with collaborators Jianghua Chen, Cees Storm, Jean-François Joanny, and Henny Zandbergen. The nanopore work at Delft is financially supported by FOM, NanoNed and NWO.

Competing interest statement

The author declares that he has no competing financial interests.

Engineering atomic and molecular nanostructures at surfaces

Johannes V. Barth[1,2], Giovanni Costantini[3] & Klaus Kern[1,3]

The fabrication methods of the microelectronics industry have been refined to produce ever smaller devices, but will soon reach their fundamental limits. A promising alternative route to even smaller functional systems with nanometre dimensions is the autonomous ordering and assembly of atoms and molecules on atomically well-defined surfaces. This approach combines ease of fabrication with exquisite control over the shape, composition and mesoscale organization of the surface structures formed. Once the mechanisms controlling the self-ordering phenomena are fully understood, the self-assembly and growth processes can be steered to create a wide range of surface nanostructures from metallic, semiconducting and molecular materials.

In his classic talk of 1959, Richard P. Feynman pointed out[1] that there is "plenty of room at the bottom". He predicted exciting new phenomena that might revolutionize science and technology and affect our everyday lives — if only we were to gain precise control over matter, down to the atomic level. The decades since then have seen the invention of the scanning tunnelling microscope that allows us to image and manipulate individual molecules and atoms[2,3]. We also have access to nanostructured materials with extraordinary functional properties, such as semiconductor quantum dots and carbon nanotubes[4,5], and a growing understanding of how structural features control the function of such small systems.

These complementary developments are different aspects of nanotechnology, which aims to create and use structures, devices and systems in the size range of about 0.1–100 nm (covering the atomic, molecular and macromolecular length scales). Because of this focus on the nanometre scale, nanotechnology might meet the emerging needs of industries that have thrived on continued miniaturization and now face serious difficulties in upholding the trend, particularly in microelectronics[6] and magnetic data storage[7]. But even if nanosystems and nanodevices with suitable performance characteristics are available, nanotechnology solutions will find practical use only if they are economically viable. We will need to develop methods for the controlled mass fabrication of functional atomic or molecular assemblies and their integration into usable macroscopic systems and devices.

The two basic approaches to creating surface patterns and devices on substrates in a controlled and repeatable manner are the 'top-down' and 'bottom-up' techniques[8] (Fig. 1). The former may be seen as modern analogues of ancient methods such as lithography, writing or stamping, but capable of creating features down to the sub-100 nm range. The sophisticated tools allowing such precision are electron-beam writing, and advanced lithographic techniques that use extreme ultraviolet or even hard X-ray radiation[9]. Methods based on electron-beam writing achieve very high spatial resolution at reasonable capital costs, but operational capacity is limited by the serial nature of the process (although electron-projection methods may overcome this limitation). The next-generation production lines used by the semiconductor industry are likely to be based on X-ray

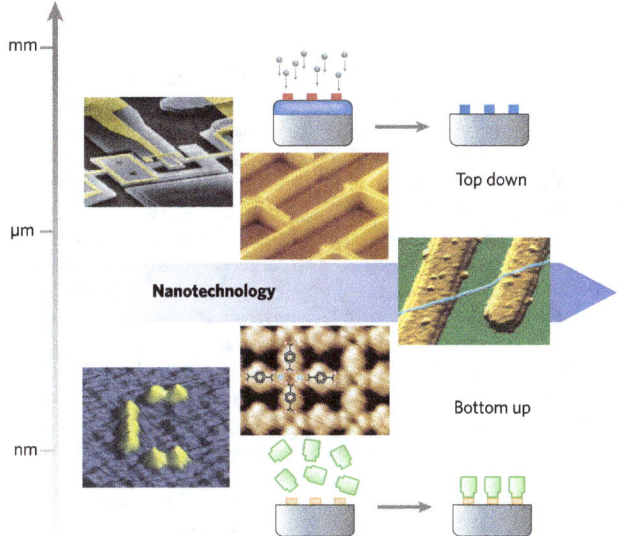

Figure 1 | Two approaches to control matter at the nanoscale. For top-down fabrication, methods such as lithography, writing or stamping are used to define the desired features. The bottom-up techniques make use of self-processes for ordering of supramolecular or solid-state architectures from the atomic to the mesoscopic scale. Shown (clockwise from top) are an electron microscopy image of a nanomechanical electrometer obtained by electron-beam lithography[92], patterned films of carbon nanotubes obtained by microcontact printing and catalytic growth[93], a single carbon nanotube connecting two electrodes[94], a regular metal-organic nanoporous network integrating iron atoms and functional molecules[78], and seven carbon monoxide molecules forming the letter 'C' positioned with the tip of a scanning tunnelling microscope (image taken from http://www.physics.ubc.ca/~stm/).

lithography, which allows parallel processing. But the upgrade will require huge investments and extensive equipment development, to deal with the need for vacuum environments, short-wavelength optics, radiation sources and so on.

[1]Institut de Physique des Nanostructures, École Polytechnique Fédérale de Lausanne, CH-1015 Lausanne, Switzerland; [2]Departments of Chemistry and Physics & Astronomy, University of British Columbia, Vancouver, British Columbia V6T 1Z4, Canada ; [3]Max-Planck-Institut für Festkörperforschung, Heisenbergstrasse 1, D-70569 Stuttgart, Germany.

When citing this review, please cite the original version as shown on the contents page of this chapter.

Considerable efforts have also been invested in developing and exploring alternative top-down patterning methods. A particularly versatile, rapid and low-cost technique is microcontact printing[10], which uses soft and hard stamps to transfer patterns with feature sizes above 100 nm onto a wide range of substrates; however, it becomes increasingly demanding for smaller feature sizes. Ultimate precision is achieved with scanning probe techniques, which are now an established (albeit cumbersome) method for the direct writing and positioning of individual atoms[3]. Prototype scanning force arrays that operate massively in parallel and thus multiply throughput have recently been developed[11], but scanning probe methods seem unlikely to find industrial use in the near future.

Top-down methods essentially 'impose' a structure or pattern on the substrate being processed. In contrast, bottom-up methods aim to guide the assembly of atomic and molecular constituents into organized surface structures through processes inherent in the manipulated system. Here, we outline how self-organized growth and self-assembly at well-defined surfaces (some of which may have been created using top-down methods) can serve as an efficient tool for the bottom-up fabrication of functional structures and patterns on the nanometre scale. We focus on atomic-level investigations and highlight what we regard as particularly informative illustrations of how this approach might lead to useful nanometre-scale surface structures. A brief introduction to the elementary processes governing surface self-ordering provides a foundation for the subsequent discussion of how these processes can be tuned in metallic, semiconducting and molecular systems to obtain surface structures with desired geometric order and well-defined shapes.

Basic concepts in surface structuring

Common to all bottom-up strategies for the fabrication of nanostructures at surfaces is that they are essentially based on growth phenomena. Atoms or molecules (or both) are deposited on the substrate (in vacuum, ambient atmosphere or solution) and nanometre-scale structures evolve as a result of a multitude of atomistic processes. This is inherently a non-equilibrium phenomenon and any growth scenario is governed by the competition between kinetics and thermodynamics. We thus use the term 'self-organized growth' to describe autonomous order phenomena mediated by mesoscale force fields or kinetic limitations in growth processes, whereas 'self-assembly' is reserved for the spontaneous association of a supramolecular architecture from its molecular constituents[12–14]. The term 'self-organization', in contrast to self-organized growth and self-assembly, is usually used in a different context, as it relates to dissipative structure formation in systems far from thermodynamic equilibrium[15] and the initial emergence of biological macromolecules[16].

The primary mechanism in the growth of surface nanostructures from adsorbed species is the transport of these species on a flat terrace (see Fig. 2), involving random hopping processes at the substrate atomic lattice[17,18]. This surface diffusion is thermally activated; that is, diffusion barriers need to be surmounted when moving from one stable (or metastable) adsorption configuration to another. As is typical for such processes, the diffusivity D — the mean square distance travelled by an adsorbate per unit time — obeys an Arrhenius law; this holds for atoms as well as rigid organic molecules[19]. If we now consider a growth experiment where atoms or molecules are deposited on a surface at a constant deposition rate F, then the ratio D/F determines the average distance that an adsorbed species has to travel to meet another adsorbate, either for nucleation of a new aggregate or attachment to an already formed island. The ratio of deposition to diffusion rate D/F is thus the key parameter characterizing growth kinetics. If deposition is slower than diffusion (large values of D/F), growth occurs close to equilibrium conditions; that is, the adsorbed species have enough time to explore the potential energy surface so that the system reaches a minimum energy configuration. If deposition is fast relative to diffusion (small D/F), then the pattern of growth is essentially determined by kinetics; individual processes, notably those leading to metastable structures, are increasingly important.

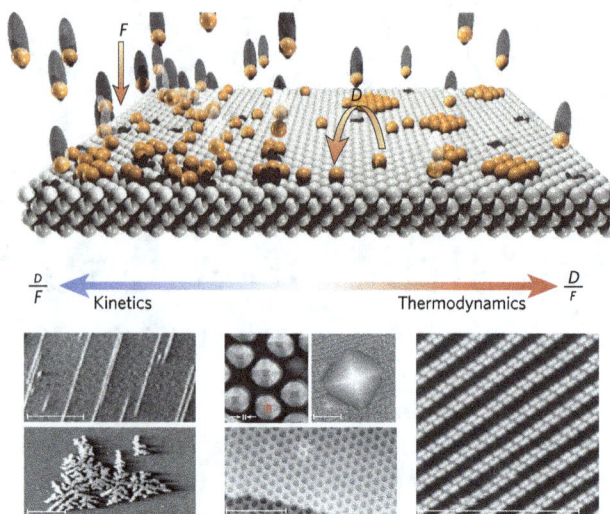

Figure 2 | Atomic-scale view of growth processes at surfaces. Atoms or molecules are deposited from the vapour phase. On adsorption they diffuse on terraces to meet other adspecies, resulting in nucleation of aggregates or attachment to already existing islands. The type of growth is largely determined by the ratio between diffusion rate D and deposition flux F. Metallic islands are controlled by growth kinetics at small D/F values. The hierarchy in the barrier of diffusing atoms can be translated into geometric order and well-defined shapes and length scales of the resulting nanostructures. The micrographs on the left-hand side show monatomic Cu chains grown on an anisotropic Pd(110) substrate (upper image) and Ag dendrites on hexagonal Pt(111) (ref. 20) (lower image). Semiconductor nanostructures are usually grown at intermediate D/F and their morphology is determined by the complex interplay between kinetics and thermodynamics. Strain effects are particularly important and can be used to achieve mesoscopic ordering. The micrographs in the centre show pyramidal and dome-shaped Ge semiconductor quantum dots grown on Si(100)[95] (upper right and upper left panels, respectively) and a boron nitride nanomesh on Rh(111)[96] (lower panel). To allow for supramolecular self-assembly based on molecular recognition, conditions close to equilibrium are required (large D/F values, or post-deposition equilibration). The micrograph on the right shows as an example a supramolecular nanograting of rod-like benzoic acid molecules on Ag(111). It consists of repulsively interacting molecular twin chains, which are stabilized by intermolecular hydrogen bonds[26]. Scale bars, 20 nm in every image.

Low-temperature growth of metal nanostructures on metal surfaces is the prototype of kinetically controlled growth methods. Metal bonds have essentially no directionality that can be used to direct interatomic interactions. Instead, kinetic control provides an elegant way to manipulate the structure and morphology of metallic nanostructures. On homogeneous surfaces, their shape and size are largely determined by the competition between the different movements the atoms can make along the surface, such as diffusion of adatoms on surface terraces, over steps, along edges and across corners or kinks. Each of these displacement modes has a characteristic energy barrier, which will to a first approximation scale with the local coordination of the diffusing atom: the diffusion of an atom over a terrace will have a lower energy barrier than diffusion along an edge or crossing of a corner, and descending an edge is often an energetically more costly process than terrace diffusion. A given material system thus has a natural hierarchy of diffusion barriers associated with these different atomistic processes. This makes it possible to shape growing aggregates by selective activation or suppression of particular diffusion processes through external growth parameters (temperature and metal deposition flux) and through the choice of a substrate with appropriate symmetry[20]. Judicious tuning of the relative importance of different diffusion processes has allowed on-demand fabrication of a host of metal nanostructures, ranging from small compact uniform clusters and large faceted islands to fractals, dendrites and atomically thin chains[17,18,20–22].

For metals, then, formation of surface structures can be controlled only by controlling the complex kinetics of the different diffusion processes at play, given the high energies associated with covalent bond formation and the limited information for spatial organization. In contrast, molecules that can participate in weak and directional non-covalent bonds may be programmed to form desired supramolecular structures[23]. The basic concepts ruling the self-assembly of three-dimensional supramolecular structures can also guide the assembly of adsorbed molecules into low-dimensional supramolecular systems that show a high degree of order on the nanometre scale. The approach does not require control over a hierarchy of activated diffusive motions, but operates near equilibrium conditions where the D/F ratio is a circumstantial parameter. Ordering may occur after deposition, and in favourable situations self-correction through the elimination of transiently formed defective structures is possible. The parameters crucial for this type of structure control are the surface mobility of molecules, their lateral interactions and their coupling to the surface atomic lattice. These depend on the chemical nature of the system and the atomic environment and symmetry of the substrate[19,24,25], all of which can be used to tune the delicate balance of lateral interactions and molecule–substrate coupling in order to steer supramolecular organization towards the desired structures[26].

If surface-supported nanostructures are to be organized on the mesoscale (10–1,000 nm), then at least some of the forces used for that purpose need to act over length scales comparable to the desired feature size; that is, they must be much longer-ranging than atomic distances. Such long-range forces can arise from various physical effects, including elastic and electrostatic interactions. Elastic forces are generally relevant to surfaces and epitaxial systems, given that atoms on a surface or in an epilayer are always under stress, even in the case of pristine surfaces or homoepitaxial systems[27]. The resultant forces typically give rise to regular two-dimensional strain relief or reconstruction patterns[21,22,28,29] with feature sizes of 2–20 nm, which can then serve as templates to control the growth of further patterns. In heteroepitaxial systems, the elastic energy associated with the inherent lattice mismatch of the materials can induce not only such lateral ordering, but also three-dimensional aggregation. 'Stranski–Krastanow growth', in which three-dimensional islands form spontaneously above a critical film thickness, is a well-established method for creating semiconductor quantum dots[30]. This nanostructure formation process is driven by thermodynamics (that is, strain relief overcompensates for the increase in surface energy associated with the transition to three-dimensional growth), but the resulting structures are usually metastable and their exact shape, size and composition result from a delicate interplay between thermodynamic and kinetic effects.

Magnetism at the physical limit

Gordon Moore observed in 1965 that improved fabrication technologies resulted in the doubling of the number of silicon field-effect transistors per unit area roughly every 18 months. 'Moore's law', which has achieved almost cult status, still describes with remarkable precision the advances in complementary metal oxide semiconductor (CMOS) technology that continue to increase information processing speeds. But the exponential growth of the information industry relies just as much on improvements in data storage, which uses small regions of ferromagnetic material with opposite magnetization to store 'zeros' and 'ones' in a hard disk. The continued downscaling of these storage domains outperforms even the stunning development of CMOS technology. During the past decade, storage density has doubled almost every 12 months and has reached 100 Gbit per square inch today. But as in the case of CMOS technologies, the drive for further miniaturization faces fundamental physical limits[6,7]. The decrease in ferromagnetic domain size is accompanied by a decrease in the magnetic anisotropy energy K, which prevents spontaneous changes in magnetization direction. For very small domains, K is comparable to thermal

energies so that thermal fluctuations can randomly flip the magnetization direction. This renders the domains superparamagnetic, with all stable magnetic order lost. The effect can be quantified by considering that for a single domain, the time for reversal of the magnetization orientation due to thermal fluctuations follows an Arrhenius law of the type $\tau = \tau_0 \exp(nK/k_B T)$ (here τ_0 is a pre-factor of the order of 10^{-9} s, n the number of atoms in the domain, k_B the Boltzmann constant). For magnetic anisotropy energies of 40 μeV per atom (a characteristic value for bulk hexagonally close-packed cobalt) and a typical stability requirement of $\tau > 10$ years, magnetically stable nanostructures thus need to contain roughly $n \approx 10^5$ atoms. In today's recording media, several hundred to a thousand of such individual domains are needed to realize one magnetic bit that can be reliably written and read, with high signal-to-noise ratio. Clearly, if miniaturization is to result in further increased storage capacities, we need to extend or avoid the superparamagnetic limit.

An obvious strategy is to develop new materials and structures with a substantially higher anisotropy energy K. A useful pointer is that K depends on spin–orbit interactions and on orbital magnetic moments, and hence on the precise atomic structure of magnetic materials[31,32]. The orbital magnetic moment m_L is particularly sensitive to the local atomic configuration and influences the magnetocrystalline anisotropy. In bulk materials m_L is largely quenched through the crystal field; but in low-dimensional nanostructured materials, the reduced symmetry of the electron wavefunctions can result in strongly anisotropic orbital magnetization that will boost the magnetocrystalline contribution to the anisotropy energy. This effect should be particularly pronounced for atomic-scale structures with constituent atoms that have a reduced average coordination, which in turn results in m_L values approaching those typically seen for free atoms[33]. The effect will be significant for structures that range in size from the single adatom to clusters composed of a few atoms to several tens of atoms at most, and it is difficult to envisage their efficient production without the use of bottom-up fabrication methods[34,35].

Judiciously used, self-organized metal growth on surfaces can produce a variety of useful nanoscale patterns with high densities in a fast, parallel process[20]. For example, the step decoration method[36,37] allows ready formation of uniform arrays of cobalt chains on a Pt(997) surface. The method makes use of the fact that step edges act as preferential nucleation sites for the deposited cobalt atoms, because of the increased coordination experienced at step sites relative to terrace sites. Growth in a well-defined temperature range then results in uniform cobalt chains of monatomic width over the entire sample, forming dense arrays of parallel one-dimensional nanowires. By adjusting the cobalt coverage on the surface and the average step spacing of the platinum surface, width and separation of the nanowires can be independently controlled. Similarly, bimetallic islands containing a Pt core and Co rim are readily obtained by Pt deposition and annealing to create the core islands, followed by Co deposition and annealing to create the rim[38].

Such surface-supported cobalt nanostructures, when examined, reveal how magnetic properties change as the size of the structures is reduced to a few atoms[39,40]. In the case of two-dimensional cobalt clusters on Pt(111), orbital moment and magnetic anisotropy energy increase markedly as the cluster size decreases (Fig. 3a). The orbital moment increases from $m_L = 0.3$ μ_B (where μ_B is the Bohr magneton) for clusters composed of about 10–15 atoms, to 0.59 μ_B for tetramers and 0.78 μ_B for trimers, and to a maximum value of 1.1 μ_B for a single adatom. The latter's orbital moment is more than seven times the bulk value. The magnetic anisotropy energy K shows a similar trend (Fig. 3a), reaching values as large as 9.3 meV for a single adatom[40]. Atomic-scale Co nanostructures may thus have K values up to two orders of magnitude larger than that of bulk hexagonal close-packed (h.c.p.) cobalt. Typical magnetic recording materials such as Co/Pt multilayers and even the permanent magnet SmCo$_5$ also have significantly lower K values ($K \approx 0.3$ meV atom^{-1} and $K \approx 1.8$ meV atom^{-1}, respectively) than the cobalt nanostructures. This pronounced size dependence of magnetic properties arises from the low coordination of Co atoms in the atomic-scale nanostructures, which can reduce d-state hybridization and the crystal field potential that is produced by the electric field of neighbouring lattice atoms. We expect that detailed insight into the effects of coordination number on both m_L and K will open new avenues for the design of nanostructures with promising magnetic properties. For example, the magnetic anisotropy of two-dimensional (2D) bimetallic islands with Pt core and Co rim is entirely determined by the rim of Co atoms at the perimeter, which accounts for extreme anisotropy energies[38].

Two-dimensional Co nanoparticles are superparamagnetic down to the lowest temperatures. In contrast, monatomic Co chains containing on average 80 atoms (Fig. 3b, c) are ferromagnetic[39] at 4.2 K. The observation of a paramagnetic response at 45 K implies the

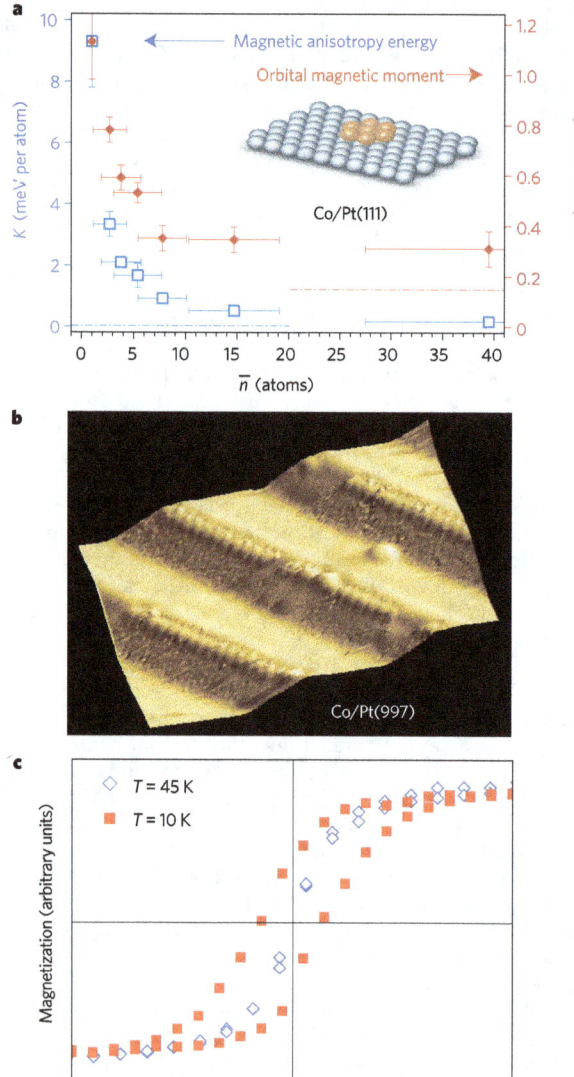

Figure 3 | Magnetism at the spatial limit. a, Magnetic anisotropy energy K (squares) and orbital magnetic moment m_L (diamonds) of Co atoms and two-dimensional clusters at the Pt(111) surface as a function of size n. The dashed lines represent the K and m_L values for bulk h.c.p. Co (blue and red, respectively)[40]. **b,** Scanning tunnelling microscopy image of monatomic cobalt chains decorating the steps of a regularly stepped platinum surface. The average distance between neighbouring chains is 2 nm. **c,** On cooling below 15 K, Co blocks become ferromagnetic, indicated by the opening up of a hysteresis loop in the magnetization curve[39].

absence of single-domain 1D ferromagnetic coupling, but the shape of the magnetization curve reveals the onset of short-range magnetic order, with spins coupling into local blocks of roughly 15 atoms. Long-range order is forbidden in infinite 1D systems at true equilibrium, yet it may exist in finite systems over relatively short timescales. In the case of the monatomic Co chain, a transition into a long-range ferromagnetically ordered state occurs at 15 K and is evident from the hysteresis in the magnetization curve (see Fig. 3c). This behaviour is due to the low coordination of the Co atoms on the vicinal Pt substrate, which results in a large anisotropy energy of 2.1 meV atom^{-1} that locks the magnetization of each spin block along the easy axis of the system. On the timescale of the experiment, ferromagnetic coupling thus effectively extends over the entire chain array[41] and gives rise to the smallest elemental magnet yet fabricated.

The magnetic behaviour seen in Co nanostructures is not only relevant for our fundamental understanding of low-dimensional magnetism, but has important implications for magnetic data storage technology. That is, the increase in magnetic anisotropy energies by more than two orders of magnitude, relative to the values seen in more traditional transition-metal systems, implies that a few hundred cobalt atoms might suffice to realize a stable magnetic bit at room temperature.

Semiconductor artificial atoms

In metallic nanostructures, because of the effects of coordination, every atom 'counts' with respect to the magnetic properties. For semiconductor materials, functional properties tend to be less sensitive to the exact number of constituent atoms: desired quantum effects already arise for structures with dimensions of 10–100 nm and containing somewhere between 10^3 and 10^6 atoms in a crystalline lattice. In this size range, the energy spectrum of electrons and holes confined in all three dimensions within these 'quantum dots' becomes discrete and in many ways similar to the spectrum of atoms[4]. Still, quantum dots are very much solid-state nanostructures, and their energy spectrum, which controls many of the physical properties of interest, can be adjusted over a wide range by tuning composition, size, lattice strain and morphology. These features make semiconductor quantum dots attractive for the design and fabrication of new electronic, magnetic and photonic devices and other functional materials.

Semiconductor quantum dots are often prepared as colloidal nanocrystals (see the review in this issue by Yin & Alivisatos, p. 664) but here we will focus only on semiconductor quantum dots supported on surfaces or embedded in solids. These systems can be prepared by using a wide range of methods, including lithography, etching and site-selective implantation[42]. But fabrication methods based on self-organized growth at surfaces are particularly attractive because they yield quantum dots with virtually no interface defects that might adversely affect performance, and because they can produce particularly small structures with widely separated energy levels that are essential for room-temperature operation. The approach can also produce high-density quantum dot structures in a fast, parallel process that is compatible with existing semiconductor technology and therefore permits mass fabrication and high levels of integration. However, to use self-organized growth effectively for quantum dot fabrication, we need detailed insight into the nucleation and growth processes involved, so as to tune the dot properties precisely and control their intrinsic statistical inhomogeneity.

As so many physical properties depend on quantum dot size and shape, it is essential to know the actual morphology of the 3D semiconductor islands that form on deposition of atoms from the gas phase, and to know how they evolve during post-growth treatments. But even though these systems have been intensely studied for more than a decade and many of their electronic and optical properties characterized, the structure of nucleated semiconductor islands and their subsequent morphological evolution remain incompletely understood. It is therefore encouraging that a common framework[43] can describe quantum dots that develop in the two most studied model

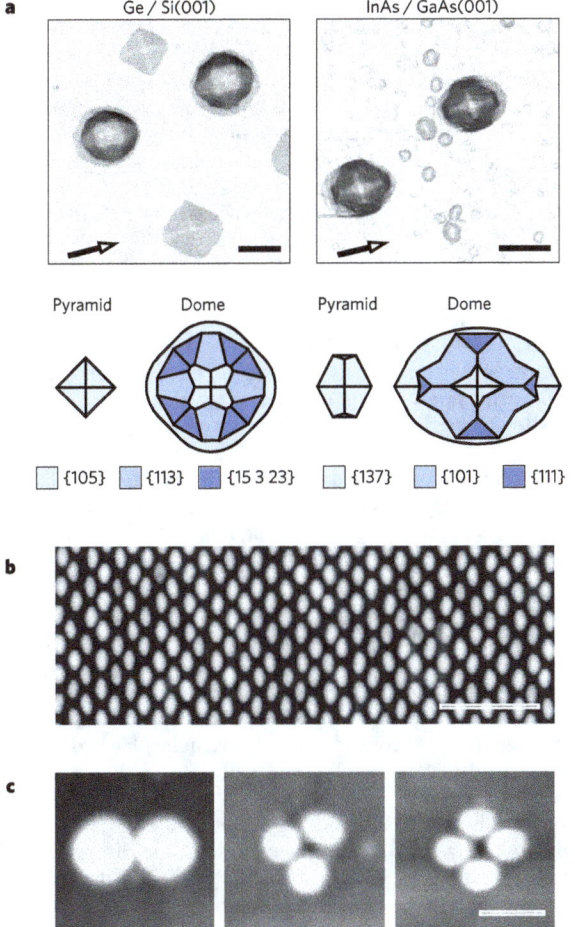

Figure 4 | Semiconductor quantum structures. a, Scanning tunnelling microscope images of pyramid and dome islands for the two main representative systems in semiconductor lattice-mismatched heteroepitaxy. The corresponding schematic structural models are also shown[43]. **b,** Atomic force topography of a regular array of InGaAs quantum dots reflecting self-organized growth on a prestructured GaAs(001) substrate[50]. **c,** Lateral quantum dot molecules grown in the InAs/GaAs(001) system. Bi-, tri- and quad-molecules can be produced by adjusting the substrate temperature and amount of deposited material[60]. Scale bars correspond to 50 nm in **a** and **c,** and to 500 nm in **b.**

considerable impact on our ability to design and engineer quantum dot structures.

In addition to controlling the properties of individual quantum dots, many of the applications so far envisaged also require precise arrangement of these structures into ordered arrays. For example, regular arrangement is obvious for systems or devices that require the addressing or coupling of individual quantum dots or the further processing of quantum dot signals, as in the case of single-electron[45], single-photon[46] and quantum computation[47] devices. Similarly, uniformity in position and spacing is critical for applications that make use of quantum dot ensembles, and where overall device performance depends on the mutual interaction between the individual dots (as in the case of cellular automata based on quantum dots[48], discussed below). But a high degree of lateral order has the additional advantage that formation of these structures usually ensures high uniformity in quantum dot properties, as statistical fluctuations are greatly suppressed if each dot experiences the same local environment during growth.

Quantum dots may be laterally organized using self-ordering processes that are mediated by elastic interactions, or using patterned substrates to direct their growth. The former approach depends on the same forces that induce the spontaneous formation of islands in the Stranski–Krastanow growth mode, and will drive the ordering of islands on length scales of the same order of magnitude as the size of the islands[49]. But it is almost impossible to obtain defect-free mesoscopic quantum dot arrangements based on this approach only. In contrast, highly regular structures are readily obtained by using electron-beam or optical lithography (top-down methods) first to create patterned substrates, which then serve as templates to direct the self-organized Stranski–Krastanow growth of three-dimensional semiconductor islands (a bottom-up method). The artificial periodic modulations of the surface are thus translated into perfect quantum dot arrays[50,51], as illustrated by the example shown in Fig. 4b. A further advantage of this approach is that it allows the resulting quantum dot structures to be connected to larger structures for integration into complex devices.

Efforts to controllably fabricate and characterize semiconductor quantum dots have mainly been driven by the desire to develop systems that take advantage of their extremely small dimensions and low power dissipation. For example, the performance of lasers can be substantially improved by using quantum dots as the active medium[52]. The tunable and discrete energy levels typical of quantum dots mean that the choice of emitted wavelength can be adjusted with unprecedented flexibility; the small active volume permits laser operation at low power, high frequency and low threshold currents that are independent of the working temperature. Information technology is another field in which the properties of quantum dots might prove attractive. One example is a cellular automaton[48] in which binary information is encoded in the configuration of charge distributed among the quantum dots and interaction between the dots provided by Coulomb interactions. Such an ensemble, with appropriately designed dot arrangement, is an essentially classical device that can reproduce the effect of wires and logic gates and implement any complex Boolean operation[53]. But quantum dots are also attracting interest for quantum computation applications, where information is encoded using quantum two-state systems ('qubits') that can be prepared in a superposition of the two states and thus enable dramatic increases in computing capabilities[54–56]. The basic building block is a two-qubit quantum gate that performs unitary operations on one qubit depending on the (quantum) state of the other[57]. Such a gate can be realized using quantum dot molecules; that is, sets of quantum-mechanically coupled quantum dots whose charge carrier wavefunctions are delocalized over the entire structure. The first semiconductor quantum dot molecules with controlled separation between the dots have been fabricated using top-down methods[58], although their sensitivity to thermal perturbations precludes any scope for use in 'real world' devices. Decreasing the size of the quantum dots should allow

systems: during the growth of Ge on Si(001) and during the growth of InAs on GaAs(001). The two types of quantum dots are both produced in the Stranski–Krastanow growth mode, with defect-free but strained 3D islands forming spontaneously on top of a thin wetting layer during lattice-mismatched heteroepitaxial growth. In both systems, only two discrete, well-defined families of islands develop: small islands that are bounded by one type of shallow facets and referred to as pyramids, and larger, multi-faceted islands that are characterized by steeper facets and referred to as domes (see Fig. 4a). When overgrowing the initially formed islands with the substrate material (Si and GaAs, respectively) to create the actual quantum dots, the capping process in both systems involves extension of the shallow facets at the expense of the steeper ones and a considerable reduction in island height. These experimental observations confirm theoretical predictions[44] that common, well-defined island shapes occur during growth and evolution, independent of the specific material system considered. It might therefore be possible to develop a common framework to explain at least qualitatively island growth and evolution for many material combinations that follow the Stranski–Krastanow growth mode. We expect the availability of such a universally applicable descriptive model to have

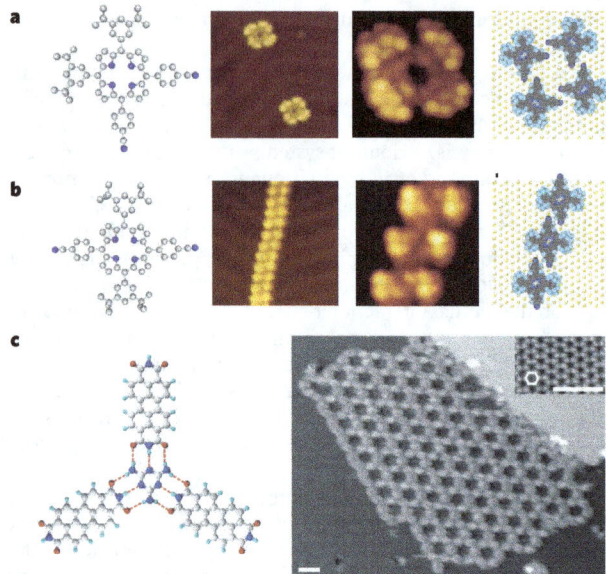

Figure 5 | Steering self-assembly of supramolecular nanostructures using hydrogen-bonding. a, b, Porphyrins substituted with two functional cyanophenyl moieties in a *cis* or *trans* configuration[67]. **a,** The *cis* species assembles in compact clusters of four molecules. **b,** With the *trans* species linear molecular chains are obtained. Mesoscopic ordering is in both cases dictated by the preferential attachment at the elbows of the chevron reconstruction of the Au(111) substrate used, visible as weak corrugation lines. Imaged area at left (right) is 20 nm^2 (5.3 nm^2). **c,** The complementary tectons perylene tetracarboxylic di-imide and melamine form a trigonal motif, where each intermolecular linkage is stabilized by three hydrogen bonds. This repeat unit gives rise to the regular nanoporous honeycomb layers fabricated on the hexagonal Ag-passivated Si(111) substrate shown in the inset[75]. Scale bars, 3 nm.

room-temperature operation. But controlled formation and precise arrangement of such small structures are challenging and likely to require self-organized growth, as illustrated by a strategy that makes use of spontaneous alignment of stacked quantum dot layers to form vertical quantum dot molecules[59]. The selective addressing of specific quantum gate parts is likely to be more feasible using laterally coupled quantum dot molecules, which have now also been produced through self-organized growth[60,61]. Figure 4c shows quantum dot molecules containing two, three and four dots; these were created in InAs/GaAs(001) using an elaborate growth–overgrowth–etching–regrowth procedure that is primarily based on self-organized growth[60]. The electronic properties of these quantum dot molecules remain to be explored, but their structural characteristics and the ability to combine them into highly ordered arrays are promising.

Supramolecular engineering

The range of functional nanometre-sized structures that can be fashioned from metallic or semiconducting materials through self-organized growth is inevitably somewhat limited, in that design and fabrication methods need to be based on the functional and structural features inherent in these materials. This contrasts with the construction of molecular nanoscale structures and patterns: the power of chemical synthesis provides access to a potentially vast range of functionally and structurally diverse building blocks ('tectons'), which can be linked through different types of relatively weak, non-covalent interactions (predominantly hydrogen bonds and metal–ligand interactions) to yield organized supramolecular architectures with tailor-made properties[14,23,62]. But although much is known about how supramolecular chemistry — the chemistry of the intermolecular non-covalent bond — can be tuned to create desired supramolecular crystals or supramolecular compounds in solution, this knowledge cannot be directly translated to guide the assembly

of adsorbed molecules into larger surface structures[63,64]. To do so, the influence of the substrate atomic lattice and substrate electronic structure on non-covalent bonds needs to be fully understood. For example, the substrate used will often alter the electronic properties of adsorbed ligands so that solution-based coordination chemistry concepts cannot be applied without appropriate modification[65]. Interactions between adsorbed molecules and their substrate may also perturb the surface-state free electron gas in metallic substrates, which in turn can influence how adsorbates are arranged on the surface[66]. More direct effects are that substrates may be reactive and chemically modify the functional moieties of the adsorbed building blocks, and the fact that topological surface features can influence interactions between adsorbed species for geometric reasons. These various effects make it obviously difficult to translate supramolecular concepts developed for crystals or solutions; but they can be used as tools for steering non-covalent interactions through the choice of templates with appropriate symmetry, surface patterns or chemical functions.

Planar molecules with extended π-systems have found particularly wide use because they tend to bond to surfaces in a flat-lying geometry, which allows functional groups at the molecular periphery to approach each other easily and to engage in non-covalent interactions. Provided the molecules retain mobility on the substrate with their functional groups not obstructed by the surface, supramolecular surface structures readily form as a result of two-dimensional self-assembly: the lateral coupling of suitably designed tectons. The tunability of supramolecular surface patterns through tecton design is illustrated in Fig. 5a, b[67]: the exact position of two cyanophenyl substituents at the periphery of a porphyrin core steers the intermolecular hydrogen bonding that provides lateral coupling and hence dictates the structure of the assemblies formed on a gold surface. Whereas the *cis* configuration gives rise to discrete clusters made up of four molecules (Fig. 5a), the *trans* configuration produces extended one-dimensional supramolecular chains (Fig. 5b). The spatial distribution of these clusters and chains is dictated by the surface pattern of the gold surface, which in the case of the Au(111) surface used results from a chevron reconstruction[28]. A similar influence on supramolecular ordering is seen in the case of 1-nitronaphthalene[68], emphasizing that patterned substrates are generally useful to guide the formation of low-dimensional molecular nanostructures[69,70].

A systematic study of the self-assembly behaviour of 4-[trans-2-(pyrid-4-yl-vinyl)]benzoic acid (PVBA) illustrates how the materials characteristics and symmetry of the substrate can affect the subtle balance between intermolecular interactions and molecule–surface interactions. The rod-like PVBA molecule, which contains a benzoic acid head group and a pyridyl tail group, self-assembles through head-to-tail hydrogen bonding[26,71]. If metallic palladium is used as substrate, molecule–substrate coupling is strong and dominates over intermolecular interactions; this prevents the formation of regular surface patterns. On close-packed noble-metal surfaces, the PVBA molecules are more mobile and able to assemble into highly regular, one-dimensional supramolecular arrangements resembling 'nanogratings' (see for example the scanning tunnelling microscope image reproduced in Fig. 2)[26]. Each stripe in the nanograting consists of two discrete chains of hydrogen-bonded PVBA molecules. The chains making up one stripe are held together through weak interchain hydrogen bonds, and the patterning of the stripes relative to each other appears to be due to interchain repulsions. The stripes are each about 1 nm wide, but the periodicity of the grating can be tuned from about 2 to 10 nm by controlling how much PVBA is deposited, or by taking advantage of the fact that they preferentially assemble at dislocation arrays of reconstructed substrates such as Au(111)[71,72]. Surface feature control at the molecular length scale is beyond the limits of current lithographic techniques. But for the self-assembled patterns to find practical use, methods need to be developed to transform the molecular arrangements into more rigid structures while retaining their precise spatial organization.

Extended 2D open network structures have been created using a number of different systems. One example is trimesic acid (TMA, $C_6H_3(COOH)_3$), in which regular dimerization of the carboxyl groups present results in an open network structure that reflects the molecule's three-fold symmetry[73,74]. This pattern is also encountered in organic TMA crystals, illustrating that in favourable cases motifs from organic solids can be replicated on surfaces. Another extended network structure is based on the classic H-bond motif of the melamine–cyanuric acid system, with perylene tetracarboxylic diimide serving as linear linker and melamine as trigonal connector. As illustrated in Fig. 5c, self-assembly through triple H-bond coupling yields a 2D bimolecular honeycomb network[75]. The high degree of order obtained on a substrate with appreciable surface corrugation (Ag-passivated Si(111)) emphasizes that even though individual intermolecular couplings may be weak, multiple weak linkages nevertheless enable stable and regular assemblies to form. Network stability is an essential feature if nanoporous surface patterns are used as templates to guide the formation of subsequent layers, as demonstrated by the formation of regular C_{60} arrays on the 2D bimolecular honeycomb arrangement[75].

Metal–ligand interactions are generally stronger than hydrogen bonds and thus result in more robust entities. Moreover, the incorporation of metal centres increases the scope of the functional properties of the nanoarchitectures and allows us to use design strategies based on metal-directed assembly[76,77]. This makes the controlled fabrication of surface structures based on metal–organic coordination appealing, but the approach can be challenging. Difficulties arise from the tendency of deposited metal atoms to interact strongly with the substrates used, in extreme cases resulting in surface reconstruction or alloying. Consequently the deposition sequence and substrate temperature have to be carefully controlled to avoid spurious effects and to achieve the formation of regular nanosystems.

We illustrate the principles underlying the interaction of organic linker molecules and transition metals at surfaces with the coordination behaviour of benzenepolycarboxylic acid species and iron (Fe) atoms on a copper Cu(100) surface. This system forms a variety of two-dimensional surface-supported open networks. The basic architectural motif of these networks depends on the relative concentrations of metal atoms and ligand molecules, with careful tuning of this ratio resulting in mononuclear metal-carboxylate clusters, 1D coordination polymers or fully connected 2D networks. The mononuclear complexes (Fig. 6a) are obtained[78] on depositing about 0.3 Fe atoms per linear terephthalate linker molecule (TPA, $C_6H_4(COO)_2$). As indicated in the scheme, a central Fe atom is coordinated to four TPA molecules through Fe-carboxylate bonds, with the resultant Fe(TPA)$_4$ complexes organized in a (6 × 6) unit cell with respect to the substrate. The complexes form a highly ordered array that covers entire terraces of the substrate, with this perfect long-range organization suggested[78,79] to result from weak hydrogen-bonding interactions between the complexes. The individual Fe centres are thus arranged in a perfect square lattice with a 15 Å periodicity. Clearly, any attempt to position large-scale arrays of single Fe atoms in a similar way using a top-down technique would be prohibitively time-consuming, if not impossible.

Increasing the amount of deposited metal to about 1 Fe atom per linker molecules results in networks of polymeric ladder structures, as illustrated in Fig. 6b for the system based on trimellate as linker[80]. This network frequently covers entire substrate terraces and constitutes a regular array of open nanocavities, each with an effective opening of (3 × 10) Å². Increasing the metal concentration further, to about two Fe atoms per linker molecule, yields fully interconnected 2D metal–organic networks with complete two-dimensional reticulation[78,80]. Examples are shown in one of the insets of Fig. 1 (using TPA as linker[78]) and in Fig. 6c (using as linker a longer analogue of TPA, 4,1′,4′,1″-terphenyl-1,4″-dicarboxylic acid or TDA[81]). Both networks are thermally robust. Because the networks 'compartmentalize' the copper substrate into nanometre-sized cavities, they can steer the

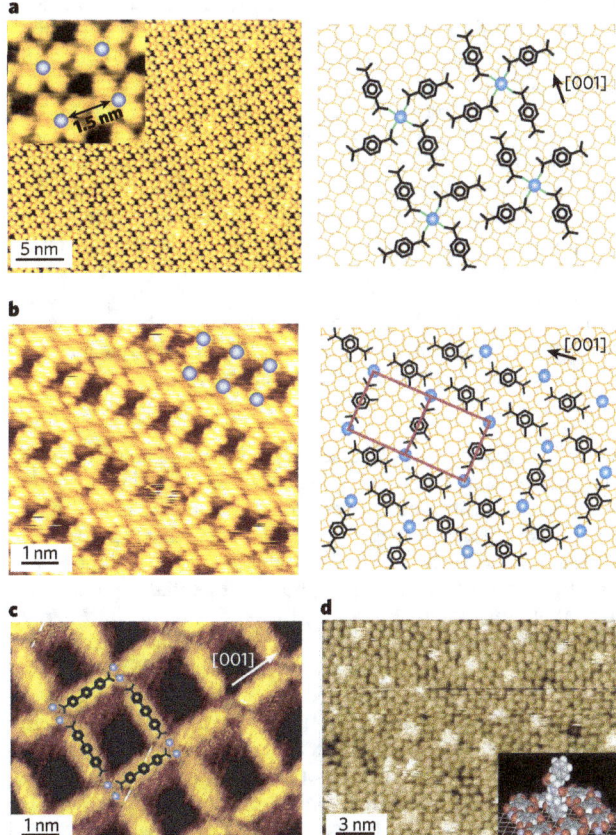

Figure 6 | Metallosupramolecular assembly of low-dimensional Fe-carboxylate coordination systems on a square Cu(100) substrate. a, The mononuclear Fe(TPA)$_4$ complexes are stabilized by metal–ligand interactions. Their perfect ordering in a (6 × 6) square array with periodicity 15 Å is mediated by substrate templating and weak intercomplex hydrogen bridges[78]. **b,** One-dimensional Fe-trimellate coordination polymer with a higher coverage ratio of Fe per tecton. In the ladder structure indicated, with a (4 × 4) repeat unit, there is a continuous 1D Fe-carboxylate linkage framing open cavities[80]. **c,** Using a linear terphenyl dicarboxylate species as linker with increased length, regular 2D reticulated coordination networks with nanometre-sized cavities are obtained[81]. **d,** Site-selective uptake of L, L-diphenylalanine peptide molecules in a 2D nanoporous Fe-carboxylate architecture[84]; a model for the dipeptide guests in an upright position is depicted in the inset (S. Stepanow, N. Lin, J.V.B. and K.K., unpublished observations).

organization of subsequently deposited molecules, as has been illustrated with C_{60} molecules[81].

As might be expected, substrate and linker symmetry play an important role in 2D supramolecular engineering of metal–organic surface structures. By replacing the Cu(100) surface with its square symmetry by the anisotropic Cu(110) surface, the 1D anisotropy of the substrate can effectively be transferred to the resulting coordination compounds. Trimesic acid with Cu and Fe on the (110) surface is found to form strictly linear metal–organic coordination chains[82]. The local linker geometry is of equal importance. For instance, whereas TMA linkers with three-fold molecular symmetry can form mononuclear Fe(TMA)$_4$ complexes that resemble their terephthalate Fe(TPA)$_4$ counterparts[83], the complexes cannot assemble into perfectly square arrays but form instead an extended coordination network with a regular arrangement of nanocavities[84]. These nanocavities provide well-defined reaction spaces and can even be used as selective receptors for biomolecules. This is as demonstrated in Fig. 6d where the L, L-diphenylalanine peptide species is used as molecular guest in a two-dimensional metal-carboxylate host-system (S. Stepanow, N. Lin, J.V.B. and K.K., unpublished observation). The design principles established for 3D carboxylate framework reticular synthesis[85,86] can thus also guide the design of 2D supramolecular metal–organic systems. That is, careful selection of suitable linker structures (in conjunction with appropriate metals) provides an effective strategy for adjusting the size of the cavities or pores present in the 2D assemblies, and the chemical functionalities lining the 'walls' of the pores are determined by the characteristics of the side groups present in the linker. Given the immense wealth and scope of reticular chemistry, a wide range of surface-supported structures with pores of different sizes and chemical characteristics should in principle be accessible. These might find use in the fabrication of patterned surface templates, for the control of host–guest chemistry involving surface structures, or in heterogeneous catalysis in which reactants can interact with the substrate and the supramolecular metal–organic surface structures.

Outlook

Self-organized growth and self-assembly at surfaces can serve as an efficient and versatile tool for creating low-dimensional nanostructures. It offers exquisite control over feature size and organization on the atomic and mesoscopic length scales. We believe that these process characteristics, in combination with the ability to produce high-density structures in a fast and parallel fashion, are essential requirements for any nanofabrication methodology that aims to contribute to the quest for further miniaturization in the microelectronics industry and elsewhere. However, even though processes that make use of self-ordering growth have already yielded systems with intriguing functional properties, many challenges still need to be addressed before such strategies find wide practical use. For example, the incorporation of nanostructures into more complex organized architectures and their effective interfacing to the macroscopic world are vital for any applications. We would expect that this can be achieved by combining bottom-up and top-down techniques, with the former providing ready access to features with sizes below 10 nm, and the latter allowing for integration of these structures into larger functional systems. This general approach should also result in new materials and devices that might find use beyond the applications traditionally targeted by miniaturization efforts, particularly when it is guided by new insights into the physics of small systems or combined with chemical[35,41,87,88] and biological[89–91] bottom-up methods. ■

1. Feynman, R. P. There's plenty of room at the bottom. *Eng. Sci.* **23**, 22–36 (1960).
2. Binnig, G. & Rohrer, H. Scanning tunneling microscopy — from birth to adolescence. *Rev. Mod. Phys.* **59**, 615–625 (1987).
3. Eigler, D. M. & Schweizer, E. K. Positioning single atoms with a scanning tunnelling microscope. *Nature* **344**, 524–526 (1990).
4. Kastner, M. A. Artificial atoms. *Phys. Today* **46**, 24–31 (1993).
5. Dekker, C. Carbon nanotubes as molecular quantum wires. *Phys. Today* **52**, 22–28 (1999).
6. Bohr, M. T. Nanotechnology goals and challenges for electronics applications. *IEEE Trans. Nanotechnol.* **1**, 56–62 (2002).
7. Thomson, D. A. & Best, J. S. The future of magnetic data storage technology. *IBM J. Res. Dev.* **3**, 311–321 (2000).
8. Gates, B. D. *et al.* New approaches to nanofabrication: molding, printing, and other techniques. *Chem. Rev.* **105**, 1171–1196 (2005).
9. Ito, T. & Okazaki, S. Pushing the limits of lithography. *Nature* **406**, 1027–1031 (2000).
10. Xia, Y. N., Rogers, J. A., Paul, K. E. & Whitesides, G. M. Unconventional methods for fabricating and patterning nanostructures. *Chem. Rev.* **99**, 1823–1848 (1999).
11. Vettiger, P. *et al.* The 'millipede'-nanotechnology entering data storage. *IEEE Trans. Nanotechnol.* **1**, 39–55 (2002).
12. Lindsey, J. S. Self-assembly in synthetic routes to devices. Biological principles and chemical perspectives: a review. *New J. Chem.* **15**, 153–180 (1991).
13. Whitesides, G. M., Mathias, J. P. & Seto, C. T. Molecular self-assembly and nanochemistry — a chemical strategy for the synthesis of nanostructures. *Science* **254**, 1312–1319 (1991).
14. Philp, D. & Stoddart, J. F. Self-assembly in natural and unnatural systems. *Angew. Chem. Int. Edn Engl.* **35**, 1154–1196 (1996).
15. Nicolis, G. & Prigogine, I. *Self-Organization in Non-Equilibrium Systems: From Dissipative Structure Formation to Order through Fluctuations* (Wiley, New York, 1977).
16. Eigen, M. Self-organization of matter and the evolution of biological macromolecules. *Naturwissenschaften* **33**, 465–523 (1971).
17. Zhang, Z. Y. & Lagally, M. G. Atomistic processes in the early stages of thin-film growth. *Science* **276**, 377–383 (1997).
18. Brune, H. Microscopic view of epitaxial growth: nucleation and aggregation. *Surf. Sci. Rep.* **31**, 121–229 (1998).
19. Barth, J. V. Transport of adsorbates at metal surfaces : From thermal migration to hot precursors. *Surf. Sci. Rep.* **40**, 75–150 (2000).
20. Röder, H., Hahn, E., Brune, H., Bucher, J. P. & Kern, K. Building one-dimensional and two-dimensional nanostructures by diffusion-controlled aggregation at surfaces. *Nature* **366**, 141–143 (1993).
21. Brune, H., Giovannini, M., Bromann, K. & Kern, K. Self-organized growth of nanostructure arrays on strain-relief patterns. *Nature* **394**, 451–453 (1998).
22. Li, J. L. *et al.* Spontaneous assembly of perfectly ordered identical-size nanocluster arrays. *Phys. Rev. Lett.* **88**, 066101 (2002).
23. Lehn, J.-M. *Supramolecular Chemistry, Concepts and Perspectives* (VCH, Weinheim, 1995).
24. Gimzewski, J. K. & Joachim, C. Nanoscale science of single molecules using local probes. *Science* **283**, 1683–1688 (1999).
25. Rosei, F. *et al.* Properties of large organic molecules at surfaces. *Prog. Surf. Sci.* **71**, 95–146 (2003).
26. Barth, J. V. *et al.* Building supramolecular nanostructures at surfaces by hydrogen bonding. *Angew. Chem. Int. Edn Engl.* **39**, 1230–1234 (2000).
27. Ibach, H. The role of surface stress in reconstruction, epitaxial growth and stabilization of mesoscopic structures. *Surf. Sci. Rep.* **29**, 193–263 (1997).
28. Barth, J. V., Brune, H., Ertl, G. & Behm, R. J. Scanning tunneling microscopy observations on the reconstructed Au(111) surface — atomic structure, long-range superstructure, rotational domains, and surface defects. *Phys. Rev. B* **42**, 9307–9318 (1990).
29. Kern, K. *et al.* Long-range spatial self-organization in the adsorbate-induced restructuring of surfaces — Cu(110)–(2x1)O. *Phys. Rev. Lett.* **67**, 855–858 (1991).
30. Teichert, C. Self-organization of nanostructures in semiconductor heteroepitaxy. *Phys. Rep.* **365**, 335–432 (2002).
31. Bruno, P. Tight-binding approach to the orbital magnetic moment and magnetocrystalline anisotropy of transition-metal monolayers. *Phys. Rev. B* **39**, 865–868 (1989).
32. van der Laan, G. Microscopic origin of magnetocrystalline anisotropy in transition metal thin films. *J. Phys. Cond. Mat.* **10**, 3239–3253 (1998).
33. Wildberger, K., Stepanyuk, V. S., Lang, P., Zeller, R. & Dederichs, P. H. Magnetic nanostructures — 4d clusters on Ag(001). *Phys. Rev. Lett.* **75**, 509–512 (1995).
34. Himpsel, F. J., Ortega, J. E., Mankey, G. J. & Willis, R. F. Magnetic nanostructures. *Adv. Phys.* **47**, 511–597 (1998).
35. Sun, S., Murray, C. B., Weller, D., Folks, L. & Moser, A. Monodisperse FePt nanoparticles and ferromagnetic FePt nanocrystal superlattices. *Science* **287**, 1989–1992 (2000).
36. Gambardella, P., Blanc, M., Brune, H., Kuhnke, K. & Kern, K. One-dimensional metal chains on Pt vicinal surfaces. *Phys. Rev. B* **61**, 2254–2262 (2000).
37. Kuhnke, K. & Kern, K. Vicinal metal surfaces as nanotemplates for the growth of low-dimensional structures. *J. Phys. Cond. Mat.* **15**, S3311–S3335 (2003).
38. Rusponi, S. *et al.* The remarkable difference between surface and step atoms in the magnetic anisotropy of two-dimensional nanostructures. *Nature Mater.* **2**, 546–551 (2003).
39. Gambardella, P. *et al.* Ferromagnetism in one-dimensional monatomic metal chains. *Nature* **416**, 301–304 (2002).
40. Gambardella, P. *et al.* Giant magnetic anisotropy of single cobalt atoms and nanoparticles. *Science* **300**, 1130–1133 (2003).
41. Gambardella, P. *et al.* Oscillatory magnetic anisotropy in one-dimensional atomic wires. *Phys. Rev. Lett.* **93**, 077203 (2004).
42. Bimberg, D., Grundmann, M. & Ledentsov, N. N. *Quantum Dot Heterostructures* (Wiley, Chichester, 1999).
43. Costantini, G. *et al.* Universal island shapes of self-organized semiconductor quantum dots. *Appl. Phys. Lett.* **85**, 5673–5675 (2004).
44. Daruka, I., Tersoff, J. & Barabasi, A. L. Shape transition in growth of strained islands. *Phys. Rev. Lett.* **82**, 2753–2756 (1999).
45. Warburton, R. J. *et al.* Optical emission from a charge-tunable quantum ring. *Nature* **405**, 926–929 (2000).
46. Michler, P. *et al.* A quantum dot single-photon turnstile device. *Science* **290**, 2282–2284 (2000).
47. Burkard, G., Loss, D. & DiVincenzo, D. P. Coupled quantum dots as quantum gates. *Phys. Rev. B* **59**, 2070–2078 (1999).
48. Cole, T. & Lusth, J. C. Quantum-dot cellular automata. *Prog. Quantum Electron.* **25**, 165–189 (2001).
49. Shchukin, V. A. & Bimberg, D. Spontaneous ordering of nanostructures on crystal surfaces. *Rev. Mod. Phys.* **71**, 1125–1171 (1999).
50. Heidemeyer, H., Denker, U., Müller, C. & Schmidt, O. G. Morphology response to strain field interferences in stacks of highly ordered quantum dot arrays. *Phys. Rev. Lett.* **91**, 196103 (2003).
51. Lee, H., Johnson, J. A., He, M. Y., Speck, J. S. & Petroff, P. M. Strain-engineered self-assembled semiconductor quantum dot lattices. *Appl. Phys. Lett.* **78**, 105–107 (2001).
52. Arakawa, Y. & Sakaki, H. Multidimensional quantum well laser and temperature-dependence of its threshold current. *Appl. Phys. Lett.* **40**, 939–941 (1982).
53. Nakajima, F., Miyoshi, Y., Motohisa, J. & Fukui, T. Single-electron AND/NAND logic circuits based on a self-organized dot network. *Appl. Phys. Lett.* **83**, 2680–2682 (2003).
54. Feynman, R. P. Simulating physics with computers. *Int. J. Theor. Phys.* **21**, 467–488 (1982).
55. Shor, P. in *Proc 35th Annu. Symp. Foundations of Computer Science* (IEEE, Los Alamitos, 1994).
56. Deutsch, D. Quantum theory, the Church–Turing principle and the universal quantum computer. *Proc. R. Soc. Lond. A* **400**, 97–117 (1985).
57. Barenco, A. *et al.* Elementary gates for quantum computation. *Phys. Rev. A* **52**, 3457–3467 (1995).
58. Schedelbeck, G., Wegscheider, W., Bichler, M. & Abstreiter, G. Coupled quantum dots fabricated by cleaved edge overgrowth: From artificial atoms to molecules. *Science* **278**, 1792–1795 (1997).
59. Bayer, M. *et al.* Coupling and entangling of quantum states in quantum dot molecules. *Science* **291**, 451–453 (2001).
60. Songmuang, R., Kiravittaya, S. & Schmidt, O. G. Formation of lateral quantum dot molecules around self-assembled nanoholes. *Appl. Phys. Lett.* **82**, 2892–2894 (2003).
61. Deng, X. & Krishnamurthy, M. Self-assembly of quantum-dot molecules: heterogeneous nucleation of SiGe islands on Si(100). *Phys. Rev. Lett.* **81**, 1473–1476 (1998).
62. Prins, L. J., Reinhoudt, D. N. & Timmerman, P. Non-covalent synthesis using hydrogen bonding. *Angew. Chem. Int. Edn Engl.* **40**, 2382 (2001).
63. Barth, J. V., Weckesser, J., Lin, N., Dmitriev, S. & Kern, K. Supramolecular architectures and nanostructures at surfaces. *Appl. Phys. A* **76**, 645 (2003).

64. Feyter, S. D. & Schryver, F. C. D. Two-dimensional supramolecular self-assembly probed by scanning tunneling microscopy. *Chem. Soc. Rev.* **32,** 139–150 (2003).

65. Lin, N., Dmitriev, A., Weckesser, J., Barth, J. V. & Kern, K. Real-time single-molecule imaging of the formation and dynamics of coordination compounds. *Angew. Chem. Int. Edn Engl.* **41,** 4779 (2002).

66. Lukas, S., Witte, G. & Wöll, C. Novel mechanism for molecular self-assembly on metal substrates : unidirectional rows of pentacene on Cu(110) produced by substrate-mediated repulsion. *Phys. Rev. Lett.* **88,** 028301 (2002).

67. Yokoyama, T., Yokoyama, S., Kamikado, T., Okuno, Y. & Mashiko, S. Selective assembly on a surface of supramolecular aggregates of controlled size and shape. *Nature* **413,** 619–621 (2001).

68. Böhringer, M. et al. Two-dimensional self-assembly of supramolecular clusters and chains. *Phys. Rev. Lett.* **83,** 324–327 (1999).

69. Otero, R. et al. One-dimensional assembly and selective orientation of Lander molecules on an O-Cu template. *Angew. Chem. Int. Edn Engl.* **43,** 2092–2095 (2004).

70. Clair, S., Pons, S., Brune, H., Kern, K. & Barth, J. V. Mesoscopic metallosupramolecular texturing through hierarchic assembly. *Angew. Chem. Int. Edn Engl.* (in the press).

71. Barth, J. V. et al. Stereochemical effects in supramolecular self-assembly at surfaces: 1-D vs. 2-D enantiomorphic ordering for PVBA and PEBA on Ag(111). *J. Am. Chem. Soc.* **124,** 7991–8000 (2002).

72. Weckesser, J., Vita, A. D., Barth, J. V., Cai, C. & Kern, K. Mesoscopic correlation of supramolecular chirality in one-dimensional hydrogen-bonded assemblies. *Phys. Rev. Lett.* **87,** 096101 (2001).

73. Dmitriev, A., Lin, N., Weckesser, J., Barth, J. V. & Kern, K. Supramolecular assemblies of trimesic acid on a Cu(100) surface. *J. Phys. Chem. B* **106,** 6907–6912 (2002).

74. Griessl, S., Lackinger, M., Edelwirth, M., Hietschold, M. & Heckl, W. M. Self-assembled two-dimensional molecular host-guest architectures from trimesic acid. *Single Molecules* **3,** 25–31 (2002).

75. Theobald, J. A., Oxtoby, N. S., Phillips, M. A., Champness, N. R. & Beton, P. H. Controlling molecular deposition and layer structure with supramolecular surface assemblies. *Nature* **424,** 1029–1031 (2003).

76. Leininger, S., Olenyuk, B. & Stang, P. J. Self-assembly of discrete cyclic nanostructures mediated by transition metals. *Chem. Rev.* **100,** 853–908 (2000).

77. Holiday, B. J. & Mirkin, C. A. Strategies for the construction of supramolecular compounds through coordination chemistry. *Angew. Chem. Int. Edn Engl.* **40,** 2022–2043 (2002).

78. Lingenfelder, M. et al. Towards surface-supported supramolecular architectures: tailored coordination assembly of 1,4-benzenedicarboxylate and Fe on Cu(100). *Chem. Eur. J.* **10,** 1913–1919 (2004).

79. Dmitriev, A. et al. Design of extended surface-supported chiral metal-organic arrays comprising mononuclear iron centers. *Langmuir* **41,** 4799–4801 (2004).

80. Dmitriev, A., Spillmann, H., Lin, N., Barth, J. V. & Kern, K. Modular assembly of two-dimensional metal-organic coordination networks at a metal surface. *Angew. Chem. Int. Edn Engl.* **41,** 2670–2673 (2003).

81. Stepanow, S. et al. Steering molecular organization and host-guest interactions using tailor-made two-dimensional nanoporous coordination systems. *Nature Mater.* **3,** 229–233 (2004).

82. Classen, T. et al. Templated growth of metal-organic coordination chains at surfaces. *Angew. Chem. Int. Edn Engl.* (in the press).

83. Messina, P. et al. Direct observation of chiral metal-organic complexes assembled on a Cu(100) surface. *J. Am. Chem. Soc.* **124,** 14000–14001 (2002).

84. Spillmann, H. et al. Hierarchical assembly of two-dimensional homochiral nanocavity arrays. *J. Am. Chem. Soc.* **125,** 10725–10728 (2003).

85. Yaghi, O. M. et al. Reticular synthesis and the design of new materials. *Nature* **423,** 705–714 (2003).

86. Kitagawa, S., Kitaura, R. & Noro, S. Functional coordination polymers. *Angew. Chem. Int. Edn Engl.* **43,** 2334–2375 (2004).

87. Joachim, C., Gimzewski, J. K. & Aviram, A. Electronics using hybrid-molecular and mono-molecular devices. *Nature* **408,** 541–548 (2000).

88. Ouyang, M. & Awscholom, D. D. Coherent spin transfer between molecularly bridged quantum dots. *Science* **301,** 1074–1078 (2003).

89. Niemeyer, C. M. Nanoparticles, proteins and nucleic acids: biotechnology meets materials science. *Angew. Chem. Int. Edn Engl.* **40,** 4128–4158 (2001).

90. Seeman, N. C. & Belcher, A. M. Emulating biology: Building nanostructures from the bottom up. *Proc. Natl Acad. Sci. USA* **99,** 6451–6455 (2002).

91. Sarikaya, M., Tamerler, C., Jen, A. K. Y., Schulten, K. & Baneyx, F. Molecular biomimetics: nanotechnology through biology. *Nature Mater.* **2,** 577–585 (2003).

92. Cleland, A. N. & Roukes, M. L. A nanometre-scale mechanical electrometer. *Nature* **392,** 160–162 (1998).

93. Kind, H. et al. Patterned films of nanotubes using microcontact printing of catalysts. *Adv. Mater.* **11,** 1285–1289 (1999).

94. Tans, S. J., Devoret, M. H., Groeneveld, R. J. A. & Dekker, C. Electron-electron correlations in carbon nanotubes. *Nature* **394,** 761–764 (1998).

95. Ross, F. M., Tromp, R. M. & Reuter, M. C. Transition states between pyramids and domes during Ge/Si island growth. *Science* **286,** 1931–1934 (1999).

96. Corso, M. et al. Boron nitride nanomesh. *Science* **303,** 217–220 (2004).

Acknowledgements K.K. thanks the many students, postdocs and scientific collaborators who have contributed to the exploration of the atomic world of surfaces and nanostructures. Special thanks go to N. Lin for his enthusiasm in advancing the concepts of supramolecular chemistry at surfaces.

Author Information Reprints and permissions information is available at npg.nature.com/reprintsandpermissions. The authors declare no competing financial interests. Correspondence and requests for materials should be addressed to K.K. (k.kern@fkf.mpg.de).

MOLECULAR MACHINES AND DEVICES

Making molecular machines work

In this review we chart recent advances in what is at once an old and very new field of endeavour — the achievement of control of motion at the molecular level including solid-state and surface-mounted rotors, and its natural progression to the development of synthetic molecular machines. Besides a discussion of design principles used to control linear and rotary motion in such molecular systems, this review will address the advances towards the construction of synthetic machines that can perform useful functions. Approaches taken by several research groups to construct wholly synthetic molecular machines and devices are compared. This will be illustrated with molecular rotors, elevators, valves, transporters, muscles and other motor functions used to develop smart materials. The demonstration of molecular machinery is highlighted through recent examples of systems capable of effecting macroscopic movement through concerted molecular motion. Several approaches to illustrate how molecular motor systems have been used to accomplish work are discussed. We will conclude with prospects for future developments in this exciting field of nanotechnology.

WESLEY R. BROWNE, BEN L. FERINGA*

Organic and Molecular Inorganic Chemistry, Stratingh Institute, University of Groningen, Nijenborgh 4, 9747 AG, Groningen, The Netherlands.

*e-mail: B.L.Feringa@rug.nl

Consider a world composed of nanometre-sized factories and self-repairing molecular machines where complex and responsive processes operate under exquisite control; where translational and rotational movement is directed with precision; a nano-world fuelled by chemical and light energy. What images come to mind? The fantastical universes described in the science fiction of Asimov and his contemporaries? To a scientist, perhaps the 'simple' cell springs more easily to mind with its intricate arrangement of organelles and enzymatic systems fuelled by solar energy (as in photosynthetic systems) or by the chemical energy stored in the molecular bonds of nucleotide triphosphates (for example, ATP)[1]. Understanding and harnessing such phenomenal biological systems provides a strong incentive to design active nanostructures that can operate as molecular machines[2], and although our current efforts to control motion at the molecular level may appear awkward compared with these natural systems, it should not be forgotten that nature has had a 4.5 billion year head start.

Biological motors[3] convert chemical energy to effect stepwise linear or rotary motion, and are essential in controlling and performing a wide variety of biological functions. Linear motor proteins are central to many biological processes including muscle contraction, intracellular transport and signal transduction, and ATP synthase, a genuine molecular rotary motor, is involved in the synthesis and hydrolysis of ATP[1,2,4]. Other fascinating examples include membrane translocation proteins, the flagella motor that enables bacterial movement[5] and proteins that can entrap and release guests through chemomechanical motion[2,3]. In recent years the development of biomolecular motors (and natural–synthetic hybrid systems) towards the construction of sensors, actuators

and transporters has seen tremendous progress[6,7]. Biological motors are important components in the fabrication of dynamic smart materials, and semi-synthetic DNA-based structures have been explored in building a variety of mechanical motor-like functions[8–10].

Although biological motors are capable of complex and intricate functions, a key disadvantage of their application *ex vivo* arises in their inherent instability and restrictions in the environmental conditions they operate in[11]. Whereas nature is capable of maintaining and repairing damaged molecular systems, such complex repair mechanisms are beyond the capabilities of current nanotechnology. By contrast, wholly synthetic systems, which can tolerate a more diverse range of conditions than biological machines, offer considerable advantages in the development of complex nanomachinery.

Furthermore the sheer excitement of being able to build completely artificial motors and machines of nanodimensions, in the bottom-up approach promulgated by Feynman[12] several decades ago, has provided the drive to attempt the synthetic challenges that such molecular machines present. A synthetic approach to the construction of molecular devices not only serves to mimic and build on the elegant systems nature has to offer but also can be used to develop systems that are not restricted to nature's small, albeit versatile, synthetic toolbox.

Central to any molecular machine is the motor component and hence it is conceivable that molecular motors and machines (Fig. 1)[13–23] will play as prominent a role in the nanotechnological revolution of the twenty-first century as their macroscopic counterparts — the steam and internal combustion engines — played in catalysing the industrial revolution of the nineteenth century.

MOLECULAR MOTION

Before surveying various approaches taken towards developing artificial molecular machines, it may be pertinent to first address a

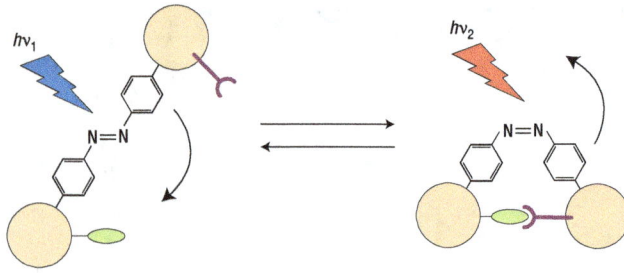

Figure 1 What makes a molecule a machine? In this example of a molecular machine we are able to switch between two molecular states (shapes) in a controlled manner as part of a repetitious mechanical cycle. An azobenzene molecule can exist in two forms (left: *trans* with the bulky groups on opposite sides of the double bond, and right: *cis* with the bulky groups on the same side). The bulky groups can be moved closer together or further apart by switching between the *cis* and *trans* forms (see also, Fig. 8b). This switching can be performed by using light of two different colours, one to go from *cis* to *trans*, the other to reverse the process.

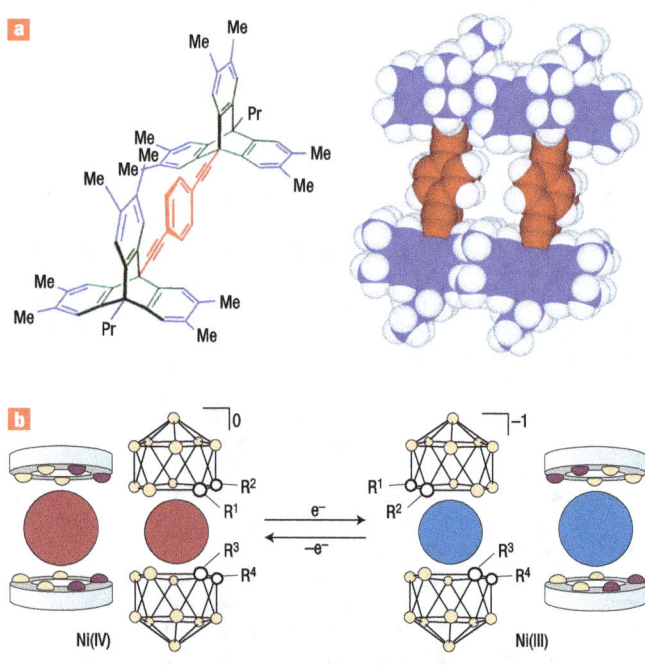

Figure 2 Examples of non-directionally controlled molecular rotors. **a**, A molecular rotor (left) where hindrance to rotation of the central phenyl ring (the rotor) is removed by the upper and lower bulky molecular end units (the stators). Sufficient spacing for the phenyl rings is generated to allow fast rotation in the solid state as illustrated for two rotor units (right). Reproduced with permission from ref. 32. Copyright (2005) National Academy of Sciences, USA. **b**, An electrochemically driven rotor where the upper and lower carborane (polyhedral clusters comprising boron and carbon atoms) moieties are bound as ligands to a nickel ion functioning as a 'ball-bearing'. Oxidation and reduction of the nickel centre (the Ni^{3+} / Ni^{4+} redox cycle) leads to rotation of the upper ligand relative to the lower ligand, changing the relative position of the alkyl groups (R^{1-4}) attached to the carborane ligands. Reprinted with permission from ref. 36. Copyright (2004) American Association of the Advancement of Science.

sometimes controversial, if somewhat semantic, aspect of this field of nanotechnology: specifically, what do we mean when we refer to molecules as machines and motors?

At the heart of every machine is its motor. *The Oxford Dictionary of English*[24] defines a motor as "a thing that imparts motion"; work as "the operation of a force in producing movement or other physical change"; and motion as "the condition of a body, when at each successive point in time it occupies a different position or orientation in space". Perhaps, a more utilitarian definition of a motor is a device that converts fuel, be it chemical, thermal or light, into kinetic energy in a controlled manner — that is, it makes things move.

A major difficulty in operating molecular machines, however, lies not in achieving motion at the molecular level but in controlling their operation, especially their directionality. Whereas the most critical design issue facing macroscopic machines is their function, in designing molecular machines making motion 'visible' is of paramount importance. It is for this reason that scientists have turned to using 'reporting' components in the design and construction of molecular machines and, more often than not, asymmetry is key to the successful demonstration of directionally controlled motion. Indeed asymmetry, as we shall see, is central to directional control in the majority of systems reported to date. In terms of size, however, bigger is not always better. Indeed, smaller molecular systems offer distinct advantages over larger assemblies in terms of synthetic chemistry, characterization and importantly in the ability to fine-tune molecular properties systematically. Finally, to translate molecular movement to macroscopic levels, many molecular motors must be able to work cooperatively.

BROWNIAN ROTORS

By using the more flexible definition of a motor proposed above, it is possible to recognize systems that make use of brownian motion and changes in chemical equilibrium as molecular machines. It is important to realize however, that although it is often expedient in understanding nanomachines and nanomotors to make direct analogies to macroscopic machines, in answering fundamental questions regarding problems associated with friction, wear, transmission, efficiency, fuel, motion and work, such facile comparisons often serve to cloud rather than simplify issues, and hence must be made with caution.

When considering a molecular motor and its operation, it should be noted that the forces that control the movement of macroscopic objects — in particular gravity — have little relevance to molecular machines of nanodimensions. In the world of the molecular machine, brownian storms rage relentlessly, and although refuge from the random brownian motion in solution can be found by resting on surfaces, even there, a molecule must contend with often significant thermally driven motion. This is not all bad news, however, as molecular dynamics is central to the operation of a molecular motor, as exemplified by the many membrane-immobilized biological motor systems. To move in such a turbulent world the molecular machine must either exploit brownian motion — as in brownian ratchets (see below) — or overcome it.

In designing motors[25] at the molecular level, random thermal brownian motion[26] must therefore be taken into consideration. Indeed, nature uses the concept of the brownian ratchet to excellent effect in the action of linear and rotary protein motors[25]. In contrast to ordinary motors, in which energy input induces motion, biological motors use energy to restrain brownian motion selectively[27]. In a brownian ratchet system the random-molecular-level motion is harnessed to achieve net directional movement, and crucially the resulting biased change in the system is not reversed but progresses in a linear or rotary fashion. The problems associated with random thermal motion of molecules have been emphasized in discussions of ratchet-type motors ("if you can't beat the chaos why not exploit it"[25]) and nanomachines ("navigators on the nanoscale have to accommodate to the brownian storms"[28]). It is in this direction that several groups have approached the challenge of controlling motion in molecular ratchets based on rotaxanes and catananes[29,30].

Molecular rotor systems that embrace thermal brownian motion are ubiquitous. Numerous groups have focused on designing systems to understand molecular motion at the molecular and submolecular level in so-called technomimetic[31] rotors (Fig. 2)[32-37]. It is important to note, however, that these systems do not allow for directional control of movement. The dynamic processes observed for molecular rotor systems in solution, on surfaces and in the solid state are the subject of a recent comprehensive review and will not be discussed in depth here[38,39].

Although the majority of studies described have focused on molecular systems in solution, recent reports on crystalline molecular rotors[32,33] (Fig. 2a) have addressed pertinent questions with regard to speed, friction and potential wear in a non-fluidic environment. Not only is fast free rotation in the solid state observed, but it was shown that, by careful consideration of the nature of the stator units and the molecular packing in the crystal lattice, the barrier to rotation can be reduced to near negligible levels. For instance, rotary motion around a carbon–carbon single bond can be extremely fast (up to 100 MHz), albeit with no control over the directionality of the rotation[33]. It is worth remembering that molecular machines and molecules in general are not rigid, but undergo changes in their shape incessantly. From these studies it is clear that the dynamics within molecules and molecular flexibility are crucial factors to take into account in the design of artificial systems[35].

Although the direction of the motion in these systems is random, important design features demonstrate how control over rotary motion might be realized. For example, in the carborane rotors[36] shown in Fig. 2b, it has been suggested that unidirectionality of rotation might be achieved by the introduction of additional asymmetry using bulky groups. Despite these advances, brownian motion presents a tremendous challenge to the construction of molecular machines. Approaches can be taken to overcome the brownian turbulence, such as immobilization of molecular machines in membranes and on surfaces (see below) and construction of machines large enough to overcome brownian motion.

BETWEEN BROWNIAN AND NEWTONIAN MOTION

With the latter approach, to be able to build larger synthetic machines, the question arises as to how large the machines need to be. Where does the boundary between the domination of macroscopic 'controllable' newtonian motion and random brownian motion lie? The recent work of Whitesides[40], Sen[41,42], Ozin, Manners[43], and co-workers, although based on non-molecular systems, provides some insight. In their multimetallic systems, a combination of a metal capable of catalysing the decomposition of hydrogen peroxide (the chemical fuel) and a relatively inert metal allow for the generation of a local oxygen gradient and/or (difference in) surface tension, which provides the stimulus to move the metallic object with limited directional control (Fig. 3).

It is clear that the size of the object is critical to both the mechanism by which movement is induced and the extent of non-brownian motion involved[42]. For objects >50 μm in size, the movement induced by oxygen evolution is related to the physical effect of bubble

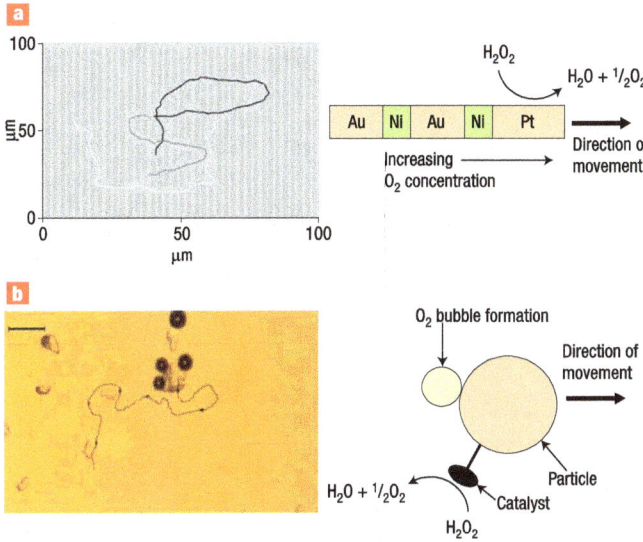

Figure 3 Chemically fuelled autonomously moving objects. The fuel applied is hydrogen peroxide. The propulsion is attributed to the decomposition of H_2O_2 to H_2O and O_2 creating a local oxygen concentration gradient and/or a difference in surface tension leading to translational movement. **a**, A striped Pt/Ni/Au/Ni/Au nanorod and the trajectory path of a nanorod spelling the letters 'PSU' in 5 wt% aq. H_2O_2. Reproduced with permission from ref. 42. Copyright (2005) Wiley. **b**, Design of an autonomous moving microparticle and the trajectory followed by the particle on H_2O_2 decomposition (the black spots indicate O_2 bubble formation). In this system a molecular catalyst is anchored to an 80 μm silica particle. Note the difference in the mechanism for movement in these systems (oxygen gradients/surface tension or pressure from bubble formation). Figure reproduced from ref. 45 by permission of The Royal Society of Chemistry.

formation, whereas for particles <1 μm in size, the motion of the particles is indistinguishable from the inherent brownian motion of the system. Between 1 and 20 μm, however, movement is driven neither by brownian motion nor bubble formation but rather by oxygen concentration gradients[41] in a manner proposed to be similar to that used by some bacteria[44]. Achieving directional control in these systems is difficult but not impossible. It was demonstrated that the directionality can be controlled magnetically by using platinum–nickel–gold rods, in which platinum acts to create the oxygen gradient (that is, uses the fuel to generate propulsion, Fig. 3a) whereas the anisotropy of a magnetic nickel layer allows for control of the direction of the movement[42].

Extension of this approach to molecular systems capable of chemically fuelled autonomous movement has been demonstrated recently, in our group, through the use of a synthetic catalyst that can effect efficient hydrogen peroxide decomposition (Fig. 3b)[45]. Despite the fact that control over directionality of the translational movement is limited thus far, this latter molecular system demonstrates that the conversion of chemical to kinetic energy, and thereby movement of a micro-object, with the concerted action of molecular-scale motors is feasible.

Although in the previous example hydrogen peroxide was used as the chemical fuel to effect motion, in molecular systems, powering motion can be achieved by other chemical fuels, pH and redox changes and of course by light[46]. Ideally the fuel used should not involve the generation of waste products, damage to the molecular machine and should allow for reversible external control, hence although chemically fuelled systems have been

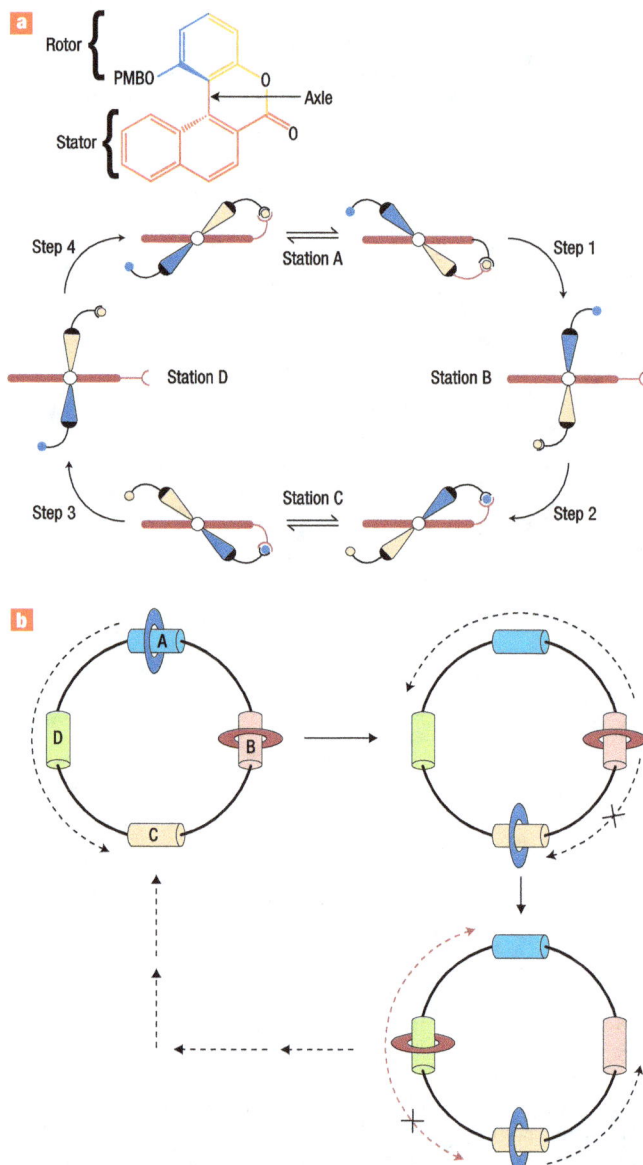

Figure 4 Rotary molecular motors in which a sequence of steps results in a full 360° unidirectional movement around the central axis. **a**, Molecular structure (station A) of a reversible, unidirectional molecular rotary motor driven by chemical energy and a schematic illustration of the rotary process. The bond-breaking processes rely on chiral non-racemic chemical fuel that discriminates between the two dynamically equilibrating helical forms, A and C. The bond-making processes between the rotor (yellow/blue) and the stator (red) use the principle that the sequential reactions are selective for either the blue or yellow parts of the rotor unit; that is, either the yellow or the blue end of the rotor can be selectively bound to the stator (red) to form A or C. Reprinted with permission from ref. 48. Copyright (2005) American Association of the Advancement of Science. **b**, A four-station [3]catenane (a molecular assembly in which one or more rings are inter-locked) in which unidirectional movement along the ring is controlled by sequential chemical and photochemical reactions. Adapted with permission from ref 49. Part of the entire rotary process is shown; at each stage one ring blocks the reverse rotation of the other ring ensuring unidirectionality over the entire cycle.

demonstrated, the major focus in recent years has been on molecular machines fuelled photo- and/or electrochemically.

In the following sections we will describe several recent examples of molecular machines that can achieve controlled use of fuel to effect motion, and highlight key examples of how some of these systems demonstrate that molecular (nanoscopic) motion can realize effective macroscopic work.

UNIDIRECTIONAL ROTARY MOLECULAR MOTORS

Biosystems frequently rely on ATP as their energy source[1], however very few examples of artificial motors that use exothermic chemical reactions to power unidirectional rotary motion have been reported to date. To build synthetic molecular rotary motors, it is apparent that three criteria must be satisfied: (i) repetitive 360° rotation, (ii) consumption of energy and, of course, (iii) control over directionality.

Kelly and co-workers have achieved limited (120°) unidirectional rotation around a single carbon–carbon bond in a modified molecular ratchet using phosgene as the chemical fuel[47]. Recently, our own group reported the unidirectional 360° rotation of a synthetic molecular motor fuelled entirely by a sequence of chemical conversions (Fig. 4a)[48]. Importantly, the sense of rotary motion is governed by the choice of chemical reagents that control the rotor movement through four distinct stations. Within each station the rotor's brownian motion relative to the stator is restricted by structural features. Although the principle of a unidirectional rotary motor driven by a chemical fuel has been demonstrated, the requirement of a sequence of non-compatible chemical steps precludes continuous rotary motion with these systems thus far.

In a multicomponent approach to a combined chemically and photochemically driven unidirectional rotary motor, Leigh and co-workers have designed a complex [3]catenane-based molecular system comprising a large ring bearing four different stations and two smaller rings (Fig. 4b)[49]. Sequential changes in position of the small rings along the four stations is achieved by applying light, chemical stimuli or heat, resulting in a unidirectional movement of the small rings along the larger one. In this process the backward brownian movement of one ring is prevented by the presence of the second ring. This system has also been engineered to make backward rotation possible by judicious choice of the components used[50].

By using light as fuel, however, continuous unidirectional rotary motion can be achieved. The photochemical step in these systems is the *cis* to *trans* isomerization of a carbon–carbon double bond (an alkene), which allows for a 180° rotation of one part of the molecule relative to another (Fig. 5a). This isomerization results in the movement of bulky groups into unstable positions, which relax thermally. The first generation of a light-driven unidirectional rotary motor[51] comprises a central alkene unit that functions as the axis of rotation, and a chiral helical structure (Fig. 5b). The orientation of the methyl (Me) group determines the most stable shape that the molecule prefers to adopt and hence dictates if clockwise or anticlockwise rotary motion of one half of the molecule (the propeller) with respect to the other (the stator) occurs. By applying light and heat, unidirectional rotation proceeds as a sequential four-step process. The first generation of this approach to a molecular motor was extended in the second-generation motors, where the stator and rotor parts are quite different in molecular structure[52], enabling additional components to be attached to either the top or the bottom half[53] and surface attachment of the stator (Fig. 5c)[54].

In order to apply these molecular motors, however, rotation at an appreciable speed — that is, 360° rotations per second — is desirable, if not essential. The structure of the second-generation motor is perfectly amenable to a whole range of structural modifications, which allow for fine tuning of the motor's properties, both photochemical and thermal. The speed of the photochemical step (*cis–trans* isomerization, Fig. 5) is of the

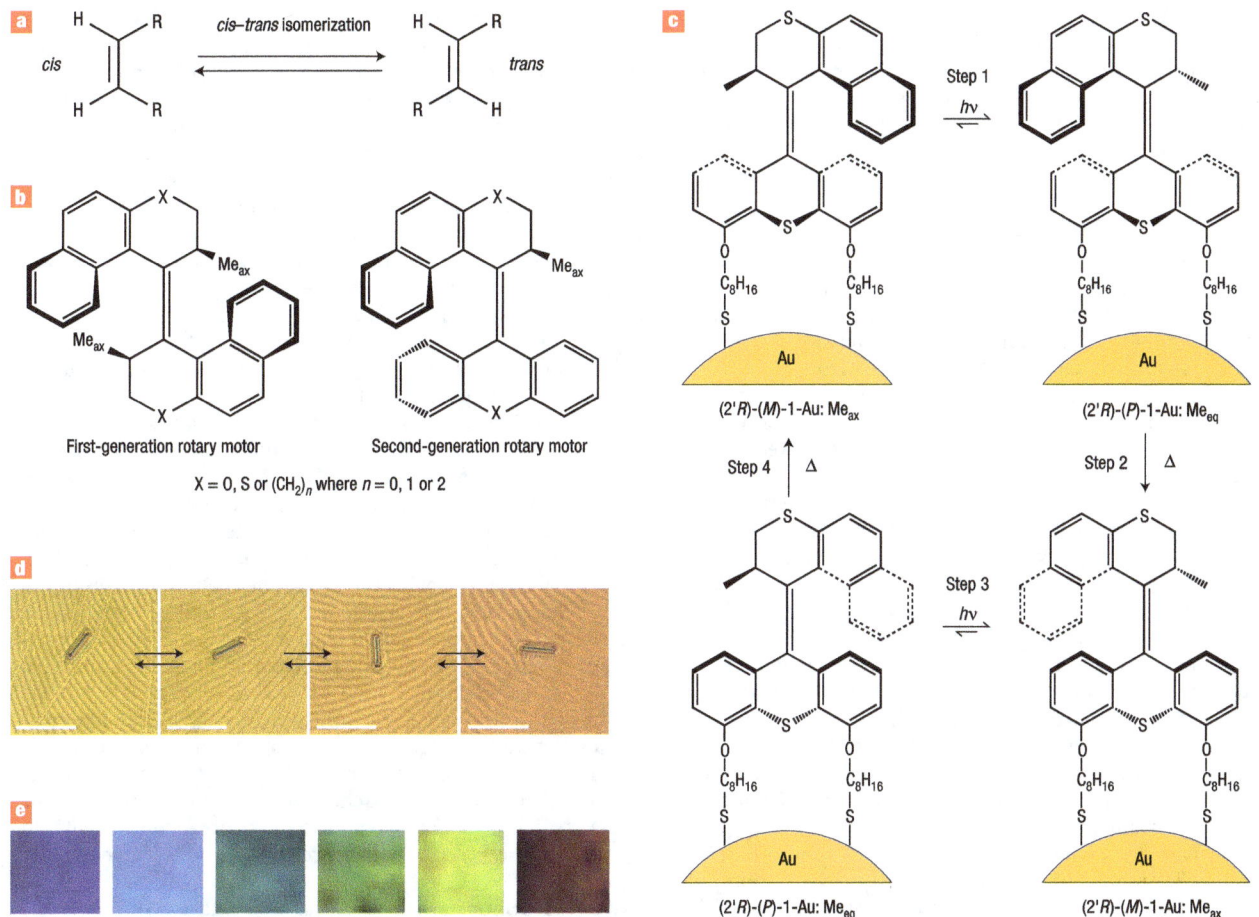

Figure 5 Light-driven unidirectional rotary motors in action. **a**, Principle of *cis–trans* isomerization around a double bond connecting two groups (R). Light absorption induces a rotary motion around the double bond, which results in a change in the relative position of the two R groups. **b**, Examples of first- and second-generation rotary motors that operate by photochemical *cis–trans* isomerization (Me_{ax} and Me_{eq} indicate axial and equatorial orientations of the methyl groups, respectively). **c**, Second-generation rotary motor immobilized on a gold surface. The first photochemical *trans*-to-*cis* isomerization forces the molecule into a highly strained *cis*-geometry. Rotary motion continues in the same direction as the consequence of a thermal helix inversion that results in a release of the strain in the molecule. Subsequent irradiation induces a *cis*-to-*trans* isomerization, which is followed by a second thermal helix inversion, again with release of conformational strain, to provide the initial state. Reproduced with permission from ref. 54. **d**, Glass rod rotating on a liquid crystal, doped with a second-generation light-driven molecular motor, during irradiation with ultraviolet light. Frames were taken at 15 s intervals and show clockwise rotations of 28° (frame 2), 141° (frame 3) and 226° (frame 4) of the rod relative to the position in frame 1. Scale bars, 50 µm. Note that heating was not involved as the thermal isomerization was efficient at room temperature. Reproduced with permission from ref. 101. **e**, Example of colour changes observed in motor-doped liquid crystals on irradiation. Liquid-crystal phase: E7(Merck) 6.16 wt% doping. Reproduced with permission from ref. 102. Copyright (2002) National Academy of Sciences, USA.

order of 1 or 2 ps (ref. 55). Hence the rate of rotation is controlled by the thermally induced relaxation of the unstable to the stable state, which requires bulky groups (for example, the methyl and phenyl rings) to pass each other. Structural modification of the bridging atoms and substituents allows for reduction of the energy barrier of the thermal steps, which has led to a 1.2-million-fold increase in the rotation speeds achievable with the second-generation motors[56]. These improved motors allow the propeller to rotate unidirectionally on irradiation at up to 80 revolutions per second at 20 °C.

LINEAR MOLECULAR MOTORS

Rotaxane-based systems have dominated the field of artificial molecular machines designed to achieve translational motion[20,57].

Typically, such systems contain a ring component that shuttles reversibly between stations on the shaft (Fig. 6) with the movement controlled by redox chemistry[58–60], pH changes[61] or light[62].

The development of increasingly sophisticated photochemically active rotaxanes has culminated recently in linear motors powered by light[63], which allow for movement of a shuttle up to 1.5 nm with a frequency of 10 kHz. These systems operate with a quantum efficiency of up to 20%. Translating the light-driven motion in these molecules into useful work remains a challenge[64], however, the recent development by Balzani, Stoddart and co-workers[65] of an autonomous photo-driven rotaxane is a significant step forward in achieving this goal.

A design for an artificial molecular muscle based on the sliding motions of the ring along the shaft of rotaxanes was also reported[66]. Two interlocked rotaxanes allow for elongation and contraction by binding different metal ions, a system reminiscent of the Ca^{2+}-driven processes that occur during muscle contraction. An example of a chemically driven system designed to mimic processive enzymes[67] has been reported recently[68] and is a significant step in the development of synthetic processive molecular machines akin to those found in nature[67].

MULTICOMPONENT MECHANICAL MACHINES

The construction of molecular machines, in which mechanical motion of different units operate in concert, for example when rotary motion of one part is coupled to linear motion of another part, requires the design of integrated multicomponent systems. A system denoted as a 'molecular elevator' composed of three rotaxane units interlocked mechanically to a platform was reported recently (Fig. 7a)[69,70]. The charge of one of the two stations is sensitive to pH changes and as a result the platform can move between two levels by adding acid or base.

Although large changes in physical properties can be brought about using *cis–trans* isomerization, the direct conversion of light energy into 'mechanical work', that is, significant motion, is a more challenging goal. Following an earlier design of molecular scissors[71], the coupling of several molecular motions was achieved in a light-powered 'molecular pedal'[72]. A change in molecular shape on *cis–trans* photoisomerization of an azobenzene unit is transmitted via a pivot point (a ferrocene unit) and a pedal-like motion of large flat zinc porphyrin units to induce a clockwise or anticlockwise rotary motion in a bound rotor guest. This system demonstrates that small changes in molecular structure can be used to drive much larger mechanical changes, remote from the point of initial mechanical motion.

Recently, the light-driven molecular motors, shown in Fig. 5, were used by Tour and co-workers to construct a prototype light-powered "nanocar"[73]. In this approach carborane 'wheels' were attached covalently to a molecular rotor 'engine' (Fig. 7b). Although this multicomponent system did not compromise the functionality of the motor unit itself, light-driven movement across a surface awaits demonstration.

MOTORS ON SURFACES

The molecular motors discussed above that convert chemical or light energy into directional rotary or linear motion all operate in solution, and although brownian motion can be overcome by building micrometre-dimension devices, another approach to overcoming brownian motion and bringing about order is to immobilize machines on a surface. However for nanomachines to be able to operate in devices or to perform useful work it will be essential that the motors do not lose functionality when immobilized. Anchoring and subsequently addressing molecular machines on surfaces is critical to the successful interfacing of nanomechanical systems with the macroscopic world.

Altitudinal molecular rotors (the rotor axle is parallel to the surface), which exhibit a strong dipole moment, have been immobilized on Au(111) surfaces[74]. Barrier-height imaging using scanning tunnelling microscopy measurements revealed that the direction of the dipole of the rotor unit turns in response to a static electric field imposed by the tip of the microscope. On the basis of molecular dynamics calculations, it was suggested that limited directionality in their motion should be possible using oscillating electrical fields normal to the surface[75]. A similar approach has been reported very recently by Miyake and co-workers[76] using lanthanide complexes, in which the large metal ion is sandwiched between two flat porphyrins, the speed of rotation of which can be modified by varying the lanthanide metal used[77].

The second-generation light-driven unidirectional molecular rotors discussed above (Fig. 5c) have been immobilized successfully on nanoparticle gold surfaces to yield an azimuthal (rotor axle normal to the surface) motor. In these systems the stator component carries two thiol-functionalized 'legs' that connect the entire motor to the surface[54]. Repetitive and unidirectional 360° rotary motion with respect to a surface was observed to occur on irradiation. The use of two attachment points per molecule prevents uncontrolled thermal rotation of the entire motor system with respect to the surface. However, this study also demonstrates the difficulties that may be encountered in such immobilization[54]. The length and nature of the connecting units (legs) is frequently a critical issue in the attachment of photoactive compounds to metal surfaces as it can affect excited-state processes dramatically[78].

The tetrathiafulvene unit has proven to be a mainstay in the design of electrochemically driven rotaxane-based molecular machines, as it can be converted between two stable states (bistable), that is, the neutral and oxidized forms, which show markedly different behaviour when interacting with other molecular components[79]. Analogous to *cis–trans* isomerization, facile reversible oxidation to the mono- or dicationic states of the tetrathiafulvene unit can be used to effect large changes in the position of rings along the shaft unit of rotaxanes[80]. These redox-switchable 'bistable' rotaxanes have been applied as monolayers in molecular switch tunnel junction devices[81,82]. Importantly, the position of the shuttle can be controlled by either reduction or oxidation but the conductance of the system is measured only as a function of the position of the shuttle on the rotaxane. In contrast to more conventional approaches to redox-switchable systems where a change in redox state is used to control conductivity, in this example the redox changes are used to move a molecular component only.

Whereas demonstration of linear and rotary motion on the nanometre scale is apparent in the synthetic molecular machines described above, realizing the summation of motion at the molecular level to achieve macroscopic motion, for instance mimicking the operation of myosin/actin-driven movement of muscles, is more challenging[83]. Nevertheless, it has been demonstrated that such translation of work can be achieved[84,85]. The approaches taken include: (i) changes in molecular properties such as surface tension, (ii) indirect macroscopic movement through amplification of molecular motion and (iii) direct translation of molecular to macroscopic motion.

MOLECULAR MOTORS AT WORK

In the individual steps of the artificial molecular motors constructed so far, the system acts as a mechanical switch. Molecular switches have found widespread application in the dynamic control of bulk materials and single-molecule properties, as well as molecular devices that take advantage of the ability to trigger functions through external signals[21]. Molecular switches can be used to modulate properties such as surface wetability[86], polymer elasticity[87], lateral pressure profile of bilayers[88], host–guest recognition[89], supramolecular organization[90], catalysis[91], colour[92], fluorescence[93], conductance[94] and enzyme activity[95–97]. The use of photochemical molecular switches has the advantage of short response times, clean and tuneable energy input, and the ability to convert an optical input into a variety of useful output signals.

It is therefore unsurprising that indirect translation of changes in molecular properties to achieve macroscopic movement has used the reversible photochemical switching of bistable molecules, such as azobenzenes, mounted on surfaces[98,99]. Photochemical

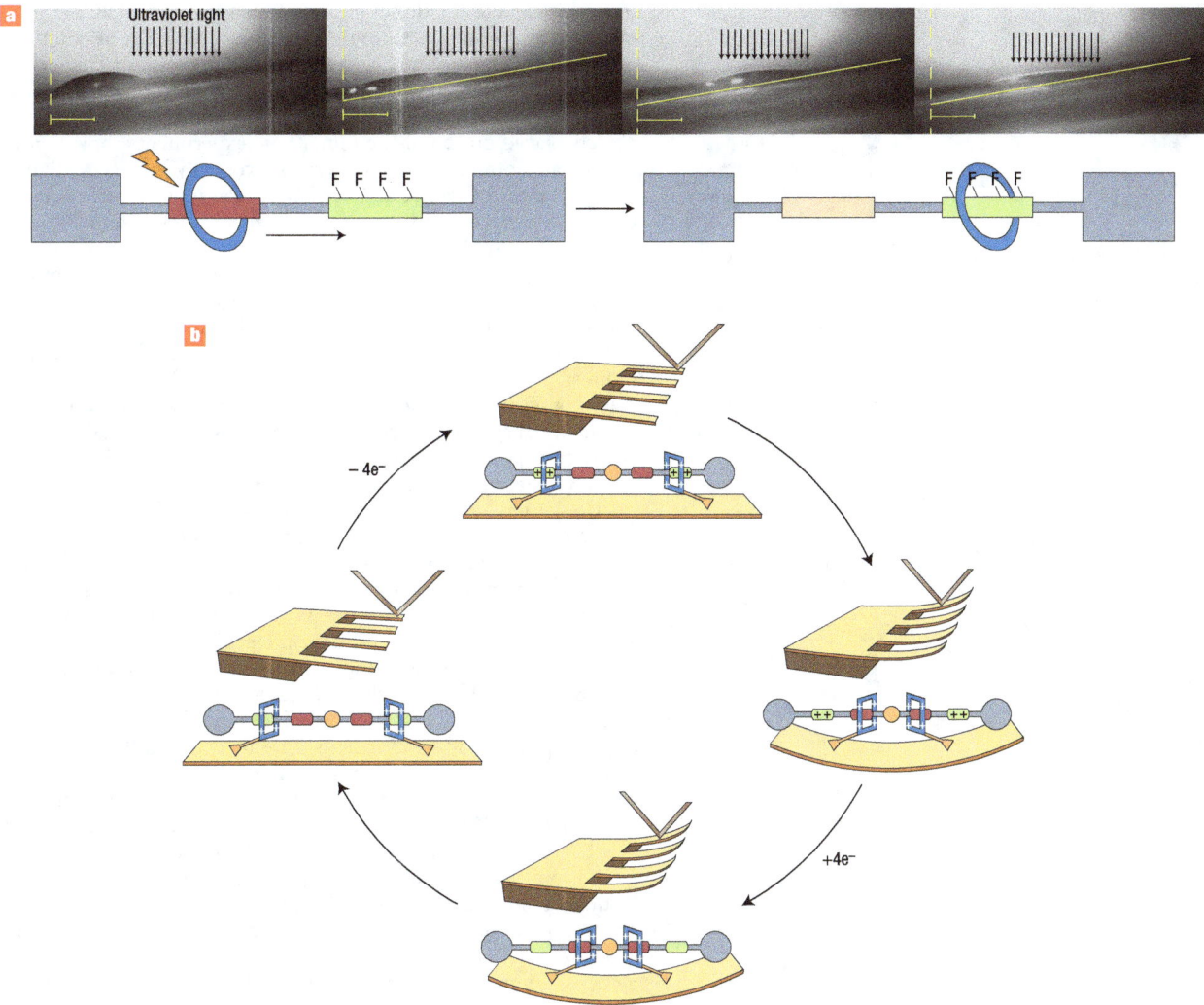

Figure 6 Synthetic molecular systems designed to achieve translational motion. These linear molecular motors are so-called rotaxanes in which one (or more) rings can move from one binding site to another along a shaft. The change in equilibrium position is triggered by an external signal. **a**, Macroscopic transport of liquids by surface-bound rotaxanes as a synthetic molecular machine. The position of the ring exposes or conceals a fluoroalkane component and concomitantly changes the surface energy. The ring is moved by shining light on the surface coated with motor molecules. When the area above the droplet of liquid is 'switched', the liquid is attracted to the irradiated zone, and in this way is transported across the surface and, in this case, up a slope (with 12° incline). Scale bars = 100 mm. Reproduced with permission from ref. 101. **b**, Design of a molecular muscle and the bending of cantilever beams by the cooperative action of several linear rotaxane motors. The system is based on a bistable [3]rotaxane structure involving a rotaxane with two positively charged rings and four stations, two of which are redox active units and can be switched between a neutral and positively charged form. The rings are connected to the surface of the cantilever via disulphide containing tethers to form a self-assembled monolayer. The rings are initially sitting on the redox-active stations in their neutral state (structure on the left). Oxidation of the redox-active stations, forces the rings to move to the inner neutral stations and results in contraction of the rotaxane, bending the beam. Adapted with permission from ref. 85. Copyright (2005) American Chemical Society.

isomerization of azobenzenes attached to surfaces was used to induce wetting/dewetting by changing the surface energy of the monolayers and allowing for the transport of liquid droplets across surfaces[100]. Similar millimetre-scale directional transport of droplets on a surface with light was achieved with a photoactive rotaxane operating as a linear nanomechanical switch (Fig. 6a)[101]. The system comprises a shuttle (the ring that moves along the shaft) on a thread with a fluoroalkane station and a photo-responsive fumaramide station, which has a high binding affinity for the shuttle. The entire system was anchored through physisorption on a modified gold surface.

Photochemical *trans–cis* isomerization of a fumaramide unit in the thread reduces the binding affinity to the shuttle drastically and, as a consequence, the equilibrium position of the shuttle changes in favour of the fluoroalkane station. The shuttle movement is therefore used to expose or conceal the fluoroalkane part, which results in a change in surface energy and, as a consequence, the movement of liquid. The collective action of a monolayer of these molecular shuttles makes them operate as a motor and has sufficient power to move a microlitre droplet of diiodomethane on a millimetre scale up a 12° incline (Fig. 6a).

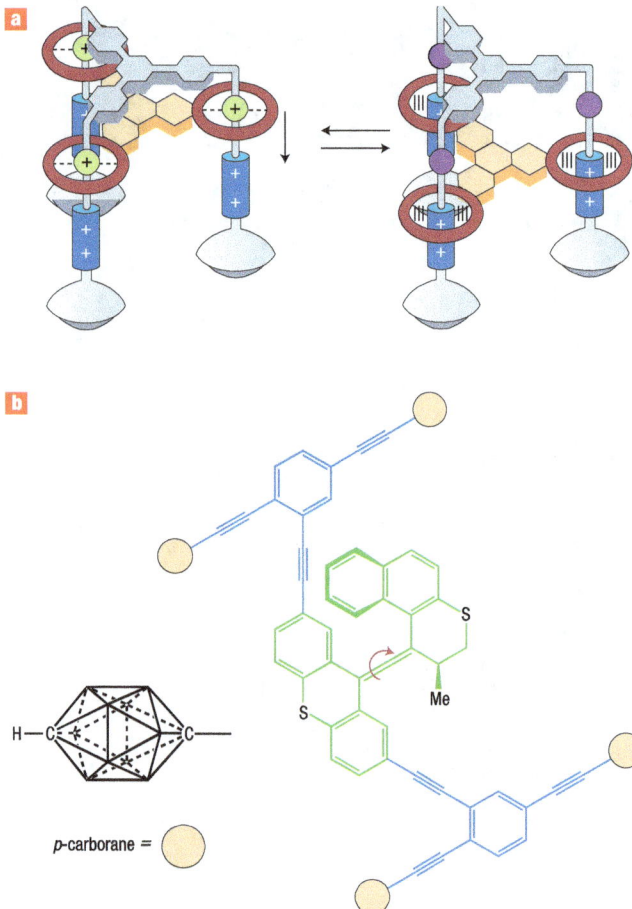

p-carborane =

Figure 7 Systems designed as multicomponent mechanical machines. **a**, A multicomponent rotaxane that operates as a molecular elevator. A platform is connected to three rotaxane units. Acid–base controlled switching in all three rotaxane units results in movement of the position of the platform relative to the rotaxane legs. Adapted with permission from ref. 70. Copyright (2006) American Chemical Society. **b**, Prototype for a light-driven molecular car. Four carborane 'wheels' are connected to a molecular frame that incorporates a light driven molecular motor. The linear molecular rods connecting the wheels act as axles and the central rotary motor component is incorporated to move the molecular car forward, by 'paddling'[73].

Although changes in molecular structure can be used to effect large changes in macroscopic properties, indirect macroscopic movement induced by changes in molecular chirality was demonstrated only recently. The reversible rotation of surface textures and microscale objects can be achieved using specially designed second-generation light-driven rotary molecular motors embedded in a liquid-crystal film (Fig. 5d)[102]. This motor has a right-handed helical structure and is very effective at inducing dynamic helical organization when applied as a chiral dopant in a liquid-crystalline film. The energy provided by light (the fuel for the motor) does not effect movement of a macroscopic object by itself. However, change in chirality of the motor is amplified through the liquid-crystalline host matrix. The result is a reorganization of the polygonal texture of the film in a clockwise or anticlockwise rotational fashion following the photochemical or thermal isomerization steps of the motor molecule. The rotation of the motor used as a dopant is 'transmitted' to the surface through the

reorganization of the liquid crystal. A surface relief of 20 nm was observed by non-contact atomic force microscopy. The orientation of the surface relief alters in response to the topology change in the embedded molecular motor. This reorganization generates a torque sufficient to achieve unidirectional rotation of microscopic objects placed on top of the film. These experiments show that a molecular motor can actually perform work. By harvesting light-energy and the collective action of molecular motors, microscopic objects that exceed the nanomotor in size by a factor of 10,000 can be rotated, with directional control.

The application of synthetic molecular motors in liquid-crystalline systems is not limited to the movement of objects. The change in handedness of the motor molecule during the rotary steps induces a reorganization of the liquid-crystalline film and induces a reversible and, importantly, tuneable change in the colour of the film (Fig. 5e)[103]. Therefore the macroscopic properties of a material can also be changed using a molecular rotary motor, and in this specific case, pixel generation with colours covering the entire visible spectrum is achieved readily.

Concerted macroscopic motion induced by changes in molecular structure is the fundamental basis of muscular movement. Hence, the possibility of a direct translation of molecular motion to macroscopic levels is, of course, an important goal. The molecular muscle developed by Stoddart and co-workers is based on linear molecular motors and uses a bistable [3]-rotaxane structure with two rings and four stations (Fig. 6b)[84,85]. Tethers attached to each ring anchor the whole rotaxane system as a self-assembled monolayer to the gold surface of microcantilever beams. The array of cantilever beams, decorated with these rotaxane molecules, shows reversible bending through sequential addition of chemical oxidants or reductants. The redox chemistry that drives the process is possible because of the presence of the redox-active units at two of the four stations. Redox-switching of the redox-active units initiates a change in the equilibrium position of the positively charged rings on the rod of the rotaxane, driven by a change in electrostatic repulsion. This change in inter-ring distance induces a change in mechanical stress on one side of the microcantilever. The combined effect of a monolayer of these bistable units operating collectively on a surface is to induce a bending of the cantilever. This demonstrates that controlled molecular motion in a wholly synthetic system can effect macroscopic movement.

In the systems developed by Gaub[104,105] Broer[106] and co-workers, photochromic azobenzenes have been used as tools to generate large reversible anisotropic changes in linear and stiff crosslinked polymer networks, respectively. The simplicity of these systems and the reversibility of the changes on irradiation with different wavelengths of light holds considerable potential for the future development of molecular-based polymer actuator materials.

MOLECULAR VALVES

That biological motors perform work and are engaged in well-defined mechanical tasks such as muscle contraction or the transport of objects is apparent in all living systems. Controlling motion using molecular switches is particularly attractive for the construction of nanomechanical valves[107]. For instance, the affinity of crown-ether type receptors to bind cations can be modulated reversibly by light when azobenzene photo-switches are incorporated in these receptors[89].

A light-actuated nanovalve derived from a channel protein was constructed in our group to control photochemical transport of solutes across a lipid bilayer. The valve consists of a channel protein modified with a photochemical active spiropyran switch[108]. The reversible molecular photochemical switch acts as a valve control for the 3 nm channel (Fig. 8a). The valve can be opened and closed with ultraviolet and visible light, respectively. This is possible because the neutral switch molecule converts to its highly polar zwitterionic form on irradiation with ultraviolet light. The hybrid protein valve is compatible with a liposome encapsulation system and allows for external photo-control of transport through the channel, demonstrated in the controlled release of an encapsulated fluorescent compound.

A different approach to photochemical valve control involves a bio-hybrid system in which movement through a channel is controlled allosterically[109]. A semisynthetic ligand-gated ion channel that can be turned on and off by ultraviolet and visible light irradiation, has been developed using an azobenzene (Fig. 1) optical switch (Fig. 8b). The azobenzene switch is attached both to the protein and the glutamate residue, which is specific for the allosteric site (a signal binding site on the protein, which regulates the operation of a separate remote functional component) responsible for closing the protein channel. The point of attachment is naturally critical to its operation, and in contrast to the previous example, the switching unit is attached to the outside of the channel rather than the inside. *Trans*-to-*cis* photoisomerization of the azobenzene unit results in a large geometric change in the molecule and, as a consequence, glutamate binds to the receptor and the channel opens. In this nanomechanical valve, several structural units and functions operate in concert to allow reversible channel gating controlled by light.

Redox processes offer an attractive alternative to light-controlled nanomechanical valves because of the possibility of integration into nano-electronic devices. A quite simple yet elegant example of such a molecular valve is based on cyclophanes (large molecular rings), where oxidation can be used to open and reduction to re-close a cavity[110]. Although the channel itself is very small when compared with the protein-based systems described above, this system does hold potential for the development of fully synthetic molecular valves.

An alternative approach to redox-controllable reversible nanovalves is found in a rotaxane-based valve system[111]. The principle involves reversible blocking of solid pores, which allows controlled release of fluorescent dyes from mesoporous (pore size 1.5–2.0 nm) silica particles.

The construction of functioning nanovalves, with movable molecular control units to regulate flow of substances, is a significant step towards the development of real nanomachines. Such nanovalves are, however, attractive components in themselves, for use in drug delivery systems with controlled release, signal transduction, sensors and nanofluidic systems, for example.

CONCLUSIONS AND OUTLOOK

The exquisite solutions nature has found to control molecular motion, evident in the fascinating biological linear and rotary motors, has served as a major source of inspiration for scientists to conceptualize, design and build — using a bottom-up approach — entirely synthetic molecular machines. The desire, ultimately, to construct and control molecular machines, fuels one of the great endeavours of contemporary science. The first primitive artificial molecular motors have been constructed and it has been demonstrated that energy consumption can be used to induce controlled and unidirectional motion. Linear and rotary molecular motors have been anchored to surfaces without loss of function — a significant step towards future nanomachines and devices. Furthermore, it has been demonstrated unequivocally that both linear and rotary motors can perform work and can

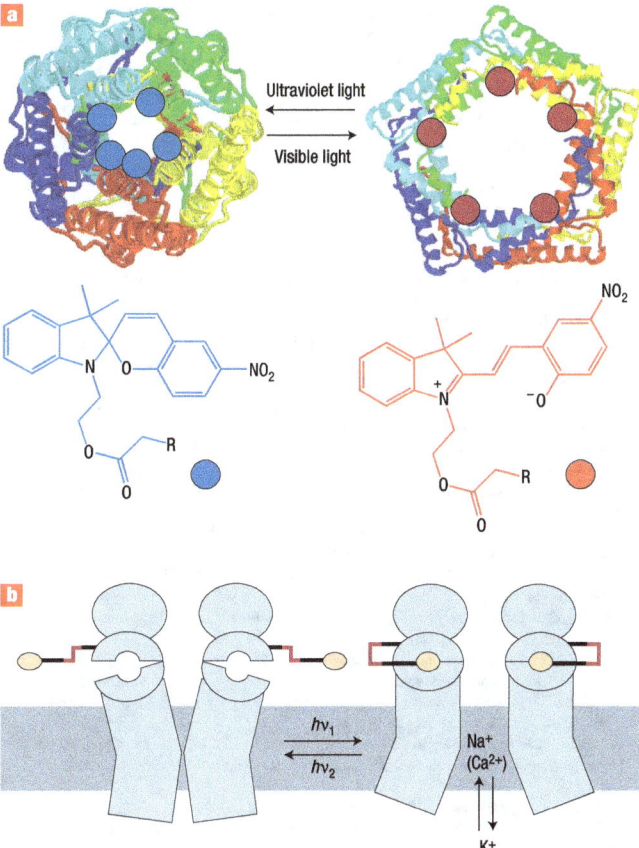

Figure 8 Two approaches to the opening and closing of nanovalves using molecular switches. **a**, A light-actuated nanovalve based on a mechano-sensitive channel protein modified with spiropyran photoswitches[108]. When ultraviolet light is shone on the protein, the molecular switch is converted from its neutral, hydrophobic, form to a charged polar form. The change in hydrophobicity in the channel results in the channel opening. Visible light reverses the process and closes the channel again. **b**, Photochemical allosteric control of a glutamate-sensitive protein channel based on the azobenzene molecular switch. In this example, the switching unit is not incorporated in the channel itself but instead is located on the outside of the channel protein. When light is shone on the azobenzene switch (see Fig. 1), the glutamate is brought into contact with a receptor site on the outside of the protein. The binding of glutamate to this site results in opening of the protein valve. Again, the process is reversed by shining light of a different colour on the protein, which moves the glutamate away from the control site. Reproduced with permission from ref. 109.

move objects. However, although the first applications of molecular motors to the control of other functions have been realized, the whole field is still very much in its infancy and offers ample opportunity in the design of nanomechanical devices.

Major challenges in the development of useful nanomachines remain, such as the development of fast and repetitive movement over longer time frames, directional movement along specified trajectories, integration of fully functional molecular motors in nanomachines and devices, catalytic molecular motors, systems that can transport cargo and so on. As complexity increases in these dynamic nanosystems, mastery of structure, function and communication across the traditional scientific boundaries will prove essential and indeed will serve to stimulate many areas of the synthetic, analytical and physical sciences. In view of the wide range of functions that biological motors play in nature and the role that macroscopic motors and machines play in daily life, the current limitation to the development and application of synthetic molecular machines and motors is perhaps only the imagination of the nanomotorists themselves.

doi:10.1038/nnano.2006.45

References

1. Berg, J. M., Tymoczko, J. L. & Stryer, L. *Biochemistry* 5th edn (W. H. Freeman, New York, 2006).
2. Kinbara, K. & Aida, T. Toward intelligent molecular machines: directed motions of biological and artificial molecules and assemblies. *Chem. Rev.* **105**, 1377–1400 (2005).
3. Schliwa, M. (ed.) *Molecular Motors* (Wiley-VCH, Weinheim, Germany 2003).
4. Boyer, P. D. Molecular motors: What makes ATP synthase spin? *Nature* **402**, 247–249 (1999).
5. Bray, D. *Cell Movements: From Molecules to Motility* (Garland, New York, 1992).
6. Hess, H. & Bachand, G. D. Biomolecular motors. *Nanotoday* **8**, 22–29 (2005).
7. Hess, H. & Vogel, V., Molecular shuttles based on motor proteins: active transport in synthetic environments. *Rev. Mol. Biotechnol.* **82**, 67–85 (2001).
8. Yan, H., Zhang, X. P., Shen, Z. Y. & Seeman, N. C. A robust DNA mechanical device controlled by hybridization topology. *Nature* **415**, 62–65 (2002).
9. Bath, J., Green, S. J. & Turberfield, A. J. A free-running DNA motor powered by a nicking enzyme. *Angew. Chem. Int. Edn* **44**, 4358–4361 (2005).
10. Alberti P. & Mergny J. L. DNA duplex-quadruplex exchange as the basis for a nanomolecular machine. *Proc. Natl Acad. Sci. USA* **100**, 1569–1573 (2003).
11. Abraham, R. T. & Tibbetts, R. S., Cell biology: Guiding ATM to broken DNA. *Science* **308**, 510–511 (2005).
12. Feynman, R. P. *The Pleasure of Finding Things Out* (Perseus Books: Cambridge, Massachusetts, 1999). *There's Plenty of Room at the Bottom* www.its.caltech.edu/~feynman
13. Davis, A. P. Synthetic molecular motors. *Nature* **401**, 120–121 (1999).
14. Harada, A. Cyclodextrin-based molecular machines. *Acc. Chem. Res.* **34**, 456–464 (2001).
15. Amendola, V., Fabbrizzi, L., Mangano, C. & Pallavicini, P. Molecular machines based on metal ion translocation. *Acc. Chem. Res.* **34**, 488–493 (2001).
16. Collin, J.-P., Dietrich-Buchecker, C., Gavina, P., Jimenez-Molero, M. C. & Sauvage, J.-P. Shuttles and muscles: linear molecular machines based on transition metals, *Acc. Chem. Res.* **34**, 477–487 (2001).
17. Sauvage, J.-P. (ed.) *Molecular Machines and Motors* (Springer, Berlin, 2001).
18. Feringa, B. L. In control of motion: from molecular switches to molecular motors. *Acc. Chem. Res.* **34**, 504–513 (2001).
19. Feringa, B. L., van Delden, R. A., Koumura, N. & Geertsema, E. M. Chiroptical molecular switches. *Chem. Rev.* **100**, 1789–1816 (2001).
20. Stoddart, J. F. Molecular machines, *Acc. Chem. Res.* **34**, 410–411 (2001).
21. Feringa, B. L. (ed.) *Molecular Switches* (Wiley-VCH, Weinheim, Germany, 2001).
22. Easton, C. J., Lincoln, S. F., Barr, L. & Onagi, H. Molecular reactors and machines: How useful are molecular mechanical devices? *Chem. Eur. J.* **10**, 3120–3128 (2004).
23. Ozin, G. A., Manners, I., Fournier-Bidoz, S. & Arsenault, A. Dream machines. *Adv. Mater.* **17**, 3011–3018 (2005).
24. Soanes, C. & Stevenson, A. (eds) *Oxford Dictionary of English* (Oxford Univ. Press, Oxford, 2005).
25. Astumian, R. D. Making molecules into motors. *Sci. Am.* **285**, 45–51 (2001).
26. Astumian, R. D. Thermodynamics and kinetics of a brownian motor. *Science* **276**, 917–922 (1997).
27. Rozenbaum, V. M., Yang, D.-Y., Lin, S. H. & Tsong, T. Y. Catalytic wheel as a brownian motor. *J. Phys. Chem. B* **108**, 15880–15889 (2004).
28. Whitesides, G. M. The once and future nanomachine. Biology outmatches futurists' most elaborate fantasies for molecular robots. *Sci. Am.* **285**, 78–84 (2001).
29. Chatterjee, M. N., Kay, E. R. & Leigh, D. A. Beyond switches: Ratcheting a particle energetically uphill with a compartmentalized molecular machine. *J. Am. Chem. Soc.* **128**, 4058–4073 (2006).
30. Siegel, J. Inventing the nanomolecular wheel. *Science* **310**, 63–64 (2005).
31. Rapenne, G. Synthesis of technomimetic molecules: towards rotation control in single molecular machines and motors. *Org. Biomol. Chem.* **3**, 1165–1169 (2005).
32. Garcia-Garibay, M. A. Crystalline molecular machines: Encoding supramolecular dynamics into molecular structure. *Proc. Natl Acad. Sci. USA* **102**, 10771–10776 (2005).
33. Khuong, T.-A. V., Zepeda, G., Ruiz, R., Khan, S. I. & Garcia-Garibay, M. A. Molecular compasses and gyroscopes: Engineering molecular crystals with fast internal rotation. *Cryst. Growth Des.* **4**, 15–18 (2004).
34. Caskey, D. C. & Michl, J. Toward self-assembled surface-mounted prismatic altitudinal rotors. A test case: trigonal and tetragonal prisms. *J. Org. Chem.* **70**, 5442–5448 (2005).
35. Horinek, D. & Michl, J. Surface-mounted altitudinal molecular rotors in alternating electric field: single-molecule parametric oscillator molecular dynamics. *Proc. Natl Acad. Sci. USA* **102**, 14175–14180 (2005).
36. Hawthorne, M. F. *et al.* Electrical or photocontrol of the rotary motion of a metallacarborane. *Science* **303**, 1849–1851 (2004).
37. Nawara, A. J., Shima, T., Hampel, F. & Gladysz, J. A. Gyroscope-like molecules consisting of PdX$_2$/PtX$_2$ rotators encased in three-spoke stators: synthesis via alkene metathesis, and facile substitution and demetalation. *J. Am. Chem. Soc.* **128**, 4962–4963 (2006).
38. Khuong, T.-A. V., Nuñez, J. E., Godinez, C. E. & Garcia-Garibay, M. A. Crystalline molecular machines: A quest toward solid-state dynamics and function. *Acc. Chem. Res.* **39**, 413–422 (2006).
39. Kottas, G. S., Clarke, L. I., Horinek, D. & Michl, J. Artificial molecular rotors. *Chem. Rev.* **105**, 1281–1376 (2005).
40. Rustem, F., Ismagilov, A. S., Bowden, N. & Whitesides, G. M. Autonomous movement and self-assembly. *Angew. Chem. Int. Edn* **41**, 652–654 (2002).
41. Paxton, W. F. *et al.* Catalytic nanomotors: Autonomous movement of striped nanorods. *J. Am. Chem. Soc.* **126**, 13424–13431 (2004).
42. Kline, T. R., Paxton, W. F., Mallouk, T. E. & Sen, A. Catalytic nanomotors: remote-controlled autonomous movement of striped metallic nanorods. *Angew. Chem. Int. Edn* **44**, 744–746 (2005).
43. Fournier-Bidoz, S., Arsenault, A. C., Manners, I. & Ozin. G. A. Synthetic self-propelled nanorotors. *Chem. Commun.* 441–443 (2005).
44. DiLuzio, W. R. *et al.* Escherichia coli swim on the right-hand side. *Nature* **435**, 1271–1274 (2005).
45. Vicario, J. *et al.* Catalytic molecular motors: Fueling autonomous movement by surface bond synthetic Manganese catalases. *Chem. Commun.* 3936–3938 (2005).
46. Ballardini, R., Balzani, V., Credi, A., Gandolfi, M. T. & Venturi, M. Artificial molecular-level machines: Which energy to make them work? *Acc. Chem. Res.* **34**, 445–455 (2001).
47. Kelly, T. R., De Silva, H. & Silva, R. A. Undirectional rotary motion in a molecular system. *Nature* **401**, 150–152 (1999).
48. Fletcher, S. P., Dumur, F., Pollard, M. M. & Feringa, B. L. A reversible, unidirectional molecular rotary motor driven by chemical energy. *Science* **310**, 80–82 (2005).
49. Leigh, D. A., Wong, J. K. Y., Dehez, F. & Zerbetto, F. Unidirectional rotation in a mechanically interlocked molecular rotor. *Nature* **424**, 174–179 (2003).
50. Hernandez, J. V., Kay, E. R. & Leigh, D. A. A reversible synthetic rotary molecular motor. *Science* **306**, 1532–1537 (2004).
51. Koumura, N. Zijlstra, R. W. J., van Delden R. A., Harada, N. & Feringa, B. L. Light-driven molecular rotor. *Nature* **401**, 152–155 (1999).
52. Feringa, B. L., van Delden, R. A. & ter Wiel, M. K. J. In control of switching, motion, and organization. *Pure Appl. Chem.* **75**, 563–575 (2003).
53. Koumura, N., Geertsema, E. M., van Gelder, M. B., Meetsma, A. & Feringa, B, L. Second generation light-driven molecular motors. Unidirectional rotation controlled by a single stereogenic center with near-perfect photoequilibria and acceleration of the speed of rotation by structural modification. *J. Am. Chem. Soc.* **124**, 5037–5051 (2002).
54. van Delden, R. A. *et al.* Unidirectional molecular motor on a gold surface. *Nature* **437**, 1337–1340 (2005).
55. Kwok, W. M. *et al.* Time-resolved resonance Raman study of S-1 *cis*-stilbene and its deuterated isotopomers. *J. Raman Spec.* **34**, 886–891 (2003).
56. Vicario, J., Walko, M., Meetsma, A. & Feringa, B. L. Fine tuning of the rotary motion by structural modification in light-driven unidirectional molecular motors. *J. Am. Chem. Soc.* **128**, 5127–5135 (2006).
57. Schalley, C. A., Beizai, K. & Vogtle, F. On the way to rotaxane-based molecular motors: Studies in molecular mobility and topological chirality. *Acc. Chem. Res.* **34**, 465–476 (2001).
58. Bissell, R. A., Cordova, E., Kaifer, A. E. & Stoddart, J. F. A chemically and electrochemically switchable molecular shuttle. *Nature* **369**, 133–137 (1994).
59. Alteri, A. *et al.* Electrochemically switchable hydrogen-bonded molecular shuttles. *J. Am. Chem. Soc.* **125**, 8644–8654 (2003).
60. Nygaard, S. *et al.* Quantifying the working stroke of tetrathiafulvalene-based electrochemically-driven linear motor-molecules. *Chem. Commun.* 144–146 (2006).
61. Lowe, J. N., Silvi, S., Stoddart, J. F., Badjic, J. D. & Credi, A. A mechanically interlocked bundle. *Chem. Eur. J.* **10**, 1926–1935 (2004).
62. Perez, E. M., Dryden, D. T. F., Leigh, D. A., Teobaldi, G. & Zerbetto, F. A generic basis for some simple light-operated mechanical molecular machines. *J. Am. Chem. Soc.* **126**, 12210–12211 (2004).
63. Brouwer, A. M. *et al.* Photoinduction of fast, reversible translational motion in a hydrogen-bonded molecular shuttle. *Science* **291**, 2124–2128 (2001).
64. Kay, E. R. & Leigh, D. A. Photochemistry: lighting up nanomachines. *Nature* **440**, 286–287 (2006).
65. Balzani, V. *et al.* Autonomous artificial nanomotor powered by sunlight. *Proc. Natl Acad. Sci. USA* **103**, 1178–1183 (2006).
66. Jimenez-Molero, C. M., Dietrich-Buchecker, C. & Sauvage, J. P. Towards artificial muscles at the nanometric level. *Chem. Commun.* 1613–1616 (2003).
67. Perkins, T. T., Li, H.-W, Dalal, R. V., Gelles, J. & Block, S. M. Forward and reverse motion of single RecBCD molecules on DNA. *Biophys. J.* **86**, 1640–1648 (2001).
68. Thordarson, P., Bijsterveld, E. J. A., Rowan, A. E. & Nolte, R. J. M. Epoxidation of polybutadiene by a topologically linked catalyst. *Nature* **424**, 915–918 (2003).
69. Badjic, J. D., Balzani, V., Credi, A., Silvi, S. & Stoddart, J. F. A molecular elevator. *Science* **303**, 1845–1849 (2004).
70. Badjic, J. D. *et al.* Operating molecular elevators. *J. Am. Chem. Soc.* **128**, 1489–1499 (2006).
71. Muraoka, T., Kinbara, K., Kobayashi, Y. & Aida, T. Light-driven open-close motion of chiral molecular scissors. *J. Am. Chem. Soc.* **125**, 5612–5613 (2003).
72. Muraoka, T., Kinbara, K. & Aida, T. Mechanical twisting of a guest by a photoresponsive host. *Nature* **440**, 512–515 (2006).
73. Morin, J.-F., Shirai, Y. & Tour, J. M. En route to a motorized nanocar. *Org. Lett.* **8**, 1713–1716 (2006).
74. Zheng, X. *et al.* Dipolar and nonpolar altitudinal molecular rotors mounted on a Au(111) surface. *J. Am. Chem. Soc.* **126**, 4540–4542 (2004).
75. Magnera, T. F. & Michl, J. Altitudinal surface-mounted molecular rotors. *Top. Curr. Chem.* **262**, 63–97 (2005).
76. Otsuki, J., Kawaguchi, S., Yamakawa, T., Asakawa, M. & Miyake, K. Arrays of double-decker porphyrins on highly oriented pyrolytic graphite. *Langmuir* **22**, 5708–5715 (2006).
77. Ikeda, M., Takeuchi, M., Shinkai, S., Tani, F. & Naruta Y. Synthesis of new diaryl-substituted triple-decker and tetraaryl-substituted double-decker lanthanum(III) porphyrins and their porphyrin ring rotational speed as compared with that of double-decker cerium(IV) porphyrins. *Bull. Chem. Soc. Jpn* **74**, 739–746 (2001).
78. Thomas, K. G., Ipe, B. I. & Sudeep, P. K. Photochemistry of chromophore-functionalized gold nanoparticles. *Pure Appl. Chem.* **74**, 1731–1738 (2002).
79. Ashton, P. R. *et al.* A three-pole supramolecular switch. *J. Am. Chem. Soc.* **121**, 3951–3957 (1999).
80. Huang, T. J. *et al.* Mechanical shuttling of linear motor-molecules in condensed phases on solid substrates. *Nano Lett.* **4**, 2065–2071 (2004).
81. Flood, A. H., Wong, E. W. & Stoddart, J. F. Models of charge transport and transfer in molecular switch tunnel junctions of bistable catenanes and rotaxanes. *Chem. Phys.* **324**, 280–290 (2006).

82. DeIonno, E., Tseng, H.-R., Harvey, D. D., Stoddart, J. F. & Heath, J. R. Infrared spectroscopic characterization of [2]rotaxane molecular switch tunnel junction devices. *J. Phys. Chem. B* **110**, 7609–7612 (2006).

83. Butt, H.-J. Towards powering nanometer-scale devices with molecular motors, single molecule engines. *Macromol. Chem. Phys.* **207**, 573–575 (2006).

84. Huang, J. *et al.* A nanomechanical device based on linear molecular motors. *Appl. Phys. Lett.* **85**, 5391–5393 (2003).

85. Liu, Y. *et al.* Linear artificial molecular muscles. *J. Am. Chem. Soc.* **127**, 9745–9759 (2005).

86. Ge, H. L. *et al.* Photoswitched wettability on inverse opal modified by a self-assembled azobenzene monolayer. *Chem. Phys. Chem.* **7**, 575–578 (2006).

87. Hugel, T., Holland, N. B., Cattani, A., Moroder, L., Seitz, M. & Gaub, H. E. Single-molecule optomechanical cycle. *Science* **296**, 1103–1106 (2002).

88. Botelho, A. V., Gibson, N. J., Thurmond, R. L., Wang, Y.& Brown, M. F. Conformational energetics of rhodopsin modulated by nonlamellar-forming lipids. *Biochemistry* **41**, 6354–6368 (2002).

89. Takeuchi, M., Ikeda, M., Sugasaki, A. & Shinkai, S. Molecular design of artificial molecular and ion recognition systems with allosteric guest responses. *Acc. Chem. Res.* **34**, 865–873 (2001).

90. de Jong, J. J. D., Lucas, L. N., Kellogg, R. M., van Esch, J. H. & Feringa, B. L. Reversible optical transcription of supramolecular chirality into molecular chirality. *Science* **304**, 278–281 (2004).

91. Sud, D., Norsten, T. B. & Branda, N. R. Photoswitching of stereoselectivity in catalysis using a copper dithienylethene complex. *Angew. Chem. Int. Edn* **44**, 2019–2021 (2005).

92. Irie, M. Diarylethenes for memories and switches. *Chem. Rev.* **100**, 1685–1716 (2000).

93. Tsivgoulis, G. M. & Lehn, J. M. Photoswitched and functionalized oligothiophenes: Synthesis and photochemical and electrochemical properties. *Chem. Eur. J.* **2**, 1399–1406 (1996).

94. Dulic, D. *et al.* One-way optoelectronic switching of photochromic molecules on gold. *Phys. Rev. Lett.* **91**, 207402 (2003).

95. Willner, I., Doron, A. & Katz, E. Gated molecular and biomolecular optoelectronic systems via photoisomerizable monolayer electrodes. *J. Phys. Org. Chem.* **11**, 546–560 (1998).

96. Willner, I. & Katz, E. Integration of layered redox proteins and conductive supports for bioelectronic applications. *Angew. Chem. Int. Edn* **39**, 1180–1218 (2000).

97. Nomura, A. M., Marnett, A. B., Shimba, N., Dotsch, V. & Craik, C. S. Induced structure of a helical switch as a mechanism to regulate enzymatic activity. *Nature Struc. Mol. Biol.* **12**, 1019–1020 (2005).

98. Furumi, S., Kidowaki, M., Ogawa, M., Nishiura, Y. & Ichimura, K. Surface-mediated photoalignment of discotic liquid crystals on azobenzene polymer films. *J. Phys. Chem. B* **109**, 9245–9254 (2005).

99. Raduge, C., Papastavrou, G., Kurth, D. G. & Motschmann, H. Controlling wettability by light: illuminating the molecular mechanism. *Eur. Phys. J. E* **10**, 103–114 (2003).

100. Oh, S. K., Nakagawa M. & Ichimura K. Photocontrol of liquid motion on an azobenzene monolayer. *J. Mater. Chem.* **12**, 2262–2269 (2002).

101. Berna, J. *et al.* Macroscopic transport by synthetic molecular machines. *Nature Mater.* **4**, 704–710 (2005).

102. Eelkema, R. *et al.* Nanomotor rotates microscale objects. A molecular motor in a liquid-crystal film uses light to turn items thousands of times larger than itself. *Nature* **440**, 163 (2006).

103. van Delden, R. A., Koumura, N., Harada, N. & Feringa, B. L. Unidirectional rotary motion in a liquid crystalline environment: Color tuning by a molecular motor. *Proc. Natl Acad. Sci. USA* **99**, 4945–4949 (2002).

104. Holland, N. B. *et al.* Single molecule force spectroscopy of azobenzene polymers: switching elasticity of single photochromic macromolecules. *Macromolecules* **36**, 2015–2023 (2003).

105. Hugel, T. *et al.* Single-molecule optomechanical cycle. *Science* **296**, 1103–1106 (2002).

106. Harris, K. D. *et al.* Large amplitude light-induced motion in high elastic modulus polymer actuators. *J. Mater. Chem.* **15**, 5043–5048 (2005).

107. Hess, H., Bachand, G. D. & Vogel, V. Powering nanodevices with biomolecular motors, *Chem. Eur. J.* **10**, 2110–2116 (2004).

108. Koçer, A., Walko, M., Meijberg, W. & Feringa B. L. A light-actuated nanovalve derived from a channel protein. *Science* **309**, 755–758 (2005).

109. Volgraf, M. *et al.* Allosteric control of an ionotropic glutamate receptor with an optical switch. *Nature Chem. Biol.* **2**, 47–52 (2006).

110. Kanazawa, H., Higuchi, M. & Yamamoto K. An electric cyclophane: Cavity control based on the rotation of a paraphenylene by redox switching. *J. Am. Chem. Soc.* **127**, 16404–16405 (2005).

111. Nguyen, T. D. *et al.* A reversible molecular valve. *Proc. Natl. Acad. Sci. USA* **102**, 10029–10034 (2005).

Acknowledgements

The Authors thank M. M. Pollard for many suggestions and reading of the manuscript.

Molecular logic and computing

Molecular substrates can be viewed as computational devices that process physical or chemical 'inputs' to generate 'outputs' based on a set of logical operators. By recognizing this conceptual crossover between chemistry and computation, it can be argued that the success of life itself is founded on a much longer-term revolution in information handling when compared with the modern semiconductor computing industry. Many of the simpler logic operations can be identified within chemical reactions and phenomena, as well as being produced in specifically designed systems. Some degree of integration can also be arranged, leading, in some instances, to arithmetic processing. These molecular logic systems can also lend themselves to convenient reconfiguring. Their clearest application area is in the life sciences, where their small size is a distinct advantage over conventional semiconductor counterparts. Molecular logic designs aid chemical (especially intracellular) sensing, small object recognition and intelligent diagnostics.

A. PRASANNA DE SILVA[1] AND SEIICHI UCHIYAMA[2]

[1]School of Chemistry and Chemical Engineering, Queen's University, Belfast BT9 5AG, Northern Ireland; [2]Graduate School of Pharmaceutical Sciences, University of Tokyo, 7-3-1 Hongo, Bunkyo-ku, Tokyo 113-0033, Japan
e-mail: a.desilva@qub.ac.uk; seiichi@mol.f.u-tokyo.ac.jp

Do you remember your first chemistry experiment? Perhaps you poured in a reagent and maybe heated a substance in a test tube. You saw that the initial substance was transformed into the product because it was distinguishable by the change of a property such as colour or physical form (Fig. 1a). This process can be viewed differently, however, through the eyes of a computer scientist (Fig. 1b) whereby the product is an 'output' of a combination of chemical and physical 'inputs'. Naturally, this analogy requires several caveats. For instance, inputs to many simple computational logic devices continually control the output. In a chemical sense, therefore, our choice of reaction needs to be reversible, otherwise each device would only provide one constant output. However, other simple computational logic devices can be latched in a given output state. Some of these have chemical analogues too[1,2].

BOOLEAN LOGIC

Although the semiconductor business brought computation to worldwide attention, computational ideas were available before transistors were fabricated and have a long history[3]. Briefly, the binary number system allows the assignment of 'true' to '1' and 'false' to '0', which is the couching of statements in mathematical terms[4]. The lasting legacy of George Boole's time in Ireland[3] was the method of manipulating these simple numbers with logic operations. Logic gates, which perform such operations, have just one input and one output in their simplest form (Fig. 2a). The NOT gate inverts any binary signal that it receives, as shown

in its truth table. The other three (YES, PASS 1 and PASS 0) are considered to be trivial by the semiconductor business, but we will see their utility later in the applications of molecular logic. 2-input logic allows more options, a few of which are shown in Fig. 2b. For example, the AND gate requires both inputs to be '1' before an output of '1' is sent. Importantly, some of these 2-input gates allow the construction of arithmetic processors, which are critical to modern computers. There is a similar history in recognizing computational concepts in biology[5,6], including human behaviour. It is educational to consider the computations required inside your head to recognize someone who approaches you. Many questions are asked about that person's face. Answers are obtained by inspection and compared with stored information, all of this being completed in the subsecond timescale. At the level of biomolecules, proteins of various kinds receive chemical inputs and output product molecules in a process similar to that shown in Fig. 1a. Sometimes, a modification of the protein itself is the output. Such operations occur in their thousands, and often in series, within each of your cells at each moment of your life.

WHY MOLECULES?

Computation with molecules is no less realistic than with well-established semiconductor materials that dominate in today's computers. Although the properties of semiconductors can be tuned by considering size effects or doping strategies, the vast array of chemical reactions that have been developed over the last 150 years allows the synthetic chemist to access a diverse range of molecular structures. In this way, the properties of molecules can be modified rationally in a systematic process, offering a much greater degree of variability and control of their properties. Moreover, assemblies of molecules (supermolecules[7]) containing multiple modules — each with its own function — are particularly promising for information handling because, for example, one module could be used to capture a chemical

When citing this review, please cite the original version as shown on the contents page of this chapter.

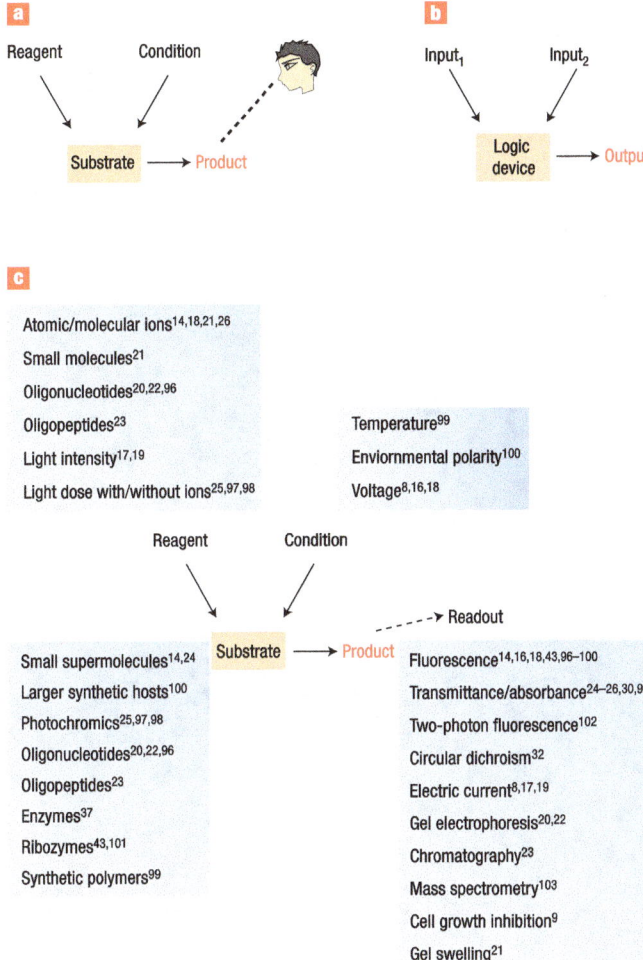

Figure 1 Chemical reactions and computational processes. **a**, A molecular substrate can be treated with a chemical reagent and/or placed under different environmental conditions to generate a product that can be observed because of a change in a physical property. **b**, In a conceptually similar process, a logic device can process a number of inputs to generate a measurable output. Various patterns of electrical inputs entering a computational logic device and producing electrical outputs in various patterns have many similarities (and some key differences) with what goes on in a typical chemical reaction. **c**, A wide range of chemical phenomena and reactions involving molecules, from masses as small as 100 Da and as large as many thousands can now be explicitly interpreted in a logical content. Various biomolecules feature prominently. Photochemical, thermal or electrochemical stimulation is common.

'input' and another would then be triggered to give an output of some kind.

The issue of connectivity is frequently raised when designing nanosized devices and this is the strength of larger semiconductor devices, where the output from one device is used to control another. It is not easy to pass the output from one molecular logic gate to serve as an input to the next because the inputs and outputs differ in nature in many instances. It must be stressed, however, that molecular computation need not follow the semiconductor blueprint. Although one aim may be to make molecular substitutes for semiconductors within existing logic devices[8], this developing field should embrace its differences and follow conceptually

different paths. Examples of this will be given at several points during this review.

One example is the integration of biological entities such as cells into synthetic logic systems in which the strengths of semiconductor devices disappear because of issues of size and material incompatibility. Cells process much information concerning internal or external chemicals in order to survive and grow[5,6], which requires intracellular species to be present at the right level in the right place, at the right time. Cell viability will be affected if these conditions are not met by certain species. Logical combinations of 'high' or 'low' levels (compared to the normal) can be arranged, via an integrated synthetic device, to release a therapeutic agent[9] or to generate a light signal for monitoring purposes (see the section on applications). The speed of molecular devices is limited (to millisecond time scales) by rates of diffusion and binding but this is tolerable when biological processes of similar or longer duration are involved. At this level of information processing within living systems, computation — by any means and of any magnitude — becomes significant.

An important consideration when using molecular logic devices is addressing. Semiconductor logic devices have wires along which electrical input signals are sent and other wires along which electrical output signals emerge. Though some molecular counterparts are beginning to be addressed in this way[8], most are addressed differently. For instance, a chemical (atomic or molecular) input signal hits the molecular device by simple diffusion in solution. The output signal emerges as fluorescence light, which can be picked up at a distance. The simplest molecular information processors are sensors and these are already loyal servants of cell biology. For instance, Tsien's sensor, so-called fura-2, measures Ca^{2+} concentration changes[10], which occur in millisecond timescales and in micrometre spaces within living cells during excitation and rest. The private life of intracellular Ca^{2+} is laid bare in a cinematic fashion. Like many other molecular sensors, fura-2 operates at the molecular level and can be addressed optically as a matter of routine. The function of related sensors has even been observed at the level of single molecules[11]. The routine optical addressing happens as follows: a microscope examines the intensity of fluorescence light in a chosen colour (510 nm) from each pixel while exciting the sample with ultraviolet light (335 nm). Furthermore, information about chemical concentrations can be gathered from nanometric spaces near membranes in a similar way by making use of hydrophobic effects to carefully position molecular devices[12]. H^+ concentrations near membranes are responsible for generating ATP (adenosine triphosphate), the energy currency in the cell. Single-device addressing[11], though elegant, is unnecessary for the present purpose if the devices are all in similar environments because they will report the same average information.

Another issue concerns the stability of molecular computational devices. The longevity of inorganic semiconductor materials contrasts with the decay of organic molecules. Yet again, however, the unique opportunities of interfacing molecular computational systems with cellular biosystems mean that such durability is not necessary in order to perform their function. Indeed, many cell imaging experiments carried out over several hours would be served more than adequately by, for instance, fluorescent molecular logic systems. The real-life success of their cousins, fluorescent sensors, was mentioned above. Nevertheless durability can be managed if needed, by reducing the intensity of exciting light and by using more efficient ways of detecting emission. Thus, the common engineering issues raised against molecular computation are not problematic when the biological and chemical arena of molecular computation is acknowledged. Therefore, one of the original, albeit controversial, aims of nanotechnology[13] — to produce active molecular devices with information processing capability — is a realistic proposition.

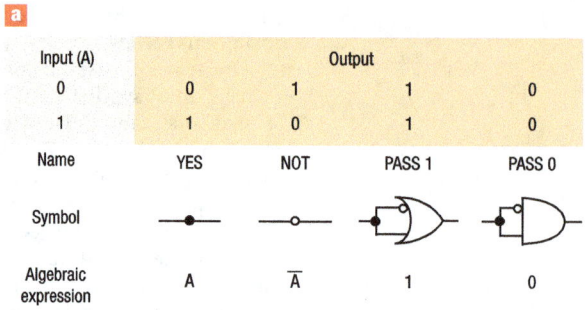

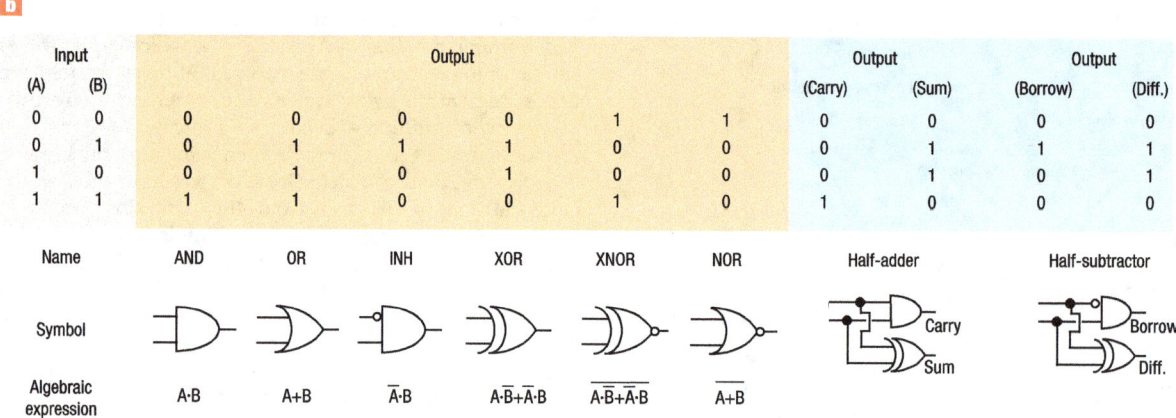

a

Input (A)	Output			
0	0	1	1	0
1	1	0	1	0
Name	YES	NOT	PASS 1	PASS 0
Symbol				
Algebraic expression	A	$\overline{A}$	1	0

b

Input		Output						Output		Output	
(A)	(B)							(Carry)	(Sum)	(Borrow)	(Diff.)
0	0	0	0	0	0	1	1	0	0	0	0
0	1	0	1	1	1	0	0	0	1	1	1
1	0	0	1	0	1	0	0	0	1	0	1
1	1	1	1	0	0	1	0	1	0	0	0
Name		AND	OR	INH	XOR	XNOR	NOR	Half-adder		Half-subtractor	
Symbol								Carry / Sum		Borrow / Diff.	
Algebraic expression		A·B	A+B	$\overline{A}$·B	A·$\overline{B}$+$\overline{A}$·B	$\overline{A·\overline{B}+\overline{A}·B}$	$\overline{A+B}$				

Figure 2 Boolean binary logic operations: truth tables, symbols and algebraic expressions. **a**, 1-input logic. **b**, 2-input logic. The half-adder is composed of AND and XOR gates running in parallel to process the two input bits so that the AND gate outputs the carry digit and the XOR gate outputs the sum digit. The half-subtractor has parallel INH and XOR gates, which output the borrow and difference digits respectively. A 3-input full-adder is composed of one half-adder passing its sum digit to a second half-adder, with the two carry digits being admitted to an OR gate. A full-subtractor is constructed in an analogous manner.

MOLECULAR LOGIC

Logic operations can be achieved by using ions to control the fluorescence of a molecule[14]. In the example shown in Fig. 3, a molecule comprises three functional units separated by spacers. At one end of the molecule is an aromatic anthracenyl ring system (shown in blue) that would normally fluoresce blue (known as a fluorophore) when it is exposed to ultraviolet light. In this system, however, the fluorescence is quenched because of a faster process, known as photoinduced electron transfer (PET)[15], in which an electron is transferred (Fig. 3a) to the anthracenyl group from either the nitrogen or oxygen atoms in the other parts of the molecule. The nitrogen group can act as a receptor for hydrogen ions (that is, acids) and the ring of oxygen atoms can be used to capture sodium cations, which are just the right size to fit snugly inside. If one of these two receptor sites (shown in green) is occupied (Fig. 3b,c), PET will still occur from the other, and no fluorescence will be observed.

If, however, both receptor sites are filled (Fig 3d), that is, H$^+$ and Na$^+$ ions are both present, PET is prevented because the electrons are now tied up binding ions and strong fluorescence is observed. The overall result is AND logic, as an output of '1' (that is, bright fluorescence) only occurs when both ion inputs are present, that is, H$^+$ and Na$^+$ values of '1'. This general structure of fluorophore–spacer$_1$–receptor$_1$–spacer$_2$–receptor$_2$, in which PET pathways are either blocked or not, lies at the heart of many molecular logic systems.

The properties of the fluorophore can be tuned as desired and each receptor can be modified to specifically recognize different guest species, in this case, G$_1$ and G$_2$, respectively. Other cases based on this molecular architecture — all with fluorescent PET parentage — will come up for discussion in the following paragraphs.

The recognition of logical content in photochemical phenomena has led to a myriad of related approaches, making molecular logic a truly general concept. Different systems can be classified according to several chemical or physical parameters (Fig. 1c). Various substrates, reagents, reaction conditions and readout modes have been commandeered. From a computing viewpoint it is clear that many kinds of inputs are used, each of which has its merits in a wider context. Biological entities contain many chemicals that can naturally serve as inputs to synthetic devices embedded therein. Electrical inputs to molecular logic gates can be supplied along wires by conventional semiconductor devices. On the other hand, optical inputs can be delivered remotely. Some of these cases featured in Fig. 1c need bulk materials, such as electrodes[8,16–19], gels[20–22], adsorptive media[23], stirred-flow reactors[24], or cuvet assemblies[25,26]. Although these approaches cannot be used for single-molecule detection, this level of sensitivity is not required for many studies and applications. In fact, only a few fluorescent logic gates have been examined at the single-molecule level[11] under carefully chosen conditions. The electrode-based examples have the advantage of being potentially in-line with semiconductor electronic technology[4]. Comprehensive collections of molecular

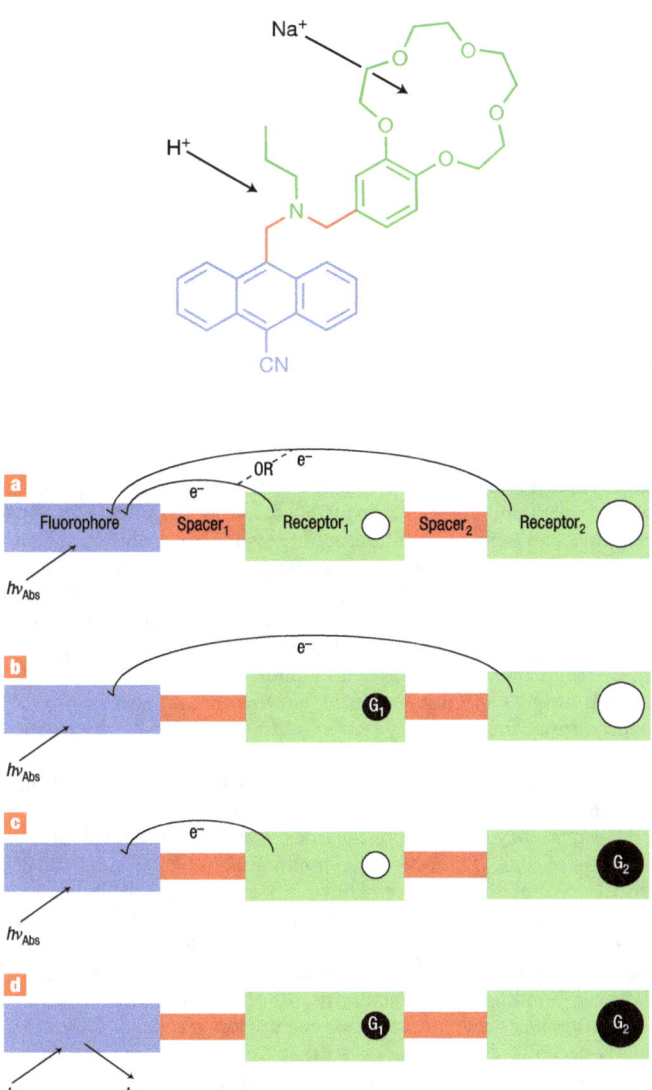

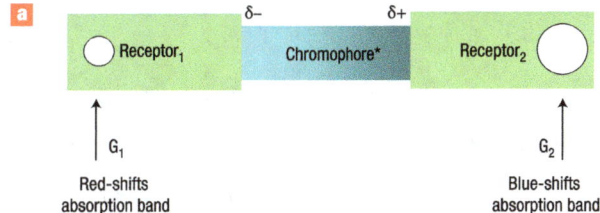

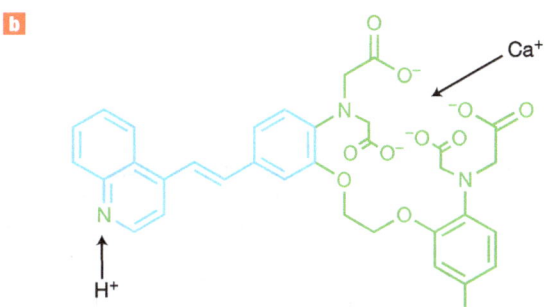

Figure 3 A molecular logic gate based on photoinduced electron transfer. Two chemical receptors (green) comprising an amine (receptor$_1$), which serves to bind H$^+$, and a benzocrown ether (receptor$_2$), which targets Na$^+$, are connected by hydrocarbon spacers (red) to each other and to a cyanoanthracene dye (blue) — a fluorophore that absorbs and emits light. **a**, The 'fluorophore–spacer$_1$–receptor$_1$–spacer$_2$–receptor$_2$' system has two possible paths of transferring an electron to the excited fluorophore — one from each unoccupied receptor — resulting in quenching of the fluorescence. **b,c**, Occupation of a receptor with its preferred guest cation blocks one pathway of electron transfer. **d**, Fluorescence results only if both receptors are occupied.

Figure 4 A molecular logic gate based on internal charge transfer. The photoactive component (chromophore) is shown in blue, and the chemically active components (receptors) are shown in green. **a**, When excited, the presence of electron-pushing and electron-pulling groups at the two ends of the 'receptor$_1$–chromophore–receptor$_2$' system causes a substantial shift of electron density from the electron-pushing group to the electron-pulling group. This causes the charge separation in the excited chromophore. When a cationic guest G$_1$ arrives at receptor$_1$, its proximity to the δ− charge causes a drop in the energy of the excited state, hence redshifting the absorption band. The opposite is true when cationic guest G$_2$ arrives at receptor$_2$ owing to the δ+ charge. **b**, The quinoline nitrogen (left) is receptor$_1$ and serves to bind H$^+$ as well as to be the electron-pulling group in the chromophore. The aniline nitrogen (right) is the electron-pushing group and is also a part of the Ca^{2+} binding amino acid receptor$_2$.

logic operations from the older literature are available[1,27–29] for the interested reader.

Although fluorescence is widely used as the output for molecular logic systems and can scale the heights of single-molecule detection, the simple effect of light absorbance should not be ignored. The ion-controlled colour change of 'receptor$_1$–chromophore–receptor$_2$' systems is illustrated in Fig. 4a and involves internal charge transfer (ICT) excited states[27]. These are electronic excited states with substantial charge separation, which arise through electron density moving from one end of the molecule to the other during excitation. The case in Fig. 4a is designed so that capture of the correct guest species by one of the receptors stabilizes the ICT state to produce a redshift in the UV–Vis absorption spectrum, whereas the other receptor/guest combination does the opposite. The design rests on the fact that receptor$_1$ is close to the negative terminal of the excited chromophore and that receptor$_2$ neighbours the positive terminal. Therefore, the presence of both guests (G$_1$ and G$_2$) leaves the absorption spectrum unchanged[30,31]. Hence, the case in Fig. 4b exemplifies this general approach to XNOR logic with absorbance output. The two amino acid units and the interspersed oxygen atoms form a three-dimensional crypt (righthand side of the green structure) of receptor$_2$, which selectively captures Ca^{2+}. The nitrogen atom within the quinoline group acts as the receptor$_1$ for H$^+$. As Fig. 2b shows, the output pattern of XNOR can be inverted to produce the pattern of XOR. As light absorbance is an inverse function of light transmittance, the case in Fig. 4b immediately becomes an XOR gate when transmittance is considered as the output.

As pointed out earlier, there are many properties that can be measured to determine the output of a molecular logic gate. In particular, monitoring changes in the rotation of the plane of polarized light in chiral molecules as a function of wavelength (a phenomenon known as optical rotatory dispersion or ORD) or in the differential absorption of circularly polarized light as a function of wavelength

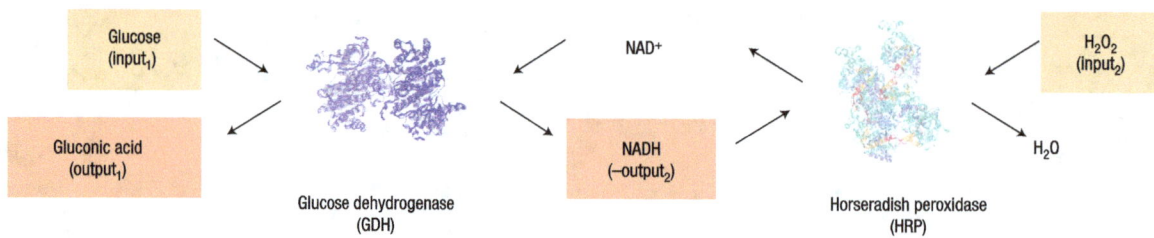

Figure 5 Molecular logic gates based on enzyme cascades. Coupling of enzymes by feeding the output of one as the input of another is a very useful way of developing molecular logic systems. While the enzymes are the devices that form the system, the various chemicals involved can be carefully declared as inputs and outputs. Starting conditions can also be set. For example, the fall in steady-state NADH concentration (output$_2$ '1') occurs if glucose is excluded (input$_1$ '0') from the enzyme system and H_2O_2 is admitted (input$_2$ '1'). This becomes INH logic (Fig. 2b). Adapted with permission from ref. 38. Copyright (2006) Wiley-VCH.

(known as circular dichroism or CD) can have advantages for reading the output. Whenever a molecular device is monitored by UV–Vis absorption spectroscopy or even fluorescence spectroscopy, excited states are invariably produced. As the energies of these excited states are comparable to the strengths of chemical bonds, this can lead to slow photochemical destruction even in the most carefully chosen cases, causing a limit on robustness. This has been considered under the 'Why molecules?' section, but laboratories keen on extending the useful readable life of molecular devices have turned to ORD and CD[32]. ORD spectra in particular occur outside the electronic absorption band as an absorption is a limited feature on the general dispersion stretching over large wavelength ranges. This is a reminder that many optical properties (for example, refractive index), unlike absorption, vary very gradually with wavelength. Thus ORD spectra can be monitored without populating excited states. As indicated above, ORD and CD properties only emerge from chiral molecules. Zhang and Zhu's work[32] on chiral photochromics can be enjoyed in this light because photochromics provide the switching ability that we require for logic operations. Indeed, photochromics can produce sharp changes in outputs such as light absorbance at a suitable wavelength when receiving light input at a different wavelength. More generally, inducing coloration of otherwise colourless photochromics with ions and light has examples spanning two decades[33–35] and has been developed more recently for logic purposes by Raymo and Giordani[36].

Biomolecule-based logic operations can be nicely illustrated with Willner's use of enzyme cascades[37,38] (Fig. 5). The enzyme glucose dehydrogenase (GDH) uses glucose as a substrate (input$_1$) in the presence of the cofactor NAD^+ to produce gluconic acid (output$_1$). Horseradish peroxidase (HRP) takes in H_2O_2 (input$_2$) to produce H_2O with the aid of NADH. These two enzymes run in parallel because the cofactor cycles between its two redox states (NAD^+ and NADH). If there is no NAD^+ initially, and if the system is allowed time to reach a steady state, the NADH concentration (as measured based on its absorption of light) drops only on the condition that glucose is absent and H_2O_2 is present. If this absorption change is taken as output$_2$, the coupled enzyme system manifests INH logic.

When glucose and H_2O_2 are both present, the NAD^+ produced by HRP's processing of H_2O_2 is taken up by GDH to process glucose so that the NADH level remains almost constant. On the other hand, gluconic acid (output$_1$, measured based on its absorption of light after reaction with hydroxylamine and ferric ion) accumulates only if glucose and H_2O_2 are both present, that is, AND logic is seen. If H_2O_2 is not available, NADH cannot be converted to NAD^+ by HRP so that GDH cannot process glucose in turn. Many native enzymes and cascades can be exploited, without any need for molecular synthesis in-house. Earlier enzyme-based logic by Conrad and Zauner was visionary but produced small signal changes[39]. Konermann's example also used a

native protein (cytochrome c) to demonstrate AND logic[40]: if present together, urea and H^+ inputs reversibly denature the protein so that an amino acid (tryptophan) fluorophore avoids being quenched by a haem porphyrin, which is located at the protein active site. Another early case by Lotan[41], though requiring synthetic modification, has an appealing design. The active pocket of the common hydrolytic enzyme α-chymotrypsin is covered or exposed by an appended group whose geometry can be switched by photoirradiation. Proflavine, a classical drug, is an inhibitor of the enzyme, but if converted to its reduced form, the inhibitory power is lost. Naturally, the enzyme is active (output '1') only when its active site is exposed and when the inhibitor is powerless. This becomes AND logic if the coding of the two inputs (enzyme and the inhibitor) is chosen to fit.

Besides enzymes, oligonucleotides have also provided lovely examples of logic. DNA-based computing, although without direct use of logic functions, has been known since 1994 (ref. 42). Stojanovic uses a DNA-based hydrolytic catalyst folded into a stem-loop shape where the catalytic activity resides[43]. An oligonucleotide with a weak link (at a ribonucleotide) is the substrate, designed to have good hydrogen-bonding complementarity with the ends of the stem-loop. The substrate carries a fluorescence energy donor and a fluorescence energy acceptor at its terminals so that we only see emission from the acceptor and not from the donor (output '0'). However, when the substrate is allowed to hybridize with the catalyst, hydrolysis occurs and the fluorescence energy donor is parted from the fluorescence acceptor. Then we only see emission from the fluorescence donor (output '1'). However, a new oligonucleotide (which will be our input) can be designed to hybridize with the catalyst by stretching it out, that is, the catalytic stem-loop is lost. Thus, when this input is present (input '1'), the substrate remains as it is and the fluorescence acceptor emission is seen (output '0'), that is, NOT logic.

We also note an interesting, though theoretical, proposal[44] to use magnetoresistance as a logic output from single molecules. Magneto-resistance is where an electrical resistance changes in response to a magnetic field, and is used in read heads of electronic computer hard drives. To extract this phenomenon from single molecules, cyclic π- electron systems connected to three metal contacts are proposed. The ring allows electron interference effects to take place. One contact carries the input electrical signal and the other two contacts serve to carry the two outputs in parallel. However, this requirement of three metal contacts to a molecule is difficult to realize at present[45].

LOGIC FOR SENSING

Sensing is an analogue experiment, with small changes of, say, fluorescence output arising from small changes of chemical analyte concentrations. Nevertheless, it is often exploited in an 'off–on' or digital manner for managerial 'go/no-go' decisions,

Box 1 Molecular logic gates based on PET

a, A selective sensor for glucosamine that binds to the diol (yellow) at one end and the ammonium group (pink) at the other. **b**, A selective sensor for D-tartaric acid, that contains two degenerate binding sites for oxygen-containing functional groups (yellow) at each end of the target molecule. **c**, A phosphorescence-based 3-input INH gate driven by H⁺, β-cyclodextrin and O₂ inputs. O₂ is the disabling input. **d**, A logic gate with several logic configurations, depending on the inputs H⁺, Zn²⁺ or Cu²⁺. **e**, A H⁺, Ca²⁺-driven AND gate. The design of the sensors in **a**, **b** and **e** are similar to that of Fig. 3. **c** is similar, but simpler as only one receptor is involved. However a phosphor, rather than a fluorophore, is present. The design in **d** is similar to that in **c**, though an excited state interaction between the two fluorophores is an extra complexity.

particularly in medical diagnostic situations. This type of analysis corresponds to 1–input YES logic gates. The opposite type, with an 'on–off' response, is 1–input NOT logic. A few recent examples are noted here. Mice everywhere will sleep easier because of the development of an 'off–on' fluorescent PET sensor for saxitoxin[46]. Samples suspected of containing this marine toxin, originating in the red tide, were tested up until now with a mouse bioassay. Similar success has been achieved with PET sensors for monitoring cellular pH[47] as well as environmentally hazardous materials such as lead[48] and mercury[49]. A nice example of a fluorescent 'on–off' sensor for copper[50] has also been developed.

From a computational standpoint, chemically irreversible systems may not be all that appealing because they do not allow for a device to be reset — that is, once the output is set, it cannot be changed. In

a broader context, however, such systems can be vitally important, especially for targets for which no receptors are known. In these cases, detection relies on the formation of chemical bonds — rather than a reversible binding event — and changes in the fluorescence of a given reagent are monitored. Among these difficult targets are chemical warfare agents such as diethylchlorophosphate. Two research groups applied fluorescent PET switching designs to this problem almost simultaneously in the same journal[51,52]. The first fluorescent PET reagents designed to produce irreversible 'switching' appeared in 1998 (ref. 53).

We have discussed AND logic gates driven by two separate chemical inputs (which can, in general, even be degenerate). Now, what if these two (or more) inputs were rigidly connected together? Whereas metal ions are not so convenient for this, many organic functionalities can be easily joined-up in chosen geometries. For instance, even a simple sugar molecule is bristling with hydroxyl units rigidly arranged in three-dimensional space. This type of multifunctional molecule abounds in living systems. No wonder, as each of these is involved in the correct step of metabolic processes only when its specific receptor comes along. If specificity was lost, any other molecule within the compartment would be taken up by the receptor resulting in chaos within the cell. If we build molecular logic gates carrying multiple receptors, correctly positioned to temporarily capture these multifunctional molecular targets, we will have sensors with optimized selectivity, with much biorelevance. This avenue is so valuable that ideas of molecular logic and computation have a good future here, if nowhere else. We illustrate this with a fluorescent sensor selective for glucosamine[54] (See Box 1a). Old[55-57] and new[58] cases showing improved selectivity for diamines (either protonated or in their neutral state) with fluorescence/absorbance output are known. The fact that these diamines include the 'molecules of death' — cadaverine and putrescine — adds to the human interest of such sensing.

The ability to discriminate between different mirror-image forms of molecules (so-called enantiomers) can also be achieved with this approach. For example[59], when the chiral AND logic gate depicted in Box 1b binds to one enantiomer of tartaric acid (known as the 'D'-form), both receptor units are occupied and a fluorescence enhancement is observed. In molecular logic terms, both inputs are '1' and so there is an output of '1', manifested as an increase in fluorescence. On the other hand, L-tartaric acid, the mirror image form of D-tartaric acid, does not fit onto the AND gate as well and so one receptor site is not bound and PET prevents fluorescence. In this latter case, one of the inputs is '0' and, therefore, the output from this AND logic gate is also '0', that is, no fluorescence enhancement. This work is underpinned by earlier studies of this system[60,61].

INTEGRATION OF MOLECULAR LOGIC

The success of semiconductor-based computing depends on physical integration of unit logic operations. Much biological information processing also clearly depends on cascades down a stream. In contrast, the physical integration of molecular logic devices described so far is difficult. For instance, H^+ and Na^+ inputs produce fluorescence output from the AND gate in Fig. 3. It is not straightforward to take such a light output from one gate and convert it to an ionic signal to serve as input for a second gate. Indeed, the physical joining-up of many electronic devices succeeds because the same 0 V or 5 V electrical signal acts as the inputs and outputs in each of these devices. This is quantitative input–output homogeneity. In fact, the first molecular logic devices were unequivocally demonstrated because their inputs and outputs were so different that there was no danger of the output feeding back into the input channel and short-circuiting it.

When inputs and outputs are quantitatively homogeneous, they must be kept apart by arranging wired conduits for them. Such conduits naturally increase the complexity of the working device. In the case of metal wiring of molecular gates, controversies have arisen concerning the metal–molecule interface[62]. So it is no wonder that progress in molecular logic integration to produce small-scale gate arrays does not involve physical interconnection of molecular gates but rather uses ideas of functional integration outlined in 1999 (ref. 63), which removes the need for molecule–molecule linking. This means that the input–output pattern of a single molecular device operating via a set of mechanisms is recognized as fitting the truth table of a multigate array. For instance, the symbol for a NOR gate (Fig. 2b) shows how it can be physically constructed by connecting the input of a NOT gate to the output of an OR gate in an electrical context. On the other hand, the same truth table is achieved if we arrange for fluorescence

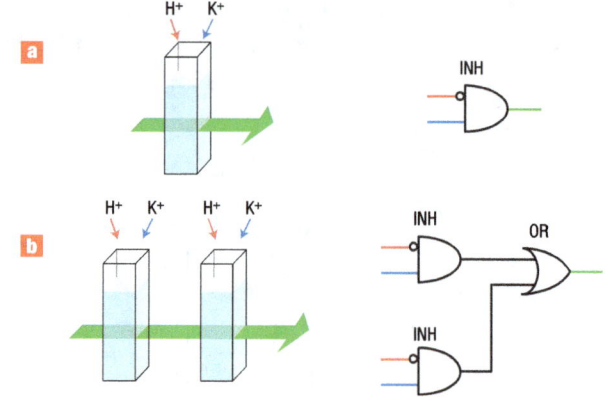

Figure 6 Integrated molecular logic gates based on colour-forming reactions. **a**, Sodium salt of $[Fe(CN)_5NO]^{2-}$ reacts rapidly with a thiol RSH to generate the purple-coloured $[Fe(CN)_5N(O)SR]^{3-}$ only if H^+ concentration (red input channel) is low and K^+ concentration (blue input channel) is high. The intensity of the purple colour is measured as the absorbance at 520 nm (green output channel). This is INH logic. **b**, Measuring the absorbance through two reaction cells in series naturally produces a result interpretable as a more complex logic gate array of two INH gates feeding their outputs into an OR gate.

quenching by two chemicals in a single molecular species[63]. Thus NOR logic with chemical inputs, as well as fluorescence output, is realized directly.

Some recent examples of functional integration include 3-input INH[64], enabled NOR[65], and NOR[66] logic. The first of these, which is displayed in Box 1c[64], gives phosphorescence (a longer-lived version of fluorescence) as the output if H^+ and β-cyclodextrin are present and if O_2 is absent. The phosphorescence emission of the bromonaphthalene unit (shown in blue) is quenched by PET, as seen previously in Fig. 3 for fluorescence. The nitrogen atom serves as the electron donor. It also serves as a receptor for H^+, in which case PET is prevented allowing phosphorescence to reassert itself. However there are two more obstacles in its path. First, the excited state of bromonaphthalene can easily collide with another copy of itself to cause mutual annihilation. This deactivation process is prevented by enveloping the bromonaphthalene unit in β-cyclodextrin, which is a donut-shaped molecule with a sufficiently large cavity. Second, the excited state of bromonaphthalene is easily deactivated by collision with molecular oxygen. So, phosphorescence is observed only when molecular oxygen is removed from the sample. As is the norm in digital electronics[4], these cases are representative of arrays of component gates (2–input AND, 2–input OR and NOT).

Another innovation leads to higher levels of integration. Chemical experiments are often conducted in one pot. Additional complexity is naturally introduced by considering arrays of pots/cells. Up to now, we distinguished inputs by their nature (that is, H^+ or Na^+) alone. Now their spatial positions (or addresses) become available as a distinguishing feature. An increase in the number of inputs means that bigger truth tables will result. These are analyzable in terms of larger arrays of logic gate components. Raymo and Giordani were the first to illustrate this idea with acid-sensitive photochromics[25,36,67]. However, let's consider Szaciłowski's recent example[26] based on a classical colour test for sulphide that many of us encountered in school chemistry laboratories. Colourless $[Fe(CN)_5NO]^{2-}$ reacts with thiols — organic sulphur compounds of structure RSH where R is a general organic group — to produce purple-coloured $[Fe(CN)_5N(O)SR]^{3-}$, which absorbs 520 nm light strongly. So our output will be the absorbance

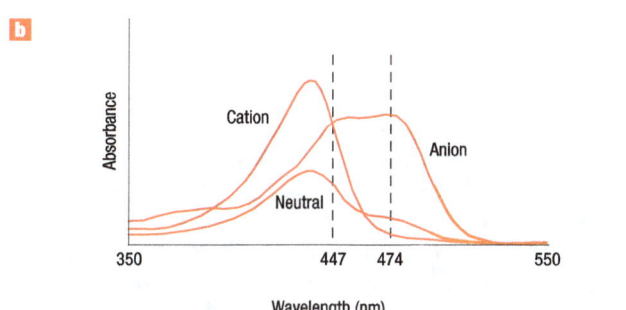

Figure 7 A moleculator based on fluorescein dye. **a**, Chemical structures of cationic, neutral, and anionic forms. Each of these can serve as the starting state of the molecular calculator, but we will focus on the neutral form (the middle structure). Equimolar doses of H⁺ or OH⁻ are the chemical inputs applied. **b**, Electronic absorption spectra of fluorescein in water. Absorbance values at chosen wavelengths (for example, 447 and 474 nm) are the outputs. The relationships between these inputs and outputs can be cast into truth tables such as those in Fig. 2b. Reproduced, with permission, from ref. 78. Copyright (2006) ACS.

at 520 nm. Fig. 6a shows the zero-dimensional case. The reaction is run in a single cell. The product appears only if the concentration of H⁺ is low and if the concentration of K⁺ is high. K⁺ stabilizes $[Fe(CN)_5N(O)SR]^{3-}$ by electrostatic attraction and drives the reaction forward. Low H⁺ concentration is important for success because the reactive form of the thiol is RS⁻. In high concentrations, H⁺ will react with RS⁻ to produce electrically neutral RSH, which is less reactive. So the overall behaviour fits INH logic.

The one-dimensional variant is a line of two cells, with a 520 nm light beam traversing them (Fig. 6b). Obviously, a high absorbance is detected if either cell generates the product. As noted above, the condition for product formation is low H⁺ concentration and high K⁺ concentration. Both cells producing a purple colour would still give a high absorbance. Only if both cells have no reaction will we get a low absorbance signal. Thus the equivalent logic gate array becomes two INH gates feeding an OR gate (Fig. 6b). Szaciłowski extends this approach to two- and three-dimensional arrays of cells to strengthen his case. He also exploits the sensitivity of the reaction to many variables (nature of sulphide derivative, light, temperature, pressure and so on) to further increase its logical complexity, though not all variables produce non-linear behaviour.

RECONFIGURABLE MOLECULAR LOGIC

Up to now we have discussed molecules that were each designed to perform a given logic operation. Much of semiconductor-based computing hardware is also based on hard-wired logic circuits of a given configuration. The ability to change the configuration of a circuit on command is a more recent innovation of some power. This naturally allows the creation of multiple logic circuits, each

optimized to carry out different tasks. Molecular logic gates can be reconfigured too. In fact, this can be done conveniently because, unlike electrons in semiconductors, many types of chemicals and many colours of light, which are all distinguishable from one another, serve as input/outputs[68]. The following paragraphs expand on this theme with specific examples.

The tried-and-tested path of fluorescence quenching by PET and its recovery by ion-binding is the basis of the action of di(anthrylmethyl)polyamine shown in Box 1d[69]. However this case differs from those discussed thus far by possessing two identical fluorescent anthracenyl units (shown in blue). When one of these is excited following light absorption, it can meet and 'stick' to the other within the excited state lifetime. The emission wavelength signature of the single anthracenyl unit is very different to that of the excited state of this dimer, which means that two fluorescence channels (at 416 nm and 520 nm respectively) can be observed. Each of these two emissions usually grows at the expense of the other so they produce different output patterns. For instance, the 520 nm emission dominates at low acidity, as the hydrophobic anthracenyl terminals crowd together in aqueous solution. The 416 nm band asserts itself at moderate acidity when the nitrogen atoms pick up protons, which cause a repulsion between the parts of the receptor chain (shown in green) and prevent the anthracenyl units from approaching each other. Thus the output at 416 nm indicates YES logic whereas 520 nm output displays NOT logic.

The logic behaviour of such PET-based anthrylmethylpolyamines is known to be reconfigurable according to the nature/level of the ion inputs[70]. Filled electron shell metal ions such as Zn²⁺ switch the fluorescence 'on' by blocking PET, whereas Cu²⁺ switches the emission 'off' (at a suitable pH). The latter effect arises because, at moderately high acidity, H⁺ will bind to the nitrogen atoms, which will block PET and therefore switch fluorescence 'on'. But Cu²⁺ ions bind strongly to the nitrogen atoms, so when they are added, they drive the H⁺ out. Unlike Zn²⁺, Cu²⁺ is coloured (blue) — it has an excited state whose energy is lower than that of the anthracene excited state. Thus the fluorescence of the anthracene is switched 'off' by transfer of its energy to the Cu²⁺ centre. In general, the level of input rarely has a reconfiguring effect, but here the logic type can be altered by increasing the acidity — at high enough acidity, binding of Cu²⁺ is prevented and fluorescence quenching is not seen. Thus the acidity serves as a reconfiguring channel.

A subtler form of logic reconfiguring arises by changing the threshold of the 'high' or 'low' input signal. The fluorescent PET system shown in Box 1e[71] shows AND logic with H⁺ and Ca²⁺ inputs provided that the H⁺ input is defined as 'high' at pH 6. Alternatively, if the H⁺ input is defined as 'high' at pH 2, the response converts to a simple PASS 0 gate where neither H⁺ or Ca²⁺ inputs revive fluorescence. At such high acidities, Ca²⁺ cannot compete for its amino acid receptor owing to protonation. The availability of carboxylic acid groups under such conditions allows intramolecular hydrogen bonding with the π-electron system of the excited fluorophore (shown in blue), which drains its energy.

There are examples of reconfiguring molecular logic by means of the excitation/observation wavelength[30,36] in the case of ICT systems relying on analyte-induced spectral shifts (Fig. 4). A wavelength shift of an absorption or emission spectral peak naturally causes a rise in intensity at one wavelength and a fall in another. Then, as discussed under the monomer and excited dimer emissions of the case in Box 1d, different logic types arise when the intensities at two chosen wavelengths are monitored as outputs. It is quite easy to arrange for such monitoring to be done all at the same time. So, multiple logic configurations can be observed simultaneously in a single molecular species[30] — something that is difficult to achieve in simple semiconductor systems, but commonly considered in quantum computation, which places much store in simultaneous processing of data. As ICT systems are the basis of molecular pH indicators found in pre-university chemistry laboratories and in flower pigments, the amazing conclusion is that aspects of quantum computers are present in your school and in your garden.

NUMBER HANDLING WITH MOLECULES

Number crunching is the public perception of computing, which any approach to molecular logic and computation needs to address. Molecular numerical operations have a variety of approaches[31,38,39,72–75]. The logic basis of numeracy has been expanded into gaming[76]. Remember your first sum? You learned that one and one is two. Implicitly you were introduced to the ascending hierarchy of numbers of zero, one and two. Importantly, your molecule-based brain learned this information for application throughout your life. A bit later, you were introduced to sum and carry digits. For example, when you tried to add seven to eight and ran out of fingers. Fifteen arose when one of tens was carried forward (to the left) and five of ones remained. The latter is the sum digit. Unlike decimal numbers from your childhood, the binary numbers in your calculator or computer deal with only zero and one, but the concept of sum and carry digits remains. The half-adder is the logic device which accomplishes this by running two binary digits into a parallel array of AND and XOR gates (Fig. 2b), where the carry digit is outputted from the AND gate and the sum digit emerges from the XOR gate.

Molecular-scale numeracy was first used in 2000 (ref. 31) when the 2-input AND gate (Box 1e) and the 2-input XOR gate (Fig. 4b), driven by H^+ and Ca^{2+}, ran in parallel. As in semiconductor-based half-adders, the AND gate output gave the carry bit while the XOR gate outputted the sum bit. Older cases of DNA molecule-based (but not molecular-scale, as gel electrophoresis is required) bit addition are known[20]. Wild's spectral hole-burning experiments for arithmetic are even older[77], but the logic operations were externally impressed on molecules rather than being inherent properties.

The recent moleculator (molecular calculator) from Margulies, Melman and Shanzer[78] is essentially pure and clever thought. Most of the previous experimental work was scattered in the literature, but Margulies *et al.* interpreted it from the vantage point of molecular logic. Their work revolves around the common fluorescent dye fluorescein, which can exist in several electrically charged forms depending on the pH value of the solution (Fig. 7a). It is shown how a full-adder and a full-subtractor (Fig. 2b) can exist within one molecular entity by integrating AND, XOR, INH and OR logic functions. The ability to observe different logic behaviours at different wavelengths (discussed in the section on reconfigurable molecular logic) and the use of inputs that are chemically identical but distinguishable to the experimentalist (known as degenerate inputs) are key to the success of this example. Let's sample this paper by discussing how the neutral state of fluorescein (the central molecule in Fig. 7a) can produce a half-subtractor (Fig. 2b). The basis of a half-adder was pointed out in the opening paragraph of this section. The half-subtractor is a similar arithmetic device, which permits subtraction of binary digits. It produces a difference digit and a borrow digit. These two are outputted by a XOR gate and an INH gate (Fig. 2b) respectively, operating in parallel.

The impact of H^+ (input$_1$) on the neutral state gives the cationic form whereas OH^- (input$_2$) gives the anionic form (Fig. 7a). Their absorption spectra (Fig. 7b) are clearly different from one another. This is an empirical observation. As discussed earlier concerning Fig. 4a, absorbance is a convenient logic output. Examining the spectra in Fig. 7b at 447 nm shows us that the absorbance is low (output '0') for the neutral form (input$_1$ '0', input$_2$ '0'). On the other hand, the absorbance is high (output '1') for the cation (input$_1$ '1', input$_2$ '0') or anion (input$_1$ '0', input$_2$ '1') form. We still need the fourth row of the

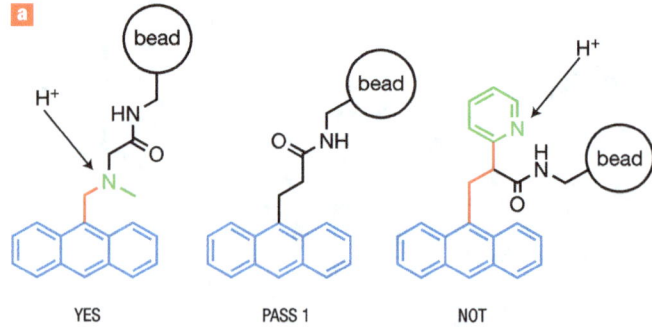

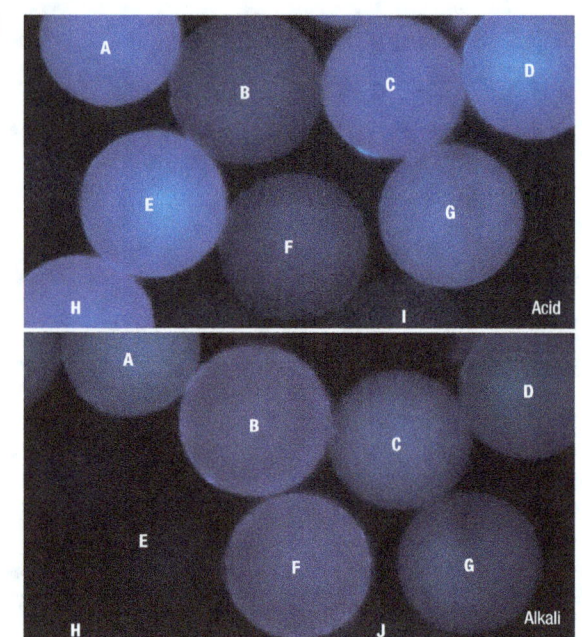

Figure 8 Molecular computational identification: materials and operation. **a**, Molecular YES, PASS 1, and NOT logic gates attached to small polymer beads. Their design is similar to that described in Fig. 3, with YES being simpler with only one receptor. PASS 1 does not possess a PET-active receptor. NOT has a receptor that permits PET only after guest binding. **b**, Fluorescence micrographs of beads with different logic tags. The H^+-driven logic type of each bead is assigned as A: PASS 1, B: NOT, C: PASS 1, D: PASS 1 + YES (1: 1), E: YES, F: NOT, G: PASS 1, H: YES, I: PASS 0. Condition: methanol:water (1:1, v/v) in the presence of acid (HCl) and alkali (NaOH) excited at 366 nm. Reproduced from ref. 89.

right-hand column of the half-subtractor truth table (Fig. 2b). Under the present regime the input condition (input$_1$ '1', input$_2$ '1') means that the neutral form of fluorescein is plied with equimolar amounts of H^+ and OH^-. This causes mutual annihilation of the two inputs, or, in chemical terms, acid–base neutralization. So it produces the same output (output '0') as the first row (input$_1$ '0', input$_2$ '0'), that is, the absorption spectrum of the fluorescein neutral form. This is XOR logic.

A similar examination of the spectra in Fig. 7b, but at 474 nm, shows us that the absorbance is high (output '1') only in the anionic form (input$_1$ '0', input$_2$ '1'). All the other three sets of input conditions produce low absorbance (output '0'), which satisfies the truth table for INH logic, which is the left-hand column of the half-subtractor

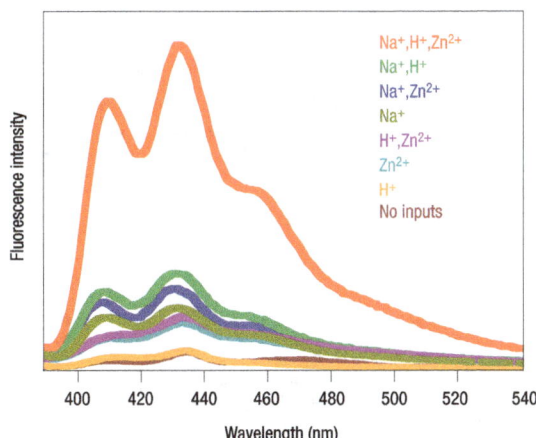

Figure 9 'Lab-on-a-molecule' and its operation. **a**, This design is similar to, but more complex than, that described in Fig. 3 as three receptors (green) are involved rather than two. **b**, Fluorescence emission spectra in water excited at 379 nm under the eight experimental conditions (colour coded). It is clear that the fluorescence is significantly stronger when the three inputs of Na$^+$, H$^+$ and Zn^{2+} are all present at sufficiently high concentrations rather than when only a pair, one or none of inputs are made available. Reproduced with permission from ref. 92. Copyright (2006) ACS.

truth table (Fig. 2b). The moleculator shows that a variety of selective receptors is not always required to demonstrate a rather complex logic array.

APPLICATIONS OF MOLECULAR LOGIC AND COMPUTATION

Molecular logic and computation has been around for 14 years and is beginning to come of age — the first applications of molecular computation are appearing. In fact, very sceptical opinions have been expressed[79,80] about this field. As a field develops, the question from sceptics changes from 'can it be done?' to 'can it do anything really useful?'. Then, new applications must address unanswered questions. Hearts and minds are won in general by applications of public interest and then a field can be said to have achieved some degree of credibility. Such applications are vital for the future of molecular logic and computation as the prognosis for solving real-world problems with DNA-based computation[42] is claimed to be poor[81].

Consider the general problem of object identification in populations. Cars have number-plates. People have faces/biometrics/passports. What of small objects in the micro/nanometric domain? These are crucial for (1) cells from patients being screened for diseases and (2) polymer bead supports used in combinatorial chemistry, which is the method of generating large families of related molecules by making many possible combinations of a given set of building

blocks. Polymer beads allow the use of the clever 'split and mix' method of generating chemical diversity in a few steps[82,83]. Briefly, separate batches of beads are appended with building blocks, A, B and C, for instance. Then these batches are mixed well and split into three batches again. Treatment of the first batch with building block D, the second with E and the third with F produces beads built up with AD, BD and CD in the first batch, AE, BE and CE in the second and AF, BF and CF in the third. The procedure can be continued to build up chains.

Semiconductor-based object identification methods, that is, RFID (radiofrequency identification) are widespread for large objects[84], such as smart keys carried in your pocket or even your pet dog (once a suitable chip is implanted). However this approach cannot be applied conveniently to the small-scale situations mentioned above. Indeed, polymer beads used in combinatorial chemistry have been combined with RFID tags for commercial purposes only by pooling large numbers of chemically identical beads into a bigger container[85,86]. Coloured fluorescent dyes have been used to provide molecular tags for beads and cells[87,88]. However, molecular emission bands are quite broad in terms of their emission range. Therefore different bands tend to overlap, making it hard to distinguish many different tags or combinations of them, that is, fluorescent colour only allows us to conveniently distinguish limited numbers of tags.

A much greater number of tags can be distinguished if we add extra capabilities to fluorescent dyes by converting them into molecular computing elements[89]. First, computer elements have diverse logic types (Fig. 2). Second, unlike their semiconductor cousins, they have a large diversity of inputs: chemical species like H$^+$, Na$^+$, glucose and so on, and physical quantities like temperature, light dose and environmental polarity. Third, the input level, which triggers the output, also builds diversity. For chemical inputs, this threshold adjustment is made by tuning the corresponding receptor binding strength in the molecular gate. Fourth, and importantly, arraying the above gate choices to produce many-valued logic systems (known in computation[90] and indicated in chemistry[91]) gives a huge diversity running into many millions. In addition to these clear advantages for tagging, the method is simple to implement. The tagged beads, for instance, are gently washed with a solution containing a low level of the chemical input (input '0') and then their output fluorescence intensities are observed. The experiment is repeated, but after washing with a solution containing a high level of the chemical input (input '1'). This allows the build-up of a truth table from which the tag's logic identity is ascertained. The method is dubbed 'wash and watch' because of this washing of the beads followed by watching for the fluorescence output.

Fig. 8 shows two fluorescence micrographs obtained with 366 nm excitation. The beads in the sample are individually tagged with one of five H$^+$-driven logic gates. Four of these are YES, PASS 1, NOT and PASS 0 (Fig. 2a). The last means that the bead carries no tag and therefore emits no fluorescence (output '0'). The fifth is a logic array PASS 1 + YES (1:1), where precursors of these two logic gates are applied in equimolar amounts to derivatize one bead. A quick look is all that is needed to identify these. Let's do this identification one by one. Bead A glows blue with equally high intensity (output '1') following an acid wash (H$^+$ input '1') and following an alkali wash (H$^+$ input '0'). So bead A carries a PASS 1 logic tag. Beads C and G are more copies of beads tagged with PASS 1 gates. Bead B is virtually invisible (output '0') following an acid wash (input '1'), but comes alive with strong blue fluorescence (output '1') following an alkali wash (input '0'). Reference to Fig. 2a reminds us that NOT logic is the assignment. The same applies to bead F. Bead E (and H) is the opposite: glowing strongly blue (output '1') in acid solution (input '1'), but switching off (output '0') its fluorescence in alkali solution (input '0'). This is the YES logic (Fig. 2a) tag. Bead I remains switched off in its fluorescence (output '0') in both acid and alkali washes, confirming its tag as PASS 0 (Fig. 2a). We kept the best for last: bead D. Fig. 8b clearly shows the significant fluorescence (output '1') in alkali solution (input '0'), whereas the fluorescence intensity doubles (output '2') following the wash (input '1'). Notably, this is not a binary 'on/off' situation but displays the two higher states of ternary logic ('2', '1' and '0'). This 2:1 output pattern can be seen to arise from the 1:1 sum of YES and PASS 1. So bead D is carrying the PASS 1 + YES (1:1). We note that all these logic gates use the same fluorophore, that is, tag diversity is seen even for a dye of one given colour. This method of molecular computational identification can be of immediate benefit in combinatorial chemistry where the use of large libraries based on polymer beads has been losing popularity.

Another area where molecular logic and computation can be applied is in intelligent diagnostics for medical purposes. We have already seen how molecular computing elements are contributing to improved sensing. Now we examine an extension of these. A patient visits a doctor. After a brief examination, blood tests are ordered. When the results are available, the doctor checks the series of parameters for those that overshoot the normal and those that undershoot. Then (s)he makes a logical combination of these outliers, in the Boolean sense, to lead to the diagnosis. For instance, a high cholesterol, a high LDL (low density lipoprotein) and a high CRP (C-reactive protein) indicates a cardiovascular problem. The typical fluorescence intensity–analyte concentration profile (response function) of a large population of identical copies of 'off–on' molecular sensors is S-shaped (or sigmoidal). Plateaux can be seen at very low and very high analyte concentrations for the 'off' and 'on' states respectively. The intermediate part of this response function is the sloping part where a small change in analyte concentration leads to a small change in fluorescence intensity, which is the sensing regime discussed in the section 'Logic for sensing'. We can assign the midpoint of the sloping part as the threshold where the 'off' state tips to the 'on' state and vice versa. The analyte concentration corresponding to this midpoint can be adjusted by controlling the analyte binding strength of the receptor unit within the sensor. We can choose/tune receptors so that only higher than normal concentrations of the analyte will elicit a fluorescence 'on' signal. By using several receptors within a multi-input AND gate, several 'high' analytes can be simultaneously detected with one fluorescence signal.

Fig. 9a shows a PET system[92] of the 'receptor$_1$–spacer$_1$–fluorophore–spacer$_2$–receptor$_2$–spacer$_3$–receptor$_3$' format which is a 3-input AND gate driven by H$^+$, Na$^+$ and Zn^{2+} inputs. Receptor$_1$ is the ring of oxygen atoms used to capture Na$^+$ and receptor$_2$ is based on the solo nitrogen used to bind H$^+$, as seen previously in Fig. 3. Receptor$_3$ is an amino acid derivative developed by Gunnlaugsson to catch Zn^{2+} rather selectively[93]. Each of these receptors can launch a PET process, in the spirit of Fig. 3 (except that the example in Fig. 3 contained two receptors instead of three), to the anthracenyl fluorophore. So each of these three receptors must be blocked by binding to its analyte present at above-threshold concentration for fluorescence to switch 'on' (Fig. 9b). This is a 'lab-on-a-molecule' as it performs three tests. It checks whether each parameter exceeds its pre-set threshold and then applies a 3-input AND logic function before releasing a 'high' fluorescence signal. Direct detection of analyte set logic patterns (cases for analyte pairs being known)[94,95] to produce 'well/ill' decisions[92] is another worthwhile capability of small molecules.

CONCLUSIONS

Designed supermolecules in the small nanometre range are highly functionalized nano-objects that can draw on two centuries of accumulated wisdom in chemistry. They display several foundational aspects of logic and computation, despite being a rudimentary mimicry of what traditional semiconductor logic does so well. All of the 1-input, 1-output and 2-input,

1-output logic operations have been demonstrated in the molecular regime. Many of these have been generalized to encompass various chemical species, absorption/emission colours and other variables, as well as various design principles. Small-scale integration, reconfigurability and numeric capability of molecular logic have all been shown. All of this proves that the field is growing well but it does not ensure its long-term vitality. This will come about when unsolved problems of a general nature are addressed. Luckily, this has now begun.

Molecular computational identification can identify small objects in large populations in biotechnologically important situations. Crucially, these objects are too small for traditional semiconductor logic to be applied. Molecular sensors based on logic ideas are already in service in intracellular situations where traditional semiconductor logic cannot go. Though not at the same stage of advancement, 'lab-on-a-molecule' systems combine sensing and logic to produce decision makers truly miniscule in size. However, the successes achieved so far are nothing compared with what is possible if fresh bright minds can be attracted into the field of molecular logic and computation. That is what this review hopes for.

doi:10.1038/nnano.2007.188

References
1. Raymo, F. M. Digital processing and communication with molecular switches. Adv. Mater. 14, 401–414 (2002).
2. Margulies, D. Felder, C. E., Melman, G. & Shanzer, A. A molecular keypad lock: a photochemical device capable of authorizing password entries. J. Am. Chem. Soc. 129, 347–354 (2007).
3. Boole, G. An Investigation of the Laws of Thought (Dover, New York, 1958).
4. Malvino, A. P. & Brown, J. A. Digital Computer Electronics 3rd edn (Glencoe, New York, 1993).
5. Bray, D. Protein molecules as computational elements in living cells. Nature 376, 307–312 (1995).
6. Williams, K. A. Three-dimensional structure of the ion-coupled transport protein NhaA. Nature 403, 112–115 (2000).
7. Lehn, J.-M. Supramolecular Chemistry (VCH, Weinheim, 1995).
8. Collier, C. P. et al. Electronically configurable molecular-based logic gates. Science 285, 391–394 (1999).
9. Amir, R. J., Popkov, M., Lerner, R. A., Barbas III, C. F. & Shabat, D. Prodrug activation gated by a molecular "OR" logic trigger. Angew. Chem. Int. Edn 44, 4378–4381 (2005).
10. Grynkiewicz, G., Poenie, M. & Tsien, R. Y. A new generation of Ca²⁺ indicators with greatly improved fluorescence properties. J. Biol. Chem. 260, 3440–3450 (1985).
11. Holman, M. W. & Adams, D. M. Using single-molecule fluorescence spectroscopy to study electron transfer. ChemPhysChem 5, 1831–1836 (2004).
12. Uchiyama, S., McClean, G. D., Iwai, K. & de Silva, A. P. Membrane media create small nanospaces for molecular computation. J. Am. Chem. Soc. 127, 8920–8921 (2005).
13. Drexler, K. E. Engines of Creation (Anchor Press, New York, 1986).
14. de Silva, A. P., Gunaratne, H. Q. N. & McCoy, C. P. A molecular photoionic AND gate based on fluorescent signalling. Nature 364, 42–44 (1993).
15. Bissell, R. A. et al. Fluorescent PET (photoinduced electron transfer) sensors. Top. Curr. Chem. 168, 223–264 (1993).
16. Komura, T., Niu, G. Y., Yamaguchi, T. & Asano, M. Redox and ionic-binding switched fluorescence of phenosafranine and thionine included in Nafion® films. Electrochim. Acta 48, 631–639 (2003).
17. Szaciłowski, K., Macyk, W. & Stochel, G. Light-driven OR and XOR programmable chemical logic gates. J. Am. Chem. Soc. 128, 4550–4551 (2006).
18. Biancardo, M., Bignozzi, C., Doyle, H. & Redmond, G. A potential and ion switched molecular photonic logic gate. Chem. Commun. 3918–3920 (2005).
19. Matsui, J., Mitsuishi, M., Aoki, A. & Miyashita, T. Molecular optical gating devices based on polymer nanosheets assemblies. J. Am. Chem. Soc. 126, 3708–3709 (2004).
20. Yurke, B., Mills Jr., A. P. & Cheng, S. L. DNA implementation of addition in which the input strands are separate from the operator strands. Biosystems 52, 165–174 (1999).
21. Schneider, H.-J., Tianjun, L., Lomadze, N. & Palm, B. Cooperativity in a chemomechanical polymer: a chemically induced macroscopic logic gate. Adv. Mater. 16, 613–615 (2004).
22. Okamoto, A., Tanaka, K. & Saito, I. DNA logic gates. J. Am. Chem. Soc. 126, 9458–9463 (2004).
23. Ashkenasy, G. & Ghadiri, M. R. Boolean logic functions of a synthetic peptide network. J. Am. Chem. Soc. 126, 11140–11141 (2004).
24. Blittersdorf, R., Müller, J. & Schneider, F. W. Chemical visualization of Boolean functions: a simple chemical computer. J. Chem. Educ. 72, 760–763 (1995).
25. Raymo, F. M. & Giordani, S. All-optical processing with molecular switches. Proc. Natl. Acad. Sci. USA 99, 4941–4944 (2002).
26. Szaciłowski, K. Molecular logic gates based on pentacyanoferrate complexes: from simple gates to three-dimensional logic systems. Chem. Eur. J. 10, 2520–2528 (2004).
27. de Silva, A. P. et al. Signaling recognition events with fluorescent sensors and switches. Chem. Rev. 97, 1515–1566 (1997).
28. Balzani, V., Credi, A. & Venturi, M. Molecular Devices and Machines (Wiley-VCH, Weinheim, 2003).
29. de Silva, A. P. & McClenaghan, N. D. Molecular-scale logic gates. Chem. Eur. J. 10, 574–586 (2004).

30. de Silva, A. P. & McClenaghan, N. D. Simultaneously multiply-configurable or superposed molecular logic systems composed of ICT (internal charge transfer) chromophores and fluorophores integrated with one- or two-ion receptors. *Chem. Eur. J.* **8,** 4935–4945 (2002).
31. de Silva, A. P. & McClenaghan, N. D. Proof-of-principle of molecular-scale arithmetic. *J. Am. Chem. Soc.* **122,** 3965–3966 (2000).
32. Zhou, Y., Zhang, D., Zhang, Y., Tang, Y. & Zhu, D. Tuning the CD spectrum and optical rotation value of a new binaphthalene molecule with two spiropyran units: mimicking the function of a molecular "AND" logic gate and a new chiral molecular switch. *J. Org. Chem.* **70,** 6164–6170 (2005).
33. Tamaki, T. & Ichimura, K. Photochromic chelating spironaphthoxazines. *J. Chem. Soc., Chem. Commun.* 1477–1479 (1989).
34. Gobbi, L., Seiler, P. & Diederich, F. A novel three-way chromophoric molecular switch: pH and light controllable switching cycles. *Angew. Chem. Int. Edn* **38,** 674–678 (1999).
35. Wojtyk, J. T. C., Kazmaier, P. M. & Buncel, E. Effects of metal ion complexation on the spiropyran–merocyanine interconversion: development of a thermally stable photo-switch. *Chem. Commun.* 1703–1704 (1998).
36. Raymo, F. M. & Giordani, S. Multichannel digital transmission in an optical network of communicating molecules. *J. Am. Chem. Soc.* **124,** 2004–2007 (2002).
37. Baron, R., Lioubashevski, O., Katz, E., Niazov, T. & Willner, I. Two coupled enzymes perform in parallel the 'AND' and 'InhibAND' logic gate operations. *Org. Biomol. Chem.* **4,** 989–991 (2006).
38. Baron, R., Lioubashevski, O., Katz, E., Niazov, T. & Willner, I. Elementary arithmetic operations by enzymes: a model for metabolic pathway based computing. *Angew. Chem. Int. Edn* **45,** 1572–1576 (2006).
39. Zauner, K.-P. & Conrad, M. Enzymatic Computing. *Biotechnol. Prog.* **17,** 553–559 (2001).
40. Deonarine, A. S., Clark, S. M. & Konermann, L. Implementation of a multifunctional logic gate based on folding/unfolding transitions of a protein. *Future Gener. Comp. Sy.* **19,** 87–97 (2003).
41. Sivan, S., Tuchman, S. & Lotan, N. A biochemical logic gate using an enzyme and its inhibitor. Part II: The logic gate. *BioSystems* **70,** 21–33 (2003).
42. Adleman, L. M. Molecular computation of solutions to combinatorial problems. *Science* **266,** 1021–1024 (1994).
43. Stojanovic, M. N., Mitchell, T. E. & Stefanovic, D. Deoxyribozyme-based logic gates. *J. Am. Chem. Soc.* **124,** 3555–3561 (2002).
44. Hod, O., Baer, R. & Rabani, E. A parallel electromagnetic molecular logic gate. *J. Am. Chem. Soc.* **127,** 1648–1649 (2005).
45. Tour, J. M. in *Stimulating Concepts in Chemistry* (eds. Vögtle, F. Stoddart, J. F. & Shibasaki, M.) 237–253 (Wiley-VCH, Weinheim, 2000).
46. Kele, P., Orbulescu, J., Gawley, R. E. & Leblanc, R. M. Spectroscopic detection of saxitoxin: An alternative to mouse bioassay. *Chem. Commun.* 1494–1496 (2006).
47. Hall, M. J., Allen, L. T. & O'Shea, D. F. PET modulated fluorescent sensing from the BF_2 chelated azadipyrromethene platform. *Org. Biomol. Chem.* **4,** 776–780 (2006).
48. Kwon, J. Y. *et al.* A highly selective fluorescent chemosensor for Pb^{2+}. *J. Am. Chem. Soc.* **127,** 10107–10111 (2005).
49. Yoon, S., Albers, A. E., Wong, A. P. & Chang, C. J. Screening mercury levels in fish with a selective fluorescent chemosensor. *J. Am. Chem. Soc.* **127,** 16030–16031 (2005).
50. Banthia, S. & Samanta, A. A two-dimensional chromogenic sensor as well as fluorescence inverter: selective detection of copper(II) in aqueous medium. *New J. Chem.* **29,** 1007–1010 (2005).
51. Dale, T. J. & Rebek Jr., J. Fluorescent sensors for organophosphorus nerve agent mimics. *J. Am. Chem. Soc.* **128,** 4500–4501 (2006).
52. Bencic-Nagale, S., Sternfeld, T. & Walt, D. R. Microbead chemical switches: An approach to detection of reactive organophosphate chemical warfare agent vapors. *J. Am. Chem. Soc.* **128,** 5041–5048 (2006).
53. de Silva, A. P., Gunaratne, H. Q. N. & Gunnlaugsson, T. Fluorescent PET (photoinduced electron transfer) reagents for thiols. *Tetrahedron Lett.* **39,** 5077–5080 (1998).
54. Cooper, C. R. & James, T. D. Selective D-glucosamine hydrochloride fluorescence signalling based on ammonium cation and diol recognition. *Chem. Commun.* 1419–1420 (1997).
55. Fages, F. *et al.* Synthesis and fluorescence emission properties of a bis-anthracenyl macrotricyclic ditopic receptor. Crystal structure of its dinuclear rubidium cryptate. *J. Chem. Soc., Chem. Commun.* 655–658 (1990).
56. de Silva, A. P. & Sandanayake, K. R. A. S. Fluorescence "off-on" signalling upon linear recognition and binding of α,ω-alkanediyldiammonium ions by 9,10-bis{(1-aza-4,7,10,13,16-pentaoxacyclooctadecyl)methyl}anthracene. *Angew. Chem. Int. Edn Engl.* **29,** 1173–1175 (1990).
57. Misumi, S. Recognitory coloration of cations with chromoacerands. *Top. Curr. Chem.* **165,** 163–192 (1993).
58. Secor, K., Plante, J., Avetta, C. & Glass, T. Fluorescent sensors for diamines. *J. Mater. Chem.* **15,** 4073–4077 (2005).
59. Zhao, J., Fyles, T. M. & James, T. D. Chiral binol-bisboronic acid as fluorescence sensor for sugar acids. *Angew. Chem. Int. Edn* **43,** 3461–3464 (2004).
60. James, T. D., Sandanayake, K. R. A. S. & Shinkai, S. Chiral discrimination of monosaccharides using a fluorescent molecular sensor. *Nature* **374,** 345–347 (1995).
61. Gray Jr., C. W. & Houston, T. A. Boronic acid receptors for α-hydroxycarboxylates: high affinity of Shinkai's glucose receptor for tartrate. *J. Org. Chem.* **67,** 5426–5428 (2002).
62. He, J. X. *et al.* Metal-free silicon-molecule-nanotube testbed and memory device. *Nature Mater.* **5,** 63–68 (2006).
63. de Silva, A. P. *et al.* Integration of logic functions and sequential operation of gates at the molecular-scale. *J. Am. Chem. Soc.* **121,** 1393–1394 (1999).
64. Mu, L. X., Wang, Y., Zhang, Z. & Jin, W. J. Room temperature phosphorescence pH switch based on photo-induced electron transfer. *Chinese Chem. Lett.* **15,** 1131–1134 (2004).
65. de Sousa, M., de Castro, B., Abad, S., Miranda, M. A. & Pischel, U. A molecular tool kit for the variable design of logic operations (*NOR, INH, EnNOR*). *Chem. Commun.* 2051–2053 (2006).
66. Cheng, P.-N., Chiang, P.-T. & Chiu, S.-H. A switchable macrocycle-clip complex that functions as a NOR logic gate. *Chem. Commun.* 1285–1287 (2005).
67. Giordani, S., Cejas, M. A. & Raymo, F. M. Photoinduced proton exchange between molecular switches. *Tetrahedron* **60,** 10973–10981 (2004).
68. Margulies, D., Melman, G. & Shanzer, A. Fluorescein as a model molecular calculator with reset capability. *Nature Mater.* **4,** 768–771 (2005).
69. Shiraishi, Y., Tokitoh, Y. & Hirai, T. A fluorescent molecular logic gate with multiply-configurable dual outputs. *Chem. Commun.* 5316–5318 (2005).
70. Alves, S. *et. al.* Open-chain polyamine ligands bearing an anthracene unit–chemosensors for logic operations at the molecular level. *Eur. J. Inorg. Chem.* 405–412 (2001).
71. Callan, J. F., de Silva, A. P. & McClenaghan, N. D. Switching between molecular switch types by module rearrangement: Ca^{2+}-enabled, H^+-driven 'off-on-off', H^+-driven YES and PASS 0 as well as H^+, Ca^{2+}-driven AND logic operations. *Chem. Commun.* 2048–2049 (2004).
72. Qu, D.-H., Wang, Q.-C. & Tian, H. A half adder based on a photochemically driven [2]rotaxane. *Angew. Chem. Int. Edn* **44,** 5296–5299 (2005).
73. Andréasson, J. *et al.* Molecule-based photonically switched half-adder. *J. Am. Chem. Soc.* **126,** 15926–15927 (2004).
74. Lederman, H., Macdonald, J., Stefanovic, D. & Stojanovic, M. N. Deoxyribozyme-based three-input logic gates and construction of a molecular full adder. *Biochemistry* **45,** 1194–1199 (2006).
75. Guo, X., Zhang, D., Zhang, G. & Zhu, D. Monomolecular logic: "half-adder" based on multistate/multifunctional photochromic spiropyrans. *J. Phys. Chem. B* **108,** 11942–11945 (2004).
76. Stojanovic, M. N. & Stefanovic, D. A deoxyribozyme-based molecular automaton. *Nature Biotechnol.* **21,** 1069–1074 (2003).
77. Wild, U. P., Bernet, S., Kohler, B. & Renn, A. From supramolecular photochemistry to the molecular computer. *Pure Appl. Chem.* **64,** 1335–1342 (1992).
78. Margulies, D., Melman, G. & Shanzer, A. A molecular full-adder and full-subtractor, an additional step toward a moleculator. *J. Am. Chem. Soc.* **128,** 4865–4871 (2006).
79. Tour, J. M. Molecular electronics. Synthesis and testing of components. *Acc. Chem. Res.* **33,** 791–804 (2000).
80. Williams, R. S., quoted in Ball, P. Chemistry meets computing. *Nature* **406,** 118–120 (2000).
81. Cox, J. C., Cohen, D. S. & Ellington, A. D. The complexities of DNA computation. *Trends Biotechnol.* **17,** 151–154 (1999).
82. Lam, K. S. *et al.* A new type of synthetic peptide library for identifying ligand-binding activity. *Nature* **354,** 82–84 (1991).
83. Furka, Á., Sebestyén, F., Asgedom, M. & Dibó, G. General-method for rapid synthesis of multicomponent peptide mixtures. *Int. J. Pept. Prot. Res.* **37,** 487–493 (1991).
84. Shepard, S. *RFID: Radio Frequency Identification* (McGraw-Hill, New York, 2005).
85. Nicolaou, K. C., Xiao, X.-Y., Parandoosh, Z., Senyei, A. & Nova, M. P. Radiofrequency encoded combinatorial chemistry. *Angew. Chem. Int. Edn Engl.* **34,** 2289–2291 (1995).
86. Moran, E. J. *et al.* Radio frequency tag encoded combinatorial library method for discovery of tripeptide-substituted cinnamic acid inhibitors of the protein tyrosine phosphatase PTP1B. *J. Am. Chem. Soc.* **117,** 10787–10788 (1995).
87. Walt, D. R. Bead-based fiber-optic arrays. *Science* **287,** 451–452 (2000).
88. Battersby, B. J., Lawrie, G. A., Johnston, A. P. R. & Trau, M. Optical barcoding of colloidal suspensions: applications in genomics, proteomics and drug discovery. *Chem. Commun.* 1435–1441 (2002).
89. de Silva, A. P., James, M. R., McKinney, B. O. F., Pears, P. A. & Weir, S. M. Molecular computational elements encode large populations of small objects. *Nature Mater.* **5,** 787–790 (2006).
90. Hayes, B. Third base. *Am. Scientist* **89,** 490–494 (2001).
91. Callan, J. F., de Silva, A. P., Ferguson, J., Huxley, A. J. M. & O'Brien, A. M. Fluorescent photoionic devices with two receptors and two switching mechanisms: applications to pH sensors and implications for metal ion detection. *Tetrahedron* **60,** 11125–11131 (2004).
92. Magri, D. C., Brown, G. J., McClean, G. D. & de Silva, A. P. Communicating chemical congregation: a molecular AND logic gate with three chemical inputs as a "lab-on-a-molecule" prototype. *J. Am. Chem. Soc.* **128,** 4950–4951 (2006).
93. Gunnlaugsson, T., Lee, T. C. & Parkesh, R. A highly selective and sensitive fluorescent PET (photoinduced electron transfer) chemosensor for Zn(II). *Org. Biomol. Chem.* **1,** 3265–3267 (2003).
94. Lankshear, M. D., Cowley, A. R. & Beer, P. D. Cooperative AND receptor for ion-pairs. *Chem. Commun.* 612–614 (2006).
95. Koskela, S. J. M., Fyles, T. M. & James, T. D. A ditopic fluorescent sensor for potassium fluoride. *Chem. Commun.* 945–947 (2005).
96. Saghatelian, A., Völcker, N. H., Guckian, K. M., Lin, V. S.-Y. & Ghadiri, M. R. DNA-based photonic logic gates: AND, NAND, and INHIBIT. *J. Am. Chem. Soc.* **125,** 346–347 (2003).
97. Gust, D., Moore, T. A. & Moore, A. L. Molecular switches controlled by light. *Chem. Commun.* 1169–1178 (2006).
98. Pina, F. *et al.* Multistate/multifunctional molecular-level systems: light and pH switching between the various forms of a synthetic flavylium salt. *Chem. Eur. J.* **4,** 1184–1191 (1998).
99. Uchiyama, S., Kawai, N., de Silva, A. P. & Iwai, K. Fluorescent polymeric AND logic gate with temperature and pH as inputs. *J. Am. Chem. Soc.* **126,** 3032–3033 (2004).
100. Leigh, D. A. *et al.* Patterning through controlled submolecular motion: rotaxane-based switches and logic gates that function in solution and polymer films. *Angew. Chem. Int. Edn* **44,** 3062–3067 (2005).
101. Penchovsky, R. & Breaker, R. R. Computational design and experimental validation of oligonucleotide-sensing allosteric ribozymes. *Nature Biotechnol.* **23,** 1424–1433 (2005).
102. Remacle, F., Speiser, S. & Levine, R. D. Intermolecular and intramolecular logic gates. *J. Phys. Chem. B* **105,** 5589–5591 (2001).
103. Remacle, F., Weinkauf, R. & Levine, R. D. Molecule-based photonically switched half and full adder. *J. Phys. Chem. A* **110,** 177–184 (2006).

Acknowledgements

We are grateful for support from Avecia, The Daiwa Anglo-Japanese Foundation, DEL, EPSRC, Invest NI (RTD COE 40), Japan Society for the Promotion of Science and Procter and Gamble. This paper is dedicated to Fraser Stoddart.

Harnessing biological motors to engineer systems for nanoscale transport and assembly

Living systems use biological nanomotors to build life's essential molecules—such as DNA and proteins—as well as to transport cargo inside cells with both spatial and temporal precision. Each motor is highly specialized and carries out a distinct function within the cell. Some have even evolved sophisticated mechanisms to ensure quality control during nanomanufacturing processes, whether to correct errors in biosynthesis or to detect and permit the repair of damaged transport highways. In general, these nanomotors consume chemical energy in order to undergo a series of shape changes that let them interact sequentially with other molecules. Here we review some of the many tasks that biomotors perform and analyse their underlying design principles from an engineering perspective. We also discuss experiments and strategies to integrate biomotors into synthetic environments for applications such as sensing, transport and assembly.

ANITA GOEL[1,2] AND VIOLA VOGEL[3]

[1]Nanobiosym Labs, 200 Boston Avenue, Suite 4700, Medford, Massachusetts 02155 USA; [2]Department of Physics, Harvard University, Massachusetts 02138, USA; [3]Department of Materials, ETH Zurich, Wolfgang Pauli Strasse 10. HCI F443, 8093 Zurich, Switzerland

e-mail: agoel@nanobiosym.com; viola.vogel@mat.ethz.ch

By considering how the biological machinery of our cells carries out many different functions with a high level of specificity, we can identify a number of engineering principles that can be used to harness these sophisticated molecular machines for applications outside their usual environments. Here we focus on two broad classes of nanomotors that burn chemical energy to move along linear tracks: assembly nanomotors and transport nanomotors.

SEQUENTIAL ASSEMBLY AND POLYMERIZATION

The molecular machinery found in our cells is responsible for the sequential assembly of complex biopolymers from their component building blocks (monomers): polymerases make DNA and RNA from nucleic acids, and ribosomes construct proteins from amino acids. These assembly nanomotors operate in conjunction with a master DNA or RNA template that defines the order in which individual building blocks must be incorporated into a new biopolymer. In addition to recognizing and binding the correct substrates (from a pool of many different ones), the motors must also catalyse the chemical reaction that joins them into a growing polymer chain. Moreover, both types of motors have evolved highly sophisticated mechanisms so that they are able not only to discriminate the correct monomers from the wrong ones, but also to detect and repair mistakes as they occur[1].

Molecular assembly machines or nanomotors (Fig. 1a) must effectively discriminate between substrate monomers that are structurally very similar. Polymerases must be able to distinguish between different nucleosides, and ribosomes need to recognize particular transfer-RNAs (t-RNAs) that carry a specific amino acid. These well-engineered biological nanomotors achieve this by pairing complementary Watson–Crick base pairs and comparing the geometrical fit of the monomers to their respective polymeric templates. This molecular discrimination makes use of the differential binding strengths of correctly matched and mismatched substrates, which is determined by the complementarity of the base-pairing between them.

Figure 1b illustrates the assembly process used by the DNA polymerase nanomotor. A template of single-stranded DNA binds to the nanomotor with angstrom-level precision, forming an open complex. The open complex can 'sample' the free nucleosides available. Binding of the correct nucleoside induces a conformational change in the nanomotor which then allows the new nucleoside to be added to the growing DNA strand[1]. The tight-fitting complementarity of shapes between the polymerase binding site and the properly paired base pair guarantees a 'geometric selection' for the correct nucleotide[2]. A similar mechanism is seen in *Escherichia coli* RNA polymerase, where the binding of an incorrect monomer inhibits the conformational change in the motor from an 'open' (inactive) to a 'closed' (active) conformation[3].

Ribosome motors carry out tasks much more complex than polymerases. Instead of the four nucleotide building blocks used by polymerases to assemble DNA or RNA, ribosomes must recognize and selectively arrange 20 amino acids to synthesize a protein. This fact alone increases the chance of errors. Nevertheless, ribosomes obviously work (and do so along the same principles of geometric fit

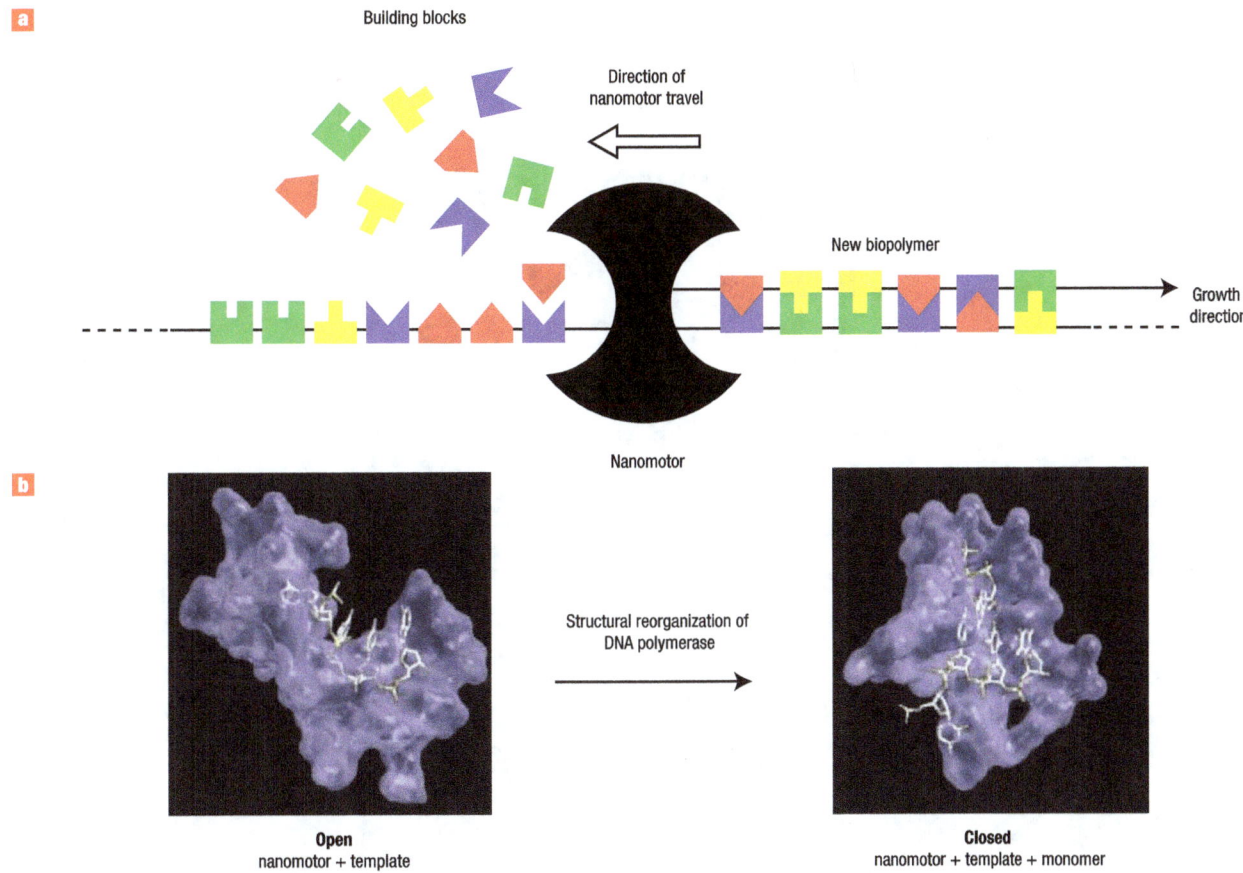

a, Building blocks

Direction of nanomotor travel

New biopolymer

Growth direction

Nanomotor

b

Structural reorganization of DNA polymerase

Open
nanomotor + template

Closed
nanomotor + template + monomer

Figure 1 Molecular discrimination during sequential assembly. **a**, The polymerase nanomotor discriminates between four different building blocks as it assembles a DNA or RNA strand complementary to its template sequence. Molecular discrimination between substrate monomers that are structurally very similar is achieved by comparing the geometrical fit of the monomers to their respective polymeric templates. **b**, The T7 DNA polymerase motor undergoes an internal structural transition from an open state (when the active site samples different nucleotides) to a closed state (when the correct nucleotide is incorporated into the nascent DNA strand). Nucleotides are added to the nascent strand one at a time. This structural transition is the rate-limiting step in the replication cycle and is thought to be dependent on the mechanical tension in the template strand[2,9,107,116,121,127,128,131]. Figure adapted from ref. 127. Copyright (2001) PNAS.

and conformational change as do polymerases) and are able to build amino acid polymers that are subsequently folded into functional proteins. But ribosomal motors can be tricked, much more easily than DNA motors, into building the 'incorrect' sequences when supplied with synthetic amino acids that resemble real ones[4].

Engineering principle no. 1: Nanomotors used in the sequential assembly of biopolymers can discriminate efficiently between similar building blocks.

The structure of molecular machines can be visualized with angstrom-level resolution using X-ray crystallography, and the sequential assembly processes they drive can be probed in real time using single-molecule techniques[5–9]. By elucidating nanomotor kinetics under load, such nanoscale techniques provide detailed insights into the single-molecule dynamics of nanomotor-driven assembly processes. Techniques such as optical and magnetic tweezers, for example, have further elucidated the polymer properties of DNA[7,10–12] and the force-dependent kinetics of molecular motors[13–18]. Single-molecule fluorescence methods such as fluorescence energy transfer, in conjunction with such biomechanical tools, are illuminating the internal conformational dynamics of these nanomotors[19–21].

As the underlying design principles of assembly nanomotors are revealed, it will become increasingly possible to use these biomachines for *ex vivo* tasks. Sequencing and PCR are two such techniques that already harness polymerase nanomotors for the *ex vivo* replication of nucleic acids. The polymerase chain reaction, or PCR, is a landmark, Nobel prize-winning technique[22] invented in the 1980s that harnessed polymerase nanomotors to amplify a very small starting sample of DNA to billions of molecules. Likewise, there are many conceivable future applications that either use assembly nanomotors *ex vivo* or mimic some of their design principles. Efforts are already under way to control these nanomotors better and thus to improve such *ex vivo* sequential assembly processes for industrial use (see, for example, the websites www.cambrios.com; www.helicosbio.com; www.nanobiosym.com; www.pacificbiosciences.com).

In contrast, current *ex vivo* methods to synthesize block copolymers rely primarily on random collisions, resulting in a wide range of length distributions and much less control over the final sequence[23]. Sequential assembly without the use of nanomotors remains limited to the synthesis of comparatively short peptides, oligonucleotides and oligosaccharides[24–26]. Common synthesizers still lack both the precision of monomer selection and the inbuilt proofreading machinery for monomer repair that nanomotors have.

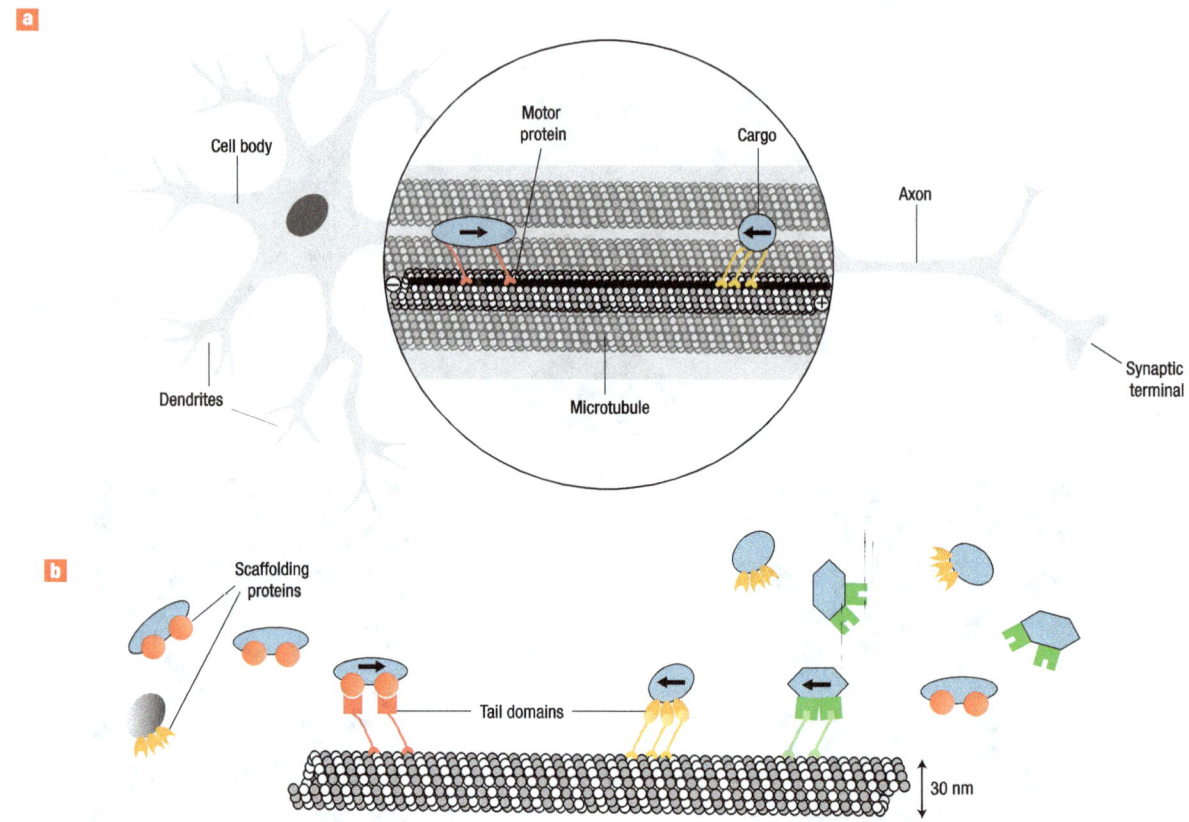

Figure 2 Motor-specific cargo transport in neurons. **a**, The axon of neurons consists of a bundle of highly aligned microtubules along which cargo is trafficked from the cell body to the synapse and vice versa. Most members of the large kinesin family (red) transport cargo towards the periphery, while other motors, including dyneins (yellow), transport cargo in the opposite direction. Motors preferentially move along a protofilament rather then side-stepping (one randomly selected protofilament is shown in dark grey). Protofilaments are assembled from the dimeric protein tubulin (white and grey spheres) which gives microtubules their structural polarity. The protofilaments then form the hollow microtubule rod. When encountering each other on the same protofilament, the much more tightly bound kinesin has the 'right of way', perhaps even forcing the dynein to step sidewise to a neighbouring protofilament[52–55]. **b**, Each member of a motor family selects its own cargo (blue shapes) through specific binding by scaffolding proteins (coloured symbols) or directly by the cargo's tail domains.

Building such copolymers with polymerase nanomotors *ex vivo* would yield much more homogeneous products of the correct sequence and precise length. Natural (for example, nanomotor-enabled) designs could inspire new technologies to synthesize custom biopolymers precisely from a given blueprint.

Ribosome motors have likewise been harnessed *ex vivo* to drive the assembly of new bio-inorganic heterostructures[27] and peptide nanowires[28,29] with gold-modified amino acids inserted into a polypeptide chain. These ribosomes are forced to use inorganically modified t-RNAs to sequentially assemble a hybrid protein containing gold nanoparticles wherever the amino acid cysteine was specified by the messenger RNA template. Such hybrid gold-containing proteins can then attach themselves selectively to materials used in electronics, such as gallium arsenide[28]. This application illustrates how biomotors could be harnessed to synthesize and assemble even non-biological constructs such as nanoelectronic components (see www.cambrios.com).

Assembly nanomotors achieve such high precision in sequential assembly by making use of three key features: (i) geometric shape-fitting selection of their building blocks (for example, nucleotides); (ii) motion along a polymeric template coupled to consumption of an energy source (for example, hydrolysis of ATP molecules); and (iii) intricate proofreading machinery to correct errors as they occur. Furthermore, nanomotor-driven assembly processes allow much more stable, precise and complex nanostructures to be engineered than can be achieved by thermally driven self-assembly techniques alone[30–32].

We should also ask whether some of these principles, which work so well at the nanoscale, could be realized at the micrometre-scale as well. Whitesides and co-workers, for example, have used simple molecular self-assembly strategies, driven by the interplay of hydrophobic and hydrophilic interactions, to assemble microfabricated objects at the mesoscale[33,34]. Perhaps the design principles used by nanomotors to improve precision and correct errors could also be harnessed to engineer future *ex vivo* systems at the nanoscale as well as on other length scales. Learning how to engineer systems that mimic the precision and control of nanomotor-driven assembly processes may ultimately lead to efficient fabrication of complex nanoscopic and mesoscopic structures.

CARGO TRANSPORT

Cells routinely use another set of nanomotors (that is, transport nanomotors) to recognize, sort, shuttle and deliver intracellular cargo along filamentous freeways to well-defined destinations, allowing molecules and organelles to become highly organized (see reviews[35–44]). This is essential for many life processes. Motor

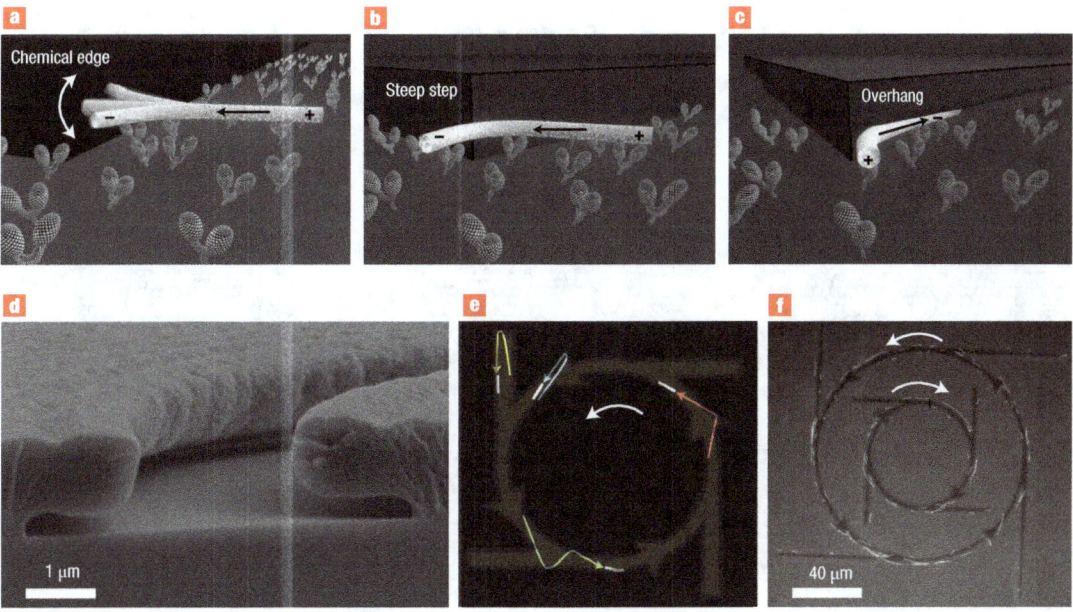

Figure 3 Track designs to guide nanomotor-driven filaments *ex vivo.* A variety of track designs have been used. **a,** A chemical edge (adhesive stripes coated with kinesin surrounded by non-adhesive areas). The filament crosses the chemical edge and ultimately falls off as it does not find kinesins on the non-adhesive areas[61]. **b,** Steep channel walls keep the microtubule on the desired path as they are forced to bend[61,65]. **c,** Overhanging walls have been shown to have the highest guidance efficiency[54]. **d,** Electron micrograph of a microfabricated open channel with overhanging walls[54]. **e,** Breaking the symmetry of micropatterns can promote directional sorting of filament movement[63,65,69,138]. The trajectories of four microtubules are shown: movement into reflector arms causes the tubule to turn around (yellow), an arrow-shaped direction rectifier allows those travelling in the desired direction to continue (red) and forces others to turn around (blue). At intersections tubules preferentially continue straight on (green). **f,** The complex microfabricated circuit analysed in **e** with open channels and overhanging walls demonstrating unidirectional movement of microtubules.

proteins transport cargo along cytoskeletal filaments to precise targets, concentrating molecules in desired locations. In intracellular transport, myosin motors are guided by actin filaments, whereas dynein and kinesin motors move along rod-like microtubules. Figure 2a illustrates how conventional kinesins transport molecular cargo along nerve axons towards the periphery, efficiently transporting material from the cell body to the synaptic region[45]. Dyneins, in contrast, move cargo in the opposite direction, so that there is active communication and recycling between both ends (see reviews[42,46]). In fact, the blockage of such bidirectional cargo transport along nerve axons can give rise to substantial neural disorders[47–50].

The long-range guidance of cargo is made possible by motors pulling their cargo along filamentous rods. Microtubules, for example, are polymerized from the dimeric tubulin into protofilaments that assemble into rigid rods around 30 nm in diameter[36]. These polymeric rods are inherently unstable: they polymerize at one end (plus) while depolymerizing from the other (minus) end, giving rise to a structural polarity. The biological advantage of using transient tracks is that they can be rapidly reconfigured on demand and in response to changing cellular needs or to various external stimuli. Highly efficient unidirectional cargo transport is realized in cells by bundling microtubules into transport highways where all microtubules are oriented in the same direction. Excessively tight bundling of microtubules, however, can greatly impair the efficiency of cargo transport, by blocking the access of motors and cargo to the microtubules in the bundle interior. Instead, microtubule-associated proteins are thought to act as repulsive polymer-brushes, thereby regulating the proximity and interactions between neighbouring microtubules[51].

Traffic control is an issue when using the filaments as tracks on which kinesin and dynein motors move in opposite directions. Although different cargoes can be selectively recognized by different members of the motor protein families and shuttled to different destinations, what happens if motors moving in opposite directions encounter each other on the same protofilament (Fig. 2b)? If two of these motors happen to run into each other, kinesin seems to have the 'right of way'. As kinesin binds the microtubule much more strongly, it is thought to force dynein to step sideways to a neighbouring protofilament[52]. Dynein shows greater lateral movement between protofilaments than kinesin[52–54] as there is a strong diffusional component to its steps[55]. When a microtubule becomes overcrowded with only kinesins, the runs of individual kinesin motors are minimally affected. But when a microtubule becomes overloaded with a mutant kinesin that is unable to step efficiently, the average speed of wild-type kinesin is reduced, whereas its processivity is hardly changed. This suggests that kinesin remains tightly bound to the microtubule when encountering an obstacle and waits until the obstacle unbinds and frees the binding site for kinesin's next step[56].

Engineering principle no. 2: Various track designs enable motors to pull their cargo along filamentous tracks, whereas others allow motors bound to micro- or nanofabricated tracks to propel the filaments which can then serve as carriers.

It is not a trivial task to engineer transport highways *ex vivo*, particularly in versatile geometries with intersections and complex shapes. Individual filaments typically allow only one-dimensional transport, as the motor-linked cargo drops off once the end of the filament is reached. Furthermore, conventional kinesin makes only a few hundred 8-nm-sized steps before dissociating from the microtubule[57,58], further limiting the use of such a system for *ex vivo* applications.

Instead of having the motors transport their cargo along filaments, motors have been immobilized on surfaces in an inverted geometry

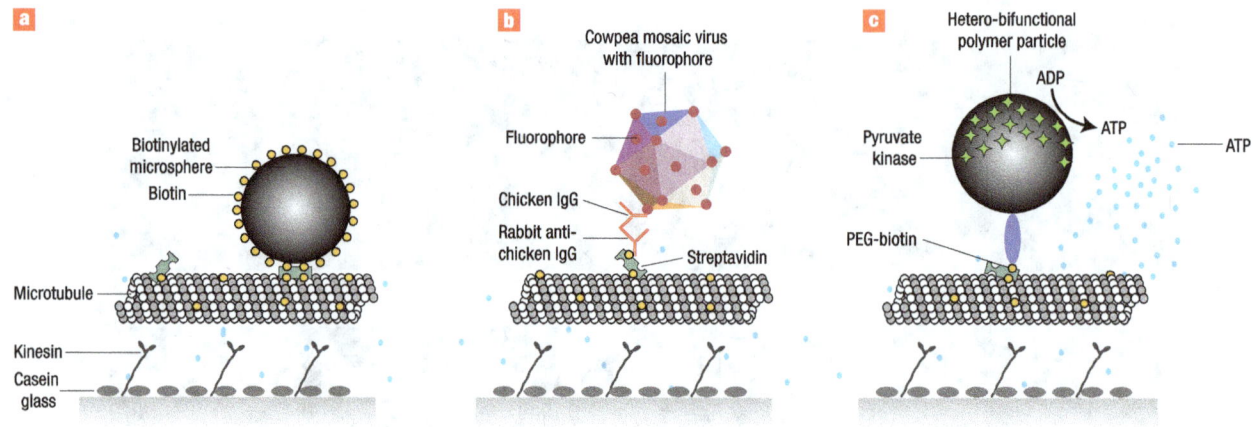

Figure 4 Selecting specific cargo by molecular recognition. A versatile toolbox exists by which synthetic and biological cargo can be coupled to microtubules. **a**, Biotinylated objects are coupled via avidin or streptavidin to biotinylated microtubules. **b**, Biological molecules, viruses[79,81] or cells can be coupled by antibody recognition. **c**, Backpacks of chemically or biologically active reagents can be shuttled around, including bioprobes[80] or tiny ATP factories[93] as shown here.

that enables the filaments to be collectively propelled forward[45]. The head domains of the kinesin and myosin motors can rotate and swivel with respect to their feet domains, which are typically bound in random orientations to the surface. These motor heads detect the structural anisotropy of the microtubules and coherently work together to propel a filament forward[59,60].

Various examples of such inverted designs for motor tracks have been engineered to guide filaments efficiently. Some of these are illustrated in Fig. 3. Inverted motility assays can be created, for example, by laying down tracks of motor proteins in microscopic stripes of chemical adhesive on an otherwise flat, protein-repellent surface, surrounded by non-adhesive surface areas. Such chemical patterns (Fig. 3a) have been explored to guide actin filaments or microtubules. The loss rate of guiding filaments increases exponentially with the angle at which they approach an adhesive/non-adhesive contact line[61]. The passage of the contact line by filaments at non-grazing angles, followed by their drop off, can be prevented by using much narrower lanes whose size is of the order of the diameter of the moving object. Such nanoscale kinesin tracks provide good guidance and have been fabricated by nanotemplating[62].

Alternatively, considerably improved guidance has been accomplished by topographic surface features (Fig. 3b). Microtubules hitting a wall are forced to bend along this obstacle and will continue to move along the wall[63–66]. The rigidity of the polymeric filaments used as shuttles thus greatly affects how tracks should be designed for optimal guidance. Whereas microtubules with a persistence length of a few millimetres can be effectively guided in channels a few micrometres wide as they are too stiff to turn around[61], the much more flexible actin filaments require channel widths in the submicrometre range[67,68]. Finally, the best long-distance guidance of microtubules has been obtained so far with overhanging walls[64,69] (Fig. 3c). The concept of topographic guidance in fact works so well that swarms of kinesin-driven microtubules have been used as independently moving probes to image unknown surface topographies. After averaging all their trajectories in the focal plane for an extended time period, the image greyscale is determined by the probability of a surface pixel being visited by a microtubule in a given time frame[70].

But how can tracks be engineered to produce *uni*directional cargo transport? All the motor-propelled filaments must move in the same direction to achieve effective long-distance transport. When polar filaments land from solution onto a motor-covered surface, however, their orientations and initial directions of movement are often

randomly distributed. Initially, various physical means, such as flow fields[71], have been introduced to promote their alignment. Strong flows eventually either force gliding microtubules to move along with the flow, or force microtubules, if either their plus or minus end is immobilized on a surface[72], to rotate around the anchoring point and along with the flow. The most universal way to control the local direction in which the filamentous shuttles are guided is to make use of asymmetric channel features. Figure 3d–f illustrates how filaments can be actively sorted according to their direction of motion by breaking the symmetry of the engineered tracks. This 'local directional sorting' has been demonstrated on surfaces patterned with open channel geometries, where asymmetric intersections are followed by dead-ended channels (that is, reflector arms), or where channels are broadened into arrow heads. Both of these topographical features not only selectively pass filaments moving in the desired direction, but can also force filaments moving in the opposite direction to turn around[65,69,73,74]. Once directional sorting has been accomplished, electric fields have been used to steer the movement of individual microtubules as they pass through engineered intersections[75,76].

In addition to using isolated nanomotors, hybrid biodevices and systems that harness self-propelling microbes could be used to drive transport processes along engineered tracks. Flagellated bacteria, for example, have been used to generate both translational and rotational motion of microscopic objects[77]. These bacteria can be attached head-on to solid surfaces, either via polystyrene beads or polydimethylsiloxane, thereby enabling the cell bodies to form a densely packed monolayer, while their flagella continue to rotate freely. In fact, a microrotary motor, fuelled by glucose and comprising a 20-μm-diameter silicon dioxide rotor, can be driven along a silicon track by the gliding bacterium *Mycoplasma*[78]. Depending on the specific application and the length scale on which transport needs to be achieved, integrating bacteria into such biohybrid devices (that work under physiological conditions) might ultimately prove more robust than relying solely on individual nanomotors.

CARGO SELECTION

To maintain intracellular contents in an inhomogeneous distribution far from equilibrium, the intracellular transport system must deliver molecular cargo and organelles on demand to precise destinations. This tight spatiotemporal control of molecular deliveries is critical for adequate cell function and survival. Molecular cargo or organelles

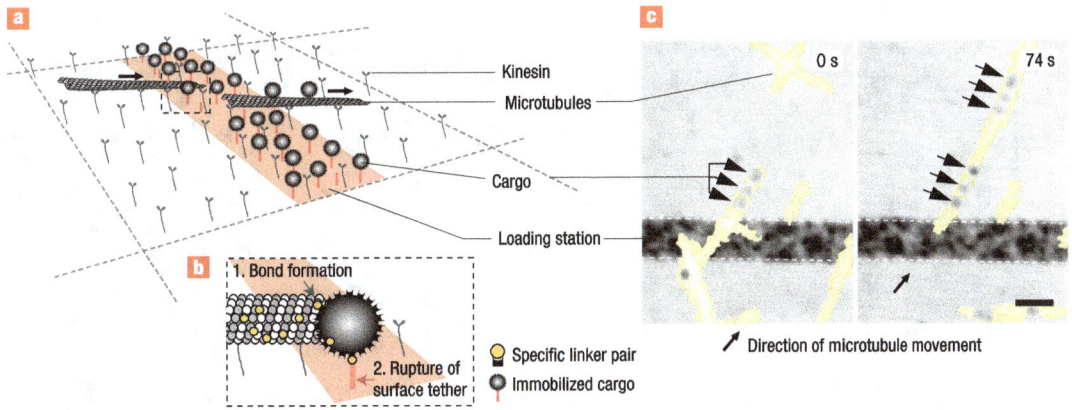

Figure 5 Cargo loading stations[93]. **a**, Stripes of immobilized cargo are fabricated by binding thiolated oligonucleotides to micropatterned lines of gold. Hybridization with complementary strands exposing antibodies at their terminal ends allows them to immobilize a versatile range of cargos that carry antibodies on their surfaces. **b**, The challenge is to tune the bond strength and valency to prevent thermal activation during cargo storage on the loading station. On collision with the shuttle (microtubule), the cargo must rapidly break off the bond it has formed with the station[88]. Fortunately, however, tensile mechanical force acting on a non-covalent bond shortens its lifetime. **c**, **d**, These concepts are used in the design of the loading stations shown here, where a microtubule moves through a stripe of immobilized gold cargo and picks up a few beads.

are typically barcoded so that they can be recognized by their specific motor protein (Fig. 4). Within cells, motors recognize cargo either from the cargo's tail domains directly, or via scaffolding proteins that link cargo to their tail domain[43].

Engineering principle no. 3: Engineered molecular recognition sites enable cargo to be selectively bonded to moving shuttles.

Although most cargo shuttled around by motors can be barcoded using the existing repertoire of biological scaffolding proteins, synthetic approaches are needed for all those *ex vivo* applications where the cargo has to be specifically linked to moving filaments. The loading and transport of biomedically relevant or engineered cargo has already been demonstrated (Fig. 4)[79–83]. Typical approaches are to tag the cargo with antibodies or to biotinylate microtubules and coat the cargo with avidin or streptavidin (Fig. 4) (for reviews, see refs 74,79), as done for polymeric and magnetic beads[84,85] (Fig. 4a), gold nanoparticles[86–88], DNA[87,89,90] and viruses[79,81] (Fig. 4b), and finally mobile bioprobes and sensors[80,81,91] (Fig.4c). However, if too much cargo is loaded onto the moving filaments and access of the propelling motors is even partially blocked, the transport velocity can be significantly impaired[92]. Finally, the binding of cargo to a moving shuttle can be used to regulate its performance. In fact, microtubules have recently been furnished with a backpack that self-supplies the energy source ATP. Cargo particles bearing pyruvate kinase have been tethered to the microtubules to provide a local ATP source[93] (Fig. 4c). The coupling of multiple motors to cargo or other scaffold materials can affect the motor performance. If single-headed instead of double-headed kinesins are used, cooperative interactions between the monomeric motors attached to protein scaffolds increase hydrolysis activity and microtubule gliding velocity[59].

At the next level of complexity, successful cargo tagging, sorting and delivery will depend on the engineering of integrated networks of cargo loading, cargo transport and cargo delivery zones. Although the construction of integrated transport circuits is still in its infancy, microfabricated loading stations have been built[88] (Fig. 5). The challenge here is to immobilize cargo on loading stations such that it is not easily detached by thermal motion, yet to allow for rapid cargo transfer to passing microtubules. By properly tuning bond strength and multivalency, and most importantly by taking advantage of the fact

that mechanical strain weakens bonds, cargo can be efficiently stored on micropatches and transferred after colliding with a microtubule[88]. Considerable fine-tuning of bond strength can be accomplished by using DNA oligomers hybridized such that the bonds are either broken by force all at once (a strong bond) or in sequence (a weak bond)[94].

As discussed above, filaments are most commonly used to shuttle molecular cargo in most emerging devices that harness linear motors for active transport. Alternatively, if the filamentous tracks could be engineered in versatile geometries, the motors themselves could be used to drag cargo coupled to the molecular recognition sites of their tail domains as in the native systems. We could thus make use of the full biological toolbox of already known or engineered scaffolding proteins that link specific motors to their respective cargoes[40,43]. So far, assemblies of microtubules organized into complex, three-dimensional patterns such as asters, vortices and networks of interconnected poles[95,96] have been successfully created in solution, and mesoscopic needles and rotating spools of microtubule bundles held together by non-covalent interactions have been engineered on surfaces[31]. All of these mesoscopic structures are uniquely related to active motor-driven motion and would not have formed purely by self-assembly without access to an energy source.

To increase the complexity of microtubule track networks, densely packed arrays of microtubules have been grown in confined spaces, consisting of open microfabricated channels with user-defined geometrical patterns[97]. The key to achieving directed transport, however, is for all microtubules within each bundle or array to be oriented in the same direction. This has been accomplished by making use of directed motility in combination with sequential assembly procedures (Fig. 6). First, microtubule seedlings have been oriented in open microfabricated and kinesin-coated channels that contain reflector arms. Once oriented by self-propelled motion, the seedlings were polymerized into mature microtubules that were confined to grow in the open channels until the channels were filled with dense networks of microtubules all oriented in the same direction[97]. Single kinesins take only a few hundred steps before they fall off, but the walking distance can be greatly increased if the cargo is pulled by more than one motor[98]. Such approaches to fabricating networks of microtubule bundles could be further expanded to engineer future devices that use either the full toolbox of native scaffolding proteins or new scaffolding proteins that target both biological and synthetic cargo.

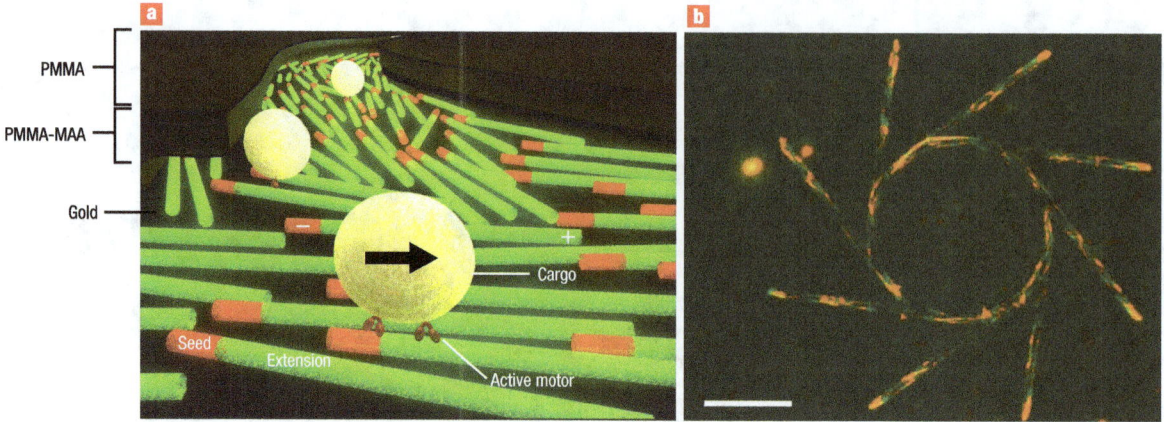

Figure 6 Filament tracks made from engineered bundles of microtubules[97]. Active transport is used to produce bundles of microtubules and confine them to user-defined geometries. **a**, Sequential assembly procedure: first, microtubule seedlings labelled in red are allowed to orient themselves in open kinesin-coated microfabricated channels that contained reflector arms. Second, and after mild fixation, the oriented seedlings are polymerized into mature microtubules through the addition of tubulin into the solution (labelled green) which preferentially binds to the plus-end (polymerizing end) of the microtubules. **b**, Fluorescence image of microtubules that have been grown in the confined space provided by the open channels until the channels were filled with dense networks of microtubules all oriented in the same direction[97]. Scale bar, 40 μm.

Nanoengineers would not be the first to harness biological motors to transport their cargo. Various pathogens are known to hijack microtubule or actin-based transport systems within host cells (reviewed in ref. 99). *Listeria monocytogenes*, for example, propels itself through the host cell cytoplasm by means of a fast-polymerizing actin filament tail[100]. Likewise, the vaccinia virus, a close relative of smallpox, uses actin polymerization to enhance its cell-to-cell spreading[101], and the alpha herpes virus hijacks kinesins to achieve long-distance transport along the microtubules of neuronal axons[102]. Signalling molecules and pathogens that cannot alter cell function and behaviour by simply passing the outer cell membrane can thus hijack the cytoskeletal highways to get transported from the cell periphery to the nucleus.

Engineering principle no. 4: By taking advantage of the existing cytoskeleton, tailored drugs and gene carriers can be actively transported to the cell nucleus.

Indeed, many viruses[37,103,104] as well as non-viral therapeutic gene carriers, such as polyethylenimine/DNA or other polymer-based gene transfer systems (that is, polyplexes)[105,106] take advantage of nanomotor-driven transport along microtubule filaments to accelerate their way through the cytoplasm towards the nucleus. Nanomotor-driven transport to the nucleus leads to a much more efficient nuclear localization than could ever be achieved by slow random diffusion through the viscous cytoplasm. Active gene carrier transport can lead to more efficient perinuclear accumulation within minutes[37,105,106]. In contrast, non-viral gene carriers that depend solely on random diffusion through the cytoplasm move much more slowly and thus have considerably reduced transfection efficiencies. Understanding how to 'hijack' molecular and cellular transport systems, instead of letting a molecule become a target for endosomal degradation[37,91], will ultimately allow the design of more efficient drug and gene carrier systems.

QUALITY CONTROL

Nanomanufacturing processes, much like macroscopic assembly lines, urgently need procedures that offer precise control over the quality of the product, including the ability to recognize and repair defects. Living systems use numerous quality control procedures to detect and repair defects occurring during the synthesis and assembly of biological nanostructures. As yet, this has not been possible in synthetic nanosystems. Many cellular mechanisms for damage surveillance and error correction rely on nanomotors. Such damage control can occur at two different levels as follows.

Engineering principle no. 5: Certain motor proteins recognize assembly mistakes and repair them at the molecular level.

DNA replication represents one of the most complex sequential assembly processes in a cell. Here the genetic information stored in the four-base code must be copied with ultra-high precision. Errors generated during replication can have disastrous biological consequences. Figure 7 illustrates the built-in mechanism used by the polymerase (DNAp) motor to repair mistakes made during the process of DNA replication[107]. When the DNAp motor misincorporates a base while replicating the template DNA strand, it slows down and switches gears from the polymerase to the exonuclease cycle. Once in exonuclease mode, it will excise the mismatched base pair and then rapidly switch back to the polymerase cycle to resume forward replication. Similar error correction mechanisms, known as 'kinetic proofreading', are conjectured to occur in RNA polymerases and ribosomal machineries[1,13,108–113].

Engineering principle no. 6: Integrated systems of motors and signalling molecules are needed to recognize and repair damage at the supramolecular level.

Nerve cells have evolved a highly regulated axonal transport system that contains an integrated damage surveillance system[114]. The traffic regulation of motors moving in opposite directions on a microtubule typically occurs in special 'turnaround' zones at the base and tip of an axon[43], but a zone for switching the organelle's direction can also be created when axonal transport is blocked at the site of nerve injury[46] (see Fig. 2). When irreparable, such blockages are often signatures of neurodegenerative diseases. For example, amyloid precursor protein[47] or tau[115] can give rise to the accumulation of protein aggregates that inhibit anterograde axonal transport, a mechanism potentially implicated in Alzheimer's disease.

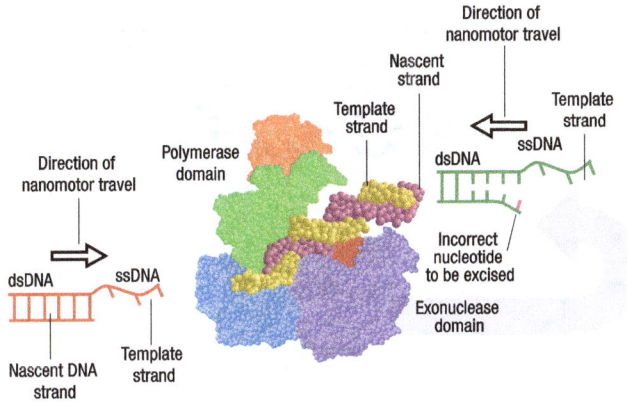

Figure 7 Quality control procedures for damage recognition and molecular repair. The DNA polymerase motor (DNAp) contains two active sites. It switches from polymerase (copying) to exonuclease (error correction) activity when it encounters a mismatched base. Mismatched bases are detected as they have weaker bonding interactions—the 'melting' temperature is lower—and this increases the chance of switching from the polymerase to the exonuclease active site[107]. In the exonuclease mode, the motor excises the incorrect base from the nascent DNA strand.

At present, there are no synthetic materials that can, in a self-regulated manner, recognize and repair defects at either the molecular or supramolecular level. Molecular recognition and repair is typically attributed to a tightly fitted stereochemical complementarity between binding partners. Nanoscale tools applied to the study of molecular recognition and repair are also elucidating the functional roles of the different structural conformations (and hence three-dimensional shapes) of the motors. For instance, the DNAp motor is in one particular conformation when it binds DNA in its copying (that is, polymerization) mode and in an entirely different conformation (that is, the exonuclease mode) when it binds DNA to proofread or excise a mistaken base from the replicated DNA strand[107]. In contrast, damage control at the supramolecular level (for example, during axonal transport) is achieved by the trafficking of signalling molecules. Deciphering the underlying engineering design principles of damage surveillance and error correction mechanisms in biological systems will inevitably allow better quality-control procedures to be integrated into nanoengineered systems.

EXTERNAL CONTROL

Engineering principle no. 7: As with macroscopic engines, external controls can regulate the performance of nanomotors on demand.

Learning how to control and manipulate the performance of nanomotors externally is another critical hurdle in harnessing nanomotors for *ex vivo* applications. By finding or engineering appropriate external knobs in the motor or its environment, its nanoscale movement can be tightly regulated, switched on and off, or otherwise manipulated on demand.

To achieve external control over the nanoscale movement of biological motors, it is important to identify the correct external parameters that can be used to control their dynamics. These external modulators of motor function ('handles') can be either naturally occurring or somehow artificially engineered into the motor to make it susceptible to a particular external control knob or regulator. Because the motion of nanomotors is typically driven by a series of conformational changes in the protein, mechanical load or strain on the motor molecule can also affect the dynamics of the motor. Nanomotors apply mechanical strain to their filaments or substrates

as they go through various internal conformational changes. This mechanical strain is intimately related to their dynamics along the substrate and hence their functional performance. Certain interstate transition rates can depend, for example[107], on the amount of intramolecular strain in the motor protein. Applying a mechanical load to a motor perturbs key mechanical transitions in the motor's kinetic pathway, and can thereby affect rates of nucleotide binding, ATP hydrolysis and product release. Single-molecule techniques are beginning to elucidate how mechanical strain on a motor protein might be used to regulate its biological functions (for example, nanoscale assembly or transport)[13,55,107,116–120].

The single-molecule dynamics of the DNA polymerase (DNAp) motor, as it converts single-stranded (ss) DNA to double-stranded (ds) DNA, has been probed, for example, through the differential elasticity of ssDNA and dsDNA (see Fig. 8). The T7 DNA polymerase motor replicates DNA at rates of more than 100 bases per second and this rate steadily decreases with mechanical tension greater than about 5 pN on the DNA template[9]. The motor can work against a maximum of about 34 pN of template tension[9]. The replication rates for the Klenow and Sequenase DNA polymerases also decrease when the ssDNA template tension exceeds 4 pN, and completely ceases at tensions greater than 20 pN (ref. 121). Likewise, single-molecule techniques have allowed direct observation of the RNA polymerase (RNAp) motor moving one base at a time[122], and occasionally pausing and even backtracking[123]. Although RNAp motors are typically five- to tenfold slower than DNAp motors, the effects of DNA template tension on their dynamics are still being investigated[6]. Similarly, ribosome motors, which translate messenger RNA (mRNA) into amino acids at roughly 10 codons per second, have been found to generate about 26.5 ± 1 pN of force[124]. The underlying design principles by which these nanomotors operate are being further elucidated by theoretical models[107,116,125–128] that describe nanomachines at a level commensurate with single-molecule data. Furthermore, these molecular assembly machines can be actively directed, driven and controlled by environmental signals[107].

Consequently, an external load or force applied to the substrate or to the motor itself can be used to slow down a motor's action or stall its movement. The stalling forces of kinesin and dynein are 6 and 1 pN, respectively[58,129]. For example, the binding of two kinesin domains to a microtubule track creates an internal strain in the motor that prevents ATP from binding to the leading motor head. In this way, the two motor domains remain out-of-phase for many mechanochemical cycles and thereby provide an efficient, adaptable mechanism for achieving highly processive movement[130]. Beyond stalling the movement of motors by a mechanical load, other types of perturbations can also influence the dynamics of molecular motors, including the stretching of substrate molecules like DNA[13]. Although this external control over nanomotors has been demonstrated in a few different contexts *ex vivo*, a rich detailed mechanistic understanding of how such external control knobs can modulate the dynamics of the molecular motor is emerging from recent work on the DNA polymerase motor[9,107,116,121,127,128,131].

Remote-controlling the local ATP concentration by the photo-activated release of caged ATP can allow a nanomotor-driven transport system to be accelerated or stopped on demand[84]. External control knobs or regulators can also be engineered into the motors. For instance, point mutations can be introduced into the gene encoding the motor protein, such that it is engineered to respond to light, temperature, pH or other stimuli[43,85]. Engineering light-sensitive switches into nanomotors enables the rate of ATPase[43,132] to be regulated, thereby providing an alternate handle for tuning the motor's speed, even while the ATP concentration is kept constant and high. When additional ATP-consuming enzymes are present in solution, the rate of ATP depletion regulates the distance the shuttles move after being activated by a light pulse and before again coming to a halt[84].

Future applications could require that instead of all the shuttles being moved at the same time, only those in precisely defined locations be activated, on demand. Some of the highly conserved residues within motors help to determine the motor's ATPase rate[43]. Introducing chemical switches near those locations might provide a handle for chemical manipulation of the motor's speed. In fact, this has already been realized for a rotary motor[132] as well as for a linear kinesin motor, where the insertion of a Ca^{2+}-dependent chemical switch makes the ATPase activity steeply dependent on Ca^{2+} concentrations[133]. In addition to caged ATP, caged peptides that block binding sites could be used to regulate the motility of such systems. Caged peptides derived from the kinesin C-terminus domain have already been used to achieve photo control of kinesin-microtubule motility[134]. Instead of modulating the rate of ATP hydrolysis, the access of microtubules to the motor's head domain can also be blocked in an environmentally controlled manner. In fact, temperature has already been shown to regulate the number of kinesins that are accessible while embedded in a surface-bound film of thermoresponsive polymers[135].

The nanomotor-driven assembly of DNA by the DNA polymerase motor provides an excellent example of how precision control over the nanomotor can be achieved by various external knobs in the motor's environment[107,116,127,128]. The DNAp motor moves along the DNA template by cycling through a given sequence of geometric shape changes. The sequence of shapes or internal states of the nanomachine can be denoted by nodes on a simple network[107,116,127,128]. As illustrated in Fig. 8, this approach elucidates how mechanical tension on a DNA molecule can precisely control (or 'tune') the nanoscale dynamics of the polymerase motor along the DNA track by coupling into key conformational changes of the motor[107].

Macroscopic knobs to precision-control the motor's movement along DNA tracks can be identified by probing how the motor's dynamics vary with each external control knob (varied one at a time). Efforts are currently under way to control even more precisely the movement of these nanomotors along DNA tracks by tightly controlling the parameters in the motor's environment (see www.nanobiosym.com). Concepts of fine-tuning and robustness could also be extended to describe the sensitivity of other nanomotors (modelled as simple biochemical networks) to various external control parameters[107]. Furthermore, such a network approach[107] provides experimentally testable predictions that could aid the design of future molecular-scale manufacturing methods that integrate nanomotor-driven assembly schemes. External control of these nanomotors will be critical in harnessing them for nanoscale manufacturing applications.

CONCLUDING REMARKS

We have reviewed several key engineering design principles that enable nanomotors moving along linear templates to perform a myriad of tasks. Equally complex biomimetic tasks have not yet been mastered *ex vivo*, either by harnessing biological motors or via synthetic analogues. Engineering insights into how such tasks are carried out by the biological nanosystems will inspire new technologies that harness nanomotor-driven processes to build new systems for nanoscale transport and assembly.

Sequential assembly and nanoscale transport, combined with features currently attributed only to biological materials, such as self-repair and healing, might one day become an integral part of future materials and bio-hybrid devices. In the near term, molecular biology techniques could be used to synthesize and assemble nanoelectronic components with more control (www.cambrios.com; see also ref. 29). Numerous proof-of-concept experiments using nanomotors integrated into synthetic microdevices have already been demonstrated (see reviews[74,136]). Among many others, these applications include stretching surface-bound molecules by moving microtubules[87,90]; probing the lifetime of a single receptor–ligand interaction via a cantilevered microtubule that acts as a piconewton

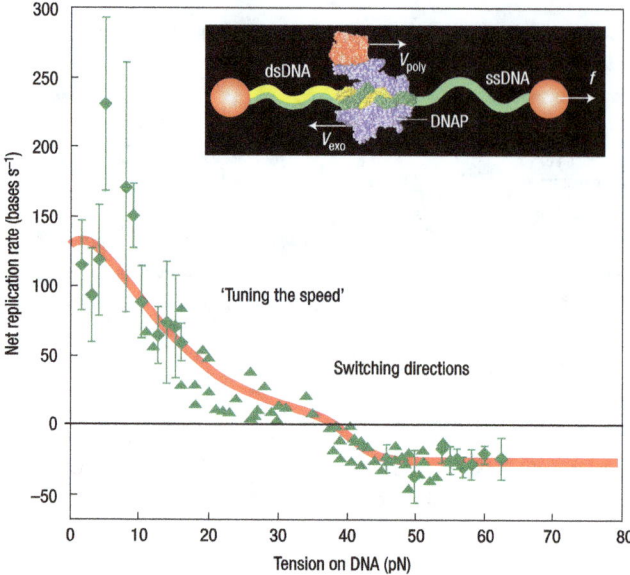

Figure 8 Precision control of nanomotors with external control 'knobs'. The net replication rate of a DNAp motor can be controlled by the mechanical tension on the DNA template strand. Single-molecule data for the motor's force-dependent velocity (two sets of data—diamonds and triangles—are shown, relating to constant force and constant extension measurements) can be described by a network model (red curve) as shown here. The change in net replication rate shows how external controls can change the dynamics of the nanomotor. This model illustrates how environmental control knobs can tune the dynamics of the nanomotor by altering the rate constants associated with its various internal transitions[106]. Tensions between 0 and 35 pN control the net replication rate, whereas tensions above 35 pN actually reverse the velocity of the nanomotor. Inset, experimental setup: a single DNA molecule is stretched between two plastic beads as the motor catalyses the conversion of single-stranded to double-stranded DNA. Figure adapted from ref. 106.

force sensor[85]; topographic surface imaging by self-propelled probes[70]; and cargo pick-up from loading stations[88] as illustrated in Fig. 5.

Although much progress is being made in the synthesis of artificial motors (see review[137]), it has been difficult, in practice, to synthesize artificial motors that come even close in performance to their natural counterparts (see review[39]). Harnessing biological motors to perform nanoscale manufacturing tasks might thus be the best near-term strategy. Although many individual nanoparts can be easily manufactured, the high-throughput assembly of these nanocomponents into complex structures is still non-trivial. At present, no *ex vivo* technology exists that can actively guide such nanoscale assembly processes. Despite advances in deciphering the underlying engineering design principles of nanomotors, many hurdles still impede harnessing them for *ex vivo* transport and sequential assembly in nanosystems. Although the use of biological nanomotors puts intrinsic constraints on the conditions under which they can be assembled and used in biohybrid devices, many of their sophisticated tasks are still poorly mimicked by synthetic analogues. Understanding the details of how these little nanomachines convert chemical energy into controlled movements will nevertheless inspire new approaches to engineer synthetic counterparts that might some day be used under harsher conditions, operate at more extreme temperatures, or simply have longer shelf lives.

Certain stages of the materials production process might one day be replaced by nanomotor-driven sequential self-assembly, allowing much more control at the molecular level. Biological motors are already being used to drive the efficient fabrication of complex nanoscopic and mesoscopic structures, such as nanowires[31] and supramolecular assemblies. Techniques for precision control of nanomotors that read DNA are also being used to engineer integrated systems for rapid DNA detection and analysis (www.nanobiosym.com). The specificity and control of assembly and transport shown by biological systems offers many opportunities to those interested in assembly of complex nanosystems. Most importantly, the intricate schemes of proofreading and damage repair—features that have not yet been realized in any manmade nanosystems—should provide inspiration for those interested in producing synthetic systems capable of similarly complex tasks.

doi:10.1038/nnano.2008.190

Published online: 27 July 2008.

References

1. Rodnina, M. V. & Wintermeyer, W. Fidelity of aminoacyl-tRNA selection on the ribosome: kinetic and structural mechanisms. *Annu. Rev. Biochem.* **70**, 415–435 (2001).
2. Kunkel, T. A. DNA replication fidelity. *J. Biol. Chem.* **279**, 16895–16898 (2004).
3. Erie, D. A., Hajiseyedjavadi, O., Young, M. C. & von Hippel, P. H. Multiple RNA polymerase conformations and GreA: control of the fidelity of transcription. *Science* **262**, 867 (1993).
4. Liu, D. R., Magliery, T. J., Pastrnak, M. & Schultz, P. G. Engineering a tRNA and aminoacyl-tRNA synthetase for the site-specific incorporation of unnatural amino acids into proteins in vivo. *Proc. Natl Acad. Sci.* **94**, 10092–10097 (1997).
5. Bustamante, C., Smith, S. B., Liphardt, J. & Smith, D. Single-molecule studies of DNA mechanics. *Curr. Opin. Struct. Biol.* **10**, 279–285 (2000).
6. Davenport, R. J., Wuite, G. J. L., Landick, R. & Bustamante, C. Single-molecule study of transcriptional pausing and arrest by *E. coli* RNA polymerase. *Science* **287**, 2497–2500 (2000).
7. Greulich, K. O. Single-Molecule Studies on DNA and RNA. *ChemPhysChem* **6**, 2459–2471 (2005).
8. Wang, M. D. *et al.* Force and velocity measured for single molecules of RNA polymerase. *Science* **282**, 902–907 (1998).
9. Wuite, G. J., Smith, S. B., Young, M., Keller, D. & Bustamante, C. Single-molecule studies of the effect of template tension on T7 DNA polymerase activity. *Nature* **404**, 103–106 (2000).
10. Smith, S. B., Cui, Y. & Bustamante, C. Overstretching B-DNA: The elastic response of individual double-stranded and single-stranded DNA molecules. *Science* **271**, 795 (1996).
11. Smith, S. B., Finzi, L. & Bustamante, C. Direct mechanical measurements of the elasticity of single DNA molecules by using magnetic beads. *Science* **258**, 1122 (1992).
12. Williams, M. C. & Rouzina, I. Force spectroscopy of single DNA and RNA molecules. *Curr. Opin. Struct. Biol* **12**, 330–336 (2002).
13. Bustamante, C., Bryant, Z. & Smith, S. B. Ten years of tension: single-molecule DNA mechanics. *Nature* **421**, 423–427 (2003).
14. Jeney, S., Stelzer, E. H., Grubmuller, H. & Florin, E. L. Mechanical properties of single motor molecules studied by three-dimensional thermal force probing in optical tweezers. *ChemPhysChem* **5**, 1150–1158 (2004).
15. Mehta, A. D. Single-molecule biomechanics with optical methods. *Science* **283**, 1689–1695 (1999).
16. Mogilner, A. & Oster, G. Polymer motors: pushing out the front and pulling up the back. *Curr. Biol.* **13**, 721–733 (2003).
17. Schnitzer, M. J., Visscher, K. & Block, S. M. Force production by single kinesin motors. *Nature Cell Biol.* **2**, 718–723. (2000).
18. Strick, T., Allemand, J. F., Croquette, V. & Bensimon, D. The manipulation of single biomolecules. *Phys. Today* **54**, 46–51 (October, 2001).
19. Ha, T. Single-molecule fluorescence methods for the study of nucleic acids. *Curr. Opin. Struct. Biol.* **11**, 287–292 (2001).
20. Kapanidis, A. N. *et al.* Initial transcription by RNA polymerase proceeds through a DNA-scrunching mechanism. *Science* **314**, 1144–1147 (2006).
21. Keller, R. A. *et al.* Single-molecule fluorescence analysis in solution. *Appl. Spectrosc.* **50**, 12A-32A (1996).
22. Mullis, K. B. *The Polymerase Chain Reaction.* Nobel Lecture (1993).
23. van Hest, J. C. M. & Tirrell, D. A. Protein-based materials, toward a new level of structural control. *Chem. Commun.* **19**, 1897–1904 (2001).
24. Fodor, S. P. *et al.* Multiplexed biochemical assays with biological chips. *Nature* **364**, 555–556 (1993).
25. Merrifield, R. B. Automated synthesis of peptides. *Science* **150**, 178–185 (1965).
26. Ratner, D. M., Swanson, E. R. & Seeberger, P. H. Automated synthesis of a protected N-linked glycoprotein core pentasaccharide. *Org. Lett.* **5**, 4717–4720 (2003).
27. Ball, P. It all falls into place. *Nature* **413**, 667–668 (2001).
28. Pavel, I. S. *Assembly of gold nanoparticles by ribosomal molecular machines* PhD thesis, Univ. Texas at Austin (2005).
29. Whaley, S. R., English, D. S., Hu, E. L., Barbara, P. F. & Belcher, A. M. Selection of peptides with semiconductor binding specificity for directed nanocrystal assembly. *Nature* **405**, 665–668 (2000).
30. Chen, H. L. & Goel, A. in *DNA Computing. Lecture Notes in Computer Science* Vol. 3384, 62–75 (Springer, Berlin/Heidelberg, 2005).
31. Hess, H. *et al.* Molecular self-assembly of 'nanowires'and 'nanospools' using active transport. *Nano Lett.* **5**, 629–633 (2005).
32. Winfree, E. & Bekbolatov, R. in *DNA Computing. Lecture Notes in Computer Science* Vol. 2943, 126–144 (Springer, Berlin/Heidelberg, 2004).
33. Choi, I. S., Bowden, N. & Whitesides, G. M. Macroscopic, hierarchical, two-dimensional self-assembly. *Angew. Chem. Intl Edn Engl.* **38**, 3078–3081 (1999).

34. Whitesides, G. M. & Boncheva, M. Beyond molecules: self-assembly of mesoscopic and macroscopic components. *Proc. Natl Acad. Sci. USA* **99**, 4769–4774 (2002).
35. Caviston, J. P. & Holzbaur, E. L. Microtubule motors at the intersection of trafficking and transport. *Trends Cell Biol.* **16**, 530–537 (2006).
36. Howard, J. *Mechanics of Motor Proteins and the Cytoskeleton* (Sinauer, Sunderland, Massachusetts, 2001).
37. Lakadamyali, M., Rust, M. & Zhuang, X. Ligands for clathrin-mediated endocytosis are differentially sorted into distinct populations of early endosomes. *Cell* **124**, 997–1009 (2006).
38. Lakadamyali, M., Rust, M. J., Babcock, H. P. & Zhuang, X. Visualizing infection of individual influenza viruses. *Proc. Natl Acad. Sci. USA* **100**, 9280–9285 (2003).
39. Månsson, A. & Linke, H. *Controlled Nanoscale Motion.* Proc. Nobel Symp. 131, Vol. 711. (Springer, Berlin, 2007).
40. Miki, H., Okada, Y. & Hirokawa, N. Analysis of the kinesin superfamily: insights into structure and function. *Trends Cell Biol.* **15**, 467–476 (2005).
41. Rust, M. J., Lakadamyali, M., Zhang, F. & Zhuang, X. Assembly of endocytic machinery around individual influenza viruses during viral entry. *Nature Struct. Mol. Biol.* **11**, 567–573 (2004).
42. Sotelo-Silveira, J. R., Calliari, A., Kun, A., Koenig, E. & Sotelo, J. R. RNA trafficking in axons. *Traffic* **7**, 508–515 (2006).
43. Vale, R. D. The molecular motor toolbox for intracellular transport. *Cell* **112**, 467–480. (2003).
44. Vallee, R. B. & Sheetz, M. P. Targeting of motor proteins. *Science* **271**, 1539–1544 (1996).
45. Vale, R. D., Reese, T. S. & Sheetz, M. P. Identification of a novel force-generating protein, kinesin, involved in microtubule-based motility. *Cell* **42**, 39–50 (1985).
46. Guzik, B. W. & Goldstein, L. S. Microtubule-dependent transport in neurons: steps towards an understanding of regulation, function and dysfunction. *Curr. Opin. Cell Biol.* **16**, 443–450 (2004).
47. Gunawardena, S. & Goldstein, L. S. Disruption of axonal transport and neuronal viability by amyloid precursor protein mutations in Drosophila. *Neuron* **32**, 389–401 (2001).
48. Gunawardena, S. & Goldstein, L. S. Cargo-carrying motor vehicles on the neuronal highway: transport pathways and neurodegenerative disease. *J. Neurobiol.* **58**, 258–271 (2004).
49. Mandelkow E, M. E. Kinesin motors and disease. *Trends Cell Biol.* **12**, 585–591 (2002).
50. Ström, A. L. *et al.* Retrograde axonal transport and motor neuron disease. *J. Neurochem.* Preprint at <http://www.ncbi.nlm.nih.gov/pubmed/18384644> (23 April 2008).
51. Mukhopadhyay, R. & Hoh, J. H. AFM force measurements on microtubule-associated proteins: the projection domain exerts a long-range repulsive force. *FEBS Lett.* **505**, 374–378 (2001).
52. Mizuno, N. *et al.* Dynein and kinesin share an overlapping microtubule-binding site. *EMBO J.* **23**, 2459–2467 (2004).
53. Vale, R. D. & Toyoshima, Y. Y. Rotation and translocation of microtubules in vitro induced by dyneins from Tetrahymena cilia. *Cell* **52**, 459–469 (1988).
54. Wang, Z., Khan, S. & Sheetz, M. P. Single cytoplasmic dynein molecule movements: characterization and comparison with kinesin. *Biophys. J.* **69**, 2011–2023 (1995).
55. Reck-Peterson, S. L. *et al.* Single-molecule analysis of dynein processivity and stepping behavior. *Cell* **126**, 335–348 (2006).
56. Seitz, A. & Surrey, T. Processive movement of single kinesins on crowded microtubules visualized using quantum dots. *EMBO J.* **25**, 267–277 (2006).
57. Coppin, C. M., Finer, J. T., Spudich, J. A. & Vale, R. D. Detection of sub-8-nm movements of kinesin by high-resolution optical-trap microscopy. *Proc. Natl Acad. Sci. USA* **93**, 1913–1917 (1996).
58. Svoboda, K., Schmidt, C. F., Schnapp, B. J. & Block, S. M. Direct observation of kinesin stepping by optical trapping interferometry. *Nature* **365**, 721–727 (1993).
59. Diehl, M. R., Zhang, K., Lee, H. J. & Tirrell, D. A. Engineering cooperativity in biomotor-protein assemblies. *Science* **311**, 1468–1471 (2006).
60. Hunt, A. J. & Howard, J. Kinesin swivels to permit microtubule movement in any direction. *Proc. Natl Acad. Sci. USA* **90**, 11653–11657 (1993).
61. Clemmens, J. *et al.* Principles of microtubule guiding on microfabricated kinesin-coated surfaces: chemical and topographic surface patterns. *Langmuir* **19**, 10967–10974 (2003).
62. Reuther, C., Hajdo, L., Tucker, R., Kasprzak, A. A. & Diez, S. Biotemplated nanopatterning of planar surfaces with molecular motors. *Nano Lett.* **6**, 2177–2183 (2006).
63. Clemmens, J., Hess, H., Howard, J. & Vogel, V. Analysis of microtubule guidance by microfabricated channels coated with kinesin. *Langmuir* **19**, 1738–1744 (2003).
64. Hess, H. *et al.* Molecular shuttles operating undercover: a new photolithographic approach for the fabrication of structured surfaces supporting directed motility. *Nano Lett.* **3**, 1651–1655 (2003).
65. Hiratsuka, Y., Tada, T., Oiwa, K., Kanayama, T. & Uyeda, T. Q. Controlling the direction of kinesin-driven microtubule movements along microlithographic tracks. *Biophys. J.* **81**, 1555–1561. (2001).
66. Moorjani, S. G., Jia, L., Jackson, T. N. & Hancock, W. O. Lithographically patterned channels spatially segregate kinesin motor activity and effectively guide microtubule movements. *Nano Lett.* **3**, 633–637 (2003).
67. Bunk, R. *et al.* Actomyosin motility on nanostructured surfaces. *Biochem. Biophys. Res. Commun.* **301**, 783–788 (2003).
68. Sundberg, M. *et al.* Actin filament guidance on a chip: toward high-throughput assays and lab-on-a-chip applications. *Langmuir* **22**, 7286–7295 (2006).
69. Clemmens, J. *et al.* Motor-protein 'roundabouts': microtubules moving on kinesin-coated tracks through engineered networks. *Lab Chip* **4**, 83–86 (2004).
70. Hess, H., Clemmens, J., Howard, J. & Vogel, V. Surface imaging by self-propelled nanoscale probes. *Nano Lett.* **2**, 113–116 (2002).
71. Stracke, R., Böhm, K. J., Burgold, J., Schacht, H.-J. & Unger, E. Physical and technical parameters determining the functioning of a kinesin-based cell-free motor system. *Nanotechnology* **11**, 52–56 (2000).
72. Brown, T. B. & Hancock, W. O. A polarized microtubule array for kinesin-powered nanoscale assembly and force generation. *Nano Lett.* **28**, 571–576 (2005).
73. Nitta, T., Tanahashi, A., Hirano, M. & Hess, H. Simulating molecular shuttle movements: towards computer-aided design of nanoscale transport systems. *Lab Chip* **6**, 881–885 (2006).
74. Vogel, V. & Hess, H. in *Lecture Notes Proceedings Nobel Symposium* Vol. 711, 367–383 (Springer, Berlin/Heidelberg, 2007).

75. Stracke, R., Bohm, K. J., Wollweber, L., Tuszynski, J. A. & Unger, E. Analysis of the migration behaviour of single microtubules in electric fields. *Biochem. Biophys. Res. Commun.* **293**, 602–609. (2002).

76. van den Heuvel, M. G., de Graaff, M. P. & Dekker, C. Molecular sorting by electrical steering of microtubules in kinesin-coated channels. *Science* **312**, 910–914 (2006).

77. Darnton, N., Turner, L., Breuer, K. & Berg, H. C. Moving fluid with bacterial carpets. *Biophys. J.* **86**, 1863–1870 (2004).

78. Hiratsuka, Y., Miyata, M., Tada, T. & Uyeda, T. Q. A microrotary motor powered by bacteria. *Proc. Natl Acad. Sci. USA* **103**, 13618–13623 (2006).

79. Bachand, G. D., Rivera, S. B., Carroll-Portillo, A., Hess, H. & Bachand, M. Active capture and transport of virus particles using a biomolecular motor-driven, nanoscale antibody sandwich assay. *Small* **2**, 381–385 (2006).

80. Hirabayashi, M. *et al.* Malachite green-conjugated microtubules as mobile bioprobes selective for malachite green aptamers with capturing/releasing ability. *Biotechnol. Bioeng.* **94**, 473–480 (2006).

81. Martin, B. D. *et al.* An engineered virus as a bright fluorescent tag and scaffold for cargo proteins: capture and transport by gliding microtubules. *J. Nanosci. Nanotechnol.* **6**, 2451–2460 (2006).

82. Muthukrishnan, G., Hutchins, B. M., Williams, M. E. & Hancock, W. O. Transport of semiconductor nanocrystals by kinesin molecular motors. *Small* **2**, 626–630 (2006).

83. Taira, S. *et al.* Selective detection and transport of fully matched DNA by DNA-loaded microtubule and kinesin motor protein. *Biotechnol. Bioeng.* **95**, 533–538 (2006).

84. Hess, H., Clemmens, J., Qin, D., Howard, J. & Vogel, V. Light-controlled molecular shuttles made from motor proteins carrying cargo on engineered surfaces. *Nano Lett.* **1**, 235–239 (2001).

85. Hess, H., Howard, J. & Vogel, V. A piconewton forcemeter assembled from microtubules and kinesins. *Nano Lett.* **2**, 1113–1115 (2002).

86. Boal, A. K., Bachand, G. D., Rivera, S. B. & Bunker, B. C. Interactions between cargo-carrying biomolecular shuttles. *Nanotechnology* **17**, 349–354 (2006).

87. Diez, S. *et al.* Stretching and transporting DNA molecules using motor proteins. *Nano Lett.* **3**, 1251–1254 (2003).

88. Brunner, C., Wahnes, C. & Vogel, V. Cargo pick-up from engineered loading stations by kinesin driven molecular shuttles. *Lab on a Chip* **7**, 1263–1271 (2007).

89. Ramachandran, S., Ernst, K. H., Bachand, G. D., Vogel, V. & Hess, H. Selective loading of kinesin-powered molecular shuttles with protein cargo and its application to biosensing. *Small* **2**, 330 (2006).

90. Dinu, C. Z. *et al.* Parallel manipulation of bifunctional DNA molecules on structured surfaces using kinesin-driven microtubules. *Small* **2**, 1090–1098 (2006).

91. Soldati, T. & Schliwa, M. Powering membrane traffic in endocytosis and recycling. *Nature Rev. Mol. Cell Biol.* **7**, 897–908 (2006).

92. Bachand, M., Trent, A. M., Bunker, B. C. & Bachand, G. D. Physical factors affecting kinesin-based transport of synthetic nanoparticle cargo. *J. Nanosci. Nanotechnol.* **5**, 718–722 (2005).

93. Du, Y. Z. *et al.* Motor protein nano-biomachine powered by self-supplying ATP. *Chem. Commun.*, 2080–2082 (2005).

94. Kufer, S. K., Puchner, E. M., Gumpp, H., Liedl, T. & Gaub, H. E. Single-molecule cut-and-paste surface assembly. *Science* **319**, 594–596 (2008).

95. Chakravarty, A., Howard, L. & Compton, D. A. A mechanistic model for the organization of microtubule asters by motor and non-motor proteins in a mammalian mitotic extract. *Mol. Biol. Cell* **15**, 2116–2132 (2004).

96. Surrey, T., Nedelec, F., Leibler, S. & Karsenti, E. Physical properties determining self-organization of motors and microtubules. *Science* **292**, 1167–1171. (2001).

97. Doot, R. K., Hess, H., Vogel, V. Engineered networks of oriented microtubule filaments for directed cargo transport. *Soft Matter* **3**, 349 - 356 (2007).

98. Beeg, J. *et al.* Transport of beads by several kinesin motors. *Biophys. J.* **94**, 532 (2008).

99. Henry, T., Gorvel, J. P. & Meresse, S. Molecular motors hijacking by intracellular pathogens. *Cell Microbiol.* **8**, 23–32 (2006).

100. Soo, F. S. & Theriot, J. A. Large-scale quantitative analysis of sources of variation in the actin polymerization-based movement of *Listeria monocytogenes*. *Biophys. J.* **89**, 703–723 (2005).

101. Rietdorf, J. *et al.* Kinesin-dependent movement on microtubules precedes actin-based motility of vaccinia virus. *Nature Cell Biol.* **2**, 992–1000 (2001).

102. Smith, G. A., Gross, S. P. & Enquist, L. W. Herpes viruses use bidirectional fast-axonal transport to spread in sensory neurons. *Proc. Natl Acad. Sci. USA* **98**, 3466–3470 (2001).

103. Döhner, K., Nagel, C.-H. & Sodeik, B. Viral stop-and-go along microtubules: taking a ride with dynein and kinesins. *Trends Microbiol.* **13**, 320–327 (2005).

104. Sodeik, B. Unchain my heart, baby let me go: the entry and intracellular transport of HIV. *Cell Biol.* **159**, 393–395 (2002).

105. Kulkarni, R. P., Wu, D. D., Davis, M. E. & Fraser, S. E. Quantitating intracellular transport of polyplexes by spatio-temporal image correlation spectroscopy. *Proc. Natl Acad. Sci. USA* **102**, 7523–7528 (2005).

106. Suh, J., Wirtz, D. & Hanes, J. Efficient active transport of gene nanocarriers to the cell nucleus. *Proc. Natl Acad. Sci. USA* **100**, 3878–3882. (2003).

107. Goel, A., Astumian, R. D. & Herschbach, D. Tuning and switching a DNA polymerase motor with mechanical tension. *Proc. Natl Acad. Sci. USA* **100**, 9699–9704 (2003).

108. Donlin, M. J., Patel, S. S. & Johnson, K. A. Kinetic partitioning between the exonuclease and polymerase sites in DNA error correction. *Biochemistry* **30**, 538–546 (1991).

109. Fersht, A. R., Knill-Jones, J. W. & Tsui, W. C. Kinetic basis of spontaneous mutation. Misinsertion frequencies, proofreading specificities and cost of proofreading by DNA polymerases of *Escherichia coli*. *J. Mol. Biol.* **156**, 37–51 (1982).

110. Hopfield, J. J. Kinetic proofreading: a new mechanism for reducing errors in biosynthetic processes requiring high specificity. *Proc. Natl Acad. Sci. USA* **71**, 4135–4139 (1974).

111. Hopfield, J. J. The energy relay: a proofreading scheme based on dynamic cooperativity and lacking all characteristic symptoms of kinetic proofreading in DNA replication and protein synthesis. *Proc. Natl Acad. Sci. USA* **77**, 5248 (1980).

112. Rodnina, M. V. & Wintermeyer, W. Ribosome fidelity: tRNA discrimination, proofreading and induced fit. *Trends Biochem. Sci.* **26**, 124–130 (2001).

113. Wang, D. & Hawley, D. K. Identification of a $3' \rightarrow 5'$ exonuclease activity associated with human RNA polymerase II. *Proc. Natl Acad. Sci. USA* **90**, 843–847 (1993).

114. Cavalli, V., Kujala, P., Klumperman, J. & Goldstein, L. S. Sunday Driver links axonal transport to damage signaling. *J. Cell Biol.* **168**, 775–787 (2005).

115. Mandelkow, E. M., Stamer, K., Vogel, R., Thies, E. & Mandelkow, E. Clogging of axons by tau, inhibition of axonal traffic and starvation of synapses. *Neurobiol. Aging* **24**, 1079–1085 (2003).

116. Goel, A., Ellenberger, T., Frank-Kamenetskii, M. D. & Herschbach, D. Unifying themes in DNA replication: reconciling single molecule kinetic studies with structural data on DNA polymerases. *J. Biomol. Struct. Dyn.* **19**, 571–584 (2002).

117. Guydosh, N. R. & Block, S. M. Backsteps induced by nucleotide analogs suggest the front head of kinesin is gated by strain. *Proc. Natl Acad. Sci. USA* **103**, 8054–8059 (2006).

118. Spudich, J. Molecular motors take tension in stride. *Cell* **126**, 242–244 (2006).

119. Vale, R. D. & Milligan, R. A. The way things move: looking under the hood of molecular motor proteins. *Science* **288**, 88–95 (2000).

120. Veigel, C., Schmitz, S., Wang, F. & Sellers, J. R. Load-dependent kinetics of myosin-V can explain its high processivity. *Nature Cell Biol.* **7**, 861–869 (2005).

121. Maier, B., Bensimon, D. & Croquette, V. Replication by a single DNA polymerase of a stretched single-stranded DNA. *Proc. Natl Acad. Sci. USA* **97**, 12002–12007 (2000).

122. Abbondanzieri, E. A., Greenleaf, W. J., Shaevitz, J. W., Landick, R. & Block, S. M. Direct observation of base-pair stepping by RNA polymerase. *Nature* **438**, 460–465 (2005).

123. Shaevitz, J. W., Abbondanzieri, E. A., Landick, R. & Block, S. Backtracking by single RNA polymerase molecules observed at near-base pair resolution. *Nature* **426**, 684–687 (2003).

124. Sinha, D. K., Bhalla, U. S. & Shivashankar, G. V. Kinetic measurement of ribosome motor stalling force. *Appl. Phys. Lett.* **85**, 4789–4791 (2004).

125. Astumian, R. D. Thermodynamics and kinetics of a Brownian motor. *Science* **276**, 917–922 (1997).

126. Bustamante, C., Keller, D. & Oster, G. The physics of molecular motors. *Acc. Chem. Res.* **34**, 412–420 (2001).

127. Goel, A., Frank-Kamenetskii, M. D., Ellenberger, T. & Herschbach, D. Tuning DNA 'strings': modulating the rate of DNA replication with mechanical tension. *Proc. Natl Acad. Sci. USA* **98**, 8485–8489 (2001).

128. Goel, A. & Herschbach, D. R. Controlling the speed and direction of molecular motors that replicate DNA. *Proc. SPIE* **5110**, 63–68 (2003).

129. Mallik, R., Carter, B. C., Lex, S. A., King, S. J. & Gross, S. P. Cytoplasmic dynein functions as a gear in response to load. *Nature* **427**, 649–652 (2004).

130. Rosenfeld, S. S., Fordyce, P. M., Jefferson, G. M., King, P. H. & Block, S. M. Stepping and stretching. How kinesin uses internal strain to walk processively. *J. Biol. Chem.* **278**, 18550–18556 (2003).

131. Andricioaei, I., Goel, A., Herschbach, D. & Karplus, M. Dependence of DNA polymerase replication rate on external forces: a model based on molecular dynamics simulations. *Biophys. J.* **87**, 1478–1497 (2004).

132. Liu, H. *et al.* Control of a biomolecular motor-powered nanodevice with an engineered chemical switch. *Nature Mater.* **1**, 173–177 (2002).

133. Konishi, K., Uyeda, T. Q. & Kubo, T. Genetic engineering of a Ca(2+) dependent chemical switch into the linear biomotor kinesin. *FEBS Lett.* **580**, 3589–3594 (2006).

134. Nomura, A., Uyeda, T. Q., Yumoto, N. & Tatsu, Y. Photo-control of kinesin-microtubule motility using caged peptides derived from the kinesin C-terminus domain. *Chem. Comm.* **1**, 3588–3590 (2006).

135. Ionov, L., Stamm, M. & Diez, S. Reversible switching of microtubule motility using thermoresponsive polymer surfaces. *Nano Lett.* **6**, 1982–1987 (2006).

136. van den Heuvel, M. G. L. & Dekker, C. Motor proteins at work for nanotechnology. *Science* **317**, 333 - 336 (2007).

137. Browne, W. R. & Feringa, B. L. Making molecular machines work. *Nature Nanotech.* **1**, 25–35 (2006).

138. Hess, H. *et al.* Ratchet patterns sort molecular shuttles. *Appl. Phys. A* **75**, 309–313 (2002).

Acknowledgements

We thank Sheila Luna, Christian Brunner and Jennifer Wilson for the artwork, and all of our collaborators who contributed thoughts and experiments. At the same time, we apologize to all authors whose work we could not cite owing to space limitations.

Correspondence and requests for materials should be addressed to A.G. or V.V.

Designed DNA molecules: principles and applications of molecular nanotechnology

Anne Condon

Abstract | Long admired for its informational role in the cell, DNA is now emerging as an ideal molecule for molecular nanotechnology. Biologists and biochemists have discovered DNA sequences and structures with new functional properties, which are able to prevent the expression of harmful genes or detect macromolecules at low concentrations. Physical and computational scientists can design rigid DNA structures that serve as scaffolds for the organization of matter at the molecular scale, and can build simple DNA-computing devices, diagnostic machines and DNA motors. The integration of biological and engineering advances offers great potential for therapeutic and diagnostic applications, and for nanoscale electronic engineering.

Rationally designed DNA
DNA sequences that have been designed, using manual or computational methods, to fold into a particular structure or perform a particular function.

In vitro selection (of DNA)
The selection of DNA strands with desired functional properties from an initial set using cycles of selection for functional properties and enrichment of strands with the desired properties.

Molecular evolution (of DNA)
Similar to *in vitro* selection of DNA, except that strands might be mutated between cycles, thereby diversifying the set of DNA strands.

The Department of Computer Science, 2366 Main Mall, University of British Columbia, Vancouver, British Columbia V6T 1Z4, Canada.
e-mail: condon@cs.ubc.ca
doi:10.1038/nrg1892
Published online 13 June 2006

The cellular role of DNA is relatively limited, perhaps because of the restrictions imposed by bonding between complementary strands. Outside the cell, however, nanoengineers are uncovering the many hidden talents of DNA. The DNA sequence is able to process information in biochemical assays, its structure is an ideal building material, and its folding pathway allows DNA to move and respond to its environment. Here, I review several innovative applications of designed DNA molecules, some underlying design principles and the prospects for future developments. The applications of new DNA-based technologies range from molecular diagnostics to protein purification and from therapeutics to the assembly of tiny electronic circuits. These uses exploit the properties of DNA at three levels — sequence only, structure (which depends on sequence) and folding pathways (which depend on both sequence and structure) — as the following examples show.

By taking their cue from the genetic code found in nature, nanoengineers are now using DNA sequences *in vitro* to direct the synthesis and evolution of molecules with new functions and reactive properties. When attached to other molecules, detection methods that exploit DNA sequences allow complementary molecules to be identified at extremely low concentrations. DNA sequences can also be used, in a non-biological setting, to control the layout of electronic components in a nanocircuit.

DNA strands can fold into stable structures that have valuable functional and material properties. Sensitive methods for detecting DNA have been extended to detect proteins and other macromolecular targets of DNA aptamers — that is, nucleic-acid molecules that have high binding specificity for their targets. Lattices and even three-dimensional (3D) objects have been self-assembled from rigid DNA structures. DNA lattices can organize proteins and nanoelectronic components on a surface with unprecedented precision, making it possible to gain a better understanding of the properties and interactions of proteins.

The folding pathway of a DNA molecule to its stable structure allows it to move and perform mechanical functions. The energy released in DNA-folding pathways has been used *in vitro* to drive motors, providing the ability to release, grab or cleave target molecules and even to control the release of drugs, depending on the outcome of certain diagnostic tests.

This review focuses on the recent developments in generating rationally designed DNA molecules and their applications. The past decade has seen huge advances in the development of design principles, particularly by exploiting DNA structure and folding pathways, but also in applications of designed DNA sequences. Together, techniques for the *in vitro* selection or molecular evolution of DNA molecules from random pools have given spectacular yields, including DNA aptamers and enzymes, which can inhibit the expression of harmful genes or disrupt protein function. Excellent reviews of selected DNA molecules can be found elsewhere[1–3]; this paper includes only examples of designed DNA molecules, or molecules that combine both selected and designed

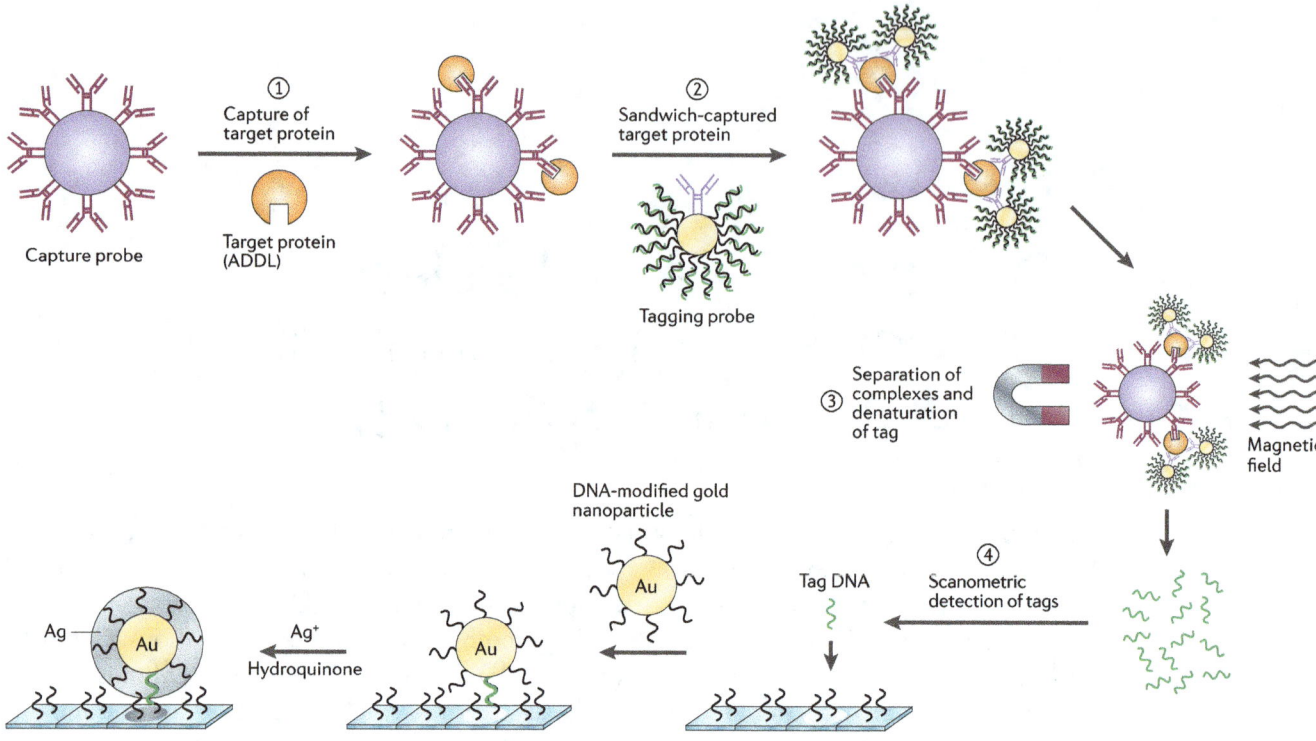

Figure 1 | Exploiting DNA sequences: sensitive molecular detection. Bio-barcode amplification allows detection of molecules, at low attomolar levels, from biological samples, such as sera or spinal fluids[8,10]. Two types of probe are used to detect a protein target, which, in the illustration, consists of amyloid-β-derived diffusible ligands (ADDLs) in human cerebrospinal fluid. Tagging probes consist of a gold nanoparticle (yellow) to which an antibody for the target (blue) and complements of DNA tags (black) are attached. DNA tags (green) are then hybridized to their complements. Capture probes consist of magnetic beads (blue circles) to which antibodies (pink) are attached. On exposure to the magnetic beads, the target, ADDL, is caught on the capture probes (step 1). It is then 'sandwiched' between capture and tagging probes, by binding to antibodies on both types of probe (step 2). The resulting complexes are separated from solution using a magnetic field and washed (step 3), at which point the tags denature from their complements. The concentration of ADDL in the sample can then be estimated on the basis of the amount of tag that is released. To this end, Georganolpoulou *et al.*[10] used scanometric DNA detection (step 4). In this method, tag DNA is captured on a glass slide, which has been spotted with DNA strands that are complementary to part of the tag. Gold nanoparticles (Au), which contain DNA strands complementary to the remaining (unhybridized) part of the DNA tag, then sandwich the tag between a slide and nanoparticle. The nanoparticles are coated with silver (Ag), and the light intensity scattered from each spot is measured. Adapted with permission from REF. 8 © (2003) American Association for the Advancement of Science and REF. 10 © (2005) National Academy of Sciences.

Fluorescence spectroscopy
The analysis of fluorescence spectra that are emitted by fluorophores when they are excited by visible light or ultraviolet rays.

DNA microarray
An array of spots on a planar surface, with copies of an ssDNA immobilized on each spot. The array can be used to detect DNA strands that are complementary to those immobilized to the surface.

features. The examples are organized by the primary property of DNA used: sequence, structure or folding pathway (although the distinctions are sometimes blurred). Overall, I hope that the reader will gain a renewed appreciation for the extraordinary abilities of DNA, and that the range of examples will stimulate ideas for further practical applications.

Innovative uses of DNA sequence

Many uses of DNA exploit its sequence properties, along with the ability of a DNA sequence to hybridize with its complement. Following a short description of DNA-sequence properties and the techniques for manipulating these molecules, I describe several applications, ranging from sensitive molecular detection to DNA computing and the assembly of nanoelectronic components. Finally, I describe some challenges and prospects for this field.

DNA sequences have three major assets: a four-base digital make-up, an ability to bind a complementary sequence with high affinity, and stability in a wide range of environmental conditions. DNA has an incredible information density — roughly 1 bit per nm^3 — that can be retrieved from an ssDNA sequence by its ability to recognize and hybridize with its complement. Hybridization is reversible and, in part, thermodynamically driven: when temperature decreases complementary ssDNA in solution tends to hybridize, and when temperature increases dsDNA tends to denature into single strands. The applications that exploit the features of DNA sequences also depend crucially on efficient technologies for the sequencing, synthesis, amplification (by PCR) and detection (by fluorescence spectroscopy and DNA microarray analysis) of DNA strands.

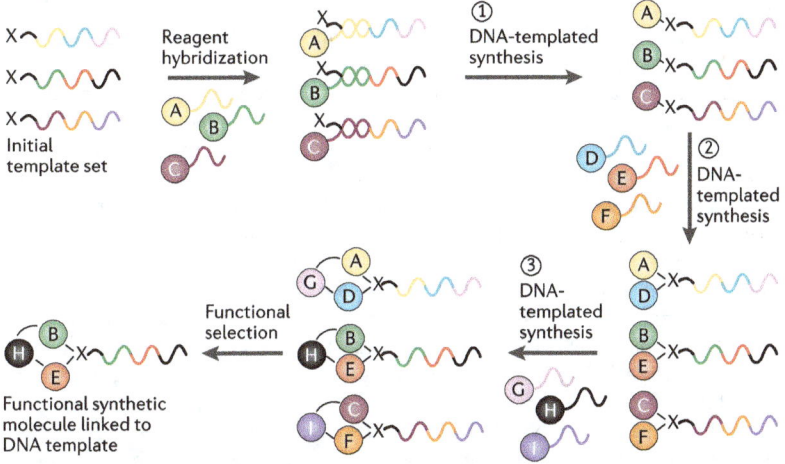

Figure 2 | Exploiting DNA sequences: DNA-templated synthesis of macrocycles. Three macrocycles, each composed of four cyclically-linked units, are synthesized[15]. The starting material consists of nine types of reagent (A–I), each of which is linked to a distinct biotinylated DNA tag (indicated by a coloured strand). A set of three DNA templates, each the concatenation of three sequences that are complementary to three distinct tags, is initially synthesized. These templates represent the three macrocycles to be synthesized. Attached to each initial template is a single unit (X; a lysine derivative) that will form part of each macrocycle. The macrocycles are then synthesized in a multi-step process, in which each step increases the length of the synthesized molecule by one. In the first illustrated synthesis step, the tags attached to reagents A, B and C (yellow, green and dark pink, respectively) hybridize to each of the three templates, bringing the X unit into proximity with the A, B and C units (one per template). The units then conjugate with the X unit by amine acylation, to produce synthesized molecules of length 2. The process is repeated (synthesis steps 2 and 3) to produce three synthesized four-unit macrocycles. The biotin present on the tags allows the macrocycles to be purified by capturing them on streptavidin beads and then releasing them. Following synthesis, one macrocycle is selected for function, and its tag reveals its identity. Adapted with permission from REF. 15 © (2001) American Association for the Advancement of Science.

Macrocycle
A cyclic molecule with large molecular weight, typically constructed from subunits. One example is porphyrin, which is a component of heme and the prosthetic group in haemoglobins and myoglobins, and is responsible for oxygen transport in the blood.

DNA code
A mapping from a set of DNA sequences to the set of monomers of other polymers; for example, the genetic code maps the set of three-letter codons to the set of amino acids.

Cytokines
Proteins that aid in the formation and development of blood cells, and are important in immune-system functioning.

DNA tags, which are sequences that are chemically attached to other molecules, represent the most successful and varied application of DNA-sequence properties. DNA tags are used in the highly sensitive detection of disease markers, the parallel synthesis of new compounds and the discovery of new reagents — as such, they have been applied in disease diagnosis and drug development. Two powerful principles unify the applications of DNA tags. First, new DNA codes[4] can direct the cheap and efficient parallel synthesis of large sets of polymers. In this case, each DNA tag is composed of ordered segments that specify the sequence of polymeric units of one molecule to be synthesized, just as a sequence of natural codons specifies the sequence of amino acids that make up a protein. Second, efficient methods for amplifying, sequencing and detecting DNA can be extended to new types of molecule using their attached DNA tags. Several applications of DNA tags are described next.

Sensitive molecular detection. Molecular-detection methods can help to identify disease markers, such as cytokines in human serum samples, and, in so doing, aid the diagnosis of some cancers and immunodeficiency-related diseases[5]. Detection methods can also identify allergens and pollutants[6]. Highly sensitive detection,

at the attomolar (10^{-18} mole) level or better, is of great practical importance: it allows more efficient screening at blood banks through the pooling of samples, and it facilitates early disease detection when treatment might be more promising and paediatric research when sample sizes are necessarily small[6].

Early DNA-based detection methods, such as immuno-PCR[6,7], involved many cumbersome steps, including PCR. Bio-barcode amplification (BCA), which has been developed by Nam and colleagues[8], is a new streamlined method that achieves highly sensitive detection at attomolar concentrations without the need for PCR amplification. In BCA, the target protein to be detected is 'sandwiched' between two probes: a tagging probe and a capture probe (FIG. 1). Both probes contain antibodies, which bind to the target protein such that the protein is sandwiched between the probes. Two key innovations are that many copies of a DNA tag are attached to the tagging probe, and that tags can be removed for detection purposes using a simple denaturing process. BCA has been successfully applied to the detection of prostate-specific antigen (an indicator of prostate cancer and breast cancer[8,9]), amyloid-β–derived diffusible ligands (ADDLs; indicators of Alzheimer disease[10]) and interleukin 2 (a cytokine protein that is involved in inflammation and immune responses in humans[11]). Georganopoulou and colleagues[10] quantified differences in ADDL levels in clinically realistic samples of cerebrospinal fluid that were taken from human tissue. Overall, they found a significantly higher (low femtomolar) concentration of ADDLs in diseased patients, compared with samples from control patients who had not been diagnosed with Alzheimer disease.

Fredriksson *et al.*[12] described another ingenious detection scheme that uses a pair of DNA aptamers, rather than antibodies, as the means of binding to the target protein. Unlike the BCA method, this 'proximal ligation' method is limited to recognizing protein targets for which DNA aptamers are known, and also requires the use of PCR. However, proximal ligation avoids washing or separation steps, and can detect a cytokine protein at subattomolar concentrations even in biological samples.

DNA-templated synthesis. In the cell, the genetic code supports both translation (whereby protein products encoded by DNA are synthesized) and evolution (whereby cycles of translation, mutation and selection produce better-adapted products in the form of evolved species). The ultimate goal of DNA-templated synthesis (DTS)[13] is to emulate these processes *in vitro* in a parallel fashion. DTS can create polymers with strong binding affinity to a target of interest, such as a disease marker or toxin (FIG. 2). DTS offers several advantages over traditional methods for developing functional molecules, including parallelism, ease of purification and mutational evolution. Rosenbaum and Liu[14] reported the first DTS of molecules that do not have a ribose backbone, which are called peptide nucleic acids (PNAs); these are valuable substitutes for DNA in environments where the backbone of DNA is prone to degradation, such as in therapeutic uses in living cells. Gartner *et al.*[15] showed

Immuno-PCR

A protein-detection method in which specially designed linker molecules capture a copy of the protein of interest and a DNA tag. The captured tags are then amplified and detected using PCR, followed by either fluorescence spectroscopy or chip-based detection methods.

Enone

An unsaturated chemical compound or functional group consisting of a conjugated system of an alkene and a ketone.

Boolean logic

An algebra for reasoning about logical values ('true' or 'false'), which includes AND, OR and NOT operations, and is also useful for computing with binary ('1' or '0') values.

that DTS can be followed by *in vitro* selection for desired functional properties — in this case, a molecule that inhibits the activity of carbonic anhydrase. Carbonic anhydrase-driven reactions, which convert carbon dioxide and water into hydrogen ions and bicarbonate, have several functions in the body (such as maintaining an acidic environment in the stomach and transporting carbon dioxide by red blood cells) and so inhibitors of such reactions have therapeutic uses.

One limitation of DTS is that synthesis can only occur in environments in which dsDNA will not denature, because binding of complementary strands directs synthesis steps. Halpin and colleagues[16–18] describe a synthesis method that overcomes this limitation by splitting templates into pools at intermediate stages of synthesis. With the pooling method, synthesis can occur at high temperatures or with organic solvents. They reported both the synthesis of a set of 1 million peptides, each comprised of 5 amino acids, and the selection from this set of a peptide with affinity for the 3-E7 antibody, which inhibits vascular endothelial growth factor.

Parallel reaction discovery. The purpose of parallel reaction-discovery systems is to efficiently test many pairs of reagents simultaneously in a single solution, to identify unknown binding pairs that might have innovative uses[19,20]. This process has been hindered by the difficulty of bringing each pair of reactants to be tested into close proximity with one another while avoiding crossreactivity among pairs of reagents that are not of interest. Another challenge is to identify those pairs that do react. DNA tags solve both of these problems: the molecules to be paired have complementary regions on their tags so that they come together by hybridization, and sequencing of the tags provides the means to identify reactive pairs. As proof of principle, Kanan *et al.*[20] used their method to discover — from 168 combinations of reagents — that an enone could be produced by an alkyne–alkene coupling reaction in the presence of a palladium catalyst.

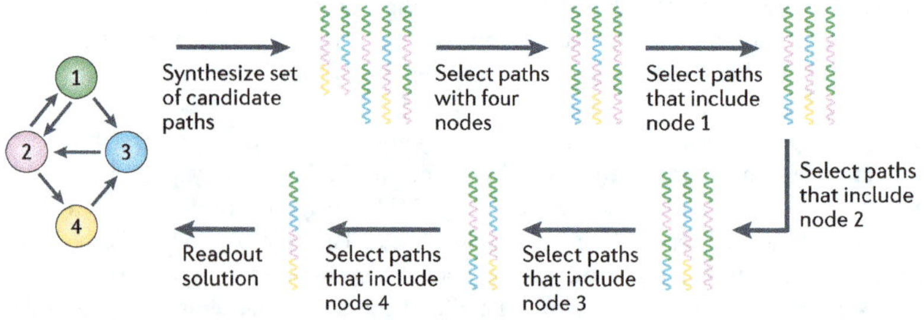

Computing with DNA. In the applications described so far, DNA sequence has an informational role. Is computation possible for information stored in DNA? In essence, computation is the process of determining an output from a list of inputs, using elementary instructions, where the available instructions depend on the context of the computation. For example, a schoolchild can determine the sum of two multiple-digit input numbers written on paper, using a sequence of single-digit additions. By contrast, a computer calculates the sum of numbers in binary format, using the principles of Boolean logic. Computation is not, however, limited to paper or silicon media. Indeed, as Adleman[21] showed (FIG. 3), DNA is an attractive medium because of its high information density. Inputs to the Adleman computation are DNA sequences that represent candidate solutions to a computational problem. The computation repeatedly prunes out incorrect solutions, so that, ultimately, the true solution is selected. This select-and-prune process is analogous to *in vitro* selection, except that sequences are selected on the basis of their information content rather than their functionality. In the largest experiment so far, Braich *et al.*[22] reported a search-and-prune computation involving more than 1 million inputs, representing potential solutions to a problem of mathematical logic. Although this approach is not competitive with conventional computers, the work has inspired many creative approaches for DNA computation that exploit structure and folding pathways, which are described in later sections.

Organization of nanoelectronic components and other materials. Single-walled carbon nanotubes (CNTs) have been used to realize tiny electronic transistors, which have potential for assembly into nanocircuits. There are several challenges to building a circuit from CNTs, however, including controlling the layout of the component CNTs and placing the wires to connect the components according to a specific pattern. DNA tags (and also tags made of PNA) offer great promise in meeting these challenges. DNA tags that are attached to CNTs can direct their assembly by binding a tag with its complement[23]. DNA strands have been used to spatially position CNTs, making it possible to assemble them into circuits[24]. In addition, ssDNA can be stretched across a surface to connect two electrodes by binding the ends of the ssDNA with complementary strands that are attached to the electrodes. Silver can then be deposited along the stretched DNA strand, to create a nanowire that connects the electrodes[25,26].

Varying the DNA strands that are wrapped around CNTs alters the electrostatic properties of the CNT–DNA hybrid in a way that depends on the diameter of the CNT[27]. DNA-wrapped CNTs also enhance the sensitivity of nanotube devices in the detection of gases[28]. Storhoff and Mirkin[29], and Niemeyer[30] review further uses of DNA for materials synthesis — for example, in the assembly of networks of gold nanoparticles or the organization of enzymes so as to speed up chemical reactions[31].

Figure 3 | Exploiting DNA sequences: DNA computation. Adleman[21] used select-and-prune parallel computation to determine whether there is a path (that is, a sequence of linked nodes) through a network, such that the path visits each node exactly once. In the illustrated example (simplified from the Adleman experiment), the input set contains a sample of five paths in a simple four-node network (shown on the left). Nodes are represented by short DNA strands, and paths are represented as concatenations of the corresponding node strands. The first round, which is accomplished by gel electrophoresis and shown on the right, selects paths that have exactly four nodes, therefore pruning two paths from the input sample. The second, third, fourth and fifth rounds prune out paths that do not visit node 1, 2, 3 or 4, in corresponding order (accomplished using DNA-sequence complementarity and magnetic-bead separation). Only one path remains at the end of the computation, revealing the solution: 1→3→2→4.

Challenges and opportunities. DTS and the involved reactions have excellent potential for use in the discovery of drugs or other new compounds. For practical applications, it will be necessary to broaden the scale and range of these methods. An adaptive approach involving repeated rounds of synthesis, in which templates of already-synthesized molecules that show some useful function could be mutated or recombined, might diversify the synthesis pool[17]. Another possibility is to integrate DNA-computing methods, which select molecules on the basis of the information content of their labels, with methods that select or organize molecules on the basis of their function.

Innovative uses of DNA structure

The diverse structures that DNA can have vastly expand its potential uses in nanotechnology and biotechnology applications. Structure is the key feature that allows the specific binding of DNA to proteins. It is now also possible to engineer rigid scaffolds from DNA. The integration of the scaffolding and binding structural properties of DNA yields two-dimensional (2D) ordered arrays of proteins, which might be useful for the detailed study of protein structure. DNA structure even supports computation, making it possible, in principle, to program the layout of arbitrarily complex 2D or 3D patterns from DNA.

DNA structure, which might involve one or more DNA molecules, is largely shaped by the hybridization of complementary intermolecular or intramolecular base pairs. The most stable structure formed by DNA molecules in solution depends on environmental factors, such as temperature, pH level and salt concentration. The double helix is just one possible conformation; structures might have branching helices that stem from junctions or loops (FIG. 4a). The stacking of adjacent base pairs causes short helices to be rigid, unlike ssDNA. Some DNA molecules enjoy further structural complexity, such as G quartets (FIG. 4b), which can fit snugly in crevices of proteins, and therefore can have high binding specificity for important protein targets. One DNA aptamer exploits a G-quartet structure to bind to thrombin and can inhibit thrombin-catalysed fibrin-clot formation *in vitro* in samples of human plasma[32].

Substantial progress has also been made recently in the design of branched and rigid shapes from DNA. These new structures support the creation of 2D scaffolds and complex patterns, and even 3D structures. Single molecules can now be arrayed with nanoscale precision on DNA scaffolds. Because many exciting developments in engineered DNA structures have been reported in the past year or so, they are the primary focus of this section.

Molecular detection. Sensitive detection of molecules can take advantage of structure as well as sequence. Li *et al.*[33] and Stavis *et al.*[34] considered the case of multiplexed detection, where the goal is to determine which of many DNA fragments, derived from pathogens, are present in solution. Instead of sequence tags, Li *et al.*[33] used branched DNA structures called nanobarcodes, with red or green fluorophores on the branch tips (FIG. 4a). Nanobarcodes are distinguishable, using fluorescence

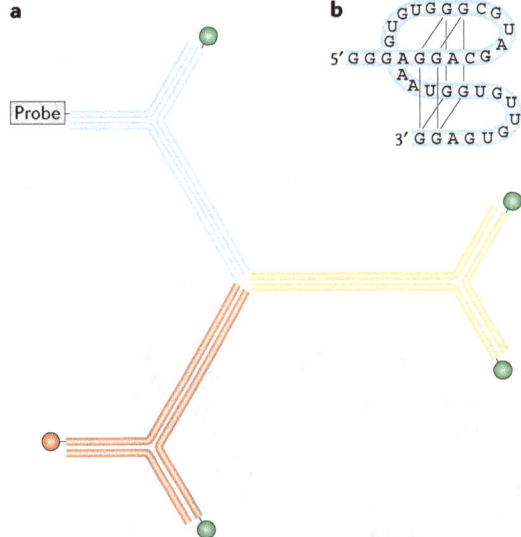

Figure 4 | **Exploiting DNA structure: branched structures. a** | The nanobarcode, which is a branched structure with fluorescent labels at the tips of its branches[33], is constructed from Y-shaped components, each formed from three DNA strands (indicated by colours). Each Y-shaped structure binds to another by complementary sticky ends. Ligation is then used to seal the nicks (at positions where distinct strands meet) in the overall structure. Nanobarcodes are distinguished by the ratio of red and green fluorophores (represented by circles) on their branches, and have been used for multiplexed detection. For each type of protein to be detected, a probe is selected that has high binding specificity for the target. A tagging probe is constructed by attaching nanobarcodes of common type (that is, ratio of red to green fluorophores) to copies of the probe. The tagging probe and an additional large capture probe (not shown) are introduced into a solution; if the target is present in the solution, it is sandwiched between the tagging and capture probes, and detected by fluorescence spectroscopy. **b** | A G-quartet structure in a nucleic-acid aptamer, which binds to thrombin (not shown) in the presence of potassium ions[81].

spectroscopy, by their ratios of red to green fluorophores. As proof of concept, nanobarcodes were used to detect the presence or absence of DNA fragments from four pathogens (*Bacillus anthracis*, *Francisella tularensis*, Ebola virus and the severe acute respiratory syndrome (SARS) coronavirus) in a prepared solution with a detection limit of 620 attomoles. Stavis *et al.*[34] showed that, by flowing nanobarcode tags through microfluidic channels, more rapid detection is possible.

DNA scaffolding. A long-term goal of Seeman and co-workers[35,36] has been to build lattices and other rigid structures from DNA in a controlled scalable fashion. DNA lattices can form a 2D scaffold on which biological macromolecules can be arrayed. DNA lattices could also bring enzymes into proximity, thereby catalysing numerous reactions on a large scale, and serve as guides for the placement of nanoelectronic components on a surface,

---

G quartets
Nucleic-acid structures in which four G nucleotides form a stable structure.

Thrombin
A protein that is involved in coagulation and anti-coagulation, and can recognize many macromolecular substrates.

Microfluidic channels
Systems of channels, with diameters that range from micrometres to millimetres, through which fluids can flow. Can be used to separate and detect DNA molecules in the fluids.

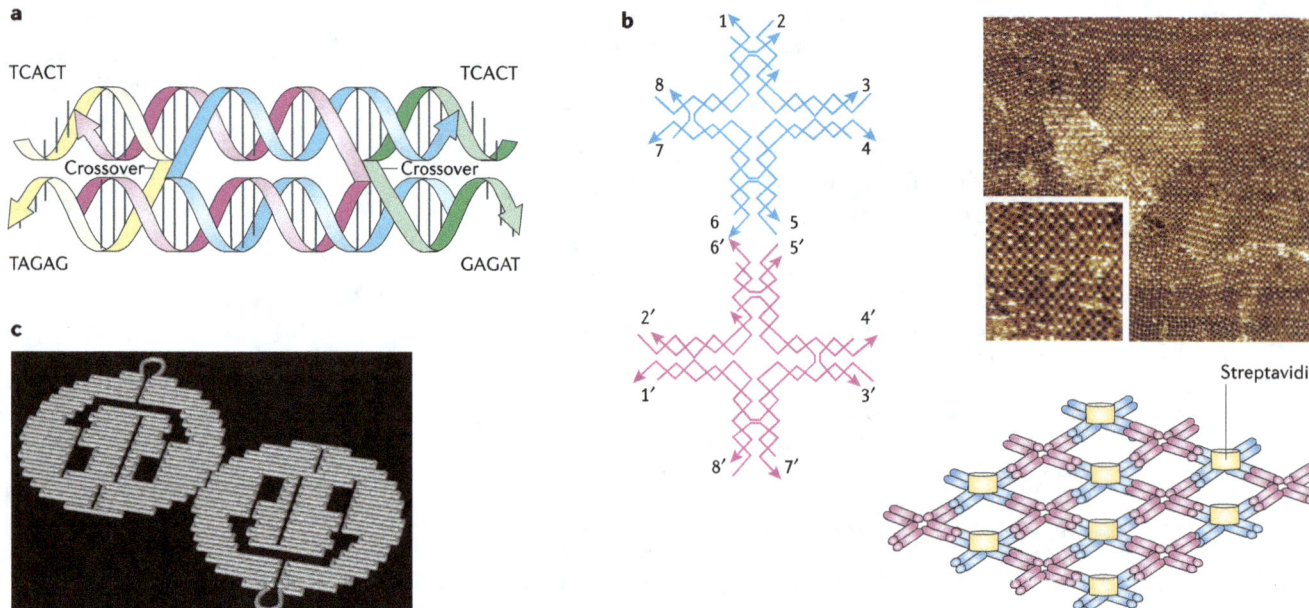

Figure 5 | Exploiting DNA structure: double-crossover structures and self assembly. a | A double-crossover (DX) molecule[37] composed of two helices with strands that cross at two points, thereby creating a rigid structure. Five-nucleotide single-stranded sequences, called sticky ends, allow DX molecules to bind to each other in a programmable fashion; that is, two DX molecules bind only if they have complementary sticky ends. **b |** Rigid junction molecules with four emanating helices, in which each of the four branches contain a single crossover[39]. This molecule was used to construct a DNA lattice, with biotinylated components, on which streptavidin was arrayed. The atomic-force microscopy image shows a resulting lattice, arrayed with streptavidin (visible as the light spots; the scale of the image is 1 μm × 1 μm and the inset is 150 nm × 150 nm). **c |** Smiley face constructed using the scaffolded origami technique[42]. A long scaffold strand can be laid out in rows, to cover the desired pattern. Rothemund used a 7,249-nucleotide DNA strand from the virus M13mp18 (a strand that is inexpensive to purchase). More than 200 short DNA strands (each about the length of a PCR primer) are used to hold the long scaffold strand in the desired pattern. The diagram shows how the long DNA strand is laid out to create the resulting pattern. Panel **a** reproduced with permission from REF. 48 © (2004) American Chemical Society. Panel **b** adapted with permission from REF. 40 © (2005) American Chemical Society. Panel **c** reproduced with permission from REF. 42 © (2006) Macmillan Publishers Ltd.

thereby forming a nanocircuit. In a 3D DNA lattice, the cavities could serve as cages in which proteins are arrayed, making it possible to uncover new properties of the macromolecules with X-ray crystallography.

Many DNA scaffolds are constructed from rectangular structures, called tiles, which are composed of numerous short DNA strands. Emanating from these tiles are one or more single strands, called sticky ends, which can bind to complementary sticky ends on other tiles. Self-assembly of large structures from small tiles ensues when many tiles bind to each other. Important challenges in tile design are ensuring that the tiles are rigid (a necessary property for scaffolding) and that adjacent assembled tiles are co-planar.

The double-crossover (DX) molecule of Fu and Seeman[37] (FIG. 5a) is an early, elegant and widely used design, from which the first 2D DNA lattices were built by Winfree et al.[38]. Other designs and applications have followed[39–41], which additionally incorporate strategies to array protein molecules in regular arrangements on the lattice. Such arrays could be used to study protein structure using electron microscopy or to construct spatially well-defined complexes involving several enzymes. Specific examples are the lattices of Yan et al.[39] and Park et al.[40], in which streptavidin binds to biotinylated DNA molecules at lattice junctions, thereby precisely controlling the arrangement of the streptavidin molecules, for example, with a distance of approximately 20 nm between adjacent molecules (FIG. 5b). Liu et al.[41] incorporated the DNA aptamer for thrombin (FIG. 4b) into a hairpin structure at regular points of a one-dimensional (1D) lattice. This approach should, in principle, allow organization of any molecule for which there is a DNA aptamer.

In a different approach, known as single-stranded origami[42], a long strand folds in an intricate way, aided by short strands that direct the folding into the desired configuration. Shih et al.[43] used this approach to construct the first rigid 3D structure, crafting an octahedron from a single long (1,669 nucleotide) strand and five shorter (30–35 nucleotide) strands. The short molecules bind so as to form appropriately placed rigid DX structures, which become edges of the octahedron. Rothemund[42] illustrated the generality of a different method, known as scaffolded origami, by constructing smiley faces, a world map and many other complex 2D patterns (FIG. 5c). To design a pattern, Rothemund uses a computational tool

Atomic-force microscopy
The analysis of the contours of a surface, which are based on measurements of Van der Waals forces between the surface and a scanning probe that is attached to a tiny cantilever.

X-ray crystallography
The analysis of the diffraction patterns of X-rays of a crystal, used to determine the structure of the molecule.

Double-crossover (DX) molecule
A type of rigid DNA rectangular-shaped molecule that is comprised of four or five DNA strands, with four ssDNA strands emanating from its corners.

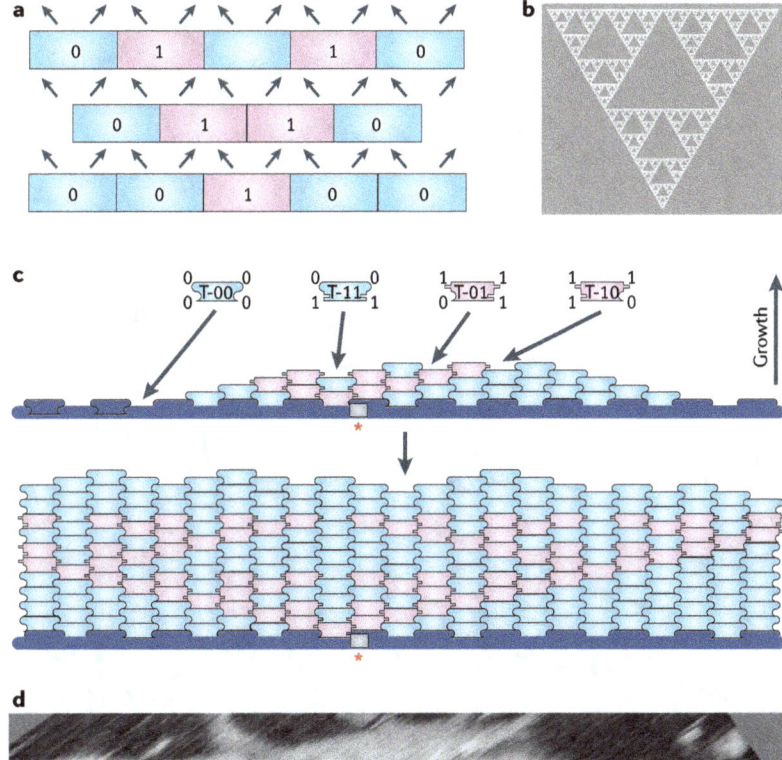

juxtaposed on adjacent rows, crossing over from one row to another in a manner resembling DX molecules. The staple strands therefore hold the pattern together rigidly when combined with the scaffold strand in solution.

Goodman *et al.*[44] gave an elegant design for a rigid 3D DNA tetrahedron that, like the octahedron, has a triangulated frame. Each tetrahedron is formed from five short strands, designed to favour the formation of tetrahedron edges in a hierarchical fashion, thereby providing faster assembly and higher yield than previous methods for the construction of 2D structures. Other new approaches to the construction of rigid DNA structures that can be replicated have been proposed by von Kiedrowski and co-workers[45–47].

Nanoelectronic assembly. Self-assembled tile structures made of DNA show promise in the construction of conductive nanowires and 2D arrays of nanocomponents. DNA tiles that are not in fact planar, but have some curvature, self-assemble into nanotubes rather than planar lattices[39,48]. Liu *et al.*[49] created a nanowire by metallizing a DNA nanotube with silver. The wire had significantly higher conductivity than wires built from dsDNA.

Many circuit subsystems, such as random-access memory and programmable logic arrays, are arranged in a 2D-array pattern. As a first step towards the construction of nanoelectronic arrays, Le *et al.*[50] assembled DNA–gold nanocomponents onto a 2D DNA array of DX molecules, with some molecules containing an extruding poly(A) ssDNA strand. Gold nanocomponents covalently linked to poly(T) strands then hybridize to the extruding poly(A) strands, resulting in a 2D array of gold nanoparticles. Such arrays could serve as a nanoscale memory, in which the electronic state of the nanoparticles could be read by scanning probes.

Algorithmic self-assembly. Winfree[51] forged a fundamental link between DNA self-assembly and computation: sticky-ended binding of tiles is, in fact, powerful enough to support general-purpose computation. To understand why this is true, consider the construction of a large jigsaw puzzle after first discarding the picture on the box cover. There is a way to make a coherent whole out of all the pieces. The 'program' for doing this is not a traditional sequence of instructions, but rather is implicitly specified by interconnection patterns (which function as sticky ends) between pieces. The DNA lattices and nanotubes described above are examples of periodic structures, the assembly of which is programmed by their sticky ends. Mao *et al.*[52] and Yan *et al.*[53] described the first aperiodic tile assembly in one dimension. A Sierpinski triangle is an elegant example of a 2D aperiodic pattern, which has been assembled using just four tiles, given a suitable starting frame, by Rothemund *et al.*[54] (FIG. 6).

Challenges and opportunities. Techniques for the scalable assembly of rigid 2D and 3D DNA structures suffer from errors that are due to inherent limitations in the specificity of sticky ends and the control of hybridization kinetics. Error-correction techniques have the potential to rectify mistakes in the self-assembly process[55,56]. DNA

Figure 6 | Construction of a two-dimensional aperiodic lattice. Four types of double-crossover (DX) molecule, distinguished by their sticky ends (**a**), represent four types of tile from which a Sierpinski triangle can self-assemble (**b**). Tiles assemble upwards from the blue frame, starting from the position marked with the red asterisk, forming patterned rows (**c**). Each tile is marked with either 0 or 1 (corresponding to the sticky ends of the DX molecule) on each of its four corners, and is added to the assembly in such a way that the mark on its lower-left corner matches the mark on the adjoining tile immediately to its left in the row below it, and similarly the mark on its lower-right corner matches the mark on the adjoining tile immediately to its right in the row below it. Crucial to the success of the construction is that the two sticky ends on the lower end of a DX molecule (tile) bind cooperatively to the upper sticky ends of the two matching DX molecules (tiles) below it; a tile with just one matching sticky end is considered to be too weak to persist. Analogously, if working on a puzzle, by moving inward from one corner of the frame and ensuring that each inserted puzzle piece properly connects with two pieces already in the puzzle, mistakes can be reduced. The proof proposed by Winfree to show that DNA self-assembly is able to perform general-purpose computation builds on the classical work of Wang[82], and relies on the cooperative binding of DNA tiles. An atomic-force microscopy image (**d**; courtesy of Erik Winfree, California Institute of Technology) shows a recognizable Sierpinski triangle fragment, with each error marked by a red cross. Red asterisks again represent starting points of self-assembly. Scale bar represents 100 nm. Reproduced with permission from REF. 54.

to plan the layout of a single long DNA strand, called a scaffold strand, which fills in the desired pattern row by row — similar to laying out yarn to create a string picture. Short staple strands are then designed, which can hybridize to segments of the scaffold strand that are

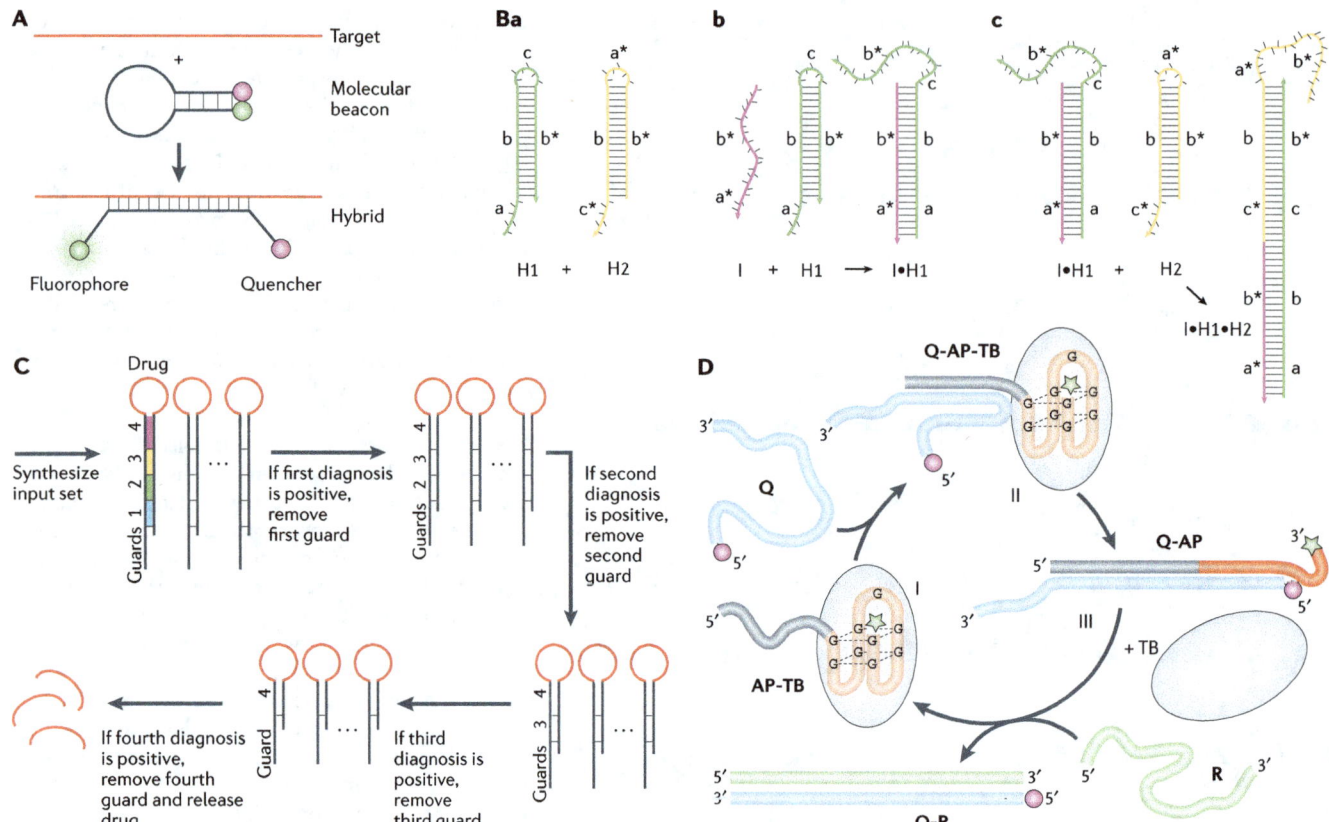

Figure 7 | **DNA-folding pathways and their uses. A** | A molecular beacon in which a stable hairpin structure, closed by a short helix, switches to a helix with emanating single-stranded ends on introduction of a strand that is complementary to the loop sequence of the hairpin. This change in structure separates the fluorescent label from the quencher, thereby producing a fluorescence signal. **B** | A folding pathway of the hybridization chain reaction[68], the initial structure of which consists of two types of hairpin, closed by long helices. **Ba** | Initially, two types of sequence fold into hairpin structures. The first type, H1, has a helix formed from complementary sequences (b and b*), a hairpin loop formed by sequence c and an unpaired sequence (a) emanating from the 5′ end of the helix. The second type, H2, has a similar helix with reverse polarity, but differs in its hairpin loop, which is the sequence a* complementary to a. **Bb** | On introduction of the target sequence (a*b*), the target binds to the unpaired 'a' region of an H1 hairpin and opens the stem through a branch migration, leaving cb* unpaired. **Bc** | This unpaired region then binds to and opens an H2 hairpin, exposing a new unpaired region (a*b*) on H2. This unpaired a*b* region is similar to the initial target sequence, forming the basis for a chain reaction that produces 'polymers' of alternating H1 and H2 components. **C** | A computation for disease diagnosis and drug delivery[71]. The inputs are all identical, consisting of a hairpin structure with a long helix, and a drug sequence encapsulated in the hairpin loop. The computation process is similar to a search-and-prune computation (FIG. 3), except that selection steps are replaced by diagnostic tests for disease indicators (high or low levels of expression of a gene) in the environment. Each positive diagnosis causes shortening of the hairpin helices, so that the drug is released if all diagnoses are positive. Therefore, in this computation, the inputs change at each step, in contrast to the search-and-prune computation of FIG. 3; this highlights the wide range of ways in which DNA-based computations can proceed. **D** | A device that binds and releases thrombin[72]. A DNA sequence contains the aptamer (AP, in red) for thrombin (TB, grey oval), concatenated with an additional ssDNA sequence (grey) and a fluorescence signal (green star). Addition of a blue sequence (a quencher, Q) causes the blue and grey regions to bind, ultimately disrupting the G-quartet structure of the thrombin aptamer and causing thrombin to be released. On addition of the green sequence (a remover, R), the blue and green sequences bind, displacing the black–red aptamer sequence through branch migration, and causing thrombin to be bound again. Further cycles are possible, by cycling addition of blue and green strands. Panel **b** reproduced with permission from REF. 68 © (2004) American Association for the Advancement of Science. Panel **c** reproduced with permission from REF. 72 © (2004) John Wiley & Sons Inc.

self-assembly of electronic components raises new research challenges, and opportunities for the design and modelling of circuit architectures[57].

Because, at present, there is limited understanding of the rules of formation of DNA tertiary structure, the development of DNA aptamers and enzymes has employed *in vitro*-selection methods rather than rational design. However, to develop larger DNA molecules with new binding properties, it is likely that a combination of rational design and selection will be effective. Combined approaches have already been used for the design of new RNA molecules[58].

Innovative uses of DNA-folding pathways

When a DNA molecule is not in its stable structure, perhaps because of a change in environmental conditions, it switches to a form that is stable in the new environment. In doing so, the molecule typically passes through a sequence of intermediate structures, called the folding pathway. The folding pathway of DNA can be put to good use in molecular detection to diagnose and respond to environmental conditions, to release energy (which can be used to drive nanoscale motors) and to execute computations.

FIGURE 7 illustrates several folding-pathway (or structure-switching) scenarios. In a typical case, DNA molecules are initially bound to each other, but have some single-stranded regions. When a new DNA molecule is introduced that can bind with a single-stranded region, it can displace base pairs by a process called branch migration. This displacement can cause the potential energy trapped in loops of the initial structure to be released, thereby providing fuel for DNA devices, such as DNA tweezers[59], and can also cause the movement of the DNA molecules. Simmel and Dittmer[60] provide a comprehensive overview of several feats of motion by DNA molecules, including unidirectional walking and movement along a track. Such devices might aid in the transport of tiny molecules in a nanoscale system. Alternatively, changes in buffer conditions, which are achieved by introducing positively charged ions, can cause a DNA helix to come apart and switch from its usual B form (in which the helix twists to the right) to the less standard Z form (in which the helix twists to the left)[61,62]. The addition of enzymes that cleave DNA also causes a change in the environment and a corresponding change in the stable structure of the collection of DNA molecules, which could be used to detect the presence of the enzymes. The following examples highlight work that directly explores the potential for applications of nanodevices that exploit folding pathways.

Molecular detection. Molecular beacons (FIG. 7A) are used to signal the presence of nucleic-acid pathogens and to monitor amplification by PCR[63,64], both in cells and *in vitro*. To detect a target DNA strand, the beacon initially forms a hairpin structure closed by a short helix, in which the loop sequence includes the complement of the target strand. One end of the sequence is linked to a fluorophore, whereas the other end is linked to a quencher. Binding of complement to target causes the hairpin to open, thereby separating the fluorophore and quencher, and causing a fluorescent signal to be emitted. Hamaguchi *et al.*[65], and Nutiu and Li[66] showed that features of molecular beacons and DNA aptamers could be integrated to obtain devices that signal on detection of proteins. For example, by incorporating the aptamer into the hairpin of the molecular beacon, the hairpin sequence can bind to the protein target of the aptamer, so that the hairpin opens and a fluorescent signal ensues. Stojanovic *et al.*[67] showed that, by combining both an aptamer for a reagent and a DNA enzyme in a single strand, it is possible to bring the enzyme and reagent together, thereby catalysing a reaction and detecting the signal.

The signal produced by molecular beacons is, at best, proportional to the amount of target to which the beacons are exposed, as each beacon can bind to, at most, one target. Dirks and Pierce[68] introduce the hybridization chain reaction (HCR) as a means to amplify the signal of an aptamer (sensor) for a DNA target (FIG. 7B). HCR reactions have the property that the molecular weight of the amplification polymers varies inversely with the concentration of the target molecule.

In a different chemical-sensing scheme, Heller *et al.*[69] show that dsDNA that is adsorbed onto a single-walled carbon nanotube can switch from the B to the Z form in the presence of cations, which are attracted to the negatively charged DNA backbone. This causes a detectable change in emission energy from the nanotube; moreover, the transition is reversible on removal of the ions. The DNA-coated nanotubes have been used to detect cations in the blood and tissues of living mammalian cells.

Diagnosis and drug release. Benenson *et al.*[70,71] designed a DNA device that performs several diagnostic tests for high or low concentrations of certain DNA molecules. The device embodies a simple 'if–then' computation: if the results of all tests are positive, then the device releases a drug (FIG. 7C). Their drug is a short antisense DNA that can bind to the mRNA transcribed from harmful genes, such as cancer-causing genes, thereby preventing their expression. The device progresses through several diagnostic states, which indicate the outcomes (positive or negative) of the diagnostic tests performed so far. These states are concretely represented by intermediate structures formed by the DNA components of the device. In the initial structure, the drug sequence is included in the loop of a hairpin closed by a long helix. On each positive diagnosis, the helix becomes shorter (through the action of an enzyme and other participating DNA strands), so that the loop sequence is completely cleaved from its closing helix and is therefore released only if all diagnoses are positive.

Binding and release of a protein target. Dittmer *et al.*[72] constructed a DNA device that can bind and release thrombin in a controlled fashion (FIG. 7D). The ability to perform such a function is an important step towards the goal of building artificial systems that can emulate the functions performed by molecules in the cell. One of the strands used in their device is the DNA aptamer for thrombin, with additional unpaired bases, called sticky ends, which do not interfere with the formation of the thrombin-binding G-quartet structure of the aptamer. This DNA aptamer is used to bind thrombin, thereby preventing it from performing other functions. To release thrombin, a second DNA strand is introduced, which binds with the sticky end of the aptamer and then displaces thrombin. A third strand can be introduced that binds with the second; this releases the thrombin-binding strand, so that it once again binds with thrombin. By alternating addition of the second and third strands, the device can repeatedly cycle between its binding and releasing functions.

Random-access memory
A means of storing digital information, in which data can be accessed and changed by directly accessing the location where they are stored; this is contrary to serial access memory, such as a tape, where it is necessary to access the data in serial order.

Programmable logic array
An array of logic gates that can be programmed to compute simple logic functions.

DNA tweezers
A device made from DNA, in which the angle subtended by two connected rigid helices can be made large or small by the use of additional DNA 'fuel' strands, thereby emulating the opening and closing of tweezers.

Challenges and opportunities. Using DNA devices *in vivo* presents formidable challenges, such as the need to power the devices with DNA (or other) fuel or to handle waste products. Dittmer *et al.*[73,74] suggest that the instructions for the synthesis of DNA machine components could be encoded in an artificial gene, and that expression of the gene could then control the workings of the device. They give a simple demonstration *in vitro*. However, the issue of waste-product disposal was not addressed in this work. Other long-term uses of DNA motors could be in the development of systems that are able to undergo synthesis[75], self-replication or evolution[76]; such systems in the cell rely on mechanisms for controlled transport and release of materials — tasks that DNA can, in principle, perform.

There are undoubtedly many other ways in which DNA-structure switching or folding pathways could be used to keep track of the states of a system; recent work in the use of structure to encode states of a molecular machine suggest some intriguing possibilities[77,78].

Conclusions

As shown by the examples above, there has been tremendous recent progress in the design and controlled use of DNA molecules. The stability and sequence properties of DNA are ideal for information storage and processing in a biological setting, and can facilitate the synthesis and discovery of new molecules. The structural properties of DNA make it effective as a rigid 2D or even 3D scaffold for the assembly of molecules and electronic components. The folding pathway of DNA allows it to work as a mechanical device and adapt to changing environmental conditions.

In turn, the process of designing structures and devices from DNA helps to advance our understanding of the properties of DNA and, more generally, of how complex biological systems work.

In the future, the ability to build devices that integrate several functions of DNA, such as the discovery of new reactions with the synthesis and directed evolution of promising reagents, could further enhance the usefulness of DNA. Additionally, the insertion of genes that encode DNA molecules or devices with diagnostic or therapeutic applications into synthetic gene networks could provide a means for expression of those devices in cells or in other diagnostic environments[79,80]. DNA might eventually do itself out of business by allowing the synthesis, discovery and control of molecules with more complex functions.

1. Breaker, R. R. Natural and engineered nucleic acids as tools to explore biology. *Nature* **432**, 838–845 (2004).
2. Emilsson, G. M. & Breaker, R. R. Deoxyribozymes: new activities and new applications. *Cell. Mol. Life Sci.* **59**, 596–607 (2002).
3. Joyce, G. F. Directed evolution of nucleic acid enzymes. *Ann. Rev. Biochem.* **73**, 791–836 (2004).
4. Brenner, S. & Lerner, R. A. Encoded combinatorial chemistry. *Proc. Natl Acad. Sci. USA* **89**, 5381–5383 (1992).
5. Janeway, C. A., Travers, P., Walport, M. & Capra, J. D. *Immunobiology: The Immune System in Health and Disease* 6th edn (Garland Science Publishing, New York, 2005).
6. Niemeyer, C. M., Adler, M. & Wacker, R. Immuno-PCR: high sensitivity detection of proteins by nucleic acid amplification. *Trends Biotechnol.* **23**, 208–216 (2005).
7. Sano, T., Smith, C. L. & Cantor, C. R. Immuno-PCR: very sensitive antigen detection by means of specific antibody–DNA. *Science* **258**, 120–122 (1992).
8. Nam, J.-M., Thaxton, C. S. & Mirkin, C. A. Nanoparticle-based bio-bar codes for the ultrasensitive detection of proteins. *Science* **301**, 1884–1886 (2003).
   Introduces the BCA method for the highly sensitive detection of proteins. Subsequent papers extend the use of the method to disease diagnosis.
9. Nam, J.-M., Stoeva, S. & Mirkin, C. A. Bio-bar-code-based DNA detection with PCR-like sensitivity. *J. Am. Chem. Soc.* **126**, 5932–5933 (2004).
10. Georganopoulou, D. G. *et al.* Nanoparticle-based detection in cerebral spinal fluid of a soluble pathogenic biomarker for Alzheimer's disease. *Proc. Natl Acad. Sci. USA* **102**, 2273–2276 (2005).
11. Nam, J.-M., Wise, A. R. & Groves, J. T. Colorimetric bio-barcode amplification assay for cytokines. *Anal. Chem.* **77**, 6985–6988 (2005).
12. Fredriksson, S. *et al.* Protein detection using proximity-dependent DNA ligation assays. *Nature Biotechnol.* **20**, 473–477 (2002).
13. Gartner, Z. J. & Liu, D. R. The generality of DNA-templated synthesis as a basis for evolving non-natural small molecules. *J. Am. Chem. Soc.* **123**, 6961–6963 (2001).
14. Rosenbaum, D. M. & Liu, D. R. Efficient and sequence-specific DNA-templated polymerization of peptide nucleic acid aldehydes. *J. Am. Chem. Soc.* **125**, 13924–13925 (2003).
15. Gartner, Z. J. *et al.* DNA-templated organic synthesis and selection of a library of macrocycles. *Science* **305**, 1601–1605 (2004).
    Illustrates the use of DTS to create a pool of molecules from which those with new functional properties might be selected.
16. Halpin, D. R. & Harbury, P. B. DNA display I. Sequence-encoded routing of DNA populations. *PLoS Biol.* **2**, 1015–1021 (2004).
17. Halpin, D. R. & Harbury, P. B. DNA display II. Genetic manipulation of combinatorial chemistry libraries for small-molecule evolution. *PLoS Biol.* **2**, e174 (2004).
    Reports a method for the synthesis, selection and evolution of new compounds in parallel from a large pool.
18. Halpin, D. R., Lee, J. A., Wrenn, S. J. & Harbury, P. B. DNA display III. Solid-phase organic synthesis on unprotected DNA. *PLoS Biol.* **2**, e175 (2004).
19. Calderone, C. T., Puckett, J. W., Gartner, Z. J. & Liu, D. R. Directing otherwise incompatible reactions in a single solution by using DNA-templated organic synthesis. *Angew. Chem. Int. Ed.* **41**, 4104–4108 (2002).
20. Kanan, M. W., Rozeman, M. M., Sakurai, K., Snyder, T. M. & Liu, D. R. Reaction discovery enabled by DNA-templated synthesis and *in vitro* selection. *Nature* **431**, 545–549 (2004).
21. Adleman, L. Molecular computation of solutions to combinatorial problems. *Science* **266**, 1021–1024 (1994).
    Introduces the principle of DNA computation by solving a tiny combinatorial problem in a test tube.
22. Braich, R. S., Chelyapov, N., Johnson, C., Rothemund, P. W. K. & Adleman, L. Solution of a 20-variable 3-SAT problem on a DNA computer. *Science* **296**, 499–502 (2002).
23. Williams, K. A., Veenhuizen, P. T. M., de la Torre, B. G., Eritja, R. & Dekker, C. Carbon nanotubes with DNA recognition. *Nature* **420**, 19–26 (2002).
24. Xin, H. & Woolley, A. T. DNA-templated nanotube localization. *J. Am. Chem. Soc.* **125**, 8710–8711 (2003).
25. Braun, E., Eichen, Y., Sivan, U. & Ben-Yoseph, G. DNA-templated assembly and electrode attachment of a conducting silver wire. *Nature* **391**, 775–778 (1998).
26. Keren, K., Berman, R. S., Buchstab, E., Sivan, U. & Braun, E. DNA-templated carbon nanotube field-effect transistor. *Science* **302**, 1380–1382 (2003).
27. Zheng, M. *et al.* Structure-based carbon nanotube sorting by sequence-dependent DNA assembly. *Science* **302**, 1545–1548 (2003).
28. Staii, C. & Johnson, A. T. DNA-decorated carbon nanotubes for chemical sensing. *Nano Lett.* **5**, 1774–1778 (2005).
29. Storhoff, J. J. & Mirkin, C. A. Programmed materials synthesis with DNA. *Chem. Rev.* **99**, 1849–1862 (1999).
30. Niemeyer, C. M. Nanoparticles, proteins, and nucleic acids: biotechnology meets materials science. *Angew. Chem. Int. Ed.* **40**, 4128–4158 (2001).
31. Niemeyer, C. M., Koehler, J. & Wuerdemann, C. DNA-directed assembly of bienzymic complexes from *in vivo* biotinylated NAD(P)H:FMN oxidoreductase and luciferase. *ChemBioChem* **02–03**, 242–245 (2002).
32. Bock, L. C., Griffin, L. C., Latham, J. A., Vermaas, E. H. & Toole, J. J. Selection of single-stranded DNA molecules that bind and inhibit human thrombin. *Nature* **355**, 564–566 (1992).
33. Li, Y. *et al.* Multiplexed detection of pathogen DNA with DNA-based fluorescence nanobarcodes. *Nature Biotechnol.* **23**, 885–889 (2005).
34. Stavis, S. M. *et al.* Detection and identification of nucleic acid engineered fluorescent labels in submicrometre fluidic channels. *Nanotechnology* **16**, S314–S323 (2005).
35. Seeman, N. C. DNA nanotechnology: novel DNA constructions. *Annu. Rev. Biophys. Biomol. Struct.* **27**, 225–248 (1998).
36. Seeman, N. C. DNA in a material world. *Nature* **421**, 427–431 (2003).
37. Fu, T.-J. & Seeman, N. C. DNA double crossover molecules. *Biochemistry* **32**, 3211–3220 (1993).
38. Winfree, E., Liu, F., Wenzler, L. A. & Seeman, N. C. Design and self-assembly of two-dimensional DNA crystals. *Nature* **394**, 539–544 (1998).
    Illustrates the principle of algorithmic self-assembly computation, through the construction of rigid 2D lattices from rigid DX molecules.
39. Yan, H., Park, S.-H., Finkelstein, G., Reif, J. H. & LaBean, T. H. DNA-templated self-assembly of protein arrays and highly conductive nanowires. *Science* **301**, 1882–1884 (2003).
    Describes the construction and use of 2D DNA lattices for the organization of proteins, and the construction of DNA nanotubes.
40. Park, S.-H. *et al.* Programmable DNA self-assemblies for nanoscale organization of ligands and proteins. *Nano Lett.* **5**, 729–733 (2005).

41. Liu, Y., Lin, C., Li, H. & Yan, H. Aptamer-directed self-assembly of protein arrays on a DNA nanostructure. *Angew. Chem. Int. Ed.* **44**, 4333–4338 (2005).

42. Rothemund, P. Folding DNA to create nanoscale shapes and patterns. *Nature* **440**, 297–302 (2006).
    Reports on the scaffolded origami technique for constructing complex 2D patterns.

43. Shih, W. M., Quispe, J. D. & Joyce, G. F. A 1.7-kilobase single-stranded DNA that folds into a nanoscale octahedron. *Nature* **427**, 618–621 (2004).
    Describes the construction of the first 3D rigid shape — an octahedron — from DNA, using short scaffold DNA strands to direct the folding of a long DNA strand.

44. Goodman, R. P. *et al.* Rapid chiral assembly of rigid DNA building blocks for molecular nanofabrication. *Science* **310**, 1661–1665 (2005).

45. von Kiedrowski, G. *et al.* Toward replicatable, multifunctional, nanoscaffolded machines. a chemical manifesto. *Pure Appl. Chem.* **75**, 609–619 (2003).

46. Scheffler, M., Dorenbeck, A., Jordan, S., Wustefeld, M. & von Kiedrowski, G. Self-assembly of trisoligonucleotidyls: the case for nanoacetylene and nano-cyclobutadiene. *Angew. Chem. Int. Ed.* **38**, 3311–3315 (1999).

47. Eckhardt, L. H. *et al.* DNA nanotechnology: chemical copying of connectivity. *Nature* **420**, 286 (2002).

48. Rothemund, P. W. K. *et al.* Design and characterization of programmable DNA nanotubes. *J. Am. Chem. Soc.* **126**, 16344–16353 (2004).

49. Liu, D., Park, S.-H., Reif, J. H. & LaBean, T. H. DNA nanotubes self-assembled from TX tiles as templates for conductive nanowires. *Proc. Natl Acad. Sci. USA* **101**, 717–722 (2004).

50. Le, J. D. *et al.* DNA-templated self-assembly of metallic nanocomponent arrays on a surface. *Nano Lett.* **4**, 2343–2347 (2004).

51. Winfree, E. On the computational power of DNA annealing and ligation. *DIMACS Ser. Discrete Math. Theoret. Comput. Sci.* **27**, 199–219 (1996).

52. Mao, C., LaBean, T. H., Reif, J. H. & Seeman, N. C. Logical computation using algorithmic self-assembly of DNA triple-crossover molecules. *Nature* **407**, 493–496 (2000).

53. Yan, H., LaBean, T. H., Feng, L. & Reif, J. H. Directed nucleation assembly of DNA tile complexes for barcode-patterned lattices. *Proc. Natl Acad. Sci. USA* **100**, 8103–8108 (2003).

54. Rothemund, P. W. K., Papadakis, N. P. & Winfree, E. Algorithmic self-assembly of DNA Sierpinski triangles. *PLoS Biol.* **2**, e424 (2004).

55. Winfree, E. & Bekbolatov, R. Proofreading tile sets: error correction for algorithmic self-assembly. *Lect. Notes Comput. Sci.* **2943**, 126–144 (2004).

56. Chen, H-L. & Goel, A. Error free self-assembly using error prone tiles. *Lect. Notes Comput. Sci.* **3384**, 62–75 (2005).

57. Dwyer, C., Lebeck, A. R. & Sorin, D. J. Self-assembled architectures and the temporal aspects of computing. *Computer* **38**, 56–64 (2005).

58. Breaker, R. Engineered allosteric ribozymes as biosensor components. *Curr. Opin. Biotechnol.* **13**, 31–39 (2002).

59. Yurke, B., Turberfield, A. J., Mills, A. P., Simmel, F. C. & Neumann, J. L. A DNA-fuelled molecular machine made of DNA. *Nature* **406**, 605–608 (2000).

60. Simmel, F. C. & Dittmer, W. U. DNA nanodevices. *Small* **1**, 284–299 (2005).

61. Jovin, T. M., Soumpasis, D. M. & McIntosh, L. P. The transition between B-DNA and Z-DNA. *Annu. Rev. Phys. Chem.* **38**, 521–558 (1987).

62. Mao, C. D., Sun, W. Q., Shen, Z. Y. & Seeman, N. C. A nanomechanical device based on the BZ transition of DNA. *Nature* **397**, 144–146 (1999).

63. Tyagi, S. & Kramer, F. R. Molecular beacons: probes that fluoresce upon hybridization. *Nature Biotechnol.* **14**, 303–308 (1996).

64. Anthony, T. & Subramaniam, V. Molecular beacons: nucleic acid hybridization and emerging applications. *J. Biomol. Struct. Dyn.* **19**, 497–504 (2001).

65. Hamaguchi, N., Ellington, A. & Stanton, M. Aptamer beacons for the direct detection of proteins. *Anal. Biochem.* **294**, 126–131 (2001).

66. Nutiu, R. & Li, Y. Structure-switching signaling aptamers. *J. Am. Chem. Soc.* **125**, 4771–4778 (2003).

67. Stojanovic, M. N., de Prada, P. & Landry, D. W. Catalytic molecular beacons. *Chembiochem* **2**, 411–415 (2001).

68. Dirks, R. M. & Pierce, N. A. Triggered amplification by hybridization chain reaction. *Proc. Natl Acad. Sci. USA* **101**, 15275–15278 (2004).

69. Heller, D. A. *et al.* Optical detection of DNA conformational polymorphism on single-walled carbon nanotubes. *Science* **311**, 508–511 (2006).

70. Benenson, Y. *et al.* Programmable and autonomous computing machine made of biomolecules. *Nature* **414**, 430–434 (2001).

71. Benenson, Y., Gil, B., Ben-Dor, U., Adar, R. & Shapiro, E. An autonomous molecular computer for logical control of gene expression. *Nature* **429**, 423–429 (2004).
    Describes the design and uses of a computational DNA device *in vitro* to control the release of a drug.

72. Dittmer, W. U., Reuter, A. & Simmel, F. C. A DNA-based machine that can cyclically bind and release thrombin. *Angew. Chem. Int. Ed.* **43**, 3550–3553 (2004).
    Presents the design of a mechanical device, built from DNA, which can bind and release thrombin.

73. Dittmer, W. U. & Simmel, F. C. Transcriptional control of DNA-based nanomachines. *Nano Lett.* **4**, 689–691 (2004).

74. Dittmer, W. U., Kempter, S., Radler, J. O. & Simmel, F. C. Using gene regulation to program DNA-based molecular devices. *Small* **1**, 709–712 (2005).

75. Liao, S. & Seeman, N. C. Translation of DNA signals into polymer assembly instructions. *Science* **306**, 2702–2704 (2004).

76. Yurke, B. & Mills, A. P. Using DNA to power nanostructures. *Gen. Prog. Evol. Mach.* **4**, 111–122 (2003).

77. Uejima, H. & Hagiya, M. Secondary structure design of multi-state DNA machines based on sequential structure transitions. *Lect. Notes Comp. Sci.* **2943**, 74–85 (2004).

78. Stojanovic, M. N. & Stefanovic, D. A deoxyribozyme-based molecular automaton. *Nature Biotechnol.* **21**, 1069–1074 (2003).

79. McDaniel, R. & Weiss, R. Advances in synthetic biology: on the path from prototypes to applications. *Curr. Opin. Biotechnol.* **16**, 476–483 (2005).

80. Weiss, R. Challenges and opportunities in programming living cells. *The Bridge* **33**, 39–46 (2003).

81. Macaya, R. F., Schultze, P., Smith, F. W., Roe, J. A. & Feigon, J. Thrombin-binding DNA aptamer forms a unimolecular quadruplex. *Proc. Natl Acad. Sci. USA* **90**, 3745–3749 (1993).

82. Wang, H. *An Unsolvable Problem on Dominoes*. *Technical Report BL-30 (II-15)* (Harvard Computation Laboratory, Cambridge, Massachusetts, 1962).

Competing interests statement
The author declares no competing financial interests.

DATABASES
The following terms in this article are linked online to:
OMIM: http://www.ncbi.nlm.nih.gov/entrez/query.fcgi?db=OMIM
Alzheimer disease | breast cancer | prostate cancer
UniProtKB: http://ca.expasy.org/sprot
interleukin 2

FURTHER INFORMATION
Anne Condon's homepage: http://www.cs.ubc.ca/~condon
Access to this links box is available online.

DNA nanomachines

We are learning to build synthetic molecular machinery from DNA. This research is inspired by biological systems in which individual molecules act, singly and in concert, as specialized machines: our ambition is to create new technologies to perform tasks that are currently beyond our reach. DNA nanomachines are made by self-assembly, using techniques that rely on the sequence-specific interactions that bind complementary oligonucleotides together in a double helix. They can be activated by interactions with specific signalling molecules or by changes in their environment. Devices that change state in response to an external trigger might be used for molecular sensing, intelligent drug delivery or programmable chemical synthesis. Biological molecular motors that carry cargoes within cells have inspired the construction of rudimentary DNA walkers that run along self-assembled tracks. It has even proved possible to create DNA motors that move autonomously, obtaining energy by catalysing the reaction of DNA or RNA fuels.

JONATHAN BATH AND ANDREW J. TURBERFIELD*

University of Oxford, Department of Physics, Clarendon Laboratory, Parks Road, Oxford OX1 3PU, UK

*e-mail: a.turberfield@physics.ox.ac.uk

The remarkable specificity of the interactions between complementary nucleotides makes DNA a useful construction material: interactions between short strands of DNA can be controlled with confidence through design of their base sequences (Box 1). The construction of branched junctions between double helices[1] makes it possible to create complex three-dimensional objects[2–5], such as the tetrahedron[5] shown in Fig. 1, by self-assembly. One way to exploit this extraordinarily precise architectural control is to use self-assembled DNA templates to position functional molecules: examples include molecular electronic circuits[6,7], near-field optical devices[8] and enzyme networks[9].

It is an obvious extension of this research to convert static DNA structures into machines. DNA is not the natural choice of material to build active structures with because it lacks the structural and catalytic versatility of proteins and RNA (for both DNA and RNA, Watson–Crick base pairing is the strongest interaction determining inter- and intramolecular interactions, but RNA has a much richer repertoire of weaker non-covalent interactions that can stabilize complex structures[10]). If we could cope with the interactions required for a three-dimensional fold we would design more competent machines made, as in nature, from RNA and proteins[11,12]. We make nanomachines from DNA because the simplicity of its structure and interactions allows us to control its assembly.

In this review we concentrate on research that is leading towards the development of synthetic molecular motors. We start by showing how DNA nanostructures can be made to switch between two states in response to molecular or environmental signals; we describe how a device can be moved along a track by operating molecular switches in the correct sequence; we finish with an account of the current state

of development of autonomous molecular motors that are inspired by the natural protein motors myosin and kinesin. Closely related work on DNA sensors and DNA-templated chemistry is described briefly in Boxes 2 and 3.

MOLECULAR SWITCHES

The simplest active DNA nanostructures are switches or actuators that can be driven between two conformations. Motion is induced by changes in temperature or ionic conditions, or by the binding of a signalling molecule, often a DNA strand.

CONFORMATION CHANGES INDUCED BY CHANGES IN ENVIRONMENT

Rotary motion can be produced by changing the twist of DNA. Double-stranded DNA with the sequence $(CG)_n$ can be flipped from the usual right-handed helix (B-DNA) to a left-handed conformation (Z-DNA). This transition is favoured by high salt concentrations and low temperatures[13]. One of the earliest nanomechanical DNA devices[14] used this transition to change the angle between two rigid DNA tiles connected by a $(CG)_{10}$ stem. Each tile carried a reporter fluorophore. Förster resonant energy transfer (FRET) between fluorophores allows sensitive measurement of their separation on a nanometre length scale: the efficiency of energy transfer, mediated by a dipole–dipole interaction, scales as the inverse sixth power of their separation[15]. When the B–Z transition was induced by an increase in ionic strength, FRET measurements showed an increase in the separation between the fluorophores consistent with the expected relative rotation of the tiles by ~3.5 turns.

Yang and co-workers converted changes in the twist of DNA into linear motion[16]. Their device consisted of a closed loop of double-stranded DNA attached to opposite arms of a four-arm Holliday junction (Box 1)[17]. A Holliday junction can migrate (isomerize) by breaking identical base pairs in one pair of opposite arms and remaking them in the other pair. A change in the

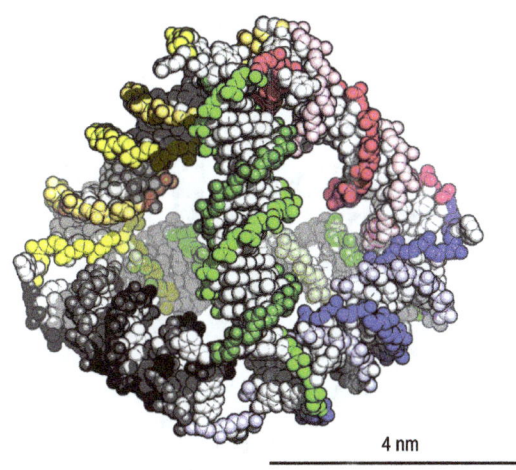

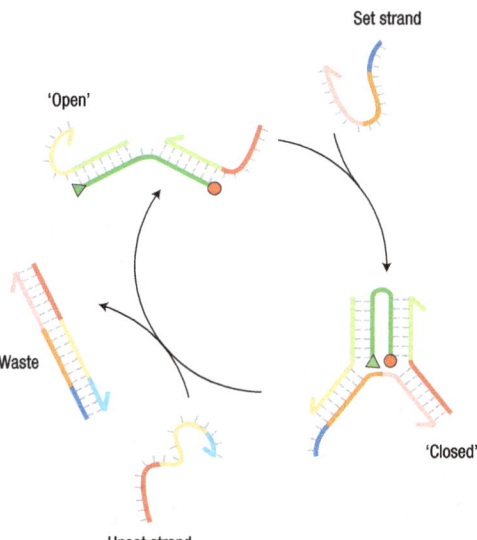

Figure 1 Self-assembly of a nanometre-scale object. The DNA tetrahedron[5] has relatively stiff double-stranded edges linked by flexible single-stranded hinges. A cargo, for example a protein[48], can be trapped in the central cavity of the tetrahedron. Mechanical devices built from DNA could be used to open the tetrahedron (R. P. Goodman, M. Heilemann, A. N. Kapenidis & A.J.T., manuscript in preparation) to control access to the cargo.

Figure 2 A DNA nanomachine driven by repeated sequential addition of DNA control strands. DNA tweezers[28] have two double-stranded arms connected by a flexible single-stranded hinge. The 'set' strand pulls the arms into a closed conformation by hybridizing to single-stranded tails at the ends of the arms. A short region of the set strand remains single-stranded even when it is hybridized to the tweezers: this region serves as a toehold that allows the 'unset' strand to hybridize to the set strand and strip it from the device, returning the tweezers to the open configuration and generating a double-stranded waste product. The state of the device can be determined by measuring the separation between donor and acceptor fluorophores (represented by the green triangle and red circle) using FRET.

conformation of the DNA within the loop was initiated by adding ethidium bromide: this intercalating dye binds between adjacent base pairs, lengthening and partially unwinding the double helix. The resulting stress was relieved by junction migration: by shortening the protruding arms of the junction the loop was allowed to lengthen without changing the total number of twists within it.

Environmentally driven changes in the conformation of single-stranded DNA can induce linear motion. Under slightly acidic conditions, a single strand with appropriately spaced cytosine bases folds into an i-motif — a compact three-dimensional structure that is held together by cytosine–protonated-cytosine base pairs[18]. In the presence of a near-complementary strand of DNA there is competition between the i-motif and an extended double helix formed by hybridization of the two strands (note that a perfectly complementary strand would itself fold into a stable structure called the G-quadruplex[19]). The i-motif device can be switched between compact and extended states by changing the pH[20,21]. Cyclic switching can be driven by an oscillating chemical reaction[22,23].

The i-motif-to-duplex transition has been made to do mechanical work. One surface of a silicon cantilever was coated with tethered cytosine-containing strands and formation of the i-motif induced a compressive surface stress that bent the cantilever[24]. The origin of the surface stress, which is also observed when complementary strands hybridize to tethered probes[25], is not understood. Electrostatic repulsion between the compact i-motifs plays a part, but the effect persists with high salt concentrations at which interactions over a distance comparable to the separation between strands are effectively screened.

It has also been suggested that the conformational change resulting from pH-dependent binding of a single strand of DNA to a duplex to form a triple-helical structure could be used as the basis of a nanomechanical actuator[26,27].

Devices such as those described above can be used to monitor and report on their environment. Box 2 describes how active DNA nanostructures are being developed as sensors, how elementary logical operations can be performed on their outputs, and how the

combination of sensors and computation might be used to create smart drug-delivery systems.

CONFORMATION CHANGES INDUCED BY SIGNALLING

Yurke and co-workers[28] constructed a pair of DNA tweezers with two rigid double-stranded arms connected at one end by a flexible single-stranded hinge (Fig. 2). In the open configuration, the arms rotate freely about the hinge. Single-stranded tails extend from the free end of both arms and serve as attachment points for a control strand — the 'set' or 'fuel' strand — that can pull the arms together by hybridizing to both of them. A short region of the set strand remains single-stranded even when hybridized to the device: it serves as a toehold for hybridization of a complementary 'unset' or 'antifuel' strand that strips the set strand from the device by branch migration (Box 1). Displacement of the set strand generates a double-stranded waste product and resets the device to its initial open configuration. The device can be driven through many cycles of operation simply by repeated sequential addition of set and unset strands. The time to half completion of a single switching operation is ~10 s at typical (micromolar) control-strand concentrations: the rate constant for toehold-mediated strand exchange is ~10^5 M^{-1} s^{-1} (ref. 29). Operation of the DNA tweezers has recently been characterized using single molecule FRET measurements[30].

Strand displacement has been used to effect conformation changes in a wide variety of systems. A number of variations on the tweezers have been reported including a device where the arms are pushed apart instead of being pulled together[31] and a three-state device[32]. Yan and co-workers[33] constructed a linear array of rigid DNA tiles in which adjacent tiles could be flipped between *cis* and *trans* conformations by stripping away and replacing control

Box 1 The building material

In **a**, the DNA backbone, shown in green, is an alternating string of deoxyribose sugars and phosphate groups whose polarity is indicated by the end-labels 5′ and 3′. One of the four bases G, A, T, C (guanine, adenine, thymine, cytosine) is attached to each sugar (bases are shown in grey). Hydrogen bonding between complementary CG and AT base pairs holds two strands of DNA together to form a right-handed double helix (formation of the double helix by base pairing between complementary strands is described as hybridization). For clarity, the helical character of DNA is omitted from the figures that follow: the backbone is represented as a coloured line and the bases (paired or unpaired) as short grey lines. Complementary sequences are indicated by light and dark shades of the same colour.

Double-stranded DNA has a persistence length of 50 nm (ref. 90). On the length scale of most devices described here (a few tens of base pairs, or ~10 nm) it can be considered to be stiff and straight. Single-stranded DNA is very flexible: its persistence length is in the order of three backbone units[91].

Enzymes can be used to break the phosphodiester backbone of DNA. A typical restriction enzyme recognizes a specific sequence of bases within double-stranded DNA and cuts both strands within the recognition sequence (black triangles indicate cuts made by the enzyme EcoRI).

In **b**, short single-stranded 'sticky ends', shown in red, can be used to join two segments of double-stranded DNA. Hybridization between complementary sticky ends leaves nicks in the backbone where the 3′ end of one strand meets the 5′ end of another. Nicks can be repaired by DNA ligase, an enzyme that catalyses formation of a phosphodiester bond between adjacent 3′ hydroxyl and 5′ phosphate groups.

A four-arm junction made from four strands of DNA (**c**) illustrates the principle that self-assembly can be controlled through design of the component strands. The interactions that hold together the four double-helical arms are encoded in the base sequences of the strands by ensuring that sections of DNA that should hybridize to each other have complementary sequences. Undesired interactions are 'designed out' by adjusting the sequences to reduce the stability of competing structures[92–94]. For a simple structure, this design process will ensure that when the component strands are annealed the desired structure will be formed in high yield. An analogous procedure was used to create the tetrahedron shown in Fig. 1 in a single assembly step[5].

A Holliday junction is a four-arm structure in which opposite arms have the same nucleotide sequences[17]. The position of the junction can migrate in a process in which base pairs in one opposite pair of arms break and are reformed in the other pair.

A short single-stranded sticky end can also serve as a toehold for strand exchange[29], as shown in **d**. A complementary strand which binds first to the toehold can displace the shorter strand of the original duplex: strand exchange proceeds by means of a random walk of the branch point separating regions hybridized to the competing strands. Strand exchange is driven by the greater stability of the duplex formed with the invading strand, which results from the additional base pairs in the toehold region. Several devices that use strand exchange to drive conformation changes are described in this review.

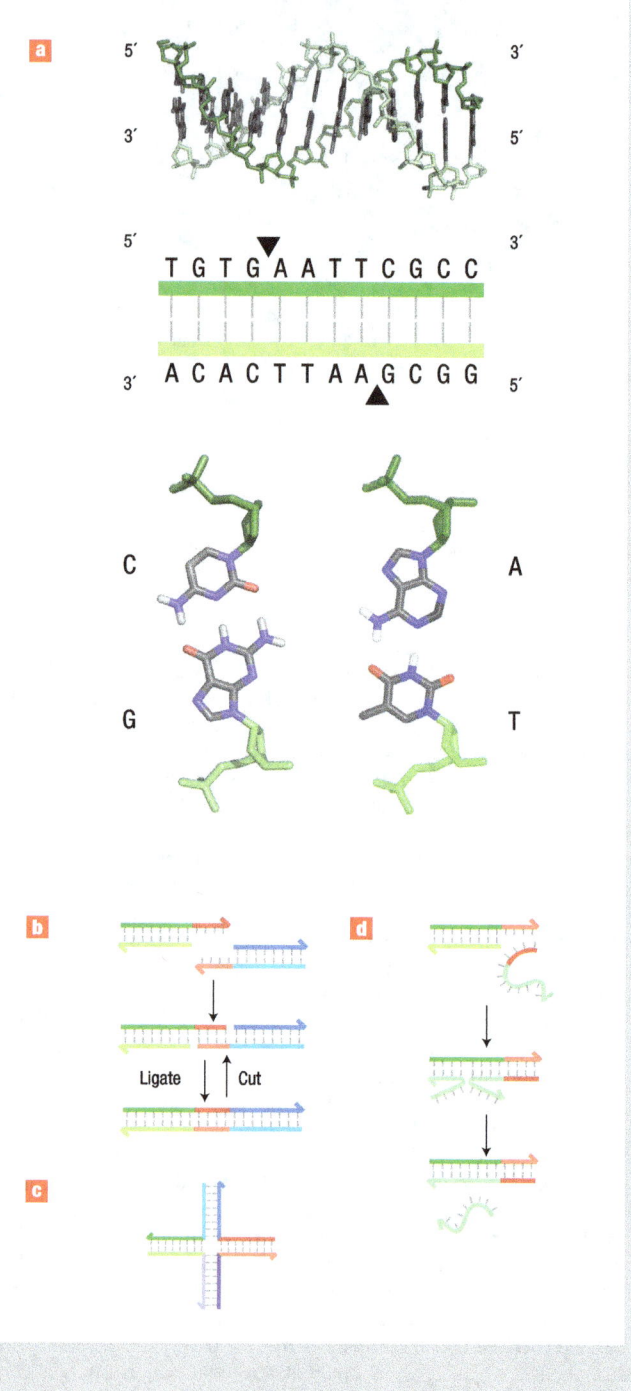

strands. Different conformations were easily distinguished by atomic force microscopy. The same device has recently been incorporated into a two-dimensional DNA array[34]. Feng *et al.*[35] created a two-dimensional array that could be switched between states with different lattice spacings and Hazarika and co-workers[36] used strand displacement to reverse the aggregation of gold nanoparticles held

together by DNA bridges[37]. A number of groups have used DNA control strands to switch a section of DNA between a single-stranded state, designed to fold into a compact G-quadruplex stabilized by hydrogen-bonded tetrads of guanine[19], and an extended duplex[38–40].

Motion of DNA devices can be triggered using RNA control strands[41]. By using a specific messenger RNA (mRNA) as a control

Box 2 Sensors that can process information

A DNA device that changes conformation in response to a chemical or physical input, and that couples the change to a useful output is a sensor. Temperature[95] and pH[20] sensors have been described. FRET changes have been used to detect the deformation of a double helix when bound by a protein and to estimate the DNA–protein interaction energy[96]. Molecular beacons[97] are routinely used to detect specific DNA sequences. Hybridization to the target opens a hairpin loop and separates fluorophores attached to the neck domains. Aptamers, which are oligonucleotides capable of binding specific targets are generated from random sequence libraries by repeated *in vitro* selection and amplification[98]. Catalytic oligonucleotides (DNA enzymes and ribozymes) can be selected using a similar strategy. DNA sensors that incorporate aptamers can be designed to detect small molecules or proteins[83,99]. Release of the blood clotting

protein thrombin from a DNA aptamer has been triggered using a DNA control strand[99-101].

DNA-based sensors can incorporate gain. An ATP-binding aptamer has been used to trigger the polymerization of DNA hairpin loops[83]. An RNA aptamer has been used to couple detection of a protein to the release and activation of a viper venom enzyme that triggers a blood-clotting cascade, generating a visible signal by precipitating polystyrene microspheres[102]. A colour-based lead sensor has been constructed from a lead-dependent DNA enzyme that cleaves the DNA linkages holding together a network of gold nanoparticles[103]. Various schemes for detecting target DNA sequences make use of DNA enzymes to amplify the input signal: the target sequence can switch on a DNA enzyme[104,105] or a ribozyme[106], or act as a primer to trigger the production of a DNA enzyme[107] by isothermal amplification[108].

Ribozymes whose activity is regulated by control strands can be used to perform logical operations[109,110]. In **a**, the DNA enzyme (light green) binds to a DNA substrate (dark green) incorporating a ribobase (white) and cuts it. A simple modification to the DNA enzyme produces a NOT gate (**b**): the DNA enzyme will not cut in the presence of strand x (blue) because its binding disrupts the active site. In **c**, further modification produces a DNA enzyme that will not cut in the absence of strand y (pink) — the substrate will only bind when y is present. Simple programs can be executed by combinations of gates[111,112].

Seelig *et al.* have built logic gates that are based on DNA hybridization rather than DNA hydrolysis[113]. Their AND gate uses hybridization of the first input strand to the gate complex to uncover a toehold that allows binding of a second input strand, releasing the output strand. Outputs from one operation can be used as inputs to another, opening the way to the creation of multilayer circuits.

In principle, molecular computation could be used to control a therapeutic device. For example, a device operating within cells could trigger programmed cell death in response to a specific set of mRNAs. Working towards this goal, Benenson *et al.* constructed a system of enzymes and DNA fragments that can release a short strand of DNA (which could have a therapeutic effect) after successful completion of a series of tests for the presence or absence of DNA signals[114]. Control of gene expression has been demonstrated using engineered riboregulators, non-coding RNA fragments that can interfere with transcription and whose interaction with the target gene can be modulated by binding of a signal RNA or small molecule[115,116].

---

strand, a tweezer device has been used to sense *in vitro* transcription[42,43]. These experiments show how DNA devices could be controlled by transcriptional circuits[44-47], and point to future applications in which the presence of a specific mRNA within a cell might trigger an event such as the release of a caged drug[48].

The maximum force exerted by hybridization of a signalling strand of DNA can be estimated as the free energy change of hybridization divided by the distance through which the hybridizing strands must move together, and is of the order of 10 pN (ref. 28). The fact that protein motors generate similar forces[49-52] encourages speculation that DNA hybridization could be used to drive DNA devices that mimic natural molecular motors.

CLOCKED WALKERS

Machines that can be driven through more than one conformational transition are capable of relatively sophisticated tasks, such as

programmable synthesis of oligomers (Box 3) or directed motion along a track. They can be controlled by means of a set of DNA instructions that determine the state of the machine by addressing each of the component devices independently. The simplest example is a 3-bit rewritable memory consisting of three single-stranded anchorages attached to a rigid double-stranded scaffold[53]. The anchorages have unique base sequences so that they can be individually addressed. Each control strand delivers an instruction to the device to change its state: set strands hybridize to straighten the anchorages and are removed by unset strands.

Shin and Pierce[54] used a similar strategy to build a device with two distinguishable feet that walks on a track with four different anchorages. Each step requires the sequential addition of two instruction strands: the first lifts the back foot from the track and the second replaces it ahead of the stationary foot (Fig. 3a). Directional movement, which is possible because the sequence of anchorages has no inversion symmetry, requires a set of eight DNA

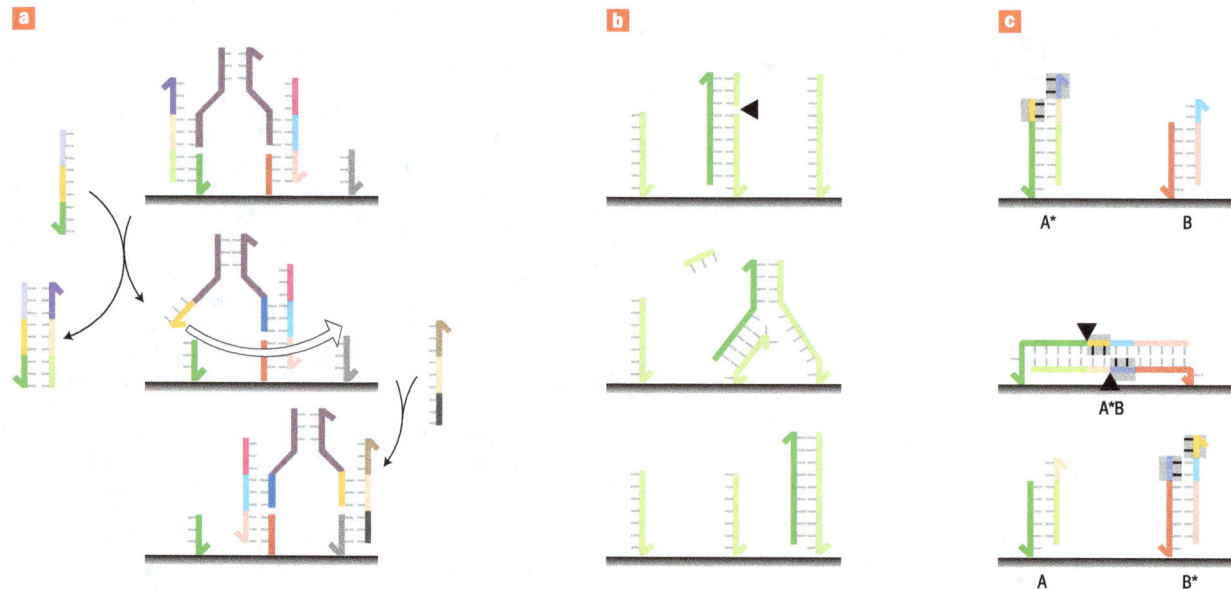

Figure 3 DNA nanomachines that execute directional stepwise movement along linear tracks. **a**, The Shin and Pierce walker[54] has two distinct feet and steps along a track that displays a sequence of four distinct single-stranded anchorages (the first three are shown). Each step is driven by sequential addition of two control strands, one that lifts the back foot from the track and one that binds it to a new anchorage ahead of the stationary foot. The single step shown here requires two instruction strands. A total of eight instruction strands are required for extended operation of the motor. **b**, Autonomous movement can be driven by enzymatic hydrolysis of the DNA[71] or RNA[70] backbone. Binding of the cargo (dark green) to an anchorage (light green) enables an enzyme to cleave the anchorage (the cleavage site is indicated by a black triangle). A short fragment of the anchorage is released, leaving the cargo with a single-stranded toehold that can bind to the intact anchorage ahead of it; the cargo can then step forward by a branch migration reaction. Destruction of the track in the wake of the cargo imposes directionality. **c**, A cargo can be passed autonomously from one anchorage to the next by a repeated cycle of enzymatic ligation and hydrolysis[76]. The cargo consists of two short fragments of DNA (shaded in grey). It is passed along a track with four distinct double-stranded anchorages. In the top panel the cargo is covalently attached to anchorage A. Anchorage B ahead of it has a sticky end that is complementary to the free end of the cargo. DNA ligase joins the two anchorages covalently with the cargo bridging the gap between them (middle panel). This creates a sequence of bases that is recognized by a restriction enzyme, which cleaves the cargo from A, leaving it covalently attached to B (bottom panel), and so on. Cleavage sites are represented by black triangles.

instructions. A similar scheme was used to roll two DNA 'gears' against each other[55]. The Sherman and Seeman walker[56] uses a different stepping mechanism: the front foot steps forward then the back foot catches up (the front foot always remains ahead of the back foot). Both walkers could step indefinitely along a track composed of a repeating sequence of anchorages provided that the free foot is never offered a choice of anchorages. For the Shin and Pierce device, which walks along a repeating sequence of four anchorages, this could be achieved by ensuring that the free foot can only bind to an anchorage adjacent to the bound foot (the sequence of anchorages could be reduced to three if the number of control strands were increased to twelve). For the inchworm mechanism of the Sherman and Seeman walker, the feet are sometimes bound to non-adjacent sites: continuous operation requires either a sequence of more than four anchorages or the constraint that the feet cannot swing past each other on the track. If the latter condition were met, an inchworm device with distinguishable feet could walk on a repeating sequence of only two anchorages, although if it did manage to reverse, it would not correct itself. Extended tracks for DNA walkers might be made from DNA nanotubes, which are straight structures with persistence lengths of many micrometres[57-60], or from arrays built on periodic[61-63] or non-periodic[64] DNA templates.

The walking devices described in this section are not true molecular motors because they cannot complete a cycle of motion without external intervention. Lack of autonomy brings the advantage of increased controllability: the device can be stopped at a desired location or reversed simply by changing the order in which instruction strands are added. However, the construction of a free-running DNA motor that does not require external intervention would be a significant accomplishment.

MOLECULAR MOTORS

The ambition to use DNA to construct autonomous motors that step along linear tracks is inspired by biological motors like myosin, kinesin and dynein that use free energy from hydrolysis of ATP to drive directional movement. Biological motors are astonishingly competent: they can be very fast, moving loads at speeds of up to 60 µm s[-1] (ref. 65), and processive, travelling distances of up to 1 µm before dissociating from their tracks[66,49].

A chemically driven molecular motor is a catalyst for the reaction of the fuel from which it obtains energy. Biological motors couple conformational changes, including track binding and unbinding and rotation amplified by lever arms, to the binding and hydrolysis of ATP and release of ADP[67]. Three different sources of energy for synthetic DNA motors have been explored: hydrolysis of the DNA backbone and of ATP (both of which involve making and breaking covalent bonds), and DNA hybridization.

MOTORS POWERED BY DNA AND RNA HYDROLYSIS

The free energy released on hydrolysis of the phosphodiester backbone of a DNA or RNA fuel can drive an autonomous device that catalyses the reaction. RNA hydrolysis can be catalysed by a limited set of DNA sequences including the '10-23' DNA

The ribosome creates proteins by linking amino acids according to instructions encoded in mRNA. Could a DNA nanomachine be used to build an oligomer specified by instructions encoded in DNA? The challenge is to read and process a sequence of instructions in the correct order and to respond by promoting the appropriate chemical reactions.

Halpin and Harbury used DNA-encoded instructions to synthesize short polypeptides. DNA instruction strands consisted of a string of unique sequence tags (codons) that specified both the identity of an amino acid and its position in the polypeptide product[117,118]. This strand remained covalently attached to the growing chain and was used to direct its path between reaction vessels, in each of which one amino acid was added. At each step, molecules with the appropriate tag for a given reaction were selected by hybridization to complementary sequence tags (anticodons) immobilized on a solid support.

Liao and Seeman constructed a reconfigurable device that can template assembly of four different DNA products[119]. At the heart of the device are two conformational switches that can be rotated by 180° by interaction with DNA instruction strands. The device promotes the ligation of rigid DNA tiles. The four states of the device offer up to four different combinations of splints that bind and hold together single-stranded sticky ends

protruding from different tiles, and thus catalyse ligation of four different tile sequences. This device does not read instructions in sequence; instead its switches are initialized independently before synthesis begins. The length of the product is limited by the length of the device.

It is possible to promote other chemical reactions simply by attaching reactants to complementary DNA strands and using hybridization to hold them together and thus increase the reaction probability[120]. Multistep synthesis can be carried out by controlling the secondary structure of the DNA template: masks that prevent reagents from coming into contact can be removed by increasing the temperature[121]. DNA-templated chemistry can also be used to assemble linear and branched organic precursors into complex structures whose connectivity is programmed by the template[122,123]. DNA devices can be used to trigger chemical reactions by changing the effective concentration of reagents in response to a change in pH[124] or to the addition of a control strand[125].

Progress in the construction of linear motors built from DNA, together with the demonstration of DNA-templated chemistry, sets the following challenge: to construct a DNA device that acts, like the ribosome, as a chemical assembler that can process an instruction tape of arbitrary length.

enzyme[68] — a short DNA strand consisting of a catalytic loop that cuts a substrate held in place by hybridization to two arms that flank the loop. This reaction was used to drive autonomous conformational changes in a tweezer-like device[69]. The arms of the tweezers are tethered together by a single strand of DNA that contains the catalytic sequence. Hybridization of a single-stranded fuel to the tether pushes the arms apart and positions two RNA bases in the fuel close to the catalytic motif of the DNA enzyme. The DNA enzyme cuts the fuel between the RNA bases to create two short fragments that dissociate spontaneously, restoring the random coil configuration of the tether, bringing the arms closer together.

Two groups have used DNA hydrolysis to power a motor that moves along a track (Fig. 3b). In both cases the track consists of repeated identical single-stranded anchorages attached to a double-stranded backbone. The cargo is a strand that can hybridize to any anchorage and, in conjunction with a catalyst, enable its cleavage. In the motor constructed by Tian and co-workers[70], each anchorage contains two RNA bases and can be cleaved when hybridized to the cargo, which contains the 10-23 catalytic domain[68]. In the motor created by Bath and co-workers[71] the cargo–anchorage duplex contains a recognition site for a genetically modified nicking restriction enzyme (Box 1), N.BbvC IB (ref. 72), that cuts only the anchorage strand. The outer segment that is cleaved from the anchorage is short enough to dissociate spontaneously from the cargo, leaving the complementary section of the cargo as a single-stranded overhang that can hybridize with the next anchorage. The cargo is transferred completely to the new anchorage by a branch migration reaction and the operating cycle can be repeated. It is important that the rate of stepping without cleavage be small, as such a step creates the possibility that after the next cleavage the cargo will step backwards and be marooned. The free energy input is provided by the hydrolysis of the DNA backbone and includes the configurational entropy of the liberated fragment. Autonomous directional motion is achieved by making cleavage of the anchorage conditional on

the presence of the cargo and rapid motion of the cargo along the track conditional on cleavage. Unidirectionality is achieved by the simple expedient of destroying the track that the cargo has passed over, although the track does not impose an initial direction on the motion unless it is prepared with the cargo at one end. At all stages in the cycle the cargo is hybridized to one or two anchorages and the stronger the interaction, the lower the probability of dissociation of the cargo from the track. Unfortunately, strong binding of the cargo also has the effect of slowing down transfer of the cargo between anchorages: the strength of the cargo–track interaction must be tuned to achieve an appropriate balance between speed of operation and processivity. Ultimately the rate is limited by the turnover of the catalyst (0.003 s^{-1} for N.BbvC IB (ref. 73) and 0.06 s^{-1} for the 10-23 ribozyme[68]). The destruction of the track after a single passage limits the potential applications of these devices.

MOTORS POWERED BY ATP HYDROLYSIS

Several designs for motors[74–76] and computing devices[77,78] involve cycles of restriction (backbone cleavage) and ligation (backbone joining). Bonds broken by a restriction enzyme are repaired by DNA ligase (Box 1), which converts one molecule of ATP to AMP for every phosphodiester bond made. The free energy put into a cycle of operation for one of these devices is thus provided by ATP hydrolysis. Yin and co-workers[76] constructed an autonomous DNA motor that passes its cargo down a track from one anchorage to the next (Fig. 3c). Its operating cycle is designed to prevent dissociation of the cargo from the track: the cargo is attached covalently, rather than by hybridization, and release from one anchorage is made conditional on attachment to the next. The track consists of a double-stranded DNA backbone to which double-stranded anchorages are attached by short single-stranded tethers. Each anchorage has a 3-base 3′ overhang. The cargo consists of two 3-base DNA fragments, one of which is ligated to each strand of one double-stranded anchorage. The track, which is intrinsically directional, consists of an infinitely repeated sequence of four anchorages A–D (although only a three-anchorage track has been tested[76]). An anchorage to which the cargo has been ligated is indicated by an asterisk, A* for instance. The operating cycle of the motor is as follows: the sticky end (overhang) of A* is complementary to that of B; the two sticky ends hybridize and are ligated by T4 DNA ligase. The A*B complex contains a sequence of base pairs that is recognized by a restriction enzyme, which cuts it to create A and B*, transferring the cargo to anchorage B. Similarly, the cargo can be transferred from B* to C, C* to D and D* to A and so on. Two restriction enzymes (PflM I and BstAP I), which have different recognition sites, are used — one for the first and last steps in the sequence and one for the middle two steps. (The four-anchorage sequence, which allows the track for this motor to be extended indefinitely, can be obtained by adding a fourth anchorage, D 5'-CTGG-3'/5'-CCAGCAG-3', to the three-anchorage track published by Yin and co-workers[76]).

The motor is unidirectional: if, say, B* is bound to A in an idling step then the B*A complex can be cut to recreate B* and A (repeating the previous forwards step) but not B and A* (a backwards step). Energy is input at each ligation step by ATP hydrolysis, and dissipated at each enzymatic cleavage (hydrolysis of the DNA backbone releases free energy). Passage of the cargo does not permanently change the track. The motor is processive: the cargo is always covalently attached to at least one, if not two, anchorages. In principle the motor is capable of indefinite unidirectional, processive, autonomous motion. In practice, this scheme is complicated by the need to use three enzymes simultaneously.

CATALYSIS OF DNA HYBRIDIZATION AS AN ENERGY SOURCE

The energy that drives most of the two-state devices and clocked walkers described in the first part of this review is provided by DNA hybridization. In general, their operating cycles involve a conformation change driven by interaction with at least one fuel strand that is later removed by hybridization with the corresponding antifuel strand. The fuel–antifuel duplex is a waste product. Both processes are driven by the decrease in free energy on forming additional base pairs, which can be substantial: the free energy change on creation of ten base pairs is comparable to that for ATP hydrolysis ($\Delta G^\circ_{ATP} = -7.7$ kcal mol^{-1} (ref. 79) and under cellular conditions $\Delta G_{ATP} \approx -14$ kcal mol^{-1}; $\Delta G^\circ_{hybridization} \approx -1.4$ kcal M^{-1} per base pair[80]).

In order to use DNA hybridization as an energy source for a free-running molecular motor, it is necessary to create both a metastable DNA fuel, whose spontaneous hybridization rate is slow compared with the cycle time of the motor, and to have a means of catalysing its hybridization. Turberfield and co-workers introduced the idea that DNA loops could be used as a fuel[81]. In order to simplify FRET measurement of reaction rates, they used two-strand loop complexes. Hairpin loops, formed by hybridization of complementary domains at either end of a single strand, are also suitable[82–84]. Figure 4 uses hairpin loops as an example to introduce the ideas of a metastable DNA fuel and of a hybridization catalyst. Two hairpins with identical necks and complementary loop domains can hybridize to produce a continuous duplex (Fig. 4). This reaction is driven by the free energy of hybridization of the loop bases (bases in the necks are hybridized both before and after the reaction, although to different partners), which more than compensates for the loss in configurational entropy associated with dimer formation.

Although hybridization can be initiated by interaction of unpaired bases in the loop domains, it is strongly hindered by closure of the necks: hybridization of the loop bases is limited by the topological impossibility of winding one loop around another to form a double helix. The reaction can be speeded up by competitive opening of at least one neck by means of an

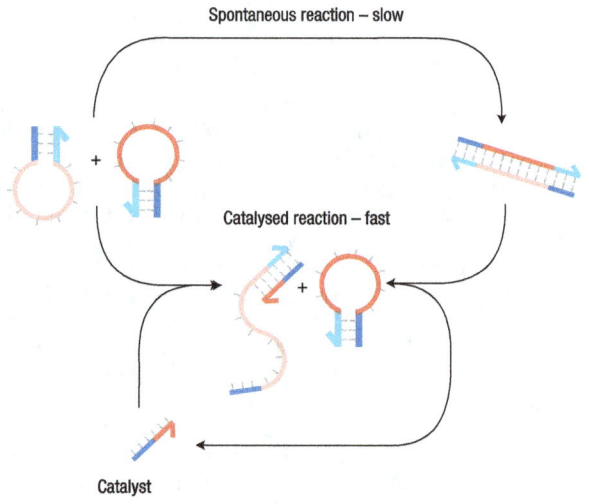

Figure 4 Hairpin loops to fuel DNA motors[81]. DNA loop complexes can be used as an energy source. The DNA hairpin has a single-stranded loop trapped by a double-stranded neck. Hybridization of complementary hairpins releases free energy of ~1.4 kcal mol^{-1} per additional base pair formed[80], but is hindered because the loop domains cannot wind round one another without opening at least one neck. The reaction can be catalysed by a short strand that binds to a single-stranded toehold, either in the loop or at the far end of the neck, and opens the hairpin by hybridizing to the neck along its entire length[81–86]. Once opened a loop interacts relatively quickly with its complement and releases the catalyst strand.

'opening strand' that contains a neck sequence. A useful opening strand also incorporates a toehold domain[81] that can initiate the opening of the neck by hybridizing either to the first few bases in the loop domain or to an unprotected single-stranded extension at the free end of the neck (Box 1). The opened loop can react more rapidly with its complement. It can, for example, thread though the loop of an unopened hairpin to facilitate hybridization in the loop domain. Complete hybridization of the hairpin with its complement displaces the opening strand (but not from an external toehold, which must be short enough to allow spontaneous dissociation). The freed opening strand can then hybridize to another loop and accelerate its reaction with its complement: it is a catalyst for loop–loop hybridization. If a single strand of DNA that undergoes a transition from random coil to duplex can be described as a machine then a hybridization catalyst was the first autonomous DNA machine[81]. (It might be helpful to restrict the use of 'machine' to the Oxford English Dictionary definition "an apparatus constructed to perform a task".)

Interactions between complementary pairs of hairpins[82,84] and of two-strand loops[81,85,86] have been studied with reference to their potential use as fuel. Dirks and Pierce have demonstrated a triggered amplification reaction[83] in which a loop-opening initiator triggers a cascade of hairpin–hairpin reactions. Seelig and co-workers[86] pointed out that the loops studied in ref. 81 can form a stable 'kissed' complex held together by partial base-pairing between loop domains without disruption of the necks. The kissed complex was purified and shown to be particularly long-lived ($t_{1/2} \approx 10^6$ s). Catalysis of the hybridization of this complex by a neck-opening strand can speed up their reaction to form a duplex by a factor of 5,000 (ref. 86). The authors point out that the kissed complex may be suitable for use as a one-component fuel.

We show in Fig. 5 an early unpublished design of our own, which demonstrated that it is possible to create a motor from a

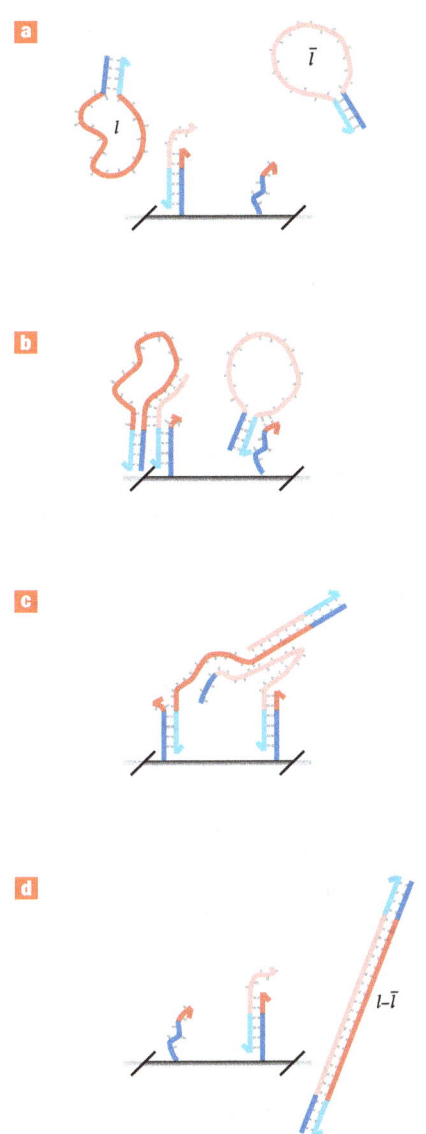

Figure 5 Scheme for a hybridization-powered molecular motor. **a**, The cargo is shown bound to one of an array of single-stranded anchorages attached to a rigid track. Movement of the cargo to the next anchorage is coupled to hybridization of complementary hairpin loops I and $\bar{I}$. **b**, An empty anchorage is designed to bind to the first few bases of the loop domain of the $\bar{I}$ hairpin (pink) and to competitively open the hairpin's neck (blue) by a strand-displacement reaction, in the same way as the hybridization catalyst shown in Fig. 4. The cargo is complementary to the anchorage sequence and has an extended toehold region designed to interact with the loop domain of the complementary hairpin I. The track-bound cargo can open an I hairpin in a strand-exchange reaction involving nucleation and migration of a four-arm Holliday junction (Box 1). In this reaction the blue domains in the hairpin neck, the cargo and the anchorage exchange partners, displacing the cargo from its initial anchorage but keeping it securely attached to the track. **c**, By forcing open the necks of complementary hairpins I and $\bar{I}$ the track-bound cargo and adjacent empty anchorage catalyse their hybridization. **d**, When the open loops hybridize to produce an I-$\bar{I}$ duplex (a waste product), the cargo is deposited on the next anchorage by strand exchange. One loop–loop reaction is thus coupled to one step down the track. This motor could be made directional by preparing all anchorages with bound (open) $\bar{I}$ hairpins then removing excess $\bar{I}$. Motion of the cargo would leave empty anchorages behind, preventing backwards steps.

hybridization catalyst. A single step of a cargo along a track is driven by hybridization of one pair of complementary hairpins. Successful operation of such a free-running hybridization-powered motor has not yet been reported.

The use of DNA hybridization as an energy source is attractive because the rates of different reactions can be tuned independently by adjusting the base sequences and concentrations of the corresponding fuel strands. The demonstration of an autonomous molecular motor made from DNA that is capable of extracting energy from a metastable DNA fuel remains a significant challenge.

OUTLOOK

Research objectives, such as those listed below, are suggested by the example of natural molecular machines, the drive to miniaturize existing devices and the possibility of implementing models of computation in molecular systems. Biological systems set challenging — perhaps impossibly high — benchmarks.

- The use of static DNA templates (Box 3) is only the first stage in the development of DNA-directed synthesis. The example of the ribosome suggests that the track down which a synthetic molecular motor moves could be used as an instruction tape to control the stepwise addition of different monomers to create an oligomer of arbitrary length and sequence.
- The direct experimental realization of a Turing machine as a molecular machine, moving on and altering a DNA instruction tape according to its internal state and transition rules, could be used as a challenge to drive the development of active DNA devices.
- DNA machines are built by rational design whereas biological machines evolve. We should explore the combination of rational design with *in vitro* selection or evolution as a way to build more competent machines.
- The simplicity of DNA is both an advantage and a drawback when making molecular machines. Where chemical versatility is required, synthetic DNA bases[87] and backbone linkages[88,89] and RNA and peptide components should be considered.
- Many DNA sensors and other active devices are inspired by potential medical applications. We should investigate the behaviour of DNA devices *in vivo* and explore techniques for introducing them into cells.

As our knowledge of how to engineer self-assembling systems improves, so the fields of DNA, RNA and peptide design will converge. As the resolution of semiconductor lithography improves, so self-assembly and top-down nanofabrication will be integrated. In constructing functional molecular devices we have the opportunity to learn about biology while imitating its processes. We have hardly begun to explore the potential of synthetic molecular machinery.

doi:10.1038/nnano.2007.104

References
1. Seeman, N. C. Nucleic acid junctions and lattices. *J. Theor. Biol.* **99**, 237–247 (1982).
2. Chen, J. H. & Seeman, N. C. Synthesis from DNA of a molecule with the connectivity of a cube. *Nature* **350**, 631–633 (1991).
3. Zhang, Y. W. & Seeman, N. C. Construction of a DNA-truncated octahedron. *J. Am. Chem. Soc.* **116**, 1661–1669 (1994).
4. Shih, W. M., Quispe, J. D. & Joyce, G. F. A 1.7-kilobase single-stranded DNA that folds into a nanoscale octahedron. *Nature* **427**, 618–621 (2004).
5. Goodman, R. P. *et al.* Rapid chiral assembly of rigid DNA building blocks for molecular nanofabrication. *Science* **310**, 1661–1665 (2005).
6. Robinson, B. H. & Seeman, N. C. The design of a biochip: a self-assembling molecular-scale memory device. *Protein Eng.* **1**, 295–300 (1987).
7. Keren, K. *et al.* Sequence-specific molecular lithography on single DNA molecules. *Science* **297**, 72–75 (2002).

8. Heilemann, M. *et al.* Multistep energy transfer in single molecular photonic wires. *J. Am. Chem. Soc.* **126,** 6514–6515 (2004).
9. Niemeyer, C. M., Koehler, J. & Wuerdermann, C. DNA-directed assembly of bienzymic complexes from in vivo biotinylated NAD(P)H:FMN oxidoreductase and luciferase. *ChemBioChem* **3,** 242–245 (2002).
10. Cate, J. H. *et al.* Crystal structure of a group I ribozyme domain: Principles of RNA packing. *Science* **273,** 1678–1685 (1996).
11. DeGrado, W. F., Summa, C. M., Pavone, V., Nastri, F. & Lombardi, A. De novo design and structural characterization of proteins and metalloproteins. *Annu. Rev. Biochem.* **68,** 779–819 (1999).
12. Chworos, A. *et al.* Building programmable jigsaw puzzles with RNA. *Science* **306,** 2068–2072 (2004).
13. Pohl, F. M. & Joyin, T. M. Salt-induced co-operative conformational change of a synthetic DNA: equilibrium and kinetic studies with poly (dG-dC). *J. Mol. Biol.* **67,** 375–396 (1972).
14. Mao, C., Sun, W., Shen, Z. & Seeman, N. C. A nanomechanical device based on the B-Z transition of DNA. *Nature* **397,** 144–146 (1999).
15. Stryer, L. & Haugland, R. P. Energy transfer: a spectroscopic ruler. *Proc. Natl Acad. Sci. USA* **58,** 719–726 (1967).
16. Yang, X., Vologodskii, A. V., Liu, B., Kemper, B. & Seeman, N. C. Torsional control of double-stranded DNA branch migration. *Biopolymers* **45,** 69–83 (1998).
17. Holliday, R. A mechanism for gene conversion in fungi. *Genet. Res.* **5,** 282–304 (1964).
18. Gehring, K., Leroy, J. L. & Gueron, M. A tetrameric DNA structure with protonated cytosine-cytosine base-pairs. *Nature* **363** 561–565 (1993).
19. Aboul-ela, F., Murchie, A. I. H. & Lilley, D. M. J. NMR study of parallel-stranded tetraplex formation by the hexadeoxynucleotide d(TG_4T). *Nature* **360,** 280–282 (1992).
20. Liu, D. & Balasubramanian, S. A proton-fuelled DNA nanomachine. *Angew. Chem. Int. Edn* **42,** 5734–5736 (2003).
21. Liu, D. *et al.* A reversible pH-driven DNA nanoswitch array. *J. Am. Chem. Soc.* **128,** 2067–2071 (2006).
22. Liedl, T. & Simmel, F. C. Switching the conformation of a DNA molecule with a chemical oscillator. *Nano Lett.* **5,** 1894–1898 (2005).
23. Liedl, T., Olapinksi, M. & Simmel, F. C. A surface-bound DNA switch driven by a chemical oscillator. *Angew. Chem. Int. Edn* **45,** 5007–5010 (2006).
24. Shu, W. *et al.* DNA molecular motor driven micromechanical cantilever arrays. *J. Am. Chem. Soc.* **127,** 17054–17060 (2005).
25. Baller, M. K. et al. A cantilever array-based artificial nose. *Ultramicroscopy* **82** 1–9 (2001).
26. Chen, Y., Lee, S.-H. & Mao, C. A DNA nanomachine based on a duplex-triplex transition. *Angew. Chem. Int. Edn* **43,** 5335–5338 (2004).
27. Brucale, M., Zuccheri, G. & Samori, B. The dynamic properties of an intramolecular transition from DNA duplex to cytosine-thymine motif triplex. *Org. Biomol. Chem.* **3,** 575–577 (2005).
28. Yurke, B., Turberfield, A. J., Mills, A. P. Jr, Simmel, F. C. & Neumann, J. L. A DNA-fuelled molecular machine made of DNA. *Nature* **406,** 605–608 (2000).
29. Yurke, B. & Mills, A. P. Jr. Using DNA to power nanostructures. *Genetic Programming and Evolvable Machines* **4,** 111–122 (2003).
30. Muller, B. K., Reuter, A., Simmel, F. C. & Lamb, D. C. Single-pair FRET characterization of DNA tweezers. *Nano Lett.* **6** 2814–2820 (2006).
31. Simmel, F. C. & Yurke, B. Using DNA to construct and power a nanoactuator. *Phys. Rev. E* **63,** 041913 (2001).
32. Simmel, F. C. & Yurke, B. A DNA-based molecular device switchable between three distinct mechanical states. *Appl. Phys. Lett.* **80,** 883–885 (2002).
33. Yan, H., Zhang, X., Shen, Z. & Seeman, N. C. A robust DNA mechanical device controlled by hybridization topology. *Nature* **415,** 62–65 (2002).
34. Ding, B. & Seeman, N. C. Operation of a DNA robot arm inserted into a 2D DNA crystalline substrate. *Science* **314,** 1583–1585.
35. Feng, L., Park, H., Reif, J. H. & Yan, H. A two-state DNA lattice switched by DNA nanoactuator. *Angew. Chem. Int. Edn* **42,** 4342–4346 (2003).
36. Hazarika, P., Ceyhan, B. & Niemeyer, C. M. Reversible switching of DNA-gold nanoparticle aggregation. *Angew. Chem. Int. Edn* **43,** 6469–6471 (2004).
37. Mirkin, C. A., Letsinger, R. L., Mucic, R. C. & Storhoff, J. J. A DNA-based method for rationally assembling nanoparticles into macroscopic materials. *Nature* **382,** 607–609 (1996).
38. Li, J. J. & Tan, W. A single DNA molecule nanomotor. *Nano Lett.* **2,** 315–318 (2002).
39. Alberti, P. & Mergny, J.-L. DNA duplex-quadruplex exchange as the basis for a nanomolecular machine. *Proc. Natl Acad. Sci. USA* **100,** 1569–1573 (2003).
40. Wang, Y. Zhang, Y. & Ong, N. P. Speeding up a single-molecule DNA device with a simple catalyst. *Phys Rev. E* **72,** 051918 (2005).
41. Zhong, H. & Seeman, N. C. RNA used to control a rotary device. *Nano Lett.* **6,** 2899–2903 (2006).
42. Dittmer, W. U. & Simmel, F. C. Transcriptional control of DNA-based nanomachines. *Nano Lett.* **4,** 689–691 (2004).
43. Dittmer, W. U., Kempter, S., Radler, J. O. & Simmel, F. C. Using gene regulation to program DNA-based molecular devices. *Small* **7,** 709–712 (2005).
44. Gardner, T. S., Cantor, C. R. & Collins, J. J. Construction of a genetic toggle switch in *Escherichia coli*. *Nature* **403,** 339–342 (2000).
45. Elowitz, M. B. & Leibler, S. A synthetic oscillatory network of transcriptional regulators. *Nature* **403,** 335–338 (2000).
46. Becskei, A. & Serrano, L. Engineering stability in gene networks by autoregulation. *Nature* **405,** 590–593 (2000).
47. Kim, J., White, K. S. & Winfree, E. Construction of an *in vitro* bistable circuit from synthetic transcriptional switches. *Mol. Syst. Biol.* **2,** 68 (2006).
48. Erben, C. M., Goodman, R. P. & Turberfield, A. J. Single-molecule protein encapsulation in a rigid DNA cage. *Angew. Chem. Int. Edn* **45,** 7414–7417 (2006).
49. Svoboda, K., Schmidt, C. F., Schnapp, B. J. & Block, S. M. Direct observation of kinesin stepping byoptical trapping interferometry. *Nature* **365,** 721–727 (1993).
50. Kuo, S. C. & Sheetz, M. P. Force of single kinesin groups measured with optical tweezers. *Science* **260,** 232–234 (1993).
51. Finer, J. T., Simmons, R. M. & Spudich, J. A. Single myosin group mechanics: piconewton forces and nanometre steps. *Nature* **368,** 113–119 (1994).
52. Ishijima, A. *et al.* Single-molecule analysis of the actomyosin motor using nano-manipulation. *Biochem. Biophys. Res. Commun.* **199,** 1057–1063 (1994).
53. Shin, J.-S. & Pierce, N. A. Rewritable memory by controllable nanopatterning of DNA. *Nano Lett.* **4** 905–909 (2004).
54. Shin, J.-S. & Pierce, N. A. A synthetic DNA walker for molecular transport. *J. Am. Chem. Soc.* **126,** 10834–10835 (2004).
55. Tian, Y. & Mao, C. A pair of DNA circles continuously rolls against each other. *J. Am. Chem. Soc.* **126,** 11410–11411 (2004).
56. Sherman, W. B. & Seeman, N. C. A precisely controlled DNA biped walking device. *Nano Lett.* **4,** 1203–1207 (2004).
57. Mitchell, J. C., Harris, J. R., Malo, J., Bath, J. & Turberfield, A. J. Self-assembly of chiral DNA nanotubes. *J. Am. Chem. Soc.* **126,** 16342–16343 (2004).
58. Rothemund, P. W. K. *et al.* Design and characterization of programmable DNA nanotubes. *J. Am. Chem. Soc.* **126,** 16344–16352 (2004).
59. Liu, D., Park, S. H., Reif, J. H. & LaBean, T. H. DNA nanotubes self-assembled from triple-crossover tiles as templates for conductive nanowires. *Proc. Natl Acad. Sci. USA* **101,** 717–722 (2004).
60. Mathieu, F. *et al.* Six-helix bundles designed from DNA. *Nano Lett.* **5,** 661–665 (2005).
61. Lubrich, D., Bath, J. & Turberfield, A. J. Design and assembly of double-crossover linear arrays of micrometre length using rolling circle replication. *Nanotechnology* **16,** 1574–1577 (2005).
62. Beyer, S., Nickels, P. & Simmel, F. C. Periodic DNA nanotemplates synthesized by rolling circle amplification. *Nano Lett.* **5,** 719–722 (2005).
63. Deng, Z., Tian, Y., Lee, S. H., Ribbe, A. E. & Mao, C. DNA-encoded self-assembly of gold nanoparticles into one-dimensional arrays. *Angew. Chem. Int. Edn* **44,** 3582–3585 (2005).
64. Rothemund, P. W. K. Folding DNA to create nanoscale shapes and patterns. *Nature* **440,** 298–302 (2006).
65. Higashi-Fujime, S. *et al.* The fastest actin-based motor protein from the green algae, *Chara*, and its distinct mode of interaction with actin. *FEBS Lett.* **375,** 151–154 (1995).
66. Howard, J., Hudspeth, A. J. & Vale, R. D. Movement of microtubules by single kinesin molecules. *Nature* **342,** 154–158 (1989).
67. Vale, R. D. & Milligan, R. A. The way things move: Looking under the hood of molecular motor proteins. *Science* **288,** 88–95 (2000).
68. Santoro, S. W. & Joyce, G. F. A general purpose RNA-cleaving DNA enzyme. *Proc. Natl Acad. Sci. USA* **94,** 4262–4266 (1997).
69. Chen, Y., Wang, M. & Mao, C. An autonomous DNA nanomotor powered by a DNA enzyme. *Angew. Chem. Int. Edn* **43,** 3554–3557 (2004).
70. Tian, Y., He, Y., Peng, Y. & Mao, C. A DNA enzyme that walks processively and autonomously along a one-dimensional track. *Angew. Chem. Int. Edn* **44,** 4355–4358 (2005).
71. Bath, J., Green, S. J. & Turberfield, A. J. A free-running DNA motor powered by a nicking enzyme. *Angew. Chem. Int. Edn* **44,** 4358–4361 (2005).
72. Heiter, D. F., Lunnen, K. D. & Wilson, G. G. Site-specific DNA-nicking mutants of the heterodimeric restriction endonuclease R.BbvCI. *J. Mol. Biol.* **348,** 631–640 (2005).
73. Bellamy, S. R. W. *et al.* Cleavage of individual DNA strands by the different subunits of the heterodimeric restriction endonuclease BbvCI. *J. Mol. Biol.* **348,** 641–653 (2005).
74. Reif, J. H. The design of autonomous DNA nanomechanical devices: Walking and rolling DNA. *Lect. Notes Comput. Sc.* **2568,** 22–37 (2003).
75. Yin, P., Turberfield, A. J., Sahu, S. & Reif, J. H. Designs for autonomous unidirectional walking DNA devices. *Lect. Notes Comput. Sc.* **3384,** 410–425 (2005).
76. Yin, P., Yan, H., Daniell, X. G., Turberfield, A. J. & Reif, J. H. A unidirectional DNA walker that moves autonomously along a DNA track. *Angew. Chem. Int. Edn* **43,** 4906–4911 (2004).
77. Benenson Y. *et al.* Programmable and autonomous computing machine made of biomolecules. *Nature* **414** 430–434 (2001).
78. Yin, P., Sahu, S., Turberfield, A. J. & Reif, J. H. Design of autonomous DNA cellular automata. *Lect. Notes Comput. Sc.* **3892,** 399–416 (2006).
79. Alberty, R. A. & Goldbert, R. N. Standard Thermodynamic Formation Properties for the Adenosine 5'-Triphosphate Series. *Biochemistry* **31,** 10610–10615 (1992).
80. SantaLucia, J. A unified view of polymer, dumbell, and oligonucleotide nearest neighbour thermodynamics. *Proc. Natl Acad. Sci. USA.* **95,**1460–1465 (1998).
81. Turberfield, A. J. *et al.* DNA fuel for free-running nanomachines. *Phys. Rev. Lett.* **90,** 118102 (2003).
82. Bois, J. S. *et al.* Topological constraints in nucleic acid hybridization kinetics. *Nucleic Acids Res.* **33,** 4090–4095 (2005).
83. Dirks, R. M. & Pierce, N. A. (2004). Triggered amplification by hybridization chain reaction. *Proc. Natl Acad. Sci. USA* **101,** 15275–15278.
84. Green, S. J., Lubrich, L. & Turberfield, A. J. DNA hairpins: fuel for autonomous DNA devices. *Biophys. J.* **91,** 2966–2975 (2006).
85. Seelig, G., Yurke, B. & Winfree, E. DNA hybridization catalysts and catalyst circuits. *Lect. Notes Comput. Sc.* **3384,** 329–343 (2005).
86. Seelig, G., Yurke, B. & Winfree, E. Catalysed relaxation of a metastable fuel. *J. Am. Chem. Soc.* **128,** 12211–12220 (2006).
87. Kool, E. T. Replacing the nucleobases of DNA with designer molecules. *Acc. Chem. Res.* **35,** 936–943 (2002).
88. Nielsen, P. E., Egholm M., Berg R. H. & Buchardt, O. Sequence-selective recognition of a DNA by strand displacement with a thymine-substituted polyamide. *Science* **254,** 1497–1500 (1991).
89. Koshkin, A. A. *et al.* Synthesis of the adenin, cytosine, guanine, 5-methylcytosine, thimine and uracil bicyclonucleotide monomers, oligomerization, and unprecedented nucleic acid recognition. *Tetrahedron* **54,** 3607–3630 (1998).
90. Smith, S. B., Finzi, L. & Bustamante, C. Direct mechanical measurements of the elasticity of single DNA molecules by using magnetic beads. *Science* **258,** 1122–1126 (1992).
91. Smith, S. B., Cui, Y. & Bustamante, C. Overstretching B-DNA: the elastic response of individual double-stranded and single-stranded DNA molecules. *Science* **271,** 795–799 (1996).
92. Seeman, N. C. De novo design of sequences for nucleic-acid structural engineering. *J. Biomol. Struc. Dyn.* **8,** 573–581 (1990).

93. Dirks, R. M., Lin, M., Winfree, E. & Pierce, N. A. Paradigms for computational nucleic acid design. *Nucleic Acids Res.* **32**, 1392–1403 (2004).
94. Goodman, R. P. NANEV: a program employing evolutionary methods for the design of nucleic acid nanostructures. *Biotechniques* **38**, 548–550 (2005).
95. Tashiro, R. & Sugiyama, H. A nanothermometer based on the different stackings of B- and Z-DNA. *Angew. Chem. Int. Edn* **42**, 6018–6020 (2003).
96. Shen, W., Bruist, M. F., Goodman, S. D. & Seeman, N. C. A protein-driven DNA device that measures the excess binding energy of proteins that distort DNA. *Angew. Chem. Int. Edn* **43**, 4750–4752 (2004).
97. Tyagi, S. & Kramer, F. R. Molecular beacons: probes that fluoresce upon hybridization. *Nature Biotechnol.* **14**, 303–308 (1996).
98. Wilson, D.W. & Szostak, J. W. In vitro selection of functional nucleic acids. *Ann. Rev. Biochem.* **68**, 611–648 (1999).
99. Dittmer, W. U., Reuter, A. & Simmel, F. C. A DNA-based machine that can cyclically bind and release thrombin. *Angew. Chem. Int. Edn* **43**, 3550–3553 (2004).
100. Rusconi, C. P. *et al.* RNA aptamers as reversible antagonists of coagulation factor IXa. *Nature* **419**, 90–94 (2002).
101. Beyer, S. & Simmel, F. C. A modular DNA signal translator for the controlled release of a protein by an aptamer. *Nucleic Acids Res.* **34**, 1581–1587 (2006).
102. Chelyapov, N. Allosteric aptamers controlling a signal amplification cascade allow visual detection of molecules at picomolar concentration. *Biochemistry* **45**, 2461–2466 (2006).
103. Liu, J. & Lu, Y. A colorimetric lead biosensor using DNA enzyme-directed assembly of gold nanoparticles. *J. Am. Chem. Soc.* **125**, 6642–6643 (2003).
104. Porta H. & Lizardi, P. M. An allosteric hammerhead ribozyme. *Biotechnology* **13**, 161–164 (1995).
105. Stojanovic, M. N., de Prada, P. & Landry, D. W. Catalytic molecular beacons. *ChemBioChem* **2**, 411–415 (2001).
106. Robertson, M. P. & Ellington, A. D. *In vitro* selection of an allosteric ribozyme that transduces analytes to amplicons. *Nature Biotechnol.* **17**, 62–66 (1999).
107. Weizmann, Y. *et al.* A virus spotlighted by an autonomous DNA machine. *Angew. Chem. Int. Edn* **45**, 7384–7388 (2006).
108. Van Ness, J., Van Ness, L. K. & Galas, D. J. Isothermal reactions for the amplification of oligonucleotides. *Proc. Natl Acad. Sci USA* **100**, 4504–4509 (2003).
109. Stojanovic, M. N., Mitchell, T. E. & Stefanovic, D. Deoxyribozyme-based logic gates. *J. Am. Chem. Soc.* **124**, 3555–3561 (2002).
110. Penchovsky, R. & Breaker, R. R. Computational design and experimental testing of oligonucleotide-sensing allosteric ribozyme. *Nature Biotechnol.* **23**, 1424–1433 (2005).
111. Stojanovic, M. N. & Stefanovic, D. Deoxyribosome-based half adder. *J. Am. Chem. Soc.* **125** 6673–6676 (2002).
112. Stojanovic, M. N. & Stefanovic, D. A deoxyribozyme-based molecular automaton. *Nature Biotechnol.* **21**, 1069–1074 (2003).
113. Seelig, G., Soloveichik, D., Zhang, D. Y. & Winfree, E. Enzyme-free nucleic acid logic circuits. *Science* **314**, 1585–1588 (2006).
114. Benenson, Y., Gil, B., Ben-Dor, U., Adar, R. & Shapiro, E. An autonomous molecular computer for logical control of gene expression. *Nature* **429**, 423–428 (2004).
115. Isaacs, F. J. *et al.* Engineered riboregulators enable post-transcriptional control of gene expression. *Nature Biotechnol.* **22**, 841–847 (2005).
116. Bayer, T. S. & Smolke, C. D. Programmable ligand-controlled riboregulators of eukaryotic gene expression. *Nature Biotechnol.* **23**, 337–343 (2005).
117. Halpin, D. R. & Harbury, P. R. DNA display I: Sequence-encoded routing of DNA populations. *PloS Biol.* **2**, 1015–1021 (2004).
118. Halpin, D. R. & Harbury, P. R. DNA display II: Genetic manipulation of combinatorial chemistry libraries for small-molecule evolution. *PloS Biol.* **2**, 1022–1030 (2004).
119. Liao, S. & Seeman, N. C. Translation of DNA signals into polymer assembly instructions. *Science* **306**, 2072–2074 (2004).
120. Gartner, Z. J. & Liu, D. R. The generality of DNA-templated synthesis as a basis for evolving non-natural small molecules. *J. Am. Chem. Soc.* **123**, 6961–6963 (2001).
121. Snyder, T. M. & Liu, D. R. Ordered multistep synthesis in a single solution directed by DNA templates. *Angew. Chem. Int. Edn* **44**, 7379–7382 (2005).
122. Gothelf, K. V., Thomsen, A., Nielsen, M., Clo, E., & Brown, R. S. Modular DNA-programmed assembly of linear and branched conjugated nanostructures. *J. Am. Chem. Soc.* **126**, 1044–1046 (2004).
123. Eckardt, L. H. *et al.* DNA nanotechnology: chemical copying of connectivity. *Nature* **420**, 286 (2002).
124. Chen, Y. & Mao, C. Reprogramming DNA-directed reactions on the basis of a DNA conformational change. *J. Am. Chem. Soc.* **126**, 13240–13241 (2004).
125. Chhabra, R., Sharma, J., Liu, Y. & Yan, H. Addressable molecular tweezers for DNA-templated coupling reactions. *Nano Lett.* **6**, 978–983 (2006).

Acknowledgements
This work was supported by the UK research councils BBSRC, EPSRC and MRC, and by the MoD through the UK Bionanotechnology IRC.

Competing financial interests
The authors declare no competing financial interests.

NANOELECTRONICS

Nanoelectronics from the bottom up

Electronics obtained through the bottom-up approach of molecular-level control of material composition and structure may lead to devices and fabrication strategies not possible with top-down methods. This review presents a brief summary of bottom-up and hybrid bottom-up/top-down strategies for nanoelectronics with an emphasis on memories based on the crossbar motif. First, we will discuss representative electromechanical and resistance-change memory devices based on carbon nanotube and core–shell nanowire structures, respectively. These device structures show robust switching, promising performance metrics and the potential for terabit-scale density. Second, we will review architectures being developed for circuit-level integration, hybrid crossbar/CMOS circuits and array-based systems, including experimental demonstrations of key concepts such as lithography-independent, chemically coded stochastic demultipluxers. Finally, bottom-up fabrication approaches, including the opportunity for assembly of three-dimensional, vertically integrated multifunctional circuits, will be critically discussed.

WEI LU[1] AND CHARLES M. LIEBER[2]

[1]Department of Electrical Engineering and Computer Science, University of Michigan, Ann Arbor, Michigan 48109, USA
[2]Department of Chemistry and Chemical Biology, and School of Engineering and Applied Sciences, Harvard University, Cambridge, Massachusetts 02138, USA
e-mail: wluee@eecs.umich.edu; cml@cmliris.harvard.edu

Over the past four decades, sustained advances in integrated circuit technologies for memory and processors have given us computers with ever more powerful processing capabilities and consumer electronics with ever increasing non-volatile memory capacity[1]. But as device features are pushed towards the deep sub-100-nm regime, the conventional scaling methods of the semiconductor industry face increasing technological and fundamental challenges[1-3]. For example, device size fluctuations may result in a large spread in device characteristics at the nanoscale, affecting key parameters such as the threshold voltage and on/off currents. The increasing costs associated with lithography equipment and operating facilities needed for traditional manufacturing might also create an economic barrier to continued increases in the capabilities of conventional processor and memory chips[1-3].

To continue the remarkably successful scaling of conventional complementary metal oxide semiconductor (CMOS) technology and possibly produce new paradigms for logic and memory, many researchers have been investigating devices based on nanostructures, and in particular carbon nanotubes (CNTs) and semiconductor nanowires[4,5]. For these nanostructures, at least one critical device dimension — the nanoscale wire diameter — is defined during the growth or chemical synthesis process and can be controlled with near-atomic-scale precision. This control, which goes beyond that achievable in top-down lithography, represents one of several key features motivating these efforts. More generally, predictable and

well-controlled nanostructure growth implies that materials with distinct chemical composition, structure, size and morphology can be assembled by design to build specific functional devices and integrated circuits[4,5]. This 'bottom-up' paradigm, analogous to the way that biology so successfully works, may prove to be a solution to the technological challenges faced by the semiconductor industry, and could open up new strategies for increasing overall device density by allowing aggressive scaling in three dimensions. Here we review unique features and opportunities of the bottom-up approach to nanoelectronics, assess where it might merge with today's top-down industry, and look at the challenges that must be met to take this area of science towards commercial applications.

NANOSCALE MEMORY ARCHITECTURE

An attractive architecture for using the strengths of the CNT and nanowire building blocks for memory and logic consists of an array of crossed wires[6-8] — the 'crossbar' — as shown in Fig. 1. The assembly of distinct nanoscale wire elements into a crossbar enables the size and function of key device features to be defined during the synthesis of the building blocks and their subsequent assembly, instead of during lithography and processing steps. Specifically, the bit size in the crossbar structure is defined by the diameters of the orthogonal nanoscale wires, the electronic characteristics of the functional element are defined by the two crossed wires (for example, coaxial core–shell materials), and the device density is governed by the density at which the CNTs and nanowires are assembled. This architecture is also attractive owing to its simplicity and the fact that scalable bottom-up techniques have already yielded 2D meshes of chemically synthesized nanowires[9-11], CNTs[12] and molecular devices[13,14].

The crossbar structure can be used for non-volatile memory applications when assembled nanowires yield a configurable junction

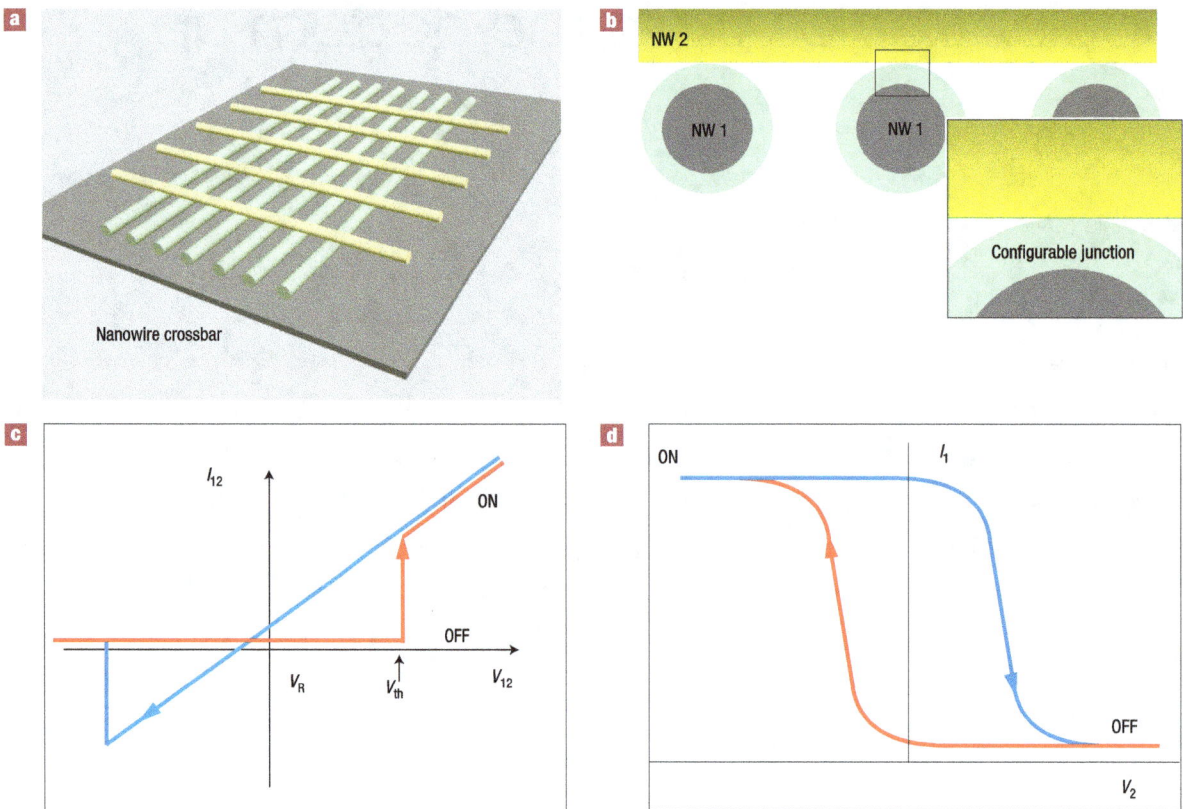

Figure 1 Crossbar memory switches. **a**, Schematic illustration of a nanowire crossbar memory. **b**, Cross-sectional view of the crossbar memory in **a** along a row-nanowire. A memory bit is represented by a configurable bistable junction formed between a pair of column- and row-wires. **c**, Schematic current–voltage (I–V) curve of a configurable junction that can be modelled as a two-terminal hysteretic resistor. Current I_{12} (voltage V_{12}) is measured (applied) across the column- and row-nanowire pair sandwiching the configurable junction. **d**, In a FET-based crossbar hysteresis is observed in the FET current (I_1) as the voltage on the control wire (V_2) is changed.

at each crosspoint (Fig. 1b). Two general cases can be pictured in which each crosspoint serves as either a resistive switching element (Fig. 1c) or a configurable field-effect transistor (FET; Fig. 1d). In the former case, each crosspoint can be configured independently into a low-impedance ('closed') state or a high-impedance ('open') state simply by controlling the applied voltage between the pair of nanowire electrodes, forming the basis of a memory bit (Fig. 1c). In the latter, each crosspoint FET is configured by the control nanowire which can switch the charge or polarization state of the junction (Fig. 1d). The specific natures of the materials used for the two nanoscale wires in the crossbar and the junction thus define the nature of the memory elements. This conceptual framework has been used for studies of crossbar memory structures based on crossed CNT electromechanical devices[12], bistable molecules[14] and nanowires[15] as discussed below.

The crossbar architecture can in principle be scaled to a memory density level of terabits per square centimetre by reducing the nanowire or CNT average pitch to 10 nm, a feat possible with advanced nanofabrication techniques[16–19] or more simply based on direct assembly[12] of the nanoscale-diameter wires themselves. Dense nanoscale memories, like their CMOS counterparts, will require logic components to write and access stored information. For example, a demultiplexer is a logic component providing address selection and interfacing the high-density memory array with external circuitry. Demultiplexers can be built into the crossbar structure with the same geometry using either nanowire transistors[20] or hysteretic

resistor-logic[14]. For more general logic operations, hybrid CMOS/crossbar structures may be required[21–23] in which the CMOS circuitry provides the needed voltage gain and peripheral functions such as input/output, coding/decoding and line driving. The cost of a hybrid approach is increased complexity of chip design and chip area overhead. Alternatively, general computing schemes could be based on a crossbar motif in which nanowire crosspoints function as locally gated FETs[8] or simple hysteretic two-terminal resistors[24,25]. In the latter, signal restoration can be achieved by latches programmed inside the crossbar, although at a cost to operation speed. Key architectural issues and opportunities for nanoelectronics are addressed in greater detail below.

MEMORY ELEMENTS

Molecule-based memories have received considerable attention partly because of the logic that a single molecule would represent the high-density limit for memory bits[6]. Negative differential resistance (NDR) and hysteretic resistance switching have been observed in molecule-based devices formed using break junctions[26], nanopores[27], scanning probes[28,29] and crossbar structures[14,30,31]. The mechanism of the observed device behaviour has been attributed to several effects, including charge-transfer-induced conformational change[32–34], electromechanical switching[35], stochastic conformational changes[28] and metal filament formation[29], where the latter is unrelated to the structures or electrical states of the molecules. This controversy about the mechanism of molecule device behaviour may be due in part to the challenges associated with probing molecular states in devices and obtaining well-controlled molecule–electrode interfaces. Performance metrics are also less satisfactory in most molecule-based memories in terms of device yield, on/off ratio, switching speed, endurance cycles and retention time than in existing commercial devices or devices based on CNTs and nanowires. A detailed discussion of molecule-based memory devices can be found in ref. 36.

In addition, a closer examination of molecule-based crossbar memories leads to the observation that the device density is limited by the size of the nanowire electrodes rather than that of the molecules, even in the smallest proof-of-concept devices (for example, around 100 molecules are present at each crosspoint for the 33-nm-pitch device in ref. 31). As a result, even though memories based on single molecules represent the ultimate density potential, this advantage may not come into play until single (or few) molecules can be reliably deposited in arrays with sub-10-nm pitch size, and even then the molecules may not be robust enough for workable memories. In the foreseeable future, we believe high-density crossbar memories will first be produced using a more reliable inorganic or solid-state medium, provided that the storage medium can be scaled in a similar way to the smallest electrodes that can be produced.

One of the first successful implementations of this concept[12] — using the nanoscale wires as both electrodes and functional elements in a crossbar architecture — used the suspended CNT motif shown in Fig. 2a. This crossbar consists of a set of parallel CNTs or nanowires on a substrate and a set of perpendicular single-walled nanotubes (SWNTs) suspended on a periodic array of supports. Each crosspoint in this structure corresponds to a device element with a SWNT suspended above a perpendicular nanoscale wire. Bistability was shown to be possible through the interplay of the elastic energy, which produces a potential energy minimum at finite separation (when the upper nanotube is freely suspended), and the attractive van der Waals energy, which creates a second energy minimum when the suspended SWNT is deflected into contact with the lower wire. These two minima correspond to well-defined OFF and ON states, respectively: the separated upper-to-lower nanotube junction resistance will be very high, whereas the contacted junction resistance will be orders of magnitude lower. Devices can be switched between these OFF and ON states by transiently charging the nanotubes to produce attractive or repulsive electrostatic forces. The viability of this concept was first demonstrated[12] in 2000, and more recently has served as the basis for non-volatile memory chips being developed by Nantero[37]. This electromechanical crossbar switch array has the potential to approach an integration level of 10^{12} elements cm^{-2} with an element operation frequency in excess of 100 gigahertz (ref. 12), although such promise remains to be demonstrated.

Very different strategies for non-volatile memory can also be achieved with the crossbar architecture by using different nanowire materials. For example, by growing nanowires with p-type crystalline silicon (c-Si) cores and amorphous silicon (a-Si) shells, we have recently observed hysteretic resistance switching at Si-nanowire/metal-nanowire crosspoints in which a-Si acts as the information storage medium (Y. Dong, G. Yu, W. Lu, M. C. McAlpine & C. M. Lieber, manuscript in preparation; Fig. 3). Amorphous Si has been extensively studied in the past for potential non-volatile memory applications in planar metal-to-metal (M2M) structures[38,39], where resistance switching was attributed to metal filament formation (retraction) inside the a-Si matrix to yield high (low) conductance ON (OFF) states[39] (Fig. 3a inset). Devices based on M/a-Si/c-Si nanowires have several attractive features that go beyond earlier planar M2M devices. First, they show intrinsic rectification: the device in the ON state behaves like a diode (Fig. 3a), which is desirable for crossbar-based circuits as it mitigates cross-talk between elements[40,41]. Indeed, the bits

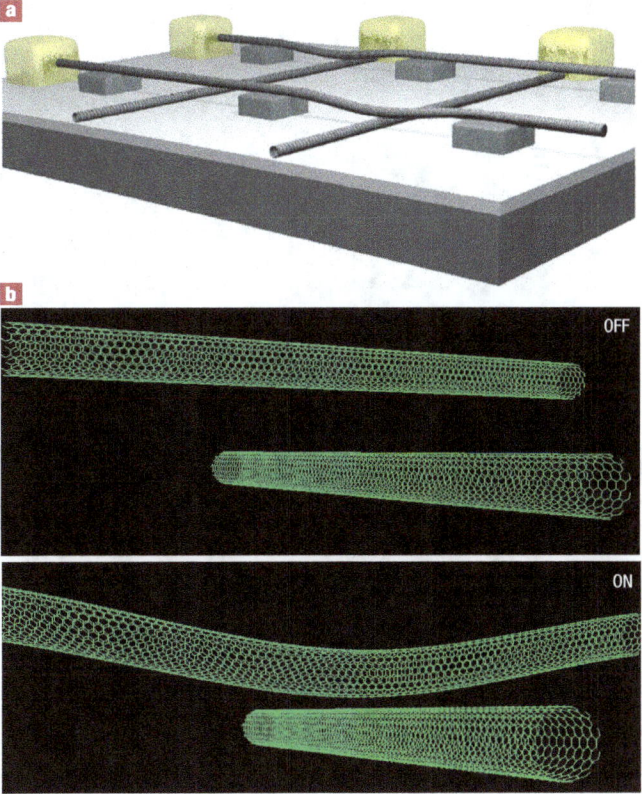

Figure 2 Electromechanical crossbar memory based on carbon nanotubes. **a**, Three-dimensional view of a suspended crossbar array showing four junctions with two elements in the ON (contact) state and two elements in the OFF (separated) state. The substrate consists of a conducting layer (for example highly doped silicon, dark grey) that terminates in a thin dielectric layer (for example SiO$_2$, light grey). The lower nanotubes are supported directly on the dielectric film, whereas the upper nanotubes are suspended by periodic inorganic or organic supports (grey blocks). Each nanotube is contacted by a metal electrode (yellow blocks). **b**, Calculated structures of the 20-nm pitch (10, 10) SWNT device element in the OFF (top) and ON (bottom) states. Reprinted with permission from ref. 12.

in multi-element one-dimensional and crossbar arrays (Fig. 3b) can be written, read and rewritten in a fully independent manner (Fig. 3c).

In addition, this a-Si/c-Si nanowire system offers scaling potential comparable to the use of molecules as the storage medium in crossbar memories. We have observed reliable resistance switching and near-100% yield in M/a-Si/c-Si nanowire devices with active device sizes down to 20 nm × 20 nm (limited by the size of the nanowire and lithography resolution), corresponding to a potential device density better than 6×10^{10} bits cm^{-2}, and substantially better than expected from previous work on planar M2M devices where micrometre-sized metal filaments precluded aggressive scaling[39,42,43]. The crossed Si–nanowire/metal–nanowire memory elements show other features that suggest substantial promise as an emerging memory technology. For example, the ON/OFF ratio is typically larger than 10^4 (in comparison, molecular[14,30,31] and chalcogenide[44] elements show ratios under 10^2). The programming current, ~0.01 mA, is much less than that reported for phase-change memories[44], 0.1–0.2 mA, and is more compatible with the minimal MOS transistor drive current desired for high-density integration, whereas the programming speed, <100 ns, is attractive for flash-memory types of applications. These attractive

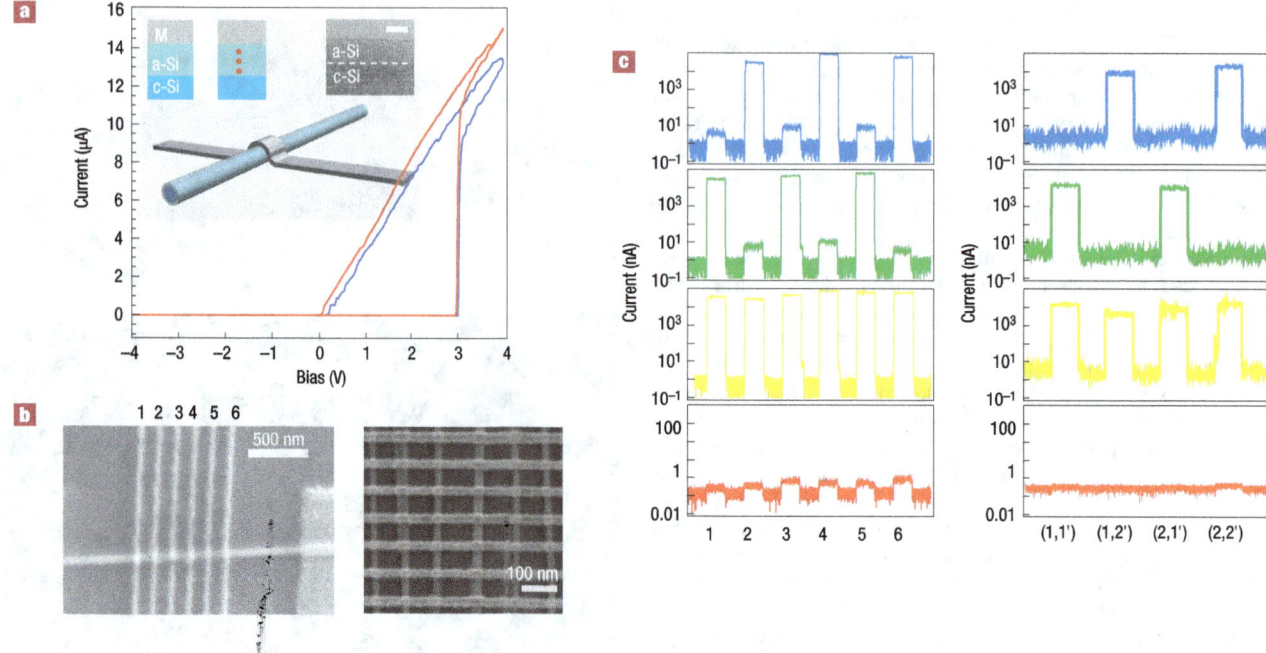

Figure 3 Hysteretic-resistor memories based on core–shell nanowires. **a**, Hysteretic resistance-switching observed at a crosspoint. Unlike the linear schematic curve shown in Fig. 1c, an asymmetric diode-like behaviour can be observed in the ON state. The initial cycle is shown in red and subsequent cycles are in blue. Main inset: schematic of the crossbar memory formed by metal wires (rows) and core–shell nanowires made of a-Si/c-Si (columns). Top left: schematic of a single memory bit showing metal and c-Si electrodes sandwiching an a-Si layer serving as the switchable medium. Top centre: formation of a metal filament inside the a-Si layer changes the memory bit from the high-resistance OFF state to the low-resistance ON state. Top right: high-resolution transmission electron microscope image of a c-Si/a-Si core–shell nanowire. Dashed line indicates the interface between the c-Si and a-Si. Scale bar, 5 nm. **b**, Left panel: scanning electron microscope (SEM) image of a 1 × 6 crossbar array fabricated on a single nanowire with 30-nm metal line width and 150-nm spacing. Right panel: SEM image of a high-density 2D crossbar memory with 100-nm pitch. The core–shell nanowires (columns) were assembled using the Langmuir–Blodgett method and the metal wires (rows) were fabricated using lithography. **c**, Memory bits that can be controllably written and read from the 1 × 6 array and a representative 2 × 2 array. Adapted from Y. Dong, G. Yu, W. Lu, M. C. McAlpine and C. M. Lieber, manuscript in preparation.

characteristics highlight the potential for such bottom-up crossbar arrays, and also suggest that further effort focused on optimizing the endurance and retention time would be worthwhile. Furthermore, the fact that the switching phenomenon is very robust in the nanowire system suggests that high-density memory devices based on the same mechanism may be obtained using commercial top-down fabrication techniques in an all-silicon-based approach. This observation highlights a key point for research in nanoelectronics: besides the ultimate device in which all components are assembled bottom-up at the molecular level, nanostructures may lead to new devices that can in turn result in suitable top-down fabrication for near-term large-scale applications.

The bottom-up approach may also lead to improved performance of memory devices based on conventional structures, or serve as a well-controlled platform for studying the memory mechanism. One such example is crystalline nanowire-based phase-change memory (PCM). PCM has sparked considerable recent interest as a potential next-generation non-volatile solid-state memory technology, and has already been shown to possess many of the necessary attributes, including high resistance contrast, the potential for multilevel storage, and better endurance and write speeds than flash memory. Detailed discussions on PCM can be found in ref. 45. In the bottom-up approach to PCM devices, crystalline nanowires serve as the channel material and can yield reduced PCM reset currents[46–49]. In a PCM, a large current is typically required to heat the channel material enough to yield the amorphous high-resistance reset state. Reduction of the reset current will result in faster amorphization with less power consumption, and allow faster memory switching speed and higher

reliability. The reset current is normally lowered by reducing the sample size, which leads to smaller heat capacity per unit length so the device can be heated to sufficiently high temperature with lower joule power[50]. By using GeTe or $Ge_2Sb_2Te_5$ nanowires, it has now been shown[46–49] that currents can be lowered to about 0.4 mA. The reset current could be further reduced by decreasing nanowire diameter, thus confirming the role of channel size in PCM devices suggested by studies on lithographically defined samples[50].

ELECTRONIC CIRCUITS

Recent development of controlled high-yield assembly of crossed nanowire p–n diodes and FETs has enabled the bottom-up approach to be used for assembly of nanoelectronic circuits[10], such as logic gates, that ultimately must be integrated together with memory arrays for read/write operations or to build stand-alone processors. For example, a two-input two-output AND logic gate was assembled from 1(p-Si) × 3(n-GaN) multiple junction arrays (Fig. 4a). The V_o–$V_{i1(i2)}$ data shows constant low V_o when the other input $V_{i2(i1)}$ is low, and nearly linear behaviour when the other input is set at high. Correspondingly, logic-0 is observed from this device when either one or both of the inputs are low, as $V_i = 0$ corresponds to a forward-biased, low-resistance p–n diode that pulls down the output. The logic-1 state is observed only when both inputs are high, as this condition corresponds to reverse-biased p–n diodes with resistances much larger than the constant resistor; that is, little voltage drop across the constant resistor and a high voltage at the output.

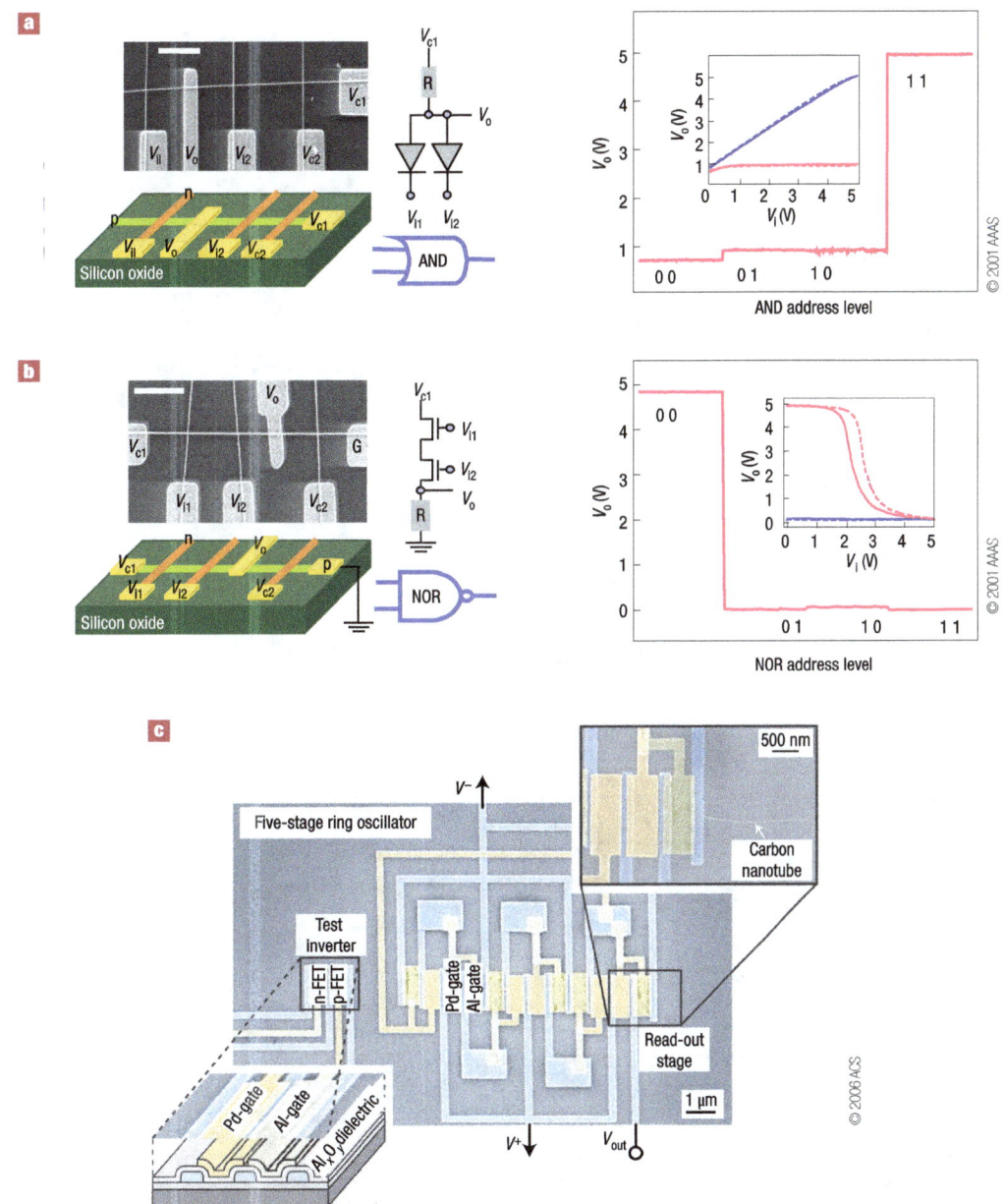

Figure 4 Nanowire and nanotube-based transistor logic. **a**, Left panel: schematic of logic AND gate constructed from a 1 × 3 crossed nanowire array. Insets: a typical SEM (scale bar, 1 μm) of the assembled AND gate and symbolic electronic circuit. Right panel: the output voltage (V_o) versus the four possible logic address level inputs from (V_{i1}, V_{i2}) at preset control (V_{c1}, V_{c2})). Inset: the input and output voltage characteristics, V_o–V_i, where the solid and dashed red (blue) lines correspond to V_o–V_{i1} and V_o–V_{i2} when the other input is 0 (1). **b**, Left panel: schematic of logic NOR gate constructed from a 1 × 3 crossed nanowire array. Insets: an example SEM (scale bar, 1 μm) and symbolic electronic circuit. Right panel: the output voltage versus the four possible logic address level inputs. Inset: the V_o–V_i relation, where the solid and dashed red (blue) lines correspond to V_o–V_{i1} and V_o–V_{i2} when the other input is 0 (1). The slope of the data shows that the device voltage gain is larger than 5. Parts **a** and **b** reprinted with permission from ref. 10. **c**, SEM image of a SWNT ring oscillator consisting of five CMOS inverter stages. Adapted with permission from ref. 51.

In addition, multi-input FET-based NOR logic gates have been studied using assembled 1(p-Si) × 3(n-GaN) crossed nanowire-FET arrays (Fig. 4b)[10]. The V_o–V_i relation of a two-input device shows constant low V_o when the other input is high, and a nonlinear response with large change in V_o when the other input is set low. The logic-0 state is observed when either one or both of the inputs are high. A logic-1 state can only be achieved when both of the transistors are on; that is, both inputs low. Analysis of the V_o–V_i data demonstrates that

these two-input NOR gates routinely have gains over five, which is a critical characteristic as it allows interconnection of arrays of logic gates without signal restoration at each stage.

More recently, hybrid bottom-up/top-down approaches have also been used to build circuits based on CNTs and nanowires, including relatively high-frequency oscillators. For example, Avouris and co-workers[51] have demonstrated a CMOS-type ring oscillator on a single long CNT that was assembled on a substrate surface. A key feature

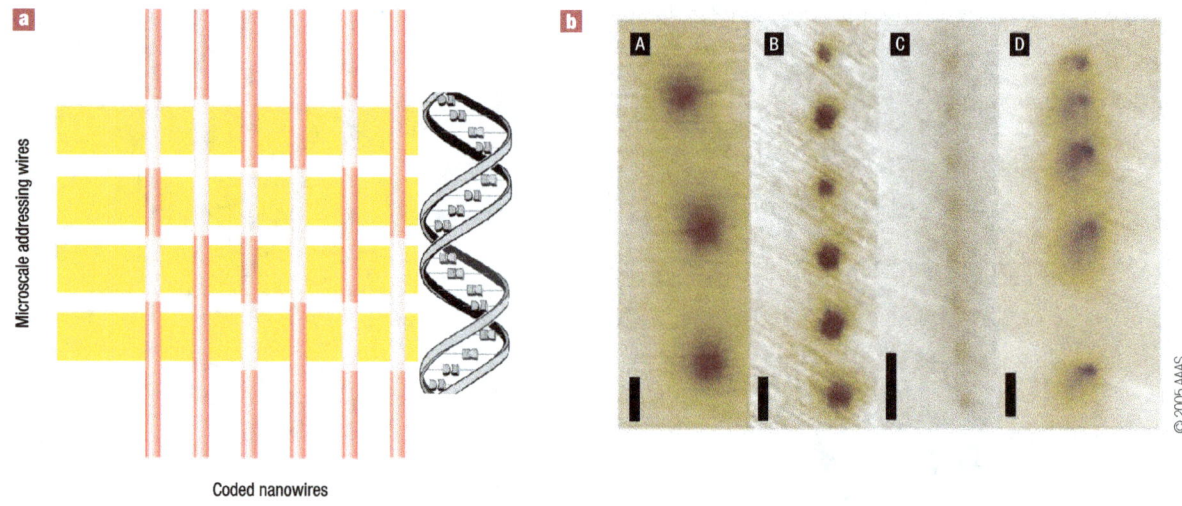

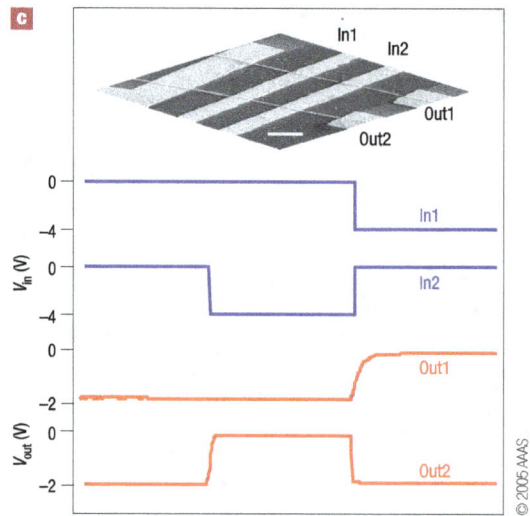

Figure 5 Mixed-scale crossbar demuxes. **a**, Schematic illustration of the mixed-scale demux. Address coding in the nanowires is achieved through modulation doping in a way analogous to gene coding in DNA molecules. The DNA and modulation are not shown to scale. **b**, Scalable synthesis of modulation-doped nanowires with a doping profile of $n^+ - (n - n^+)_N$ for (A) $N = 3$, (B) $N = 6$, (C) $N = 8$ and (D) $N = 5$. Here $n^+(n)$ means heavily (lightly) doped segments and N is the number of repeat units. The results were obtained using a scanning-gate microscopy technique with the lightly doped n-segments shown as the black regions. Both the spacing and length of the doped segments can be controlled during nanowire growth. Scale bars, 1 μm. **c**, Top: SEM image of a 2 × 2 mixed-scale demux configured using two modulation-doped silicon nanowires as outputs (Out1 and Out2) and two gold metal gates as inputs (In1 and In2). Scale bar, 1 μm. Bottom: plots of input (blue) and output (red) voltages for the 2 × 2 demux. Parts **b** and **c** reprinted with permission from ref. 54.

of this work is that it combines the unique electronic properties of single-walled CNTs with controlled top-down lithography to pattern different work-function metals on the CNT so that both n-type and p-type regions can be realized as required for a CMOS device (Fig. 4c). Notably, in this five-stage device, which consists of five p-type and five n-type FETs, signal analysis showed resonances at up to 52 MHz that are consistent with stable oscillation of the device. This combination of strategies could be particularly advantageous for near-term nanoelectronic development, although ultimately hybrid approaches face many of the same constraints that limit conventional top-down processing used in industry today. Furthermore, the one-dimensional nature of nanowires and nanotubes indicates that circuits are more easily built along the long axis of a single nanoscale wire.

This offers both design advantages and challenges: on the one hand, the nanowire or nanotube naturally serves as local interconnects and eliminates additional wiring; on the other hand, the success of these devices depends critically on the development of new architectures, including crossbar-motif logic architectures to take full advantage of the unparalleled material and structural control offered by the bottom-up approach.

ADDRESSING NANOARRAYS

Addressing large numbers of memory bits inside a crossbar array will require demultiplexers (demuxes) that allow a relatively small number of control wires to access the large number of nanoscale wires selectively inside the memory array. In an ideal demux, *n* pairs of microscale control wires can be used to select 2^n nanowires; that is, 10 pairs of control wires can specifically address each of the $2^{10} = 1,024$ nanowires serving as the columns (or rows) of a 1 Mbit crossbar memory. By using a pair of demuxes serving respectively as row and column selectors, all 2^{2n} crosspoints (memory bits) inside the memory array can be selectively addressed by $2n$ pairs of control wires.

The first prototype crossbar demux was based on the transistor function of semiconductor nanowires[20] in which control wires serve as gate electrodes that cross nanowires extending into and forming the columns or rows of the memory array. 'Address coding' of the nanowires inside the demux was obtained by surface modification during the 'programming' process at specific crosspoints to give them a lower threshold voltage V_T than the unmodified ones[15,20]. The coded nanowire demux effectively forms a NOR plane: the nanowire will be turned off if any one of the control wires crossing it at the modified positions is in the high-voltage '1' state.

Coding specific crosspoints through lithography, however, presents a significant manufacturing challenge for large demux arrays and limits the ultimate scaling advantage of assembled nanowire and CNT building blocks. An intriguing approach that overcomes this lithography barrier involves stochastic demuxes formed using modulation-doped semiconductor nanowires. Threshold voltage modulation and hence address coding is obtained at the different crosspoints between gate wires and the nanowire channel, which may be either heavily or lightly doped (Fig. 5a)[52]. In this approach, the manufacturing cost of lithography and functionalizing specific crosspoints to obtain address coding is exchanged for that of creating modulation-doped nanowires during growth; that is, information is encoded during nanowire synthesis, not by lithography during circuit fabrication. Interestingly, this concept of coding during nanowire synthesis has many analogies with biology where the information is encoded during the synthesis of the linear sequence of bases in a replicated DNA molecule (Fig. 5a).

The stochastic demux scheme is particularly suitable for crossbars based on chemically grown nanowires, because the position registry is generally lost during the assembly using flow-alignment[53] or Langmuir–Blodgett[11] techniques, resulting in arrays of aligned albeit randomly positioned nanowires. Stochastic demultiplexing based on coded nanowires was recently demonstrated by Yang *et al.*[54], where modulation doping along the nanowire axis was achieved with full control of the size, spacing and number of the modulated regions during the growth process (Fig. 5b). Crossed nanowire/microscale control wire demux arrays formed by randomly positioned coded-nanowires were shown to follow predictions[52,54] (Fig. 5c), and thus demonstrate that it is possible to construct a unique demux that is independent of the constraints of lithography used in conventional top-down systems.

In electromechanical and resistive-switching crossbar memories, the memory bits are formed by hysteresis resistors or diodes[14]. By permanently setting specific crosspoints in these resistor arrays to the 'closed' state during the configuration process, wired-OR and AND logic (although with a reduced voltage margin) can be obtained by control wires crossing the same nanowire at the closed points. Using the wired-OR and AND logic, row and column demuxes may be built using mixed-scale crossbars consisting of nanowires leading to the memory array and microscale control wires from the CMOS circuitry. The crosspoints in the mixed-scale crossbar demuxes need to be configured 'permanently' at manufacturing time or during configuration, and should not be affected by the write/erase operations of the memory array. This may require deposition of a different switching medium at the demuxes that shows larger switching threshold voltages compared with inside the memory array. Furthermore, linear resistor logic suffers from limited voltage margin: the voltage difference between the selected and non-selected nanowires may not be large enough to

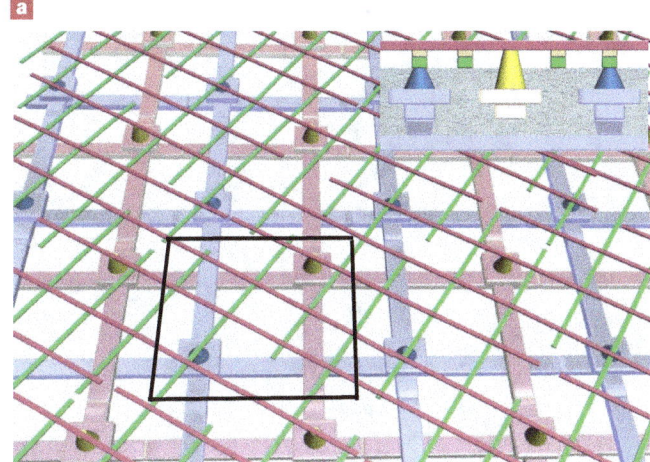

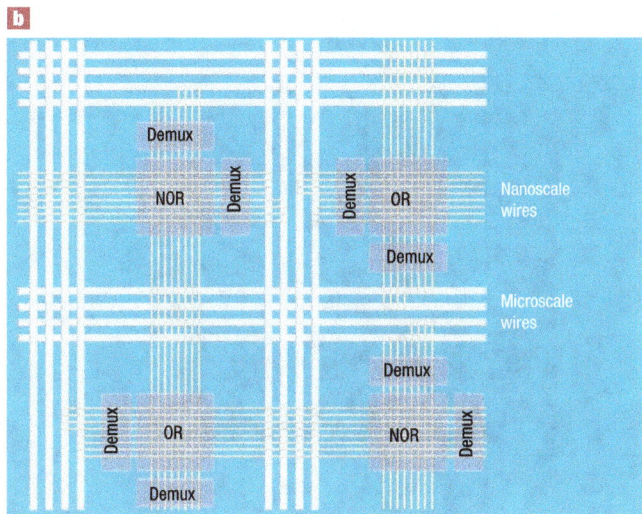

Figure 6 Hybrid crossbar circuits. **a**, Hybrid crossbar/CMOS system based on the CMOL approach. Two sets of pins with different heights separately connect the row and column nanowires in the crossbar with the CMOS circuitry underneath. **b**, Schematic of the array-based system architecture. Each node in the array is a crossbar-based memory or logic device with its own address demux.

complete the specific write/erase function. To improve the voltage margin, several techniques have been proposed using CMOS coding to eliminate the worst-case scenarios[41,55]. However, considering the limited geometry and functions available in the crossbar resistor-logic, it is reasonable to expect workable demuxes to be in the form of hybrid resistor-crossbar/transistor structures in which the transistor circuitry provides the more difficult functions such as signal gain, restoration, inversion and impedance matching.

HYBRID TO NANO ARCHITECTURES

A hybrid crossbar/CMOS circuit takes advantage of both worlds: the ultra-high device density offered by the crossbar structure and the flexibility offered by the CMOS circuitry. But tradeoffs have to be made: larger crossbar proportion results in higher device density with greater circuit design and fabrication challenges, whereas larger CMOS proportion results in increased circuit functionality at a cost of increased chip area overhead. One hybrid crossbar/

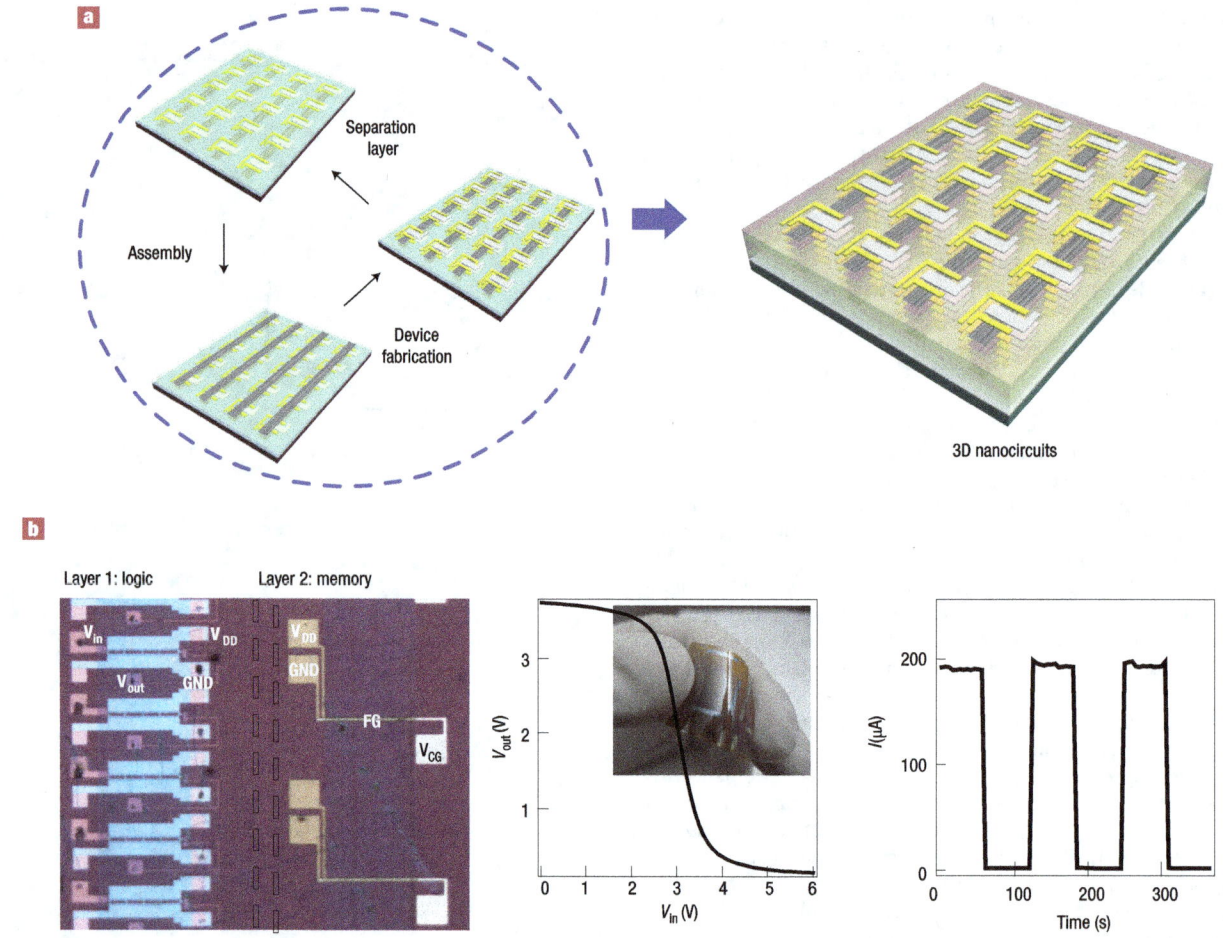

Figure 7 Assembly and integration. **a**, Schematic of three-dimensional integration of multifunctional devices through a layer-by-layer process. **b**, Optical image (left) of nanowire-inverters (layer 1) and floating gate memory (layer 2) on Kapton substrate; d.c. characteristics of the nanowire inverter (centre); switching characteristics of the nanowire floating-gate memory (right). Inset to the centre panel shows functional devices on flexible Kapton substrate. V_{CG}, V_{FG} and V_{DD} are the control gate, floating gate and supply voltages respectively. Reprinted with permission from ref. 59.

CMOS approach that has received considerable attention is the CMOS/nanowire/molecular hybrid (known as CMOL) structure[21,22].

In the CMOL circuit (Fig. 6a), the CMOS-to-nanowire interface is provided by distributing two sets of via-pins uniformly over the whole circuit area. The two sets of pins have different heights so that the taller (shorter) set is connected to the top (bottom) nanowire electrodes in the crossbar array from the CMOS cells underneath. By rotating the crossbar at a small angle (determined by the relative pitch size of the crossbar and CMOS) with respect to the CMOS layer, each nanowire in the crossbar array can be connected to exactly one CMOS cell. In this approach, the most difficult functions — inversion, gain and demultiplexing — are moved into the CMOS layer, using the nanowires solely for data storage (in the case of memories), or wired-OR logic (in the case of logic circuits) and signal routing. A density advantage can still be maintained because only $2n$ CMOS cells are used to control n^2 memory bits; however, the viability of processing on top of a high-performance CMOS structure on which considerable process development/cost has been expended needs to be further scrutinized.

Disregarding practical issues, preliminary studies have shown that CMOL memory with 32-nm CMOS technology could store 1 terabit of data in a 2 cm × 2 cm device[22]. In another hybrid crossbar/CMOS proposal in which even more functions are shifted to CMOS circuitry and the crossbar is used only for data routing[23], easier device fabrication and lower power dissipation are achieved at the cost of lower performance metrics including speed, density and defect-tolerance capabilities. Given these advances, the question seems to have shifted from whether hybrid cross/CMOS circuits are needed to how to implement these structures, and how to optimize the architecture by properly dividing functions between the two classes of components.

Once memory and logic devices (circuits) are implemented, array-based system architectures can be made by interconnecting the crossbar devices such that the output from one crossbar array forms the input of the other (Fig. 6b)[8,21]. The system architecture resembles the nano-architecture at the lower device level: It addresses a similar set of issues such as signal restoration and address demuxing, with the exception that each node in the array is now replaced by a functional memory or logic device. Moreover, the large connectivity of the 2D crossbar arrays provides the possibility to map neuromorphic networks onto distributed crossbar networks in a way that was not possible with CMOS-based approaches, with the potential to achieve artificial neural circuits with comparable density but orders-of-magnitude faster communication speed compared with the cerebral cortex[56].

DEVICE ASSEMBLY TECHNIQUES

To create large-scale crossbar memory and logic devices from a bottom-up approach, effective assembly techniques are required. Using fluidic assisted alignment, Duan et al.[15] and Huang et al.[10] have demonstrated prototype crossbar circuits based on crossed nanowire devices. Further progress has also been made on larger hierarchically patterned arrays of aligned and crossed nanowires at centimetre scales[11,57]. In one study[57], a monolayer of aligned nanowires with controlled spacing was first produced through the Langmuir–Blodgett approach, followed by the formation of crossed-wire structures by depositing a second aligned layer at right angles to the first. Photolithography was then used to define a pattern over the entire substrate surface, setting the array dimensions and array pitch. Finally, nanowires outside the patterned areas were removed and metal interconnects could be deposited by conventional lithography.

New assembly methods are still needed for use on a wafer scale where bottom-up and top-down approaches might feasibly be merged, for example, to produce hybrid memory chips. On the other hand, the simple parallel wire structures make the crossbar-based devices suitable for certain top-down techniques, particularly imprint lithography, which offers the potential for high throughput and low-cost fabrication of metal nanowires[16,17,58]. A related approach termed superlattice nanowire pattern transfer (SNAP)[18] uses selective etching of a superlattice edge and subsequent metal deposition to transfer extremely dense parallel arrays of metal nanowires. Using the SNAP technique, a 160-kilobit crossbar memory structure has been demonstrated with silicon (made by selective etching) and titanium wire arrays as the bottom and top electrodes, and a monolayer of bistable rotaxane molecules as the storage medium[31]. The pitch size of the crossbar array was 33 nm, corresponding to a density of 10^{11} bits cm^{-2}, more than 30 times as high as that of the state-of-the-art DRAM or flash devices[1].

Despite these advances the ultimate potential of the bottom-up approach will not be reached unless material-independent assembly methodologies can be developed. One promising approach involves a new dry deposition strategy that enables oriented and patterned assembly of nanowires with controlled density and alignment on substrates from silicon to plastics[59]. The overall process involves optimized growth of designed nanowire material and patterned transfer of nanowires directly from a growth substrate to a second device substrate by means of contact printing (Fig. 7a). Key features include the ability to print aligned nanowires on a wafer scale and to control the density of aligned nanowires through the transfer process. Moreover, this method is readily adaptable to 3D structures. Specifically, repeating the transfer process has enabled up to ten layers of active nanowire field-effect devices to be assembled, and also a bilayer structure consisting of logic in layer 1 and non-volatile memory in layer 2 (ref. 59). This type of capability, and the ability to assemble distinct types of materials almost at will, offers one of the strongest arguments for using the bottom-up approach in the future.

FUTURE PROSPECTS

The ever-growing demand for new nanoscale memories has led to great progress in disciplines including materials growth, assembly/fabrication techniques and new circuit paradigms. It is reasonable to expect bottom-up structures to make their debut in commercial applications as memory devices (which have less stringent requirements) instead of logic devices, although only as effective large-scale and high-throughput assembly processes become available.

In the meantime, many scientific challenges need to be addressed, and we will gain from a critical examination of the diverse approaches mentioned here. For example, can molecules in the nanoscale crossbar memories be replaced by a more reliable solid-state-based switching medium that can maintain the density advantage but offer better performance metrics? Our initial studies on nanoscale resistive switching devices using a-Si as the switching medium seem to affirm this approach.

It is also likely that the next-generation non-CMOS memory device will still have substantial CMOS components — built by hybrid bottom-up/top-down structures at the physical device level, and operated with crossbar/CMOS structures at the architecture level. For example, a memory chip might be composed of imprinted nanoscale metallic crossbars, solid-state switching medium and CMOS logic circuitry that provide addressing, inversion and gain. The daunting task of producing these devices will only be overcome through collaboration between chemists, physicists and electrical and computer engineers.

Further into the future, we expect the bottom-up approach to open many unique approaches for nanoelectronics. Such areas include the development of 3D multifunctional nanoelectronic and hybrid nanoelectronic/biological systems[60] where the capability of assembling CNT and nanowire building blocks in multiple active layers, independent of material or substrate, provides a new way to consider building the future.

doi:10.1038/nmat2028

References

1. International Technology Roadmap for Semiconductors 2005 edn, available online at <http://public.itrs.net/>.
2. Frank, D. J. et al. Device scaling limits of Si MOSFETs and their application dependencies. Proc. IEEE 89, 259–288 (2001).
3. Likharev, K. K. in Nano and Giga Challenges in Microelectronics (eds Greer, J., Korkin, A. & Labanowski, J.) 27–68 (Elsevier, Amsterdam, 2003).
4. McEuen, P. L., Fuhrer, M. S. & Park, H. Single-walled carbon nanotube electronics. IEEE Trans. Nanotechnol. 1, 78–85 (2002).
5. Lieber, C. M. Nanoscale science and technology: building a big future from small things. Mater. Res. Soc. Bull. 28, 486–491 (2003).
6. Heath, J. R., Kuekes, P. J., Snider, G. S. & Williams, R. S. A defect-tolerant computer architecture: opportunities for nanotechnology. Science 280, 1716–1721 (1998).
7. Stan, M. R., Franzon, P. D., Goldstein, S. C., Lach, J. C. & Ziegler, M. M. Molecular electronics: from devices and interconnect to circuits and architecture. Proc. IEEE 91, 1940–1957 (2003).
8. DeHon, A. Array-based architecture for FET-based, nanoscale electronics. IEEE Trans. Nanotechnol. 2, 23–32 (2003).
9. Cui, Y. & Lieber, C. M. Functional nanoscale electronic devices assembled using silicon nanowire building blocks. Science 291, 851–853 (2001).
10. Huang, Y. et al. Logic gates and computation from assembled nanowire building blocks. Science 294, 1313–1317 (2001).
11. Whang, D., Jin, S., Wu, Y. & Lieber, C. M. Large-scale hierarchical organization of nanowire arrays for integrated nanosystems. Nano Lett. 3, 1255–1259 (2003).
12. Rueckes, T. et al. Carbon nanotube-based nonvolatile random access memory for molecular computing. Science 289, 94–97 (2000).
13. Collier, C. P. et al. Electronically configurable molecular-based logic gates. Science 285, 391–394 (1999).
14. Chen, Y. et al. Nanoscale molecular-switch crossbar circuits. Nanotechnology 14, 462–468 (2003).
15. Duan, X. F., Huang, Y. & Lieber, C. M. Nonvolatile memory and programmable logic from molecule-gated nanowires. Nano Lett. 2, 487–490 (2002).
16. Zankovych, S., Hoffmann, T., Seekamp, J., Bruch, J. U. & Torres, C. M. S. Nanoimprint lithography: challenges and prospects. Nanotechnology 12, 91–95 (2001).
17. Chou, S. Y., Krauss, P. R. & Renstrom, P. J. Imprint lithography with 25-nanometer resolution. Science 272, 85–87 (1996).
18. Melosh, N. A. et al. Ultrahigh-density nanowire lattices and circuits. Science 300, 112–115 (2003).
19. Brueck, S. R. J. in International Trends in Applied Optics (eds Guenther, A. H. & Holst, G. C.) 85–110 (SPIE, Bellingham, Washington, 2002).
20. Zhong, Z., Wang, D., Cui, Y., Bockrath, M. W. & Lieber, C. M. Nanowire crossbar arrays as address decoders for integrated nanosystems. Science 302, 1377–1379 (2003).
21. Strukov, D. B. & Likharev, K. K. CMOL FPGA: a reconfigurable architecture for hybrid digital circuits with two-terminal nanodevices. Nanotechnology 16, 888–900 (2005).
22. Strukov, D. B. & Likharev, K. K. Prospects for terabit-scale nanoelectronic memories. Nanotechnology 16, 137–148 (2005).
23. Snider, G. S. & Williams, R. S. Nano/CMOS architectures using a field-programmable nanowire interconnect. Nanotechnology 18, 035204 (2007).
24. Snider, G. S. Computing with hysteretic resistor crossbars. Appl. Phys. A. 80, 1165–1172 (2005).
25. Kuekes, P. J., Stewart, D. R. & Williams, R. S. The crossbar latch: logic value storage, restoration, and inversion in crossbar circuits. J. Appl. Phys. 97, 034301 (2003).
26. Reed, M. A., Zhou, C., Muller, C. J., Burgin, T. P. & Tour, J. M. Conductance of a molecular junction. Science 278, 252–254 (1997).
27. Reed, M. A., Chen, J., Rawlett, A. M., Price, D. W. & Tour, J. M. Molecular random access memory cell. Appl. Phys. Lett. 78, 3735–3737 (2001).

28. Donhauser, Z. J. *et al.* Conductance switching in single molecules through conformational changes. *Science* **292**, 2303–2307 (2001).

29. Lau, C. N., Stewart, D. R., Williams, R. S. & Bockrath, M. Direct observation of nanoscale switching centers in metal/molecule/metal structures. *Nano Lett.* **4**, 569–572 (2004).

30. Collier, C. P. *et al.* A [2]catenane-based solid state electronically reconfigurable switch. *Science* **289**, 1172–1175 (2000).

31. Green, J. E. *et al.* A 160-kilobit molecular electronic memory patterned at 10^{11} bits per square centimetre. *Nature* **445**, 414–417 (2007).

32. Seminario, J. M., Zacarias, A. G. & Tour, J. M. Theoretical study of a molecular resonant tunneling diode. *J. Am. Chem. Soc.* **122**, 3015–3020 (2000).

33. Di Ventra, M., Pantelides, S. T. & Lang, N. D. First-principles calculation of transport properties of a molecular device. *Phys. Rev. Lett.* **84**, 979–982 (2000).

34. Chen, J. *et al.* Room-temperature negative differential resistance in nanoscale molecular junctions. *Appl. Phys. Lett.* **77**, 1224–1226 (2000).

35. Flood, A. H., Stoddart, J. F., Steuerman, D. W. & Heath, J. R. Whence molecular electronics? *Science* **306**, 2055–2056 (2004).

36. Waser, R. & Aono, M. Nanoionics-based resistive switching memories. *Nature Mater.* **6**, 833–840 (2007).

37. Nantero; <http://www.nantero.com/index.html>.

38. Owen, A. E., Lecomber, P. G., Hajto, J., Rose, M. J. & Snell, A. Switching in amorphous devices. *Int. J. Electron.* **73**, 897–906 (1992).

39. Jafar, M. & Haneman, D. Switching in amorphous-silicon devices. *Phys. Rev. B* **49**, 13611–13615 (1994).

40. Scott, J. C. Is there an immortal memory? *Science* **304**, 62 (2004).

41. Kuekes, P. J., Robinett, W. & Williams, R. S. Improved voltage margins using linear error-correcting codes in resistor-logic demultiplexers for nanoelectronics. *Nanotechnology* **16**, 1419–1432 (2005).

42. Hu, J., Branz, H. M., Crandall, R. S., Ward, S. & Wang, Q. Switching and filament formation in hot-wire CVD p-type a-Si:H devices. *Thin Solid Films* **430**, 249–252 (2003).

43. Avila, A. & Asomoza, R. Switching in coplanar amorphous hydrogenated silicon devices. *Solid State Electron.* **44**, 17–27 (2000).

44. Goronkin, H. & Yang, Y. High-performance emerging solid-state memory technologies. *Mater. Res. Soc. Bull.* **29**, 805–808 (2004).

45. Wuttig, M. & Yamada, N. Phase-change materials for rewriteable data storage. *Nature Mater.* **6**, 824–832 (2007).

46. Jung, Y., Lee, S.-H., Ko, D.-K. & Agarwal, R. Synthesis and characterization of $Ge_2Sb_2Te_5$ nanowires with memory switching effect. *J. Am. Chem. Soc* **128**, 14026–14027 (2006).

47. Lee, S.-H., Ko, D.-K., Jung, Y. & Agarwal, R. Size-dependent phase transition memory switching behavior and low writing currents in GeTe nanowires. *Appl. Phys. Lett.* **89**, 223116 (2006).

48. Yu, D., Wu, J., Gu, Q. & Park, H. Germanium telluride nanowires and nanohelices with memory-switching behavior. *J. Am. Chem. Soc.* **128**, 8148–8149 (2006).

49. Meister, S. *et al.* Synthesis and characterization of phase-change nanowires. *Nano Lett* **6**, 1514–1517 (2006).

50. Lankhorst, M. H. R., Ketelaars, B. W. S. M. M. & Wolters, R. A. M. Low-cost and nanoscale non-volatile memory concept for future silicon chips. *Nature Mater.* **4**, 347–352 (2005).

51. Chen, Z. *et al.* An integrated logic circuit assembled on a single carbon nanotube. *Science* **311**, 1735 (2006).

52. DeHon, A., Lincoln, P. & Savage, J. E. Stochastic assembly of sublithographic nanoscale interfaces. *IEEE Trans. Nanotechnol.* **2**, 165–174 (2003).

53. Huang, Y., Duan, X. F., Wei, Q. Q. & Lieber, C. M. Directed assembly of one-dimensional nanostructures into functional networks. *Science* **291**, 630–633 (2001).

54. Yang, C., Zhong, Z. & Lieber, C. M. Encoding electronic properties by synthesis of axial modulation-doped silicon nanowires. *Science* **310**, 1304–1307 (2005).

55. Kuekes, P. J. *et al.* Resistor-logic demultiplexers for nanoelectronics based on constant-weight codes. *Nanotechnology* **17**, 1052–1061 (2006).

56. Likharev, K. K. & Strukov, D. B. in *Introducing Molecular Electronics* (eds Cuniberti, G., Fagas, G. & Richter, K.) 447–477 (Springer, Berlin, 2005).

57. Jin, S. *et al.* Scalable interconnection and integration of nanowire devices without registration. *Nano Lett.* **4**, 915–919 (2004).

58. Jung, G. Y. *et al.* Fabrication of multi-bit crossbar circuits at sub-50 nm half-pitch by using UV-based nanoimprint lithography. *J. Photopolym. Sci. Technol.* **18**, 565–570 (2005).

59. Javey, A., Nam, S. W., Friedman, R. S., Yan, H. & Lieber, C. M. Layer-by-layer assembly of nanowires for three-dimensional, multifunctional electronics. *Nano Lett.* **7**, 773–777 (2007).

60. Patolsky, F. *et al.* Detection, stimulation, and inhibition of neuronal signals with high-density nanowire transistor arrays. *Science* **313**, 100–104 (2006).

Acknowledgements

We thank many of our colleagues at the University of Michigan and Harvard University, particularly T. Rueckes, Y. Dong, G. Yu, Y. Chen, Z. Zhong, X. Duan, Y. Huang and S. Y. Jo for assistance, and A. DeHon and P. Lincoln for critical discussions. W.L. acknowledges support by the National Science Foundation (CCF-0621823). C.M.L. acknowledges support by Air Force Office of Scientific Research, Defense Advanced Research Projects Agency, Intel, and Samsung Electronics.

The emergence of spin electronics in data storage

Electrons have a charge and a spin, but until recently these were considered separately. In classical electronics, charges are moved by electric fields to transmit information and are stored in a capacitor to save it. In magnetic recording, magnetic fields have been used to read or write the information stored on the magnetization, which 'measures' the local orientation of spins in ferromagnets. The picture started to change in 1988, when the discovery of giant magnetoresistance opened the way to efficient control of charge transport through magnetization. The recent expansion of hard-disk recording owes much to this development. We are starting to see a new paradigm where magnetization dynamics and charge currents act on each other in nanostructured artificial materials. Ultimately, 'spin currents' could even replace charge currents for the transfer and treatment of information, allowing faster, low-energy operations: spin electronics is on its way.

CLAUDE CHAPPERT[1,2]*, ALBERT FERT[2,3] AND FRÉDÉRIC NGUYEN VAN DAU[2,3]

[1]Institut d'Electronique Fondamentale, CNRS, UMR8622, 91405 Orsay, France
[2]Université Paris Sud, 91405 Orsay, France
[3]Unité Mixte de Physique CNRS-Thales, 91767 Palaiseau, France
*e-mail: claude.chappert@ief.u-psud.fr

The interdependence between magnetization and charge transport is not a new story. For instance, anisotropic magnetoresistance (AMR), which links the value of the resistance to the respective orientation of magnetization and current, was first observed in 1856 by William Thomson, but its amplitude is weak (up to a few per cent variation in resistance on changing the relative orientation of magnetization and current). Nevertheless, the introduction by IBM in 1991 of a magnetoresistive read head based on AMR was a major technological step forward[1]. In hard disk drives (HDD), the head flies at constant height above the magnetic domains that define the 'bits' in the recording medium, and senses the spatial variations of the stray magnetic field of these domains. The old ring-shaped magnetic head, invented in 1933 by Eduard Schuller (AEG) for tape recording, measured a magnetic flux, and so was approaching its sensitivity limit with the reduction of the head dimensions. In the IBM head (Fig. 1), the AMR sensor is combined with a 'ring' element still used for writing, and directly senses the magnetic field through its influence on the magnetization orientation in the head. The AMR head had a relative 'magnetoresistance' (hereafter referred to as $\Delta R/R = (R_{max} - R_{min})/R_{min}$) of the order of only 1%, but this was enough to increase the growth rate of HDD storage areal density from 25% per year, its value since nearly the introduction of HDD in 1957, up to 60% per year.

FUNDAMENTALS OF SPIN ELECTRONICS

The physics behind today's fast expansion of spin electronics has also been known for a long time. A cornerstone is the 'two currents' conduction concept proposed by Mott[2] and used by Fert and Campbell[3,4] to explain specific behaviours in the conductivity of the ferromagnetic metals Fe, Ni, Co and their alloys. In such 'itinerant ferromagnets' both the $4s$ and $3d$ electron bands contribute to the density of states at the Fermi level E_F. Because of the strong exchange interaction favouring parallel orientation of electron spins, the 'spin-up' and 'spin-down' $3d$ bands are shifted in energy. This band splitting creates the imbalance between numbers n_{up} and n_{down} of $3d$ electrons that is at the origin of the ferromagnetic moment ($\mu \approx -(n_{up} - n_{down})\mu_B$/atom, where μ_B is the Bohr magneton), whereas the conduction is dominated by the unsplit $4s$ band, the $4s$ electrons having a much higher mobility. However, the spin-conserving s-to-d transitions are the main source of s-electron scattering. This has two chief consequences for transport: the spin-imbalanced density of states for $3d$ electrons at E_F results in strongly spin-dependent scattering probabilities (Fermi 'Golden Rule'), and between two spin-flip scattering events an electron can undergo many scattering events that keep the same spin direction. Thus at the limit where spin-flip scattering events are negligible, conduction happens in parallel through two spin channels that have very different conductivities.

Important length scales can be discussed in the diffusive transport model. The spin-dependent scattering probability results in very different mean free paths λ_{up} and λ_{down}, or equivalently relaxation times τ_{up} and τ_{down}. In usual thin metallic layers they scale from a few nanometres to a few tens of nanometres, with highly variable $\lambda_{down}/\lambda_{up}$ ratios: some impurities have strongly spin-dependent cross-sections, so that in Ni, for example, the ratio $\lambda_{down}/\lambda_{up}$ can reach 20 for

When citing this review, please cite the original version as shown on the contents page of this chapter.

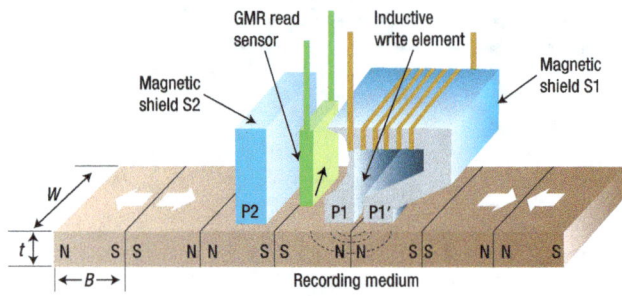

Figure 1 Magnetoresistive head for hard-disk recording. Schematic structure of the magnetoresistive head introduced by IBM for its hard disk drives in 1991. A magnetic sensor based on anisotropic magnetoresistance (left) is added to the inductive 'ring-type' head (right) still used for writing. The distances P1–P1′ and P1–P2 between the pole pieces of the magnetic shields S1 and S2 define respectively the 'write' and 'read' gaps, on which depends the minimum length B of the magnetic domains. W is the track width and t is the thickness of the recording medium. Note that in today's hard disk recording, W and B are of the order of 100 nm and 30 nm respectively, but with a different arrangement of head and domains in 'perpendicular recording'[1].

Co impurities or decrease to 0.3 with Cr doping[4]. Another important length scale is the spin-conserving drift length projected along one direction, called the spin diffusion length L_{sf} (sf denotes spin flip). It is generally much larger than the mean free path[5].

GIANT MAGNETORESISTANCE AND THE SPIN-VALVE HEAD

The founding step of spin electronics, which triggered the discovery of the giant magnetoresistance[6,7], was actually to build magnetic multilayers with individual thicknesses comparable to the mean free paths, so that evidence could be seen for spin-dependent electron transport. The principle is schematized in Fig. 2 for the simplest case of a triple-layer film of two identical ferromagnetic layers F1 and F2 sandwiching a non-magnetic metal spacer layer M, when the current circulates 'in plane' (cf. Fig. 2b). We assume $\lambda_{up}^F \gg \lambda_{down}^F$, with $\lambda_{up}^F > t_F > \lambda_{down}^F$, for the thickness t_F of the magnetic layer and $t_M \ll \lambda_M$ for the thickness t_M of the spacer layer. When the two magnetic layers are magnetized parallel (P), the spin-up electrons can travel through the sandwich nearly unscattered, providing a conductivity shortcut and a low resistance. On the contrary, in the antiparallel (AP) case, both spin-up and spin-down electrons undergo collisions in one F layer or the other, giving rise to a high resistance. The relative magnetoresistance $\Delta R/R = (R_{AP} - R_P)/R_P$ can reach 100% or more in multilayers with a high number of F/M periods. It was already 80% in the Fe/Cr multilayer of the original discovery[6], hence the name of giant magnetoresistance (GMR). Elaborate theories must of course take into account other effects such as interfacial scattering and quantum confinement of the electrons in the layers[8,9]. The GMR is an outstanding example of how structuring materials at the nanoscale can bring to light fundamental effects that provide new functionalities. And indeed the amplitude of the GMR immediately triggered intense research, soon achieving the definition of the spin-valve sensor[10–12]. In its simplest form, the spin valve is just a trilayer film of the kind displayed in Fig. 2a in which one layer (for example, F2) has its magnetization pinned along one orientation. The rotation of the free F1 layer magnetization then 'opens' (in P configuration) or 'closes' (in AP configuration) the flow of electrons, acting as a sort of valve.

The standard spin valve shows magnetoresistance values of about 5–6%. More complex spin-valve stacks, with layer thicknesses

controlled at the atomic scale and ultra-low surface roughness, are optimized to favour specular reflection of electrons towards the active part so that the magnetoresistance reaches about 20%. Following the introduction in 1997 by IBM of the spin-valve sensor (Fig. 2b) to replace the AMR sensor in magnetoresistive HDD read heads, the growth rate for storage areal density immediately increased up to 100% per year. Together, the sequential introduction of the magnetoresistance and spin-valve head, by providing a sensitive and scalable read technique, contributed to increase the raw HDD areal recording density by three orders of magnitude (from ~0.1 to ~100 Gbit in⁻²) between 1991 and 2003. This jump forward opened the way both to smaller HDD form factors (down to 0.85-inch disk diameter) for mobile appliances such as ultra-light laptops or portable multimedia players, and to unprecedented drive capacities (up to a remarkable 1 terabyte) for video recording or backup. HDDs are now replacing tape in at least the first tiers of data archival strategies, for which they provide faster random access and higher data rates. However, the areal density growth rate started to slow down after 2003, when other problems joined the now limiting spin-valve head.

Attempts were also made to develop solid-state magnetic storage. Provided that the free layer magnetization of the spin valve is constrained to take only the two opposite orientations of an easy magnetization axis, arrays of patterned spin-valve elements can be used to store binary information with resistive read-out. Spin-valve solid-state memories were indeed developed[13]. But the planar geometry of the spin-valve sensor intrinsically limits the integration into high-density nanoelectronics, and the low metallic resistance and $\Delta R/R$ value around 10% are not well adapted to CMOS electronics. For read heads also, the 'current in plane' spin-valve geometry is a limitation to the downscaling. First, it is easy to see that the integration of the planar element of Fig. 2b between the magnetic shields of the read head of Fig. 1 strongly limits the minimum dimension of the read gap between the P1 and P2 poles, a crucial requirement for reducing the bit length: indeed, two thick insulating layers are needed on either part of the sensor. The 'current perpendicular to plane' (CPP) geometry of Fig. 2c is much better for this purpose, as the spin-valve sensor can be directly connected to the magnetic shields. This configuration is also more favourable for reducing the track width.

Simple geometrical arguments based on the average electron propagation direction also lead us to expect higher magnetoresistance values for a spin valve in the CPP configuration. The fundamental study of CPP GMR was indeed extremely productive in terms of the new concepts of spin injection and spin accumulation[14]. The basic principle is displayed in Fig. 3. Again we assume that $\tau_{up} \gg \tau_{down}$ in the ferromagnetic metal. So when a current flows from a ferromagnetic layer (F) to a non-magnetic layer (N), away from the interface the current densities j_{up} and j_{down} must be very different on the ferromagnetic side, and equal on the non-magnetic side. The necessary adjustment requires that, in the area near the interfaces, more electrons from the spin-up channel flip their spins. This occurs through an 'accumulation' of spin-up electrons, that is, a splitting of the E_{Fup} and E_{Fdown} Fermi energies, which induces spin-flips and adjusts the incoming and outgoing spin fluxes. The spin-accumulation decays exponentially on each side of the interface on the scale of the respective spin diffusion lengths L_{sf}^F and L_{sf}^N. In this spin accumulation zone, the spin polarization of the current decreases progressively going from the magnetic conductor to the non-magnetic one, so that a spin-polarized current is 'injected' into the non-magnetic metal up to a distance that can reach a few hundreds of nanometres, well beyond the ballistic range. This concept has been extended to more complex interfaces between metals and semiconductors[15] (Fig. 3c). Likewise, the concept applies when an interfacial resistance such as a Shottky or insulating barrier exists[16]. The spin-injection effect has also been demonstrated for

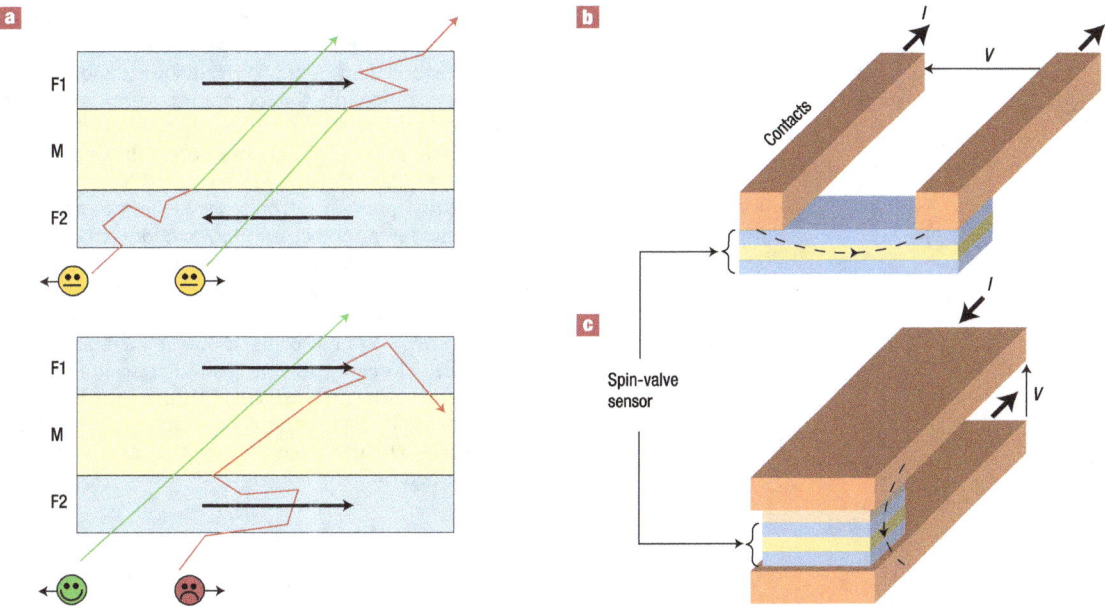

Figure 2 The spin valve. **a**, Schematic representation of the spin-valve effect in a trilayer film of two identical ferromagnetic layers F1 and F2 sandwiching a non-magnetic metal spacer layer M, the current circulating in plane. When the two magnetic layers are magnetized parallel (lower scheme), the spin-up electrons (spin antiparallel to the magnetization) can travel through the sandwich nearly unscattered, providing a conductivity shortcut and a low resistance. In contrast, in the antiparallel case (top scheme) both spin-up and spin-down electrons undergo collisions in either F1 or F2, giving rise to a higher overall resistance. **b**, Schematic arrangement of the 'current in plane' spin-valve sensor in a read head. **c**, Schematic arrangement of the 'current perpendicular to plane' spin-valve sensor in a read head. In both configurations, the recording medium travels parallel to the front face of the sensor.

planar geometries[17], and proposed for use in three-terminal devices such as the spin transistor[18]. We will come back to this point later.

If the magnetoresistance ratio is larger in the CPP geometry, exceeding 20%, CPP GMR samples are much more difficult to fabricate[19]. In particular, quantitative measurements require either superconducting contacts[20] or long multilayered wires[21] to ensure current lines perpendicular to the layers. And the resistance and magnetoresistance still remain too small for optimal application to read heads[22]. Considerable industrial research is going on (see for instance ref. 23).

MAGNETIC TUNNEL JUNCTION

Another big step forward came from replacing the non-magnetic metallic spacer layer M of the spin valve by a thin (~1–2 nm) non-magnetic insulating layer, thus creating a magnetic tunnel junction (MTJ). In that configuration the electrons travel from one ferromagnetic layer to the other by a tunnel effect, which conserves the spin (Fig. 4a). Again, this is not a new story, as it was proposed by Jullière[24] in 1975, but its practical realization with a high magnetoresistance (up to 30% at 4.2 K) had to wait until 1995 once considerable progress had been made in deposition and nanopatterning techniques[25,26]. The first MTJs used an amorphous Al_2O_3 insulating layer between ferromagnetic metal layers: the tunnel magnetoresistance (TMR) of such stacks reached a limit around 70% at room temperature. Much higher effects were later obtained with a single-crystal MgO barrier[27,28]. Within such a barrier, the tunnelling current is carried by evanescent waves of several well-defined symmetries, which, at least for high-quality interfaces, have an interfacial connection in the metals only to Bloch waves of the same symmetry at the Fermi level[29,30]. In the typical case of MgO(001) between Co electrodes for instance,

the decay is much slower and the transmission higher for the evanescent waves of symmetry Δ_1, so that the TMR comes from the specific high spin polarization of the states of symmetry Δ_1 in a zone of the Fermi surface of Co near [001]. The MgO barrier is thus 'active' in selecting a symmetry of high spin polarization, leading to latest record values of $\Delta R/R = 1010\%$ at 5 K, 500% at room temperature (Fig. 4b)[31]. There is no sharp symmetry selection in amorphous barriers such as Al_2O_3, which explains the much lower TMR ratios achieved.

The magnetic tunnel junction is clearly a CPP 'vertical' device, with a magnetic behaviour similar to a spin valve but magnetoresistance values up to two orders of magnitude higher. It is stable up to reasonable breakdown voltages (above 1 V), and the equipment suppliers rapidly developed reliable techniques to scale down its dimensions to well below 100 nm. So its development had an immediate impact on storage applications. Indeed, a TMR read head was commercialized by Seagate[32] in 2005 (Fig. 4c), providing a higher sensitivity. This may prove to be a short-lived option: MTJs have an intrinsic high resistance (with resistance–area products above 1 Ω cm²), and with further downscaling it will become difficult to maintain a high signal-to-noise ratio with an increasing sensor resistance. Then CPP spin valves or degenerate MTJs could become more favourable[22].

MAGNETIC RANDOM ACCESS MEMORY

The 'high' MTJ resistance is actually more adapted to nanoelectronics. The 1995 publications started a race to develop the magnetic random access memory, or MRAM[13]. Figure 5a shows the principle of this magnetic solid state memory, in the basic 'cross point' architecture. The binary information 0 and 1 is recorded on the two opposite orientations of the magnetization of the free layer along its easy magnetization axis. The MTJs are

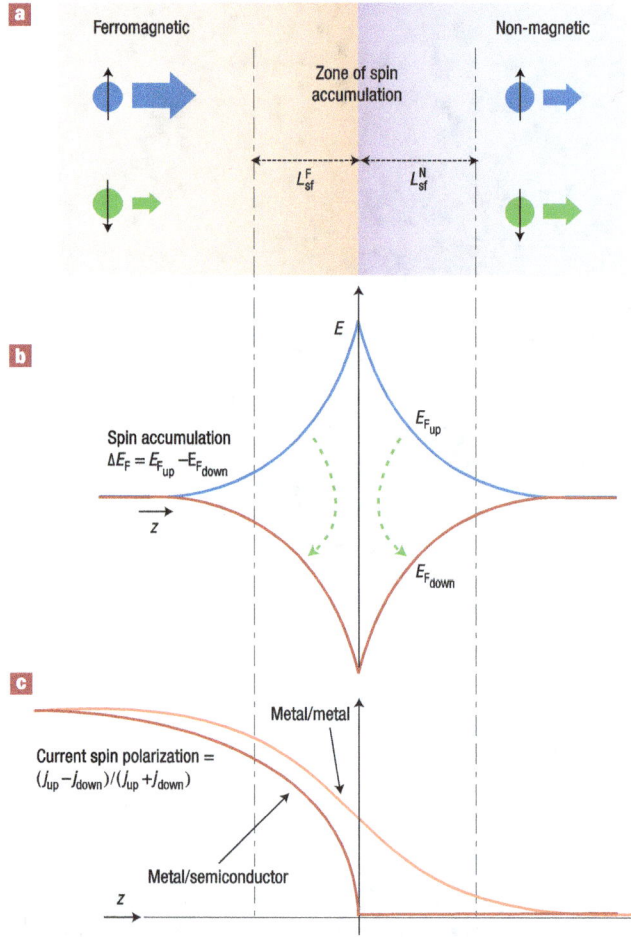

Figure 3 Spin accumulation. Schematic representation of the spin accumulation at an interface between a ferromagnetic metal and a non-magnetic layer, adapted from ref. 132. **a**, Spin-up and spin-down current far from an interface between the ferromagnetic and nonmagnetic conductors (outside the spin-accumulation zone). L^F_{sf} and L^N_{sf} are, respectively, the spin diffusion lengths in the ferromagnetic and non-magnetic layers. **b**, Splitting of the Fermi levels $E_{F_{up}}$ and $E_{F_{down}}$ at the interface. The dashed green arrows symbolize the transfer of current between the two channels by the unbalanced spin flips caused by the out-of-equilibrium spin-split distribution, which governs the depolarization of the electron current between the left and the right. With the current in the opposite direction, there is an inversion of the spin accumulation and opposite spin flips, which polarizes the current across the spin-accumulation zone. **c**, Variation of the current spin polarization when there is an approximate balance between the spin flips on both sides (metal/metal) and when the spin flips on the left side are predominant (metal/semiconductor, for example). The current densities for spin-up and spin-down electrons are j_{up} and j_{down}, respectively.

connected to the crossing points of two perpendicular arrays of parallel conducting lines. For writing, current pulses are sent through one line of each array, and only at the crossing point of these lines is the resulting magnetic field high enough to orient the magnetization of the free layer. For reading, the resistance between the two lines connecting the addressed cell is measured. In principle, this cross point architecture promises very high densities. In practice, the amplitude of magnetoresistance remains too low for fast, reliable reading because of the unwanted current paths as well as the direct one through the addressed cell. So,

realistic cells add one transistor per cell, resulting in more complex 1T/1MTJ cell architectures such as the one represented in Fig. 5b. Several demonstrator circuits were rapidly presented by most leading semiconductor companies, culminating with the first MRAM product, a 4-Mbit stand-alone memory[33] commercialized by Freescale in 2006 (Fig. 5c), and voted 'Product of the Year' by *Electronics Products Magazine* in January 2007.

The MRAM potentially combines key advantages such as non-volatility, infinite endurance and fast random access (down to 5 ns read/write time[34]) that make it a likely candidate for becoming the 'universal memory', one of the chief aims of nanoelectronics. Such a memory is able to provide data/code (Flash, ROM) and execution (DRAM, SRAM) storage using a single memory technology on the same die. Moreover, in June 2007 Freescale introduced a new version able to work in the expanded temperature range of –40 °C to 105 °C, thus qualifying for military and space applications where the MRAM will also benefit from the intrinsic resistance to radiation of magnetic storage.

NANOMAGNETISM

Progress in spin electronics cannot be separated from the development of 'nanomagnetism'. In particular, the engineering of magnetic properties at atom level in multilayers was developed in parallel with GMR and helped to make it possible. Improved knowledge of the role of interface effects, and the use of the layer thickness as a parameter, led to the development of artificial magnetic materials with finely tuned new properties. This had a direct impact on spin storage.

The magnetic storage of binary information requires the engineering of an energy barrier between two opposite orientations of the magnetization, able to stop thermally excited reversals[35]. This 'magnetic anisotropy' has several competing origins. The strongest one is usually the shape anisotropy due to the dipole–dipole magnetic interaction, which induces the well-known in-plane easy magnetization of thin films. But the main effect used in recording is the magnetocrystalline anisotropy, an atomic effect correlated to the symmetry of the immediate atomic environment. The interface anisotropy, initially proposed by Néel[36], takes advantage of the break in translational symmetry at an interface to generate giant magnetic anisotropies, able to overcome the shape anisotropy and induce a stable perpendicular magnetization axis (PMA) in ultrathin films and multilayers. This PMA was first observed in 1967 on single-atomic-layer films[37], and achieved in 1985 in Co/Pd multilayer[38] and Au/Co/Au films[39] with more practical thicknesses. New materials could even be predicted from *ab initio* calculations[40]. PMA is now used for recording media in the 'perpendicular' HDD introduced by Seagate, Hitachi and Toshiba in 2005–06, which helped to restore the present 40% growth rate after the slow-down experienced in 2003–04.

Exchange bias is another crucial effect linked to an interface, here between a ferromagnetic and antiferromagnetic layer: the antiferromagnetic layer has no net magnetic moment that could be sensitive to an applied field, but may retain a large magnetic anisotropy, which, transferred to the ferromagnetic layer through interfacial exchange interaction, contributes to stabilizing the orientation of its magnetization. This is also an old story[41], but with progress in interface control[42] it gained wide application in the spin valve and magnetic tunnel junctions for pinning the magnetization of the reference magnetic layer. And it can also be used in a new kind of MRAM[43], or to fight thermal excitations of magnetic nanoparticles[44] for future storage.

The growing structural quality of interfaces has led to the spin-dependent quantum confinement of the electrons in metallic ultrathin layers. Together with interfacial band hybridization this

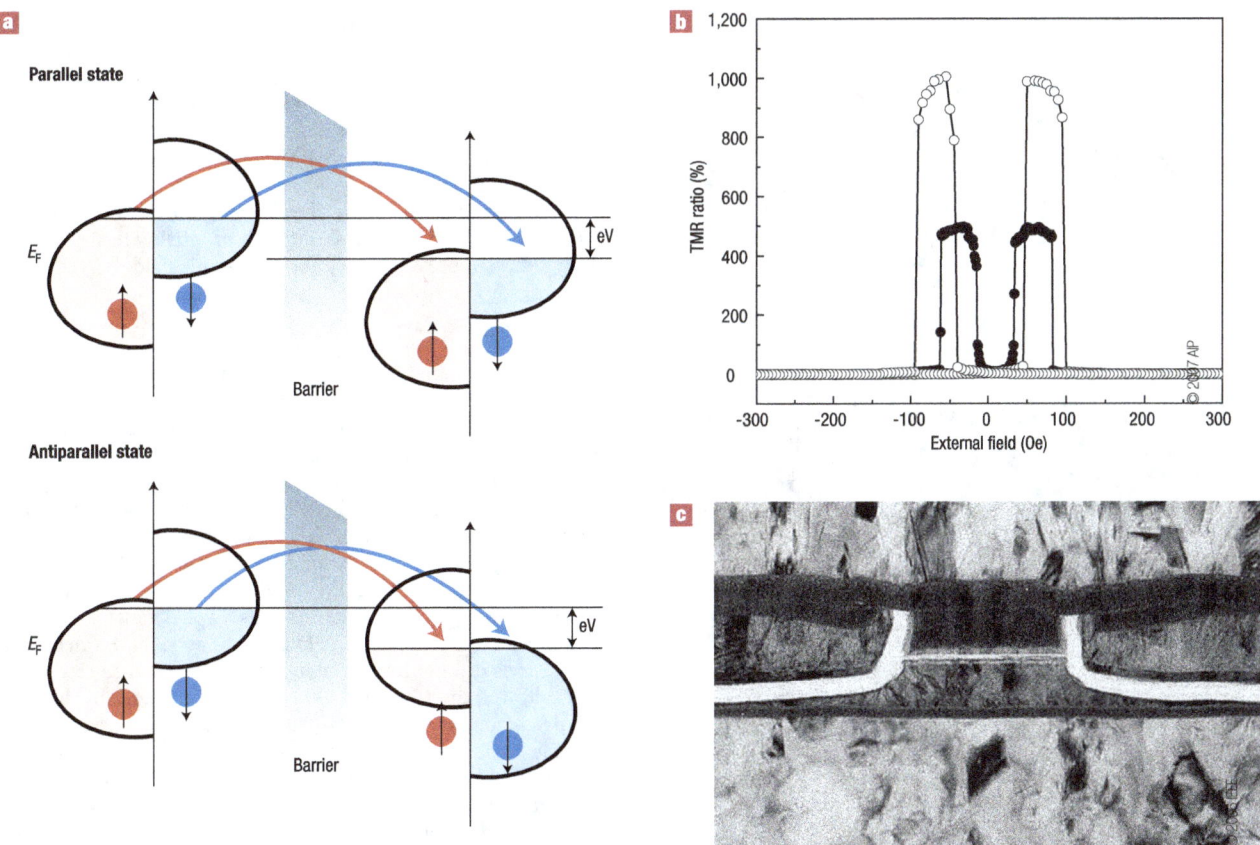

Figure 4 The magnetic tunnel junction. **a**, Schematic representation of the tunnel magnetoresistance in the case of two identical ferromagnetic metal layers separated by a non-magnetic amorphous insulating barrier such as Al_2O_3. The tunnelling process conserves the spin. When electron states on each side of the barrier are spin-polarized, then electrons will more easily find free states to tunnel to when the magnetizations are parallel (top picture) than when they are antiparallel (bottom picture). **b**, Record high magnetoresistance TMR = $(R_{max} - R_{min})/R_{min}$ for the magnetic stack $(Co_{25}Fe_{75/80}B_{20}$ (4 nm)/MgO (2.1 nm)/$(Co_{25}Fe_{75/80}B_{20}$ (4.3 nm) annealed at 475 °C after growth, measured at room temperature (filled circles) and at 5 K (open circles). Reprinted with permission from ref. 31. **c**, Transmission electron microscope cross-section of a TMR read head from Seagate. Reprinted with permission from ref. 32. The tunnel junction stack appears vertically at the centre of the picture, with the tunnel barrier at the level of the thin white horizontal line. The thick bent lines on both sides are the insulating layers between top and bottom contacts. The two thick light grey layers on top and bottom are the magnetic pole pieces (see Fig. 1). The track width of the TMR element is typically 90–100 nm.

now makes it possible to control the exchange interaction between ferromagnetic layers separated by a non-magnetic layer[45–49]. This, for instance, enables synthetic antiferromagnets (SAF) to be built, these being trilayers where two ferromagnetic layers are kept magnetized antiparallel by exchange through a non-magnetic spacer. Depending on their exact structure and properties, such SAFs can provide either improved non-volatility for constant writing field[50], or more reliable writing in solid state devices[51], or simply a weak sensitivity to applied field and minimal stray field when the net magnetic moment is brought to zero. They are thus used now as recording media for longitudinal HDD products[1] as well as in Freescale's MRAM product. And they are ubiquitous in all pinned layers of spin-valve sensors and MRAM cells where they help the exchange bias and minimize the interlayer dipole–dipole interaction.

So at the beginning of this century outstanding progress had been made both in designing the magnetic properties through atomic engineering, and in understanding and controlling the spin-dependent electron transport. And materials exist that allow thermally stable magnetic particles to be produced down to sizes of a few nanometres[52], seemingly opening a bright future for a high-density spin storage.

WRITING IS THE PROBLEM

The use of a magnetic field to write the information still remained a considerable limitation. This can easily be understood. Let us assume that information is stored in the form of the magnetization orientation of a nanoparticle of volume V. The energy barrier fighting the thermal excitations is given by KV, where K is the anisotropy constant per unit volume. Non-volatility — usually defined by a maximum error rate, for example 10^{-9}, over a 10-year period — is obtained when $KV > 50$–$60k_BT$, where k_B is the Boltzmann constant and T the temperature. So reducing V requires a corresponding increase in K, but then the writing field increases proportionally with K, whereas the power available to create it decreases as the dimensions are downscaled.

The problem is well known in HDD, where it was recently postponed by first introducing the SAF longitudinal media and then changing to perpendicular recording. The future also promises heat-assisted magnetic recording (HAMR)[1], where the magnetic field is helped by local heating of the media temporarily reducing the energy barrier, as in magneto-optical recording. But the fundamental limit is still there, together with other problems linked to the rotating nature

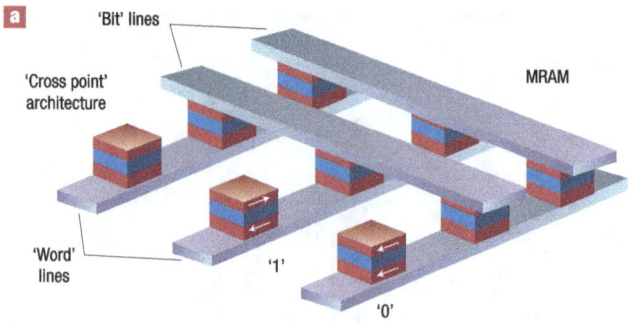

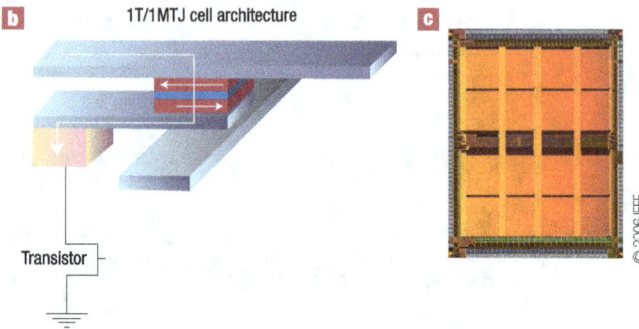

Figure 5 Magnetic random access memory. **a**, Principle of MRAM, in the basic cross-point architecture. The binary information 0 and 1 is recorded on the two opposite orientations of the magnetization of the free layer of magnetic tunnel junctions (MTJ), which are connected to the crossing points of two perpendicular arrays of parallel conducting lines. For writing, current pulses are sent through one line of each array, and only at the crossing point of these lines is the resulting magnetic field high enough to orient the magnetization of the free layer. For reading, the resistance between the two lines connecting the addressed cell is measured. **b**, To remove the unwanted current paths around the direct one through the MTJ cell addressed for reading, the usual MRAM cell architecture has one transistor per cell added, resulting in more complex 1T/1MTJ cell architecture such as the one represented here. **c**, Photograph of the first MRAM product, a 4-Mbit stand-alone memory commercialized by Freescale in 2006. Reprinted with permission from ref. 33.

of HDD storage such as mechanical tracking, slow access time (hardly reduced in several decades and still a few milliseconds) and energy consumption. The future markets of HDD, in particular through the competition with Flash storage, will depend on how such problems are solved. For instance, HDD currently maintains an areal density growth rate of roughly 40%, in line with Moore's law defining the minimum cell size of Flash. But meanwhile multi-level Flash cells have been introduced (2 bits per cell), and huge efforts have succeeded in reducing their cost.

In MRAM the writing problem was immediately worse than in HDD, as the conducting lines have much smaller dimensions, with a strong limitation in current density around 10^7 A cm^{-2} due to electromigration. Also, it is not possible in a very large-scale integration circuit to include an optimized 'ring-like' ferromagnetic circuit to 'channel' the magnetic induction to the magnetic media. A magnetic channelling was developed for MRAM[53], but the effect is limited (to a factor of about two) and requires costly fabrication steps. Finally, when approaching the downscaling limits the unavoidable distribution of writing parameters, coupled to the large stray fields in such densely packed arrays, leads to spreading program errors. Freescale researchers elegantly solved this reliability problem by

replacing the standard ferromagnetic free layer with a SAF layer, written using a spin-flop process[33,51,54], and this opened the way to the first MRAM product. But this is at the expense of higher writing currents (around 10 mA), and clearly limits the achievable densities and the downscaling. As for HDD, one solution could be heat-assisted recording (TAS-RAM)[43,55]. It was also proposed that writing could be assisted by microwave excitation at the ferromagnetic resonance frequency of the free layer[56–58], a technique that could also be useful for hard disks. But such promising techniques do not completely suppress the need for a magnetic field.

SPIN TRANSFER — A NEW ROUTE FOR WRITING MAGNETIC INFORMATION

The hoped for breakthrough for spin storage was provided by the prediction[59,60] in 1996 that the magnetization orientation of a free magnetic layer could be controlled by direct transfer of spin angular momentum from a spin-polarized current. In 2000, the first experimental demonstration that a Co/Cu/Co CPP spin-valve nanopillar can be reversibly switched by this 'spin-transfer effect' between its low (parallel) and high (antiparallel) magnetoresistance states was presented[61]. The concept of spin transfer actually dates back to the 1970s, with the prediction[62] and observations[63] of domain-wall dragging by currents. Spin-transfer effects had also been predicted[64] for MTJs as early as 1989. Somehow, those predictions did not immediately trigger the intense research work that followed the 1996 and 2000 publications, possibly because the required fabrication technologies were not mature enough.

The principle of 'spin-transfer torque' (STT) writing in nanopillars is shown schematically in Fig. 6a for the usual case of $3d$ ferromagnetic metals (for example Co) in a spin-valve structure with a non-magnetic metal spacer (for example Cu). Let us consider a 'thick' ferromagnetic layer F1, whereas the ferromagnetic layer F2 and the spacer M are 'thin' (compared with the length scales of spin-polarized transport[65]). F1 and F2 are initially magnetized along different directions. A current of s electrons flowing from F1 to F2 will acquire through F1, acting as a spin polarizer, an average spin polarization approximately along the magnetization of F1. When the electrons reach F2, the $s–d$ exchange interaction quickly aligns the average spin moment along the magnetization of F2. In the process, the s electrons have lost a transverse spin angular momentum, which, because of the total angular momentum conservation law, is 'transferred' to the magnetization of F2. This results in a torque tending to align F2 magnetization towards the spin moment of the incoming electrons, and thus towards the magnetization of F1. Because the loss of transverse spin momentum happens over a very short distance (around 1 nm), the torque is an interfacial effect, more efficient on a thin layer. But a more important result is that the amplitude of the torque per unit area is proportional to the injected current density, so that the writing current decreases proportionally to the cross-sectional area of the structure. With today's advances in nanotechnologies and the easy access to sizes below 100 nm, this represents an important advantage of spin transfer over field-induced writing.

A realistic treatment of the effect[65] includes both quantum effects at the interfaces (spin-dependent transmission of Bloch states) and diffusive transport theory (spin-accumulation effects), and the dynamical behaviour can be studied through a modified Landau–Lifshitz–Gilbert equation describing the damped precession of magnetization in the presence of STT and thermal excitations[65,66]. The principle of STT writing of a MRAM cell is shown in Fig. 6b. Electrons flowing from the thick 'polarizing' layer to the thin free layer favour a parallel orientation of the magnetizations: if the initial state is antiparallel, then beyond a threshold current density j_{C+} the free layer will switch. When the electrons flow from the free to the polarizing layer, it can be shown that the effective spin moment injected in the free layer is opposed to the magnetization of

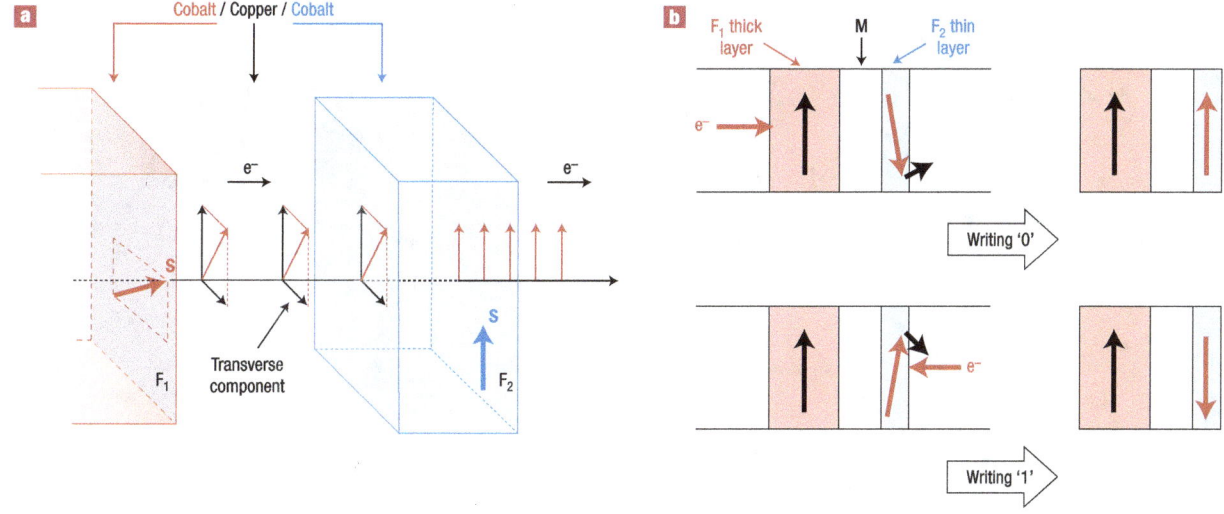

Figure 6 Spin-transfer switching. **a**, Principle of the STT effect, for a typical case of a Co(F1)/Cu/Co(F2) trilayer pillar. A current of s electrons flowing from left to right will acquire through F1 (assumed to be thick and acting as a spin polarizer) an average spin moment along the magnetization of F1. When the electrons reach F2, the s–d exchange interaction quickly aligns the average spin moment along the magnetization of F2. To conserve the total angular momentum, the transverse spin angular momentum lost by the electrons is transferred to the magnetization of F2, which senses a resulting torque tending to align its magnetization towards F1. **b**, Principle of STT writing of a MRAM cell: reversing the current flowing through the cell will induce either parallel or antiparallel orientation of the two ferromagnetic layers F1 and F2.

the polarizing layer, writing an antiparallel configuration beyond a threshold current density j_C.

Since the first observation on Co/Cu/Co trilayers, STT writing has been achieved on many different stacks, including exchange-bias pinned layers and SAF layers[67]. It also works with tunnel junctions[68], and in particular with MgO tunnel barriers[69]. Moreover, the threshold current densities are becoming nearly compatible with NMOS transistor output as predicted by the International Technology Roadmap for Semiconductors. And in the near future, TAS and STT writing modes could be combined for an even smaller switching current.

Companies have already presented several demonstrations of 'spin-RAM'[70,71]. As shown in Fig. 7, the cell structure has now become extremely simple, opening the way to high densities. Is it enough to compete with NAND Flash on mass data storage in standalone memories? In terms of areal density, Flash still offers a smaller multi-bit cell, and three-dimensional (3D) stacking has been announced[72]. But MRAM has other advantages, such as potentially infinite endurance (compared with ~10^5 cycles for a Flash) and potential for sub-nanosecond operation[66,73,74], that make it competitive as universal memory.

PERSPECTIVES

One grave limitation to ultra-high-density spin-MRAM is the requirement of one transistor per cell (1T/1MTJ). The cross-point memory architecture provides a way of reaching very high densities[75], lower fabrication costs and a potential for 3D stacking of several recording layers. Intermediate cell structures such as 1T/4MTJ[76] have also been proposed. Multi-level cell operation was also recently achieved in TAS-RAM[77]. But in all cases increased density is obtained at the expense of smaller signal amplitude and thus much slower read.

One step to fight these limitations, at least partially, would be to gain at least one order of magnitude in the amplitude of the magnetoresistance, moving towards a true 'current switch' with

$\Delta R/R$ values comparable, for example, to those of phase-change RAM (PCRAM). This could be achieved by replacing metallic ferromagnetic layers with 100% spin-polarized conductors such as half-metallic oxides[78] or Heussler alloys[79,80], or diluted magnetic semiconductors (DMS)[81,82]. DMS furthermore open the way to new effects such as tunnelling anisotropic magnetoresistance[83,84], and, in the long term, to quantum dot storage devices[85]. Another promising concept is that of the 'spin filter'[86] where tunnelling happens through a ferromagnetic barrier: the transmission varies exponentially with the square root of the barrier height, which itself depends on electron spin direction versus barrier magnetization. Such developments would of course also benefit HDD. Another approach is to build a three-terminal device that can produce both the transistor effect and magnetoresistance in a single magnetic device[87–89].

All these ideas show promising results at low temperature, but depend on materials issues such as obtaining Curie temperatures well above room temperature, mastering a complex stoichiometry (oxides, Heussler alloys) at an interface, or maintaining the fabrication thermal budget compatible with the CMOS process, though these new materials usually require specific high-temperature growth. Moreover, three-terminal devices can so far provide only very small currents (microamperes at most), below CMOS compatibility levels, and independent writing of one or two magnetic layers may prove difficult to scale down even using spin transfer.

Future magnetic mass storage could come instead from domain wall devices, in an approach conceptually close to that of the former 'bubble' magnetic memories[90]. File architecture for mass storage does not require random access to a single bit or word as proposed in spin-RAM, but can accommodate random access only to large sectors in which binary information is written or read sequentially, such as in HDD. A chain of domain walls in a magnetic stripe can indeed represent such a sector storing binary information, but a simple application of a uniform magnetic field would immediately destroy it by annihilating the reverse domain. It was first proposed[91,92] to use an oscillating magnetic field, uniform over the whole chip, acting on a specific domain wall circuit behaving as a shift register: the global

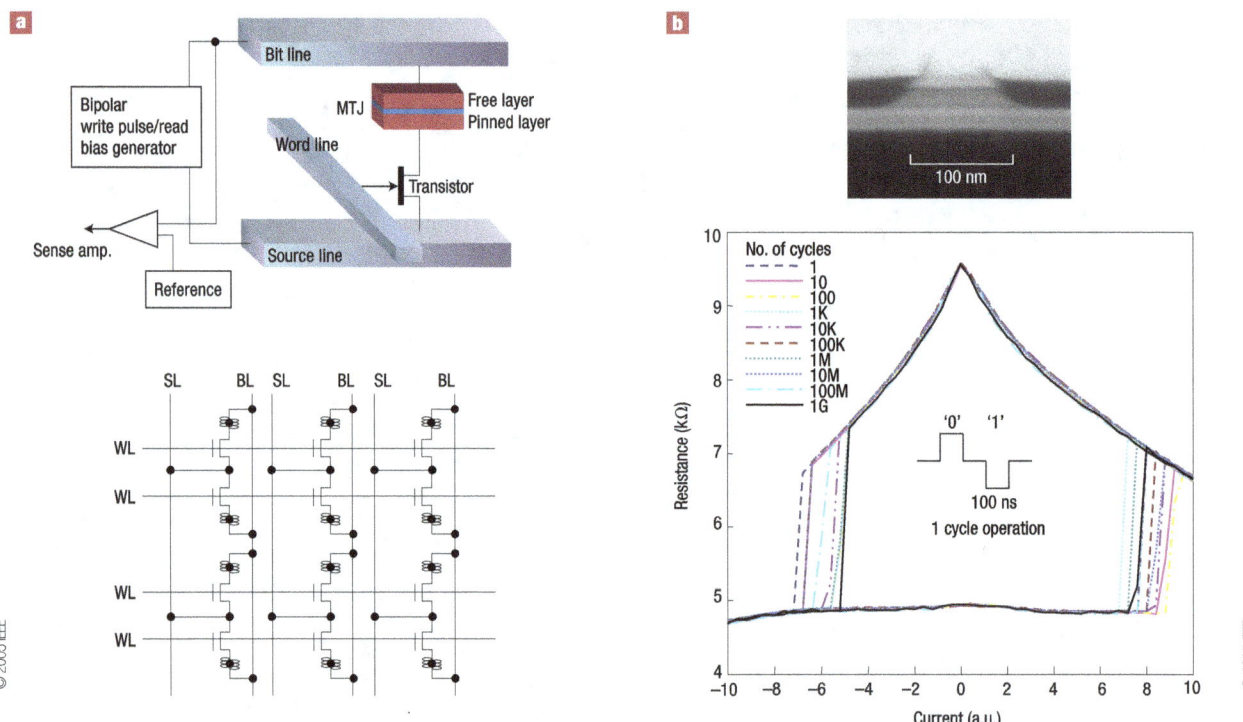

Figure 7 The spin-RAM. **a**, Schematic architecture of a spin-RAM; upper panel, scheme of the memory cell, and lower panel, tentative architecture of the cell array. Reprinted with permission from ref. 70. **b**, Resistance versus current hysteresis loop of a spin-RAM cell. Reprinted with permission from ref. 71. The different colours show the evolution of the loop after an increasing number (up to 1 G = 10⁹) of writing cycles (100 ns pulses of successively positive and negative currents, see image). This demonstrates excellent stability. TEM image: TMR device size 100 nm × 50 nm; free layer CoFe (1.0 nm) / NiFe (2.0 nm); tunnel barrier MgO (1.0 nm).

character of the applied field makes downscaling and 3D stacking achievable[93]. Even more promising for very-high-density solid-state integration, a current injected in the magnetic stripe applies the same pressure to all domain walls along the direction of electron travel, propagating the walls simultaneously at the same speed, without losing information. This mimics the fast passing of bits in front of the head in HDD recording, but here there is no moving part (hence increased ruggedness) and addressing a sector would be done by CMOS electronics with microsecond access times. This scheme thus opens the way to very compact 'storage track memory devices'[94], and could also be used in a new kind of MRAM[95] (Fig. 8). But many advances are needed before its practical implementation.

The theory mixes spin transfer torque (electrons crossing a domain wall transfer spin angular momentum to the non-uniform magnetization in the wall) and mechanical momentum transfer (electrons are reflected from narrow walls), leading to a domain wall propagation controlled with a current density beyond a threshold[96,97]. The effect has indeed been observed in metals magnetized in plane[98,99] and perpendicular to plane[100], or in DMS[101] (with orders of magnitude lower currents but much higher resistance and other problems). However, even qualitative agreement between experience and theory is not straightforward[102,103].

On the experimental aspect, crucial progress has to be made in reliably controlling domain wall propagation with low current densities. First non-volatile reliable trapping of the domain walls must be provided at dedicated positions in the 'data sector'. Most works use simple notches[104,105], but more elaborate pinning profiles have also been proposed[106,107]. But not much has been published on the evolution of the thermal stability of the pinning when downscaling the dimensions. Besides, even in a parallel stripe and even more at

artificial notches with a distribution of patterning defects, domain walls can take different structures that are close in energy, and may even change structure under a current pulse[105,108–110]. This is a potential problem for reliable operation, because the parameters of current-induced propagation should depend on the wall structure. Last but not least, the threshold current density necessary to depin a domain wall from a trap, or even to start motion in a parallel stripe, is still too high for applications (above a few 10⁶ A cm⁻²). Great progress has recently been made by using the natural radiofrequency dynamics of a domain wall pinned in a potential well[111–113]. Note, however, that if the current density is still too high, the cross-section of a thin-film stripe is intrinsically small so the threshold current is already much smaller than is available from the smallest CMOS transistors, a key asset for developing low-power data-storage devices (Fig. 8b). Finally, the achievable domain wall speed is also a crucial issue for fast data rates[114–117]. For instance, with maximum domain wall speed around 100 m s⁻¹ and an on-track linear density of one domain wall every 200 nm, already very demanding values, the data rate would be around 0.5 Gbit s⁻¹. This would be in line with today's HDD data rates, but is far from being demonstrated.

In the long term, even higher densities could be reached by replacing domain walls by smaller magnetization vortices[118], in a trend similar to that from bubble to Bloch line memories[90].

We have tried in this review to share our amazement at the outstanding progress in spin electronics over the past two decades, under the convergence of a chain of scientific breakthroughs and technology advances. If giant magnetoresistance opened the way in 1988 to control transport through magnetization, spin transfer now allows magnetization to be controlled through transport, closing the loop for a new paradigm, from magnetic recording to spin storage.

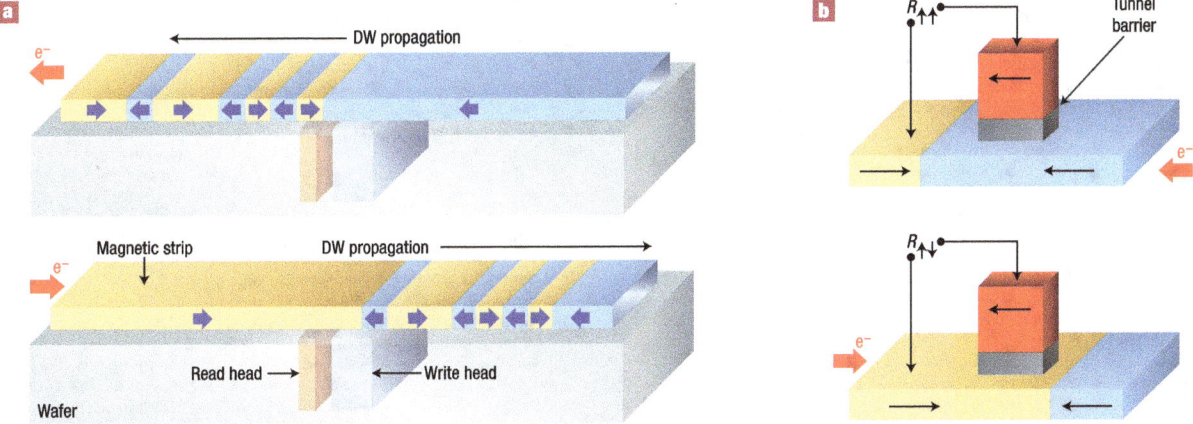

Figure 8 Domain wall storage devices. Examples of storage devices using current-induced domain wall (DW) propagation. **a**, In the concept first proposed by Parkin[94], the binary information is stored by a chain of domain walls in a magnetic stripe. An electrical current in the stripe, by applying the same pressure to all the domain walls, moves them simultaneously at the same speed for a sequential reading (or writing) at fixed read and write heads. A reverse current can move the domain walls in the opposite direction for resetting, or in an alternative solution the domain walls might turn on a loop. This mimics the fast passing of bits in front of the head in HDD recording, but here there is no moving part and addressing a sector would be done by CMOS electronics at microsecond access times. The initial scheme[94] proposes to store data in vertical stripes: this would open the way to very compact high capacity 'storage track memory devices'. Other schemes now propose multilayers of in-plane domain tracks, which would be easier to fabricate. **b**, Scheme of a MRAM cell using domain wall propagation from one stable position to another on either side of a magnetic tunnel junction (ref. 95).

Traditional hard-disk recording has gained orders of magnitude in storage capacity, thus entering the consumer electronics markets. And the MRAM magnetic solid-state memory is now in production, although as yet only for niche markets.

But the intrinsic speed and endurance of magnetic recording, together with the potential of the spin-RAM to work with CMOS-compatible electrical parameters, could open the way to applications where data storage would not be the primary objective although non-volatility would still be a key asset. Along this line, it has been proposed that logic calculations through magnetic interactions could be performed, in magnetic quantum cellular automata[119,120], or in domain wall logic[92], for low-power massively parallel logic operations under a uniform cyclical magnetic field. MTJs can also be used for logic calculation, either directly[121] (a nice idea but one whose practical realization is uncertain) or by a dense integration of MTJs into CMOS logic circuits[122,123] where they bring instant ON/OFF, run time re-programmability, and overall improved operation safety. Magnetism would thus enter the realm of the CPU. But a great step forward would be to realize three-terminal spin electronic devices that would enable a complete programmable logic function to be packed into a single nanodevice. Let us assume a source–gate–drain device where the magnetization of magnetic source and drain could be independently controlled, injecting spin-polarized electrons into a channel of spin-dependent transmission that could also be controlled by a gate voltage. Such a device has multiple inputs to control a multi-level output, realizing a logic function that can be programmed through the non-volatile magnetic configurations. The first proposition of this kind[18], despite recent progress in injecting spin polarization into semiconductors[124], has not yet been achieved in practice. New concepts are being proposed[125–127], and more could be realized, for instance, with molecules[128–132]. In the longer term, the use of spin injection and spin currents[21,133] may lead to the development of 'spin logic' devices[134].

Ultimately, 'magnetic' writing will again become a problem in much smaller and more complex devices, and new routes will have to be found. As in nanoelectronics, zero-current, gate-voltage-controlled writing would be ideal. Preliminary results have recently been obtained by using interfacial coupling with piezoelectric or even multiferroic materials[135–137], or through electric field control of ferromagnetism in DMS[138,139]. In an even more futuristic approach, switching by spin currents only (no charge currents) has been announced[140], in a pioneering step towards nanoelectronics using spin currents only. As a whole, finding solutions to the magnetic writing problem may prove to be a key issue on the way to future spin electronics, as it has been for the past evolution of magnetic recording.

doi:10.1038/nmat2024

References

1. Moser, A. *et al.* Magnetic recording: advancing into the future. *J. Phys. D* **35**, R157–R167 (2002).
2. Mott, N. Electrons in transition metals. *Adv. Phys.* **13**, 325–422 (1964).
3. Fert, A. & Campbell, I. A. Two-current conduction in nickel. *Phys. Rev. Lett.* **21**, 1190–1192 (1968).
4. Fert, A. & Campbell, I. Electrical resistivity of ferromagnetic nickel and iron based alloys. *J. Phys. F* **6**, 849–871 (1976).
5. Fert, A., Duvail, J. & Valet, T. Spin relaxation effects in the perpendicular magnetoresistance of magnetic multilayers. *Phys. Rev. B* **52**, 6513–6521 (1995).
6. Baibich, M. N. *et al.* Giant magnetoresistance of (001)Fe/(001)Cr magnetic superlattices. *Phys. Rev. Lett.* **61**, 2472–2475 (1988).
7. Binasch, G., Grünberg, P., Saurenbach, F. & Zinn, W. Enhanced magnetoresistance in layered magnetic structures with antiferromagnetic interlayer exchange. *Phys. Rev. B* **39**, 4828–4830 (1989).
8. Levy, P. M. & Mertig, I. in *Spin Dependent Transport in Magnetic Nanostructures* (eds Maekawa, S. & Shinjo, T.) Ch. 2, 47-112 (CRC, Boca Raton, 2002).
9. Fert, A., Barthélémy, A. & Petroff, F. in *Nanomagnetism: Ultrathin Films, Multilayers and Nanostructures* (eds Mills, D. M. & Bland, J. A. C.) Ch. 6 (Elsevier, Amsterdam, 2006).
10. Grünberg, P. Magnetic field sensor with ferromagnetic thin layers having magnetically antiparallel polarized components. US patent 4,949,039 (1990).
11. Dieny, B. *et al.* Magnetoresistive sensor based on the spin valve effect. US patent 5,206,590 (1993).
12. Dieny, B. *et al.* Giant magnetoresistance in soft ferromagnetic multilayers. *Phys. Rev. B* **43**, 1297–1300 (1991).
13. Daughton, J. M. Magnetic tunneling applied to memory. *J. Appl. Phys.* **81**, 3758–3763 (1997).
14. Valet, T. & Fert, A. Theory of the perpendicular magnetoresistance in magnetic multilayers. *Phys. Rev. B* **48**, 7099–7113 (1993).
15. Schmidt, G., Ferrand, D., Molenkamp, L. W., Filip, A. T. & van Wees, B. J. Fundamental obstacle for electrical spin injection from a ferromagnetic metal into a diffusive semiconductor. *Phys. Rev. B* **62**, R4790–R4793 (2000).
16. Fert, A. & Jaffrès, H. Conditions for efficient spin injection from a ferromagnetic metal into a semiconductor. *Phys. Rev. B* **64**, 184420 (2001).
17. Jedema, F. J., Filip, A. T. & van Wees, B. J. Electrical spin injection and accumulation at room temperature in an all-metal mesoscopic spin valve. *Nature* **410**, 345–348 (2001).
18. Datta, S. & Das, B. Electronic analog of the electro-optic modulator. *Appl. Phys. Lett.* **56**, 665–667 (1990).
19. Gijs, M. A. M., Lenczowski, S. K. J. & Giesbers, J. B. Perpendicular giant magnetoresistance of microstructured Fe/Cr magnetic multilayers from 4.2 to 300 K. *Phys. Rev. Lett.* **70**, 3343–3346 (1993).

20. Bass, J. & Pratt, W. P. Current-perpendicular (CPP) magnetoresistance in magnetic metallic multilayers. *J. Magn. Magn. Mater.* **200**, 274–289 (1999).
21. Fert, A. & Piraux, L. Magnetic nanowires. *J. Magn. Magn. Mater.* **200**, 338–358 (1999).
22. Takagishi, M. *et al.* The applicability of CPP-GMR heads for magnetic recording. *IEEE Trans. Magn.* **38**, 2277–2282 (2002).
23. Childress, J. *et al.* Fabrication and recording study of all-metal dual-spin-valve CPP read heads. *IEEE Trans. Magn.* **42**, 2444–2446 (2006).
24. Jullière, M. Tunneling between ferromagnetic films. *Phys. Lett. A* **54**, 225–226 (1975).
25. Moodera, J. S., Kinder, L. R., Wong, T. M. & Meservey, R. Large magnetoresistance at room temperature in ferromagnetic thin film tunnel junctions. *Phys. Rev. Lett.* **74**, 3273–3276 (1995).
26. Miyazaki, T. & Tezuka, N. Giant magnetic tunneling effect in Fe/Al_2O_3/Fe junction. *J. Magn. Magn. Mater.* **139**, L231-L234 (1995).
27. Parkin, S. S. P. *et al.* Giant tunnelling magnetoresistance at room temperature with MgO (100) tunnel barriers. *Nature Mater.* **3**, 862–867 (2004).
28. Yuasa, S., Nagahama, T., Fukushima, A., Suzuki, Y. & Ando, K. Giant room-temperature magnetoresistance in single-crystal Fe/MgO/Fe magnetic tunnel junctions. *Nature Mater.* **3**, 868–871 (2004).
29. Butler, W. H., Zhang, X., Schulthess, T. C. & MacLaren, J. M. Spin-dependent tunneling conductance of Fe/MgO/Fe sandwiches. *Phys. Rev. B* **63**, 054416 (2001).
30. Mathon, J. & Umerski, A. Theory of tunneling magnetoresistance of an epitaxial Fe/MgO/Fe(001) junction. *Phys. Rev. B* **63**, 220403 (2001).
31. Lee, Y. M., Hayakawa, J., Ikeda, S., Matsukura, F. & Ohno, H. Effect of electrode composition on the tunnel magnetoresistance of pseudo-spin-valve magnetic tunnel junction with a MgO tunnel barrier. *Appl. Phys. Lett.* **90**, 212507 (2007).
32. Mao, S. *et al.* Commercial TMR heads for hard disk drives: characterization and extendibility at 300 gbit/in^2. *IEEE Trans. Magn.* **42**, 97–102 (2006).
33. Engel, B. *et al.* A 4-Mb toggle MRAM based on a novel bit and switching method. *IEEE Trans. Magn.* **41**, 132–136 (2005).
34. DeBrosse, J. *et al.* A high-speed 128-kb MRAM core for future universal memory applications. *IEEE J. Solid-State Circ.* **39**, 678–683 (2004).
35. Brown, W. F. Thermal fluctuations of a single-domain particle. *Phys. Rev.* **130**, 1677–1686 (1963).
36. Néel, L. Anisotropie superficielle et surstructures d'orientation magnétique. *J. Phys. Rad.* **15**, 225–239 (1954).
37. Gradmann, U. & Müller, J. Flat ferromagnetic, epitaxial 48Ni/52Fe(111) films of few atomic layers. *Phys. Status Solidi B* **27**, 313–324 (1968).
38. Carcia, P. F., Meinhaldt, A. D. & Suna, A. Perpendicular magnetic anisotropy in Pd/Co thin film layered structures. *Appl. Phys. Lett.* **47**, 178–180 (1985).
39. Chappert, C., Renard, D., Beauvillain, P. & Renard, J. Ferromagnetism of very thin films of nickel and cobalt. *J. Magn. Magn. Mater.* **54–57**, 795–796 (1986).
40. Daalderop, G. H. O., Kelly, P. J. & den Broeder, F. J. A. Prediction and confirmation of perpendicular magnetic anisotropy in Co/Ni multilayers. *Phys. Rev. Lett.* **68**, 682–685 (1992).
41. Meiklejohn, W. H. & Bean, C. P. New magnetic anisotropy. *Phys. Rev.* **102**, 1413–1414 (1956).
42. Nogues, J. *et al.* Exchange bias in nanostructures. *Phys. Rep.* **422**, 65–117 (2005).
43. Prejbeanu, I. *et al.* Thermally assisted switching in exchange-biased storage layer magnetic tunnel junctions. *IEEE Trans. Magn.* **40**, 2625–2627 (2004).
44. Skumryev, V. *et al.* Beating the superparamagnetic limit with exchange bias. *Nature* **423**, 850–853 (2003).
45. Grünberg, P., Schreiber, R., Pang, Y., Brodsky, M. B. & Sowers, H. Layered magnetic structures: evidence for antiferromagnetic coupling of Fe layers across Cr interlayers. *Phys. Rev. Lett.* **57**, 2442–2445 (1986).
46. Majkrzak, C. F. *et al.* Observation of a magnetic antiphase domain structure with long-range order in a synthetic Gd-Y superlattice. *Phys. Rev. Lett.* **56**, 2700–2703 (1986).
47. Parkin, S. S. P., More, N. & Roche, K. P. Oscillations in exchange coupling and magnetoresistance in metallic superlattice structures: Co/Ru, Co/Cr, and Fe/Cr. *Phys. Rev. Lett.* **64**, 2304–2307 (1990).
48. Bruno, P. & Chappert, C. Oscillatory coupling between ferromagnetic layers separated by a nonmagnetic metal spacer. *Phys. Rev. Lett.* **67**, 1602–1605(1991).
49. Bruno, P. Theory of interlayer magnetic coupling. *Phys. Rev. B* **52**, 411–439 (1995).
50. Margulies, D. T., Berger, A., Moser, A., Schabes, M. E. & Fullerton, E. E. The energy barriers in antiferromagnetically coupled media. *Appl. Phys. Lett.* **82**, 3701–3703 (2003).
51. Savchenko, L., Engel, B. N., Rizzo, N. D., Deherrera, M. F. & Janesky J. A. Method of writing to scalable magnetoresistance random access memory element. US patent 6,545,906B1 (2003).
52. Weller, D. *et al.* High K_u materials approach to 100 Gbits/in^2. *IEEE Trans. Magn.* **36**, 10–15 (2000).
53. Durlam, M. *et al.* Low power 1 Mbit MRAM based on 1T1MTJ bit cell integrated with copper interconnects. *Symp. VLSI Techn. Dig.*, 158–161 (2002).
54. Worledge, D. C. Spin flop switching for magnetic random access memory. *Appl. Phys. Lett.* **84**, 4559–4561 (2004).
55. Daughton, J. M. & Pohm, A. V. Design of Curie point written magnetoresistance random access memory cells. *J. Appl. Phys.* **93**, 7304–7306 (2003).
56. Rizzo, N. D. & Engel, B. N. MRAM write apparatus and method. US patent 6,351,409 (2002).
57. Thirion, C., Wernsdorfer, W. & Mailly, D. Switching of magnetization by nonlinear resonance studied in single nanoparticles. *Nature Mater.* **2**, 524–527 (2003).
58. Nembach, H. T. *et al.* Microwave assisted switching in a $Ni_{81}Fe_{19}$ ellipsoid. *Appl. Phys. Lett.* **90**, 062503 (2007).
59. Slonczewski, J. Current-driven excitation of magnetic multilayers. *J. Magn. Magn. Mater.* **159**, L1-L7 (1996).
60. Berger, L. Emission of spin waves by a magnetic multilayer traversed by a current. *Phys. Rev. B* **54**, 9353–9358 (1996).
61. Albert, F. J., Katine, J. A., Buhrman, R. A. & Ralph, D. C. Spin-polarized current switching of a Co thin film nanomagnet. *Appl. Phys. Lett.* **77**, 3809–3811 (2000).
62. Berger, L. Prediction of a domain-drag effect in uniaxial, non-compensated, ferromagnetic metals. *J. Phys. Chem. Solids* **35**, 947–956 (1974).
63. Freitas, P. P. & Berger, L. Observation of *s–d* exchange force between domain walls and electric current in very thin Permalloy films. *J. Appl. Phys.* **57**, 1266–1269 (1985).
64. Slonczewski, J. C. Conductance and exchange coupling of two ferromagnets separated by a tunneling barrier. *Phys. Rev. B* **39**, 6995–7002 (1989).
65. Stiles, M. & Miltat, J. in *Spin Dynamics in Confined Magnetic Structures III* (eds Hillebrands, B. & Thiaville, A.) (Springer, Berlin, 2006)
66. Sun, J. Z. Spin–current interaction with a monodomain magnetic body: a model study. *Phys. Rev. B* **62**, 570–578 (2000).
67. Ralph, D. & Buhrman, R., in *Concepts in Spintronics* (ed. Maekawa, S.) (Oxford Univ. Press, 2006)
68. Huai, Y., Albert, F., Nguyen, P., Pakala, M. & Valet, T. Observation of spin-transfer switching in deep submicron-sized and low-resistance magnetic tunnel junctions. *Appl. Phys. Lett.* **84**, 3118–3120 (2004).
69. Hayakawa, J. *et al.* Current-induced magnetization switching in MgO barrier based magnetic tunnel junctions with CoFeB/Ru/CoFeB synthetic ferrimagnetic free layer. *Jpn. J. Appl. Phys.* **45**, L1057–L1060 (2006).
70. Hosomi, M. *et al.* Novel nonvolatile memory with spin torque transfer magnetization switching: spin-ram. *IEDM Tech. Dig.* 459–462 (2005).
71. Kawahara, T. *et al.* 2Mb spin-transfer torque RAM (SPRAM) with bit-by-bit bidirectional current write and parallelizing-direction current read. *ISSCC Dig. Tech. Papers*, 480–481 (2007).
72. Jung, S. *et al.* Three dimensionally stacked NAND Flash memory technology using stacking single crystal Si layers on ILD and TANOS structure for beyond 30 nm node. *IEDM Tech. Dig.*, 1–4 (2006).
73. Ito, K., Devolder, T., Chappert, C., Carey, M. J. & Katine, J. A. Micromagnetic simulation of spin transfer torque switching combined with precessional motion from a hard axis magnetic field. *Appl. Phys. Lett.* **89**, 252509 (2006).
74. Devolder, T., Chappert, C. & Ito, K. Sub-ns spin-transfer switching: compared benefits of free layer biasing and pinned layer biasing. *Phys. Rev. B* **75**, 224430 (2007).
75. Sakimura, N. *et al.* A 512 kb cross-point cell MRAM. *ISSCC Dig. Tech. Papers*, 278–279 (2003).
76. Tanizaki, H. *et al.* A high-density and high-speed 1T-4MTJ MRAM with voltage offset self-reference sensing scheme. *Asian Solid-State Circuits Conf. Dig. Tech. Papers*, 303–306 (2006).
77. Leuschner, R. *et al.* Thermal select MRAM with a 2-bit cell capability for beyond 65 nm technology node. *IEDM Tech. Dig.*, 1–4 (2006).
78. Bowen, M. *et al.* Nearly total spin polarization in $La_{2/3}Sr_{1/3}MnO_3$ from tunnelling experiments. *Appl. Phys. Lett.* **82**, 233–235 (2003).
79. Ishikawa, T. *et al.* Spin-dependent tunneling characteristics of fully epitaxial magnetic tunneling junctions with a full-Heusler alloy Co_2MnSi thin film and a MgO tunnel barrier. *Appl. Phys. Lett.* **89**, 192505 (2006).
80. Marukame, T., Ishikawa, T., Matsuda, K., Uemura, T. & Yamamoto, M. High tunnel magnetoresistance in fully epitaxial magnetic tunnel junctions with a full-Heusler alloy $Co_2Cr_{0.6}Fe_{0.4}Al$ thin film. *Appl. Phys. Lett.* **88**, 262503 (2006).
81. Chiba, D., Sato, Y., Kita, T., Matsukura, F. & Ohno, H. Current-driven magnetization reversal in a ferromagnetic semiconductor (Ga,Mn)As/GaAs/(Ga,Mn)As tunnel junction. *Phys. Rev. Lett.* **93**, 216602 (2004).
82. Elsen, M. Spin transfer experiments on (Ga,Mn)As/(In,Ga)As/(Ga,Mn)As tunnel junctions. *Phys. Rev. B* **73**, 035303 (2006).
83. Gould, C. Tunneling anisotropic magnetoresistance: a spin-valve-like tunnel magnetoresistance using a single magnetic layer. *Phys. Rev. Lett.* **93**, 117203 (2004).
84. Gould, C., Schmidt, G. & Molenkamp, L. W. Tunneling anisotropic magnetoresistance-based devices. *IEEE Trans. Electron Dev.* **54**, 977–983 (2007).
85. Enaya, H., Semenov, Y. G., Kim, K. W. & Zavada, J. M. Electrical manipulation of nonvolatile spin cell based on diluted magnetic semiconductor quantum dots. *IEEE Trans. Electron Dev.* **54**, 1032–1039 (2007).
86. LeClair, P. *et al.* Large magnetoresistance using hybrid spin filter devices. *Appl. Phys. Lett.* **80**, 625–627 (2002).
87. Monsma, D. J., Lodder, J. C., Popma, T. J. A. & Dieny, B. Perpendicular hot electron spin-valve effect in a new magnetic field sensor: the spin-valve transistor. *Phys. Rev. Lett.* **74**, 5260–5263 (1995).
88. van Dijken, S., Jiang, X. & Parkin, S. S. P. Room temperature operation of a high output current magnetic tunnel transistor. *Appl. Phys. Lett.* **80**, 3364–3366 (2002).
89. Hehn, M., Montaigne, F. & Schuhl, A. Hot-electron three-terminal devices based on magnetic tunnel junction stacks. *Phys. Rev. B* **66**, 144411 (2002).
90. Hubert, A. & Schäfer, R. *Magnetic Domains* (Springer, Berlin, 1998).
91. Allwood, D. A. *et al.* Submicrometer ferromagnetic NOT gate and shift register. *Science* **296**, 2003–2006 (2002).
92. Allwood, D. A. *et al.* Magnetic domain-wall logic. *Science* **309**, 1688–1692 (2005).
93. Cowburn, R. P. & Allwood, D. A. Multiple layer magnetic logic memory device. UK patent GB2,430,318A (2007).
94. Parkin, S. S. P. Shiftable magnetic shift register and method using the same. US patent 6,834,005B1 (2004).
95. Cros, V., Grollier, J., Munoz Sanchez, M., Fert, A. & Nguyen Van Dau, F. Spin electronics device. Patent WO 2006/064022 (2006).
96. Tatara, G. & Kohno, H. Theory of current-driven domain wall motion: spin transfer versus momentum transfer. *Phys. Rev. Lett.* **92**, 086601 (2004).
97. Li, Z. & Zhang, S. Domain-wall dynamics and spin-wave excitations with spin-transfer torques. *Phys. Rev. Lett.* **92**, 207203 (2004).
98. Grollier, J. *et al.* Switching a spin valve back and forth by current-induced domain wall motion. *Appl. Phys. Lett.* **83**, 509 (2003).
99. Yamaguchi, A. *et al.* Real-space observation of current-driven domain wall motion in submicron magnetic wires. *Phys. Rev. Lett.* **92**, 077205 (2004).
100. Ravelosona, D., Lacour, D., Katine, J. A., Terris, B. D. & Chappert, C. Nanometer scale observation of high efficiency thermally assisted current-driven domain wall depinning. *Phys. Rev. Lett.* **95**, 117203 (2005).
101. Yamanouchi, M., Chiba, D., Matsukura, F. & Ohno, H. Current-induced domain-wall switching in a ferromagnetic semiconductor structure. *Nature* **428**, 539–542 (2004).
102. Thiaville, A., Nakatani, Y., Miltat, J. & Suzuki, Y. Micromagnetic understanding of current-driven domain wall motion in patterned nanowires. *Europhys. Lett.* **69**, 990–996 (2005).

103. Piechon, F. & Thiaville, A. Spin transfer torque in continuous textures: Semiclassical Boltzmann approach. *Phys. Rev. B* **75**, 174414 (2007).

104. Himeno, A. *et al.* Dynamics of a magnetic domain wall in magnetic wires with an artificial neck. *J. Appl. Phys.* **93**, 8430–8432 (2003).

105. Hayashi, M. *et al.* Dependence of current and field driven depinning of domain walls on their structure and chirality in permalloy nanowires. *Phys. Rev. Lett.* **97**, 207205 (2006).

106. Allwood, D. A., Xiong, G. & Cowburn, R. P. Domain wall diodes in ferromagnetic planar nanowires. *Appl. Phys. Lett.* **85**, 2848–2853 (2004).

107. Faulkner, C. C. *et al.* Artificial domain wall nanotraps in $Ni_{81}Fe_{19}$ wires. *J. Appl. Phys.* **95**, 6717–6719 (2004).

108. Klaui, M. *et al.* Direct observation of domain-wall configurations transformed by spin currents. *Phys. Rev. Lett.* **95**, 026601 (2005).

109. Klaui, M. *et al.* Current-induced vortex nucleation and annihilation in vortex domain walls. *Appl. Phys. Lett.* **88**, 232507 (2006).

110. He, J., Li, Z. & Zhang, S. Current-driven vortex domain wall dynamics by micromagnetic simulations. *Phys. Rev. B* **73**, 184408 (2006).

111. Saitoh, E., Miyajima, H., Yamaoka, T. & Tatara, G. Current-induced resonance and mass determination of a single magnetic domain wall. *Nature* **432**, 203–206 (2004).

112. Thomas, L. *et al.* Oscillatory dependence of current-driven magnetic domain wall motion on current pulse length. *Nature* **443**, 197–200 (2006).

113. Thomas, L. *et al.* Resonant amplification of magnetic domain-wall motion by a train of current pulses. *Science* **315**, 1553–1556 (2007).

114. Nakatani, Y., Thiaville, A. & Miltat, J. Faster magnetic walls in rough wires. *Nature Mater.* **2**, 521–523 (2003).

115. Lim, C. K. *et al.* Domain wall displacement induced by subnanosecond pulsed current. *Appl. Phys. Lett.* **84**, 2820–2822 (2004).

116. Hayashi, M. *et al.* Current driven domain wall velocities exceeding the spin angular momentum transfer rate in permalloy nanowires. *Phys. Rev. Lett.* **98**, 037204. (2007).

117. Yamanouchi, M., Chiba, D., Matsukura, F., Dietl, T. & Ohno, H. Velocity of domain-wall motion induced by electrical current in the ferromagnetic semiconductor (Ga,Mn)As. *Phys. Rev. Lett.* **96**, 096601 (2006).

118. Kasai, S., Nakatani, Y., Kobayashi, K., Kohno, H. & Ono, T. Current-driven resonant excitation of magnetic vortices. *Phys. Rev. Lett.* **97**, 107204 (2006).

119. Cowburn, R. P. & Welland, M. E. Room temperature magnetic quantum cellular automata. *Science* **287**, 1466–1468 (2000).

120. Imre, A. *et al.* Majority logic gate for magnetic quantum-dot cellular automata. *Science* **311**, 205–208 (2006).

121. Ney, A., Pampuch, C., Koch, R. & Ploog, K. H. Programmable computing with a single magnetoresistive element. *Nature* **425**, 485–487 (2003).

122. Black, W. C. J. & Das, B. Programmable logic using giant-magnetoresistance and spin-dependent tunneling devices. *J. Appl. Phys.* **87**, 6674–6679 (2000).

123. Zhao, W. *et al.* Integration of Spin-RAM technology in FPGA circuits. *Proc. ICSICT* 799–802 (2006).

124. Min, B., Motohashi, K., Lodder, C. & Jansen, R. Tunable spin-tunnel contacts to silicon using low-work-function ferromagnets. *Nature Mater.* **5**, 817–822 (2006).

125. Hall, K. C., Lau, W. H., Gundogdu, K., Flatte, M. E. & Boggess, T. F. Nonmagnetic semiconductor spin transistor. *Appl. Phys. Lett.* **83**, 2937–2939 (2003).

126. Hall, K. C. & Flatte, M. E. Performance of a spin-based insulated gate field effect transistor. *Appl. Phys. Lett.* **88**, 162503 (2006).

127. Tanaka, M. & Sugahara, S. MOS-based spin devices for reconfigurable logic. *IEEE Trans. Electron Dev.* **54**, 961–976 (2007).

128. Pasupathy, A. N. *et al.* The Kondo effect in the presence of ferromagnetism. *Science* **306**, 86–89 (2004).

129. Sahoo, S., Kontos, T., Schonenberger, C. & Surgers, C. Electrical spin injection in multiwall carbon nanotubes with transparent ferromagnetic contacts. *Appl. Phys. Lett.* **86**, 112109 (2005).

130. Hueso, L. E. *et al.* Transformation of spin information into large electrical signals using carbon nanotubes. *Nature* **445**, 410–413 (2007).

131. Romeike, C., Wegewijs, M. R., Ruben, M., Wenzel, W. & Schoeller, H. Charge-switchable molecular magnet and spin blockade of tunneling. *Phys. Rev. B* **75**, 064404 (2007).

132. Fert, A., George, J., Jaffres, H. & Mattana, R. Semiconductors between spin-polarized sources and drains. *IEEE Trans. Electron Dev.* **54**, 921–932 (2007).

133. Kimura, T., Hamrle, J. & Otani, Y. Estimation of spin-diffusion length from the magnitude of spin-current absorption: multiterminal ferromagnetic/nonferromagnetic hybrid structures. *Phys. Rev. B* **72**, 014461 (2005).

134. Dery, H., Dalal, P., Cywinski, L. & Sham, L. J. Spin-based logic in semiconductors for reconfigurable large-scale circuits. *Nature* **447**, 573–576 (2007).

135. Khomskii, D. Multiferroics: Different ways to combine magnetism and ferroelectricity. *J. Magn. Magn. Mater.* **306**, 1–8 (2006).

136. Zavaliche, F. *et al.* Electric field-induced magnetization switching in epitaxial columnar nanostructures. *Nano Lett.* **5**, 1793–1796 (2005).

137. Zhao, T. *et al.* Electrical control of antiferromagnetic domains in multiferroic $BiFeO_3$ films at room temperature. *Nature Mater.* **5**, 823–829 (2006).

138. Chiba, D., Matsukura, F. & Ohno, H. Electric-field control of ferromagnetism in (Ga,Mn)As. *Appl. Phys. Lett.* **89**, 162505 (2006).

139. Wunderlich, J. *et al.* Coulomb blockade anisotropic magnetoresistance effect in a (Ga,Mn)As single-electron transistor. *Phys. Rev. Lett.* **97**, 077201 (2006).

140. Kimura, T., Otani, Y. & Hamrle, J. Switching magnetization of a nanoscale ferromagnetic particle using nonlocal spin injection. *Phys. Rev. Lett.* **96**, 037201 (2006).

Acknowledgements

C.C. acknowledges support from the EU Specific Support Action WIND (IST 033658). The authors also benefit from EU contracts Spinswitch (MRTN-CT-2006-035327) and Nanospin (STREP FET 015728). Correspondence and requests for materials should be addressed to C.C.

Nanoionics-based resistive switching memories

Many metal–insulator–metal systems show electrically induced resistive switching effects and have therefore been proposed as the basis for future non-volatile memories. They combine the advantages of Flash and DRAM (dynamic random access memories) while avoiding their drawbacks, and they might be highly scalable. Here we propose a coarse-grained classification into primarily thermal, electrical or ion-migration-induced switching mechanisms. The ion-migration effects are coupled to redox processes which cause the change in resistance. They are subdivided into cation-migration cells, based on the electrochemical growth and dissolution of metallic filaments, and anion-migration cells, typically realized with transition metal oxides as the insulator, in which electronically conducting paths of sub-oxides are formed and removed by local redox processes. From this insight, we take a brief look into molecular switching systems. Finally, we discuss chip architecture and scaling issues.

RAINER WASER[1,2]* AND MASAKAZU AONO[3,4]

[1]Institut für Werkstoffe der Elektrotechnik 2, RWTH Aachen University, 52056 Aachen, Germany
[2]Institut für Festkörperforschung/CNI—Center of Nanoelectronics for Information Technology, Forschungszentrum Jülich, 52425 Jülich, Germany
[3]Nanomaterials Laboratories, National Institute for Material Science, 1-1 Namiki, Tsukuba, Ibaraki 305-0044, Japan
[4]ICORP/Japan Science and Technology Agency, 4-1-8 Honcho, Kawaguchi, Saitama 332-0012, Japan

*e-mail: r.waser@fz-juelich.de

Memory concepts that have been recently pursued range from spin-based memories (magnetoresistive random access memories, or MRAM for short, and related ideas), in which a magnetic field is involved in the resistance switching, to phase-change RAM (PCRAM), in which thermal processes control a phase transition in the switching material from the amorphous to the crystalline state. Yet another class of resistive switching phenomena is based on the electrically stimulated change of the resistance of a metal–insulator–metal (MIM) memory cell, usually called resistance switching RAM, or RRAM for short. The 'M' in MIM denotes any reasonably good electron conductor, often different for the two sides, and the 'I' stands for an insulator, often ion-conducting material. Typically, an initial electroforming step such as a current-limited electric breakdown is induced in the virgin sample. This step preconditions the system which can subsequently be switched between a conductive ON state and a less conductive OFF state. The necessity of this initial step and its mechanism strongly depend on the switching class, as will be described.

Starting with the report on oxide insulators by Hickmott[1] in 1962, a huge variety of materials in a MIM configuration have been reported to show hysteretic resistance switching. In general, the 'I' in MIM can be one of a wide range of binary and multinary oxides and higher chalcogenides as well as organic compounds, and the

'M' stands for a similarly large variety of metal electrodes including electron-conducting non-metals. A first period of high research activity up to the mid-1980s has been comprehensively reviewed elsewhere[2–4]. The current period started in the late 1990s, triggered by Asamitsu et al.[5], Kozicki et al.[6] and Beck et al.[7].

Before we turn to the basic principles of these switching phenomena, we need to distinguish between two schemes with respect to the electrical polarity required for resistively switching MIM systems. Switching is called unipolar (or symmetric) when the switching procedure does not depend on the polarity of the voltage and current signal. A system in its high-resistance state (OFF) is switched ('set') by a threshold voltage into the low-resistance state (ON) as sketched in Fig. 1a. The current is limited by the compliance current of the control circuit. The 'reset' into the OFF state takes place at a higher current and a voltage below the set voltage. In this respect, PCRAMs show unipolar switching (without compliance current in this case). In contrast, the characteristic is called bipolar (or antisymmetric) when the set to an ON state occurs at one voltage polarity and the reset to the OFF state on reversed voltage polarity (Fig. 1b). The structure of the system must have some asymmetry, such as different electrode materials or the voltage polarity during the initial electroforming step, in order to show bipolar switching behaviour. In both characteristics, unipolar and bipolar, reading of the state is conducted at small voltages that do not affect the state.

CLASSIFICATION OF SWITCHING MECHANISMS

In conjunction with the discussion of conceivable switching mechanisms, we must address the question of the geometrical location of the switching event in a MIM structure. With respect to the cross-section of the electrode pad, the switching to the ON state is typically reported as a confined, filamentary effect rather than a homogeneously distributed one, leading to a resistance that is independent of pad size. In planar MIM structures, filaments along the surface are observed

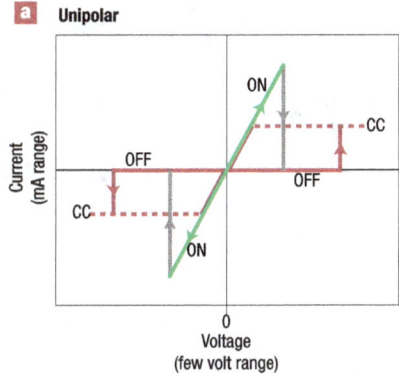

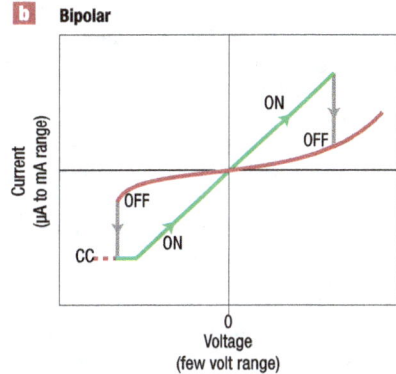

Figure 1 Classification of the switching characteristics in a voltage sweeping experiment. Depending on the specific system, the curves vary considerably. The purpose of these sketches is to differentiate between the two possible switching directions. Dashed lines indicate that the real voltage at the system will differ from the control voltage because of the compliance current (CC) in action. **a**, Unipolar switching. The set voltage is always higher than the voltage at which reset takes place, and the reset current is always higher than the CC during set operation. **b**, Bipolar switching. The set operation takes place on one polarity of the voltage or current, and the reset operation requires the opposite polarity. In some systems, no CC is used.

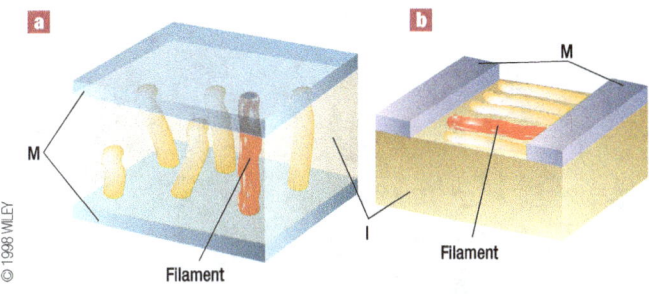

© 1998 WILEY

Figure 2 Sketch of filamentary conduction in MIM structures. Redrawn with modifications from ref. 4. **a**, Vertical stack configuration. **b**, Lateral, planar configuration. The red tube indicates the filament responsible for the ON state.

(Fig. 2). For both cases, in the perpendicular direction (along the path between the electrodes), evidence for interface effects are more frequently described than bulk switching effects.

Conceivable mechanisms for the resistive switching in MIM systems often consist of a combination of physical and/or chemical effects. As a first approach, however, they can be classified according to whether the dominant contribution comes from a thermal effect, an electronic effect, or an ionic effect. Caution must be exercised, because in many reports the switching mechanism has not yet been elucidated or suggestions are based on little experimental and theoretical evidence. Unfortunately, papers often do not report full details of the sample preparation and electrode deposition, the polarity of voltage applied during electroforming and during the first voltage pulse or voltage sweep, the electrode pad-size dependence of the current, the temperature dependence of the electrical response, the yield and the statistics in the cell-to-cell data, or the shift of characteristics on repeated cycling. As a consequence, in many cases we cannot make a comparison of the results and a reasonable assignment of a switching mechanism. Therefore, we attempt a coarse-grained classification of conceivable mechanisms and try to assign selected examples.

A typical resistive switching based on a thermal effect shows a unipolar characteristic. It is initiated by a voltage-induced partial dielectric breakdown in which the material in a discharge filament

is modified by Joule heating. Because of the compliance current, only a weak conductive filament with a controlled resistance is formed. This filament may be composed of the electrode metal transported into the insulator, carbon from residual organics[4] or decomposed insulator material such as sub-oxides[8]. During the reset transition, this conductive filament is again disrupted thermally because of high power density of the order of 10^{12} W cm^{-3} generated locally, similar to a traditional household fuse but on the nanoscale. Hence, we refer to this mechanism as the fuse–antifuse type. One candidate out of many is NiO, first reported in the 1960s[9]. Recently, the filamentary nature of the conductive path in the ON state has been confirmed for NiO (ref. 10) and TiO$_2$ (ref. 11). Cells based on Pt/NiO/Pt thin films have been successfully integrated into CMOS (complementary metal oxide semiconductor) technology to demonstrate non-volatile memory operation[12]. A critical parameter for this unipolar switching effect seems to be the value of the compliance current. In fact, it has recently been demonstrated that a TiO$_2$ thin film shows bipolar switching, and that this can be changed to unipolar switching characteristics by setting the compliance current to a larger value[13].

Electronic charge injection and/or charge displacement effects can be considered as another origin of resistive switching. One possibility is the charge-trap model[14], in which charges are injected by Fowler–Nordheim tunnelling at high electric fields and subsequently trapped at sites such as defects or metal nanoparticles in the insulator. This modifies the electrostatic barrier character of the MIM structure and hence its resistance, resembling the gate–channel resistance in a Flash field-effect transistor (FET). For example, gold nanoclusters incorporated in either polymeric[15,16] or inorganic insulator films[17] can be trapping sites. In a modified model, trapping at interface states is thought to affect the adjacent Schottky barrier at various metal/semiconducting perovskite interfaces[18–20]. A similar mechanism has been reported for ZnSe–Ge heterojunctions[21].

Another possible model is the insulator–metal transition (IMT), in which electronic charge injection acts like doping to induce an IMT in perovskite-type oxides such as (Pr,Ca)MnO$_3$ (refs 5,22,23) and SrTiO$_3$:Cr (ref. 24). A generic model by Rozenberg *et al.*[25] has recently been extended to bipolar switching[26].

Finally, a model based on ferroelectricity has been proposed by Esaki[27] and theoretically described by Kohlstedt *et al.*[28, 29]. Here, an ultrathin ferroelectric insulator is assumed whose ferroelectric polarization direction influences the tunnelling current through the insulator.

In this review, we will focus on MIM systems in which ionic transport and electrochemical redox reactions provide the essential mechanism for bipolar resistive switching. It is this area in which nanoelectronics[30] becomes intimately connected to nanoionics[31]. One class in this category relies on the sequence of the following processes: the oxidation of an electrochemically active electrode metal such as Ag; the drift of the mobile Ag^+ cations in the ion-conducting layer; their discharge at the (inert) counterelectrode leading to a growth of Ag dendrites, which form a highly conductive filament in the ON state of the cell[32]. When the polarity of the applied voltage is reversed, an electrochemical dissolution of the conductive bridges takes place, resetting the system into the OFF state. Instead of silver, it is possible to use copper and other metals in the moderate area of the standard electrochemical potential series. A second class in this category operates through the migration of anions, typically oxygen ions, towards the anode (better described by the migration of oxygen vacancies towards the cathode), a subsequent change of the stoichiometry, and a valence change of the cation sublattice associated with a modified electronic conductivity. We will outline the main similarities and differences between the two classes, summarize the current state of knowledge and technology, and sketch future work in this area.

REDOX PROCESSES INDUCED BY CATION MIGRATION

Silver growth from heated argentite (α-Ag_2S) was reported more than 400 years ago[33]. Extensive studies following the first study of electric conduction of Ag_2S by Faraday in the early nineteenth century[34] established the thermodynamical theory of ionic conduction which is a key characteristic of a solid electrolyte. A galvanic cell was used to control stoichiometric deviation of Ag^+ cations in Ag_2S crystals[35], and this revealed a mechanism for the silver whisker growth.

By using solid electrolytes in which conduction is due to metal cations, the formation and annihilation of a metal filament in the MIM system can be controlled. To achieve the bipolar switching behaviour, the MIM system consists of an electrode made from an electrochemicall active metal, a solid electrolyte as an ion-conducting 'I' layer, and a counter electrode made from an inert metal. Switching behaviour due to silver dendrite formation and annihilation was first reported by Hirose in 1976 using Ag-photodoped amorphous As_2S_3 as 'I' of a MIM system with a lateral structure[36]. Kozicki et al. succeeded in developing the MIM system in a vertical configuration by using GeSe as the ion conductor and applied this to making non-volatile memory[6]. Recently, resistive switching GeSe structures as small as 20 nm have been fabricated (Fig. 3).

In these systems, metal cations in the ionic conductors migrate towards the cathode made of inert materials and are reduced there. The reduced metal atoms form a metal filament which grows towards the anode to turn on the switch. As the anode is made of electrochemically active material, metal atoms of the anode are oxidized and dissolved into the ionic conductor, maintaining the number of metal cations for continuous electrochemical deposition. In the case of Ag_2S, the following chemical reaction occurs at the anode and the cathode:

$$Ag^+(Ag_2S) + e^- \xrightleftharpoons[\text{Oxidation}]{\text{Reduction}} Ag \qquad (1)$$

On changing the polarity of the bias voltage, metal atoms dissolve at the edge of the metal filament, eventually annihilating the filament so that the switch is turned off. Because the chemical reaction ideally does not cause any damage to the MIM system, the switch may in principle be expected to work indefinitely[6].

Solid electrolyte can be used as one of the 'metal' electrodes of the MIM system. Terabe et al. developed this type of MIM switch using an electronic and ionic mixed conductor for one of the electrodes[32]. In this case, a vacuum nano-gap is used as an insulator layer, and a metal

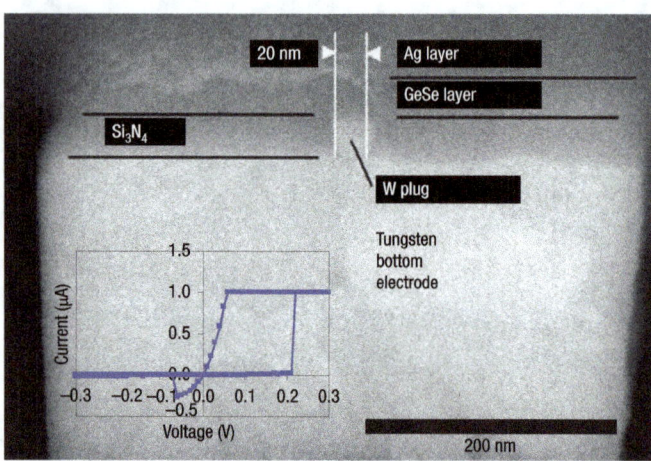

Figure 3 Cross-section of a vertical type of MIM switch using Ag^+ conducting solid electrolyte. A silver filament is electrochemically formed in the GeSe layer to turn on the switch. The cross-section is shown for the device for which the inset I–V curve was recorded. Reprinted with permission from ref. 44.

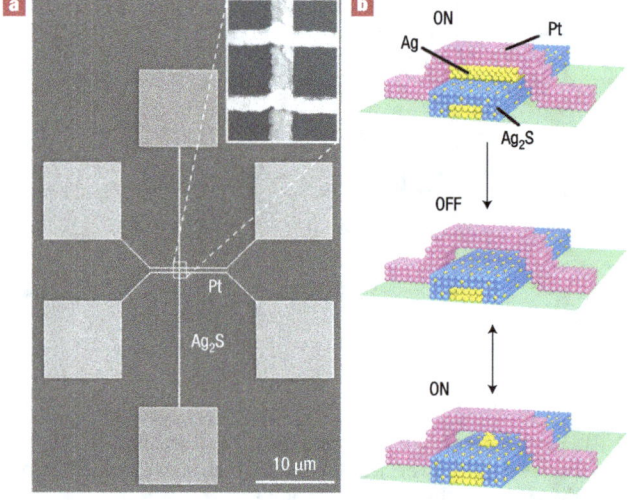

Figure 4 Scanning electron micrograph of an atomic switch and its operating mechanism[32]. Silver atoms precipitated from the Ag_2S electrode make a bridge in a vacuum gap of 1 nm between the two electrodes. From ref. 32.

filament grows in the gap to bridge the mixed conductor electrode and the counterelectrode (Fig. 4). Because the switching is caused by an electrochemical reaction, increasing a switching bias voltage shortens the switching time exponentially[37]. Reducing the size of a metal filament also produces faster switching.

In both types of MIM system, the switching effect is most easily found for material systems that are cation conductors for the redox active species (such as Ag^+ in Ag_2S) or, alternatively, compounds with high solubility of the redox active cation (such as Ag^+ in $GeSe_x$). The switches can be operated with smaller bias voltage, such as 0.2 V, which may be preferable for memory devices with low power consumption[6,38].

Some applications, such as non-volatile switches in reconfigurable large-scale integrated systems (LSIs), require higher threshold bias

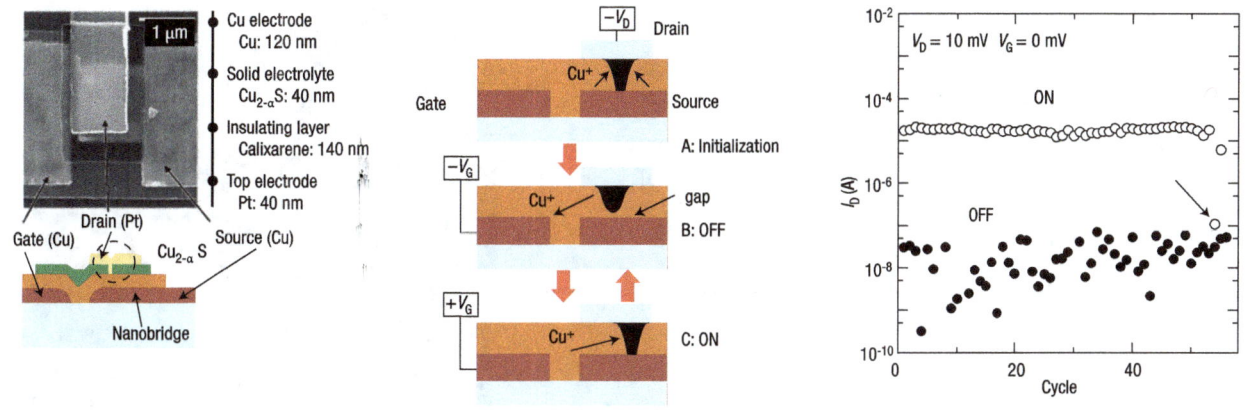

Figure 5 Three-terminal solid electrolyte switch. Electrochemical reaction for formation and annihilation of a metal filament between the source and the drain electrodes is controlled by the gate voltage. Reprinted with permission from ref. 47.

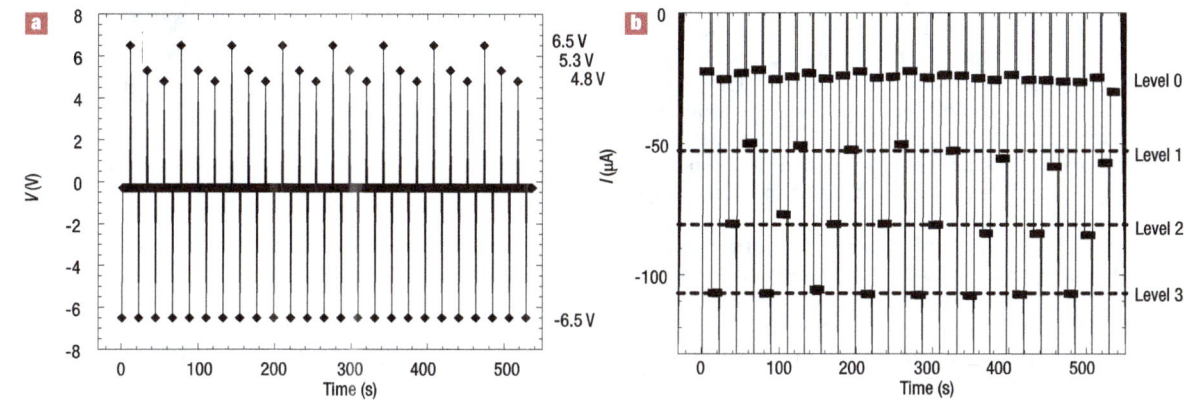

Figure 6 Multilevel switching in a Cr-doped $SrZrO_3$ MIM cell operated at 77 K. **a**, Voltage stimulation. **b**, Current response. Reprinted with permission from ref. 7. By applying voltage pulses of appropriate levels, the system can be set to three different ON state resistances (levels 1 to 3). Together with the OFF state (level 0) this represents a two-bit memory cell.

voltages than that of CMOS devices[38]. Because the switching bias voltage is mainly determined by the activation energy for the chemical reaction and the ionic diffusion constant, the operating bias voltage can be tuned by choice of materials for MIM systems, especially the ion-conducting material. For instance, Ta_2O_5 with Cu cations has been reported[39] to have an operating bias voltage of 2 V. Therefore, switches using a variety of solid electrolyte materials such as (Zn,Cd)S, WO_3, and SiO_2 are being investigated[40–42]. It is interesting to note that systems such as Ag/(Zn,Cd)S/Pt, which were initially attributed to a completely different category[43], are now shown to fall into this cation-migration class[40]. The state of the art with respect to non-volatile memories is demonstrated by Ag/GeSe$_x$/W cells integrated into 90-nm CMOS technology[44,45]. In this work, an active matrix concept with one access transistor per switching cell has been used and a prototype integration density of 2 Mbit has been achieved.

Making use of an electrochemical reaction makes it possible to configure three-terminal devices in which switching can be controlled by the gate electrode. Gate-controlled formation and annihilation of a metal filament was first demonstrated by using (liquid) electrolyte[46], and it has been confirmed with fully solid-state three-terminal devices

(Fig. 5)[47]. Such devices, in which the control line is separated from the conduction line, widen the possibility of practical use of the solid electrolyte switches.

Although the electrochemistry of the cation-migration-based resistive switching cells is reasonably well understood, some questions remain open: for example, the microscopic nature of the cation conduction paths; the impact of thermal effects, details of the electrode reactions in particular for the OFF switching, as well as the interrelationship of the electrolyte nature, the cell geometry and the morphology of the metal dendrites on the nanoscale. In specific systems, such as the GeSe$_x$ electrolyte, the role of defects that are also known to affect phase-change alloys has not been clarified yet[48].

REDOX PROCESSES INDUCED BY ANION MIGRATION

In many oxides, in particular in transition metal oxides, oxygen ions defects, typically oxygen vacancies, are much more mobile than cations. If the cathode blocks ion exchange reactions during an electroforming process, an oxygen-deficient region starts to build and to expand towards the anode. Transition metal cations accommodate this deficiency by trapping electrons emitted from the cathode. In the case of TiO_2 or titanates, for example, this reduction reaction

$$ne^- + Ti^{4+} \rightarrow Ti^{(4-n)+} \tag{2}$$

is equivalent to filling the Ti $3d$ band. The reduced valence states of the transition metal cations which are generated by this electrochemical process typically turn the oxide into a metallically conducting phase, such as $TiO_{2-n/2}$ for approximately $n > 1.5$. This 'virtual cathode' moves towards the anode and will finally form a conductive path[49]. At the anode, the oxidation reaction may lead to the evolution of oxygen gas, according to

$$O_O \rightarrow V_O^{\cdot\cdot} + 2e^- + 1/2O_2 \tag{3}$$

where $V_O^{\cdot\cdot}$ denotes oxygen vacancies with a double positive charge with respect to the regular lattice and O_O represents an oxygen ion on a regular site according to the Kröger–Vink notation. As an alternative to reaction (3), the anode or the material nearby may be oxidized. The electroforming conditions depend on the MIM system. Macroscopic single crystals typically require some 100 V for several hours, whereas for thin films the first switching cycle at about 1 V may be sufficient. The total charge has been found to control the electroforming[50]. Once the electroforming is completed, the bipolar switching obviously takes place through local redox reactions between the virtual cathode and the anode, by forming or breaking the conductive contact. Depending on the charge transfer during the switching, the resistance of the system can be established at intermediate levels, which might help in creating multibit storage in future memory cells (Fig. 6)[7].

As in the case of the electrochemical metallization process, R_{ON} is usually not pad-size dependent, indicating a filamentary switching in MIM structures, as reported, for example, for nanocrystalline Ta_2O_5 and Nb_2O_5 thin films[51], VO_2 thin films[52], TiO_2 thin films[11], nanoscale confined TiO_2 (ref. 53) and epitaxial $SrZrO_3$:Cr thin films[54]. Use of a conductive-tip atomic force microscope (C-AFM) technique has shown the conductive filaments to be identical to dislocations in the case of undoped $SrTiO_3$ single crystals and thin films[55]. Figure 7 shows that the conductivity enhancement by several orders of magnitude is confined to a region 1–2 nm wide at the exit of a dislocation. By repeated scanning with a suitably biased AFM tip, the dislocation can be made to switch between an ON and an OFF state. In the ON state, the conductivity shows metallic behaviour, in accordance with first principles calculations[55]. At the surface of ultrathin epitaxial $SrTiO_3$ films, entire areas can be reversibly switched between an ON and OFF state by C-AFM (Fig. 8). For these films also, high-resolution studies reveal the filamentary nature of the conductivity and their possible correlation to dislocations[56]. In the case of other materials, such as $(La,Sr)MnO_3$ thin films, the conductivity is confined to boundary regions between islands of about 100 nm diameter[57].

Lateral MIM configurations allow for the observation of filaments along their extension between the electrodes. In undoped $SrTiO_3$ single crystals, the formation of conductive filaments within a network of dislocations has been verified by combining electrocoloration studies by optical microscopy with C-AFM scans (Fig. 9)[55]. Infrared thermal microscopy of a Cr-doped $SrTiO_3$ single crystal during a current load of 5 mA confirms the confinement of the current path and shows a 'hot spot' near the electrode, where the virtual cathode touches the anode (Fig. 10)[58].

It must be noted that this class of ion-migration-based switching effects is much less well understood than the switching induced by cation migration that was described in the previous section. Open questions remain about the microscopic details of the ion transport, the defect structure and electronic charge transport properties of the

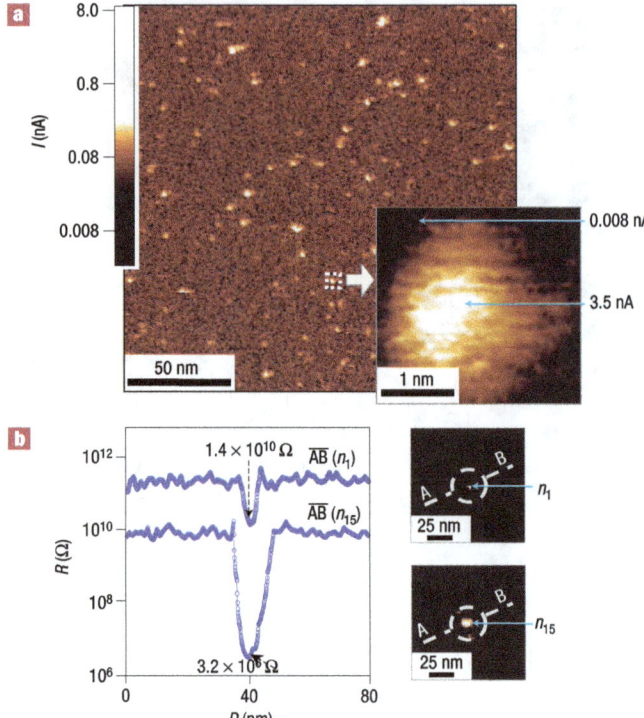

Figure 7 Conductance of individual dislocations in $SrTiO_3$. **a**, A conductivity map of the surface of an undoped $SrTiO_3$ single crystal as recorded by C-AFM after modest thermal reduction. **b**, Line scan across the selected spot (D denoting the distance along AB) showing the dynamic range of the conductance increase as a result of the application of a negative voltage to the AFM tip. Right: conductivity maps of the selected spot before and after electroformation. From ref. 55.

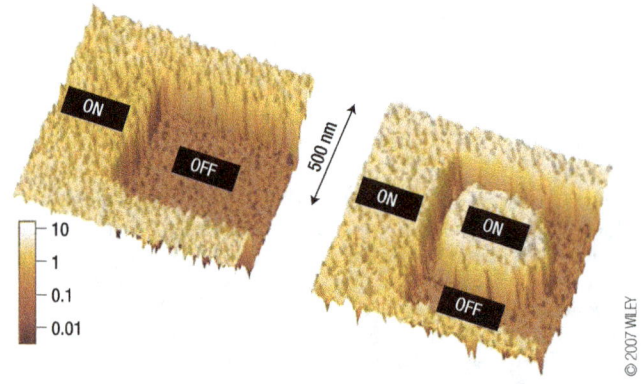

Figure 8 Area-wide switching of an epitaxial 10-nm $SrTiO_3$ thin film by C-AFM. The left scan is produced with a tip voltage of −6 V and subsequently with +6 V in the inner area; the right scan is subsequently scanned with −6 V within the inner ON-state area. Reprinted with permission from ref. 56.

conductive channels formed, details of the electrochemical redox reactions involved, and so forth. In several cases, it is not even clear what ions are involved in the process and whether the system falls into the cation- or anion-migration class.

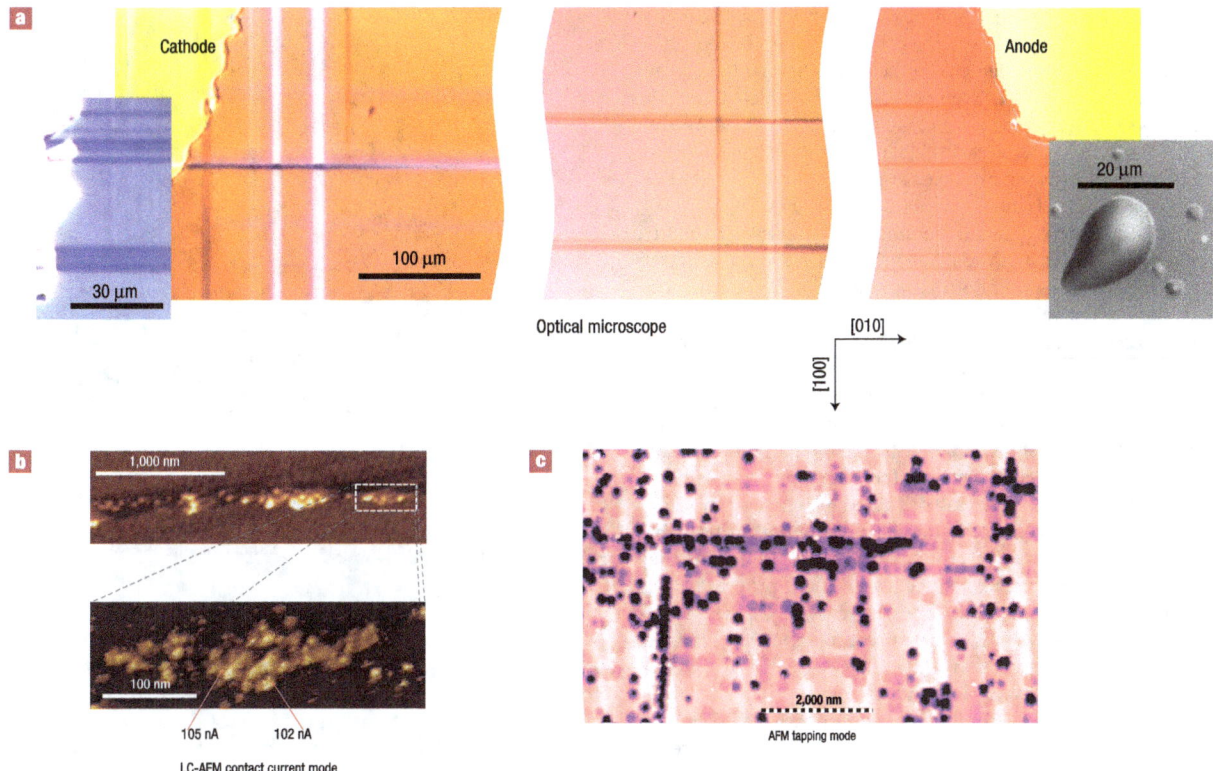

Figure 9 Filamentary structure induced by electroformation in an undoped SrTiO₃ single crystal. From ref. 55. **a**, Optical micrograph of the filamentary structure created in the skin region of a thermally pre-reduced crystal by electroformation between planar gold electrodes. Segments from left to right: near the cathode, in the central region, and near the anode. Clearly visible is the orthogonal network along the crystallographic [100] direction of the crystal. Insets show the possible fine structure of filaments at the cathode (left) and gas bubble that have developed under the anode metal (right). **b**, High resolution of the scene as recorded by C-AFM at a location between the electrodes, where filaments have terminated at the surface. Measurements support the filamentary character of conductance and the fine structure of the high-conductivity spots. **c**, Etch pits of the same kind of crystal give an example of the distribution of disl ocations crossing the surface, highlighting the natural tendency to agglomerate along crystallographic directions.

RESISTIVELY SWITCHING SYSTEMS INVOLVING ORGANIC MOLECULES

Resistive switching has been seen in a large variety of MIM systems in which the 'I' layer is represented by organic molecules or polymers of typically 30 nm to >1 μm thickness[59–61]. The characteristics of the organic compounds include redox activity, the formation of a charge-transfer complex, and the formation of donor–acceptor couples. A recent review[62] summarizes the literature comprehensively and classifies the conceivable switching mechanisms, but in many cases the database is too weak to draw definite conclusions. Note that in several studies an aluminium top electrode has been used that is deposited onto the organic layer in an *ex situ* deposition step. Recent control experiments indicate that the organic layer is not always essential for the switching. Instead the switching event seems to take place in a thin Al₂O₃ layer inevitably formed between the organic layer and the aluminium metal, as found for rose Bengal[63], for polyethylenedioxythiophene (PEDOT)[64], and for Cu-tetracyano-quinodimethane (Cu-TCNQ)[65]. In the latter case, a thin aluminium oxide/hydroxide layer was suggested as the conducting layer for copper ions in a cation-migration-based electrochemical switch.

In another class of potential molecular memories, the 'I' layer in a MIM system is a monomolecular film or even a single molecule contacted by a metal tip on the nanoscale. The aim of these molecular electronic studies is to make use of processes such as molecular redox, molecular configuration and conformation changes, or molecular electronic excitations, as well as molecular spin properties that may

affect the electron transfer coefficient[66]. In one of the experiments, however, in which a redox process within a specific catenane molecule was originally attributed to the resistive switching of a monolayer[67], a control experiment using electronically inactive alkanoic acids revealed a very similar *I–V* behaviour[68], and a conceivable mechanism based on the electrochemistry of the oxide layers formed at the metal electrodes has been suggested[69].

These examples show that great care must be exercised in attributing mechanistic models to observed switching events and that many critical control experiments are required to obtain an microscopic understanding. Certainly, specific organic molecules have advantageous electronic properties. But the inherent characteristics of these molecules are easily masked by the electrode materials and the experimental boundary conditions. The redox activity of viologen molecules unveiled in an electrochemical *in situ* experiment using scanning tunnelling microscopy (STM)[70,71] and of specific oligophenylene molecules in mechanically controllable break-junction experiments[72] are some of of the rare examples in which inherent molecular properties are observed.

CHIP ARCHITECTURE, RELIABILITY, SCALING AND OUTLOOK

In a random access memory (RAM) the storage cells are organized in a matrix. Along the rows and columns of the matrix, there are write and read lines, respectively, which are connected to electronic line drives and sense amplifiers in the periphery of the matrix[30].

Resistively switching storage cells may be organized in a passive cross-bar matrix, just connecting the word and bit lines at each node. The detailed circuit requirements depend on the type (unipolar or bipolar) of switch. Alternatively, in an active matrix, there is a select transistor at each node which decouples the storage cell if it is not addressed. This concept significantly reduces crosstalk and disturb signals in the matrix. Although a passive matrix can, in principle, be fabricated on a $(4/n)F^2$ scheme per cell (where F denotes the minimum feature size of the fabrication technology and n the number of memory layers in a multilayer stack), the cells in an active matrix require somewhat more space, of the order of 4 to $8F^2$. Both passive matrices[38,73,74] and active matrices[12,45] have been realized in prototype RRAMs. In passive arrays the storage cells need to incorporate diodes in series with the switchable resistors in order to avoid signal bypasses by cells in the ON state. For oxide-based unipolar cells, a sandwich concept has been proposed to integrate the diode function[75]. In the case of bipolar cells, serial elements with a Zener diode or varistor characteristic are required. In addition, for medium and large passive matrices the interconnect resistances must be taken into account, requiring dedicated reference schemes[76].

Resistive switching cells offer application opportunities that go beyond mere high-density memory devices. In particular, they can be used as reconfigurable switches in field-programmable gate-array (FPGA) type logic too. About 10 years ago, the Hewlett-Packard research labs developed a prototype computer, called the Teramac, entirely built from conventional, CMOS-based FPGAs, interconnected through several hierarchical levels by a 'fat tree'[77]. They replaced the traditional computer programming based on von Neumann architecture by reconfiguration of the look-up matrices in the FPGAs of the Teramac. The concept proved to be efficient and, as a particular benefit, highly defect-tolerant because during the configuration phase routes could be made around all defects. Defect tolerance is one of the essential demands on any future computer as the defect rate will inherently increase with decreasing feature sizes. The far-reaching idea of the Teramac project has been to replace the complex cells in the conventional FPGAs by nanoscale two-terminal resistive switches. In recent years, considerable progress has been made in resistively switching matrices (described in this review) and in the further elaboration of the architecture concept[78–81].

Compared with several other emerging memory concepts, the RRAM concept including its different variants is immature. As a consequence, the performance and reliability, and in particular the microscopic mechanism of processes limiting the reliability, have not yet been studied in detail. First results look promising. Switching times of under 10 ns have been reported for individual oxide cells[82]. In the case of cation-migration-induced processes, fatigue of the switching hysteresis of high-density RRAM prototypes did not occur within 10^6 write cycles and 10^{12} read cycles[12,83]. A data retention time of over 10 years has been extrapolated, for example, for Ag/GeSe$_x$ RRAMs[83]. Yet these are just first results. Once we fully understand the particular switching mechanisms, we will need detailed studies of reliability, including a thorough investigation of all conceivable failure mechanisms, and optimization steps based on these.

For the further evolution of nanoelectronics, the question of inherent physical limits to scaling as well as possible technological barriers to scaling is most important. One ultimate limit will be given by the tunnelling distance between neighbouring cells as well as the leakage current from the word and bit line, again potentially dominated by tunnelling[84]. Another limit is obviously set by the lateral extension of the switching area, typically the cross-section of the switching filament as described above. These limits depend on the class of the resistive switching mechanism. The scaling of thermally switching cells will depend on the specific heat capacities and thermal conductivities of the materials involved in the cell, limiting the scaling by the onset of thermal cross-talk as in PCRAMs[48]. Cells based on electrical effects depend on the size of electrostatic barriers controlled

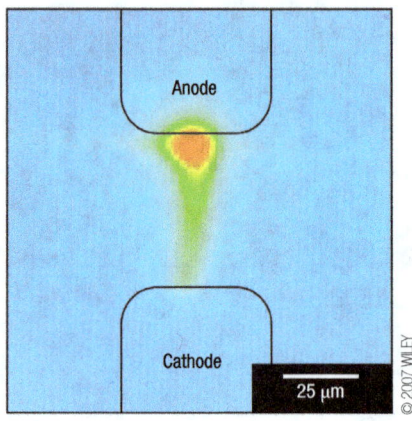

Figure 10 Infrared thermal micrograph of a planar Cr-doped SrTiO$_3$ single-crystal cell. The cell has a current of 5 mA at an applied voltage of 30 V. In the colour scale, blue and red represent room temperature and elevated temperature, respectively. From ref. 58.

by the shift of the charges involved[85]. Ionic switching cells may be limited in scaling by the random diffusion of migrating ions involved as well as the metal atoms, in the case of cation-migration-induced redox switching. They may also be limited by poor uniformity due to inherent impurities in the materials. If this is the case, then obviously the downscaling may be limited by the requirements of data retention. Proper selection of materials for their diffusion rates and precisely defined material interfaces on the atomic scale may limit this problem to the point that it falls beyond the ultimate tunnelling limit. In this respect, ion-migration-induced redox-type switching might offer huge potential for future high-density non-volatile memories.

Technologically, the scaling of the RRAMs will be determined by the fabrication of efficient and reliable electrode contacts and interconnects within the matrix. Although great progress has been made in recent years[74,78], the structures are still far from any tunnelling limit. Another obvious limit to scaling of the chip size will be the size of the periphery circuit and, for active matrices, the size of the access transistors within the matrix. Concepts have been proposed in which the resistive switching cross-bar matrix is slightly rotated against the array of CMOS cells underneath, lifting the alignment constraints considerably[80,86]. This approach can be used to continue the downscaling of the resistive switching cells without having to shrink the access transistors.

Much research effort is still needed to explore the potential of the resistive switching effect in general, and ion-migration-based redox effects in particular, and to exploit this potential to its limits. Questions requiring further attention include a deeper understanding of the microscopic mechanism of the switching, the process and material optimization, the effects limiting the reliability, all aspects of fabrication technology, and the guidelines for scaling.

doi:10.1038/nmat2023

References
1. Hickmott, T. W. Low-frequency negative resistance in thin anodic oxide films. *J. Appl. Phys.* **33**, 2669–2682 (1962).
2. Dearnaley, G., Stoneham, A. M. & Morgan, D. V. Electrical phenomena in amorphous oxide films. *Rep. Prog. Phys.* **33**, 1129–1191 (1970).
3. Oxley, D. P. Electroforming, switching and memory effects in oxide thin films. *Electrocomponent Sci. Technol. UK* **3**, 217–224 (1977).
4. Pagnia, H. & Sotnik, N. Bistable switching in electroformed metal-insulator-metal devices. *Phys. Status Solidi* **108**, 11–65 (1988).
5. Asamitsu, A., Tomioka, Y., Kuwahara, H. & Tokura, Y. Current switching of resistive states in magnetoresistive manganites. *Nature* **388**, 50–52 (1997).
6. Kozicki, M. N., Yun, M., Hilt, L. & Singh, A. Applications of programmable resistance changes in metal-doped chalcogenides. *Pennington NJ USA: Electrochem. Soc.* 298–309 (1999).

7. Beck, A., Bednorz, J. G., Gerber, C., Rossel, C. & Widmer, D. Reproducible switching effect in thin oxide films for memory applications. *Appl. Phys. Lett.* **77,** 139–141 (2000).
8. Chudnovskii, F. A., Odynets, L. L., Pergament, A. L. & Stefanovich, G. B. Electroforming and switching in oxides of transition metals: the role of metal-insulator transition in the switching mechanism. *J. Solid State Chem.* **122,** 95–99 (1996).
9. Bruyere, J. C. & Chakraverty, B. K. Switching and negative resistance in thin films of nickel oxide. *Appl. Phys. Lett.* **16,** 40–43 (1970).
10. Kim, D. C. *et al.* Electrical observations of filamentary conductions for the resistive memory switching in NiO films. *Appl. Phys. Lett.* **88,** 202102 (2006).
11. Choi, B. J. *et al.* Resistive switching mechanism of TiO_2 thin films grown by atomic-layer deposition. *J. Appl. Phys.* **98,** 033715 (2005).
12. Baek, I. G. *et al.* Highly scalable nonvolatile resistive memory using simple binary oxide driven by asymmetric unipolar voltage pulses. *IEDM Tech. Digest,* 587–590 (2005).
13. Jeong, D. S., Schroeder, H. & Waser, R. Coexistence of bipolar and unipolar resistive switching behaviors. *Electrochem. Solid-State Lett.* **10,** G51-G53 (2007).
14. Simmons, J. G. & Verderber, R. R. New conduction and reversible memory phenomena in thin insulating films. *Proc. R. Soc.Lond. A* **301,** 77–102 (1967).
15. Ouyang, J. Y., Chu, C. W., Szmanda, C. R., Ma, L. P. & Yang, Y. Programmable polymer thin film and non-volatile memory device. *Nature Mater.* **3,** 918–922 (2004).
16. Bozano, L. D. *et al.* Organic materials and thin-film structures for cross-point memory cells based on trapping in metallic nanoparticles. *Adv. Funct. Mater.* **15,** 1933–1939 (2005).
17. Guan, W. *et al.* Fabrication and charging characteristics of MOS capacitor structure with metal nanocrystals embedded in gate oxide. *J. Phys. D* **40,** 2754–2758 (2007).
18. Sawa, A., Fujii, T., Kawasaki, M. & Tokura, Y. Interface resistance switching at a few nanometer thick perovskite manganite active layers. *Appl. Phys. Lett.* **88,** 232112 (2006).
19. Fujii, T. *et al.* Hysteretic current–voltage characteristics and resistance switching at an epitaxial oxide Schottky junction $SrRuO_3/SrTi_{0.99}Nb_{0.01}O_3$. *Appl. Phys. Lett.* **86,** 012107 (2005).
20. Lee, D. *et al.* in *Proc. Non-Volatile Memory Technology Symposium* (ed. Campbell, K.) 89–93 (IEEE, Piscataway, New Jersey, 2006).
21. Hovel, H. J. & Urgell, J. J. Switching and memory characteristics of ZnSe–Ge heterojunctions. *J. Appl. Phys.* **42,** 5076–5083 (1971).
22. Fors, R., Khartsev, S. I. & Grishin, A. M. Giant resistance switching in metal-insulator-manganite junctions: evidence for Mott transition. *Phys. Rev. B* **71,** 045305 (2005).
23. Kim, D. S., Kim, Y. H., Lee, C. E. & Kim, Y. T. Colossal electroresistance mechanism in a $Au/Pr_{0.7}Ca_{0.3}MnO_3/Pt$ sandwich structure: evidence for a Mott transition. *Phys. Rev. B* **74,** 174430 (2006).
24. Meijer, G. I. *et al.* Valence states of Cr and the insulator-to-metal transition in Cr-doped SrTiO3. *Phys. Rev. B* **72,** 155102 (2005).
25. Rozenberg, M. J., Inoue, I. H. & Sanchez, M. J. Nonvolatile memory with multilevel switching: a basic model. *Phys. Rev. Lett.* **92,** 178302 (2004).
26. Rozenberg, M. J., Inoue, I. H. & Sanchez, M. J. Strong electron correlation effects in nonvolatile electronic memory devices. *Appl. Phys. Lett.* **88,** 033510 (2006).
27. Esaki, L., Laibowitz, R. B. & Stiles, P. J. Polar Switch. *IBM Tech. Discl. Bull.* **13,** 2161 (1971).
28. Kohlstedt, H., Pertsev, N. A., Contreras, J. R. & Waser, R. Theoretical current–voltage characteristics of ferroelectric tunnel junctions. *Phys. Rev. B* **72,** 125341 (2005).
29. Tsymbal, E. Y. & Kohlstedt, H. Tunneling across a ferroelectric. *Science* **313,** 181–183 (2006).
30. Waser, R. *Nanoelectronics and Information Technology* 2nd edn (Wiley-VCH, Weinheim, 2003).
31. Maier, J. Nanoionics: ion transport and electrochemical storage in confined systems. *Nature Mater.* **4,** 805–818 (2005).
32. Terabe, K., Hasegawa, T., Nakayama, T. & Aono, M. Quantized conductance atomic switch. *Nature* **433,** 47–50 (2005).
33. Ercker, L. *Treatise on Ores and Assaying* (1547) (transl. Sisco, A. G. & Smith, C. S., Univ. Chicago, 1951), p. 177.
34. Faraday, M. *Phil. Trans. R. Soc. Lond.* **123,** 507–522 (1833).
35. Wagner, C. Physical chemistry of ionic crystals involving small concentrations of foreign substances. *J. Phys. Chem.* **57,** 738–742 (1953).
36. Hirose, Y. & Hirose, H. Polarity-dependent memory switching and behaviour of Ag dendrite in Ag-photodoped amorphous As_2S_3 films. *J. Appl. Phys.* **47,** 2767–2772 (1976).
37. Tamura, T. *et al.* Switching property of atomic switch controlled by solid electrochemical reaction. *Jpn. J. Appl. Phys.* **45,** L364-L366 (2006).
38. Kaeriyama, S. *et al.* A nonvolatile programmable solid-electrolyte nanometer switch. *IEEE J. Solid-State Circuits USA* **40,** 168–176 (2005).
39. Sakamoto, T. *et al.* A Ta_2O_5 solid-electrolyte switch with improved reliability. *VLSI Technol. Digest Tech. Pap.* (in the press).
40. Zheng-Wang *et al.* Resistive switching mechanism in $Zn_xCd_{1-x}S$ nonvolatile memory devices. *IEEE Electron Dev. Lett.* **28,** 14–16 (2007).
41. Kozicki, M. N., Gopalan, C., Balakrishnan, M. & Mitkova, M. A low-power nonvolatile switching element based on copper-tungsten oxide solid electrolyte. *IEEE Trans. Nanotechnol.* **5,** 535–544 (2006).
42. Schindler, C., Puthen Thermadam, S. C., Kozicki, R. &Waser, M. N. Bipolar and unipolar resistive switching in Cu-doped SiO_2. *IEEE Trans. Electron Dev.* (in the press).
43. van-der-Sluis, P. Non-volatile memory cells based on $Zn_xCd_{1-x}S$ ferroelectric Schottky diodes. *Appl. Phys. Lett.* **82,** 4089–4091 (2003).
44. Kund, M. *et al.* Conductive bridging RAM (CBRAM): an emerging non-volatile memory technology scalable to sub 20 nm. *IEDM Tech. Digest,* 754–757 (2005).
45. Dietrich, S. *et al.* A nonvolatile 2-Mbit CBRAM memory core featuring advanced read and program control. *IEEE J.Solid-State Circuits* **42,** 839–845 (2007).
46. Xie, F. Q., Nittler, L., Obermair, C. & Schimmel, T. Gate-controlled atomic quantum switch. *Phys. Rev. Lett.* **93,** 128303 (2004).
47. Banno, N., Sakamoto, T., Hasegawa, T., Terabe, K. & Aono, M. Effect of ion diffusion on switching voltage of solid-electrolyte nanometer switch. *Jpn. J. Appl. Phys.* **45,** 3666–3668 (2006).
48. Wuttig, M. & Yamada, N. Phase change materials for rewriteable data storage. *Nature Mater.* **6,** 824–832 (2007).
49. Baiatu, T., Waser, R. & Hardtl, K. H. DC electrical degradation of perovskite-type titanates. III. A model of the mechanism. *J. Am. Ceram. Soc.* **73,** 1663–1673 (1990).
50. Watanabe, Y. *et al.* Current-driven insulator-conductor transition and nonvolatile memory in chromium-doped $SrTiO_3$ single crystals. *Appl. Phys. Lett.* **78,** 3738–3740 (2001).
51. Pinto, R. Filamentary switching and memory action in thin anodic oxides. *Phys. Lett. A* **35,** 155–156 (1971).
52. Beaulieu, R. P., Sulway, D. V. & Cox, C. D. The detection of current filaments in VO_2 thin-film switches using the scanning electron microscope. *Solid-State Electron.* **3,** 428–429 (1973).
53. Ogimoto, Y., Tamia, Y., Kawasaki, M. & Tokura, Y. Resistance switching memory device with a nanoscale confined current path. *Appl. Phys. Lett.* **90,** 143515 (2007).
54. Rossel, C., Meijer, G. I., Bremaud, D. & Widmer, D. Electrical current distribution across a metal-insulator-metal structure during bistable switching. *J. Appl. Phys.* **90,** 2892–2898 (2001).
55. Szot, K., Speier, W., Bihlmayer, G. & Waser, R. Switching the electrical resistance of individual dislocations in single-crystalline $SrTiO_3$. *Nature Mater.* **5,** 312–320 (2006).
56. Szot, K., Dittmann, R., Speier, W. & Waser, R. Nanoscale resistive switching. *Phys. Status Solidi* **1,** R86–R88 (2007).
57. Chen, X., Wu, N., Strozier, J. & Ignatiev, A. Spatially extended nature of resistive switching in perovskite oxide thin films. *Appl. Phys. Lett.* **89,** 063507 (2006).
58. Janousch, M. *et al.* Role of oxygen vacancies in Cr-doped $SrTiO_3$ for resistance-change memory. *Adv. Mater.* **19,** 2232–2235 (2007).
59. Pender, L. F. & Fleming, R. J. Memory switching in glow discharge polymerized thin films. *J. Appl. Phys.* **46,** 3426–3431 (1975).
60. Potember, R. S., Poehler, T. O. & Cowan, D. O. Electrical switching and memory phenomena in Cu-TCNQ thin films. *Appl. Phys. Lett.* **34,** 405–407 (1979).
61. Bandyopadhyay, A. & Pal, A. J. Large conductance switching and memory effects in organic molecules for data-storage applications. *Appl. Phys. Lett.* **82,** 1215–1217 (2003).
62. Scott, J. C. & Bozano, L. D. Nonvolatile memory elements based on organic materials. *Adv. Mater.* **19,** 1452–1463 (2007).
63. Karthauser, S. *et al.* Resistive switching of rose bengal devices: a molecular effect? *J. Appl. Phys.* **100,** 094504 (2006).
64. Colle, M., Buchel, M. & de-Leeuw, D. M. Switching and filamentary conduction in non-volatile organic memories. *Org. Electron.* **7,** 305–312 (2006).
65. Kever, T., Boettger, U., Schindler, Ch. & Waser, R. On the origin of bistable resistive switching in Cu:TCNQ. *Appl. Phys. Lett.* **91,** 083506 (2007).
66. Feringa, B. L. *Molecular Switches* (Wiley-VCH, Weinheim, 2001).
67. Collier, C. P. *et al.* A [2]catenane-based solid state electronically reconfigurable switch. *Science* **289,** 1172–1175 (2000).
68. Stewart, D. R. *et al.* Molecule-independent electrical switching in Pt/organic monolayer/Ti devices. *Nano Lett.* **4,** 133–136 (2004).
69. Blackstock, J. J. *et al.* Internal structure of a molecular junction device: chemical reduction of PtO_2 by Ti. *J. Phys. Chem. C* **111,** 16–20 (2007).
70. Li, Z. *et al.* Two-dimensional assembly and local redox-activity of molecular hybrid structures in an electrochemical environment. *Faraday Disc.* **131,** 121–143 (2005).
71. Li, Z., Pobelov, I., Han, B., Wandlowski, T., Blaszczyk, A. & Mayor, M. Conductance of redox-active single molecular junctions: an electrochemical approach. *Nanotechnology* **18,** 1–8 (2007).
72. Lörtscher, E., Ciszek, J. W., Tour, J. & Riel, H. Reversible and controllable switching of a single-molecule junction. *Small* **2,** 973–977 (2006).
73. Wu, W. *et al.* One-kilobit cross-bar molecular memory circuits at 30-nm half-pitch fabricated by nanoimprint lithography. *Appl. Phys. A* **80,** 1173–1178 (2005).
74. Green, J. E. *et al.* A 160-kilobit molecular electronic memory patterned at 10^{11} bits per square centimetre. *Nature* **445,** 14–17 (2007).
75. Lee, M. J. *et al.* A low-temperature grown oxide diode as a new switch element for high-density, nonvolatile memories. *Adv. Mater.* **19,** 73–76 (2007).
76. Mustafa, J. & Waser, R. A novel reference scheme for reading passive resistive crossbar memories. *IEEE Trans. Nanotechnol.* **5,** 687–691 (2006).
77. Heath, J. R., Kuekes, P. J., Snider, G. S. & Williams, R. S. A defect-tolerant computer architecture: opportunities for nanotechnology. *Science* **280,** 1716–1721 (1998).
78. Snider, G., Kuekes, P., Hogg, T. & Williams, R. S. Nanoelectronic architectures. *Appl. Phys. A* **80,** 1183–1195 (2005).
79. DeHon, A., Randy Huang, & Wawrzynek, J. Stochastic spatial routing for reconfigurable networks. *Microprocessors Microsyst.* **30,** 301–318 (2006).
80. Likharev, K. K. & Strukov, D. B. in *Introducing Molecular Electronics. Lecture Notes in Physics* Vol. **680** (eds Cuniberti, G., Richter, K. & Fagas, G.) 447–477 (Springer, Berlin, 2006).
81. Lu, W. & Lieber, C. M. Nanoelectronics from the bottom up. *Nature Mater.* **6,** 841–850 (2006).
82. Ignatiev, A. *et al.* Resistance switching in perovskite thin films. *Phys. Stat. Sol. B* **243,** 2089–2097 (2006).
83. Honigschmid, H. *et al.* A non-volatile 2Mbit CBRAM memory core featuring advanced read and program control. *VLSI Circuits Symp. Tech. Digest,* 110–11(2006).
84. Zhirnov, V. V., Cavin-R-K-III, Hutchby, J. A. & Bourianoff, G. I. Limits to binary logic switch scaling—a gedanken model. *Proc. IEEE USA* **91,** 1934–1939 (2003).
85. Cavin, R. K., Zhirnov, V. V., Herr, D. J. C., Alba Avila, & Hutchby, J. Research directions and challenges in nanoelectronics. *J.Nanoparticle Res.* **8,** 841–858 (2006).
86. Snider, G. S. & Williams, R. S. Nano/CMOS architectures using a field-programmable nanowire interconnect. *Nanotechnology* **18,** 1–11 (2007).

Acknowledgements

We thank J. G. Bednorz (IBM Research, Zurich), U-In Chung, I. G. Baek and S. O. Park (Samsung Electronics), Y. Zhang (Intel, Santa Clara), R. Bruchhaus (Qimonda, Munich), V. Zhirnov (SRC), and K. Szot and R. Dittmann (Research Center Jülich) for valuable comments.

Correspondence and requests for materials should be addressed to R.W.

Technology and metrology of new electronic materials and devices

Scaling of the metal oxide semiconductor (MOS) field-effect transistor has been the basis of the semiconductor industry for nearly 30 years. Traditional materials have been pushed to their limits, which means that entirely new materials (such as high-κ gate dielectrics and metal gate electrodes), and new device structures are required. These materials and structures will probably allow MOS devices to remain competitive for at least another ten years. Beyond this timeframe, entirely new device structures (such as nanowire or molecular devices) and computational paradigms will almost certainly be needed to improve performance. The development of new nanoscale electronic devices and materials places increasingly stringent requirements on metrology.

ERIC M. VOGEL

The University of Texas at Dallas, Department of Electrical Engineering, 2601 North Floyd Road, Richardson, Texas 75083, USA; previously at the National Institute of Standards and Technology, Semiconductor Electronics Division, 100 Bureau Drive, MS 8120, Gaithersburg, Maryland 2089, USA.

e-mail: eric.vogel@utdallas.edu

For over 30 years, the planar silicon metal oxide semiconductor (MOS) field-effect transistor (FET) has been the basis of integrated circuits[1,2]. In 1970, integrated circuits contained thousands of MOSFETs, each having dimensions of tens of micrometres. Today's chips contain almost one billion MOSFETs, each having physical dimensions of tens of nanometres. The exponential increase of device density (Moore's Law[3]) and scaling of device dimension into the nanotechnology regime (see Fig. 1) have resulted in the vast improvements observed in numerous electronic devices, from PCs and mobile phones to control systems found in automobiles and jet airplanes. Although considerable innovation and investment was required to realize this rate of dimensional scaling, until recently very little has changed in the materials and design of the basic MOSFET. In the future, a variety of new materials and device structures will be required to continue MOSFET scaling[1,2,4–10]. Furthermore, as silicon MOSFET technology approaches its limits[11], entirely new device structures and computational paradigms will be required to replace and augment traditional MOSFETs. These possible emerging technologies span the realm from transistors made of silicon nanowires to devices made of nanoscale molecules.

Whether one is considering future nanoscale planar silicon MOSFETs or an emerging technology to replace the MOSFET, the electronic properties of all of these nanodevices are extremely susceptible to small perturbations in properties such as dimension, structure, roughness and defects, which means there is a significant need for precise metrology. Furthermore, the insertion of a variety of new materials into the conventional MOSFET and radically new materials and devices for potential emerging replacement technologies (for example, molecular and spin) present further challenges for

metrology. Although the devices and materials being considered for future MOS technologies and beyond are broad, the overarching challenges to metrology remain. The ability to measure the physical, chemical and electronic properties that control final device electrical characteristics will be crucial to the research and development of future electronic devices.

TECHNOLOGY

TRADITIONAL MOS

As illustrated in Fig. 2a, the traditional MOSFET consists of a silicon substrate, a highly doped polysilicon gate electrode, a gate dielectric of SiO_2, and doped source and drain. In 1970, the MOSFET channel length was approximately 10 μm, the SiO_2 gate dielectric thickness was approximately 100 nm, and the operating voltage was approximately 10 V (refs 2,12). For over 30 years, the device density and speed of integrated circuits has been increased by reducing the MOSFET lateral dimension as shown in Fig. 1. To maintain, or improve, device performance when reducing channel length, most other device parameters need to be scaled: the substrate doping must be increased, the depth of the source/drain junction must be decreased, the supply voltage must be decreased and the capacitance of the SiO_2 gate dielectric must be increased (that is, SiO_2 thickness must be decreased).

Although there have been some changes of materials in the past 30 years (for example, moving from aluminium to highly doped polysilicon gate electrodes in the 1970s, the addition of lightly doped drain extensions in the 1980s, the addition of nitrogen to the SiO_2 gate dielectric to form silicon oxynitride in the late 1990s), the basic structure and materials of the traditional MOSFET have seen very little change. Today's MOSFET has an effective channel length of approximately 30 nm, a silicon oxynitride gate dielectric thickness of approximately 1.2 nm and a supply voltage of approximately 1.1 V. Although it may be possible to manufacture traditional MOS devices with a thinner layer of silicon oxynitride, tunnelling leakage current larger than that associated with the 1.2 nm gate dielectric (≥ 10 A cm^{-2}) results in unacceptably high power dissipation (≥ 10 W)

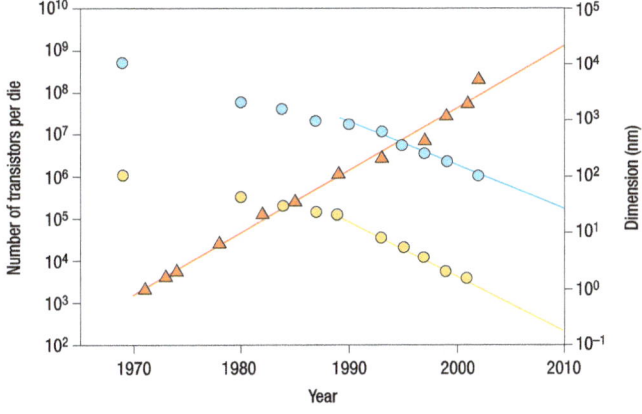

Figure 1 Moore's Law and scaling of transistor dimensions. The number of transistors (red triangles), transistor minimum lateral feature size (blue circles), and transistor minimum physical oxide thickness (yellow circles) as a function of time (data extracted from the Intel web site). The exponential increase in the number of transistors on an integrated circuit (Moore's Law[3]) requires the lateral dimensions of the transistor to scale downward. In order to improve the performance (speed) of the transistor, the physical oxide thickness must scale at approximately the same rate. Exponential fits to the data are shown as lines in the figure.

for the integrated circuit. Furthermore, because of high electric fields, the highly doped polysilicon no longer behaves as a metal and exhibits significant depletion, further limiting MOSFET performance. Owing to the fundamental limits associated with traditional MOSFET materials, new materials ('materials-limited MOS') and/or structures ('non-classical MOS') will be required to realize future improvements in MOSFET performance.

MATERIALS-LIMITED MOS

Insulators with high dielectric constant (for example, HfO_2) are being developed to replace silicon oxynitrde[9]. High-κ dielectrics allow a high capacitance with a thicker film so as to reduce the direct tunnelling leakage current. However, these dielectrics have a large number of technological problems, perhaps the worst of which is a generally poor interface with silicon. A wide variety of metals (for example, Ru, Ta, Pt, W, N and Si binaries and ternaries) are being considered as possible replacements to highly doped polysilicon[13,14]. The work function of the metal (minimum energy needed to remove an electron from the metal to vacuum) is critical in defining the threshold or turn-on voltage of the FET. Metal gate electrodes that have a work function similar to highly doped n-type silicon and highly doped p-type silicon must be found. The metal must have low resistance to current flow to ensure the speed of the device and be thermodynamically stable during subsequent processes, which can reach temperatures of ~1000 °C. Finally, the metal must not degrade the electrical properties of the underlying high-κ dielectric.

New materials for the substrate are also being developed and produced to directly improve the speed of the MOSFET without necessarily decreasing device dimensions. The most widely studied substrate material to date is strained silicon, formed either by growing silicon on top of relaxed $Si_{1-x}Ge_x$ or by introducing strain through the side using $Si_{1-x}Ge_x$ source/drains[2]. The smaller lattice constant of silicon as compared to $Si_{1-x}Ge_x$ results in strain and pronounced mobility enhancement for both holes and electrons. Because of their high mobility, III–V materials (for example,

InGaAs, InSb) on a silicon platform are also being considered as future substrate materials[7].

NON-CLASSICAL MOS

To extend beyond the performance improvements found with dimensional scaling and new materials for planar MOSFETs, non-classical MOSFET structures are also being considered. The simplest is that of silicon-on-insulator (SOI), as shown in Fig. 2b, where the active silicon is on a layer of thick silicon dioxide, which decreases the parasitic junction capacitance from the source/drain to the substrate[5,10], thereby increasing the speed of the FET. In partially depleted (PD) SOI, the silicon depletion layer is less than the thickness of the silicon, whereas in fully depleted (FD) SOI, the silicon depletion layer is greater than the thickness of the silicon. FDSOI exhibits nearly ideal drain current–gate voltage characteristics (ideal sub-threshold slope) because the back substrate has little control over the channel.

A more radical non-classical device structure is the tri-gate or finFET as illustrated in Fig. 2c (refs 2,5,15–17). These devices exhibit good short-channel behaviour because the wrap-around gate can strongly control the channel but they pose a host of technological problems. For example, a high-quality gate oxide must be formed on an etched sidewall, the wire of silicon must typically be undoped to maintain a correct and controlled threshold voltage, and the fin height and width must be scaled appropriately[17].

BEYOND MOS

It is likely that MOSFET dimensional scaling with changes in both materials and device structure will enable necessary improvements in device performance well into the next decade. However, at some point, new 'beyond-MOS' devices will be needed to improve performance[11]. The devices being considered for beyond MOS can be roughly organized into four classifications: confined-dimension, molecular, strongly correlated and spin[1,18,19].

Confined-dimension devices include memories based on nanocrystals[20], thin semiconductors for resonant tunnelling diodes[21], quantum dots for cellular automata[22,23], carbon nanotube[24] or bottom-up nanowire transistors[25,26], and single-electron transistors[27]. The basis of molecular electronic devices is the behaviour of the electronic properties of single molecules or monolayers[28]. The potential for molecular electronics originates from its nanoscale dimension, the possibility of synthesizing very specific electronic properties, and the promise of fabrication through self-assembly. Although numerous molecular devices have been demonstrated, there is still significant controversy concerning the repeatability of measurements and the understanding of charge transport[29–31]. Forming ohmic but non-interacting metallic contacts to molecules is a specific problem.

Devices based on strongly correlated materials include ferroelectric memories[32] and ferromagnetic logic devices. Ferroelectric materials such as $(Ba,Sr)TiO_3$ exhibit memory function through spontaneous changes in dipole moment with the application of an electric field. Ferromagnetic devices formed of materials such as Fe, Ni, and Co have computational states based on local ferromagnetic orientation. The foundation of spin-based devices such as the spin MOSFET[33,34], spin-torque transistor[35], and spin-gain transistor[36] is the use of electron spin as the computational state. These devices are of interest because of their inherent low power dissipation as compared with charge-based devices. However, such devices are still in their infancy.

Silicon-based MOSFETs are approaching device densities of 10^9 devices per chip (with ten year reliability), lateral dimensions of 10 nm, and gate delays of 1 ps. Finding a technology with substantially better capabilities to simply replace the MOSFET in conventional architectures is highly unlikely[37]. Instead, these beyond-MOS technologies are more likely to be used as a modification to the MOS platform or in wholly different architectures (for example,

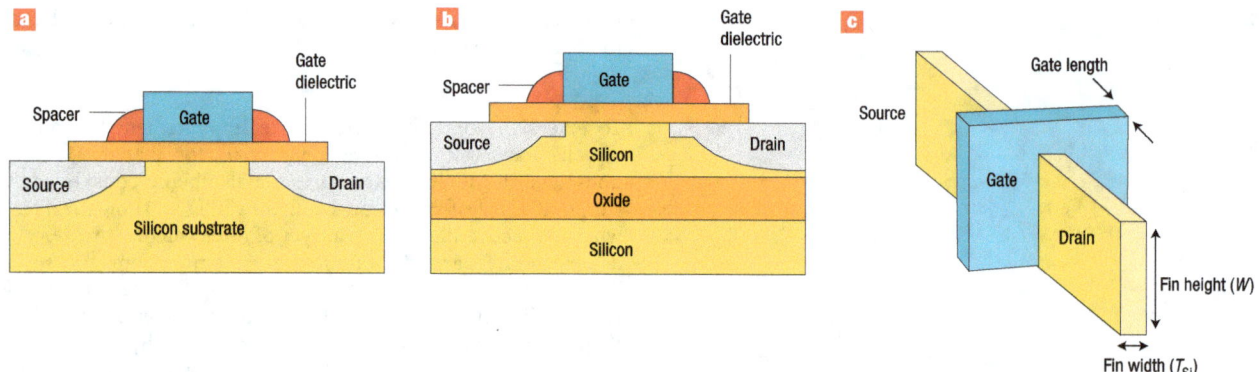

Figure 2 Illustrations of silicon transistors. **a,** A traditional n-channel MOSFET uses a highly doped n-type polysilicon gate electrode, a highly doped n-type source/drain, a p-type substrate, and a silicon dioxide or oxynitride gate dielectric. **b,** A silicon-on-insulator (SOI) MOSFET is similar to the traditional MOSFET except the active silicon is on a thick layer of silicon dioxide. This electrical isolation of the silicon reduces parasitic junction capacitance and improves device performance. **c,** A finFET is a three-dimensional version of a MOSFET. The gate electrode wraps around a confined silicon channel providing improved electrostatic control of the channel electrons.

quantum[38,39], defect tolerant[40] and biologically inspired[41]) aimed at applications that complement MOS. For example, one potential implementation of a quantum computer involves the use of dopants in silicon and single-electron transistors[38]. Quantum computers are exponentially more efficient than binary computers at applications such as searching a non-sorted database and number factorization. However, the probabilistic nature of quantum computing limits its applicability to more general computing.

METROLOGY REQUIREMENTS

Given the wide breadth of materials and devices being considered for future MOS and beyond, it is impossible to cover every measurement challenge adequately and it is not the purpose of this review to discuss all possible measurement tools and methods. However, there are important overarching measurement challenges that pervade most of these seemingly disparate technologies. The following describes these themes by illustrating the sensitivity of device behaviour to properties such as dimension, structure, and composition, and gives examples of solutions to these measurement challenges.

NANOMETRE DIMENSION

Perhaps the most fundamental measurement challenge for future MOSFETs and emerging devices is that of dimension, both horizontal (that is, critical dimension or linewidth) and vertical (film thickness). By 2010, the most recent edition of the International Technology Roadmap for Semiconductors (ITRS)[1] predicts that state-of-the-art MOSFETs in manufacturing will have physical gate lengths of approximately 20 nm controlled to within approximately 2 nm. In order to precisely measure a difference of 2 nm, the precision (3σ) of the critical dimension measurement must be much smaller — 0.37 nm (in the order of one atom) — by 2010. By 2020, the required precision drops to 0.12 nm. This is even more critical for devices such as finFETs or nanowire FETs that have extremely narrow channels. Figure 3 shows the drain current versus gate voltage characteristics for a finFET with different silicon fin widths. The off-state current ($V_g = 0$ V) varies by almost two orders of magnitude for a 2 nm change in the fin width (approximately one order of magnitude for every two atomic layers). This extreme sensitivity of electrical properties to nanoscale dimension illustrates the importance of dimensional metrology. Scanning electron

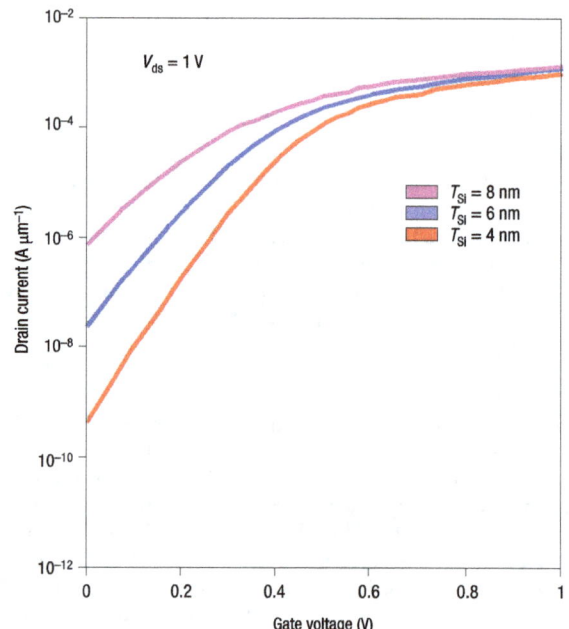

Figure 3 Simulation of a finFET. Drain current versus gate voltage characteristics of a finFET (illustrated in Fig. 2) for different values of fin width (T_{Si}) as defined in Fig. 2. The increase of the characteristic slope at low gate voltages with decreasing T_{Si} is due to increased electrostatic control of the gate over carriers in the channel. The strong sensitivity of the current at 0 V on T_{Si} (almost one order of magnitude in current for every 2 nm change in T_{Si}) illustrates the need for highly precise dimensional measurements with precision much less than 1 nm. Courtesy of T.-J. King, Univ. California, Berkeley.

microscopy (SEM), transmission electron microscopy (TEM) and atomic force microscopy are being developed to meet these needs[42–44]. As the dimensions of future devices approach the atomic scale, sample preparation, which can alter the dimension to be measured, becomes increasingly important.

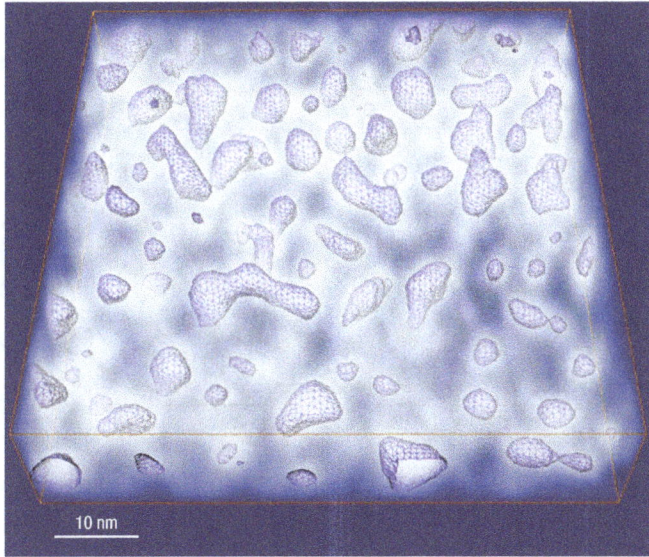

Figure 4 The structure of silicon quantum dots. Tomographic reconstruction of the silicon plasmon signal at 17 eV, visualized by volume rendering (white 'fog') and an iso-surface at fixed threshold (blue shapes) for silicon quantum dots embedded in SiO$_2$. In modelling and predicting the behaviour of devices based on silicon quantum dots, it is largely assumed that the quantum dots are perfectly spherical. These results indicate that the silicon dots are not perfectly spherical, illustrating the need for three-dimensional structural measurements at the nanoscale. Courtesy of A. Yurtserver and D. Muller, Cornell Univ.

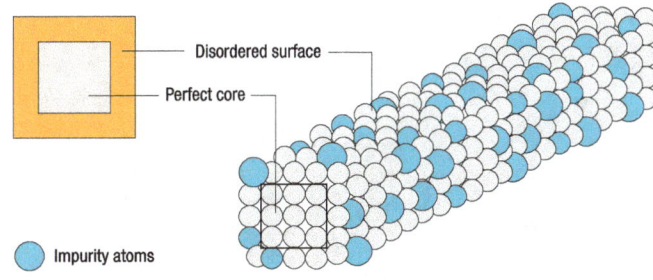

Figure 5 The importance of compositional metrology. Schematic illustration of a nanowire with surface disorder induced by shell doping. The mobility of electrons in silicon nanowires is a critical parameter that strongly impacts its measured electrical behaviour. Numerical results by Zhong[61] have suggested that doping of an outer shell of a silicon nanowire while keeping the core of the nanowire unperturbed may result in an increase in the electron mobility (in contrast to bulk silicon). There are currently no techniques capable of mapping the composition of nanowires at the spatial resolutions necessary to experimentally verify this result. Reprinted with permission from ref. 61. Copyright 2006 American Chemical Society.

The vertical dimension (thickness) of materials being considered for future devices is also critical[1, 45–47]. The thickness of silicon dioxide in today's MOSFET is approximately 1.2 nm controlled to within 0.04 nm[1,47]. This means that the precision (3σ) of the critical dimension measurement must be 0.0048 nm. Similar requirements

are found for a wide variety of future devices and materials. FDSOI FETs will have silicon thicknesses approaching 15 nm and will require 0.0075 nm measurement precision by 2010. Resonant tunnelling diodes will have extremely thin quantum wells and self-assembled monolayers for molecular electronics have thicknesses in the order of nanometres.

Indirect techniques such as spectroscopic ellipsometry, X-ray reflectivity, and capacitance-based measurements are being used and developed to measure thickness in this regime[1,45–48]. These indirect techniques typically require some *a priori* knowledge of the fundamental properties and constants of the materials being measured. The breadth of materials being considered for future devices and the fact that their properties may change in the nanoscale complicates these indirect measurements. Furthermore, techniques available today to measure thickness to within 0.04 nm typically require large spot sizes or sample areas to achieve enough signal. For example, spectroscopic ellipsometers have a measurement diameter from micrometres to millimetres and capacitance–voltage measurements are typically made on test structures measuring tens of micrometres. Therefore, these techniques provide a measure of film thickness averaged over the probed sample. As device dimensions continue to scale into the nanometre regime, the assumption that thickness determined from large probes is relevant to the electrical device of interest (for example, a MOSFET with 30 nm channel length) is likely to become less valid. For example, the strain associated with a nanoscale MOSFET may have a strong impact on the growth and thickness of a dielectric. The value of thickness obtained from a large-area capacitor or optical measurement may not be relevant to the technology or device of interest. As the dimensions of future devices approach the atomic scale, technologically relevant measures of dimension become increasingly important.

THREE-DIMENSIONAL STRUCTURE

It is not only the lateral and vertical dimension of a nanodevice that is important or even necessarily relevant. The complete structure in multiple dimensions is critical for many devices and materials. For planar MOSFETs, line edge roughness (LER), line width roughness (LWR) and surface roughness is critical. By 2010, the ITRS predicts that LWR must be less than 1.4 nm and that the precision (3σ) of the LWR measurement must be 0.28 nm, in the order of one atom[1].

Furthermore, many of the devices being considered do not possess simple planar geometries. The exact shape, rather than simple dimension, of a nanowire has been shown to strongly influence its final electrical properties[49,50]. The thickness of a dielectric or film surrounding a nanowire is likely to be non-uniform. It is largely assumed that silicon quantum dots embedded in SiO$_2$ (for use in optical and memory applications) are spherical. However, as shown in Fig. 4, recent plasmon tomography measurements indicate that silicon particles have complex morphologies and high surface to volume ratios rather than the commonly assumed near-spherical structures. These complex morphologies would affect quantum-confined excitons and interface density of states, directly impacting final device properties[51,52]. A model that attempts to predict the electrical properties of devices based on spherical silicon quantum dots may be incorrect. Techniques such as TEM holography and tomography, aberration correction SEM, and STM have shown promise for measuring structure at the dimensions of interest[53–59].

ATOMIC COMPOSITION

The electrical properties of many devices are sensitive to small changes in composition or even the placement of small numbers of atoms. By 2010, the distribution of dopant atoms within the conventional MOSFET must be characterized with 3 nm spatial resolution and 4% accuracy[1]. Characterizing the dopant distribution of nanowire devices is even more critical[60]. For example, it has been theoretically shown that shell-doping of nanowires, as illustrated in Fig. 5, results in enhanced electron mobility[61]. Rigorously confirming this prediction would require measurement of dopant profiles on the atomic scale.

Beyond dopants, another issue with bottom-up nanowires is that small amounts of metal catalyst used to grow the nanowires can migrate into the nanowire and affect the electrical properties[62]. It has also been shown that small numbers of bonded water molecules can change the conduction properties of carbon nanotubes from p-type to n-type[63,64] and the conductivity of single molecules has been shown to be strongly affected by single charged surface atoms[65]. One approach to quantum information processing proposes using single phosphorous atoms to make a quantum computer[38]. All of these technologies and issues point to the need for significant improvements in compositional metrology.

New characterization technology and understanding is needed for the problems described above, namely, characterizing compositions approaching single atoms with spatial resolution approaching angstroms. Common dopant profiling techniques such as secondary ion mass spectroscopy (SIMS) are not possible on these three-dimensional structures. Techniques for two-dimensional dopant profiling, such as scanning capacitance microscopy, are not easily performed on nanowires because the effective tip size is larger than the nanowire, but development is ongoing[66,67]. Techniques such as annular dark-field scanning transmission electron microscopy (ADF-STEM) have shown some success in detecting individual atoms[67-70]. ADF-STEM is a Z-contrast technique — the intensity scattered by an atom is approximately proportional to its atomic number (Z) squared. Achieving atomic resolution requires extremely sensitive cross-sectional sample preparation (~5-nm-thick samples with thickness variation less than the contrast associated with one atom). Further development of compositional metrology will be required for future electronic devices.

ELECTRONIC STRUCTURE AND PROPERTIES
Electronic density of states, dielectric function, bandgap, work function and density of electrically active defects are perhaps the properties most directly related to electrical-device behaviour. For example, future MOSFETs will likely have a complex stack of materials in which the density of states, work function and defect density will be critical. The metal gate electrode may consist of a thick top contact metal gate electrode and a thin layer of metal used to control effective work function[14]. It has been shown that, in the nanoscale regime, the thickness of this thin metal layer is a key parameter defining effective work function as measured from device characteristics[14]. The gate dielectric will consist of a high-permittivity material (such as HfO_2) on a very thin layer of SiO_2 to reduce the interfacial defect density near the substrate. The band offsets between these dielectrics, the substrate material and the gate electrode determine not only the tunnel current through the gate stack, but also the threshold voltage of the device[71,72]. Electrically active defects at the interfaces of these materials degrade device reliability as well as the effective mobility of carriers in the substrate[73-75].

Similar issues are present in beyond-MOS devices. An understanding of the local density of states of the molecules comprising the self-assembled monolayer is critical to understanding transport through the molecules[76]. The energy offsets and electrically active defects between the molecules and its electrodes are perhaps even more critical[30,76]. The bandgap and electronic structure (semiconducting versus metallic) of carbon nanotubes are critical to their electrical behaviour[77-80]. A low electrically active defect density is critical to achieving high negative differential resistance in resonant tunnelling diodes[81] and to charge offset stability in single electron devices[82]. Scattering is an important problem limiting spintronic devices[83].

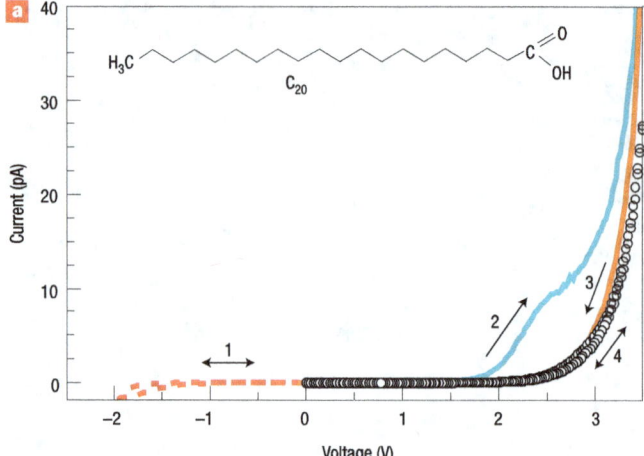

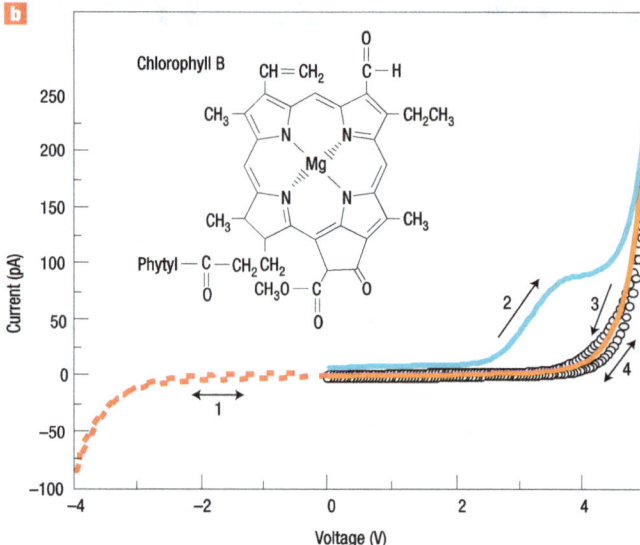

Figure 6 Does the molecule matter? Shown are typical experimental hysteretic current–voltage curves of Al/AlO$_x$/molecule/Ti/Al planar crossbar devices incorporating Langmuir–Blodgett organic monolayers of **a**, eicosanoic acid (C$_{20}$) and **b**, chlorophyll-B. The fact that the above hysterisis affect is observed for two very different molecular species suggests the importance of the interface of the metal electrodes with the molecule. The results illustrate the importance of performing compositional and structural measurements on the final device of interest since the entire device (molecule and electrodes) is critical to its final electrical properties. Reprinted with permission from ref. 30. Copyright (2005) Springer Science and Business Media.

Techniques such as photoelectron spectroscopy and inverse photoelectron spectroscopy have been used to map the valence and electron density of states in thin dielectric films[72,84]. One- and two-photon photoelectron spectroscopy has also been used to determine the electronic structure of self-assembled monolayers around the Fermi level[76]. These techniques are primarily applicable to planar thin films and monolayers and cannot typically be used to map the spatial dependence of properties. Scanned probe techniques such as scanning Kelvin probe microscopy[66] and versions of scanning tunnelling microscopy can provide information on local electronic structure[85].

However, the size of the probe tip is a limiting factor. Techniques such as fluorescence, Raman spectroscopy and Rayleigh scattering have been used to measure the bandgap of carbon nanotubes[79,80,86,87]. The ability to measure the electronic density of states, band offsets, and electrically active defect density as a function of energy and space provides valuable information for engineering the final electrical properties of MOS and beyond devices.

PROPERTIES OF DEVICE STRUCTURES

Measurement of electronic materials prior to their placement in a device structure is an important part of developing a new electronic device. However, the nanoscale dimension and the processing necessary to fabricate the device is critical to its final physical and electrical properties. For example, straining silicon by depositing it on $Si_{1-x}Ge_x$ is being used for enhancing carrier mobility. However, the final strain found in the device is determined not only by the lattice mismatch between the epitaxial silicon and $Si_{1-x}Ge_x$, but also by many other materials and processing steps associated with the final device such as sidewalls, capping layers and thermal processes[2,88]. Measuring properties such as composition, strain and structure within the final nanoscale device structure is crucial.

The final device structure can also dramatically change the intended electronic properties of the material of interest. For example, the interaction between the electrodes and molecules in molecular electronics is critical to the transport properties of the final device. The current–voltage characteristics of crossbar device structures containing eicosanoic acid and chlorophyll B, respectively, are shown in Figs 6a and 6b (ref. 30). Eicosanoic acid is a simple molecule and is expected to act as a passive film with simple tunnelling current–voltage behaviour. The hysterisis observed for this simple molecule and its qualitatively similar characteristics to chlorophyll B (a very different molecular species) suggests that the final electrical properties are determined not only by the molecules, but also by the interaction of the molecule with its electrodes. Techniques such as inelastic tunnelling spectroscopy, which probe vibrational modes of the molecules through measurement of the device current–voltage behaviour,

have been developed to determine which molecular species and bonds are present in the final device structure[31]. Techniques such as backside-incidence Fourier-transform infrared spectroscopy have been used to investigate the interaction of top-metallization with molecular monolayers[30]. The above molecular electronics example illustrates the importance of performing characterization of the full integrated device structure.

Electrical test structures can be used to characterize the physical properties of nanoscale materials. For example, electrically active defects in the gate dielectric of the MOSFET are at densities ($\sim 10^{10}$ cm^{-2}) that typically cannot be measured using physical characterization techniques. Methods such as charge pumping have been developed which permit sensitive extraction of both the energy and depth dependence of these defects[75]. This technique determines the electrically active defect density by measuring the amount of current associated with the capture of free carriers induced by applying periodic pulses to the gate of a FET. By comparing the depth and energy dependence of the extracted electrically active defect density from transistor measurements with physical measurements of composition and structure, important insights on the physical origin of the defects can be elucidated. This approach of integrated technology and measurement will be increasingly critical to future technological development (Box 1).

As another example, the ability of an injected spin-polarized current to retain its spin polarization (spin injection efficiency) is an important parameter in spin-based devices such as lateral spin valves. The spin injection efficiency of a spin valve can be deduced through measurement of the voltage induced by spin accumulation using separate detectors fabricated with the spin valve[89]. It has been shown that the exact configuration of the detector affects whether the measured value is due only to the signal of interest from the channel of the spin valve or includes other spurious affects[89]. The extreme sensitivity of the electrical properties of nanoscale devices on their physical properties can be exploited to precisely characterize many properties of nanoscale materials. However, care must be taken to ensure that the test structure and analysis of these indirect electrical measurements provide meaningful properties and do not

Box 1 Integrating metrology with technology development

The metrology requirements described in this review suggest that a single characterization tool is needed that can measure the location (to within 0.001 nm) and type of every atom within a three-dimensional device structure of interest (even if the structure is buried below 100 nm of material) and provide information on other parameters of interest such as electronic density of states and strain as a function of position. It is apparent that there will not be a characterization tool available to meet all of these challenges in time to have an impact on the development of new electronic materials and devices required by the semiconductor industry. Nevertheless, development continues at a rapid pace without the ability to physically measure every parameter of interest. For example, the semiconductor industry is developing uniaxial strain engineering through heteroexpitaxy or capping layers to enhance MOSFET performance[1,2,90]. As there are currently no metrology tools available that can measure strain magnitudes of interest within the nanoscale MOSFET, researchers are calibrating strain levels by applying known stress to a transistor while measuring drive current.

Furthermore, every characterization method has strengths and weaknesses. There is typically not one characterization method that provides all of the information necessary to describe a given property of a material. The complexity and dimensions of the materials and devices being considered often result in different measured values for nominally the same property of a given material. For example, the thickness of ultrathin silicon dioxide measured using transmission electron microscopy (TEM), spectroscopic ellipsometry (SE) and capacitance–voltage (C–V) can give results varying by as much as 20% (ref. 47). The thickness measured using TEM depends on sample preparation as well as the methodology used to define the interface. The thickness measured using SE and C–V depends dramatically on the model assumed in data interpretation[91].

Although developing a single metrology capability to provide increasingly accurate measurements is critical, it is also extremely important to develop a fundamental understanding of the interrelationship between measurements from many different physical characterization tools and measurements of the technology of interest. This approach may be the only way to obtain the information necessary to develop new electronic materials and devices.

include spurious effects that are not directly associated with the electronic material of interest.

OUTLOOK

The silicon transistor has been a technological and economic engine for over three decades. Metrology has been an important infrastructure enabling the exponential increase of device density and exponential decrease of device dimension. New nanoscale materials and devices will be required to sustain this technological revolution but the extreme sensitivity of the electronic properties of these devices to their nanoscale physical properties and the insertion of radically new materials present significant challenges to metrology. The ability to measure the physical, chemical and electronic properties of these new materials and devices will become even more critical in continuing the advance of the MOS and finding technologies to extend or replace it.

doi: 10.1038/nnano.2006.142

References

1. *International Technology Roadmap for Semiconductors* (Semiconductor Industry Association, 2005); www.ltrs.net
2. Thompson, S. E. *et al.* In search of "forever," continued transistor scaling one new material at a time. *IEEE Trans. Semiconduct. M.* **18**, 26–36 (2005).
3. Moore, G. Cramming more components onto integrated circuits. *Electronics* **38**, 114–117 (1965).
4. Chang, L. L. *et al.* Extremely scaled silicon nano-CMOS devices. *Proc. IEEE* **91**, 1860–1873 (2003).
5. Chang, L. L. *et al.* Moore's law lives on: ultra-thin body SOI and FinFET CMOS transistors look to continue Moore's law for many years to come. *IEEE Circuits Device.* **19**, 35–42 (2003).
6. Chang, L. L., Ieong, M. & Yang, M. CMOS circuit performance enhancement by surface orientation optimization. *IEEE Trans. Electron. Dev.* **51**, 1621–1627 (2004).
7. Chau, R., Datta, S. & Majumdar, A. Opportunities and Challenges of III–V Nanoelectronics for Future High-Speed, Low-Power Logic Applications. *IEEE CSIC Symposium Technical Digest* 17–20 (2005).
8. Taur, Y. *et al.* CMOS scaling into the nanometer regime. *Proc. IEEE* **85**, 486–504 (1997).
9. Wilk, G. D., Wallace, R. M. & Anthony, J. M. High-kappa gate dielectrics: Current status and materials properties considerations. *J. Appl. Phys.* **89**, 5243–5275 (2001).
10. Wong, H. S. P., Frank, D. J., Solomon, P. M., Wann, C. H. J. & Welser, J. J. Nanoscale CMOS. *Proc. IEEE* **87**, 537–570 (1999).
11. Muller, D. A. A sound barrier for silicon? *Nature Mater.* **4**, 645–647 (2005).
12. Sah, C. T. Evolution Of The MOS-Transistor — from Conception To VLSI. *Proc. IEEE* **76**, 1280–1326 (1988).
13. Song, S. C. *et al.* Integration issues of high-k and metal gate into conventional CMOS technology. *Thin Solid Films* **504**, 170–173 (2006).
14. Jha, R., Lee, J., Majhi, P. & Misra, V. Investigation of work function tuning using multiple layer metal gate electrodes stacks for complementary metal-oxide-semiconductor applications. *Appl. Phys. Lett.* **87**, 223503 (2005).
15. Hisamoto, D. *et al.* FinFET — a self-aligned double-gate MOSFET scalable to 20 nm. *IEEE Trans. Electron. Dev.* **47**, 2320–2325 (2000).
16. Xiong, S. Y. & Bokor, J. Sensitivity of double-gate and finFET devices to process variations. *IEEE Trans. Electron. Dev.* **50**, 2255–2261 (2003).
17. Anil, K. G., Henson, K., Biesemans, S. & Collaert, N. Layout density analysis of finFETs. *Proc. 33rd ESSDERC* 139–142 (2003).
18. Hutchby, J. A., Bourianoff, G. L., Zhirnov, V. V. & Brewer, J. E. Emerging research memory and logic technologies. *IEEE Circuits Device.* **21**, 47–51 (2005).
19. Zhirnov, V. V., Hutchby, J. A., Bourianoff, G. I. & Brewer, J. E. Emerging research logic devices. *IEEE Circuits Device.* **21**, 37–46 (2005).
20. Hanafi, H. I., Tiwari, S. & Khan, I. Fast and long retention-time nano-crystal memory. *IEEE Trans. Electron. Dev.* **43**, 1553–1558 (1996).
21. Reed, M. A., Frensley, W. R., Matyi, R. J., Randall, J. N. & Seabaugh, A. C. Realization Of A 3-Terminal Resonant Tunneling Device: The Bipolar Quantum Resonant Tunneling Transistor. *Appl. Phys. Lett.* **54**, 1034–1036 (1989).
22. Lent, C. S. & Isaksen, B. Clocked molecular quantum-dot cellular automata. *IEEE Trans. Electron. Dev.* **50**, 1890–1896 (2003).
23. Lent, C. S., Isaksen, B. & Lieberman, M. Molecular quantum-dot cellular automata. *J. Am. Chem. Soc.* **125**, 1056–1063 (2003).
24. Chen, J., Klinke, C., Afzali, A. & Avouris, P. Self-aligned carbon nanotube transistors with charge transfer doping. *Appl. Phys. Lett.* **86**, 123108 (2005).
25. Chung, S. W., Yu, J. Y. & Heath, J. R. Silicon nanowire devices. *Appl. Phys. Lett.* **76**, 2068–2070 (2000).
26. Cui, Y., Zhong, Z. H., Wang, D. L., Wang, W. U. & Lieber, C. M. High performance silicon nanowire field effect transistors. *Nano Lett.* **3**, 149–152 (2003).
27. Chen, R. H., Korotkov, A. N. & Likharev, K. K. Single-electron transistor logic. *Appl. Phys. Lett.* **68**, 1954–1956 (1996).
28. Reed, M. A. Molecular-scale electronics. *Proc. IEEE* **87**, 652–658 (1999).
29. Reed, M. A. Molecular electronics: Back under control. *Nature Mater.* **3**, 286–287 (2004).
30. Richter, C. A., Stewart, D. R., Ohlberg, D. A. A. & Stanley Williams, R. Electrical characterization of Al/AlO_x/molecule/Ti/Al devices. *Appl. Phys. A* **80**, 1355–1362 (2005).
31. Wang, W. Y., Lee, T. H. & Reed, M. A. Electronic transport in molecular self-assembled monolayer devices. *Proc. IEEE* **93**, 1815–1824 (2005).
32. Arimoto, Y. & Ishiwara, H. Current status of ferroelectric random-access memory. *MRS Bull.* **29**, 823–828 (2004).
33. Datta, S. & Das, B. Electronic Analog Of The Electrooptic Modulator. *Appl. Phys. Lett.* **56**, 665–667 (1990).
34. Zutic, I., Fabian, J. & Das Sarma, S. Spintronics: Fundamentals and applications. *Rev. Mod. Phys.* **76**, 323–410 (2004).
35. Bauer, G. E. W., Brataas, A., Tserkovnyak, Y. & van Wees, B. J. Spin-torque transistor. *Appl. Phys. Lett.* **82**, 3928–3930 (2003).
36. Nikonov, D. E. & Bourianoff, G. I. Spin gain transistor in ferromagnetic semiconductors: The semiconductor Bloch-equations approach. *IEEE Trans. Nanotechnol.* **4**, 206–214 (2005).
37. Zhirnov, V. V., Cavin, R. K., Hutchby, J. A. & Bourianoff, G. I. Limits to binary logic switch scaling — a Gedanken model. *Proc. IEEE* **91**, 1934–1939 (2003).
38. Kane, B. E. A silicon-based nuclear spin quantum computer. *Nature* **393**, 133–137 (1998).
39. Shor, P. W. Polynomial-time algorithms for prime factorization and discrete logarithms on a quantum computer. *SIAM J. Comp.* **26**, 1484–1509 (1997).
40. Heath, J. R., Kuekes, P. J., Snider, G. S. & Williams, R. S. A defect-tolerant computer architecture: Opportunities for nanotechnology. *Science* **280**, 1716–1721 (1998).
41. Sarpeshkar, R. Analog versus digital: Extrapolating from electronics to neurobiology. *Neural Comput.* **10**, 1601–1638 (1998).
42. Diebold, A. C. Metrology technology for the 70-nm node: Process control through amplification and averaging microscopic changes. *IEEE Trans. Semiconduct. M.* **15**, 169–182 (2002).
43. Diebold, A. C. & Joy, D. A critical analysis of techniques and future CD metrology needs. *Solid State Technol.* **46**, 63 (2003).
44. Marchman, H. M. & Griffith, J. E. in *Handbook of Silicon Semiconductor Metrology* (ed. Diebold, A. C.) (Marcel Dekker, New York, 2001).
45. Diebold, A. C. *et al.* Characterization and production metrology of gate dielectric films. *Mat. Sci. Semicon. Proc.* **4**, 3–8 (2001).
46. Diebold, A. C. *et al.* Thin dielectric film thickness determination by advanced transmission electron microscopy. *Microscopy Microanal.* **9**, 493–508 (2003).
47. Ehrstein, J. *et al.* A comparison of thickness values for very thin SiO_2 films by using ellipsometric, capacitance-voltage, and HRTEM measurements. *J. Electrochem. Soc.* **153**, F12-F19 (2006).
48. Tompkins, H. H. & McGahan, W. A. *Spectroscopic Ellipsometry and Reflectometry* (Academic Press, New York, 1999).
49. Guo, J., Wang, J., Polizzi, E., Datta, S. & Lundstrom, M. Electrostatics of nanowire transistors. *IEEE Trans. Nanotechnol.* **2**, 329–334 (2003).
50. Vashaee, D. *et al.* Electrostatics of nanowire transistors with triangular cross sections. *J. Appl. Phys.* **99**, 054310 (2006).
51. Fonoberov, V. A., Pokatilov, E. P., Fomin, V. M. & Devreese, J. T. Photoluminescence of tetrahedral quantum-dot quantum wells. *Physica E* **26**, 63–66 (2005).
52. Tokura, Y., Sasaki, S., Austing, D. G. & Tarucha, S. Excitation spectra and exchange interactions in circular and elliptical quantum dots. *Physica B* **298**, 260–266 (2001).
53. Bals, S., Van Tendeloo, G. & Kisielowski, C. A new approach for electron tomography: Annular dark-field transmission electron microscopy. *Adv. Mater.* **18**, 892 (2006).
54. Cowley, J. M. Off-axis STEM or TEM holography combined with four-dimensional diffraction imaging. *Microscopy Microanal.* **10**, 9–15 (2004).
55. Cumings, J., Zettl, A., McCartney, M. R. & Spence, J. C. H. Electron holography of field-emitting carbon nanotubes. *Phys. Rev. Lett.* **88**, 056804 (2002).
56. Joy, D. C. The aberration corrected SEM. *AIP Conference Proceedings* **788**, 535–542 (2005).
57. Kim, M. J., Wallace, R. M. & Gnade, B. E. HRTEM for nano-electronic materials research. *Characterization and Metrology for ULSI Technology* **788**, 558–564 (2005).
58. Shekhawat, G. S. & Dravid, V. P. Nanoscale imaging of buried structures via scanning near-field ultrasound holography. *Science* **310**, 89–92 (2005).
59. Fallahi, P. *et al.* Imaging a single-electron quantum dot. *Nano Lett.* **5**, 223–226 (2005).
60. Garcia-Gutierrez, D. I. *et al.* Study of two-dimensional B doping profile in Si fin field-effect transistor structures by high angle annular dark field in scanning transmission electron microscopy mode. *J. Vac. Sci. Tech. B* **24**, 730–738 (2006).
61. Zhong, J. X. & Stocks, G. M. Localization/quasi-delocalization transitions and quasi-mobility-edges in shell-doped nanowires. *Nano Lett.* **6**, 128–132 (2006).
62. Hannon, J. B., Kodambaka, S., Ross, F. M. & Tromp, R. M. The influence of the surface migration of gold on the growth of silicon nanowires. *Nature* **440**, 69–71 (2006).
63. Cao, J., Wang, Q. & Dai, H. Electron transport in very clean, as-grown suspended carbon nanotubes. *Nature Mater.* **4**, 745–749 (2005).
64. Na, P. S. *et al.* Investigation of the humidity effect on the electrical properties of single-walled carbon nanotube transistors. *Appl. Phys. Lett.* **87** (2005).
65. Piva, P. G. *et al.* Field regulation of single-molecule conductivity by a charged surface atom. *Nature* **435**, 658–661 (2005).
66. Bonnell, D. A. & Shao, R. Local behavior of complex materials: scanning probes and nano structure. *Current Opin. Solid St. M.* **7**, 161–171 (2003).
67. Castell, M. R., Muller, D. A. & Voyles, P. M. Dopant mapping for the nanotechnology age. *Nature Mater.* **2**, 129–131 (2003).
68. Voyles, P. M., Grazul, J. L. & Muller, D. A. Imaging individual atoms inside crystals with ADF-STEM. *Ultramicroscopy* **96**, 251–273 (2003).
69. Voyles, P. M., Muller, D. A., Grazul, J. L., Citrin, P. H. & Gossmann, H. J. L. Atomic-scale imaging of individual dopant atoms and clusters in highly n-type bulk Si. *Nature* **416**, 826–829 (2002).
70. Voyles, P. M., Muller, D. A. & Kirkland, E. J. Depth-dependent imaging of individual dopant atoms in silicon. *Microscopy Microanal.* **10**, 291–300 (2004).
71. Sayan, S. *et al.* Band alignment issues related to HfO_2/SiO_2/p-Si gate stacks. *J. Appl. Phys.* **96**, 7485–7491 (2004).
72. Sayan, S. *et al.* Valence and conduction band offsets of a ZrO_2/SiO_xN_y/n-Si CMOS gate stack: A combined photoemission and inverse photoemission study. *Phys. Status Solidi B* **241**, 2246–2252 (2004).

73. Kerber, A. *et al.* Charge trapping in SiO₂/HfO₂ gate dielectrics: Comparison between charge-pumping and pulsed I-D.-V.-G. *Microelectron. Eng.* **72,** 267–272 (2004).

74. Han, J. P. *et al.* Asymmetric energy distribution of interface traps in n- and p-MOSFETs with HfO₂ gate dielectric on ultrathin SiON buffer layer. *IEEE Electron. Dev. Lett.* **25,** 126–128 (2004).

75. Heh, D. *et al.* Spatial distributions of trapping centers in HfO₂/SiO₂ gate stacks. *Appl. Phys. Lett.* **88,** 152907 (2006).

76. Zangmeister, C. D., Robey, S. W., van Zee, R. D., Yao, Y. & Tour, J. M. Fermi level alignment and electronic levels in "molecular wire" self-assembled monolayers on Au. *J. Phys. Chem. B* **108,** 16187–16193 (2004).

77. Kim, P., Odom, T. W., Jin-Lin, H. & Lieber, C. M. Electronic density of states of atomically resolved single-walled carbon nanotubes: van Hove singularities and end states. *Phys. Rev. Lett.* **82,** 1225–1228 (1999).

78. Reich, S., Thomsen, C. & Ordejon, P. Electronic band structure of isolated and bundled carbon nanotubes. *Phys. Rev. B* **65,** 155411 (2002).

79. Sfeir, M. Y. *et al.* Probing electronic transitions in individual carbon nanotubes by Rayleigh scattering. *Science* **306,** 1540–1543 (2004).

80. Wildoer, J. W. G., Venema, L. C., Rinzler, A. G., Smalley, R. E. & Dekker, C. Electronic structure of atomically resolved carbon nanotubes. *Nature* **391,** 59–62 (1998).

81. Lake, R., Brar, B., Wilk, G. D., Seabaugh, A. & Klimeck, G. in *Compound Semiconductors 1997 Institute of Physics Conference Series* 617–620 (1998).

82. Zimmerman, N. M., Huber, W. H., Fujiwara, A. & Takahashi, Y. Excellent charge offset stability in a Si-based single-electron tunneling transistor. *Appl. Phys. Lett.* **79,** 3188–3190 (2001).

83. Hong, C. *et al.* Spin-polarized reflection in a two-dimensional electron system. *Appl. Phys. Lett.* **86,** 32113 (2005).

84. Sayan, S. *et al.* Band alignment issues related to HfO₂/SiO₂/p-Si gate stacks. *J. Appl. Phys.* **96,** 7485–7491 (2004).

85. Bussmann, E., Zheng, N. & Williams, C. C. Single-electron manpulation to and from SiO₂ surface by electrostatic force microscopy. *Appl. Phys. Lett.* **86,** 163109 (2005).

86. Bachilo, S. M. *et al.* Structure-assigned optical spectra of single-walled carbon nanotubes. *Science* **298,** 2361–2366 (2002).

87. Odom, T. W., Jin-Lin, H., Kim, P. & Lieber, C. M. Atomic structure and electronic properties of single-walled carbon nanotubes. *Nature* **391,** 62–64 (1998).

88. Diebold, A. Introduction of stress requires stress metrology methods. *Solid State Technol.* **48,** 59 (2005).

89. Ku, J., Chang, J., Han, S., Ha, J. & Eom, J. Electrical spin injection and accumulation in ferromagnetic/Au/ferromagnetic lateral spin valves. *J. Appl. Phys.* **99,** 08H705 (2006).

90. Lin, H. N. *et al.* Correlating drain-current with strain-induced mobility in nanoscale strained CMOSFETs. *IEEE Electron. Dev. Lett.* **27,** 659–661 (2006).

91. Richter, C. A., Hefner, A. R. & Vogel, E. M. A comparison of quantum-mechanical capacitance-voltage simulators. *IEEE Electron. Dev. Lett.* **22,** 35–37 (2001).

Acknowledgments

The author acknowledges the support of the Jonsson School of Engineering at the University of Texas at Dallas, the NIST Office of Microelectronics Programs, and the NIST Semiconductor Electronics Division. Contribution of the National Institute of Standards and Technology is not subject to US copyright. The author would like to thank Curt Richter, Steve Knight, David Seiler and Erik Secula for careful reading of the manuscript.

Competing financial interests

The author declares no competing financial interests.

Carbon-based electronics

The semiconductor industry has been able to improve the performance of electronic systems for more than four decades by making ever-smaller devices. However, this approach will soon encounter both scientific and technical limits, which is why the industry is exploring a number of alternative device technologies. Here we review the progress that has been made with carbon nanotubes and, more recently, graphene layers and nanoribbons. Field-effect transistors based on semiconductor nanotubes and graphene nanoribbons have already been demonstrated, and metallic nanotubes could be used as high-performance interconnects. Moreover, owing to the excellent optical properties of nanotubes it could be possible to make both electronic and optoelectronic devices from the same material.

PHAEDON AVOURIS*, ZHIHONG CHEN AND VASILI PEREBEINOS

IBM T. J. Watson Research Center, Yorktown Heights, New York, 10598, USA
*e-mail: avouris@us.ibm.com

In the last few decades we have witnessed dramatic advances in electronics that have found uses in computing, communications, automation and other applications that affect just about every aspect of our lives. To a large extent these advances have been the result of the continuous miniaturization or 'scaling' of electronic devices, particularly of silicon-based transistors, that has led to denser, faster and more power-efficient circuitry. Obviously, however, this device scaling and performance enhancement cannot continue forever; a number of limitations in fundamental scientific as well as technological nature place limits on the ultimate size and performance of silicon devices.

The realization of the approaching limits has inspired a worldwide effort to develop alternative device technologies. Some approaches involve moving away from traditional electron transport-based electronics: for example, the development of spin-based devices. Another approach, on which we focus here, maintains the operating principles of the currently used devices, primarily that of the field-effect transistor, but replaces a key component of the device, the conducting channel, with carbon nanomaterials such as one-dimensional (1D) carbon nanotubes (CNT) or two-dimensional (2D) graphene layers, which have superior electrical properties[1]. Furthermore, semiconducting carbon nanotubes are direct bandgap materials providing an ideal system to study optics and optoelectronics in one dimension and explore the possibility of basing both electronics and optoelectronic technologies on the same material.

In this article we examine the electronic structure, electrical transport and optoelectronic properties of CNTs, with a focus on the physical phenomena involved. We discuss briefly the emerging area of study involving single graphene layers and narrow graphene nanoribbons (GNRs). We analyse the switching mechanism and characteristics of single-CNT field-effect transistors (FETs), GNR FETs and efforts towards device integration. We also describe the principles of simple CNT optoelectronic devices such as electroluminescent light emitters and photodetectors.

ELECTRONIC STRUCTURE

Graphite is a well-known allotropic form of carbon composed of layers of sp^2-bonded carbon atoms in a honeycomb arrangement (Fig. 1a). The different carbon layers in graphite interact weakly, primarily by van der Waals forces. This interaction produces a small valence and conduction band overlap of about 40 meV, which makes graphite overall a semi-metal. The electronic structure of graphene, an individual layer of graphite, was first discussed by P. R. Wallace in 1947 (ref. 2). There are about 3 million layers in a millimetre thickness of graphite and the experimental study of graphene has been thwarted by the difficulty of isolating and studying this single atomic layer.

Recently, however, graphene became the object of intense experimental study when it was realized that single layers, or a few layers, could be produced relatively easily by mechanical exfoliation of graphite[3], or by heating SiC (ref. 4). Figure 1b shows the peculiar single-particle band structure of this 2D material. The linear dispersion at low energies makes the electrons and holes in graphene mimic relativistic particles that are described by the Dirac relativistic equation for particles with spin 1/2, and they are usually referred to as Dirac Fermions. Their dispersion, $E_{2D} = \hbar v_F \sqrt{k_x^2 + k_y^2}$, is analogous to that of photons, $E_k = \hbar c k$, but with the velocity of light c replaced by $v_F \approx 10^6$ m s^{-1}, the Fermi velocity. Thus, electrons and holes in graphene have zero effective mass and a velocity that is about 300 times slower than that of light. This linear dispersion relationship also means that quasi-particles in graphene display properties quite different to those observed in conventional three-dimensional materials, which have parabolic dispersion relationships. For example, graphene displays an anomalous quantum Hall effect and half-integer quantization of the Hall conductivity[5,6]. The quantum Hall effect in graphene can be observed even at room temperature[7].

The electronic structure of CNTs is usually discussed on the basis of the band structure of graphene. The CNT is thought of as being formed by the rolling of a piece of a ribbon of graphene to form a seamless cylinder. To a large extent, the remarkable electrical properties of carbon nanotubes have their origins in the unusual electronic structure of graphene. The rolling process forming the nanotube and the resulting nanotube structure are specified by a pair of integers (n,m) defining the chiral vector

When citing this review, please cite the original version as shown on the contents page of this chapter.

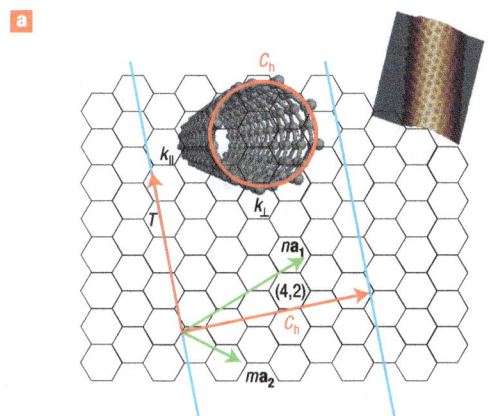

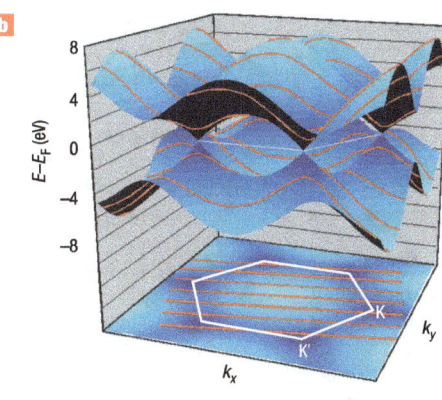

Figure 1 The structure of graphene and carbon nanotubes. **a**, The carbon atoms in a single sheet of graphene are arranged in a honeycomb lattice. A nanotube can be formed by rolling a ribbon of graphene along a chiral vector, C_h, defined by two integers, such as the (4,2) chiral vector shown here. The insets show the definitions of $k_\perp$ and $k_\parallel$ (left), and a scanning tunnelling microscope image (right) of a single-walled nanotube. **b**, The band structure (top) and Brillouin zone (bottom) of graphene. The valence band (which is of π-character) and the conduction band (π*-character) touch at six points that lie at the Fermi energy, but only two of these points — the K and K' points — are inequivalent. At these Dirac points, the density-of-states is zero, so graphene can be considered as a zero-gap semiconductor. At low energies, the dispersion is linear, determined by the conical sections involving the K and K' points. The quantization of the circumferential momentum, $k_\perp$, leads to the formation of a set of discrete energy sub-bands for each nanotube (red parallel lines). The relation of these lines to the band structure of graphene determines the electronic structure of the nanotube. If the lines pass through the K or K' points, the nanotube is a metal; if they do not (as in **b**), the nanotube is a semiconductor. Specifically, (n,n) nanotubes (armchair tubes) are always metallic, and (n,m) nanotubes with $n−m = 3j$, where $j = 1,2,3…$, are nearly metallic with a small, curvature-induced gap that has a $1/d^2$ dependence. Tubes with $n−m ≠ 3j$ are semiconductors.

$C_h = n\mathbf{a_1} + m\mathbf{a_2}$ that describes the circumference of the nanotube ($C_h = \pi d_{CNT}$), where $\mathbf{a_1}$ and $\mathbf{a_2}$ are the unit vectors of the graphene honeycomb lattice. The periodic boundary conditions around the circumference of a nanotube require that the component of the momentum along the circumference[8], $k_\perp$, is quantized: $C_h k_\perp = 2\pi\nu$ where ν is a non-zero integer. On the other hand, electron motion along the length of the tube is free and $k_\parallel$ is a continuous variable. As explained in Fig. 1a this quantization leads to the formation of metallic and semiconducting nanotubes. In the simple tight-binding model the bandgap of semiconducting nanotubes E_g is given by $E_g = 4\hbar v_F/3d_{CNT} = \gamma(2R_{C–C}/d_{CNT})$, where γ is the hopping matrix element (~3 eV), $R_{c–c}$ is the C–C bond length and d_{CNT} is the CNT diameter[8]. Further details on this subject can be found in recent review articles[9,10].

Whereas 2D graphene is a semi-metal, electrons (and holes) can be further confined by forming narrow ribbons, for example by quantizing k_x or k_y. This confinement should open a gap and make the GNR a finite gap semiconductor[4,11–13]. The confinement gap is expected to be inversely proportional to the width W of the GNR, given approximately by: $\Delta E_C \approx 2\pi\hbar v_F/3W$ (refs 13–17). We should note here the differences in confinement quantization in CNTs and GNRs. In a 1D quantum box of width W, quantization requires that $k_\perp = n\pi/W$, thus the allowed energy states are spaced as π/W. In a circular box of circumference C, on the other hand, the requirement is: $k_\perp = 2n\pi/C$ and the states are spaced apart as $2\pi/C$. Thus, qualitatively we expect a larger confinement gap in a CNT than in a GNR with the same confinement dimension. In addition to the difference in the gap value, the electronic states in GNRs are not degenerate, whereas those of CNTs are doubly degenerate. This is due to the difference in the boundary conditions: in a GNR the wavefunction has to vanish at the edges, whereas in a CNT the wavefunction is periodic in the circumference direction.

The above picture of the nanotube electronic structure is a single-electron model that accounts well for many of the CNT ground states properties. Interactions between electrons, however, can be important and can modify, among other properties, the predicted

semiconducting bandgaps and can affect the nature of the excited states of the CNTs[18].

NANOTUBE ELECTRICAL PROPERTIES

Individual nanotubes, like macroscopic structures, can be characterized by a set of electrical properties — resistance, capacitance and inductance — which arise from the intrinsic structure of the nanotube and its interaction with other objects. Electrical transport inside the CNTs is affected by scattering by defects and by lattice vibrations that lead to resistance, similar to that in bulk materials. However, the 1D nature of the CNT and their strong covalent bonding drastically affects these processes. Scattering by small angles is not allowed in a 1D material, only forward and backward motion of the carriers. Most importantly the 1D nature of the CNT leads to a new type of quantized resistance related to its contacts with three-dimensional (3D) macroscopic objects such as the metal electrodes[19,20]. The confinement of the electrons in the CNT around its circumference produces a small number of discrete states (modes) that overlap the continuous states of the metal electrodes. This mismatch of the number of states that can transport the current in the CNT and the electrodes leads to a quantized contact resistance, R_Q. The size of the resistance is determined by the number of modes, M, in the CNT that have energies lying between the Fermi levels of the electrodes: $R_Q = h/(2e^2M)$. For a metallic CNT, $M=2$ so that $R_Q = h/4e^2 = 6.45$ kΩ.

Of course, as well as this quantum resistance there are other forms of contact resistance such as that attributable to the presence of Schottky barriers at metal–semiconducting nanotube interfaces, of which we will speak later, and 'parasitic' resistance, which is simply due to bad contacts. When the only resistance present is the quantum resistance, transport in the CNT is ballistic — that is, no carrier scattering or energy dissipation takes place in the body of the CNT. This is obviously an important transport regime that is uniquely accessible to CNT conductors. The length over which a CNT can behave as a ballistic conductor depends on its structural perfection, temperature and the size of the driving electric field. In general, ballistic transport can be achieved over lengths typical of modern scaled electronic devices, that is ≤100 nm. At the other extreme, in long CNTs, or at high bias, many scattering collisions can take place and the so-called diffusive limit of transport that is typical of conventional conductors is reached. In this limit the carriers have a finite mobility. However, in CNTs this can be very high — as much as 1,000 times higher than in bulk silicon.

The intrinsic electronic structure of a CNT also leads to a capacitance that is related to its density-of-states — that is, how its energy states are distributed in energy — and it is independent of electrostatics. This quantum capacitance, C_Q, is small — of the order of 10^{-16} F μm^{-1} (ref. 21). In addition to C_Q, a CNT incorporated in a structure has an electrostatic capacitance, C_G, which arises from its coupling to surrounding conductors and as such depends on the device geometry and dielectric structure. In a typical metal oxide semiconductor FET (MOSFET) and in a planar graphene FET, $C_G \approx 1/t_{ins}$, where t_{ins} is the thickness of the gate insulator. In single-CNT FETs, geometry leads to a weaker variation, $C_G \approx 1/\ln(t_{ins})$. C_G and C_Q are coupled in series — that is, $1/C_{total} = 1/C_G + 1/C_Q$ — and therefore the smaller capacitance should dominate. In most experimental CNTFETs, C_G is smaller than C_Q, but in a highly miniaturized CNTFET with a high dielectric-constant insulator, $C_Q \lesssim C_G$ and therefore C_Q can dominate the total capacitance and determine the performance of the device.

Finally, CNTs have inductance, which is a resistance to any changes in the current flowing through them[22]. Again, there is a quantum and a classical contribution. The quantum inductance, usually referred to as kinetic inductance, L_K, is the resistance to the change of the kinetic energy of the electrons of the CNT. It leads to electron velocities that lag in phase with respect to the external driving field and is proportional to the density of states of the CNT. Classical self-inductance depends on the CNT diameter, geometry of the structure and the magnetic permeability of the medium. The total inductance is the sum of the two values, so that the larger inductance, L_K, dominates ($L_K \approx 16$ nH μm^{-1}, $L_C \approx 1$ nH μm^{-1}). In response to an a.c. signal, a CNT behaves like a transmission line owing to its inductance.

SCATTERING MECHANISMS AND TRANSPORT

As we already indicated above, CNTs are unique materials in terms of their long elastic mean free path, which is of the order of a micrometre. In metallic CNTs in particular, the symmetry of the band structure forbids backscattering as long as the carrier energies are below the second sub-band[23]. Long-range Coulomb scattering is ineffective, but a strong, short-ranged potential can lead to backscattering. In general, as the diameter of the CNT increases, the influence of a defect decreases because the influence of the CNT wavefunction is diluted[24]. Transport in graphene has recently been proposed as an example of the Klein paradox[25]: unimpeded penetration of relativistic particles through high and wide potential barriers[26]. Thus, unlike conventional 2D systems where strong disorder leads to Anderson localization, the transparency of the barriers (at least for some angles) in graphene could lead to efficient percolation of localized regions even for E_F close to the Dirac point.

Because elastic scattering in CNTs is weak, inelastic scattering processes determine their transport properties. These processes depend on the energy (applied bias) of the carriers. At low temperatures and low bias, only low-energy acoustic phonons can scatter the electrons, which results in an inverse temperature dependence of the carrier mobility in semiconducting CNTs[27,28], unlike in bulk materials, where acoustic phonon scattering typically leads to a $\sim 1/T^5$ temperature dependence of mobility[29]. The stronger

temperature dependence in three dimensions is due to phase-space restrictions and the occurrence of low-angle scattering[29]. In carbon nanotubes, acoustic phonon scattering is predominantly a backscattering event, and the phase-space is nearly temperature independent. Only a small fraction of phonons in the vicinity of the zone centre and zone boundary can effectively participate in the scattering. This is the result of the energy and momentum conservation law requirements and the mismatch in the phonon sound velocity and electron band velocity. The low-field mobility is very high in carbon nanotubes, even at room temperature[30]. This is unlike many other materials, such as III–V semiconductors, which have a very high mobility at low temperatures, but a substantially degraded one at room temperature. The main reason for the uniquely high mobility in carbon nanotubes is the very weak electron–acoustic-phonon coupling and the very large optical phonon energy of about 200 meV.

In addition to the low-energy acoustic phonons, electron (or hole) scattering by the radial breathing mode (RBM) is important in the low bias regime. The RBM phonon energy is inversely proportional to the tube diameter[8], and its energy is comparable to the thermal energy at room temperature for tubes in the diameter range of $d_{CNT} = 1.5$–2.0 nm, which are of interest for electronic applications. As the acoustic mean free path is very long — of the order of a micrometre at room temperature — electrons can be accelerated up to the RBM energy not only thermally, but also by an applied bias of a few V cm^{-1}.

The 1D nature of the electronic states of CNTs leads to van-Hove singularities in the density of states, which in turn are responsible for the non-monotonic dependence of the mobility on band filling[28,30,31]. As the band is filled — by the applied gate potential — the mobility initially increases owing to the loss of available final states for the scattered particles. Once the Fermi level reaches a higher-energy sub-band, an additional scattering channel opens and again lowers the mobility. The inverse diameter dependence of the effective mass and of the electron–phonon coupling strength leads to a quadratic diameter dependence of the mobility[27], which is confirmed experimentally[28].

Unlike acoustic phonon scattering, optical phonon scattering is very strong in carbon nanotubes; optical phonons contract and elongate the C–C bond length and lead to a strong modulation of the electronic structure. However, for electrons to emit an optical phonon, their energies must be larger than the optical phonon energy. This can only be achieved under high bias conditions. Such scattering processes were first observed in metallic tubes[32–34] and later in semiconducting tubes[35]. In metallic tubes, the current was found to saturate at about 25 μA owing to the short optical phonon mean-free-path, of the order of 10–20 nm, whereas in semiconducting CNTs, a velocity saturation was observed[35] in accord with earlier theoretical predictions[27,36].

As the carrier energy increases further, other inelastic processes can take place, in particular, impact excitation[37,38]. The analogous process in bulk semiconductors is impact ionization, where a high-energy electron can lose its energy by scattering to a lower energy state and impact exciting an electron–hole pair. In low-dimensional materials like carbon nanotubes, the electron–hole interaction is very strong, which leads to the formation of excitons with large binding energies (a few tenths of an eV)[39–41]. As a result, in CNTs, impact excitation primarily produces excitons[38]. The impact excitation process is governed by the Coulomb interaction, which is very strong in 1D systems, and indeed calculations suggest that impact excitation processes in CNTs are much more efficient (about four orders of magnitude stronger) than in conventional bulk semiconductors[38].

Unlike in 3D bulk materials, the longitudinal momentum along the tube axis of the produced excitons is nearly zero, whereas the angular momentum is finite. Furthermore, the difference between the impact excitation and the impact ionization rates, neglecting the strong electron–hole interaction in the produced electron–hole pair,

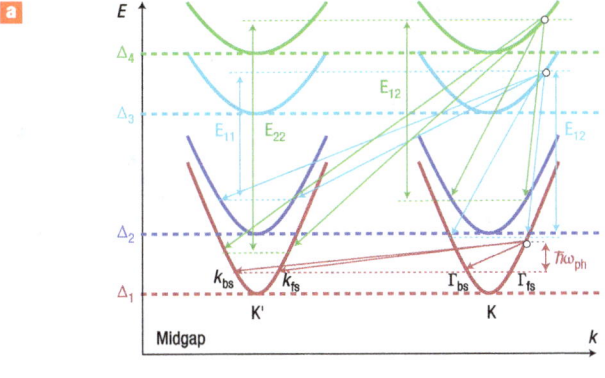

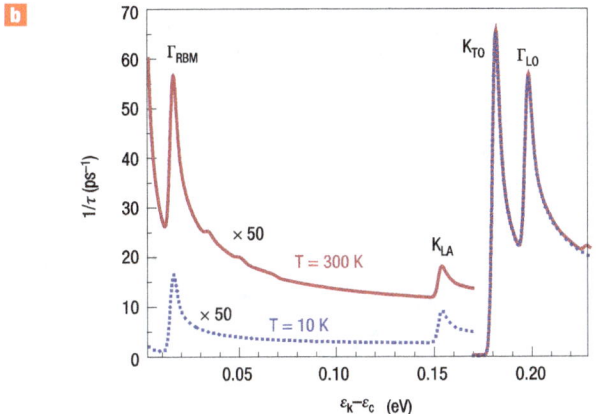

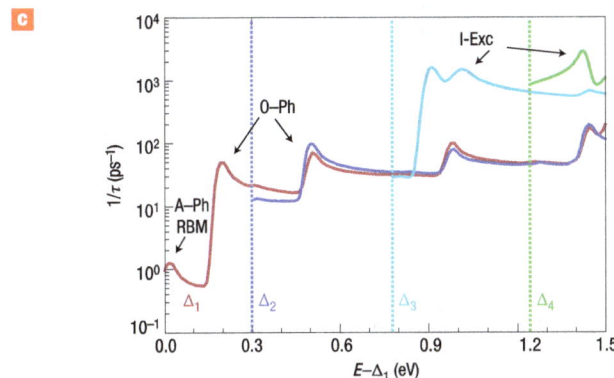

Figure 2 Inelastic scattering in carbon nanotubes. **a**, Schematic illustration of the intra-sub-band (Γ) and inter-sub-band (K) phonon scattering mechanisms (red) and electron impact excitation (blue and green curves) for the first four conduction bands. The different conduction band edges are labelled as Δ_i and the resulting electronic excitations are denoted as E_{ij}. Subscripts bs and fs stand for the back and forward scattering. **b**, Calculated phonon scattering rate for a (25,0) nanotube showing weak acoustic phonon scattering and strong optical phonon scattering. **c**, Calculated inelastic scattering rate for a (19,0) nanotube over a wide carrier energy range. Different colours correspond to the scattering rates of electrons in bands with different circumferential angular momentum. The vertical lines show the bottoms of the conduction bands 2 (blue), 3 (cyan) and 4 (green) with respect to the fundamental band edge Δ_1. Some of the characteristic peaks in the scattering, due to the longtitudinal (LA) acoustic phonons (A-Ph), radial breathing mode (RBM), longitudinal (LO) and transverse (TO) optical phonons (O-Ph) and impact electronic excitation (I-Exc), are labelled. In **b** and **c** the electron scattering rate is shown as a function of the excess energy of the electron above the first conduction-band minimum.

is much higher in carbon nanotubes than in other studied systems. The energy necessary for the electronic excitation is provided by the 'hot' carriers. Conservation of circumferential angular momentum $k_\perp$ plays a critical role in determining the threshold energy, E_{th}, for the onset of impact excitation in CNTs. The carriers are accelerated by the field, but they also lose energy to phonons, particularly to high-energy optical phonons and electronic transitions. This problem can be treated by solving the corresponding Boltzmann equation[38]. It is found that the exciton production rate P (per unit carrier) varies exponentially with the applied field F as $P \approx \exp(-E_{th}/eF\lambda_{op})$, where E_{th} is the excitation threshold and λ_{op} (~20–40 nm) is the electron mean free path due to optical phonon scattering. The energy of the optical phonons is usually efficiently dissipated into the heat bath provided by the substrate. However, in the case of suspended nanotubes, evidence for the existence of a non-equilibrium optical phonon excitation has been presented. This phonon excitation degrades the carrier mobility in CNTs[42,43]. The effect was included in the Boltzmann model by having the optical phonons at different temperatures, T_{op}, while keeping the other phonons at ambient temperature. The resulting exciton production rate could well be fitted by an exponential dependence with an effective temperature, T_{eff} (ref. 38).

In summary, the inelastic scattering rates determining transport properties of carbon nanotubes vary by four orders of magnitude depending on the energy of the electrons and their angular momentum (sub-band index) as shown in Fig. 2. The weakest is the acoustic (primarily RBM) phonon scattering, which has linear temperature dependence. The optical phonon scattering rate, which is two orders of magnitude stronger, is nearly temperature independent. Finally, another two orders of magnitude stronger than the optical phonon scattering is impact excitation. It can only be observed at very large local fields of above ~3×10^4 V cm^{-1}, at which higher energy CNT sub-bands can be populated with the charge carriers. At higher fields CNT dielectric breakdown takes place[44,45].

ELECTRICAL SWITCHING OF CARBON NANOTUBES

Whereas metallic, particularly multiwalled, CNTs offer the possibility of use as high-performance interconnects in very-large-scale integrated systems[46], transport in semiconducting CNTs can be switched 'on' and 'off', which means that semiconducting CNTs can be used as a basis of novel transistors.

The flow of electricity in a semiconductor requires some kind of activation (for example, heating or light absorption) to get the charge carriers over the gap, or modulation of the gap by some external influence such as an applied strain, electric field or magnetic field. The most common mode of switching, which forms the basis of modern microelectronic industry, is by means of an external electric field, in the so-called field-effect transistor. The idea of the FET is quite old, but it was not until 1960 when John Atalla of Bell Laboratories demonstrated the modern form of the FET, the MOSFET. The basic structure involves a channel made of a semiconductor, typically silicon connected to two electrodes: a source (S) and a drain (D). An insulating thin film, usually SiO$_2$, separates the channel and the source and drain from a third electrode called the gate (G). By applying a voltage at this gate electrode with respect to the source (V_g), we can modulate (switch) the conductance of the semiconducting channel (Fig 3a). The carriers (electrons or holes) travelling from the source to the drain encounter a material- and structure-dependent energy barrier in the bulk of the semiconductor. This barrier is in the conduction band for an electron (n-type semiconductor), or in the valence band for a hole (p-type). The electric field generated by the biased gate, depending on its direction, lowers or raises the barrier and thus changes the conductivity. For example, for electron conduction, application of a positive V_g lowers the barrier, whereas a negative V_g raises it.

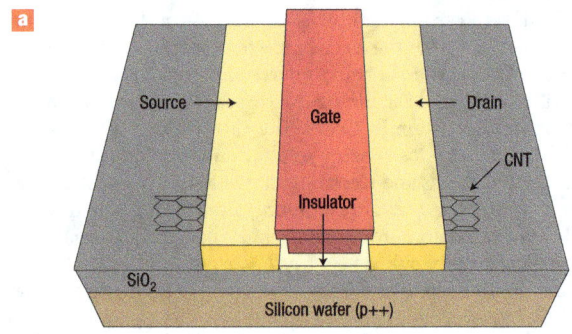

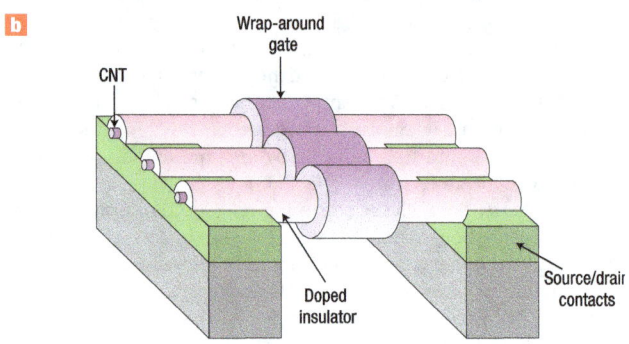

Figure 3 Designs of carbon nanotube field-effect transistors. **a**, Schematic of a top-gated carbon nanotube field-effect transistor. **b**, Schematic of an array of nanotube transistors with wrap-around gates and doped gate extensions (courtesy of P. M. Solomon).

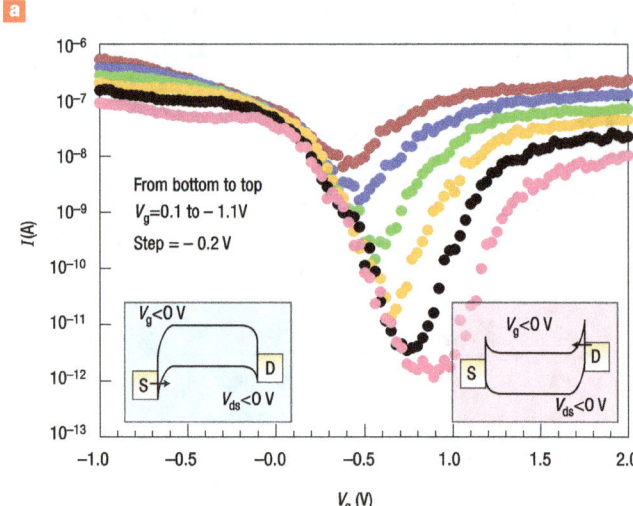

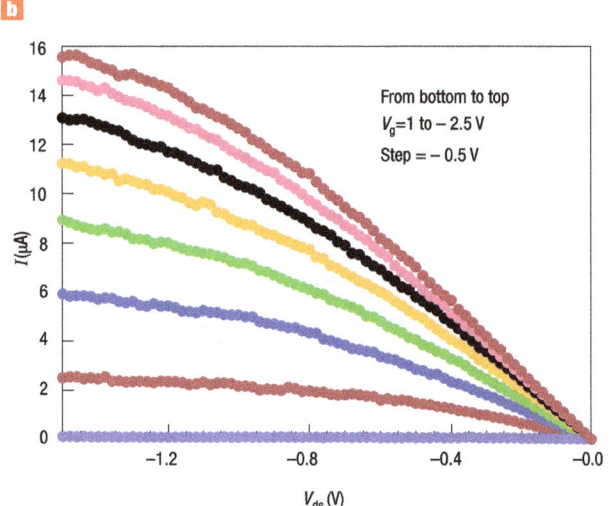

Figure 4 Performance characteristics for a single nanotube transistor. **a**, Ambipolar transfer characteristics (current versus gate voltage): drain bias increases from −0.1 V to −1.1 V in −0.2 V steps. Left inset: schematic of the band structure of a Schottky barrier semiconducting CNT in a FET under negative gate bias. Holes are injected from the source (S). Right inset: under positive gate bias electrons are injected from the drain (D). **b**, Output characteristics (current versus drain bias): gate voltage changes from +1 V (off state) to −2.5 V (on state) in −0.5 V steps.

The first CNTFETs were reported in 1998[47,48]. In a CNTFET the role of the channel is played by one (or more) semiconducting nanotubes. In Fig. 3 we show schematically two different CNTFET structures: a top-gated CNTFET and an array of CNTFETs with 'wrap-around' gates. Typical current–voltage characteristics of a CNTFET are shown in Fig. 4. The atomic and electronic structures of the nanotube give it a number of unique advantages as a FET channel. First, its small diameter (1–2 nm) allows optimum coupling between the gate and the channel, that being the ability of the gate to control the potential of the channel. This is particularly true for the wrap-around gate configuration shown in Fig. 3b. This strong coupling makes the CNT the ultimate 'thin-body' semiconductor system and allows the devices to be made shorter while avoiding the dreaded 'short-channel effects'[49], which basically involve the loss of control of the device by the gate field. The fact that all bonds in the CNT are satisfied and the surface is smooth also has important implications. Scattering by surface states and roughness, which plagues conventional FETs, especially at high V_g values, is absent. Of course, the key advantage is the low scattering in the CNT and the high mobility of the FET channel.

CNTFETs have a number of other differences with conventional MOSFETs. The profile of the energy bands of the CNTFET at a given drain bias, V_{ds}, is determined by V_g and the capacitance of the FET. As we discussed above there are two contributions to the capacitance: C_G and C_Q coupled in series. Now in conventional devices, $C_G < C_Q$ and C_G is the controlling term, but in very small (scaled down) CNTFETs, $C_G \approx C_Q$ or even $C_G > C_Q$ and the quantum capacitance can be dominant. The quantum capacitance has an important consequence on how the gate voltage affects the band profile (energy barrier) of the CNT. In a conventional MOSFET the ability of the gate to control the potential in the channel is limited once V_g exceeds the threshold voltage, V_{th}

(ref. 49). Increasing the gate voltage above this value does not change the bands any more because the increased charge pulled into the channel by an increase in V_g pushes the bands back up by electrostatic repulsion. In a C_Q-controlled device, however, the gate can still retain control of the potential in the CNT channel. Higher CNT conduction bands can now be pulled by the gate below the Fermi level of the S, thus contributing to the current[50].

The above discussion on CNTFETs assumed the existence of only one, bulk barrier in the motion of the carriers as in MOSFETs. In MOSFETs, in addition to the channel, the source, drain and gate are also made of heavily doped Si and the contacts are ohmic. This is not generally true for the CNT–metal contacts used in CNT electronics[51–54]. Metals like Au, Ti, Pd and Al are used for the source and drain electrodes. The different work functions of the metal and the CNT lead to transfer of charge at their interface. The resulting interface dipole produces an energy barrier, the so-called Schottky barrier. The alignment of the Fermi levels of the metal and CNT, and therefore the Schottky barrier height, depend on their respective work functions (Φ), the CNT bandgap and the details of chemical bonding at the interface.

There are two Schottky barriers in a FET; one at the source and another at the drain as shown schematically in the insets of Fig. 4a. As long as one of the barriers is much higher than the other, the FET operates as a unipolar device; that is, it transports one type of carrier: electrons, or holes. For example, a high Φ metal such as Pd could be used to form a nearly barrierless contact for holes[55] (valence band close to the metal Fermi level, E_F), that is to optimize a p-type CNTFET operation, but electron injection at the other end would then experience the maximum barrier of E_g. Correspondingly, a low work-function metal, for example, Al, will optimize electron transport, but will inhibit hole transport. However, typically, carrier transport through the metal-CNT interface is dominated by quantum mechanical tunnelling through the Schottky barrier rather than thermally-activated — thermionic — emission over the Schottky barrier[53,56,57] and, therefore, the thickness of the Schottky barrier becomes critical. In 1D systems like CNTs, the intrinsic screening is weak and the distribution of the potential is determined by the screening provided by the nearby metallic gate[51], and also decreases with increasing d_{CNT}. Therefore, we should expect that, in a thin gate oxide CNTFET, both Schottky barriers can become thin enough to allow, depending on the bias, the injection of either electrons or holes, or of both carriers simultaneously. Such a type of transistor is called ambipolar[52,53,58]. The insets of Fig. 4a show how the application of different gate voltages (V_g) brings about the injection of electrons or holes. We can now understand the $I-V_g$ characteristics of the ambipolar CNTFET shown in Fig. 4a. For a strongly negative or positive V_g, holes or electrons are injected, respectively. However, for V_g values in the vicinity of the current minimum, both electrons and holes contribute to the current, and at $V_g = V_{ds}/2$, the two currents are of equal size[53,58].

Schottky barriers in a CNTFET impact the characteristics of both the 'on' and 'off' states[59]. An important parameter in this regime is the inverse sub-threshold slope, S, that measures the efficiency with which the gate switches the channel, and is defined as $S = (d\log_{10}I/dV_g)^{-1}$. In a transistor with ohmic source and drain contacts (as in a conventional Si MOSFET), S is limited by thermionic emission over the channel and is $\approx k_BT/q$ — that is, ~60 meV per decade at 300 K. However, in a transistor with Schottky barriers dominating its transport, S is significantly higher, about 100–150 mV per decade for oxide thickness $t(SiO_2) \lesssim 10$ nm. It is not only the 'on' current of the transistor that is important. The current in the 'off' state is equally important. Maintaining a low leakage current to keep both the passive power at a minimum and a reasonable I_{on}/I_{off} ratio ($\geq 10^4$ is typically desired in logic applications) is critical. Unipolar CNTFETs have I_{on}/I_{off} ratios in the range of 10^5–10^7.

The fact that Schottky barriers exist at CNT–metal interfaces does not imply that we have to be content with ambipolar behaviour and its implications. There are ways to further screen the Schottky barrier and get closer to a MOSFET-like bulk-switched device. Two different approaches have already been demonstrated. In one case, a double-gated CNT was used where the gates near the contacts selectively thinned the Schottky barriers, while a central gate was independently used to switch the bulk of the CNTFET (ref. 60). In this way, ambipolar behaviour and excess leakage current was eliminated and a sub-threshold slope of 60 mV per decade was achieved. As well as this electrostatic approach, chemical approaches have also been successful[61–63]. Substitutional doping of single-walled nanotubes using the traditional type of dopants such as B or P atoms has not been successful so far. This could be due to the large strain these impurities

introduce to the CNT lattice. A different type of doping is produced by atoms or molecules that chemically adsorb on the CNT through a charge-transfer mechanism[61,62]. By selectively doping the contact regions of the CNTFET and using a central gate for switching, a bulk-like behaviour can be achieved. A further improvement of the sub-threshold slope of CNTFETs to values below the thermal limit was achieved, taking advantage of band-to-band tunnelling as a way of filtering the energy distribution of the electrons in CNTs. In this way S values of 40 mV per decade were obtained at room temperature[64].

Through the years a number of different CNTFET designs have been implemented, attempting to optimize their performance. For example, high-k materials have been used as gate insulators[65], wrap-around gates have been used to increase the coupling of the gate and the CNT channel, self-aligned designs and contact extensions to improve the device capacitance[61,66], and channel lengths have been scaled down to 18 nm (ref. 67).

In general, the key advantage of CNTFETs over Si MOSFETs in logic applications is their much lower capacitance of ~10 aF (for a $d_{CNT} = 1$ nm, $L = 10$ nm, $t_{ox} = 5$ nm device) and their somewhat lower operating voltage. Furthermore, the small size of the CNTs allows the fabrication of aligned arrays with high packing density. However, careful design and engineering are needed to make sure that the wire interconnect capacitance, typically ~0.2 aF nm^{-1}, does not overwhelm the capacitance of the CNT. It takes only about 50 nm of wiring for the interconnect to limit the capacitance of the 10 aF device. An area of clear advantage of CNTFETs is their lower switching energy per logic transition. The dynamic switching energy of a device is given by: $1/2(C_{dev}+C_{wire})V^2$, where C_{dev} and C_{wire} are the device and wiring capacitance contributions, respectively. To minimize the switching energy, minimum-sized devices, and interconnects should be used as well as the minimum supply voltage. Then the CNT can have a considerable advantage, of up to a factor of six, because of its smaller intrinsic capacitance and size[68].

ALTERNATING CURRENT PERFORMANCE OF CNTFETS

Although most of the work on CNTFETs has concentrated so far on their d.c. properties, the a.c. properties are technologically most relevant. Theoretically, it is predicted that a short nanotube operating in the ballistic regime, and the quantum capacitance limit should be able to provide gain in the THz range[69]. However, directly measuring the a.c. performance of a nanostructure such as a single nanotube is very difficult as the input impedance of a CNTFET is much higher than 50 Ω and the capacitance is typically in the aF range. A different approach is to operate the CNTFET as a mixer or rectifier[70,71–73]. Because of the nonlinear $I-V$ characteristics of the transistor, an a.c. signal applied to the source becomes rectified and leads to a measurable d.c. current change. Using this approach, it was recently demonstrated that it is possible to obtain mixing up to a frequency of 50 GHz (ref. 72). As the authors pointed out, this is by no means the limit of performance of CNTFETs. The observed behaviour was determined by the parasitic capacitances of their system. This again underscores the need for a global optimization of a circuit before the unique properties of the nanotube can be effectively utilized.

NANOTUBE INTEGRATED CIRCUITS

The fabrication and evaluation of CNT-based devices has advanced beyond single devices to include logic gates[65,74–77] and, more recently more complex structures such as ring oscillators have been fabricated[78]. To build these circuits the energy efficient CMOS (complementary-MOS) architecture is preferred. This involves pairs of n- and p-type transistors. CNTs, where the valence and conduction bands are mirror images of each other (equal effective masses for

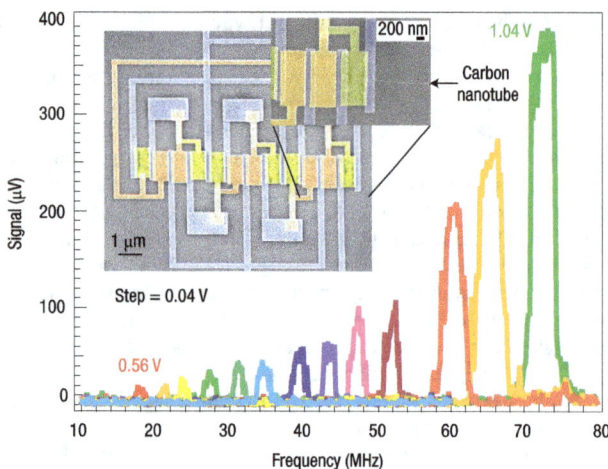

Figure 5 Ring-oscillator circuit based on a single nanotube. Scanning electron microscope image of the single nanotube, 5-stage ring-oscillator. The graph shows how the ringing frequency changes as the supply voltage is increased from 0.56 V to 1.04 V in 0.04 V steps.

electrons and holes) are ideally suited for such applications. The two types (p- and n-) of FETs can be made by doping the CNTs as discussed above. However, controlling doping in nanoscale devices is difficult. Fluctuations in the number and position of the dopants can have a profound effect on device performance. It has been shown recently, however, that the ambipolar behaviour of an undoped CNT can be successfully utilized to implement the CMOS architecture. A CMOS inverter is formed with a pair of p- and n- transistors operating under conditions where one FET is fully turned on while the other one is off. The ambipolar characteristics of a CNFET provide a pair of p- and n-type branches; however, the leakage branch of one type always has a comparable current level to the main branch (on-current) of the other. Therefore, obtaining control of the threshold voltage in an ambipolar device is not only important to match the desired supply voltage window, but is also critical to achieve a pair of p- and n-type characteristics which are suitable for CMOS logic.

For a given undoped CNFET with a fixed energy gap and oxide thickness, tuning the work function of the gate metal is the only way to control the threshold voltage. The gate work function acts like an extra voltage source in addition to the applied gate voltage. When the work functions are properly selected, the two characteristics can be relatively shifted towards each other and this leads to a distinguishable on-state in one and an off-state in the other. Fig. 5 shows scanning electron microscope image of the most complex structure so far based on a single carbon nanotube. It is a ring oscillator using Pd gates for p-FETs and Al gates for n-FETs. The frequency response from the ring oscillator shows a strong dependence on the supply voltage. A 72 MHz frequency is measured for a supply voltage of about 1 V. The small measured signal is due to the mismatch between the high output impedance of the nanotube circuit and the low input impedance of the measurement set-up. This impedance mismatch is a general problem of nanocircuits that needs to be addressed.

SWITCHING OF GRAPHENE NANORIBBONS

Having zero gap, graphene cannot be used directly in applications such as FETs for logic applications as the transistor cannot be turned completely off. Although the density of states at the Dirac point is

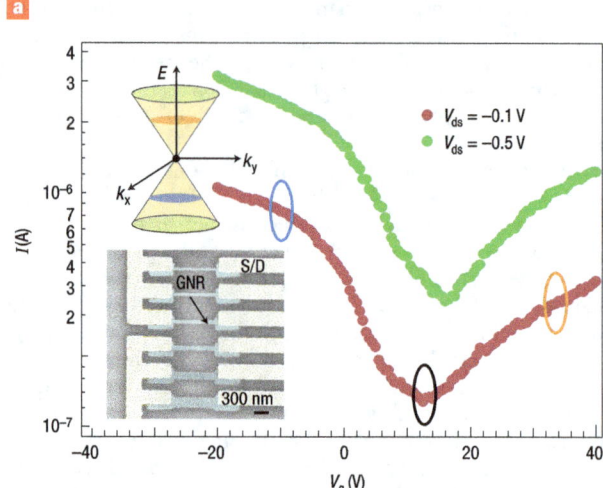

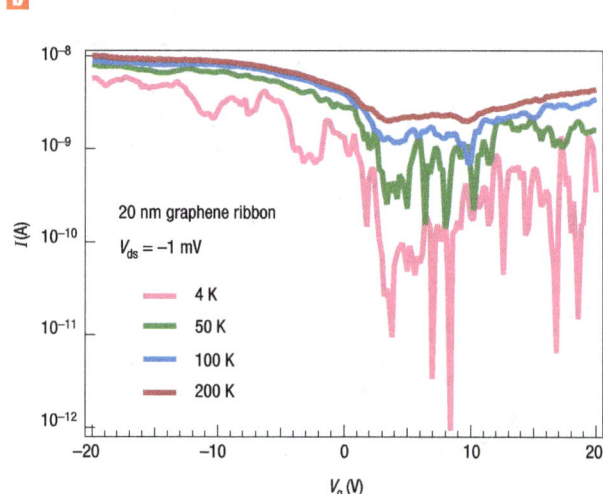

Figure 6 Graphene nanoribbon transistors. **a**, Current versus gate voltage characteristics of a 20 nm width graphene nanoribbon FET at room temperature. $V_{ds} = -0.1$ V (red) and -0.5 V (green). The blue, black and orange states represent, in the current versus gate voltage curves, the hole branch, minimum current, and electron branch, respectively. The insets show linear energy dispersion around the Dirac point, and the scanning electron microscope image of wired graphene nanoribbons of variable width produced by electron-beam lithography and etching. The widths of the ribbons are 20 nm (top), 30 nm, 40 nm, 50 nm, 100 nm and 200 nm (bottom). **b**, Temperature dependence of the current versus gate voltage characteristics of the 20 nm graphene nano-ribbon FET. $V_{ds} = -1$ mV. $T = 4$ K, 50 K, 100 K and 200 K.

zero, graphene has a minimum conductivity when the Fermi level is aligned with it that is found experimentally to be of the order of e^2/h (ref. 5,79–83). However, as we already discussed earlier, in addition to the 2D confinement in the plane of graphene, the graphene electrons (or holes) can be further confined by forming narrow ribbons and thus open a gap[4,11–13].

Very recently, electron-beam lithography and etching techniques have been used to introduce confinement to the 2D graphene[11,12]. The dependence of the transport properties on the ribbon width has been studied, and the energy gap that was opened up was indeed found to be inversely proportional to the ribbon width[12]. The smallest ribbon

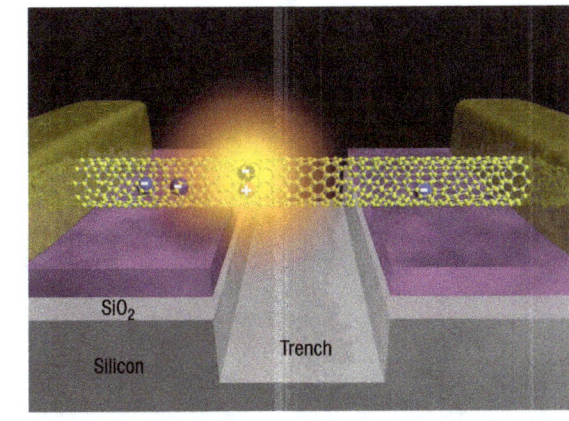

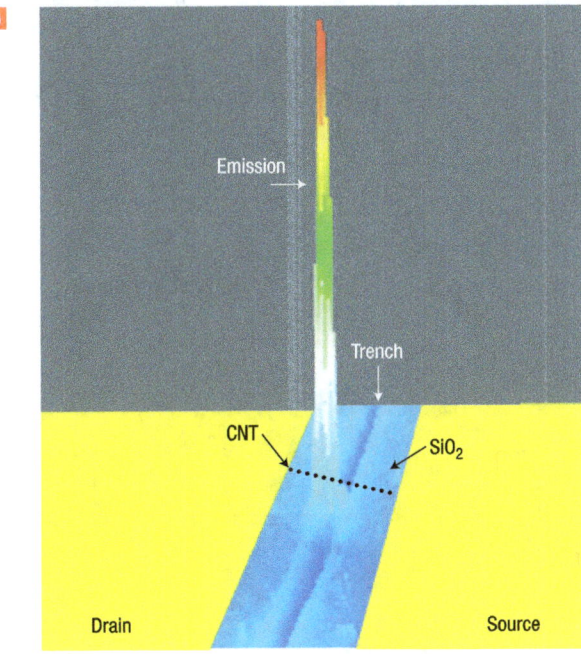

Figure 7 Light emission from a nanotube. **a,** Schematic of a modified ('trenched'), back-gated nanotube transistor used to produce a sudden change in the potential along the nanotube. **b,** Optical image of the trench and of the light emitted at its edge.

or zigzag, but a mixture of such edges may co-exist along the length of the graphene ribbon.

The possible existence of zigzag edges in graphene ribbons leads to localized edge states at the Fermi level[14,84]. Further complications could arise from the fact that cutting a ribbon leaves behind carbon dangling bonds that need to be saturated. Depending on the nature of the saturating group, the bonding and the potential at the edges are different from those in the bulk of the GNR. The combination of these factors could lead to a degradation of the mobility of the carriers by edge scattering, while the role of contact barriers will increase as scaling of the dimensions of the GNR proceeds. A general resistivity increase trend has been experimentally observed when the ribbon gets smaller than about 50 nm (ref. 11). More thorough studies are needed to find out the key contributors to this effect and possible ways to preserve the high mobility nature of the 2D graphene.

As in the case of CNTs, further confinement from two dimensions to quasi-zero-dimensions has recently been achieved on graphene[13]. A combined Coulomb blockade and confinement gap in a 10 nm island reaches $\sim 10k_BT$ at room temperature. This large gap makes it possible to operate it as a single electron transistor at room temperature. However, the challenge still remains of how to gain precise control of the graphene size and shape to obtain reproducible characteristics among different devices.

NANOTUBE OPTOELECTRONIC DEVICES

Electrically or optically generated electron and hole carriers in semiconductors can recombine by a variety of different mechanisms (direct or indirect recombination). In most cases, the recombination energy will be released as heat (phonons), but a fraction of the recombination events may involve the emission of a photon. This light-emission process is called 'electroluminescence' and is extensively used to produce solid-state light sources such as light emitting diodes (LEDs).

In order to fabricate LEDs, or other electroluminescent devices, one must generate and bring together significant populations of electrons and holes. Conventionally, this is achieved at the interface between a hole-doped and an electron-doped semiconductor (a p–n junction). As we have seen above, however, in ambipolar CNTFETs, by applying the appropriate biases, both electrons and holes can be simultaneously injected from the source and drain of the CNTFET. The quasi-1D character of the CNT confines the two types of carriers, which are driven towards each other. Indeed, radiative recombination in an ambipolar CNTFET was reported in ref. 85. Although the mechanism of that emission is similar to that of an LED, there is an important difference: the CNT is not doped so there is no clear p–n junction. As a result, the light does not originate from a fixed point along the CNT, but its origin can be translated by simply changing V_g, which determines the local potential in a long CNT device[85,86]. The overall emission intensity is maximized when the hole and electron currents become equal, which as we already indicated, occurs when $V_g = V_{ds}/2$. The emitted light is, as expected, polarized along the tube axis and the spectrum is, within the resolution of the measurements, the same as the photoluminescence spectrum of nanotubes with the same diameter.

In addition to this 'mobile' emission, localized electroluminescence is also observed from particular spots on a SWNT even under unipolar transport conditions. As light generation by radiative recombination requires that both types of carriers are present, we must conclude that at these spots electron–hole pairs are actively generated. Our studies showed that light-generating spots include a variety of inhomogeneities such as trapped charges in the gate insulator of the FET and interfaces between materials with different dielectric constants. In general these inhomogeneities produce voltage drops (electrical resistance) along the CNT and

reproducibly defined by lithography is about 20 nm, shown in Fig. 6a, and induces the opening of a confinement gap of about 30 meV (ref. 11). At room temperature, this is too small a gap to have an observable effect on modulation of the current by the gate voltage, as shown in Fig. 6b. The GNR channel exhibits resistivity modulation by the gate voltage, which causes a significant Fermi-level shift (see inset of Fig. 6a). However, when the temperature is sufficiently lowered, the confinement gap starts to impede carrier injection and the device shows a clear on/off ratio as a normal semiconductor (Fig. 6b). Strong current fluctuations are observed at low temperatures for narrow ribbons, which is an indication of electron–electron interactions. Interestingly, no dependence on cutting angle of the graphene ribbon was observed experimentally[12]. This is likely the result of the present limitations of lithography rather than an intrinsic feature of graphene ribbon. State-of-the-art lithographic technology does not allow the formation of a uniquely defined ribbon edge, for example, armchair generate large, local electric fields[87]. These fields can accelerate the carriers to an energy that allows them to generate electron–hole pairs through an intra-nanotube impact-excitation process. This excitation mechanism offers a number of possibilities for new applications. For example, more intense and brighter light sources can be produced because unipolar currents in CNTs can be higher than ambipolar currents, carrier multiplication can take place in the high fields and the emission is more localized. Furthermore, the excitation is not limited by dipole selection rules as photoexcitation is, and a number of different states can be populated.

This approach to generating high densities of excitations in low-dimensional systems can be used to resolve problems of current interest such as their mutual interactions and boson condensation. Recent work illustrates one way the impact excitation may be implemented in an LED[37]. A CNTFET was modified as shown in Fig. 7 to induce a sharp discontinuity in the potential; that is, to produce a waterfall-like energy landscape along the CNT channel. Electrons that are accelerated at the discontinuity can acquire enough energy to exceed the impact-excitation threshold and generate a bright light emitter with a yield that is about 1,000 times higher than that produced by recombination in an ambipolar CNTFET (Fig. 7)[37]. Another use of the localized electroluminescence can be as a much-needed analytical tool for identifying defects in either the CNT itself or the gate insulator. Parallel measurements of the mobile and localized electroluminescence of the CNTFETs and their I–V characteristics can provide unique information regarding the effect of specific defects and inhomogeneities on the electrical transport in CNTs.

Photoconductivity is the reverse process of electroluminescence, with optical radiation producing electron–hole pairs that are separated by the applied field. Single CNT photoconductivity was reported for the first time in 2003 (ref. 88). An example of such a measurement is shown in Fig. 8 (ref. 89). The resonant excitation of a CNT generates an electric current and can be used as a nanosized photodetector, a photo switch, or as a spectroscopic tool. Alternatively, in the open-circuit configuration the device generates a photovoltage. Internal fields such as those formed at Schottky contacts, or defect sites can also separate photogenerated electron–hole pairs and can be used to image such sites and determine the band bending in an open-circuit configuration[90–92]. Thus, a CNTFET device can be used as a transistor, a light emitter or a light detector[18]. Choosing between these different modes of operation only requires changes in the electrical inputs.

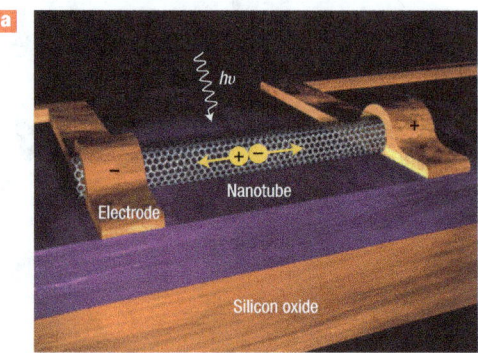

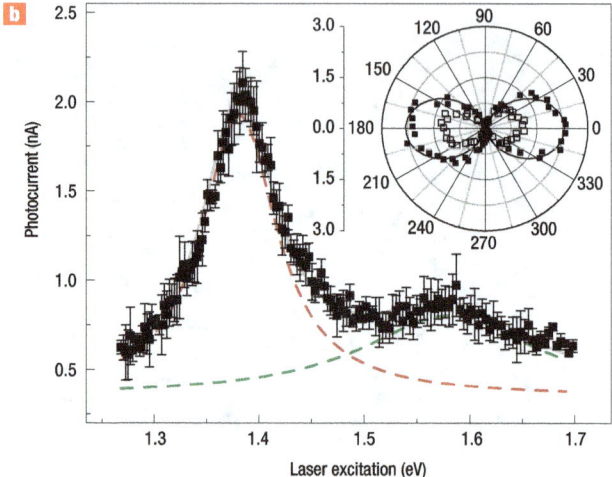

Figure 8 Photoconductivity with a nanotube. **a**, Schematic of a photoconductivity experiment. **b**, Induced photocurrent in a CNTFET as a function of the energy of the incident light. The strong peak corresponds to the second allowed exciton transition of the nanotube (E_{22} excitation); the weaker peak corresponds to the simultaneous excitation of one quantum of the G-phonon mode of the nanotube. The inset shows that both transitions are polarized along the nanotube axis.

THE FUTURE

Carbon nanotubes have provided us with an ideal model system to study electrical and optical phenomena on the nanometre scale. 1D materials with their exotic properties, long the realm of theoretical studies, are now open to experimentation. New states of condensed matter such as the Luttinger-Tomonaga liquid can be studied. Graphene is a novel, covalent 2D system that has already been found to exhibit a number of unique phenomena such as anomalous quantum Hall effect, Klein paradox and so forth. There is no doubt that in the future we will continue to obtain new information on the physics of the nanoscale through the study of nanotubes and graphene. Nanotube and graphene research is also teaching us how to handle and process nanomaterials and develop nanotechnology in general. In terms of direct technological applications, we focused here, because of limited space, only on electronic switching and light emission/detection. Nanotubes offer the potential of very fast (THz) transistors, ultimately scaled logic devices, and simpler and cheaper self-assembly based fabrication. In addition, transistors with properly functionalized CNTs can and are already being used as sensitive and selective chemical and biosensors. CNT-based nano-light sources and detectors may allow intra-chip optical communications and individual molecule level spectroscopy. The excellent electrical conduction of

metallic CNTs may eventually allow the development of electronic systems where both active devices and interconnects are based on the same material — CNTs. Further integration to include optics could lead to a unified electronic — optoelectronic technology.

Graphene nanoribbons have field-switching capabilities, unlike graphene, in which the field switching is limited owing to its semi-metallic nature. The limited I_{on}/I_{off} of graphene field-effect devices may inhibit their application in computer logic, but may be appropriate for RF applications. Evidence, however, has been presented recently that bilayer graphene can provide a sizable and field-tuneable bandgap and as such may provide the basis of novel field-effect electronic and optoelectronic devices[93,94]. A gate-controlled p–n junction has also been demonstrated in graphene, and used to further explore the quantum Hall effect[95,96]. The large electronic and spin coherence lengths of both materials could also lead to quantum interference and spintronic devices. Spin-polarized injection can be achieved using manganite electrodes. The weak spin-orbit coupling and short transit times of carriers in nanotubes tend to preserve the spin polarization and can lead to high $\Delta V/V$ ratios, where V is the bias voltage and ΔV its variation when the magnetic configuration is changed[97]. Spin transport and spin precession over micrometre-scale distances have also been demonstrated in single graphene sheets at room temperature[98].

How soon do we expect to see these developments? What is the bottleneck in the development of nanotubes, graphene and indeed in any high-end nanotechnology? The main hurdle is our current inability to produce large amounts of identical nanostructures. Nanotubes come in many sizes and structures and the same is true of many other nanostructures. For example, there is no reliable way to directly produce a single CNT type such as will be needed in a large integrated system, and patterning graphene is limited by the resolution of current lithographic techniques. However, promising signs that these problems will be circumvented soon are appearing. Already, separation of single-type nanotubes from mixtures has been achieved, and direct growth of graphene is possible. Similarly encouraging results are being reported in the area of self-assembly of CNT devices.

doi:10.1038/nnano.2007.300

Published online: 30 September 2007.

References

1. Avouris, P. Electronics with carbon nanotubes. *Phys. World* **20**, 40–45 (March 2007).
2. Wallace, P. R. The band theory of graphite. *Phys. Rev.* **71**, 622–634 (1947).
3. Novoselov, K. S. et al. Electric field effect in atomically thin carbon films. *Science* **306**, 666–669 (2004).
4. Berger, C. et al. Electronic confinement and coherence in patterned epitaxial graphene. *Science* **312**, 1191–1196 (2006).
5. Novoselov, K. S. et al. Two-dimensional gas of massless Dirac fermions in graphene. *Nature* **438**, 197–200 (2005).
6. Zhang, Y., Tan, Y.-W., Stormer, H. L. & Kim, P. Experimental observation of the quantum Hall effect and Berry's phase in graphene. *Nature* **438**, 201–204 (2005).
7. Novoselov, K. S. et al. Room-temperature quantum Hall effect in graphene. *Science* **315**, 1379 (2007).
8. Saito, R., Dresselhaus, G., & Dresselhaus, M. S. (eds.) Physical properties of carbon nanotubes. (Imperial College Press, London, 1998)
9. Anantram, M. P. & Leonard, F. Physics of carbon nanotube electronic devices. *Rep. Prog. Phys.* **69**, 507–561 (2006).
10. Charlier, J.-C., Blase, X., & Roche, S. Electronic and transport properties of nanotubes. *Rev. Mod. Phys.* **79**, 677–733 (2007).
11. Chen, Z., Lin, Y.-M., Rooks, M. J., & Avouris, P. Graphene nano-ribbon electronics. Preprint at <http://arxiv.org/abs/cond-mat/0701599> (2007).
12. Han, M. Y., Özyilmaz, B., Zhang, Y., & Kim, P. Energy band gap engineering of graphene nanoribbons. *Phys. Rev. Lett.* **98**, 206805 (2007).
13. Geim, A. K. & Novoselov, K. S. The rise of graphene. *Nature Mater.* **6**, 183–191 (2007).
14. Nakada, K., Fujita, M., Dresselhaus, G., & Dresselhaus, M. S. Edge state in graphene ribbons: Nanometer size effect and edge shape dependence. *Phys. Rev. B* **54**, 17954–17961 (1996).
15. Son, Y.-W., Cohen, M. L., & Louie, S. G. Energy gaps in graphene nanoribbons. *Phys. Rev. Lett.* **97**, 216803 (2006).
16. Barone, V., Hod, O. & Scuseria, G. E. Electronic structure and stability of semiconducting graphene nanoribbons. *Nano Lett.* **6**, 2748–2754 (2006).
17. White, C. T., Li, J., Gunlycke, D. & Mintmire, J. W. Hidden one-electron interactions in carbon nanotubes revealed in graphene nanostrips. *Nano Lett.* **7**, 825–830 (2007).
18. Avouris, P. et al. Carbon nanotube optoelectronics. *Physica Status Solidi B* **243**, 3197–3203 (2006).
19. Landauer, R. Spatial variation of currents and fields due to localized scatterers in metallic conduction. *IBM J. Res. Dev.* **32**, 306 (1988).
20. Büttiker, M. Symmetry of electrical conduction. *IBM J. Res. Dev.* **32**, 317 (1988).
21. Ilani, S., Donev, L. A. K., Kindermann, M. & McEuen, P. L. Measurement of the quantum capacitance of interacting electrons in carbon nanotubes. *Nature Phys.* **2**, 687–691 (2006).
22. Burke, P. An RF circuit model for carbon nanotubes. *IEEE Trans. Nanotechnol.* **2**, 55–58 (2003).
23. Ando, T. & Nakanishi, T. Impurity scattering in carbon nanotubes — absence of backscattering. *J. Phys. Soc. Jpn.* **67**, 1104–1113 (1998).
24. White, C. T. & Todorov, T. N. Carbon nanotubes as long ballistic conductors. *Nature* **393**, 240–242 (1998).
25. Klein, O. An introduction to Kaluza–Klein theories. *Z. Phys.* **37**, 895 (1926).
26. Katsnelson, M. I., Novoselov, K. S. & Geim, A. K. Unconventional quantum Hall effect and Berry's phase of 2π in bilayer graphene. *Nature Phys.* **2**, 177–180 (2006).
27. Perebeinos, V., Tersoff, J. & Avouris, P. Electron-phonon interaction and transport in semiconducting carbon nanotubes. *Phys. Rev. Lett.* **94**, 086802 (2005).
28. Zhou, X., Park, J-Y, Huang, S., Liu, J. & McEuen, P. L. Band structure, phonon scattering, and the performance limit of single-walled carbon nanotube transistors. *Phys. Rev. Lett.* **95**, 146805 (2005).
29. Ashcroft, N. W. & Mermin, N. D. *Solid State Physics* (Saunders, 1976).
30. Durkop, T., Getty, S. A., Cobas, E., & Fuhrer, M. S. Extraordinary mobility in semiconducting carbon nanotubes. *Nano Lett.* **4**, 35–39 (2004).
31. Perebeinos, V., Tersoff, J. & Avouris, P. Mobility in semiconducting carbon nanotubes at finite carrier density. *Nano Lett.* **6**, 205–208 (2006).
32. Yao, Z., Kane, C. L. & Dekker, C. High-field electrical transport in single-wall carbon nanotubes. *Phys. Rev. Lett.* **84**, 2941–2944 (2000).

33. Javey, A. et al. High-field quasiballistic transport in short carbon nanotubes. *Phys. Rev. Lett.* **92**, 106804 (2004).
34. Park, J. Y. et al. Electron-phonon scattering in metallic single-walled carbon nanotubes. *Nano Lett.* **4**, 517–520 (2004).
35. Chen, Y.-F. & Fuhrer, M. S. Electric-field-dependent charge-carrier velocity in semiconducting carbon nanotubes. *Phys. Rev. Lett.* **95**, 236803 (2005).
36. Pennington, G. & Goldsman, N. Semiclassical transport and phonon scattering of electrons in semiconducting carbon nanotubes. *Phys. Rev. B* **68**, 045426 (2003).
37. Chen, J. et al. Bright infrared emission from electrically induced excitons in carbon nanotubes. *Science* **310**, 1171–1174 (2005).
38. Perebeinos, V. & Avouris, P. Impact excitation by hot carriers in carbon nanotubes. *Phys. Rev. B* **74**, 121410R (2006).
39. Ando, T. J. Excitons in carbon nanotubes. *J. Phys. Soc. Jpn* **66**, 1066–1073 (1997).
40. Spataru, C. D., Ismail-Beigi, S., Benedict, L. X. & Louie, S. G. Excitonic effects and optical spectra of single-walled carbon nanotubes. *Phys. Rev. Lett.* **92**, 077402 (2004).
41. Perebeinos, V., Tersoff, J. & Avouris, P. Scaling of excitons in carbon nanotubes. *Phys. Rev. Lett.* **92**, 257402 (2004).
42. Pop, E. et al. Negative differential conductance and hot phonons in suspended nanotube molecular wires. *Phys. Rev. Lett.* **95**, 155505 (2005).
43. Lazzeri, M., Pisanec, S., Mauri, F., Ferrari, A. C. & Robertson J. Electron transport and hot phonons in carbon nanotubes. *Phys. Rev. Lett.* **95**, 236802 (2005).
44. Collins, P. G., Arnold, M. S. & Avouris, P. Engineering carbon nanotubes and nanotube circuits using electrical breakdown. *Science* **292**, 706–709 (2001).
45. Collins, P. G. Hersam, M., Arnold, M., Martel, R. & Avouris, P. Current saturation and electrical breakdown in multiwalled carbon nanotubes. *Phys. Rev. Lett.* **86**, 3128–3131 (2001).
46. Naeemi, A., Sarvati, R. & Meindl, J. D. Performance comparison between carbon nanotube and copper interconnects for GSI. *IEDM Digest* 699–702 (2004).
47. Tans, S. J., Verschueren, A. R. M. & Dekker, C. Room-temperature transistor based on a single carbon nanotube. *Nature* **393**, 49–52 (1998).
48. Martel, R., Schmidt, T, Shea, H. R., Hertel, T. & Avouris, P. Single- and multi-wall carbon nanotube field-effect transistors. *Appl. Phys. Lett.* **73**, 2447–2449 (1998).
49. Sze, S. M. *Physics of Semiconductor Devices*. (Wiley, New York, 1981).
50. Appenzeller, J., Knoch, J., Radosavljević, M. & Avouris, P. Multimode transport in Schottky-barrier carbon-nanotube field-effect transistors. *Phys. Rev. Lett.* **92**, 226802 (2004).
51. Léonard, F. & Tersoff, J. Role of Fermi-level pinning in nanotube Schottky diodes. *Phys. Rev. Lett.* **84**, 4693–4696 (2000).
52. Martel, R. et al. Ambipolar electrical transport in semiconducting single-wall carbon nanotubes. *Phys. Rev. Lett.* **87**, 256805 (2001).
53. Heinze, S. et al. Carbon nanotubes as Schottky barrier transistors. *Phys. Rev. Lett.* **89**, 106801 (2002).
54. Appenzeller, J. et al. Field-modulated carrier transport in carbon nanotube transistors. *Phys. Rev. Lett.* **89**, 126801 (2002).
55. Javey, A. et al. Ballistic carbon nanotube field-effect transistors. *Nature* **424**, 654–657 (2003).
56. Appenzeller, J., Radosavljevi, M., Knoch, J. & Avouris, P. Tunneling versus thermionic emission in one-dimensional semiconductors. *Phys. Rev. Lett.* **92**, 048301 (2004).
57. Chen, Z., Appenzeller, J. Knoch, J., Lin, Y-M & Avouris, P. The role of metal-nanotube contact in the performance of carbon nanotube field-effect transistors. *Nano Lett.* **5**, 1497–1502 (2005).
58. Radosavljevic, M., Heinze, S., Tersoff, J. & Avouris, P. Drain voltage scaling in carbon nanotube transistors. *Appl. Phys. Lett.* **83**, 2435–2437 (2003).
59. Avouris, P. Carbon nanotube electronics. *Proc. IEEE* **91**, 1772–1784 (2003).
60. Lin, Y.-M., Appenzeller, J., Knoch, J. & Avouris, P. High-performance carbon nanotube field-effect transistor with tunable polarities. *IEEE Trans. Nanotechnol.* **4**, 481–489 (2005).
61. Chen, J., Klinke, C., Afzali, A. & Avouris, P. Self-aligned carbon nanotube transistors with charge transfer doping. *Appl. Phys. Lett.* **86**, 123108 (2005).
62. Klinke, C., Chen, J., Afzali, A. & Avouris, P. Charge transfer induced polarity switching in carbon nanotube transistors. *Nano Lett.* **5**, 555–558 (2006).
63. Javey, A. et al. High performance n-type carbon nanotube field-effect transistors with chemically doped contacts. *Nano Lett.* **5**, 345–348 (2005).
64. Appenzeller, J., Lin, Y.-M., Knoch, J. & Avouris, P. Band-to-band tunneling in carbon nanotube field-effect transistors. *Phys. Rev. Lett.* **93**, 196805 (2004).
65. Javey, A. et al. High-k dielectrics for advanced carbon nanotube transistors and logic gates. *Nature Mater.* **1**, 241–246 (2002).
66. Javey, A. et al. Self-aligned ballistic molecular transistors and electrically parallel nanotube arrays. *Nano Lett.* **4**, 1319–1322 (2004).
67. Seidel, R. V. et al. Sub-20 nm short channel carbon nanotube transistors. *Nano Lett.* **5**, 147–150 (2005).
68. Solomon, P. M. in *Future Trends in Microelectronics: Up the Nano Creek* (eds Luryi, S., Xu, J. M., & Zaslavsky, A.) 212–223 (Wiley, New York, 2007).
69. Castro, L. C. et al. Method for predicting fₜ for carbon nanotube FETs. *IEEE Trans. Nanotechnol.* **4**, 699–704 (2005).
70. Frank, D. J. & Appenzeller, J. High frequency response in carbon nanotube field-effect transistors. *IEEE Electron. Device Lett.* **25**, 34–36 (2004).
71. Li, S, D., Yu, Z., Yen, S-F, Tang, W. C. & Burke, P. J. Carbon nanotube transistor operation at 2.6 GHz. *Nano Lett.* **4**, 753–756 (2004).
72. Rosenblatt, S., Lin, H., Sazonova, V., Tiwari, S. & McEuen, P. L. Mixing at 50GHz using a single-walled carbon nanotube transistor. *Appl. Phys. Lett.* **87**, 153111–15313 (2005).
73. Bethoux, J-M. et al. An 8-GHz fₜ carbon nanotube field-effect transistor for gigahertz range applications. *IEEE Electron. Device Lett.* **27**, 681–683 (2006).
74. Bachtold, A., Hadley, P., Nakanishi, T. & Dekker, C. Logic circuits with carbon nanotube transistors. *Science* **294**, 1317–1320 (2001).
75. Derycke, V., Martel, R., Appenzeller, J. & Avouris, P. Carbon nanotube inter- and intramolecular logic gates. *Nano Lett.* **1**, 453–456 (2001).
76. Liu, X., Lee, C., Zhou, C. & Han, J. Carbon nanotube field-effect inverters. *Appl. Phys. Lett.* **79**, 3329–3331 (2001).

77. Javey, A., Wang, Q., Ural, A., Li, Y. & Dai, H. Carbon nanotube transistor arrays for multistage complementary logic and ring oscillators. *Nano Lett.* **2**, 929–932 (2002).

78. Chen, Z. *et al.* An integrated logic circuit assembled on a single carbon nanotube. *Science* **311**, 1735 (2006).

79. Katsnelson, M. I. Minimal conductivity in bilayer graphene. *Eur. Phys. J. B* **51**, 157 (2006).

80. Gusynin, V. P. & Sharapov, S. G. Unconventional integer quantum hall effect in graphene. *Phys. Rev. Lett.* **95**, 146801 (2005).

81. Peres, N. M. R., Guines, F. & Neto, A. H. C. Electronic properties of disordered two-dimensional carbon. *Phys. Rev. B* **73**, 125411 (2006).

82. Tworzydo, J., Trauzettel, B., Titov, M., Rycerz, A. & Beenakker, C. W. J. Sub-Poissonian shot noise in graphene. *Phys. Rev. Lett.* **96**, 246802 (2006).

83. Ziegler, K., Robust transport properties in graphene. *Phys. Rev. Lett.* **97**, 266802 (2006).

84. Niimi, Y. *et al.* Scanning tunneling microscopy and spectroscopy of the electronic local density of states of graphite surfaces near monoatomic step edges. *Phys. Rev. B* **73**, 085421 (2006).

85. Misewich, J. A. *et al.* Electrically induced optical emission from a carbon nanotube FET. *Science* **300**, 783–786 (2003).

86. Freitag, M. *et al.* Mobile ambipolar domain in carbon-nanotube infrared emitters. *Phys. Rev. Lett.* **93**, 076803 (2004).

87. Freitag, M. *et al.* Electrically excited, localized infrared emission from single carbon nanotubes. *Nano Lett.* **6**, 1425–1433 (2006).

88. Freitag, M. *et al.* Photoconductivity of single carbon nanotubes. *Nano Lett.* **3**, 1067–1071 (2003).

89. Qiu, X., Freitag, M., Perebeinos, V. & Avouris, P. Photoconductivity spectra of single carbon nanotubes: Implications on the nature of their excited states. *Nano Lett.* **5**, 749–752 (2005).

90. Balasubramanian, K *et al.* Photoelectronic transport imaging of individual semiconducting carbon nanotubes. *Appl. Phys. Lett.* **84**, 2400–2402 (2004).

91. Freitag, M. *et al.* Imaging of the Schottky barriers and charge depletion in carbon nanotube transistors. *Nano Lett.* **7**, 2037–2042 (2007).

92. Freitag, M. *et al.* Scanning photovoltage microscopy of potential modulations in carbon nanotubes. *Appl. Phys. Lett.* **91**, 031101 (2007).

93. Castro, E. V. *et al.* Biased bilayer graphene: semiconductor with a gap tunable by electric field effect. Preprint at < http://arxiv.org/abs/cond-mat/0611342 > (2006).

94. Ohta, T. *et al.* Controlling the electronic structure of bilayer graphene. *Science* **313**, 951–954 (2006).

95. Williams, J. R., DiCarlo, L. & Marcus, C. M. Quantum Hall effect in a gate-controlled p-n junction of graphene. *Science* **317**, 638–641 (2007).

96. Abanin, D. A. & Levitov, L. S. Quantized transport in graphene p-n junctions in a magnetic field. *Science* **317**, 641–643 (2007).

97. Hueso, L. E. *et al.* Transformation of spin information into large electrical signals using carbon nanotubes. *Nature* **445**, 410–413 (2007).

98. Tombros, N., Jozsa, C., Popinciuc, M., Jonkman, H. T. & van Wees, B. J. Electronic spin transport and spin precession in single graphene layers at room temperature. *Nature* **448**, 571–574 (2007).

Electron transport in molecular junctions

Building an electronic device using individual molecules is one of the ultimate goals in nanotechnology. To achieve this it will be necessary to measure, control and understand electron transport through molecules attached to electrodes. Substantial progress has been made over the past decade and we present here an overview of some of the recent advances. Topics covered include molecular wires, two-terminal switches and diodes, three-terminal transistor-like devices and hybrid devices that use various different signals (light, magnetic fields, and chemical and mechanical signals) to control electron transport in molecules. We also discuss further issues, including molecule–electrode contacts, local heating- and current-induced instabilities, stochastic fluctuations and the development of characterization tools.

N. J. TAO

Department of Electrical Engineering
Arizona State University, Tempe, Arizona 85287

e-mail: nongjian.tao@asu.edu

Although molecular electronics has been proposed as an alternative to silicon in post-CMOS devices, molecules with unique functions may have applications that are complementary to the silicon-based microelectronics. To date, many molecules with wonderful electronic properties have been identified and more with desired properties are being synthesized in chemistry labs. In addition to electronic properties, many molecules possess rich optical, magnetic, thermoelectric, electromechanical and molecular recognition properties, which may lead to new devices that are not possible using conventional materials or approaches (Fig. 1).

Despite there being many unsolved issues in molecular electronics, important and solid advances have been made over the past decade. These advances include: (1) demonstration of simple molecular device functions; (2) development of many different experimental approaches for measuring electron transport in single molecules and theoretical methods for describing the electron transport properties; (3) emergence of new characterization techniques that help to bridge theories and experiments; (4) hybrid devices, such as molecular sensors, which have been actively pursued parallel to the efforts in pure electronic devices.

This paper will provide an overview of some of these advances and discuss remaining issues that must be solved in order to reach the dream of functional devices. Molecular electronics is a diverse and rapidly growing field. Having limited space and references, we have inevitably overlooked important work in preparing this review. Fortunately, there are many excellent reviews covering various aspects of molecular electronics, which will amend these deficiencies (see, for example, reviews published within the past three years[1–3]).

MOLECULAR WIRES

A wire-shaped molecule that can efficiently transport charge has been actively pursued as it may provide interconnections for molecular devices. The interest can also be traced back to its close relevance to charge transfer in biological systems, such as photosynthesis and DNA, as well as to conducting polymers. Such a molecule is often referred to as a molecular wire, although the definition is still a subject of debate[4]. To date, many wire-shaped molecules have been studied, which can be loosely divided into two categories: saturated and conjugated chains.

SATURATED CHAINS

A well-studied molecular system is the alkane consisting of saturated C–C bonds terminated by linkers that can bind to electrodes[5–11]. These molecules have large gaps between their highest occupied molecular orbitals (HOMO) and lowest unoccupied molecular orbitals (LUMO) and are considered to be poorly conducting wires. However, they serve as a model system for testing both experimental techniques and theoretical calculations. Experiments have found that the conductance (G) of alkanes decreases exponentially with molecular length (L), and can be described by $G = A\exp(-\beta L)$, where A is a constant and β is a decay constant varying between ~0.7–0.9 Å^{-1} (Fig. 2a). Similar decay has also been found in peptide chains[12]. The exponential decay, together with characteristic current–voltage (I–V) curves and temperature independence[7,13], suggests electron tunnelling as the conduction mechanism for these molecules. Although the exponential decay of the conductance with length is similar in most experiments, comparison of the absolute conductance values of different experiments is not straightforward. This is partly because some experiments measure many molecules and others measure single molecules. Even in the case of single-molecule measurements, there may be two or more conductance values for the same molecule resulting from different molecule–electrode contact geometries[13–17].

CONJUGATED MOLECULES

Because of the rapid decrease in tunnelling rate with distance, in most experiments the conductance is too small to be measured for alkane chains longer than 2–3 nm. For long-distance charge transport, conjugated molecules with alternating double and single bonds or delocalized π electrons are better candidates. These molecules have much smaller HOMO–LUMO gaps compared to alkanes and should therefore transport charge more efficiently. However, for short conjugated molecules, off-resonance tunnelling is still believed to be the conduction mechanism. With increasing

When citing this review, please cite the original version as shown on the contents page of this chapter.

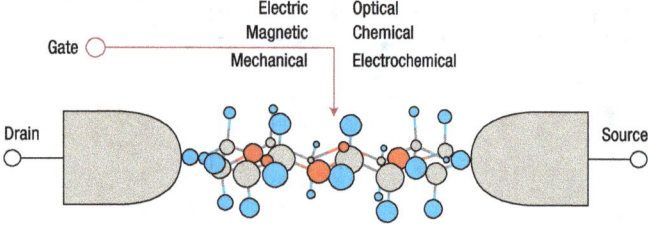

Figure 1 Illustration of a single molecule attached to two electrodes as a basic component in molecular electronics. Electron transport through the molecule may be controlled electrically, magnetically, optically, mechanically, chemically and electrochemically, leading to various potential device applications. To reach the ultimate goal in device applications, experimental techniques to fabricate such an electrode–molecule–electrode junction, and theoretical methods to describe the electron transport properties must be developed. Both are challenging tasks, but rapid advances have been made in recent years.

length, it is thought that tunnelling is replaced by hopping in which charges hop from one site to the next along the molecule. Hopping is thermally activated and has weaker length-dependence than the tunnelling process[4].

Conjugated molecules have been studied by measuring the charge transfer rate through the molecules bridged between an acceptor and a donor group[4], or between a redox group and a bulk electrode[18]. These experiments have established that: (1) charge transport can occur over a much greater distance in conjugated molecules than in alkanes; (2) a crossover from a tunnelling regime to a hopping regime occurs when the length of the molecule is increased.

Direct conductance measurements of conjugated molecules have been reported using different approaches, and interesting phenomena have been observed[19–22], although systematic studies into the length dependence of the conductance of single conjugated molecules bridged between two electrodes are relatively rare. One such measurement has been made in carotenoid polyenes with different lengths using a combined break junction and statistical analysis approach (Fig. 2b)[23]. The measured β is 0.22 ± 0.04 Å^{-1}, which is in close agreement with the value obtained from first-principles simulations (0.22 ± 0.01 Å^{-1}) (Fig. 2a). The small value of β demonstrates that conductance drops off slowly with chain length, confirming that carotenoid conjugated chains are relatively good molecular 'wires'. Charge transport in oligothiophenes with three and four repeating units have been measured, and the longer molecule was found to be more conductive than the shorter one[24]. This unusual length-dependent conductance was attributed to a smaller HOMO–LUMO gap and a closer position of the HOMO to the Fermi levels of the probing electrodes for the longer molecule. This conclusion is supported by the ultraviolet–visible absorption, electrochemical measurements, and also by the dependence of the conductance on the electrochemical gate, which shifts the HOMO relative to the Fermi levels.

Compared to charge transfer rate measurements, direct conductance measurements are still at an early stage. For example, charge transfer rate experiments have established the important role of the electron coupling to the nuclear coordinates of the molecules and the surrounding solvent molecules[25]. The rate measurements have also revealed the crossover from tunnelling to hopping[26]. These effects have yet to be studied and confirmed in the direct single-molecule conductance measurements. Finally it is important to note that charge transfer rate and direct conductance methods measure relevant but different quantities. Direct conductance measurement of

a molecule requires the molecule to bridge two electrodes that are connected to a power source, forming a continuous circuit. Although the two electrodes act like a donor and acceptor, the energy levels of the electrodes are continuous bands, whereas the energy levels of the donor and acceptors in the charge transfer measurements are discrete. A relationship between charge transfer rate and conductance has been worked out by Niztan[27], which requires certain parameters such as the coupling strength between the molecules and the two electrodes to be known.

TWO-TERMINAL DEVICES

Device applications demand molecules to work not only like a wire but also an active element that can perform one or a set of controllable functions. Silicon-based electronics relies critically on three-terminal devices, such as transistors. However, owing to the difficulty of placing a third gate electrode close to a molecule, most studies in molecular electronics to date have focused on two-terminal devices. Nevertheless, two-terminal devices can exhibit interesting and important behaviour that is useful for applications. Examples include rectification[28], negative differential resistance[29] and conductance switching effects[30].

MOLECULAR DIODES

A diode or rectifier is an important component in electronics that allows an electric current to flow in one direction, but blocks it in the opposite direction. Inspired by the visionary work by Aviram and Ratner[28], building diodes using single molecules has been pursued by many groups[31]. The basic structure of early molecular diodes consisted of a donor and an acceptor separated by a σ bridge, with σ being some saturated covalent bond linking the donor and acceptor and providing a tunnelling barrier between them. In this donor–σ–acceptor structure, the diode behaviour was expected to occur as a consequence of different thresholds at positive and negative bias voltages. Such Aviram–Ratner molecular diodes have been demonstrated using Langmuir–Blodgett films[32] and block copolymers[33] sandwiched between two planar electrodes. When using self-assembly to build a diode, the orientation of the molecules must be controlled. One way to achieve this is to synthesize molecules terminating in a thiol at each end, both of which are protected with different groups[34]. One protection group is removed first, allowing the thiol at that end to bind to one electrode, followed by removal of the second protection group allowing the other end to link to another electrode.

The experiments described above involve a large number of molecules. The single-molecule diode has been demonstrated in a molecule consisting of two weakly coupled conjugated units, using a mechanically controlled break-junction method[35,36] (Fig. 3). The two conjugated units are different — one is fluorized and the other one is not — and this breaks the symmetry of the molecular junction. When sweeping the bias voltage, the energy levels of both units are shifted relative to each other. Whenever an unoccupied level passes by an occupied one, an additional transport channel opens up and the current increases by a certain amount — one step. The height of each step depends on the bias polarity, which gives rise to the diode-like I–V curve results as shown in Fig. 3.

The presence of donor and acceptor groups introduces non-symmetry to the molecular junctions. More generally, one may expect rectification to arise from non-symmetric molecular junctions that do not necessarily contain a pair of donor and acceptor groups[37]. Indeed, diode-like I–Vs have been observed by using different molecule–electrode contacts on the two ends[38,39], non-symmetric molecules[34,40,41], or both[42]. However, non-symmetry alone is not sufficient to achieve a larger diode effect if the bias voltage drops entirely at the molecule–electrode interfaces[43]. Another type of molecular diode was proposed based on bias-polarity-dependent structural changes in molecules[44]. This structural-change-induced function is completely different from the principles of semiconductor diodes.

NDR EFFECT

A region of decreasing current with increasing voltage in the I–V curve characteristic of a device is referred to as negative differential resistance (NDR). The NDR effect was first discovered in the Esaki diode, which opened a new avenue in many niche applications such as low-power memory. In an Esaki diode the current initially increases with the bias as electrons tunnel through the p–n junction barrier because the conduction band on the n-side is aligned with the valence band on the p-side. As voltage increases further the conduction and valence bands become misaligned and the current drops, resulting in NDR. Other devices that exhibit NDR are semiconductor heterostructures in which electrons tunnel through a quantum well formed by two tunnel barriers. The current increases initially when the energy level of one side of the quantum well lines up with an energy level of the quantum well. After reaching a maximum at resonance, the current decreases and gives rise to NDR. The demonstration of NDR in molecular devices[29,45] has inspired much interest in the effect[46-55].

The physical basis of NDR is interband tunnelling in the Esaki diode, and resonant tunnelling in the semiconductor quantum-well heterostructures. Although the observations of NDR in scanning tunnelling microscopy (STM) have been attributed to narrow features in the local density of states of the tip apex atom, which is analogous to that in semiconductor heterostructures[45], the mechanisms of NDR in molecular devices in general are likely to involve different processes. In the most widely studied nitro-substituted oligo(phenylene ethynylene) (OPE), a conjugated molecule consisting of three benzene rings linked by single–triple–single bonds, the NDR was attributed to the electrochemical reduction of the NO_2 group[29]. Excellent correlation was observed between the electrochemical potentials and the first negative threshold potentials for electric conduction of OPE derivatives[56]. Further evidence of the correlation between the electrochemical reduction and the NDR-like effect was found in the electron transport measurement of single nitro-OPE molecules in an electrochemical cell[48]. These results support the theories that NDR involves charging (reduction) and/or structural change[50,57-59]. NDR or NDR-like behaviour has been reported in several other molecular systems[51-55] and evidence shows that this NDR-like behaviour involves chemical reactions[60] and dissociations[54,61]. Although it is questionable whether an irregular or transient peak in the I–V curve should be referred to as NDR, the characteristic decrease in current with increasing voltage must be robust and reproducible for device applications, and a better understanding of the mechanisms is crucial to further development.

TWO-TERMINAL MOLECULAR SWITCHES

A two-terminal device in which the conductance can be reversibly switched between two states with an applied voltage is useful for molecular memory and logic devices[30,62,63]. One class of such devices is based on mechanically interlocked bistable complexes, such as catenanes and rotaxanes using Langmuir–Blodgett films[64]. Whereas the voltage-gated configurational change of the molecules was proposed as the switching mechanism[30], recent experiments suggest alternative mechanisms[65]. Despite complications, the rich switching properties of catenanes and rotaxanes upon photo-, electrochemical and electrical stimulation make the molecules interesting candidates for molecular devices[64].

Other systems that could be candidates for molecular switches are OPE derivatives[62], which are known for their NDR effect discussed above. A recent study using three different methods — STM, crossed-wire junction and magnetic-bead junction — has revealed reversible switching behaviour in the system providing strong evidence for a two-terminal molecular switch[66] (Fig. 4). It is believed that charge

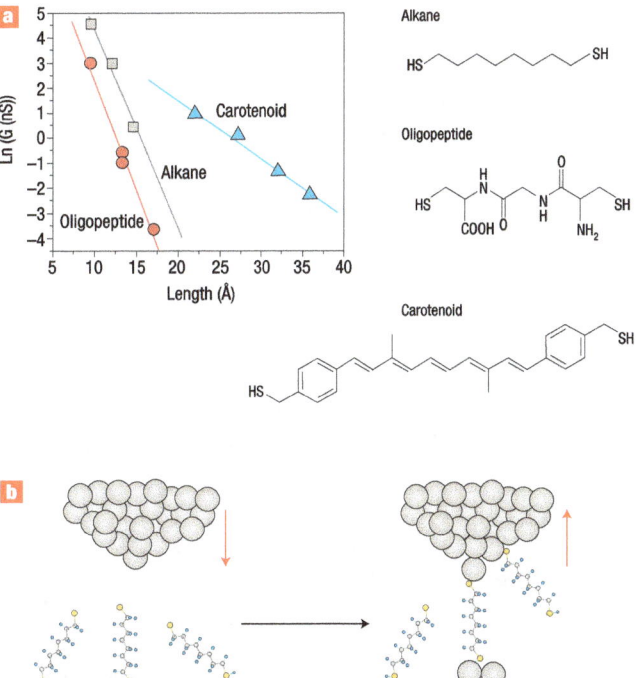

Figure 2 Length dependence of conductance G for saturated chains (alkanes[13] and peptides[12]) and conjugated molecules (carotenoids[23]). **a**, The conductance in each system decreases exponentially with the length but with a different slope. **b**, Experimental technique used to measure single molecule conductance[13]. An Au electrode is moved into contact with another Au electrode, which is covered with sample molecules terminated with proper linkers. As the first electrode is pulled away again the bridging molecules break contact with one of the two electrodes. Each breakage is revealed as a step in the conductance trace. Statistical analysis of the individual traces allows determination of the conductance of the molecule.

trapping, structural changes in the molecules and the interplay between them are likely to be responsible for the switching phenomenon[59]. A different switching mechanism based on bias-driven changes in the molecule/electrode hybridization or bond angle has also been proposed[67].

Two-terminal molecular switches have been reported in well-defined STM experiments involving current-induced structural changes in molecules adsorbed on surfaces[68]. A two-state switching has been reported in Zn(II)–porphyrin, which is due to electric-field-induced structural change in the molecules[68]. This type of experiment will continue to provide basic knowledge for building devices.

THREE-TERMINAL DEVICES

Two-terminal devices are interesting and can be useful, but a three-terminal field-effect transistor (FET)-like device is highly desired. Theoretical models have predicted that the conductance of a single molecule can be modulated with a gate electrode in a fashion similar to a conventional FET[69,70]. Experimental demonstration of this FET behaviour in single molecules has, however, been difficult because it requires the gate electrode to be placed close to the molecule for effective gate control. Several different gate configurations have been proposed and demonstrated in recent years.

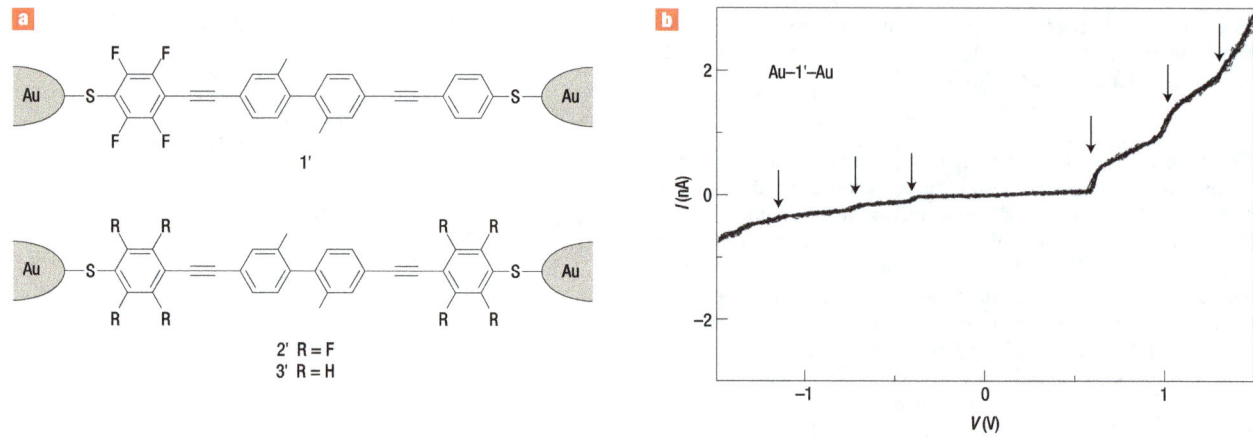

Figure 3 a, A molecular diode consisting of a single molecule (1') covalently linked to two Au electrodes of a mechanically controlled break junction. The molecule consists of two weakly coupled conjugated units, one is fluorized and the other is not, which breaks the symmetry of the molecular junction. Symmetric molecules, 2' and 3', each consisting of two identical conjugated units, serve as controls. **b**, The *I–V* curve for 1' showing rectification for the Au–1'–Au junction, where the arrows point to stepwise increases in the current. Reprinted with permission from ref. 36. Copyright (2005) National Academy of Sciences, USA.

BACK GATE

One possible gate configuration is the back gate approach, which consists of a pair of electrodes — source and drain — on a solid substrate, such as Si or Al substrate (Fig. 5a). The Si substrate is heavily doped to serve as a back gate, and a thin oxide layer separates the Si or Al from the source and drain electrodes. This configuration has been used to demonstrate FET-like behaviour in carbon nanotubes[71]. For most molecules, the source–drain separation has to be a few nm or less. To create such a small separation, unconventional techniques, such as electromigration[72,73] electrochemical[74,75] and other novel methods[21,76–81], have been developed. Using the back gate approach, interesting single-electron transistor[72,73,78,82] and Kondo effects[72,73] have been observed. The back gate approach has allowed researchers to study various phenomena in single molecules or a small number of molecules, but the need for sophisticated fabrication facilities and the low yield of the devices have prevented it from becoming a wider-spread technique. Research in this direction will benefit from better structural characterization of the molecular junctions formed by the various techniques. Finally these back-gate single-molecule transistors are so far largely in the single-electron transistor regime. Large and continuous field modulation like the conventional FET has yet to be demonstrated.

ELECTROCHEMICAL GATE

Several groups have recently demonstrated true FET-like modulation in single molecules using an electrochemical gate (Fig. 5b)[83–87], in which the gate voltage is applied between the source and a gate electrode inserted in the electrolyte[88–90]. Since the gate voltage falls across the double layers at the electrode-electrolyte interfaces, the effective gate thickness is of the order of a few solvated ions, which results in a large gate field. This large field has allowed electrochemists to reversibly switch redox molecules between oxidized and reduced states. Most of these gate effects observed to date are rather small. A recent study of perylene tetracarboxylic diimide (PTCDI), a redox molecule, found that the current through the molecule can be reversibly varied over nearly three orders of magnitude (Fig. 6)[84]. In terms of the energy diagram (Fig. 6c), decreasing the gate voltage shifts the LUMO of the redox molecule towards the Fermi levels of the electrodes causing a large increase in the current through

the molecule. To fully understand the electrochemical gate effect, a theory must be developed that considers coupling of the electrons with the vibration modes of the molecules and the surrounding solvent molecules. Both are known to be important in the electron transfer theories of redox molecules[25].

OTHER GATES

The goal of regulating current through a single molecule may be achieved by a chemical binding event (Fig. 5c)[91]. Such a concept has been demonstrated in self-assembled monolayers of electron donor, tetramethyl xylyl dithiol (TMXYL). Conductance measurements by STM show that TMXYL switches from insulating to ohmic behaviour as a result of reaction with an electron acceptor, tetracyanoethylene (TCNE)[92]. Reversible chemical-binding-induced conductance changes have also been observed in the study of Cu^{2+} and Ni^{2+} binding to oligopeptides using an STM break-junction approach[41]. In general, the chemical gate mechanism may involve many processes, including field effect, charge transfer, chemical reaction and structural changes.

A field-controlled conductance change in styrene-derived molecules on hydrogen terminated Si(100) has been observed using an ultrahigh vacuum STM, where the gate field originates from a fixed point charge of a nearby dangling bond[93]. Although this is a two-terminal experiment, the atomic-resolution STM provides a clean demonstration of single-molecule field-effect concept. We note that the charge could be ions or dopants, which is also the principle of electrochemical gating that controls charge surrounding the molecule by regulating the distribution of ions.

HYBRID DEVICES

Whereas most current efforts have focused on electrical properties of molecules analogous to conventional microelectronic devices, it has become apparent that molecules often possess unique properties that have no parallel in conventional materials. These unique molecular features include optical, magnetic, mechanical and thermoelectric properties. The interplay of these properties and the electrical properties of the molecules promise new hybrid devices with functions that cannot be achieved in Si-based devices[94]. We will discuss some of these efforts below.

MOLECULAR ELECTROMECHANICAL DEVICES

Electron transport in molecules is often accompanied by a large structural change or mechanical movement in the molecules[95]. This phenomenon is well known in chemistry and electron transfer studies, and is believed to play an important role in the molecular switches and NDR effects, discussed above. The interplay between the electrical and mechanical properties in bulk materials is responsible for piezoelectric, piezoresistive and other important effects. Finding analogous effects in single molecules is exciting. It has been shown that electron tunnelling through a C_{60} molecule increases rapidly when the molecule is pressed by an STM tip[96]. More recently, a conducting AFM-method has been applied to study the effect of a stretching force on the conductance of single molecules[24]. In the case of octanedithiol, the conductance decreases by 10–20% before the molecular junction breaks down. This piezoresistive-like effect is primarily due to the elongation of the molecular junction. However, a much greater stretching-induced decrease in the conductance was observed in oligothiophene and attributed to the change in the HOMO–LUMO gap[24].

Interesting electromechanical properties that do not exist in bulk materials have also been reported in molecular systems[96-99]. For example, novel nanomechanical phenomena have been reported in a single C_{60} transistor, in which the centre of mass of the C_{60} molecule oscillates between the source and drain electrodes[97]. The oscillation, with a frequency of about 1.2 THz, is manifested in the transport current owing to the coupling between the mechanical oscillation and the single-electron hopping process. The oscillation frequency is in good agreement with a simple theoretical estimate based on van der Waals and electrostatic interactions between C_{60} molecules and gold electrodes. The C_{60} in such a Au–C_{60}–Au junction may be driven into oscillation by a current via an inelastic tunnelling process in which the electrons transfer energy to the molecule and cause the molecule to oscillate[100].

Examples discussed above involve relatively small structural changes or mechanical motions. Molecules are known to undergo large structural changes and rotational and translational motions, which have been proposed for molecular machines, including motors, shuttles and actuators (see ref. 101 for review) This line of research is also exciting because of the inspiration from biological systems, which achieve many functions through mechanical motions. Examples of molecular machines have been demonstrated in solution and solid phases. The recent trend of studying molecular machines on surfaces represents an important step towards practical application. The ultimate task is the construction of individually wired molecular machines, or wireless machines that can perform programmed functions.

MOLECULAR OPTOELECTRONICS

Controlling current in a molecule with light may lead to optoelectronic devices. One mechanism is through light-induced conformational changes in the molecule. Optical switching of photochromic dithienylethene from the conducting to the insulating state has been observed when illuminating the molecule with visible light[102]. A similar finding has been reported on single dithienylethene derivatives using an STM break-junction technique[103]. The switching process in the molecules is found to be due to a swapping of the HOMO and LUMO during the conformational change[104], which is different from the traditional semiconductor-based optoelectronic mechanism in which the photocurrent depends on the photo-induced excess carriers. Current in molecular junctions may also be controlled or induced by light through other mechanisms, such as adiabatic pumping[105], optical resonance activation[106] and strong-field optical control[107]. The opposite phenomenon is light emission as a result of current in a molecular junction. Such current-induced light emission in molecular systems[108-110] has been observed by STM. The high spatial

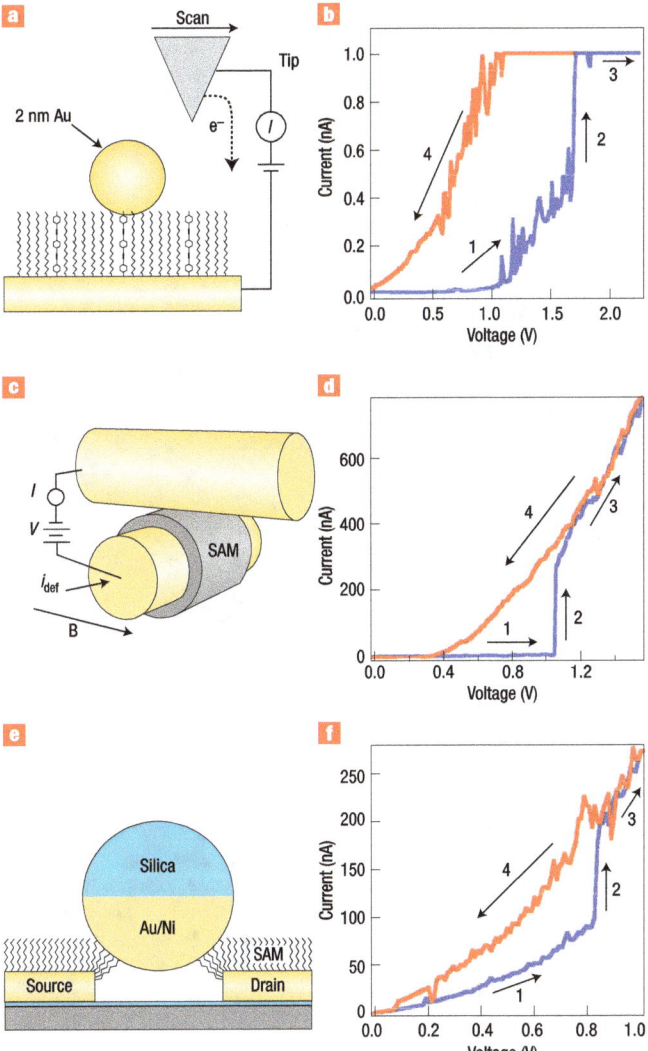

Figure 4 Demonstration of two-state molecular switching behaviour in three different platforms. **a, b**, Diagram and I–V curve of STM experiment, where current through an OPE molecule is measured between a Au substrate and an STM tip. **c, d**, Diagram and I–V curve of crossed-wire tunnelling junction, in which a self-assembled monolayer (SAM) of the molecule is sandwiched between two cross wires. The separation between the wires is controlled by passing a current, i_{def}, through a wire in a magnetic field, B. **e, f**, Diagram and and I–V curve of a magnetic bead junction. Note that two molecular junctions are formed in this method, one between the source and the bead and the other between the bead and the drain. Each of the I–V curves shows an abrupt increase (labelled 2) in the current when the voltage is swept positively (blue curves labelled 1, 2 and 3) and the current returns to the original value gradually when the voltage is swept negatively (red curves labelled 4). Reprinted with permission from ref. 66.

resolution of the STM enabled the demonstration of variations in light-emission spectra from different parts of the molecule[110].

Quenching of photo-excited molecules by metal electrodes can profoundly affect the optoelectronic switching of the molecules[111]. In a study by Dulic et al.[102] only one-way switching was observed, which was attributed to quenching of the molecules. To reduce the interaction between the molecule and the metal substrate, an oxide layer can be

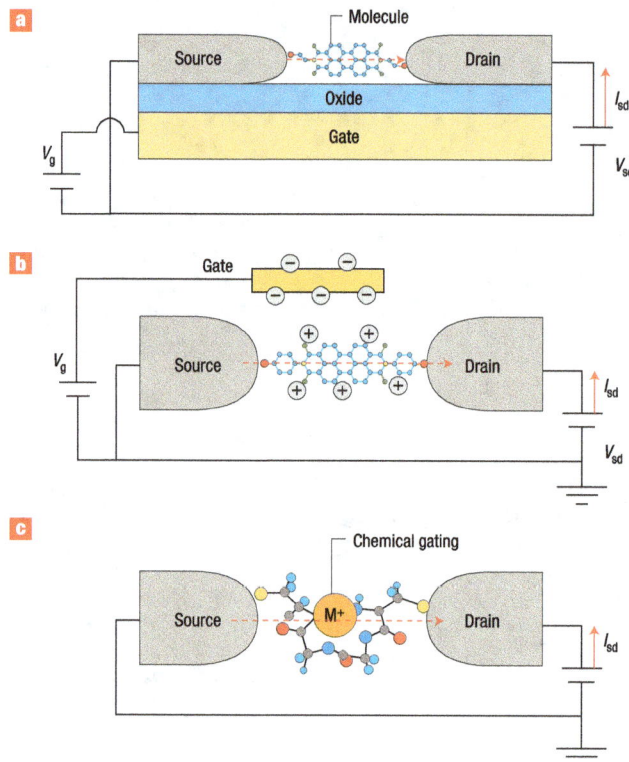

Figure 5 Controlling current through a molecule using different gates. **a**, Back gate: a molecule attached to source and drain electrodes on an oxidized metal or heavily doped Si gate (substrate). **b**, Electrochemical gate: a molecule bridged between source and drain electrodes in an electrolyte in which a gate field is applied by a third electrode inserted in the electrolyte. For proper potential control, an additional electrode, called the counter electrode, is often used (not drawn). **c**, Chemical gate: current through the molecule is controlled via a reversible chemical event, such as binding, reaction, doping or complexation.

placed between them[110]. A linker group between the photochromic centre and the electrodes may also help to retain the photoswitching function of molecules[103]. Another important consideration is irreversible photobleaching of the molecules. Recent studies show that metal clusters can be used like single molecules to create stable light-emitting diodes and logic gates[112].

MOLECULAR SPINTRONICS

Spintronics involves manipulating the spin rather than the charge of the electron in electronic devices, giving them potential applications in information storage and processing. The ability to control electron spins in molecules leads to molecular spintronics. An important advantage of molecular spintronics is weak spin-orbit and hyperfine interactions in organic molecules, which may help to preserve electron spin polarization over a much greater distance than in conventional semiconductor systems. Several theoretical works have calculated spin-polarized transport in molecules[113,114], including recent works[115,116] that combine density functional theory and non-equilibrium Green function methods. These theories differ from each other, but they all predict a large magnetoresistance effect in single molecules bridged between two ferromagnetic electrodes.

Experimental demonstrations of coherent spin-transport and magnetoresistance have been reported in several molecular materials,

including photo-excited electrons across conjugated molecular bridges[117] and conjugated organic semiconductors[118]. Recently, the spin-polarized electron tunnelling through a barrier consisting of a self-assembled monolayer of octanethiol between two magnetic electrodes has been measured, and the work demonstrates that spin-conserving transport in molecular devices is possible for low-energy electrons[119]. These experiments involve a large number of molecules whereas most theories usually consider a single molecule. Single-molecule spintronics will allow a closer comparison between theory and experiments. Recently, large magnetoresistive effects have been observed in a single-molecule junction of C_{60} bridged between two Ni electrodes[120]. More single-molecule spintronics studies are anticipated.

MOLECULAR SENSORS

Electrically wiring a single molecule into an electronic circuit also allows one to read chemical and biological information of the molecule electronically. For example, single base-pair mismatches[121,122], protein–DNA interactions and oxidative damage may be detected and analysed by simply measuring the conductance of a DNA molecule[123]. The binding of a guest species to a single host molecule has been studied by measuring the conductance of the host molecule wired to two electrodes[41]. Just like molecular electronic devices, a single-molecule sensor device may not be necessary in practice, and molecular detection using a nanowire[124] presents an approach that may find applications in the near future. However, the ability to detect a single molecule or single-molecule property represents the ultimate limit of analytical science, and study of single-molecule events will provide a unique opportunity to understand molecular recognition processes.

ISSUES AND CHALLENGES

The examples discussed above are merely a small fraction of the tremendous advances achieved over the past several years. These studies have revealed rich and wonderful electronic, mechanical, optoelectronic, magnetic and molecular recognition functions that may serve as individual device elements. Synergetic integration of various different functions obviously requires a new architecture or system-level design and fabrication. Whereas system-level discussion is omitted in the present paper, architecture design and individual device development must work hand in hand. On an individual device level, there are still key issues that need to be resolved and understood. Some of these issues are discussed below.

MOLECULE–ELECTRODE INTERFACES

To create a useful device, molecules must be electrically wired to the outside world reliably by electrodes. It is known that the conductance of a molecule is sensitive, not only to the nature of the chemical bonds[125] between the molecule and probing electrodes, but also to the atomic-scale details of the molecule–electrode contact geometry[126]. Precise control of the contact geometry down to the atomic scale has been a difficult challenge. A widely used method is to attach molecules terminated with thiol groups to gold electrodes. This approach has many advantages, but also undesired effects, such as high metal-atom mobility and large polarization field at the metal–molecule interface. Researchers have looked into other types of contacts, involving C–C (refs 42, 127), C–Si (ref. 128) and other bonds[129,130]. Although the molecule–electrode contact has a profound effect on the conductance, it does not necessarily mask the intrinsic switching properties of molecules[131]. So an alternative approach is to build devices by taking advantage of the intrinsic properties of the molecules rather than the absolute conductance values.

STOCHASTIC FLUCTUATIONS

Unwanted random telegraphic switching of conductance has been observed in metal–molecule–metal junctions[132–134]. Two mechanisms have been proposed: molecular motion due to conformational change, and bond-fluctuation due to the molecules tethered to the gold surface becoming attached or detached in a random manner. Similar telegraphic switching behaviours have long been reported in the conductance measurements of many nanoscale systems, including in submicrometre Si-metal-oxide field-effect transistors[135]. These random fluctuations between two or several discrete levels are believed to be the origin of the well-known 1/f noise in electronic devices[136]. More recent examples include conducting polymer nanojunctions, in which a single or a few polymer strands are sandwiched between two electrodes[137]. The telegraphic noises in these systems have been attributed to charge trapping in defect states. It is interesting to note that similar fluctuations have been observed in optical spectroscopies of single molecules. These stochastic fluctuations are not necessarily a fundamental limit in device applications as deterministic devices have been demonstrated, but they must be understood and controlled.

CURRENT-INDUCED INSTABILITY AND LOCAL HEATING

Electromigration and local heating are known to be important considerations in the design of conventional electronics. It is natural to ask how important this effect is in molecular electronics. Current-induced instability and local heating arise from energy exchanged between electrons and phonons[138]. In a nanoscale junction the inelastic electron mean free path is often large compared with the junction size, so that each electron should release only a small fraction of its energy during transport in the junction. However, substantial effects can still arise due to the large current density in the nanojunction. An experimental approach has been reported to determine the local temperature in single molecules by measuring the force required to detach molecules bound to electrodes[139]. Since the detachment is thermally activated, the force is sensitive to the local temperature of the molecule–electrode contact, which provides an estimate of the effective local temperature of the molecule.

CHARACTERIZATION TOOLS

To understand the above issues, structural and spectroscopic characterization tools are highly desired. STM has played, and will continue to play, an important role in the understanding of electron transport in single molecules because of its capability for both microscopy and tunnelling spectroscopy. As STM requires well-defined surfaces, such as single crystals, it is not suitable for many devices or samples. Atomic force microscopy (AFM) is another useful structural characterization tool. Although it typically has lower resolution than STM, a combined force and conductance measurement with AFM provides unique information about the bonding nature of molecule–electrode contacts[24]. Ballistic electron microscopy[140] is also capable of revealing structural information about molecules on electrodes. Another structural characterization tool is X-ray, which has been shown to provide angstrom-resolution structural information of a silicon–molecule–mercury junction[141]. Surface spectroscopy techniques, including photoemission spectroscopy, have been powerful in determining the electronic properties of molecule–electrode interfaces and relative alignments of the HOMO and LUMO to the Fermi level of the electrodes[22]. However, most of the spectroscopy techniques are difficult to apply to molecules sandwiched between two electrodes. Inelastic electron tunnelling spectroscopy[142,143] is an exception, but it requires low temperatures to resolve low-energy vibration modes. Optical techniques, such as infrared and

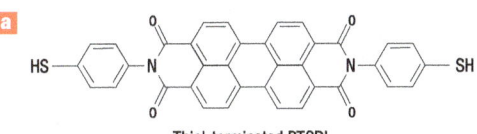
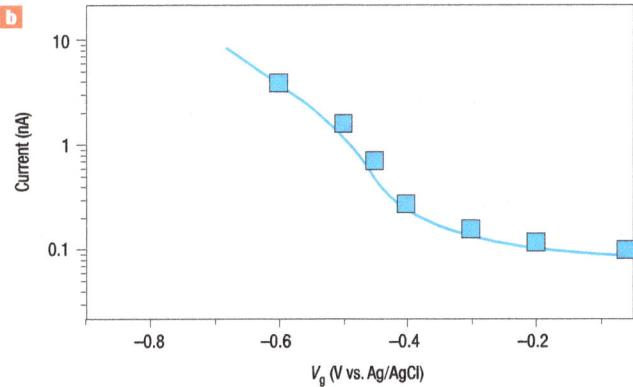
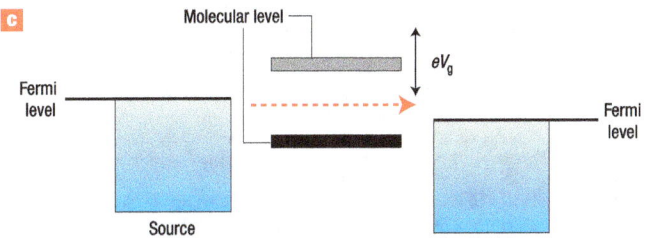

Figure 6 Controlling electron transport through a redox molecule, PTCDI, by switching the redox state of the molecule with an electrochemical gate. **a**, Structural illustration of PTCDI. **b**, Source–drain current versus gate voltage (V_g with respect to an Ag/AgCl reference electrode) for a dithiol-terminated PTCDI covalently bound to Au source and drain electrodes in 0.1 M NaClO$_4$ aqueous solution. The squares were experimental data determined from the peak position of the conductance histograms and the solid line was obtained by recording the current while continuously sweeping the gate voltage[84]. **c**, Adjusting the gate voltage shifts the molecular energy levels relative to the source and drain Fermi energy levels, which gives rise to changes in the transport current through the molecule.

Raman spectroscopy, have shown promise in probing the chemical properties of molecules in junctions at room temperature[144].

THEORETICAL CHALLENGES

Recent advances in theoretical development[145] are also substantial, but are beyond the scope of this review. The gap between theoretical calculations and experimental data is decreasing for many relatively simple molecular systems where off-resonance tunnelling dominates. Further development to include realistic treatment of molecules, electrodes and the atomic-scale details of the molecule–electrode interface is critical. Owing to the lack of atomic-scale details of most experiments, the recent treatment of a range of different molecule–electrode contact geometries is an important step for a fair comparison between theories and experiments[14,15]. Non-equilibrium Greens functions and density functional theory approaches provide reasonable descriptions of systems involving coherent tunnelling but the incorporation of strong electronic correlations and electron–vibration couplings also requires further development. Sophisticated and successful theories have been developed to describe strong electron–vibration and electron–solvent couplings in electron transfer phenomena in molecules. Extending this to electron transport is critical for understanding the electrochemical gate experiments and to bridge the gap between electron transfer theories developed by the chemistry community and electron transport theories developed by the physics and engineering communities.

CONCLUDING REMARKS

Molecular electronics has been one of the fastest growing fields and has brought together scientists and engineers from different disciplines. Much has been learned about the fundamental electron transport in molecules, and niche applications in selected areas are emerging, but many of the remaining challenges are still formidable. Although it is always difficult to predict the future, it is certain that research will help to bridge hard electronics to the soft molecular world, and the field will continue to grow.

doi: 10.1038/nnano.2006.130

References

1. Cuniberti, G., Fagas, G. & Richter, K. *Introducing Molecular Electronics*. (Springer, Berlin and Heidelberg, 2005).
2. Selzer, Y. & Allara, D. L. Single-Molecule Electrical Junctions. *Ann. Rev. Phys. Chem.* **57,** 593–623 (2006).
3. Joachim, C. & Ratner, M. A. Molecular Electronics Special Feature: Molecular electronics: Some views on transport junctions and beyond. *Proc. Natl Acad. Sci. USA* **102,** 8801–8808 (2005).
4. Weiss, E. A., Wasielewski, M. R. & Ratner, M. A. Molecules as wires: Molecule-assisted movement of charge and energy. *Top. Curr. Chem.* **257,** 103–133 (2005).
5. Slowinski, K., Chamberlain, R. V., Miller, C. J. & Majda, M. Through-bond and chain-to-chain coupling. Two pathways in electron tunneling through liquid alkanethiol monolayers on mercury electrodes. *J. Am. Chem. Soc.* **119,** 11910–11919 (1997).
6. Wold, D. J. & Frisbie, C. D. Formation of metal-molecule-metal tunnel junctions: Microcontacts to alkanethiol monolayers with a conducting AFM tip. *J. Am. Chem. Soc.* **122,** 2970–2971 (2000).
7. Wang, W., Lee, T. & Reed, M. A. Mechanism of electron conduction in self-assembled alkanethiol monolayer devices. *Phys. Rev. B* **68,** 035416 (2003).
8. Haiss, W. *et al.* Thermal gating of the single molecule conductance of alkanedithiols. *Faraday Discuss.* **131,** 253–264 (2006).
9. Xu, B. Q. & Tao, N. J. Measurement of single molecule conductance by repeated formation of molecular junctions. *Science* **301,** 1221–1223 (2003).
10. Venkataraman, L. *et al.* Single-molecule circuits with well-defined molecular conductance. *Nano Lett.* **6,** 458–462 (2006).
11. Cui, X. D. *et al.* Reproducible measurement of single-molecule conductivity. *Science* **294,** 571–574 (2001).
12. Xiao, X., Xu, B. & Tao, N. Conductance titration of single-peptide molecules. *J. Am. Chem. Soc.* **126,** 5370–5371 (2004).
13. Li, X. *et al.* Conductance of single alkanedithiols: Conduction mechanism and effect of molecule-electrode contacts. *J. Am. Chem. Soc.* **128,** 2135–2141 (2006).
14. Hu, Y. B., Zhu, Y., Gao, H. J. & Guo, H. Conductance of an ensemble of molecular wires: A statistical analysis. *Phys. Rev. Lett.* **95,** 156803 (2005).
15. Muller, K. H. Effect of the atomic configuration of gold electrodes on the electrical conduction of alkanedithiol molecules. *Phys. Rev. B* **73,** 045403 (2006).
16. Lee, M. H., Speyer, G. & Sankey, O. F. Electron transport through single alkane molecules with different contact geometries on gold. *Phys. Status Solidi B* **243,** 2021–2029 (2006).
17. Ishizuka, K. *et al.* Effect of molecule-electrode contacts on single-molecule conductivity of mu-conjugated system measured by scanning tunnelling microscopy under ultrahigh vacuum. *Japn. J. Appl. Phys. 1* **45,** 2037–2040 (2006).
18. Chidsey, C. E. D. Free energy and temperature dependence of electron transfer at the metal–electrolyte interface. *Science* **251,** 919–922 (1991).
19. Bumm, L. A. *et al.* Are single molecular wires conducting? *Science* **271,** 1705–1707 (1996).
20. Kergueris, C. *et al.* Electron transport through a metal-molecule-metal junction. *Phys. Rev. B* **59,** 12505–12513 (1999).
21. Zhitenev, N. B., Meng, H. & Bao, Z. Conductance of small molecular junctions. *Phys. Rev. Lett.* **88,** 226801. (2002).
22. Kim, B., Beebe, J. M., Jun, Y., Zhu, X. Y. & Frisbie, C. D. Correlation between HOMO alignment and contact resistance in molecular junctions: Aromatic thiols versus aromatic isocyanides. *J. Am. Chem. Soc.* **128,** 4970–4971 (2006).
23. He, J. *et al.* Electronic decay constant of carotenoid polyenes from single-molecule measurements. *J. Am. Chem. Soc.* **127,** 1384–1385 (2005).
24. Xu, B., Li, X., Xiao, X., Sakaguchi, H. & Tao, N. Electromechanical and conductance switching properties of single oligothiophene molecules. *Nano Lett.* **5,** 1491–1495 (2005).
25. Kuznetsov, A. M. & Ulstrup, J. *Electron Transfer in Chemistry and Biology. An Introduction to the Theory.* (Wiley, Chichester, 1999).
26. Weiss, E. A. *et al.* Conformationally gated switching between superexchange and hopping within oligo-p-phenylene-based molecular wires. *J. Am. Chem. Soc.* **127,** 11842–11850 (2005).
27. Nitzan, A. A relationship between electron-transfer rates and molecular conduction. *J. Phys. Chem. A* **105,** 2677–2679 (2001).
28. Aviram, A. & Ratner, M. Molecular rectifiers. *Chem. Phys. Lett.* **29,** 277–283 (1974).
29. Chen, J., Reed, M. A., Rawlett, A. M. & Tour, J. M. Large on-off ratios and negative differential resistance in a molecular electronic device. *Science* **286,** 1550–1552 (1999).
30. Collier, C. P. *et al.* Electronically configurable molecular-based logic gates. *Science* **285,** 391–394 (1999).
31. Metzger, R. M. Unimolecular electrical rectifiers. *Chem. Rev.* **103,** 3803–3834 (2003).
32. Martin, A. S., Sambles, J. R. & Ashwell, G. J. Molecular rectifier. *Phys. Rev. Lett.* **70,** 218–221 (1993).
33. Ng, M.-K. & Yu, L. Synthesis of amphiphilic conjugated diblock oligomers as molecular diodes. *Angew. Chem. Int. Edn* **41,** 3598–3601 (2002).
34. Jiang, P., Morales, G. M., You, W. & Yu, L. Synthesis of diode molecules and their sequential assembly to control electron transport. *Angew. Chem. Int. Edn* **43,** 4471–4475 (2004).
35. Smit, R. H. M. *et al.* Measurement of the conductance of a hydrogen molecule. *Nature* **419,** 906–909 (2002).
36. Elbing, M. *et al.* A single-molecule diode. *Proc. Natl Acad. Sci. USA* **102,** 8815–8820 (2005).
37. Liu, R., Ke, S. H., Yang, W. T. & Baranger, H. U. Organometallic molecular rectification. *J. Chem. Phys.* **124,** 024718 (2006).
38. Reichert, J. *et al.* Driving current through single organic molecules. *Phy. Rev. Lett.* **88,** 176804 (2002).
39. Kushmerick, J. G., Whitaker, C. M., Pollack, S. K., Schull, T. L. & Shashidar, R. Tuning current rectification across molecular junctions. *Nanotechnology* **15,** S489–S493 (2004).
40. Chabinyc, M. L. *et al.* Molecular rectification in a metal-insulator-metal junction based on self-assembled monolayers. *J. Am. Chem. Soc.* **124,** 11730–11736 (2002).
41. Xiao, X., Xu, B. & Tao, N. Changes in the conductance of single peptide molecules upon metal-ion binding. *Angew. Chem. Int. Edn* **43,** 6148–6152 (2004).
42. McCreery, R. *et al.* Molecular rectification and conductance switching in carbon-based molecular junctions by structural rearrangement accompanying electron injection. *J. Am. Chem. Soc.* **125,** 10748–10758 (2003).
43. Mujica, V., Ratner, M. A. & Nitzan, A. Molecular rectification: why is it so rare? *Chem. Phys.* **281,** 147–150 (2002).
44. Troisi, A. & Ratner, M. A. Conformational molecular rectifiers. *Nano Lett.* **4,** 591–595 (2004).
45. Xue, Y. Q. *et al.* Negative differential resistance in the scanning-tunneling spectroscopy of organic molecules. *Phys. Rev. B* **59,** R7852-R7855 (1999).
46. Fan, F. R. F. *et al.* Charge transport through self-assembled monolayers of compounds of interest in molecular electronics. *J. Am. Chem. Soc.* **124,** 5550–5560 (2002).
47. Rawlett, A. M. *et al.* Electrical measurements of a dithiolated electronic molecule via conducting atomic force microscopy. *Appl. Phys. Lett.* **81,** 3043–3045 (2002).
48. Xiao, X., Nagahara, L. A., Rawlett, A. M. & Tao, N. Electrochemical gate-controlled conductance of single oligo(phenylene ethynylene)s. *J. Am. Chem. Soc.* **127,** 9235–9240 (2005).
49. Le, J. D., He, Y., Hoye, T. R., Mead, C. C. & Kiehl, R. A. Negative differential resistance in a bilayer molecular junction. *Appl. Phys. Lett.* **83,** 5518–5520 (2003).
50. Kiehl, R. A., Le, J. D., Candra, P., Hoye, R. C. & Hoye, T. R. Charge storage model for hysteretic negative-differential resistance in metal-molecule-metal junctions. *Appl. Phys. Lett.* **88,** 172102 (2006).
51. Wassel, R. A. C., Grace, M., Fuierer, R. R., Feldheim, D. L. & Gorman, C. B. Attenuating negative differential resistance in an electroactive self-assembled monolayer-based junction. *J. Am. Chem. Soc.* **126,** 295–300 (2004).
52. Zeng, C., Wang, H., Wang, B., Yang, J. & Hou, J. G. Negative differential-resistance device involving two C_{60} molecules. *Appl. Phys. Lett.* **77,** 3595–3597 (2000).
53. Guisinger, N. P., Greene, M. E., Basu, R., Baluch, A. S. & Hersam, M. C. Room temperature negative differential resistance through individual organic molecules on silicon surfaces. *Nano Lett.* **4,** 55–59 (2004).
54. Salomon, A., Arad-Yellin, R., Shanzer, A., Karton, A. & Cahen, D. Stable room-temperature molecular negative differential resistance based on molecule-electrode interface chemistry. *J. Am. Chem. Soc.* **126,** 11648–11657 (2004).
55. Gaudioso, J. & Ho, W. Steric turnoff of vibrationally mediated negative differential resistance in a single molecule. *Angew. Chem. Int. Edn* **40,** 4080–4082 (2001).
56. Fan, F. R. F. *et al.* Electrons are transported through phenylene-ethynylene oligomer monolayers via localized molecular orbitals. *J. Am. Chem. Soc.,* **126,** 2568–2573 (2004).
57. Karzazi, Y., Cornil, J. & Brédas, J. L. Negative differential resistance behavior in conjugated molecular wires incorporating spacers: A quantum-chemical description. *J. Am. Chem. Soc.* **123,** 10076 - 10084 (2001).
58. Emberly, E. G. & Kirczenow, G. Current-driven conformational changes, charging, and negative differential resistance in molecular wires. *Phys. Rev. B* **64,** 125318 (2001).
59. Galperin, M., Ratner, M. A. & Nitzan, A. Hysteresis, switching, and negative differential resistance in molecular junctions: A polaron model. *Nano Lett.* **5,** 125–130 (2005).
60. He, J. & Lindsay, S. M. On the mechanism of negative differential resistance in ferrocenylundecanethiol self-assembled monolayers. *J. Am. Chem. Soc.* **127,** 11932–11933 (2005).
61. Pitters, J. L. & Wolkow, R. A. Detailed studies of molecular conductance using atomic resolution scanning tunneling microscopy. *Nano Lett.* **6,** 390–397 (2006).
62. Reed, M. A., Chen, J., Rawlett, A. M., Price, D. W. & Tour, J. M. Molecular random access memory cell. *Appl. Phys. Lett.* **78,** 3735–3737 (2001).
63. Cai, L. T. *et al.* Reversible bistable switching in nanoscale thiol-substituted oligoaniline molecular junctions. *Nano Lett.* **5,** 2365–2372 (2005).
64. Moonen, N. N. P., Flood, A. H., Fernandez, J. M. & Stoddart, J. F. Towards a rational design of molecular switches and sensors from their basic building blocks. *Top. Curr. Chem.* **262,** 99–132 (2005).
65. Stewart, D. R. *et al.* Molecule-independent electrical switching in Pt/organic monolayer/Ti devices. *Nano Lett.* **4,** 133–136 (2004).
66. Blum, A. S. *et al.* Molecularly inherent voltage-controlled conductance switching. *Nature Mater.* **4,** 167–172 (2005).
67. Keane, Z. K., Ciszek, J. W., Tour, J. M. & Natelson, D. Three-terminal devices to examine single-molecule conductance switching. *Nano Lett.* **6,** 1518–1521 (2006).
68. Qiu, X. H., Nazin, G. V. & Ho, W. Mechanisms of reversible conformational transitions in a single molecule. *Phys. Rev. Lett.* **93,** 196806 (2004).
69. Di Ventra, M., Pantelides, S. T. & Lang, N. D. The benzene molecule as a molecular resonant-tunneling transistor. *Appl. Phys. Lett.* **76,** 3448–3450 (2000).

70. Damle, P., Rakshit, T., Paulsson, M. & Datta, S. Current-voltage characteristics of molecular conductors: two versus three terminal. *IEEE T. Nanotechnol.* **1,** 145–1153 (2002).

71. Tans, S. J., Verschueren, A. R. M. & Dekker, C. Room temperature transistor based on a single carbon nanotube. *Nature* **393,** 49–52 (1998).

72. Park, J. *et al.* Coulomb blockade and the Kondo effect in single-atom transistors. *Nature* **417,** 722–725 (2002).

73. Liang, W. J., Shores, M., Bockrath, M., Long, J. R. & Park, H. Kondo resonance in a single-molecule transistor. *Nature* **417,** 725–729 (2002).

74. Li, C. Z., Bogozi, A., Huang, W. & Tao, N. J. Fabrication of stable metallic nanowires with quantized conductance. *Nanotechnology* **10,** 221–223 (1999).

75. Morpurgo, A. F., Marcus, C. M. & Robinson, D. B. Controlled fabrication of metallic electrodes with atomic separation. *Appl. Phys. Lett.* **14,** 2082 (1999).

76. Lee, J.-O. *et al.* Absence of strong gate effects in electrical measurements on phenylene-based conjugated molecules. *Nano Lett.* **3,** 113–117 (2003).

77. Luber, S. M. *et al.* Nanometre spaced electrodes on a cleaved AlGaAs surface. *Nanotechnology* **16,** 1182–1185 (2005).

78. Kubatkin, S. *et al.* Single-electron transistor of a single organic molecule with access to several redox states. *Nature* **425,** 698–701 (2003).

79. Ghosh, S. *et al.* Device structure for electronic transport through individual molecules using nanoelectrodes. *Appl. Phys. Lett.* **87,** 233509 (2005).

80. Champagne, A. R., Pasupathy, A. N. & Ralph, D. C. Mechanically adjustable and electrically gated single-molecule transistors. *Nano Lett.* **5,** 305–308 (2005).

81. Dadosh, T. *et al.* Measurement of the conductance of single conjugated molecules. *Nature* **436,** 677–680 (2005).

82. Chae, D. H. *et al.* Vibrational excitations in single trimetal-molecule transistors. *Nano Lett.* **6,** 165–168 (2006).

83. Haiss, W. *et al.* Redox state dependence of single molecule conductivity. *J. Am. Chem. Soc.* **125,** 15294–15295 (2003).

84. Xu, B., Xiao, X., Yang, X., Zang, L. & Tao, N. Large gate modulation in the current of a room temperature single molecule transistor. *J. Am. Chem. Soc.* **127,** 2386–2387 (2005).

85. Chen, F. *et al.* A molecular switch based on potential-induced changes of oxidation state. *Nano Lett.* **5,** 503–506 (2005).

86. Albrecht, T., Guckian, A., Ulstrup, J. & Vos, J. G. Transistor-like behavior of transition metal complexes. *Nano Lett.* **5,** 1451–1455 (2005).

87. Li, Z. *et al.* Two-dimensional assembly and local redox-activity of molecular hybrid structures in an electrochemical environment. *Faraday Discuss.* **131,** 121–143 (2006).

88. Tao, N. J. Probing potential-tuned resonant tunneling through redox molecules with scanning tunneling mircoscopy. *Phys. Rev. Lett.* **76,** 4066–4069 (1996).

89. Gittins, D. I., Bethell, D., Schiffrin, D. J. & Nichols, R. J. A nanometre-scale electronic switch consisting of a metal cluster and redox-addressable groups. *Nature* **408,** 67–69 (2000).

90. Tran, E., Rampi, M. A. & Whitesides, G. M. Electron transfer in a Hg-SAM//SAM-Hg junction mediated by redox centers. *Angew. Chem., Int. Edn* **43,** 3835–3839 (2004).

91. Mujica, V., Nitzan, A., Datta, S., Ratner, M. A. & Kubiak, C. P. Molecular wire junctions: Tuning the conductance. *J. Phys. Chem. B* **107,** 91–95 (2003).

92. Kasibhatla, B. S. T. *et al.* Reversibly altering electronic conduction through a single molecule by a chemical binding event. *J. Phys. Chem. B* **107,** 12378–12382 (2003).

93. Piva, P. G. *et al.* Field regulation of single-molecule conductivity by a charged surface atom. *Nature* **435,** 658–661 (2005).

94. Joachim, C., Gimzewski, J. K. & Aviram, A. Electronics using hybrid-molecular and mono-molecular devices. *Nature* **408,** 541–548 (2000).

95. Troisi, A. & Ratner, M. A. Conformational molecular rectifiers. *Nano Lett.* **4,** 591–595 (2004).

96. Joachim, C. & Gimzewski, J. K. An electromechanical amplifier using a single molecule. *Chem. Phys. Lett.* **265,** 353–357 (1997).

97. Park, H. *et al.* Nanomechanical oscillations in a single-C-60 transistor. *Nature* **407,** 57–60 (2000).

98. Cao, J., Wang, Q. & Dai, H. Electromechanical Properties of Metallic, Quasimetallic, and semiconducting carbon nanotubes under stretching. *Phys. Rev. Lett.* **90,** 157601 (2003).

99. Minot, E. D. *et al.* Tuning carbon nanotube band gaps with strain. *Phys. Rev. Lett.* **90,** 156401 (2003).

100. Kaun, C. C. & Seideman, T. Current-driven oscillations and time-dependent transport in nanojunctions. *Phys. Rev. Lett.* **94,** 226801 (2005).

101. Brwone, W. R. & Feringa, B. *Nature Nanotech.* **1,** 25–35 (2006).

102. Dulic, D. *et al.* One-way optoelectronic switching of photochromic molecules on gold. *Phys. Rev. Lett.* **91,** 207402 (2003).

103. He, J. *et al.* Switching of a photochromic molecule on gold electrodes: single-molecule measurements. *Nanotechnology* **16,** 695–702 (2005).

104. Li, J., Speyer, G. & Sankey, O. F. Conduction switching of photochromic molecules. *Phys. Rev. Lett.* **93,** 248302 (2004).

105. Moskalets, M. & Büttiker, M. Adiabatic quantum pump in the presence of external ac voltages. *Phys. Rev. B* **69,** 205316 (2004).

106. Galperin, M. & Nitzan, A. Current-induced light emission and light-induced current in molecular-tunneling junctions. *Phys. Rev. Lett.* **95,** 206802 (2005).

107. Lehmann, J., Camalet, S., Kohler, S. & Hänggi, P. Laser controlled molecular switches and transistors. *Chem. Phys. Lett.* **368,** 282–288 (2003).

108. Flaxer, E., Sneh, O. & Chesnovsky, O. Molecular light-emission induced by inelastic electron-tunneling. *Science* **262,** 2012–2014 (1993).

109. Berndt, R. *et al.* Photon-emission at molecular resolution induced by a scanning tunneling microscope. *Science* **262,** 1425–1427 (1993).

110. Qiu, X. H., Nazin, G. V. & Ho, W. Vibrationally resolved fluorescence excited with submolecular precision. *Science* **299,** 542–546 (2003).

111. Buker, J. & Kirczenow, G. Two-probe theory of scanning tunneling microscopy of single molecules: Zn(II)-etioporphyrin on alumina. *Phys. Rev. B* **72,** 205338 (2005).

112. Lee, T. H., Gonzalez, J. I., Zheng, J. & Dickson, R. M. Single-molecule optoelectronics. *Acc. Chem. Res.* **38,** 534–541 (2005).

113. Emberly, E. G. & Kirczenow, G. Molecular spintronics: spin-dependent electron transport in molecular wires. *Chem. Phys.* **281,** 311–324 (2002).

114. Pati, R., Senapati, L., Ajayan, P. M. & Nayak, S. K. First-principles calculations of spin-polarized electron transport in a molecular wire: Molecular spin valve. *Phys. Rev. B* **68,** 100407 (2003).

115. Rocha, A. R. *et al.* Towards molecular spintronics. *Nature Mater.* **4,** 335–339 (2005).

116. Waldron, D., Haney, P., Larade, B., MacDonald, A. & Guo, H. Nonlinear spin current and magnetoresistance of molecular tunnel junctions. *Phys. Rev. Lett.* **96,** 166804 (2006).

117. Ouyang, M. & Awschalom, D. D. Coherent spin transfer between molecularly bridged quantum dots. *Science* **301,** 1074–1078 (2003).

118. Xiong, Z. H., Wu, D., Vardeny, Z. V. & Shi, J. Giant magnetoresistance in organic spin-valves. *Nature* **427,** 821–824 (2004).

119. Petta, J. R., Slater, S. K. & Ralph, D. C. Spin-dependent transport in molecular tunnel junctions. *Phys. Rev. Lett.* **93,** 136601 (2004).

120. Pasupathy, A. N. *et al.* The Kondo Effect in the Presence of Ferromagnetism. *Science* **306,** 86–89 (2005).

121. Boon, E. M., Ceres, D. M., Drummond, T. G., Hill, M. G. & J. K. Barton, J. K. Mutation detection by electrocatalysis at DNA-modified electrodes. *Nature Biotechnol.* **18,** 1096–1100 (2000).

122. Hihath, J., Xu, B., Zhang, P. & Tao, N. Study of single-nucleotide polymorphisms by means of electrical conductance measurements. *Proc. Natl Acad. Sci. USA* **102,** 16979–16983 (2005).

123. Porath, D., Cuniberti, G. & Di Felice, R. Charge transport in DNA-based devices. *Top. Curr. Chem.* **237,** 183 (2004).

124. Cui, Y., Wei, Q., Park, H. & Lieber, C. M. Nanowire nanosensors for highly sensitive and selective detection of biological and chemical species. *Science* **293,** 1289–1292 (2001).

125. Xue, Y. & Ratner, M. A. End group effect on electrical transport through individual molecules: A microscopic study. *Phys. Rev. B* **69,** 085403 (2004).

126. Basch, H., Cohen, R. & Ratner, M. A. Interface geometry and molecular junction conductance: Geometric fluctuation and stochastic switching. *Nano Lett.* **5,** 1668–1675 (2005).

127. Guo, X. F. *et al.* Covalently bridging gaps in single-walled carbon nanotubes with conducting molecules. *Science* **311,** 356–359 (2006).

128. Stewart, M. P. *et al.* Direct covalent grafting of conjugated molecules onto Si, GaAs, and Pd surfaces from aryldiazonium salts. *J. Am. Chem. Soc.* **126,** 370–378 (2004).

129. McGuiness, C. L. *et al.* Molecular self-assembly at bare semiconductor surfaces: Preparation and characterization of highly organized octadecanethiolate monolayers on GaAs(001). *J. Am. Chem. Soc.* **128,** 5231–5243 (2006).

130. Tulevski, G. S., Myers, M. B., Hybertsen, M. S., Steigerwald, M. L. & Nuckolls, C. Formation of catalytic metal-molecule contacts. *Science* **309,** 591–594 (2005).

131. Li, X. L. *et al.* Controlling charge transport in single molecules using electrochemical gate. *Faraday Discuss.* **131,** 111–120 (2006).

132. Donhauser, Z. J. *et al.* Conductance switching in single molecules through conformational changes. *Science* **292,** 2303–2307 (2001).

133. Ramachandran, G. K. *et al.* A bond-fluctuation mechanism for stochastic switching in wired molecules. *Science* **300,** 1413–1416 (2003).

134. Wassel, R. A., Fuierer, R. R., Kim, N. & Gorman, C. B. Stochastic variation in conductance on the nanometer scale: A general phenomenon. *Nano Lett.* **3,** 1617–1620 (2003).

135. Ralls, K. S. *et al.* Discrete resistance switching in submicrometer silicon inversion layers: individual interface traps and low-frequency (1/f?) noise. *Phys. Rev. Lett.* **52,** 228–231 (1984).

136. Dutta, P. & Horn, P. M. Low-frequency fluctuations in solids: 1/f noise. *Rev. Mod. Phys.* **53,** 497–516 (1981).

137. He, H. X. *et al.* A conducting polymer nanojunction switch. *J. Am. Chem. Soc.* **123,** 7730–7731 (2001).

138. Chen, Y.-C., Zwolak, M. & Di Ventra, M. Local heating in nanoscale conductors. *Nano Lett.* **3,** 1691 (2003).

139. Huang, Z. F., Xu, B. Q., Chen, Y. C., Di Ventra, M. & Tao, N. J. Measurement of current-induced local heating in a single molecule junction. *Nano Lett.* **6,** 1240–1244 (2006).

140. Li, W. J. *et al.* Ballistic electron emission microscopy studies of Au/molecule/n-GaAs diodes. *J. Phys. Chem. B* **109,** 6252–6256 (2005).

141. Lefenfeldt, M. *et al.* Direct structural observation of a molecular junction by high-energy x-ray reflectometry. *Proc. Natl Acad. Sci. USA* **103,** 2541–2545 (2006).

142. Kushmerick, J. G. *et al.* Vibronic contributions to charge transport across molecular junctions. *Nano Lett.* **4,** 639–642 (2004).

143. Wang, W. Y., Lee, T., Kretzschmar, I. & Reed, M. A. Inelastic electron tunneling spectroscopy of an alkanedithiol self-assembled monolayer. *Nano Lett.* **4,** 643–646 (2004).

144. McCreery, R. L. Analytical challenges in molecular electronics. *Anal. Chem.* **78,** 3490–3497 (2006).

145. Seideman, T. & Guo, H. Quantum transport and current-triggered dynamics in molecular tunnel junctions. *J. Theor. Comp. Chem.* **2,** 439–458 (2003).

Acknowledgements

I thank Hong Guo, Fang Chen and Josh Hihath for critical reading of the manuscript and NSF, DOE, VW and DARPA for support.

Molecular spintronics using single-molecule magnets

A revolution in electronics is in view, with the contemporary evolution of the two novel disciplines of spintronics and molecular electronics. A fundamental link between these two fields can be established using molecular magnetic materials and, in particular, single-molecule magnets. Here, we review the first progress in the resulting field, molecular spintronics, which will enable the manipulation of spin and charges in electronic devices containing one or more molecules. We discuss the advantages over more conventional materials, and the potential applications in information storage and processing. We also outline current challenges in the field, and propose convenient schemes to overcome them.

LAPO BOGANI AND WOLFGANG WERNSDORFER

Institut Néel, CNRS & Université Joseph Fourier, BP 166, 25 Avenue des Martyrs, 38042 GRENOBLE Cedex 9, France

e-mail: wolfgang.wernsdorfer@grenoble.cnrs.fr

The contemporary exploitation of electronic and spin degrees of freedom is a particularly promising field both at fundamental and applied levels[1]. This discipline, called spintronics, has already seen the passage from fundamental physics to technological devices in a record time of ten years, and holds great promise for the future[2]. Spintronic systems exploit the fact that the electron current is composed of spin-up and spin-down carriers, which carry information encoded in their spin state and interact differently with magnetic materials. Information encoded in spins persists when the device is switched off, can be manipulated without using magnetic fields and can be written using low energies, to cite just a few advantages of this approach. New efforts are now directed to obtaining spintronic devices that preserve and exploit quantum coherence, and fundamental investigations are shifting from metals to semiconducting[2,3] and organic materials[4]. The latter is currently used in applications such as organic light-emitting diode (OLED) displays and organic transistors. The concomitant trend towards ever-smaller electronic devices, having already passed the nanoscale, is driving electronics to its ultimate molecular-scale limit[5,6], which offers the possibility of exploiting quantum effects.

In this context, a new field of molecular spintronics is emerging that combines the ideas and the advantages of spintronics and molecular electronics[7]. The key point is the creation of molecular devices using one or a few magnetic molecules. Compounds of the single-molecule magnet (SMM) class[8,9] seem particularly attractive: their magnetization relaxation time is extremely long at low temperatures, reaching years below 2 K (ref. 8). These systems, combining the advantages of the molecular scale with the properties of bulk magnetic materials, look attractive for high-density information storage and also, due to their long coherence times[10,11], for quantum computing[12–14]. Moreover, their molecular nature leads to appealing quantum effects of the static and dynamic magnetic properties[15–17]. The rich physics behind the magnetic behaviour produces interesting effects such as negative differential conductance and complete current suppression[18,19], which could be used in electronics. In addition, specific functions (for

example, switchability with light, electric field and so on) could be directly integrated into the molecule.

This progress article aims to point out the potential of SMMs in molecular spintronics, as recently revealed by experimental and theoretical work, and to delineate future challenges and opportunities. We first show the unique chemical and physical properties of SMMs and then present three basic molecular schemes that demonstrate the state of the art of the emerging field of molecular spintronics. We propose possible solutions to existing problems, and particular emphasis is laid on the motivations and expected results. In the sideboxes, we summarize the quantum properties of SMMs (Box 1) and the basics of molecular electronics (Box 2).

CHARACTERISTICS OF SINGLE-MOLECULE MAGNETS

SMMs possess the correct chemical characteristics to overcome several problems associated with molecular junctions. They consist of an inner magnetic core with a surrounding shell of organic ligands[8,9] that can be tailored to bind them on surfaces or into junctions[20–23] (Fig. 1). To strengthen magnetic interactions between the core ions, SMMs often have delocalized bonds, which can enhance their conducting properties. SMMs come in a variety of shapes and sizes and allow selective substitutions of the ligands to alter the coupling to the environment[8,9,20,21,24]. It is also possible to exchange the magnetic ions, thus changing the magnetic properties without modifying the structure or coupling[25]. Although grafting SMMs on surfaces has already led to important results, even more spectacular results will emerge from the rational design and tuning of SMM-based junctions.

From a physical viewpoint, SMMs combine the classical macroscale properties of a magnet to the quantum properties of a nanoscale entity. They offer crucial advantages over magnetic nanoparticles in that they are perfectly monodisperse and can be studied in crystals. They display an impressive array of quantum effects (Box 1), ranging from quantum tunnelling of the magnetization[15,16] to Berry-phase interference[17] and quantum coherence[11], with important consequences for the physics of spintronic devices. Although the magnetic properties of SMMs can be affected when depositing on surfaces or between leads[24], these systems remain a

Box 1 Dynamics of SMMs

SMMs are nanoscale magnetic molecules that exhibit slow relaxation of the magnetization at low temperatures[8,9]. They often crystallize into high-quality crystals, thus allowing for a structural determination and rationalization of their properties. The magnetic properties are described using a spin hamiltonian, which takes the form $\mathcal{H} = DS_z^2 + E(S_x^2 - S_y^2) + g\mu_B\mu_0 \mathbf{S}\cdot\mathbf{H}$, where S_x, S_y and S_z are the spin components, D and E are the magnetic anisotropy constants, and the term $g\mu_B\mu_0\mathbf{S}\cdot\mathbf{H}$ describes the Zeeman energy associated with an applied magnetic field $\mathbf{H}$. Excited spin states and higher-order anisotropy terms are neglected for simplicity. For SMMs, $D < 0$ and an easy axis of magnetization is present along S_z. The energy potential takes the form of a double well, with the levels $S_m = \pm S$ having the lowest energy and a potential barrier in between. To reverse the magnetization, the spin has to overcome a potential energy barrier $\Delta = DS_z^2$ by climbing up and down all the $(2S+1)$ energy levels (Fig. B1a). The relaxation time will thus follow a thermally activated law, increasing exponentially as T is lowered. When the relaxation time is long at low temperatures, the system displays superparamagnetic behaviour with the opening of a hysteresis cycle (Fig. B1b).

Owing to their molecular nature, SMMs exhibit some striking quantum effects that affect the magnetization reversal mechanism. One of the most prominent is quantum tunnelling of the magnetization[8,9,15–17]. When $\mathbf{H}$ is applied along z, levels with $S_m < 0$ are shifted up and levels with $S_m > 0$ are shifted down. Levels of positive and negative quantum numbers will then cross at $\mathbf{H} = 0$ and at certain resonance fields. Transverse terms, containing S_x or S_y, turn the crossings into an avoided level-crossing and quantum tunnelling between the two wells can occur (Fig. B1a). The availability of this additional tunnelling mechanism for the relaxation of the magnetization produces the characteristic steps at the resonance fields in the hysteresis curve of SMMs (Fig. B1b).

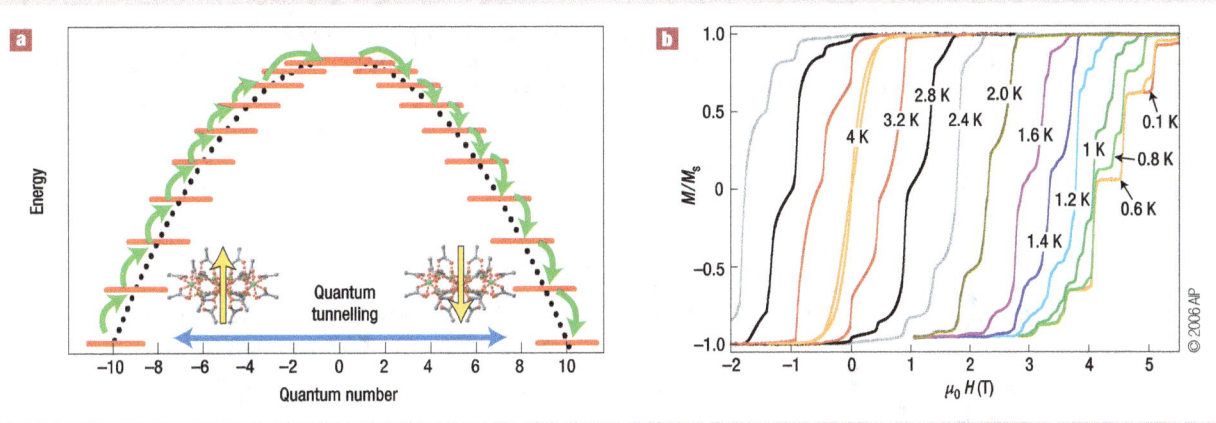

Figure B1 Dynamics of the magnetization in SMMs. **a**, Schematic representation of the energy landscape of a SMM with a spin ground state $S = 10$. The magnetization reversal can occur via quantum tunnelling between energy levels (blue arrow) when the enery levels in the two wells are in resonance. Phonon absorption (green arrows) can also excite the spin up to the top of the potential energy barrier with the quantum number $M = 0$, and phonon emission descends the spin to the second well. **b**, Hysteresis loops of single crystals of $[Mn_{12}O_{12}(O_2CCH_2C(CH_3)_3)_{16}(CH_3OH)_4]$ SMM at different temperatures and a constant field sweep rate of 2 mT s^{-1} (data from ref. 65). The loops exhibit a series of steps, which are due to resonant quantum tunnelling between energy levels. As the temperature is lowered, there is a decrease in the transition rate due to reduced thermally assisted tunnelling. The hysteresis loops become temperature-independent below 0.6 K, demonstrating quantum tunnelling at the lowest energy levels.

step ahead of non-molecular nanoparticles, which have large size and anisotropy distributions.

MOLECULAR SPIN-TRANSISTOR

The first scheme we consider is a magnetic molecule attached between two non-magnetic electrodes. One possibility is to use a scanning tunnelling microscope (STM) tip as the first electrode and the conducting substrate as the second one (Fig. 2a). So far, only few atoms on surfaces have been probed in this way, revealing interesting Kondo effects[26] and single-atom magnetic anisotropies[27]. The next scientific step is to pass from atoms to molecules to observe richer physics and to modify the properties of the magnetic objects. Although isolated SMMs on gold have been obtained[20–23], the rather drastic experimental requirements for studying them by STM — that is, very low temperatures and high magnetic fields — have not yet been achieved. The first theoretical work predicted that quantum tunnelling of the magnetization (Box 1) is detectable via the electric current flowing through the molecule[28], thus enabling the readout of the quantum dynamics of a single molecule.

Another possibility concerns break-junction devices[29], which contain a gate electrode in addition to source and drain electrodes. Such a three-terminal transport device, called a molecular spin-transistor, is a single-electron transistor with non-magnetic electrodes and a single magnetic molecule as the island. The current passes through the magnetic molecule through the source and drain electrodes, and the electronic transport properties are tuned using a gate voltage V_g (Fig. 2b). Similarly to molecular electronics (Box 2), two experimental regimes can be distinguished, depending on the coupling between molecule and electrodes.

WEAK-COUPLING LIMIT

In the weak-coupling limit, charging effects dominate the transport. Transport takes place when a molecular orbital is in resonance with the Fermi energy E_F of the leads (Fig. B2), and electrons can then tunnel through the energy barrier into the molecular level and out

Box 2 Molecular electronics

The energy spectrum of molecules and semiconducting quantum dots (QDs), are characterized by discrete electronic energy levels (Fig. B2). As much of the physics is the same whether a molecule or a QD is placed in the junction, the two are often used interchangeably. Although it is often complicated to know the spacing of all levels, many properties are determined by the energies of the highest occupied molecular orbital (HOMO) and the lowest unoccupied molecular orbital (LUMO).

Molecules can be weakly or strongly coupled to the electrodes, with low or high perturbations of molecular states by the presence of the electrodes (Fig. B2a–c). Physically, this is depicted by an energy barrier between the contacts and the molecule, whose height determines the mixing between the wavefunctions of the molecule and of the electrodes. When the energy broadening of the levels due to hybridization (Γ) is smaller than the charging energy U of the molecule ($\Gamma < U$), the molecule is weakly coupled to the leads, whereas for $\Gamma > U$ strong coupling is obtained (Fig. B2a–c).

For $\Gamma < U$ and at temperatures T comparable to the energy between the Fermi energy E_F of the leads and the HOMO or LUMO, electron transport occurs by emptying the HOMO or filling the LUMO. At low T, electron transport is blocked (Fig. B2d)

except for V_g values that bring the molecular levels in resonance with E_F (Fig. B2e) or for bias voltages that bring E_F to the energy of molecular levels (Fig. B2f). This regime, called Coulomb-blockade, leads to large regions of blocked conductance, forming characteristic Coulomb diamonds in differential conductance plots[30,31]. Vibrational and spin excitation levels can also lead to resonance levels and appear inside the diamonds, making transport measurements essentially a form of spectroscopy (Fig. B2g).

For $\Gamma > U$, the molecular wavefunctions are not appropriate eigenstates of the system, and are replaced by hybrid states. Hybridization shifts both the LUMO and HOMO closer to E_F and broadens the electronic states. Because electrons are delocalized between electrodes and the molecule, the system can be described by an occupancy fraction of the new HOMO and LUMO levels. When an unpaired electron occupies the HOMO, electrons moving to and from the molecule reverse the spin of the unpaired electron and a screening of the spin occurs, in analogy to the well-known Kondo effect[37–40]. Below a certain Kondo temperature T_K, the screening leads to a zero-bias conductance resonance, which is associated with the entangled state of electrons in the leads and in the molecule.

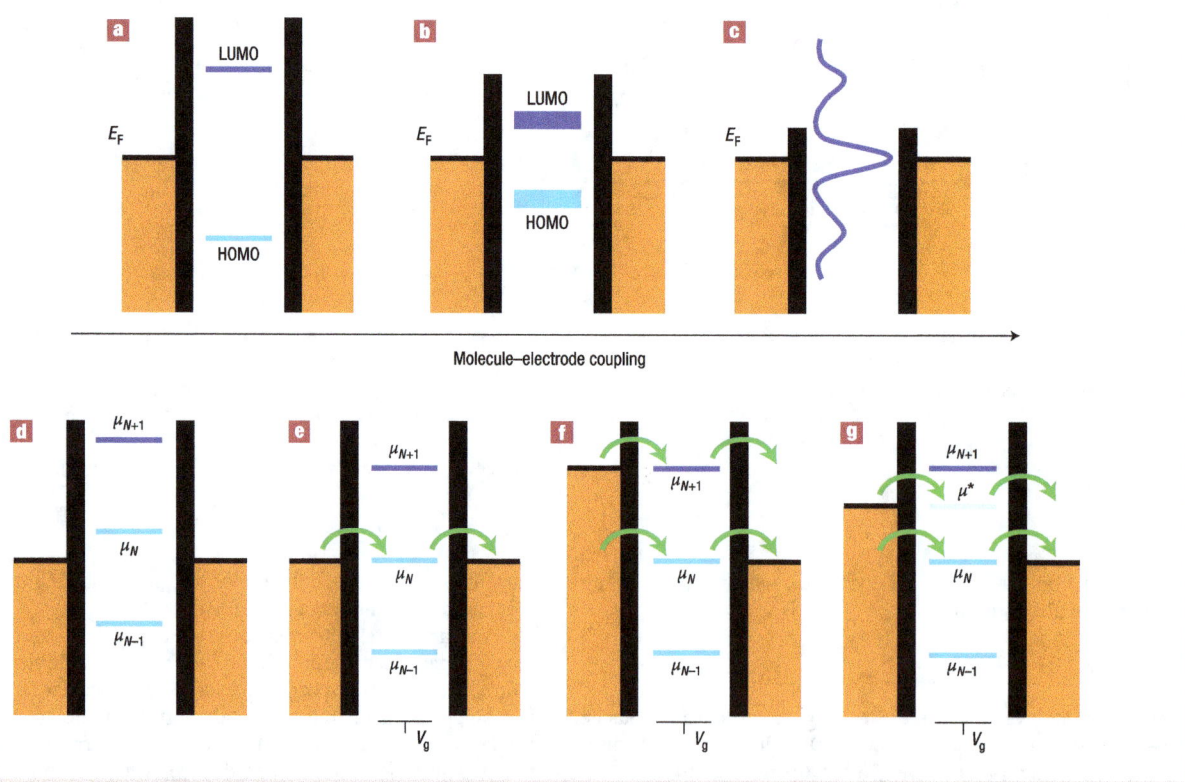

Figure B2 Schematic representation of the basic processes of non-magnetic molecular electronics. **a–c,** Change of the molecular levels on varying the transparency of the tunnel barriers (black) between the electrodes and the molecule. The HOMO (light blue) and LUMO (deep blue) are well defined in the weak-coupling regime (**a**), become broader and closer to E_F as coupling is increased (**b**) and eventually give rise to a semicontinuous multipeaked structure for very strong coupling (**c**). In this case, and when an unpaired spin characterizes the ground state of the QD, a screening of the spin occurs, in analogy to the well-known Kondo effect in solids containing magnetic impurities[37–40]. **d–g,** Processes giving rise to Coulomb diamonds; μ_N, μ_{N-1} and μ_{N+1} denote the electronic energy levels of the QD. Out of resonance (**d**) electrons cannot pass through the device. They can tunnel through the barrier only if the molecular levels are put in resonance with E_F using V_g (**e**). At higher bias voltage, two or more electronic levels (**f**) can contribute to the electron transport. Additional energy levels (μ^*; **g**), for example vibrational levels, excited electronic or nuclear spin levels, can also induce electronic transport through the QD.

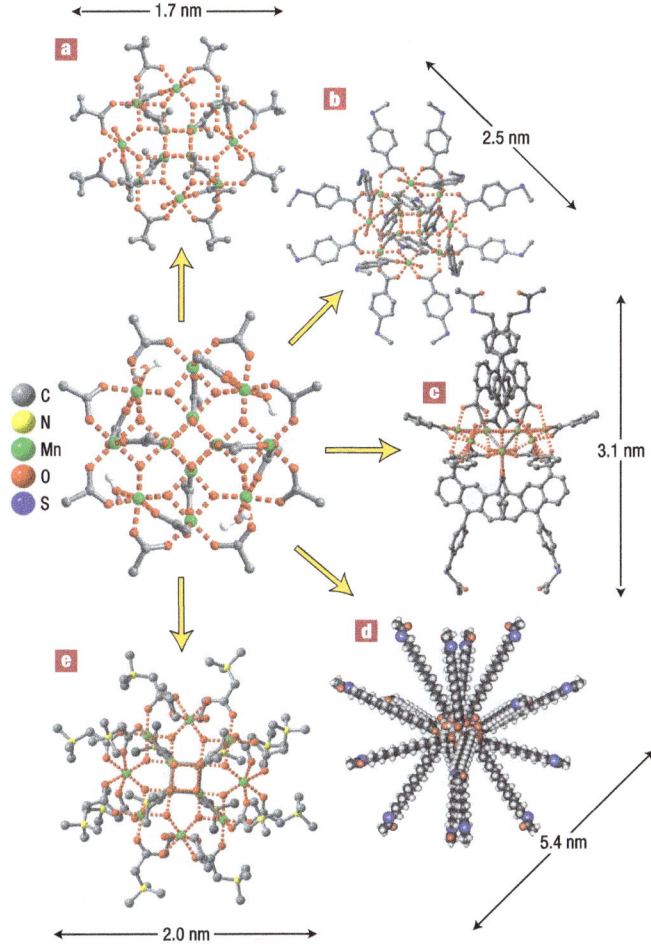

Figure 1 Representative examples of the peripheral functionalization of the outer organic shell of the $[Mn_{12}O_{12}(CH_3COO)_{16}(H_2O)_4]$ SMM (centre). Different functionalizations used to graft the SMM to surfaces are displayed: **a**, $[Mn_{12}O_{12}(C(CH_3)_3COO)_{16}(H_2O)_4]$. **b**, $[Mn_{12}O_{12}(p\text{-}CH_3S\text{-}C_6H_4\text{-}COO)_{16}(H_2O)_4]$. **c**, $[Mn_{12}O_{12}(O_2CC_6H_5)_8(1,8\text{-dicarboxyl-}10\text{-}(4\text{-acetylsulphanylmethyl-phenyl})\text{-}$ anthracene-1,8-dicarboxylic acid)$_4(H_2O)_4]$. **d**, $[Mn_{12}O_{12}(CH_3OS(CH_2)_{15}COO)_{16}(H_2O)_4]$. **e**, $[Mn_{12}O_{12}((CH_3)_3NCH_2COO)_{16}(H_2O)_4]^{16+}$. All structures are determined by X-ray crystallography, except **d**, which is a model structure. Solvent molecules have been omitted.

into the drain electrode. The resonance condition is obtained by shifting the energy levels with V_g and Coulomb-blockade diamonds appear in differential conductivity maps as a function of source–drain voltage and V_g (ref. 30).

The experimental realization of this scheme has been achieved by substituting the acetate molecules of the $[Mn_{12}O_{12}(CH_3COO)_{16}(H_2O)_4]$ SMM with thiol-containing ligands (Fig. 1), which bind the SMM to the gold electrodes with strong and reliable covalent bonds[18]. An alternative route is to use short but weak binding ligands[19]: in both cases the peripheral groups act as tunnel barriers and help conserve the magnetic properties of the SMM in the junction. As the electron transfer involves the charging of the molecule, we must consider, in addition to the neutral state, the magnetic properties of the negatively and positively charged species. This introduces an important difference with respect to the homologous measurements on diamagnetic molecules, where the assumption is often made that charging of the

molecule does not significantly alter the internal degrees of freedom[31]. Because crystals of the charged species can be obtained, SMMs permit direct comparison between spectroscopic transport measurements and more traditional characterization methods. In particular, magnetization measurements, electron paramagnetic resonance, and neutron spectroscopy can provide energy-level spacings and anisotropy parameters[8,9]. In the case of the $[Mn_{12}O_{12}(CH_3COO)_{16}(H_2O)_4]$ SMM, positively charged clusters possess a lower anisotropy barrier[32]. As revealed by the first Coulomb-blockade measurements, the presence of these states is fundamental for explaining transport through the clusters[18,19]. Negative differential conductance was found that might be due to the magnetic characteristics of SMMs.

Studies in magnetic fields showed the first evidence of spin-transistor properties[19]. Degeneracy at zero field and nonlinear behaviour of the excitations as a function of field are typical of tunnelling via a magnetic molecule. In these first studies, the lack of a hysteretic response can be due, besides environmental effects[24], to the alternation of the molecules during the grafting procedure, to the population of excited states with lower energy barriers, or it might also be induced by the source–drain voltage scan performed at each field value.

Theoretical investigations in the weak-coupling regime predict many interesting effects. For example, a direct link between shot noise measurements and the detailed microscopic magnetic structure of SMMs has been proposed[33], allowing the connection of structural and magnetic parameters to the transport features and therefore a characterization of SMMs using transport measurements. This opens the way to rational design of SMMs for spintronics, and to test the physical properties of related compounds. The first step in this direction has already been made by comparing the expected response of chemically related SMMs[34]. Note that this direct link cannot be established for nanoparticles or quantum-dots (QDs) because they do not possess a unique chemical structure.

A complete theoretical analysis as a function of the angle between the easy axis of magnetization and the magnetic field showed that the response persists whatever the orientation of the SMM in the junction, and that even films of SMMs should retain many salient properties of single-molecule devices[35,36].

STRONG-COUPLING LIMIT

For strong electronic coupling between the molecule and the leads, higher-order tunnel processes become important, leading to the Kondo effect[37–40] (Box 2). This regime has been attained using paramagnetic molecules containing one[41] or two magnetic centres[42], but remains elusive for SMMs.

The first mononuclear magnetic molecule investigated (Fig. 2c) is a Co(II) ion bound by two terpyridine ligands, TerPy, attached to the electrodes with chemical groups of variable length[41]. The system with the longer alkyl spacer, due to a lower transparency of the barrier, displays Coulomb-blockade diamonds, which are characteristic for the weak-coupling regime, but no Kondo peak. Experiments conducted as a function of magnetic field reveal the presence of excited states related to spin excitations, in agreement with the effective spin $S = 1/2$ state usually attributed to Co^{2+} ions at low temperatures. However, a Landé factor $g = 2.1$ is found, which is unexpected for Co^{2+} ions having strong spin-orbit coupling and high magnetic anisotropy — this point therefore needs further investigation. The same complex with the thiol directly connected to the TerPy ligand (Fig. 2d) shows strong coupling to the electrodes, with exceptionally high Kondo temperatures of around 25 K (ref. 41).

Additional physical effects of considerable interest were obtained using a simple molecule containing two magnetic centres[42]. This molecule, the divanadium molecule $[(N,N',N''\text{-trimethyl-}1,4,7\text{-}$ triazacyclononane)$_2V_2(CN)_4(\mu\text{-}C_4N_4)]$ (Fig. 2e) was directly grafted to the electrodes, so as to have the highest possible transparency[42]. The

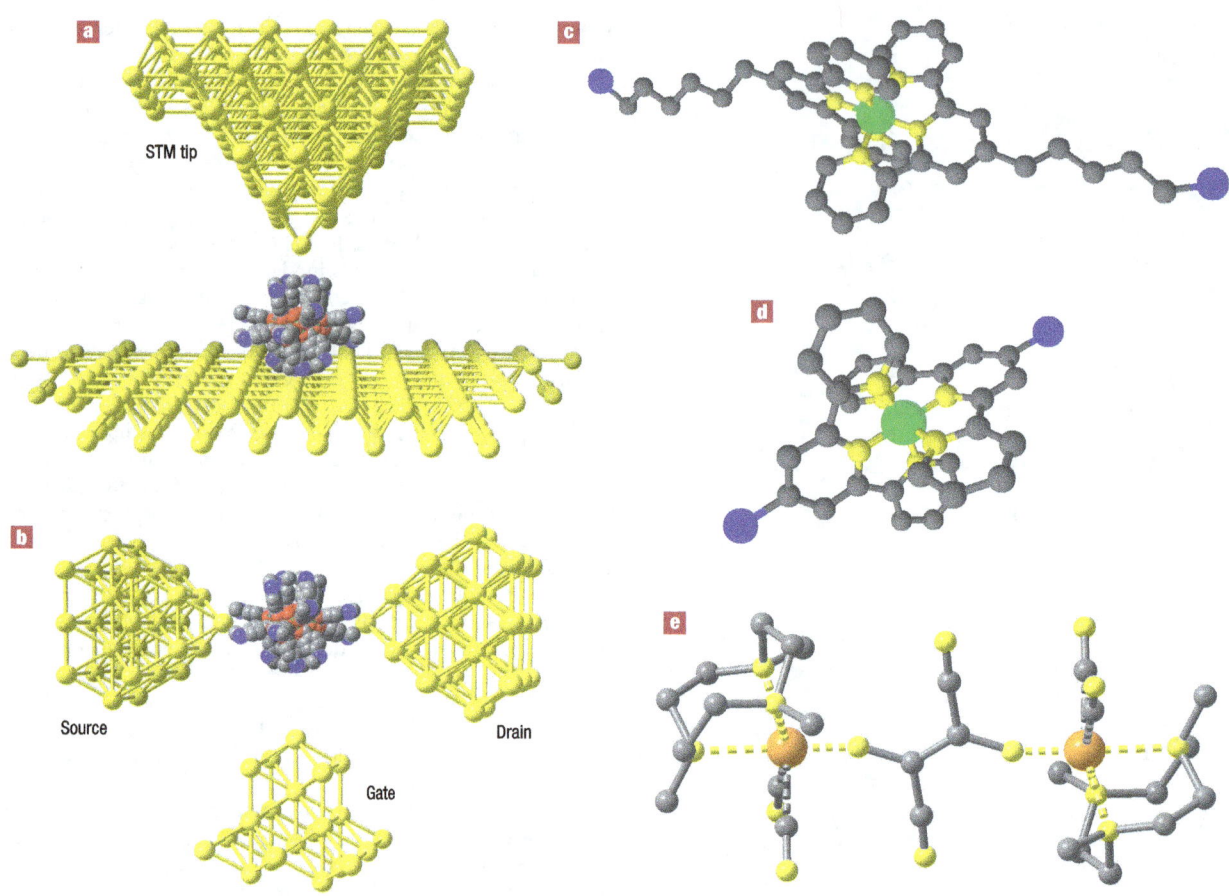

Figure 2 Transport experiments on SMMs. **a**, Schematic showing the use of an STM tip to perform transport experiments on surface-grafted SMMs. **b**, Schematic of SMM-based molecular transistors, in which a gate voltage can modulate transport. **c**, [Co(TerPy)$_2$] mononuclear magnet molecule with alkyl spacers, permitting transport in the weakly coupled regime[41]. **d**, [Co(TerPy)$_2$] without spacers, showing strong coupling and the Kondo effect[41]. **e**, Divanadium [(N,N',N''-trimethyl-1,4,7-triazacyclononane)$_2$V$_2$(CN)$_4$ (μ-C$_4$N$_4$)] molecular magnet showing the Kondo effect only in the charged state[42]. The colour code is the same as for Fig. 1, except for Co atoms (green) and V atoms (orange).

molecule can be tuned, using V_g, into two differently charged states. The neutral state, due to antiferromagnetic coupling between the two magnetic centres, has $S = 0$, whereas the positively charged state has $S = 1/2$. Kondo features are found, as expected[37–40], only for the state in which the molecule has a non-zero spin moment. This nicely demonstrates that magnetic molecules with multiple centres and antiferromagnetic interactions enable the Kondo effect to be switched on and off, depending on their charge state. The Kondo temperature is again exceptionally high, exceeding 30 K, and its characterization as a function of V_g indicates that not only spin but also orbital degrees of freedom have an important role in the Kondo resonance of single molecules. Molecular magnets, in which spin–orbit interaction can be tuned without altering the structure[9,25], are appealing for investigating this physics further.

The Kondo temperatures observed in the two cases[41,42] are much higher than those obtained for QDs and carbon nanotubes[37–40], and are extremely encouraging. The study of the superparamagnetic transition of SMMs while in the Kondo regime thus seems achievable, possibly leading to an interesting interplay of the two effects. To observe the Kondo regime one might start with small SMMs[25,43], with core states more affected by the proximity of the leads, and use short and strongly bridging ligands to connect SMMs to the electrodes[20,41].

Theoretical investigations have explored the rich physics of this regime[44,45], revealing that the Kondo effect should even be visible in SMMs with $S > 1/2$ (ref. 45). This is in contrast to expectations for a system with an anisotropy barrier, where the blocked spin should hinder co-tunnelling processes. However in SMMs, the presence of a transverse anisotropy induces a Kondo resonance peak[45]. The observation of this new physical phenomenon should be possible because of the tunability of SMMs, allowing a rational choice of the physical parameters governing the tunnelling process: low-symmetry transverse terms are particularly useful, because selection rules apply for high-symmetry terms.

The first theoretical predictions argued that the Kondo effect should be present only for half-integer spin molecules. However, the particular quantum properties of SMMs allow the Kondo effect to occur even for integer spins. In addition, the presence of the so-called Berry-phase interference[17], a geometrical quantum-phase effect, can produce not only one Kondo resonance peak, but a series of peaks as a function of applied magnetic field[46]. These predictions demonstrate how the molecular nature of SMMs and the quantum effects they exhibit differentiate them from inorganic QDs and nanoparticles, and should permit the observation of otherwise prohibited phenomena.

MOLECULAR SPIN-VALVE

A molecular spin-valve (SV)[7] is similar to a spin transistor but contains at least two magnetic elements (Fig. 3a,b). SVs change

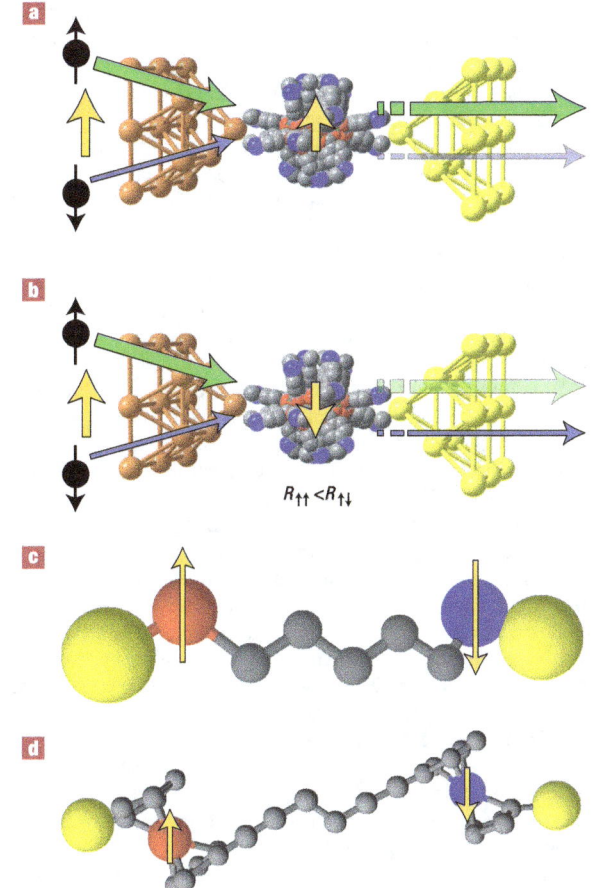

Figure 3 Spin valves based on molecular magnets. Yellow arrows represent the magnetization direction. **a**, Parallel configuration of the magnetic source electrode (orange) and molecular magnetization, with diamagnetic drain electrode (yellow). Spin-up majority carriers (thick green arrow) are not affected by the molecular magnetization, whereas the spin-down minority carriers (thin blue arrow) are partially reflected back. **b**, Antiparallel configuration: majority spin-up electrons are only partially transmitted by the differently polarized molecule, whereas the minority spin-down electrons pass unaffected. Assuming that the spin-up contribution to the current is larger in the magnetic contact, this configuration has higher resistance than that of the previous case. **c**, Theoretical schematic of a spin-valve configuration with non-magnetic metal electrodes[7] and **d**, proposed molecular magnet [$(C_5H_5)Co(C_5H_4-CCCH_2CH_2CC-C_5H_4)Fe(C_5H_5)$] between gold electrodes: a conjugated molecule bridges the cobaltocene (red) and ferrocene (blue) moieties[51].

their electrical resistance for different mutual alignments of the magnetizations of the electrodes and of the molecule, analogous to a polarizer–analyser set-up. Non-molecular devices are already used in hard-disk drives, owing to the giant- and tunnel-magnetoresistance effects. Good efficiency has already been demonstrated for organic materials[4], molecular SVs are therefore actively sought after[47,48]. As only few examples of molecular SVs exist[49,50], the fundamental physics behind these devices remains largely unexplored, and is likely to be the focus of considerable attention in the near future. The simplest SV consists of a diamagnetic molecule in between two magnetic leads, which can be metallic or semiconducting. The first experiments sandwiched a C_{60} fullerene between Ni electrodes, and exhibited a very large negative magnetoresistance effect[49]. Another interesting possibility is to use carbon nanotubes connected with

magnetic half-metallic electrodes transforming spin information into large electrical signals[50].

A SMM-based SV can have one or two magnetic electrodes (Fig. 3a,b), or the molecule can possess two magnetic centres in between two non-magnetic leads (Fig. 3c,d), in a scheme reminiscent of early theoretical models of SVs[7]. Molecules with two magnetic centres connected by a molecular spacer are well known in molecular magnetism, and a double-metallocene junction has been studied theoretically[51]. This seems a good choice, as the metallocenes leave the d-electrons of the metals largely unperturbed.

Theory indicates that, when using SMMs, the contemporary presence, at high bias, of large currents and slow relaxation will form a physically interesting regime[52,53]. Only spins parallel to the molecular magnetization can flow through the SMM and the current will display, for a time equivalent to the relaxation time, a very high spin polarization. For large currents this process can lead to a selective drain of spins with one orientation from the source electrode, thus transferring a large amount of magnetic moment from one lead to the other. This phenomenon, due to a sole SMM, has been named giant spin amplification[52], and offers a convenient way to read the magnetic state of the molecule. The switching of the device seems more complicated, at first sight, involving a two-step process that includes the application of a magnetic field and the variation of the bias voltage. However, it has recently been suggested that the spin-polarized current itself can be sufficient to switch the magnetization of a SMM[54]. The switching can be detected in the current as a step if both leads are magnetic and have parallel magnetization, or as a sharp peak for the antiparallel configuration.

MOLECULAR MULTIDOT DEVICES

Even more interesting phenomena can be obtained by transport measurements through a controlled number of molecules or multiple centres inside the same molecule[55]. The double-metallocene molecule[51] (Fig. 3d) can already be considered as a multidot system with level crossings giving rise to negative differential conductance[51]. The possibilities offered by the chemistry of SMMs seem particularly promising here[11,56,57]. Only one experiment, up to now, has considered the chemical fabrication of multidot devices[58], obtaining a controlled sequence of CdSe QDs connected by p-benzenedithiol molecular bridges (Fig. 4a), in a process reminiscent of the solid-phase synthesis of peptides[59]. Ultrafast pump–probe magneto-optical experiments, probing the spin transfer through the π-conjugated molecular bridge, indicate that the process is coherent and that the bridging molecule acts as a quantum-information bus. Coupled SMMs or spin-transfer molecular compounds[56,57] use similar molecular bridges, and we can thus expect analogous performances for spin transfer in bridged SMMs. With small decoherence[11–14], this should open the way to solid-state quantum computation protocols.

A simpler situation concerns a double-dot (Fig. 4b) with three terminals, where the current passes through a non-magnetic quantum conductor (quantum wire, nanotube, molecule or QD). The magnetic molecule is only weakly coupled to the non-magnetic conductor but its spin can influence the transport properties, permitting readout of the spin state with minimal back-action. Several mechanisms can be exploited to couple the two systems. One appealing way is to use a carbon nanotube as a detector of the magnetic flux variation, possibly using a nanoSQUID[60]. Another possibility involves the indirect detection of the spin state through electrometry. Indeed, a non-magnetic quantum conductor at low temperatures behaves as a QD for which charging processes become quantized, giving rise to Coulomb blockade and the Kondo effect depending on the coupling to the leads (Box 2). Any slight change in the electrostatic environment (controlled by the gate) can induce a shift of the Coulomb diamonds of the device, leading to a conductivity variation of the QD at constant

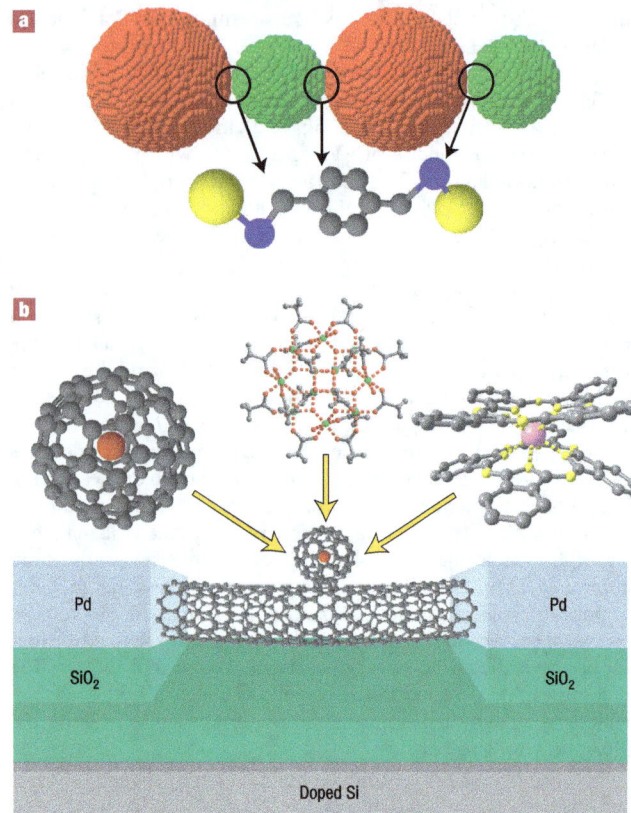

Figure 4 Multidot devices. **a**, Chemically bridged multi-QDs forming chains in which coherent spin transfer can be achieved and followed optically. The p-C$_6$H$_4$-(CH$_2$SH)$_2$ molecule, depicted below the chain, bridges the inorganic QDs and acts as a molecular buffer. **b**, Magnetic molecules proposed for grafting on suspended carbon nanotubes connected to Pd electrodes (form left to right): a C$_{60}$ fullerene including a rare-earth atom[64], the [Mn$_{12}$O$_{12}$(C(CH$_3$)$_3$COO)$_{16}$(H$_2$O)$_4$] SMM[20] and the rare-earth-based double-decker [Tb(phtalocyanine)$_2$] SMM[25]. The gate voltage of the double-dot device is obtained by a doped Si substrate covered by a SiO$_2$ insulating layer.

V_g. QDs are therefore accurate electrometers. When the QD is coupled, even weakly, with a magnetic object, owing to the Zeeman energy the spin flip at non-zero field induces a change of the electrostatic environment of the QD. This 'magneto-Coulomb' effect[61], enables detection of the magnetization reversal of the molecule.

Other routes of multidot devices use weak exchange or dipole coupling between the magnetic molecule and the QD[56]. It is interesting to probe these effects as a function of the number of trapped electrons, because odd or even numbers of electrons should lead to different couplings.

The main advantage of these double-dot devices is that the weak coupling between the SMM and the quantum conductur ensures preservation of the intrinsic properties of the SMM. Because coupling is small, these devices will enable non-destructive readout of the spin states.

OUTLOOK

In conclusion, SMM-based molecular spintronics is an emerging field full of fascinating challenges for theoreticians, experimental physicists and chemists alike. These devices are exciting experimental playgrounds for testing quantum effects at the single-molecule level.

The unique properties of SMMs will soon lead to the design of molecules for specific transport characteristics using the flexibility of supramolecular chemistry.

SMM-based systems might not find their way into everyday electronics, because of their low operational temperatures. However, modulating the response of spintronic devices using quantum effects, besides its fundamental importance, can lead to important consequences for applications. As molecular-based storage devices are becoming feasible[62], SMMs offer the possibility to magnetically store information at a molecular level. The absence of conformational changes that usually destroy coherence, as in molecular machines[62], makes the development of SMM-based quantum computation devices possible. This is particularly true when considering that the coherence time of SMMs is extremely long, compared with that of metallic and semiconducting materials. Another important advantage of molecular spintronics is the weak spin-orbit and hyperfine interactions of organic molecules, which help preserve electron-spin polarization over a much greater distance than in conventional semiconductor systems. Because a spin-polarized current, passing through a SMM, should allow the reading and reversing of its magnetization, such devices can be a breakthrough for quantum information processing. Last but not least, specific functions (for example, switchability with light, electric field and so on) could be directly integrated into the molecule.

In the perspective of molecular spintronics, some new SMMs look most promising. Rare-earth-based double-decker molecules, for example, offer a family of isostructural SMMs with tunable anisotropy[25]. The conjugated structure of the porphyrine ligands is particularly suitable for electron transport and, owing to their symmetric disposition around the rare-earth element, the relevant magnetic parameters can be calculated or extracted from electron paramagnetic resonance and optical data[24]. Because these systems have very strong couplings between electron and nuclear spins, they exhibit exotic characteristics that can be used as fingerprints of the presence of the molecule in the junctions or on the nanotubes[63]. Magnetic atoms imprisoned inside fullerenes[64] also look very attractive. They offer a clean environment around an otherwise isolated atom that can be nicely modelled and compared to transport measurements of empty fullerenes[49]. Finally, we would like to mention that there is a vast amount of trivial-looking magnetic molecules in the chemistry literature, which represent a precious unexploited resource for molecular spintronics.

doi:10.1038/nmat2133

References
1. Wolf, S. A. et al. Spintronics: a spin-based electronics vision for the future. Science 294, 1488–1495 (2001).
2. Awshalom, D. D. & Flatté, M. M. Challenges for semiconductor spintronics. Nature Phys. 3, 153–159 (2007).
3. Dery, H., Dalal, P., Cywiński, Ł. & Sham, L. J. Spin-based logic in semiconductors for reconfigurable large-scale circuits. Nature 447, 573–576 (2007).
4. Xiong, Z. H., Wu, D., Valy Vardeny, Z. & Shi, J. Giant magnetoresistance in organic spin-valves. Nature 427, 821–824 (2004).
5. Ratner, M. A. (ed.) Special feature on molecular electronics. Proc. Natl Acad. Sci. 102, 8800–8837 (2005).
6. Tao, N. J. Electron transport in molecular junctions. Nature Nanotech. 1, 173–181 (2006).
7. Sanvito, S. & Rocha, A. R. Molecular spintronics: The art of driving spin through molecules. J. Comput. Theor. Nanosci. 3, 624–642 (2006).
8. Christou, G., Gatteschi, D., Hendrickson, D. N. & Sessoli, R. Mater. Res. Soc. Bull. 25, 66–71 (2000).
9. Gatteschi, D., Sessoli, R. & Villain, J. Molecular Nanomagnets (Oxford Univ. Press, New York, 2007).
10. Carretta, S. et al. Quantum oscillations of the total spin in a heterometallic antiferromagnetic ring: Evidence from neutron spectroscopy. Phys. Rev. Lett. 98, 167401 (2007).
11. Ardavan, A. et al. Will spin-relaxation times in molecular magnets permit quantum information processing? Phys. Rev. Lett. 98, 057201 (2007).
12. Leuenberger, M. N. & Loss, D. Quantum computing in molecular magnets. Nature 410, 789–793 (2001).
13. Troiani, F. et al. Molecular engineering of antiferromagnetic rings for quantum computation. Phys. Rev. Lett. 94, 207208 (2005).
14. Lehmann, J., Gaita-Ariño, A., Coronado, E. & Loss, D. Spin qubits with electrically gated polyoxometalate molecules. Nature Nanotech. 2, 312–317 (2007).

15. Friedman, J. R., Sarachik, M. P., Tejada, J. & Ziolo, R. Macroscopic measurement of resonant magnetization tunneling in high-spin molecules. *Phys. Rev. Lett.* **76,** 3830 (1996).

16. Thomas, L. *et al.* Macroscopic quantum tunnelling of magnetization in a single crystal of nanomagnets. *Nature* **383,** 145–147 (1996).

17. Wernsdorfer, W. & Sessoli, R. Quantum phase interference and parity effects in magnetic molecular clusters. *Science* **284,** 133–135 (1999).

18. Heersche, H. B. *et al.* Electron transport through single Mn_{12} molecular magnets. *Phys. Rev. Lett.* **96,** 206801 (2006).

19. Jo, M.-H. *et al.* Signatures of molecular magnetism in single-molecule transport spectroscopy. *Nano. Lett.* **6,** 2014–2020 (2006).

20. Cornia, A. *et al.* Preparation of novel materials using SMMs. *Struct. Bond.* **122,** 133–161 (2006).

21. Fleury, B. *et al.* a new approach to grafting a monolayer of oriented Mn12 nanomagnets on silicon. *Chem. Comm.* 2020–2022 (2005).

22. Naitabdi, A., Bucher, J.-P., Gerbier, Ph., Rabu, P. & Drillon, M. Self-assembly and magnetism of Mn_{12} nanomagnets on native and functionalized gold surfaces. *Adv. Mater.* **17,** 1612–1616 (2005).

23. Coronado, E. *et al.* Polycationic Mn_{12} single-molecule magnets as electron reservoirs with *S*>10 ground states. *Angew. Chem. Int. Ed.* **43,** 6152–6156 (2004).

24. Bogani, L. *et al.* Magneto-optical investigations of nanostructured materials based on single-molecule magnets monitor strong environmental effects. *Adv. Mater.* **19,** 3906–3911 (2007).

25. Ishikawa, N., Sugita, M., Ishikawa, T., Koshihara, S. & Kaizu, Y. Mononuclear lanthanide complexes with a long magnetization relaxation time at high temperatures: A new category of magnets at the single-molecular level. *J. Phys. Chem. B* **108,** 11265–11271 (2004).

26. Wahl, P. *et al.* Exchange interaction between single magnetic adatoms. *Phys. Rev. Lett.* **98,** 056601 (2007).

27. Hirjibehedin, C. F. *et al.* Large magnetic anisotropy of a single atomic spin embedded in a surface molecular network. *Science* **317,** 1199–1202 (2007).

28. Kim, G.-H. & Kim, T.-S. Electronic transport in single molecule magnets on metallic surfaces. *Phys Rev. Lett.* **92,** 137203 (2004).

29. Park, H., Kim, A. K. L., Alivisatos, A. P., Park J. & McEuen, P. L. Fabrication of metallic electrodes with nanometer separation by electromigration. *Appl. Phys. Lett.* **75,** 301–303 (1999).

30. Hanson, R., Kouwenhoven, L. P., Petta, J. R., Tarucha S. & Vandersypen L. M. K. Spins in few-electron quantum dots. *Rev. Mod. Phys.* **79,** 1217–1265 (2007).

31. Kouwenhoven, L. P. *et al.* in *Mesoscopic Electron Transport* (eds Sohn, L. L., Kouwenhoven, L. P. & Schön, G.) 105–214 (Series E 345, Kluwer, 1997).

32. Chakov, N. E., Soler, M., Wernsdorfer, W., Asbboud, K. A. & Christou, G. Single-molecule magnets: Structural characterization, magnetic properties, and ¹⁹F NMR spectroscopy of a Mn₁₂ family spanning three oxidation levels. *Inorg. Chem.* **44,** 5304–5321 (2005).

33. Romeike, C., Wegewijs, M. R. & Schoeller, H. Spin quantum tunneling in single molecule magnets: Fingerprints in transport spectroscopy of current and noise. *Phys. Rev. Lett.* **96,** 196805 (2006).

34. Romeike, C. *et al.* Charge-switchable molecular nanomagnet and spin-blockade tunneling. *Phys Rev. B* **75,** 064404 (2007).

35. Erste, F. & Timm, C. Cotunneling and nonequilibrium magnetization in magnetic molecular monolayers. *Phys Rev. B* **75,** 195341 (2007).

36. Timm, C. Tunneling through magnetic molecules with arbitrary angle between easy axis and magnetic field. *Phys Rev. B* **76,** 014421 (2007).

37. Goldhaber-Gordon, D. *et al.* Kondo effect in a single-electron transistor. *Nature* **391,** 156–159 (1998).

38. Cronenwett, S. M., Oosterkamp, T. H. & Kouwenhoven, L. P. A tunable Kondo effect in quantum dots. *Science* **281,** 540–544 (1998).

39. van der Wiel, W. G. *et al.* The Kondo effect in the unitary limit. *Science* **289,** 2105–2108 (2000).

40. Nygard, J., Cobden, D. H. & Lindelof, P. E. Kondo physics in carbon nanotubes. *Nature* **408,** 342–346 (2000).

41. Park, J. *et al.* Coulomb blockade and the Kondo effect in single-atom transistors. *Nature* **417,** 722–725 (2002).

42. Liang, W., Shores, M. P., Bockrath, M. & Long, J. R. Kondo resonance in a single-molecule transistor. *Nature* **417,** 725–729 (2002).

43. Aromi, G. & Brechin, E. K. Synthesis of 3d metallic single-molecule magnets. *Struct. Bond.* **122,** 1–67 (2006).

44. Romeike, C., Wegewijs, M. R., Hofstetter W. & Schoeller, H. Kondo-transport spectroscopy of single molecule magnets. *Phys. Rev. Lett.* **97,** 206601 (2006).

45. Romeike, C., Wegewijs, M. R., Hofstetter, W. & Schoeller, H. Quantum-tunneling-induced kondo effect in single molecular magnets. *Phys. Rev. Lett.* **96,** 196601 (2006).

46. Leuenberger, M. N. & Mucciolo, E. R. Berry-phase oscillations of the Kondo effect in single-molecule magnets. *Phys. Rev. Lett.* **97,** 126601 (2006).

47. Waldron, D., Haney, P., Larade, B., MacDonald, A. & Guo, H. Nonlinear spin current and magnetoresistance of molecular tunnel junctions. *Phys. Rev. Lett.* **96,** 166804 (2006).

48. Rocha, A. R. *et al.* Towards molecular spintronics. *Nature Mater.* **4,** 335–339 (2005).

49. Pasupathy, A. N. *et al.* The Kondo effect in the presence of ferromagnetism. *Science* **306,** 86–89 (2004).

50. Hueso L. E. *et al.* Transformation of spin information into large electrical signals using carbon nanotubes. *Nature* **445,** 410–413 (2007).

51. Liu, R., Ke, S.-H., Baranger, H. U. &Yang, W. Negative differential resistance and hysteresis through an organometallic molecule from molecular-level crossing. *J. Am. Chem. Soc.***128,** 6274–6275 (2006).

52. Timm, C. & Elste, F. Spin amplification, reading, and writing in transport through anisotropic magnetic molecules. *Phys. Rev. B* **73,** 235304 (2006).

53. Elste, F. & Timm, C. Transport through anisotropic magnetic molecules with partially ferromagnetic leads: Spin-charge conversion and negative differential conductance. *Phys. Rev. B* **73,** 235305 (2006).

54. Misiorny, M. & Barnaś, J. Magnetic switching of a single molecular magnet due to spin-polarized current. *Phys. Rev. B* **75,** 134425 (2007).

55. van der Wiel, W. G. *et al.* Electron transport through double quantum dots. *Rev. Mod. Phys.* **75,** 1–22 (2003).

56. Wernsdorfer, W., Aliaga-Alcalde, N., Hendrickson, D. N. & Christou G. Exchange-biased quantum tunneling in a supramolecular dimer of single-molecule magnets. *Nature* **416,** 406–409 (2002).

57. Affronte, M. *et al.* Linking rings through diamines and clusters: Exploring synthetic methods for making magnetic quantum gates. *Angew. Chem. Int. Ed.* **44,** 6496–6490 (2005).

58. Ouyang, M. & Awshalom, D. Coherent Spin-transfer between molecularly bridged quantum dots. *Science* **301,** 1074–1076 (2003).

59. Merrifield, B. *Solid Phase Synthesis* Available at <http://nobelprize.org/nobel_prizes/chemistry/laureates/1984/merrifield-lecture.html>.

60. Cleuziou, J.-P., Wernsdorfer, W., Bouchiat, V., Ondarçuhu, T. & Monthioux, M. Carbon nanotube superconducting quantum interference device. *Nature Nanotech.* **1,** 53–59 (2006).

61. Shimada, H., Ono, K & Otuka, Y. Driving the single-electron device with a magnetic field. *J. Appl. Phys.* **93,** 8259–8264 (2003).

62. Green, J. E. *et al.* A 160-kilobit molecular electronic memory patterned at 10^{11} bits per square centimeter. *Nature* **445,** 414–418 (2007).

63. Ishikawa, N., Sugita, M. & Wernsdorfer, W. Quantum tunneling of magnetization in lanthanide single-molecule magnets: Bis(phthalocyaninato)terbium and bis(phthalocyaninato)dysprosium anions. *Angew. Chem. Int. Ed.* **44,** 2931–2935 (2005).

64. Kasumov, A. *et al.* Proximity effect in a superconductor-metallofullerene-superconductor molecular junction. *Phys. Rev. B* **72,** 033414 (2005).

65. Wernsdorfer, W., Murugesu, M. & Christou, G. Resonant tunneling in truly axial symmetry Mn_{12} single-molecule magnets: Sharp crossover between thermally assisted and pure quantum tunneling. *Phys. Rev. Lett.* **96,** 057208 (2006).

Acknowledgements

This work is partially financed by ANR-PNANO, Contract MolSpintronics No. ANR-06-NANO-27 and by EC-RTN QUEMOLNA Contract No. MRTN-CT-2003-504880. L.B. acknowledges E.U. support through the EIF-041565-'MoST' Marie Curie fellowship.

NANOPHOTONICS

Light in tiny holes

C. Genet[1] & T. W. Ebbesen[1]

The presence of tiny holes in an opaque metal film, with sizes smaller than the wavelength of incident light, leads to a wide variety of unexpected optical properties such as strongly enhanced transmission of light through the holes and wavelength filtering. These intriguing effects are now known to be due to the interaction of the light with electronic resonances in the surface of the metal film, and they can be controlled by adjusting the size and geometry of the holes. This knowledge is opening up exciting new opportunities in applications ranging from subwavelength optics and optoelectronics to chemical sensing and biophysics.

A hole in a screen is probably the simplest optical element possible, and was an object of curiosity and technological application long before it was scientifically analysed. A pinhole was at the heart of the camera obscura used by the Flemish painters in the sixteenth century to project an image (albeit upside down) onto their canvases. It was in the middle of the seventeenth century that Grimaldi first described diffraction from a circular aperture[1], contributing to the foundation of classical optics. Despite their apparent simplicity and although they were much larger than the wavelength of light, such apertures remained the object of scientific study and debates for centuries thereafter, as an accurate description and experimental characterization of their optics turned out to be extremely difficult.

In the twentieth century, the interest naturally shifted to subwavelength holes as the technology evolved towards longer wavelengths of the electromagnetic spectrum. With the rising importance of microwave technology in the war effort of the 1940s, Bethe treated the diffractive properties of an idealized subwavelength hole, that is, a hole in a perfectly conducting metal screen of zero thickness[2] (Box 1). His predictions, notably that the optical transmission would be very weak, became the reference for issues associated with the miniaturization of optical elements and the development of modern characterization tools beyond the diffraction limit, such as the scanning near-field optical microscope (SNOM), which typically has a small aperture in the metal-coated tip as the probing element[3].

In this context, the report of the extraordinary transmission phenomenon through arrays of subwavelength holes milled in an opaque metal screen[4] generated considerable interest because it showed that orders of magnitude more light than Bethe's prediction could be transmitted through the holes. This has since stimulated much fundamental research and promoted subwavelength apertures as a core element of new optical devices. Central to this phenomenon is the role of surface waves such as surface plasmons (SP), which are essentially electromagnetic waves trapped at a metallic surface through their interaction with the free electrons of the metal[5,6] (Box 2). This combination of surface waves and subwavelength apertures is what distinguishes the enhanced transmission phenomenon from the idealized Bethe treatment and gives rise to the enhancement. Moreover, modern nanofabrication techniques allow us to tailor the dynamics of this combination by structuring the surface at the subwavelength scale. This opens up a wealth of possibilities and applications from chemical sensors to atom optics.

We will review here the present understanding of the transmission through subwavelength apertures in metal screens, starting for the sake of clarity with simple isolated holes and ending with arrays. As we will see, SPs play an essential role at optical wavelengths in all the considered structures. Applications such as tracking single molecule fluorescence in biology, enhanced vibrational spectroscopy of molecular monolayers and ultrafast photodetectors for optoelectronics illustrate the broad implications for science and technology.

Single apertures

Figure 1a shows a single hole milled in a free-standing Ag film, characterized by both the diameter of the hole and its depth. When Bethe considered such a system, he idealized the structure by assuming that the film was infinitely thin and that the metal was a perfect conductor. With these assumptions, he derived a very simple expression for the transmission efficiency η_B (normalized to the aperture area)[2]:

$$\eta_B = 64(kr)^4/27\pi^2 \qquad (1)$$

where $k = 2\pi/\lambda$ is the norm of the wavevector of the incoming light of wavelength λ, and r is the radius of the hole. It is immediately apparent that η_B scales as $(r/\lambda)^4$ and that therefore we would expect the optical transmission to drop rapidly as λ becomes larger than r, as shown in Box 1. In addition, the transmission efficiency is further attenuated exponentially if the real depth of the hole is taken into account[7]. This exponential dependence reflects the fact that the light cannot propagate through the hole if $\lambda > 4r$, whereupon the transmission becomes a tunnelling process. The cutoff condition $\lambda > 4r$ is of course a first approximation and in real situations the cutoff occurs at longer wavelengths when the finite conductivity is taken into account[8] (see Box 1).

Bethe also predicted that the light would diffract as it emerges from the hole in an angular pattern that depends on the orientation relative to the polarization of the incident light[2]. If the diffraction pattern is scanned along the direction of the incoming polarization the intensity should be constant (like a spherical wave in a plane) while in the perpendicular direction, the intensity decreases with increasing angle (the angular dependence is a $\cos^2\theta$ function, typical of a dipole emission pattern).

The increasing use of SNOM and interest in the extraordinary transmission phenomenon have stimulated experimental[9–11] and theoretical[12–16] studies, the results of which challenge Bethe's predictions. In particular, it has become possible to measure the transmission and diffraction from a single subwavelength aperture in a metallic film at optical wavelengths[9–11]. Angular measurements at

[1]ISIS, Université Louis Pasteur and CNRS (UMR7006), 8 allée G. Monge, 67000 Strasbourg, France.

When citing this review, please cite the original version as shown on the contents page of this chapter.

the exit of subwavelength apertures have revealed that the light diffracts less than expected[9,10]. Similarly, the transmitted light can have unexpected features[10]. The simple circular aperture of Fig. 1a has a transmission spectrum with a peak as shown in Fig. 1b not predicted

---

When light scatters through apertures, it diffracts at the edges. In the subwavelength regime, Bethe was able to give a theoretical description of the diffraction of light at a given wavelength λ through a circular hole of radius $r \ll \lambda$ in the idealized situation of an infinitely thin and perfect metal sheet. He has shown that the transmission $T(\lambda)$ scales uniformly with the ratio of r to λ to the power of four, as described in equation (1) and schematically shown below in Box 1 Fig. 1.

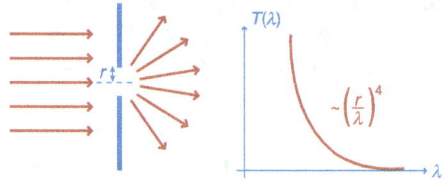

Box 1 Figure 1 | Diffraction and typical transmission spectrum of visible light through a subwavelength hole in an infinitely thin perfect metal film.

However, a real aperture is characterized by a depth and therefore has waveguide properties. The transmission of light through such a guide is very different from the propagation of light in empty space. The confined space of the waveguide essentially modifies the dispersion relation of the electromagnetic field. The lateral dimensions of the waveguide define the wavelength at which light can no longer propagate through the aperture. This wavelength is known as the cutoff wavelength λ_c. When the incident wavelength $\lambda > \lambda_c$ the transmission is exponentially small, characterizing the non-propagating regime as shown in Box 1 Fig. 2. With real metals, the cutoff wavelength cannot be sharply defined because one goes continuously from propagative to evanescent regime as the wavelength increases.

There is a straightforward relationship between the cross-section of the waveguide and λ_c. However, one should take into account that λ_c for an aperture in a real metal is increased by taking the skin-depth into account, reflecting the penetration of the electromagnetic field inside the walls of the metal waveguide. It is possible to control and even to eliminate cutoff wavelengths even when the lateral dimensions are much smaller than λ, by playing with more complex geometries. While simple apertures are always characterized by the existence of cutoff wavelengths, an annular hole, for example, which resembles a coaxial cable, has no cutoff wavelength and is always propagating. The polarization of the incident light is also an important parameter, and with non-cylindrical waveguides, the transmission can be made extremely polarization sensitive. A striking illustration is provided by a slit. Here, for incident polarization parallel to the long axis, the transmission can be made subwavelength, as soon as the short dimension of the rectangle is smaller than λ. However, for the perpendicular polarization, no matter how narrow the guide is, the light always propagates through it. This allows for many possibilities in the choice of geometry depending on the application.

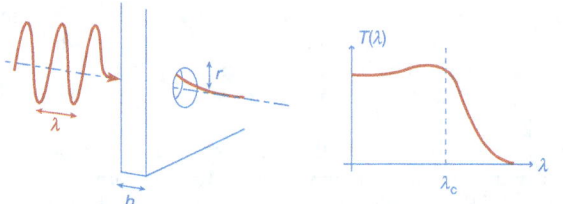

Box 1 Figure 2 | A cylindrical waveguide with a radius r much smaller than the wavelength λ of the incident electromagnetic field milled in a metal film of thickness h. The exponentially decreasing tail represents the attenuation of the subwavelength regime. A transmission spectrum can reveal the different propagating and evanescent regimes.

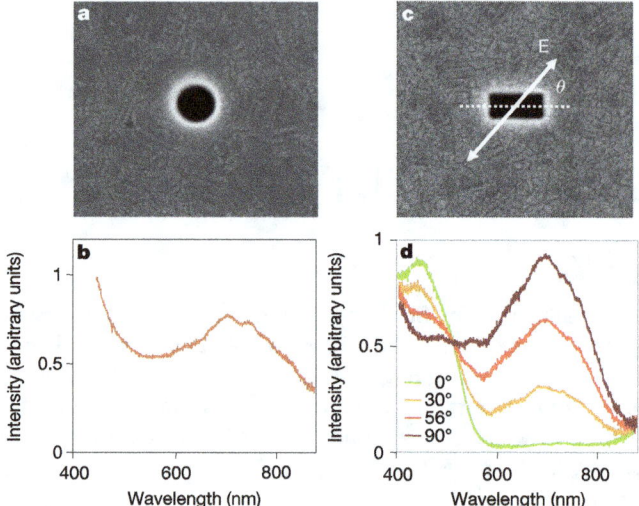

Figure 1 | Optical transmission properties of single holes in metal films. The holes were milled in suspended optically thick Ag films illuminated with white light. **a,** A circular aperture and **b,** its transmission spectrum for a 270 nm diameter in a 200-nm-thick film. **c,** A rectangular aperture and **d,** its transmission spectrum as a function of the polarization angle θ for the following geometrical parameters: 210 nm $\times$ 310 nm, film thickness 700 nm. Figure adapted from ref. 10, with permission.

by equation (1) or by other conventional theories[2,7]. Similar measurements can be made on a rectangular hole (Fig. 1c) where the spectrum becomes sensitive to the incident light polarization as can be seen in Fig. 1d. The appearance of such resonant peaks can be understood as the excitation of SP modes at the edges of the hole, a type known as localized SP that has been confirmed by theoretical studies[14]. By aligning the incoming polarization on either the short or long axis of the rectangular hole, we can selectively excite the corresponding localized SPs (Fig. 1d). Such behaviour is very reminiscent of elongated metal particles, the colours of which are also determined by localized SPs. Whereas the localized SP modes are defined by the lateral dimensions of the aperture, theoretical studies have shown that in addition to such SP modes[14] other resonant modes defined along the depth of the hole might also be present and contribute to the transmission signal[12]. Further experimental studies on this issue at optical wavelengths are necessary.

Bethe's theory describes the transmission as a smooth decreasing function of the wavelength, as given by equation (1) and shown in Box 1, whereas the experiments discussed above reveal the presence of a resonance superimposed on a smooth background, thus providing an enhancement at the resonant wavelengths. In all the structures presented in this review, it is always the presence of some type of resonance that leads to transmission enhancement. This reveals yet again that Bethe's theory is too idealized to treat situations where surface modes are involved and where propagating or evanescent modes can additionally be excited inside the hollow aperture[12], thereby significantly underestimating the transmission efficiency. We define the transmission as being extraordinary when it is so enhanced that the transmission efficiency η is larger than 1, in other words when the flux of photons per unit area emerging from the hole is larger than the incident flux per unit area. As we shall see in the next sections, η can be much larger than one for certain aperture structures under appropriate conditions.

For experimental reasons, it is very difficult to quantify η for a single hole. As was pointed out above, the emission pattern from a singe aperture in a real metal is not isotropic and therefore the absolute transmission can only be determined by measuring the absolute intensity over all angles and then summing the data. This remains an experimental challenge. As we shall see in the section on optimizing

At the interface separating a dielectric with a permittivity ε_d and a metal with a permittivity ε_m, SPs can be resonantly excited by the coupling between free surface charges of the metal and the incident electromagnetic field. Such a mode is characterized by a surface wave vector that obeys the following dispersion relation:

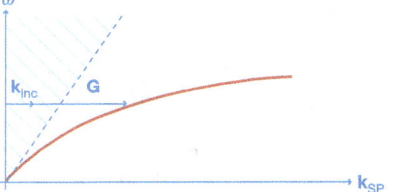

Box 2 Figure 1 | SP dispersion relation. The dotted line corresponds to the light line. The hatched sector of propagating waves does not overlap with the evanescent sector below the light line that fully contains the SP dispersion relation. $\mathbf{k}_{inc}$ is the transverse component of the incident wave vector and $\mathbf{G}$ corresponds to the momentum needed to couple to the SP mode in the evanescent sector.

$$k_{SP} = \frac{\omega}{c}\sqrt{\frac{\varepsilon_m \varepsilon_d}{\varepsilon_m + \varepsilon_d}}$$

Here, ω is the pulsation of the electromagnetic field and c the velocity of light in vacuum. Provided that the real part of ε_m is smaller than $-\varepsilon_d$, this wave vector has positive real and imaginary parts. The latter corresponds to the propagation length of the surface wave before it is damped inside the metal, and can be tens of micrometres at the smooth surface of noble metals, such as Au or Ag, at optical wavelengths. The real part of k_{SP} is plotted in Box 2 Fig. 1. It is always below the light line that separates free-space photons from evanescent ones. This implies immediately that such a mode is evanescent and therefore cannot be excited directly by freely propagating light. A given additional momentum $\mathbf{G}$ is needed to go from the propagating sector where the wave vector $\mathbf{k}_{inc}$ of the incident light falls to the evanescent one where SP modes exist. This is expressed in the simple resonance condition $\mathbf{k}_{SP} = \mathbf{k}_{inc} + \mathbf{G}$, which is a function of the incident pulsation and incident angle θ.

One way to provide the missing momentum $\mathbf{G}$ necessary for coupling incoming light to SPs is to use a periodic array. In one dimension for instance, it can be shown that $\mathbf{G}$ is related to multiples of $2\pi/a_0$ where a_0 is the period of the structure. This is the origin of the optical resonant behaviour of the array, because only when:

$$k_{SP} = k_0 \sin\theta + i \times \frac{2\pi}{a_0}$$

does light couple to SPs (i is an integer). The electromagnetic wave is then trapped momentarily on the surface, giving rise to the transmission peaks. The array generates a complex band structure, as schematically shown in Box 2 Fig. 2. At every multiple of π/a_0 (Brillouin zone edges), SPs are back-reflected so strongly that they cannot propagate any more. Bandgaps appear in the SP dispersion relation, corresponding to stationary waves and high field enhancements.

It should be noted that, when illuminated, non-periodic structures such as single holes, sharp edges, particles and so on can generate localized SP modes. This is possible when the dimensions of the defects are smaller than the wavelength of the incident field, generating a broad spectrum of $\mathbf{G}$ vectors (stemming from the spatial Fourier spectrum of the particular defect) in which a solution to the coupling condition $\mathbf{k}_{SP} = \mathbf{k}_{inc} + \mathbf{G}$ can be found. The coupling efficiency is dependent on the particular profile of the defect.

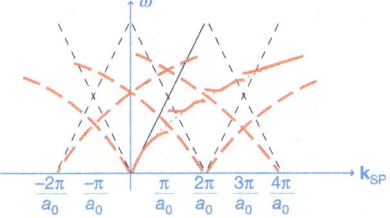

Box 2 Figure 2 | SP band structure on a periodic array.

subwavelength apertures below, the best transmission signals are obtained in noble metals such Au and Ag. To obtain detectable resonances at visible wavelengths, the dimensions of the holes should be of the order of 150 to 300 nm and the films not much thicker than 200 to 300 nm.

Small apertures are routinely used in SNOM tips to explore and to map with subwavelength resolution the electromagnetic field in the immediate vicinity of a surface[3]. More recently, tiny apertures have been implemented in fluorescence correlation spectroscopy[17–19], a powerful technique for the study of the diffusion and reaction of single fluorescent biomolecules in which the information is derived from the analysis of the statistical fluctuations of individual molecules as they move through a small volume. Traditionally, the volume is defined by the focal point of a laser beam, that is, about 1 μm^3, which puts a limit on the upper concentration that can be used while still observing statistical fluctuations. By using small apertures in metal (see Fig. 2)[18,19], the analysed volume has been reduced by a factor of 1,000, allowing one to study molecular events at nearly millimolar concentrations—closer to biological conditions. In addition, such structures give rise to other benefits: the localized SP fields increase the excitation rate of the molecules in its vicinity[10,14,19], the emission pattern is potentially directional[9,10] and the branching ratios from the fluorescent state are affected[19]. All these can lead to an increase in the detected signal, rendering fluorescence correlation spectroscopy ever more useful as a tool for biology.

Single apertures surrounded by periodic corrugations

With modern nanofabrication techniques it is possible to modify the optical properties of a single aperture by sculpting the surrounding material at the scale of the wavelength[20–28]. Such modifications give rise to much higher transmission than single holes at selected wavelengths and in addition, novel lensing effects including beaming can be induced by texturing the output surface of the aperture—as discussed next.

When a single aperture is surrounded by circular corrugations as shown in Fig. 3a, the periodic structure acts like an antenna to couple the incident light into SPs at a given λ. As a consequence the electromagnetic fields at the surface become intense above the aperture, resulting in very high transmission efficiencies and a well-defined spectrum (Fig. 3a). Here the resonant wavelength is mainly determined by the periodicity of the grooves, which provides the necessary momentum and energy-matching conditions (as explained in Box 2). The resonance is, however, slightly more red-shifted than the period owing to the interaction with the light directly transmitted through the hole. This should be considered in tuning the structure to be bright at a desired wavelength. When such a structure is milled in a metal like Ag, the value of η can be much larger than one[21]. Again, absolute quantification is difficult, but compared to a bare single hole of the same dimensions the transmission gain can be an order of magnitude at resonant wavelengths[20]. This, as we shall see below, has important applications.

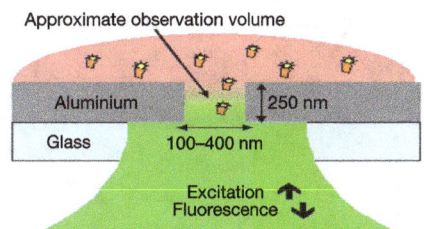

Figure 2 | Schematic diagram of the fluorescence correlation spectroscopy in a single hole. The fluorescence of individual molecules is collected as they diffuse through the observation volume defined by the hole in the metal film. The fluorescence is collected from the same side as the incident excitation. Courtesy of J. Wenger.

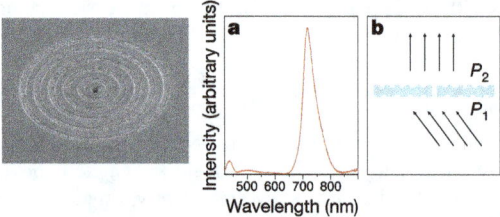

Figure 3 | Optical properties of single apertures surrounded by periodic corrugations. a, Transmission spectrum of a single hole surrounded by periodic corrugations (left) prepared by focused ion beam (hole diameter 300 nm, period 650 nm). **b,** Schematic illustration of redirecting beam by single-slit aperture surrounded by grooves of different periodicity on the input (P_1) and output (P_2) surfaces.

If the output surface surrounding the aperture is also corrugated, a surprisingly narrow beam can be generated, having a divergence of less than a few degrees[20], which is far smaller than that of the single apertures discussed earlier. This is because the light emerging from the hole couples to the periodic structure of the exit surface and to the modes existing in the grooves—which in turn scatter the surface waves into freely propagating light[22–24]. This then interferes with the light that has travelled directly through the hole generating the focused beam. A variant of the double-sided bull's-eye structure is a slit with parallel grooves on both sides of the film, which in addition disperses light spatially according to wavelength[20,28]. Such double-sided structures act as a novel kind of optical element[20,23–25]. They can have a focal plane like a lens but at the same time have other unusual features. For instance, by having grooves with different periods on either side of the film next to a slit, the direction of the output beam can be made independent of the input beam, unlike conventional lenses or gratings, suggesting many practical applications (Fig. 3b). This ability to redirect the beam stems from the way input and output corrugations act like two separate independent gratings connected by a pinhole.

The antenna capacity of the corrugations to concentrate the photons at the tiny central aperture also opens up other technological possibilities such as a bright subwavelength spot for ultradense optical-data storage and nonlinear phenomena[29–31]. Of paramount importance to modern optical telecommunication are photodetectors that can translate an optical signal into an electrical one and thereby convert the flow of information being carried through the telecom network into a displayable signal on the screen. Such photodetectors must therefore be as fast as possible to handle the large amount of data flow. Typically the operating speed of a photodetector scales inversely with the size of its photoelectrical element, but the size cannot be made too small because then it would no longer collect enough photons. To circumvent this problem, an ultrafast photode-

tector has been realized that elegantly combines a very small photoelectrical element with a bull's-eye antenna structure (shown in Fig. 4) that collects and concentrates the incoming photons[32]. This combination illustrates well the potential benefit of plasmonic devices for optoelectronics.

Hole arrays

Periodic arrays of holes in an opaque metal film have so far been the structures that have found the most applications owing to the simplicity with which their spectral properties can be tuned and scaled. Among other things, they can act as filters for which the transmitted colour can be selected by merely adjusting the period. As we saw in the previous section (and Box 2), periodic metallic structures can convert light into SPs by providing the necessary momentum conservation for the coupling process. It is therefore not surprising that periodic arrays of holes such as those shown in Fig. 5 can give rise to the extraordinary transmission phenomenon[4] where the transmission spectrum contains a set of peaks with η larger than one even when the individual holes are so small that they do not allow propagation of light (Fig. 5). Hole arrays have been characterized in great detail both theoretically[33–59] and experimentally[58–79]. As in the case of single holes surrounded by periodic grooves[22], the process can be divided into three steps: the coupling of light to SPs on the incident surface, transmission through the holes to the second surface and then re-emission from the second surface. At the peak transmissions, standing SP waves are formed on the surface (Box 2). The intensity of SP electromagnetic fields above each hole compensates for the otherwise inefficient transmission through each individual hole (Box 1).

If we apply the momentum-matching conditions discussed in Box 2 to a two-dimensional triangular array shown in Fig. 5, we can show that the peak positions λ_{max} at normal incidence are given in a first approximation by:

$$\lambda_{max} = \frac{P}{\sqrt{\frac{4}{3}(i^2 + ij + j^2)}} \sqrt{\frac{\varepsilon_m \varepsilon_d}{\varepsilon_m + \varepsilon_d}} \qquad (2)$$

where P is the period of the array, ε_m and ε_d are respectively the dielectric constants of the metal and the dielectric material in contact with the metal and i, j are the scattering orders of the array. Because equation (2) does not take into account the presence of the holes and the associated scattering losses, it neglects the interference that gives rise to a resonance shift[42,43]. As a consequence, it predicts peak positions at wavelengths slightly shorter than those observed experimentally, as can be seen in Fig. 5.

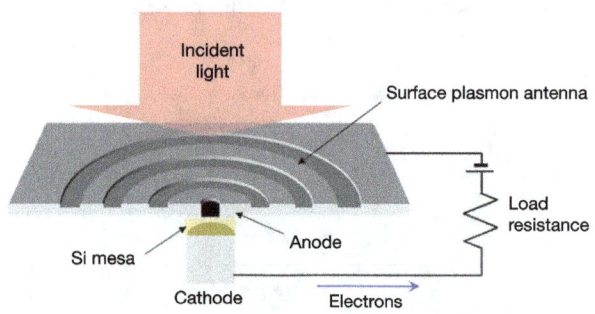

Figure 4 | Ultrafast miniature photodetector. This device consists of a small Si photoelectric element and a SP antenna (reproduced from ref. 32 with permission). The incoming light is harvested by the periodic structure surrounding the central hole, which then transmits it to the underlying photodetector.

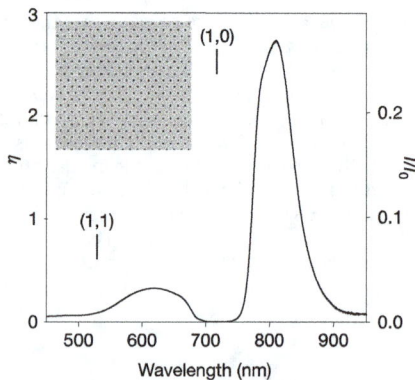

Figure 5 | Transmission spectrum of hole arrays. The triangular hole array was milled in a 225-nm-thick Au film on a glass substrate with an index-matching liquid on the air side (hole diameter 170 nm, period 520 nm). The transmission spectrum is measured at normal incidence using collimated white light. The inset shows the image of the actual array. I/I_o is the absolute transmission of the array and η is the same transmission but normalized to the area occupied by the holes.

Implicit in the resonance conditions defined by equations such as equation (2) are the symmetry relations of the array. Therefore the SPs generated in the array will propagate along defined symmetry axes with their own polarization depending on the (i, j) number of the mode. This results in a rich polarization behaviour that can be revealed in particular under focused light illumination[79].

We emphasize that both surfaces on either side of the holes can sustain SP modes offset from each other by the difference in ε_d of the material in immediate contact with the metal surface (typically glass and air), as predicted by equation (2). Hence, the transmission spectrum of asymmetric structures contains two sets of peaks, each set belonging to one of the surfaces. In many applications, hole arrays of a finite size are used for practical reasons. If the arrays contain small numbers of holes then the periodicity is not well defined and the contribution from the edges becomes significant, changing the spectrum and leading to unusual re-emission patterns[33].

One interesting feature of hole arrays is the fact that each hole on the output surface acts like a new point source for the light. Therefore, if a plane wave (that is, a collimated beam) impinges on the input surface, then a plane wave is reconstructed through classical interference as the light travels away from the output surface. Naturally, because the array is also a grating, the transmission gives rise to different diffraction orders depending on the wavelength to period ratio. For the longest-wavelength (1,0) peak shown in Fig. 5, only the 0th order diffraction is formed, because $\lambda > P$. When $\lambda < P$, higher diffraction orders gradually appear as the wavelength becomes shorter.

The shape and dimensions of the holes in an array do influence its transmission spectrum[64,65]. For instance, in the case of non-propagative apertures, switching from circular to rectangular holes changes the spectrum as a result of the simultaneous change in both the localized SP mode associated with each individual hole and the cutoff wavelength (the wavelength above which the aperture no longer allows light propagation; see Box 1). Nevertheless, the spectrum is dominated by the SP modes because of the periodicity of the array[65]. If the transmission peak falls below the cutoff, its intensity drops exponentially with the depth of the hole and hence the film thickness (or hole depth) is a critical parameter in these structures[33]. It should be noted that arrays of slits have more complex transmission spectra than do hole arrays because the slits can be made propagative under the appropriate polarization (Box 1). As a consequence, the transmission spectra typically contain the signature of both cavity modes in the slits (often at wavelengths that equal twice the slit depth divided by an integer) and SP modes, owing to the slit periodicity[35,48–54,61].

Hole arrays have many applications, from optical elements to sensors for chemistry and biology. For instance, the array acts like a tunable filter because the wavelength selectivity of the array transmission can be adjusted simply by changing the period, as predicted by equation (2) and illustrated in Fig. 6. The letters 'hv' are obtained by fabricating a periodic dimple array in which some of the dimples are milled through to form holes, which in turn reveal the spectral signature of the array.

The combination of the large electromagnetic fields generated by the SPs on the hole arrays, their sensitivity to the dielectric medium in contact with the surface (equation (2)) and the simplicity of the arrays have spurred efforts to use them to detect molecules and to enhance spectroscopic signals (fluorescence, Raman and so on)[80–89]. In this perspective, the enhanced infrared molecular vibrational spectroscopy exemplifies well the usefulness of the hole arrays for chemistry and biology[80]. Arrays of square holes in a Ni film with a periodicity tuned to the infrared were prepared and modified with Cu oxide to induce the catalytic transformation of methanol to formaldehyde on the surface. When such an array with a single adsorbed molecular layer on surface is placed in a Fourier-transform infrared apparatus, the infrared vibrational absorption spectrum (Fig. 7) that is extracted is at least 100 times stronger than if the apparatus had probed a molecular layer on an inert dielectric substrate[80,87]. The large signal enhancement is due to the fact that when the light is trapped momentarily on the surface in terms of SPs, its interaction time with the molecules increases and therefore so does the probability of the absorption. Note that absorption enhancement of electronic transitions is also observed in the visible part of the spectrum, although the enhancement factor is then only a factor of ten, owing to the shorter SP lifetime on the surface at optical wavelengths[90]. Needless to say, such results are extremely promising for studying molecular monolayers and surface reactions by static or time-resolved spectroscopy because this method provides a much better signal-to-noise ratio and is relatively easy to implement.

Other considerations for optimizing apertures

So far we have discussed mainly the broad features associated with tiny holes in opaque metal films. The unique properties of these holes being related to the presence of SPs, they are in turn very much dependent on geometric factors and the properties of the metal.

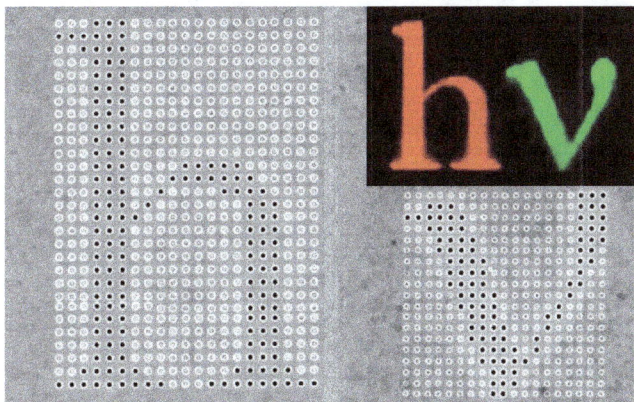

Figure 6 | Holes in a dimple array generating the letters 'hv' in transmission. An array of dimples is prepared by focused-ion-beam milling an Ag film. Some of the dimples are milled through to the other side so that light can be transmitted. When this structure is illuminated with white light, the transmitted colour is determined by the period of the array. In this case the periods were chosen to be 550 and 450 nm respectively to achieve the red and green colours.

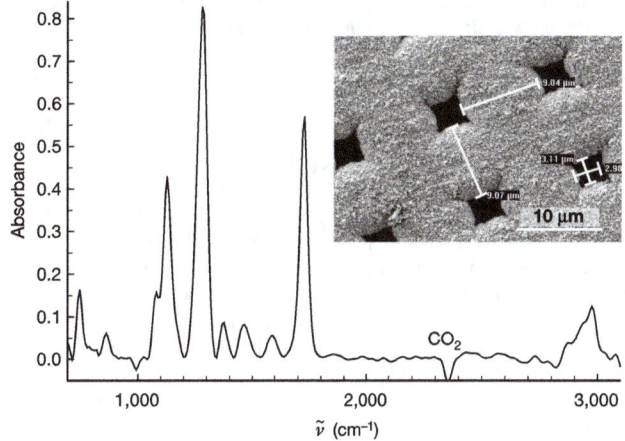

Figure 7 | Infrared enhanced vibrational spectra. Vibrational absorption spectrum of formaldehyde (CH_2O) monolayer adsorbed on a Ni hole array covered with Cu oxide (adapted from ref. 81 with permission). Note that absorbance of 0.8 implies that 84% of the incident light is absorbed by the molecular monolayer. The Ni hole array here acts like the antenna, trapping the light momentarily on the surface and therefore increasing the likelihood of absorption by the CH_2O.

The choice of the metal depends on the wavelength to be used because the dielectric constants of metals are wavelength dependent. Ideally, the dielectric constant of the metal should have a high absolute value for the real part and a small imaginary part that determines the absorption into the metal. This combination gives rise to high SP fields at the surface and minimizes the losses. Therefore Ag is ideally suited to obtain high transmission in the visible part of the spectrum, while above 600 nm Au is even better because it suffers little oxidation. In the infrared, metals such as Ni or Cu can also be used.

Interestingly, high transmittivity through structures similar to those described above but scaled to the microwave region of the spectrum have also been reported where SPs are not considered to exist[91–95]. This is explained by the fact that when a metal surface is corrugated, it is the effective dielectric constant that is modified rather than the bulk metal[37] (this is similar to the way the wetting properties of a material are changed by nanostructuring). As a consequence, SP-like surface waves (also known as 'spoof plasmons') are formed, which enhances the transmittivity[94,95]. More generally, it is essential to trap the electromagnetic wave in the vicinity of the aperture to observe enhanced transmission and its related phenomena. We therefore expect that surface waves such as surface phonon polaritons can also be used. There has been some discussion on whether SPs are involved in the optical transmission of aperture structures but a recent analysis has confirmed the key role of SPs[96,97], in agreement with the vast majority of studies.

The geometrical factors that influence the optical properties of the holes are numerous: symmetry of the structure, the aspect ratios and shape of the holes, aperture area, profile of the corrugations and so forth. These variables determine the electromagnetic field distribution on the surface, the propagation dynamics of the surface waves and their scattering efficiencies, and, the in-plane and out-of-plane coupling to light. The depth of the holes (or thickness of the metal film) is important for several reasons. If the films are too thin, they are partially transparent to the incident light and no holes are necessary to achieve significant transmission, especially if the surfaces are in addition resonantly corrugated[98,99]. To obtain a large contrast between the aperture brightness and the surrounding metal, the metal must be opaque (optically thick), which implies that the film thickness must be several times the skin-depth of the metal. The skin-depth is the penetration depth in the metal at which incoming light intensity has been reduced by $1/e$. Typical skin-depths are of the order of 20 nm for noble metals in the visible spectrum, so film thicknesses of the order of 200 nm are appropriate at optical wavelengths. Even in such thick films, the surface modes on either side of a metal film can also interact via the holes, split and give rise to modes with new energies. Such an effect is especially visible in hole arrays and disappears as the film thickness increases and the modes on either side are decoupled[33,37].

There are many ways to fabricate aperture structures, depending on the scale of interest. For the optical regime, the techniques of choice are: focused ion beam, electron beam lithography and photolithography. The latter two involve several steps but are particularly useful for large-scale structures. The focused ion beam technique, in which the sample is milled by focused ion bombardment, is ideally suited for texturing the metal surface, for instance when preparing grooves around an aperture. Finally, care is needed in preparing the metal films because their quality is an important parameter in the optical properties of the structure.

Potential applications

Surface-wave-activated holes in metal films are finding applications well beyond the illustrations given in the above paragraphs. In the field of opto-electronics for instance, studies are being carried out to extract more light from light-emitting devices[100]. The metal electrodes of such devices, which are normally a source of loss, can be structured with holes to help extract the light from the diode. The need for ever-smaller features on electronic chips is pushing photo-

lithography to use shorter wavelengths, with the associated increased costs and complications. The use of extraordinary optical transmission could perhaps circumvent this problem by using SP-activated lithography masks, which allow subwavelength features in the near-field and high throughput[101–103].

The combination of molecules and holes is another promising area of application, whether for the realization of devices or for the spectroscopic purposes illustrated above. The high optical contrast of SP-activated holes, their small sizes and their simplicity make them ideal candidates for integration on biochips as sensing elements. As in all SP-enhanced phenomena, both the input and output optical fields can be strengthened, with the additional feature that the structure can potentially focus the signal towards a detector. For the purpose of making SP-active devices, the transmission of hole arrays can be switched by controlling the refractive index of molecular materials either electrically[104] or optically up to terahertz speeds[105].

Finally, subwavelength holes might find use in quantum and atom optics. For instance, hole arrays are promising tools in the study of the physical nature—quantum versus classical—of SPs as collective excitations when implemented in quantum entanglement experiments[106,107]. It has been shown theoretically that the extraordinary transmission phenomenon can also be expected for matter waves involving ultracold atoms[108] such as those used in Bose–Einstein condensates. This presents opportunities to create optical elements to manipulate atoms and control their direction.

The potential of the optics of tiny holes in metal screens lies in the contrast between the strong opacity of the metal and the aperture, combined with the fact that the metal allows for high local field enhancements. In addition, the properties of these apertures can be tailored by structuring the metal with modern nanofabrication techniques. The simplicity of the structures and their ease of use should further expand their application in a variety of areas and lead to unsuspected developments.

1. Grimaldi, F.-M. in *Physico-mathesis de Lumine, Coloribus, et Iride, Aliisque Sequenti Pagina Indicatis* 9 (Bologna, 1665).
2. Bethe, H. A. Theory of diffraction by small holes. *Phys. Rev.* **66**, 163–182 (1944).
3. Betzig, E. & Trautman, J. K. Near-field optics: microscopy, spectroscopy, and surface modification beyond the diffraction limit. *Science* **257**, 189–194 (1992).
4. Ebbesen, T. W., Lezec, H. J., Ghaemi, H. F., Thio, T. & Wolff, P. A. Extraordinary optical transmission through sub-wavelength hole arrays. *Nature* **391**, 667–669 (1998).
5. Ritchie, R. H. Plasma losses by fast electrons in thin films. *Phys. Rev.* **106**, 874–881 (1957).
6. Barnes, W. L., Dereux, A. & Ebbesen, T. W. Surface plasmon subwavelength optics. *Nature* **424**, 824–830 (2003).
7. Roberts, A. Electromagnetic theory of diffraction by a circular aperture in a thick, perfectly conducting screen. *J. Opt. Soc. Am. A* **4**, 1970–1983 (1987).
8. Gordon, R. & Brolo, A. Increased cut-off wavelength for a subwavelength hole in a real metal. *Opt. Express* **13**, 1933–1938 (2005).
9. Obermüller, C. & Karrai, K. Far-field characterization of diffracting apertures. *Appl. Phys. Lett.* **67**, 3408–3410 (1995).
10. Degiron, A., Lezec, H. J., Yamamoto, N. & Ebbesen, T. W. Optical transmission properties of a single subwavelength aperture in a real metal. *Opt. Commun.* **239**, 61–66 (2004).
11. Yin, L. *et al.* Surface palsmons at single nanoholes in Au films. *Appl. Phys. Lett.* **85**, 467–469 (2004).
12. Garcia-Vidal, F. J., Moreno, E., Porto, J. A. & Martin-Moreno, L. Transmission of light through a single rectangular hole. *Phys. Rev. Lett.* **95**, 103901 (2005).
13. Chang, C.-W., Sarychev, A. K. & Shalaev, V. M. Light diffraction by a subwavelength circular aperture. *Laser Phys. Lett.* **2**, 351–355 (2005).
14. Popov, E. *et al.* Surface plasmon excitation on a single subwavelength hole in a metallic sheet. *Appl. Opt.* **44**, 2332–2337 (2005).
15. Webb, K. J. & Li, J. Analysis of transmission from small apertures in conducting films. *Phys. Rev. B* **73**, 033401 (2006).
16. Garcia de Abajo, F. J. Light transmission through a single cylindrical hole in a metallic film. *Opt. Express* **10**, 1475–1484 (2002).
17. Magde, D., Elson, E. & Webb, W. W. Thermodynamic fluctuations in a reacting system - measurement by fluorescence correlation spectroscopy. *Phys. Rev. Lett.* **29**, 705–707 (1972).
18. Levene, M. J. *et al.* Zero-mode waveguides for single molecule analysis at high concentrations. *Science* **299**, 682–686 (2003).
19. Rignault, H. *et al.* Enhancement of single-molecule fluorescence detection in subwavelength apertures. *Phys. Rev. Lett.* **95**, 117401 (2005).

Nature | VOL 445 | 4 January 2007 | www.nature.com/nature

20. Lezec, H. J. et al. Beaming light from a subwavelength aperture. *Science* **297**, 820–822 (2002).

21. Thio, T., Pellerin, K. M., Linke, R. A., Lezec, H. J. & Ebbesen, T. W. Enhanced light transmission through a single subwavelength aperture. *Opt. Lett.* **26**, 1972–1974 (2001).

22. Degiron, A. & Ebbesen, T. W. Analysis of the transmission process through single apertures surrounded by periodic corrugations. *Opt. Express* **12**, 3694–3700 (2004).

23. Martin-Moreno, L., Garcia-Vidal, F. J., Lezec, H. J., Degiron, A. & Ebbesen, T. W. Theory of highly directional emission from a single subwavelength aperture surrounded by surface corrugations. *Phys. Rev. Lett.* **90**, 167401 (2003).

24. Garcia-Vidal, F. J., Lezec, H. J., Ebbesen, T. W. & Martin-Moreno, L. Multiple paths to enhance optical transmission through a subwavelength slit. *Phys. Rev. Lett.* **90**, 213901 (2003).

25. Garcia-Vidal, F. J., Martin-Moreno, L., Lezec, H. J. & Ebbesen, T. W. Focusing light with a single subwavelength aperture flanked by surface corrugations. *Appl. Phys. Lett.* **83**, 4500–4502 (2003).

26. Yu, L.-B. et al. Physical origin of directional beaming from a subwavelength slit. *Phys. Rev. B* **71**, 041405(R) (2005).

27. Ishi, T., Fujikata, J. & Ohashi, K. Large optical transmission through a single subwavelength hole associated with a sharp-apex grating. *Jpn J. Appl. Phys.* **44**, L170–L172 (2005).

28. Sun, Z. & Kim, H. K. Refractive transmission of light and beam shaping with metallic nano-optics lenses. *Appl. Phys. Lett.* **85**, 642–644 (2004).

29. Gbur, G., Schouten, H. F. & Visser, T. D. Achieving superresolution in near-field optical data readout systems using surface plasmons. *Appl. Phys. Lett.* **87**, 191109 (2005).

30. Fujikata, J. et al. Surface plasmon enhancement effect and its application to near-field optical recording. *Trans. Magn. Soc. Jpn* **4**, 255–259 (2004).

31. Nahata, A., Linke, R. A., Ishi, T. & Ohashi, K. Enhanced nonlinear optical conversion from a periodically nanostructured metal film. *Opt. Lett.* **28**, 423–425 (2003).

32. Ishi, T., Fujikata, J., Makita, K., Baba, T. & Ohashi, K. Si nano-photodiode with a surface plasmon antenna. *Jpn J. Appl. Phys.* **44**, L364–L366 (2005).

33. Degiron, A., Lezec, H. J., Barnes, W. L. & Ebbesen, T. W. Effects of hole depth on enhanced light transmission through subwavelength hole arrays. *Appl. Phys. Lett.* **81**, 4327–4329 (2002).

34. Bravo-abad, J. et al. How light emerges from an illuminated array of subwavelength holes. *Nature Phys.* **2**, 120–123 (2006).

35. Porto, J. A., Garcia-Vidal, F. J. & Pendry, J. B. Transmission resonances on metalllic gratings with very narrow slits. *Phys. Rev. Lett.* **83**, 2845–2848 (1999).

36. Strelniker, Y. M. & Bergman, D. J. Optical transmission through metal films with a subwavelength hole array in the presence of a magnetic field. *Phys. Rev. B* **59**, R12763 (1999).

37. Martin-Moreno, L. et al. Theory of extraordinary optical transmission through subwavelength hole arrays. *Phys. Rev. Lett.* **86**, 1114–1117 (2001).

38. Popov, E., Neviere, M., Enoch, S. & Reinisch, R. Theory of light transmission through subwavelength periodic hole arrays. *Phys. Rev. B* **62**, 16100 (2000).

39. Barbara, A., Quemerais, P., Bustarret, E. & Lopez-Rios, T. Optical transmission through subwavelength metallic gratings. *Phys. Rev. B* **66**, 161403(R) (2002).

40. Baida, F. I. & Van Labeke, D. Light transmission by subwavelength annular aperture arrays in metallic films. *Opt. Commun.* **209**, 17–22 (2002).

41. Sarychev, A. K., Podolskiy, V. A., Dykhne, A. M. & Shalaev, V. M. Resonance transmittance through a metal film with subwavelength holes. *IEEE J. Quant. Elect.* **38**, 956–963 (2002).

42. Sarrazin, M., Vigneron, J. P. & Vigoureux, J.-M. Role of Wood anomalies in optical properties of thin metallic films with a bidimensional array of subwavelength holes. *Phys. Rev. B* **67**, 085415 (2003).

43. Genet, C., van Exter, M. P. & Woerdman, J. P. Fano-type interpretation of red shifts and red tails in hole array transmission spectra. *Opt. Commun.* **225**, 331–336 (2003).

44. Zayats, A. V., Salomon, L. & de Fornel, F. How light gets through periodically nanostructured metal films: a role of surface polaritonic crystals. *J. Microsc.* **210**, 344–349 (2003).

45. Lalanne, P., Rodier, J. C. & Hugonin, J. P. Surface plasmons of metallic surfaces perforated by nanohole arrays. *J. Opt. Pure Appl. Opt.* **7**, 422–426 (2005).

46. Lomakin, V. & Michielssen, E. Enhanced transmission through metallic plates perforated by arrays of subwavelength holes and sandwiched between dielectric slabs. *Phys. Rev. B* **71**, 235117 (2005).

47. Müller, R., Malyarchuk, V. & Lienau, C. Three-dimensional theory on light-induced near-field dynamics in a metal film with a periodic array of nanoholes. *Phys. Rev. B* **68**, 205415 (2003).

48. Takakura, Y. Optical resonance in a narrow slit in a thick metallic screen. *Phys. Rev. Lett.* **86**, 5601–5603 (2001).

49. Shipman, S. P. & Venakides, S. Resonant transmission near nonrobust periodic slab modes. *Phys. Rev. E* **71**, 026611 (2005).

50. Shen, J. T., Catrysse, P. B. & Fan, S. Mechanism for designing metallic metamaterials with a high index of refraction. *Phys. Rev. Lett.* **94**, 197401 (2005).

51. Xie, Y., Zakharian, A. R., Moloney, J. V. & Mansuripur, M. Transmission of light through slit apertures in metallic films. *Opt. Express* **12**, 6106–6121 (2004).

52. Lee, K. G. & Park, Q.-H. Coupling of surface plasmon polaritons and light in metallic nanoslits. *Phys. Rev. Lett.* **95**, 103902 (2005).

53. Marquier, F., Greffet, J.-J., Collin, S., Pardo, F. & Pelouard, J. L. Resonant transmission through metallic film due to coupled modes. *Opt. Express* **13**, 70–76 (2005).

54. Skigin, D. C. & Depine, R. A. Transmission resonances of metallic compound gratings with subwavelength slits. *Phys. Rev. Lett.* **95**, 217402 (2005).

55. Kim, K. Y., Cho, Y. K., Tae, H. S. & Lee, J.-H. Light transmission along dispersive plasmonic gap and its subwavelength guidance characteristics. *Opt. Express* **14**, 320–330 (2006).

56. Liu, W.-C. & Tsai, D. P. Optical tunnelling effect of surface plasmon polaritons and localized surface plasmon resonance. *Phys. Rev. B* **65**, 155423 (2005).

57. Garcia de Abajo, F. J., Saenz, J. J., Campillo, I. & Dolado, J. S. Site and lattice resonances in metallic hole arrays. *Opt. Express* **14**, 7–18 (2006).

58. Chang, S.-H., Gray, S. K. & Schatz, G. C. Surface plasmon generation and light transmission by isolated nanoholes and arrays of nanoholes in thin metal films. *Opt. Express* **13**, 3150–3165 (2005).

59. Bravo-Abad, J., Garcia-Vidal, F. J. & Martin-Moreno, L. Resonant transmission of light through finite chains of subwavelength holes in a metallic film. *Phys. Rev. Lett.* **93**, 227401 (2005).

60. Ghaemi, H. F., Thio, T., Grupp, D. E., Ebbesen, T. W. & Lezec, H. J. Surface plasmons enhance optical transmission through subwavelength holes. *Phys. Rev. B* **58**, 6779–6782 (1998).

61. Sun, Z., Jung, Y. S. & Kim, H. K. Role of surface plasmons in the optical interaction in metallic gratings with narrow slits. *Appl. Phys. Lett.* **83**, 3021–3023 (2003).

62. Barnes, W. L., Murray, W. A., Dintinger, J., Devaux, E. & Ebbesen, T. W. Surface plasmon polaritons and their role in the enhanced transmission of light through periodic arrays of sub-wavelength holes in a metal film. *Phys. Rev. Lett.* **92**, 107401 (2004).

63. Prikulis, J., Hanarp, P., Olofsson, L., Sutherland, D. & Kall, M. Optical spectroscopy of nanometric holes in thin gold films. *Nano Lett.* **4**, 1003–1007 (2004).

64. Klein Koerkamp, K. J., Enoch, S., Segerink, F. B., van Hulst, N. F. & Kuipers, L. Strong influence of hole shape on extraordinary transmission through periodic arrays of subwavelength holes. *Phys. Rev. Lett.* **92**, 183901 (2004).

65. Degiron, A. & Ebbesen, T. W. The role of localized surface plasmon modes in the enhanced transmission of periodic subwavelength apertures. *J. Opt. Pure Appl. Opt.* **7**, S90–S96 (2005).

66. Gordon, R. et al. Strong polarization in the optical transmission through elliptical nanohole arrays. *Phys. Rev. Lett.* **92**, 037401 (2004).

67. Ye, Y.-H. & Zhang, J.-Y. Enhanced light transmission through cascaded metal films perforated with periodic hole arrays. *Opt. Lett.* **30**, 1521–1523 (2005).

68. Krasavin, A. V. et al. Polarization conversion and "focusing" of light propagating through a small chiral hole in a metallic screen. *Appl. Phys. Lett.* **86**, 201105 (2005).

69. Wang, Q.-J., Li, J.-Q., Huang, C.-P., Zhang, C. & Zhu, Y.-Y. Enhanced optical transmission through metal films with rotation-symmetrical hole arrays. *Appl. Phys. Lett.* **87**, 091105 (2005).

70. Ropers, C. et al. Femtosecond light transmission and subradiant damping in plasmonic crystals. *Phys. Rev. Lett.* **94**, 113901 (2005).

71. Dogariu, A., Thio, T., Wang, L. J., Ebbesen, T. W. & Lezec, H. J. Delay in light transmission through small apertures. *Opt. Lett.* **26**, 450–452 (2001).

72. Halté, V., Benabbas, A., Guidoni, L. & Bigot, J.-Y. Femtosecond dynamics of the transmission of gold arrays of subwavelength holes. *Phys. Status Solidi (b)* **242**, 1872–1876 (2005).

73. Dechant, A. & Elezzabi, A. Y. Femtosecond optical pulse propagation in subwavelength metallic slits. *Appl. Phys. Lett.* **84**, 4678–4680 (2004).

74. Kwak, E.-S. et al. Surface plasmon standing waves in large-area subwavelength hole arrays. *Nano Lett.* **5**, 1963–1967 (2005).

75. Kim, D. S. et al. Microscopic origin of surface-plasmon radiation in plasmonic band-gap nanostructures. *Phys. Rev. Lett.* **91**, 143901 (2003).

76. Egorov, D., Dennis, B. S., Blumberg, G. & Haftel, M. I. Two-dimensional control of surface plasmons and directional beaming from arrays of subwavelength apertures. *Phys. Rev. B* **70**, 033404 (2004).

77. Chyan, J. Y., Chang, C. A. & Yeh, J. A. Development and characterization of a broad-bandwidth polarization-insensitive subwavelength optical device. *Nanotechnology* **17**, 40–44 (2006).

78. Schouten, H. F. et al. Plasmon-assisted two-slit transmission: Young's experiment revisited. *Phys. Rev. Lett.* **94**, 053901 (2005).

79. Altewischer, E., van Exter, M. P. & Woerdman, J. P. Polarization analysis of propagating surface plasmons in a subwavelength hole array. *J. Opt. Soc. Am. B* **20**, 1927–1931 (2003).

80. Williams, S. M. et al. Use of the extraordinary infrared transmission of metallic subwavelength arrays to study the catalyzed reaction of methanol to formaldehyde on copper oxide. *J. Phys. Chem. B* **108**, 11833–11837 (2004).

81. Brolo, A. G. et al. Enhanced fluorescence from arrays of nanoholes in a gold film. *J. Am. Chem. Soc.* **127**, 14936–14941 (2005).

82. Liu, Y., Bishop, J., Williams, L., Blair, S. & Herron, J. Biosensing based upon molecular confinement in a metallic nanocavity arrays. *Nanotechnology* **15**, 1368–1374 (2004).

83. Brolo, A. G., Gordon, R., Leathem, B. & Kavanagh, K. L. Surface plasmon sensor based on the enhanced light transmission through arrays of nanoholes in gold films. *Langmuir* **20**, 4813–4815 (2004).

84. Moran, C. E., Steele, J. M. & Halas, N. J. Chemical and dielectric manipulation of plasmonic band gap of metallodielectric arrays. *Nano Lett.* **4**, 1497–1500 (2004).

85. Stark, P. R. H., Halleck, A. E. & Larson, D. N. Short order nanohole arrays in metals for highly sensitive probing of local indices of refraction as the basis for a highly multiplexed biosensor technology. *Methods* **37**, 37–47 (2005).

86. Brolo, A. G., Arctander, E., Gordon, R., Leathem, B. & Kavanagh, K. L. Nanohole-enhanced Raman scattering. *Nano Lett.* **4**, 2015–2018 (2004).

87. Williams, S. M. *et al.* Scaffolding for nanotechnology: extraordinary infrared transmission of microarrays for stacked sensors and surface spectroscopy. *Nanotechnology* **15**, S495–S503 (2004).

88. Coe, J. V. *et al.* Extra IR transmission with metallic arrays of subwavelength holes. *Anal. Chem.* **78**, 1385–1389 (2006).

89. Rindzevicius, T. *et al.* Plasmonic sensing characteristics of single nanometric holes. *Nano Lett.* **5**, 2335–2339 (2005).

90. Dintinger, J., Klein, S. & Ebbesen, T. W. Molecule–surface plasmon interactions in hole arrays: enhanced absorption, refractive index changes and all-optical switching. *Adv. Mat.* **18**, 1267–1270 (2006).

91. Gomez Rivas, J., Schotsch, C., Haring Bolivar, P. & Kurz, H. Enhanced transmission of Thz radiation through subwavelength holes. *Phys. Rev. B* **68**, 201306(R) (2003).

92. Shou, X., Agrawal, A. & Nahata, A. Role of metal thickness on the enhanced transmission properties of a periodic array of subwavelength apertures. *Opt. Express* **13**, 9834–9840 (2005).

93. Lockyear, M. J., Hibbins, A. P. & Sambles, J. R. Surface-topography-induced enhanced transmission and directivity of microwave radiation through a subwavelength circular metal aperture. *Appl. Phys. Lett.* **84**, 2040–2042 (2004).

94. Pendry, J. B., Martin-Moreno, L. & Garcia-Vidal, F. J. Mimicking surface plasmons with structured surfaces. *Science* **305**, 847–848 (2004).

95. Garcia-Vidal, F. J., Martin-Moreno, L. & Pendry, J. B. Surfaces with holes in them: new plasmonic metamaterials. *J. Opt. Pure Appl. Opt.* **7**, S97–S101 (2005).

96. Lalanne, P. & Hugonin, J. P. Interaction between optical nano-objects at metallo-dielectric interfaces. *Nature Phys.* **2**, 551–556 (2006).

97. Visser, T. D. Surface plasmons at work? *Nature Phys.* **2**, 509–510 (2006).

98. Gruhlke, R., Hod, W. & Hall, D. Surface-plasmon cross coupling in molecular fluorescence near a corrugated thin film. *Phys. Rev. Lett.* **56**, 2838–2841 (1986).

99. Bonod, N., Enoch, S., Li, L., Popov, E. & Nevière, M. Resonant optical transmission through thin metallic films with and without holes. *Opt. Express* **11**, 482–490 (2003).

100. Liu, C., Kamaev, V. & Vardeny, Z. V. Efficency enhancement of an organic light-emitting diode with a cathode forming two-dimensional periodic hole array. *Appl. Phys. Lett.* **86**, 143501 (2005).

101. Srituravanich, W., Fang, N., Sun, C., Luo, Q. & Zhang, X. Plasmonic nanolithography. *Nano Lett.* **4**, 1085–1088 (2004).

102. Luo, X. & Ishihara, T. Sub-100nm photolithography based on plasmon resonance. *Jpn J. Appl. Phys.* **43**, 4017–4021 (2004).

103. Shao, D. B. & Che, S. C. Surface-plasmon-assisted nanoscale photolithography by polarized light. *Appl. Phys. Lett.* **86**, 253107 (2005).

104. Kim, T. J., Thio, T., Ebbesen, T. W., Grupp, D. E. & Lezec, H. J. Control of optical transmission through metals perforated with subwavelength hole arrays. *Opt. Lett.* **24**, 256–258 (1999).

105. Dintinger, J., Robel, I., Kamat, P. V., Genet, C. & Ebbesen, T. W. Terahertz all-optical molecule-plasmon modulation. *Adv. Mater.* **18**, 1645–1648 (2006).

106. Altewisher, E., van Exter, M. P. & Woerdman, J. P. Plasmon-assisted transmission of entangled photons. *Nature* **418**, 304–306 (2002).

107. Fasel, S. *et al.* Energy-time entanglement preservation in plasmon-assisted light transmission. *Phys. Rev. Lett.* **94**, 110501 (2005).

108. Moreno, E., Fernandez-Dominguez, A. I., Cirac, I. J., Garcia-Vidal, F. J. & Martin-Moreno, L. Resonant transmission of cold atoms through subwavelength apertures. *Phys. Rev. Lett.* **95**, 170406 (2005).

Acknowledgements Our research was supported by the European Community, Network of Excellence PLASMO-NANO-DEVICES, STREP SPP, the ANR grant COEXUS, the CNRS, and the French Ministry of Higher Education and Research.

Author Information Reprints and permissions information is available at www.nature.com/reprints. The authors declare no competing financial interests. Correspondence should be addressed to T.W.E. (ebbesen@isis-ulp.org).

Nano-optics from sensing to waveguiding

The design and realization of metallic nanostructures with tunable plasmon resonances has been greatly advanced by combining a wealth of nanofabrication techniques with advances in computational electromagnetic design. Plasmonics — a rapidly emerging subdiscipline of nanophotonics — is aimed at exploiting both localized and propagating surface plasmons for technologically important applications, specifically in sensing and waveguiding. Here we present a brief overview of this rapidly growing research field.

SURBHI LAL[1,3], STEPHAN LINK[2,3] AND NAOMI J. HALAS[1,2,3,*]

[1]Department of Electrical and Computer Engineering, [2]Department of Chemistry and [3]Laboratory for Nanophotonics, Rice University, 6100 Main Street, Houston, Texas 77005-1892, USA

*e-mail: halas@rice.edu

Plasmonics is a subfield of nanophotonics that is concerned primarily with the manipulation of light at the nanoscale, based on the properties of propagating and localized surface plasmons. Plasmons are the collective oscillations of the electron gas in a metal or semiconductor. Optical waves can couple to these electron oscillations in the form of propagating surface waves or localized excitations, depending on the geometry. Although all conductive materials, such as metals, support plasmons, the coinage metals (that is, copper, silver and gold) have been most closely associated with the field of plasmonics as their plasmon resonances lie closer to the visible region of the spectrum, allowing plasmon excitation by standard optical sources and methods. The field of plasmonics is based on exploiting plasmons for a variety of tasks, by designing and manipulating the geometry of metallic structures, and consequently their plasmon-resonant properties.

In quantum theory, a plasmon is a quasiparticle that results from the quantization of plasma oscillations interacting with a photon. Despite their origins in quantum mechanics, the properties of plasmons can be described rigorously by classical electrodynamics. Surface plasmons are supported by structures at all length scales and are certainly not limited to quantum confined systems. For thin metal films, for example, surface plasmons are the electromagnetic waves that propagate along metallic–dielectric interfaces[1,2]. They can exist at any interface and frequency range where the real dielectric constants of the media constituting the interface are of opposite signs. Small noble metal particles, with dimensions from a few up to several hundred nanometres, support localized surface plasmon oscillations that create large electromagnetic fields at the nanoparticle surface[2-4].

The plasmon resonance frequency, determined by the frequency-dependent dielectric function of the metal and the dielectric constant of the surrounding medium, is strongly dependent on the size and shape of the nanostructure[5-8]. Although both the surface plasmon resonance for a thin film and the localized surface plasmon resonance supported on isolated nanoparticles have been of great interest in the scientific community, the resonances supported by single nanoparticles in particular have received considerable attention owing to numerous significant advances in nanoparticle synthesis. Wet chemical synthesis methods have now made it possible to fabricate plasmonic nanoparticles having a variety of shapes (for example spheres[9], triangles[10], prisms[11], rods[12] and cubes[13]) with controllable sizes and narrow size distributions. Furthermore, metallic–dielectric and metallic–metallic core–shell nanoparticles and mixed metallic–alloy nanoparticles with different shapes (for example, nanoshells[14-16] and nanorice[17]) have been prepared. Fabrication of this large variety of different structures makes a variety of applications possible, as the spectral position of the surface plasmon resonance depends on both the shape and the size of the nanoparticle. For spherical nanoparticles, a resonance of the oscillating electrons with an incident optical wave occurs when the negative of the real part of the particle's dielectric constant equals twice the value of the dielectric constant of the medium (see theoretical section below). However, for non-spherical particle shapes, the electron oscillation is non-isotropic and localized either along the principal axes[18] or at the edges and corners of the nanoparticle[19] (or both), leading to an additional shape-dependent depolarization and splitting of the surface plasmon resonance into several modes (such as longitudinal and transverse modes for nanorods or symmetric and antisymmetric modes in nanoshells[20]). A large variety of structures have been synthesized and characterized whose plasmon resonances may be varied over the entire visible to mid-infrared part of the electromagnetic spectrum (see Fig. 1). In contrast to this explosion of plasmonic nanoparticles of various shapes and sizes, the controlled assembly and integration of nanoparticles into plasmonic materials and devices remains a challenge at present. Fabrication strategies using clean room and wet chemical techniques are leading to numerous advances in nanoparticle assembly and integration methods.

Within this rapidly developing and highly multidisciplinary field, several key research directions have been emerging with the potential for robust and commercializable technological applications. These range from enhanced sensing and spectroscopy for chemical identification and detection of biomolecules or biological agents, near-field optics and scanning microscopy using metallic probe tips, to signal propagation with metal-based waveguides. Broadly, one can distinguish between two main areas based on the properties of plasmons: sensing applications based on either the refractive-index dependence of the plasmon resonance or the amplification of the optical field near the nanostructure; and manipulation and guiding of light using plasmonic waveguides. This review focuses specifically on

When citing this review, please cite the original version as shown on the contents page of this chapter.

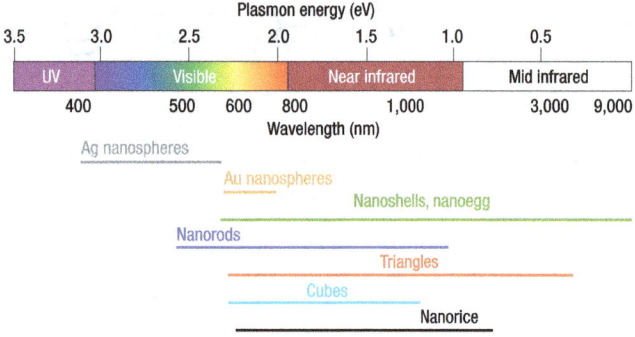

Figure 1 Nanoparticle resonances. A range of plasmon resonances for a variety of particle morphologies.

the plasmonic properties of nanostructures for sensing applications, and on plasmonic waveguides. We first start by reviewing some theoretical methods used throughout the field of plasmonics. For the many preparation methods developed for metallic nanostructures we refer the interested reader to the cited literature.

THEORETICAL BACKGROUND

As a first approximation, the plasmon resonance for a small spherical metallic nanoparticle can be understood by a simple Drude free-electron model, assuming that the positively charged metal atoms are fixed in place and that the valence electrons are dispersed throughout a solid sphere of overall positive charge. In the quasi-static limit, where the wavelength of light is much larger than the size of the particle, the force exerted by the electromagnetic field of the incident light moves all the free electrons collectively. Using the boundary condition that the electric field is continuous across the surface of the sphere, the static polarizability can be expressed as

$$\alpha = 4\pi R^3 \, \frac{\varepsilon - \varepsilon_{\mathrm{m}}}{\varepsilon + 2\varepsilon_{\mathrm{m}}},$$

where R is the sphere radius, ε is the complex dielectric function of the metal and ε_{m} is the dielectric constant of the embedding medium. The polarizability shows a resonance when the denominator is minimized, which occurs when the magnitude of the real part of the complex dielectric function, $\varepsilon_{\mathrm{real}}$, is $-2\varepsilon_{\mathrm{m}}$. This resonance condition for the polarizability leads to a strong extinction of light at the plasmon resonance frequency. Within this free-electron description, plasmons can be thought of as collective oscillations of the conduction-band electrons induced by an interacting light wave.

To determine the plasmon-resonant properties of arbitrarily shaped particles, solutions to Maxwell's equations must be obtained. For nanoparticles with a spherical symmetry, Mie scattering theory[21,22] provides a rigorous solution that describes well the optical spectra of spheres of any size. To determine the extinction spectrum of a nanoparticle using Mie theory, the electromagnetic fields of the incident wave, scattered wave and the wave inside the particle are expressed as the sum of a series of vector spherical harmonic basis functions. The electromagnetic fields must then satisfy Maxwell's boundary conditions of continuity at the junction between the nanoparticle and the embedding medium. Mie theory is exact and accounts for field-retardation effects that become significant for particles whose size is comparable to the wavelength of light. A

complete description of the Mie scattering problem can be found in the classic book by Craig Bohren and Donald Huffman[3].

For non-spherical geometries, brute-force computational methods, such as the discrete dipole approximation, DDA (ref. 23), and the finite-difference time domain, FDTD (refs 24,25), are widely used. In the DDA, the scattering and absorption properties of arbitrarily shaped nanostructures are calculated by approximating the complete nanostructure as a finite array of polarizable point dipoles. In response to an external field, the points acquire a dipole moment, and the scattering properties are then calculated as the interaction of a finite number of closely spaced dipoles.

The FDTD is a computational method based on numerically evaluating the temporal evolution of electromagnetic fields using Maxwell's equations. Because Maxwell's equations relate time-dependent changes in the electric (magnetic) fields to spatial variations in the magnetic (electric) fields, FDTD programs implement an explicit time-marching algorithm for solving Maxwell's curl equations, usually on an offset cartesian spatial grid. Modelling a new system is therefore reduced to grid generation, instead of deriving geometry-specific equations. In addition, the time-marching aspect of the FDTD method enables direct observations to be made of both near- and far-field values of the electromagnetic fields at any time during the simulation. With these 'snap shots', the time evolution of the electromagnetic fields can be calculated directly. The FDTD has become an extremely powerful technique for modelling nanostructures with complex shapes as well as arrangements of multiple nanostructures.

Although DDA and FDTD correctly predict the spectral response of arbitrarily shaped nanostructures, they provide little physical insight into the nature and origin of a plasmon. Thus, it is difficult to rationally design nanostructures with predictable plasmon resonances based on DDA and FDTD. Plasmon hybridization (PH) theory is a more intuitive method for calculating the plasmon resonances of complex nanostructures. The PH model is a mesoscale electromagnetic analogue of the molecular-orbital theory used to predict how atomic orbitals interact to form molecular orbitals. Plasmon hybridization theory separates complex nanoparticle geometries into simpler constituent parts and then calculates how the plasmon resonances of the elementary parts interact with each other to generate the hybridized plasmon modes of the composite nanostructure. For example, the plasmon resonances of a silica core with a gold shell, called a nanoshell, can be explained as the interacting plasmons of a solid spherical gold particle and a spherical cavity inside a bulk block of gold (see Fig. 2a). The plasmon resonances of nanoshells can then be understood as the interaction or hybridization of the sphere and cavity plasmons. This hybridization leads to a higher-energy antisymmetric plasmon and a lower-energy symmetric plasmon. The symmetric plasmon has a larger dipole moment and couples easily with light giving rise to plasmon absorption (see Fig. 2a). The tunability of the nanoshell plasmon resonance (Fig. 2b) arises from the combination of silica-particle size (cavity plasmon) and gold-shell thickness (interaction distance). Thinner (thicker) shells have a stronger (weaker) interaction, leading to larger (smaller) energy splitting of the symmetric and antisymmetric plasmon modes. Therefore PH theory provides an elegant and more intuitive plasmon description and has become a useful guide in the engineering of metallic nanostructures with predictable resonances when applied to other complex geometries with interacting plasmons.

PLASMONIC SENSING

REFRACTIVE-INDEX PLASMON SENSING

Chemical sensing based on surface plasmon resonances (SPRs) can be accomplished in a variety of ways. Historically, SPR sensing was

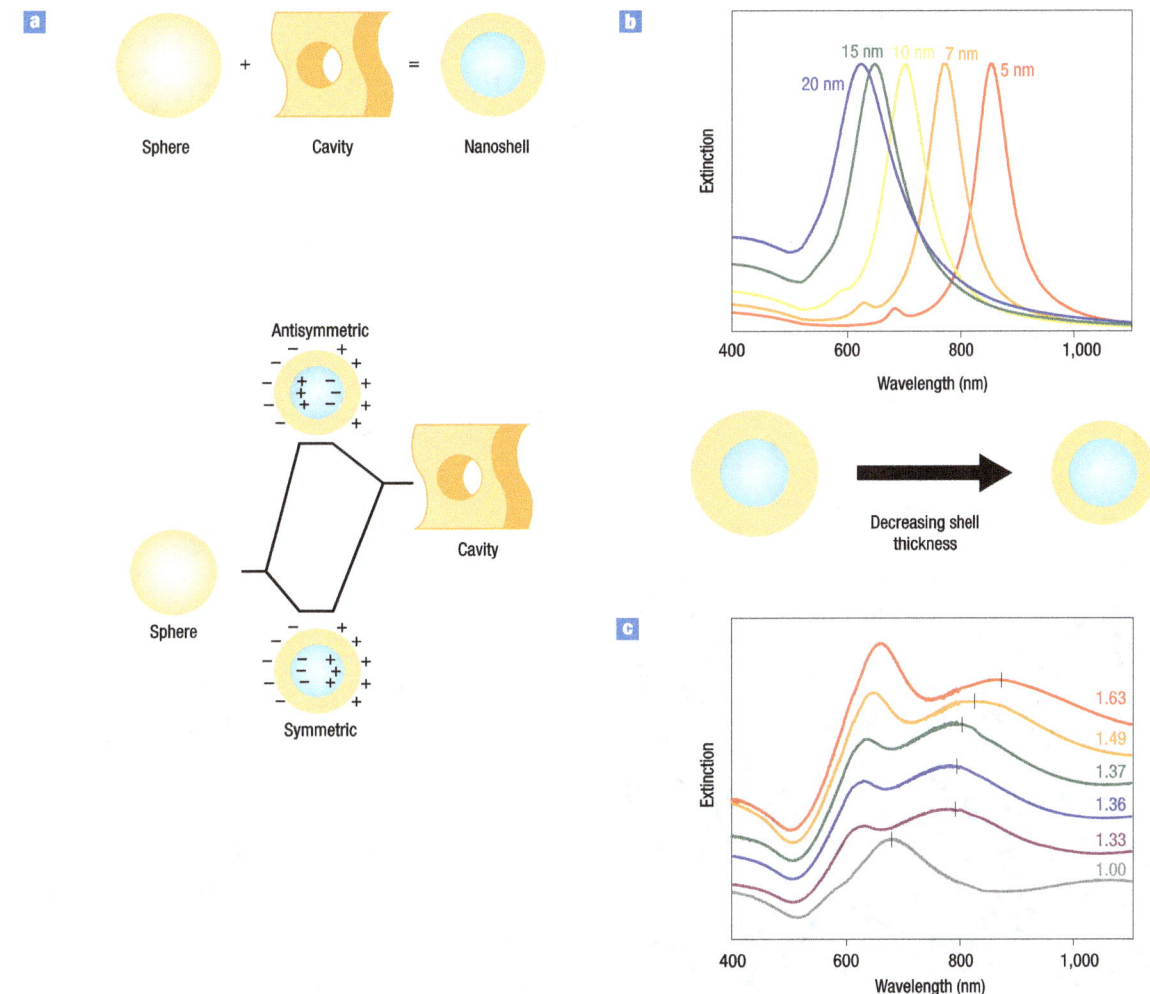

Figure 2 Tunability of nanoshells. **a**, Plasmon hybridization. Nanoshell plasmons can be understood as a hybridization of a sphere and a cavity plasmon. The schematic of the plasmon hybridization shows the charge distribution forming symmetric and antisymmetric plasmon resonances. **b**, Plasmon resonances of a 120-nm-diameter silica core coated with varying thicknesses of gold shell. Note the blue shift in the plasmon resonance as the size of the gold layer increases from 5 nm to 20 nm. **c**, Surface plasmon resonance (SPR) shifts of a nanoshell's resonance in solutions with different refractive indices varying from 1.00 to 1.63.

first performed using propagating surface plasmons on continuous metal films that had been chemically functionalized. Modifications in the chemical environment due to binding of molecules to a functionalized film were monitored as a change in the incidence angle required for surface plasmon excitation in an evanescent coupling geometry[26]. Similarly, metallic nanoparticles that support the excitation of a localized SPR are also highly sensitive to the environment, because the resonance frequency depends on the dielectric constant of the local medium (for example see Fig. 2c and refs 4,7). In nanoparticle-based SPR sensing, changes in the dielectric constant of the medium surrounding the metallic nanostructure are detected by measuring the shift of the SPR absorbance maximum (see Fig. 2c and refs 26,27). This enables the detection of molecules with different dielectric properties at the nanoparticle surface. Although the SPR response is not chemically specific, this disadvantage has been overcome by detection techniques using metal particles conjugated to, for example, antibodies or DNA, allowing for that selectivity of target-receptor molecules through binding specificities[28,29]. For example, the highly robust biotin–streptavidin

labelling scheme has been used extensively for plasmon sensing in biological systems. Although SPR spectroscopy using nanoparticles is less sensitive at present than its thin-film-based analogue, it has the advantages of portability (that is, the technique can be miniaturized) and affordability[28] ($5,000 versus $300,000).

LOCAL FIELD ENHANCEMENT

When an electromagnetic wave interacts with a roughened metallic surface or a nanoparticle, the electromagnetic fields near the surface are greatly enhanced relative to the incident electromagnetic field. This phenomenon is due to two processes. The first is the 'lightning rod' effect, conventionally described as the crowding of the electric field lines at a sharp metallic tip. The second process is the excitation of localized surface plasmons at the metal surface and is responsible for the amplification of fluorescence and second-harmonic generation from thin metal films[30]. For metal nanoparticles often both processes are involved in creating the localized enhanced near field. Although this electromagnetic enhancement effect has been exploited in several spectroscopy techniques, it is mainly the amplification of the

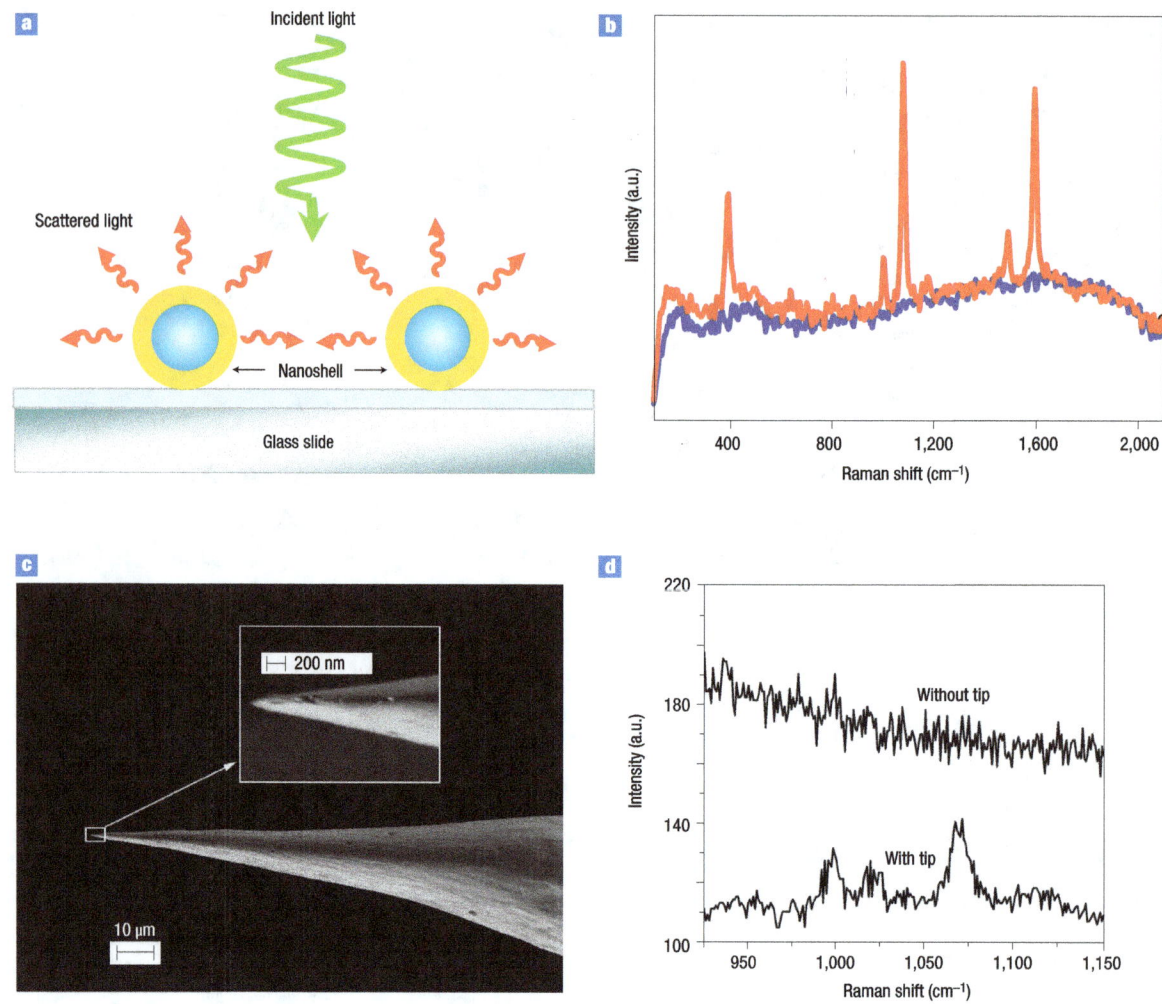

Figure 3 Surface-enhanced Raman spectroscopy. **a**, Schematic sample geometry for nanoparticle-based SERS. **b**, Raman spectrum of *p*-mercaptoaniline collected with no nanoshells (blue) and SERS spectra with nanoshells (red). **c**, Scanning electron microsopy image of a typical silver tip used for TERS. The inset shows a zoom in of the tip apex, which has a radius of curvature sharper than 50 nm. **d**, Tip-enhanced Raman spectra from a benzenethiol self-assembled monolayer on a planar gold surface. Spectra were collected before and after the approach of the tip with an exposure time of 10 seconds per frame. Parts **c** and **d** reprinted with permission from ref. 79. Copyright (2007) ACS.

otherwise very weak Raman signal that has triggered the greatest amount of interest. Surface-enhanced Raman spectroscopy (SERS) is a very attractive spectroscopic method because the Raman signal, unlike fluorescence, contains detailed information derived from molecular structure that may be useful in chemical identification. It was discovered accidentally, and then pursued aggressively by numerous groups, before a deeper understanding of how plasmon properties could be manipulated to maximize signal amplification began to emerge. At present this topic is becoming an important field in plasmonics, with a concentrated effort in developing and optimizing SERS substrates for practical use. Surface-enhanced Raman spectrocopy was first discovered for molecules attached to roughened silver electrodes[31,32] as the molecules showed a large increase in their scattered Raman intensity with typical enhancement factors of 10^6. It is now well understood that these large enhancement factors are due to electromagnetic enhancement caused by plasmon excitation, and 'chemical effects' — a broader term that encompasses all molecule–metal interactions that may lead to an enhanced SERS signal. The relative contributions of electromagnetic and chemical effects in numerous metal–molecule systems are still a very active topic of discussion and ongoing research.

Single-molecule spectroscopy[33,34] has revealed that the SERS ensemble enhancement factor of 10^6 is actually due to a distribution of enhancement factors because some molecule–substrate combinations show much smaller or no enhancements, whereas others can reach values of up to 10^{14} for dye molecules on aggregated gold and silver nanoparticles. In this context, nanoparticle aggregates can serve as efficient nanoantennas that harvest and focus the incident light, resulting in greatly enhanced electromagnetic fields in the direct nanoscale proximity of the nanoparticles. Work in this area has led to the improvement of SERS substrates, making SERS a chemical-specific sensing technique surpassing the capabilities achieved in single-molecule fluorescence spectroscopy.

It is now well understood that maximizing the SERS signal requires a combination of factors. First, the nanostructure or nanoparticle substrate must have a strong plasmon resonance. Nanostructures and nanoparticle assemblies with nanoscale gaps have been found to be the most efficient substrates owing to junction plasmons created in the gaps (see following section). Next, because SERS excitation is a near-field phenomenon and the near field decays exponentially with distance away from the nanoparticle surface[35], it is important that the analyte molecule is located within the enhanced fringing field of the nanoparticle surface or in the highly enhanced field region in the gaps between adjacent nanoparticles. Finally, a strong correlation has been observed between the SERS excitation wavelength and the SPR maximum[36,37]. Maximum enhancement is achieved under resonant excitation conditions, which requires that the excitation laser is tuned near the plasmon resonance of the nanoparticle substrate or, alternatively, that the plasmon resonance of the substrate is tuned near the excitation-laser wavelength.

JUNCTION PLASMONS

Aggregated colloidal particles show localized areas of intense local fields, or 'hot spots', that give rise to the largest enhancement factors and make single-molecule SERS possible. However, the heterogeneity of 'hot spots' in SERS substrates consisting of nanoparticle aggregates makes quantitative measurements unreliable and optimization difficult. For the past few years there has been a great emphasis on rationally designing substrate geometries to achieve large enhancement factors. A variety of shapes and geometries have been explored as SERS substrates, including metal island films[38], large silver and gold colloids[39], silver triangle arrays[26], silver and gold nanoshells[36] and fractal silver films[40]. In many applications where single-molecule detection is not required, substrates are optimized in such a way as to achieve the largest enhancement factors and the densest coverage of the detected adsorbate molecules. Another approach for optimizing SERS substrates is the fabrication of nanoscale structures with controllable nanoscale gaps between two or more nanoparticles.

The simplest geometry of a structure with a nanoscale gap is a pair of closely spaced nanoparticles, usually referred to as a nanoparticle dimer. For spherical nanoparticles, we can invoke PH to explain theoretically the dimer plasmon as the hybrid plasmon of two interacting sphere plasmons[41]. The dimer plasmon is polarized[42–45] and for polarization perpendicular to the interparticle axis of the dimer, the hybridized plasmon is slightly blueshifted with respect to the single-particle plasmon. For incident polarization parallel to the dimer axis, there is a strong hybridization of the sphere plasmons, and the symmetric plasmon shows a continuous redshift as the gap between the spheres decreases. This has been observed experimentally for well-defined dimers made by electron-beam lithography[42–44]. In addition to the strong redshift of the plasmon resonance, there is an accumulation of charge in the gap between the two nanoparticles leading to a large enhancement of the near field and an increase in the far-field extinction. It is important to point out that the enhancement in the junction is much larger than the sum of the enhancements for individual nanospheres[36,43,45].

NANOSHELL-BASED SUBSTRATES

The plasmon resonances of nanoshells can be tuned over a wide spectral range by changing the geometrical dimensions of the core and shell thickness (see Fig. 2b). Additional tunability of the plasmon resonance can be achieved through asymmetric nanoshell geometries such as nanorice[17], a prolate spheroidal core–shell particle or a nanoegg[46] — a nanoshell with an offset core. The nanoshell plasmons furthermore give rise to large near-field enhancements when they are excited resonantly. Nanoshell synthesis methods have been optimized and PH has successfully been applied to rationalize the plasmon properties of single nanoshells, nanoshell dimers, clusters[47], nanorice[17] and a nanoegg[46]. This makes nanoshells a well-characterized and highly reproducible substrate with large SERS enhancements and an ideal candidate for investigating the electromagnetic origins of SERS (refs 48,49).

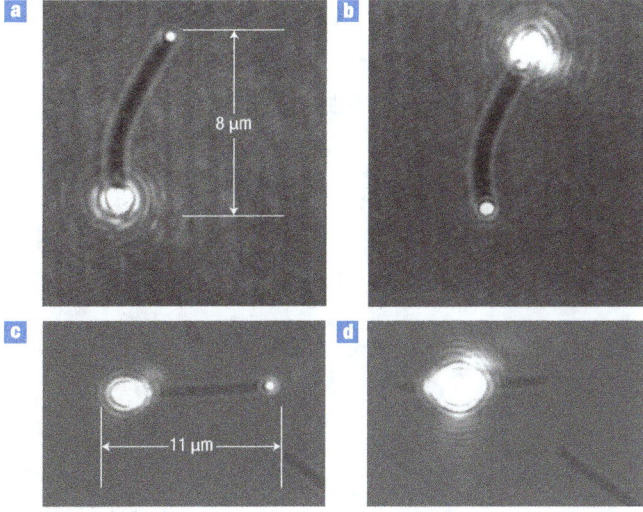

Figure 4 Micrographs showing the spatial sensitivity of launching plasmons. **a**, Nanowire with excitation at the bottom end. **b**, The same nanowire when excited from the top end. **c**, A different nanowire excited at the left end. **d**, The same nanowire with a laser positioned at the middle of the nanowire. Notice that the plasmon is not excited in this geometry. Figure reproduced with permission from ref. 67. Copyright (2006) ACS.

Enhancement factors of around 10^9 have been achieved with SERS even for non-resonant molecules such as para-mercaptoaniline (pMA) on gold nanoshells (see Fig. 3a,b and ref. 48). When the single nanoshell plasmon is resonantly excited by the pump laser, a linear response of the SERS signal with nanoshell density was observed indicating that individual nanoshells are responsible for the observed SERS signal. Nanoshell dimers and arrays have also produced large enhancement factors of more than 10^8 over a large interrogation area[36,50]. Nanoshell arrays furthermore support hybrid junction plasmons that can be tuned into the infrared region of the spectrum, which extends the usefulness of nanoshell-based substrates to measurements of surface-enhanced infrared absorption, SEIRA (ref. 50). Finally, nanoshells have been used to characterize the effect of metals (gold versus silver)[15,51], surface defects[51] and morphology on the SERS response. Thus, nanoshells with their well-characterized nanoenvironments are ideal substrates for sensitive chem–bio detection using SERS.

TIP-ENHANCED RAMAN SPECTROSCOPY

Junction plasmons or 'hot spots' in nanoparticle aggregates and assemblies are small, and it is often difficult to place the SERS analyte into the hot spot. An alternative approach that uses nanosized metallic or metal-coated tips to produce an enhanced optical field at the tip apex has been developed for SERS detection[52,53]. The metal tip is scanned over the analyte molecules, and the SERS signal is recorded as a function of tip position using an experimental set-up similar to near-field scanning optical microscopy. Tip-enhanced Raman spectrocopy or TERS offers enhanced SERS signals together with high spatial resolution beyond the diffraction limit[54,55].

Initial TERS experiments were performed with a tapered metal tip, which was brought close to the analyte molecules illuminated in a diffraction-limited laser spot[56]. The spatial resolution of these measurements depends mainly on the diameter of the tip apex. This imposes a severe limitation on TERS because the scattered fields from the tip decrease sharply as the tip size is reduced.

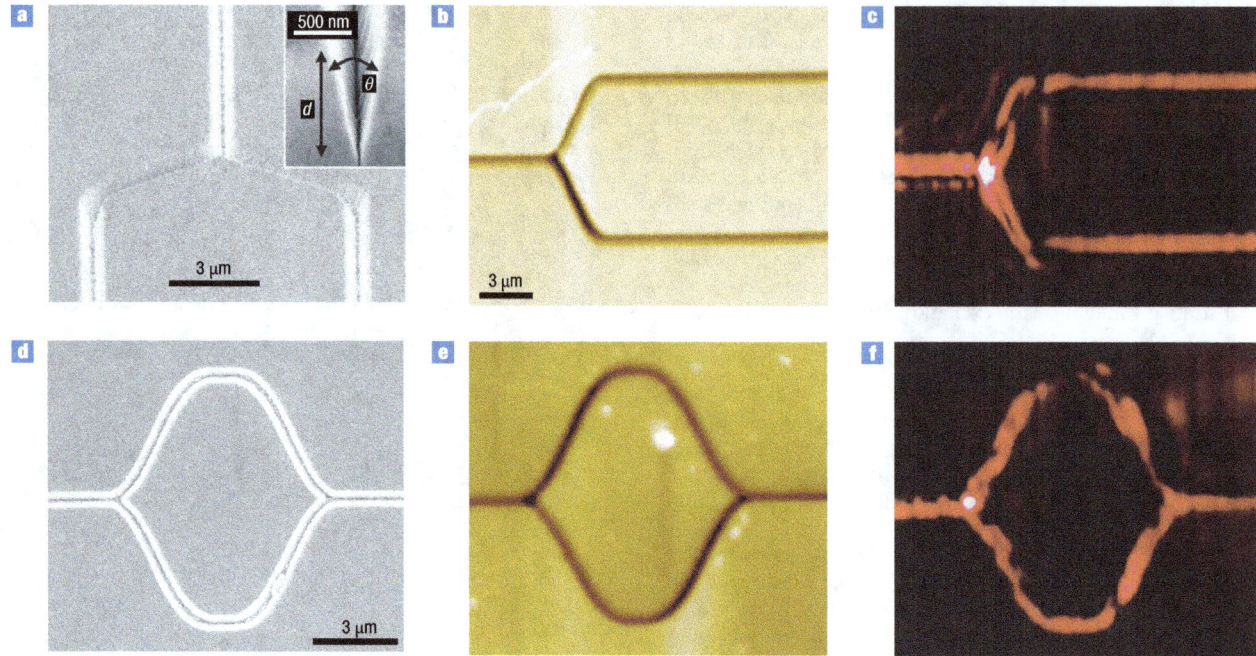

Figure 5 Y-splitter and Mach–Zehnder interferometer for plasmons[78]. **a**, Scanning electron microscope image, along with, **b**, topographical and, **c**, scanning near-field optical microscopy images of a Y-splitter. The inset in **a** is an image showing a typical groove profile of the channel waveguide. **d–f**, Scanning electron microscope (**d**), topographical (**e**) and scanning near-field optical microscope (**f**) images of the Mach–Zehnder interferometer. Copyright (2006) *Nature*.

A tip coated with a granular silver film has been successfully used to overcome this limitation because only one silver nanosphere near the apex of the tip acts as a nanoantenna to excite surface plasmons and to create enhanced local fields. Both of these geometries rely on the 'lightning rod' effect for field enhancement. A second TERS geometry is based on junction plasmons created between a metallic tip and a flat metallic surface that also serves as the analyte substrate[52]. Figure 3c,d shows an SEM image of a silver tip used for TERS along with the signal enhancement obtained when the tip is close to benzenethiol molecules on an optically flat gold film.

Enhancement factors of 10^6–10^8 under non-resonant excitation conditions have been reported for benzenethiol molecules adsorbed on optically flat gold and platinum surfaces probed with a gold tip placed one nanometre from the surface[52]. Owing to the high spatial resolution of TERS and the large field enhancements in easily accessible junctions between the tip and metal surface, TERS promises to be a powerful technique for real-time chem–bio sensing with nanoscale spatial resolution.

PLASMONIC WAVEGUIDES

As demand for computer processing speed increases, on-chip optical data transfer promises to greatly enhance current electronic computation schemes. However, for an optical technology to be feasible, light must be confined and routed in dimensions smaller than its wavelength to allow for sufficient miniaturization of chips. Surface plasmon waveguides offer the possibility of propagating electromagnetic radiation much like an optical fibre does, but using metallic waveguide structures that are an order of magnitude smaller[57,58]. The diffraction of classical optics constrains the dimension of an optical fibre to be larger than about $\lambda/2$ (where λ is the wavelength, giving dimensions of about 750 nm for telecommunications applications), whereas plasmonic devices are

only limited in their size by the dimensions of the input–output coupler and the transmitting medium. Plasmonic devices are therefore ideal candidates for next-generation nanoscale electronic applications such as light-on-a-chip.

Plasmon propagation has been explored for a variety of structures, such as patterned planar metal films, metal stripes, nanowires, nanoparticle chains, and metal–insulator–metal (MIM) waveguides. Two-dimensional plasmon propagation at the metal–air interface of a planar metal film was achieved by introducing height modulations. Scattering of the plasmons at holes or ordered nanoparticle arrays acting as local defects enables guiding of the plasmons. A propagation length of 10 µm has been observed for a system consisting of silica nanostructures deposited on a silica substrate and overcoated with a 70-nm-thick silver film[59]. Ordered arrangements of these surface modulations can be used to create functional elements like Bragg mirrors[59] or focusing lenses[60]. Focusing lenses were fabricated by milling a curved line of closely spaced holes into a metal film. The authors of ref. 60 elegantly demonstrated the focusing of their nanohole array by coupling a plasmon in a metal-stripe waveguide.

A particularly simple geometry of a plasmon waveguide is a long, thin metal stripe sandwiched between two insulator layers, that is, an insulator–metal–insulator (IMI) geometry[61]. Plasmon propagation lengths of several micrometres in a 200-nm-wide stripe have been observed. The propagation lengths depend strongly on the width of the metal stripe[62,63], and it has been shown both theoretically[64] and experimentally[65] that, as the stripe waveguide becomes narrower, the number of guided modes decreases. At a certain stripe width, no more guided modes exist, and thus propagation on the stripe is very limited. For micrometre-wide stripes however, the propagating plasmon modes can reach attenuation lengths of a centimetre at telecommunication wavelengths in the near infrared. The IMI waveguide geometry also includes metal stripes on an insulator substrate surrounded by air. It is worth mentioning this asymmetric geometry separately because the plasmon modes excited at the metal–air interface suffer radiative losses in addition to ohmic heating. These additional losses cause shorter propagation distances compared with symmetric IMI structures[63,64]. The same trend of decreasing propagation length with decreasing metal stripe width is also observed for the air–metal–insulator structures, confirming that, for metal-stripe waveguides, there is a trade-off between confinement and propagation length. Although the metal-stripe waveguide can, in general, support long propagation lengths for large stripe widths, it is not as useful for the purpose of subwavelength confinement.

In addition to lithographically prepared metal stripes, which are several hundred nanometres wide, plasmon propagation has also been reported for chemically prepared silver and gold nanorods or nanowires with transverse and longitudinal dimensions of less than 100 nm and around 4 μm, respectively. Light was coupled in at one end of the rod and observed to emerge from the other, proving that plasmons did in fact propagate along the rod[66,67] (see Fig. 4). More recent experiments were carried out with even longer and narrower silver wires (25 nm wide and about 50 μm long)[68]. Plasmon propagation was visualized by fluorescence markers along the nanowire and terminated after a propagation distance of about 15 μm. Similar to light propagation in optical fibres, plasmon propagation is not limited to a straight line, but electromagnetic radiation contained in the plasmon can be bent and split using appropriate plasmonic structures[67,69]. For example, a metallic wire branching into more than one path enables a plasmon to continue propagating down both paths of the wire, similar to an optical beamsplitter. These observations show the great opportunity to manipulate plasmons in the same way that light has been manipulated in optics, but using much smaller components.

The interparticle coupling responsible for the giant local field enhancement at the junctions between nanoparticles can also be exploited for plasmon propagation. Linear chains of nanoparticles can act as a plasmonic waveguide[70] because the metal nanoparticles couple to each other by means of the optical near field when irradiated with light[71,72]. Using a near-field scanning optical microscope to locally excite plasmons on one end of a nanoparticle chain, plasmon propagation and waveguiding has been observed along the nanoparticle chain by directly measuring the energy transport[73]. It is anticipated that fabricating nanoparticle chains with complex architectures could be far superior to conventional top-down approaches such as electron-beam lithography[74].

The waveguide geometry consisting of an insulator with a metal cladding has recently stimulated increased interest. Theoretical studies suggested that these MIM structures could offer superior subwavelength confinement for plasmon waveguides. Although IMI structures (for example, metal stripes sandwiched between two insulators or gaps and grooves in a planar film) generally supported longer propagation lengths along the waveguide, the MIM structures confine the plasmon to the insulator part with far superior ability to achieve subwavelength confinement[75]. Planar-multilayered MIM systems, furthermore, support both plasmonic modes and photonic modes (normal propagating electromagnetic modes)[76,77]. In Ag/Si$_3$N$_4$/Ag layered MIMs, light can be coupled into and out of the waveguide structures using slit openings in the metal, enabling broadband propagation of electromagnetic energy over distances of several micrometres. Similarly, V-shaped grooves milled into a gold film support propagating plasmon modes localized at the bottom of the groove. Propagation lengths of 100 μm at a wavelength of 1,500 nm have been realized in metal grooves[78]. The groove parameters (depth, d, which is about 1.1–1.3 μm, and groove angle, θ, about 25°) can be optimized to enable low-loss single-mode propagation at the bottom of the groove at telecom wavelengths. The shape of the grooves can be designed

to function as complex optical components, such as Y-splitters and Mach–Zehnder interferometers. Figure 5 shows SEM images, along with topographical and near-field optical images of a Y-splitter (Fig. 5a–c) and a Mach–Zehnder interferometer (Fig. 5d–f) that can be used to realize plasmonic components in an integrated optical circuit. Inspired by the success of these plasmonic waveguides, more complicated plasmonic elements are being actively explored, such as Y-splitters, interferometers and ring resonators[78], with the ultimate goal of building all plasmonic-based active devices that include plasmonic light sources, waveguides and detectors.

SUMMARY

In conclusion, the rapidly growing subdiscipline of nanophotonics known as plasmonics has yielded unique phenomena based on light–metal interactions that may give rise to new, useful applications and technologies. With applications ranging from 'light-on-a-chip' to 'lab-on-a-chip', the future of this field looks exceptionally bright and promising. We have no doubt that useful technologies will emerge from the development of plasmonics for a wide range of applications.

doi:10.1038/nphoton.2007.223

References

1. Raether, H. R. *Surface Plasmons on Smooth and Rough Surfaces and on Gratings* (Springer, New York, 1988).
2. Maier, S. *Plasmonics: Fundamentals and Applications* (Springer, New York, 2007).
3. Bohren, C. F. & Huffman, D. R. *Absorption and Scattering of Light by Small Particles* (Wiley, New York, 1983).
4. Kreibig, U. & Vollmer, M. *Optical Properties of Metal Clusters* (Springer, Berlin, 1995).
5. Grady, N. K., Halas, N. J. & Nordlander, P. Influence of dielectric function properties on the optical response of plasmon resonant metallic nanoparticles. *Chem. Phys. Lett.* **399**, 167–171 (2004).
6. Kelly, K. L., Coronado, E., Zhao, L. L. & Schatz, G. C. The optical properties of metal nanoparticles: The influence of size, shape, and dielectric environment. *J. Phys. Chem. B* **107**, 668–677 (2003).
7. Mock, J. J., Smith, D. R. & Schultz, S. Local refractive index dependence of plasmon resonance spectra from individual nanoparticles. *Nano Lett.* **3**, 485–491 (2003).
8. Underwood, S. & Mulvaney, P. Effect of the solution refractive index on the color of gold colloids. *Langmuir* **10**, 3427–3430 (1994).
9. Grabar, K. C., Freeman, R. G., Hommer, M. B. & Natan, M. J. Preparation and characterization of Au colloid monolayers. *Anal. Chem.* **67**, 735–743 (1995).
10. Hulteen, J. C. & Van Duyne, R. P. Nanosphere lithography: A materials general fabrication process for periodic particle array surfaces. *J. Vac. Sci. Technol. A* **13**, 1553–1558 (1995).
11. Bastys, V., Pastoriza-Santos, I., Rodríguez-González, B., Vaisnoras, R. & Liz-Marzán, L. M. Formation of silver nanoprisms with surface plasmons at communication wavelengths. *Adv. Funct. Mater.* **16**, 766–773 (2006).
12. Nikoobakht, B. & El-Sayed, M. A. Preparation and growth mechanism of gold nanorods (NRs) using seed-mediated growth method. *Chem. Mater.* **15**, 1957–1962 (2003).
13. Sun, Y. G. & Xia, Y. N. Shape-controlled synthesis of gold and silver nanoparticles. *Science* **298**, 2176–2179 (2002).
14. Oldenburg, S. J., Averitt, R. D., Westcott, S. L. & Halas, N. J. Nanoengineering of optical resonances. *Chem. Phys. Lett.* **288**, 243–247 (1998).
15. Jackson, J. B. & Halas, N. J. Silver nanoshells: Variations in morphologies and optical properties. *J. Phys. Chem. B* **105**, 2743–2746 (2001).
16. Sun, Y., Mayers, B. & Xia, Y. Metal nanostructures with hollow interiors. *Adv. Mater.* **15**, 641–646 (2003).
17. Wang, H., Brandl, D. W., Le, F., Nordlander, P. & Halas, N. J. Nanorice: A hybrid plasmonic nanostructure. *Nano Lett.* **6**, 827–832 (2006).
18. Landes, C. F., Link, S., Mohamed, M. B., Nikoobakht, B. & El-Sayed, M. A. Some properties of spherical and rod-shaped semiconductor and metal nanocrystals. *Pure Appl. Chem.* **74**, 1675–1692 (2002).
19. McLellan, J. M., Siekkinen, A. R. & Xia, Y. The SERS activity of a supported Ag nanocube strongly depends on its orientation relative to laser polarization. *Nano Lett.* **7**, 1013–1017 (2007).
20. Prodan, E. & Nordlander, P. Plasmon hybridization in spherical nanoparticles. *J. Chem. Phys.* **120**, 5444–5454 (2004).
21. Mie, G. Articles on the optical characteristics of turbid media, especially colloidal metal solutions. *Ann. Phys. (Leipz.)* **25**, 377–445 (1908).
22. Aden, A. L. & Kerker, M. Scattering of electromagnetic waves from two concentric spheres. *J. App. Phys.* **22**, 1242–1246 (1951).
23. Draine, B. T. & Flatau, P. J. Discrete-dipole approximation for scattering calculations. *J. Opt. Soc. Am. A* **11**, 1491–1499 (1994).
24. Futamata, M., Maruyama, Y. & Ishikawa, M. Local electric field and scattering cross section of Ag nanoparticles under surface plasmon resonance by finite difference time domain method. *J. Phys. Chem. B* **107**, 7607–7617 (2003).
25. Oubre, C. & Nordlander, P. Optical properties of metallodielectric nanostructures calculated using the finite difference time domain method. *J. Phys. Chem. B* **108**, 17740–17747 (2004).
26. Hayes, C. L. & Van Duyne, R. P. Plasmon-sampled surface-enhanced Raman excitation spectroscopy. *J. Phys. Chem. B* **107**, 7426–7433 (2003).

27. Liao, H., Nehl, C. L. & Hafner, J. H. Biomedical applications of plasmon resonant metal nanoparticles. *Nanomedicine* **1**, 201–208 (2006).

28. Haes, A. & Van Duyne, R. P. A unified view of propagating and localized surface plasmon resonance biosensors. *Anal. Bioanal. Chem.* **379**, 920–930 (2004).

29. Taton, T. A., Mirkin, C. A. & Letsinger, R. L. Scanometric DNA array detection with nanoparticle probes. *Science* **289**, 1757–1760 (2000).

30. Novotny, L. & Hecht, B. *Principles of Nano-Optics* (Cambridge Univ. Press, Cambridge, 2006).

31. Jeanmaire, D. L. & Van Duyne, R. P. Surface raman spectroelectrochemistry: Part 1. Heterocyclic, aromatic and aliphatic amines adsorbed on the anodized silver electrodes. *J. Electroanal. Chem.* **84**, 1–20 (1977).

32. Albrecht, M. G. & Creighton, J. A. Anomalously intense Raman spectra of pyridine at a silver electrode. *J. Am. Chem. Soc.* **99**, 5215–5217 (1977).

33. Kneipp, K. *et al.* Single molecule detection using surface enhanced Raman scattering. *Phys. Rev. Lett.* **78**, 1667–1670 (1997).

34. Nie, S. & Emory, S. R. Probing single molecules and single nanoparticles by surface-enhanced Raman scattering. *Science* **275**, 1102–1106 (1997).

35. Lal, S., Grady, N. K., Goodrich, G. P. & Halas, N. J. Profiling the near field of a plasmonic nanoparticle with Raman-based molecular rulers. *Nano Lett.* **6**, 2338–2343 (2006).

36. Talley, C. E. *et al.* Surface-enhanced Raman scattering from individual Au nanoparticles and nanoparticle dimer substrates. *Nano Lett.* **5**, 1569–1574 (2005).

37. McFarland, A. D., Young, M. A., Dieringer, J. A. & Van Duyne, R. P. Wavelength-scanned surface-enhanced Raman excitation spectroscopy. *J. Phys. Chem. B* **109**, 11279–11285 (2005).

38. Jennings, C. & Aroca, R. Surface-enhanced Raman scattering from copper and zinc phthalocyanine complexes by silver and indium island films. *Anal. Chem.* **56**, 2033–2035 (1984).

39. Michaels, A. M., Jiang, J. & Brus, L. Ag nanocrystal junctions as the site for surface-enhanced Raman scattering of single rhodamine 6G molecules. *J. Phys. Chem. B* **104**, 11965–11971 (2000).

40. Drachev, V. P. *et al.* Adaptive silver films for surface-enhanced Raman spectroscopy of biomolecules. *J. Raman Spectrosc.* **36**, 648–656 (2005).

41. Nordlander, P., Oubre, C., Prodan, E., Li, K. & Stockman, M. I. Plasmon hybridization in nanoparticle dimers. *Nano Lett.* **4**, 899–903 (2004).

42. Rechberger, W. *et al.* Optical properties of two interacting gold nanoparticles. *Opt. Comm.* **220**, 137–141 (2003).

43. Atay, T., Song, J.-H. & Nurmikko, A. V. Strongly interacting plasmon nanoparticle pairs: From dipole-dipole interaction to conductively coupled regime. *Nano Lett.* **4**, 1627–1631 (2004).

44. Gunnarsson, L. *et al.* Confined plasmons in nanofabricated single silver particle pairs: Experimental observations of strong interparticle interactions. *J. Phys. Chem. B* **109**, 1079–1087 (2005).

45. Romero, I., Aizpurua, J., Bryant, G. W. & de Abajo, F. J. G. Plasmons in nearly touching metallic nanoparticles: Singular response in the limit of touching dimers. *Opt. Express* **14**, 9988–9999 (2006).

46. Wang, H. *et al.* Symmetry-breaking in individual plasmonic nanoparticles. *Proc. Natl Acad. Sci. USA* **103**, 10856–10860 (2006).

47. Brandl, D. W., Mirin, N. A. & Nordlander, P. Plasmon modes of nanosphere trimers and quadrumers. *J. Phys. Chem. B* **110**, 12302–12310 (2006).

48. Jackson, J. B. & Halas, N. J. Surface-enhanced Raman scattering on tunable plasmonic nanoparticle substrates. *Proc. Natl Acad. Sci.* **101**, 17930–17935 (2004).

49. Jackson, J. B., Westcott, S. L., Hirsch, L. R., West, J. L. & Halas, N. J. Controlling the surface enhanced Raman effect via the nanoshell geometry. *Appl. Phys. Lett.* **82**, 257–259 (2003).

50. Wang, H., Le, F., Kundu, J., Nordlander, P. & Halas, N. J. Nanoshell arrays as multifunctional surface enhanced Raman/IR spectroscopic substrates. *Angew. Chem. Int. Edn* (in the press).

51. Wang, H. *et al.* Controlled texturing modifies the surface topography and plasmonic properties of Au nanoshells. *J. Phys. Chem. B* **109**, 11083–11087 (2005).

52. Zhang, W. *et al.* Nanoscale roughness on metal surfaces can increase tip-enhanced Raman scattering by an order of magnitude. *Nano Lett.* **7**, 1401–1405 (2007).

53. Neacsu, C. C., Dreyer, J., Behr, N. & Raschke, M. B. Scanning-probe Raman spectroscopy with single-molecule sensitivity. *Phys. Rev. B* **73**, 193406 (2006).

54. Pettinger, B. Tip-enhanced Raman spectroscopy (TERS). *Top. Appl. Phys.* **103**, 217–240 (2006).

55. Verma, P., Inouye, Y. & Kawata, S. *Surface-enhanced Raman Scattering: Physics and Applications – Topics in Applied Physics* (eds Kneipp, K., Moskovits, M. & Kneipp, M.) 241–262 (Springer, Berlin, 2006).

56. Domke, K. F., Zhang, D. & Pettinger, B. Toward Raman fingerprints of single dye molecules at atomically smooth Au(111). *J. Am. Chem. Soc.* **128**, 14721–14727 (2006).

57. Maier, S. A. *et al.* Plasmonics - A route to nanoscale optical devices. *Adv. Mater.* **13**, 1501–1505 (2001).

58. Barnes, W. L., Dereux, A. & Ebbesen, T. W. Surface plasmon subwavelength optics. *Nature* **424**, 824–830 (2003).

59. Dittlbacher, H. *et al.* Fluorescence imaging of surface plasmon fields. *Appl. Phys. Lett.* **80**, 404–406 (2002).

60. Yin, L. *et al.* Subwavelength focusing and guiding of surface plasmons. *Nano Lett.* **5**, 1399–1402 (2005).

61. Charbonneau, R., Berini, P., Berolo, E. & Lisicka-Shrzek, E. Experimental observation of plasmon-polariton waves supported by a thin metal film of finite width. *Opt. Lett.* **25**, 844–846 (2000).

62. Krenn, J. R. *et al.* Non-diffraction-limited light transport by gold nanowires. *Europhys. Lett.* **60**, 663–669 (2002).

63. Lamprecht, B. *et al.* Surface plasmon propagation in micrsoscale metal stripes. *App. Phys. Lett.* **79**, 51–53 (2001).

64. Zia, R., Selker, M. D. & Brongersma, M. L. Leaky and bound modes of surface plasmon waveguides. *Phys. Rev. B* **71**, 165431 (2005).

65. Zia, R., Schuller, J. A. & Brongersma, M. L. Near-field characterization of guided polariton propagation and cutoff in surface plasmon waveguides. *Phys. Rev. B* **74**, 165415 (2006).

66. Dickson, R. M. & Lyon, L. A. Unidirectional plasmon propagation in metallic nanowires. *J. Phys. Chem. B* **104**, 6095–6098 (2000).

67. Sanders, A. W. *et al.* Observation of plasmon propagation, redirection, and fan-out in silver nanowires. *Nano Lett.* **6**, 1822–1826 (2006).

68. Graff, A., Wagner, D., Ditlbacher, H. & Kreibig, U. Silver nanowires. *Euro. Phys. J. D* **34**, 263–269 (2005).

69. Knight, M. W. *et al.* Nanoparticle-mediated coupling of light into a nanowire. *Nano Lett.* **7**, 2346–2350 (2007).

70. Quinten, M., Leitner, A., Krenn, J. R. & Aussenegg, F. R. Electromagnetic energy transport via linear chains of silver nanoparticles. *Opt. Lett.* **23**, 1331–1333 (1998).

71. Maier, S. A., Brongersma, M. L., Kik, P. G. & Atwater, H. A. Observation of near-field coupling in metal nanoparticle chains using far-field polarization spectroscopy. *Phys. Rev. B* **65**, 193408 (2002).

72. Maier, S. A., Kik, P. G. & Atwater, H. A. Observation of coupled plasmon-polariton modes in Au nanoparticle chain waveguides of different lengths: Estimation of waveguide loss. *Appl. Phys. Lett.* **81**, 1714–1716 (2002).

73. Maier, S. A. *et al.* Local detection of electromagnetic energy transport below the diffraction limit in metal nanoparticle plasmon waveguides. *Nature Mater.* **2**, 229–232 (2003).

74. Lin, S., Li, M., Dujardin, E., Girard, C. & Mann, S. One-dimensional plasmon coupling by facile self-assembly of gold nanoparticles into branched chain networks. *Adv. Mater.* **17**, 2553–2559 (2005).

75. Zia, R., Selker, M. D., Catrysse, P. B. & Brongersma, M. I. Geometries and materials for subwavelength surface plasmon modes. *J. Opt. Soc. Am. A* **21**, 2442–2446 (2004).

76. Dionne, J. A., Sweatlock, L. A., Atwater, H. A. & Polman, A. Plasmon slot waveguides: Towards chip-scale propagation with subwavelength-scale localization. *Phys. Rev. B* **73**, 035407 (2006).

77. Dionne, J. A., Lezec, H. J. & Atwater, H. A. Highly confined photon transport in subwavelength metallic slot waveguides. *Nano Lett.* **6**, 1928–1932 (2006).

78. Bozhevolnyi, S. I., Volkov, V. S., Devaux, E., Laluet, J.-Y. & Ebbesen, T. W. Channel plasmon subwavelength waveguide components including interferometers and ring resonators. *Nature* **440**, 508–511 (2006).

79. Zhang, W., Yeo, B. S., Schmid, T. & Zenobi, R. Single molecule tip-enhanced Raman spectroscopy with silver tips. *J. Phys. Chem. C* **111**, 1733–1738 (2007).

Acknowledgements

S. Lal acknowledges useful discussions with J. Britt Lassiter and Nathaniel K. Grady. This work was supported by the Robert A. Welch Foundation grants C-1664 and C-1220, and the Department of Defense Multidisciplinary University Research Initiative (MURI) grant W911NF-04-01-0203. Additional funding was provided by grants from the Air Force Office of Scientific Research, The National Science Foundation and National Aeronautics and Space Administration.

Semiconductor quantum light sources

Lasers and LEDs have a statistical distribution in the number of photons emitted within a given time interval. Applications exploiting the quantum properties of light require sources for which either individual photons, or pairs, are generated in a regulated stream. Here we review recent research on single-photon sources based on the emission of a single semiconductor quantum dot. In just a few years remarkable progress has been made in generating indistinguishable single photons and entangled-photon pairs using such structures. This suggests that it may be possible to realize compact, robust, LED-like semiconductor devices for quantum light generation.

ANDREW J. SHIELDS

Toshiba Research Europe Limited, 260 Cambridge Science Park, Cambridge CB4 0WE, UK

e-mail: andrew.shields@crl.toshiba.co.uk

Applying quantum light states to photonic applications allows functionalities that are not possible using 'ordinary' classical light. For example, carrying information with single photons provides a means to test the secrecy of optical communications, which could soon be applied to the problem of sharing digital cryptographic keys[1,2]. Although secure quantum-key-distribution systems based on weak laser pulses have already been realized for simple point-to-point links, true single-photon sources would improve their performance[3]. Furthermore, quantum light sources are important for future quantum-communication protocols, such as quantum teleportation[4]. Here quantum networks sharing entanglement could be used to distribute keys over longer distances or through more complex topologies[5].

A natural progression would be to use photons for quantum-information processing, as well as communication. In this regard it is relatively straightforward to encode and manipulate quantum information on a photon. On the other hand, single photons do not interact strongly with each another, which is a prerequisite for a simple photon logic gate. In linear-optics quantum computing[6,7] (LOQC) this problem is solved using projective measurements to induce an effective interaction between the photons. Here, triggered sources of single photons and entangled pairs are required both as the qubit carriers, and as auxiliary sources to test the successful operation of the gates. Although the component requirements for LOQC are challenging, they have recently been relaxed significantly by new theoretical schemes[7]. Quantum light states will also probably become increasingly important for various types of precision optical measurement[8].

For these applications, light sources, which generate pure single-photon states 'on demand' in response to an external trigger signal, are preferred. Key performance measures for such a source are the efficiency, defined as the fraction of photons collected in the experiment or application per trigger, and the second-order correlation function at zero delay (see text box). The second-order correlation function at zero delay is essentially a measure of the two-photon rate compared with a classical source, with random emission times, of the same average intensity. In order to construct applications involving more than one photon, it is also important that photons emitted from the source (at different times), as well as those from different sources, are otherwise indistinguishable.

In the absence of a convenient triggered single-photon source, most experiments in quantum optics rely on nonlinear optical processes for generating quantum light states. Optically pumping a crystal with a $\chi^{(2)}$ nonlinearity has a finite probability of generating a pair of lower-energy photons through parametric down conversion. This may be used to prepare photon pairs with time-bin entanglement[9], entangled polarizations[10,11], or alternatively single-photon states 'heralded' by the other photon in the pair[12]. A $\chi^{(3)}$ nonlinearity in a semiconductor has also been used to generate entangled pairs[13]. As these nonlinear processes occur randomly, there is always a finite probability of generating two pairs that increases with pump power. As double pairs degrade the fidelity of quantum optical gates, the pump laser power must be restricted to reduce the rate of double pairs to an acceptable level, which has a detrimental effect on the efficiency of the source[14]. This means that although down-conversion sources continue to be highly successful in demonstrating few-photon quantum optical gates, scaling to large numbers may be problematic. Solutions have been proposed based on switching multiple sources[15], or storing photons in a switched fibre loop[16].

Preferably the quantum light source should generate exactly one single photon, or entangled pair, per excitation trigger pulse. This may be achieved using the emission of a single quantum system. After relaxation, a quantum system is, by definition, no longer excited and therefore unable to re-emit. Photon antibunching — the tendency of a quantum source to emit photons separated in time — was first demonstrated in the resonance fluorescence of a low-density vapour of sodium atoms[17], and subsequently for a single ion[18].

Quantum dots are often referred to as 'artificial atoms', as their electron motion is quantized in all three spatial directions, resulting in a discrete energy-level spectrum, like that of an atom. They provide a quantum system, which can be grown within robust, monolithic semiconductor devices and can be engineered to have a wide range of desired properties. In the following, recent progress towards the realization of semiconductor technology for quantum photonics is reviewed. An excellent account of the early work can be found in ref. 19. Space restrictions limit discussion of work on other quantized systems. For this we refer the reader to the comprehensive review in ref. 20.

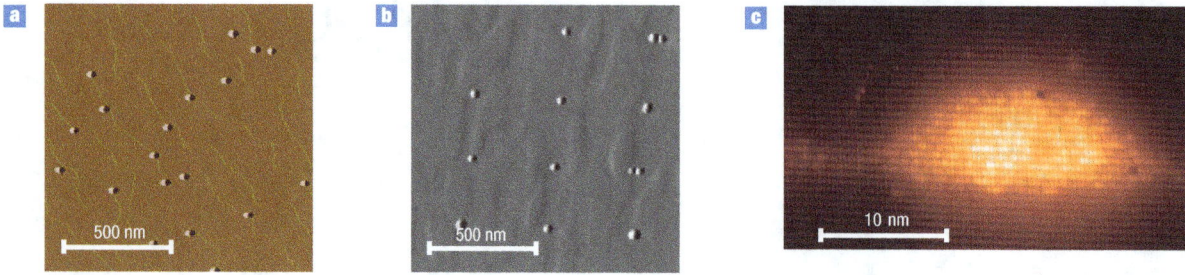

Figure 1 Self-assembled quantum dots. **a**, Image of a layer of InAs/GaAs self-assembled quantum dots recorded with an atomic-force microscope. Each blob corresponds to a dot with typical lateral diameters of 20–30 nm and a height of 4–8 nm. **b**, Atomic-force-microscope image of a layer of InAs quantum dots whose locations have been seeded by a matrix of nanometre-sized pits patterned onto the wafer surface. Under optimal conditions, up to 60% of the etch pits contain a single dot. Reproduced with permission from ref. 23. Copyright (2006) JSAP. **c**, Cross-sectional scanning-tunnelling-microscope image of an InAs dot inside a GaAs device. Image courtesy of P. Koenraad, Eindhoven.

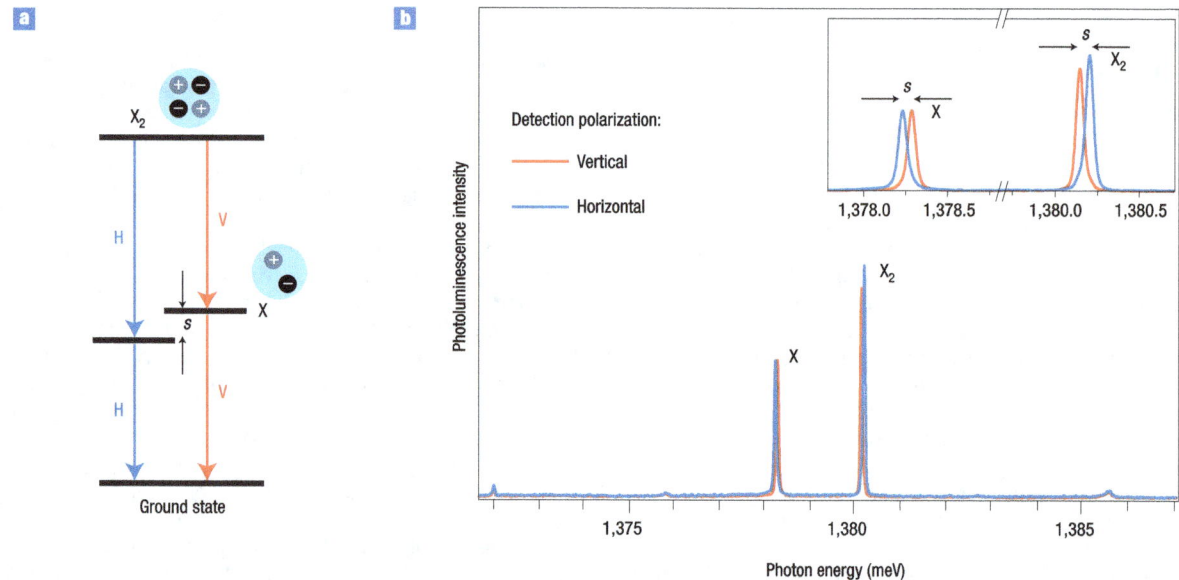

Figure 2 Optical spectrum of a quantum dot. **a**, Schematic of the biexciton cascade of a quantum dot. **b**, Typical photoluminescence spectrum of a single quantum dot showing sharp line emission due to the biexciton, X_2, and exciton, X, photon emitted by the cascade. The inset shows the polarization splitting of the transitions originating from the spin splitting of the exciton levels.

OPTICAL PROPERTIES OF SINGLE QUANTUM DOTS

Nanoscale quantum dots with good optical properties can be fabricated using a natural-growth mode of strained-layer semiconductors[21]. When InAs is deposited on GaAs, it initially grows as a strained two-dimensional sheet, but beyond some critical thickness, tiny islands like those shown in Fig. 1a form in order to minimize the surface strain. Overgrowth of the islands leads to the coherent incorporation of $In_xGa_{1-x}As$ dots into the crystal structure of the device, as can be seen in the cross-sectional image of Fig. 1c. The most intensively studied are small InAs dots on GaAs emitting at wavelengths around 900–950 nm at low temperatures, which can be conveniently measured with low-noise silicon single-photon detectors.

A less desirable feature of the self-organizing technique is that the dots form at random positions on the growth surface. However, considerable progress has been made recently to control the dot position within the device structure (Fig. 1b) by patterning nanometre-sized pits on the growth surface[22,23].

As InGaAs has a lower-energy bandgap than GaAs, the quantum dot forms a potential trap for electrons and holes. If sufficiently small, the dot contains just a few quantized levels in the conduction and valence bands, each of which holds two electrons or holes of opposite spin. Illumination by a picosecond laser pulse excites electrons and holes, which rapidly relax to the lowest-lying energy states either side of the bandgap. A quantum dot can thus capture two electrons and two holes to form the biexciton state, which decays by a radiative cascade, as shown schematically in Fig. 2a. One of the trapped electrons

Box 1 Photon-correlation measurements

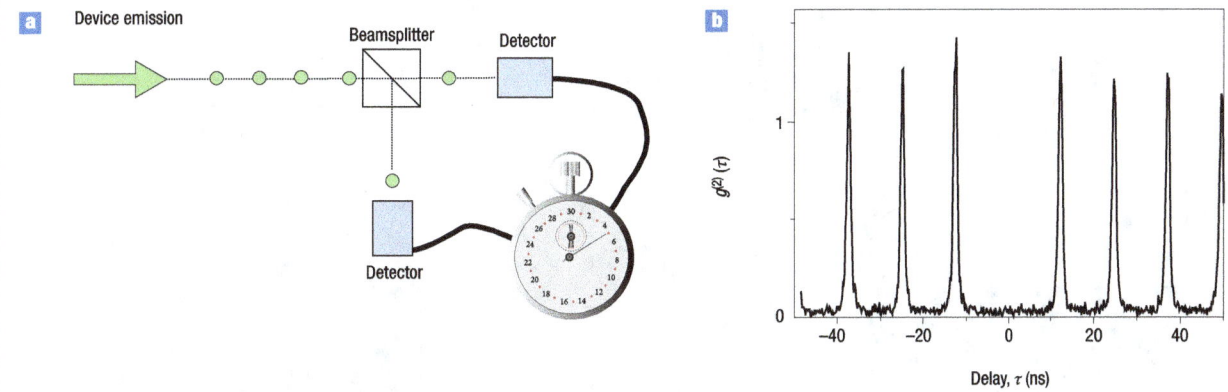

Figure B1 Measuring the correlation. **a,** Schematic of the set-up used for photon-correlation measurements. **b,** Second-order correlation function of the exciton emission of a single dot in a pillar microcavity. Reproduced with pemission from ref. 41. Copyright (2005) OSA.

The photon statistics of light can be studied by means of the second-order correlation function, $g^{(2)}(\tau)$, which describes the correlation between the intensity of the light field, I, with that after a delay τ and is given by[100]:

$$g^{(2)}(\tau) = \frac{\langle I(t)I(t+\tau)\rangle}{\langle I(t)\rangle^2} .$$

This function can be measured directly using the Hanbury-Brown and Twiss[101] interferometer, comprising a 50:50 beamsplitter and two single-photon detectors, shown in the figure. For delays much less than the average time between detection events (that is, for low intensities), the distribution in the delays between clicks in each of the two detectors is proportional to $g^{(2)}(\tau)$.

For a continuous light source with random emission times, such as an ideal laser or LED, $g^{(2)}(\tau) = 1$. It shows there is no correlation in the emission time of any two photons from the source. A source for which $g^{(2)}(\tau = 0) > 1$ is described as 'bunched' as there is an enhanced probability of two photons being emitted within a short time interval. Photons emitted by quantum light sources are typically 'antibunched', ($g^{(2)}(\tau = 0) < 1$) and tend to be separated in time.

Of particular interest in communication and computing systems are pulsed light sources, for which the emission occurs at times defined by an external clock. In this case $g^{(2)}(\tau)$ consists of a series of peaks separated by a clock period. For an ideal single-photon source, the peak at zero time delay is absent, $g^{(2)}(\tau = 0) = 0$; as the source cannot produce more than one photon per excitation period, clearly the two detectors cannot fire simultaneously.

The figure shows $g^{(2)}(\tau)$ recorded for resonant pulsed optical excitation of the X emission of a single quantum dot in a pillar microcavity. Notice the almost complete absence of the peak at zero delay: the definitive signature of a single-photon source. The noise seen at $\tau = 0$ demonstrates that the rate of two-photon emission is more than 50 times less than that of an ideal laser with the same average intensity. The bunching behaviour observed for the finite delay peaks is explained by intermittent trapping of a charge carrier in the dot[102]. This trace was taken for quasi-resonant laser excitation of the dot, which avoids creating carriers in the surrounding semiconductor. For higher-energy laser excitation, the suppression in $g^{(2)}(0)$ is typically reduced indicating occasional two-photon pulses due to emission from the layers surrounding the dot, but this can be minimized with careful sample design.

recombines with one of the holes and generates a first photon (called the biexciton photon, X_2). This leaves a single electron–hole pair in the dot (the exciton state), which subsequently also recombines to generate a second (exciton, X) photon. The biexciton and exciton photons have distinct energies, as can be seen in the low temperature photoluminescence spectrum of Fig. 2a, owing to the different Coulomb energies of their initial and final states. Often a number of other weaker lines can also be seen owing to recombination of charged excitons, which form intermittently when the dot captures an extra electron or hole[24]. Larger quantum dots, with several confined electron and hole levels, have a richer optical signature owing to the large number of exciton complexes that can be confined.

High-resolution spectroscopy reveals that the X_2 and X transitions of a dot are in fact both doublets with linearly polarized components

parallel to the [110] and [1–10] axes of the semiconductor crystal, labelled here H and V, respectively[25,26]. The origin of this polarization splitting is an asymmetry in the electron–hole exchange interaction of the dot, which produces a splitting of the exciton spin states. The asymmetry derives from an elongation of the dot along one crystal axis and inbuilt strain in the crystal. It mixes the exciton eigenstates of a symmetric dot, with total azimuthal spin $J_z = +1$ and −1, into symmetric and antisymmetric combinations, which couple to two H- or two V-polarized photons, respectively, as shown in Fig. 2.

The exciton state of the dot has a typical lifetime of about one nanosecond, which is due purely to radiative decay. As this is much longer than the duration of the exciting laser pulse, or the lifetime of the photo-excited carrier population in the surrounding semiconductor, only one X photon can be emitted

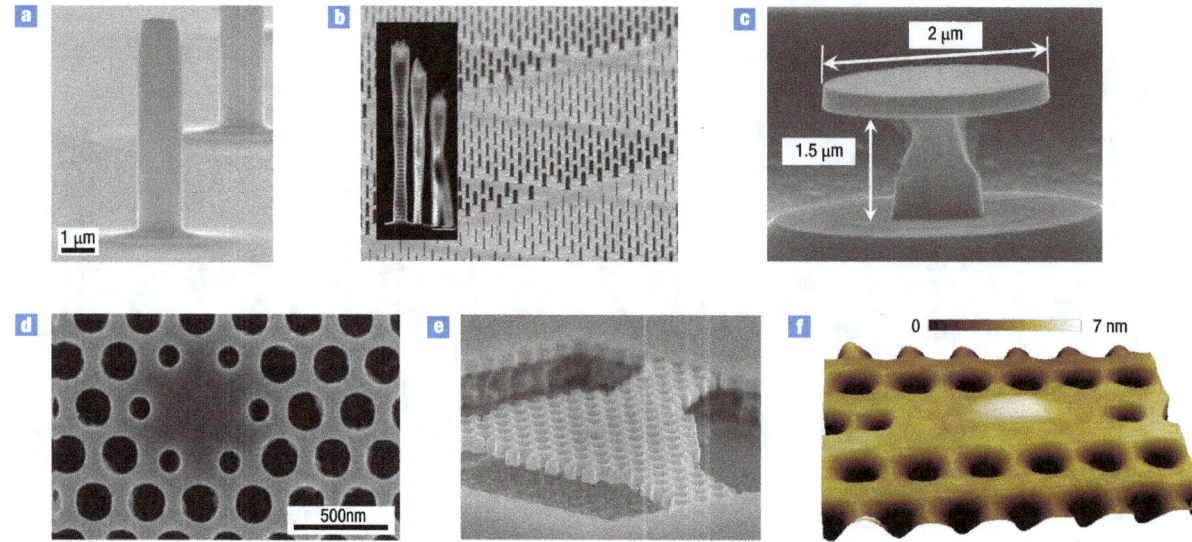

Figure 3 Scanning-electron-microscope images of semiconductor cavities. **a,b,** Pillar microcavities. **c,** microdisks. **d–f,** Photonic-bandgap defect cavities. The structures were fabricated at: **a,** University of Würzburg; **b,c,e,** CNRS-LPN (UPR-20); **d,** Univ. Cambridge; **f,** UCSB/ETHZ. (Image sources and permissions: **a,** Ref. 56. **c,** Ref. 51, copyright (2005) APS. **d,** Ref. 47, copyright (2006) AIP. **e,f,** Ref. 48)

per laser pulse. This can be proved, as first reported[27] by Peter Michler, Atac Imamoglu and their colleagues in Santa Barbara, by measuring the second-order correlation function, $g^{(2)}(\tau)$ of the exciton photoluminescence[28,29] (text box). In fact each of the exciton complexes of the dot generates at most one photon per excitation cycle, which also allows single-photon emission from the biexciton or charged exciton transitions[30].

Cross-correlation measurements[31–33] between the X and X_2 photons confirm the time correlation expected for the cascade in Fig. 2a, that is, the X photon follows the X_2 one. Indeed the shape of the cross-correlation function for both continuous-wave and pulsed excitation can be accurately described with a simple rate-equation model and the experimentally measured X and X_2 decay rates[34].

SEMICONDUCTOR MICROCAVITIES

A major advantage of using self-assembled quantum dots for single-photon generation is that they can be easily incorporated into cavities using standard semiconductor growth and processing techniques. Cavity effects are useful for directing the emission from the dot into an experiment or application, as well as for modifying the photon-emission dynamics[35,36]. Purcell[37] predicted enhanced spontaneous emission from a source in a cavity when the source energy coincides with that of the cavity mode, owing to the greater density of optical states into which it can emit. For an ideal cavity, in which the emitter is located at the maximum of the electric field with its dipole aligned with the local electric field, the enhancement in decay rate is given by the Purcell factor, $F_p = (3/4\pi^2)\,(\lambda/n)^3\,Q/V$, where λ is the wavelength of the emission, n is the refractive index, Q is the quality factor (a measure of the time a photon is trapped in the cavity) and V is the effective mode volume. Thus high photon-collection efficiency, and simultaneously fast radiative decay, requires small cavities with highly reflecting mirrors and a high degree of structural perfection. Furthermore, without controlling the location of the dot in the cavity, as discussed below, it may be difficult to achieve the full enhancement predicted by the Purcell formula.

Figure 3 shows images of some of the single-quantum-dot cavity structures that have proved most successful. Pillar microcavities, formed by etching cylindrical pillars into semiconductor Bragg mirrors placed either side of the dot layer, have shown large Purcell enhancements and have a highly directional emission profile, thus making good single-photon sources[38–41]. Purcell factors of around six have been measured directly[40,41], through the rate of cavity-enhanced radiative decay compared with that of a dot without a cavity, implying a coupling to the cavity mode of $\beta = (F_p-1)/F_p > 83\%$, if we assume the leaky modes are unaffected by the cavity. However, the experimentally determined photon-collection efficiency, which is a more pertinent parameter for applications, is typically around 10%, due to the fact that not all the cavity mode can be coupled into an experiment and that there is scattering of the mode by the rough pillar edges. We can expect that the photon-collection efficiency will increase with improvements to the processing technology or new designs of microcavity.

Another means of forming a cavity is to etch a series of holes in a suspended slab of semiconductor, so as to form a lateral variation in the refractive index, which creates a forbidden energy gap for photonic modes in which light cannot propagate[42]. Photons can then be trapped in a central irregularity in this structure: usually an unetched portion of the slab. Such photonic-bandgap defect cavities have been fabricated in silicon with Q values approaching 10^6 (refs 43 and 44). High-quality active cavities have also been demonstrated in GaAs containing InAs quantum dots[45–48]. A radiative lifetime of 86 ps, corresponding to a Purcell factor of $F_p \approx 12$, has been reported[47], and very recently a lifetime of 60 ps was measured for a cavity in the strong-coupling regime[48].

If the Q-value is sufficiently large, the system enters the strong-coupling regime, where the excitation oscillates coherently between an exciton in the dot and a photon in the cavity. The spectral signature of strong coupling, an anticrossing between the dot line and the cavity mode, has been observed for quantum dots in pillar microcavities[49], photonic-bandgap defect cavities[50], microdisks[51] and microspheres[52]. It has been demonstrated for atom cavities that strong coupling allows the deterministic generation of single photons[53,54]. Single-photon

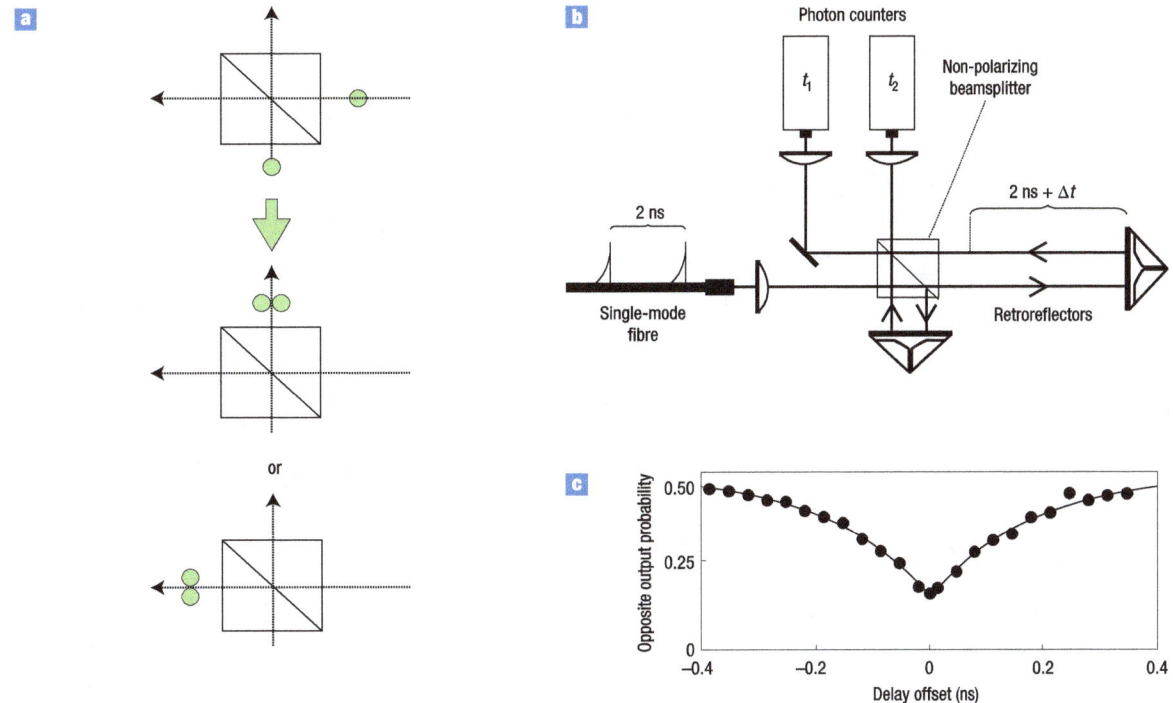

Figure 4 Two-photon interference. **a**, If the two photons are indistinguishable, the two outcomes resulting in one photon in either arm interfere destructively. This results in the two photons always exiting the beamsplitter together. **b**, Schematic of an experiment using two photons emitted successively from a quantum dot. **c**, Experimental data showing suppression of the co-incidence rate in **b** when the delay between input photons is zero owing to two-photon interference[60]. Copyright (2002) *Nature*, courtesy of Y. Yamamoto, Stanford University.

sources in the strong-coupling regime can be expected to have very-high extraction efficiencies and to be time-bandwidth limited[55]. Encouragingly, single-photon emission has been reported recently for a dot in a strongly coupled pillar microcavity[56].

Another interesting recent development is the ability to locate a single quantum dot within the cavity, as this ensures the largest possible coupling and removes background emission, as well as other undesirable effects due to other dots in the cavity. Above we discussed techniques to control the dot position on the growth surface. The other approach is to position the cavity around the dot. One technique combines microphotoluminescence spectroscopy to locate the dot position, with *in situ* laser photolithography to pattern markers on the wafer surface[57]. An alternative involves growing a vertical stack of dots so that their location can be revealed by scanning the wafer surface[58], as shown in Fig. 3f. Recently this technique has allowed larger coupling energies for a single dot in a photonic-bandgap defect cavity[48].

PHOTON INDISTINGUISHABILITY

Cavity effects are important for rendering different photons from the source indistinguishable, which is essential for many applications in quantum information. When two identical photons are incident simultaneously on the opposite input ports of a 50:50 beamsplitter, they will always exit through the same output port[59], as shown schematically in Fig. 4a. This occurs because of destructive interference in the probability amplitude of the final state in which one photon exits through each output port. The amplitude of the case where both photons are reflected exactly cancels with that where both are transmitted, due to the $\pi/2$ phase change on reflection, provided the two photons are entirely identical.

Two-photon interference of two single photons, emitted successively from a quantum dot in a weakly coupled pillar microcavity, was first reported by the Stanford group[60]. Figure 4b shows a schematic of their experiment. Notice the reduction of the co-incidence count rate measured between detectors in either output port, when the two photons are injected simultaneously (Fig. 4c). The dip does not extend completely to zero, indicating that the two photons sometimes exit the beamsplitter in opposite ports. The measured reduction in the co-incidence rate at zero delay of 69% implies an overlap for the single-photon wavepackets of 0.81, after correcting for the imperfect single-photon visibility of the interferometer. Two-photon interference dips of 66% and 75% have been reported by Bennett *et al.*[61] and Vauroutsis *et al.*[62] Similar results have been obtained for a single dot in a photonic-bandgap defect cavity[63].

This two-photon interference visibility is limited by the finite coherence time of the photons emitted by the quantum dot[64], which renders them distinguishable. The depth of the dip in Fig. 4c depends on the ratio of the radiative decay time to the coherence time of the dot, that is $R = 2\tau_{\text{decay}}/\tau_{\text{coh}}$. When this ratio is equal to unity, the coherence time is limited by radiative decay and the source has perfect two-photon interference. The most successful approach thus far has been to extend τ_{coh} by resonant optical excitation of the dot and reduce τ_{decay} using the Purcell effect in a pillar microcavity, to give values of $R \approx 1.5$. In the future, higher visibilities may be achieved with a larger Purcell enhancement, using a single-dot cavity in the strong-coupling regime or with electrical gating described in the next section.

A source of indistinguishable single photons was used by Fattal *et al.*[65,66] to generate entanglement between post-selected pairs. This involves simply rotating the polarization of one of the photons incident

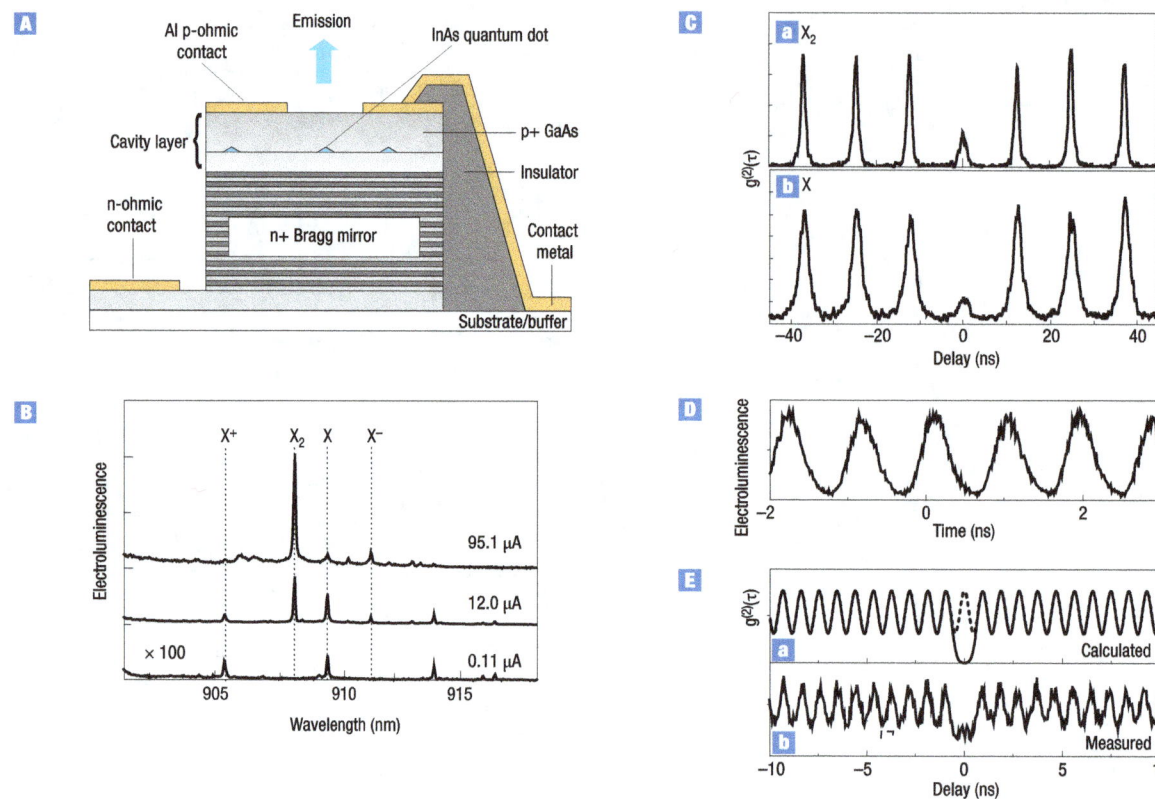

Figure 5 Electrically driven single-photon emission. **A**, Schematic of a single-photon LED. **B**, Electroluminescence spectra of the device. Notice the spectra are dominated by the exciton X and biexciton X_2 lines, which have linear and quadratic dependence on drive current, respectively. Other weak lines are due to charged excitons. **C**, Second-order correlation function recorded for the exciton (**a**) and biexciton (**b**) emission lines. **D**, Time-resolved electroluminescence from a device operating with a 1.07 GHz repetition rate. **E**, Modelled (**a**) and measured (**b**) second-order correlation function of the biexciton electroluminescence at 1.07 GHz. (Adapted with permission from ref. 71, copyright (2005) AIP, and ref. 73, copyright (2005) APS.)

on the final beamsplitter in Fig. 4a by 90°. By post-selecting the results where the two photons arrive at the beamsplitter at the same time and where there is one photon in each output arm (labelled 1 and 2), the measured pairs should correspond to the Bell state:

$$\Psi^- = \frac{1}{\sqrt{2}} \left(|H_1 V_2 > - |V_1 H_2 > \right). \qquad (1)$$

Note that only if the two photons are indistinguishable, and thus the entanglement is only in the photon polarization, can the two terms in equation (1) interfere. Analysis of the density matrix published by Fattal et al.[65] reveals a fidelity of the post-selected pairs to the state in equation (1) of 0.69, beyond the classical limit of 0.5. This source of entangled pairs differs in an important way from that based on the biexciton cascade described below. Post-selection implies that the photons are destroyed when this scheme succeeds. This is a problem for some quantum-information applications such as LOQC, but could be usefully applied to quantum key distribution[65].

SINGLE-PHOTON LEDS

An early proposal for an electrical single-photon source by Kim et al.[67] was based on etching a semiconductor heterostructure that had a Coulomb blockade. However, the light emission from this etched structure was too weak to allow the second-order correlation function to be studied. Recently, encouraging progress has been

made towards the realization of a single-photon source based on quantizing a lateral electrical-injection current[68,69]. However, the most successful approach so far has been to integrate self-assembled quantum dots into conventional p-i-n doped junctions.

In the first report of electrically driven single-photon emission by Yuan et al.[70], the electroluminescence of a single dot was isolated by forming a micrometre-diameter emission aperture in the opaque top contact of the p-i-n diode. Figure 5A shows an improved emission-aperture single-photon LED after Bennett et al.[71], which incorporates an optical cavity formed between a high-reflectivity Bragg mirror and the semiconductor–air interface in the aperture. This structure forms a weak cavity, which enhances the measured collection efficiency tenfold compared with devices without a cavity[72].

Single-photon pulses are generated by exciting the diode with a train of short voltage pulses. The second-order correlation function $g^{(2)}(\tau)$ of either the X or X_2 electroluminescence (Fig. 5C) shows the suppression of the zero-delay peak indicative of single-photon emission[71]. The finite rate of multiphoton pulses is due mostly to background emission from layers other than the dot, which is also seen for non-resonant optical excitation. Such electrical contacts also allow the temporal characteristics of the single-photon source to be tailored. By applying a negative bias to the diode between the electrical injection pulses, Bennett et al.[73] reduced the jitter in the photon-emission time to less than 100 ps. This allowed the repetition rate of the single-photon source to be increased to

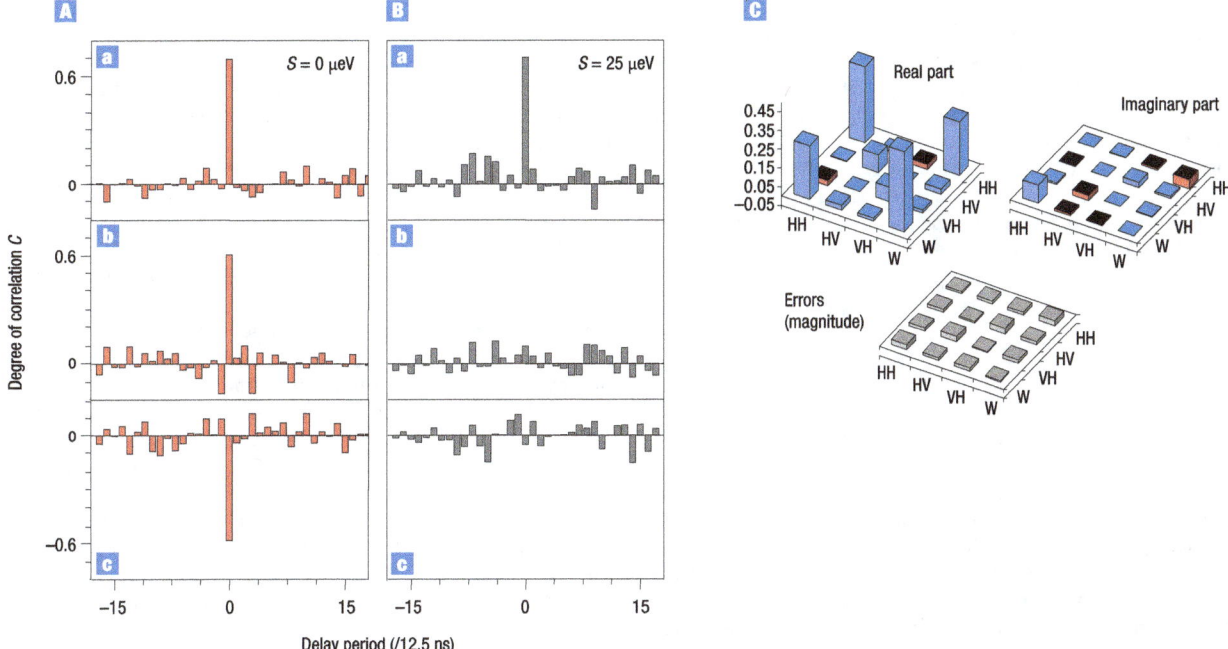

Figure 6 Generation of entangled photons by a quantum dot. **A**, Degree of correlation measured for a dot with exciton polarization splitting $S = 0$ µeV in linear (**a**), diagonal (**b**) and circular (**c**) polarization bases as a function of the delay between the X and X_2 photons (in units of the repetition cycle). The correlation is defined as the rate of co-polarized pairs minus the rate of cross-polarized pairs divided by the total rate. Notice that the values at finite delay show no correlation, as expected for pairs emitted in different laser excitation cycles. More interesting are the peaks close to zero time delay, corresponding to X and X_2 photons emitted from the same cascade. The presence of strong correlations for all three types of measurement for the dot with zero exciton splitting can only be explained if the X and X_2 polarizations are entangled. **B**, Degree of correlations measured for the dot in **A** subject to an in-plane magnetic field so as to produce an exciton polarization splitting of $S = 25$ µeV. Notice that the correlation in diagonal and circular bases has vanished, indicating only classical correlations at finite splitting. **C**, Two-photon density matrix of the device emission in **A** along with errors calculated from the errors on the correlation measurements. The strong off-diagonal terms appear owing to entanglement. (**a**, **c**, Adapted with permission from ref. 92. Copyright (2006) IOP.)

1.07 GHz (Fig. 5D) while retaining good single-photon emission characteristics (Fig. 5E). Such electrical gating could provide a technique for producing time-bandwidth-limited single-photons from quantum dots.

Another promising approach is to aperture the current flowing through the device[74,75]. This is achieved by growing a thin AlAs layer within the intrinsic region of the p-i-n junction and later exposing the mesa to wet oxidation in a furnace, converting the AlAs layer around the outer edge of the mesa to insulating AlO_x. By careful control of the oxidation time, a micrometre-diameter conducting aperture can be formed within the insulating ring of AlO_x. Such structures have the advantage of exciting just a single dot within the structure, thereby reducing the amount of background emission. The oxide annulus also confines the optical mode laterally within the structure, potentially allowing high photon-extraction efficiency.

Altering the nanostructure or materials that comprise the quantum dot allows considerable control over the emission wavelength and other characteristics. Most of the experimental work done so far has concentrated on small InAs quantum dots emitting at around 900–950 nm, as these have well understood optical properties and can be detected with low-noise silicon single-photon detectors. On the other hand the shallow confinement potentials of this system mean they emit only at low temperatures. At shorter wavelengths optically pumped single-photon emission has been demonstrated at around 350 nm using GaN/AlGaN (ref. 76), 500 nm using CdSe/ZnSSe (ref. 77) and 682 nm using

InP–GaInP (ref. 78) quantum dots. GaN/AlGaN and CdSe/ZnSSe quantum dots have been shown to operate at 200 K.

It is very important for quantum communications to develop sources at longer wavelengths in the fibre-optic transmission bands at 1.3 µm and 1.55 µm. This may be achieved using InAs/GaAs heterostructures by depositing more InAs to form larger quantum dots. These larger dots offer deeper confinement potentials than those at 900 nm and thus often emit at room temperature[79]. Optically pumped single-photon emission at telecom wavelengths has been achieved using a number of techniques to prepare low densities of longer-wavelength dots, including a bimodal-growth mode in molecular-beam epitaxy to form low densities of large dots[80], ultralow-growth-rate molecular-beam epitaxy[81] and metal–organic chemical vapour deposition[82]. Recently, the first electrically driven single-photon source at a telecom wavelength has been demonstrated[83].

GENERATION OF ENTANGLED PHOTONS

By collecting both the X_2 and X photons emitted by the biexciton cascade, a single quantum dot may also be used as a source of photon pairs. Polarization-correlation measurements on these pairs revealed that the two photons were classically correlated with the same linear polarization[84–86]. This occurs because the cascade can proceed by means of one of two intermediate exciton spin states, as described above and shown in Fig. 2a, one of which couples to two H- and the other to two V-polarized photons. The emission is thus a statistical mixture of $|H_{X_2}H_X\rangle$ and $|V_{X_2}V_X\rangle$, although exciton spin scattering during the cascade (discussed below) ensures there are also some cross-polarized pairs.

The spin splitting[87,88] of the exciton state of the dot distinguishes the H- and V-polarized pairs and prevents the emission of entangled pairs predicted by Benson *et al.*[89] If this splitting could be removed, the H and V components would interfere in appropriately designed experiments. The emitted two-photon state should then be written as a superposition of HH and VV, which can be recast in either the diagonal (spanned by D, A) or circular (σ^+, σ^-) polarization bases, that is:

$$\Phi^+ = \frac{1}{\sqrt{2}} \left(|H_{X2} H_X> + |V_{X2} V_X> \right)$$
$$= \frac{1}{\sqrt{2}} \left(|D_{X2} D_X> + |A_{X2} A_X> \right)$$
$$= \frac{1}{\sqrt{2}} \left(|\sigma^+_{X2} \sigma_X> + |\sigma^+_{X2} \sigma_X> \right). \qquad (2)$$

Equal weighting of the HH and VV terms assumes the source to be unpolarized, as indicated by experimental measurements.

Equation (2) suggests that, for zero exciton spin splitting, the biexciton cascade generates entangled photon pairs, similar to those seen for atoms[90]. Entanglement of the X or X_2 photons was recently observed experimentally for the first time by Stevenson, Young and co-workers[91,92], using two different schemes to cancel the exciton spin splitting. An alternative approach by Akopian *et al.*[93] using dots with finite exciton splitting, post-selects photons emitted in a narrow spectral band where the two polarization lines overlap.

The exciton spin splitting depends on the exciton-emission energy, tending to zero for InAs dots emitting close to 1.4 eV and then inverting for higher emission energies[94,95]. These correspond to shallow quantum dots for which the carrier wavefunctions extend into the barrier material reducing the electron–hole exchange. Zero splitting can be achieved by either careful control of the growth conditions to achieve dots emitting close to the desired energy, or by annealing samples emitting at lower energies[94]. The exciton spin splitting may be continuously tuned by applying a magnetic field in the plane of the dot[96]. It has been observed that the signatures of entanglement then appear only when the exciton spin splitting is close to zero[91]. Other promising schemes to tune the exciton spin splitting are now emerging, including application of strain[97] and an electric field[98,99].

Figure 6A plots polarization correlations reported by Young *et al.*[92] for a dot with zero exciton spin splitting (achieved by control of the growth conditions). Pairs emitted in the same cascade (that is, with zero delay) show a very striking positive correlation (co-polarization) measuring in either, rectilinear or diagonal bases and anticorrelation (cross-polarization) when measuring in a circular basis. This is exactly the behaviour expected for the entangled state of equation (2). In contrast, a dot with finite splitting shows polarization correlation for the rectilinear basis only, with no correlation for diagonal or circular measurements (Fig. 6B). The strong correlations seen for all three bases in Fig. 6A could not be produced by any classical light source or mixture of classical sources and is proof that the source generates entangled photons. The measured[92] two-photon density matrix (Fig. 6C) projects onto the expected $1/\sqrt{2}$ ($|H_{X2} H_X> + |V_{X2} V_X>$) state with a fidelity (that is, a probability) of 0.702 ± 0.022, exceeding the classical limit (0.5) by 9 standard deviations.

Two processes contribute to the 'wrongly' correlated pairs, which impair the fidelity of the entangled photon source. The first of these is due to background emission from layers in the sample other than the dot. This background emission, which is unpolarized and dilutes the entangled photons from the dot, limited the fidelity observed in the first report[91] of triggered entangled photon pairs from a quantum dot

and has been subsequently reduced using a better sample design[92]. The second mechanism, which is an intrinsic feature of the dot, is exciton spin scattering during the biexciton cascade. It is interesting that this process does not seem to depend strongly on the exciton spin splitting. It may be reduced by suppressing the scattering using resonant excitation or alternatively by using cavity effects to reduce the time required for the radiative cascade.

OUTLOOK

The past several years have seen remarkable progress in quantum light generation using semiconductor devices. However, despite considerable progress many challenges still remain. The structural integrity of cavities must continue to improve, thereby enhancing quality factors. This, combined with the ability to reliably position single dots within the cavity, will further enhance photon-collection efficiencies and the Rabi energy in the strong-coupling regime. It is also important to realize all the benefits of these cavity effects in more practical electrically driven sources. Meanwhile bandstructure engineering of the quantum dots will allow a wider range of wavelengths to be accessed for both single and entangled photon sources, as well as structures that can operate at higher temperatures. Techniques for fine tuning the characteristics of individual emitters will also be important.

One of the most interesting aspects of semiconductor quantum optics is that we may be able to use quantum dots not only as quantum light emitters, but also as the logic and memory elements, which are required in quantum information processing. Although LOQC is scalable theoretically, quantum computing with photons would be much easier with a useful single-photon nonlinearity. Such nonlinearity may be achieved with a quantum dot in a cavity in the strong-coupling regime. Encouragingly, strong coupling of a single quantum dot with various types of cavity has already been observed in the spectral domain. Eventually it may even be possible to integrate photon emission, logic, memory and detection elements into single semiconductor chips to form a photonic integrated circuit for quantum information processing.

doi:10.1038/nphoton.2007.46

References
1. Gisin, N., Ribordy, G., Tittel, W. & Zbinden, H. Quantum cryptography. *Rev. Mod Physics* **74**, 145–195 (2001).
2. Dusek, M., Lutkenhaus, N. & Hendrych, M. *Progress in Optics* Vol. 49 (ed. Wolf, E.) Ch. 5 (Elsevier, The Netherlands, 2006).
3. Waks, E. *et al.* Secure communication: Quantum cryptography with a photon turnstile. *Nature* **420**, 762 (2002).
4. Bouwmeester, D. *et al.* Experimental quantum teleportation. *Nature* **390**, 575–579 (1997).
5. Briegel, H.-J., Dür, W., Cirac, J. I. & Zoller, P. Quantum repeaters: The role of imperfect local operations in quantum communication. *Phys. Rev. Lett.* **81**, 5932–5935 (1998).
6. Knill, E., Laflamme, R. & Milburn, G. J. A scheme for efficient quantum computation with linear optics. *Nature* **409**, 46–52 (2001).
7. Kok, P., Munro, W. J., Nemoto, K., Ralph, T. C., Dowling, J. P. & Milburn, G. J. Linear optical quantum computing. *Rev. Mod. Phys.* **79**, 135 (2007).
8. Giovannetti, V., Lloyd, S. & Maccone, L. Quantum-enhanced measurements: Beating the standard quantum limit. *Science* **306**, 1330–1336 (2004).
9. Brendel, J., Gisin, N., Tittel, W. & Zbinden, H. Pulsed energy-time entangled twin-photon source for quantum communication. *Phys. Rev. Lett.* **82**, 2594–2597 (1999).
10. Shih, Y. H. & Alley, C. O. New type of Einstein-Podolsky-Rosen-Bohm experiment using pairs of light quanta produced by optical parametric down conversion. *Phys. Rev. Lett.* **61**, 2921–2924 (1988).
11. Ou, Z. Y. & Mandel, L. Violation of Bell's inequality and classical probability in a two-photon correlation experiment. *Phys. Rev. Lett.* **61**, 50–53 (1988).
12. Fasel, S. *et al.* High quality asynchronous heralded single-photon source at telecom wavelength. *New J. Phys.* **6**, 163 (2004).
13. Edamatsu, K., Oohata, G., Shimizu, R. & Itoh, T. Generation of ultraviolet entangled photons in a semiconductor. *Nature* **431**, 167–170 (2004).
14. Scarani, V., de Riedmatten, H., Marcikic, I., Zbinden, H. & Gisin, N. *Eur. Phys. J. D* **32**, 129–138 (2005).
15. Migdall, A., Branning, D. & Castelletto, S. Tailoring single-photon and multiphoton probabilities of a single-photon on-demand source. *Phys. Rev. A* **66**, 053805 (2002).
16. Pittman, T. B., Jacobs, B. C. & Franson, J. D. Single-photons on pseudodemand from stored parametric down-conversion. *Phys. Rev. A* **66**, 042303 (2002).
17. Kimble, H. J., Dagenais, M. & Mandel, L. Photon antibunching in resonance fluorescence. *Phys. Rev. Lett.* **39**, 691–695 (1977).

18. Diedrich, F. & Walther, H. Nonclassical radiation of a single stored ion. *Phys. Rev. Lett.* **58,** 203–206 (1987).
19. Michler P. *et al.* in *Single Quantum Dots* 315 (Springer, Berlin, 2003).
20. Lounis, B. & Orrit, M. Single photon sources. *Rep. Prog. Phys.* **68,** 1129–1179 (2005).
21. Bimberg, D., Grundmann, M. & Ledentsov, N. N. *Quantum Dot Heterostructures* (Wiley, Chichester, 1999).
22. Song, H. Z. *et al.* Site-controlled photoluminescence at telecommunication wavelength from InAs/InP quantum dots. *Appl. Phys. Lett.* **86,** 113118 (2005).
23. Atkinson P. *et al.* Site control of InAs quantum dots using ex-situ electron-beam lithographic patterning of GaAs substrates. *Jpn J. Appl. Phys.* **45,** 2519–2521 (2006).
24. Landin, L., Miller, M. S., Pistol, M.-E., Pryor, C. E. & Samuelson, L. Optical studies of individual InAs quantum dots in GaAs: Few-particle effects. *Science* **280,** 262–264 (1998).
25. Gammon, D., Snow, E. S., Shanabrook, B. V., Katzer, D. S. & Park, D. Fine structure splitting in the optical spectra of single GaAs quantum dots. *Phys. Rev. Lett.* **76,** 3005–3008 (1996).
26. Kulakovskii, V. D. *et al.* Fine structure of biexciton emission in symmetric and asymmetric CdSe/ZnSe single quantum dots. *Phys. Rev. Lett.* **82,** 1780–1783 (1999).
27. Michler, P. *et al.* A quantum dot single-photon turnstile device. *Science* **290,** 2282–2285 (2000).
28. Santori, C., Pelton, M., Solomon, G., Dale, Y. & Yamamoto, Y. Triggered single-photons from a quantum dot. *Phys. Rev. Lett* **86,** 1502–1505 (2001).
29. Zwiller, V. *et al.* Single quantum dots emit single-photons at a time: Antibunching experiments. *Appl. Phys. Lett.* **78,** 2476–2478 (2001).
30. Thompson, R. M. *et al.* Single-photon emission from exciton complexes in individual quantum dots. *Phys. Rev. B* **64,** 201302 (2001).
31. Moreau, E. *et al.* Quantum cascade of photons in semiconductor quantum dots. *Phys. Rev. Lett.* **87,** 183601 (2001).
32. Regelman, D. V. *et al.* Semiconductor quantum dot: A quantum light source of multicolor photons with tunable statistics. *Phys. Rev. Lett.* **87,** 257401 (2001).
33. Kiraz, A. *et al.* Photon correlation spectroscopy of a single quantum dot. *Phys. Rev. B* **65,** 161303 (2002).
34. Shields, A. J., Stevenson, R. M., Thompson, R., Yuan, Z. & Kardynal, B. *Nano-Physics and Bio-Electronics* (Elsevier, Amsterdam, 2002).
35. Vahala, K. J. Optical microcavities. *Nature* **424,** 839–846 (2003).
36. Barnes, W. L. *et al.* Solid-state single-photon sources: Light collection strategies. *Euro. Phys. J. D* **18,** 197–210 (2002).
37. Purcell E. Spontaneous emission probabilities at radio frequencies. *Phys. Rev.* **69,** 681 (1946).
38. Moreau, E. *et al.* Single-mode solid-state single-photon source based on isolated quantum dots in pillar microcavities. *Appl. Phys. Lett.* **79,** 2865–2867 (2001).
39. Pelton, M. *et al.* Efficient source of single-photons: A single quantum dot in a micropost microcavity. *Phys. Rev. Lett.* **89,** 233602 (2002).
40. Vučković, J., Fattal, D., Santori, C., Solomon, G. S. & Yamamoto, Y. Enhanced single-photon emission from a quantum dot in a micropost microcavity. *Appl. Phys. Lett.* **82,** 3596–3598 (2003).
41. Bennett, A. J., Unitt, D., Atkinson P., Ritchie D. A. & Shields A. J. High performance single-photon sources from photo-lithographically defined pillar microcavities. *Opt. Express* **13,** 50–55 (2005).
42. Yablonovitch, E. Inhibited spontaneous emission in solid state physics and electronics. *Phys. Rev. Lett.* **58,** 2059–2062 (1987).
43. Song, B.-S., Noda, S., Asano, T. & Akahane, Y. Ultra-high-*Q* photonic double-heterostructure nanocavity. *Nature Mater.* **4,** 207–210 (2005).
44. Notomi, M. *et al.* Ultrahigh-*Q* photonic crystal nanocavities realized by the local width modulation of a line defect. *Appl. Phys. Lett.* **88,** 041112 (2006).
45. Kress, A. *et al.* Manipulation of the spontaneous emission dynamics of quantum dots in two-dimensional photonic crystals. *Phys. Rev. B* **71,** 241304 (2005).
46. Englund, D. *et al.* Controlling the spontaneous emission rate of single quantum dots in a two-dimensional photonic crystal. *Phys. Rev. Lett.* **95,** 013904 (2005).
47. Gevaux, D. G. *et al.* Enhancement and suppression of spontaneous emission by temperature tuning InAs quantum dots to photonic crystal cavities. *Appl. Phys. Lett.* **88,** 131101 (2006).
48. Hennessy *et al.* Quantum nature of a strongly coupled single quantum dot-cavity system. *Nature* **445,** 896–899 (2007).
49. Reithmaier, J. P. *et al.* Strong coupling in a single quantum dot–semiconductor microcavity system. *Nature* **432,** 197–200 (2004).
50. Yoshie, T. *et al.* Vacuum Rabi splitting with a single quantum dot in a photonic crystal nanocavity. *Nature* **432,** 200–203 (2004).
51. Peter, E. *et al.* Exciton-photon strong-coupling regime for a single quantum dot embedded in a microcavity. *Phys. Rev. Lett.* **95,** 067401 (2005).
52. Le Thomas, N. *et al.* Cavity QED with semiconductor Nanocrystals. *Nano Lett.* **6,** 557–561 (2006).
53. Kuhn, A., Hennrich, M. & Rempe, G. Determinsitic single photon source for distributed quantum networking. *Phys. Rev. Lett.* **89,** 067901 (2002)
54. McKeever, J. *et al.* Deterministic generation of single photons from one atom trapped in a cavity. *Science* **303,** 1992–1994 (2004).
55. Cui, G. & Raymer, M. G. Quantum efficiency of single photon sources in cavity-QED strong-coupling regime. *Opt. Express* **13,** 9660–9665 (2005).
56. Press, D. *et al.* Photon antibunching from a single quantum dot-microcavity system in the strong coupling regime. <http://arxiv.org/abs/quant-ph/0609193> (2006).
57. Lee, K. H. *et al.* Registration of single quantum dots using cryogenic laser photolithography. *Appl. Phys. Lett.* **88,** 193106 (2006).
58. Badolato A. *et al.* Deterministic coupling of single quantum dots to single nanocavity modes. *Science* **308,** 1158–1161 (2005).
59. Hong, C. K., Ou, Z. Y. & Mandel, L. Measurement of subpicosecond time intervals between two photons by interference. *Phys. Rev. Lett.* **59,** 2044–2046 (1987).
60. Santori, C., Fattal, D., Vučković, J., Solomon, G. S. & Yamamoto, Y. Indistinguishable photons from a single-photon device. *Nature* **419,** 594–597 (2002).
61. Bennett, A. J., Unitt, D., Atkinson, P., Ritchie, D. A. & Shields, A. J. Influence of exciton dynamics on the interference of two photons from a microcavity single-photon source. *Opt. Express* **13,** 7772–7778 (2005).
62. Varoutsis, S. *et al.* Restoration of photon indistinguishability in the emission of a semiconductor quantum dot. *Phys. Rev. B* **72,** 041303 (2005).
63. Laurent, S. *et al.* Indistinguishable single photons from a single quantum dot in two-dimensional photonic crystal cavity. *Appl. Phys. Lett.* **87,** 163107 (2005).
64. Kammerer. C. *et al.* Interferometric correlation spectroscopy in single quantum dots. *Appl. Phys. Lett.* **81,** 2737–2739 (2002).
65. Fattal, D., Diamanti, E., Inoue, K. & Yamamoto, Y. Quantum teleportation with a quantum dot single-photon source. *Phys. Rev. Lett.* **92,** 037904 (2004).
66. Fattal, D. *et al.* Entanglement formation and violation of Bell's inequality with a semiconductor single-photon source. *Phys. Rev. Lett.* **92,** 037903 (2004).
67. Imamoglu, A. & Yamamoto, Y. Turnstile device for heralded single photons: Coulomb blockade of electron and hole tunneling in quantum confined p-i-n heterojunctions. *Phys. Rev. Lett.* **72,** 210–213 (1994).
68. Cecchini, M. *et al.* Surface acoustic wave-driven planar light-emitting device. *Appl. Phys. Lett.* **85,** 3020–3022 (2005).
69. Gell, J. R. *et al.* Surface acoustic wave driven luminescence from a lateral p-n junction. *Appl. Phys. Lett.* **89,** 243505 (2006).
70. Yuan, Z. *et al.* Electrically driven single-photon source. *Science* **295,** 102–105 (2002).
71. Bennett, A. J. *et al.* A microcavity single-photon emitting diode. *Appl. Phys. Lett.* **86,** 181102 (2005).
72. Abram, I., Robert, I. & Kuszelewicz, R. Spontaneous emission control in semiconductor microcavities with metallic or Bragg mirrors. *IEEE. J. Quant. Elect.* **34,** 71–76 (1998).
73. Bennett, A. J. *et al.* Electrical control of the uncertainty in the time of single-photon emission events. *Phys. Rev. B* **72,** 033316 (2005).
74. Ellis, D., Bennett, A. J., Shields, A. J., Atkinson, P. & Ritchie, D. A. Electrically addressing a single self-assembled quantum dot. *Appl. Phys. Lett.* **88,** 133509 (2006).
75. Lochman, A. Electrically driven single quantum dot polarised single photon emitter. *Electron. Lett.* **42,** 774–775 (2006).
76. Kako, S. *et al.* A gallium nitride single-photon source operating at 200 K. *Nature Mater.* **5,** 887–892 (2006).
77. Sebald, K. *et al.* Single-photon emission of CdSe quantum dots at temperatures up to 200 K. *Appl. Phys. Lett.* **81,** 2920–2922 (2002).
78. Aichele, T., Zwiller, V. & Benson O. Visible single-photon generation from semiconductor quantum dots. *New J. Phys.* **6,** 90 (2004).
79. Le Ru, E. C., Fack, J. & Murray, R. Temperature and excitation density dependence of the photoluminescence from annealed InAs/GaAs quantum dots. *Phys. Rev. B* **67,** 245318 (2003).
80. Ward, M. B. On-demand single-photon source for 1.3 μm telecom fiber. *Appl. Phys. Lett.* **86,** 201111 (2005).
81. Zinoni, C. *et al.* Time-resolved and antibunching experiments on single quantum dots at 1300nm. *Appl. Phys. Lett.* **88,** 131102 (2006).
82. Miyazawa, T. *et al.* Single-photon generation in the 1.55-μm optical-fiber band from an InAs/InP quantum dot. *Jpn J. Appl. Phys.* **44,** L620-622 (2005).
83. Ward, M. B. *et al.* Electrically driven telecommunication wavelength single-photon source. *Appl. Phys. Lett.* **90,** 063512 (2007).
84. Stevenson, R. M. Quantum dots as a photon source for passive quantum key encoding. *Phys. Rev. B* **66,** 081302 (2002).
85. Santori, C., Fattal, D., Pelton, M., Solomon, G. S. & Yamamoto, Y. Polarization-correlated photon pairs from a single quantum dot. *Phys. Rev. B* **66,** 045308 (2002).
86. Ulrich, S. M., Strauf, S., Michler, P., Bacher, G. & Forchel, A. Triggered polarization-correlated photon pairs from a single CdSe quantum dot. *Appl. Phys. Lett.* **83,** 1848–1850 (2003).
87. van Kesteren, H. W., Cosman, E. C., van der Poel, W. A. J. A. & Foxon C. T. Fine structure of excitons in type-II GaAs/AlAs quantum wells. *Phys. Rev. B* **41,** 5283–5292 (1990).
88. Blackwood, E., Snelling, M. J., Harley, R. T., Andrews, S. R. & Foxon, C. T. B. Exchange interaction of excitons in GaAs heterostructures. *Phys. Rev. B* **50,** 14246–14254 (1994).
89. Benson, O., Santori, C., Pelton, M. & Yamamoto, Y. Regulated and entangled photons from a single quantum dot. *Phys. Rev. Lett.* **84,** 2513–2516 (2000).
90. Aspect, A., Grangier, P. & Roger, G. Experimental tests of realistic local theories via Bell's theorem. *Phys. Rev. Lett.* **47,** 460–463 (1981).
91. Stevenson, R. M. *et al.* A semiconductor source of triggered entangled photon pairs. *Nature* **439,** 179–182 (2006).
92. Young, R. J. *et al.* Improved fidelity of triggered entangled photons from single quantum dots. *New J. Phys.* **8,** 29 (2006).
93. Akopian, N. *et al.* Entangled photon pairs from semiconductor quantum dots. *Phys. Rev. Lett.* **96,** 130501 (2006).
94. Young, R. J. *et al.* Inversion of exciton level splitting in quantum dots. *Phys. Rev. B* **72,** 113305 (2005).
95. Seguin, R. *et al.* Size-dependent fine-structure splitting in self-organized InAs/GaAs quantum dots. *Phys. Rev. Lett.* **95,** 257402 (2005).
96. Stevenson, R. M. *et al.* Magnetic-field-induced reduction of the exciton polarisation splitting in InAs quantum dots. *Phys. Rev. B* **73,** 033306 (2006).
97. Seidl, S., Kroner, M., Högele, A. & Karrai, K. Effect of uniaxial stress on excitons in a self-assembled quantum dot. *Appl. Phys. Lett.* **88,** 203113 (2006).
98. Gerardot, B. D. *et al.* Manipulating exciton fine-structure in quantum dot with a lateral electric field. < http://arxiv.org/abs/cond-mat/0608711> (2006).
99. Kowalik, K. *et al.* Influence of an in-plane electric field on exciton fine structure in InAs-GaAs self-assembled quantum dots. *Appl. Phys. Lett.* **86,** 041907 (2005).
100. Walls, D. F. & Milburn, G. J. *Quantum Optics* (Springer, Berlin, 1994).
101. Hanbury Brown, R. & Twiss, R. Q. A new type of interferometer for use in radio astronomy. *Phil. Mag.* **45,** 663 (1954).
102. Santori, C. *et al.* Submicrosecond correlations in photoluminescence from InAs quantum dots. *Phys. Rev. B* **69,** 205324 (2004).

Acknowledgements

The author would like to thank Mark Stevenson, Robert Young, Anthony Bennett and Martin Ward for their comments during the preparation of the manuscript and the UK Department of Trade and Industry 'Optical Systems for Digital Age', Engineering and Physical Sciences Research Council and European Commission Future and Emerging Technologies arm of the 1st programme for supporting research on quantum light sources.

Biomimetics of photonic nanostructures

Biomimetics is the extraction of good design from nature. One approach to optical biomimetics focuses on the use of conventional engineering methods to make direct analogues of the reflectors and anti-reflectors found in nature. However, recent collaborations between biologists, physicists, engineers, chemists and materials scientists have ventured beyond experiments that merely mimic what happens in nature, leading to a thriving new area of research involving biomimetics through cell culture. In this new approach, the nanoengineering efficiency of living cells is harnessed and natural organisms such as diatoms and viruses are used to make nanostructures that could have commercial applications.

ANDREW R. PARKER[1,2] AND HELEN E. TOWNLEY[3]

[1]Department of Zoology, The Natural History Museum, Cromwell Road, London SW7 5BD, UK; [2]Department of Biological Sciences, University of Sydney, NSW 2006, Australia; [3]Department of Zoology, University of Oxford, South Parks Road, Oxford OX1 3PS, UK

e-mail: andrew.parker@green.ox.ac.uk; helen.townley@zoo.ox.ac.uk

Three centuries of research, beginning with Hooke and Newton, have revealed a diversity of optical devices at the submicrometre scale in nature[1]. These include one-dimensional multilayer reflectors, two-dimensional diffraction gratings and three-dimensional liquid crystals. In 2001 the first photonic crystal in an animal was identified[2] and since then the scientific effort in this subject has accelerated. Now we know of a variety of two-dimensional and three-dimensional photonic crystals in nature, including some designs not encountered previously in physics.

However, some of the optical nanostructures found in nature have such an elaborate architecture at the nanoscale that we simply cannot copy them using current engineering techniques or, if they can be copied, the effort involved is so great that commercial-scale manufacture would never be cost-effective. An alternative approach is to exploit the fact that plants and animals can make these designs very efficiently (for example, ref. 3). Therefore we can let nature manufacture the devices for us through cell-culture techniques.

Animal cells are about 10 μm in size and plant cells can have sizes of up to about 100 μm, so they are both of a suitable scale for nanostructure production. Although the overall size of the structures produced by a single cell is about the size of the cell itself, these structures can involve sub-structures on a smaller scale (typically of the order of 100 nm). The success of cell culture depends on the species and on the type of cell. Insect cells, for instance, can be cultured at room temperature, whereas an incubator is required for mammalian cells. It is also necessary, where solid media are involved, to establish a culture medium that the cells can adhere to in order for them to develop to the stage where they can make photonic devices.

This article will first describe progress in the conventional engineering approach to optical biomimetics, and then discuss the more recent cell-culture approach at greater length.

ENGINEERING ANTI-REFLECTORS AND IRIDESCENT DEVICES

Some insects benefit from anti-reflective surfaces, either on their eyes to see in low-light conditions, or on their wings to reduce surface reflections from transparent structures for the purpose of camouflage. Anti-reflective surfaces are found, for instance, on the corneas of moth and butterfly eyes[4] and on the transparent wings of hawkmoths[5]. These surfaces consist of cylindrical nodules with rounded tips arranged in a hexagonal array with a periodicity of around 240 nm (Fig. 1a). Effectively they introduce a gradual refractive index profile at an interface between chitin (a polysaccharide with a refractive index, n, of 1.54 that is often embedded in a proteinaceous matrix) and air ($n = 1.0$), and reduce reflectivity by a factor of about ten.

This 'moth-eye structure' was first reproduced at its correct scale by crossing three gratings at 120° using lithographic techniques, and has been used as an anti-reflective surface on glass windows in Scandinavia[6]. Here, plastic sheets bearing the anti-reflector were attached to each interior surface of triple glazed windows using refractive-index-matching glue to provide a significant difference in reflectivity. Today the moth-eye structure can be made extremely accurately using electron-beam etching[7] (Fig. 1c), and is used commercially on solid plastic and other lenses.

A different type of anti-reflective device, in the form of a sinusoidal grating of 250 nm periodicity, was discovered on the cornea of a 45 million-year-old fly preserved in amber[8] (Fig. 1b). This type of grating is particularly useful where light is incident at a range of angles (within a single plane, perpendicular to the grooves of the grating), as demonstrated by a model made in a photoresist using lithographic methods[8]. Consequently it has been used on the surfaces of solar panels, providing a 10% increase in energy capture by reducing the reflected portion of sunlight[9]. Again, this device is embossed onto plastic sheets using holographic techniques.

Many birds, insects (particularly butterflies and beetles), fishes and lesser-known marine animals exploit photonic nanostructures on their surfaces to make their colour change with viewing angle (iridescence) and/or appear 'metallic'. These visual effects appear more pronounced than those produced by pigments and are used to attract the attention of potential mates or to startle predators.

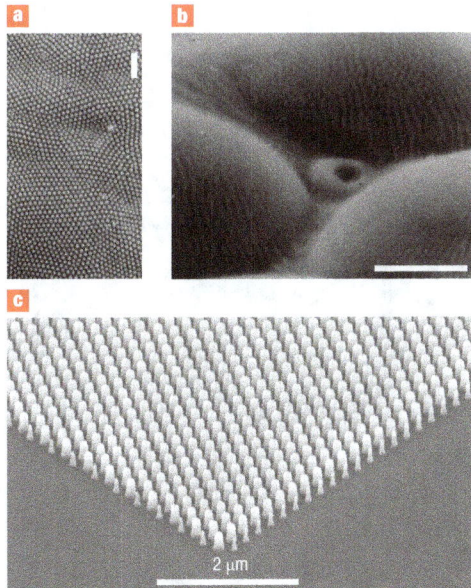

Figure 1 Natural and fabricated anti-reflective surfaces. **a**, Scanning electron micrograph of an anti-reflective surface from the eye of a moth. Domains corresponding to individual cells are evident in this image Scale bar = 1 μm. **b**, Anti-reflector with ridges on three facets on the eye of a 45 million-year-old fly (dolichopodid). Scale bar = 3 μm. Courtesy of P. Mierzejewski. **c**, This biomimetic replica of a moth eye[7] was fabricated with ion-beam etching. Although the replica is accurate, it can only be fabricated over areas of a few micrometres across, whereas a natural device has the potential to be larger as the size is limited by the number of appropriate cells present. Scale bar = 2 μm. Copyright (2006) IEEE.

An obvious application for such visually attractive and optically sophisticated devices is within the anti-counterfeiting industry[10]. Although much of the work in this area is not published, researchers are developing devices with different levels of sophistication, from effects that are discernable by the eye to fine-scale optical characteristics (such as polarization and angular properties) that can be read only by specialized detectors.

However, new research — such as the replication of an iridescent beetle cuticle by Vigneron *et al.*[11] — aims to exploit these devices in the cosmetics (including packaging), paint, printing and clothing industries. Interestingly, the range of refractive indices available in artificial structures is much larger than those found in natural reflectors: *n* can be as high as 4 for a semiconductor, compared with a maximum of around 1.8 for natural materials. This could make these structures considerably more useful for applications, even if the natural structure geometry is maintained.

Original work exploiting the reflectors found in nature involved copying the design but not the size, with reflectors being scaled-up to work at longer wavelengths. For example, the *Morpho* butterflies — a group of large, usually blue, butterflies from Central and South America — typically possess optical structures with a 'Christmas tree' profile. A microwave analogue of this structure, which could be used as an antenna or as an anti-reflection coating for radar, has been manufactured using rapid prototyping. The layers in these devices are about 1 mm thick, compared with 100 nm in the butterflies. However, techniques are now available to manufacture natural reflectors at their true size.

Nanostructures causing iridescence include photonic crystals and unusually sculpted three-dimensional architectures. Photonic crystals are materials with an ordered subwavelength structure that can control the propagation of light, only allowing certain wavelengths (or ranges of wavelengths) to pass through the crystal, just as atomic crystals only allow electrons with energies in certain ranges to propagate[12]. Examples include opal (a hexagonal or square array of 250 nm spheres) and inverse opal (a hexagonal array of similar sized and shaped holes in a solid matrix). Hummingbird feather barbs provide an example of unusually sculpted three-dimensional architectures, with the iridescence often caused by variations in porosity. Such devices have been mimicked using aqueous-based layering techniques[13].

The greatest diversity of three-dimensional architectures can be found in butterfly scales, which can include micro-ribs with nanoridges, concave multilayered pits, blazed gratings and randomly punctuated nanolayers[14–16]. The cuticles of many beetles contain chiral films that produce iridescent effects with circular or elliptical polarization properties[17]. These have been replicated in titania for specialized coatings (Fig. 2). The titania mimic can be nanoengineered for a wide range of resonant wavelengths; the lowest so far is a pitch of 60 nm for a circular Bragg resonance at 220 nm in a Sc_2O_3 film (I. Hodgkinson, personal communication).

Biomimetic work on the photonic crystal fibres of the *Aphrodita* sea mouse is underway. The sea mouse contains spines (tubes), the walls of which are packed with smaller tubes that are 500 nm in diameter and have varying internal diameters (50–400 nm). These provide a bandgap in the red region, and will be manufactured by an extrusion technique. Larger glass tubes, arranged proportionally, will be heated and pulled through a drawing tower until they reach the correct dimensions.

Analogues of blue *Morpho* butterfly (Fig. 3a) scales have also been manufactured. *Morpho* wings typically contain two layers of scales — a quarter-wave stack to generate colour and another layer above it to scatter the light. Early copies of the *Morpho* wing did not have this scattering layer: rather, the quarter-wave stack was deposited onto a roughened substrate, and the resulting roughness in the stack scattered the light[18]. Nevertheless, these devices closely matched the butterfly wings, with the colour changing only slightly as the angle was varied over 180°. This effect, which is actually quite difficult to achieve, is useful for various types of optical filter.

Recently, focused ion beam chemical vapour deposition (FIB-CVD) has been used to make more accurate reproductions of the *Morpho* Christmas tree structures[19] (Fig. 3c), but this method is not suitable for low-cost mass production, so the nanostructures produced are currently limited to high-cost items, such as filters for flat-panel displays. Recently the scales of butterfly wings have been used as templates for replicating these structures in ZnO (ref. 20). This technique involves sacrificing a scale to make each replica, but the enhanced optical properties of the replacement materials could make that profitable.

CELL CULTURE TECHNIQUES AND EARLY SUCCESSES WITH DIATOMS

An alternative approach to optical biomimetics involves harnessing the actual processes inside cells that make natural optical devices, rather than trying to use conventional fabrication techniques. Below we discuss the use of diatoms and viruses to make optical nanostructures, and in the section "Lessons from cell engineering" we discuss an alternative manufacturing approach based on the emulation of natural engineering processes.

Current work on cell-culture techniques centres on butterfly scales. The cells that make the butterfly scales are identified in chrysalises, dissected and plated out. The individual cells are then separated, kept alive in culture and prompted to manufacture scales through the addition of growth hormones. We have cultured blue *Morpho* butterfly scales in the lab that have optical and structural

Figure 2 Replicating the iridescent cuticle of a beetle. **a**, The Manuka (scarab) beetle exploits chiral films in a manner similar to a liquid crystal to make its cuticle iridescent. **b**, Biomimetic replicas (each around 2 cm²) made of titania have different colours depending on the pitch of the film. The colour of the replica varies with angle in the same way as the cuticle of the beetle. The circular polarization properties of the replica and the beetle are also the same. **c**, A scanning electron micrograph of the chiral reflector in the cuticle of the beetle and **d**, the replica. Scale bar = 400 nm. Reproduced with permission from ref 17. Copyright (2005) Taylor and Francis.

characteristics that are identical to those found in natural scales. The cultured scales could be embedded in a polymer or mixed into a paint, where they may float to the surface and self-align. There are also applications in optical sensors for vapours[21]. However, if cultured scales are to be used in real-world applications, further work is needed to increase the level of production and to find better ways of harvesting the scales from laboratory equipment.

It is, however, much easier to work with single-celled organisms, such as diatoms. A diatom is a photosynthetic micro-organism and its cell wall (called the frustule) is made of pectin, a polysaccharide, impregnated with silica. The frustule contains pores (Fig. 4a) and slits that give the protoplasm in the cell access to the external environment. There are more than 100,000 different species of diatoms, which are generally 20–200 µm in diameter or length, but can be up to 2 mm long. Diatoms have been proposed to build photonic devices directly in three-dimensions[22]. The biological function of the optical effect (Fig. 4b) is at present unknown but it may affect light collection by the diatom. This type of photonic device can be made in silicon using a deep photochemical etching technique that was initially developed by Lehmann and co-workers[23] (see also Fig. 4c). However, it is possible to exploit the fact that the number of diatoms grows exponentially with time — each individual can give rise to 100 million descendents in a month.

Unlike most manufacturing processes, diatoms achieve a high degree of complexity and hierarchical structure under mild physiological conditions. Importantly, the size of the pores does not scale with the size of the cell, thus maintaining the pattern. Fuhrmann *et al.*[22] showed that the presence of these pores in the silica cell wall of the diatom *Coscinodiscus granii* means that the frustule can be regarded as a photonic crystal slab waveguide. Furthermore, they present models to show that light may be coupled into the waveguide and give photonic resonances in the visible spectral range.

The silica surface of the diatom is amenable to simple chemical functionalization (Fig. 4d,e). An interesting example of this uses a DNA-modified diatom template for the control of nanoparticle assembly[24]. Gold particles were coated with DNA complementary to that bound to the surface of the diatom. Subsequently, the gold particles were bound to the diatom surface via the sequence specific DNA interaction. Using this method, up to seven layers were added showing how a hierarchical structure could be built onto the template.

Porous silicon is known to luminesce in the visible region of the spectrum when irradiated with ultraviolet light[25]. This photoluminescence (PL) emission from the silica skeleton of diatoms was exploited by DeStafano[26] in the production of an optical sensor. It was shown that the PL of *Thalassiosira rotula* is strongly dependent on the surrounding environment.

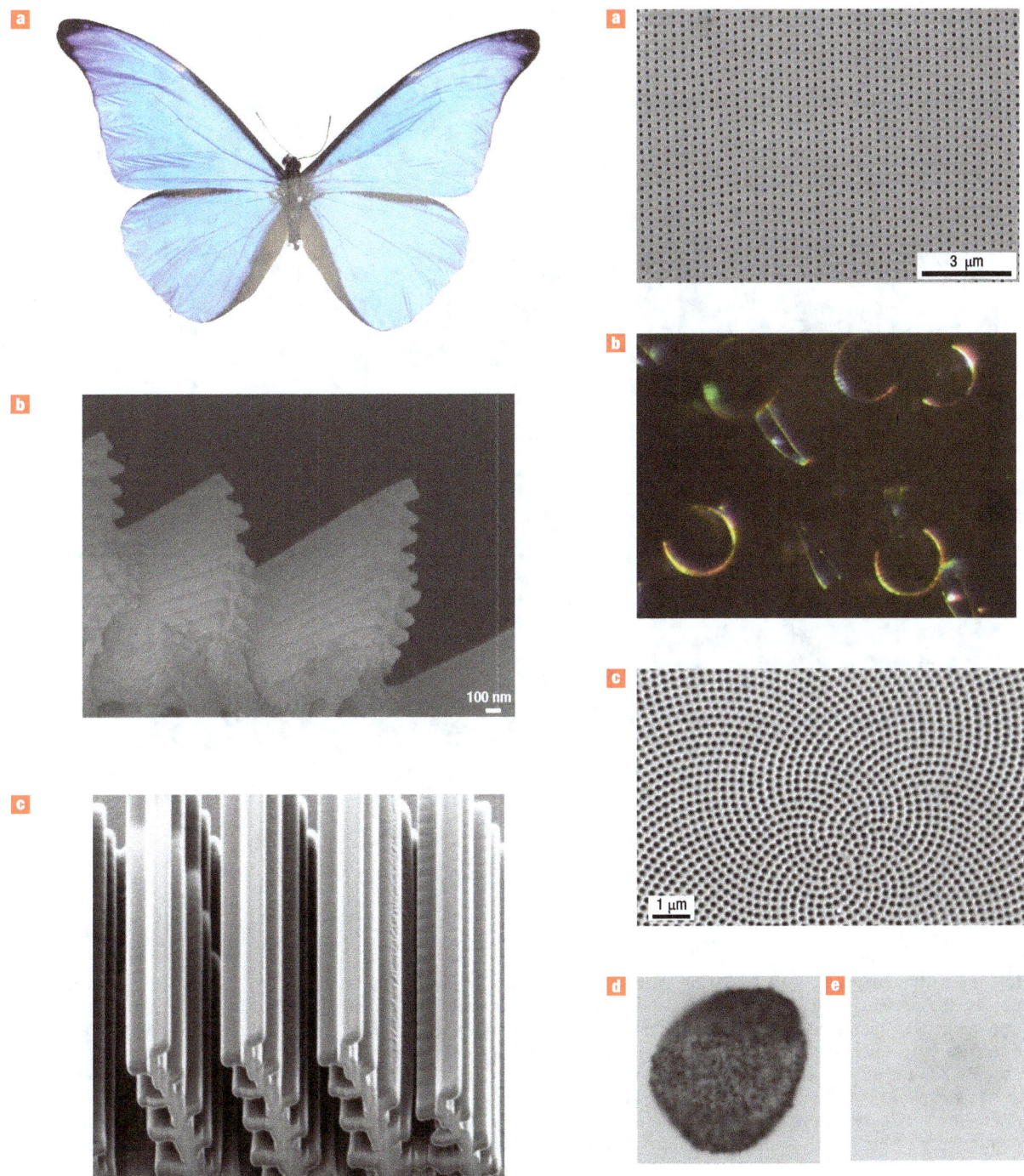

Figure 3 The iridescent wings of a *Morpho* butterfly. **a**, The complex nanoarchitecture found in the wings of this *Morpho* gives them a distinctive iridescent blue colour and presents significant challenges to researchers trying to make replicas. **b**, Scanning electron micrographs of the structures in the scales of the wing that reflect blue light. **c**, A mimic fabricated with the FIB-CVD method. Both structures give a wavelength peak at around 440 nm at an angle of about 30°. However, the replica can only be made to cover areas in the order of micrometres, whereas the butterfly wings are several centimetres across. Reproduced with permission from ref. 19. Copyright (2004) Japanese Society of Applied Physics.

Figure 4 The periodic structures found in the walls of some diatoms could have useful optical properties for applications. **a**, A scanning electron micrograph of the girdle band (which form the side wall of the frustule) of *Coscinodiscus granii*. **b**, The iridescent effect of several isolated girdle bands. **c**, A scanning electron micrograph of a replica of a diatom frustule made by deep photochemical etching. This method has been patented for photonic crystal applications. Reproduced with permission from ref. 46. Copyright (2006) Royal Society. **d,e** The surface of the diatom can also be chemically modified for specific applications. For example, silanization of the surface followed by treatment with a heterobifunctional crosslinker allows an antibody to be attached. These modified diatoms can be used to reveal the presence of a secondary antibody (**d**), whereas non-modified diatoms cannot (**e**).

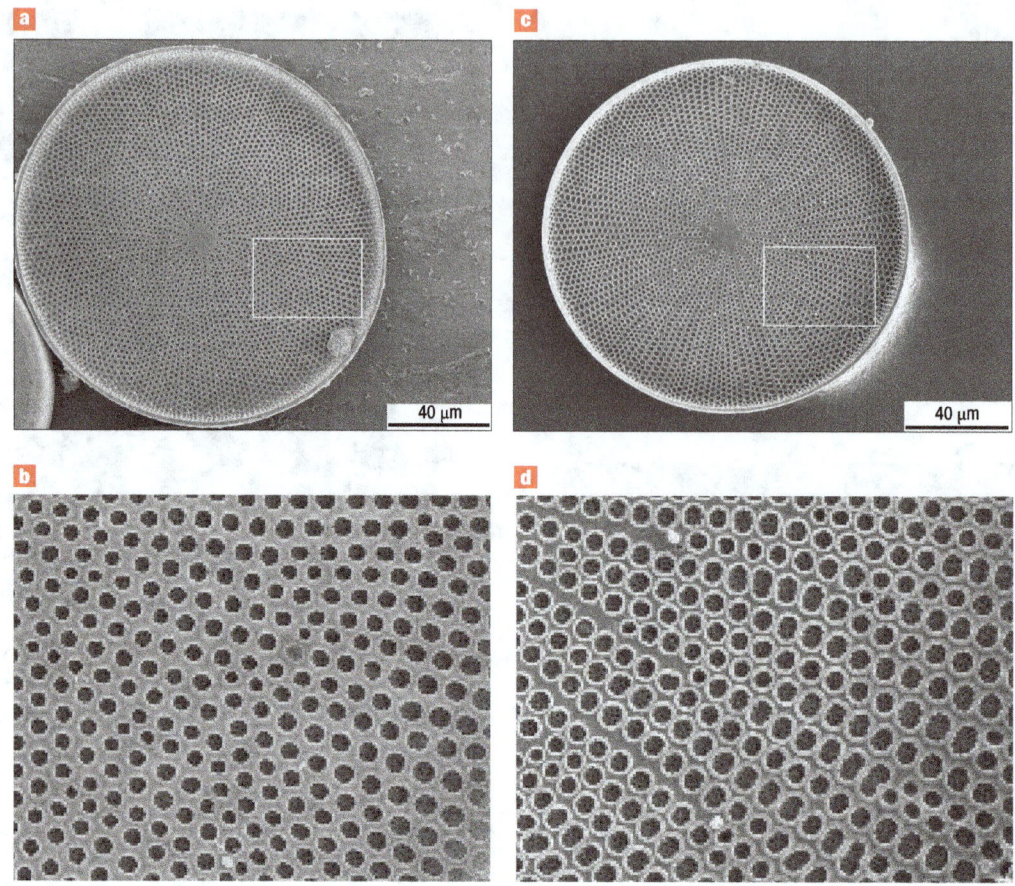

Figure 5 Made-to-measure structures in cell walls. **a,b**, Scanning electron micrographs showing the pore pattern in the silica-based cell wall of the diatom *Coscinodiscus wailesii* grown under ordinary conditions. **c,d**, If the growth takes place in the presence of nickel sulphate, the pores become much larger. The influence of the environmental conditions on the physical properties of the walls could be exploited in applications. For instance, an optical sensor could detect the change in the pore size from their effect on light, thus providing information about the environmental conditions.

Both the optical intensity and peaks are affected by gases and organic vapours. In the presence of NO_2, acetone and ethanol, the photoluminescence was quenched because these substances attract electrons from the silica skeleton of the diatoms and hence quench the PL. On the other hand, substances that donate electrons, such as xylene and pyridine, had the opposite effect, and increased PL intensity almost ten times. Both quenching and enhancements were reversible as soon as the atmosphere was replaced by air.

The silica inherent to diatoms does not provide the optimum chemistry/refractive index for many applications. Sandhage *et al.*[27] have devised an inorganic molecular conversion reaction that preserves the size, shape and morphology of the diatom whilst changing its composition. They perfected a displacement reaction to convert biologically derived silica structures such as frustules into new compositions. Magnesium was shown to convert SiO_2 diatoms by a vapour phase reaction at 900 °C to MgO of identical shape and structure, with a liquid Mg_2Si by-product. Similarly when diatoms were exposed to titanium fluoride gas, the titanium displaced the silicon, yielding a diatom structure made up entirely of titanium dioxide; a material used in some commercial solar cells. Recently, this group extended its work on silica diatoms to work at lower temperatures (650 °C; ref 28).

An alternative approach is to 'hijack' the process that deposits the silica into the frustule so that another material is deposited instead. Rorrer *et al.*[29] have used this *in vivo* approach to incorporate germanium directly into the frustule. Using a two-stage cultivation process the photosynthetic marine diatom *Nitzschia frustulum* was shown to assimilate soluble germanium and fabricate Si–Ge oxide nanostructured composite materials. As germanium is a semiconductor, these structures could have applications in electronics, optoelectronics, photonics, thin-film displays and solar cells.

Porous glasses impregnated with organic dye molecules are promising solid media for tuneable lasers and nonlinear optical devices, luminescent solar concentrators, gas sensors and active waveguides. Biogenic porous silica has an open sponge-like structure and its surface is naturally OH-terminated. Hildebrand and Palenik[30] have shown that Rhodamine B and 6G are able to stain diatom silica *in vivo*, and have determined that the dye treatment could survive the harsh acid treatment needed to remove the surface organic layer from the silica frustule.

Now, attention is beginning to turn to coccolithophores — single-celled marine algae that are also abundant in marine environments. These cells secrete calcitic photonic crystal frustules that, like diatoms, can take a diversity of forms[31].

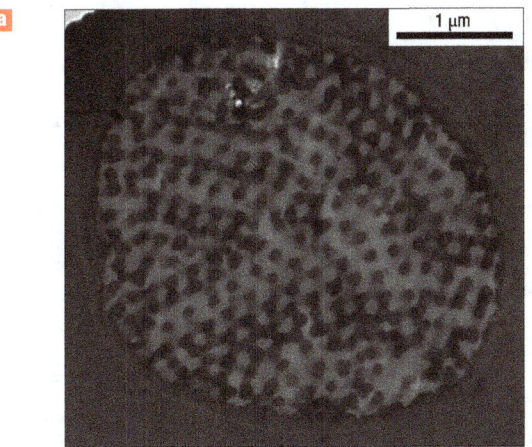 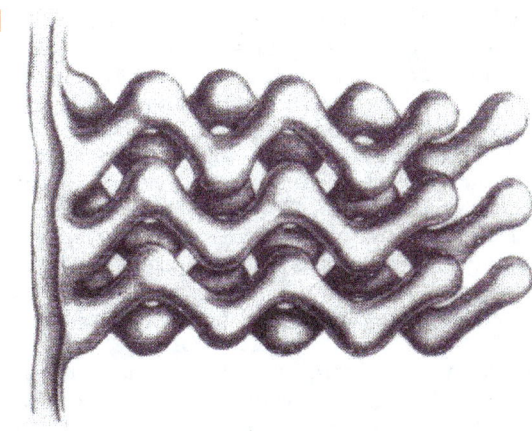

Figure 6 The basic elements of cell engineering. **a**, A transmission electron micrograph of the inverse-opal-type photonic crystal of the weevil *Metapocyrtus* sp. **b**, Illustration of tubular christae in mitochondria from the chloride cell of sardine larvae as an example of animal intracellular membrane form. Evidence suggests that pre-existing internal cell structures play a role in the manufacture of natural nanostructures. If these internal structures can be altered, the cell will produce different nanostructures. Reproduced with permission from ref. 47.

BUILDING OPTICAL DEVICES WITH IRIDOVIRUSES

Viruses are infectious particles made up of the viral genome packaged inside a protein capsid. The iridovirus family comprises a diverse array of large (120–300 nm in diameter) viruses with icosahedral symmetry. The viruses replicate in the cytoplasm (the watery matrix) of insect cells. Within the infected cell the virus particles form a paracrystalline array that causes Bragg refraction of light. This property has largely been considered aesthetic to date but now iridoviruses are being used to create biophotonic crystals. These can be used for the control of light, with researchers at the Wright–Patterson Air Force Base in the US undertaking large-scale virus production and purification, as well as exploring ways to change the properties of the crystals by manipulating the surfaces of the iridoviruses. This work could have applications in a broad range of optical technologies, ranging from sensors to waveguides[32].

Virus nanoparticles, specifically *Chilo* and *Wiseana* Invertebrate Iridovirus, have been used as building blocks for iridescent nanoparticle assemblies. Virus particles were assembled *in vitro*, yielding films and monoliths with optical iridescence arising from multiple Bragg scattering from close-packed crystalline structures of the iridovirus. Bulk viral assemblies were prepared by centrifugation followed by the addition of glutaraldehyde, a crosslinking agent, and long-range assemblies were prepared using a cell design that forced virus assembly within a confined geometry followed by crosslinking.

In addition, virus particles were used as core substrates in the fabrication of nanostructures that comprise a dielectric core surrounded by a metallic shell. More specifically, a gold shell was assembled around the viral core by attaching small gold nanoparticles to the virus surface using inherent chemical functionality of the protein capsid[33]. These gold nanoparticles then acted as nucleation sites for electroless deposition of gold ions from solution. Such nanoshells could be manufactured in large quantities, and provide cores with a narrower size distribution and smaller diameters (below 80 nm) than currently used for silica. These investigations demonstrated that direct harvesting of biological structures, rather than biochemical modification of protein sequences, is a viable route to create unique optically active materials.

LESSONS FROM CELL ENGINEERING

Where cell culture is concerned it is enough to know that cells do make optical nanostructures, which can be farmed appropriately. In the future, however, it may be possible to emulate the natural engineering processes ourselves by reacting the same concentrations of chemicals under the same environmental conditions, perhaps with the help of purpose built nanomachinery.

For the Manuka beetle we already have some insights into the influence of micro-environmental conditions on the growth of the 'liquid crystal' reflectors in its cuticle. The elongated chitin molecules that make up the reflector gradually self-assemble into a liquid crystal arrangement, during drying (as the molecules become closer to each other)[34]. Here, the precise shape of the chitin molecules becomes important. Chitin has also been shown to form submicrometre spheres, such as those that make up the opal structure in a weevil's scales[35], simply by varying the pH[36].

To date, however, the process that is best understood is the formation of the frustule in diatoms. The frustule is formed by the controlled precipitation of silica within a specialized membrane vesicle called the silica deposition vesicle (SDV). Once inside the SDV, silicic acid is converted into silica particles, each measuring approximately 50 nm in diameter, and these particles then aggregate to form larger blocks. The silica is deposited in a pattern that appears to be defined by the presence of organelles, such as mitochondria, that are spaced at regular intervals along the cytoplasmic side of the SDV[37]. These organelles are thought to physically prevent the silica reaching specific parts of the silica cell wall, thus leaving a space. This process is very fast, presumably because conditions are optimal for the synthesis of amorphous silica. The involvement of the cell organelles as obstacles during the silica deposition process results in a final species-specific cell wall with intricate architecture.

The mechanism whereby diatoms use intracellular components to dictate the final pattern of the frustule may provide a route for directed evolution. Growth of *Skeletonema costatum* in sublethal concentrations of mercury and zinc[38] results in cells with swollen organelles, dilated membranes, and membrane-bound cavities. Frustule abnormalities have also been reported in *Nitzschia liebethrutti* grown in the presence of mercury and tin[39]. Both metals resulted in a reduction in the length-to-width ratios of the diatoms, fused pores and a reduction in the number of pores per frustule. These abnormalities were thought to arise from enzyme disruption, either at the silica deposition site or at the nuclear level. We grew *Coscinodiscus wailesii* in sublethal concentrations of nickel and observed an increase in the size of the pores (Fig. 5a,b) and a change in the phospholuminescent properties of the frustule, which could have applications in sensing.

Other structures in cells called 'trans-Golgi-derived vesicles' are known to manufacture three-dimensional photonic crystals in coccolithophores[40]. The organelles in cells therefore appear to precisely control (photonic) crystal growth (of calcium carbonate in coccolithophores) and packing (of silica in the diatoms)[41,42]. Indeed, Ghiradella[14] suggests that structures within the cell play a role in the development of some butterfly scales, as does Overton[43], reporting that microtubules and microfibrils also play an important role. Studies of the diversity of photonic crystals produced by different cells also suggest such internal cell structures are used as moulds and nanomachinery (Fig. 6).

Indeed, the same few designs are found again and again within highly unrelated species, suggesting that the basic eukaryote cell contains an array of pre-existing structures that can be called upon to play a role in the manufacture of complex photonic nanostructures in any taxon[44]. If the same moulds, scaffolds, templates or machinery are used each time, it is not surprising that the photonic devices show similarities. Equally, if similar physical processes are involved, such as molecular self-assembly, then similar nanoarchitectures could be expected to reoccur[44].

The ultimate goal in the field of optical biomimetics, therefore, could be to replicate such nanomachinery and/or to provide conditions under which, if the correct ingredients are supplied in suitable quantities, the optical nanostructures will self-assemble with precision. DNA machines are already made by self-assembly[45], so this goal is not unrealistic.

doi:10.1038/nnano.2007.152

Published online: 3 June 2007.

References

1. Parker, A. R. 515 million years of structural colour. *J. Opt. A* **2**, R15–28 (2000).
2. Parker, A. R., McPhedran, R. C., McKenzie, D. R., Botten, L. C. & Nicorovici, N. A. P. Aphrodite's iridescence. *Nature* **409**, 36–37 (2001).
3. Gordon R. & Parkinson, J. Potential roles for diatomists in nanotechnology. *J. Nanosci. Nanotech.* **5**, 51–56 (2005).
4. Miller, W. H., Moller, A. R. & Bernhard, C. G. in *The Functional Organisation of the Compound Eye* (ed. Bernhard C. G.). 21–33 (Pergamon Press, Oxford, 1966).
5. Yoshida, A., Motoyama, M., Kosaku, A. & Miyamoto, K. Antireflective nanoprotuberance array in the transparent wing of a hawkmoth *Cephanodes hylas*. *Zool. Sci.* **14**, 737–741 (1997).
6. Gale, M. Diffraction, beauty and commerce. *Phys. World* **2**, 24–28 (October, 1989).
7. Boden, S. A. & Bagnall, D. M. in *Proc. 4th World Conference on Photovoltaic Energy Conversion*, Hawaii 1358–1361 (IEEE, 2006).
8. Parker, A. R., Hegedus, Z. & Watts, R. A. Solar-absorber type antireflector on the eye of an Eocene fly (45Ma) *Proc. R. Soc. Lond. B* **265**, 811–815 (1998).
9. Beale, B. Fly eye on the prize. *The Bulletin* 46–48 (25 May 1999).
10. Berthier *et al.* Anticounterfeiting using biomimetic polarization effects. *Appl. Phys. A* **86**, 123 (2007).
11. Vigneron, J. P. *et al.* Spectral filtering of visible light by the cuticle of metallic woodboring beetles and microfabrication of a matching bioinspired material. *Phys. Rev. E* **73**, 041905 (2006).
12. López, C. Three-dimensional photonic band-gap materials: semiconductors for light *J. Opt. A* **8**, R1–14 (2006).
13. Cohen, R. E., Zhai, L., Nolte, A. & Rubner, M. F. pH gated porosity transitions of polyelectrolyte multilayers in confined geometries and their applications as tunable Bragg reflectors. *Macromolecules* **37**, 6113 (2004).
14. Ghiradella, H. Structure and development of iridescent butterfly scales: lattices and laminae. *J. Morph.* **202**, 69–88 (1989).
15. Vukusic, P., Sambles J. R., Lawrence C. R. & Wootton, R. J. Structural colour: Now you see it — now you don't. *Nature* **410**, 36 (2001).
16. Berthier, S. *Iridescence: The Physical Colors of Insects* (Springer, New York, 2006).
17. DeSilva, L. *et al.* Natural and nanoengineered chiral reflectors: structural colour of manuka beetles and titania coatings. *Electromagnetics* **25**, 391–408 (2005).
18. Kinoshita, S., Yoshioka, S., Fujii, Y. & Okamoto, N. Photophysics of structural color in the *Morpho* butterflies. *Forma* **17**, 103–121 (2002).
19. Watanabe, K., Hoshino, T., Kanda, K., Haruyama, Y. & Matsui, S. Brilliant blue observation from a *Morpho*-butterfly-scale quasi-structure. *Jpn. J. Appl. Phys.* **44**, L48–L50 (2005).
20. Zhang, W. *et al.* Biomimetic zinc oxide replica with structural color using butterfly (*Ideopsis similis*) wings as templates. *Bioinspir. Biomim.* (submitted).
21. Potyrailo, R. A. *et al. Morpho* butterfly wing scales demonstrate highly selective vapour response. *Nature Photonics* **1**, 123–128 (2007).
22. Fuhrmann, T., Landwehr, S., El Rharbi-Kucki, M. & Sumper, M. Diatoms as living photonic crystals. *Appl. Phys. B* **78**, 257–260 (2004).
23. Lehmann, V. On the origin of electrochemical oscillations at silicon electrodes. *J. Electrochem. Soc.* **143**, 1313 (1993).
24. Rosi, N. L., Thaxton, C. S. & Mirkin, C. A. Control of nanoparticle assembly by using DNA-modified diatom templates. *Angew. Chem. Int. Edn* **43**, 5500–5503 (2004).
25. Cullis, A. G., Canham, L. T. & Calcott, P. D. J. The structural and luminescence properties of porous silicon. *J. Appl. Phys.* **82**, 909–965 (1997).
26. De Stefano, L., Rendina, I., De Stefano, M., Bismuto, A. & Maddalena, P. Marine diatoms as optical chemical sensors. *Appl. Phys. Lett.* **87**, 233902 (2005).
27. Sandhage, K. H. *et al.* Novel, bioclastic route to self-assembled, 3D, chemically tailored meso/nanostructures: Shape-preserving reactive conversion of biosilica (diatom) microshells. *Adv. Mater.* **14**, 429–433 (2002).
28. Bao, Z. Chemical reduction of three-dimensional silica micro-assemblies into microporous silicon replicas *Nature* **446**, 172–175 (2007).
29. Rorrer, G. L. *et al.* Biosynthesis of silicon-germanium oxide nanocomposites by the marine diatom *Nitzschia frustulum*. *J. Nanosci. Nanotechnol.* **5**, 41–49 (2004).
30. Hildebrand, M. & Palenik, B. *Investigation into the Optical Properties of Nanostructured Silica from Diatoms*. Grant report (2003); www.stormingmedia.co.uk
31. Quintero-Torres, R., Aragon, J. L., Torres, M., Estrada, M. & Gros, L. Strong far field coherent scattering of UV radiation by holococcolithophores. *Phys. Rev. E* **74** (2006).
32. Juhl, S. B. *et al.* Assembly of Wisean Iridovirus: Viruses for photonic crystals. *Adv. Funct. Mater.* **16**, 1086–1094 (2002).
33. Radloff, C., Vaia, R. A., Brunton, J., Bouwer, G. T. & Ward, V. K. Metal nanoshell assembly on a virus bioscaffold. *Nano Lett.* **5**, 1187–1191 (2005).
34. Giraud-Guille, M. M., Besseau, L. & Martin, R. Liquid crystalline assemblies of collagen in bone and in vitro systems. *J. Biomech.* **36**, 1571–1579 (2003).
35. Parker, A. R., Welch, V. L., Driver, D. & Martini, N. An opal analogue discovered in a weevil. *Nature* **426**, 786–787 (2003).
36. Murray, S. B. & Neville, A. C. The role of pH, temperature and nucleation in the formation of cholesteric liquid crystal spherulites from chitin and chitosan. *Int. J. Biol. Macromol.* **22**, 137 (1998).
37. Schmid, A. M. M. Aspects of morphogenesis and function of diatom cell walls with implications for taxonomy. *Protoplasma* **181**, 43–60 (1994).
38. Smith, M. A. The effect of heavy metals on the cytoplasmic fine structure of Skeletonema costatum (Bacillariophyta). *Protoplasma* **116**, 14–23 (1983).
39. Saboski, E. Effects of mercury and tin on frustular ultrastructure of the marine diatom *Nitzschia liebethrutti*. *Water Air Soil Poll.* **8**, 461–466 (1977).
40. Corstjens, P. L. A. M. & Gonzales, E. L. Effects of nitrogen and phosphorus availability on the expression of the coccolith-vesicle v-ATPase (subunit C) of Pleurochrysis (Haptophyta). *J. Phycol.* **40**, 82–87 (2004).
41. Klaveness D. & Paasche E. in *Biochemistry and Physiology of Protozoa* 2nd edn, Vol. 1 (eds Hutner, S. H. & Levandowsky, M.) (Academic Press, New York, 1979).
42. Klaveness, D. & Guillard, R. R. L. The requirement for silicon in *Synura petersenii* (Chrysophyceae). *J. Phycol.* **11**, 349–355 (1975).
43. Overton, J. Microtubules and microfibrils in morphogenesis of the scale cells of *Ephestia kuhniella*. *J. Cell Biol.* **29**, 293–305 (1966).
44. Parker, A. R. Conservative photonic crystals imply indirect transcription from genotype to phenotype. *Rec. Res. Develop. Entomol.* **5**, 1–10 (2006).
45. Bath, J. & Turberfield. A. J. DNA nanomachines *Nature. Nanotech.* **2**, 275–284 (2007).
46. Parker, G. J. *et al.* Highly engineered mesoporous structures for optical processing. *Phil. Trans. R. Soc. Lond. A* **364**, 189–199 (2006).
47. Threadgold, L. T. *The Ultrastructure of the Animal Cell* (Pergamon Press, Oxford, 1967).

Acknowledgements

This work was funded by the Royal Society (University Research Fellowship), the Australian Research Council, Framework 6 of the European Union, and RCUK (Basic Technology Grant).

Competing financial interests

The authors declare no competing financial interests.

NANOBIOTECHNOLOGY AND NANOMEDICINE

Nanoparticle therapeutics: an emerging treatment modality for cancer

Mark E. Davis*, Zhuo (Georgia) Chen‡ and Dong M. Shin‡

Abstract | Nanoparticles — particles in the size range 1–100 nm — are emerging as a class of therapeutics for cancer. Early clinical results suggest that nanoparticle therapeutics can show enhanced efficacy, while simultaneously reducing side effects, owing to properties such as more targeted localization in tumours and active cellular uptake. Here, we highlight the features of nanoparticle therapeutics that distinguish them from previous anticancer therapies, and describe how these features provide the potential for therapeutic effects that are not achievable with other modalities. While large numbers of preclinical studies have been published, the emphasis here is placed on preclinical and clinical studies that are likely to affect clinical investigations and their implications for advancing the treatment of patients with cancer.

Nanoparticle therapeutics are typically particles comprised of therapeutic entities, such as small-molecule drugs, peptides, proteins and nucleic acids, and components that assemble with the therapeutic entities, such as lipids and polymers, to form nanoparticles (FIG. 1). Such nanoparticles can have enhanced anticancer effects compared with the therapeutic entities they contain. This is owing to more specific targeting to tumour tissues via improved pharmacokinetics and pharmacodynamics, and active intracellular delivery (FIG. 1). These properties depend on the size and surface properties (including the presence of targeting ligands) of the nanoparticles.

In this Review, we first briefly discuss the key properties of nanoparticles and how they differ from other types of cancer drugs. Next, we summarize current clinical uses of first-generation nanoparticle therapeutics and describe how newer, experimental nanoparticle therapeutics differ from first-generation therapeutics. Finally, we discuss what is on the near horizon for cancer therapy that use nanoparticles. Although an enormous amount of research is ongoing in this area, most will not be translatable to the clinic. Some of the main obstacles include the use of immunostimulatory components, the use of components that have barriers to large-scale current good manufacturing practice (cGMP) production and/ or hurdles in the development of well-defined chemistry, manufacturing and controls assays. A limited number of nanoparticle systems have reached a clinical application (or are about to reach this status), and information

is becoming available to begin to understand some of the issues of moving these experimental systems into humans. Thus, our emphasis here is on issues involving the translation of these experimental nanoparticle therapeutics into the clinic and their application in clinical settings. Although there are several experimental approaches utilizing nanoparticles that can be affected by external stimulation, we focus our attention on systemically administered nanoparticles that do not require external stimuli.

Key properties of anticancer nanoparticles

Nanoparticle size. It is currently thought that the diameter of nanoparticle therapeutics for cancer should be in the range of 10–100 nm. The lower bound is based on the measurement of sieving coefficients for the glomerular capillary wall, as it is estimated that the threshold for first-pass elimination by the kidneys is 10 nm (diameter)[1]. The upper bound on size is not as well defined at this time. The vasculature in tumours is known to be leaky to macromolecules. The lymph system of tumours in mouse models is poorly operational and macromolecules leaking from the blood vessels accumulate — a phenomenon known as "enhanced permeability and retention (EPR) effect"[2]. Numerous lines of evidence suggest that this phenomenon is also operational in humans. It has been shown that entities in the order of hundreds of nanometre in size can leak out of the blood vessels and accumulate within tumours. However, large

*Chemical Engineering, California Institute of Technology, Pasadena, California 91125, USA.
‡Winship Cancer Institute, Emory School of Medicine, Atlanta, Georgia 30322, USA.
Correspondence to M.E.D.
e-mail: mdavis@cheme.caltech.edu
doi:10.1038/nrd2614*

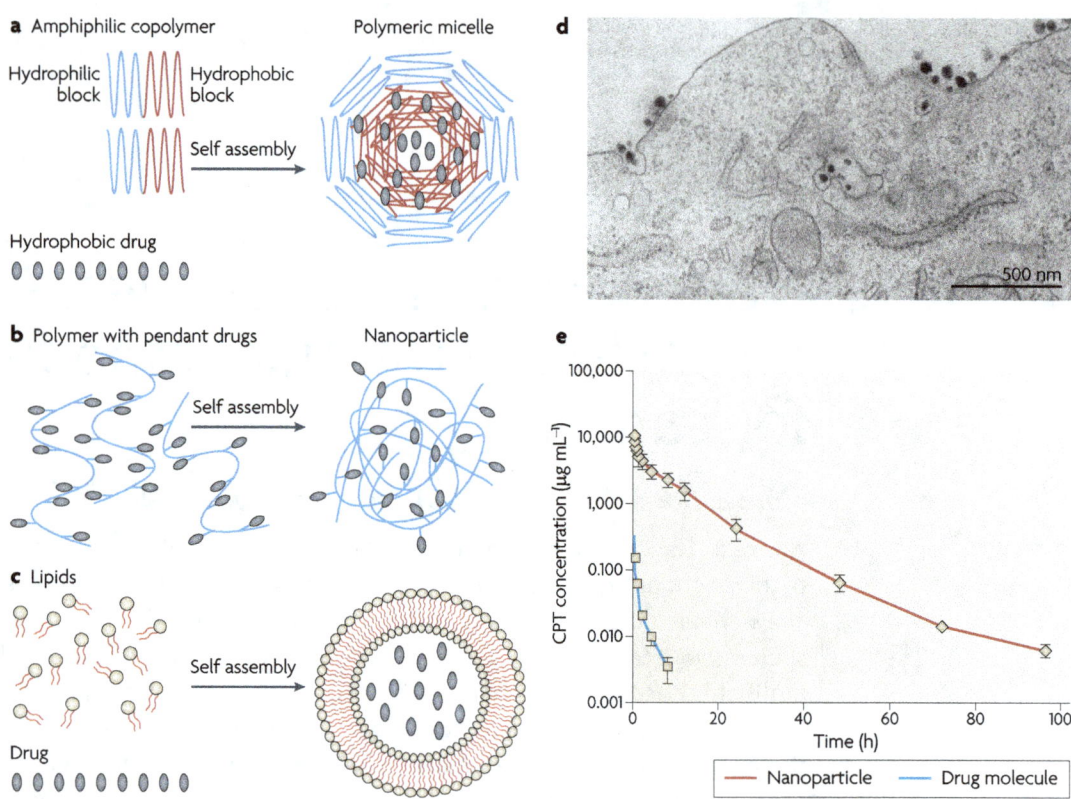

Figure 1 | Major classes of nanoparticles that are in clinical trials and some of their properties. a | Nanoparticles formed from therapeutic entities and block copolymers that can form polymeric micelles. **b |** Nanoparticles that form with polymer–drug conjugates. **c |** Nanoparticles formed of liposomes. **d |** Nanoparticles can enter cells via endocytosis. This figure shows a transmission electron micrograph of nanoparticles at the surface of a cancer cell, entering the cell and within endocytic vesicles. **e |** Nanoparticles can have extended pharmacokinetics over the therapeutic entity alone. The data are from a small-molecule drug (CPT) and a nanoparticle containing CPT in rats. Panel (**e**) is adapted with permission from REF. 94 © Springer (2006).

macromolecules or nanoparticles could have limited diffusion in the extracellular space[3]. Experiments from animal models suggest that sub-150 nm, neutral or slightly negatively charged entities can move through tumour tissue[4]. Additionally, recent data show that nanoparticles in the 50–100 nm size range that carry a very slight positive charge can penetrate throughout large tumours following systemic administration[5]. Thus, well-designed nanoparticles in the 10–100 nm size range and with a surface charge either slightly positive or slightly negative should have accessibility to and within disseminated tumours when dosed into the circulatory system. If this size range is correct, then these nanoparticles will be restricted from exiting normal vasculature (that requires sizes less than 1–2 nm); however they will still be able to access the liver, as entities up to 100–150 nm in diameter are able to do so.

Nanoparticle surface properties. Nanoparticles have high surface-to-volume ratios when compared with larger particles, and so control of their surface properties is crucial to their behaviour in humans (for example, see REF. 6). The ultimate fate of nanoparticles within the body can be determined by the interactions of nanoparticles with

their local environment, which depends on a combination of size and surface properties. Nanoparticles that are sterically stabilized (for example by polyethylene glycol (PEG) polymers on their surface) and have surface charges that are either slightly negative or slightly positive tend to have minimal self–self and self–non-self interactions. Also, the inside surface of blood vessels and the surface of cells contain many negatively charged components, which would repel negatively charged nanoparticles. As the surface charge becomes larger (either positive or negative), macrophage scavenging is increased and can lead to greater clearance by the reticuloendothelial system. Thus, minimizing nonspecific interactions via steric stabilization and control of surface charge helps to prevent nanoparticle loss to undesired locations. However, the complete removal of nonspecific interactions is not currently possible, and so there is always some particle loss; the key is to minimize these interactions as much as possible.

If nanoparticle loss could be avoided, it would be expected that the distribution of the nanoparticles within a mammal would be uniform if no size restrictions existed on the basis of thermodynamic considerations. However, there are numerous size-restricted locations within the body that would create non-uniformity. For example, the brain is protected by the blood–brain barrier, which has severe size and surface property limitations for entrance. By understanding the size and surface property requirements for reaching specified sites within the body, localization of nanoparticles to these sites can be accomplished.

Nanoparticle-targeting ligands. The addition of targeting ligands that provide specific nanoparticle–cell surface interactions can play a vital role in the ultimate location of the nanoparticle. For example, nanoparticles can be targeted to cancer cells if their surfaces contain moieties such as small molecules, peptides, proteins or antibodies. These moieties can bind with cancer cell-surface receptor proteins, such as transferrin receptors, that are known to be increased in number on a wide range of cancer cells[7]. These targeting ligands enable nanoparticles to bind to cell-surface receptors and enter cells by receptor-mediated endocytosis (FIG. 1). Recent work comparing non-targeted and targeted nanoparticles (lipid-based[8] or polymer-based[9]) has shown that the primary role of the targeting ligands is to enhance cellular uptake into cancer cells rather than increasing the accumulation in the tumour.

Distinguishing features of nanoparticle therapeutics for cancer. Nanoparticles can be tuned to provide long or short circulation times by careful control of size and surface properties. Also, they can be directed to specific cell types within target organs (for example, hepatocytes versus Kupffer cells in the liver[10]). While other types of cancer therapeutics such as molecular conjugates (for example, antibody–drug conjugates) can also meet these minimum specifications, targeted nanoparticles have at least five features that distinguish them from other therapeutic modalities for cancer.

First, nanoparticles can carry a large payload of drug entity and protect it from degradation. For example, a 70 nm nanoparticle can contain approximately 2,000 small interfering RNA (siRNA) molecules[11], whereas antibody conjugates carry fewer than ten[12]. These high payload amounts can also be achieved with other drug types such as small-molecule or peptide drugs. Furthermore, nanoparticle payloads are located within the particle, and their type and number do not affect the pharmacokinetic properties and biodistribution of the nanoparticles. This is unlike molecular conjugates in which the type and number of therapeutic entities conjugated to the targeting ligand (such as an antibody) significantly modifies the overall properties of the conjugate.

Second, the nanoparticles are sufficiently large to contain multiple targeting ligands that can allow multivalent binding to cell-surface receptors[13]. Nanoparticles have two parameters for tuning the binding to target cells: the affinity of the targeting moiety and the density of the targeting moiety. The multivalency effects can lead to high effective affinities when using arrangements of low-affinity ligands[13–15]. Thus, the repertoire of molecules that can be used as targeting agents is greatly expanded as many low-affinity ligands that are not sufficient for

use as molecular conjugates can now be attached on nanoparticles to create higher affinity via multivalent binding to cell-surface receptors (FIG. 2).

Third, nanoparticles are sufficiently large to accommodate multiple types of drug molecules. Numerous therapeutic interventions can be simultaneously applied with a nanoparticle in a controlled manner. As mentioned in the first point, the fact that the pharmacokinetic properties of the nanoparticle are not modified by the amount of the therapeutic also holds true with multiple types of the therapeutics being combined together within the nanoparticles.

Fourth, the release kinetics of drug molecules from nanoparticles can be tuned to match the mechanism of action. For example, topoisomerase I inhibitors such as the camptothecin-based chemotherapeutic drugs are reversible binders of the enzyme. So, the mechanism of action for camptothecin-based drugs on the topoisomerase I enzyme suggests enhanced efficacy with prolonged exposure of the drug[16], making a slow release from the nanoparticles most desirable. With siRNA, the gene inhibition kinetics are greatly influenced by cell-cycle times[17,18], and for uses in cancer there may not be a need for slow release of the therapeutic agent.

Fifth, nanoparticles could have the potential to bypass multidrug resistance mechanisms that involve cell-surface protein pumps (for example, glycoprotein P), as they enter cells via endocytosis (FIG. 1 and discussed below). Overall, it seems that controlled combination of these features through nanoparticle design could minimize side effects of anticancer drugs while enhancing efficacy, and clinical results are emerging that suggest that this promise is starting to be realized. Nanoparticle types and results from the clinic are summarized below.

Nanoparticles as anticancer agents

Nanoscaled systems for systemic cancer therapy and their latest stage of development are summarized in TABLE 1. We have included PEG-containing proteins and PEG-conjugated small molecules, which, as single molecules in solution, can be defined as nanoscale therapeutics or as nanoparticles if they have some degree of polymer–polymer interaction to give assembled entities with more than one polymer chain contained within.

Liposomes (~100 nm and larger) carrying chemotherapeutic small-molecule drugs have been approved for cancer since the mid-1990s, and are mainly used to solubilize drugs, leading to biodistributions that favour higher uptake by the tumour than the free drug[19]. However, liposomes do not provide control for the time of drug release, and in most cases do not achieve effective intracellular delivery of the drug molecules[19], therefore limiting their potential to be useful against multidrug-resistant cancers.

A representative example shown in TABLE 1 is Doxil (Ortho Biotech), a PEG-liposome containing the cytotoxic drug doxorubicin. Doxil was originally approved for the treatment of AIDS-related Kaposi's sarcoma and is now approved for use in ovarian cancer and multiple myeloma. This agent circulates in the body as a nanoparticle and has

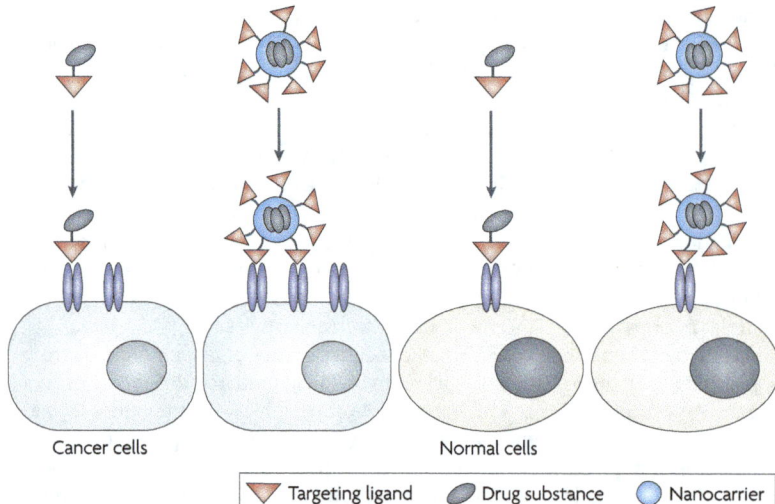

Cancer cells   Normal cells

Targeting ligand   Drug substance   Nanocarrier

Figure 2 | **Nanoparticles with numerous targeting ligands can provide multivalent binding to the surface of cells with high receptor density.** When the surface density of the receptor is low on normal cells, then a molecular conjugate with a single targeting agent and a targeted nanoparticle can compete equally for the receptor as only one ligand–receptor interaction may occur. However, when there is a high surface density of the receptor on cancer cells (for example, the transferrin receptor), then the targeted nanoparticle can engage numerous receptors simultaneously (multivalency) to provide enhanced interactions over the one ligand–one receptor interaction that would occur with a molecular conjugate.

Clearance
This is the volume of blood/plasma cleared of the drug per time. Lower clearances are indicative of higher circulation times.

Neutropaenia
Neutropaenia, usually induced by chemotherapy, is a myelosuppression that involves mainly the neutrophil lineage of white blood cells. Severe (grade 3 or 4) neutropaenia with infection is life-threatening, which should be prevented by treatment with growth factors. For more information see the Common Toxicity Criteria at the US National Cancer Institute web site (see Further information).

with nanoparticles such as Doxil, XYOTAX (CT-2103) and IT-101. The longer circulation times of the nanoparticles compared with the free drug alone can improve tumour uptake (for example, see REF. 23). Moreover, polymeric micelles (sub-100 nm) have been shown to accumulate more readily in tumours than the larger liposomes[24]. Additionally, movement of a particle throughout a tumour is also size dependent as described earlier. It is speculated that nanoparticles that are smaller than 100 nm but larger than 10 nm (to avoid renal clearance) will be optimal for tumour penetration. Therefore, careful control of size will be important to the pharmacokinetics, biodistribution, tumour accumulation and tumour penetration of the nanoparticle therapeutic.

Some of the nanoparticles that are now in clinical testing also have mechanisms to control the release of the drug, as discussed below with IT-101. These methodologies are based on cleavage of a chemical bond between the particle and the drug by hydrolysis; by enzymes that are located within and outside cells, for example, lysozymes, esterases; or by enzymes that are located only within cells, for example, cathepsin B.

Clinical trials using non-targeting nanoparticles. Doxil has been used in the clinic for over two decades and it has been discussed extensively elsewhere[25–27]. Numerous other liposomes containing drugs such as irinotecan and SN-38 are currently in clinical trials (ClinicalTrials.gov; see Further information).

ABI-007 (Abraxane; Abraxis Bioscience/AstraZeneca), an albumin-bound nanoparticle of paclitaxel (~120 nm in mean diameter) was developed to retain the therapeutic benefits of paclitaxel but eliminate the toxicities associated with the emulsifier Cremophor EL in the paclitaxel formulation (Taxol) and its generic equivalents[28]. The maximally tolerated dose (MTD) of Abraxane was approximately 70–80% higher than that reported for Taxol. In a Phase III study of 454 patients with metastatic breast cancer given Abraxane (260 mg per m^2) or Taxol (175 mg per m^2) intravenously every 3 weeks, response rates were significantly greater in patients treated with Abraxane than those receiving Taxol[29]. Despite the increased dose of paclitaxel in the Abraxane group, the incidence of grade 4 neutropaenia was significantly lower with Abraxane than with Taxol (9% versus 22%; $p = 0.001$). Pharmacokinetic assessments also showed that paclitaxel clearance and volume of distribution were higher for Abraxane than for Taxol. Clearance was 13 litres per hour per m^2 for Abraxane versus 14.76 litres per hour per m^2 for Taxol ($p = 0.048$)[28]. Distribution was 663.8 litres per m^2 for Abraxane versus 433.4 litres per m^2 for Taxol ($p = 0.04$)[28]. These differences in pharmacokinetic properties may be associated with the higher intratumoral concentrations observed with Abraxane compared with the equivalent dose of Taxol. However, it should be noted that it is not clear whether or not the nanoparticles dissolve when infused into the patient, and there are indications that they do (see below in final section). Thus, the clinical benefit from Abraxane is probably not due to its functioning as a nanoparticle but to other factors such as the removal of Cremophor EL from the formulation that causes toxicities of its own[30].

a half-life ~100-times longer than free doxorubicin (see below). Its primary advantage in the clinic is the reduction in cardiotoxicity over that of doxorubicin[20,21].

However, such nanoscaled systems have also shown that unwanted attributes can manifest themselves. For example, although Doxil has been shown to have reduced cardiotoxicity compared with free doxorubicin, it also has skin toxicity that is not observed with the drug alone[22]. Newer nanoparticle systems, as defined by a higher degree of multifunctionality (incorporating features such as slow release and/or targeting ligands), have enhanced features such as reduced toxicity without the emergence of other toxicities (as with Doxil) compared with the initial approved products, and some of these attributes are described below.

Nanoparticles without targeting ligands

TABLE 2 compares several nanoparticle-based therapeutics with the drug molecules that they carry. The types of particles include liposomes, polymer micelles and polymer-based nanoparticles.

In each case — for example, doxorubicin compared with the doxorubicin-carrying nanoparticles SP1049C, NK911 and Doxil — the nanoparticle alters the pharmacokinetic properties of the drug molecule. Circulation half-lives are listed in TABLE 2, although they are difficult to compare because different models are used for their determination. Clearance is a common pharmacokinetic parameter that is readily available from clinical data, and it is a better indicator of differences in circulation times among the therapeutics. Dramatically reduced clearances have been obtained

Table 1 | **Nanoscaled systems for systemic cancer therapy**

Platform	Latest stage of development	Examples
Liposomes	Approved	DaunoXome, Doxil
Albumin-based particles	Approved	Abraxane
PEGylated proteins	Approved	Oncospar, PEG-Intron, PEGASYS, Neulasta
Biodegradable polymer–drug composites	Clinical trials	Doxorubicin Transdrug
Polymeric micelles	Clinical trials	Genexol-PM*, SP1049C, NK911, NK012, NK105, NC-6004
Polymer–drug conjugate-based particles	Clinical trials	XYOTAX (CT-2103), CT-2106, IT-101, AP5280, AP5346, FCE28068 (PK1), FCE28069 (PK2), PNU166148, PNU166945, MAG-CPT, DE-310, Pegamotecan, NKTR-102, EZN-2208
Dendrimers	Preclinical	Polyamidoamine (PAMAM)
Inorganic or other solid particles	Preclinical (except for gold nanoparticle that is clinical)	Carbon nanotubes, silica particles, gold particles (CYT-6091)

*Approved in South Korea. PEG, polyethylene glycol.

Several nanoscaled therapeutics based on PEGylated proteins have been approved or are in clinical trials. PEGylation has been applied to various proteins including enzymes, cytokines and monoclonal antibody Fab fragments. PEGylation provides a means to increase protein solubility, reduce immunogenicity, prevent rapid renal clearance (due to the increased size of conjugates) and prolong plasma half-life[31]. PEG-L-asparaginase (Oncospar; Enzon) was approved by the US Food and Drug Administration in 1994 to treat acute lymphoblastic leukaemia[32,33]. Although free-drug L-asparaginase depletes asparagine and is active against acute lymphoblastic leukaemia and lymphoma, it frequently induces a hypersensitivity reaction and antibody production that leads to its premature clearance from the circulation. Phase I studies with Oncospar showed an increased plasma half-life compared with the naked enzyme and a reduced frequency of hypersensitivity reaction[34]. A subsequent Phase II study showed a partial response and a reduced hypersensitivity reaction in some Oncospar-treated patients with refractory lymphoma[35,36]. PEG–recombinant arginine deaminase (PEG–rhArg) has been assessed either as a single agent or in combination with 5-fluorouracil[37,38]. Weekly intramuscular injection of PEG–rhArg showed a clinical activity in hepatocellular carcinoma with the achievement of low arginine concentrations of <2 μM[38].

PEG has also been linked to biological response modifiers such as interferon-α (IFN-α) and recombinant granulocyte colony-stimulating factor (G-CSF)[39,40]. Two PEGylated IFN-α conjugates, PEGASYS (Roche) for IFN-α2a and PEGINTRON (Schering) for IFN-α2b, have shown clinically superior antiviral activity compared with free IFN-α and are approved for hepatitis C therapy[41]. These PEG–IFN-α conjugates have been shown to be effective for the treatment of melanoma and renal cell carcinoma[42–44], and are currently being tested in other solid tumours. Because of the prolonged half-lives of PEG–IFN-α conjugates, they can be given by subcutaneous injection once every 12 weeks instead of three-times per week for free IFNs.

Because of the success of PEGylated proteins, it is not surprising that PEG polymers have also been conjugated with small-molecule drugs to create nanoparticles. Pegamotecan is a camptothecin conjugated with linear PEG and has a molecular mass of ~40,000 Daltons. Two Phase I trials were completed with Pegamotecan (weekly dosing[45] and every 3-week dosing[46]) as was a Phase II trial[47]; however Enzon is no longer pursuing this conjugate. Other PEG-conjugated chemotherapeutics are currently in Phase I trials for advanced solid tumours, including PEG–irinotecan (NKTR-102) and PEG–SN-38 (EZN-2208) (ClinicalTrials.gov; see Further information). The PEGylated small molecules have increased circulation times relative to the free drug[45,46], thus providing the potential for greater tumour accumulation via the EPR effects.

In addition to altering the pharmacokinetics of the therapeutic, nanoscaled systems can provide other functions. For example, PG–paclitaxel conjugates that link high loading levels of paclitaxel (37% wt/wt) with a biodegradeable polymer, poly-L-glutamic acid (PG) have been in numerous clinical trials including Phase III, and XYOTAX (also called CT-2103) is clinically the most advanced polymer–small-molecule conjugate used for systemic administration. Paclitaxel is released from the polymer to a small extent by slow hydrolysis (up to 14% over 24 hours), but is released to a greater extent following lysosomal cathepsin B degradation of the polymer backbone after endocytic uptake[48]. EPR-mediated tumour targeting and enhanced efficacy of PG–paclitaxel has been observed in many preclinical tumour models, with improvement of the safety profile due to both decreased exposure to normal tissue and improved drug solubility[49,50]. Phase I/II studies showed a significant number of partial responses or stable disease in patients with non-small-cell lung cancer (NSCLC), renal cell carcinoma, mesothelioma or paclitaxel-resistant ovarian cancer[51]. Severe side effects included neutropaenia and peripheral neuropathy, which are classical paclitaxel-associated toxicities. In a randomized Phase III trial, PG–paclitaxel was compared with gemcitabine or vinorelbine as a first-line therapy for performance status 2 patients with NSCLC[52]. Patients receiving the conjugate showed significantly reduced side effects when compared with control patients, most of whom received gemcitabine, but the nanoparticle conjugate failed to show significance for overall improved survival in comparison with either of the non-nanoparticle drugs. However, there was a greater increase in survival for women treated with PG–paclitaxel compared with men[53]. Such enhanced activities in female patients might correlate with oestrogen levels, as oestrogen has been shown to increase the expression of cathepsin B[54]. A definitive trial is now ongoing to compare PG–paclitaxel and free paclitaxel (175 mg per m²) as a first-line therapy for women with NSCLC. A PG–camptothecin (CT-2106) nanoparticle has also been tested in Phase I/II clinical trials[45,46,55].

Table 2 | Comparison of pharmacokinetics (human) of small-molecule drugs with nanoparticle therapeutics

Name	Formulation	Diameter (nm)	$t_{1/2}$ (h)	Clearance (ml/min•kg)	Comments	Refs
Doxorubicin (DOX)	0.9% NaCl	NA	0.8	14.4	Small-molecule drug	24
SP1049C	Pluronic micelle + DOX	22–27	2.4	12.6	Micelle nanoparticle	24
NK911	PEG–Asp micelle + DOX	40	2.8	6.7	Micelle nanoparticle	24
Doxil	PEG–liposome + DOX	80–90	84.0	0.02	PEGylated liposome nanoparticle with long circulation	24
Taxol (paclitaxel)	Cremophor EL	NA	21.8 (20.5)	3.9 (9.2)	Small-molecule drug	24 (28)
Genexol-PM	PEG–PLA micelle + paclitaxel	20–50	11.0	4.8	Micelle nanoparticle	24
Abraxane	Albumin + paclitaxel	120*	21.6	6.5	Albumin nanoparticle before injection; status in vivo unknown	28
XYOTAX	PG + paclitaxel	Unknown	70–120	0.07–0.12	Polymer nanoparticle	23
Camptosar (prodrug of SN-38)	0.9% NaCl	NA	11.7	5.8	Small-molecule prodrug	95
LE-SN-38	Liposome + SN-38	Unknown	7–58	3.5–13.6	Liposome nanoparticle	97
Topotecan (camptothecin analogue)	0.9% NaCl	NA	3.0	13.5	Small-molecule drug	96
CT-2106	PG + camptothecin	Unknown	65–99	0.44	Polymer nanoparticle	98
IT-101	Cyclodextrin-containing polymer + camptothecin	30–40	38	0.03	Polymer nanoparticle with extended circulation times	66

*May dissolve upon exposure to blood. NA, not applicable; PEG, polyethylene glycol; PEG–PLA, block copolymer of PEG and poly(L-lactic acid); PG, polyglutamic acid; SN-38, 7-ethyl-10-hydroxycamptothecin.

Haematological toxicity
This includes suppression of red blood cells, white blood cells or platelet counts, and is usually induced by chemotherapeutic agents. Grade 4 toxicity, including severe anaemia, leucopaneia or thrombocytopaenia, requires immediate intervention to prevent life-threatening conditions. For more information see the Common Toxicity Criteria at the US National Cancer Institute web site (see Further information).

Cardiotoxicity
Cardiotoxicity is a toxicity that affects the heart functions. It includes arrhythmia, cardiac pumping dysfunction and eventually heart failure when it develops in severity. Grade 4 (severe) cardiotoxicity is associated with the life-threatening condition of arrhythmia or heart failure. For more information see the Common Toxicity Criteria at the US National Cancer Institute web site (see Further information).

Other polymer–small-molecule conjugates have completed several clinical programmes. For example, N-(2-hydroxypropyl)methacrylamide (HPMA) polymer conjugates demonstrated EPR-mediated tumour targeting[56], and improved antitumour efficacy in animal studies[57]. HPMA-polymer-Gly-Phe-Leu-Gly-doxorubicin (PK1; FCE28068) entered Phase I studies with a dosing schedule of once every 3 weeks at increasing doses up to a MTD of 320 mg per m[2] (doxorubicin equivalent), which is fourfold to fivefold higher than the usual dose of doxorubicin[58]. The dose-limiting toxicities were neutropaenia and mucositis, but the cumulative doses reached 1,680 mg per m[2] without observation of cardiotoxicity. Antitumour activity was observed in anthracycline-resistant breast cancer, NSCLC and colorectal cancer. A clinical pharmacokinetic study showed prolonged plasma circulation[58,59]. The HPMA conjugates of doxorubicin (PK1), paclitaxel (PNU166945), camptothecin (MAG-CPT) and platinum-containing drugs (AP5280 and AP5346) have been tested in the clinic, but at this time only the platinum-containing conjugate AP5346 remains in clinical development. The other conjugates have either shown unacceptable toxicity or disappointing efficacy[60].

NK911, a polymeric micelle nanoparticle, was designed for the enhanced delivery of doxorubicin. NK911 was given intravenously to patients with solid tumours every 3 weeks in a Phase I trial[61]. The starting dose was 6 mg per m[2] (doxorubicin equivalent) with dose escalation. A total of 23 patients were enrolled and a MTD of 67 mg per m[2] and a dose-limiting toxicity of neutropaenia were observed. Among these 23 patients, a partial response was observed in a patient with pancreatic cancer. The recommended Phase II dose was determined to be 50 mg per m[2] every 3 weeks.

SP1049C, a novel anticancer agent containing doxorubicin and two non-ionic pluronic block co-polymers, was designed to increase efficacy compared with doxorubicin. In a Phase I study, the pharmacokinetic profile of SP1049C showed a slower clearance than has been reported for free doxorubicin[62]. A subsequent Phase II study was conducted with SP1049C. Among 19 eligible patients with adenocarcinoma of the oesophagus, nine partial responses (47%) and eight stable diseases (42%) were achieved[63]. Haematological toxicities (grade 3 to 4) included neutropaenia (62%) and anaemia (5%), resulting in nine patients (43%) having the dose reduced to 55mg per m[2]. Grade 1 cardiotoxicity was seen in four (19%) patients. The investigators concluded that SP1049C appears to be active in monotherapy in this group of patients and combination studies with other active agents are recommended.

Genexol-PM[64,65], a polymeric micelle containing paclitaxel, has just received approval in South Korea and represents the first such nanoparticle therapeutic to be approved for the treatment of cancer. Genexol-PM is currently in Phase II trials in the United States. In a completed Phase I trial, of the 21 patients treated there were three partial responses (14%) and two of these patients were refractory to prior taxane therapy[64]. Similar to Abraxane, Genexol-PM does not require the use of pre-medications that are normally required with Taxol. A Phase II trial with patients with metastatic breast cancer has been reported for Genexol-PM administered at 300 mg per m² every 3 weeks[65]. Of the 39 patients that were evaluated, there were 5 complete responses (13%), 19 partial responses (49%), 13 stable diseases (33%) and 2 with progressive diseases (5%). However, because of grade 3 neuropathy, 17 patients had to have dose reductions.

Some of the newer multifunctional nanoparticles, such as IT-101, a conjugate of camptothecin and a cyclodextrin-based polymer, can have greatly extended circulation times (shown both in animals and in humans[66]), enter tumour cells and allow slow release of the drug[67]. Initial results from a Phase I clinical trial showed that several patients receiving IT-101, at doses compatible with a high quality of life, had long-term stable disease (approximately 1 year and more)[66]. These results and others are suggestive that agents shown to be active against tumours that are resistant to the drug via multidrug-resistant-mediated pump mechanisms[67] in animal experiments, may also be active in drug-resistant tumours in humans (discussed below).

Multifunctional nanoparticles such at IT-101 can provide for significantly reduced side effects without the generation of new toxicities compared with the drug that is contained within them. Because of the low side-effect profile of IT-101, it will be tested in a Phase II clinical trial in ovarian cancer as maintenance therapy (ClinicalTrials.gov; see Further information). IT-101 will be dosed in women who normally 'wait and watch' for disease progression to occur after chemotherapy to test whether or not the time to disease progression can be prolonged. If successful, this will open a new paradigm for nanoparticle therapeutics based on their potential to provide low toxicity and high efficacy.

Nanoparticles with targeting ligands
The nanoparticles listed in TABLE 2 utilize passive targeting to reach tumours. That is, it is thought that the leaky vasculature of tumours allows the nanoparticles to extravasate, whereas the normal vasculature does not (this property is involved in the altered biodistribution of nanoparticles compared with the drug molecule). Ultimately, active targeting via the inclusion of a targeting ligand on the nanoparticles is envisioned to provide the most effective therapy. TABLE 3 lists the few ligand-targeted therapeutics that are either approved or in (was in) the clinic.

PK2 (FCE28069), a HPMA-polymer-Gly-Phe-Leu-Gly-doxorubicin conjugate that also contains the sugar galactosamine, was the first ligand-targeted nanoparticle to reach the clinic. The galactose-based ligand was used to target the asialoglycoprotein receptor (ASGPR), which is expressed on hepatocytes, in the hope that its high expression is retained on primary liver cancer cells.

However, as ASGPR is also expressed on healthy hepatocytes, the targeted nanoparticles accumulated in normal liver cells as well as in the tumour. In a Phase I trial, PK2 had a MTD of 160 mg per m² (doxorubicin equivalent dose), with dose-limiting toxicities that are typical of anthracyclines[69]. Of the 23 patients with primary hepatocellular carcinoma, two had progressive diseases, lasting 26 and 47 months, a third showed a reduction of tumour volume, and 11 had stable diseases[70]. Concentrations of drug in liver were 15–20% of the administered dose after 24 hours[70] and the concentrations in the tumour were 12–50-fold higher than would have been achieved through free doxorubicin.

At present, the only targeted nanoparticles in the clinic are MBP-426, which contains the cytotoxic platinum-based drug oxaliplatin in a liposome; SGT-53, a liposome containing a plasmid coding for the tumour suppressor p53; and CALAA-01, a polymer–siRNA composite. MCC-465, a targeted doxorubicin-containing liposome, does not appear to be currently in use. These nanoparticles all target the transferrin receptor, which is known to be upregulated in many types of cancer[7].

Additionally, targeted nanoparticles can have active mechanisms for the intracellular release of the therapeutic agent. For example, CALAA-01 is a targeted nanoparticle that has high drug (siRNA) payload per targeting ligand, proven multivalent binding to cancer cell surfaces and an active drug (siRNA) release mechanism that is triggered upon the recognition of intracellular localization by pH decline below a value of 6.0 (which occurs in the endocytic pathway)[5,9,11,71].

As mentioned above, recent work comparing non-targeted and targeted nanoparticles (lipid-based[8] or polymer-based[9] nanoparticles) has shown that the primary role of the targeting ligands is to enhance cellular uptake into cancer cells and to minimize the accumulation in normal tissue. This behaviour suggests that the colloidal properties of nanoparticles will determine their biodistribution, whereas the targeting ligand serves to increase the intracellular uptake in the target tumour. If this turns out to be the case in general, then the targeting ligand affinity and surface density on the nanoparticle will determine cellular uptake. Recently, Zhou et al. showed that the affinity–density relationships for nanoparticles can be determined[72]. The density of single-chain Fv antibody fragments was important for nanoparticle uptake and high affinities were not necessary if high densities were used. The results clearly demonstrated that high-density, low-affinity antibody fragments can provide uptake into cancer cells that is not increased when the affinity is increased. Although the specific affinities and densities are probably a function of cell-surface receptor densities, this work illustrates the importance of multivalency of the targeted nanoparticles discussed here.

Targeting efflux-pump-mediated resistance?
A major clinical obstacle that limits the efficacy of cancer therapeutics is the resistance of cancer cells to a multitude of chemotherapeutic and biological agents, known as multidrug resistance (MDR)[73,74]. MDR can

---

Neuropathy
Neuropathy is a disorder of the nervous system that includes dysfunction of cranial, motor and sensory nerves. When grade 3 or 4 neuropathy is developed, it can significantly jeopardize normal functions. Severe neuropathy includes paralysis, paraesthesia and disabling cognitive impairments. For more information see the Common Toxicity Criteria at the US National Cancer Institute web site (see Further information).

Table 3 | Examples of ligand-targeted therapeutic agents

Name	Targeting agent	Therapeutic agent	Status	Comments	Refs
Gemtuzumab ozogamicin (Mylotarg; UCB/Wyeth)	Humanized anti-CD33 antibody	Calicheamicin	Approved	Antibody–drug conjugate	99
Denileukin diftitox (Ontak; Ligand Pharmaceuticals/Eisai)	Interleukin 2	Diphtheria toxin fragment	Approved	Fusion protein of targeting agent and therapeutic protein	100
Ibritumomab tiuxetan (Zevalin; Cell Therapeutics)	Mouse anti-CD20 antibody	^{90}Yttrium	Approved	Antibody–radioactive element conjugate	101
Tositumomab (Bexxar; GlaxoSmithKline)	Mouse anti-CD20 antibody	^{131}Iodine	Approved	Antibody–radioactive element conjugate	101
FCE28069 (PK2)	Galactose	Doxorubicin	Phase I (stopped)	Small-molecule targeting agent conjugated to polymer nanoparticle	102
MCC-465	F(ab')$_2$ fragment of human antibody GAH	Doxorubicin	Phase I	Liposome nanoparticle containing antibody fragment targeting agent	103
MBP-426	Transferrin	Oxaliplatin	Phase I	Liposome nanoparticle containing human transferrin protein targeting agent	104,105
SGT-53	Antibody fragment to transferrin receptor	Plasmid DNA with p53 gene	Phase I	Liposome nanoparticle containing antibody fragment targeting agent	106
CALAA-01	Transferrin	Small interfering RNA	Phase I	Polymer-based nanoparticle containing human transferrin protein targeting agent	71,107

be caused by physiological barriers (non-cellular-based mechanisms), or by alterations in the biological and biochemical characteristics of cancer cells (cellular mechanisms). In the first case, non-cellular drug resistance mechanisms can be due to poorly vascularized tumour regions that greatly reduce drug access to the tumour tissues and thus protect cancerous cells from drug-induced cytotoxicity. Furthermore, high interstitial pressure and low microvascular pressure may also impede extravasation of drug molecules. In the second case, resistance of tumours to therapeutic intervention can be due to cellular mechanisms, including alteration of specific enzyme systems for drug metabolism, reduction of apoptotic activity, induction of the cellular repair system, mutation of the drug target, or increasing drug efflux in tumour cells.

Among these mechanisms, changes in the drug-efflux pump are the best known and most extensively investigated. The discovery that P-glycoprotein (P-gp) mediates active efflux of chemotherapeutic drugs from tumour cells initiated this line of reasoning[75,76]. P-gp is a product of the *MDR1* gene and is a 170-kDa transmembrane glycoprotein that functions as a transporter or efflux pump (it is one of the ATP-binding cassette (ABC) proteins) that removes drug out of cells. To date, numerous inhibitors of ABC transporters (including P-gp) have been investigated as potential anticancer agents[77]; however, the results have been disappointing.

Alternative strategies for overcoming drug resistance could be based on systems that allow selective drug accumulation in tumour tissues, tumour cells or even compartments of tumour cells without increased systemic toxicity. This might be provided by nanoparticle-based drugs because they enter cells by endocytosis (FIGS 1,3). Also, by choosing an appropriate nanoparticle formulation, it is possible to protect a drug from the acidic (or other degrading entities such as nucleases for oligonucleotides) microenvironment it encounters before entering into tumour cells. For example, doxorubicin-loaded poly(alkyl cyanoacrylate) (PACA) nanoparticles are able to penetrate cells without being recognized by P-gp[78,79]. Using PACA nanoparticles, it has been demonstrated that the MDR of P388 leukaemia cells in culture was partially overcome. PEG-coated PACA nanoparticles were prepared from a poly(PEGcyanoacrylate-co-hexadecyl cyanoacrylate) co-polymer[79]. In murine cancer models, these nanoparticles circulated longer in the blood stream, whereas their uptake by the liver was reduced. An increased accumulation of the drug in tumour tissue was observed when the drug was administered in the form of PEG-coated PACA nanoparticles[80]. Additionally, Schluep *et al.* showed that IT-101 administered systemically in mice can overcome P-gp resistance in mouse tumour xenografts[67]. This was further evaluated in mice bearing six different xenografts — LS 174T and HT29 colorectal cancer; H1299 NSCLC;

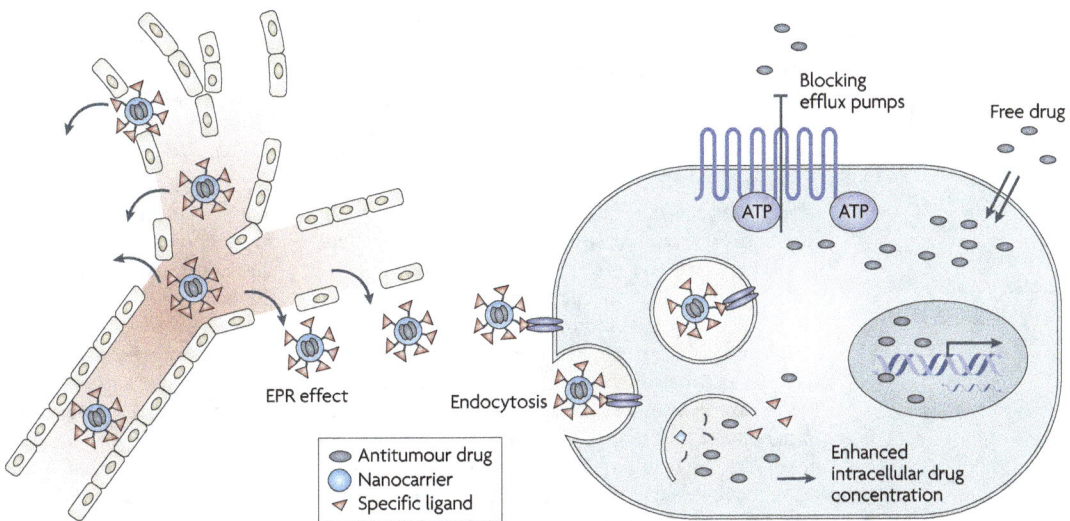

Figure 3 | **Nanoparticles can overcome surface efflux pump mediated drug resistance.** Efflux-pump mediated resistance is based on the rapid elimination of free drug that enters the cell. Drugs that inhibit efflux pumps are in development, but have had limited success to date. Nanoparticle agents are designed to utilize the enhanced permeability and retention (EPR) effect to exit blood vessels in the tumour, to target surface receptors on tumour cells, and to enter tumour cells by endocytosis before releasing their drug payloads. This method of delivery allows for high intracellular drug concentrations that can overcome efflux-pump mediated drug resistance.

H69 small-cell lung cancer; Panc-1 pancreatic cancer; MDA-MB-231 breast cancer — and one disseminated xenograft (TC71-luc Ewing's sarcoma). In all cases, a single treatment cycle of three weekly doses of IT-101 resulted in significant antitumour effects. Complete tumour regression was observed in some of the animals bearing H1299 tumours, and in the majority of the animals with disseminated Ewing's sarcoma[67]. Furthermore, IT-101 was shown to be effective in a number of tumours (for example, HT29 colorectal tumours), which are resistant to treatment with irinotecan[67]. This is consistent with the hypothesis that polymeric drug conjugates may be able to overcome certain kinds of MDR (FIG. 3).

As strategies using targeted nanoparticles take advantage of their binding to cell-surface receptors that are then endocytosed, this approach has also been applied to overcome drug resistance. Folate-receptor-targeted, pH-sensitive polymeric micelles containing doxorubicin[81], transferrin-conjugated paclitaxel nanoparticles[82] and transferrin-ligated liposomes containing oxaliplatin[83] all exhibited greater antitumour activity than the respective free drugs in drug-resistant mouse models. Overall, these studies show that nanoparticles of different types that contain small-molecule drugs and targeting moieties can outperform non-targeted nanoparticles.

There is no conclusive proof that nanoparticles have bypassed surface pump mediated resistance in humans, although clinical results that report partial responses in patients[65,68,84] who had previously failed drug therapies are suggestive of this possibility. For example, patients who were refractory to prior taxane therapy (for example, NSCLC with paclitaxel/carboplatin and ovarian cancer with paclitaxel/carboplatin and paclitaxel regimens) experienced objective responses after treatments with Genexol-PM, a polymeric micelle nanoparticle containing paclitaxel, or XYOTAX (CT-2103), a PG–paclitaxel conjugate[64,68].

Achievements and future challenges

Nanoparticles provide opportunities for designing and tuning properties that are not possible with other types of therapeutics, and as more clinical data become available (see also reviews on polymer–drug conjugates[85] and polymeric micelles[86] for further clinical data), the nanoparticle approach should improve further as the optimal properties are elucidated. As illustrated by the agents in TABLES 1–3, nanoparticle-based therapeutics are evolving, and newer, more sophisticated multifunctional nanoparticles are reaching the clinic. Results from these trials are already fuelling enthusiasm for this type of therapeutic modality.

Some of the important features of nanoparticles observed in preclinical studies have been confirmed in humans, such as extended pharmacokinetic data with sub-100 nm particles[66]. As discussed above, there are also suggestions that pump-mediated MDR might be overcome, and side effects significantly reduced (without the emergence of new side effects) while providing improved efficacy. These combined features may allow for new therapeutic strategies such as maintenance therapy.

Although there are numerous positive features of nanoparticle therapeutics for cancer, there are also issues of concern. First, while size can provide useful features such as large payloads and accommodation of multiple targeting ligands, it also can be a detriment.

At present, it remains unknown how nanoparticles move through tumour tissue once they have localized into the tumour area. Tumour penetration is important and especially so when the nanoparticles are designed to carry the drug molecules into the cancer cells before release. Much further work is required to understand how nanoparticles function in humans as early claims of some nanoparticle function are now being called into question. For example, with Abraxane there is evidence to suggest that the proposed nanoparticle delivery may not be the true mechanism leading to enhanced amounts of drug in tumours[87].

Second, there are valid concerns about nanoparticle toxicity, as little is known about how nanoscale entities behave in humans. The size and surface properties of nanoparticles can give them access to locations that are not available to larger particles. Surface properties also affect biodistribution through mechanisms such as nonspecific binding to proteins in the blood, removal by macrophages and by causing local disturbances in barriers that would otherwise limit their access. An example of the latter phenomenon was recently published: in this study neutral and slightly negatively charged nanoparticles did not alter the integrity of the blood–brain barrier in rats, whereas highly charged nanoparticles did regardless of whether they were positively or negatively charged[88]. Studies of this type suggest that further work is necessary in order to fully define the biocompatibility of nanoparticles in humans. The careful analysis of toxicities of nanostructures in animal models revealed no detrimental effects for some (for example, silica coated magnetic 50 nm particles[89]) but toxicity for others (for example, carbon nanotubes[90]). As expected, the size and surface properties of the nanoparticles dictate their behaviour and more data are necessary in order to develop understanding of their structure–property relationships. Nevertheless, some nanoparticles described in this article have passed rigorous toxicity testing for regulatory approvals and have years of experience in humans. Although each new nanostructure will need to be tested, there is good reason to believe that nanoparticles can ultimately be used in humans as effective systemic medicines and imaging agents[91,92]. As more biocompatibility data become available, further understanding of what is needed to tune the size and surface properties of nanoparticles to provide safety will aid the creation of new, more effective nanomedicines for systemic use[93].

Third, there are important commercial and regulatory challenges to be tackled with the emerging generation of more complex nanoparticles, in part owing to their multicomponent nature. Such nanoparticles are likely to be difficult and expensive to manufacture at large scale with appropriate quality. However, some highly complex nanoparticles have reached the clinic. For example, CALAA-01 is a four-component system that assembles into a highly multifunctional, targeted nanoparticle that contains siRNA. This multicomponent system is now in clinical studies and this example shows that complex nanoparticles can be manufactured at cGMP and satisfy regulatory requirements, at

least for the initiation of Phase I trials. It remains to be seen whether nanoparticles of this complexity can reach the market. In addition to supporting the cost of development, intellectual property costs could be higher because so many components are needed to create the nanoparticle and each might have multiple intellectual property licenses for use. Given these barriers to commercialization, it seems that lack of sufficient financial support could be an issue, and in addition it is likely that approved products will be expensive because of these issues.

There are also numerous efforts focused on combining imaging and therapeutic agents within the same particle. Although there are situations in which this combination might be useful, there are numerous others were this would not. For example, imaging is not necessary every time therapy is administered, especially if this is daily or even more frequently. To carry along an expensive imaging agent and not use it is not a particularly good idea with the rising costs of medicines. Additionally, there are significant regulatory and developmental issues that make the concept of commercializing a combined agent daunting. An alternative, appropriate methodology for these situations is to create individual nanoparticles for imaging and therapy in which the size and surface properties are essentially the same between the two nanoparticle types. The closest analogue to this combination is antibodies possessing a radionucleotide for imaging and another for therapy. As the size and surface properties define the biodistribution, the imaging agent should localize similarly to the therapeutic agent. This is a general strategy that is achievable using nanoparticles that can use numerous types of therapeutics and imaging modalities. One can imagine nanoparticle imaging agents that provide information on intracellular targets. The molecular target of the disease could be verified to exist in a patient before treatment, and as the observation was made via a nanoparticle with the same size and surface properties of the therapeutic particle, the therapy would be expected to reach the target. This combination will allow personalized medicine in the sense that treatment does not have to occur until the target is known to exist in the patient. Also, follow-up imaging can be performed to verify that the target has been reached and that the therapy is working.

There is no doubt that nanoparticle therapeutics with increasing multifunctionality will exist in the future. As newer and more complex nanoparticle systems appear, better methodologies to define biocompatibility will need to be created, especially those that can assess intracellular biocompatibility. While the details of issues regarding scale-up and cGMP production are not often discussed, sophisticated nanoparticles such as CALAA-01 show that efforts towards overcoming cGMP and regulatory hurdles is progressing. Although many challenges exist for the translation of nanoparticles that are currently research tools into approved products for patients, their potential advantages should drive their successful development, and the continuing emergence of a new class of anticancer therapies.

1. Venturoli, D. & Rippe, B. Ficoll and dextran vs. globular proteins as probes for testing glomerular permselectivity: effects of molecular size, shape, charge and deformability. *Am. J. Physiol.* **288**, F605–F613 (2005).
2. Matsumura, Y. & Maeda, H. A new concept of macromolecular therapies in cancer chemotherapy: mechanism of tumortropic accumulation of proteins and the antitumor agent SMANCS. *Cancer Res.* **6**, 6387–6392 (1986).
3. Dreher, M. R. *et al.* Tumor vascular permeability, accumulation, and penetration of macromolecular drug carriers. *J. Natl Cancer Inst.* **98**, 335–344 (2006).
4. Nomura, T., Koreeda, N., Yamashita, F., Takakura, Y. & Hashida, M. Effect of particle size and charge on the disposition of lipid carriers after intratumoral injection into tissue-isolated tumors. *Pharm Res.* **15**, 128–132 (1998).
5. Hu-Lieskovan, S., Heidel, J. D., Bartlett, D. W., Davis, M. E. & Triche, T. J. Sequence-specific knockdown of EWS-FLI1 by targeted, nonviral delivery of small interfering RNA inhibits tumor growth in a murine model of metastatic Ewing's sarcoma. *Cancer Res.* **65**, 8984–8992 (2005).
6. Chen, M. Y. *et al.* Surface properties, more than size, limiting convective distribution of virus-sized particles and viruses in the central nervous system. *J. Neurosurg.* **103**, 311–319 (2005).
7. Gatter, K. C. *et al.* Transferrin receptors in human tissues: their distribution and possible clinical relevance. *J. Clin. Pathol.* **36**, 539–545 (1983).
8. Kirpotin, D. B. *et al.* Antibody targeting of long-circulating lipidic particles does not increase tumor localization but does increase internalization in animal models. *Cancer Res.* **66**, 6732–6740 (2006).
9. Bartlett, D. W., Su, H., Hildebrandt, I. J., Weber, W. A. & Davis, M. E. Impact of tumor-specific targeting on the biodistribution and efficacy of siRNA nanoparticles measured by multimodality *in vivo* imaging. *Proc. Natl Acad. Sci. USA* **104**, 15549–15554 (2007).
10. Popielarski, S. R., Hu-Lieskovan, S., French, S. W., Triche, T. J. & Davis, M. E. A nanoparticle-based model delivery system to guide the rational design of gene delivery to the liver. 2. *In vitro* and *in vivo* uptake results. *Bioconjug. Chem.* **16**, 1071–1080 (2005).
11. Bartlett, D. W. & Davis, M. E. Physicochemical and biological characterization of targeted, nucleic acid-containing nanoparticles. *Bioconjug. Chem.* **18**, 456–468 (2007).
12. Song, E. *et al.* Antibody mediated *in vivo* delivery of small interfering RNAs via cell-surface receptors. *Nature Biotech.* **23**, 709–717 (2005).
13. Hong, S. *et al.* The binding avidity of a nanoparticle-based multivalent targeted drug delivery platform. *Chem. Biol.* **14**, 107–115 (2007).
14. Montet, X., Funovics, M., Montet-Abou, K., Weissleder, R. & Josephson, L. Multivalent effects of RGD peptides obtained by nanoparticle display. *J. Med. Chem.* **49**, 6087–6093 (2006).
15. Carlson, C. B., Mowery, P., Owen, R. M., Dykhuizen, E. C. & Kiessling, L. L. Selective tumor cell targeting using low-affinity, multivalent interactions. *ACS Chem. Biol.* **2**, 119–127 (2007).
16. Pommier, Y. Camptothecins and topoisomerase I: a foot in the door. Targeting the genome beyond topoisomerase I with camptothecins and novel anticancer drugs: importance of DNA replication, repair and cell cycle check points. *Curr. Med. Chem. Anticancer Agents* **4**, 429–434 (2004).
17. Bartlett, D. W. & Davis, M. E. Insights into the kinetics of siRNA-mediated gene silencing from live-cell and live-animal bioluminescent imaging. *Nucl. Acid Res.* **34**, 322–333 (2006).
18. Bartlett, D. W. & Davis, M. E. Effect of siRNA nuclease stability on the *in vitro* and *in vivo* kinetics of siRNA-mediated gene silencing. *Biotech. Bioeng.* **97**, 909–921 (2007).
19. Zamboni, W. C. Liposomal, nanoparticle, and conjugated formulation of anticancer agents. *Clin. Cancer Res.* **11**, 8230–8234 (2005).
20. Rahman, A. M., Yusuf S. W. & Ewer, M. S. Anthracycline-induced cardiotoxicity and the cardiac-sparing effect of liposomal formulation. *Int. J. Nanomedicine* **2**, 567–583 (2007).
21. Batist, G. Cardiac safety of liposomal anthracyclines. *Cardiovasc. Toxicol.* **7**, 72–74 (2007).
22. Uziely, B. *et al.* Liposomal doxorubicin: antitumor activity and unique toxicities during two complementary Phase I studies. *J. Clin. Oncol.* **13**, 1777–1785 (1995).
23. Boddy, A. V. *et al.* A phase I and pharmacokinetic study of paclitaxel poliglumex (XYOTAX), investigating both 3-weekly and 2-weekly schedules. *Clin. Cancer Res.* **11**, 7834–7840 (2005).
24. Sutton, D., Nasongkla, N., Blanco, E. & Gao, J. Functionalized micellar systems for cancer targeted drug delivery. *Pharm. Res.* **24**, 1029–1046 (2007).
25. O'Brien, M. E. Single-agent treatment with pegylated liposomal doxorubicin for metastatic breast cancer. *Anticancer Drugs* **19**, 1–7 (2008).
26. Green, A. E. & Rose P. G. Pegylated liposomal doxorubicin in ovarian cancer. *Int. J. Nanomedicine* **1**, 229–239 (2006).
27. Chanan-Khan, A. A. & Lee, K. Pegylated liposomal doxorubicin and immunomodulatory drug combinations in multiple myeloma: rationale and clinical experience. *Clin. Lymphoma Myeloma* **7** (Suppl. 4), S163–S169 (2007).
28. Sparreboom, A. *et al.* Comparative preclinical and clinical pharmacokinetics of a Cremophor-free, nanoparticle albumin-bound paclitaxel (ABI-007) and paclitaxel formulated in Cremophor (Taxol). *Clin. Cancer Res.* **11**, 4136–4143 (2005).
29. Gradishar, W. J. *et al.* Phase III trial of nanoparticle albumin-bound paclitaxel compared with polyethylated castor oil-based paclitaxel in women with breast cancer. *J. Clin Oncol.* **23**, 7794–7803 (2005).
30. Henderson, I. C. & Bhatia, V. Nab-paclitaxel for breast cancer: a new formulation with an improved safety profile and greater efficacy. *Expert Rev. Anticancer Ther.* **7**, 919–943 (2007).
31. Harris, J. M. & Chess, R. B. Effect of peglyation on pharmaceuticals. *Nature Rev. Drug Discov.* **2**, 214–221 (2003).
32. Fuertges, F. & Abuchowski, A. The clinical efficacy of poly(ethylene glycol) modified proteins. *J. Control. Release* **11**, 139–148 (1990).
33. Graham, M. L. Pegaspargase: a review of clinical studies. *Adv. Drug Deliv. Rev.* **55**. 1293–1302 (2003).
34. Ho, D. H. *et al.* Clinical pharmacology of polyethylene glycol–asparaginase. *Drug Metab. Disposit.* **14**, 349–352 (1986).
35. Kurtzberg, J., Moore, J. O. Scudiery, D. & Franklin, A. A phase II study of polyethylene glycol (PEG) conjugated L-asparaginase in patients with refractory acute leukaemias. *Proc. Am. Assoc. Cancer Res.* **29**, 213 (1988).
36. Abshire, T. C., Pollock, B. H., Billett, A. L., Bradley, P. & Buchanan, G. R. Weekly polyethylene glycol conjugated-asparaginase compared with biweekly dosing produces superior induction remission rates in childhood relapsed acute lymphoblastic leukemia: a Pediatric Oncology Group Study. *Blood* **96**, 1709–1715 (2000).
37. Cheng, P. N. *et al.* Pegylated recombinant human Arginase (rhArg-peg 5000Mw) has *in vitro* and *in vitro* anti-proliferative potential and apoptotic activities in human hepatocellular carcinoma (HCC). *J. Clin. Oncol.* **26** (Suppl. 16), 3179 (2005).
38. Delman, K. A. *et al.* Phase I/II trial of pegylated arginine deiminase (ADI-PEG20) in unresectable hepatocellular carcinoma. *J. Clin. Oncol.* **26** (Suppl. 16), 4139 (2005).
39. Molineux, G. The design and development of pegfilgrastim (PEG-rmetHuG-CSF, Neulasta). *Curr. Pharm. Des.* **10**, 1235–1244 (2004).
40. Tanaka, H. Satake-Ishikawa, R., Ishikawa M., Matsuki, S. & Asano, K. Pharmacokinetics of recombined human granulocyte colony-stimulating factor conjugated to polyethylene glycol in rats. *Cancer Res.* **51**, 3710–3714 (1991).
41. Wang, Y. S. *et al.* Structural and biological characterisation of pegylated recombinant interferon α2b and its therapeutic implications. *Adv. Drug Del. Rev.* **54**, 547–570 (2002).
42. Fiaherty, L., Heilbrun, L., Marsack, C. & Vaishampayan U. N. Phase II trial of pegylated interferon (Peg-Intron) and thalidomide (Thal) in pretreated metatastic malignant melanomal. *J. Clin. Oncol.* **23** (Suppl. 16), 7562 (2005).
43. Bukowski, R. *et al.* Pegylated interferon α-2b treatment for patients with solid tumors: a phase I/II study. *J. Clin. Oncol.* **20**, 3841–3849 (2002).
44. Holmes, F. A. *et al.* Comparable efficacy and safety profiles of once-per-cycle pegfilgrastim and daily injection filgrastim in chemotherapy-induced neutropenia: a multicenter dose-finding study in women with breast cancer. *Ann. Oncol.* **13**, 903–909 (2002).
45. Posey, J. A. 3rd *et al.* Phase I study of weekly polyethylene glycol-camptothecin in patients with advanced solid tumors and lymphoma. *Clin. Cancer Res.* **11**, 7866–7871 (2005).
46. Rowinsky, E. K. *et al.* A phase I and pharmacokinetic study of pegylated camptothecin as a 1-hour infusion every 3 weeks in patients with advanced solid malignancies. *J. Clin. Oncol.* **21**, 148–157 (2003).
47. Scott, L. C. *et al.* A phase II study of pegylated-camptothecin (pegamotecan) in the treatment of locally advanced and metastatic gastric and gastro-oesophageal junction adnenocarcinoma. *Cancer Chemother. Pharmacol.* 9 April 2008 (doi:10.1007/s00280-008-0746-2).
48. Shaffer, S. A. *et al.* Proteolysis of Xyotax™ by lysosomal cathepsin B; metabolic profiling in tumor cells using LC-MS. *Eur. J. Cancer* **38** (Suppl. 7), S129 (2002).
49. Singer, J. W. *et al.* Poly-(L)-glutamic acid-paclitaxel (CT-2103) [XYOTAX], a biodegradable polymeric drug conjugate: characterization, preclinical pharmacology, and preliminary clinical data. *Adv. Exp. Med. Biol.* **519**, 81–99 (2003).
50. Singer, J. W. *et al.* Paclitaxel poliglumex (XYOTAX; CT-2103) [XYOTAX™]: an intracellularly targeted taxane. *Anticancer Drugs* **16**, 243–254 (2005).
51. Todd, R. *et al.* Phase I and pharmacological study of CT-2103, a poly(L-glutamic acid)-paclitaxel conjugate. *Proc. AACR-NCI-EORTC 12th Int. Conf. Mol. Targets Cancer Therapeut. Discov. Develop. Clin. Valid.* Abstract 439 (2001).
52. Langer, C. J. *et al.* Paclitaxel poliglumex (PPX)/ carboplatin vs. Paclitaxel/carboplatin for the treatment of PS2 patients with chemotherapy-naive advanced non-small cell lung cancer (NSCLC): a phase III study. *J. Clin. Oncol.* **23** (Suppl. 16), LBA7011 (2005).
53. Socinski, M. & Ramalingham, S. XYOTAX in NSCLC and other solid tumors. Emerging evidence on biological differences between men and women: is gender-specific therapy warranted? *Chemotherapy Foundation web site* [online], <http://www.chemotherapyfoundationsymposium.org/meeting_archives/meetingarchives_tcf2005_main.html> (2005).
54. Kremer, M., Judd, J., Rifkin, B., Auszmann, J. & Oursler, M. J. Estrogen modulation of osteoclast lysosomal-enzyme secretion. *J. Cell. Biochem.* **57**, 271–279 (1995).
55. Homsi, J. *et al.* Phase I trial of poly-L-glutamate camptothecin (CT-2106) administered weekly in patient with advanced solid malignancies. *Clin. Cancer Res.* **13**, 5855–5861 (2007).
56. Seymour, L. W. *et al.* Tumouritropism and anticancer efficacy of polymer-based doxorubicin prodrugs in the treatment of subcutaneous murine B16F10 melanoma. *Br. J. Cancer* **70**, 636–641 (1994).
57. Duncan, R. *et al.* Preclinical evaluation of polymer-bound doxorubicin. *J. Control. Release* **19**, 331–346 (1992).
58. Vasey, P. *et al.* Phase I clinical and pharmacokinetic study of PKI (HPMA copolymer doxorubicin) first member of a new class of chemotherapeutics agents: drug–polymer conjugates. *Clin. Cancer Res.* **5**, 83–94 (1999).
59. Thomson, A. H. *et al.* Population pharmacokinetics in phase I drug development: a phase I study of PK1 in patients with solid tumors. *Br. J. Cancer* **81**, 99–107 (1999).
60. Duncan, R. Designing polymer conjugates as lysomsomotropic nanomedicines. *Biochem. Soc. Trans.* **35**, 56–60 (2007).
61. Matsumura, Y. *et al.* Phase I clinical trial and pharmacokinetic evaluation of NK911, a micelle-encapsulated doxorubicin. *Br. J. Cancer.* **91**, 1775–1781 (2004).
62. Danson, S. *et al.* Phase I dose escalation and pharmacokinetic study of pluronic polymer-bound doxorubicin (SP1049C) in patients with advanced cancer. *Br. J. Cancer.* **11**, 2085–2091 (2004).
63. Armstrong, A. *et al.* SP1049C as first-line therapy in advanced (inoperable or metatastic) adenocarcinoma of the oesophagus: a phase II window study. *J. Clin. Oncol.* **24**, 4080 (2006).
64. Kim, T. Y. *et al.* Phase I and pharmacokinetic study of Genexol-PM, a cremophor-free, polymeric micelle-formulated paclitaxel, in patients with advanced malignancies. *Clin. Cancer Res.* **10**, 3708–3716 (2004).
65. Lee K. S. *et al.* Multicenter phase II study of a cremophor-free polymeric micelle-formulated paclitaxel in patients with metastatic breast cancer. *J. Clin. Oncol.* **24**, (Suppl. 18), 10520 (2006).

66. Yen, Y. *et al.* First-in-human phase I trial of a cyclodextrin-containing polymer-camptothecin nanoparticle in patients with various solid tumors. *J. Clin. Oncol.* **25** (Suppl. 18), 14078 (2007).

67. Schluep, T. *et al.* Preclinical efficacy of the camptothecin–polymer conjugate IT-101 in multiple cancer models. *Clin. Cancer Res.* **12**, 1606–1614 (2006).

68. Nemunaitis, J. *et al.* Phase I study of CT-2103, a polymer-conjugated paclitaxel, and carboplatin in patients with advanced solid tumors. *Cancer Invest.* **23**, 671–676 (2005).

69. Seymour, L. W. *et al.* Hepatic drug targeting: Phase I evaluation of polymer bound doxorubicin. *J. Clin. Oncol.* **20**, 1668–1676 (2002).

70. Julyan, P. J. *et al.* Preliminary clinical study of the distribution of HPMA copolymer-doxorubicin bearing galactosamine. *J. Control. Release* **57**, 281–290 (1999).

71. Heidel, J. D. *et al.* Administration in non-human primates of escalating intravenous doses of targeted nanoparticles containing ribonucleotide reductase subunit M2 siRNA. *Proc. Natl Acad. Sci. USA* **104**, 5715–5721 (2007).

72. Zhou, Y. *et al.* Impact of single-chain Fv antibody fragment affinity on nanoparticle targeting of epidermal growth factor receptor-expressing tumor cells. *J. Mol. Biol.* **371**, 934–947 (2007).

73. Gottesman, M. M. Mechanisms of cancer drug resistance. *Annu. Rev. Med.* **53**, 615–627 (2002).

74. Gottesman, M. M., Fojo, T. & Bates, S. E. Multidrug resistance in cancer: role of ATP-dependent transporters. *Nature Rev. Cancer* **2**, 48–58 (2002).

75. Modok, S., Mellor, H. R. & Callaghan, R. Modulation of multidrug resistance efflux pump activity to overcome chemoresistance in cancer. *Curr. Opin. Pharmacol.* **6**, 350–354 (2006).

76. Nobili, S., Landini, I., Giglioni, B. & Mini, E. Pharmacological strategies for overcoming multidrug resistance. *Curr. Drug Targets* **7**, 861–879 (2006).

77. Thomas, H. & Coley, H. M. Overcoming multidrug resistance in cancer: an update on the clinical strategy of inhibiting p-glycoprotein. *Cancer Control* **10**, 159–165 (2003).

78. Pepin, X. *et al.* On the use of ion-pair chromatography to elucidate doxorubicin release mechanism from polyalkylcyanoacrylate nanoparticles at the cellular level. *J. Chromatogr. B Biomed. Sci. Appl.* **702**, 181–191 (1997).

79. Vauthier, C., Dubernet, C., Chauvierre, C., Brigger, I. & Couvreur, P. Drug delivery to resistant tumors: the potential of poly(alkyl cyanoacrylate) nanoparticles. *J. Control. Release* **93**, 151–160 (2003).

80. Peracchia, M. T. *et al.* Stealth PEGylated polycyanoacrylate nanoparticles for intravenous administration and splenic targeting. *J. Control. Release* **60**, 121–128 (1999).

81. Lee, E. S., Na, K. & Bae, Y. H. Doxorubicin loaded pH-sensitive polymeric micelles for reversal of resistant MCF-7 tumor. *J. Control. Release* **103**, 405–418 (2005).

82. Sahoo, S. K., Ma, W. & Labhasetwar, V. Efficacy of transferrin-conjugated paclitaxel-loaded nanoparticles in a murine model of prostate cancer. *Int. J. Cancer* **112**, 335–340 (2004).

83. Suzuki, R. *et al.* Effective anti-tumor activity of oxaliplatin encapsulated in transferrin-PEG-liposome. *Int. J. Pharm.* **346**, 143–150 (2008).

84. Northfelt, D. W. *et al.* Efficacy of pegylated-liposomal doxorubicin in the treatment of AIDS-related Kaposi's sarcoma after failure of standard chemotherapy. *J. Clin. Oncol.* **15**, 653–659 (1997).

85. Li, C. & Wallace, S. Polymer–drug conjugates: recent development in clinical oncology. *Adv. Drug Del. Rev.* **60**, 886–898 (2008).

86. Matsumura, Y. Poly(amino acid) micelle nanocarriers in preclinical and clinical studies. *Adv. Drug Del. Rev.* **60**, 899–914 (2008).

87. Gardner, E. R. *et al.* Randomized crossover pharmacokinetic study of solvent-based paclitaxel and nab-paclitaxel. *Clin. Cancer Res.* **14**, 4200–4205 (2008).

88. Lockman, P. R., Koziara, J. M., Mumper, R. J. & Allen, D. D. Nanoparticle surface charges alter blood–brain barrier integrity and permeability. *J. Drug Target.* **12**, 635–641 (2004).

89. Kim, J. S. *et al.* Toxicity and tissue distribution of magnetic nanoparticles in mice. *Toxicol. Sci.* **89**, 338–347 (2006).

90. Salvador-Morales, C. Complement activation and protein adsorption by carbon nanotubes. *Mol. Immunol.* **43**, 193–201 (2006).

91. Wang, X., Yang, L., Chen, Z. G. & Shin, D. M. Application of nanotechnology in cancer therapy and imaging. *CA Cancer J. Clin.* **58**, 97–110 (2008).

92. Qian, X. *et al.* In vivo tumor targeting and spectroscopic detection with surface-enhanced Raman nanoparticle tags. *Nature Biotech.* **26**, 83–90 (2008).

93. Cho, K., Wang X., Nie, S., Chen Z.G. & Shin, D. M. Therapeutic nanoparticles for drug delivery in cancer. *Clin. Cancer Res.* **14**, 1310–1316 (2008).

94. Schluep, T. *et al.* Pharmacokinetics and biodistribution of the camptohecin-polymer conjugate IT-101 in rats and tumor-bearing mice. *Cancer Chemother. Pharmacol.* **57**, 654–662 (2006).

95. Pfizer. CAMPTOSAR U.S. Physician Prescribing Information. *Pfizer web site* [online], < www.pfizer. com/files/products/uspi_camptosar.pdf > (2008).

96. Herben, V. M., ten Bokkel Huinink, W. W. & Beijnen, J. H. Clinical pharmacokinetics of topotecan. *Clin. Pharmacokinet.* **31**, 85–102 (1996).

97. Kraut, E. H. *et al.* Final results of a phase I study of liposome encapsulated SN-38 (LE-SN38): safety, pharmacogenomics, pharmacokinetics, and tumor response. *J. Clin. Oncol.* **23** (Suppl. 16), 2017 (2005).

98. Daud, A. *et al.* Phase I trial of CT-2106 (polyglutamated camptothecin) administered weekly in patients with advanced solid tumor malignancies. *J. Clin. Oncol.* **24** (Suppl. 18), 2015 (2006).

99. Bross, P. F. *et al.* Approval summary: gemtuzumab ozogamicin in relapsed acute myeloid leukemia. *Clin. Cancer Res.* **7**, 1490–1496 (2001).

100. Kawakami, K., Nakajima, O., Morishita, R. & Nagai, R. Targeted anticancer immunotoxins and cytotoxic agents with direct killing moieties. *ScientificWorldJournal* **6**, 781–790 (2006).

101. Allen, T. M. Ligand-targeted therapeutics in anticancer therapy. *Nature Rev. Cancer* **2**, 750–763 (2002).

102. Duncan, R. Polymer conjugates as anticancer nano-medicines. *Nature Rev. Cancer* **6**, 688–701 (2006).

103. Matsumura, M. *et al.* Phase I and pharmacokinetic study of MCC-465, a doxorubicin (DXR) encapsulated in PEG immunoliposome, in patients with metastatic stomach cancer. *Ann. Oncol.* **15**, 517–525 (2004).

104. MedBiopharm Co., Ltd. Safety study of MBP-426 (liposomal oxaliplatin suspension for injection) to treat advanced or metastatic solid tumors. *ClinicalTrials.gov web site* [online], < http://www.clinicaltrials.gov/ct/ show/NCT00355888 > (2008).

105. Phan, A. *et al.* Open label phase I study of MBP-426, a novel formulation of oxaliplatin, in patients with advanced or metastatic solid tumors. *Proc. AACR-NCI-EORTC Int. Conf. Mol. Targets Cancer Therapeut. Discov. Develop. Clin. Valid.* Abstract C115 (2007).

106. SynerGene Therapeutics, Inc. Safety Study of Infusion of SGT-53 to Treat Solid Tumors. *ClinicalTrials.gov web site* [online], < http://www.clinicaltrials.gov/ct2/show/ NCT00470613 > (2008).

107. Calando Pharmaceuticals. Safety Study of CALAA-01 to Treat Solid Tumor Cancers. *ClinicalTrials.gov web site* [online], < http://www.clinicaltrials.gov/ct/show/ NCT00689065 > (2008).

Acknowledgements
This article is partially supported (Z.C. and D.M.S.) by a grant from the US National Cancer Institute, Center for Cancer Nanotechnology of Excellence (U54 CA119338).

FURTHER INFORMATION
ClinicalTrials.gov: http://clinicaltrials.gov
US National Cancer Institute:
http://ctep.cancer.gov/reporting/ctc_archive.html

ALL LINKS ARE ACTIVE IN THE ONLINE PDF

Neuroscience nanotechnology: progress, opportunities and challenges

Gabriel A. Silva

Abstract | Nanotechnologies exploit materials and devices with a functional organization that has been engineered at the nanometre scale. The application of nanotechnology in cell biology and physiology enables targeted interactions at a fundamental molecular level. In neuroscience, this entails specific interactions with neurons and glial cells. Examples of current research include technologies that are designed to better interact with neural cells, advanced molecular imaging technologies, materials and hybrid molecules used in neural regeneration, neuroprotection, and targeted delivery of drugs and small molecules across the blood–brain barrier.

Self-assembly
The self-organization of molecules into supermolecular structures. Self-assembly is triggered by specific chemical or physical variables, such as a change in temperature or concentration, and reflects an energy minimization process.

Nanodevices and nanomaterials can interact with biological systems at fundamental, molecular levels with a high degree of specificity. By taking advantage of this unique molecular specificity, these nanotechnologies can stimulate, respond to and interact with target cells and tissues in controlled ways to induce desired physiological responses, while minimizing undesirable effects. Applications of nanotechnology in basic and clinical neuroscience are only in the early stages of development, partly because of the complexities associated with interacting with neural cells and the mammalian nervous system. Despite this, an impressive body of research is emerging that hints at the potential contributions these technologies could make to neuroscience research.

This review discusses the basic concepts associated with nanotechnology and its current applications in neuroscience. The first section attempts to answer the questions concerning what nanotechnology is and what it encompasses. The second section gives an overview of the main areas of neuroscience nanotechnology research. Specifically, it discusses neuronal adhesion and growth, interfacing and stimulating neurons at a molecular level, imaging and manipulating neurons and glia using functionalized quantum dots, approaches for functional neural regeneration, approaches for neuroprotection and nanotechnologies for crossing the blood–brain barrier (BBB). The third section discusses some of the unique challenges encountered when applying nanotechnology to neural cells and the nervous system, and the tremendous impact this technology might have on neuroscience research. The fourth section attempts to define the roles of neuroscientists in advancing neuroscience nanotechnology. The final

section speculates on the results that might derive from current applications of nanotechnology and the types of applications that might have the earliest impact in neuroscience.

What is nanotechnology?

Nanotechnologies are technologies that use engineered materials or devices with the smallest functional organization on the nanometre scale (that is, one billionth of a metre) in at least one dimension, typically ranging from 1 to ~100 nanometres. This implies that some aspect of the material or device can be manipulated and controlled by physical and/or chemical means at nanometre resolutions, which results in functional properties that are unique to the engineered technology and not shown by its constituent elements. For example, DNA self-assembly can yield DNA nanotube supermolecular structures that form electrically conducting nanowires that have potential for use in nanoelectronic devices[1]. In addition, boron-doped silicon nanowires can function as real-time ultrahigh sensitive detectors for biological and chemical factors[2]: nanowires modified with biotin specifically and accurately detect picomolar concentrations of streptavidin. Nanotechnologies are therefore primarily defined by their functional properties, which determine how they interact with other disciplines. Although the chemical and/or physical make up of a nanomaterial or device is important in the overall technological process, it is secondary to their engineering and functional properties. Considering the two examples described above, DNA does not have the intrinsic capacity to function as an electrically conducting nanowire, and neither boron nor silicon can detect specific chemical

Departments of Bioengineering and Ophthalmology, and the Neurosciences Program, University of California, San Diego, UCSD Jacobs Retina Center 0946, 9415 Campus Point Drive, La Jolla, California 92037-0946, USA. e-mail: gsilva@ucsd.edu
doi:10.1038/nrn1827

Box 1 | Synthetic approaches in nanotechnology

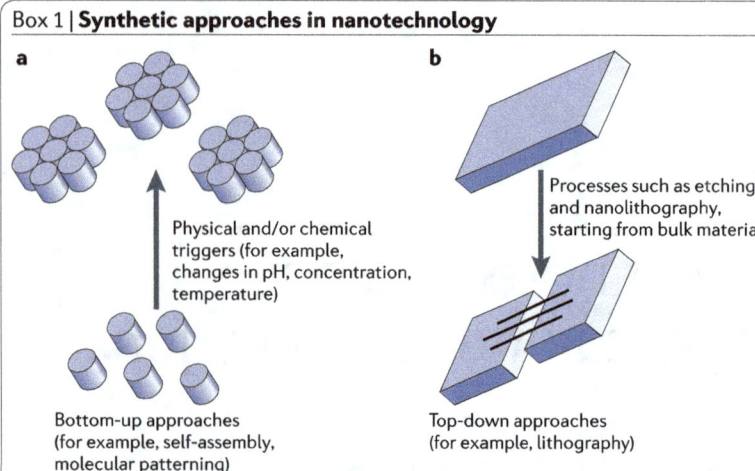

a Physical and/or chemical triggers (for example, changes in pH, concentration, temperature)

Bottom-up approaches
(for example, self-assembly, molecular patterning)

b Processes such as etching and nanolithography, starting from bulk material

Top-down approaches
(for example, lithography)

Different approaches used in the synthesis of nanomaterials and nanodevices can accommodate solid, liquid, and/or gaseous precursor materials. In general, most of these techniques can be classified as bottom-up and top-down approaches, and strategies that have elements of both. Bottom-up approaches (panel **a**) start with one or more defined molecular species, which undergo certain processes that result in a higher-ordered and -organized structure. Examples of bottom-up approaches include systems that self-assemble, a process that is triggered by a local change in a chemical or physical condition. Related techniques include templating and scaffolding methods, such as biomineralization, which rely on backbone structures to support and guide the nucleation and growth of a nanomaterial. Top-down approaches (panel **b**) begin with a bulk material that incorporates nanoscale details, such as nanolithography and etching techniques. Specific examples include dip-pen nanolithography (in which specific molecules are deposited into desired configurations) and electrostatic atomic force nanolithography (in which molecules are moved around to form desired structures). In all cases, the resultant structures have novel engineered chemical and/or physical properties that the original constituent materials do not have. These emergent properties allow controlled interactions of the nanomaterial or device with its target system.

Bottom-up technologies
Materials or devices engineered from constituent elements such as specific molecules that are organized into higher-order functional structures.

Top-down technologies
Materials or devices that are engineered from a bulk material. The various forms of lithography are examples of top-down engineering approaches.

Lithography
The process of producing patterns in bulk materials. The most common forms of lithography are those associated with the production of semiconductor integrated circuits.

species. However, DNA nanowires and boron-doped silicon have gained these novel functional properties. These technologies are described by the new functional outcomes of bringing these components together, not because they had to contain DNA, or boron or silicon. Other materials or devices can be made using the same building blocks with different functional or engineered outcomes, and both electrically conducting nanowires and high-sensitivity chemical sensors can be produced using other chemistry and synthetic approaches. In this functional definition of nanotechnology, it is implicit that this is not a new area of science *per se*, and that the interdisciplinary convergence of basic fields (such as chemistry, physics, mathematics and biology) and applied fields (such as materials science and the various areas of engineering) contributes to the functional outcomes of the technology. In this framework, nanotechnology can be regarded as an interdisciplinary pursuit that involves the design, synthesis and characterization of nanomaterials and devices that have the types of property discussed above. In particular, this engineering definition of nanotechnology is what sets it apart from chemistry. Chemistry is an integral component of nanotechnology, but the two are not synonymous, a point that is a potential source of confusion. Chemistry involves the manipulation of matter at nanometre scales

resulting in chemical products with specific intrinsic properties (for example, a defined melting point, pK_a or charge distribution) that affect how they interact with their environments. As emphasized above, nanotechnologies are primarily defined in terms of their extrinsic engineered functional properties (of which the intrinsic chemistry could be one of many contributory factors that define these properties).

From the neuroscientist's perspective, the most important aspect of nanotechnologies is the application of these technologies to neuroscience questions and challenges[3,4]. There are two key types of nanotechnology application in neuroscience: 'platform nanotechnologies' that can be readily adapted to address neuroscience questions; and 'tailored nanotechnologies' that are specifically designed to resolve a particular neurobiological issue. This is also the case for other areas of biology and medicine. Platform nanotechnologies use materials or devices with unique physical and/or chemical properties that can potentially have wide-ranging applications in different fields. These technologies have considerable promise, and scientists and engineers are constantly looking for new ways to explore the potential of platform nanotechnologies. Tailored nanotechnologies begin with a well-defined biological question, and are developed to specifically address that issue. Owing to the inherent complexity of biological systems in general, and the nervous system in particular, the tailored approach often results in highly specialized technologies that are designed to interact with their target systems in sophisticated and well-defined ways, and so will be better suited to tackle the particular problem than a generic platform technology. However, because tailored nanotechnologies are highly specialized, their broader application to other biological systems could be limited. In many cases, the application of particular nanotechnologies in neuroscience has been derived from what was originally a platform technology, and some tailored technologies can be modified to address other (but usually related) scientific questions. From a synthesis standpoint, nanotechnologies can be classified as variations of bottom-up technologies, such as self-assembly, or top-down technologies, such as lithographic methods (BOX 1). Many nanotechnologies combine aspects of both strategies.

Examples of current work

This section reviews applications of nanotechnology in basic and clinical neuroscience (TABLE 1). It is organized according to the type of application, and compares different nanotechnology approaches applied to similar or related neurobiological or physiological objectives. In general, the discussion provides more breadth than depth to give the reader a broad sense of ongoing research.

The section is divided into two subsections: applications of nanotechnology in basic neuroscience; and applications of nanotechnology in clinical neuroscience. Applications of nanotechnology in basic neuroscience include those that investigate molecular, cellular and physiological processes (FIG. 1). This first subsection discusses three specific areas. First, nanoengineered materials and

Table 1 | Applications of nanotechnologies in neuroscience

Nanotechnology	Applications	Refs
Basic neuroscience		
Molecular deposition and lithographic patterning of neuronal-specific molecules with nanometre resolution	Study of cellular communication and signalling; test systems for drugs and other molecules	5–10
Atomic force microscopy (measures molecularly functionalized surfaces)	Interact, record and/or stimulate neurons at the molecular level	11–13
Functionalized quantum dots	High-resolution spatial and temporal imaging; molecular dynamics and tracking	14–19
Clinical neuroscience		
Self-assembling peptide amphiphile nanofibre networks	Neuronal differentiation from progenitor cells; neural regeneration	20
Derivatives of hydroxyl-functionalized fullerenes (fullerenols)	Neuroprotection mediated by limiting the effects of free radicals following injury	23–26
Poly(ethylene glycol) and polyethylenimine nanogels; poly(butylcyanoacrylate) nanoparticles	Transport of drugs and small molecules across the blood–brain barrier	27–36

aimed at limiting and reversing neuropathological disease states (FIG. 2). This subsection discusses nanotechnology approaches designed to support and/or promote the functional regeneration of the nervous system[20,21]; neuroprotective strategies, in particular those that use fullerene derivatives[22–26]; and nanotechnology approaches that facilitate the delivery of drugs and small molecules across the BBB[27–38].

As with any classification scheme, there might be a sense of forced categorization when classifying nanotechnologies, as some nanotechnologies can be used in more than one area. For example, all three of the basic nanotechnology applications can also contribute to an understanding of neuropathophysiology; and all three of the clinical nanotechnology applications can also increase our understanding of basic molecular and cellular neurobiology. But, for the most part, basic neuroscience applications primarily concern themselves with understanding basic molecular and cellular mechanisms without necessarily considering their potential clinical implications, whereas clinical neuroscience applications are designed to primarily target disease events, and make use of basic molecular and cellular neurobiology only when necessary.

approaches for promoting neuronal adhesion and growth to understand the underlying neurobiology of these processes or to support other technologies designed to interact with neurons *in vivo* (for example, coating of recording or stimulating electrodes)[5–10]. Second, nanoengineered materials and approaches for directly interacting, recording and/or stimulating neurons at a molecular level[11–13]. Third, imaging applications using nanotechnology tools, in particular, those that focus on chemically functionalized semiconductor quantum dots[14–19]. Applications of nanotechnology in clinical neuroscience include research

Applications in basic neuroscience. Molecular deposition and lithographic patterning of neuronal-specific molecules with nanometre resolutions[5–10] are an extension of micropatterning approaches[39–44]. The deposition of proteins and other molecules that promote and support neuronal adhesion and growth on surfaces that do not support these processes enables the geometric selective patterning and growth of neurons (for example, controlled neurite extension). This allows the study of cellular communication and signalling, and provides a test system for investigating the effects of drugs and other molecules. The ability to control this process at nanometre as

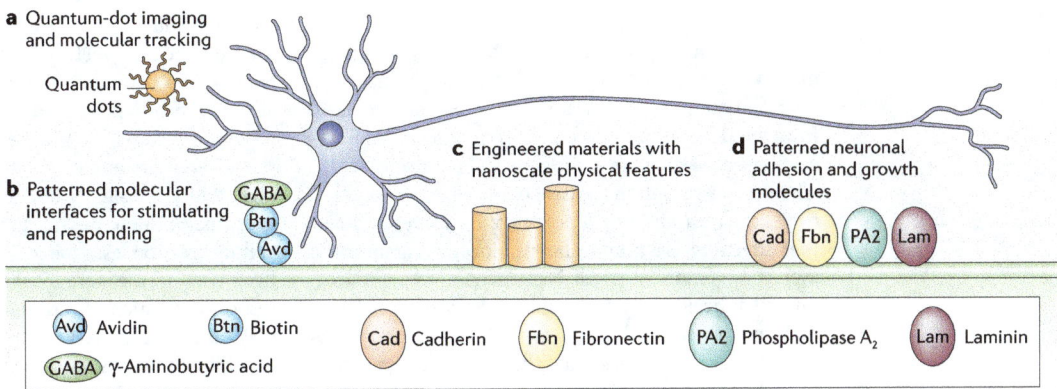

Figure 1 | Applications of nanotechnologies in basic neuroscience. Nanomaterials and nanodevices that interact with neurons and glia at the molecular level can be used to influence and respond to cellular events. In all cases, these engineered technologies allow controlled interactions at cellular and subcellular scales. **a** | Chemically functionalized fluorescent quantum dot nanocrystals used to visualize ligand–target interactions. **b** | Surfaces modified with neurotransmitter ligands to induce controlled signalling. For example, GABA (γ-aminobutyric acid) was immobilized, via an avidin–biotin linkage, to different surfaces to stimulate neurons in predictable (that is, patterned) ways. **c** | Engineered materials with nanoscale physical features that produce ultrastructural morphological changes. **d** | Surfaces and materials functionalized with different neuronal-specific effector molecules, such as cadherin and laminin, to induce controlled cellular adhesion and growth.

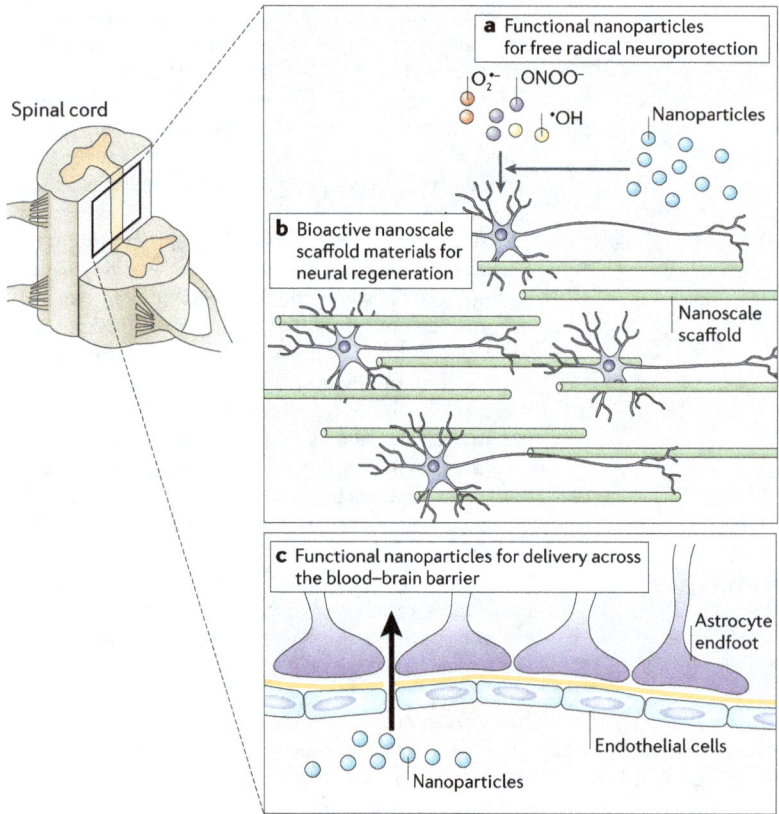

Spinal cord

a Functional nanoparticles for free radical neuroprotection

$O_2^{\cdot-}$ ONOO$^-$

$^\cdot$OH

Nanoparticles

b Bioactive nanoscale scaffold materials for neural regeneration

Nanoscale scaffold

c Functional nanoparticles for delivery across the blood–brain barrier

Astrocyte endfoot

Endothelial cells

Nanoparticles

Figure 2 | Applications of nanotechnology in clinical neuroscience.
Nanotechnology can be used to limit and/or reverse neuropathological disease
processes at a molecular level or facilitate and support other approaches with this
goal. **a** | Nanoparticles that promote neuroprotection by limiting the effects of free
radicals produced following trauma (for example, those produced by CNS secondary
injury mechanisms). **b** | The development and use of nanoengineered scaffold
materials that mimic the extracellular matrix and provide a physical and/or bioactive
environment for neural regeneration. **c** | Nanoparticles designed to allow the
transport of drugs and small molecules across the blood–brain barrier.

the growth of cerebellar neurons[5]. These studies suggest
that bioactive ultrathin layers could coat electrodes
designed for long-term implants to promote cell adhesion
and limit immune responses. In a different approach,
electrodes coated with nanoporous silicon increased
neurite outgrowth from PC12 cells, which are clonally
derived neuronal precursor cells, compared with uncoated
electrodes, and decreased glial responses, thereby limiting
the insulating effects of the glial scar[10]. Coating electrodes
with ultrathin bioactive layers might have other advantages, for example, limiting the increase in the thickness
of the electrodes and thereby minimizing local trauma
due to their insertion and resultant cellular responses.
Other research has shown the effects of nanoscale
physical features on neuronal behaviour. Substantia nigra
neurons cultured on silicon dioxide (SiO_2) surfaces with
different nanoscale topographies had differential cell
adhesion properties[6]. Neurons cultured on surfaces
with physical features (that is, surface roughness)
between 20–70 nm adhered and grew better than neurons
cultured on surfaces with features <10 nm or >70 nm.
These neurons also had normal morphologies and
normal production of tyrosine hydroxylase, a marker of
metabolic activity.

Another emerging area of neuroscience nanotechnology is materials and devices that have been designed to
interact, record and/or stimulate neurons at the molecular level[11–13]. Recent research has demonstrated the
feasibility of functionalizing mica or glass tethered with
the inhibitory neurotransmitter GABA (γ-aminobutyric
acid) and its analogue muscimol (5-aminomethyl-
3-hydroxyisoxazole) through biotin–avidin binding
interactions[12,13] (FIG. 1). The functional integrity of the
bound version of the neurotransmitter, which *in vivo*
functions not as a bound ligand but as a diffusible
messenger across the synaptic cleft, was shown electro-
physiologically *in vitro* by eliciting an agonist response
to cloned GABA$_A$ (GABA type A) and GABA$_C$ receptors in *Xenopus* oocytes[12]. Such sophisticated systems,
although still in the early conceptual and testing phases,
may provide powerful molecule-based platforms for
testing drugs and neural prosthetic devices. At present,
all neural prostheses (including neural retinal prostheses)
rely on micron-scale features and cannot interact with
the nervous system in a controlled way at the molecular
level, which is a significant disadvantage. Other research
is focusing on achieving nanoscale measurements of
cellular responses. Atomic force microscopy (AFM) has
been used to measure local nanometre morphological
responses to micro-electrode array stimulation of neuroblastoma cells[11]. AFM is a technique that, among other
capabilities, allows the measurement of height changes
in the topography of a surface (for example, a living cell
or a synthetic material) with nanometre resolution[45,46]; in
essence, it is a nanoscale cantilever that measures surface
topologies at the atomic level. This technique can measure cross-sectional changes in cell height (between 100
and 300 nm), which are produced by biphasic pulses at
a frequency of 1 Hz, thereby providing information on
ultrafine morphological changes to electrical stimuli in
neurons that cannot be achieved by other technologies.

Atomic force microscopy
(AFM). Scanning probe
microscopy that uses a sharp
probe moving over the surface
of a sample to measure
topographic spatial
information.

opposed to micron resolutions enables the investigation
of how neurons respond to anisotropic physical and
chemical cues. Micron-scale patterning can provide a
functional boundary for controlling and influencing
cellular behaviour, but ultimately the neuron detects a
stimulus (or stimuli in the case of multiple signals) that,
because of the (relatively large) micron-scale resolution of
the patterning, is a homogeneous bioactive signal that is
averaged over the entire cell. Nanotechnology approaches
present subcellular stimuli that can vary from one part
of the neuron to another. For example, photolithography
and layer-by-layer self-assembly have been used to pattern
phospholipase A$_2$, which promotes neuronal adhesion,
on a background of poly(diallyldimethylammonium
chloride) (PDDA)[8]. This approach facilitates nanoscale
patterning at resolutions that can yield complex functional architectures that are tailored to the needs of a
particular experiment. Layer-by-layer self-assembly has
also been used on silicon rubber to pattern alternating
laminin and poly-D-lysine or fibronectin/poly-D-lysine
ultrathin layers, which are 3.5–4.4 nm thick, that support

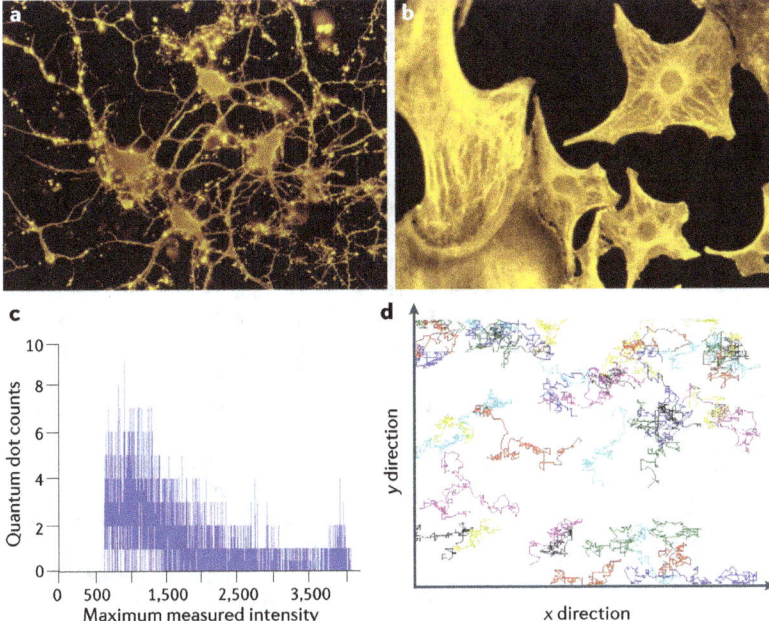

Figure 3 | The quantum dot toolbox. Fluorescent quantum dots are nanoscale particles that can be chemically functionalized by attaching a large variety of biological molecules to their outer surface (for example, antibodies, peptides and trophic factors). This allows specific molecular interactions in both live and fixed target cells, which can be visualized at high resolutions by taking advantage of the unique optical properties of quantum dots, such as their prolonged photo-stability (that is, minimal photobleaching), large excitation absorption spectra and extremely narrow emission spectra; specific examples of applications of quantum dots in neuroscience can be found in REFS 14–19. **a** | Primary rat cortical neurons labelled with conjugates of quantum dots and anti-β-tubulin III antibody. β-tubulin is a neuronal-specific intermediate filament protein and so serves as a neuronal marker. **b** | Primary rat astrocytes labelled with quantum dot–anti-glial fibrillary acidic protein (GFAP) antibody conjugates. GFAP is a glial specific intermediate filament protein. **c** | One of the main advantages of quantum dot nanotechnology is that qualitative observations as well as quantitative data can be obtained, which provide detailed molecular and biophysical information about the biological system being investigated. For example, by using computational and morphometric tools, individual quantum dots can be counted across a sample image to yield information on the distribution and number of ligand–target interactions. This graph shows the number of quantum dots with a particular intensity. Such quantitative measures could be used in measuring the expression level of cellular markers, such as β-tubulin in **a** or GFAP in **b**, which are conjugated to the quantum dots. **d** | Quantum dots can also be used to carry out single-particle tracking of ligand–target pairs, such as tracking the motion of a receptor in a cell membrane. Illustration of the trajectory of a field of 55 quantum dots undergoing Brownian diffusion, with individual trajectories corresponding to individual quantum dots. The small size, photochemistry and bioactivity of functionalized quantum dots provide an extensive new toolbox for investigating molecular and cellular processes in neurons and glia. Images and data courtesy of the Silva Research Group, University of California, San Diego, California, USA.

Photobleaching
The progressive loss of fluorescence signal intensity due to exposure to light. This can result in a decreased signal-to-noise ratio.

An area of nanotechnology that holds significant promise for probing the details of molecular and cellular processes in neural cells is functionalized semiconductor quantum dot nanocrystals[14–19] (FIG. 1). Quantum dots are nanometre-sized particles comprising a heavy metal core of materials such as cadmium–selenium or cadmium telluride, with an intermediate unreactive zinc sulphide shell and an outer coating composed of selective bioactive molecules tailored to a particular application.

The physical nature of quantum dots gives them unique and highly stable fluorescent optical properties that can be changed by altering their chemistry and physical size. Quantum dots can be tagged with fluorescent proteins of interest using different chemical approaches similar to fluorophore immunocytochemistry. However, quantum dots have significant advantages compared with other fluorescent techniques. Quantum dots undergo minimal photobleaching and, because they have broad absorption spectra but narrow emission spectra[47–50], can have much higher signal-to-noise ratios, which result in dramatically improved signal detection. In addition, quantum dots can be used for single-particle tracking of target molecules in live cells, such as tracking ligand–receptor dynamics in the cell membrane[51–53] (FIG. 3). Quantum dot labelling of both fixed and live cells is well established, and has been used in a wide variety of cell types — mostly in vitro or in situ[54–64], with some examples in vivo[65,66]. Despite the growing literature on the uses of quantum dots in the study of various cell types, their application in the labelling of neurons[14–19] and glia[19] has been slower to develop, and care must be taken to validate labelling methods that are specific for neural cells, as methods for labelling other cell types are not necessarily suitable for neural cells[19]. Recent research has illustrated the potential of this technology in neuroscience. The real-time dynamics of glycine receptors in spinal neurons have been tracked and analysed using single-particle tracking over periods of seconds to minutes[15]. These investigators characterized the dynamics of glycine receptor diffusion, which differed as a function of the spatial localization of the receptors relative to the synapse depending on whether they were in synaptic, perisynaptic or extrasynaptic regions. In another example, immobilized quantum dots that were conjugated with β-nerve growth factor (βNGF) were shown to interact with TrkA receptors in PC12 cells and to regulate their differentiation into neurons in a controlled way[18]. This could provide new tools for studying neuronal signalling processes. Although most applications of quantum dots in neuroscience have taken place ex vivo, in vivo microangiography of mouse brains has been achieved using serum that has been labelled with quantum dots[14]. New functionalization and labelling methods of quantum dots have been developed and subsequently tested using labelled AMPA (α-amino-3-hydroxy-5-methyl-4-isoxazole propionic acid) neurotransmitter receptors[16], and by measuring the cytotoxicity of hippocampal neurons[17]. Although quantum-dot nanotechnology, when used correctly, has little cytotoxic effects on cells in vitro (so that experimental results are not affected), their in vivo applications present different challenges due to the possibility of local and systemic toxicity. To address these issues, the safety of using quantum dots with both neural cells and other cell types is an active area of research[67–69].

Applications in clinical neuroscience. Applications of nanotechnology that are intended to limit and reverse neurological disorders by promoting neural regeneration and achieving neuroprotection are active areas of

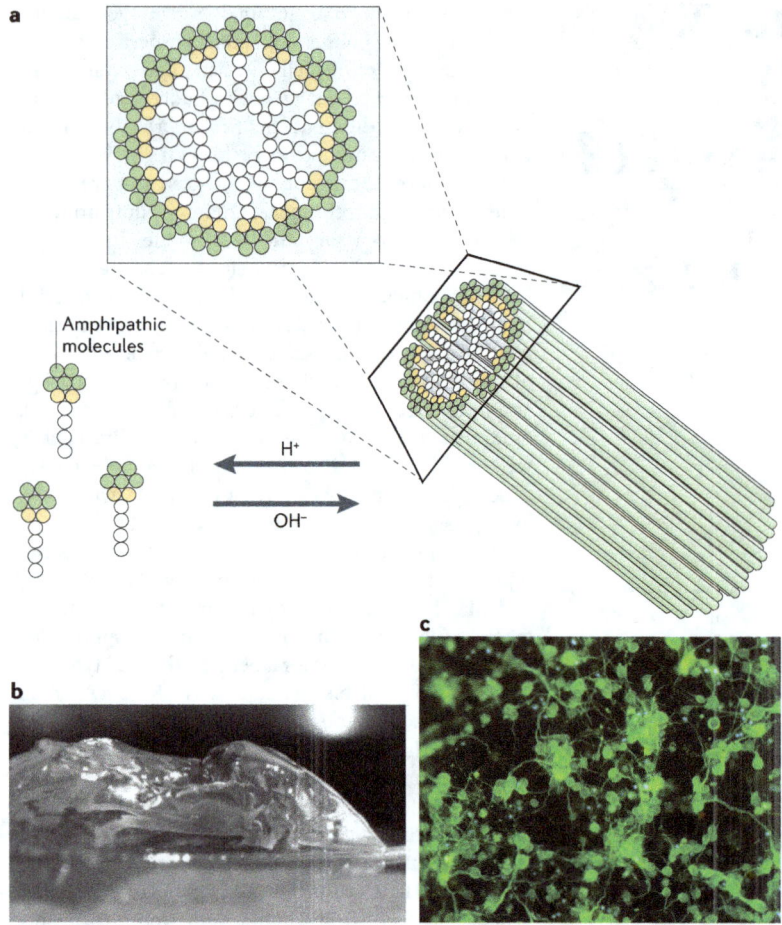

a

Amphipathic
molecules

H+

OH-

b

c

Figure 4 | **Example of an engineered nanomaterial for neural regeneration.**
Engineered nanomaterials enable the highly specific induction of controlled cellular
interactions that can promote desired neurobiological effects. **a** | Peptide-amphiphile
molecules, which consist of a hydrophilic peptide head group (green circles) and a
hydrophobic carbon tail (white circles) joined by a peptide spacer region (yellow
circles), can be coaxed to self-assemble into elongated micelles to produce a dense
nanofibre matrix[99,100]. Under physiological conditions, the self-assembly process traps
the surrounding aqueous environment and macroscopically produces a self-
supporting gel in which neural progenitor cells and stem cells can be encapsulated. In
this way, the growth and differentiation of neural progenitor cells and stem cells can
be controlled. **b** | An example of a peptide-amphiphile nanogel on a 12 mm glass
coverslip. **c** | The surface of the nanofibres consists of laminin-derived, neuronal-
specific pentapeptides, which are encountered by the encapsulated cells at high
concentrations, resulting in robust differentiation into neurons while suppressing
astrocyte differentiation[21]. The cells are stained for the neuronal marker β-tubulin III
(green) and the astrocyte marker glial fibrillary acidic protein (none is present). All
nuclei were stained with a nonspecific nuclear Hoescht stain. Images courtesy of the
Stupp laboratory, Northwestern University, Evanston, Illinois, USA.

Absorption spectra
The range of wavelengths over
which a molecule, such as a
fluorophore, or a nanoparticle,
such as a quantum dot, are
energetically excited.

this type of work is PLLA scaffolds with an ultrastructure
consisting of cast PLLA fibres, which have diameters of
50–350 nm and porosity of ~85%[20]. The scaffolds were
constructed using liquid–liquid phase separation by
dissolving PLLA in tetrahydrofuran (THF) rather than
casting them on glass. When cultured in the scaffolds,
neonatal mouse cerebellar progenitor cells were able to
extend neurites and differentiate into mature neurons. A
fundamentally different approach for the development
of a nanomaterial that promotes and supports neural
regeneration is the self-assembly of nanofibre networks
composed of peptide-amphiphile molecules[21] (FIG. 4).
On exposure to physiological ionic conditions, peptide-
amphiphile molecules, which consisted of a hydrophobic
carbon tail and a hydrophilic peptide head group, self-
assembled into a dense network of nanofibres. This
trapped the surrounding water molecules and formed
a weak self-supporting gel at the macroscopic level. The
hydrophilic peptide head groups, which formed the
outside of the fibres, consisted of the bioactive laminin-
derived peptide IKVAV, which promotes neurite sprout-
ing and growth[80–83]. Encapsulation of neural progenitor
cells from embryonic mouse cortex in the nanofibre networks resulted in fast and robust neuronal differentiation
(30% and 50% of neural progenitor cells differentiated
into neurons at 1 and 7 days *in vitro*, respectively), with
minimal astrocytic differentiation (1% and 5% of neural
progenitor cells differentiated into astrocytes at 1 and 7
days *in vitro*, respectively). This approach could therefore promote neuronal differentiation at an injury site
while potentially limiting the effects of reactive gliosis
and glial scarring, which are ubiquitous neuropathological disease processes.

Applications of nanotechnologies for neuroprotection have focused on limiting the damaging effects
of free radicals generated after injury, which is a key
neuropathological process that contributes to CNS
ischaemia, trauma and degenerative disorders[84–88].
Fullerenols, which are derivatives of hydroxyl-func-
tionalized fullerenes (molecules composed of regular
arrangements of carbon atoms[89–93]), have been shown
to have antioxidant properties. They also function as
free radical scavengers, which can lead to a reduction
in the extent of excitotoxicity and apoptosis induced by
glutamate, NMDA (*N*-methyl-D-aspartate), AMPA and
kainate[23–26]. Fullerenol-mediated neuroprotection has
been shown *in vitro* and *in vivo*. Fullerenol limits excitotoxicity and apoptosis of cultured cortical neurons
in vitro, and delays the onset of motor degeneration
in vivo in a mouse model of familial amyotrophic lateral sclerosis. The neuroprotective effect of fullerenols
might be partly mediated by inhibition of glutamate
receptors, as they had no effect on $GABA_A$ or taurine
receptors. They also lowered glutamate-induced eleva-
tions in intracellular calcium, which is an important
mechanism of neuronal excitotoxicity[23–26].

Another clinically relevant area of intense research
is the design of functionalized nanoparticles that can
be administered systemically and deliver drugs and
small molecules across the BBB[27–36]. This is a major
clinical objective for the treatment of a wide range

research. The development of nanoengineered scaffolds
that support and promote neurite and axonal growth
are evolving from tissue engineering approaches based
on the manipulation of bulk materials. Examples of
micron-scale tissue engineering include poly-(L-lactic)
acid (PLLA) and other synthetic hydrogels that have
engineered microscale features, and scaffolds derived
from naturally occurring materials such as collagen[70–79].
One example of a nanoengineered system derived from

of neurological disorders. To achieve this, various materials and synthetic approaches are being investigated. Oligonucleotides have been delivered in gels of crosslinked poly(ethylene glycol) and polyethylenimine[28]. Charge differences in the electrostatic forces between the gel and spontaneously negatively charged oligonucleotides provide a reversible delivery mechanism that can be used for shuttling molecules across the BBB and then releasing them from the delivery system. Neuropeptides (such as enkephalins), the NMDA receptor antagonist MRZ 2/576, and the chemotherapeutic drug doxorubicin have been absorbed onto the surface of poly(butylcyanoacrylate) nanoparticles coated with polysorbate 80 (REFS 31,33,34,37,38). The polysorbate on the surface of the nanoparticles adsorbs apolipoprotein B and apolipoprotein E from the blood, and the nanoparticles are taken up by brain capillary endothelial cells via receptor-mediated endocytosis[38]. Nanoparticles that target tumours in the CNS may be a particularly important application of this technology due to the high morbidity and mortality associated with often aggressive neoplasms in the physically confined spaces of the cranium and spinal canal.

Challenges and opportunities

The challenges associated with nanotechnology applications in neuroscience are numerous, but the impact it can have on understanding how the nervous system works, how it fails in disease and how we can intervene at a molecular level is significant. Ultimately, the challenges and opportunities presented by nanotechnology stem from the fact that this technology provides a way to interact with neural cells at the molecular level, which has both positive and negative aspects. The ability to exploit drugs, small molecules, neurotransmitters and neural developmental factors offers the potential to tailor technologies to particular applications. For example, neural developmental factors, such as the cadherins, laminins and bone morphometric protein families, as well as their receptors, can be manipulated in new ways. Nanotechnology offers the capacity to take advantage of the functional specificity of these molecules by incorporating them into engineered materials and devices to have highly targeted effects. The laminins, for example, are large multi-domain trimeric proteins composed of α, β and γ chains, of which 12 isoforms are known[82,83,94,95]. The isoforms contain different bioactive peptide sequences, which have varying affinities for specific cell types and can induce different effects. For example, the laminin 1 isoform, which is the most studied laminin, contains at least 48 different short peptide sequences that promote neuronal adhesion and neurite outgrowth, and some of these peptides (25 of 48 tested) have such effects on specific types of neuron[82]. This degree of molecular specificity, which is conferred not only by laminins but by many other signalling molecules that are important in the development and function of the nervous system, can be used to design highly selective nanotechnologies. Indeed, any desired cellular signalling pathway can be targeted using this approach.

This discussion also suggests the main technical challenges that are encountered when using nanotechnology applications in neuroscience: the need for greater specificity; multiple induced physiological functions; and minimal side effects. Greater specificity of interactions with target cells and tissues will result in more significant and specific physiological effects, which should also reduce undesirable and deleterious side effects that are induced by the technology. Another important challenge is the requirement for technologies that are able to multitask, carrying out a diverse set of specific cellular and physiological functions, such as targeting multiple receptors or ligands. This is particularly important when attempting to address multi-dimensional CNS disorders that are the result of numerous interdependent molecular and biochemical events (for example, secondary injury following traumatic brain injury or spinal cord injury). At present, synthetic and engineering processes are not advanced enough to allow nanotechnologies that have been designed to interact with the nervous system to fully meet these criteria.

From a biological perspective, the most significant successes of nanotechnology applications in neuroscience will be those that appreciate a detailed understanding of neurobiology and take advantage of the known (and unknown) molecular details. As suggested above, the main challenge is the ability to design and use more sophisticated technologies that are able to carry out highly targeted and specific functions while minimizing nonspecific interactions. To achieve this, both the design and engineering aspects of nanotechnology as well as our understanding of the underlying neurobiology are crucial. This, in turn, will require more interaction between neuroscientists and physical scientists such as chemists and materials scientists. This is not a trivial issue as the scientific language and culture between different disciplines can vary considerably. This is an increasing challenge for interdisciplinary science, which requires people with different training, skills and conceptions of how science should be conducted coming together first to understand and agree on a common challenge or problem, and then to agree on how to address that issue. The next section discusses further the role of neuroscientists in the development and application of nanotechnology.

The above discussion pertains to all applications of nanotechnology to neuroscience. Applications of nanotechnology to the nervous system *in vivo* present additional challenges. In particular, the inherent complexity of the CNS, as well as its difficult and anatomically restrictive nature, poses a unique set of obstacles. Cellular heterogeneity and multi-dimensional cellular interactions (for example, spatial and temporal summation of postsynaptic potentials) underlie the nervous system's anatomical and functional 'wiring' that is the basis of its extremely complex information processing. Nanotechnologies designed to interact with CNS cells and processes *in vivo* must take this complexity into consideration, if only to avoid disrupting it. Failure to do so may result in unforeseen and unacceptable 'side effects' in the nervous system and/or other physiological systems. A significant challenge in

Emission spectra
The range of wavelengths over which a molecule, such as a fluorophore, or a nanoparticle, such as a quantum dot, emit light.

Synaptic, perisynaptic or extrasynaptic regions
Areas where neurotransmitter receptors cluster at, near, or outside the synapse, respectively.

TrkA receptors
A family of proto-oncogene receptors found throughout the central and peripheral nervous system that bind β-nerve growth factor, which results in downstream signalling effects.

in vivo applications of nanotechnology is that they are designed to physically interact with neural cells at cellular and subcellular levels, but ultimately aim at engaging functional interactions at a systemic level, which usually involves large groups of interacting neurons and glia. At present, there are only a few applications of this type; nonetheless, although technically and conceptually challenging, these types of application could have a significant impact on clinical neuroscience. However, there is still a tremendous amount of (exciting) work to be done.

Apart from physiological complexity, the second main consideration for *in vivo* applications of nanotechnology is that they must consider the highly anatomically restrictive nature of the CNS. The structures of the CNS are well protected from mechanical and physical injury, and are immunologically privileged behind the BBB and blood–retina barrier, which have unique molecular and cellular environments. Nanotechnologies designed for *in vivo* applications must be efficiently delivered with minimal disruption to these structures before it can carry out its primary function. This will surely present significant technical challenges. Similarly, extreme care must be taken to understand and avoid potential safety pitfalls, including both systemic and local side effects associated with the delivery and primary function of the applied technology — an issue that is unique to *in vivo* nanotechnology[96–98]. As mentioned above, investigating the safety of nanotechnologies is an active and important area of research[67–69]. Despite all these challenges, the applications of nanotechnology both *in vivo* and *ex vivo* offer tremendous opportunities for understanding normal physiology and for developing therapies.

The role of the neuroscientist

Neuroscientists have a unique role in developing nanotechnologies. Neuroscientists — both researchers and clinicians — need to identify potential applications of nanotechnology in neuroscience and neurology to maximize their impact. Scientists with other specialties can develop powerful platform technologies and even provide neuroscience-specific examples, but it is only with direct input from and in partnership with neuroscientists that broad neurophysiological and clinical applications can be properly formulated and addressed. This requires highly interdisciplinary collaborations with consideration of the requirements of both parties. Some neuroscientists

develop neuroscience nanotechnologies for their particular purposes, most likely geared towards specific questions or objectives for their research. But most neuroscientists find that they lack the expertise and resources needed to design, synthesize and characterize sophisticated nano-engineered materials or devices. It would be unrealistic to assume that we as neuroscientists have knowledge and skills in these areas that are equivalent to those of chemists or materials scientists who have devoted their careers to the synthetic aspects of these technologies. However, chemists and material scientists do not have the comprehensive training in neurobiology, neurophysiology and neuropathology required to fully appreciate and exploit the potential of nanotechnology in neuroscience. Therefore, it is crucial that different disciplines are able to communicate with each other using a common technical language, which is not a trivial issue — a nucleus means different things to a physicist and a cell biologist. To this end, it is important for neuroscientists who wish to pursue the development of nanotechnology to educate themselves across disciplines. To envision new applications of nanotechnology, it is necessary to understand what has already been accomplished and what can be achieved with this technology.

Future directions

Applications of nanotechnology to neuroscience are already having significant effects, which will continue in the foreseeable future. Short-term progress has benefited *in vitro* and *ex vivo* studies of neural cells, often supporting or augmenting standard technologies. These advances contribute to both our basic understanding of cellular neurobiology and neurophysiology, and to our understanding and interpretation of neuropathology. Although the development of nanotechnologies designed to interact with the nervous system *in vivo* is slow and challenging, they will have significant, direct clinical implications. Nanotechnologies targeted at supporting cellular or pharmacological therapies or facilitating direct physiological effects *in vivo* will make significant contributions to clinical care and prevention. The reason for the tremendous potential that nanotechnology applications can have in biology and medicine in general and neuroscience in particular stems from the capacity of these technologies to specifically interact with cells at the molecular level.

1. Liu, D., Park, S. H., Reif, J. H. & LaBean, T. H. DNA nanotubes self-assembled from triple-crossover tiles as templates for conductive nanowires. *Proc. Natl Acad. Sci. USA* **101**, 717–722 (2004).
2. Cui, Y., Wei, Q., Park, H. & Lieber, C. M. Nanowire nanosensors for highly sensitive and selective detection of biological and chemical species. *Science* **293**, 1289–1292 (2001).
3. Silva, G. A. Nanotechnology approaches for the regeneration and neuroprotection of the central nervous system. *Surg. Neurol.* **63**, 301–306 (2005).
4. Silva, G. A. Small neuroscience: the nanostructure of the central nervous system and emerging nanotechnology applications. *Curr. Nanosci.* **3**, 225–236 (2005).
5. Ai, H. *et al.* Biocompatibility of layer-by-layer self-assembled nanofilm on silicone rubber for neurons. *J. Neurosci. Methods* **128**, 1–8 (2003).
6. Fan, Y. W. *et al.* Culture of neural cells on silicon wafers with nano-scale surface topograph. *J. Neurosci. Methods* **120**, 17–23 (2002).
7. Kim, D. H., Abidian, M. & Martin, D. C. Conducting polymers grown in hydrogel scaffolds coated on neural prosthetic devices. *J. Biomed. Mater. Res. A* **71**, 577–585 (2004).
8. Mohammed, J. S., DeCoster, M. A. & McShane, M. J. Micropatterning of nanoengineered surfaces to study neuronal cell attachment *in vitro. Biomacromolecules* **5**, 1745–1755 (2004).
9. Kramer, S. *et al.* Preparation of protein gradients through the controlled deposition of protein–nanoparticle conjugates onto functionalized surfaces. *J. Am. Chem. Soc.* **126**, 5388–5395 (2004).
10. Moxon, K. A. *et al.* Nanostructured surface modification of ceramic-based microelectrodes to enhance biocompatibility for a direct brain–machine interface. *IEEE Trans. Biomed. Eng.* **51**, 881–889 (2004).
    A good example of electrode modification using nanoengineered materials to improve neuronal adhesion while limiting the effects of contaminating cells to improve the functionality of the electrodes.
11. Shenai, M. B. *et al.* A novel MEA/AFM platform for measurement of real-time, nanometric morphological alterations of electrically stimulated neuroblastoma cells. *IEEE Trans. Nanobioscience* **3**, 111–117 (2004).
12. Vu, T. Q. *et al.* Activation of membrane receptors by a neurotransmitter conjugate designed for surface attachment. *Biomaterials* **26**, 1895–1903 (2005).
    An excellent example of a nanoengineered cell-signalling platform technology designed to selectively and controllably stimulate target neurons by immobilizing neurotransmitter molecules on a surface.

13. Saifuddin, U. *et al*. Assembly and characterization of biofunctional neurotransmitter-immobilized surfaces for interaction with postsynaptic membrane receptors. *J. Biomed. Mater. Res. A* **66**, 184–191 (2003).

14. Levene, M. J., Dombeck, D. A., Kasischke, K. A., Molloy, R. P. & Webb, W. W. *In vivo* multiphoton microscopy of deep brain tissue. *J. Neurophysiol.* **91**, 1908–1912 (2004).

15. Dahan, M. *et al*. Diffusion dynamics of glycine receptors revealed by single-quantum dot tracking. *Science* **302**, 442–445 (2003).
    First major work showing the application of functionalized quantum dots to target neurons in the investigation of a specific neurophysiological process. The authors took advantage of the physical properties of quantum dots to achieve single-particle tracking of glycine receptors.

16. Howarth, M., Takao, K., Hayashi, Y. & Ting, A. Y. Targeting quantum dots to surface proteins in living cells with biotin ligase. *Proc. Natl Acad. Sci. USA* **102**, 7583–7588 (2005).

17. Fan, H. *et al*. Surfactant-assisted synthesis of water-soluble and biocompatible semiconductor quantum dot micelles. *Nano. Lett.* **5**, 645–648 (2005).

18. Vu, T. Q. *et al*. Peptide-conjugated quantum dots activate neuronal receptors and initiate downstream signaling of neurite growth. *Nano. Lett.* **5**, 603–607 (2005).

19. Pathak, S., Cao, E., Davidson, M., Jin, S.-H. & Silva, G. A. Quantum dot applications in neuroscience: new tools for probing neurons and glia. *J. Neurosci.* (in the press).

20. Yang, F. *et al*. Fabrication of nano-structured porous PLLA scaffold intended for nerve tissue engineering. *Biomaterials* **25**, 1891–1900 (2004).

21. Silva, G. A. *et al*. Selective differentiation of neural progenitor cells by high-epitope density nanofibers. *Science* **303**, 1352–1355 (2004).
    First major work describing the application of a nanoengineered material designed specifically for neural regeneration that had multiple, functional properties, including *in vitro* self-assembly under physiological conditions and preferential differentiation of neurons over astrocytes.

22. Dugan, L. L. *et al*. Fullerene-based antioxidants and neurodegenerative disorders. *Parkinsonism Relat. Disord.* **7**, 243–246 (2001).

23. Dugan, L. L., Gabrielsen, J. K., Yu, S. P., Lin, T. S. & Choi, D. W. Buckminsterfullerenol free radical scavengers reduce excitotoxic and apoptotic death of cultured cortical neurons. *Neurobiol. Dis.* **3**, 129–135 (1996).

24. Dugan, L. L. *et al*. Carboxyfullerenes as neuroprotective agents. *Proc. Natl Acad. Sci. USA* **94**, 9434–9439 (1997).

25. Jin, H. *et al*. Polyhydroxylated C_{60}, fullerenols, as glutamate receptor antagonists and neuroprotective agents. *J. Neurosci. Res.* **62**, 600–607 (2000).

26. Dugan, L. L. *et al*. Fullerene-based antioxidants and neurodegenerative disorders. **7**, 243–246 (2001).
    A good example of the application of fullerene-derived nanoparticles intended to protect the CNS from free radical toxicity following injury. This and related research have shown promising initial results *in vivo*.

27. Lockman, P. R., Mumper, R. J., Khan, M. A. & Allen, D. D. Nanoparticle technology for drug delivery across the blood–brain barrier. *Drug Dev. Ind. Pharm.* **28**, 1–13 (2002).
    Reviews current strategies for delivering drugs and other small molecules across the BBB.

28. Vinogradov, S. V., Batrakova, E. V. & Kabanov, A. V. Nanogels for oligonucleotide delivery to the brain. *Bioconjug. Chem.* **15**, 50–60 (2004).

29. Koziara, J. M., Lockman, P. R., Allen, D. D. & Mumper, R. J. *In situ* blood–brain barrier transport of nanoparticles. *Pharm. Res.* **20**, 1772–1778 (2003).

30. Olbrich, C., Gessner, A., Kayser, O. & Muller, R. H. Lipid–drug-conjugate (LDC) nanoparticles as novel carrier system for the hydrophilic antitrypanosomal drug diminazenediaceturate. *J. Drug Target.* **10**, 387–396 (2002).

31. Schroeder, U., Sommerfeld, P., Ulrich, S. & Sabel, B. A. Nanoparticle technology for delivery of drugs across the blood–brain barrier. *J. Pharm. Sci.* **87**, 1305–1307 (1998).

32. Kreuter, J. *et al*. Direct evidence that polysorbate-80-coated poly(butylcyanoacrylate) nanoparticles deliver drugs to the CNS via specific mechanisms requiring prior binding of drug to the nanoparticles. *Pharm. Res.* **20**, 409–416 (2003).

33. Kreuter, J. Nanoparticulate systems for brain delivery of drugs. *Adv. Drug Deliv. Rev.* **47**, 65–81 (2001).

34. Alyaudtin, R. N. *et al*. Interaction of poly(butylcyanoacrylate) nanoparticles with the blood–brain barrier *in vivo* and *in vitro*. *J. Drug Target.* **9**, 209–221 (2001).

35. Brigger, I. *et al*. Poly(ethylene glycol)-coated hexadecylcyanoacrylate nanospheres display a combined effect for brain tumor targeting. *J. Pharmacol. Exp. Ther.* **303**, 928–936 (2002).

36. Garcia-Garcia, E. *et al*. A relevant *in vitro* rat model for the evaluation of blood–brain barrier translocation of nanoparticles. *Cell. Mol. Life Sci.* **62**, 1400–1408 (2005).

37. Gelperina, S. E. *et al*. Toxicological studies of doxorubicin bound to polysorbate 80-coated poly(butyl cyanoacrylate) nanoparticles in healthy rats and rats with intracranial glioblastoma. *Toxicol. Lett.* **126**, 131–141 (2002).
    Describes a practical example of the delivery of the anticancer drug doxorubicin across the BBB *in vivo* and illustrates the significant potential these technologies have for drug delivery to the CNS.

38. Kreuter, J. *et al*. Apolipoprotein-mediated transport of nanoparticle-bound drugs across the blood–brain barrier. *J. Drug Target.* **10**, 317–325 (2002).

39. Park, T. H. & Shuler, M. L. Integration of cell culture and microfabrication technology. *Biotechnol. Prog.* **19**, 243–253 (2003).

40. Oliva, A. A. Jr, James, C. D., Kingman, C. E., Craighead, H. G. & Banker, G. A. Patterning axonal guidance molecules using a novel strategy for microcontact printing. *Neurochem. Res.* **28**, 1639–1648 (2003).

41. Andersson, H. & van den Berg, A. Microfabrication and microfluidics for tissue engineering: state of the art and future opportunities. *Lab. Chip* **4**, 98–103 (2004).

42. Branch, D. W., Wheeler, B. C., Brewer, G. J. & Leckband, D. E. Long-term maintenance of patterns of hippocampal pyramidal cells on substrates of polyethylene glycol and microstamped polylysine. *IEEE Trans. Biomed. Eng.* **47**, 290–300 (2000).
    An excellent example of classical microscale-patterning approaches for controlling the growth and connections of patterned neurons.

43. Wheeler, B. C., Corey, J. M., Brewer, G. J. & Branch, D. W. Microcontact printing for precise control of nerve cell growth in culture. *J. Biomech. Eng.* **121**, 73–78 (1999).

44. Chang, J. C., Brewer, G. J. & Wheeler, B. C. Modulation of neural network activity by patterning. *Biosens. Bioelectron.* **16**, 527–533 (2001).

45. Hansma, H. G., Kasuya, K. & Oroudjev, E. Atomic force microscopy imaging and pulling of nucleic acids. *Curr. Opin. Struct. Biol.* **14**, 380–385 (2004).

46. Santos, N. C. & Castanho, M. A. An overview of the biophysical applications of atomic force microscopy. *Biophys. Chem.* **107**, 133–149 (2004).

47. West, J. L. & Halas, N. J. Engineered nanomaterials for biophotonics applications: improving sensing, imaging, and therapeutics. *Annu. Rev. Biomed. Eng.* **5**, 285–292 (2003).

48. Murphy, C. J. Optical sensing with quantum dots. *Anal. Chem.* **74**, 520A–526A (2002).

49. Chan, W. C. & Nie, S. Quantum dot bioconjugates for ultrasensitive nonisotopic detection. *Science* **281**, 2016–2018 (1998).

50. Vanmaekelbergh, D. & Liljeroth, P. Electron-conducting quantum dot solids: novel materials based on colloidal semiconductor nanocrystals. *Chem. Soc. Rev.* **34**, 299–312 (2005).

51. Saxton, M. J. & Jacobson, K. Single-particle tracking: applications to membrane dynamics. *Annu. Rev. Biophys. Biomol. Struct.* **26**, 373–399 (1997).

52. Bonneau, S., Cohen, L. D. & Dahan, M. A multiple target approach for single quantum dot tracking. *IEEE Int. Symp. Biomed Imaging*, Arlington, Virginia, USA, 15–18 April, 664–667 (2004).

53. Bonneau, S., Dahan, M. & Cohen, L. D. Single quantum dot tracking based on perceptual grouping using minimal paths in a spatiotemporal volume. *IEEE Trans. Image Process.* **14**, 1384–1395 (2005).

54. Arya, H. *et al*. Quantum dots in bio-imaging: revolution by the small. *Biochem. Biophys. Res. Commun.* **329**, 1173–1177 (2005).

55. Mansson, A. *et al*. *In vitro* sliding of actin filaments labelled with single quantum dots. *Biochem. Biophys. Res. Commun.* **314**, 529–534 (2004).

56. Lidke, D. S. *et al*. Quantum dot ligands provide new insights into erbB/HER receptor-mediated signal transduction. *Nature Biotechnol.* **22**, 198–203 (2004).

57. Jaiswal, J. K., Goldman, E. R., Mattoussi, H. & Simon, S. M. Use of quantum dots for live cell imaging. *Nature Methods* **1**, 73–78 (2004).

58. Wu, X. *et al*. Immunofluorescent labeling of cancer marker Her2 and other cellular targets with semiconductor quantum dots. *Nature Biotechnol.* **21**, 41–46 (2003).

59. Watson, A., Wu, X. & Bruchez, M. Lighting up cells with quantum dots. *Biotechniques* **34**, 296–300, 302–303 (2003).

60. Tokumasu, F. & Dvorak, J. Development and application of quantum dots for immunocytochemistry of human erythrocytes. *J. Microsc.* **211**, 256–261 (2003).

61. Ness, J. M., Akhtar, R. S., Latham, C. B. & Roth, K. A. Combined tyramide signal amplification and quantum dots for sensitive and photostable immunofluorescence detection. *J. Histochem. Cytochem.* **51**, 981–987 (2003).

62. Jaiswal, J. K., Mattoussi, H., Mauro, J. M. & Simon, S. M. Long-term multiple color imaging of live cells using quantum dot bioconjugates. *Nature Biotechnol.* **21**, 47–51 (2003).

63. Gao, X. & Nie, S. Molecular profiling of single cells and tissue specimens with quantum dots. *Trends Biotechnol.* **21**, 371–373 (2003).

64. Chan, W. C. *et al*. Luminescent quantum dots for multiplexed biological detection and imaging. *Curr. Opin. Biotechnol.* **13**, 40–46 (2002).

65. Akerman, M. E., Chan, W. C., Laakkonen, P., Bhatia, S. N. & Ruoslahti, E. Nanocrystal targeting *in vivo*. *Proc. Natl Acad. Sci. USA* **99**, 12617–12621 (2002).

66. Michalet, X. *et al*. Quantum dots for live cells, *in vivo* imaging, and diagnostics. *Science* **307**, 538–544 (2005).

67. Braydich-Stolle, L., Hussain, S., Schlager, J. & Hofmann, M. C. *In vitro* cytotoxicity of nanoparticles in mammalian germ-line stem cells. *Toxicol. Sci.* **88**, 412–419 (2005).

68. Lovric, J. *et al*. Differences in subcellular distribution and toxicity of green and red emitting CdTe quantum dots. *J. Mol. Med.* **83**, 377–385 (2005).

69. Voura, E. B., Jaiswal, J. K., Mattoussi, H. & Simon, S. M. Tracking metastatic tumor cell extravasation with quantum dot nanocrystals and fluorescence emission-scanning microscopy. *Nature Med.* **10**, 993–998 (2004).

70. Stang, F., Fansa, H., Wolf, G. & Keilhoff, G. Collagen nerve conduits — assessment of biocompatibility and axonal regeneration. *Biomed. Mater. Eng.* **15**, 3–12 (2005).

71. Gamez, E. *et al*. Photofabricated gelatin-based nerve conduits: nerve tissue regeneration potentials. *Cell Transplant.* **13**, 549–564 (2004).

72. Ma, W. *et al*. CNS stem and progenitor cell differentiation into functional neuronal circuits in three-dimensional collagen gels. *Exp. Neurol.* **190**, 276–288 (2004).

73. Park, K. I., Teng, Y. D. & Snyder, E. Y. The injured brain interacts reciprocally with neural stem cells supported by scaffolds to reconstitute lost tissue. *Nature Biotechnol.* **20**, 1111–1117 (2002).

74. Balgude, A. P., Yu, X., Szymanski, A. & Bellamkonda, R. V. Agarose gel stiffness determines rate of DRG neurite extension in 3D cultures. *Biomaterials* **22**, 1077–1084 (2001).

75. Dillon, G. P., Yu, X., Sridharan, A., Ranieri, J. P. & Bellamkonda, R. V. The influence of physical structure and charge on neurite extension in a 3D hydrogel scaffold. *J. Biomater. Sci. Polym. Ed.* **9**, 1049–1069 (1998).

76. Katayama, Y. *et al*. Coil-reinforced hydrogel tubes promote nerve regeneration equivalent to that of nerve autografts. *Biomaterials* **27**, 505–518 (2006).

77. Tsai, E. C., Dalton, P. D., Shoichet, M. S. & Tator, C. H. Matrix inclusion within synthetic hydrogel guidance channels improves specific supraspinal and local axonal regeneration after complete spinal cord transection. *Biomaterials* **27**, 519–533 (2006).

78. Levesque, S. G., Lim, R. M. & Shoichet, M. S. Macroporous interconnected dextran scaffolds of controlled porosity for tissue-engineering applications. *Biomaterials* **26**, 7436–7446 (2005).

79. Yu, T. T. & Shoichet, M. S. Guided cell adhesion and outgrowth in peptide-modified channels for neural tissue engineering. *Biomaterials* **26**, 1507–1514 (2005).

80. Tashiro, K. *et al.* A synthetic peptide containing the IKVAV sequence from the A chain of laminin mediates cell attachment, migration, and neurite outgrowth. *J. Biol. Chem.* **264**, 16174–16182 (1989).

81. Nomizu, M. *et al.* Structure–activity study of a laminin α1 chain active peptide segment Ile-Lys-Val-Ala-Val (IKVAV). *FEBS Lett.* **365**, 227–231 (1995).

82. Powell, S. K. *et al.* Neural cell response to multiple novel sites on laminin-1. *J. Neurosci. Res.* **61**, 302–312 (2000).

83. Tunggal, P., Smyth, N., Paulsson, M. & Ott, M. C. Laminins: structure and genetic regulation. *Microsc. Res. Tech.* **51**, 214–227 (2000).

84. Blass, J. P. Cerebrometabolic abnormalities in Alzheimer's disease. *Neurol. Res.* **25**, 556–566 (2003).

85. Gagliardi, R. J. Neuroprotection, excitotoxicity and NMDA antagonists. *Arq. Neuropsiquiatr.* **58**, 583–588 (2000).

86. Lo, E. H., Dalkara, T. & Moskowitz, M. A. Mechanisms, challenges and opportunities in stroke. *Nature Rev. Neurosci.* **4**, 399–415 (2003).

87. Mahadik, S. P. & Mukherjee, S. Free radical pathology and antioxidant defense in schizophrenia: a review. *Schizophr. Res.* **19**, 1–17 (1996).

88. Mishra, O. P. & Delivoria-Papadopoulos, M. Cellular mechanisms of hypoxic injury in the developing brain. *Brain Res. Bull.* **48**, 233–238 (1999).

89. Gust, D., Moore, T. A. & Moore, A. L. Photochemistry of supramolecular systems containing C60. *J. Photochem. Photobiol. B* **58**, 63–71 (2000).

90. Zhang, J., Albelda, M. T., Liu, Y. & Canary, J. W. Chiral nanotechnology. *Chirality* **17**, 404–420 (2005).

91. Fortina, P., Kricka, L. J., Surrey, S. & Grodzinski, P. Nanobiotechnology: the promise and reality of new approaches to molecular recognition. *Trends Biotechnol.* **23**, 168–173 (2005).

92. Scott, L. T. Methods for the chemical synthesis of fullerenes. *Angew. Chem. Int. Ed. Engl.* **43**, 4994–5007 (2004).

93. Segura, J. L. & Martin, N. New concepts in tetrathiafulvalene chemistry. *Angew. Chem. Int. Ed. Engl.* **40**, 1372–1409 (2001).

94. Cheng, Y. S., Champliaud, M. F., Burgeson, R. E., Marinkovich, M. P. & Yurchenco, P. D. Self-assembly of laminin isoforms. *J. Biol. Chem.* **272**, 31525–31532 (1997).

95. Hohenester, E. & Engel, J. Domain structure and organisation in extracellular matrix proteins. *Matrix Biol.* **21**, 115–128 (2002).

96. Giles, J. Size matters when it comes to safety, report warns. *Nature* **430**, 599 (2004).

97. Giles, J. Nanotechnology: what is there to fear from something so small? *Nature* **426**, 750 (2003).

98. Oberdorster, G., Oberdorster, E. & Oberdorster, J. Nanotoxicology: an emerging discipline evolving from studies of ultrafine particles. *Environ. Health Perspect.* **113**, 823–839 (2005).

99. Hartgerink, J. D., Beniash, E. & Stupp, S. I. Peptide-amphiphile nanofibers: a versatile scaffold for the preparation of self-assembling materials. *Proc. Natl Acad. Sci. USA* **99**, 5133–5138 (2002).

100. Hartgerink, J. D., Beniash, E. & Stupp, S. I. Self-assembly and mineralization of peptide-amphiphile nanofibers. *Science* **294**, 1684–1688 (2001).

Acknowledgements
This work was supported by the Whitaker Foundation, Arlingon, Virginia, USA, and the Stein Clinical Research Institute at the University of California, San Diego, USA. Quantum dots were kindly provided free of charge by Quantum Dot Corporation.

Competing interests statement
The author declares no competing financial interests.

DATABASES
The following terms in this article are linked online to:
Entrez Gene: http://www.ncbi.nlm.nih.gov/entrez/query.fcgi?db=gene
AMPA receptors | apolipoprotein B | apolipoprotein E | GABA$_A$ receptor | NMDA receptors

FURTHER INFORMATION
The National Nanotechnology Initiative:
http://www.nano.gov
Silva's laboratory: http://www.silva.ucsd.edu
American Academy of Nanomedicine: http://www.aananomed.org
Nano Science and Technology Institute Nanotechnology to Neuroscience symposium: http://www.nsti.org/Nanotech2006/symposia/Nanotech_Neurology.html
Access to this interactive links box is free online.

The potential and challenges of nanopore sequencing

Daniel Branton[1], David W Deamer[2], Andre Marziali[3], Hagan Bayley[4], Steven A Benner[5], Thomas Butler[6], Massimiliano Di Ventra[7], Slaven Garaj[8], Andrew Hibbs[9], Xiaohua Huang[10], Stevan B Jovanovich[11], Predrag S Krstic[12], Stuart Lindsay[13], Xinsheng Sean Ling[14], Carlos H Mastrangelo[15], Amit Meller[16], John S Oliver[17], Yuriy V Pershin[7], J Michael Ramsey[18], Robert Riehn[19], Gautam V Soni[16], Vincent Tabard-Cossa[3], Meni Wanunu[16], Matthew Wiggin[20] & Jeffery A Schloss[21]

A nanopore-based device provides single-molecule detection and analytical capabilities that are achieved by electrophoretically driving molecules in solution through a nano-scale pore. The nanopore provides a highly confined space within which single nucleic acid polymers can be analyzed at high throughput by one of a variety of means, and the perfect processivity that can be enforced in a narrow pore ensures that the native order of the nucleobases in a polynucleotide is reflected in the sequence of signals that is detected. Kilobase length polymers (single-stranded genomic DNA or RNA) or small molecules (e.g., nucleosides) can be identified and characterized without amplification or labeling, a unique analytical capability that makes inexpensive, rapid DNA sequencing a possibility. Further research and development to overcome current challenges to nanopore identification of each successive nucleotide in a DNA strand offers the prospect of 'third generation' instruments that will sequence a diploid mammalian genome for ~$1,000 in ~24 h.

When a small (~100 mV) voltage bias is imposed across a nanopore in a membrane separating two chambers containing aqueous electrolytes, the resulting ionic current through the pore can be measured with standard electrophysiological techniques. Bearing in mind that the opening and closing of many biological channels depends on relatively small peptide moieties physically blocking the channel, one of us (Deamer) at the

University of California (Santa Cruz) and George Church of Harvard (Cambridge, MA, USA) (personal communication) independently proposed that if a strand of DNA or RNA could be electrophoretically driven through a nanopore of suitable diameter, the nucleobases would similarly modulate the ionic current through the nanopore. Subsequently, Deamer, Branton and colleagues[1] demonstrated that single-stranded DNA (ssDNA) and RNA molecules can be driven through a pore-forming protein and detected by their effect on the ionic current through this nanopore (**Fig. 1a**). This system used the *Staphylococcus aureus* toxin, α-hemolysin, the use of which as a biosensor had been pioneered by Bayley and his colleagues[2]. The Bayley group[3] has since shown that an α-hemolysin pore is remarkably stable and remains functional at close to the boiling point of water. Because the inside diameter of the α-hemolysin pore is barely as large as the diameter of a single nucleic acid strand, the results of Deamer, Branton and colleagues[1] showed that a nanopore can locally unravel a coiled nucleic acid so that its nucleotides are translocated through the pore in strictly single-file, sequential order. Because the current of ions through the nanopore is partially blocked by the translocating molecule, each translocating molecule produces a readily detected reduction of the ionic current relative to that which flows through the open, unblocked pore. Given this fact, Deamer, Branton and colleagues[1] hypothesized that if each nucleotide in the polymer produced a characteristic modulation of the ionic current during its passage through the nanopore, the sequence of current modulations would reflect the sequence of bases in the polymer.

To test this hypothesis, two groups (Deamer[4], Meller and Branton[5]) investigated the current modulations caused by several different RNA and ssDNA polynucleotides. These experiments showed that pore current is blocked to a substantially greater degree by polyC RNA than by

[1]Department of Molecular and Cell Biology, Harvard University, Cambridge, Massachusetts 02138, USA. [2]Department of Chemistry and Biochemistry, University of California, Santa Cruz, California 95064, USA. [3]Department of Physics and Astronomy, University of British Columbia, Vancouver, British Columbia V6T 1Z1, Canada. [4]Department of Chemical Biology, Oxford University, Oxford OX1 3TA, UK. [5]Foundation for Applied Molecular Evolution, Gainesville, Florida 32604, USA. [6]Department of Physics, University of Washington, Seattle, Washington 98195, USA. [7]Department of Physics, University of California at San Diego, La Jolla, California 92093, USA. [8]Department of Physics, Harvard University, Cambridge, Massachusetts 02138, USA. [9]Electronic BioSciences, San Diego, California 92121, USA. [10]Department of Bioengineering, University of California at San Diego, La Jolla, California 92093, USA. [11]Microchip Biotechnologies Inc., Dublin, California 94568, USA. [12]Oak Ridge National Laboratory, Oak Ridge, Tennessee 37831, USA. [13]Departments of Physics and Chemistry and the Biodesign Institute, Arizona State University, Tempe, Arizona 85287, USA. [14]Department of Physics, Brown University, Providence, Rhode Island 02912, USA. [15]Electrical Engineering and Computer Science, Case Western Reserve University, Cleveland, Ohio 44106, USA. [16]Biomedical Engineering, Boston University, Boston, Massachusetts 02215, USA. [17]NABsys, Inc., Providence, Rhode Island 02906, USA. [18]Department of Chemistry, University of North Carolina, Chapel Hill, North Carolina 27599, USA. [19]Department of Physics, North Carolina State University, Raleigh, North Carolina 27695, USA. [20]Department of Biochemistry, University of British Columbia, Vancouver, British Columbia V6T 1Z3, Canada. [21]National Human Genome Research Institute, National Institutes of Health, Bethesda, Maryland 20892, USA. Correspondence should be addressed to D.B.(dbranton@harvard.edu).

Published online 9 October 2008; doi:10.1038/nbt.1495

polyA RNA, whereas other experiments with RNA molecules containing 30 As followed by 70 Cs revealed that the transition from polyA to polyC segments within a single RNA molecule is readily detectable (Deamer[4]). Such easily measured distinctions between purine and pyrimidine ribonucleotides were unfortunately not as clear for deoxyribonucleotides (Meller and Branton[5]). In fact, the current level differences that had been observed with RNAs turned out to be a reflection of base stacking and other secondary structural differences between polyA and polyC

oligomers (Deamer[4]), and further measurements using various DNA homopolymers revealed only small ion-current differences (~5% or less) between deoxypurine and deoxypyrimidine oligomers (Meller and Branton[5]). Single nucleotide discrimination could not be achieved because the ion-current blockades were found to be the consequence of the ~10–15 nucleotides (rather than any single nucleotide) that occupy the membrane-spanning domain of the α-hemolysin pore (Meller and Branton[6]; **Fig. 1a**).

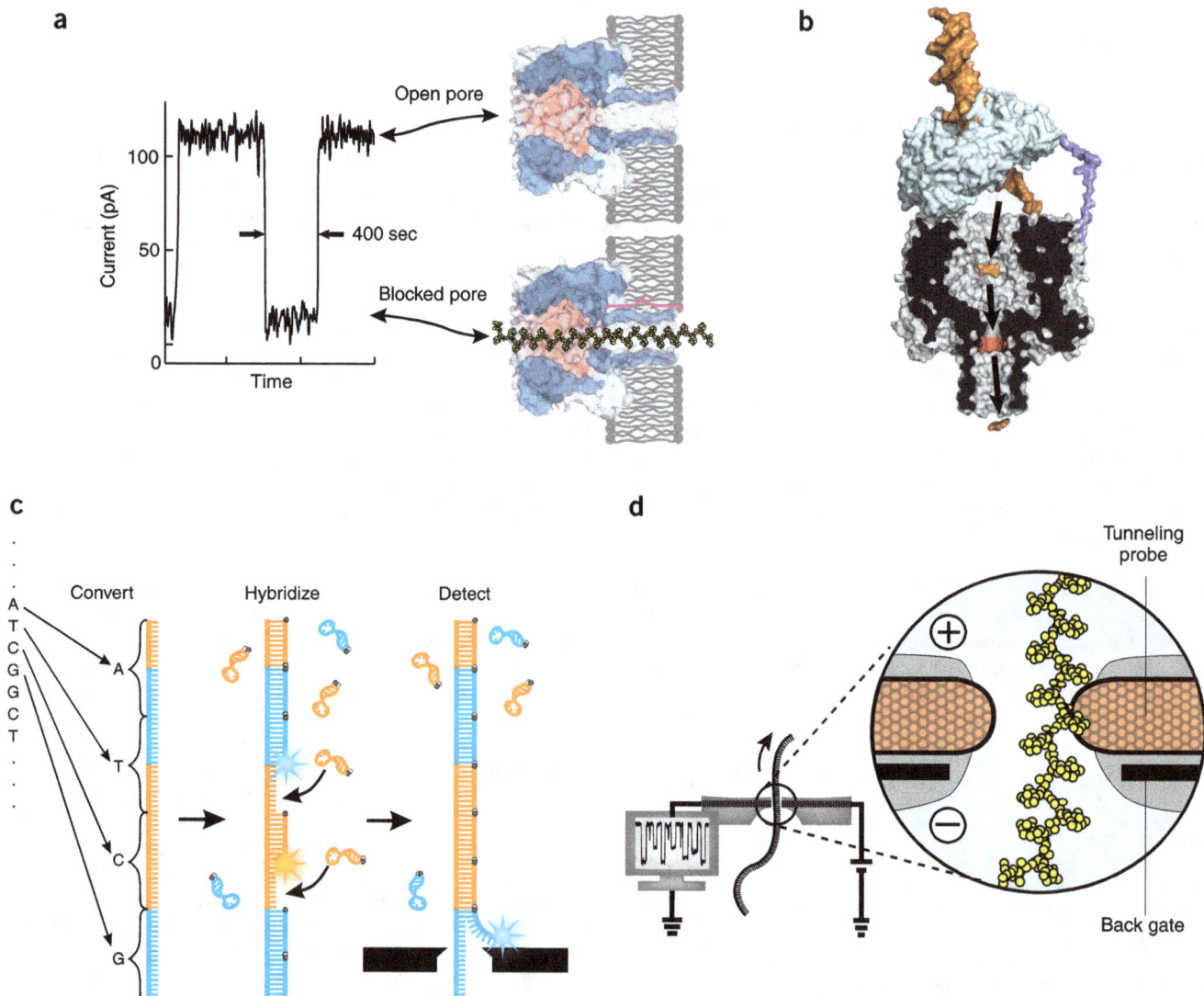

Figure 1 Approaches to nanopore sequencing. (**a**) Strand-sequencing using ionic current blockage. A typical trace of the ionic current amplitude (left) through an α-hemolysin pore clearly differentiates between an open pore (top right) and one blocked by a strand of DNA (bottom right) but cannot distinguish between the ~12 nucleotides that simultaneously block the narrow transmembrane channel domain (red bracket). (**b**) Exonuclease-sequencing by modulation of the ionic current. An exonuclease (pale blue) attached to the top of an α-hemolysin pore through a genetically encoded (deep blue), or chemical, linker sequentially cleaves dNMPs (gold) off the end of a DNA strand (in this case, one strand of a double-stranded DNA). A dNMP's identity (A, T, G or C) is determined by the level of the current blockade it causes when driven into an aminocyclodextrin adaptor (red) lodged within the pore. After a few milliseconds, the dNMP is released and exits on the opposite side of the bilayer. (**c**) Nanopore sequencing using synthetic DNA and optical readout. Each nucleotide in the target DNA that is to be sequenced is first converted into a longer DNA strand composed of pairs of two different code-units (colored orange and blue for illustration); each code-unit is a 12-base-long oligomer. After hybridizing the converted DNA with molecular beacons that are complementary to the code units, these beacons are stripped off using a nanopore. The sequence of the original DNA is read by detecting the discrete short-lived photon-bursts as each oligo is stripped. (**d**) Strand-sequencing using transverse electron currents. DNA is driven through a nanopore functionalized with embedded emitter and collector tunneling probes (orange) and a back gate (black). The amplitude of the tunneling currents that traverse through the nucleotides is expected to differentiate each nucleobase as the DNA is electrophoretically driven through the pore (arrow).

Although these early nanopore experiments disappointed naive expectations of an easy path to inexpensive DNA sequencing, they demonstrated the extraordinary single-molecule sensitivity of nanopores and stimulated many successful applications that have led to a growing literature comprising both theoretical and experimental studies related to the nanopore analysis of nucleic acids[7–10] (Deamer and Branton[7], Marziali[8], Wanunu and Meller[10]). Since the first demonstrations that an electric field can drive even kilobase lengths of ssDNA molecules through nanopores, the prospect of an inexpensive direct physical route to massive sequencing capacity has stimulated nanopore research using either protein pores in a lipid bilayer (Bayley[11]) and, more recently, fabricated nanopores in solid-state (Branton[12]), or plastic materials[13]. Indeed, under the auspices of the US National Human Genome Research Institute (NHGRI) funding program that was initiated in 2004 with the aim of reducing the cost of genome sequencing to ~$1,000 in 10 years, several grants have been made to nanopore sequencing platforms[14] (http://grants.nih.gov/grants/guide/rfa-files/RFA-HG-04-003.html.

Although nanopores are the basis for several important single-molecule applications (ref.9, Wanunu and Meller[10]), substantial lengths of DNA have yet to be sequenced with a nanopore. In view of the demonstrable progress and cost reductions with sequencing by synthesis[15,16], is continued research toward nanopore sequencing justified? This perspective presents current views on the promises of and the challenges for the development of nanopore sequencing, with a goal of engaging the broader community of scientists and engineers to propose solutions.

Justifying nanopore sequencing development

One of the most compelling advantages of nanopore sequencing is the prospect of inexpensive sample preparation requiring minimal chemistries or enzyme-dependent amplification. Furthermore, a nanopore sensor eliminates the need for nucleotides and polymerases or ligases during readout. Thus, the costs of nanopore sequencing, be it by direct strand sequencing or by one of the other methods discussed here, are projected to be far lower than ensemble sequencing by the Sanger method, or any of the recently commercialized massively parallel approaches (454, Roche, Basel; Solexa, Illumina, San Diego; SOLiD, Applied Biosystems, Foster City, CA, USA/Agencourt, Beverly, MA, USA; HelioScope, Helicos, Cambridge, MA, USA[17]). Unlike these approaches, the ideal nanopore sequencing approach would not require the use of purified fluorescent reagents and would use unamplified genomic DNA, thus eliminating enzymes, cloning and amplification steps.

The components of an ideal commercial sequencing system using electrical measurements would consist of a disposable detector chip containing an array of nanopores having the required integrated microfluidics and electronic probes; and a bench-top instrument, or portable system, that controls the fluidics and electronic elements of the chip and processes the raw sequence data. Assuming one chip will be used to sequence a sample with the complexity of a human genome, the cost of sequencing a complete human genome will be the cost of preparing the genomic DNA from a biological sample (e.g., blood), the amortized cost of the instrument, and the cost of one disposable detector chip.

Nanopore sequencing could in principle achieve a sixfold sequence coverage with less than 1 μg of genomic DNA (<10^6 copies of the target genome extracted from <10^6 cells). But the concentration-limited rate at which a nanopore can capture the diffusing DNA molecules from the source volume (Meller and Bayley[18]) will probably require ~10^8 copies of the target genome to provide adequate throughput from the 25–50 μl volume of source material required to feed an array of nanopores. Approximately 10^8 copies of the target genome corresponds to ~700 μg of human diploid genomic material, which can be directly obtained

without amplification by using commercially available kits for the isolation, purification and concentration of genomic DNA. Such kits can obtain ~1,000 μg of purified, high molecular weight (>50,000 base-pair fragments) genomic DNA from ~20 ml of blood at a cost that is likely to be <$40/sample (QIAamp DNA Blood Maxi Kit: http://www1.qiagen.com/Products/).

For direct strand sequencing in a nanopore, the diploid mammalian genome, consisting of 6×10^9 base pairs, would be fragmented into 50,000 base-pair lengths and dissociated into ssDNA (e.g., by high pH). The extremely long reads of ~50,000 bases that may be possible with nanopore methods should greatly simplify the genome assembly process. If nanopores indeed enable minimal sample processing and obviate the need for labeling, the cost of such sequencing would be dominated by the cost of the disposable chip and the amortized cost of the instrument, which is estimated to total <$1,000 per mammalian genome. But although these cost and read-length forecasts of nanopore technology are exceptionally promising, several key technological challenges must be addressed before nanopore sequencing can be implemented.

Challenges to developing nanopore sequencing

Several different approaches to using nanopores for base recognition and resolution are being considered. Those examined below are not intended to form an exhaustive list of such approaches but instead illustrate the major challenges common to most of these efforts.

Measurement of ionic current blockades as ssDNA is driven through a biopore or a solid-state pore. Although several experiments have clearly demonstrated that modulations of ionic current during translocation of RNA or DNA strands can be used to discriminate between polynucleotides (Deamer[4], Meller and Branton[5], Bayley[19]), none of the natural or manmade nanopore structures used to date has had the appropriate geometry to detect the features of only one nucleotide at a time while the polymer is translocating through the pore. None of these nanopores has channels shorter than ~5 nm and, because at least 10–15 nucleotides of ssDNA extend through a channel of this length, all of these nucleotides together contribute to the ionic current blockade (Meller and Branton[6]; **Fig. 1a**). Even an 'infinitely short' channel would not achieve the required resolution, as the high electric field region that determines the electrical 'read' region of the channel[20–23] will extend for approximately one channel diameter to each side of the channel. Because the channel diameter needs to be large enough to translocate ssDNA (~1.5 nm), the ~3 nm (2×1.5 nm) electrical 'read' region of an 'infinitely short' channel puts a fundamental restriction on the spatial resolution that can be achieved when using only current blockade measurements. Furthermore, single-stranded polynucleotides are translocated at average rates that approach ~1 nucleotide/μs when driven through a nanopore by a ~150 mV bias. Resolving single bases with small pA currents will require a means of slowing the translocation so that the time that each base occupies the nanopore detector is ≥1 msec, and possibly larger (Deamer and Branton[7]).

Alternatively, even though a nanopore cannot yet resolve the single bases separated by ~0.4 nm in a DNA strand, coarser-grained current-blockage information could be used to infer sequence by using a nanopore in conjunction with sequencing by hybridization[24]. The original concept of sequencing by hybridization contemplates aligning hybridization probes of known sequences to derive the sequence of an unknown ssDNA strand[24]. For de novo sequencing by hybridization, both the location and the number of probes bound to the long, unknown DNA strand that is to be sequenced must be known. Sequencing by hybridization alone does not provide this information. But current blockade measurements from a nanopore can readily distinguish the passage of ssDNA from the passage of double-stranded DNA (dsDNA) in a pore that is large enough to translocate dsDNA[25]. Because a nanopore is able to discriminate between ssDNA and dsDNA, it may be able to detect and resolve the location and number of oligonucleotide probes that are hybridized to a long translocating ssDNA. Thus, if the standard routines for sequencing by hybridization[24] could be enhanced by nanopore-derived information regarding the location and number of double strand regions (that is, bound oligonucleotide probes) on the DNA strands that are to be sequenced, *de novo* sequencing should be possible. This is the basic concept of hybridization-assisted nanopore sequencing (HANS; Ling[26]). Current research on the HANS method faces two challenges. Can a nanopore determine the location of a hybridized probe with sufficient accuracy to enhance sequencing by hybridization? What length of DNA sequence can be reliably reconstructed, given the practical limitations of detecting bound probes and locating them precisely on the DNA strand that is to be sequenced?

Measurement of ionic current blockades from individual nucleotides sequentially cleaved off the end of a DNA strand and driven through a biopore. At the time Keller and colleagues[27] recognized that it might be possible to sequence single molecules of DNA by identifying the deoxynucleoside monophosphates (dNMPs) released by an exonuclease from the end of a DNA or RNA chain, there was no obvious way to identify individual unlabeled bases after their release. Recent work (Bayley[28,29]) indicates that unlabeled bases can be identified by α-hemolysin when fitted with an aminocyclodextrin adaptor (Bayley[28]), and methods have now been developed to covalently attach cyclodextrins within the lumen of the α-hemolysin pore (Bayley[29]). On the basis of this work, Oxford Nanopore Technologies (Oxford, UK) has recently succeeded in covalently attaching the aminocyclodextrin adaptor within the α-hemolysin pore (**Fig. 1b**). When a dNMP is captured and driven through the α-hemolysin-aminocyclodextrin pore in a lipid bilayer membrane, the ionic current through the pore is reduced to one of four levels, each of which reflects which of the four dNMPs—A, T, G or C—is translocating. Furthermore, because all four of the ionic current blockage levels are easily distinguishable from the current that flows through the open, unblocked pore, the current traces can provide an accurate count of the total number of dNMPs that have been translocated through the α-hemolysin-aminocyclodextrin pore. For sequencing, it will now be important to assure that 100% of the exonuclease-released dNMPs are captured in the pore and efficiently expelled on the opposite side of the membrane. Because this approach uses the nanopore to identify the released dNMPs, rather than identifying the bases of an intact DNA strand, the strictly single-file, sequential passage of the bases that a nanopore can enforce is lost. It will therefore be especially important to demonstrate that the sequence of independently read dNMPs reflects the order in which the bases are cleaved from the DNA (that is, no overtaking or double counting). Finally, the choice and attachment of the exonuclease to the nanopore must be considered. A genetic construct in which the nuclease and α-hemolysin genes are spliced together might be used or the nuclease might be chemically attached to assure delivery of the released dNMPs into the nanopore. The enzyme should be processive and, for low noise detection, active in high salt. Preferably, the enzyme should digest dsDNA, which is readily produced from genomic DNA and easy to handle.

Nanopore sequencing using converted targets and optical readout. Another readout modality in development for nanopore-based sequencing converts the sequence information of DNA into a two-color scheme that is then optically read (Soni and Meller[30], Meller[31]). Whereas attachment of fluorescent probes to each and every nucleotide in DNA is

difficult, methods are available to systematically encode and substitute each and every nucleotide in the genome with a specific permutation of two different 12-mer oligos (A and B), concatenated in a specific order (AB, BA, AA, BB) that reflects and encodes the nucleotide sequence of the unknown DNA[32] (**Fig. 1c**). This converts the quaternary DNA code of A, T, G and C into a binary code in which each base is represented by a pair of 12-mer oligos (A and B). An automated, massively-parallel process developed by Lingvitae (Oslo; http://www.lingvitae.com/DPTutorial.php) currently requires ~24 h for the conversion of a complete human genome into a DNA mixture consisting of fragments, each corresponding to a 24-bp segment of the original genome. Work is currently underway to develop inexpensive error-free conversion of longer segments of the original genome and to greatly reduce the conversion time. The conversion process does introduce an extra biochemical step, which is not ideal, but it side-steps some of the challenges faced by other approaches and thus simplifies the subsequent sequencing readout.

For readout, this converted DNA mixture is then hybridized with a mixture containing two different 'molecular beacons' (Public Health Research Institute, Newark, NJ, USA; http://www.molecular-beacons.org/Introduction.html), each of which is a 12-mer oligo designed to complement either A or B. When free in solution, the molecular beacons produce only a very low background fluorescence because of self-quenching (**Fig. 1c**). Similarly, when hybridized to the converted DNA, the molecular beacons produce only low background fluorescence because the universal quencher at one end of each beacon is in close proximity to the fluorophore of its nearest neighbor (**Fig. 1c**). Because the beacons do fluoresce briefly as they are stripped off the complementary converted DNA strand, readout is performed by sequentially stripping off the fluorescent 12-mer oligos one at a time by driving the converted DNA strand through a <2-nm-diameter nanopore (that is, a pore diameter that strips off the complementary, fluorescently labeled 12-mer oligos (Branton[33])). The original DNA sequence is obtained by determining the color sequence of the photon bursts, where each pair of two successive bursts corresponds to a specific base. With high-density nanopore arrays (Wanunu and Meller[34]), optical readout can facilitate massive parallelism, and a high resolution electron-multiplying charge-coupled device camera could be used to probe thousands of nanopores simultaneously. Because the nanopores require no on-chip electrical contacts, surface modification, or mechanisms to regulate the translocation process, improved nanofabrication methods may make it possible to develop such nanopores in very high density arrays. Nevertheless, at this time, fabricating high-density arrays of 1.7- to 2-nm-diameter nanopores remains a substantial challenge.

Measurement of transverse tunneling currents or capacitance as ssDNA is driven through a solid-state nanopore with embedded probes. It has been proposed that tunneling currents through nucleobases that are driven through a nanopore articulated with tunneling probes may be able to distinguish among the four nucleobases of ssDNA (**Fig. 1d**)[35–39] (Di Ventra[35,38]). Single bases should be resolved because it is the transverse tunneling current from an emitter probe tip of ≤1-nm diameter that generates the nucleobase-identifying signal rather than the nucleotide occupancy through the entire length of the nanopore channel. Although simulations of attainable base contrast when using tunneling measurements for nucleobase identification have presented encouraging but differing insights into the challenges this approach must address[35,38,40–43] (Di Ventra[35,38,40,43]), the ability of a scanning tunneling microscope (STM) to reveal the atomic scale features of matter is well established[44].

As in a STM, electron tunneling currents can be in the nano-ampere range with appropriate probes[37,45,46] (Lindsay[46]). The nanoamp electron currents would make it possible to read the nucleotides at a greater speed than is possible with the pico-ampere ionic currents that flow through a <3 nm-diameter nanopore. Although this approach using only robust solid-state components and electrical measurements may ultimately be the least expensive and fastest way of sequencing a genome, four major challenges must be addressed (Di Ventra[43]). First, the voltage bias and solution conditions that optimize contrast between the bases must be determined and maintained to provide unambiguous nucleobase identification; it is difficult to predict beforehand exactly what the electronic response of the detector will be to the different DNA bases, particularly in a fluid system such as is envisioned here. Second, the device must provide a mechanism to assure that each base will assume a reproducible orientation and position on the collector probe while it is being interrogated; tunneling currents are exponentially sensitive to atomic scale changes of orientations and distances. Third, unidirectional translocation of the DNA must be controlled so that each nucleobase remains between the tunneling probes at least 0.1 msec to sample over inevitable noise and molecular motion; this translocation rate will assure that each nucleotide is sampled over a time period that is two orders of magnitude longer than that required for a state-of-the-art preamplifier[47] to sense nanoamp currents. And fourth, it remains to be shown whether the transverse current measurements can provide sufficient contrast to not only discriminate between the bases, but also provide a signal characteristic of the gaps between bases that could be used to distinguish each base from the next base in the unknown DNA sequence.

The use of single-walled carbon nanotubes has been proposed (Golovchenko and Branton, unpublished data) as a means of addressing the second and third challenges above—and possibly even the first challenge—if the carbon nanotube were to be appropriately functionalized[37]. Nanotubes bind and orient nucleobases in a specific manner[48] and the binding activation enthalpies per base lie in a range that can be modulated by temperature, ionic strength, or a voltage bias so as to control the DNA as it slides on the nanotube (unpublished data).

Another inventive solution to the challenge of identifying each base using transverse tunneling currents is to form base-specific hydrogen bonds between chemically modified metal electrodes and the nucleobases in the molecule that is to be sequenced. Ohshiro and Umezawa[45] showed that in a STM whose metallic probe is modified with thiol derivatives of adenine, guanine, cytosine, or uracil, tunneling is greatly enhanced between a sample nucleobase and its complementary nucleobase modified metallic probe. Using a cytosine-modified probe, they demonstrated base identification and electrical signals able to distinguish between TTTTTTTTGTTTTTTTTT and TTTTTTTTGGTTTTTTTTT. Their work has led Lindsay and colleagues[46] to propose a nanopore reader bearing pairs of two chemically functionalized probes, one probe of each pair able to couple to the nucleotide's phosphate moiety while the other probe base-pairs with the nucleobase (**Fig. 2**). The nucleobases can be identified by the current-distance responses as the DNA moves through the nanopore and past the reader, rather than the tunneling current in a static configuration. The functional groups on each of four such readers— A, C, G or T—would be designed to form a hydrogen-bonded path when the cognate base is translocated through the nanopore between the pair of probes (Lindsay[46]). Four such readers would be needed to generate a complete sequence, each one reading a duplicate strand. Synchronizing the translocation of four duplicate strands through four readers will pose a major challenge for this approach.

Electrostatic DNA detection and sequencing based on a metal-oxide-silicon capacitor incorporated into the nanopore has also been proposed[49–51]. Using the electron beam of a transmission electron microscope[49,52], a nanopore is fabricated in a membrane consisting of two layers of doped silicon, separated by a 5-nm-thick insulating SiO_2 layer. As DNA is translocated through the pore, variation of the electrostatic

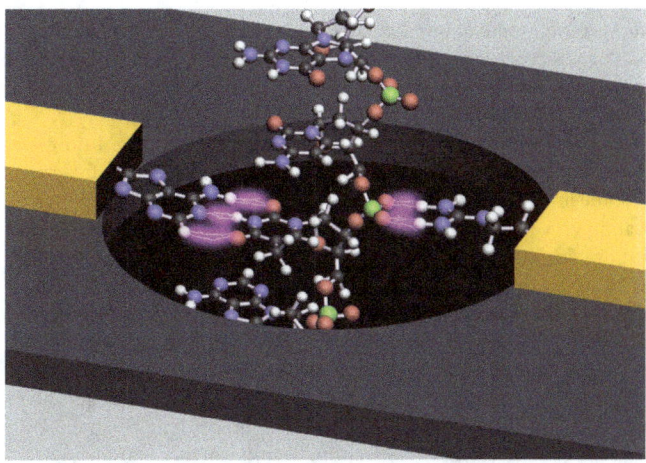

Figure 2 A nanopore reader with chemically functionalized probes. As a strand of DNA emerges from a nanopore, a 'phosphate grabber' on one functionalized electrode and a 'base reader' on the other electrode form hydrogen bonds (light blue ovals) to complete a transverse electrical circuit through each nucleotide as it is translocated through the nanopore.

potential in the pore polarizes the capacitor and voltage fluctuations on the two silicon layers are measured. Simulation results demonstrated that A, C, G and T give distinct capacitance signals and that the instrument can, in principle, resolve single-base substitutions in a DNA strand[53]. In an early trial of this approach, a voltage signal associated with DNA translocation was detected with one such device, but the time resolution was inadequate to distinguish between nucleotides[49]. The control of DNA velocity and orientation during translocation is also a major challenge for this approach.

Achieving the promise of long reads

One of the compelling potential advantages of nanopores for sequencing is the promise of long reads. Because the nanopore sensor reads molecules sequentially, base by base, as they thread through the pore, its fundamental strength is that the accuracy of a base call at one instance in time does not depend on the prior history of the system. In principle, the length of DNA that could be read with a nanopore is limited only by the practicalities of avoiding shearing during sample preparation and of limitations yet to be explored with respect to capture and threading of exceptionally long molecules through individual pores. To date, it has been demonstrated that lengths of ssDNA on the order of 25 kb have been threaded through biopores (Meller and Branton, unpublished data) and up to 5.4-kb lengths of ssDNA have been threaded through solid-state nanopores (ref.25, Branton[54]). A unique feature and promise of nanopore technologies is, therefore, that if a detection scheme is developed that allows reading of a few bases on the fly during unidirectional translocation of the DNA strand through a pore, then the extension of the technology from reading a few bases to reading thousands of bases should be straightforward. Although the expected accuracy of the read is yet unknown, insertions, deletions and other sequence errors will not compromise the read-length as de-phasing is not an issue in independent single-molecule reads. Sufficient averaging (high sequence coverage) could then reach any desired level of accuracy, as long as sequence errors are random rather than systematically sequence or position dependent.

Furthermore, given the high throughput available and anticipated in short reads from current next-generation instruments, it may be that nanopores will play a role in providing an assembly scaffold of very long reads at low accuracy to facilitate assembly of short read sequences. A hybrid combination of low accuracy long reads and high coverage, high accuracy short reads may be one path to inexpensive and rapid *de novo* sequencing. Ultimately, both of these two classes of data could be collected from nanopores.

Considering the central importance of long reads to the future of sequencing methods, additional work needs to be undertaken to determine the limitations of nanopores in capturing and sequentially translocating very long ssDNA fragments. Very high throughput detection of short single-stranded oligomers (<50 nucleotides) can be achieved (as shown by the groups of Branton and Meller[5] and Meller[55]), and for these the measured concentration-normalized capture rate constant in α-hemolysin (Meller and Branton[18]) is ~5.8 oligomers (sec μM)[-1]. Because the capture rate depends on the solution molarity and because the molar concentration (or concentration of fragment ends) must be limited to reasonably low wt/vol concentrations of long fragments to avoid excessive viscosities, it remains to be seen whether ~50-kb ssDNA fragments can be captured and threaded through small nanopores at reasonable rates. Although several publications using 3- to 6-nm-diameter pores show a reasonable number of capture translocations per minute with native 3- to 10-kb or kbp fragments of native ssDNA or dsDNA when the source chamber concentrations are in the range of 10–20 nM[25,56,57], the precise capture rate constants were not determined. In addition, although full-length λ-DNA (48 kbp) has also been captured and translocated through nanopores[51,58,59] (Branton[58]), achieving high coverage of such long reads might be most efficiently achieved by using the recently demonstrated trapping and recapture ability of a nanopore[57]. The discovery and accompanying theory that show how the same molecule that has been translocated all the way through a nanopore can be recaptured and interrogated multiple times are particularly relevant to implementing accurate sequencing. If the initial passage of an individual molecule provides an incomplete or poor quality read-out, real-time software could drive that molecule back to be re-sequenced multiple times without having to resample the entire genome.

Controlling DNA motion and translocation in a nanopore

The high speed at which DNA is translocated through nanopores[4,5,56] (Deamer[4], Meller and Branton[5]) holds the promise of ultra-fast sequencing; but the rate at which unconstrained DNA moves through these pores is also the Achilles' heel of many approaches because it implies unattainable measurements of very small currents. At 120 mV, DNA is typically translocated through an α-hemolysin pore at a rate of ~1–20 μs per nucleotide (Meller[60]). This pushes the detector bandwidth requirements to the MHz region which precludes the measurement of pico-ampere steps in ion current.

The situation is worsened by diffusion as the DNA is electrophoretically driven through the pore. Stochastic DNA motion, which is reflected in the broad distribution of transit times in both experimental[1,4–6,25,52,54,56,58,61] (Deamer and Branton[1], Deamer[4], Meller and Branton[5,6], Branton[54,58]) and theoretical studies[19,62–68], can, as indicated above, generate uncertainty in the number of bases that have passed through the nanopore. Furthermore, nonspecific interactions between the translocating DNA and the nanopore's surface may be dominated by discontinuous stick-slip phenomena[69]. Variability in the nature and frequency of interactions can give rise to non-Poisson distributions of escape times (Wiggin and Marziali[70]; Marziali[71], Wiggin, Tabard-Cossa and Marziali[72]; Meller and Wanunu[73]), such that the translocation time for two identical molecules can differ by orders of magnitude[1,4–6,51,56,61,72] (Deamer and Branton[1]; Deamer[4]; Meller and Branton[5,6]; Wiggin, Tabard-Cossa and Marziali[72]). If some of the nucleotides in a DNA strand slip between the probing elements of a nanopore in time periods that are substantially less than the average, these fast-translocating nucleotides may be missed.

Thus, a key challenge to DNA sequencing with nanopores is to find methods to slow down and control DNA translocation and reduce the fluctuations in translocation kinetics because of pore-surface interactions. DNA translocation speeds can be reduced somewhat by decreasing temperature (Meller and Branton[5], Meller[74]), or increasing solvent viscosity[75], but these methods do not reduce the variations in the translocation dynamics because of DNA-pore interactions[70–73,76] (Wiggin and Marziali[70]; Marziali[71], Wiggin, Tabard-Cossa and Marziali[72]; Meller and Wanunu[70]). Substantial reductions of the translocation rate can be achieved with processive DNA enzymes[77–79] (Akeson and Meller[77]; Deamer and Akeson[78]) which limit the translocation rate by binding to the DNA strand and preventing it from moving into the narrow confines of the pore faster than the enzyme processing rate; or by successive unzipping of DNA oligos, which then becomes the rate-limiting step for the translocation process (Soni and Meller[30], Meller[31], Branton[33]). These processing rates are typically on the time scale of a few milliseconds per base and can be controlled through ion concentrations[78–80] (Deamer and Akeson[78]), temperature and the voltage bias through the nanopore.

Ultimately, eliciting a distinct electrical signal from the space between bases to provide a clear count of the number of bases that are translocated would be ideal. Such signals would greatly facilitate further analysis of translocation kinetics and base dwell time distributions so that the detection system developers can determine the required bandwidth and performance specifications of their systems. But until such signals are available, a detailed understanding of, and methods to control, the kinetics of DNA strand translocation through a narrow pore need to be obtained. Fabricating nanopores provides the opportunity for generating nanopores with tailored surface properties that could both regulate DNA-pore surface interaction (Wanunu and Meller[81]) and reduce noise (Branton[82]; Tabard-Cossa, Wiggin and Marziali[83]). Ultimately, a combination of methods to control translocation rate and DNA-pore interaction will need to be coupled to high-bandwidth, low-noise detection to achieve the fast sequence analysis that is the promise of many nanopore approaches.

Biopore stability and fabrication of solid-state pores

The hemolysin heptamer, which until now has been the usual protein that is used to form biopores in lipid bilayers, is remarkably stable (Bayley[3]). The primary instability therefore arises from the support, typically a fluid lipid bilayer, which is difficult and time consuming to set up.

Bayley and his colleagues[84] have demonstrated that a bilayer encapsulated between two thin layers of agarose with a single inserted α-hemolysin pore is sufficiently stable to be sealed in Teflon film and stored for weeks before use. They also show that a single α-hemolysin pore can be introduced in each element of an array of such bilayers using agarose-tipped plastic or glass probes (Bayley[85,86]). Another approach to stabilizing bilayers is to use nano-scale, rather than micron-scale, apertures. Bilayers across 100- to 1,000-nm-diameter apertures at the end of glass capillaries coated with a specially formulated silanizing agent have been shown to be stable for over two weeks[87].

Very stable, functionally useful solid-state nanopores can be fabricated in silicon nitride, silicon oxide or metal oxides, using ion beam sculpting[56], e-beam drilling (Ling[88]) and atomic layer deposition (Branton[82]), but generating arrays of a large number of uniform solid-state nanopores with diameters in the 1.5- to 2.0-nm range remains a daunting task, particularly for academic research laboratories that cannot afford commercial production facilities. Articulated nanopores with buried nanotube probes for tunneling measurements have been realized, but the current research-scale fabrication methods are so tedious, slow and manpower-expensive they often cannot be used to provide even the limited number of such nanopores required for research-scale development.

There is little doubt that the accelerating rate of discovery in the field of nano-scale electronics and the proven ability of the electronics community to develop mass-production strategies for high-value components will be able to master the nano-scale science required to fabricate massive nanopore arrays. But until such time as nanopore sequencing in any form is shown to be feasible and valuable, nanopore sequencing researchers face the challenge of using only research-scale facilities rather than those that are to be found, or could be developed, in a specialized, mass-production plant.

For some nanopore applications, the ultimate stable pore is likely to be a hybrid between a solid-state pore and α-hemolysin. This might involve producing a ~5-nm pore in a synthetic membrane such as silicon nitride, then capturing an α-hemolysin heptamer in the pore in the absence of a lipid bilayer. Should this prove possible, the resulting nanopore is likely to be both highly reproducible and indefinitely stable.

Conclusions

Several advantages are offered by nanopore sequencing if it can be achieved. The most important are minimal sample preparation, sequence readout that does not require nucleotides, polymerases or ligases, and the potential of very long read-lengths (>10,000–50,000 nt). It follows that a successful nanopore sequencing device will provide a tremendous reduction in costs and might well achieve the $1,000 per mammalian genome goal set by the US National Institutes of Health. The instrument itself will be relatively inexpensive, and the time required for sixfold coverage could be as little as one day if 100 nanopores having the required integrated microfluidics and electronic probes can be fabricated into each sequencing chip. But important challenges remain. A substantial short-term challenge is to slow DNA translocation from microseconds per base to milliseconds, and several recent studies indicate that this can be achieved by using DNA-processing enzymes. If a future instrument incorporates the hemolysin heptamer, it will also be necessary to establish a stable support of some kind. Again, there is recent progress toward this end, though in the longer term it seems likely that synthetic solid-state nanopores will be preferred for a commercial instrument. Electronic sensing based on either tunneling probes or a capacitor is being tested for its ability to detect a DNA strand during translocation, but whether this is possible remains to be demonstrated. A continuing concern is that stochastic motion of the DNA molecule in transit will increase signal noise in such a sensor, thereby reducing the potential for single-base resolution. All that said, the advantages of nanopore sequencing are so attractive that work will continue unless a fundamental limitation is discovered. So far, no such limitation has emerged, and the progress toward the goal of fast, inexpensive nanopore sequencing has been both impressive and encouraging.

AUTHOR CONTRIBUTIONS
D.B. wrote this review, with additions and editorial assistance from D.W.D., A.Marziali. and H.B. S.A.B., T.B., M.D., S.G., A.H., X.H., S.B.J., P.S.K., S.L., X.S.L., C.H.M., A.Meller., J.S.O., Y.V.P., J.M.R., R.R., G.V.S., V.T.-C., M.Wanunu. and M.Wiggin. contributed some of the text and read drafts of the manuscript for accuracy. J.A.S. proposed the idea for the review and read the manuscript for accuracy.

Published online at http://www.nature.com/naturebiotechnology/
Reprints and permissions information is available online at http://npg.nature.com/reprintsandpermissions/

1. Kasianowicz, J.J., Brandin, E., Branton, D. & Deamer, D.W. Characterization of individual polynucleotide molecules using a membrane channel. *Proc. Natl. Acad. Sci. USA* **93**, 13770–13773 (1996).
2. Braha, O. *et al.* Designed protein pores as components for biosensors. *Chem. Biol.* **4**, 497–505 (1997).
3. Kang, X.F., Gu, L.-Q., Cheley, S. & Bayley, H. Single protein pores containing molecular adapters at high temperatures. *Angew. Chem. Int. Edn Engl.* **44**, 1495–1499 (2005).
4. Akeson, M., Branton, D., Kasianowicz, J.J., Brandin, E. & Deamer, D.W. Microsecond time-scale discrimination among polycytidylic acid, polyadenylic acid, and polyuridylic acid as homopolymers or as segments within single RNA molecules. *Biophys. J.* **77**, 3227–3233 (1999).
5. Meller, A., Nivon, L., Brandin, E., Golovchenko, J. & Branton, D. Rapid nanopore discrimination between single oligonucleotide molecules. *Proc. Natl. Acad. Sci. USA* **97**, 1079–1084 (2000).
6. Meller, A., Nivon, L. & Branton, D. Voltage-driven DNA translocations through a nanopore. *Phys. Rev. Lett.* **86**, 3435–3438 (2001).
7. Deamer, D.W. & Branton, D. Characterization of nucleic acids by nanopore analysis. *Acc. Chem. Res.* **35**, 817–825 (2002).
8. Nakane, J.J., Akeson, M. & Marziali, A. Nanopore sensors for nucleic acid analysis. *J. Phys. Condens. Matter* **15**, R1365–R1393 (2003).
9. Healy, K. Nanopore-based single-molecule DNA analysis. *Nanomedicine* **2**, 459–481 (2007).
10. Wanunu, M. & Meller, A. Single-molecule analysis of nucleic acids and DNA-protein interactions using nanopores. in *Single-Molecule Techniques: A Laboratory Manual* (eds. Selvin, P. & Ha, T.J.), p 395–420 (Cold Spring Harbor Laboratory Press, Cold Spring Harbor, NY, 2008).
11. Tobkes, N., Wallace, B.A. & Bayley, H. Secondary structure and assembly mechanism of an oligomeric channel protein. *Biochemistry* **24**, 1915–1920 (1985).
12. Li, J. *et al.* Ion-beam sculpting at nanometre length scales. *Nature* **412**, 166–169 (2001).
13. Harrell, C.C. *et al.* Resistive-pulse DNA detection with a conical nanopore sensor. *Langmuir* **22**, 10837–10843 (2006).
14. Schloss, J.A. How to get genomes at one ten-thousandth the cost. *Nat. Biotechnol.* **26**, 1113–1116 (2008).
15. Shendure, J. & Hanlee, J. Next-generation DNA sequencing. *Nat. Biotechnol.* **26**, 1135–1145 (2008).
16. Rothberg, J.M. & Leamon, J. The development and impact of 454 sequencing. *Nat. Biotechnol.* **26**, 1117–1124 (2008).
17. Mardis, E.R. The impact of next-generation sequencing technology on genetics. *Trends Genet.* **24**, 133–141 (2008).
18. Meller, A. & Branton, D. Single molecule measurements of DNA transport through a nanopore. *Electrophoresis* **23**, 2583–2591 (2002).
19. Ashkenasy, N., Sanchez-Quesada, J., Bayley, H. & Ghadiri, M.R. Recognizing a single base in an individual DNA strand: a step toward DNA sequencing in nanopores. *Angew. Chem. Int. Ed.* **44**, 1401–1404 (2005).
20. Aksimentiev, A., Heng, J.B., Timp, G. & Schulten, K. Microscopic kinetics of DNA translocation through synthetic nanopores. *Biophys. J.* **87**, 2086–2097 (2004).
21. Aksimentiev, A. & Schulten, K. Imaging α-hemolysin with molecular dynamics: ionic conductance, osmotic permeability, and the electrostatic potential map. *Biophys. J.* **88**, 3745–3761 (2005).
22. Muthukumar, M. & Kong, C.Y. Simulation of polymer translocation through protein channels. *Proc. Natl. Acad. Sci. USA* **103**, 5273–5278 (2006).
23. Liu, H., Qian, S. & Bau, H.H. The effect of translocating cylindrical particles on the ionic current through a nanopore. *Biophys. J.* **92**, 1164–1177 (2007).
24. Drmanac, R. *et al.* Sequencing by hybridization (SBH): advantages, achievements, and opportunities. *Adv. Biochem. Eng. Biotechnol.* **77**, 75–101 (2002).
25. Fologea, D. *et al.* Detecting single stranded DNA with a solid state nanopore. *Nano Lett.* **5**, 1905–1909 (2005).
26. Ling, X.S., Bready, B. & Pertsinidis, A. Hybridization-assisted nanopore sequencing of nucleic acids. US patent application no. 2007 0190542 (2007).
27. Jett, J.H. *et al.* High-speed DNA sequencing: an approach based upon fluorescence detection of single molecules. *J. Biomol. Struct. Dyn.* **7**, 301–309 (1989).
28. Astier, Y., Braha, O. & Bayley, H. Toward single molecule DNA sequencing: direct identification of ribonucleoside and deoxyribonucleoside 5′-monophosphates by using an engineered protein nanopore equipped with a molecular adapter. *J. Am. Chem. Soc.* **128**, 1705–1710 (2006).
29. Wu, H.-C., Astier, Y., Maglia, G., Mikhailova, E. & Bayley, H. Protein nanopores with covalently attached molecular adapters. *J. Am. Chem. Soc.* **129**, 16142–16148 (2007).
30. Soni, G.V. & Meller, A. Progress toward ultrafast DNA sequencing using solid-state nanopores. *Clin. Chem.* **53**, 1996–2001 (2007).
31. Lee, J.W. & Meller, A. Rapid DNA sequencing by direct nanoscale reading of nucleotide bases on individual DNA chains. in *Perspectives in Bioanalysis* vol. 2, (ed. Mitchelson, K.), 245–263 (Elsevier, Oxford, UK, 2007).
32. Lexow, P. Sequencing method using magnifying tags. US Patent 6,723,513 B2 (2004).
33. Sauer-Budge, A.F., Nyamwanda, J.A., Lubensky, D.K. & Branton, D. Unzipping kinetics of double-stranded DNA in a nanopore. *Phys. Rev. Lett.* **90**, 2381011–2381014 (2003).
34. Kim, M.J., Wanunu, M., Bell, D.C. & Meller, A. Rapid fabrication of uniformly sized nanopores and nanopore arrays for parallel DNA analysis. *Adv. Mater.* **18**, 3149–3153 (2006).
35. Zwolak, M. & Di Ventra, M. Electronic signature of DNA nucleotides via transverse transport. *Nano Lett.* **5**, 421–424 (2005).
36. Zikic, R. *et al.* Characterization of the tunneling conductance across DNA bases. *Phys. Rev. E* **74**, 011919 (2006).
37. Meunier, V. & Krstic, P.S. Enhancement of the transverse conductance in DNA nucleotides. *J. Chem. Phys.* **128**, 041103 (2008).
38. Lagerqvist, J., Zwolak, M. & Di Ventra, M. Fast DNA sequencing via transverse electronic transport. *Nano Lett.* **6**, 779–782 (2006).
39. Zhang, X.-G., Krstic, P.S., Zikic, R., Wells, J.C. & Fuentes-Cabrera, M. First-principles

transversal DNA conductance deconstructed. *Biophys. J.* **91**, L04–L06 (2006).

40. Lagerqvist, J., Zwolak, M. & Di Ventra, M. Influence of the environment and probes on rapid DNA sequencing via transverse electronic transport. *Biophys. J.* **93**, 2384–2390 (2007).

41. Meng, S., Maragakis, P., Papaloukas, C. & Kaxiras, E. DNA nucleoside interaction and identification with carbon nanotubes. *Nano Lett.* **47**, 45–50 (2007).

42. Xu, M., Endres, R.G. & Arakawa, Y. The electronic properties of DNA bases. *Small* **3**, 1539–1543 (2007).

43. Zwolak, M. & DiVentra, M. Physical approaches to DNA sequencing and detection. *Rev. Mod. Phys.* **80**, 141–165 (2008).

44. Golovchenko, J. The tunneling microscope: a new look at the atomic world. *Science* **232**, 48–53 (1986).

45. Ohshiro, T. & Umezawa, Y. Complementary base-pair-facilitated electron tunneling for electrically pinpointing complementary nucleobases. *Proc. Natl. Acad. Sci. USA* **103**, 10–14 (2006).

46. He, J., Lin, L., Zhang, P. & Lindsay, S. Identification of DNA base-pairing via tunnel-current decay. *Nano Lett.* **7**, 3854–3858 (2007).

47. Michel, B., Novotny, L. & Durig, U. Low-temperature compatible I–V converter. *Ultramicroscopy* **42–44**, 1647–1652 (1992).

48. Hughes, M.E., Brandin, E. & Golovchenko, J.A. Optical absorption of DNA-carbon nano-tube structures. *Nano Lett.* **7**, 1191–1194 (2007).

49. Heng, J.B. *et al.* Beyond the gene chip. *Bell Labs Tech. J.* **10**, 5–22 (2005).

50. Gracheva, M.E. *et al.* Simulation of the electric response of DNA translocation through a semiconductor nanopore-capacitor. *Nanotechnology* **17**, 622–633 (2006).

51. Sigalov, G., Comer, J., Timp, G. & Aksimentiev, A. Detection of DNA sequences using an alternating electric field in a nanopore capacitor. *Nano Lett.* **8**, 56–63 (2008).

52. Storm, A.J. *et al.* Fast DNA translocation through a solid-state nanopore. *Nano Lett.* **7**, 1193–1197 (2005).

53. Gracheva, M.E., Aksimentiev, A. & Leburton, J.-P. Electrical signatures of single-stranded DNA with single base mutations in a nanopore capacitor. *Nanotechnology* **17**, 3160–3165 (2006).

54. Fologea, D., Brandin, E., Uplinger, J., Branton, D. & Li, J. DNA conformation and base number simultaneously determined in a nanopore. *Electrophoresis* **28**, 3186–3192 (2007).

55. Mathe, J., Visram, H., Viasnoff, V., Rabin, Y. & Meller, A. Nanopore unzipping of individual DNA hairpin molecules. *Biophys. J.* **87**, 3205–3212 (2004).

56. Li, J., Gershow, M., Stein, D., Brandin, E. & Golovchenko, J. DNA molecules and configurations in a solid-state nanopore microscope. *Nat. Mater.* **2**, 611–615 (2003).

57. Gershow, M. & Golovchenko, J.A. Recapturing and trapping single molecules with a solid state nanopore. *Nature Nanotechnology* **2**, 775–779 (2007).

58. Chen, P. *et al.* Probing single DNA molecule transport using fabricated nanopores. *Nano Lett.* **4**, 2293–2298 (2004).

59. Smeets, R.M.M. *et al.* Salt dependence of ion transport and DNA translocation through solid-state nanopores. *Nano Lett.* **6**, 89–95 (2006).

60. Meller, A. Dynamics of polynucleotide transport through nanometer-scale pores. *J. Phys. Condens. Matt.* **15**, R581–R607 (2003).

61. Heng, J.B. *et al.* Sizing DNA using a nanometer-diameter pore. *Biophys. J.* **87**, 2905–2911 (2004).

62. Lubensky, D.K. & Nelson, D.R. Driven polymer translocation through a narrow pore. *Biophys. J.* **77**, 1824–1838 (1999).

63. Chern, S.S., Cardenas, A.E. & Coalson, R.D. Three-dimensional dynamic Monte Carlo simulations of driven polymer transport through a hole in a wall. *J. Chem. Phys.* **115**, 7772–7782 (2001).

64. Loebl, H.C., Randel, R., Goodwin, S.P. & Matthai, C.C. Simulation studies of polymer translocation through a channel. *Phys. Rev. E.* **67**, 041913–041911 - 041913–041915 (2003).

65. Matysiak, S., Montesi, A., Pasquali, M., Kolomeisky, A.B. & Clementi, C. Dynamics of polymer translocation through nanopores. Theory meets experiment. *Phys. Rev. Lett.* **96**, 118103 (2006).

66. Huopaniemi, I., Luo, K., Ala-Nissila, T. & Ying, S.C. Langevin dynamics simulations of polymer translocation through nanopores. *J. Chem. Phys.* **125**, 124901 (2006).

67. Chen, P. & Li, C.M. Nanopore unstacking of single-stranded DNA helices. *Small* **3**, 1204–1208 (2007).

68. Zhao, X., Payne, C.M., Cummings, P. & Lee, J.W. Single stranded DNA molecules translocation through nanoelectrode gaps. *Nanotechnology* **18**, 424018 (2007).

69. Cheikh, C. & Koper, G. Influence of the stick-slip transition on the electrokinetic behavior of nanoporous material. *Physica A* **373**, 21–28 (2007).

70. Nakane, J., Wiggin, M. & Marziali, A. A nanosensor for transmembrane capture and identification of single nucleic acid molecules. *Biophys. J.* **87**, 615–621 (2004).

71. Tropini, C. & Marziali, A. Multi-nanopore force spectroscopy for DNA analysis. *Biophys. J.* **92**, 1632–1637 (2007).

72. Wiggin, M.W., Tropini, C.T., Tabard-Cossa, V. Jetha, N.N., & Marziali, A. Non-exponential kinetics of DNA escape from α-hemolysin nanopores. *Biophys. J.* published online, doi:101529/biophysj.108.137760 (5 September 2008).

73. Wanunu, M., Chakrabarti, B., Mathe, J., Nelson, D.R. & Meller, A. Orientation-dependent interactions of DNA with an α-hemolysin channel. *Phys. Rev. E* **77**, 031904 (2008).

74. Mathe, J., Aksimentiev, A., Nelson, D.R., Schulten, K. & Meller, A. Orientation discrimination of single-stranded DNA inside the α-hemolysin membrane channel. *Proc. Natl. Acad. Sci. USA* **102**, 12377–12382 (2005).

75. Fologea, D., Uplinger, J., Thomas, B., McNabb, D.S. & Li, J. Slowing DNA translocation in a solid-state nanopore. *Nano Lett.* **5**, 1734–1737 (2005).

76. Payne, C.M., Zhao, X., Vlcek, L. & Cummings, P. Molecular dynamics simulation of ss-DNA translocation between copper nanoelectrodes incorporating electrode charge dynamics. *J. Phys. Chem. B* **112**, 1712–1717 (2008).

77. Hornblower, B. *et al.* Single-molecule analysis of DNA-protein complexes using nanopores. *Nat. Methods* **4**, 315–317 (2007).

78. Benner, S. *et al.* Sequence-specific detection of individual DNA polymerase complexes in real time using a nanopore. *Nat. Nanotechnol.* **2**, 718–724 (2007).

79. Cockroft, S.L., Chu, J., Amorin, M. & Ghadiri, M.R. A single-molecule nanopore device detects DNA polymerase activity with single-nucleotide resolution. *J. Am. Chem. Soc.* **130**, 818–820 (2008).

80. Joyce, C.M. & Steitz, T.A. Function and structure relationships in DNA-polymerases. *Annu. Rev. Biochem.* **63**, 777–822 (1994).

81. Wanunu, M. & Meller, A. Chemically-modified solid-state nanopores. *Nano Lett.* **7**, 1580–1585 (2007).

82. Chen, P. *et al.* Atomic layer deposition to fine-tune the surface properties and diameters of fabricated nanopores. *Nano Lett.* **4**, 1333–1337 (2004).

83. Tabard-Cossa, V., Trivedi, D., Wiggin, M., Jetha, N.N. & Marziali, A. Noise analysis and reduction in solid-state nanopores. *Nanotechnology* **18**, 305505–305510 (2007).

84. Kang, X.F., Cheley, S., Rice-Ficht, A.C. & Bayley, H. A storable encapsulated bilayer chip containing a single protein nanopore. *J. Am. Chem. Soc.* **129**, 4701–4705 (2007).

85. Holden, M.A. & Bayley, H. Direct introduction of single protein channels and pores into lipid bilayers. *J. Am. Chem. Soc.* **127**, 6502–6503 (2005).

86. Holden, M.A., Jayasinghe, L., Daltrop, O., Mason, A. & Bayley, H. Direct transfer of membrane proteins from bacteria to planar bilayers for rapid screening by single-channel recording. *Nat. Chem. Biol.* **2**, 314–318 (2006).

87. White, R.J. *et al.* Single ion-channel recordings using glass nanopore membranes. *J. Am. Chem. Soc.* **129**, 11766–11775 (2007).

88. Storm, A.J., Chen, J.H., Ling, X.S., Zandbergen, H.W. & Dekker, C. Fabrication of solid-state nanopores with single-nanometre precision. *Nat. Mater.* **2**, 537–541 (2003).

Atomic force microscopy as a multifunctional molecular toolbox in nanobiotechnology

With its ability to observe, manipulate and explore the functional components of the biological cell at subnanometre resolution, atomic force microscopy (AFM) has produced a wealth of new opportunities in nanobiotechnology. Evolving from an imaging technique to a multifunctional 'lab-on-a-tip', AFM-based force spectroscopy is increasingly used to study the mechanisms of molecular recognition and protein folding, and to probe the local elasticity, chemical groups and dynamics of receptor–ligand interactions in live cells. AFM cantilever arrays allow the detection of bioanalytes with picomolar sensitivity, opening new avenues for medical diagnostics and environmental monitoring. Here we review the fascinating opportunities offered by the rapid advances in AFM.

DANIEL J. MÜLLER[1] AND YVES F. DUFRÊNE[2]

[1]Biotechnology Center, Technische Universität Dresden, Tatzberg 47-51, D-01307 Dresden, Germany; [2]Unité de Chimie des Interfaces, Université Catholique de Louvain, Croix du Sud 2/18, B-1348 Louvain-la-Neuve, Belgium

e-mail: daniel.mueller@biotec.tu-dresden.de; yves.dufrene@uclouvain.be

Nanobiotechnology — the characterization, design and application of biological systems at the nanometre scale — is a rapidly evolving area at the crossroads of nanoscience, biology and engineering[1,2]. One fascinating challenge in nanobiotechnology is the bottom-up design and operation of nanoscale machines and motors made up of supramolecular systems[3,4]. The molecular components of these systems are typically cellular machineries such as haemolysin[5], porin[6], myosin[7], kinesin[8], RNA polymerase[9] and ATP synthase[10]. These machines can be set in motion in a controlled manner to work as bio-inspired nanoscale valves, engines, motors and shuttles.

Progress in nanobiotechnology strongly relies on the development of scanning probe techniques, particularly atomic force microscopy (AFM)[11,12]. For the first time, AFM techniques are allowing us to observe and manipulate biomolecular machinery in its native environment (Fig. 1). The principle of AFM is to raster-scan a sharp tip over the sample surface and to probe interaction forces with piconewton sensitivity. AFM can image surfaces in buffer solution with outstanding signal-to-noise ratio, offering a means of observing single biomolecules without the need for fixation or staining. Cellular functions are governed ultimately by the dynamic character of their working parts, which makes them difficult to understand solely from static structural information. Time-lapse AFM imaging directly observes biological specimens at work, thus providing a unique tool to approach the relationship between their structure and function. Moreover, using the AFM probe as a 'lab-on-a-tip' enables us to probe simultaneously the structure and specific biological, chemical and physical parameters of the cell's machinery.

Molecular interactions make up the basic language of all biological processes. They determine how molecules assemble into complex functional structures and switch these cellular machineries 'on' or 'off'. In addition, molecular interactions control the way that cells communicate with each other, and form higher-ordered organisms. Remarkably, AFM can measure interactions between and within single biomolecules, giving us clues on how cells direct adhesion between themselves. Such information is important for understanding tissue development, tumour metastasis, bacterial infection and an almost uncountable number of medical and biotechnological questions. AFM can also be used to determine the energy landscape of molecular recognition events and protein folding pathways. AFM is clearly emerging as a powerful, multifunctional nanoscale tool, flourishing in modern biological and medical laboratories, and opening up exciting new possibilities for nanobiotechnologists (Fig. 1).

OBSERVING CELLULAR MACHINERIES AT WORK

With current progress in AFM instrumentation, sample preparation and imaging, researchers can now routinely obtain topographs of biological specimens in physiological conditions[13–15]. The ability to visualize native biomolecules without the need for staining or fixation, together with superior signal-to-noise ratio, makes AFM a complementary tool to optical microscopy and even to X-ray and electron crystallography.

So far, the best spatial resolution obtained using AFM on corrugated cell surfaces is of the order of 10 nm. By contrast, AFM allows us to observe isolated cell membranes adsorbed onto flat supports with subnanometre resolution (Fig. 2). Prominent examples include the voltage-dependent ion channels of the mitochondrial membrane[16], ion-driven rotors of F_oF_1-ATP synthases[17,18], communication channels (Fig. 2a)[19], and G-protein coupled receptors (GPCRs) (Fig. 2b)[20], all of which cover key functions of the human body and are the targets of a large number of therapeutic drugs. The example of GPCRs demonstrates impressively that their temporary assembly with other GPCRs adapts their functions to the needs of the biological cell[21]. This is the case for many of the other cell machineries, and it is of outstanding interest to characterize factors that determine how they assemble into larger functional units. High-resolution topographs of light-harvesting complexes[22] provide insights into the molecular

When citing this review, please cite the original version as shown on the contents page of this chapter.

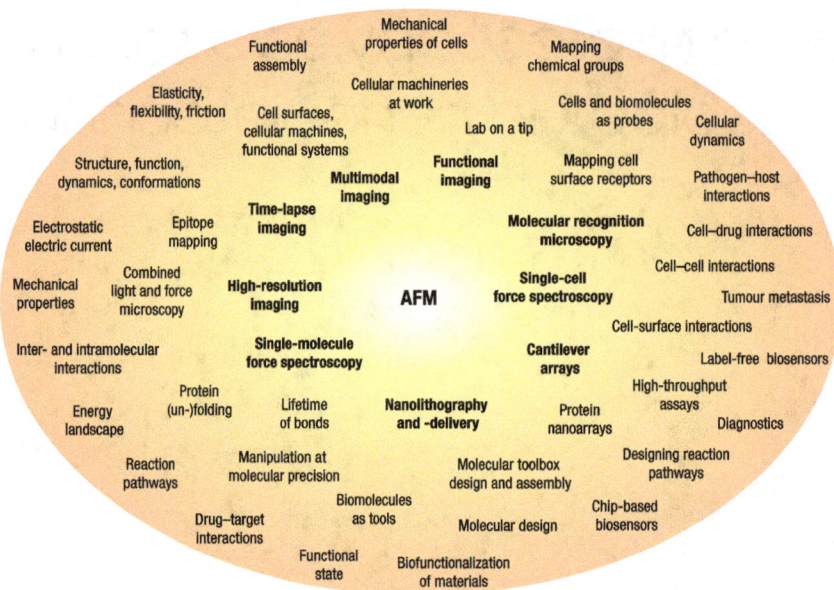

Figure 1 AFM techniques to characterize and manipulate biological systems on the nanometre scale. Originally invented to image the topography of surfaces, AFM has evolved into a multifunctional tool that makes it possible to characterize biological cells and their components with unprecedented resolution. AFM techniques allow single-molecule analyses and an understanding of various biochemical, physical and chemical properties and interactions in the cell.

rearrangements of the photosynthetic membrane (Fig. 2c). Simply tuning the light intensity optimizes the assembly of light-harvesting complexes for their conversion of light into electric energy. Similar resolution could be achieved on tubulin protofilaments that assemble into microtubuli, which are structures involved in cellular processes such as trafficking, migration and cell division. The anti-cancer agent taxol inhibits the polymerization of tubulin into microtubules and thus inhibits the division of cancer cells. AFM allows us to observe directly the action of taxol in slowing down the structural transitions of tubulin protofilaments by several orders of magnitude[23] (Fig. 2d). In the future, there is no doubt that AFM will equip most modern biological laboratories to characterize the assembly of the different functional components of the cell at nanoscopic resolution.

Watching cellular nanomachines at work provides a fascinating and direct way to access their structure–function relationships. Since the early days, time-lapse AFM has been used to observe dynamic processes ranging from the atomic, molecular and more microscopic scales[24]. In pioneering studies, time-lapse AFM was applied to observe directly the activity of RNA polymerase[25], the phospholipase-induced bilayer degradation[26] and the reversible closure of the pores of a bacterial surface layer[13]. Notably, AFM has been applied to characterize factors that trigger the function of the cell's machinery. For instance, by developing ultrafast small cantilevers of 10–30 μm in length, it was possible to monitor the formation of molecular chaperone complexes (GroEL/ES) in real time[27]. Addition of the 'biofuel' Mg-adenosine triphosphate (ATP) triggered the formation of these complexes and the fast cantilevers were able to measure their lifetimes. Developing even faster AFM cantilevers and feedback loops revealed the conformational states and lifetimes of GroEL, which were reversibly switching from the open to the closed state[28]. In another study[19], AFM revealed the reversible closure of communication channels in response to the ligand Ca^{2+}. In the presence of taurine — an aminosulphonate compound — a pH change again closes this channel, but apparently by a different gating mechanism (Fig. 2e)[29]. Whereas in the presence of Ca^{2+} the gap junction hemichannels moved their subunits centrally

towards closing the channel entrance, the pH-induced closure twisted the subunits like a camera iris.

AFM is evolving from an imaging technique to a multifunctional toolbox. Conversion of the AFM stylus to a 'lab-on-a-tip', by means of specific surface modifications, makes simultaneous imaging and probing possible for a variety of additional traits such as antigen recognition, flexibility, elasticity or electric current[12,30]. This approach complements real-time structural observations with multifunctional data that are difficult to reveal by conventional approaches. In pioneering work, pH- and voltage-dependent conformational changes were demonstrated for the channel-forming protein OmpF porin[13]. The closure mechanism discovered for OmpF, whereby the flexible extracellular loops could reversibly gate the channel entrance, was later reported to gate other porins from the same family as well. Using the AFM stylus as a lab-on-a-tip has allowed the observation of individual OmpF porins at subnanometre resolution and the detection of the electrostatic field established by their channels[31]. This field was thought to tune the channel selectivity for charged ions. Recently, a conducting high-resolution AFM probe was introduced to detect electric currents in the femtoampere range, such as those gated by membrane proteins[32]. In the future, this approach may complement the 'patch clamp technique', which basically measures ion currents of single membrane channels. A modified AFM probe may simultaneously measure and locate ion channel currents of a given membrane protein and characterize functional related conformations at (sub)nanometre resolution.

SINGLE-MOLECULE MANIPULATION

Applying a mechanical load to molecules forces bonds formed through interactions within or between the molecules. As the externally applied force overcomes the stability of a bond it breaks. Gaub[33] and Lee[34] introduced the use of single-molecule force spectroscopy (SMFS) to probe the strength of receptor–ligand bonds (Fig. 3a). These experiments initiated many others which all demonstrated

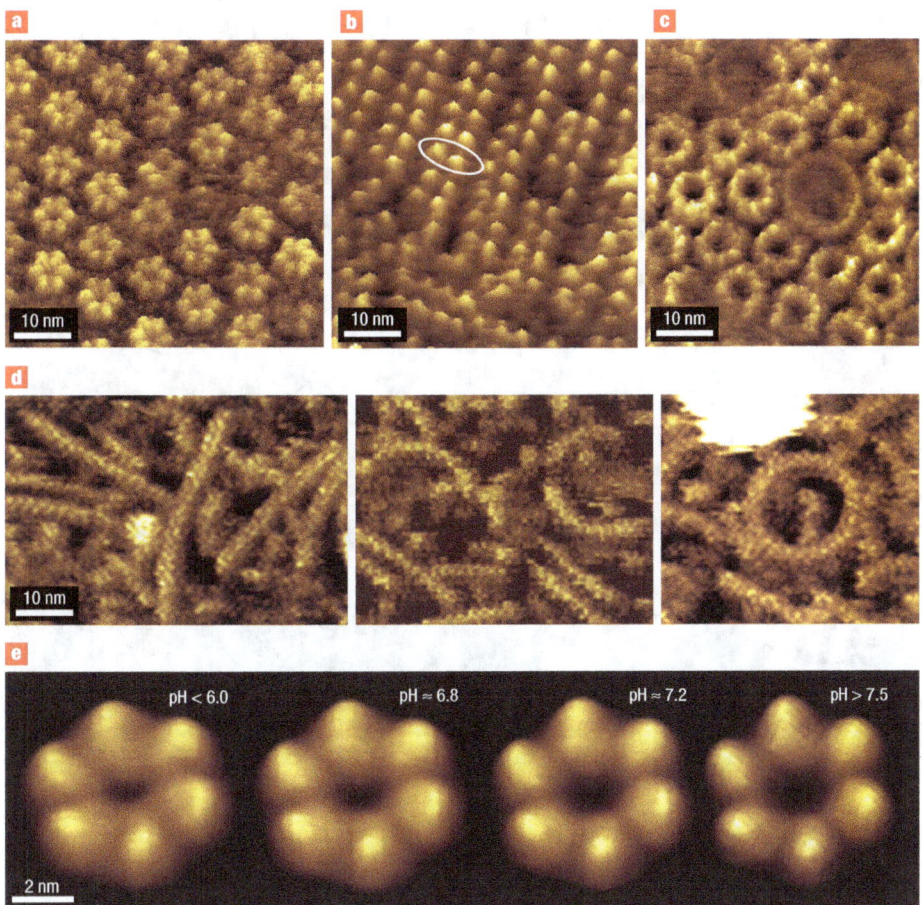

Figure 2 Observing the individuality of cellular machines at high resolution. **a**, Human communication channels known as gap junctions form hexameric pores. **b**, Bovine rhodopsin, the visual pigment of the eye, assembles into rows of dimers (circled in white). **c**, Assembly of light-harvesting II (small doughnuts) and·light-harvesting I complexes (large doughnuts, surrounding the reaction centre) change with intensity of the incident light to optimize the collection of photons and their conversion into electric energy. **d**, The anti-cancer drug taxol slows down the tubulin hydrolysis of straight protofilaments to form rings of 22 nm in diameter (far right panel). **e**, In the presence of aminosulphonate compounds, human communication channels change from a closed (pH < 6.0) to an open state (pH > 7.5).

that the forces probed by SMFS reflect interactions established within or between molecules. Providing direct access to these inter- and intramolecular interactions, SMFS has produced answers to a number of biologically and medically pertinent questions. These include unravelling of protein folding and unfolding mechanisms, receptor–ligand interactions, and ligand-binding interactions that can switch a protein's functional state.

The first protein unfolded by SMFS, and probably the best-studied protein so far, is immunoglobulin titin[35]. In humans, the titin filament represents an adjustable molecular spring in muscle sarcomeres, the repeating unit of myofibrils of striated muscles. About 90% of titin is made of immunoglobulin domains, which provide mechanical elasticity to the filament[36]. SMFS showed that simply applying a mechanical pulling force to both peptide ends induced the fully reversible unfolding of an oligomeric titin construct. Recording the applied force over the stretching distance revealed a characteristic sawtooth-like pattern of force peaks. Each force peak of this pattern reflected the unfolding of a single immunoglobulin molecule and the sequence of force peaks described the unfolding pathways of all immunoglobulin molecules within the oligomeric titin construct. These unfolding pathways, predefined by the mechanical pulling

direction using SMFS, mimicked a physiological process of an (over-)stretched sarcomere. However, most other proteins unfold in complex three-dimensional trajectories[37]. Mechanically stretching proteins at different sites showed that their resistance to unfolding strongly depends on the pulling direction[38,39], suggesting that certain unfolding directions are preferred over others. To explore individual trajectories of unfolding pathways systematically, a group led by Rief[40] stretched and unfolded single green fluorescent proteins (GFP) in different directions. Other SMFS applications characterized the properties of cellular machineries that show functional related elastic properties such as nanosprings mediating protein–protein interactions[41], coiled coils being involved in structural tasks of the cell[42], spider silk[43] and elastomeric polyproteins[44].

Dynamic SMFS (DFS) probes molecular bonds at different force-loads (applied force over time), thereby allowing the approximation of the transition state and kinetic rate of the bond's energy barrier[45]. Such energy barriers form the energy landscape used to describe reaction pathways of protein unfolding, folding, binding and function[37]. Operated in the force-clamp mode, SMFS applies a constant pulling force to a molecular bond and measures the time until the bond breaks[46,47]. Unfolding steps of the clamped protein increase the

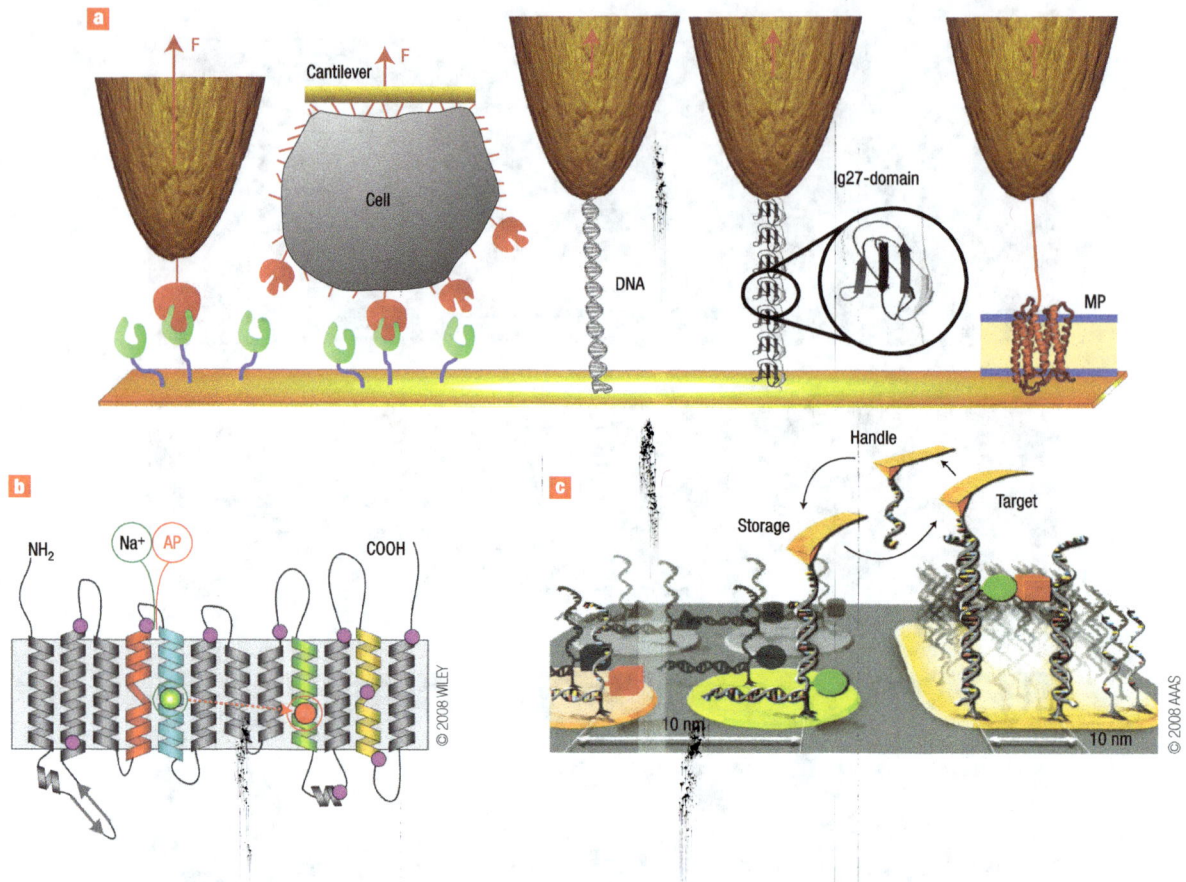

Figure 3 Single molecule manipulation, control and design. **a**, Applying AFM to probe interaction forces (F) of single biomolecules. These examples measure ligand–receptor interactions in their isolated form (left) and embedded in their cellular environment (probe replaced by a biological cell); stretching of a DNA molecule; unfolding of Ig27-titin; and unfolding of a membrane protein (MP). **b**, SMFS can detect and locate interactions (circles) on the structure of membrane proteins (here proton/sodium antiporter NhaA from *E. coli*). A ligand (Na⁺ ion) or inhibitor (AP, 2-aminoperimidine) binding to the ligand-binding site (green circle) establishes different interactions activating (green circle) or deactivating (green and red circles) the antiporter (composed of 12 transmembrane helices). **c**, Single molecules can be mechanically assembled by picking up from discrete storage sites with a DNA oligomer at the AFM probe and depositing them at a target site with nanoscopic precision. Reproduced with permission from refs 52, 53 and 63.

length of the unfolded polypeptide and the sequence of these steps describes the protein's unfolding pathway. These single-molecule analyses offer exciting opportunities in biomedicine, particularly for studying the destabilization and misfolding of proteins involved in neurodegenerative diseases such as Alzheimer's, Parkinson's, Creutzfeldt–Jakob disease or diabetes[48]. We anticipate that in the future, SMFS will increasingly be applied to help understand how changing external conditions can guide proteins along different reaction pathways leading to their malfunction.

In contrast to water-soluble proteins, membrane proteins unfold in sequential steps with each force peak in the SMFS spectra denoting a structural unfolding intermediate[49]. Each unfolding intermediate denotes an energy barrier, with their sequence reflecting the unfolding pathway chosen by the membrane protein[50]. Using SMFS, researchers can observe how changes in temperature, pH, electrolyte and mutations can direct membrane proteins to choose certain unfolding pathways over others and thus to populate trajectories in their energy landscape differently. Binding of a ligand establishes a molecular interaction that can be detected by SMFS and located precisely on the membrane protein structure[51]. Interactions that activate the functional state of a membrane protein are different from those established on inhibitor-binding (Fig. 3b). Thus, SMFS spectra recording these interactions can identify whether a ligand or an inhibitor has bonded[52]. Moreover, once

the characteristic SMFS spectrum has been assigned to the membrane proteins' functional state, conclusions can be drawn about possible interactions and the functional implications of a molecular compound. This possibility of using SMFS to screen the molecular interactions established on drug binding opens up fascinating avenues. Recent work using DFS has tracked how these interactions change the energy landscape of the membrane protein and thus the properties of certain structural regions relative to others[53]. These insights suggest that in the future, SMFS may be used to locate the binding of a drug to a target and to understand the mechanisms by which the drug functionally modulates the target[54].

Specific molecules or cells attached to the AFM cantilever can be used as reference probes for further single molecule manipulations. A prominent example is provided by oligomeric titin constructs engineered to sandwich a given protein. The known reference peaks of the SMFS spectra assigned to the unfolding of immunoglobulin titin molecules allow newly occurring force peaks to be attributed to the unfolding of the sandwiched protein[55]. This principle of sandwiching a protein between well-characterized ones is becoming a standard tool to characterize water-soluble proteins by SMFS. Whereas SMFS can detect the interactions of isolated receptors and ligands (Fig. 3a), single-cell force spectroscopy (SCFS) attaches a living cell to a cantilever to detect interactions of single-cell

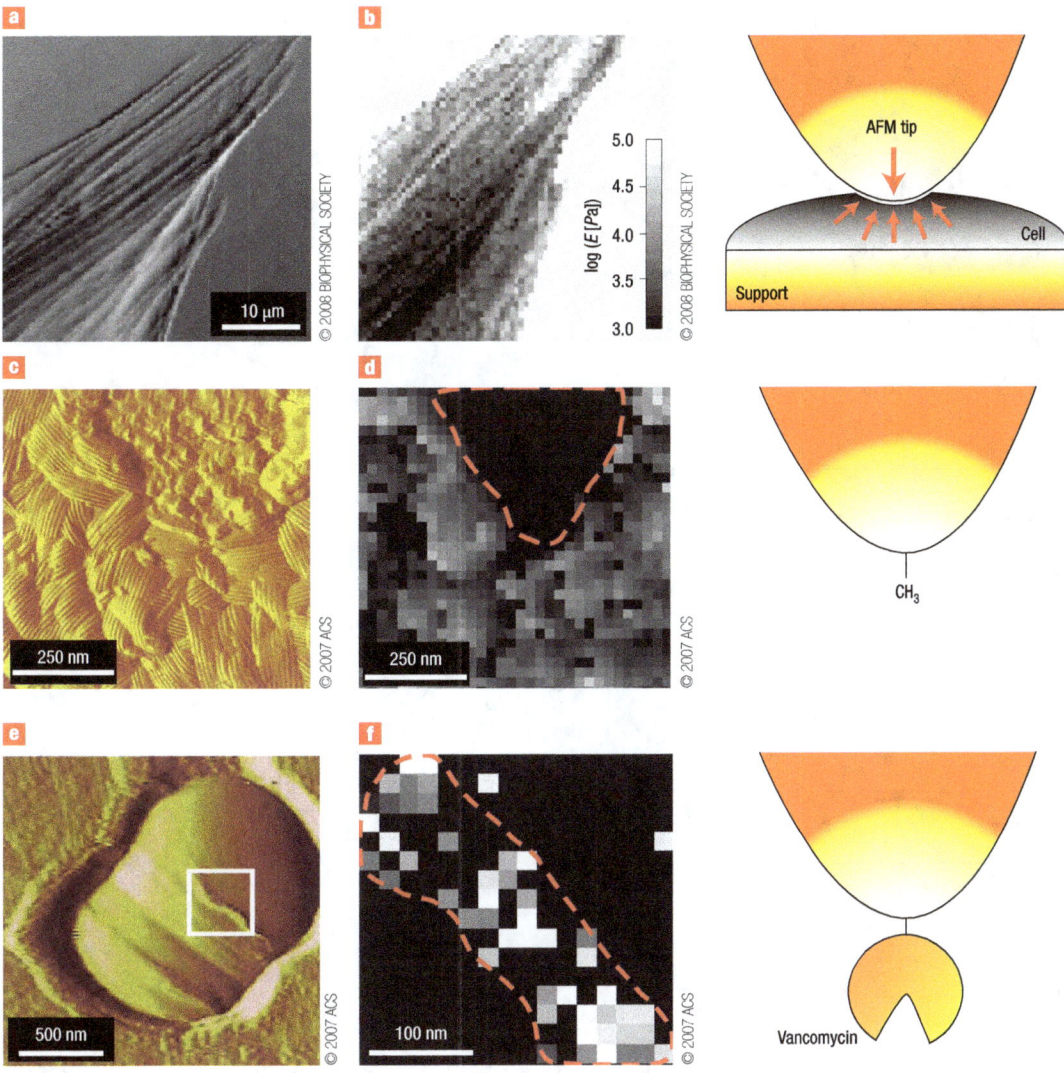

Figure 4 Nanoscale functional imaging of single live cells. **a,b,** Mapping cellular elasticity. AFM deflection image of a living fibroblast (**a**), and the corresponding elasticity map (**b**), obtained through nanomechanical force measurements (far right schematic) and showing the increasing stiffness of the fibrous structures. Reproduced with permission from ref. 65. **c,d,** Chemical force microscopy. AFM deflection image (**c**), and adhesion force map (**d**) recorded on the fungus *Aspergillus fumigatus* using a methyl-terminated tip (far right schematic), demonstrating that the nanostructured surface is globally hydrophobic (hydrophobins), with some hydrophilic nanopatches (polysaccharides, see area within dashed red line). Reproduced with permission from ref. 76. **e,f,** Imaging single receptors. AFM deflection image of *Lactococcus lactis* showing a well-defined division septum (**e**) and adhesion force map recorded with a vancomycin-terminated tip (far right schematic) on the septum region (**f**) (highlighted by the white box in **e**). Each bright pixel in the septum region (red line) reflects the detection of a single vancomycin binding site. Reproduced with permission from ref. 89.

surface receptors while they are embedded in their native cellular environment (Fig. 3a). This approach is of particular importance, as receptor–ligand interactions can be modulated by the cell's functional state. The accuracy of SCFS can detect adhesion of a cell to a support or to another cell and can resolve the contribution of individual receptors[56]. This provides the platform to study molecular adhesion mechanisms of living cells ranging from bacteria and plants to eukaryotes and prokaryotes. SCFS is of particular interest to cell biologists for characterizing cell–cell interactions at molecular resolution[57], as well as for medical researchers working in tumour metastasis[58] and for biotechnologists addressing cell adhesion to biofunctionalized surfaces[59,60].

AFM-based lithographies, particularly dip-pen nanolithography, offer a means of creating arrays of stable biomolecules with nanoscale resolution. They can present considerable advantages for biosensing applications, including short diffusion times, parallel detection of multiple targets and the requirement for only tiny amounts of sample[61,62]. In a pioneering work, Gaub's group established the first molecular toolbox in which they pick up single biomolecules with the AFM probe and drop them to places needed (Fig. 3c). Each pick and drop cycle can be monitored by fluorescence microscopy and by the characteristic force spectra during each pick-up and drop-off[53]. The ability to tinker with single biomolecules with nanoscopic precision allows the study of their structural and functional interactions. As a manipulating tool, AFM offers new ways to characterize biological matter from the cellular to the molecular level. Once characterized, each biomolecular system can easily become a potential probe that adds to the toolbox. A new era of nanobiotechnology has begun.

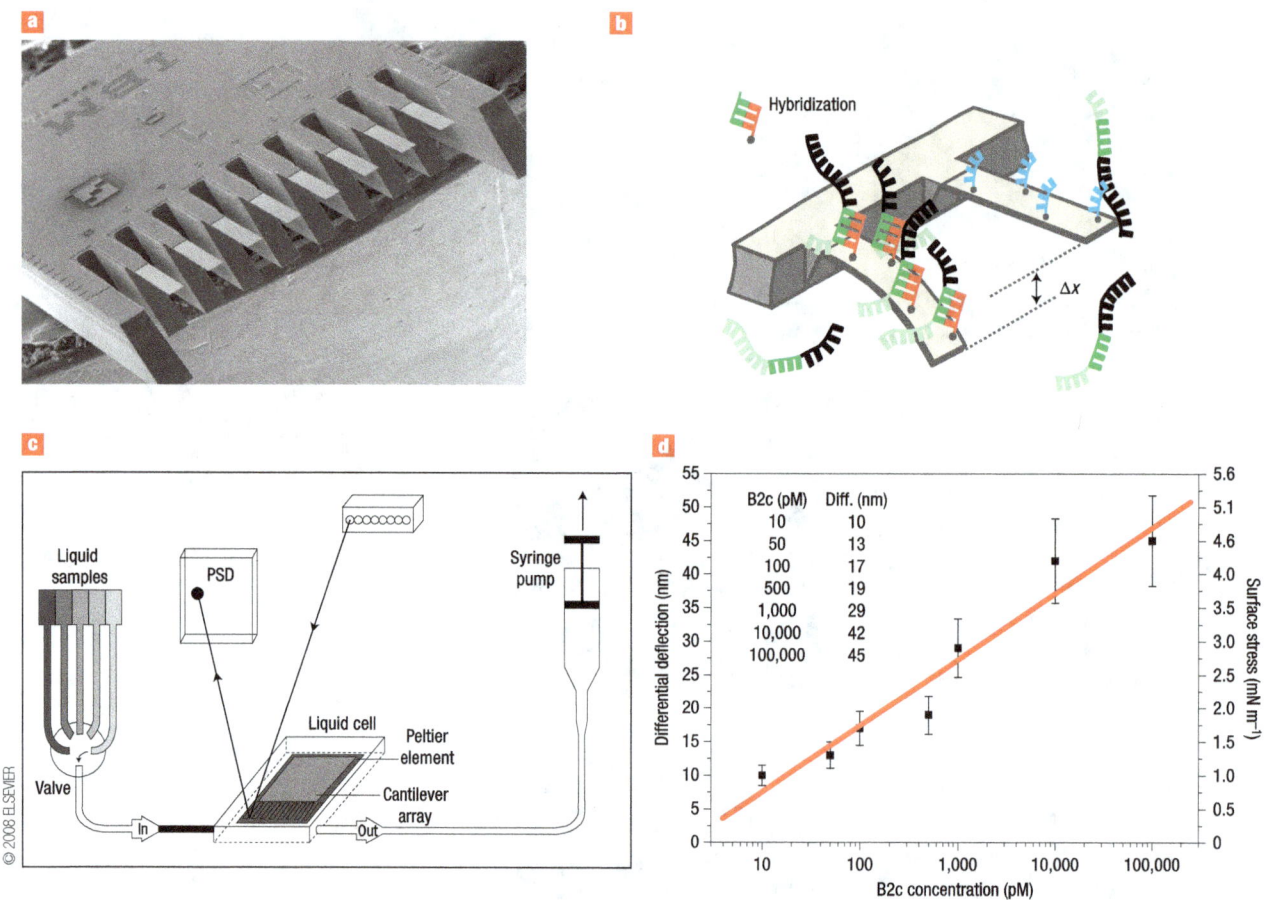

Figure 5 Microfabricated cantilever arrays as label-free biosensors. **a**, Cantilever array showing eight cantilevers that can be individually biofunctionalized. Courtesy of M. Hegner. **b**, Principle of differential readout using sensor and reference cantilevers. Two different probe oligonucleotides (in red and blue) are tethered to the two cantilevers. Nanomechanical bending of the sensor cantilever occurs on hybridization with the complementary oligonucleotides (in green). The cantilever that does not respond is used as a reference, leading to the differential readout (Δx). **c**, Set-up of the device. A laser beam is deflected at the end of a cantilever (see arrow) and its position is detected with a position-sensitive detector (PSD) to measure cantilever bending. Target solutions or buffer are circulated by means of a syringe pump. **d**, Example of differential cantilever response as a function of the concentration of complementary oligonucleotides B2c. Reproduced with permission from ref. 98 and 99.

NANOSCALE FUNCTIONAL IMAGING OF LIVE CELLS

Imaging the local properties of cells and localizing receptors on their surface is a basic challenge in cell biology and microbiology. AFM force spectroscopy may be applied to probe the elastic properties of the cell which play an important role in critical cellular functions, including migration, division and shape. The unique ability to detect and to map the cellular elasticity at the resolution of a few tens of nanometres provides information complementary to that obtained using other techniques such as magnetic or optical tweezers. Such nanoscale measurements involve analysing force–distance curves to provide quantitative information on the elasticity of the cell surface and of the cytoskeleton (that is, Young's modulus)[64]. Interestingly, two-dimensional mapping of elasticity can be obtained by recording arrays of curves over the cell surface (Fig. 4a and b)[65]. In this way, the mechanical properties of a variety of cells including fibroblasts[65,66], platelets[67], kidney cells[68], metastatic cancer cells[69], diatoms[70,71] and yeast cells[72] have been measured.

Of particular interest in medicine is the possibility of monitoring the changes in elasticity on incubation of the cells with drugs. For instance, Rotsch and Radmacher[65] investigated how various drugs that

disrupt or stabilize actin or microtubule networks affect the elasticity of cells. Disaggregation of F-actin resulted in a loss of cell rigidity. However, treatment with drugs such as colchicine, colcemide or taxol that target microtubules yielded no effect, leading to the conclusion that the actin network mainly determines the elastic properties of living cells. More recently, Cross and co-workers[69] measured the stiffness of live metastatic cancer cells taken from the body fluids of patients with suspected lung, breast and pancreas cancer. Cancer cells were substantially softer than benign cells, indicating that nanomechanical analyses can distinguish cancerous cells from normal ones even when they have similar shapes. This suggests that AFM-based nanomechanical measurements have a strong potential in biomedicine for the detection of diseases such as cancer[73].

Non-covalent interactions, such as hydrophobic and electrostatic interactions, play essential roles in biology because they promote crucial events such as protein folding and cell adhesion. Chemical force microscopy (CFM)[74,75], in which AFM tips are modified with specific chemical groups, can now be used to map the spatial arrangement of chemical groups of live cells with nanoscale resolution[76–79]. In the first such study, Dague et al.[76] used methyl-terminated tips to map hydrophobic groups on the human opportunistic pathogen

Aspergillus fumigatus (Fig. 4c,d). Native spores were shown to be strongly hydrophobic, owing to the presence of 'hydrophobins', a family of small hydrophobic proteins. This supported the notion that hydrophobic forces may promote the adhesion of the pathogen to surfaces and tissues. Notably, using a temperature-controlled stage, progressive changes of surface hydrophobicity and structure could be monitored in real time upon spore germination[77]. Similarly, CFM was applied to demonstrate that treatment of mycobacteria with different antibiotics causes a sharp decrease in cell surface hydrophobicity, reflecting the removal of cell surface mycolic acids[78,79]. Thus, CFM provides unique opportunities to resolve the distribution of chemical groups on live cells as they grow or interact with drugs.

Understanding the specific interactions between ligand and receptor molecules represents a great challenge in current life science research and is critical for developing nanobiotechnology applications. During recent years, SMFS with biologically modified tips[80] has enabled researchers to measure the molecular forces and the dynamics of a variety of receptor–ligand interactions, including those associated with avidin/streptavidin[33,34], antibodies[81], DNA[82] and cell adhesion proteins[59]. DFS can provide estimates of kinetic and energetic parameters of the unbinding process such as the dissociation rate constants that characterize the kinetic stability of the bond. The use of spatially resolved SMFS[84] or dynamic recognition force mapping[85] makes it possible to map the distribution of individual receptors on cell surfaces. This is particularly relevant in current membrane biology because membrane proteins and lipids are known to compartmentalize into nanodomains ('lipid rafts'). In this context, Chtcheglova et al.[86] applied dynamic recognition force mapping to microvascular endothelial cells to demonstrate that cadherins, which are proteins involved in homophilic cell-to-cell adhesion, are organized into nanodomains ranging from 10 to 100 nm in diameter. The method has also made it possible to quantify both the binding kinetics and the distribution of growth factor receptors on human microvascular endothelial cells[87]. In microbiology, spatially resolved SMFS has been used to map the nanoscale distribution of single cell adhesion proteins[88] and antibiotic binding sites[89] on live bacteria (Fig. 4e,f), providing insight into pathogen–host and pathogen–drug interactions. In the future, we anticipate that the nanoscale functional imaging of live cells by AFM will find important applications in many fields, particularly for understanding the molecular organization of cell membranes and cell walls, the molecular basis of cellular interactions, and the modes of action of various drugs.

LABEL-FREE CANTILEVER BIOSENSORS

Another exciting application of AFM is to use microfabricated cantilever arrays as label-free biomolecular sensors (Fig. 5). Cantilever nanosensors have important advantages over other biomolecular techniques. First, sample preconditioning such as labelling and amplification is not required. Second, specific biomolecules can be detected with picomolar (~10 pM) sensitivity, which is comparable to current standard gene-chip technology. Third, cantilever arrays are not expensive because they are microfabricated by standard low-cost silicon technology. Fourth, cantilevers are highly suitable for parallelization into arrays, allowing the simultaneous detection of multiples bioanalytes within minutes.

The general idea behind AFM-based nanosensors is to translate specific biomolecular recognition directly into nanomechanical motion. For this purpose, receptor molecules, such as DNA, membranes, proteins, peptides and antibodies, are immobilized on the cantilevers (Fig. 5a), which are then incubated with specific ligands in a liquid environment. Specific molecular recognition reactions can be detected by monitoring either the resonance frequency shift (dynamic mode) or the cantilever bending (static mode). In the dynamic mode, the cantilever is used as a kind of

microbalance to detect very small mass changes (<1 pg)[90]. The cantilever is oscillated, and mass changes on the functionalized cantilever surface are derived from shifts in the frequency of vibration. This dynamic method was used to detect viable eukaryotic cells (fungal spores), with a sensitivity of 10^3 colony-forming units per ml, opening up exciting opportunities for medical and environmental diagnostics, and for monitoring food and water quality[91,92]. Recently, Burg et al.[93] introduced microchannels into cantilevers, thereby making it possible to weigh single bacterial cells and submonolayers of adsorbed proteins in water with subfemtogram resolution.

In the static mode, only one side of the cantilever is functionalized with receptor molecules. A surface stress is generated on biomolecular recognition of the ligand and this bends the cantilever (Fig. 5b). The bending is usually detected optically, by monitoring the deflection of a laser beam (Fig. 5c). Because measurement of the deflection of a single cantilever can be misleading[90], it is essential to use multiple cantilevers with one reference cantilever (Fig. 5a,b). This parallel differential readout increases the sensor sensitivity and eliminates the influence of external factors such as thermal drift and unspecific interactions. In addition to providing an internal reference sensor, arrays of cantilevers enable multiple independent detection experiments to be performed simultaneously within minutes[94].

In the past few years, the general applicability of cantilever nanosensors has been demonstrated for probing DNA hybridization[94,95] and detecting medically important biomarkers, such as prostate-specific antigens[96], cardiac biomarker proteins[97], DNA-binding proteins[98] and mRNA markers for cancer progression[99]. In an elegant study, the biosensitivity of label-free cantilever-array sensors could be increased by orders of magnitude (~10 pM, which is comparable to current standard gene-chip technology) to detect mRNA biomarkers without amplification steps in total RNA derived from human or rat cell lines (Fig. 5d)[99]. The proposed nanosensors may become important medical diagnosis tools, leading to fast and reliable detection of biomarkers that reveal disease risk, progression or therapy response. In another recent report, DNA, RNA, protein or combinations thereof could be detected in parallel on a single cantilever array[100]. Thus, micromechanical cantilever arrays allow monitoring transcription and translation without creating artefacts, giving a more accurate picture of gene expression in either a healthy or a diseased cell.

PERSPECTIVES

AFM is revolutionizing nanobiotechnological research. Within a decade, this unconventional microscope has evolved into a multifunctional toolbox. For the first time, single cellular machineries can be watched at work at a spatial resolution of ~1 nm, and their folding and functional state explored. The technique has also been used to probe cell surface properties and to elucidate the molecular mechanisms of cell adhesion. Microfabricated cantilever arrays represent a powerful approach for the label-free, highly sensitive detection of bioanalytes. These AFM-based nanoscale analyses offer exciting opportunities in biomedicine. For instance, AFM should help refine our understanding of the destabilization of proteins involved in neurodegenerative diseases and the mechanisms by which drugs bind to and modulate the functional state of cell membrane and cell wall targets. It should also be very useful for characterizing cell–cell interactions involved in diseases such as cancer, and pathogen–host interactions. In diagnosis, nanosensors should be increasingly used for detecting biomarkers in blood samples.

But first there are technological challenges to be solved. At the molecular scale, most biological processes are much faster than the time required to record a high-resolution AFM image (~30 seconds). Nevertheless, remarkable advances have been made in developing fast scanning AFM. Improved time resolution (down to milliseconds) has allowed a number of cellular machineries to be followed in a way that was not previously accessible[27,101,102].

The full potential of AFM in cell biology will be seen when combined with advanced light microscopy and spectroscopy techniques. This multi-faceted approach will let cellular structures be identified by fluorescence microscopy, while they are imaged at high resolution using AFM. Traditionally, the resolution of a light-focusing microscope is limited by the wavelength of light (λ), and is usually around 200 nm. In stimulated emission depletion microscopy, however, the resolution is no longer strictly limited by λ, meaning cellular structures with dimensions of a few tens of nanometres can be resolved[103]. Combining this technique with AFM will permit the imaging, manipulation and probing of biological matter from microscopic down to nanoscopic scales, thus helping us to cross borders currently limited by conventional microscopy techniques. This combined approach will allow us to observe and manipulate biological structures at an almost unlimited resolution (~1 nm) and simultaneously elucidate their sophisticated functions.

Advances in SMFS methodology will shed new light on the structures, functional states and reaction pathways of an increasing number of membrane proteins, which form targets for drug development in pharmacology[54]. In particular, malfunction, destabilization and misfolding of membrane proteins are involved in major pathologies such as neuropathological diseases[48]. Water-soluble proteins have been characterized much more frequently by SMFS, and most technological developments have been established based on standardized oligomeric protein constructs such as titin or fibronectin. These proteins are now well-studied references for the characterization of new proteins by SMFS. But the continuously growing number of publications shows that we are just beginning to understand the molecular interactions driving the structure and function relationship of these proteins and that the future will bring insights that we never expected.

First, SMFS experiments could detect, locate and differentiate interactions occurring upon ligand- and inhibitor-binding that switch the functional state of proteins. SMFS-based technologies could potentially dissect the contribution of different interactions such as van der Waals, hydrophobic, hydrogen and electrostatic interactions, revealing the spatial and temporal sequence of interactions that guide the binding of a drug to a protein and finally modulate the protein function. In the distant future it may even be possible to detect these interactions acting within and between cellular machineries without the need to unfold as we currently do.

Meanwhile, progress in molecular recognition SMFS and imaging will allow routine measuring of biomolecular interactions in biosystems ranging from single molecules to tissues. Decreasing the size of cantilevers should improve the force resolution, thereby permitting smaller forces to be measured, while finer nanotube tips may help improve the lateral resolution. Advances in nanodeposition techniques, such as dip-pen nanolithography[62], will enable nanoarrays with almost unlimited chemical and biochemical complexity to be created. These structures will be useful for designing new biological chips and nanomaterials, and will help elucidate surface binding events in single particles (proteins, cells)[62].

There are many open challenges in biosensing technology. High-throughput, multi-content sensing assays are needed, and these should be achievable using devices with thousands of cantilevers working in parallel. Another exciting challenge will be to develop assays that can be applied in biomedical laboratories for routine analysis. This will involve producing cantilever assays at low cost and making them easy to personalize and to use.

doi: 10.1038/nnano.2008.100

References

1. Niemeyer, C. M. & Mirkin, C. A. *Nanobiotechnology: Concepts, Applications and Perspectives* (Wiley-VCH, Weinheim, 2004).
2. *Nanobiotechnology: Report of the National Nanotechnology Initiative Workshop*, October 2003; available at <http://www.nano.gov>.
3. Shirai, Y., Morin, J. F., Sasaki, T., Guerrero, J. M. & Tour, J. M. Recent progress on nanovehicles. *Chem. Soc. Rev.* **35**, 1043–1055 (2006).
4. Balzani, V., Credi, A. & Venturi, M. Molecular devices and machines. *Nano Today* **2**, 18–25 (April 2007).
5. Bayley, H. & Cremer, P. S. Stochastic sensors inspired by biology. *Nature* **413**, 226–230 (2001).
6. Panchal, R. G., Smart M. L., Bowser D. N., Williams D. A. & Petrou S. Pore-forming proteins and their application in biotechnology. *Curr. Pharm. Biotechnol.* **3**, 99–115 (2002).
7. Kitamura, K., Tokunaga, M., Iwane, A. H. & Yanagida, T. A single myosin head moves along an actin filament with regular steps of 5.3 nanometres. *Nature* **397**, 129–134 (1999).
8. Hess, H. *et al.* Molecular shuttles operating undercover: A new photolithographic approach for the fabrication of structured surfaces supporting directed motility. *Nano Lett.* **3**, 1651–1655 (2003).
9. Wang, M. D. *et al.* Force and velocity measured for single molecules of RNA polymerase. *Science* **282**, 902–907 (1998).
10. Soong, R. K. *et al.* Powering an inorganic nanodevice with a biomolecular motor. *Science* **290**, 1555–1558 (2000).
11. Binnig, G., Quate, C. F. & Gerber, C. Atomic force microscope. *Phys. Rev. Lett.* **56**, 930–933 (1986).
12. Gerber, C. & Lang, H. P. How the doors to the nanoworld were opened. *Nature Nanotech.* **1**, 3–5 (2006).
13. Engel, A. & Müller, D. J. Observing single biomolecules at work with the atomic force microscope. *Nature Struct. Biol.* **7**, 715–718 (2000).
14. Dufrêne, Y. F. Using nanotechniques to explore microbial surfaces. *Nature Rev. Microbiol.* **2**, 451–60 (2004).
15. Müller, D. J. *et al.* Single-molecule studies of membrane proteins. *Curr. Opin. Struct. Biol.* **16**, 489–495 (2006).
16. Hoogenboom, B. W., Suda, K., Engel, A. & Fotiadis, D. The supramolecular assemblies of voltage-dependent anion channels in the native membrane. *J. Mol. Biol.* **370**, 246–255 (2007).
17. Seelert, H., Poetsch A., Dencher, N. A., Engel, A. & Müller, D. J. Proton powered turbine of a plant motor. *Nature* **405**, 418–419 (2000).
18. Pogoryelov, D. *et al.* The c15 ring of the Spirulina platensis F-ATP synthase: F1/F0 symmetry mismatch is not obligatory. *EMBO Rep.* **6**, 1040–1044 (2005).
19. Müller, D. J., Hand, G. M., Engel, A. & Sosinsky, G. Conformational changes in surface structures of isolated Connexin26 gap junctions. *EMBO J.* **21**, 3598–3607 (2002).
20. Fotiadis, D. *et al.* Atomic-force microscopy: Rhodopsin dimers in native disc membranes. *Nature* **421**, 127–128 (2003).
21. Park, P. S., Filipek, S., Wells, J. W. & Palczewski, K. Oligomerization of G protein-coupled receptors: past, present, and future. *Biochemistry* **43**, 15643–15656 (2004).
22. Scheuring, S. & Sturgis, J. N. Chromatic adaptation of photosynthetic membranes. *Science* **309**, 484–487 (2005).
23. Elie-Caille, C. *et al.* Straight GDP-tubulin protofilaments form in the presence of taxol. *Curr. Biol.* **17**, 1765–1770 (2007).
24. Drake, B. *et al.* Imaging crystals, polymers, and processes in water with the atomic force microscope. *Science* **243**, 1586–1588 (1989).
25. Kasas, S. *et al.* Escherichia coli RNA ploymerase activity observed using atomic force microscopy. *Biochemistry* **36**, 461–468 (1997).
26. Grandbois, M., Clausen-Schaumann, H. & Gaub, H. Atomic force microscope imaging of phospholipid bilayer degradation by phospholipase A2. *Biophys. J.* **74**, 2398–2404 (1998).
27. Viani, M. B. *et al.* Probing protein–protein interactions in real time. *Nature Struct. Biol.* **7**, 644–647 (2000).
28. Yokokawa, M. *et al.* Fast-scanning atomic force microscopy reveals the ATP/ADP-dependent conformational changes of GroEL. *EMBO J.* **25**, 4567–4576 (2006).
29. Yu, J., Bippes, C. A., Hand, G. M., Müller, D. J. & Sosinsky, G. E. Aminosulfonate modulated pH-induced conformational changes in connexin26 hemichannels. *J. Biol. Chem.* **282**, 8895–8904 (2007).
30. Frederix, P. L. *et al.* Atomic force bio-analytics. *Curr. Opin. Chem. Biol.* **7**, 641–647 (2003).
31. Philippsen, A. *et al.* Imaging the electrostatic potential of transmembrane channels: Atomic probe microscopy on OmpF porin. *Biophys. J.* **82**, 1667–1676 (2002).
32. Frederix, P. L. T. M. *et al.* Assessment of insulated conductive cantilevers for biology and electrochemistry. *Nanotechnology* **16**, 997–1005 (2005).
33. Moy, V. T., Florin, E. L. & Gaub, H. E. Intermolecular forces and energies between ligands and receptors. *Science* **266**, 257–259 (1994).
34. Lee, G. U, Kidwell, D. A. & Colton, R. J. Sensing discrete streptavidin-biotin interactions with atomic force microscopy. *Langmuir* **10**, 354–357 (1994).
35. Rief, M., Gautel, M., Oesterhelt, F., Fernandez, J. M. & Gaub, H. E. Reversible unfolding of individual titin immunoglobulin domains by AFM. *Science* **276**, 1109–1112 (1997).
36. Smith, M. L. *et al.* Force-induced unfolding of fibronectin in the extracellular matrix of living cells. *PLoS Biol.* **5**, e268 (2007).
37. Wolynes, P. G., Onuchic, J. N. & Thirumalai, D. Navigating the folding routes. *Science* **267**, 1619–1620 (1995).
38. Brockwell, D. J. *et al.* Pulling geometry defines the mechanical resistance of a beta-sheet protein. *Nature Struct. Biol.* **10**, 731–737 (2003).
39. Carrion-Vazquez, M. *et al.* The mechanical stability of ubiquitin is linkage dependent. *Nature Struct. Biol.* **10**, 738–743 (2003).
40. Dietz, H., Berkemeier, F., Bertz, M. & Rief, M. Anisotropic deformation response of single protein molecules. *Proc. Natl Acad. Sci. USA* **103**, 12724–12728 (2006).
41. Lee, G. *et al.* Nanospring behaviour of ankyrin repeats. *Nature* **440**, 246–249 (2006).
42. Schwaiger, I., Sattler, C., Hostetter, D. R. & Rief, M. The myosin coiled-coil is a truly elastic protein structure. *Nature Mater.* **1**, 232–235 (2002).
43. Becker, N. *et al.* Molecular nanosprings in spider capture-silk threads. *Nature Mater.* **2**, 278–283 (2003).
44. Cao, Y. & Li, H. Polyprotein of GB1 is an ideal artificial elastomeric protein. *Nature Mater.* **6**, 109–114 (2007).
45. Evans, E. Probing the relation between force—lifetime—and chemistry in single molecular bonds. *Annu. Rev. Biophys. Biomol. Struct.* **30**, 105–128 (2001).

46. Oberhauser, A. F., Hansma, P. K., Carrion-Vazquez, M. & Fernandez, J. M. Stepwise unfolding of titin under force-clamp atomic force microscopy. *Proc. Natl Acad. Sci. USA* **98**, 468–472 (2001).
47. Fernandez, J. M. & Li, H. Force-clamp spectroscopy monitors the folding trajectory of a single protein. *Science* **303**, 1674–1678 (2004).
48. Dobson, C. M. Protein folding and misfolding. *Nature* **426**, 884–890 (2003).
49. Kedrov, A., Janovjak, H., Sapra, K. T. & Müller, D. J. Deciphering molecular interactions of native membrane proteins by single-molecule force spectroscopy. *Annu. Rev. Biophys. Biomol. Struct.* **36**, 233–260 (2007).
50. Janovjak, H., Sapra, K. T., Kedrov, A. & Müller, D. J. From valleys to ridges: Exploring the dynamic energy landscape of single membrane proteins. *Chem. Phys. Chem.* (in the press); doi:10.1002/cphc.200700662.
51. Kedrov, A., Krieg, M., Ziegler, C., Kuhlbrandt, W. & Müller, D. J. Locating ligand binding and activation of a single antiporter. *EMBO Rep.* **6**, 668–674 (2005).
52. Kedrov, A., Ziegler, C. & Müller, D. J. Differentiating ligand and inhibitor interactions of a single antiporter. *J. Mol. Biol.* **362**, 925–932 (2006).
53. Kedrov, A., Appel, M., Baumann, H., Ziegler, C. & Muller, D. J. Examining the dynamic energy landscape of an antiporter upon inhibitor binding. *J. Mol. Biol.* **375**, 1258–1266 (2008).
54. Müller, D. J., Wu, N. & Palczewski, K. Vertebrate membrane proteins: Structure, function and insights from biophysical approaches. *Pharmacol. Rev.* **60**, 43–78 (2008).
55. Best, R. B., Li, B., Steward, A., Daggett, V. & Clarke, J. Can non-mechanical proteins withstand force? Stretching barnase by atomic force microscopy and molecular dynamics simulation. *Biophys. J.* **81**, 2344–2356 (2001).
56. Benoit, M., Gabriel, D., Gerisch, G. & Gaub, H. E. Discrete interactions in cell adhesion measured by single-molecule force spectroscopy. *Nature Cell Biol.* **2**, 313–317 (2000).
57. Krieg, M. et al. Tensile forces govern germ layer organization during gastrulation. *Nature Cell Biol.* **10**, 429–436 (2008).
58. Fierro, F. A. et al. BCR/ABL expression of myeloid progenitors increases b1-integrin mediated adhesion to stromal cells. *J. Mol. Biol.* **377**, 1082–1093 (2008).
59. Li, F., Redick, S. D., Erickson, H. P. & Moy, V. T. Force measurements of the alpha5beta1 integrin–fibronectin interaction. *Biophys. J.* **84**, 1252–1262 (2003).
60. Taubenberger, A. et al. Revealing early steps of alpha2beta1 integrin-mediated adhesion to collagen type I by using single-cell force spectroscopy. *Mol. Biol. Cell* **18**, 1634–1644 (2007).
61. Tinazli, A., Piehler, J., Beuttler, M., Guckenberger, R. & Tampe, R. Native protein nanolithography that can write, read and erase. *Nature Nanotech.* **2**, 220–225 (2007).
62. Salaita, K., Wang, Y. H. & Mirkin, C. A. Applications of dip-pen nanolithography. *Nature Nanotech.* **2**, 145–155 (2007).
63. Kufer, S. K., Puchner, E., Gumpp, H., Liedl, T. & Gaub, H. E. Single molecule cut and paste. *Science* **319**, 594–596 (2008).
64. Weisenhorn, A. L., Khorsandi, M., Kasas, S., Gotzos, V. & Butt, H.-J. Deformation and height anomaly of soft surfaces studied with an AFM. *Nanotechnology* **4**, 106–113 (1993).
65. Rotsch, C. & Radmacher, M. Drug-induced changes of cytoskeletal structure and mechanics in fibroblasts: An atomic force microscopy study. *Biophys. J.* **78**, 520–535 (2000).
66. Rotsch, C., Jacobson, K. & Radmacher M. Dimensional and mechanical dynamics of active and stable edges in motile fibroblasts investigated by using atomic force microscopy. *Proc. Natl Acad. Sci. USA* **96**, 921–926 (1999).
67. Radmacher, M., Fritz, M., Kacher, C. M., Cleveland, J. P. & Hansma, P. K. Measuring the viscoelastic properties of human platelets with the atomic force microscope. *Biophys. J.* **70**, 556–567 (1996).
68. Matzke, R., Jacobson, K. & Radmacher, M. Direct, high-resolution measurement of furrow stiffening during division of adherent cells. *Nature Cell Biol.* **3**, 607–610 (2001).
69. Cross, S. E, Jin, Y. S., Rao, J. & Gimzewski, J. K. Nanomechanical analysis of cells from cancer patients. *Nature Nanotech.* **2**, 780–783 (2007).
70. Higgins, M. J., Sader, J. E., Mulvaney, P. & Wetherbee, R. Probing the surface of living diatoms with atomic force microscopy: The nanostructure and nanomechanical properties of the mucilage layer. *J. Phycol.* **39**, 722–734 (2003).
71. Francius, G., Tesson, B., Dague, E., Martin-Jézéquel, V. & Dufrêne, Y. F. Nanostructure and nanomechanics of live *Phaeodactylum tricornutum* morphotypes *Env. Microbiol.* **10**, 1344–1356 (2008).
72. Touhami, A., Nysten, B. & Dufrêne, Y. F. Nanoscale mapping of the elasticity of microbial cells by atomic force microscopy. *Langmuir* **19**, 4539–4543 (2003).
73. Suresh, S. Biomechanics and biophysics of cancer cells. *Acta Biomater.* **3**, 413–438 (2007).
74. Frisbie, C. D., Rozsnyai, L. F., Noy, A., Wrighton, M. S. & Lieber, C. M. Functional-group imaging by chemical force microscopy. *Science* **265**, 2071–2074 (1994).
75. Noy, A. Chemical force microscopy of chemical and biological interactions. *Surf. Interf. Anal.* **38**, 1429–1441 (2006).
76. Dague, E. et al. Chemical force microscopy of single live cells. *Nano Lett.* **7**, 3026–3030 (2007).
77. Dague, E., Alsteens, D., Latgé, J. P. & Dufrêne, Y. F. High-resolution cell surface dynamics of germinating *Aspergillus fumigatus* conidia. *Biophys. J.* **94**, 656–660 (2008).
78. Alsteens, D., Dague, E., Rouxhet, P. G., Baulard, A. R. & Dufrêne, Y. F. Direct measurement of hydrophobic forces on cell surfaces using AFM. *Langmuir* **23**, 11977–11979 (2007).
79. Alsteens, D. et al. Organization of the mycobacterial cell wall: a nanoscale view. *Pflugers Arch. Eur. J. Physiol.* **456**, 117–125 (2008).
80. Hinterdorfer, P. & Dufrêne, Y. F. Detection and localization of single molecular recognition events using atomic force microscopy. *Nat. Methods* **3**, 347–355 (2006).
81. Hinterdorfer, P., Baumgartner, W., Gruber, H. J., Schilcher, K. & Schindler, H. Detection and localization of individual antibody–antigen recognition events by atomic force microscopy. *Proc. Natl Acad. Sci. USA* **93**, 3477–3481 (1996).
82. Lee, G. U., Chrisey, L. A. & Colton, R. J. Direct measurement of the forces between complementary strands of DNA. *Science* **266**, 771–773 (1994).
83. Fritz, J., Katopidis, A. G., Kolbinger, F. & Anselmetti, D. Force-mediated kinetics of single P-selectin/ligand complexes observed by atomic force microscopy. *Proc. Natl Acad. Sci. USA* **95**, 12283–12288 (1998).
84. Ludwig, M., Dettmann, W. & Gaub, H. E. Atomic force microscope imaging contrast based on molecular recognition. *Biophys. J.* **72**, 445–448 (1997).
85. Raab, A. et al. Antibody recognition imaging by force microscopy. *Nature Biotechnol.* **17**, 902–905 (1999).
86. Chtcheglova, L. A., Waschke, J., Wildling, L., Drenckhahn, D. & Hinterdorfer, P. Nano-scale dynamic recognition imaging on vascular endothelial cells. *Biophys. J.* **93**, L11–L13 (2007).
87. Lee, S., Mandic, J. & Van Vliet K. J. Chemomechanical mapping of ligand-receptor binding kinetics on cells. *Proc. Natl Acad. Sci. USA* **104**, 9609–9614 (2007).
88. Dupres, V. et al. Nanoscale mapping and functional analysis of individual adhesins on living bacteria. *Nature Meth.* **2**, 515–520 (2005).
89. Gilbert, Y. et al. Single-molecule force spectroscopy and imaging of the Vancomycin/D-Ala-D-Ala interaction. *Nano Lett.* **7**, 796–801 (2007).
90. Lang, H. P., Hegner, M., Meyer, E. & Gerber, C. Nanomechanics from atomic resolution to molecular recognition based on atomic force microscopy technology. *Nanotechnology* **13**, R29–R36 (2002).
91. Nugaeva, N. et al. Micromechanical cantilever array sensors for selective fungal immobilization and fast growth detection. *Biosens. Bioelectron.* **21**, 849–856 (2005).
92. Nugaeva, N. et al. An antibody-sensitized microfabricated cantilever for the growth detection of *Aspergillus niger* spores. *Microsc. Microanal.* **13**, 13–17 (2007).
93. Burg, T. P. et al. Weighing of biomolecules, single cells and single nanoparticles in fluid. *Nature* **446**, 1066–1069 (2007).
94. McKendry, R. et al. Multiple label-free biodetection and quantitative DNA-binding assays on a nanomechanical cantilever array. *Proc. Natl Acad. Sci. USA* **99**, 9783–9788 (2002).
95. Fritz, J. et al. Translating biomolecular recognition into nanomechanics. *Science* **288**, 316–318 (2000).
96. Wu, G. H. et al. Bioassay of prostate-specific antigen (PSA) using microcantilevers. *Nature Biotechnol.* **19**, 856–860 (2001).
97. Arntz, Y. et al. Label-free protein assay based on a nanomechanical cantilever array. *Nanotechnology* **14**, 86–90 (2003).
98. Huber, F., Hegner, M., Gerber, C., Guntherodt, H. J. & Lang, H. P. Label free analysis of transcription factors using microcantilever arrays. *Biosens. Bioelectron.* **21**, 1599–1605 (2006).
99. Zhang, J. et al. Rapid and label-free nanomechanical detection of biomarker transcripts in human RNA. *Nature Nanotech.* **1**, 214–220 (2006).
100. Huber, F. et al. Analyzing gene expression using combined nanomechanical cantilever sensors. *J. Phys.: Conf. Ser.* **61**, 450–453 (2007).
101. Ando, T., Kodera, N., Takai, E., Maruyama, D., Saito, K. & Toda, A. A high-speed atomic force microscope for studying biological macromolecules. *Proc. Natl Acad. Sci. USA* **98**, 12468–12472 (2001).
102. Humphris, A. D. L., Hobbs, J. K. & Miles, M. J. Ultrahigh-speed scanning near-field optical microscopy capable of over 100 frames per second. *Appl. Phys. Lett.* **83**, 6–8 (2003).
103. Willig, K. I. et al. Nanoscale resolution in GFP-based microscopy. *Nature Meth.* **3**, 721–723 (2006).

Acknowledgements

We acknowledge support from the National Foundation for Scientific Research (FNRS), the Région wallonne, the Université catholique de Louvain (Fonds Spéciaux de Recherche), the Federal Office for Scientific, Technical and Cultural Affairs (Interuniversity Poles of Attraction Programme), the Research Department of Communauté Française de Belgique (Concerted Research Action), the Deutsche Forschungsgemeinschaft (DFG), the European Union and the Free State of Saxony. Y.F.D. is a Research Associate of the FNRS.
Correspondence and requests for materials should be addressed to D.J.M. or Y.F.D.

Immunological properties of engineered nanomaterials

Most research on the toxicology of nanomaterials has focused on the effects of nanoparticles that enter the body accidentally. There has been much less research on the toxicology of nanoparticles that are used for biomedical applications, such as drug delivery or imaging, in which the nanoparticles are deliberately placed in the body. Moreover, there are no harmonized standards for assessing the toxicity of nanoparticles to the immune system (immunotoxicity). Here we review recent research on immunotoxicity, along with data on a range of nanotechnology-based drugs that are at different stages in the approval process. Research shows that nanoparticles can stimulate and/or suppress the immune responses, and that their compatibility with the immune system is largely determined by their surface chemistry. Modifying these factors can significantly reduce the immunotoxicity of nanoparticles and make them useful platforms for drug delivery.

MARINA A. DOBROVOLSKAIA* AND SCOTT E. McNEIL

Nanotechnology Characterization Laboratory, Advanced Technology Program, SAIC-Frederick, NCI-Frederick, 1050 Boyles St., Bldg. 469, Frederick, Maryland 21702

*e-mail: marina@mail.nih.gov; ncl@ncifcrf.gov

A number of pharmaceutical products that involve nanotechnology have already been approved for clinical use, and many others are at different stages of preclinical development. Indeed, it is estimated that approximately 240 nano-enabled products entered pharmaceutical 'pipelines' in 2006 (ref. 1) (Table 1). At present, as the use of nanomaterials in biomedical products is relatively new and no formal evaluation guidelines have been established, the strategies adopted for preclinical studies on these products resemble those undertaken for other pharmaceuticals, including the way immunotoxicity is assessed[2–7].

Nanoparticles can have many different shapes and chemical compositions, which means that their properties — such as solubility and surface chemistry — can be engineered to make them suitable for a specific biomedical application. Research has shown that nanoparticles can stimulate and/or suppress immune responses by binding to proteins in the blood. Bound proteins determine particle uptake by various cells of the immune system and influence how nanoparticles interact with other blood components[8–10]. Assessing the immunotoxicity of nanobased biomedical products involves the evaluation of all these properties.

The influence of size, solubility and surface modification on the biocompatibility of nanoparticles and their use in biological applications is well known[11]. However, the effects of nanoparticle properties on the immune system are still being explored, and immunological studies of various nanopreparations generally fall into one of two categories: (1) responses to nanoparticles that are specifically modified to stimulate the immune system (for example, vaccine carriers) or (2) undesirable side-effects of other nanoparticles.

(For this review, we will follow the nanoparticle definition endorsed by the ASTM International[12] (Table 2).)

Available preclinical safety data indicates that nanoparticles are not intrinsically more immunotoxic than 'conventional' drugs (Table 3). Furthermore, reformulation of conventional drugs to include nanotechnology-derived carriers often leads to reduced systemic (whole body) immunotoxicity compared with that of unmodified formulations. For example, nanoparticles may keep drugs away from blood cells and normal tissues, releasing them only at targeted sites. Nanoparticles may also decrease the immunotoxicity of a drug by improving their solubility. Research suggests that the physico-chemical properties (size, shape, charge, surface groups, and so on) of the nanoparticles determine their immunotoxicity.

Numerous studies have attempted to address the toxicity issues associated with different nanoparticle administration routes. However, few definite trends can be established. This is probably because of the lack of a comprehensive evaluation of: (1) the composition, size, architecture and zeta potential (surface charge) of the nanoparticles; (2) the density, thickness and stability of the surface coating under physiological conditions; (3) a thorough comparison of (1) and (2) across various animal models; (4) variability in raw materials used to make these particles and differences in manufacturing procedures; and (5) thorough immunological characterization.

Nanoparticles can be administered via nasal, oral, intraocular, pulmonary (lungs) and other routes. Here, we will focus on the systemic route, which is common for a broad variety of engineered nanoparticles, and attempt to summarize their interaction with the immune system. We outline key considerations for preclinical immunological characterization of nanomaterials intended for biomedical applications.

BLOOD CONTACT PROPERTIES

Early studies show that modifying the surface of nanomaterials[13–17] alters the nanoparticle/biological fluid interface and influences the

When citing this review, please cite the original version as shown on the contents page of this chapter.

Table 1 Nanomaterials in the pharmaceutical pipeline. The products described here were summarized from ref. 1 and from the public domains of the relevant company's website as indicated.

Class of Nanomaterial	Application	Examples
Dendrimers	Drug delivery, imaging agents	VivaGel (SPL7013)
Fullerenes	Drug delivery, therapeutics	Antioxidant fullerene
Lipsomes	Drug delivery	Doxil*
Superparamagnetic particles	Drug delivery, imaging agents	Ferumoxitol, Ferumoxtran-10, TNT-Anti-Ep-CAM†
Colloidal gold particles	Drug delivery, therapeutics	Aurimune, Aurothol
Polymeric Nanoparticles	Drug delivery, therapeutics, imaging	Basulin, Contramid, MRX-951
Gold nanoshells	Photo-thermal tumour ablation	AuroLase‡
Quantum dots	Imaging, diagnostics	QDot-800 conjugate
Self-assembling peptide/protein particles	Drug delivery, therapeutics	Abraxane

*www.orthobiotech.com; †www.tritonbiosystems.com; ‡www.nanospectra.com.

way nanoparticles interact with blood constituents. Furthermore, plasma protein binding is important in determining the *in vivo* organ distribution and clearance of certain nanotechnology-derived drug carriers[8,10]. Nanoparticles whose surfaces are not modified to prevent adsorption of opsonins (blood serum proteins that signal cells to ingest the particles) were reportedly removed from the bloodstream within seconds by macrophages — cells that stimulate other immune cells and remove foreign materials by ingesting them through a process known as a phagocytosis[9].

Protein binding to the surface of nanoparticles depends on particle surface characteristics[18–21], composition and the method of preparation[22,23]. When polymeric nanoparticles, iron oxide particles, liposomes and carbon nanotubes were examined for bound proteins, the most abundant ones were albumin, apolipoprotein, immunoglobulins, complement and fibrinogen[19–21,24–30]. It has been suggested that adsorption of plasma proteins depends primarily on nanoparticle surface hydrophobicity or charge[19,28]. For example, when polyhexadecylcyanoacrylate nanoparticles were more densely coated with polyethylene glycol (PEG), less proteins were bound[28]. Whereas it has been reported that protein adsorption decreases with increasing surface charge for several polymer-based nanoparticles[31], the opposite trend has also been shown[19]. This disparity probably reflects the difficulty in making isolated changes to one physico-chemical parameter. For example, altering surface charge, which often involves addition of different functional groups on the particle surface, may also affect the surface hydrophobicity. More exact studies, which vary only one parameter at a time, are required to pin down the crucial structural factor(s) for protein binding.

Another common preclinical safety assessment examines the interaction of nanoparticles with red blood cells and platelets — cells involved in blood coagulation. Below are examples of how surface modification can improve nanoparticle–blood

compatibility. *In vitro* exposure of human whole blood to polyvinyl chloride resin particles increased blood coagulation time, aggregation and adhesion of platelets to these particles[32]. However, the same particles coated with PEG affected neither platelet counts nor coagulation[32]. Similarly, folate-coated and PEG-coated gadolinium nanoparticles did not cause platelet aggregation or activate neutrophils (a type of white blood cell)[33]. When PEGylated and non-PEGylated cetyl alcohol/polysorbate nanoparticles were compared, haemolysis (disruption of red blood cells), platelet aggregation and coagulation were found to be time- and dose-dependant and were greater for non-PEGylated particles[34]. Not all surface modifications aimed at increasing particle solubility can improve their compatibility with plasma coagulation. Water-soluble fullerene derivatives, for example, accelerated clotting times in tumour-bearing mice[35].

The biochemical cascade that removes pathogens, known as the complement system, is divided into two pathways. The classical complement pathway is activated by antigen–antibody complexes, whereas the alternative pathway is antibody independent. Nanoliposomes (Table 3) and engineered carbon nanotubes can activate the complement system[29]. Intriguingly, both single-walled (SWNTs) and double-walled (DWNTs) carbon nanotubes stimulate the classical pathway but only DWNTs have been reported to trigger the alternative pathway. Despite a clear difference in protein adsorption by SWNTs and DWNTs, the mechanism of this selective complement activation remains unknown[29].

In summary, the above data suggest that nanomaterial structure and composition may be engineered to minimize effects on biochemical and cellular components of the blood. Understanding particle structure–activity relationships at the level of interaction with blood cellular and protein components will facilitate the development of a new generation of biocompatible drug carriers.

Table 2 Definitions. The terms summarized in this table are widely applicable in the literature, yet their use is sometimes controversial. The definitions provided here have been summarized from references 12,72–74.

Terminology	Definition
Ultrafine particle	A particle ranging in size from approximately 0.1 µm (100 nm) to 0.001 µm (1 nm).
Nanoparticle	A sub-classification of ultrafine particle with lengths in two or three dimensions greater than 0.001 µm (1 nm) and smaller than about 0.1 µm (100 nm) and which may or may not exhibit a size-related intensive property.
Endocytosis	A complex and highly regulated process of macromolecule and particle internalization by cells; includes two subcategories: phagocytosis and pinocytosis.
Phagocytosis	Actin-dependent endocytic mechanism restricted to 'professional' phagocytes: macrophages, dendritic cells and neutrophils; also defined as 'cell eating', uptake of large particles; subcategories include Fcγ receptor-, complement receptor- and mannose receptor-mediated phagocytosis.
Pinocytosis	Endocytic mechanism for the cellular uptake of fluids and solutes, which is further sub-categorized into macropinocytosis (endocytic vesicle (EV) size >1 µm); clathrin-mediated endocytosis (EV size ~120 nm); caveolin-mediated endocytosis (EV size ~60 nm) and clathrin- and caveolin-independent endocytosis (EV size ~90 nm); only macropinocytosis is actin-dependent, the three other pathways are actin-independent.

Table 3 Products in the drug development pipeline. These nanoparticle formulations were selected from ref. 1 with preference given to late-stage preclinical, clinical, and approved products (many are from the Nanotechnology Characterization Lab database). The selected conventional pharmaceuticals have long clinical-use records. This is not an exhaustive list, but many nanotechnology-based drugs are associated with fewer adverse immune reactions than conventional drugs. NA: not applicable; NAD: no available data; NSC: non-small cell; i.v.: intravenous; subc.: subcutaneous; i.m.: intramuscular; NDA: new drug application; TNF: tumour necrosis factor; RBC: red blood cell; BM: bone marrow; NK: natural killer cell; EPO: erythropoietin.

Product	Type of material	Size (nm)	Indication	Route of administration	Phase	Company	Interactions with immune system
Nanotechnology-based products							
Doxil	PEGylated liposome	75–80*	Metastatic ovarian cancer	i.v.	Approved	OrthoBiotech	Complement activation[101]
Abraxane	Albumin-bound paclitaxel particles	130[102]	NSC lung cancer, breast cancer, others	i.v.	Approved	Abraxis Oncology	Hypersensitivity reactions, weaker and less frequent than in paclitaxel[101]
Basulin	Polyglutamate polymer dotted with insulin	20–50†	Type I diabetes	subc.	Phase II	Flamel Technologies	NAD
Biovant	Nanosized calcium phosphate	NAD	Vaccine component	i.m.	Phase I completed	BioSante Pharmaceuticals	Adjuvant properties Elicits positive immune response to DNA-based and protein antigens stronger and with less hypersensitivity reactions than traditional aluminum hydroxide†
Cyclosert-camptothecin	Cyclodextrin nanoparticles	3–60†	Metastatic solid tumours	i.v.	Phase I	Insert Therapeutics	NAD
INGN-401	Liposome	NAD	Metastatic lung cancer	i.v.	Phase I	Introgen	NAD
Combidex (Ferumoxtran-10)	Iron oxide nanoparticles	17–20[103] 26.1–27.5[104]	Tumour imaging	i.v.	NDA filed	Advanced Magnetics	No haematotoxicity, minor skin hypersensitivity reactions[103]
CYT-6091 (Aurimmune)	TNFα-bound PEGylated colloidal gold particles	33[105]	Solid tumours	i.v.	Phase I	CytImmune Sciences	Decreases systemic TNF toxicity, targets tumours, but not healthy tissues†
MRX-951	Branching block copolymer self-assembling nanoparticles	40–70[106]	Oncology	i.v.	Preclinical	ImaRx Therapeutics	No bone marrow toxicity even with repeated doses[104]
TNT-Anti-Ep-CAM	Polymer-coated iron oxide	NAD	Solid tumours	i.v.	Preclinical	Triton Biosystems	NAD
CYT-21001 (Auritol)	Taxol and TNFα-bound colloidal gold	26†	Solid tumours	i.v.	Preclinical	CytImmune	NAD
AuroLase	Gold nanoshell	150*	Head and neck cancer	i.v.	Preclinical	Nanospectra Biosciences	Non toxic to RBC and platelets*
Dendrimer–Magnevist complex	PAMAM dendrimer	9.8[107]	MRI imaging agent	i.v.	Preclinical	Dendritic Nanotechnologies	No interactions seen *in vitro* with RBCs, BM, platelets, coagulation, macrophage, NK and leukocyte functions[107]
Bioconjugated nanoparticles	Luminescent quantum dots encapsulated with ABC triblock copolymer	10–15[108]	Cancer	subc.	Preclinical	Emory-Georgia Tech Nanotechnology Center	Accumulation in spleen; no data available about toxicity to immune cells[106]
DF1	Dendritic fullerene	4.7–4.9*	Chemoprotection	i.v.	Preclinical	Carbon Nanotechnology	Inhibit collagen-induced platelet aggregation; no toxicity to RBC and BM *in vitro**
C3	Fullerene derivative	NAD	Chemoprotection	i.v.	Preclinical	Carbon Nanotechnology	Inhibit collagen-induced platelet aggregation; no toxicity to RBC and BM in *vitro**
Biotechnology-derived and small molecule pharmaceuticals							
Infliximab	Neutralizing TNFα antibody	NA	Ankylosing spondilytis, arthritis, Crohn's disease,	i.v.	Approved	Centrocor	Reduces inflammation; side effects include tuberculosis, invasive fungal infections and rarely hepatosplenic T-cell lymphomas[101]

Table 3 continued

Product	Type of material	Size (nm)	Indication	Route of administration	Phase	Company	Interactions with immune system
Epogen/Procrit	Recombinant human erythropoietin alpha	NA	Anaemia	i.v. subc.	Approved	Amgen/OrthoBiotech	Elevate and maintain RBC; side effects include pure red cell aplasia due to formation of EPO neutralizing antibodies[101]
Roferon-A	Recombinant human interferon alpha2	NA	Chronic hepatitis C AIDS-related Kaposhi sarcoma, melanoma	subc. i.m.	Approved	Hoffman La-Roche	Can cause or aggravate autoimmune or infectious disorders[101]
Doxorubicin HCl/ Adriamycin RDF	Doxorubicin salt	NA	Cancer	i.v.	Approved	Bedford labs/ Pharmacia&Upjohn	Myelosuppression[101]
Paclitaxel/Taxol		NA	Cancer	i.v.	Approved	SuperGen/BMS	Bone marrow suppression (neutropenia); anaphylaxis and severe hypersensitivity reactions[101]
Neoprofen	Ibuprofen lysine	NA	Patent ductus arteriosus	i.v.	Approved	Ovation	Inhibit platelet aggregation[101]

*Nanotechnology Characterization Lab, unpublished data.
†Information obtained from the public domain of the respective company's website.

IMMUNOSTIMULATORY PROPERTIES

Biological therapeutics are often immunostimulatory, that is, the immune system efficiently recognizes them as foreign substances and mounts a multilevel immune response against them. The molecular structure, architecture of folding motifs, degradation byproducts, formulation, package purity and stability of a biotechnology-derived pharmaceutical have all been shown to solicit an immune response on administration[36]. Nanomaterials that trigger the immune response are expected to do so in an equally complex process. The immunostimulatory properties of nanoparticles discussed here include their antigenicity, adjuvant properties, inflammatory responses and the mechanisms through which nanoparticles are recognized by the immune system.

ANTIGENICITY

Biotechnology-derived pharmaceuticals can cause a specific antibody response (antigenicity). Antibodies, also called immunoglobulin, are specialized proteins produced by plasma B cells in response to an antigen (any foreign material). The immune response to a composite nanoparticle-based drug potentially involves antibodies for both the particles and the surface groups. Antibodies generated in response to the nanoparticles themselves will only affect the efficacy of particle-based products. However, the immune response to nanoparticles functionalized with growth factors, receptors or other biological molecules could include formation of neutralizing antibodies that will recognize both the particle and the endogenous (body's own) molecules, with clinical implications similar to those observed with biotechnology-derived therapeutics[37,38].

There are few studies on the antigenic properties of nanoparticles. An early study on polyamidoamine (PAMAM) dendrimers did not reveal overt antigenicity of generation 3, 5, and 7 amino-terminated dendrimers[39]. Data on antigenicity of other dendrimers are not available. There are several studies on the immunogenic properties of nanosized gold for biomedical applications[40,41] but none of them report particle-specific antibody generation. To date, only two studies describe specific anti-nanoparticle immunoglobulin

formation: the first describes polyclonal C_{60}-specific antibodies with a subpopulation cross-reacting with the C_{70} fullerene[42] and the second reports C_{60}-specific monoclonal antibodies[43]. Two other studies did not observe anti-C_{60} specific immunoglobulins even when used with complete Freund's adjuvant — an oil-in-water emulsion of killed mycobacteria that stimulates cell-mediated immunity in animals[44,45]. This discrepancy in experimental results may be partly explained by the different choices of fullerene derivatives or by variation in the genetic background of the animal models. It is not currently clear if C_{60}, when used as a drug carrier, induces formation of C_{60}-specific antibodies. A variety of factors, including particle surface properties, functional groups and the genetic susceptibility and autoimmune disease history of patients receiving C_{60} may ultimately affect the systemic antigenicity of this drug carrier.

ADJUVANT PROPERTIES

Adjuvants are agents added to a vaccine to augment immune responses toward antigens. A number of studies describe the use of nanoparticles as adjuvants. For example, polymethylmethacrylate nanoparticles induced long-lasting antibody titres (the quantity of substance required to produce a reaction) in HIV2 whole virus vaccine in mice. The antibody response was 100 times higher than when a traditional aluminium sulphate (alum) adjuvant was used[46]. Similarly, when immunized with rabies glycoprotein, lipid-coated polysaccharide nanoparticles induced four times more immunoglobulin G than when immunized with alum adjuvant[47]. Amino-terminated generation-5 PAMAM dendrimers also enhanced anti-ovalbumin immunoglobulin G and immunoglobulin M levels[48]. Immunization of animals with both complete antigens and haptens — small molecules that can elicit an immune response only when attached to a large carrier such as a nanoparticle or a protein — conjugated to the surface of colloidal gold particles generated higher levels of specific antibodies than immunization of the same antigens with classical adjuvants[49]. Furthermore, the amount of antigen required to achieve a high antibody response was an order of magnitude lower than for immunization with Freund's adjuvant[49]. Injection of C_{60}

derivatives into mice resulted in an approximately 20-fold increase in anti-ovalbumin antibodies[45].

Adjuvant properties of another C_{60} derivative were weaker than those of classical Freund's and alum adjuvants during the first immunization, but increased significantly after the sixth booster dose[44]. Although the mechanisms of nanoparticle adjuvanticity are not completely understood, some studies suggest that nanoparticles can enhance antigen uptake and/or stimulate antigen-presenting cells. For example, poly-lactic-glycolic acid (PLGA) and modified lipid nanoparticles enhanced the immune response towards the lowly immunogenic cancer antigen MUC1 and plasmid DNA, respectively[50,51]. In both cases, this was attributed to the particles' ability to induce the maturation of dendritic cells (DCs) — cells that detect and capture pathogens that enter the body. Other mechanisms and nanoparticle effects on vaccine efficacy are reviewed elsewhere[52].

In summary, nanoparticles can be engineered to serve as vaccine carriers and adjuvants. Some studies suggest that nanoparticles can exacerbate allergic reactions[45] and, therefore, careful preclinical characterization is required. Comprehensive structure–activity relationship studies are necessary to further understand the critical parameters that determine the antigenic and adjuvant properties of nanoparticles.

INFLAMMATORY RESPONSES

In an inflammation, cells of the immune system are activated. The cells recognize the pathogens (or any invading foreign substance) and trigger an inflammatory response, which involves secretion of signalling molecules — known as cytokines — to attract more cells to destroy the foreign substance. Immune cells recognize nanoparticles by their surface properties and core composition and mount an inflammatory response in a similar way (Fig. 1a). However, many of the molecular events are still poorly understood. A systematic examination of a range of nanoparticle sizes and surface charges within each class of engineered nanomaterials is required to quantify how these parameters influence the inflammatory response.

In general, cationic (positively-charged) particles are more likely to induce inflammatory reactions than anionic (negatively-charged) and neutral species. For example, anionic generation-4.5 PAMAM dendrimers did not cause human leukocytes (white blood cells) to secrete cytokines[53] but cationic liposomes induced secretion of cytokines such as TNF, IL-12 and IFNγ (ref. 54). Systemic administration of another cationic nanoliposome alone or in combination with bacterial DNA did not induce cytokine production but increased the expression of DC surface markers, CD80/CD86 (ref. 51), which are important in the inflammatory response. Liposomes combined with bacterial DNA stimulated greater DC maturation than the DNA alone[51]. The immunostimulatory properties of nanoliposomes may benefit cancer treatment because they have shown anti-tumour effects in animal cancer models[55,56]. Further, PLGA-based nanoparticles loaded with oligonucleotides induced greater cytokine production and T-cell proliferation than the oligonucleotide alone[57]. Carboxyfullerene nanoparticles were seen to enhance the ability of neutrophils (another type of white blood cell) to destroy bacteria in a mouse model infected with *Streptococcus pyogenes*[58].

Trace impurities within the nanomaterial formulation can also frequently induce an inflammatory response. Early studies suggest that carbon nanotubes induce inflammatory reactions[59], but a more recent study shows that they don't when they are purified[60]. Similarly, purified iron oxide and gold nanoparticles do not induce cytokine secretion[61,62]. This shows that material purity and stability under physiological conditions can affect toxicity.

Another consideration in the inflammatory response is maintaining the Th1/Th2 response — the inflammatory reaction triggered by Th cells that direct and activate other immune cells such as B and T cells and macrophages to secrete different cytokines. This response is important for protecting against cancer cells and pathogens and to avoid hypersensitivity (undesirable and exaggerated immune response) reactions. Several studies have addressed the influence of nanoparticles on Th1 and Th2 responses. Large (>1 μm) industrialized particles induced the Th1 response, whereas smaller ones (<500 nm) were associated with Th2 (ref. 63). In contrast, some small engineered nanoparticles such as 500 nm PLGA[57,64], 270 nm PLGA[65], 80 nm and 100 nm nanoemulsions[66,67], 95 nm and 112 nm PEG–PHDA nanoparticles[68], and 123 nm dendrosome[69] induced the Th1 response. Other engineered particles (for example, 5-nm generation-5 PAMAM dendrimers) do not cause overt inflammatory reactions *in vivo*, but weakly induce Th2 cytokine production and enhanced immunoglobulin production[48,70]. More structure–activity relationship studies are required to understand how size, surface modification and charge of engineered particles influence the Th1/Th2 balance. A detailed understanding of the properties that determine the Th response to biomedical nanoparticles will allow us to engineer nanotechnology-derived drug carriers that can protect the immune system and prevent hypersensitivity reactions.

Particle stimulation of adaptive (acquired) immunity has also been described. For example, small (<100 nm) polystyrene particles promoted CD8 and CD4 T-cell responses and were associated with higher antibody levels than larger (>500 nm) particles[52]. Understanding the mechanisms requires further investigation, and is important for nanovaccine formulation development[52].

Using nanotechnology, particles can potentially be engineered to possess certain properties so they resemble pathogens and are dealt with in the body in a similar way. For example, key molecules that alert the immune system to pathogen infection are the Toll-like receptor (TLR) ligands. Modifying particles with TLR ligands essentially creates a pathogen-like particle that can be recognized by DCs. When TLR-modified nanoparticles are loaded with antigens, they can stimulate DCs to activate T cells that are antigen specific. Such regulated delivery of an antigen to the DCs can control the outcome of any immune response. In one study, DCs that received TLR-modified PLGA nanoparticles containing tetanus toxoid (a non-toxic derivative of a bacterial toxin) and MUC1 lipopeptide (a cancer vaccine candidate) induced significantly higher T-cell proliferation than control cells that were treated with the tetanus antigen alone or antigen and TLR ligands in solution[50]. In another study, modifying the particle surface with mannose (a sugar molecule) stimulated uptake of the particle in macrophages by mannose receptors[71], which are also specific for pathogen uptake and removal.

Liposome polycation plasmid DNA nanoparticles that stimulate the immune system is also thought to involve TLR-dependent signalling[51]. In contrast to Fcγ, mannose and complement receptors, which are discussed below, TLRs are not involved in phagocytosis (uptake of material), but rather regulate phagosome (membrane bound vesicle) maturation and initiation of phagosome-mediated inflammatory responses[72]. A better understanding of how modified particles stimulate TLR signals will allow drug designs that balance the intended immune responses (for example, improved vaccine efficacy) with exaggeration of inflammatory disorders.

UPTAKE MECHANISMS

Mammalian cells have five ways to internalize macromolecules and/or nanoparticles: phagocytosis (via mannose receptor-, complement receptor-, Fcγ receptor, and scavenger receptor-mediated pathways), macropinocytosis, clathrin-mediated, caveolin-mediated, and clathrin/caveolin-independent endocytosis[72–74]

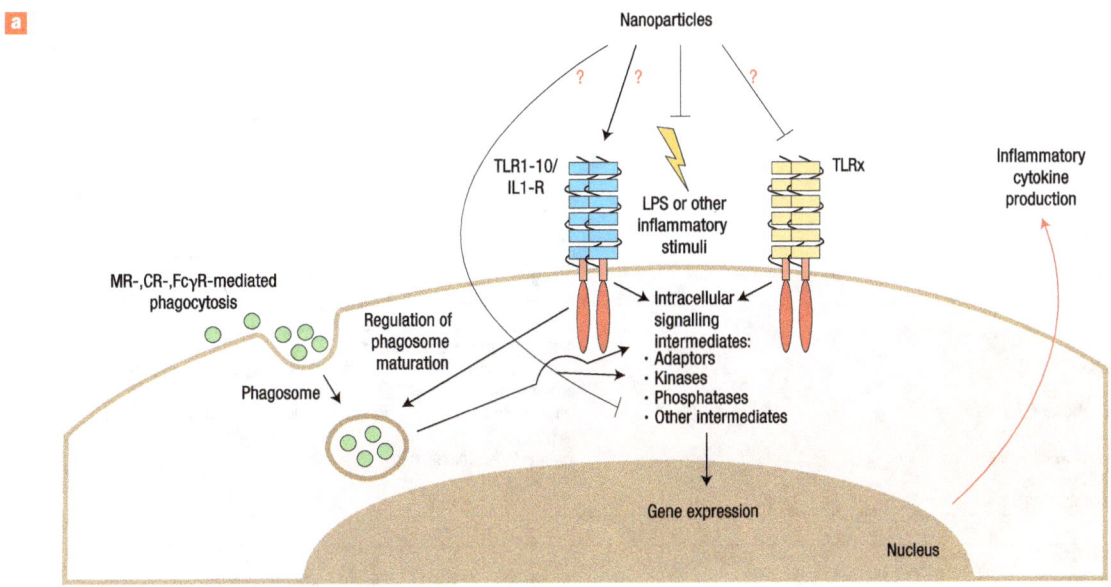

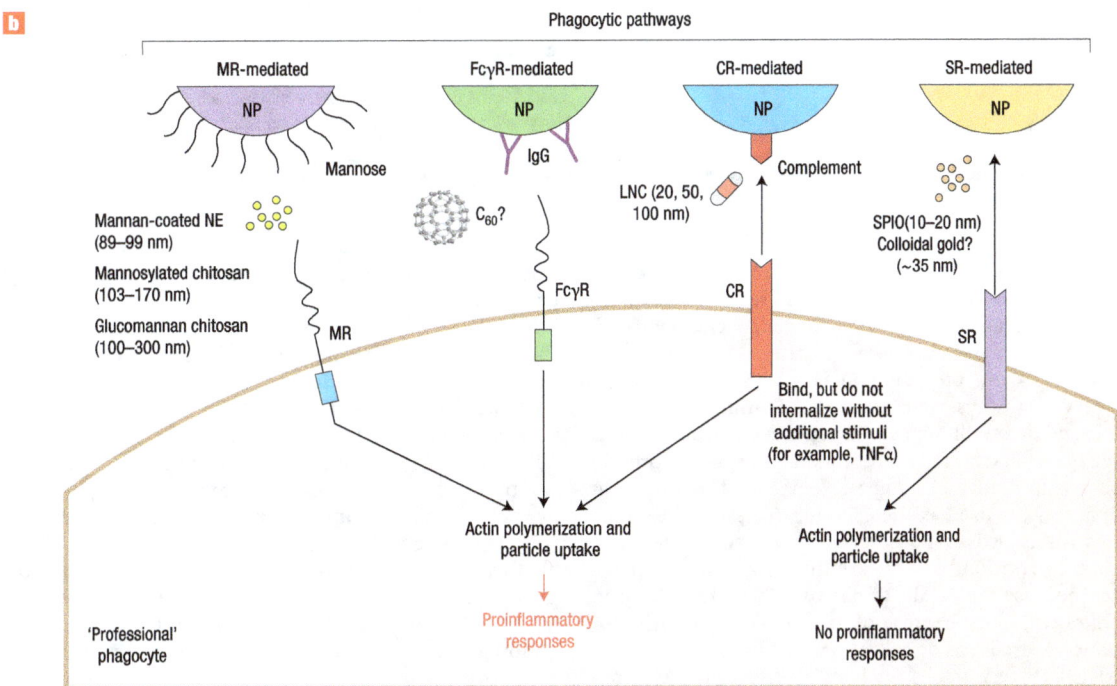

Figure 1 Cell uptake of materials and the different pathways. **a**, Cell uptake of nanoparticles by mannose receptor- (MR-), complement receptor- (CR-), and immunoglobulin Fcγ receptor- (FcγR-)mediated pathways elicit inflammatory responses and may affect toll-like receptor (TLR) signalling, which in turn affects various intracellular signalling intermediates. **b**, nanoparticles (NP) taken up via MR, FcγR and CR pathways elicit an immune response but it is possible to engineer nanoparticles for uptake via non-inflammatory pathways such as the scavenger receptor (SR-) mediated pathway. IL1-R: interleukin 1 receptor; LPS: lipopolysaccharide; NE: nanoemulsion; LNC: lipid nanocapsule; IgG: immunoglobulin G.

(see Table 2 and Fig. 1b). Each pathway involves a unique set of receptors and acts on particular types of particles. This discussion is limited to phagocytosis used by macrophages to eliminate pathogens and apoptotic cells. Below we briefly discuss the terminology and summarize some of the studies on particle uptake via phagocytic receptors.

Traditionally, the term 'phagocytosis' describes the uptake of particulate materials that are >0.5 μm (ref. 74). Other endocytic

Table 4 Preclinical evaluation of nanomaterials intended for use in biomedical applications. This table summarizes the key questions to be addressed during early phase preclinical characterization. CFU-GM: colony-forming unit granulocyte macrophage; TDAR: T cell-dependent antibody response; endotoxin: gram-negative bacteria cell-wall component lipopolysaccharide.

Assay category	Questions to address
In vitro	
Haemolysis	Do nanoparticles* change integrity of red blood cells?
Platelet aggregation	Do nanoparticles* interfere with cellular components of the blood coagulation cascade?
Coagulation time	Do nanoparticles* cause changes in coagulation factors' function?
Complement activation	Do nanoparticles* activate the complement system?
CFU-GM	Do nanoparticles cause myelosupression (toxicity to bone marrow precursors)?
Leukocyte proliferation	Do nanoparticles* have adverse effects on leukocyte proliferative responses?
Uptake by macrophages	Are nanoparticles* internalized by specialized phagocytes?
Cytokine Induction	Do nanoparticles* activate immune cells to elicit cytokine production or interfere with that caused by known immunogens?
Nitric oxide production	Do nanoparticles* induce oxidative stress? Indirect test for potential endotoxin contamination.
Cytotoxicity of natural killer cells	Do nanoparticles* interfere with the ability of natural killer cells to recognize and kill tumour target cells?
Endotoxin contamination	Pyrogen contamination test
Microbial contamination	Sterility test
Viral/mycoplasma contamination	Sterility test
In vivo	
Single-dose toxicity study	
Standard toxicity tests	These tests aim at answering the following questions:
Blood chemistry	Do nanoparticles* cause toxicity to immune cells and organs?
Haematology	Are there any indications for additional toxicity studies?
Histopathology	Additional toxicity studies are conducted on a case-by-case basis using weight-of-
Gross pathology	evidence approach.
TDAR	This test is recognized for its high predictability for human models
Host resistance studies	These tests are recommended for (1) testing the potential effects that particles might
Evaluation of cell-mediated immunity	have on host resistance towards pathogens and tumour cells and (2) to check for contact sensitization and delayed type hypersensitivity reactions
Repeated dose toxicity study	
Immunogenicity	Do nanoparticles* elicit particle specific immune response?

* 'Nanoparticles' refers to a given nanotechnology concept rather than the generalization for a given class of nanomaterials or to a nanoparticle in general.

pathways are associated with solutes and smaller particles[73]. Based on the size of endocytic vesicles, some reports categorize nanoparticle internalization as receptor-mediated endocytosis[52]. Nanoparticles, which represent a very diverse class of materials, can interact with plasma proteins and be engineered to target specific delivery pathways. For example, 50-nm polystyrene particles pre-incubated in plasma are internalized by Kupffer cells (liver macrophages) via scavenger receptors and the plasma protein fetuin was shown to mediate this uptake[75]. When 'function', that is the entry pathway, is used as the criterion instead of 'size', the term 'phagocytosis' is also a valid description. For the purposes of preclinical characterization, we will use the terminology 'uptake by phagocytes' because the details of nanoparticle internalization are largely unknown, likely because of the limited capabilities for visualizing nanomaterials inside cells.

Highly sensitive techniques such as transmission electron microscopy and scanning electron microscopy detect only electron-dense materials such as metallic nanoparticles but cannot easily image polymers, dendrimers, liposomes and other 'soft' materials similar in size to cellular organelles. Because reliable visualization is difficult, most studies use indirect assays such as luminol-based detection or flow cytometry to examine phagocytosis. The former is a sensitive assay based on an increase in luminescence upon luminol oxidation in phagolysosomes, but it can be limited by particle quenching of the luminol signal when it is inside the cell. Flow cytometry assays for phagocytosis, on the other hand, cannot distinguish between particles attached to the cell membrane and internalized particles.

Conclusions based on studies that use fluorescently labelled nanoparticles are problematic because of (1) the unknown

stability of fluorophore nanoparticles under physiological conditions and (2) the lack of controls for assessing fluorophore effects on nanoparticle properties. Nevertheless, studies show that ~90-nm mannose-coated nanoemulsions, ~130-nm mannose–chitosan nanoparticles and ~200-nm gluco–mannan particles are internalized by macrophages via a mannose receptor-mediated phagocytic pathway[66,76,77] and 20-, 50- and 100-nm lipid nanocapsules are internalized via a complement receptor-mediated pathway[78]. Data also suggest that fullerene derivatives may be internalized via the Fcγ receptor route[42,43] (Fig. 1b). Phagocytosis of microbial pathogens and nanoparticles via these pathways results in inflammatory cytokine production. Of interest are the studies with 10–20-nm superparamagnetic iron oxide nanoparticles (SPION) and ~35-nm colloidal gold nanoparticles, which demonstrate internalization via scavenger receptors[62,79]. Scavenger receptor-mediated uptake is a 'classical' pathway macrophages use to eliminate apoptotic cells and is not accompanied by pro-inflammatory cytokine secretion[74]. Indeed, neither ferumoxtran-10 (dextran-coated SPION) nor gold colloids induce inflammatory cytokines[61,62].

Although known to be influenced by particle size, shape, charge, and surface modification, the exact mechanism of colloidal nanoparticle uptake is still not well understood. Surface modifications with PEG, poloxamer and poloxamine polymers have been used as particle 'camouflage' to prevent phagocytosis[80]. The phagocytic index, evaluated by particle number per cell, was shown to increase with lipid nanocapsule size 20 nm<50 nm<100 nm (ref. 77) and PEG5000-PHDCA particle size 80 nm<171 nm<243 nm (ref. 81). It was also shown that the uptake increases with zeta potential, that is, −3.8 mV

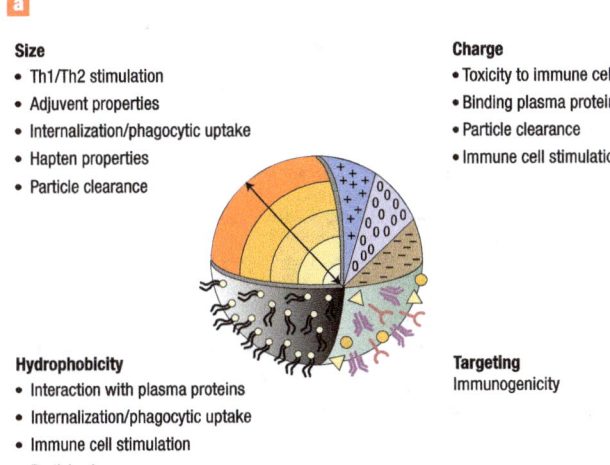

a

Size
- Th1/Th2 stimulation
- Adjuvent properties
- Internalization/phagocytic uptake
- Hapten properties
- Particle clearance

Charge
- Toxicity to immune cells
- Binding plasma proteins
- Particle clearance
- Immune cell stimulation

Hydrophobicity
- Interaction with plasma proteins
- Internalization/phagocytic uptake
- Immune cell stimulation
- Particle clearance

Targeting
Immunogenicity

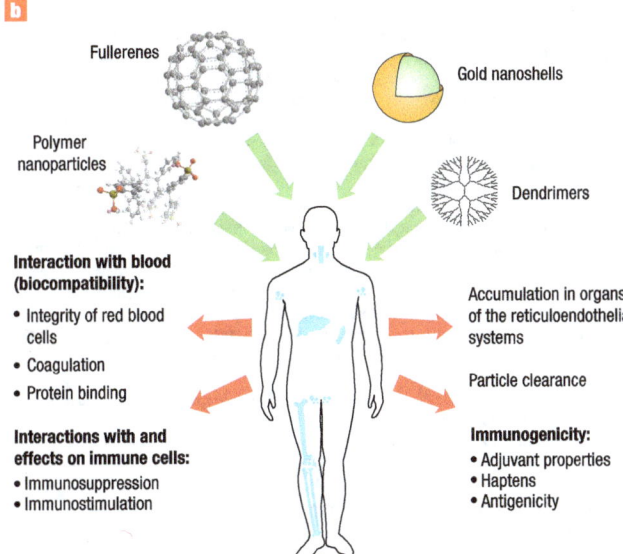

b

Fullerenes

Gold nanoshells

Polymer nanoparticles

Dendrimers

Interaction with blood (biocompatibility):
- Integrity of red blood cells
- Coagulation
- Protein binding

Interactions with and effects on immune cells:
- Immunosuppression
- Immunostimulation

Accumulation in organs of the reticuloendothelial systems

Particle clearance

Immunogenicity:
- Adjuvant properties
- Haptens
- Antigenicity

Figure 2 Nanoparticle properties determine their interaction with the immune system. **a**, The effect of nanoparticle size, charge, hydrophobicity and targeting on immunotoxicity. **b**, Some nanoparticles (shown schematically) can trigger certain immune responses as listed here. A characterization scheme for nanoparticles intended for biomedical applications must include testing for these responses. Such tests may exclude a potentially harmful drug candidate from the development pipeline and inform future studies relevant to the immunomodulatory properties of nanoparticles.

(size 169 nm)< −10.6 mV (171 nm)<− 14.8 mV (172 nm) (ref. 81) and −0.1 mV (200 nm) < +40 mV (200 nm) (ref. 82). Of great interest is a recent study using large particles (>500 nm), which shows that shape decisively determines phagocytosis initiation, whereas particle size primarily impacts phagocytosis completion[83]. More studies are required to conclude whether this holds for nanoparticles as well. In general, cationic or anionic and large particles are internalized more rapidly than smaller and neutral particles. Furthermore, the pathway for nanoparticle uptake is defined by surface groups or by opsonins, which are proteins that bind to the particles in the blood stream.

Phagosome-mediated processing and presentation of nanoparticles may differ from that of 'canonical' antigens. Certain biodegradable nanoparticles can be taken up through conventional pathogen-specific routes and can stimulate inflammatory reactions just like pathogens[66,76–78]. More mechanistic studies are required to understand how the immune system manages non-biodegradable components of nanoparticles (for example, metallic cores). Many questions remain regarding processing of multi-component and multi-functional nanoparticles. Are the individual components (the coating, core, and so on) stable inside the phagosome or do they separate? Are the biodegradable and non-biodegradable components processed together or individually? It is also interesting to know if nanoparticles will change the hydrolytic environment of maturing phagosomes — an important prerequisite for efficient antigen processing and presentation of MHC (major histocompatibility complex) class II proteins. The fundamental question, however, is whether manipulation of nanoparticle surfaces can direct particle-bound antigens to a particular pathway for more efficient processing, presentation and clearance of the antigens. This is also a central question for nanovaccine development. Challenges of vaccine delivery by nanotechnology-derived carriers are reviewed elsewhere[52].

IMMUNOSUPPRESSIVE PROPERTIES

Immunosuppression is a down-regulation of the immune response. Undesirable immunosuppression may result in an impaired ability to cope with cancer and infections, whereas pharmaceutical down-regulation of the immune response may be beneficial in the treatment of inflammatory disorders. Some examples of nanoparticle-related immunosuppression are highlighted below.

Generation-3.5 PAMAM dendrimers conjugated to glucosamine strongly inhibited induction of inflammatory cytokines and chemokines in human macrophages and DCs exposed to bacterial endotoxin[84]. Haematology studies of these conjugates did not reveal any overt toxicity, suggesting that the immunosuppressive effects of dendrimer–glucosamine conjugates can potentially treat and prevent scar tissue formation[84]. Amino-terminated generation-5 PAMAM dendrimers modified with 2-hydrohyhexyl groups suppressed inflammatory cytokine secretion *in vitro* and *in vivo*, and were protective against endotoxin-induced sepsis — the fatal overwhelming immune response towards bacterial lipopolysaccharide[85].

Fullerene derivatives can quench the production of nitric oxide by macrophages because they scavenge free radicals like superoxide dismutase — the enzyme that protects against superoxide-induced damage to cells[86]. Polymerized lipid nanoparticles that bind preferentially to selectins (cell adhesion molecules) on inflammation-activated endothelial cells attenuated both peribronchial inflammation and airway hyper-reactivity induced by an allergen. It is thought that the nanoparticles prevent eosinophils (a type of lymphocyte) from binding to the targeted selectins, which in turn decreases local inflammation and prevents tissue damage[87]. Similarly, cholesteryl-butyrate-conjugated solid lipid nanoparticles reduced neutrophil adhesion to inflammation-activated endothelial cells and are currently considered for treating colon ulcers[88]. PLGA nanoparticles (100–200 nm in diameter) functionalized with betamethasone reduced inflammation in experimentally induced arthritis in rats[89].

All these studies demonstrate that nanoparticles may be engineered to be immunosuppressive for use as anti-inflammatory therapeutics. Concerns may arise, however, about undesirable immunosuppression. Will administration of nanoparticles designed for cancer therapy or imaging interfere with the body's immune protection against cancerous cells and bacterial and viral pathogens? Tests for undesirable immunosuppression are an important part of preclinical nanoparticle evaluation. The great potential of biomedical engineering is that particles may be manipulated to reduce/eliminate immunosupression and yet retain their therapeutic potential[90].

CHALLENGES IN IMMUNOTOXICITY TESTS

Immunotoxicological analysis of new molecular entities is not a straightforward process, and there is no universal guide for immunotoxicity. For example, whereas European authorities mandate the application of immune function tests as an initial screen for immunotoxicity in sub-chronic rodent toxicity studies, the US Food and Drug Administration recommends such tests only when previous evidence indicates a need. Japanese guidelines also follows the as-needed approach[91–94]. According to an inter-laboratory survey of both European- and US-based pharmaceutical companies, an overwhelming amount of immunotoxicity studies (91%) were based on case-by-case reviews. In 55% of cases, decisions were made on the basis of a change in one or more parameter in routine toxicity tests such as lymphoid organ weight, cellular composition or histopathology[95].

As nanotechnology-based products are relatively new drug candidates[1], there are currently no agreed-upon guideline for assessing their immunotoxicity. However, data strongly suggest that nanoparticles do interact with the immune system (Fig. 2). In an attempt to provide a framework of clinical concerns for various aspects of immunotoxicity, we have suggested in Table 4 the important parameters to be addressed during an initial evaluation of nanotechnology-derived pharmaceuticals (composition and intended application of a product will define other considerations).

The main challenge in immunological studies of nanomaterials is choosing an experimental approach that is free of false-positive or false-negative readouts. The majority of the standard immunotoxicological methods are applicable to nanomaterials. However, as nanoparticles represent physically and chemically diverse materials, the classical methods cannot always be applied without modification, and novel approaches may be required. For example, many nanoparticles absorb in the UV–Vis range and some particles may catalyse enzyme reactions or quench fluorescent dyes commonly used as detection reagents in various end-point or kinetic assays. These and other methodological challenges in preclinical evaluation of nanoparticles are reviewed in detail elsewhere[96].

Both 'classical' and novel imunotoxicological assessments of nanomaterials clearly need a scrupulous stepwise validation, standardization, and demonstration of their physiological relevance. Industry, academics, and federal agencies are now collaborating to identify critical parameters in nanoparticles characterization and to establish acceptance criteria for nanomaterial-specific assays[97–100].

doi:10.1038/nnano.2007.223

References

1. Powers, M. Nanomedicine and nano device pipeline surges 68%. *NanoBiotech News* 1–69 (4 January 2006).
2. *Note for guidance on repeated-dose toxicity* (Committee for Proprietary Medicinal Products, London, 2000).
3. *Preclinical Safety Evaluation of Biotechnology — Derived Pharmaceuticals* (FDA/CBER/CDER, 1997).
4. *Developing Medical Imaging Drug and Biological Products. Part1: Conducting Safety Assessments* (FDA/CBER/CDER, 2004).
5. *Immunotoxicity Studies for Human Pharmaceuticals* (FDA/CDER, 2006).
6. *Immunotoxicology Evaluation of Investigational New Drugs* (FDA/CDER, 2002).
7. *Immunotoxicity Testing Guidance* (FDA/CDRH, 1999).
8. Leu, D. *et al.* Distribution and elimination of coated polymethyl [2–14C]methacrylate nanoparticles after intravenous injection in rats. *J. Pharm. Sci.* 73, 1433–1437 (1984).
9. Gref, R. *et al.* Biodegradable long-circulating polymeric nanospheres. *Science* 263, 1600–1603 (1994).
10. Goppert, T. M. & Muller, R. H. Polysorbate-stabilized solid lipid nanoparticles as colloidal carriers for intravenous targeting of drugs to the brain: comparison of plasma protein adsorption patterns. *J. Drug Target.* 13, 179–187 (2005).
11. McNeil, S. E. Nanotechnology for the biologist. *J. Leukoc. Biol.* 78, 585–594 (2005).
12. *Standard Terminology Relating to Nanotechnology* E 2456-06 (ASTM International, 2006).
13. Alivisatos, A. P., Gu, W. & Larabell, C. Quantum dots as cellular probes. *Annu. Rev. Biomed. Eng.* 7, 55–76 (2005).
14. Bala, I., Hariharan, S. & Kumar, M. N. PLGA nanoparticles in drug delivery: the state of the art. *Crit. Rev. Ther. Drug Carrier Syst.* 21, 387–422 (2004).
15. Bulte, J. W. & Kraitchman, D. L. Monitoring cell therapy using iron oxide MR contrast agents. *Curr. Pharm. Biotechnol.* 5, 567–584 (2004).
16. Fiorito, S. *et al.* Toxicity and biocompatibility of carbon nanoparticles. *J. Nanosci. Nanotechnol.* 6, 591–599 (2006).
17. Gupta, A. K. & Gupta, M. Cytotoxicity suppression and cellular uptake enhancement of surface modified magnetic nanoparticles. *Biomaterials* 26, 1565–1573 (2005).
18. Xu, Y. & Du, Y. Effect of molecular structure of chitosan on protein delivery properties of chitosan nanoparticles. *Int. J. Pharm.* 250, 215–226 (2003).
19. Gessner, A., Lieske, A., Paulke, B. & Muller, R. Influence of surface charge density on protein adsorption on polymeric nanoparticles: analysis by two-dimensional electrophoresis. *Eur. J. Pharm. Biopharm.* 54, 165–170 (2002).
20. Gessner, A., Lieske, A., Paulke, B. R. & Muller, R. H. Functional groups on polystyrene model nanoparticles: influence on protein adsorption. *J. Biomed. Mater. Res. A* 65, 319–326 (2003).
21. Gessner, A. *et al.* Nanoparticles with decreasing surface hydrophobicities: influence on plasma protein adsorption. *Int. J. Pharm.* 196, 245–249 (2000).
22. Luck, M. *et al.* Plasma protein adsorption on biodegradable microspheres consisting of poly(D,L-lactide-co-glycolide), poly(L-lactide) or ABA triblock copolymers containing poly(oxyethylene). Influence of production method and polymer composition. *J. Control. Release* 55, 107–120 (1998).
23. Goppert, T. M. & Muller, R. H. Protein adsorption patterns on poloxamer- and poloxamine-stabilized solid lipid nanoparticles (SLN). *Eur. J. Pharm. Biopharm.* 60, 361–372 (2005).
24. Luck, M. *et al.* Identification of plasma proteins facilitated by enrichment on particulate surfaces: analysis by two-dimensional electrophoresis and N-terminal microsequencing. *Electrophoresis* 18, 2961–2967 (1997).
25. Diederichs, J. E. Plasma protein adsorption patterns on liposomes: establishment of analytical procedure. *Electrophoresis* 17, 607–611 (1996).
26. Gessner, A., Paulke, B. R., Muller, R. H. & Goppert, T. M. Protein rejecting properties of PEG-grafted nanoparticles: influence of PEG-chain length and surface density evaluated by two-dimensional electrophoresis and bicinchoninic acid (BCA)-proteinassay. *Pharmazie* 61, 293–297 (2006).
27. Gref, R. *et al.* 'Stealth' corona-core nanoparticles surface modified by polyethylene glycol (PEG): influences of the corona (PEG chain length and surface density) and of the core composition on phagocytic uptake and plasma protein adsorption. *Colloid. Surf. B* 18, 301–313 (2000).
28. Peracchia, M. T. *et al.* Visualization of in vitro protein-rejecting properties of PEGylated stealth polycyanoacrylate nanoparticles. *Biomaterials* 20, 1269–1275 (1999).
29. Salvador-Morales, C. *et al.* Complement activation and protein adsorption by carbon nanotubes. *Mol. Immunol.* 43, 193–201 (2006).
30. Thode, K. *et al.* Determination of plasma protein adsorption on magnetic iron oxides: sample preparation. *Pharm. Res.* 14, 905–910 (1997).
31. Luck, M. *et al.* Analysis of plasma protein adsorption on polymeric nanoparticles with different surface characteristics. *J. Biomed. Mater. Res.* 39, 478–485 (1998).
32. Balakrishnan, B., Kumar, D. S., Yoshida, Y. & Jayakrishnan, A. Chemical modification of poly(vinyl chloride) resin using poly(ethylene glycol) to improve blood compatibility. *Biomaterials* 26, 3495–3502 (2005).
33. Oyewumi, M. O. *et al.* Comparison of cell uptake, biodistribution and tumor retention of folate-coated and PEG-coated gadolinium nanoparticles in tumor-bearing mice. *J. Control. Release* 95, 613–626 (2004).
34. Koziara, J. M. *et al.* Blood compatibility of cetyl alcohol/polysorbate-based nanoparticles. *Pharm. Res.* 22, 1821–1828 (2005).
35. Wang, J. *et al.* Antioxidative function and biodistribution of [Gd@C82(OH)22]n nanoparticles in tumor-bearing mice. *Biochem. Pharmacol.* 71, 872–881 (2006).
36. Chamberlain, P. & Mire-Sluis, A. R. An overview of scientific and regulatory issues for the immunogenicity of biological products. *Dev. Biol.* 112, 3–11 (2003).
37. Swanson, S. J., Ferbas, J., Mayeux, P. & Casadevall, N. Evaluation of methods to detect and characterize antibodies against recombinant human erythropoietin. *Nephron Clin. Pract.* 96, c88–c95 (2004).
38. Kivisakk, P., Alm, G. V., Fredrikson, S. & Link, H. Neutralizing and binding anti-interferon-beta (IFN-beta) antibodies. A comparison between IFN-beta-1a and IFN-beta-1b treatment in multiple sclerosis. *Eur. J. Neurol.* 7, 27–34 (2000).
39. Roberts, J. C., Bhalgat, M. K. & Zera, R. T. Preliminary biological evaluation of polyamidoamine (PAMAM) Starburst dendrimers. *J. Biomed. Mater. Res.* 30, 53–65 (1996).
40. Tomii, A. & Masugi, F. Production of anti-platelet-activating factor antibodies by the use of colloidal gold as carrier. *Jpn. J. Med. Sci. Biol.* 44, 75–80 (1991).
41. Kreuter, J. Nanoparticles as adjuvants for vaccines. *Pharm. Biotechnol.* 6, 463–472 (1995).
42. Chen, B. X. *et al.* Antigenicity of fullerenes: antibodies specific for fullerenes and their characteristics. *Proc. Natl Acad. Sci. USA* 95, 10809–10813 (1998).
43. Braden, B. C. *et al.* X-ray crystal structure of an anti-Buckminsterfullerene antibody fab fragment: biomolecular recognition of C(60). *Proc. Natl. Acad. Sci. U. S. A.* 97, 12193–12197 (2000).
44. Masalova, O. V. *et al.* Immunostimulating effect of water-soluble fullerene derivatives—perspective adjuvants for a new generation of vaccine. *Dokl. Akad. Nauk* 369, 411–413 (1999).
45. Andreev, S. M. *et al.* Immunogenic and allergenic properties of fulleren conjugates with aminoacids and proteins. *Dokl. Biochem.* 370, 4–7 (2000).
46. Stieneker, F., Kreuter, J. & Lower, J. High antibody titres in mice with polymethylmethacrylate nanoparticles as adjuvant for HIV vaccines. *AIDS* 5, 431–435 (1991).

47. Castignolles, N. *et al.* A new family of carriers (biovectors) enhances the immunogenicity of rabies antigens. *Vaccine* **14**, 1353–1360 (1996).

48. Rajananthanan, P., Attard, G. S., Sheikh, N. A. & Morrow, W. J. Evaluation of novel aggregate structures as adjuvants: composition, toxicity studies and humoral responses. *Vaccine* **17**, 715–730 (1999).

49. Dykman, L. A. *et al.* Immunogenic properties of colloidal gold. *Izv. Akad. Nauk Biol.* **1**, 86–91 (2004).

50. Diwan, M. *et al.* Biodegradable nanoparticle mediated antigen delivery to human cord blood derived dendritic cells for induction of primary T cell responses. *J. Drug Target.* **11**, 495–507 (2003).

51. Cui, Z., Han, S. J., Vangasseri, D. P. & Huang, L. Immunostimulation mechanism of LPD nanoparticle as a vaccine carrier. *Mol. Pharm.* **2**, 22–28 (2005).

52. Xiang, S. D. *et al.* Pathogen recognition and development of particulate vaccines: does size matter? *Methods* **40**, 1–9 (2006).

53. http://ncl.cancer.gov/120406.pdf

54. Tan, Y., Li, S., Pitt, B. R. & Huang, L. The inhibitory role of CpG immunostimulatory motifs in cationic lipid vector-mediated transgene expression in vivo. *Hum. Gene Ther.* **10**, 2153–2161 (1999).

55. Dileo, J. *et al.* Lipid-protamine-DNA-mediated antigen delivery to antigen-presenting cells results in enhanced anti-tumor immune responses. *Mol. Ther.* **7**, 640–648 (2003).

56. Whitmore, M. M., Li, S., Falo, L. Jr & Huang, L. Systemic administration of LPD prepared with CpG oligonucleotides inhibits the growth of established pulmonary metastases by stimulating innate and acquired antitumor immune responses. *Cancer Immunol. Immunother.* **50**, 503–514 (2001).

57. Diwan, M., Elamanchili, P., Cao, M. & Samuel, J. Dose sparing of CpG oligodeoxynucleotide vaccine adjuvants by nanoparticle delivery. *Curr. Drug Deliv.* **1**, 405–412 (2004).

58. Tsao, N. *et al.* Inhibition of group A streptococcus infection by carboxyfullerene. *Antimicrob. Agents Chemother.* **45**, 1788–1793 (2001).

59. Shvedova, A. A. *et al.* Unusual inflammatory and fibrogenic pulmonary responses to single-walled carbon nanotubes in mice. *Am. J. Physiol. Lung C.* **289**, L698–L708 (2005).

60. Pulskamp, K., Diabate, S. & Krug, H. F. Carbon nanotubes show no sign of acute toxicity but induce intracellular reactive oxygen species in dependence on contaminants. *Toxicol. Lett.* **168**, 58–74 (2007).

61. Muller, K. *et al.* Effect of ultrasmall superparamagnetic iron oxide nanoparticles (Ferumoxtran-10) on human monocyte-macrophages in vitro. *Biomaterials* **28**, 1629–1642 (2007).

62. Shukla, R. *et al.* Biocompatibility of gold nanoparticles and their endocytotic fate inside the cellular compartment: a microscopic overview. *Langmuir* **21**, 10644–10654 (2005).

63. van Zijverden, M. & Granum, B. Adjuvant activity of particulate pollutants in different mouse models. *Toxicology* **152**, 69–77 (2000).

64. Lutsiak, M. E., Kwon, G. S. & Samuel, J. Biodegradable nanoparticle delivery of a Th2-biased peptide for induction of Th1 immune responses. *J. Pharm. Pharmacol.* **58**, 739–747 (2006).

65. Chong, C. S. *et al.* Enhancement of T helper type 1 immune responses against hepatitis B virus core antigen by PLGA nanoparticle vaccine delivery. *J. Control. Release* **102**, 85–99 (2005).

66. Cui, Z. & Mumper, R. J. Coating of cationized protein on engineered nanoparticles results in enhanced immune responses. *Int. J. Pharm.* **238**, 229–239 (2002).

67. Cui, Z. *et al.* Strong T cell type-1 immune responses to HIV-1 Tat (1–72) protein-coated nanoparticles. *Vaccine* **22**, 2631–2640 (2004).

68. de Kozak, Y. *et al.* Intraocular injection of tamoxifen-loaded nanoparticles: a new treatment of experimental autoimmune uveoretinitis. *Eur. J. Immunol.* **34**, 3702–3712 (2004).

69. Balenga, N. A. *et al.* Protective efficiency of dendrosomes as novel nano-sized adjuvants for DNA vaccination against birch pollen allergy. *J. Biotechnol.* **124**, 602–614 (2006).

70. Rajananthanan, P., Attard, G. S., Sheikh, N. A. & Morrow, W. J. Novel aggregate structure adjuvants modulate lymphocyte proliferation and Th1 and Th2 cytokine profiles in ovalbumin immunized mice. *Vaccine* **18**, 140–152 (1999).

71. Cui, Z., Hsu, C. H. & Mumper, R. J. Physical characterization and macrophage cell uptake of mannan-coated nanoparticles. *Drug Dev. Ind. Pharm.* **29**, 689–700 (2003).

72. Blander, J. M. & Medzhitov, R. On regulation of phagosome maturation and antigen presentation. *Nat. Immunol.* **7**, 1029–1035 (2006).

73. Conner, S. D. & Schmid, S. L. Regulated portals of entry into the cell. *Nature* **422**, 37–44 (2003).

74. Aderem, A. & Underhill, D. M. Mechanisms of phagocytosis in macrophages. *Annu. Rev. Immunol.* **17**, 593–623 (1999).

75. Nagayama, S. *et al.* Fetuin mediates hepatic uptake of negatively charged nanoparticles via scavenger receptor. *Int. J. Pharm.* **329**, 192–198 (2007).

76. Kim, T. H. *et al.* Mannosylated chitosan nanoparticle-based cytokine gene therapy suppressed cancer growth in BALB/c mice bearing CT-26 carcinoma cells. *Mol. Cancer. Ther.* **5**, 1723–1732 (2006).

77. Cuna, M. *et al.* Development of phosphorylated glucomannan-coated chitosan nanoparticles as nanocarriers for protein delivery. *J. Nanosci. Nanotechnol.* **6**, 2887–2895 (2006).

78. Vonarbourg, A. *et al.* Evaluation of pegylated lipid nanocapsules versus complement system activation and macrophage uptake. *J. Biomed. Mater. Res. A* **78**, 620–628 (2006).

79. von Zur Muhlen, C. *et al.* Superparamagnetic iron oxide binding and uptake as imaged by magnetic resonance is mediated by the integrin receptor Mac-1 (CD11b/CD18): Implications on imaging of atherosclerotic plaques. *Atherosclerosis* **193**, 101–111 (2007).

80. Vonarbourg, A., Passirani, C., Saulnier, P. & Benoit, J. P. Parameters influencing the stealthiness of colloidal drug delivery systems. *Biomaterials* **27**, 4356–4373 (2006).

81. Fang, C. *et al.* In vivo tumor targeting of tumor necrosis factor-alpha-loaded stealth nanoparticles: effect of MePEG molecular weight and particle size. *Eur. J. Pharm. Sci.* **27**, 27–36 (2006).

82. Kwon, Y. J., Standley, S. M., Goh, S. L. & Frechet, J. M. Enhanced antigen presentation and immunostimulation of dendritic cells using acid-degradable cationic nanoparticles. *J. Control. Release* **105**, 199–212 (2005).

83. Champion, J. A. & Mitragotri, S. Role of target geometry in phagocytosis. *Proc. Natl Acad. Sci. USA.* **103**, 4930–4934 (2006).

84. Shaunak, S. *et al.* Polyvalent dendrimer glucosamine conjugates prevent scar tissue formation. *Nat. Biotechnol.* **22**, 977–984 (2004).

85. Cromer, J. R. *et al.* Functionalized dendrimers as endotoxin sponges. *Bioorg. Med. Chem. Lett.* **15**, 1295–1298 (2005).

86. Chen, Y. W., Hwang, K. C., Yen, C. C. & Lai, Y. L. Fullerene derivatives protect against oxidative stress in RAW 264.7 cells and ischemia-reperfused lungs. *Am. J. Physiol. Reg. I.* **287**, R21–R26 (2004).

87. John, A. E. *et al.* Discovery of a potent nanoparticle P-selectin antagonist with anti-inflammatory effects in allergic airway disease. *Faseb J.* **17**, 2296–2298 (2003).

88. Dianzani, C. *et al.* Cholesteryl butyrate solid lipid nanoparticles inhibit adhesion of human neutrophils to endothelial cells. *Br. J. Pharmacol.* **148**, 648–656 (2006).

89. Higaki, M. *et al.* Treatment of experimental arthritis with poly(D, L-lactic/glycolic acid) nanoparticles encapsulating betamethasone sodium phosphate. *Ann. Rheum. Dis.* **64**, 1132–1136 (2005).

90. Sundstrom, J. B. *et al.* Magnetic resonance imaging of activated proliferating rhesus macaque T cells labeled with superparamagnetic monocrystalline iron oxide nanoparticles. *J. Acquir. Immune Defic. Syndr.* **35**, 9–21 (2004).

91. Putman, E., van der Laan, J. W. & van Loveren, H. Assessing immunotoxicity: guidelines. *Fundam. Clin. Pharmacol.* **17**, 615–626 (2003).

92. Putman, E. *et al.* Assessment of the immunotoxic potential of human pharmaceuticals: a workshop report. *Drug Inf. Journal* **36**, 417–427 (2002).

93. Snodin, D. J. Regulatory immunotoxicology: does the published evidence support mandatory nonclinical immune function screening in drug development? *Regul. Toxicol. Pharmacol.* **40**, 336–355 (2004).

94. Haley, P. J. Guidelines on immunotoxicity testing of new chemical entities; consideration of a conundrum. *Newsletter of the Society of Toxicology* 3–5 (Fall, 2002).

95. Dean, J. H., Hincks, J. R. & Remandet, B. Immunotoxicology assessment in the pharmaceutical industry. *Toxicol. Lett.* **102–103**, 247–255 (1998).

96. Patri, A. K., Dobrovolskaia, M. A., Stern, S. T. & McNeil, S. E. in *Nanotechnology for Cancer Therapy* 105–138 (CRC Press, Boca Raton, 2006).

97. http://www.fda.gov/nanotechnology/

98. http://ncl.cancer.gov

99. http://www.astm.org

100. http://www.iso.org

101. http://online.factsandcomparisons.com/index.aspx/

102. http://www.fda.gov/cder/foi/

103. Sharma, R. *et al.* Safety profile of ultrasmall superparamagnetic iron oxide ferumoxtran-10: phase II clinical trial data. *J. Magn. Reson. Imaging* **9**, 291–294 (1999).

104. Enochs, W. S., Harsh, G., Hochberg, F. & Weissleder, R. Improved delineation of human brain tumors on MR images using a long-circulating, superparamagnetic iron oxide agent. *J. Magn. Reson. Imaging* **9**, 228–232 (1999).

105. Paciotti, G. F. *et al.* Colloidal gold: a novel nanoparticle vector for tumor directed drug delivery. *Drug Deliv.* **11**, 169–183 (2004).

106. Unger, E. C. *et al.* Nanoparticle drug delivery systems *The Drug Delivery Companies Report* 2001/02 61–63 (2001).

107. http://ncl.cancer.gov/120406.pdf

108. Gao, X. *et al.* In vivo cancer targeting and imaging with semiconductor quantum dots. *Nat. Biotechnol.* **22**, 969–976 (2004).

Acknowledgements

We thank Nancy Rice, Serguei Kozlov, Banu Zolnik, Jennifer Hall and Anil Patri for helpful discussion and suggestions. We are grateful to Allen Kane for assistance with illustrations. This project has been funded in whole or in part with federal funds from the National Cancer Institute, National Institutes of Health, under contract N01-CO-12400. The content of this publication does not necessarily reflect the views or policies of the Department of Health and Human Services, nor does mention of trade names, commercial products, or organizations imply endorsement by the US Government.

Competing financial interests

Authors declare no competing financial interest.

Correspondence and requests for materials should be addressed to M.A.D.

Injectable nanocarriers for biodetoxification

Hospitals routinely treat patients suffering from overdoses of drugs or other toxic chemicals as a result of illicit drug consumption, suicide attempts or accidental exposures. However, for many life-threatening situations, specific antidotes are not available and treatment is largely based on emptying the stomach, administering activated charcoal or other general measures of intoxication support. A promising strategy for managing such overdoses is to inject nanocarriers that can extract toxic agents from intoxicated tissues. To be effective, the nanocarriers must remain in the blood long enough to sequester the toxic components and/or their metabolites, and the toxin bound complex must also remain stable until it is removed from the bloodstream. Here, we discuss the principles that govern the use of injectable nanocarriers in biodetoxification and review the pharmacological performance of a number of different approaches.

JEAN-CHRISTOPHE LEROUX

Canada Research Chair in Drug Delivery, Faculty of Pharmacy, University of Montreal, P.O. Box 6128 Downtown Station, Montreal, Quebec, Canada, H3C 3J7
e-mail: Jean-Christophe.Leroux@umontreal.ca

Acute intoxications, either accidental or intentional, constitute a major public health problem worldwide[1]. Drugs account for about 40% of toxic exposures in humans, and a significant number of deaths are associated with overdoses of analgesics, antidepressants, sedatives/hypnotics/antipsychotics, stimulants and cardiovascular drugs. Unfortunately, antidotes are limited to a relatively small number of agents. For the most severe intoxications, treatments are largely based on general measures of intoxication support, such as administration of activated charcoal, gastric emptying, whole bowel irrigation, correction of electrolyte disturbances and removal of toxins through extracorporeal procedures[2]. Orally administered activated charcoal adsorbs and eliminates drugs/metabolites that are still present or being secreted in the gastrointestinal tract. Although this technique is valuable, it can only be used on conscious patients. On the other hand, whole bowel irrigation and haemodialysis are reserved for eliminating specific life-threatening toxins[3]. For instance, haemodialysis (which involves removing substances from the blood by passing the blood through a semi-permeable membrane in a bedside dialysis machine) is particularly suited for drugs or metabolites that are water soluble, have a low volume of distribution (V_d) (that is, they do not distribute to a large extent to tissues/organs), a molecular weight of less than 500 g mol^{-1} and low plasma protein binding[3].

One emerging strategy for managing overdose involves injecting nanosized particulate carriers (<1 μm) to reduce the free drug concentration in the body by acting as a sink for the toxin (Fig. 1). The injected nanocarriers that are either in the circulatory system or have diffused in the peripheral organs extract the drug from the intoxicated tissues and then exit the body via the kidneys or liver. Nanosized carriers can take the form of liposomes, nanoemulsions,

nanoparticles and macromolecules. Owing to their high specific surface area and adjustable composition/surface properties, which can be manipulated to optimize uptake and circulation time, several of these carriers can function as detoxifiers. The systems used in biodetoxification usually share the same characteristics as those used in drug delivery, with the exception that the affinity of the toxic agent to the carrier should be very high to ensure rapid and efficient removal of toxins from the peripheral tissues. Here, we examine the principles and pharmacological performance of four main carrier systems currently studied as sequestering agents.

In toxicity reversal, several parameters of the toxic agent, such as its molecular weight, ionization constant, affinity for blood proteins, V_d, half-life, toxicological profile and the presence of active metabolites, must be considered. As shown in Table 1, most drugs involved in poisoning are weak bases that are characterized by a large V_d, high protein binding and the presence of active metabolites. A large V_d may complicate the detoxification procedure, especially if the transfer rate of the toxins from the tissues to the blood is slow. Similarly, when drugs bind to blood proteins, the extraction efficiency is lowered because less drug is available for capture[4]. The potential toxicity of metabolites is also an important parameter to consider. By the time an intoxicated patient is admitted to the emergency ward, a substantial amount of the drug may have been converted into active metabolites. For example, upon oral absorption, 40% of amitriptyline – an antidepressant – is metabolized by the liver into its active demethylated form, nortriptyline[5]. Finally, close attention should be paid to the delay between drug/chemical intake and administration of the antidote. In most laboratory settings, the nanocarrier is administered prior to or within minutes after exposure to the toxic agent. This almost never occurs in practice as patients are often treated hours after the onset of symptoms.

When used as detoxifiers, injectable nanocarriers should meet a number of criteria of which innocuousness, circulation time and uptake capacity are of paramount importance. In principle, the injected carrier must remain in the blood long enough for the toxic agent to be extracted sufficiently from the peripheral

When citing this review, please cite the original version as shown on the contents page of this chapter.

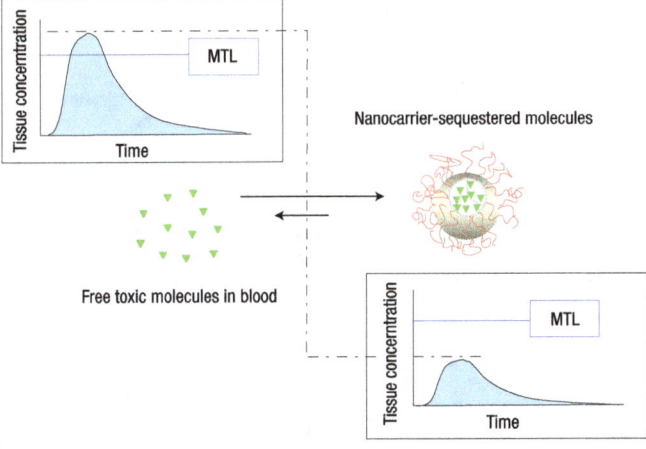

Nanocarrier-sequestered molecules

Free toxic molecules in blood

Figure 1 Treating drug overdose and chemical poisoning with nanocarriers. The ingestion of a toxic dose of a chemical results in an elevation of its tissue concentrations above the minimum toxic level (MTL; blue line). This toxic concentration is maintained until the chemical is eliminated from the tissue by diffusion and/or metabolism, resulting in a decrease of tissue levels (upper curve). The sequestration of the toxin by circulating nanocarriers allows the redistribution of the chemical from the peripheral tissues into the blood compartment. This reduces tissue exposure to the toxic compound, bringing its concentration below the MTL at a faster rate (lower curve). Note: sequestration of the toxic molecules by the nanocarrier can also take place directly in the tissues.

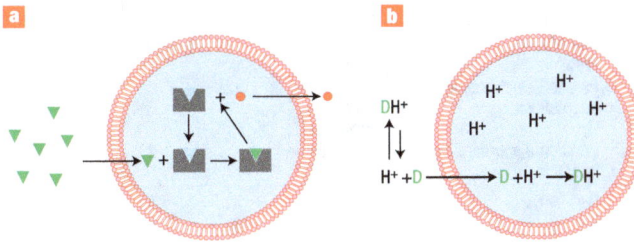

Figure 2 Schematic representation of two vesicular nanostructures used in detoxification. Nanocarriers can contain an enzyme (grey) within the vesicle (blue), which converts the toxic agent (green triangles) into an inactive product (red circles) (a), or they can maintain a transmembrane pH-gradient between the inside (blue) and outside of the vesicles (b). In the latter case, the unionized toxic agent (D) diffuses down the pH gradient into the vesicle interior where it is trapped in an ionized form (DH^+ for a weak base). Diffusion continues until the internal buffering capacity is overwhelmed.

tissues. The circulating carrier should also be stable enough to avoid rapid release of the sequestered drug back into the tissues. The circulation time of a colloid, for instance, depends on its hydrodynamic volume, shape and surface properties. Coating nanocarriers with hydrophilic, flexible polymers such as polyethylene glycol (PEG) can slow down their clearance by the immune system and improve their half-lives in the blood. For spherical colloids, maximum circulation times are obtained for those with diameters between 50–200 nm (ref. 6). Very small colloids (<8 nm) are excreted by the kidneys[7] and/or rapidly accumulate in the liver, whereas large particles (>200 nm) are

subjected to major uptake by the spleen[8]. Nanocarriers (and their toxic cargos) are generally eliminated from the bloodstream within 24 h and mostly end up in the liver where the toxic compound is metabolized. Fortunately, apart from a few compounds, most drugs rarely cause liver injury upon acute poisoning[9].

To rapidly bring tissue concentrations below the toxic threshold, the overdosed agent should be extensively and rapidly sequestered. When using oil-based nanostructures as detoxifiers, the oil (that is, lipids) needs to be carefully selected to maximize compatibility with the toxin[9]. Unfortunately, because many hydrophobic and amphiphilic drugs are poorly soluble in injectable oils that are approved for human applications[10], the nanocarrier dose required to extract the toxic agent is important. Given the non-repetitive nature of the treatment, infusing high amounts of lipids may be feasible in the context of detoxification. However, injecting large doses of carrier (>1 g kg⁻¹ or >5 ml kg⁻¹ for a typical 20%-lipid emulsion) can slow down the detoxification process because this increases the time during which the antidote is given.

The partition coefficient, P — which is a measure of a compound's differing solubility in two solvents — for the oil phase is not the sole parameter governing the uptake of toxic agents by oil-based nanostructures because amphiphilic compounds that possess hydrophilic and hydrophobic properties can adsorb at the oil/water interface. Adsorption depends on specific surface area which, in turn, depends on particle size. It has been demonstrated that the extraction capacity often increases with decreasing particle size[11,12]. Furthermore, in the case of amphiphilic charged drugs, adsorption at the interface can be enhanced by adding to the nanocarrier an oppositely charged component that interacts electrostatically with the toxic agent[13]. Chemically modifying the nanocarrier with specific functional groups, such as electron-deficient aromatic rings that bind to compounds with π-electron-rich aromatic rings[14], can also increase drug uptake and improve extraction.

An alternative strategy to optimize extraction is to create an elevated concentration gradient between the inside and outside of the nanocarrier. This can be achieved by encapsulating an enzyme that degrades the toxic agent into water-filled vesicular structures (Fig. 2a)[15]. As the toxin diffuses into the carrier and is metabolized by the enzymes, more toxic compounds can be pumped into the carrier. This detoxification measure requires the toxic agent to freely permeate the vesicle membranes and the entrapped enzyme to remain active for at least a few hours while circulating in the blood.

Another approach, which is simpler but only applicable to ionizable drugs (this includes weak bases or acids, Table 1), involves sequestering the toxic agent into nanosized vesicles by creating a transmembrane pH gradient. This concept is similar to the urinary pH manipulation technique used by clinicians to accelerate excretion of ionizable drugs from the kidneys. The neutral form of low-molecular-weight weak acids and bases can permeate vesicle membranes at much faster rates than their ionized forms. If a vesicle exhibits a pH gradient (acidic or basic for weak bases or acids, respectively), the unionized compound diffuses down its concentration gradient into the vesicle interior where it is subsequently ionized and trapped (Fig. 2b)[16]. The diffusion of the toxic agent's neutral form will continue until the interior buffering capacity is overwhelmed. This extraction process is very efficient, even for molecules that are highly protein-bound[4].

Several colloidal carriers have been investigated for detoxification applications over the past two decades (Table 2). These systems have sizes ranging from a few nanometres (polymers) to half a micrometre (emulsions in parenteral nutrition). The following section provides an overview of their pharmacological performance as sequestering agents.

Table 1 Pharmaceutical agents commonly involved in overdoses. Most drugs do not have specific antidotes. They are generally ionizable, possess a large volume of distribution (V_d), bind avidly to blood proteins and are often degraded into pharmacologically active metabolites. Compiled from refs 1,9,48.

Drug	Class	Base/acid	Half-life (h)	Protein binding (%)	V_d (l kg^{-1})	Active metabolites	Specific antidote
Acetaminophen	Analgesic	-	1–3	8–43	0.8–1	No	Acetylcysteine
Aspirin	Analgesic	Acid	2–4.5	50–80	0.1–0.3	Yes	No
Fentanyl	Analgesic, narcotic	Base	2–7	80	4	No	Naloxone
Methadone	Analgesic, narcotic	Base	15–59	80–85	1–8	No	Naloxone
Amitriptyline	Antidepressant	Base	9–25	95	8	Yes	No
Bupropion	Antidepressant	Base	14	82–88	19–21	Yes	No
Doxepine	Antidepressant	Base	8–15	80	6–8	Yes	No
Venlafaxine	Antidepressant	Base	3–7	27	7.5	Yes	No
Diltiazem	Cardiovascular drug	Base	4–6	77–85	3–7	Yes	No
Verapamil	Cardiovascular drug	Base	2–8	90	4.7	Yes	No
Metformin	Antidiabetic drug	Base	2–6	Negligible	650	No	No
Valproic acid	Anticonvulsant	Acid	5–20	80–95	0.1–0.5	Yes	No
Phenobarbital	Anticonvulsant	Acid	80–120	20–50	0.5–0.9	No	No
Alprazolam	Anti-anxiety drug	-	12–15	80	0.9–1.2	No	Flumazenil
Haloperidol	Antipsychotic	Base	20	90	18–30	Yes	No
Quetiapine	Antipsychotic	Base	6	83	10	No	No
Diphenhydramine	Antihistamine	Base	2–8	78	5	No	No

Table 2 Non-immune nanocarriers under investigation for biodetoxification. These carriers have sizes ranging from a few to several hundred nanometres. With the exception of polymers, most of the systems described so far in the literature to treat drug overdose have been prepared from natural or synthetic lipids.

System	Composition	Diameter (nm)	Mechanism*	Model toxin/drug	Reference
Liposomes	POPC/Chol/ DPPE-PEG	100	Enzymatic degradation	Paraoxon	19
	POPC/Chol/DPPE-PEG	100	Enzymatic degradation	Diisopropylfluoro-phosphate	21
	DOPC/Chol ± DSPE-PEG	1600	Chelation	^{239}Pu-phytate	49
				^{238}Pu-citrate	
	DOPC/Chol/DSPE-PEG	100–1,600	Chelation	^{238}Pu-phytate	28
	DMPC/DOPG	40–45	Partition/electrostatic interactions	Amitriptyline	50
	SPC/Chol	200	pH gradient	Haloperidol	4
	SPC/Chol	180	pH gradient	Amitriptyline	23
Nanoemulsions	Soybean oil/egg PC	430	Partition	Bupivacaine	29,30
	Ethyl butyrate/fatty acid/poloxamer	15–40	Partition/electrostatic interactions	Bupivacaine	13
	SPC/MCT/HSA-EO	118	Partition	Bupivacaine	11
	PC/MCT/HSA-PEG ± lauric acid ± DSPE-PEG	60–90	Partition/electrostatic interactions	Haloperidol Docetaxel Paclitaxel	33
	SPC/MCT/HSA-PEG ± lauric acid	60–90	Partition/electrostatic interactions	Amitriptyline	23
Nanocapsules	Hexadecane/polysiloxane-silicate	98	Partition	Quinoline Bupivacaine	35
	PS-80/ethyl butyrate/PC/polysiloxane-silicate	30–300	Partition	Quinoline	12
Nanospheres	Magnetite	8	Chelation	Uranyl(2+) cation	51
Polymers	γ-cyclodextrin derivative	~1	Hydrophobic/electrostatic interactions	Rocuronium Vecuronium	41 52
	Dinitrobenzesulfonyl chitosan	n.a.	π–π interactions	Amitriptyline	14,43

POPC = palmitoyloleoylphosphatidylcholine, Chol = cholesterol, DPPE = dipalmitoylphosphatidylethanolamine, DSPE = distearoylphosphatidylethanolamine, DOPC = dioleoylphosphatidylcholine, PG = phosphatidylglycerol, DMPC = dimyristoylphosphatidylcholine, DOPG = dioleoylphosphatidylglycerol, PC = phosphatidylcholine, MCT = medium chain triglycerides, SPC = soybean phosphatidylcholine, HSA-PEG = poly(ethylene glycol)(660)-12-hydroxystearate, PS-80 = polysorbate 80, n.a. = not available.

*In this table, all reported electrostatic interactions occur between a negatively-charged colloid and a positively charged drug.

LIPOSOMES

Liposomes are spherical vesicles that possess one or more concentric phospholipid bilayer membrane that delimit aqueous compartments. They have been extensively studied for the treatment of intoxications due to organophosphorus agents (OPs), which are toxic agents commonly found in agriculture pesticides. The first use of liposomes as antidotes for OPs was a follow-up to the work of Way and co-workers wherein resealed red blood cells served as vesicles to encapsulate the enzymes rhodanese and organophosphorus acid anhydrolase (OPAA) which degrade cyanide[17] and OPs[18], respectively (Fig. 2a). The approach was later refined by entrapping OPAA in neutral long-circulating PEGylated liposomes[19-21]. Compared with red

blood cells, liposomes are advantageous because they are built from non-human-derived material, can undergo large-scale production and exhibit a greater shelf-life. In mice, liposomal OPAA was found be quite efficient in detoxifying OPs, but only when administered in prevention (that is, prior to intoxication)[19]. Unfortunately, a substantial loss of protection against OP-induced mortality was observed when the antidote was given after the injection of the OP — a situation more likely to happen under real conditions of intoxication[20]. Although these data confirmed the therapeutic value of liposomal OPAA, they also revealed how important timing is in reversing intoxications.

As pointed out previously, transmembrane pH gradients can help take up low-molecular-weight weak acids or bases from physiological media. In an elegant study, Mayer et al.[22] demonstrated

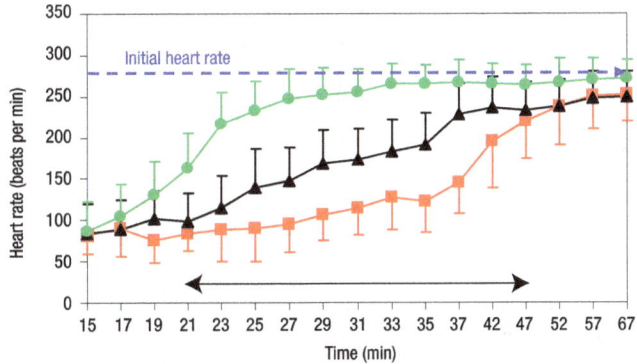

Figure 3 Heart-rate recovery after intoxication and addition of a nanocarrier. Overdose of amitriptyline — an antidepressant — elevates heart rates and brings about deleterious effects on the heart (cardiotoxicity). Isolated rat hearts were infused for 12 min with amitriptyline (17 μM) to cause intoxication and subsequently perfused with pH 7.4 buffer (red squares), pH 7.4 spherulites (black triangles), or pH 3.0 gradient spherulites (green circles) from 15 to 37 min. Perfusion of pH 3.0 gradient spherulites resulted in swift recovery of heart rate to its initial value because the nanocarrier extracted amitriptyline from the heart tissue and the protonated drug was sequestered within the vesicle aqueous core. The black arrows indicate the time during which the difference in heart beats between the pH 3.0 spherulite and pH 7.4 buffer group was statistically significant (p<0.01). Reproduced with permission from ref. 23. Copyright (2007) Elsevier.

that stealth (long-circulating) liposomes with an internal pH of 4, when administered prior to injection of the anticancer drug doxorubicin, captured the drug *in vivo* and decreased its toxicity while maintaining the drug's anti-tumour potency. The pH gradient was relatively stable with a decrease of only 1.5 units over 20 h following injection and this allowed doxorubicin to be sequestered *in situ* at clinically relevant doses of liposomes. Although the aim of this study was to show that pre-treatment with empty liposomes could improve the pharmacokinetic profiles of drugs, it revealed their potential as detoxifying agents. Along those lines, pH-gradient spherulites — a type of multilamellar liposome made from uniformly spaced concentric bilayers — were investigated to counteract an overdose of amitriptyline[23], a potentially cardiotoxic antidepressant. Isolated hearts were first perfused with amitriptyline at a concentration causing cardiotoxicity. Subsequent infusion of pH-gradient spherulites resulted in swift recoveries of heart functions (Fig. 3). These preliminary data are promising because the spherulite concentration in this investigation could be readily achieved *in vivo*[24].

The chelating agent diethylene triamine pentaacetic acid (DTPA) — a molecule that binds cations — is used to decontaminate individuals who have been exposed to toxic heavy metals such as ytterbium[25] and plutonium[26]. DTPA-Pu/Yb complexes are stable, soluble and readily eliminated in the urine. Incorporating DTPA into liposomes was shown to increase its half-life and promote its deposition into tissues such as the liver and bone where heavy metals tend to accumulate and exert their toxicity. Fattal and co-workers[27,28] prepared uncoated and PEGylated liposomes containing DTPA ranging from 100 to 1,600 nm and assessed their Pu decorporation capacities. The encapsulated DTPA exhibited a 3- to 90-fold decrease in clearance with the longest circulation time achieved with small stealth liposomes. This was accompanied by substantial DTPA deposition in the liver regardless of the liposome formulation and comparatively higher levels in the bone compared to the free

chelator. Liposomal DTPA improved the Pu excretion, thereby reducing the total Pu burden 30 days after toxic exposure[28].

NANOEMULSIONS

Liposomes have proven to be effective antidotes for amphiphilic compounds that can be inactivated by encapsulated enzymes or actively trapped within their aqueous compartments, but they may not be ideal for highly hydrophobic and poorly or non-ionizable molecules. Under these circumstances, colloidal systems such as nanoemulsions (that is, nanosized droplets of oil dispersed in an aqueous phase), where drug uptake mostly relies on a favourable partition coefficient for oil droplets, may be more appropriate. In some cases, ionizable drugs may exhibit a high affinity for oils or oil/water interfaces and can therefore be extracted by nanoemulsions.

Intralipid — a nutritional supplement — is a soybean oil-in-water emulsion (430 nm) that is stabilized with egg phosphatidylcholine lipid and is commonly injected as a source of triglycerides for individuals who cannot ingest fats orally. This emulsion was evaluated as a detoxifier for hydrophobic drugs such as bupivacaine, a local anaesthetic associated with occasional but severe and potentially lethal cardiotoxicity. In animals, it was found that the infusion of Intralipid immediately after the injection of lethal bupivacaine doses increased survival[29,30]. Case reports documenting the efficacy of Intralipid in humans experiencing anaesthetic-induced cardiotoxicity have also been published[31,32]. Considering the low affinity of bupivacaine for Intralipid ($\log P_{emulsion/plasma}<10$), the positive effect of the treatment could be partially attributed to the large dose of lipids injected (several grams of triglycerides per kilogram), which favoured drug partition to the oil phase.

Recently, the effect of pre- and post-dosing of PEGylated tricaprylin emulsions (another triglyceride based emulsion) on the pharmacokinetics and biodistribution of docetaxel (a model non-ionisable anticancer drug) was addressed[33]. The injection of the emulsion (500 mg kg^{-1}) 20 min after docetaxel administration produced a rapid drug sequestration in the blood pool. Furthermore, after uptake by the emulsion, the drug was mainly redirected to the liver and spleen, which are the main organs of colloid deposition. These findings clearly illustrate that nanoemulsions can extract drugs that have already been distributed to peripheral tissues.

NANOPARTICLES

Emulsions are particularly attractive in the biomedical field because they can be prepared from generally recognized-as-safe excipients. However, they often face inherent formulation issues arising from thermodynamic instability, which can lead to the coalescence of oil droplets over time or their disassembly in the bloodstream. Moreover, their long-term stability is also limited by difficulties in obtaining dry formulations for prolonged storage. This is particularly relevant in the context of detoxification because turnover is expected to be low. To enhance stability, nanoemulsions can be coated with a hard polymeric shell, leading to the formation of nanocapsules. The shell renders the carrier more robust and controls drug uptake kinetics[34]. Underhill et al.[35] prepared hexadecane-filled polysiloxane/silicate nanocapsules and assessed their ability to sequester bupivacaine and quinoline. *In vitro*, the nanocapsules rapidly removed these two drugs from a normal saline solution. Nonetheless, the low biodegradability of their polymeric shell and their interaction with blood components — rupturing red blood cells and delaying clotting time — would hamper their use in the clinic.

PEGylation of these nanocapsules was, however, found to improve blood compatibility[36].

An interesting concept based on injectable magnetic nanospheres was recently introduced to remove deleterious compounds. Magnetic nanospheres were functionalized with ligands that recognize a particular toxin[37]. Once bound to the ligand on the carrier, the toxin is removed using a magnetic filter unit. Such a system is still in the early development stage, and no data other than *in vitro* extraction results are available so far.

MACROMOLECULAR CARRIERS

Water-soluble macromolecules have been investigated as nanomedicines for more than three decades to increase the circulation time of bound drugs and improve their site-specific delivery[38]. Whole antibodies and antibody fragments represent one of the most studied classes of macromolecular carriers. Their first documented therapeutic human application dates back to the 1970's and concerned the treatment of digitalis (a cardiovascular drug) intoxication[39]. Since then, the concept has been applied with some success to several drugs (for example, amitriptyline) and toxins (for example, colchicine). This detoxification procedure, which is potentially very powerful owing to the high specificity and affinity of the antibody–antigen interaction, nonetheless, faces important limitations. In order to produce specific high affinity antibodies directed towards the toxic compound, the latter should have an immunogenic character, which is not always the case. Moreover, each antibody or antibody fragment can neutralize a limited number of toxic molecules, making this approach appropriate mostly for compounds that are toxic at very low doses.

Non-immune macromolecules such as cyclodextrins (cyclic oligosaccharides), can also reverse the pharmacological effect of drugs. For example, Sugammadex is a novel γ-cyclodextrin based molecule that forms an exceptionally stable 1:1 complex with the neuromuscular blocking agent, rocuronium (association constant ~10^7 M^{-1})[40]. Neuromuscular blocking drugs are used to relax muscles during surgery. After completion of the surgery, these drugs are often neutralized with a pharmacological agent to accelerate recovery from neuromuscular blockade. In Sugammadex, the dextrose units of cyclodextrin were modified to better accommodate the rocuronium and enhance the electrostatic interactions between them. The intravenous injection of Sugammadex was shown to deplete the free rocuronium in plasma and enhance its urinary excretion[41]. Several clinical studies have clearly established the remarkable efficacy of Sugammadex at quickly reversing the neuromuscular block without inducing serious adverse events[42].

Recently, oligochitosan, a linear biodegradable copolymer of *N*-acetyl-D-glucosamine and D-glucosamine (1150 g mol^{-1}), was studied as a detoxifier for amitriptyline. The polymer was modified with dinitrobenzenesulfonyl groups to selectively bind amitriptyline via π–π interactions. The functionalized polymer alone seemed inert because it did not affect blood clotting *in vitro*[14]. Perfusion of the amitriptyline-polymer complex in isolated hearts reduced the drug's cardiotoxicity, whereas the unmodified polymer had no effect[43]. This study proves that the amitriptyline binding to the chitosan derivative prevents the drug from diffusing into the heart tissue. However, the potential liver toxicity of the dinitrobenzylsulfonide moiety, which could arise at the high doses that are required for biodetoxification, remains to be investigated.

CONCLUSION AND PERSPECTIVES

Since liposomes were first proposed as a means of treating poisoning almost 35 years ago[26], tremendous progress has been made in perfecting nanocarriers for biodetoxification applications. Yet, only a single system developed so far has reached the clinical stage, partly because combining properties such as biocompatibility, long circulation time, stability and high extraction efficacy is not trivial. Recent advances in the field of nanotechnology may be exploited to successfully attain this goal. For instance, instability problems commonly encountered with liposomes can be circumvented by engineering nanosized vesicles from biodegradable multiblock polymers[44] or shell crosslinked nanocages[45]. On the other hand, the affinity between the drug and nanocarrier can be enhanced by using molecular imprinting techniques[46]. Polymeric matrices are imprinted with a template (which could, for example, be made from a drug) and are then washed away. This leaves vacant sites that can rebind the imprinted molecule with high specificity and affinity. In the future, nanocarriers may also be used to sequester high-molecular-weight hydrophilic toxins. Indeed, reverse polymeric micelles from hyperbranched and star-shape polymers can be tailored to take up macromolecules in their hydrophilic inner core[47].

doi:10.1038/nnano.2007.339

References

1. Watson, W. A. *et al.* 2004 Annual report of the American Association of Poison Control Centers Toxic Exposure Surveillance System. *Am. J. Emerg. Med.* **23**, 589–666 (2005).
2. Mokhlesi, B., Leiken, J. B., Murray, P. & Corbridge, T. C. Adult toxicology in critical care - Part I: general approach to the intoxicated patient. *Chest* **123**, 577–592 (2003).
3. Zimmerman, J. L. Poisonings and overdoses in the intensive care unit: general and specific management issues. *Crit. Care Med.* **31**, 2794–2801 (2003).
4. Babu Dhanikula, A., Lafleur, M. & Leroux, J. C. Characterization and in vitro evaluation of spherulites as sequestering vesicles with potential application in drug detoxification. *Biochim. Biophys. Acta* **1758**, 1787–1796 (2006).
5. Breyer-Pfaff, U. The metabolic fate of amitriptyline, nortiptyline and amitriptylinoxide in man. *Drug Metab. Rev.* **36**, 723–746 (2004).
6. Allen, T. M. & Cullis, P. R. Drug delivery systems: entering the mainstream. *Science* **303**, 1818–1822 (2004).
7. Yamaoka, T., Tabata, Y. & Ikada, Y. Distribution and tissue uptake of poly(ethylene glycol) with different molecular weights after intravenous administration to mice. *J. Pharm. Sci.* **83**, 601–606 (1994).
8. Moghimi, S. M., Hunt, A. C. & Murray, J. C. Long-circulating and target-specific nanoparticles: theory to practice. *Pharmacol. Rev.* **53**, 283–318 (2001).
9. Olson, K. R. Poisoning and drug overdose. (Lange Medical Books / McGraw-Hill, New York, 2007).
10. Rossi, J. & Leroux, J. C. in *Role of Lipid Excipients in Modifying Oral and Parenteral Drug Delivery* (ed. K. M. Wasan) 88–123 (John Wiley & Sons, Hoboken, 2006).
11. Morey, T. E. *et al.* Treatment of local anesthetic-induced cardiotoxicity using drug scavenging nanoparticles. *Nano Lett.* **4**, 757–759 (2004).
12. Jovanovic, A. V., Underhill, R. S., Bucholz, T. L. & Duran, R. S. Oil core and silica shell nanocapsules: toward controlling the size and the ability to sequester hydrophobic compounds. *Chem. Mater.* **17**, 3375–3383 (2005).
13. Varshney, M. *et al.* Pluronic microemulsions as nanoreservoirs for extraction of bupivacaine from normal saline. *J. Am. Chem. Soc.* **126**, 5108–5112 (2004).
14. Lee, D. W. & Baney, R. H. Oligochitosan derivatives bearing electron-deficient aromatic rings for adsorption of amitryptiline: implications for drug detoxification. *Biomacromolecules* **5**, 1310–1315 (2004).
15. Walde, P. & Ichikawa, S. Enzymes inside lipid vesicles: preparation, reactivity and applications. *Biomol. Eng.* **18**, 143–177 (2001).
16. Cullis, P. R. *et al.* Influence of pH gradients on the transbilayer transport of drugs, lipids, peptides and metal ions into large unilamellar vesicles. *Biochim. Biophys. Acta* **1331**, 187–211 (1997).
17. Leung, P. *et al.* Encapsulation of thiosulfate:cyanide sulfurtransferase by mouse erythrocytes. *Toxicol. Appl. Pharmacol.* **83**, 101–107 (1986).
18. Pei, L., Petrikovics, I. & Way, J. L. Antagonism of the lethal effects of paraoxon by carrier erythrocytes containing phosphotriesterase. *Fundam. Appl. Toxicol.* **28**, 209–214 (1995).
19. Petrikovics, I. *et al.* Antagonism of paraoxon intoxication by recombinant phosphotriesterase encapsulated within sterically stabilized liposomes. *Toxicol. Appl. Pharmacol.* **156**, 56–63 (1999).
20. Petrikovics, I. *et al.* Comparing therapeutic and prophylactic protection against the lethal effect of paraoxon. *Toxicol. Sci.* **77**, 258–262 (2004).
21. Petrikovics, I. *et al.* Long circulating liposomes encapsulating organophosphorus acid anhydrolase in diisopropylfluorophosphate antagonism. *Toxicol. Sci.* **57**, 16–21 (2000).
22. Mayer, L. D., Reamer, J. & Bally, M. B. Intravenous pretreatment with empty pH gradient liposomes alters the pharmacokinetics and toxicity of doxorubicin through in vivo active drug encapsulation. *J. Pharm. Sci.* **88**, 96–102 (1999).
23. Babu Dhanikula, A., Lamontagne, D. & Leroux, J. C. Rescue of amitriptyline-intoxicated hearts with nanosized vesicles. *Cardiovasc. Res.* **74**, 480–486 (2007).
24. Simard, P., Hoarau, D., Khalid, M. N., Roux, E. & Leroux, J. C. Preparation and in vivo evaluation of PEGylated spherulite formulations. *Biochim. Biophys. Acta* **1715**, 37–48 (2005).
25. Blank, M. L., Cress, E. A., Byrd, B. L., Washburn, L. C. & Snyder, F. Liposomal encapsulated Zn-DTPA for removing intracellular ^{169}Yb. *Health Phys.* **39**, 913–920 (1980).
26. Rahman, Y. E., Rosenthal, M. W. & Cerny, E. A. Intracellular plutonium: removal by liposome-encapsulated chelating agents. *Science* **180**, 300–302 (1973).

27. Phan, G. et al. Pharmacokinetics of DTPA entrapped in conventional and long-circulating liposomes of different size for plutonium decorporation. *J. Controlled Release* **110**, 177–188 (2005).

28. Phan, G. et al. Enhanced decorporation of plutonium by DTPA encapsulated in small PEG-coated liposomes. *Biochimie* **88**, 1843–1849 (2006).

29. Weinberg, G. L., VadeBoncouer, T., Ramaraju, G. A., Garcia-Amaro, M. F. & Cwik, M. J. Pretreatment of resuscitation with a lipid infusion shifts the dose-response to bupivacaine-induced asystole in rats. *Anesthesiology* **88**, 1071–1075 (1998).

30. Weinberg, G. L., Ripper, R., Feinstein, D. L. & Hoffman, W. Lipid emulsion infusion rescues dogs from bupivacaine-induced cardiac toxicity. *Reg. Anesth. Pain Med.* **28**, 198–202 (2003).

31. Rosenblatt, M. A., Abel, M., Fischer, G. W., Itzkovich, C. J. & Eisenkraft, J. B. Successful use of a 20% lipid emulsion to resuscitate a patient after a presumed bupivacaine-related cardiac arrest. *Anesthesiology* **105**, 217–221 (2007).

32. Foxall, G., McCahon, R., Lamb, J., Hardman, J. G. & Bedforth, N. M. Levobupivacaine-induced seizures and cardiovascular collapse treated with Intralipid. *Anesthesia* **62**, 516–518 (2007).

33. Babu Dhanikula, A. et al. Long circulating lipid nanocapsules for drug detoxification. *Biomaterials* **28**, 1248–1257 (2007).

34. Joncheray, T. J. et al. Electrochemical and spectroscopic characterization of organic compound uptake in silica core-shell nanocapsules. *Langmuir* **22**, 8684–8689 (2006).

35. Underhill, R. S. et al. Oil-filled silica nanocapsules for lipophilic drug uptake: implications for drug detoxification therapy. *Chem. Mater.* **14**, 4919–4925 (2002).

36. Jovanovic, A. V. et al. Surface modification of silica core-shell nanocapsules: biomedical implications. *Biomacromolecules* **7**, 945–949 (2006).

37. Mertz, C. J. et al. In vitro studies of functionalized magnetic nanospheres for selective removal of a simulant biotoxin. *J. Magn. Magn. Mater.* **293**, 572–577 (2005).

38. Duncan, R. Polymer conjugates as anticancer nanomedicines. *Nat. Rev. Cancer* **6**, 688–701 (2006).

39. Bateman, D. N. Digoxin-specific antibody fragments. *Toxicol. Rev.* **23**, 135–143 (2004).

40. Bom, A. et al. A novel concept of reversing neuromuscular blok: chemical encapsulation of rocuronium bromide by a cyclodextrin-based synthetic host. *Angew. Chem. Int. Edn* **41**, 266–270 (2002).

41. Sorgenfrei, I. F. et al. Reversal of rocuronium-induced neuromuscular block by the selective relaxant binding agent sugammadex. *Anesthesiology* **104**, 667–674 (2006).

42. Naguib, M. Sugammadex: another milestone in clinical neuromuscular pharmacology. *Anesth. Analg.* **104**, 575–581 (2007).

43. Lee, D. W. et al. Aromatic-aromatic interaction of amitriptyline: implication of overdosed drug detoxification. *J. Pharm. Sci.* **94**, 373–381 (2005).

44. Discher, D. E. & Eisenberg, A. Polymer vesicles. *Science* **297**, 967–973 (2002).

45. O'Reilly, R. K., Hawker, C. J. & Wooley, K. L. Cross-linked block copolymer micelles: functional nanostructures of great potential and versatility. *Chem. Soc. Rev.* **35**, 1068–1083 (2006).

46. Mayes, A. G. & Whitecombe, M. J. Synthetic strategies for the generation of molecularly imprinted organic polymers. *Adv. Drug Deliv. Rev.* **57**, 1742–1778 (2005).

47. Jones, M. C. et al. Self-assembled nanocages for hydrophilic guest molecules. *J. Am. Chem. Soc.* **128**, 14599–14605 (2006).

48. Lacy, C. F., Armstrong, L. L., Goldman, M. P. & Lance, L. L. *Drug Information Handbook* edn 14 (Lexi-Comp, Hudson, 2006).

49. Phan, G. et al. Targeting of diethylene triamine pentaacetic acid encapsulated in liposomes to rat liver: an effective strategy to prevent bone deposition and increase urine elimination of plutonium in rats. *Int. J. Radiat. Biol.* **80**, 413–422 (2004).

50. Fallon, M. S. & Chauhan, A. Sequestration of amitriptyline by liposomes. *J. Colloid Interface Sci.* **300**, 7–19 (2006).

51. Wang, L. et al. A biocompatible method of decorporation: bisphosphonate-modified magnetite nanoparticles to remove uranyl ions from blood. *J. Am. Chem. Soc.* **128**, 13358–13359 (2006).

52. Suy, K. et al. Effective reversal of moderate rocuronium- or vecuronium-induced neuromuscular block with sugammadex, a selective relaxant binding agent. *Anesthesiology* **106**, 283–288 (2007).

Acknowledgements

The Canada Research Chair program and the Natural Sciences and Engineering Research Council of Canada (NanoIP program) are acknowledged for their financial support.
Correspondence should be addressed to J.-C. Leroux.

SELECTED APPLICATIONS

Applications of dip-pen nanolithography

The ability to tailor the chemical composition and structure of a surface at the sub-100-nm length scale is important for studying topics ranging from molecular electronics to materials assembly, and for investigating biological recognition at the single biomolecule level. Dip-pen nanolithography (DPN) is a scanning probe microscopy-based nanofabrication technique that uniquely combines direct-write soft-matter compatibility with the high resolution and registry of atomic force microscopy (AFM), which makes it a powerful tool for depositing soft and hard materials, in the form of stable and functional architectures, on a variety of surfaces. The technology is accessible to any researcher who can operate an AFM instrument and is now used by more than 200 laboratories throughout the world. This article introduces DPN and reviews the rapid growth of the field of DPN-enabled research and applications over the past several years.

KHALID SALAITA, YUHUANG WANG, AND CHAD A. MIRKIN*

Department of Chemistry and International Institute for Nanotechnology, Northwestern University, 2145 Sheridan Road, Evanston, Illinois 60208–3113.
*e-mail: chadnano@northwestern.edu

Developing the ability to control the shape, size, and chemical composition of structures on the 1–100 nm length scale is an exciting challenge because it opens new research possibilities in a number of fields ranging from electronics to medical diagnostics. There are currently a wide variety of techniques for generating nanoscale architectures. Among these, scanning probe microscope-based lithographies (SPLs) offer both ultrahigh resolution and in situ imaging capabilities, and as such, these techniques are actively used as nanofabrication tools[1–3]. In some cases, scanning probe methods have even enabled researchers to manipulate matter at the individual atom or molecule level[4]. However, most SPL techniques rely on the elimination or modification (through oxidation, etching, shaving, or grafting) of a passivating layer and subsequent adsorption of the patterning material from solution[2,3,5,6]. These indirect, often surface-'destructive' approaches have a fundamental limitation with respect to the types and number of materials that can be patterned within a nano- or microscopic field of view. In addition, indirect SPL methods are limited in throughput because of difficulties associated with parallelization[3].

Dip-pen nanolithography (DPN)[7–9] is a novel scanning probe-based tool that uses an 'ink'-coated AFM tip to pattern a surface (Fig. 1). Unlike the other SPL methods, DPN is a direct-write 'constructive' lithographic tool that allows soft and hard materials to be printed from scanning probe tips onto a surface with high registration and sub-50-nm resolution. Importantly, there is no premodification of the surface through energy delivery prior to the constructive and additive ink delivery event. Because of its direct-write capability and soft-matter compatibility, DPN can be used to deposit multiple compounds, sequentially or in parallel, precisely

and exclusively where they are needed. There is no need to expose the patterning substrate to harsh conditions such as ultraviolet, ion- or electron-beam irradiation, or even non-polar solvents, thus avoiding cross-contamination or damage. These capabilities make DPN a unique and highly desirable tool for depositing biological and other soft matter on a variety of surfaces.

The first demonstration of DPN was carried out with alkanethiol molecules[7]. These molecules could be delivered to a Au surface with 15 nm resolution, and the deposited structures were stable, crystalline and indistinguishable from self-assembled monolayers (SAMs) grown in bulk solutions[11,12]. These results disproved a 1995 claim that alkanethiol molecules could not be deposited onto a gold surface using a scanning probe tip[13] and resulted in the invention of DPN[7]. During subsequent years DPN has been developed to pattern a wide variety of inks, including small organic molecules[7,11,14–21], polymers[22–27], DNA[28,29], proteins[30–33], peptides[34–36], colloidal nanoparticles[37], metal ions[38,39] and sols[40,41]. The range of patterning substrates has also been expanded to include many insulating, semiconducting and metallic substrates[17,20,24,42]. Many of these early developments have been reviewed by Ginger et al.[8] and will not be discussed here.

Because of the unique capabilities of DPN, many research groups are using this tool to address important scientific questions inherent to nanoscale structures, as well as making fundamental contributions to the development of the DPN technique. Research efforts have ranged from theoretical[43–50] and experimental[51–59] studies elucidating the basic mechanisms of molecular diffusion and transport in DPN, to the development of novel methods of ink delivery such as thermal DPN (tDPN) of metals[10] and the introduction of active and multifunctional tips[60–62].

This review will highlight the latest developments in the DPN technique with an emphasis on the unique advantages of DPN as an enabling research tool. The review is divided into three sections: (1) DPN for fabricating biological nanoarrays, (2) templated hierarchical assembly of materials, and (3) extending the capabilities of DPN.

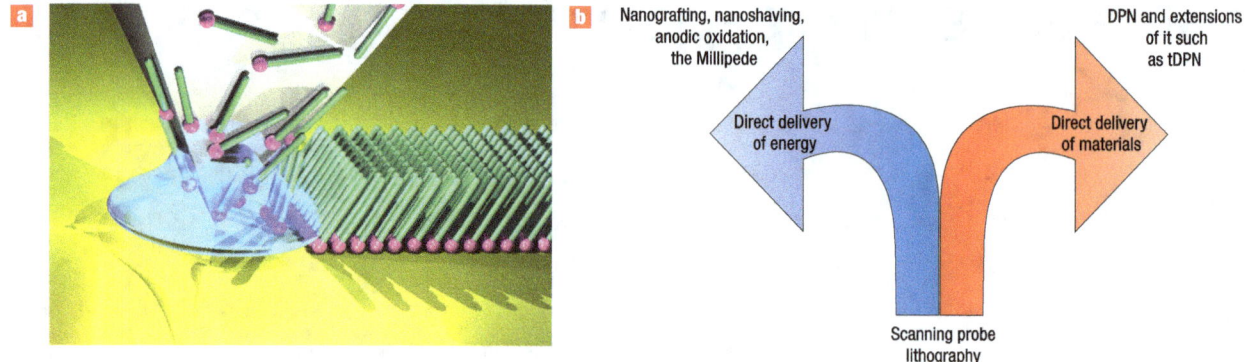

Figure 1 Dip-pen nanolithography (DPN). **a**, Schematic of DPN. An alkanethiol ink-coated AFM tip is brought into contact with a gold substrate. Molecules diffuse from the tip to the surface through or over a water meniscus and assemble in the wake of the tip path to form a stable nanostructure. **b**, Diagram showing the fundamental difference between the different types of scanning probe lithography processes. All systems developed prior to DPN involved the delivery of energy to a surface to chemically or physically modify it. For molecular printing technologies, the necessity to do this prior to an ink adsorption event presents daunting challenges in terms of parallelization and multiplexing (the use of multiple inks simultaneously). DPN involves the direct delivery of the ink to a surface, so one or more inks can be confined to different tips and many chemically pristine nanostructures can be made simultaneously. Variants of DPN such as thermal DPN[10] use energy to facilitate ink flow rather than to chemically or physically change the underlying substrate.

DPN FOR FABRICATING BIOLOGICAL NANOARRAYS

Nanoarrays of biological molecules are fundamentally exciting for a number of reasons. First, nanoarrays can hold 10^4–10^5 more features than conventional microarrays and, as a result, a larger number of targets can be screened more rapidly in an individual experiment. Second, the total area occupied by a fixed number of targets can be dramatically reduced, which translates into vastly decreased sample volumes and allows for detection of a smaller number of target molecules for a given analyte concentration. This can lead to significantly lower limits of detection (orders of magnitude from a copy number standpoint) than could otherwise be achieved with microarray technology, however, one of the drawbacks of the nanoarray is that it lowers the analyte concentration range that can be probed because device saturation occurs over a more narrow range with small feature sizes, so one must choose applications of the nanoarray where this is not a major concern. Third, nanoarrays can be used to address important fundamental questions pertaining to biomolecular recognition, since biorecognition is inherently a nano- rather than a micro- or macroscopic phenomenon.

Because of these important advantages, nanoarrays of biological molecules have the potential to become a revolutionary new platform for performing biological studies. Using DPN, such nanoarrays can be routinely fabricated, although massive multiplexing capabilities are yet to be demonstrated. Notably, a variety of biological molecules, including DNA[28,29], peptides[34–36], proteins[30–33,63], viruses[64–66] and bacteria[67] have been patterned using direct-write or indirect adsorption approaches.

These DPN-enabled nanoarrays have also been used for biological studies. For example, Lee et al. used nanoarrays of the anti-p24 antibody to screen for the human immunodeficiency HIV-1 virus (HIV-1) p24 antigen in serum samples[63]. In this work, the antibody nanoarrays were fabricated using DPN-patterned 16-mercaptohexadecanoic acid (MHA) dot features as templates for antibody immobilization. On capturing p24, the anti-p24 features increased in height by 2.3 ± 0.6 nm (measured with AFM). This height increase can be further amplified by sandwiching the captured p24 protein with anti-p24-functionalized gold nanoparticle probes (Fig. 2). Importantly, this report demonstrates that nanoarray-based assays can exceed the detection limit of conventional enzyme-linked immunosorbent assays by orders of magnitude. In this work the antibody nanoarrays are fabricated using DPN-patterned MHA dot features as a template for antibody immobilization.

In contrast to the MHA-directed immobilization in the previous example, Kang et al. used DPN to directly print integrin $\alpha_v\beta_3$ nanoarrays[33]. These nanoarrays provide a platform for investigating the molecular interaction between integrin $\alpha_v\beta_3$ and vitronectin cell adhesion protein. With protein recognition and binding, AFM topography measurements indicated a 30 ± 5 nm height increase. Using bovine serum albumin nanoarrays as a control, it was further confirmed that the patterned integrin proteins retain their biological selectivity after surface immobilization[33].

As illustrated in the previous examples, biomolecular DPN-based publications have yet to show sub-40-nm spatial resolution, and consequently, controlling the orientation or position of individual proteins is just beyond the reach of DPN. However, the critical dimension of viruses is at the ~20–200 nm length scale, and DPN serves as an ideal tool to investigate single virus particle control. As a step towards developing virus particle immobilization chemistry, both De Yoreo[64] and Smith et al.[65] have immobilized collections of cowpea mosaic virus (CPMV) particles onto DPN-generated chemical templates. Although those approaches use different chemical attachment schemes, both require the genetic modification of the CPMV virus capsid with thiol groups, and neither were able to demonstrate single-particle control. Recently, Vega et al. reported that individual tobacco mosaic viruses (TMV), which have an anisotropic tube shape (300 × 18 nm), can be immobilized by nano-affinity templates[66]. By optimizing the dimensions of the DPN-generated template and using a metal-ion-based coupling approach, they showed that single TMV particles could be isolated and oriented on MHA templates with rectangular feature dimensions of 350 × 100 nm spaced 1 μm apart. This is the first step towards using DPN to create specifically tailored surfaces, by positioning individual biological structures, in order to investigate biorecognition and virus-cell infectivity processes.

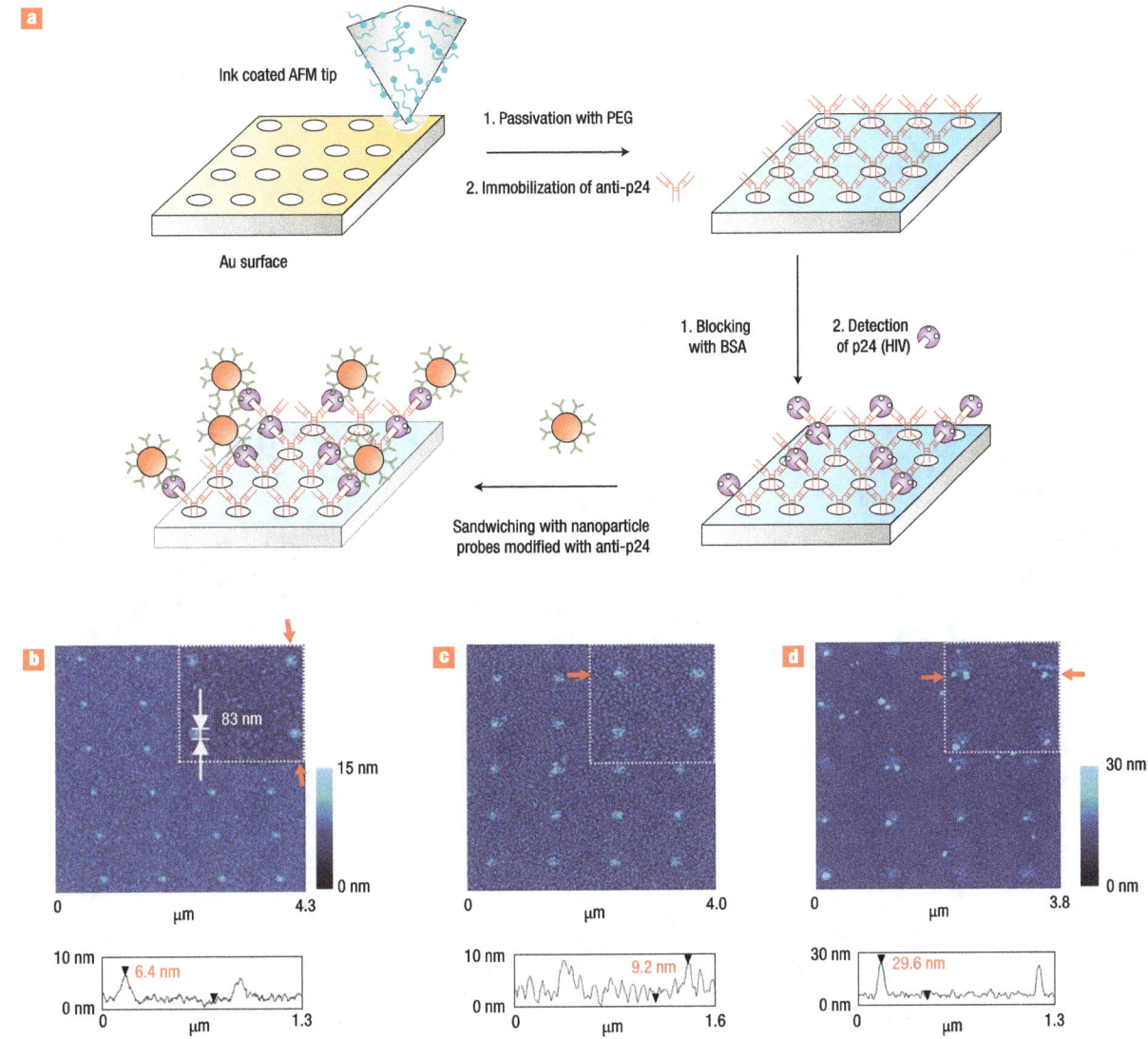

Figure 2 Detection of HIV-1 p24 antigen using a nanoarray of anti-p24 antibody. **a**, DPN-generated nanoarrays of MHA are used to immobilize anti-p24 antibodies on a Au substrate. The bare Au regions were passivated with 11-mercaptoundecyl-tri(ethylene glycol) (PEG), and nonspecific protein adsorption was minimized by blocking with bovine serum albumin (BSA). The presence of HIV-1 p24 in patient plasma is probed by measuring the height profile of the anti-p24 antibody array. The height increase signal as a result of specific antigen–antibody binding is further amplified by a nanoparticle sandwich assay. **b–d**, AFM images that demonstrate the detection of HIV-1 p24 antigen. **b**, Anti-p24 IgG protein nanoarray. Topography trace of adsorbed anti-p24 IgG on MHA. **c**, After p24 binding to anti-p24 IgG, a height increase for the IgG features is observed. **d**, p24 detection after amplification with anti-p24 IgG coated gold nanoparticles (20 nm). Reproduced with permission from ref. 63. Copyright (2004) ACS.

A variety of enzymes, patterned using a DPN-based approach, have been used to direct localized surface reactions. Enzymes are attractive inks because of the range, high yield and specificity of catalysed reactions that can be performed under very mild conditions. Chilkoti and co-workers first demonstrated that DPN can be used to deliver DNase I (a nonspecific endonuclease that digests double-stranded DNA into nucleotide fragments) onto a DNA modified Au substrate[68]. By measuring the height of the patterned regions before and after activation of the DNase I enzyme with Mg^{2+}, they demonstrated that the enzyme was biologically active and that it could catalyse the digestion of a DNA monolayer. Another approach for using enzymes in nanoscale patterning is the application of the enzyme to chemically patterned regions of a substrate to initiate specific reactions. For example, Kaplan *et al.*

demonstrated that horseradish peroxidase (HRP) can be used to polymerize the aromatic ring of 4-aminothiophenol and tyrosine ethyl ester hydrochloride to form conducting wires[69]. In a recent report[70], they further showed that HRP could be used to polymerize caffeic acid nanostructures that were deposited onto an amino-terminated monolayer. Interestingly, surface-immobilized enzymes can be used to generate metallic structures using the byproducts of catalysis. For example, Willner and co-workers demonstrated that glucose oxidase (GOx) and alkaline phosphatase (AlkPh) can be used to selectively reduce Au and Ag metal salts to form metallic wires[71]. Here DPN was used to generate the biologically active lines of both GOx and AlkPh in the same nanoscopic field of view.

The above examples provide a glimpse into the effectiveness of DPN for generating biological nanoarrays. Most of these

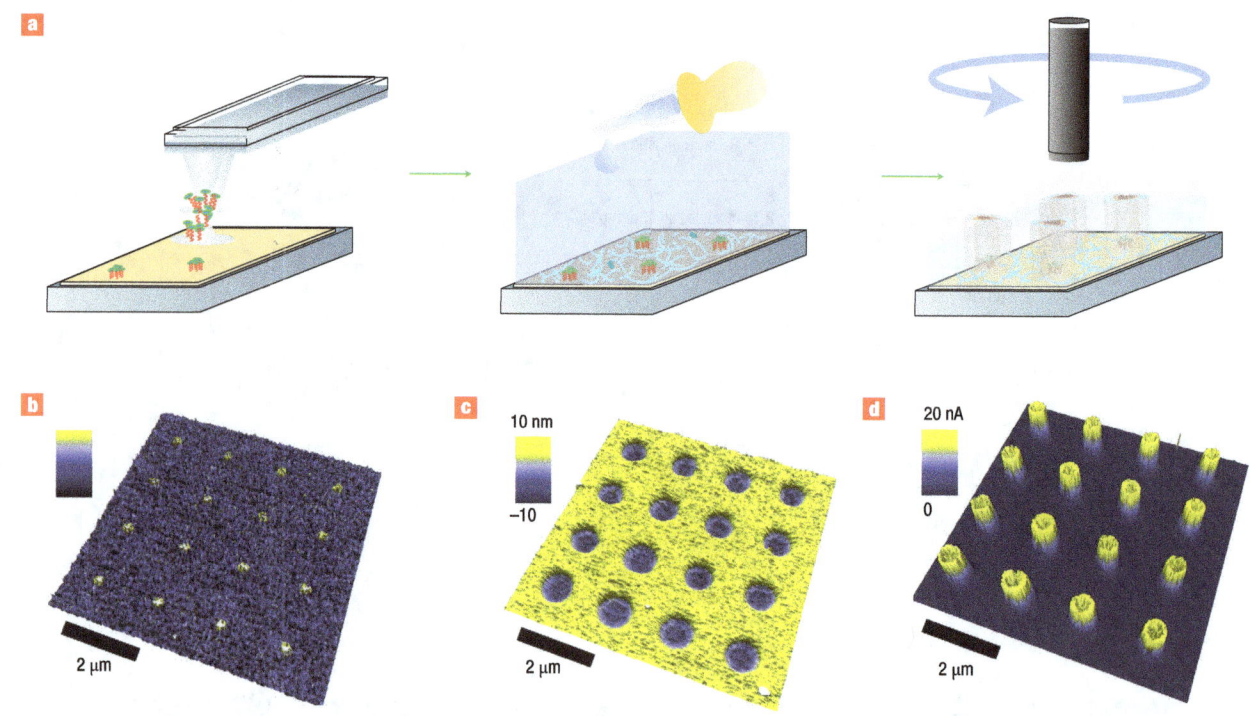

Figure 3 Monolayer templates defined by DPN are used to screen phase separation in polymer blend films. **a**, A scheme depicting the approach. DPN is used to generate nanoarrays of MHA onto an Au substrate and then a 20–60-nm-thick polymer blend of P3HT/PS is spin-coated onto this patterned surface. The resulting film morphology is subsequently characterized by AFM and optical microscopy. **b**, AFM lateral force microscopy image of MHA dot template. **c**, AFM height image (10 nm z-scale). **d**, Conductive AFM image (20 nA z-scale) of the P3HT/PS film showing distinct domains in the deposited polymer blend on top of the MHA patterns. Recessed areas with high conductivity correspond to the semiconducting P3HT-rich domains. Reproduced with permission from ref. 72. Copyright (2005) ACS.

nanoarrays consist of many features made of the same protein and this platform has allowed researchers to investigate biological recognition processes and is designed to manipulate structures such as viruses at the single particle level. It is worth noting that arrays have been fabricated that contain features made of different proteins[30,31], but applications for them are just beginning to be explored. These combined capabilities will be critical in shedding light on the roles of cooperativity and multivalency in viral infection studies, which are typically missed because of population averaging. The micro- and nano-environment of a cell plays a critical role in determining its fate and many of its functions. By interfacing cells with patterned substrates that precisely present biologically active proteins and extracellular-matrix components we can take a step towards mimicking aspects of the environment encountered by cells. As such, DPN is positioned to make a big impact in the study of important processes ranging from cell–cell adhesion to the mechanisms of cell migration. Nonetheless, one of the biggest challenges in performing experiments with live cells is the inherent heterogeneity found from sample to sample. Subsequently, the throughput limitation of single-pen DPN becomes a bottleneck in such experiments. The recent advent of massively parallel capabilities (which will be discussed later) addresses this limitation, and opens the door to performing a statistically significant number of cell-surface experiments in an accelerated fashion.

TEMPLATED HIERARCHICAL ASSEMBLY OF MATERIALS

DPN offers a rapid method of generating chemical templates with tailored chemical composition and structure, which provides

a powerful way of investigating nanoscale phenomena in fields ranging from crystallization to phase separation, and from magnetic information storage to nanoscale electronics. Surfaces can be directly modified to present almost any chemically stable functional group with sub-50-nm lateral resolution. What is particularly exciting here is that combinatorial libraries containing a range of pattern sizes, shapes, and chemical compositions can be systematically varied and routinely produced to probe physical and chemical phenomena at the micro- to nanometre length scales.

For example, Ginger *et al.* used DPN as a rapid prototyping tool to generate and study nanoscale structures in thin composite polymer films[72]. They demonstrated that DPN provides a means to study heterogeneous nucleation on surfaces by screening size-dependent interactions between polymer films and surface features. The study was performed by spin-coating 20–60-nm-thick polymer blends of poly-3-hexylthiophene/polystyrene (P3HT/PS) onto MHA dot nucleation sites generated by DPN with sizes that were systematically varied from 50 to 750 nm (Fig. 3). The domain size of the P3HT phase was found to correspond to the size of the MHA nucleation site as well as the relative ratio of P3HT/PS. In another example, Ivanisevic and co-workers investigated the patterning of polyelectrolytes on SiO_x surfaces[22,73]. The assembly of polymers on templated patterns can be performed over large areas using parallel DPN. We recently demonstrated this capability by investigating the assembly of polyelectrolyte multilayer organic thin films on an array of 67,000 MHA dot features (200 nm diameter)[74].

DPN-defined chemical templates of various sizes, shapes, and compositions have been used to screen the directed assembly of

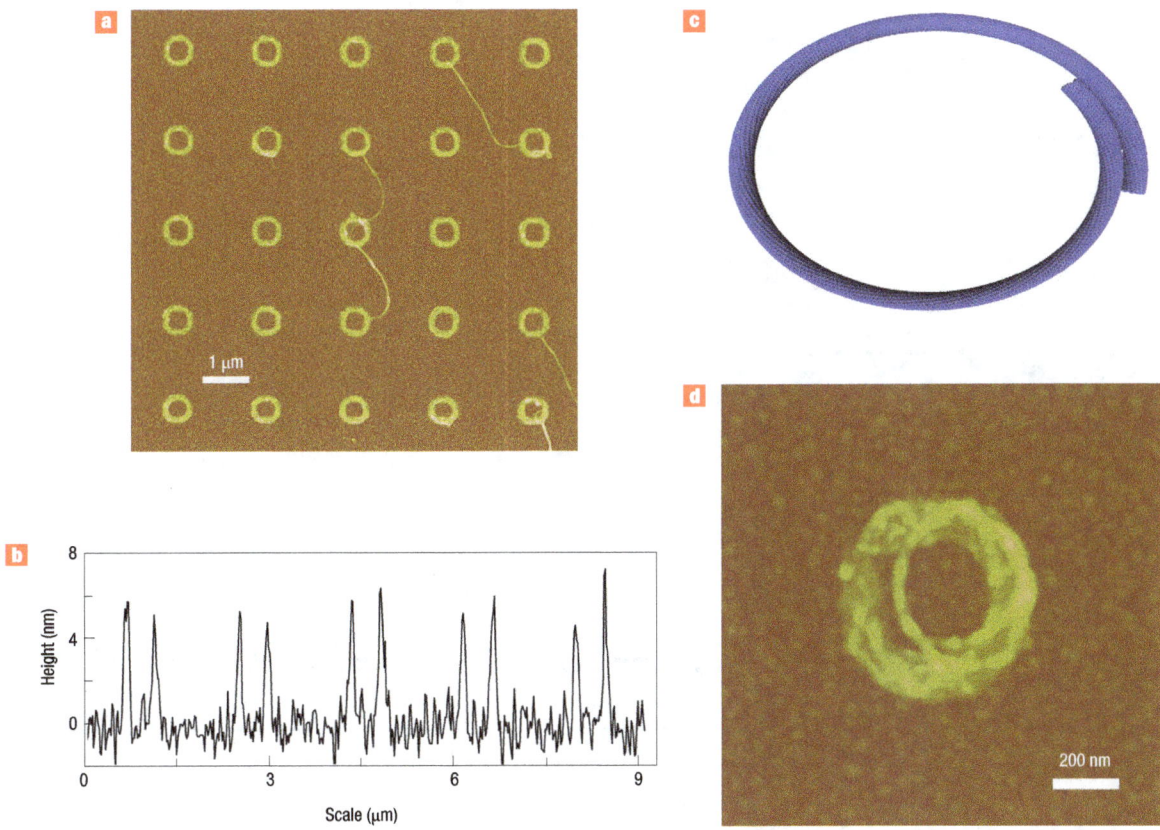

Figure 4 SWNTs assembled into rings on chemical templates. AFM tapping mode topographic images (**a**) and height profiles (**b**) of SWNT rings in a 5 x 5 array are shown. Height profile indicates that multiple SWNTs stack to form rings with relatively uniform thicknesses. Note that some SWNTs are not constrained within the ring templates and instead bridge multiple ring sites. **c**, A molecular model of a coiled SWNT. **d**, A zoom-in view of one SWNT ring. Reproduced with permission from ref. 76. Copyright (2006) National Academy of Sciences USA.

single-walled carbon nanotubes (SWNTs)[75,76] and other nanoscale building blocks[77–80]. Hong and co-workers initially reported that SWNTs could be assembled along line features of chemical templates (such as imidazole) in a solution through electrostatic interactions between SWNTs and the chemical templates[75]. In related but different work, we found that SWNTs can be assembled and manipulated using templates of MHA defined by DPN[76]. In this work, unusual, and in some cases highly strained, SWNT structures, such as dots, rings, arcs and letters, could be fabricated easily from the appropriate template (Fig. 5). Such precise control of the SWNT structure arose from an interesting phenomena occurring at the boundaries between the MHA features and the passivating 1-octadecanethiol (ODT) SAM, where the SWNTs are strongly attracted to the carboxylic groups of the MHA. Interestingly, SWNTs with lengths greater than the dimensions of the MHA feature align along the boundary of the pre-generated nano-feature in order to maximize SWNT/COOH interactions and compensate for the tension arising from SWNT bending. Taking advantage of the high-resolution and rapid prototyping capabilities of DPN, we have carried out a series of comparisons between experiments and theoretical calculations to illustrate and quantify the driving forces for this directed-assembly approach. Notably, a mathematical relationship has been developed that explains the assembly process quantitatively[76]. This relationship balances the geometrically weighted interactions between the

SWNTs and the two different SAMs that are required to overcome solvent–SWNT interactions and to have an effect on assembly. Because there is no specific chemical bonding required, this technique remains effective for the directed assembly of other nanoscale building blocks such as nanowires and nanoparticles. Notably, DPN-patterned negatively charged chemical templates (such as MHA) also allow one to reversibly capture positively charged inorganic V_2O_5 nanowires dispersed in a solution[77].

The examples described above highlight how DPN-generated chemical templates can be used as screening tools to investigate the bottom-up assembly of polymer composites and nanoscale building blocks. In another case, DPN was found useful in bridging the gap between top-down lithography and self-assembly by integrating DNA recognition into macroscopically addressable electrode junctions. In particular, Chung *et al.* used DPN to directly deposit oligonucleotide probes on lithographically defined electrode nanogaps[29]. The subsequent capture of complementary DNA-functionalized gold nanoparticles in the nanogap bridges the two electrodes and switches the sensor from an electrically insulating state to an electrically conducting one, which can be used in a simple 'on/off' readout scheme. This work illustrates how DPN can be used to interface DNA-directed nanoparticle assembly with conventional microfabrication techniques to produce primitive tunnel-junction circuits. Indeed, many proposed micro- and nanoscale biosensing schemes could benefit from such a means of directly depositing

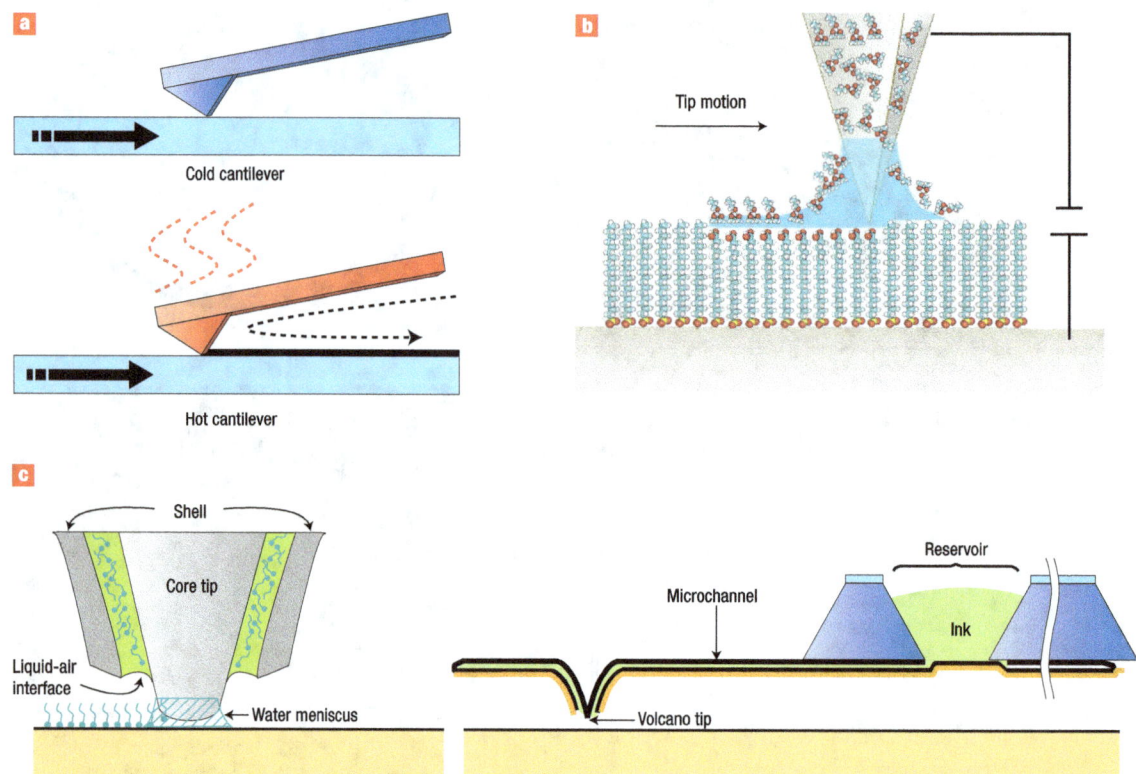

Figure 5 Examples of variants of DPN. **a**, Thermal DPN, which uses a heated AFM cantilever whose tip is coated with a solid 'ink'. When the tip is hot enough, the ink melts and flows onto the substrate. No deposition occurs when the tip is cold, so imaging without any unintended deposition is possible. **b**, Electro pen nanolithography on an octadecyltrichlorosilane (OTS)-coated surface. The terminal methyl group of the OTS is converted to a reactive COOH-terminated surface (OTSox) by applying a voltage between the conducting AFM tip and the conducting silicon substrate in a humid environment. 'Ink' molecules are delivered from the 'inked tip' to the reactive OTSox surface, forming a second layer in the same sweep. No second layer is formed on the methyl-terminated regions. **c**, Writing mechanism of the Nano Fountain Pen. Molecular ink fed from the reservoir forms a liquid–air interface at the annular aperture of the volcano tip. Molecules are transferred by diffusion from the interface to a substrate and a water meniscus is formed by capillary condensation.

ligand/receptor molecules onto specific substrate locations with the high resolution and registration offered by DPN, especially the many recent systems involving nanowire[81] and nanotube[82] arrays.

Another attractive feature of DPN is its ability to observe molecular processes *in situ*. For example, DPN was recently used to monitor the initiation and kinetics of biopolymer crystallization and growth on cleaved mica substrates[24]. This work built upon previous research showing that DPN could be used as a tool to study monolayer growth processes by serially delivering adsorbate to a gold surface[12]. In particular, poly-DL-lysine hydrobromide crystallized epitaxially into triangular single crystals when delivered to a freshly cleaved mica substrate during a DPN experiment[24]. The crystallization process could be studied and kinetically controlled on the nanometre to many-micrometre length scales. Such crystallization morphology was not observed when poly-DL-lysine was deposited with macroscopic or microscopic patterning tools. It is believed that the 'tapping mode' used in the deposition process was responsible for the nanoscale crystal nucleation observed[24].

Phase separation behaviour has been observed in DPN experiments with binary ink mixtures. For example, when mixtures of alkanethiols (such as 3-mercaptopropanoic acid and 1-hexadecanethiol) are assembled onto a bulk Au surface at certain mixing ratios, nanoscale domains form randomly on the surface[83,84]. However, when binary mixtures of alkanethiols are deposited

by DPN, phase-separated domains form in which the exterior predominantly comprises the more hydrophobic ink molecules and the interior comprises the hydrophilic ink[55]. This process has been used to improve the resolution of the technique and realize features even smaller than those attainable with conventional DPN. One ink can be used as a sacrificial material whereas the other can be used as an active component to control subsequent template-driven assembly processes[55]. Indeed, it has been shown recently that by depositing alkanethiols by DPN, it is possible to create multicomponent soft matter and solid-state structures that can be removed selectively[85].

Although most of the examples reported thus far describe chemical patterning of inks bearing simple functional groups (–COOH, –NH$_2$, –SH and so on), it is important to note that a wide variety of chemicals and chemistry have been found to be compatible with DPN. For example, Reinhoudt and co-workers recently demonstrated that complex supramolecular host–guest chemistry could be site-specifically directed using DPN[17,18]. In this example, a variety of adamantyl-modified guest molecules with different valencies were directly deposited onto a β-cyclodextrin SAM with sub-100-nm resolution. Because each β-cyclodextrin molecule has a specific recognition site, the SAM functions as a molecular 'printboard' that allows the formation of kinetically stable high-resolution patterns of guest molecules[86,87]. Vansco *et al.* have shown that amine-terminated poly(amidoamine) dendrimers can be directly deposited onto chemically activated ester SAMs[88]. The use of a reactive SAM as a patterning substrate for DPN will not only widely expand the library of ink-substrate combinations, but may also allow site-specific molecular interactions and chemistries to be explored.

EXTENDING THE CAPABILITIES OF DPN

So far we have reviewed some of the exciting work where DPN has played an important role as an enabling research tool. To extend the basic capabilities of DPN beyond the scope of single silicon nitride cantilever tips, a number of research groups, including our own, are involved in further development of the technology. These efforts include both the introduction of novel variants of the DPN technique and the development of high-throughput massively parallel DPN, as outlined below.

VARIANTS OF DPN

A number of variants of DPN have emerged with the introduction of novel multifunctional tips that are more wear resistant[89], can carry more ink[60,90–92] or can be individually actuated[61,93,94]. In addition, advances with tips that use external heating[10] or an electric field[95–97] to drive physical or chemical transformation of the ink molecules have been reported. These innovative variations have significantly expanded the capabilities of DPN.

For example, Sheehan and co-workers demonstrated that by thermally heating a tip, continuous indium lines as thin as 80 nm can be directly deposited onto glass or silicon surfaces (tDPN, Fig. 5a)[10]. An advantage of this approach is that the rate of ink diffusion can be tuned by changing the tip temperature. In an earlier paper, they claimed that solid inks, such as octadecylphosphonic acid (melting point 100 °C) are immobile at room temperature, but can be deposited through the use of tDPN[98]. However, we have recently shown that conventional DPN also works for octadecylphosphonic acid and other high melting point compounds, provided they have some solubility in the water meniscus[99]. Indeed, solubility is one of the key factors in controlling transport rate, and therefore the commercial systems use humidity control systems to influence ink transport rates. tDPN is useful for transporting metallic and certain other materials that are solids at room temperature and do not have appreciable solubility in a carrier solvent.

By combining DPN with electrochemistry, Ocko et al. introduced a new surface modification approach termed electro pen nanolithography (EPN), in which a substrate is electrochemically modified and then rapidly coupled to ink material transported from the tip[96]. This is related to the earlier work by Sagiv involving tip-induced local electro-oxidation of surface-immobilized molecules with vinyl groups[6]. By using the water meniscus both as an ink transport media and as an electrochemical cell for the oxidization of an octadecyltrichlorosilane SAM on a SiO_x surface, Ocko and co-workers showed that multiple chemical functionalities (such as a terminal SH or NH_2) could be generated in a one-step process. Significantly, this technique provides for relatively fast patterning (10 μm s^{-1}) and the construction of multilayered three-dimensional patterns (Fig. 5b).

To facilitate tip inking and to increase the quantity of adsorbed ink, novel tip modifications often have to be introduced. For example, AFM tips have been fabricated out of or coated with poly(dimethylsiloxane) (PDMS), which allows for increased ink loading[90,92,100]. Another strategy, demonstrated by Espinosa et al. integrates microfluidic channels into the AFM cantilevers to fabricate 'nanofountain pen' (NFP) probe tips capable of depositing alkanethiols onto Au surfaces with sub-50-nm resolution (Fig. 5c)[60,91]. Such fluidic ink delivery systems ensure a constant supply of ink to the surface, and keep ink molecules solvated, which can be a critical issue in maintaining the biological activity and tertiary structure of many proteins. Both nano- and micropipettes have been used as scanning probe tips to deliver reagents directly to a substrate[101–104], but the implementation of microfabricated delivery channels in the context of DPN experiment should lead to a more reliable and versatile form of nanolithography.

PARALLELIZATION OF DPN

Parallelization of the DPN technique is driven by the need for a lithographic tool that is not only capable of high-resolution patterning, but also has high throughput and massive multiplexing capabilities. Such a tool, if broadly accessible, will open up many applications important to biologists and materials scientists. These include the screening of cell-surface interactions, viral infectivity at the single-particle/single-cell level, and the ability to fabricate nanoscale photonic and electronic devices[105,106].

Throughput limitations are a problem inherent to the field of scanning probe microscopy and lithography in general. They are arguably the single largest barrier in the development of useful production-scale techniques based upon SPLs. Several groups have sought to address this problem through the use of parallel arrays of cantilevers. Most notably, the Quate group at Stanford has fabricated both linear and two-dimensional arrays of cantilevers for both imaging and lithographic applications[107–110]. They have successfully demonstrated that up to 50 conductive AFM tips in a linear array could be used to perform anodic oxidation of silicon surfaces[107–110]. For the last decade, IBM has been investigating the use of two-dimensional tip arrays for ultrahigh-density data storage applications. The approach is based on heating a cantilever tip in contact with a thin polymer film and subsequently melting the polymer above its glass transition temperature to create pits[107]. Cantilevers write data bits by heating, and data can be read by measuring the amount of heat dissipation from each tip. Arrays with as many as 32 × 32 individually addressable cantilevers, known as 'the Millipede', have been used[111–115], and up to 4,024 cantilevers have been fabricated[111]. Although these early demonstrations are impressive, it is important to note that none of these parallel approaches involve molecular transport in a lithographic process (Fig. 1b). Moreover, the challenges and design constraints for making a functional array are completely different when the focus is to deliver of energy, rather than material, to an underlying substrate.

Increasing the throughput of DPN has taken on two different approaches, one with individually actuated tips and the other with passive 'duplicating' tips. Independent tip actuation has been previously demonstrated using piezoelectric, thermoelectric, and electrostatic mechanisms[61,93,94]. Earlier demonstrations of active DPN pen arrays were based on thermally actuated tips[93]. Such tips are composed of two different materials with different thermal expansion coefficients, which leads to bending when they are heated. Using the thermal actuation approach, Liu's group and our group demonstrated the 'on/off' actuation and deposition of ODT with an array of ten cantilevers[94]. The thermal bimetallic actuation is robust, simple and effective; however, its main limitation is the undesirable effects of heat on the ink material being deposited. To address this problem, electrostatically actuated DPN probe arrays were fabricated and used to pattern ODT[61]. Parallel actuation was demonstrated with an array of ten cantilevers, which were used to generate 30–40-nm ODT lines on a Au substrate. Electrostatic actuation is an attractive approach because of reduced probe heating and reduced tip-to-tip cross talk[61].

Independently actuated tips afford the flexibility to generate chemically complex patterns. Alternatively, a simpler approach

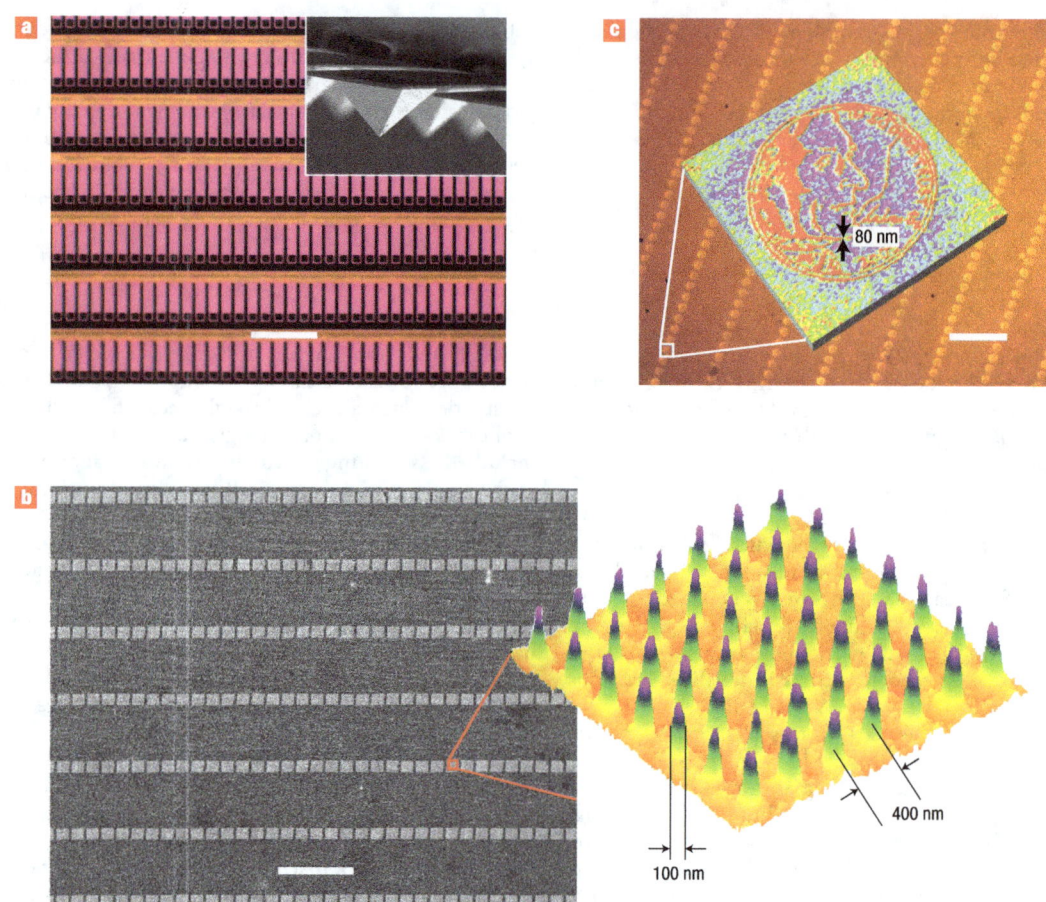

Figure 6 DPN patterning with 55,000 AFM cantilevers in parallel. **a**, Optical micrograph of a small portion of the two-dimensional array of cantilevers used for patterning. The inset shows an SEM image of the tips. **b**, SEM image of a portion of an 88,000,000 gold dot array (40 × 40 within each block) on an oxidized silicon substrate. On the right-hand side is a representative AFM topographical image of part of a block. **c**, Optical micrograph of a representative region of the substrate where ~55,000 sophisticated features were patterned. Each of the circular features is a miniaturized replica of the face of the 2005 US five cent coin. The coin bears a picture of Thomas Jefferson, who helped develop the polygraph, a letter duplicator that relies on an array of pens. Scale Bars: 100 μm. Reproduced with permission from ref. 119. Copyright (2006) Wiley.

to increasing the throughput of DPN is to use large arrays of passive probes[11,116]. In collaboration with the Liu group we have used both commercial and custom-fabricated cantilevers in linear array formats to perform DPN in parallel over a distance of one centimetre[116,117]. The resulting molecular patterns were developed in solid-state Au structures using wet chemical etching techniques and characterized using optical and electron microscopy. Notably, our group, in collaboration with NanoInk and Liu, has fabricated large two-dimensional arrays of cantilevers, which comprise as many as 1.2 million cantilevers on a 3 inch wafer[8]. Liu has also reported a novel 'spring-on-tip' design that could potentially enable higher-density tip arrays to be fabricated as the cantilever in such a design is incorporated within a pyramidal tip[118].

A working two-dimensional pen array for DPN, however, has been demonstrated only recently, and it required several innovations to make it operate in a flexible and highly controllable manner[119]. Again, it is important to note that IBM's Millipede system[111-115] was developed to deliver energy not material to a surface, and although potentially suitable for data storage applications, its design is not optimum for DPN because the tip feedback is based on heat dissipation, which drastically affects molecular transport[58]. The

first functional two-dimensional pen array for DPN incorporates a novel pen-array architecture that involves low aspect ratio pyramidal tips and curved cantilevers. In addition, a robust but simple and straightforward gravity-driven alignment protocol is used to engage the array during writing. These innovations allow a user to simultaneously align ~55,000 cantilevers with a substrate and subsequently pattern over square centimetre areas with sub-100-nm resolution (Fig. 6). Such arrays have been used to pattern small molecules like ODT and bio-relevant structures like phospholipid membranes[120]. With this approach, the throughput of DPN has been increased by more than four orders of magnitude, and as many as 450,000,000 sub-100-nm features have been patterned in less than 30 min. Remarkably, very sophisticated structures can be made using a dot matrix approach (Fig. 6c). The massive increase in throughput will undoubtedly have an impact on increasing the accessibility and lowering the cost of fabricating nanostructured materials on surfaces. Importantly, this passive 'wire-free' approach for massive parallelization of DPN has emphasized the fundamental difference between molecule-scale nanolithography and thermomechanical data storage. Indeed, for data storage, each tip must be electrically addressable, and therefore, extremely impressive but complex systems such as IBM's Millipede are required.

For molecular-deposition based nanolithographies such as DPN, relatively simple array architectures and more user-accessible approaches are possible. However, one must design and construct systems that are compatible with the wide array of inks and substrates required for each intended application (for example, combinatorial libraries of proteins, oligonucleotides, polymers and solid-state materials). Although it is conceivable that approaches based upon the Millipede could be used for materials deposition, this has not been shown to date, and if it can be accomplished, it is likely that the range of inks one can use will be limited to structures that can withstand the thermal changes that must accompany device operation. In addition, lithographic control will have to take into account the temperature-dependent diffusion coefficients of the inks. Finally, in doing so, it is critical that the end device is not so complicated that it prohibits researchers from wanting to use it. The passive pen-array approach is an important first step toward the development of a massively multiplexed molecule-based printing technology that would be widely accessible to anyone with a modest understanding of how to operate an AFM-based instrument.

With the advent of DPN, a dichotomy has emerged in the field of scanning probe lithography (Fig. 1b). Techniques can be classified as either direct and 'constructive' or indirect and 'destructive'. Constructive processes such as DPN directly deliver molecules or materials that are to be patterned. Destructive processes such as anodic oxidation[108] and nanografting[121] typically deliver energy to a surface that is often intended to damage the underlying structure, regardless of whether or not there is a subsequent molecular deposition or adsorption event. These functional differences limit where these techniques can be used, the degree of parallelization possible, and ultimately the types of applications that can be considered for each technique. Prior to DPN, patterning more than the field of view of a conventional AFM ($\sim 100 \times 100$ µm) by SPL processes involved the delivery of energy rather than materials or molecules directly to a surface, even if the intent was to eventually pattern a surface with a new chemical entity. The approach used in parallelization and ultimately application depends greatly on the modality of the patterning technique. For this reason, advances from the indirect 'destructive' approaches cannot simply be implemented in the direct-deposition 'constructive' approaches. This is painfully realized when considering that a technology like the Millipede, which was developed for polymer hole fabrication and annealing, relies on arrays with pens that maintain feedback using thermoresistive sensing and operate at ~ 350 °C. If such an approach to parallelization was applied to a technology like DPN, there would be no control over ink transport and hence feature size.

SUMMARY AND OUTLOOK

DPN is a unique and highly versatile tool that can routinely pattern a wide range of soft materials on a variety of substrates at the sub-50-nm to many-micrometre length scale. Indeed, it is the only nanofabrication tool that uniquely combines high-resolution direct-write capabilities with soft-matter compatibility. Several unconventional lithographic tools such as microcontact printing and nanoimprint lithography have been developed with the primary intention of competing with the tools used by the semiconductor industry for microelectronic applications[1]. Although DPN is finding applications in certain areas that impact the semiconductor industry such as mask fabrication, inspection and repair, thus far, its unique attributes have positioned it on a different path aimed largely at exploiting the properties and applications of nanopatterned soft materials. Over the past several years, DPN has played a pivotal role as an enabling research tool

in areas as diverse as biological recognition to materials assembly. It is opening up the field of nanoarrays and allowing researchers to explore single-particle binding and positioning events.

One of the greatest limitations in nanoscience and nanotechnology involves the lack of efficient methods for macroscopically addressing and manipulating individual nanostructures within the context of an integrated device or material. This limitation is realized in the area of nanoelectronics where researchers have become quite proficient at making nanostructures with unusual and often useful properties, but they still rely on very crude methods for aligning and integrating them within the context of a more sophisticated structure. Once these structures have been assembled, in many cases (for example, in chemical and biological sensors) there is still the problem of chemically functionalizing the individual components within the integrated structure. DPN and more sophisticated variations of DPN provide a solution to this problem, especially because researchers can take advantage of the wide range of chemistries compatible with this technique. The technique becomes even more attractive when considering the recent technological advances that have moved DPN from a low throughput but custom single-pen process to one with massively parallel, high-throughput and large-area capabilities. Taken together, the advances in DPN technology point towards the development of a 'desktop fab' for the custom fabrication of biological chips, nanomaterials, combinatorial chemical templates, and tools important for the semiconductor industry.

On the hardware side, addressable tip inking is the next challenge in developing a suite of DPN-related tools. Such capabilities will enable nanoarrays with almost unlimited chemical and biochemical complexity to be created. Researchers will be able to use these structures to control materials assembly in a very sophisticated manner and open up the field of multivalency in understanding single particle (for example, large proteins, cells, viruses, and spores) surface binding events. There are many possibilities for addressing this challenge including the use of inkjet printing technology[122] and ink wells[123,124] to coat the tips, which are sharp on the nanoscopic scale but separated on the microscopic length scale. On the scientific side, we are just beginning to learn how to use these new technologies. When one considers that virtually every material when miniaturized to the sub-100-nm length scale has new properties, DPN becomes an extraordinarily useful tool to study the fundamental consequences of miniaturization.

doi: 10.1038/nnano.2007.39

References

1. Gates, B. D. et al. New approaches to nanofabrication: Molding, printing, and other techniques. Chem. Rev. 105, 1171–1196 (2005).
2. Tseng, A. A., Notargiacomo, A. & Chen, T. P. Nanofabrication by scanning probe microscope lithography: A review. J. Vac. Sci. Tech. B 23, 877–894 (2005).
3. Kramer, S., Fuierer, R. R. & Gorman, C. B. Scanning probe lithography using self-assembled monolayers. Chem. Rev. 103, 4367–4418 (2003).
4. Eigler, D. M. & Schweizer, E. K. Positioning single atoms with a scanning tunnelling microscope. Nature 344, 524–526 (1990).
5. Liu, S., Maoz, R. & Sagiv, J. Planned nanostructures of colloidal gold via self-assembly on hierarchically assembled organic bilayer template patterns with in-situ generated terminal amino functionality. Nano Lett. 4, 845–851 (2004).
6. Maoz, R., Cohen, S. R. & Sagiv, J. Nanoelectrochemical patterning of monolayer surfaces. Toward spatially defined self-assembly of nanostructures. Adv. Mater. 11, 55–61 (1999).
7. Piner, R. D., Zhu, J., Xu, F., Hong, S. H. & Mirkin, C. A. "Dip-pen" nanolithography. Science 283, 661–663 (1999).
8. Ginger, D. S., Zhang, H. & Mirkin, C. A. The evolution of dip-pen nanolithography. Angew. Chem. Int. Edn 43, 30–45 (2004).
9. Mirkin, C. A., Piner, R. & Hong, S. Methods using scanning probe microscope tips and products therefor or produced thereby. US patent 2002063212; International patent 2000041213.
10. Nelson, B. A., King, W. P., Laracuente, A. R., Sheehan, P. E. & Whitman, L. J. Direct deposition of continuous metal nanostructures by thermal dip-pen nanolithography. Appl. Phys. Lett. 88, 033104 (2006).
11. Hong, S. H., Zhu, J. & Mirkin, C. A. Multiple ink nanolithography: Toward a multiple-pen nano-plotter. Science 286, 523–525 (1999).
12. Hong, S. H., Zhu, J. & Mirkin, C. A. A new tool for studying the in situ growth processes for self-assembled monolayers under ambient conditions. Langmuir 15, 7897–7900 (1999).

13. Jaschke, M. & Butt, H.-J. Deposition of organic material by the tip of a scanning force microscope. *Langmuir* **11,** 1061–4 (1995).
14. Zhang, Y., Salaita, K., Lim, J. H., Lee, K. B. & Mirkin, C. A. A massively parallel electrochemical approach to the miniaturization of organic micro- and nanostructures on surfaces. *Langmuir* **20,** 962–968 (2004).
15. Zhang, Y., Salaita, K., Lim, J. H. & Mirkin, C. A. Electrochemical whittling of organic nanostructures. *Nano Lett.* **2,** 1389–1392 (2002).
16. Vesper, B. J. *et al.* Surface-bound porphyrazines: Controlling reduction potentials of self-assembled monolayers through molecular proximity/orientation to a metal surface. *J. Am. Chem. Soc.* **126,** 16653–16658 (2004).
17. Bruinink, C. M. *et al.* Supramolecular microcontact printing and dip-pen nanolithography on molecular printboards. *Chem. Eur. J.* **11,** 3988–3996 (2005).
18. Auletta, T. *et al.* Writing patterns of molecules on molecular printboards. *Angew. Chem. Int. Edn* **43,** 369–373 (2004).
19. Zhou, H. L., Li, Z., Wu, A. G., Wei, G. & Liu, Z. G. Direct patterning of Rhodamine 6G molecules on mica by dip-pen nanolithography. *Appl. Surf. Sci.* **236,** 18–24 (2004).
20. Kooi, S. E., Baker, L. A., Sheehan, P. E. & Whitman, L. J. Dip-pen nanolithography of chemical templates on silicon oxide. *Adv. Mater.* **16,** 1013–1016 (2004).
21. Ivanisevic, A., McCumber, K. V. & Mirkin, C. A. Site-directed exchange studies with combinatorial libraries of nanostructures. *J. Am. Chem. Soc.* **124,** 11997–12001 (2002).
22. Nyamjav, D. & Ivanisevic, A. Properties of polyelectrolyte templates generated by dip-pen nanolithography and microcontact printing. *Chem. Mater.* **16,** 5216–5219 (2004).
23. Su, M., Aslam, M., Fu, L., Wu, N. Q. & Dravid, V. P. Dip-pen nanopatterning of photosensitive conducting polymer using a monomer ink. *Appl. Phys. Lett.* **84,** 4200–4202 (2004).
24. Liu, X. G. *et al.* The controlled evolution of a polymer single crystal. *Science* **307,** 1763–1766 (2005).
25. Lim, J. H. & Mirkin, C. A. Electrostatically driven dip-pen nanolithography of conducting polymers. *Adv. Mater.* **14,** 1474–1477 (2002).
26. Noy, A. *et al.* Fabrication of luminescent nanostructures and polymer nanowires using dip-pen nanolithography. *Nano Lett.* **2,** 109–112 (2002).
27. Qin, L. D., Park, S., Huang, L. & Mirkin, C. A. On-wire lithography. *Science* **309,** 113–115 (2005).
28. Demers, L. M., Ginger, D. S., Park, S. J., Li, Z., Chung, S. W. & Mirkin, C. A. Direct patterning of modified oligonucleotides on metals and insulators by dip-pen nanolithography. *Science* **296,** 1836–1838 (2002).
29. Chung, S. W. *et al.* Top-down meets bottom-up: Dip-pen nanolithography and DNA-directed assembly of nanoscale electrical circuits. *Small* **1,** 64–69 (2005).
30. Lee, K. B., Lim, J. H. & Mirkin, C. A. Protein nanostructures formed via direct-write dip-pen nanolithography. *J. Am. Chem. Soc.* **125,** 5588–5589 (2003).
31. Lim, J. H. *et al.* Direct-write dip-pen nanolithography of proteins on modified silicon oxide surfaces. *Angew. Chem. Int. Edn* **42,** 2309–2312 (2003).
32. Lee, K. B., Park, S. J., Mirkin, C. A., Smith, J. C. & Mrksich, M. Protein nanoarrays generated by dip-pen nanolithography. *Science* **295,** 1702–1705 (2002).
33. Lee, M. *et al.* Protein nanoarray on Prolinker™ surface constructed by atomic force microscopy dip-pen nanolithography for analysis of protein interaction. *Proteomics* **6,** 1094–1103 (2006).
34. Cho, Y. & Ivanisevic, A. TAT peptide immobilization on gold surfaces: A comparison study with a thiolated peptide and alkylthiols using AFM, XPS, and FT-IRRAS. *J. Phys. Chem. B* **109,** 6225–6232 (2005).
35. Cho, Y. & Ivanisevic, A. SiO_x surfaces with lithographic features composed of a TAT peptide. *J. Phys. Chem. B* **108,** 15223–15228 (2004).
36. Jiang, H. Z. & Stupp, S. I. Dip-pen patterning and surface assembly of peptide amphiphiles. *Langmuir* **21,** 5242–5246 (2005).
37. Gundiah, G. *et al.* Dip-pen nanolithography with magnetic Fe_2O_3 nanocrystals. *Appl. Phys. Lett.* **84,** 5341–5343 (2004).
38. Ding, L., Li, Y., Chu, H. B., Li, X. M. & Liu, J. Creation of cadmium sulfide nanostructures using AFM dip-pen nanolithography. *J. Phys. Chem. B* **109,** 22337–22340 (2005).
39. Li, J. Y., Lu, C. G., Maynor, B., Huang, S. M. & Liu, J. Controlled growth of long gan nanowires from catalyst patterns fabricated by "dip-pen" nanolithographic techniques. *Chem. Mater.* **16,** 1633–1636 (2004).
40. Fu, L., Liu, X. G., Zhang, Y., Dravid, V. P. & Mirkin, C. A. Nanopatterning of "hard" magnetic nanostructures via dip-pen nanolithography and a sol-based ink. *Nano Lett.* **3,** 757–760 (2003).
41. Su, M., Liu, X. G., Li, S. Y., Dravid, V. P. & Mirkin, C. A. Moving beyond molecules: Patterning solid-state features via dip-pen nanolithography with sol-based inks. *J. Am. Chem. Soc.* **124,** 1560–1561 (2002).
42. Agarwal, G., Naik, R. R. & Stone, M. O. Immobilization of histidine-tagged proteins on nickel by electrochemical dip pen nanolithography. *J. Am. Chem. Soc.* **125,** 7408–7412 (2003).
43. Jang, J., Schatz, G. C. & Ratner, M. A. Capillary force on a nanoscale tip in dip-pen nanolithography. *Phys. Rev. Lett.* **90,** 156104 (2003).
44. Lee, N. K. & Hong, S. H. Modeling collective behavior of molecules in nanoscale direct deposition processes. *J. Chem. Phys.* **124,** 114711–114715 (2006).
45. Ahn, Y., Hong, S. & Jang, J. Growth dynamics of self-assembled monolayers in dip-pen nanolithography. *J. Phys. Chem. B* **110,** 4270–4273 (2006).
46. Manandhar, P., Jang, J., Schatz, G. C., Ratner, M. A. & Hong, S. Anomalous surface diffusion in nanoscale direct deposition processes. *Phys. Rev. Lett.* **90,** 115505 (2003).
47. Jang, J. Y., Schatz, G. C. & Ratner, M. A. How narrow can a meniscus be? *Phys. Rev. Lett.* **92,** 085504 (2004).
48. Jang, J. K., Schatz, G. C. & Ratner, M. A. Capillary force in atomic force microscopy. *J. Chem. Phys.* **120,** 1157–1160 (2004).
49. Jang, J. Y., Schatz, G. C. & Ratner, M. A. Liquid meniscus condensation in dip-pen nanolithography. *J. Chem. Phys.* **116,** 3875–3886 (2002).
50. Cho, N., Ryu, S., Kim, B., Schatz, G. C. & Hong, S. H. Phase of molecular ink in nanoscale direct deposition processes. *J. Chem. Phys.* **124,** 024714 (2006).
51. Sheehan, P. E. & Whitman, L. J. Thiol diffusion and the role of humidity in "dip pen nanolithography". *Phys. Rev. Lett.* **88,** 156104–156107 (2002).
52. Weeks, B. L., Noy, A., Miller, A. E. & De Yoreo, J. J. Effect of dissolution kinetics on feature size in dip-pen nanolithography. *Phys. Rev. Lett.* **88,** 255505 (2002).
53. Peterson, E. J., Weeks, B. L., De Yoreo, J. J. & Schwartz, P. V. Effect of environmental conditions on dip pen nanolithography of mercaptohexadecanoic acid. *J. Phys. Chem. B* **108,** 15206–15210 (2004).
54. Schwartz, P. V. Molecular transport from an atomic force microscope tip: A comparative study of dip-pen nanolithography. *Langmuir* **18,** 4041–4046 (2002).
55. Salaita, K., Amarnath, A., Maspoch, D., Higgins, T. B. & Mirkin, C. A. Spontaneous "phase separation" of patterned binary alkanethiol mixtures. *J. Am. Chem. Soc.* **127,** 11283–11287 (2005).
56. Hampton, J. R., Dameron, A. A. & Weiss, P. S. Double-ink dip-pen nanolithography studies elucidate molecular transport. *J. Am. Chem. Soc.* **128,** 1648–1653 (2006).
57. Hampton, J. R., Dameron, A. A. & Weiss, P. S. Transport rates vary with deposition time in dip-pen nanolithography. *J. Phys. Chem. B* **109,** 23118–23120 (2005).
58. Rozhok, S., Piner, R. & Mirkin, C. A. Dip-pen nanolithography: What controls ink transport? *J. Phys. Chem. B* **107,** 751–757 (2003).
59. Rozhok, S., Sun, P., Piner, R., Lieberman, M. & Mirkin, C. A. AFM study of water meniscus formation between an AFM tip and NaCl substrate. *J. Phys. Chem. B* **108,** 7814–7819 (2004).
60. Moldovan, N., Kim, K. H. & Espinosa, H. D. Design and fabrication of a novel microfluidic nanoprobe. *J. Microelectromech. Syst.* **15,** 204–213 (2006).
61. Bullen, D. & Liu, C. Electrostatically actuated dip pen nanolithography probe arrays. *Sens. Actuators A* **125,** 504–511 (2006).
62. Wang, X. F. & Liu, C. Multifunctional probe array for nano patterning and imaging. *Nano Lett.* **5,** 1867–1872 (2005).
63. Lee, K. B., Kim, E. Y., Mirkin, C. A. & Wolinsky, S. M. The use of nanoarrays for highly sensitive and selective detection of human immunodeficiency virus type 1 in plasma. *Nano Lett.* **4,** 1869–1872 (2004).
64. Cheung, C. L. *et al.* Fabrication of assembled virus nanostructures on templates of chemoselective linkers formed by scanning probe nanolithography. *J. Am. Chem. Soc.* **125,** 6848–6849 (2003).
65. Smith, J. C. *et al.* Nanopatterning the chemospecific immobilization of cowpea mosaic virus capsid. *Nano Lett.* **3,** 883–886 (2003).
66. Vega, R. A., Maspoch, D., Salaita, K. & Mirkin, C. A. Nanoarrays of single virus particles. *Angew. Chem. Int. Edn* **44,** 6013–6015 (2005).
67. Rozhok, S. *et al.* Methods for fabricating microarrays of motile bacteria. *Small* **1,** 445–451 (2005).
68. Hyun, J., Kim, J., Craig, S. L. & Chilkoti, A. Enzymatic nanolithography of a self-assembled oligonucleotide monolayer on gold. *J. Am. Chem. Soc.* **126,** 4770–4771 (2004).
69. Xu, P. & Kaplan, D. L. Nanoscale surface patterning of enzyme-catalyzed polymeric conducting wires. *Adv. Mater.* **16,** 628–633 (2004).
70. Xu, P., Uyama, H., Whitten, J. E., Kobayashi, S. & Kaplan, D. L. Peroxidase-catalyzed in situ polymerization of surface orientated caffeic acid. *J. Am. Chem. Soc.* **127,** 11745–11753 (2005).
71. Basnar, B., Weizmann, Y., Cheglakov, Z. & Willner, I. Synthesis of nanowires using dip-pen nanolithography and biocatalytic inks. *Adv. Mater.* **18,** 713–718 (2006).
72. Coffey, D. C. & Ginger, D. S. Patterning phase separation in polymer films with dip-pen nanolithography. *J. Am. Chem. Soc.* **127,** 4564–4565 (2005).
73. Yu, M., Nyamjav, D. & Ivanisevic, A. Fabrication of positively and negatively charged polyelectrolyte structures by dip-pen nanolithography. *J. Mater. Chem.* **15,** 649–652 (2005).
74. Lee, S. W., Sanedrin, R. G., Oh, B. K. & Mirkin, C. A. Nanostructured polyelectrolyte multilayer organic thin films generated via parallel dip-pen nanolithography. *Adv. Mater.* **17,** 2749–2753 (2005).
75. Rao, S. G., Huang, L., Setyawan, W. & Hong, S. H. Large-scale assembly of carbon nanotubes. *Nature* **425,** 36–37 (2003).
76. Wang, Y. *et al.* Controlling the shape, orientation, and linkage of carbon nanotube features with nano affinity templates. *Proc. Natl Acad. Sci. USA* **103,** 2026–2031 (2006).
77. Myung, S., Lee, M., Kim, G. T., Ha, J. S. & Hong, S. Large-scale "surface-programmed assembly" of pristine vanadium oxide nanowire-based devices. *Adv. Mater.* **17,** 2361–2364 (2005).
78. Liu, X. G., Fu, L., Hong, S. H., Dravid, V. P. & Mirkin, C. A. Arrays of magnetic nanoparticles patterned via "dip-pen" nanolithography. *Adv. Mater.* **14,** 231–234 (2002).
79. Demers, L. M., Park, S.-J., Taton, T. A., Li, Z. & Mirkin, C. A. Orthogonal assembly of nanoparticles building blocks on dip-pen nanolithographically generated templates of DNA. *Angew. Chem. Int. Edn* **40,** 3071–3073 (2001).
80. Demers, L. M. & Mirkin, C. A. Combinatorial templates generated by dip-pen nanolithography for the formation of two-dimensional particle arrays. *Angew. Chem. Int. Edn* **40,** 3069–3071 (2001).
81. Zheng, G. F., Patolsky, F., Cui, Y., Wang, W. U. & Lieber, C. M. Multiplexed electrical detection of cancer markers with nanowire sensor arrays. *Nature Biotechnol.* **23,** 1294–1301 (2005).
82. Chen, R. J. *et al.* Noncovalent functionalization of carbon nanotubes for highly specific electronic biosensors. *Proc. Natl Acad. Sci. USA* **100,** 4984–4989 (2003).
83. Stranick, S. J., Parikh, A. N., Tao, Y. T., Allara, D. L. & Weiss, P. S. Phase-separation of mixed-composition self-assembled monolayers into nanometer-scale molecular domains. *J. Phys. Chem.* **98,** 7636–7646 (1994).
84. Imabayashi, S., Hobara, D., Kakiuchi, T. & Knoll, W. Selective replacement of adsorbed alkanethiols in phase-separated binary self-assembled monolayers by electrochemical partial desorption. *Langmuir* **13,** 4502–4504 (1997).
85. Salaita, K. S., Lee, S. W., Ginger, D. S. & Mirkin, C. A. DPN-generated nanostructures as positive resists for preparing lithographic masters or hole arrays. *Nano Lett.* **6,** 2493–2498 (2006).
86. Onclin, S., Ravoo, B. J. & Reinhoudt, D. N. Engineering silicon oxide surfaces using self-assembled monolayers. *Angew. Chem. Int. Edn* **44,** 6282–6304 (2005).
87. Mulder, A. *et al.* Molecular printboards on silicon oxide: Lithographic patterning of cyclodextrin monolayers with multivalent, fluorescent guest molecules. *Small* **1,** 242–253 (2005).
88. Degenhart, G. H., Dordi, B., Schonherr, H. & Vancso, G. J. Micro- and nanofabrication of robust reactive arrays based on the covalent coupling of dendrimers to activated monolayers. *Langmuir* **20,** 6216–6224 (2004).
89. Kim, K. H. *et al.* Novel ultrananocrystalline diamond probes for high-resolution low-wear nanolithographic techniques. *Small* **1,** 866–874 (2005).

90. Wang, X. F. *et al.* Scanning probe contact printing. *Langmuir* **19**, 8951–8955 (2003).
91. Kim, K. H., Moldovan, N. & Espinosa, H. D. A nanofountain probe with sub-100 nm molecular writing resolution. *Small* **1**, 632–635 (2005).
92. Zhang, H., Elghanian, R., Amro, N. A., Disawal, S. & Eby, R. Dip pen nanolithography stamp tip. *Nano Lett.* **4**, 1649–1655 (2004).
93. Wang, X. F., Bullen, D. A., Zou, J., Liu, C. & Mirkin, C. A. Thermally actuated probe array for parallel dip-pen nanolithography. *J. Vac. Sci. Tech. B* **22**, 2563–2567 (2004).
94. Bullen, D. *et al.* Parallel dip-pen nanolithography with arrays of individually addressable cantilevers. *Appl. Phys. Lett.* **84**, 789–791 (2004).
95. Li, Y., Maynor, B. W. & Liu, J. Electrochemical AFM "dip-pen" nanolithography. *J. Am. Chem. Soc.* **123**, 2105–2106 (2001).
96. Cai, Y. G. & Ocko, B. M. Electro pen nanolithography. *J. Am. Chem. Soc.* **127**, 16287–16291 (2005).
97. Unal, K., Frommer, J. & Wickramasinghe, H. K. Ultrafast molecule sorting and delivery by atomic force microscopy. *Appl. Phys. Lett.* **88**, 183105/1–183105/3 (2006).
98. Sheehan, P. E., Whitman, L. J., King, W. P. & Nelson, B. A. Nanoscale deposition of solid inks via thermal dip pen nanolithography. *Appl. Phys. Lett.* **85**, 1589–1591 (2004).
99. Huang, L., Chang, Y.-H., Kakkassery, J. J. & Mirkin, C. A. Dip-pen nanolithography of high-melting-temperature molecules. *J. Phys. Chem. B* **110**, 20756–20758 (2006).
100. Zou, J. *et al.* A mould-and-transfer technology for fabricating scanning probe microscopy probes. *J. Micromech. Microeng.* **14**, 204–211 (2004).
101. Lewis, A. *et al.* Fountain pen nanochemistry: Atomic force control of chrome etching. *Appl. Phys. Lett.* **75**, 2689–2691 (1999).
102. Ying, L. M. *et al.* The scanned nanopipette: A new tool for high resolution bioimaging and controlled deposition of biomolecules. *Phys. Chem. Chem. Phys.* **7**, 2859–2866 (2005).
103. Bruckbauer, A. *et al.* Writing with DNA and protein using a nanopipet for controlled delivery. *J. Am. Chem. Soc.* **124**, 8810–8811 (2002).
104. Bruckbauer, A. *et al.* Multicomponent submicron features of biomolecules created by voltage controlled deposition from a nanopipet. *J. Am. Chem. Soc.* **125**, 9834–9839 (2003).
105. Sniadecki, N., Desai, R. A., Ruiz, S. A. & Chen, C. S. Nanotechnology for cell-substrate interactions. *Ann. Biomed. Eng.* **34**, 59–74 (2006).
106. Haynes, C. L. & Van Duyne, R. P. Nanosphere lithography: A versatile nanofabrication tool for studies of size-dependent nanoparticle optics. *J. Phys. Chem. B* **105**, 5599–5611 (2001).
107. Lutwyche, M. *et al.* 5×5 2D AFM cantilever arrays a first step towards a terabit storage device. *Sens. Actuators A* **73**, 89–94 (1999).
108. Minne, S. C. *et al.* Centimeter scale atomic force microscope imaging and lithography. *Appl. Phys. Lett.* **73**, 1742–1744 (1998).
109. Minne, S. C., Manalis, S. R., Atalar, A. & Quate, C. F. Independent parallel lithography using the atomic force microscope. *J. Vac. Sci. Tech. B* **14**, 2456–2461 (1996).
110. Minne, S. C., Manalis, S. R. & Quate, C. F. Parallel atomic force microscopy using cantilevers with integrated piezoresistive sensors and integrated piezoelectric actuators. *Appl. Phys. Lett.* **67**, 3918–3920 (1995).
111. Despont, M., Drechsler, U., Yu, R., Pogge, H. B. & Vettiger, P. Wafer-scale microdevice transfer/interconnect: Its application in an AFM-based data-storage system. *J. Microelectromech. Syst.* **13**, 895–901 (2004).
112. Eleftheriou, E. *et al.* Millipede - a MEMS-based scanning-probe data-storage system. *IEEE Trans. Magnetics* **39**, 938–945 (2003).
113. Vettiger, P. *et al.* The "Millipede" - nanotechnology entering data storage. *IEEE Trans. Nanotechnol.* **1**, 39–55 (2002).
114. King, W. P. *et al.* Design of atomic force microscope cantilevers for combined thermomechanical writing and thermal reading in array operation. *J. Microelectromech. Syst.* **11**, 765–774 (2002).
115. Vettiger, P. *et al.* The "Millipede" - more than one thousand tips for future afm data storage. *IBM J. Res. Develop.* **44**, 323–340 (2000).
116. Zhang, M. *et al.* A mems nanoplotter with high-density parallel dip-pen manolithography probe arrays. *Nanotechnology* **13**, 212–217 (2002).
117. Salaita, K. *et al.* Sub-100 nm, centimeter-scale, parallel dip-pen nanolithography. *Small* **1**, 940–945 (2005).
118. Wang, X. F., Vincent, L., Bullen, D., Zou, J. & Liu, C. Scanning probe lithography tips with spring-on-tip designs: Analysis, fabrication, and testing. *Appl. Phys. Lett.* **87**, 054102 (2005).
119. Salaita, K. *et al.* Massively parallel dip-pen nanolithography with 55,000-pen two-dimensional arrays. *Angew. Chem. Int. Edn* **45** (2006).
120. Lenhert, S., Sun, P., Wang, Y., Mirkin, C. A. & Fuchs, H. Massively parallel dip-pen nanolithography of heterogeneous supported phospholipid multilayer patterns. *Small* **3**, 71–75 (2007).
121. Liu, G. Y., Xu, S. & Qian, Y. L. Nanofabrication of self-assembled monolayers using scanning probe lithography. *Acc. Chem. Res.* **33**, 457–466 (2000).
122. Calvert, P. Inkjet printing for materials and devices. *Chem. Mater.* **13**, 3299–3305 (2001).
123. Rosner, B. *et al.* Active probes and microfluidic ink delivery for dip pen nanolithography. *Proc. SPIE: BioMEMS Nanotechnol.* **5275**, 213–222 (2004).
124. Rosner, B. *et al.* Functional extensions of dip pen nanolithography: Active probes and microfluidic ink delivery. *Smart Mater. Struct.* **15**, S124–S130 (2006).

Acknowledgements

C.A.M. acknowledges the Air Force Office of Scientific Research, Defense Advanced Research Projects Agency, Army Research Office, National Science Foundation, and NIH through a Director's Pioneer Award for support of this work.

Competing Financial Interests

The authors declare that they have no competing financial interests.

Biosensing with plasmonic nanosensors

Recent developments have greatly improved the sensitivity of optical sensors based on metal nanoparticle arrays and single nanoparticles. We introduce the localized surface plasmon resonance (LSPR) sensor and describe how its exquisite sensitivity to size, shape and environment can be harnessed to detect molecular binding events and changes in molecular conformation. We then describe recent progress in three areas representing the most significant challenges: pushing sensitivity towards the single-molecule detection limit, combining LSPR with complementary molecular identification techniques such as surface-enhanced Raman spectroscopy, and practical development of sensors and instrumentation for routine use and high-throughput detection. This review highlights several exceptionally promising research directions and discusses how diverse applications of plasmonic nanoparticles can be integrated in the near future.

JEFFREY N. ANKER, W. PAIGE HALL,
OLGA LYANDRES, NILAM C. SHAH, JING ZHAO
AND RICHARD P. VAN DUYNE*

Chemistry Department, Northwestern University, 2145 Sheridan Road, Evanston, Illinois 60208-3113, USA

*e-mail: vanduyne@northwestern.edu

The mirror-like quality of smooth noble-metal films changes markedly when the metal is separated into particles that are smaller than the wavelength of light[1–3]. Light incident on the nanoparticles induces the conduction electrons in them to oscillate collectively with a resonant frequency that depends on the nanoparticles' size, shape and composition. As a result of these LSPR modes, the nanoparticles absorb and scatter light so intensely that single nanoparticles are easily observed by eye using dark-field (optical scattering) microscopy[2–6]. For example, a single 80-nm silver nanosphere scatters 445-nm blue light with a scattering cross-section of 3×10^{-2} μm^2, a millionfold greater than the fluorescence cross-section of a fluorescein molecule, and a thousandfold greater than the cross-section of a similarly sized nanosphere filled with fluorescein to the self-quenching limit[4]. This phenomenon enables noble-metal nanoparticles to serve as extremely intense labels for immunoassays[4,6,7], biochemical sensors[8–13] and surface-enhanced spectroscopies[14–22]. Plasmonic nanoparticles also have significant potential as nanoscale optical switches, waveguides, light sources, microscopes and lithographic tools[23–25]. A review by Zhang and co-workers on page 435 of this issue highlights the use of plasmonic structures for subwavelength imaging and lithography[26]. Unlike fluorophores, plasmonic nanoparticles do not blink or bleach, providing a virtually unlimited photon budget for observing molecular binding over arbitrarily long time intervals.

The shape of the nanoparticle extinction and scattering spectra, and in particular the peak wavelength λ_{max}, depends on nanoparticle composition, size, shape, orientation and local dielectric environment[27,28]. The LSPR can be tuned during fabrication by controlling these parameters with a variety of chemical syntheses[29–32] and lithographic techniques[33–36]. For example, an array of triangular nanoprisms can be fabricated by vapour-depositing metal through the triangular spaces in a close-packed monolayer of colloidal nanospheres, a method known as nanosphere lithography (NSL)[33]. The height of these nanoprisms is precisely controlled by the deposition time and rate, and the lateral width and interparticle spacing are adjusted by varying the diameter of the nanospheres in the mask. In addition, the nanoprisms can be converted into hemispherical particles by thermal annealing, and multi-angle deposition provides for a wide variety of structures. Figure 1 shows the extinction spectra for a series of silver nanoprisms and nanohemispheres made by NSL with λ_{max} varying between 426 and 782 nm. The spectrum of silver nanoparticles can be tuned from 380 nm to at least 6 μm by varying their size and shape, with increasing aspect ratios and sizes producing nanoparticles with redshifted LSPR spectra[27].

Although silver and gold are the most commonly used materials, LSPR is theoretically possible in any metal, alloy or semiconductor with a large negative real dielectric constant and small imaginary dielectric constant. Other materials, such as aluminium, potentially offer advantages in refractive index sensitivity, different surface chemistries, and resonances into the ultraviolet, where many organic molecules absorb light. Some plasmonic materials, such as copper, have an absorbing oxide layer that strongly damps the LSPR; this must be removed to restore a narrow resonance linewidth. In the case of copper, the copper oxide layer can be stripped by using glacial acetic acid[37].

In addition to serving as brightly coloured spatial labels in immunoassays[4,6,7] and cellular imaging[38–40], plasmonic nanoparticles also act as transducers that convert small changes in the local refractive index into spectral shifts in the intense nanoparticle extinction and scattering spectra. Most organic molecules have a higher refractive index than buffer solution; thus, when they bind to nanoparticles, the local refractive index increases, causing the extinction and scattering spectrum to redshift. Molecular binding can be monitored in real time with high sensitivity by using simple and inexpensive transmission spectrometry, which measures extinction, the sum of absorption and scattering[3,8–10,41,42]. For example, Yonzon and co-workers studied the real-time binding of conconavalin A to mannose-functionalized nanoparticles[42]. The real-time LSPR-shift assay had a similar response to a commercial surface plasmon resonance (SPR) instrument (based on propagating plasmons in a thin gold film), but demonstrated less

When citing this review, please cite the original version as shown on the contents page of this chapter.

interference from bulk refractive index. In addition, LSPR possesses greater spatial resolution, both lateral and normal, when compared with SPR. The ultimate lateral spatial resolution is achieved with single nanoparticles.

LSPR-shift assays can also be used for ultrasensitive quantification of proteins. For example, an LSPR-shift assay was used to measure concentrations of amyloid-derived diffusible ligands (ADDLs), a neurotoxin that is thought to be important in the pathology of Alzheimer's disease[43–45], at concentrations down to 100 fM (ref. 9). In the same study, elevated concentrations of ADDLs were detected in brain extract and cerebrospinal fluid of an Alzheimer's patient in comparison with an ageing control[9]. Similarly elevated levels of ADDLs were observed in a population of 15 Alzheimer's patients and 15 ageing controls by using a 'bio-barcode' assay that uses plasmonic nanoparticle labels[46].

The remainder of this review describes the three most significant challenges for plasmonic biosensors: first, pushing sensitivity towards the single-molecule detection limit; second, combining LSPR with molecular identification techniques such as surface-enhanced Raman spectroscopy (SERS) and laser desorption ionization mass spectrometry; and third, the practical development of sensors and instrumentation for routine and high-throughput detection.

IMPROVING THE DETECTION LIMIT

The LSPR spectral shift ($\Delta\lambda$) in response to changes in refractive index is approximately described as

$$\Delta\lambda \approx m(n_{\mathrm{adsorbate}} - n_{\mathrm{medium}})(1 - e^{-2d/l_d}) \qquad (1)$$

where m is the sensitivity factor (in nm per refractive index unit (RIU)), $n_{\mathrm{adsorbate}}$ and n_{medium} are the refractive indices (in RIU) of the adsorbate and medium surrounding the nanoparticle, respectively, d is the effective thickness of the adsorbate layer (in nm), and l_d is the electromagnetic field decay length (in nm)[3]. LSPR shifts are maximized by optimizing the nanoparticle characteristics, m and l_d, as well as the Δn due to molecular adsorption. The former requires careful selection of nanoparticle size, shape and composition, and the latter can be achieved with the use of larger molecules and

resonant labels. In this review we describe how LSPR sensitivity can be enhanced by improving instrumental resolution, using novel nanofabrication techniques, shifting from particle arrays to single-nanoparticle sensors, and carefully choosing resonant labels that couple with the nanoparticle plasmon resonance.

INSTRUMENTAL RESOLUTION

Advances in LSPR instrumentation and analysis made by Dahlin and co-workers have allowed the detection of LSPR spectral shifts with a noise level of less than 5×10^{-4} nm and less than 5×10^{-6} extinction units for 2-s acquisitions[10]. We term this high-resolution LSPR (HR-LSPR) spectroscopy. Using plasmonic nanohole arrays functionalized with biotin, Dahlin et al. measured neutravidin binding in real time with a detection limit of less than 0.1 ng cm^{-2} (ref. 10). This level of precision, comparable to that of commercial SPR sensors, provides a signal-to-noise ratio (S/N) of ~2,000 for typical (molecular mass ~60 kDa) protein-binding reactions.

This improvement in resolution results from collecting more light by using photodiode arrays with high saturation levels to reduce shot noise, and from the use of improved fitting algorithms to calculate λ_{max} in real time. By collecting large numbers of photons, it is possible to determine the LSPR peak position (centroid) with much greater accuracy than either the grating resolution of the spectrometer (a few nanometres for commonly used fibre-coupled portable spectrometers, depending on the slit width and gratings) or the LSPR spectral linewidth full-width at half-maximum (FWHM), which is typically on the order of 100 nm for silver nanoprism arrays. This spectral super-resolution is analogous to spatial super-resolution used to track single fluorescent molecules to nanometre resolution (far better than either the pixel spacing or the diffraction-limited spot size)[47,48]. The use of photodiode-array spectrometers with high saturation capacities, of order 10 billion photoelectrons per pixel per second, has therefore greatly enhanced LSPR resolution. This high resolution enables the detection of weaker signals, improving the resolution of real-time binding kinetics experiments. An important challenge will be to use HR-LSPR spectroscopy to observe new information about the dynamics of protein conformational changes that cause subtle changes in peak wavelength and shape[49]. Another challenge will be to apply HR-LSPR spectroscopy to more complex systems such as tracking

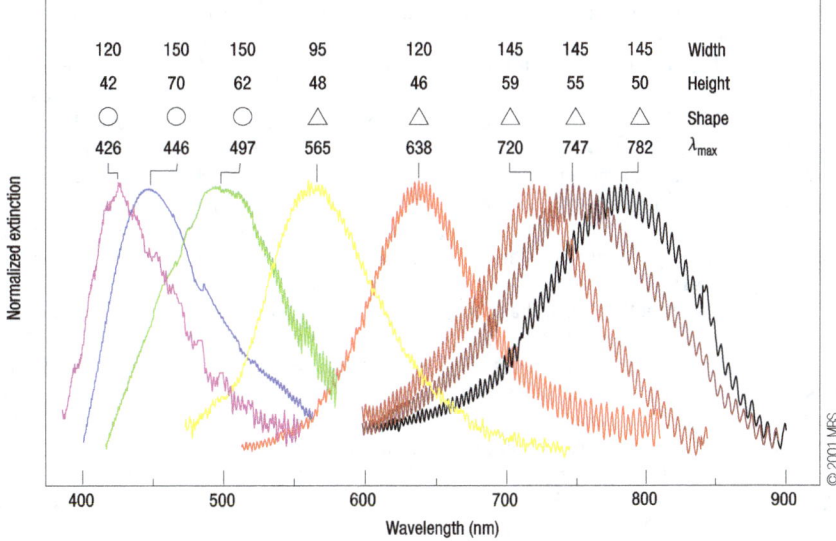

120	150	150	95	120	145	145	145	Width
42	70	62	48	46	59	55	50	Height
○	○	○	△	△	△	△	△	Shape
426	446	497	565	638	720	747	782	λ_{max}

© 2001 MRS

Figure 1 Effect of size and shape on LSPR extinction spectrum for silver nanoprisms and nanodiscs formed by nanosphere lithography. The high-frequency signal on the spectra is an interference pattern from the reflection at the front and back surfaces of the mica. Reprinted with permission from ref. 132.

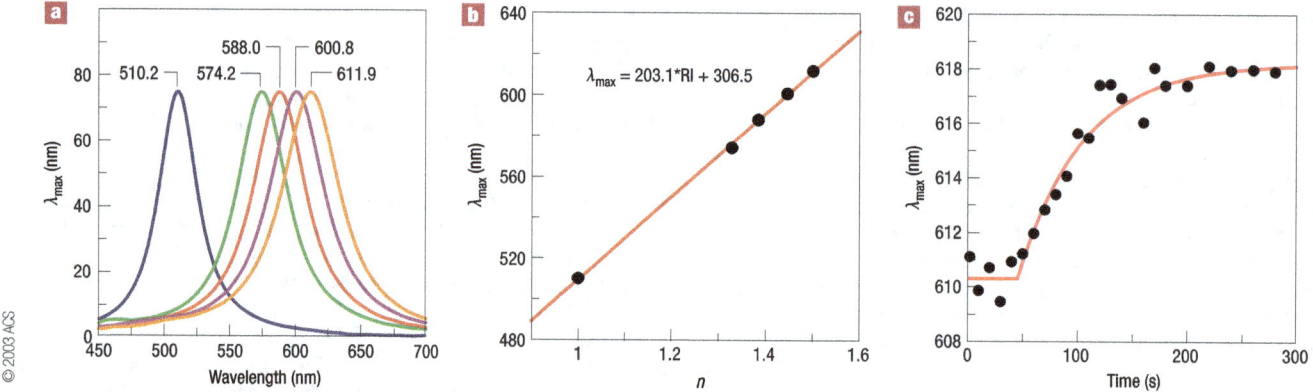

Figure 2 Single-nanoprism LSPR. **a**, Resonant Rayleigh scattering spectrum from a single silver nanoparticle in various solvent environments (left to right): nitrogen, methanol, propan-1-ol, chloroform and benzene. **b**, Plot depicting the linear relationship between the solvent refractive index n and the LSPR λ_{max}; the regression equation is $\lambda = 203.1n + 306.5$. **c**, Monitoring the real-time adsorption of octanethiol (1 mM) onto a single nanoparticle. At this concentration, the rate constant is estimated to be 0.017 s^{-1}. Reprinted with permission from ref. 11.

the binding and changes in membrane composition of intracellular systems based on single-nanoparticle spectra. Intracellular HR-LSPR spectroscopy and imaging of single nanoparticles will require the use of instruments with high spectral, spatial and temporal resolutions. In addition to inherent tradeoffs between spectral sensitivity and spatial/temporal resolution, sensitivity will be affected by background cellular scattering and extinction. Choosing methods to reduce or account for these background signals will be an important aspect of this work.

INCREASING THE SENSITIVITY OF LSPR

The sensitivity of LSPR sensors to local changes in refractive index is highly dependent on plasmon characteristics such as spectral linewidth, extinction intensity and electromagnetic-field strength and decay length. Understanding how nanoparticle characteristics affect these LSPR properties and optimizing nanoparticle design are a significant focus of current plasmonics research. The sensitivity of LSPR sensors continues to improve as nanoparticle designs and nanofabrication methods advance.

In general, increasing the aspect ratio (width/height) of NSL-fabricated nanoparticles results in redshifts in λ_{max}, higher m values and longer electromagnetic field decay lengths. In addition, theoretical modelling has shown that sharp nanoparticle features give rise to hot-spots in the electromagnetic fields that increase the sensitivity to local refractive index[50] and amplify surface-enhanced spectroscopies[14–22]. The effective electromagnetic decay length l_d can be tuned by changing the thickness and lateral dimensions of arrays of nanoprisms[42].

Alternative sensing modalities that use gold nanohole arrays and composite crystalline-metal nanostructures have been shown to exhibit even greater sensitivity in response to refractive index changes. Hicks and co-workers developed a plasmonic structure termed "film over nanowell" by first using reactive ion etching through an NSL sphere mask and removing the spheres to form a nanowell array, and then vapour-depositing a silver film over the nanowells[51]. The structure exhibited plasmon peaks with a FWHM as narrow as 35 nm (0.13 eV), and good refractive index sensitivity ($m = 538$ nm RIU^{-1} or 2 eV RIU^{-1}) with an overall figure of merit (FOM), m/FWHM (where both are measured in eV), of 14.5, surpassing the typical FOM for nanoprism arrays of about 3. Using a similar nanostructure, Henzie and co-workers reported a FOM of more than 23 for ordered arrays of nanoholes in gold fabricated by soft interference lithography[52]. The high FOM for these arrays, which exhibit a moderate m value

of ~300 nm RIU^{-1} (0.75 eV RIU^{-1}) is derived primarily from the narrow plasmon linewidth of 14.5 nm (0.032 eV) FWHM. Creatively designed nanostructures such as these demonstrate that the sensitivity of LSPR-based sensors continues to improve as advances are made in nanofabrication techniques.

In addition to increasing m values by controlling nanoparticle size, shape and composition, LSPR sensitivity can be improved by narrowing spectral linewidths and increasing extinction. Polydispersity in nanoparticle size and shape, especially those based on wet synthetic techniques, significantly broadens LSPR extinction bands. The next section will focus on how these issues can be overcome with the use of single-nanoparticle spectroscopy.

SINGLE-NANOPARTICLE SENSORS

Most LSPR spectroscopy has been performed with large ensembles of nanoparticles. However, each nanoparticle in the ensemble could potentially serve as an independent sensor. Indeed, the Van Duyne and Klar labs recently demonstrated an LSPR-shift assay with single nanoparticles[11,12,53]. Single-nanoparticle sensors offer improved absolute detection limits (total number of molecules detected) and also enable higher spatial resolution in multiplexed assays. In addition, single nanoparticles with especially narrow bandwidths can be selected from a field of view to provide better S/N resolution. Finally, single nanoparticles have promising applications for measurements in solution, or inside cells and tissues where fixed arrays are unable to penetrate[38,39,54–57]. Figure 2a,b shows the shift in LSPR scattering spectrum from a single silver nanoparticle fixed to a glass coverslip in solvents with increasing refractive index. Figure 2c shows the real-time assembly of an octanethiol self-assembled monolayer (SAM) from a 1 mM ethanolic solution onto a single silver nanoparticle. It was estimated that a fully assembled SAM covering a single silver nanoparticle with a surface area of 14,000 nm^2 has 60,000 molecules on its surface. At saturation, octanethiol produced a shift of ~8 nm, for an average shift of 10^{-4} nm per molecule. A single exponential fit yielded a rate constant of 0.017 s^{-1}, and an initial binding rate of ~1,000 molecules s^{-1}.

Although single nanoparticles enable sensitive chemical detection with high spatial resolution, nanoparticle size must be selected with care to ensure sufficient signal intensity for LSPR-shift assays. Smaller nanoparticles are advantageous in protein labelling and cellular imaging because their smaller surface area reduces non-specific interactions and enables more targeted binding. In addition, smaller particles produce more confined electromagnetic fields so they are more sensitive to single molecules, which occupy a greater portion of their sensing region. However, the nanoparticle absorbance and scattering depend strongly on nanoparticle size, scaling with nanoparticle volume for absorbance and with volume squared for scattering. Absorbance and scattering cross-sections become comparable when the nanospheres are about 60 nm in diameter for silver nanospheres and 80 nm in diameter for gold nanospheres[6,58], and nanoparticles less than 20 nm in diameter are challenging to observe. Several exciting developments have enabled smaller gold nanoparticles to be observed over extended periods without bleaching[54,55,58,59]. These techniques have primarily been used to track protein migration inside cells. However, the techniques could also be used to detect binding and conformational changes by using LSPR spectral shifts.

Confocal reflection imaging allows the detection of nanoparticles with diameters as small as 15 nm inside single cells and 5 nm in solution[55]. To measure the spectra from smaller nanoparticles, it becomes necessary either to rely on interference methods to amplify the scattering signal, or to measure absorbance, which provides a stronger signal than scattering for small particles[58]. The simplest method to detect nanoparticle absorbance is to track transmission by using bright-field imaging. With bright-field imaging and digitally enhanced contrast, single 10-nm nanoparticles can be tracked at video rate inside cells[59,60]. However, the S/N for a weakly absorbing particle is limited by noise on the majority of light that is transmitted.

To detect nanoparticle absorbance directly instead of monitoring transmission, one can monitor the temperature increase caused by optically heating the nanoparticles and their surroundings (that is, photothermal imaging)[54,61] or one can monitor the pressure wave created during rapid heating with photoacoustic imaging[62,63]. Photothermal imaging provides high-resolution nanoparticle tracking, whereas photoacoustic imaging has been used for imaging in tissues.

To perform photothermal imaging, a modulated pump laser beam is used to heat the nanoparticles, and a probe beam is used to detect the deflection caused by the thermally induced change in the refractive index. Using photothermal microscopy, single 5-nm gold nanoparticles have been tracked in living cells at video rate (see Fig. 3)[54]. In addition, the absorption spectrum of single 2-nm gold nanoparticles was acquired by measuring photothermal absorption[54]. At 37 °C, a temperature increase of 1 K typically leads to a 10^{-4} change in the refractive index of water; at higher temperatures this thermorefractive coefficient increases. The use of photothermal imaging is extremely promising for following protein trafficking over arbitrarily long time intervals with the use of individual nanoparticle labels. It is also expected that combining photothermal image tracking with LSPR spectroscopy will enable high-resolution information on molecular conformations and interactions to be obtained in the near future.

In addition to passive detection, photothermal effects can be used to heat plasmonic nanoparticles selectively for cancer therapy. Recently, researchers used gold nanoparticles to target cancer cells that overexpressed epidermal growth factor receptor (EGFR) and were thereby able to differentiate between cancer cells and normal cells. The nanoparticles were then used as photothermal agents to kill cancer cells selectively when illuminated by 514-nm light that was near the λ_{max} of the gold nanospheres[64]. Loo and co-workers used a similar photothermal therapy technique, but they used gold nanoshells with near-infrared resonances and used a near-infrared laser to kill nanoshell-labelled breast cancer cells that overexpressed the HER2 receptor[40]. Larson and co-workers achieved even greater photothermal selectivity by observing that highly proliferative cells express sufficient EGFR to cause receptor clustering and nanoparticle aggregation, resulting in an absorbance and scattering

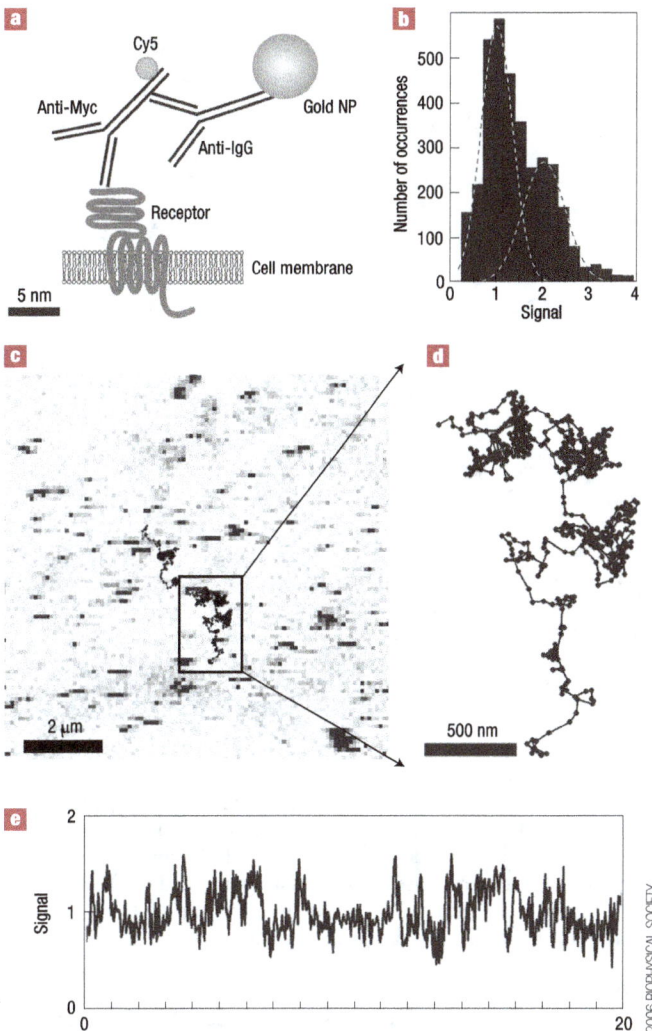

Figure 3 Photothermal imaging of gold nanoparticles in a living cell. **a**, Biological construct. NP, nanoparticle. **b**, Histogram of the signals for 5-nm gold nanoparticles detected on COS7 cells. **c**, Photothermal image of a portion of a transfected COS7 cell labelled with 5-nm gold nanoparticles detected with $S/N \approx 30$. Moving nanoparticles during the raster scan of the sample produce characteristic stripe signals. The lines are trajectories recorded with single receptors diffusing on the cell membrane. **d**, Zoom onto one recorded trajectory. **e**, Time trace of the signal amplitude calculated from the triangulation process while tracking one of the nanoparticles. Increased signal fluctuations correspond to the nanoparticle in a fast diffusive state. Reprinted with permission from ref. 54.

redshift. They used anti-EGFR-conjugated gold nanospheres with individual resonances of ~532 nm to target cancer cells, but used a 700-nm laser to selectively heat and kill only cells labelled with nanoparticle aggregates[65].

Photoacoustic imaging uses a similarly modulated laser source to heat nanoparticles but detects the light absorption by monitoring the resultant acoustic waves. The method offers the advantages of a deep optical penetration depth using near-infrared-resonant plasmonic nanoparticles, minimal interference from optical scattering, high signal contrast, and simple integration with conventional ultrasound devices[62]. In addition, multiple functionalities can be integrated

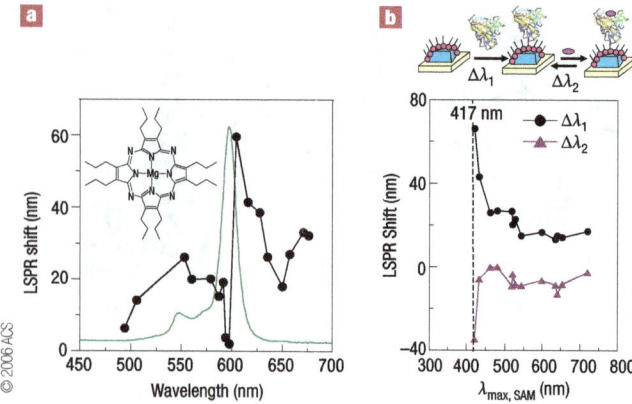

Figure 4 Wavelength-dependent LSPR shifts induced by resonant molecules. **a**, Comparison of LSPR shifts induced by a monolayer of MgPz adsorption on silver nanoparticles with the LSPR of bare silver nanoparticles (black line with dots). Inset: molecular structure of MgPz. Reprinted with permission from ref. 70. The green line is the solution absorption spectrum of MgPz. **b**, Schematic representation of CYP101 immobilized on a silver nanobiosensor followed by binding of camphor, and plots of LSPR shifts against $\lambda_{max,SAM}$ (LSPR of SAM-functionalized nanoparticles), where $\Delta\lambda_1$ is the shift on binding of CYP101 and $\Delta\lambda_2$ is the shift on binding of camphor. The vertical black dotted line denotes the molecular resonance of substrate-free CYP101. Reprinted with permission from ref. 72.

into the photoacoustic nanoparticle platform, including magnetic properties. Hybrid magnetic–plasmonic nanoparticles can be magnetically pulled or rotated to modulate optical signals and increase contrast[66,67]. They have also been used as combined photoacoustic, dark-field and magnetic-resonance-imaging contrast agents as well as vehicles for photothermal cancer therapy[65].

METHODS TO ENHANCE THE LSPR SHIFT

The previous sections have described methods of improving the sensitivity of nanoparticle structures and spectrometers. A complementary approach to increasing sensitivity, potentially towards the single-molecule limit, is to increase the effective change in refractive index per molecular binding event. This shift can be increased in at least three ways. First, larger molecules produce larger shifts roughly in proportion to the mass of the molecule[50,68,69]. Thus, proteins and macromolecules produce larger molecular shifts per molecule than small SAM molecules. Second, chromophores that absorb visible light couple strongly with the LSPR of nanoparticles to produce surprisingly large shifts[70,71] and can be used to detect small molecules binding to protein receptors[72]. Third, pairs of nanoparticles that are separated by less than about 2.5 particle radii show plasmonic coupling and marked spectral shifts[73,74]. The latter two amplification methods are both based on plasmon resonance coupling, described in more detail below.

Recent work pioneered by Van Duyne and co-workers explored the interaction between the molecular resonance of a chromophore and the plasmon resonance of a nanoparticle[70–72]. When chromophores adsorb onto silver nanoparticles, the resulting LSPR shift depends strongly on the spectral overlap between the molecular absorbance and the LSPR[70–72]. Figure 4a shows the LSPR shift of triangular silver nanoparticles induced by a monolayer of [2,3,7,8,12,13,17,18-octakis (propyl)porphyrazinato]magnesium (II) (MgPz). A drastic change in the LSPR shift was observed when the LSPR of silver nanoparticles was tuned through the molecular resonance of MgPz at 598 nm. For the non-resonant case, when the base nanoparticle λ_{max} is significantly blueshifted or redshifted from the molecular absorbance, a redshift

of about 20 nm is observed on molecular adsorption. By contrast, when the base nanoparticle λ_{max} overlaps directly with the molecular resonance, an unusually small redshift of 2 nm is observed. When the base nanoparticle LSPR is redshifted by 6 nm from the molecular absorbance, an unusually large redshift of ~60 nm occurs; that is, threefold greater than the non-resonant case. This exquisite sensitivity to molecular absorption, in which a 6-nm difference in overlap results in a 60-nm shift in extinction, paves the way to the development of highly sensitive sensors based on shifts in molecular absorbance in response to analyte concentration.

The coupling between chromophore and LSPR nanoprism arrays was used to detect the binding of camphor (molecular mass 154.24 g mol^{-1}) to the haem-containing cytochrome P450cam protein (CYP101)[72]. When camphor binds to CYP101, it displaces the water coordinated with the haem iron, which results in a 26-nm blueshift in the Soret absorption band, from 417 nm to 391 nm (refs 75, 76). Nanoparticles with varying initial λ_{max} values were exposed to CYP101 proteins, resulting in a wavelength-dependent LSPR shift on CYP101 binding (as shown in Fig. 4b, black line with dots). When these CYP101 functionalized nanoparticles were incubated with camphor, camphor binding induced a wavelength-dependent blueshift in the LSPR. For nanoparticles with an LSPR tuned close to the CYP101 resonance, an amplified LSPR redshift as large as 67 nm was obtained for protein binding, and a 37-nm blueshift for camphor binding. It should be noted that on average only 500 CYP101 molecules fit onto a nanoprism with a surface area of 14,000 nm^2, and that camphor binds to CYP101 at a ratio of 1:1; the average shift per camphor molecule is therefore 0.07 nm. This remarkably large shift in response to a small molecule opens up the possibility of detecting a wide range of analytes with the use of plasmonic nanoparticles coupled to indicator dyes. Combining indicator dyes with LSPR nanosensors allows weak changes in molecular absorption intensity or wavelength to be transduced into spectral shifts in the intense LSPR scattering and absorbance signals, thus greatly improving the sensitivity. A challenge will be to select and optimize the nanoparticle LSPR to match the dye resonance and to control for potential interferants.

In addition to coupling with resonant dyes, nanoparticles show a distance-dependent coupling with other particles or surfaces displaying plasmon resonance. Labelling proteins and ligands with nanoparticles thus provides a way to explore conformational changes and binding interactions[77]. Characterizing these dynamic protein behaviours is essential for understanding cellular activities. LSPR sensitivity can be used to detect changes in the extension of single molecules by sandwiching molecules between two nanoparticles. Such plasmonic molecular ruler structures are extremely sensitive to distance-dependent plasmonic coupling between the nanoparticle pairs[13,78–82], enabling high-resolution monitoring of molecular conformation[73,74,79,83]. The magnitude and direction of the spectral shift depend on the orientation of the nanoparticle pair with respect to the polarization axis of the incident light[79,80]. With unpolarized light, when a nanoparticle label binds to a fixed nanoparticle sensor to form a nanoparticle pair, the scattering intensity increases markedly (by about fourfold compared with a single particle) and a large spectral redshift is observed, about 75 nm for 40-nm gold nanoparticles[74,83,84]. This shift depends on the distance between nanoparticles, decreasing approximately exponentially with distance[79]. For 40-nm gold particles, the resonance shifts by about 10 nm per nanometre separation for small separations. This distance dependence is roughly proportional to particle radius, so smaller particles will be more sensitive to small changes in plasmonic spacing, although with a reduced dynamic range.

Alivisatos, Liphardt and co-workers have developed a molecular ruler to monitor the separation between single pairs of nanoparticles[74]. A molecular ruler was used to detect the hybridization of DNA oligonucleotides complementary to the single-stranded DNA (ssDNA). Gold nanospheres, 40 nm in diameter, were functionalized with streptavidin and immobilized on glass slides. The immobilized nanoparticles were then exposed to nanoparticles functionalized with biotinylated ssDNA, allowing the anchored nanoparticles to capture the ssDNA-functionalized nanoparticles and form pairs (see Fig. 5a). This binding event caused an immediate redshift and an increase in scattering intensity of the immobilized nanoparticles as a result of plasmon resonance coupling between the nanoparticles. The scattering spectra of single nanoparticles and nanoparticle pairs were monitored by dark-field microscopy (shown in Fig. 5b). Because the ssDNA chain is relatively flexible, introducing the complementary DNA to the ssDNA linkers causes hybridization to a more extended form, pushing the nanoparticles apart and resulting in a blueshift in the scattering spectrum of the nanoparticle pair (shown in Fig. 5c). By continuously monitoring the spectrum of nanoparticle pairs, one can observe the dynamics of DNA hybridization. Figure 5d shows the shift in the scattering spectra peak over time. Initially the data show 0.9-nm fluctuations in the peak position, corresponding to the extension/bending of ssDNA strands. On the addition of the complementary DNA, the scattering peak blueshifted by $\Delta\lambda = 2.1$ nm; and then remained relatively constant, indicating the formation of a DNA duplex.

Plasmonic rulers have been used to study changes in DNA persistence length in sodium chloride solutions, decrease in DNA chain length resulting from DNA nuclease activity, and DNA bending and cleavage due to single *EcoRV* restriction enzymes[73]. Very recently, plasmon resonance coupling has been applied to molecular imaging *in vivo* in cancer cells on the basis of EGFR clustering[56].

COUPLING MOLECULAR IDENTIFICATION TO LSPR SPECTROSCOPY

The LSPR-shift assay is a general technique for measuring binding affinities and rates from any molecule that changes the local refractive index. Specificity is achieved for known analytes by using molecular recognition elements such as antibodies in conjunction with hydrophilic SAMs and blocking agents to decrease non-specific binding. Specificity can also be provided by specific dye–analyte interactions in chromophore-coupled plasmonic nanoparticles, or by specific analyte-dependent conformational changes in molecules holding together pairs of plasmonic nanoparticles in plasmonic rulers, especially if the interactions can be modulated reversibly. However, LSPR-shift assays are not well suited for the identification of unknown molecules. Thus, an important challenge is to integrate complementary analytical techniques with LSPR-shift assays so that molecules can be first characterized with LSPR and then identified with a second technique. Fortunately, the plasmon-enhanced absorbance and local electromagnetic fields make LSPR substrates ideal for two molecular identification techniques, SERS and laser desorption ionization mass spectrometry.

SURFACE-ENHANCED RAMAN SPECTROSCOPY

Raman spectroscopy is a highly specific technique used to detect and identify molecules on the basis of their unique vibrational energy levels and corresponding Raman fingerprints. In Raman scattering, photons are scattered inelastically, either losing energy (Stokes shift) or gaining energy (anti-Stokes shift) equal to the molecular vibration of the probed material. When molecules adsorb onto a plasmonic nanoparticle or move to within a few nanometres of its surface, the local electromagnetic fields around the nanoparticle can enhance the Raman scattering by a factor of 10^6–10^8 for an ensemble of molecules[14,85], and by as much as 10^{14}–10^{15} for single molecules[86–88]. This phenomenon, known as surface-enhanced Raman scattering, results in a highly specific and sensitive method for molecular identification[89].

Two principal enhancement mechanisms are generally thought to explain the large SERS signals[90]. First, an electromagnetic

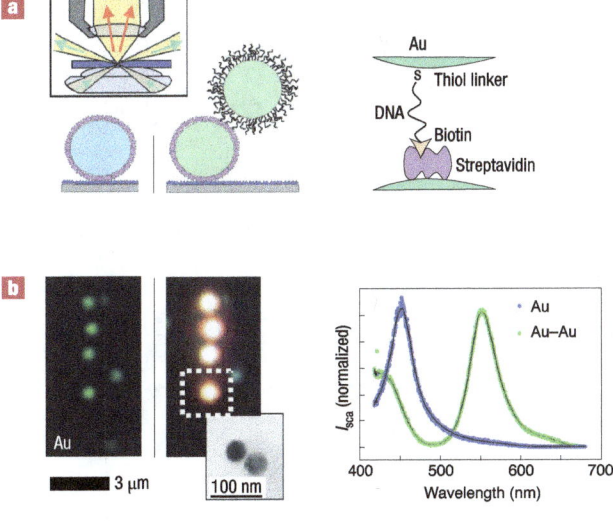

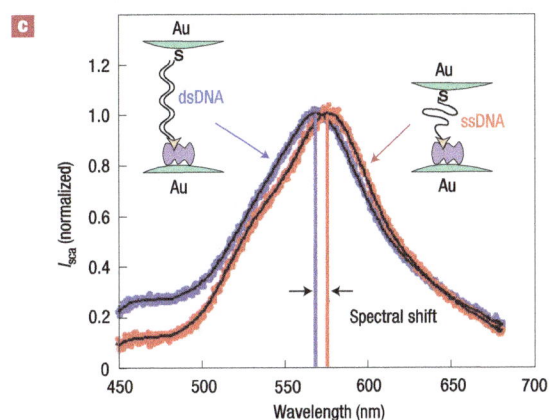

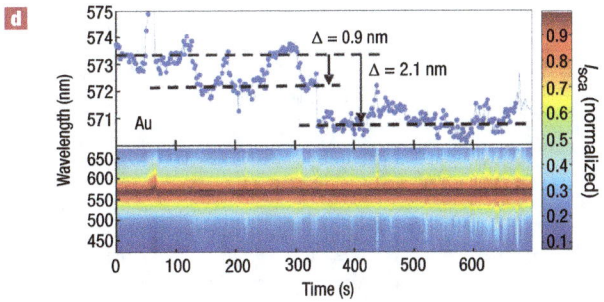

Figure 5 Molecular plasmonic ruler. **a**, Schematic illustration of nanoparticle functionalization and immobilization. Inset: principle of transmission dark-field microscopy. **b**, Dark-field images of single gold nanoparticles (green, left), and gold nanoparticle pairs (orange, right). Inset: representative transmission electron microscopy image of a nanoparticle pair. Right: representative scattering spectra of single gold nanoparticles and nanoparticle pairs. **c**, Example of a spectral shift between a gold nanoparticle pair connected with single-stranded DNA (red) and double-stranded DNA (blue). **d**, Spectral position as a function of time after the addition of complementary DNA. The scattered intensity (I_{sca}) is shown colour-coded on the bottom; the peak position obtained by fitting each spectrum is traced on the top. Discrete states are observed, indicated by horizontal dashed lines. Reprinted with permission from ref. 74.

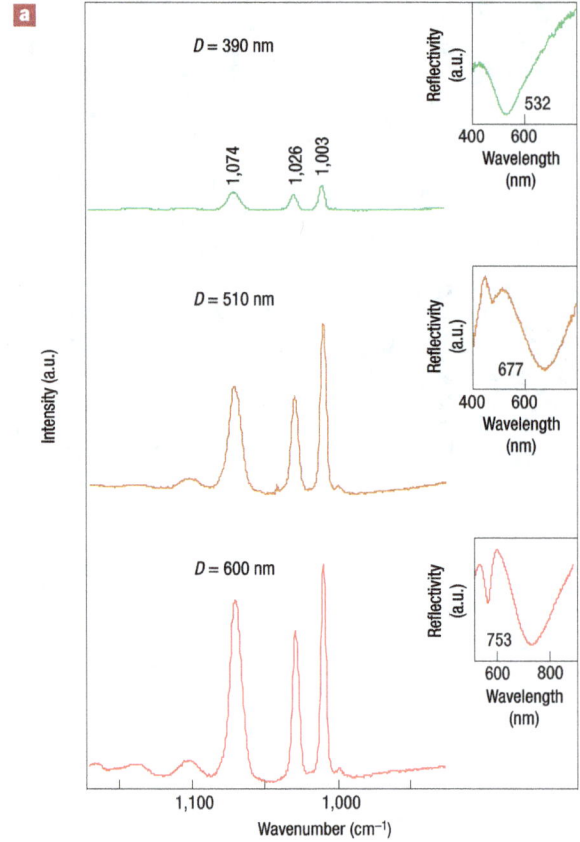

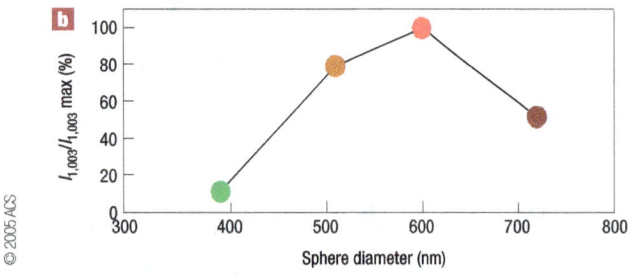

Figure 6 Tuning the LSPR to maximize the SERS signal. **a**, SERS spectrum of benzenethiol on AgFONs with varying nanosphere diameters and corresponding resonances: at 532 nm, sphere diameter $D = 390$ nm (green), at 677 nm, $D = 510$ nm (orange), and at 753 nm, $D = 600$ nm (red). The reflection spectrum is shown in the insets, with minimal reflection corresponding to maximum LSPR-induced absorbance and scattering. **b**, Relative peak heights of the 1,003 cm^{-1} SERS peak (in-plane ring deformation mode) as a function of nanosphere diameter. Reprinted with permission from ref. 93.

enhancement factor arises because LSPR modes in the metal nanoparticles focus the incident light energy at the nanoparticle surface and also increase the density of states at Stokes-shifted wavelengths. The electromagnetic enhancement factor is typically 10^5–10^8 for nanoprisms, but simulations indicate that it can be as large as 10^{13} for structures such as arrays of nanoprism dimers[91]. Second, chemical enhancement factors arise from changes in the molecular electronic state or resonant enhancements from either existing molecular excitations or newly formed charge transfer states. When

molecules are excited by light resonant with an electronic absorbance, an additional enhancement factor of ~10^1–10^3 is observed in comparison with non-resonant excitation. Single-molecule SERS has been reported for such resonantly excited dyes on a small percentage of 'hot' nanoparticles with enhancement factors of up to 10^{14}–10^{15} in total[86–88].

The electromagnetic enhancement in SERS results from Raman excitation and emission coupling with the nanoparticle LSPR modes. Because the LSPR can be easily tuned by changing the size and shape of nanoparticles, LSPR sensors are ideal candidates for a complementary molecular identification platform with SERS. The enhancement is greatest when the LSPR λ_{max} falls between the excitation wavelength and the wavelength of the scattered photon[85]. For a particular Raman band, the optimal condition is achieved when the LSPR λ_{max} equals the excitation wavelength (in absolute wavenumbers) minus one-half the Stokes shift of the band.

The complementary nature of LSPR and SERS was demonstrated by tuning LSPR substrates to optimize SERS signals. SERS-active substrates were fabricated by drop-coating a layer of nanospheres on a glass or metal substrate and subsequently depositing a 200-nm-thick silver film over nanospheres (AgFON). By varying the diameter of the nanospheres, the LSPR λ_{max} can be tuned to couple with the 750-nm excitation wavelength used in the experiments. Although AgFON substrates have broader LSPR resonances than NSL nanoprism arrays, they provide a large surface area available for binding and detection[92]. Because 200-nm films are not transparent, LSPR spectra are measured in reflectance mode with the minimum spectral reflectivity corresponding to LSPR λ_{max}. With increasing sphere diameter, the LSPR λ_{max} redshifts, making it more appropriate for near-infrared excitation.

To study the dependence of SERS on LSPR wavelength, the reflectivity and SER spectra of a monolayer of benzenethiol were acquired on AgFONs fabricated with 200-nm silver deposited over nanospheres 390, 510 and 600 nm in diameter (Fig. 6a). Figure 6b shows the SERS intensity of the 1,003 cm^{-1} benzenethiol band as a function of sphere diameter acquired with 750-nm laser illumination. The optimal LSPR resonance for the 1,003 cm^{-1} SERS band of benzenethiol is estimated to be 779 nm (750 nm minus half of the 1,003 cm^{-1} Stokes shift), close to the experimental λ_{max} of 753 nm observed for the 600-nm nanosphere AgFON. As expected, the 600-nm nanosphere AgFON produced the greatest intensity for the 1,003 cm^{-1} peak. Because each Raman band experiences a slightly different enhancement, the LSPR of the SERS-active surface should be optimized for the middle of the desired Raman spectral range. Zhang and co-workers used similar LSPR-optimized SERS substrates to detect dipicolinic acid, an anthrax biomarker. They demonstrated a limit of detection of ~2.6×10^3 for anthrax spores, which is below the infectious dose[93]. The limit of detection for anthrax was further improved by a factor of two when the AgFON surface was functionalized with an atomic layer of alumina to increase the stability and binding affinity of dipicolinic acid[94].

The complementary nature of LSPR spectral-shift assays and SERS molecular identification was also demonstrated with the anti-dinitrophenyl immunoassay system[95]. Binding of the anti-dinitrophenyl ligand to 2,4-dinitrobenzoic acid was quantified by LSPR shift measurements, and SERS was used to verify the identity of the adsorbed molecules. This is a general approach that can be applied to a wide range of molecules attached to LSPR sensors, including isotopically labelled molecules that cannot be differentiated by LSPR[87].

LASER DESORPTION IONIZATION MASS SPECTROMETRY

Mass spectrometry (MS) is another molecular identification technique that can be combined with LSPR spectroscopy. Interfacing LSPR sensors with MS is in its beginning stages, but it promises to provide

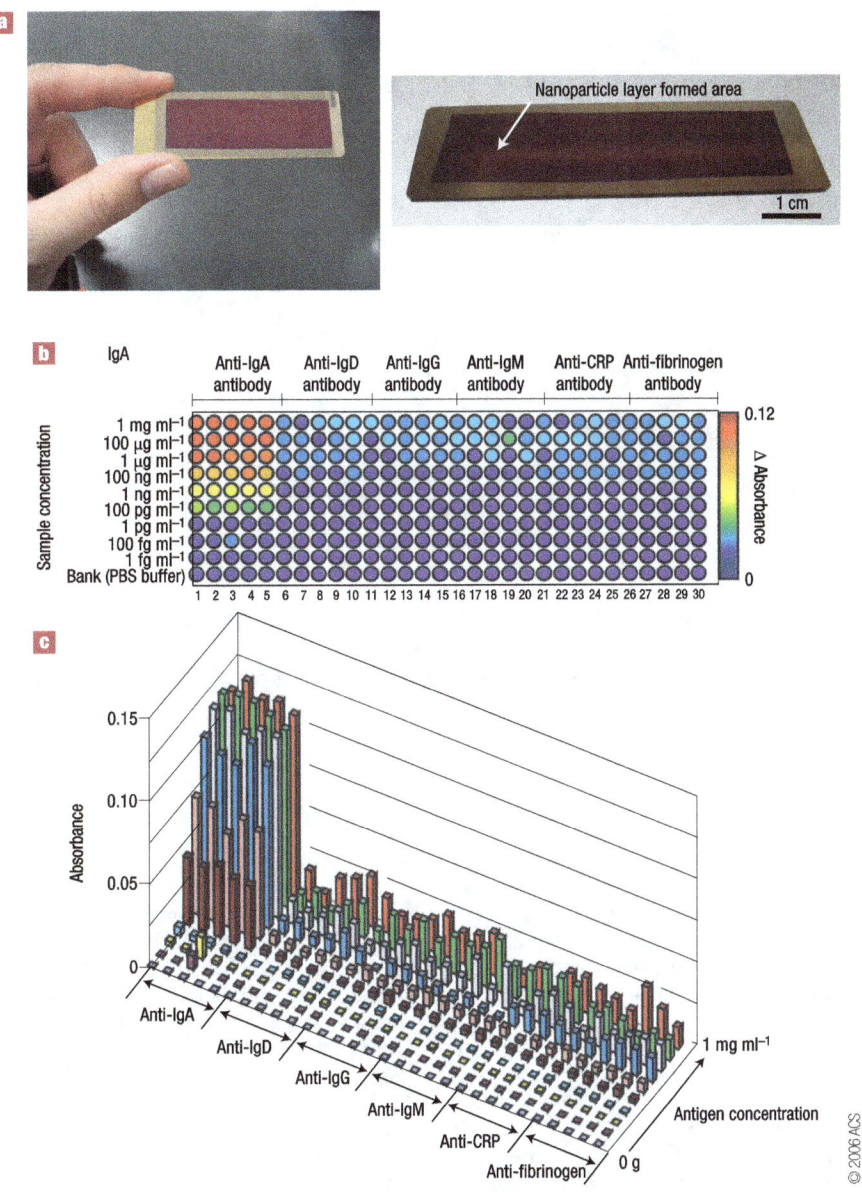

Figure 7 Uniform biochip for multiplexed LSPR detection. **a**, Photograph of the LSPR-based nanochip with dimensions 20 mm × 60 mm. The nanochip structure consists of silica nanospheres deposited on a flat gold film and coated with a gold overlayer. **b,c**, Absorbance measurements at each spot of the multiarray nanochip for binding of immunoglobulin A (IgA) to six different types of antibodies. **b**, The antibodies were immobilized on the chip, resulting in a total of 300 spots separated by 1 mm (to prevent cross-contamination). **c**, Different concentrations of antigen were incubated for 30 min and LSPR absorption spectra were then acquired with a fibre-coupled ultraviolet–visible spectrometer (USB-200_UV_vis) in a reflection geometry. Detection was linear up to 1 µg ml⁻¹ with a limit of detection of 100 pg ml⁻¹ for all of the proteins. Reprinted with permission from ref. 120.

information on unknown molecules adsorbed on LSPR sensor arrays as well as a mechanism for laser desorption ionization based on plasmon-assisted photothermal heating and photoionization.

Early work with continuous films of silver, gold and aluminium demonstrated a remarkable enhancement in metal-ion and peptide signal when the film was illuminated by visible light at the critical angle needed to excite a propagating surface plasmon resonance[96–98]. Recent work by Chen and co-workers has demonstrated a similar plasmon-assisted laser desorption ionization by using LSPR in nanoparticles[99,100]. Gold nanorods were fabricated by electrochemically depositing gold in nanoporous alumina templates. When the nanorods were

optically excited by a 532-nm Nd–YAG laser near their resonant absorbance wavelength, molecular desorption and ionization were greatly enhanced in comparison with continuous films. As expected, the MS intensity depended on the length of the nanorods and the polarization of the incident light.

Plasmonic nanoparticles have also been used to enhance MS imaging, although the enhancement mechanisms have not yet been elucidated. Altelaar and co-workers recently demonstrated that sputtering gold nanoparticles onto tissue sections improved the signal strength and image quality of images from both secondary ion mass spectrometry and ultraviolet matrix-assisted laser desorption

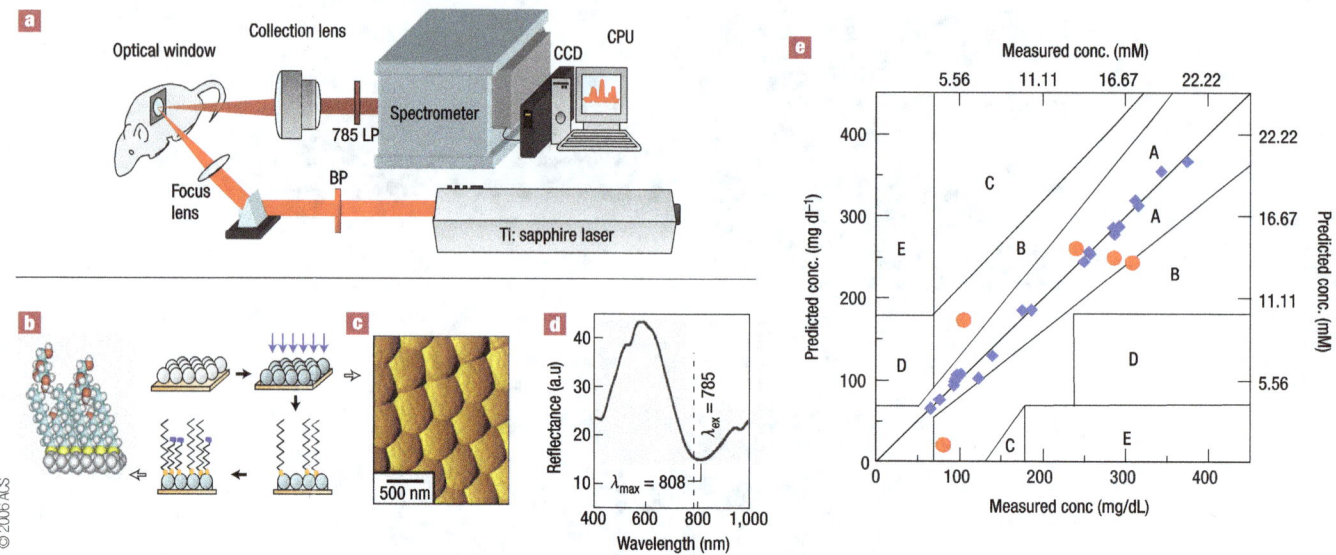

© 2006 ACS

Figure 8 The first *in vivo* SERS implantable glucose sensor. **a**, Experimental setup used for *in vivo* SERS measurements in rats. CCD, charge-coupled device; BP, band-pass filter; LP, long-pass filter. **b**, Fabrication and functionalization of SERS-active surfaces: formation of a nanosphere mask, silver deposition resulting in formation of the silver film over nanospheres (AgFON) surface, incubation in decanethiol, and incubation in mercaptohexanol. **c**, Atomic-force micrograph of a typical AgFON surface. **d**, Reflection spectrum of AgFON optimized for *in vivo* experiments. The LSPR λ_{max} is at 808 nm, slightly red of the laser excitation wavelength λ_{ex} of 785 nm. **e**, Quantitative detection of glucose *in vivo* is achieved by calibrating the sensor with an over-the-counter electrochemical glucose sensor. The calibration and prediction data points are plotted on the Clarke error grid, a standard metric for evaluating the performance of glucose sensors, with all the points falling in the A and B ranges corresponding to proper medical treatment. Reprinted with permission from ref. 123.

ionization mass spectrometry[101]. In addition, nanoparticles can also improve MS analysis by separating and concentrating analytes by centrifugation[102,103]. An important challenge will be to integrate these separate developments in MS with advances in plasmonic nanoparticle targeting and sensing to enable targeted MS imaging and multimodal chemical analysis.

PRACTICAL IMPLEMENTATIONS

The promise of LSPR sensors for measuring molecular concentrations[8,9], binding kinetics[10,11,42], dye absorbance spectra[70–72] and conformational changes[49,73,74] has now been firmly established. However, transforming these early sensor designs into robust, practical tools for a wider community is an important challenge that requires the development of uniform, well characterized, highly stable substrates and instruments. A few recent examples are presented to highlight how these goals are being pursued.

Because LSPR-based sensors are by design very sensitive to changes in the characteristics of nanoparticles, uniformity in nanoparticle size, shape and composition improve sensitivity and reliability. Uniform nanoparticles can be produced by using various techniques including nanosphere lithography[33], electron-beam lithography[35,36] and chemical synthesis[29–32]. Separation techniques such as electrophoresis can also be used to fractionate a polydisperse suspension into monodisperse subpopulations[104,105]. In addition to uniformity, it is also important to develop sensors that are stable against solvent annealing[106] and photothermal annealing if high laser intensities are used[107]. To overcome these issues, particles can be pretreated in ethanol, water and bovine serum albumin to pre-anneal them[108]. The nanoparticles can also be protected by using controlled surface chemistries or by encapsulating them in inert layers such as silica or alumina[94,108]. These surface chemistries stabilize the sensing platform and provide different affinities, permitting the detection of a wider variety of analytes.

Although some molecules can be detected on bare nanoparticle arrays and bare FONs, many important analytes, such as glucose, have a low affinity for the bare metal surface. In addition, oxidation of the metal surface can limit sensor stability and shelf life[109,110]. Furthermore, bare nanoparticle surfaces offer limited possibilities for isolating the analyte from complex mixtures. To overcome these limitations, several methods have been used to functionalize substrates for improved affinity and selectivity. Detection is usually facilitated by the use of various coatings ranging from simple alkanes to complex macrocyclic molecules[111,112]. These molecules are anchored to the noble-metal surface by a thiol group and form a SAM. The SAMs serve many functions including stabilizing the nanoparticles, preventing non-specific binding, selectively binding the analyte of interest, and bringing the analyte to within a few nanometres of the nanostructured surface. In addition, antibodies can be immobilized on the SAM to facilitate protein detection. Research is also being conducted to find alternative surface chemistries to thiol-based SAMs, which often rearrange over timescales ranging from hours to months[113,114], and suffer defects due to thermal desorption[115] and photooxidation[116,117]. One method entails encapsulating the nanoparticles in a conformal layer of alumina or titania by using either atomic layer deposition or solution-phase sol–gel deposition. This technique protects the nanoparticle from annealing and broadens the scope of LSPR-based sensors by offering new chemical functionalities. For example, alumina selectively adsorbs polar compounds and is extremely stable against oxidation and high temperatures. Recently, alumina-functionalized AgFON substrates have been used in the quantitative detection of an anthrax biomarker, calcium dipicolinate. The alumina platform also markedly increased the shelf-life of the SERS sensor (more than a year at last count)[94]. The alumina layer can be further functionalized with self-assembled monolayers by using carboxylic-acid-terminated groups[118] and silane chemistry[119].

Introducing LSPR sensors to a wider proteomics community requires the development of substrates that interface well with high-throughput instruments. Recently, a promising multi-arrayed LSPR biochip sensor for proteomics was developed by Endo and co-workers[120,121]. This biochip provides rapid, label-free detection of protein concentration in small sample volumes. The biochip is fabricated in three steps: first a gold film is deposited onto a glass substrate, then silica nanospheres are coated onto the gold film, and finally another layer of gold is deposited onto the nanospheres and the underlying gold substrate. To detect proteins, a (4,4′-dithiodibutyric acid) SAM was formed on the core-shell array. Subsequently, six different antibodies were immobilized on the SAM by using a nanolitre dispensing system. Figure 7a is a photograph of the multi-array sensor with a uniform LSPR reflection spectrum across the chip. Different concentrations of specific antigens were dispensed onto the surface and the change in the absorbance at λ_{max} was recorded for each spot. Figure 7b,c shows the change in absorbance for varying concentrations of analyte. The limit of detection for this sensor is 100 pg ml^{-1}, and the sensor response scales linearly with concentration up to 1 µg ml^{-1}. This biochip can potentially be used in a variety of applications including cancer diagnosis and microorganism detection.

In addition to developing high-throughput assays with LSPR nanosensors, another important practical challenge is to develop implantable LSPR-enabled SERS sensors. Such sensors will enable the spectroscopic detection of important analytes *in vivo* without requiring enzyme-coupled detection schemes. The first *in vivo* SERS glucose detection was recently developed in the Van Duyne laboratory (see Fig. 8)[122,123]. AgFON substrates were functionalized with a mixed SAM consisting of decanethiol (DT) and mercaptohexanol (MH). DT/MH has dual hydrophobic and hydrophilic properties; it selectively partitions glucose from interfering analytes and brings glucose closer to the nanostructured surface. The reversibility of the DT/MH AgFON sensor was demonstrated by alternately exposing the sensor to 0 and 100 mM aqueous glucose solutions (pH ≈ 7). The difference spectra demonstrate reversible partitioning and departitioning. Quantitative detection was demonstrated both *in vitro* and *in vivo* in a rat by using a partial least-squares chemometric analysis. The sensors provide stable readings in bovine plasma for at least 10 days, and show a response rate to glucose pulses of less than 30 s (ref. 122). A similarly rapid response was measured on an AgFON that had been implanted in a rat for 5 h and subsequently removed and tested in a flow cell[123]. Figure 8e depicts the *in vivo* calibration and validation models of a DT/MH-functionalized AgFON substrate implanted into a rat, demonstrating good agreement between SERS and a commercial glucose meter.

The same DT/MH-functionalized AgFON sensor also can detect other analytes, such as lactate. Both glucose and lactate SERS signals were monitored while alternately injecting 100 mM glucose and 100 mM lactate solutions into the flow cell and rinsing the surface with phosphate-buffered saline between each step. The results indicate that both analytes partition and departition successfully from the DT/MH-functionalized SAM[124].

Finally, an important challenge for detection *in vivo* is to develop biocompatible sensors to limit toxicity and biofouling. Studies are underway to test the long-term biocompatibility of the glucose/lactate sensor by using immunohistochemical techniques on tissues near the site of implantation. The use of different surface chemistries, such as a titanium overlayer, is also being explored to improve the biocompatibility of the sensor.

FUTURE DIRECTIONS

In addition to continuing to push forward on the above challenges, there are great opportunities for optimizing single-nanoparticle sensitivity by characterizing and controlling shape. It is known that the corners of nanoparticles and spaces between nanoparticles create the largest electromagnetic enhancements and are thus most important in LSPR and SERS experiments. New techniques for the rapid characterization of large numbers of individual nanoparticles with LSPR[11,38,125,126], SERS[16], fluorescence[22], atomic force microscopy and electron microscopies[16,127] will improve our understanding of how shape features relate to signal enhancements in LSPR and SERS. We also expect that single nanoparticles will increasingly be used as local light sources and chemical sensors in apertureless scanning near-field optical microscopy and chemical imaging[128,129]. The plasmonic nanoparticles serve as antennas that focus the incident light energy into a region ~20 nm in diameter defined by the nanoparticle size and shape, enabling $\lambda/500$ resolution in the information-rich 10-µm band[129]. Nanoparticles can also be used for high-resolution chemical imaging of surfaces with tip-enhanced Raman scattering (TERS)[130,131]. Although currently reported TERS enhancement factors are on the order of 10^3–10^6, further optimization is expected to yield enhancement factors at least as high as bulk nanoparticle substrates (10^6–10^8) and potentially as high as 10^{14} when including contributions from resonantly absorbing molecules[86,87].

Many practical challenges remain before LSPR biosensors can reach their full potential as analytical tools for chemical quantification, the characterization of binding kinetics, the detection of conformational changes, and molecular identification. However, progress in these areas is continuing rapidly. As novel sensing techniques are developed to increase the instrumental resolution and nanoparticle spectral shift, as fabrication techniques mature and nanoparticles become more uniform and stable, and as more diverse and complementary techniques are integrated, the future for plasmonic biosensors looks bright indeed.

doi:10.1038/nmat2162

References

1. Faraday, M. The Bakerian Lecture: experimental relations of gold (and other metals) to light. *Phil. Trans. R. Soc.* **147**, 145–181 (1857).
2. Eustis, S. & El-Sayed, M. A. Why gold nanoparticles are more precious than pretty gold: Noble metal surface plasmon resonance and its enhancement of the radiative and nonradiative properties of nanocrystals of different shapes. *Chem. Soc. Rev.* **35**, 209–217 (2006).
3. Willets, K. A. & Van Duyne, R. P. Localized surface plasmon resonance spectroscopy and sensing. *Annu. Rev. Phys. Chem.* **58**, 267–297 (2007).
4. Schultz, S., Smith, D. R., Mock, J. J. & Schultz, D. A. Single-target molecule detection with nonbleaching multicolor optical immunolabels. *Proc. Natl Acad. Sci. USA* **97**, 996–1001 (2000).
5. Bohren, C. F. & Huffman, D. R. *Absorption and Scattering of Light by Small Particles* (Wiley, New York, 1983).
6. Yguerabide, J. & Yguerabide, E. E. Light-scattering submicroscopic particles as highly fluorescent analogs and their use as tracer labels in clinical and biological applications. I. Theory. *Anal. Biochem.* **262**, 137–156 (1998).
7. Nam, J. M., Thaxton, C. S. & Mirkin, C. A. Nanoparticle-based bio-bar codes for the ultrasensitive detection of proteins. *Science* **301**, 1884–1886 (2003).
8. Yonzon, C. R. *et al.* Towards advanced chemical and biological nanosensors — an overview. *Talanta* **67**, 438–448 (2005).
9. Haes, A. J., Chang, L., Klein, W. L. & Van Duyne, R. P. Detection of a biomarker for Alzheimer's Disease from synthetic and clinical samples using a nanoscale optical biosensor. *J. Am. Chem. Soc.* **127**, 2264–2271 (2005).
10. Dahlin, A. B., Tegenfeldt, J. O. & Hook, F. Improving the instrumental resolution of sensors based on localized surface plasmon resonance. *Anal. Chem.* **78**, 4416–4423 (2006).
11. McFarland, A. D. & Van Duyne, R. P. Single silver nanoparticles as real-time optical sensors with zeptomole sensitivity. *Nano Lett.* **3**, 1057–1062 (2003).
12. Raschke, G. *et al.* Biomolecular recognition based on single gold nanoparticle light scattering. *Nano Lett.* **3**, 935–938 (2003).
13. Elghanian, R., Storhoff, J. J., Mucic, R. C., Letsinger, R. L. & Mirkin, C. A. Selective colorimetric detection of polynucleotides based on the distance-dependent optical properties of gold nanoparticles. *Science* **277**, 1078–1081 (1997).
14. Jeanmaire, D. L. & Van Duyne, R. P. Surface Raman spectroelectrochemistry. Part I. Heterocyclic, aromatic, and aliphatic amines adsorbed on the anodized silver electrode. *J. Electroanal. Chem. Interface Electrochem.* **84**, 1–20 (1977).
15. Haynes, C. L., Yonzon, C. R., Zhang, X. & Van Duyne, R. P. Surface-enhanced Raman sensors: Early history and the development of sensors for quantitative biowarfare agent and glucose detection. *J. Raman Spectrosc.* **36**, 471–484 (2005).
16. Dieringer, J. A. *et al.* Surface enhanced Raman spectroscopy: New materials, concepts, characterization tools, and applications. *Faraday Discuss.* **132**, 9–26 (2006).
17. Haller, K. L. *et al.* Spatially resolved surface enhanced second harmonic generation: Theoretical and experimental evidence for electromagnetic enhancement in the near infrared on a laser microfabricated Pt surface. *J. Chem. Phys.* **90**, 1237–1252 (1989).

18. Yang, W. H., Hulteen, J. C., Schatz, G. C. & Van Duyne, R. P. A surface-enhanced hyper-Raman and surface-enhanced Raman scattering study of *trans*-1,2-bis(4-pyridyl)ethylene adsorbed onto silver film over nanosphere electrodes. Vibrational assignments: Experiment and theory. *J. Chem. Phys.* **104**, 4313–4323 (1996).
19. Jensen, T. R., Van Duyne, R. P., Johnson, S. A. & Maroni, V. A. Surface-enhanced infrared spectroscopy: A comparison of metal island films with discrete and non-discrete surface plasmons. *Appl. Spectrosc.* **54**, 371–377 (2000).
20. Moskovits, M. Surface-enhanced spectroscopy. *Rev. Mod. Phys* **57**, 783–826 (1985).
21. Aslan, K., Lakowicz, J. R., Szmacinski, H. & Geddes, C. D. Enhanced ratiometric pH sensing using SNAFL-2 on silver island films: Metal-enhanced fluorescence sensing. *J. Fluoresc.* **15**, 37–40 (2005).
22. Chen, Y., Munechika, K. & Ginger, D. S. Dependence of fluorescence intensity on the spectral overlap between fluorophores and plasmon resonant single silver nanoparticles. *Nano Lett.* **7**, 690–696 (2007).
23. Sundaramurthy, A. *et al.* Toward nanometer-scale optical photolithography: Utilizing the near-field of bowtie optical nanoantennas. *Nano Lett.* **6**, 355–360 (2006).
24. Ozbay, E. Plasmonics: Merging photonics and electronics at nanoscale dimensions. *Science* **311**, 189–193 (2006).
25. Atwater, H. The promise of plasmonics. *Sci. Am.* **296**, 56–63 (2007).
26. Zhang, X. & Liu, Z. Superlenses to overcome the diffraction limit. *Nature Mater.* **7**, 435-441 (2008).
27. Jensen, T. R., Duval Malinsky, M., Haynes, C. L. & Van Duyne, R. P. Nanosphere lithography: Tunable localized surface plasmon resonance spectra of silver nanoparticles. *J. Phys. Chem. B* **104**, 10549–10556 (2000).
28. Jensen, T. R. *et al.* Nanosphere lithography: Effect of the external dielectric medium on the surface plasmon resonance spectrum of a periodic array of silver nanoparticles. *J. Phys. Chem. B* **103**, 9846–9853 (1999).
29. Jin, R. *et al.* Controlling anisotropic nanoparticle growth through plasmon excitation. *Nature* **425**, 487–490 (2003).
30. Nikoobakht, B. & El-Sayed, M. A. Preparation and growth mechanism of gold nanorods (NRs) using seed-mediated growth method. *Chem. Mater.* **15**, 1957–1962 (2003).
31. Burda, C., Chen, X., Narayanan, R. & El-Sayed, M. A. Chemistry and properties of nanocrystals of different shapes. *Chem. Rev.* **105**, 1025–1102 (2005).
32. Wang, H., Brandl, D. W., Le, F., Nordlander, P. & Halas, N. J. Nanorice: A hybrid plasmonic nanostructure. *Nano Lett.* **6**, 827–832 (2006).
33. Haynes, C. L. & Van Duyne, R. P. Nanosphere lithography: A versatile nanofabrication tool for studies of size-dependent nanoparticle optics. *J. Phys. Chem. B* **105**, 5599–5611 (2001).
34. Haynes, C. L., McFarland, A. D., Smith, M. T., Hulteen, J. C. & Van Duyne, R. P. Angle-resolved nanosphere lithography: Manipulation of nanoparticle size, shape, and interparticle spacing. *J. Phys. Chem. B* **106**, 1898–1902 (2002).
35. Hicks, E. M. *et al.* Controlling plasmon line shapes through diffractive coupling in linear arrays of cylindrical nanoparticles fabricated by electron beam lithography. *Nano Lett.* **5**, 1065–1070 (2005).
36. Barbillon, G. *et al.* Electron beam lithography designed chemical nanosensors based on localized surface plasmon resonance. *Surf. Sci.* **601**, 5057–5061 (2007).
37. Chan, G. H., Zhao, J., Hicks, E. M., Schatz, G. C. & Van Duyne, R. P. Plasmonic properties of copper nanoparticles fabricated by nanosphere lithography. *Nano Lett.* **7**, 1947–1952 (2007).
38. Lee, K. J., Nallathamby, P. D., Browning, L. M., Osgood, C. J. & Xu, X. H. N. In vivo imaging of transport and biocompatibility of single silver nanoparticles in early development of zebrafish embryos. *ACS Nano* **1**, 133–143 (2007).
39. Xu, X. H. N., Brownlow, W. J., Kyriacou, S. V., Wan, Q. & Viola, J. J. Real-time probing of membrane transport in living microbial cells using single nanoparticle optics and living cell imaging. *Biochemistry* **43**, 10400–10413 (2004).
40. Lin, A. *et al.* Nanoshell-enabled photonics-based imaging and therapy of cancer. *Technol. Cancer Res. Treat.* **3**, 33–40 (2004).
41. Stuart, D. A., Haes, A. J., McFarland, A. D., Nie, S. & Van Duyne, R. P. Refractive-index-sensitive, plasmon-resonant-scattering, and surface-enhanced Raman-scattering nanoparticles and arrays as biological sensing platforms. *Proc. SPIE – Int. Soc. Opt. Eng.* **5327**, 60–73 (2004).
42. Yonzon, C. R. *et al.* A comparative analysis of localized and propagating surface plasmon resonance sensors: The binding of concanavalin A to a monosaccharide functionalized self-assembled monolayer. *J. Am. Chem. Soc.* **126**, 12669–12676 (2004).
43. Lacor, P. N. *et al.* Abeta oligomer-induced aberrations in synapse composition, shape and density provide a molecular basis for loss of connectivity in Alzheimer's disease. *J. Neurosci.* **27**, 796–807 (2007).
44. Gong, Y. *et al.* Alzheimer's Disease-affected brain: Presence of oligomeric Aβ ligands (ADDLs) suggests a molecular basis for reversible memory loss. *Proc. Natl Acad. Sci. USA* **100**, 10417–10422 (2003).
45. Hardy, J. & Selkoe, D. J. The amyloid hypothesis of Alzheimer's disease: Progress and problems on the road to therapeutics. *Science* **297**, 353–356 (2002).
46. Georganopoulou, D. G. *et al.* Nanoparticle-based detection in cerebral spinal fluid of a soluble pathogenic biomarker for Alzheimer's disease. *Proc. Natl Acad. Sci. USA* **102**, 2273–2276 (2005).
47. Thompson, R. E., Larson, D. R. & Webb, W. W. Precise nanometer localization analysis for individual fluorescent probes. *Biophys. J.* **82**, 2775–2783 (2002).
48. Yildiz, A. *et al.* Myosin V walks hand-over-hand: Single fluorophore imaging with 1.5-nm localization. *Science* **300**, 2061–2065 (2003).
49. Hall. W. P. *et al.* A calcium-modulated plasmonic switch. *J. Am. Chem. Soc.* **130**, 5836–5837 (2008).
50. Haes, A. J., Zou, S., Schatz, G. C. & Van Duyne, R. P. Nanoscale optical biosensor: Short range distance dependence of the localized surface plasmon resonance of noble metal nanoparticles. *J. Phys. Chem. B* **108**, 6961–6968 (2004).
51. Hicks, E. M. *et al.* Plasmonic properties of film over nanowell surfaces fabricated by nanosphere lithographpy. *J. Phys. Chem. B* **109**, 22351–22358 (2005).
52. Henzie, J., Lee, M. H. & Odom, T. W. Multiscale patterning of plasmonic metamaterials. *Nature Nanotech.* **2**, 549–554 (2007).
53. Van Duyne, R. P., Haes, A. J. & McFarland, A. D. Nanoparticle optics: Sensing with nanoparticle arrays and single nanoparticles. *Proc. SPIE* **5223**, 197–207 (2003).
54. Lasne, D. *et al.* Single nanoparticle photothermal tracking (SNaPT) of 5-nm gold beads in live cells. *Biophys. J.* **91**, 4598–4604 (2006).
55. Shotton, D. M. Confocal scanning optical microscopy and its applications for biological specimens. *J. Cell Sci.* **94**, 175–206 (1989).
56. Aaron, J. *et al.* Plasmon resonance coupling of metal nanoparticles for molecular imaging of carcinogenesis in vivo. *J. Biomed. Optics* **12**, 034007- (2007).
57. Xu, X. H. N., Chen, J., Jeffers, R. & Kyriacou, S. Direct measurement of sizes and dynamics of single living membrane transporters using nano-optics. *Nano Lett.* **2**, 175–182 (2002).
58. Dijk, M. A. *et al.* Absorption and scattering microscopy of single metal nanoparticles. *J. Phys. Chem. Chem. Phys.* **8**, 3486–3495 (2006).
59. Geerts, H., de Brabander, M. & Nuydens, R. Nanovid microscopy. *Nature* **351**, 765–766 (1991).
60. De Brabander, M., Nuydens, R., Geuens, G., Moeremans, M. & De Mey, J. The use of submicroscopic gold particles combined with video contrast enhancement as a simple molecular probe for the living cell. *Cell Motil. Cytoskel.* **6**, 105–113 (1986).
61. Link, S. & El-Sayed, M. A. Shape and size dependence of radiative, non-radiative and photothermal properties of gold nanocrystals. *Int. Rev. Phys. Chem.* **19**, 409–453 (2000).
62. Agarwal, A. *et al.* Targeted gold nanorod contrast agent for prostate cancer detection by photoacoustic imaging. *J. Appl. Phys.* **102**, 064701 (2007).
63. Mallidi, S., Larson, T., Aaron, J., Sokolov, K. & Emelianov, S. Molecular specific optoacoustic imaging with plasmonic nanoparticles. *Opt. Express* **15**, 6583–6588 (2007).
64. El-Sayed, I. H., Huang, X. & El-Sayed, M. A. Selective laser photo-thermal therapy of epithelial carcinoma using anti-EGFR antibody conjugated gold nanoparticles. *Cancer Lett.* **239**, 129–135 (2006).
65. Larson, T. A., Bankson, J., Aaron, J. & Sokolov, K. Hybrid plasmonic magnetic nanoparticles as molecular specific agents for MRI/optical imaging and photothermal therapy of cancer cells. *Nanotechnology* **18**, 1–8 (2007).
66. Aaron, J. S. *et al.* Increased optical contrast in imaging of epidermal growth factor receptor using magnetically actuated hybrid gold/iron oxide nanoparticles. *Opt. Express* **14**, 12930–12943 (2006).
67. Liu, G. L., Lu, Y., Kim, J., Doll, J. C. & Lee, L. P. Magnetic nanocrescents as controllable surface-enhanced Raman scattering nanoprobes for biomedical imaging. *Adv. Mater.* **17**, 2683–2688 (2005).
68. Haes, A. J., Zou, S., Schatz, G. C. & Van Duyne, R. P. Nanoscale optical biosensor: The long range distance dependence of the localized surface plasmon resonance of noble metal nanoparticles. *J. Phys. Chem. B* **108**, 109–116 (2004).
69. Whitney, A. V. *et al.* Localized surface plasmon resonance nanosensor: A high-resolution distance-dependence study using atomic layer deposition. *J. Phys. Chem. B* **109**, 20522–20528 (2005).
70. Haes, A. J., Zou, S., Zhao, J., Schatz, G. C. & Van Duyne, R. P. Localized surface plasmon resonance spectroscopy near molecular resonances. *J. Am. Chem. Soc.* **128**, 10905–10914 (2006).
71. Zhao, J. *et al.* Interaction of plasmon and molecular resonances for rhodamine 6G adsorbed on silver nanoparticles. *J. Am. Chem. Soc.* **129**, 7647–7656 (2007).
72. Zhao, J. *et al.* Resonance surface plasmon spectroscopy: Low molecular weight substrate binding to cytochrome P450. *J. Am. Chem. Soc.* **128**, 11004–11005 (2006).
73. Reinhard, B. M., Sheikholeslami, S., Mastroianni, A., Alivisatos, A. P. & Liphardt, J. Use of plasmon coupling to reveal the dynamics of DNA bending and cleavage by single EcoRV restriction enzymes. *Proc. Natl Acad. Sci. USA* **104**, 2667–2672 (2007).
74. Sonnichsen, C., Reinhard, B. M., Liphardt, J. & Alivisatos, A. P. A molecular ruler based on plasmon coupling of single gold and silver nanoparticles. *Nature Biotechnol.* **23**, 741–745 (2005).
75. Lipscomb, J. D. & Gunsalus, I. C. Structural aspects of active-site of cytochrome P-450Cam. *Drug Metab. Dispos.* **1**, 1–5 (1973).
76. Sligar, S. G. Coupling of spin, substrate, and redox equilibria in cytochrome P450. *Biochemistry* **15**, 5399–5406 (1976).
77. Ingber, D. E. Cellular mechanotransduction: putting all the pieces together again. *FASEB J.* **20**, 811–827 (2006).
78. Su, K. H. *et al.* Interparticle coupling effects on plasmon resonances of nanogold particles. *Nano Lett.* **3**, 1087–1090 (2003).
79. Jain, P. K., Huang, W. & El-Sayed, M. A. On the universal scaling behavior of the distance decay of plasmon coupling in metal nanoparticle pairs: a plasmon ruler equation. *Nano Lett.* **7**, 2080–2088 (2007).
80. Rechberger, W. *et al.* Optical properties of two interacting gold nanoparticles. *Opt. Commun.* **220**, 137–141 (2003).
81. Jensen, T. R., Schatz, G. C. & Van Duyne, R. P. Nanosphere lithography: Surface plasmon resonance spectrum of a periodic array of silver nanoparticles by UV-vis extinction spectroscopy and electrodynamic modeling. *J. Phys. Chem. B* **103**, 2394–2401 (1999).
82. Fromm, D. P., Sundaramurthy, A., Schuck, P. J., Kino, G. & Moerner, W. E. Gap-dependent optical coupling of single 'bowtie' nanoantennas resonant in the visible. *Nano Lett.* **4**, 957–961 (2004).
83. Reinhard, B. M., Siu, M., Agarwal, H., Alivisatos, A. P. & Liphardt, J. Calibration of dynamic molecular rulers based on plasmon coupling between gold nanoparticles. *Nano Lett.* **5**, 2246–2252 (2005).
84. Liu, G. L. *et al.* A nanoplasmonic molecular ruler for measuring nuclease activity and DNA footprinting. *Nature Nanotech.* **1**, 47–52 (2006).
85. McFarland, A. D., Young, M. A., Dieringer, J. A. & Van Duyne, R. P. Wavelength-scanned surface-enhanced Raman excitation spectroscopy. *J. Phys. Chem. B* **109**, 11279–11285 (2005).
86. Nie, S. & Emory, S. R. Probing single molecules and single nanoparticles by surface-enhanced Raman scattering. *Science* **275**, 1102–1106 (1997).
87. Dieringer, J. A., Ii, R. B. L., Scheidt, K. A. & Van Duyne, R. P. A frequency domain existence proof of single-molecule surface-enhanced Raman spectroscopy. *J. Am. Chem. Soc.* **129**, 16249–16256 (2007).
88. Kneipp, K. *et al.* Single molecule detection using surface-enhanced Raman scattering (SERS). *Phys. Rev. Lett.* **78**, 1667–1670 (1997).
89. Van Duyne, R. P. in *Chemical and Biochemical Applications of Lasers* (ed. Moore, C. B.) Vol. 4, 101–184 (Academic, New York, 1979).
90. Zhao, L., Jensen, L. & Schatz, G. C. Pyridine-Ag 20 cluster: A model system for studying surface-enhanced Raman scattering. *J. Am. Chem. Soc.* **128**, 2911–2919 (2006).
91. Zou, S. & Schatz, G. C. Silver nanoparticle array structures that produce giant enhancements in electromagnetic fields. *Chem. Rev. Lett.* **403**, 62–67 (2005).

92. Dick, L. A., McFarland, A. D., Haynes, C. L. & Van Duyne, R. P. Metal film over nanosphere (MFON) electrodes for surface-enhanced Raman spectroscopy (SERS): Improvements in surface nanostructure stability and suppression of irreversible loss. *J. Phys. Chem. B* **106**, 853–860 (2002).

93. Zhang, X., Young, M. A., Lyandres, O. & Van Duyne, R. P. Rapid detection of an anthrax biomarker by surface-enhanced Raman spectroscopy. *J. Am. Chem. Soc.* **127**, 4484–4489 (2005).

94. Zhang, X., Zhao, J., Whitney, A. V., Elam, J. W. & Van Duyne, R. P. Ultrastable substrates for surface-enhanced Raman spectroscopy: Al₂O₃ overlayers fabricated by atomic layer deposition yield improved anthrax biomarker detection. *J. Am. Chem. Soc.* **128**, 10304–10309 (2006).

95. Yonzon, C. R., Zhang, X. & Van Duyne, R. P. Localized surface plasmon resonance immunoassay and verification using surface-enhanced Raman spectroscopy. *Proc. SPIE – Int. Soc. Opt. Eng.* **5224**, 78–85 (2003).

96. Owega, S., Lai, E. P. C. & Mullett, W. M. Laser desorption ionization of gramicidin S on thin silver films with matrix isolation in surface plasmon resonance excitation. *J. Photochem. Photobiol. A* **119**, 123–135 (1998).

97. Owega, S., Lai, E. P. C. & Bawagan, A. D. O. Surface plasmon resonance-laser desorption/ionization-time-of-flight mass spectrometry. *Anal. Chem.* **70**, 2360–2365 (1998).

98. Lee, I., Callcott, T. A. & Arakawa, E. T. Laser-induced surface-plasmon desorption of dye molecules from aluminum films. *Anal. Chem.* **64**, 476–478 (1992).

99. Chen, L. C., Ueda, T., Sagisaka, M., Hori, H. & Hiraoka, K. Visible laser desorption/ionization mass spectrometry using gold nanorods. *J. Phys. Chem. C* **111**, 2409–2415 (2007).

100. Chen, L. C., Yonehama, J., Ueda, T., Hori, H. & Hiraoka, K. Visible-laser desorption/ionization on gold nanostructures. *J. Mass Spectrom.* **42**, 346–353 (2007).

101. Altelaar, A. F. M. *et al.* Gold-enhanced biomolecular surface imaging of cells and tissue by SIMS and MALDI mass spectrometry. *Anal. Chem.* **78**, 734–742 (2006).

102. McLean, J. A., Stumpo, K. A. & Russell, D. H. Size-selected (2–10 nm) gold nanoparticles for matrix assisted laser desorption ionization of peptides. *J. Am. Chem. Soc.* **127**, 5304–5305 (2005).

103. Huang, Y. F. & Chang, H. T. Nile Red-adsorbed gold nanoparticle matrixes for determining aminothiols through surface-assisted laser desorption/ionization mass spectrometry. *Anal. Chem.* **78**, 1485–1493 (2006).

104. Xu, X. *et al.* Size and shape separation of gold nanoparticles with preparative gel electrophoresis. *J. Chromatogr. A* **1167**, 35–41 (2007).

105. Qin, W. J. & Yung, L. Y. L. Nanoparticle-DNA conjugates bearing a specific number of short DNA strands by enzymatic manipulation of nanoparticle-bound DNA. *Langmuir* **21**, 11330–11334 (2005).

106. Duval Malinsky, M., Kelly, L., Schatz, G. C. & Van Duyne, R. P. Chain length dependence and sensing capabilities of the localized surface plasmon resonance of silver nanoparticles chemically modified with alkanethiol self-assembled monolayers. *J. Am. Chem. Soc.* **123**, 1471–1482 (2001).

107. Huang, W., Qian, W. & El-Sayed, M. A. Photothermal reshaping of prismatic Au nanoparticles in periodic monolayer arrays by femtosecond laser pulses. *J. Appl. Phys.* **98**, 114301 (2005).

108. Whitney, A. V., Elam, J. W., Stair, P. C. & Van Duyne, R. P. Toward a thermally robust operando surface-enhanced Raman spectroscopy substrate. *J. Phys. Chem. C* **111**, 16827–16832 (2007).

109. Von Raben, K. U., Chang, R. K., Laube, B. L. & Barber, P. W. Wavelength dependence of surface-enhanced Raman scattering from Ag colloids with adsorbed CN complexes, SO₃²⁻, and pyridine. *J. Phys. Chem. B* **88**, 5290–5296 (1984).

110. Fornasiero, D. & Grieser, F. Analysis of the visible absorption and SERS excitation spectra of silver sols. *J. Chem. Phys.* **87**, 3213–3217 (1987).

111. Carron, K., Peitersen, L. & Lewis, M. Octadecylthiol-modified surface-enhanced Raman-spectroscopy substrates — a new method for the detection of aromatic-compounds. *Environ. Sci. Technol.* **26**, 1950–1954 (1992).

112. Love, J. C., Estroff, L. A., Kriebel, J. K., Nuzzo, R. G. & Whitesides, G. M. Self-assembled monolayers of thiolates on metals as a form of nanotechnology. *Chem. Rev.* **105**, 1103–1169 (2005).

113. Han, J. W. Temporal stability of thiophene self-assembled monolayers on Au (111). *Mol. Cryst. Liq. Cryst.* **464**, 205–209 (2007).

114. Jeong, Y., Lee, C., Ito, E., Hara, M. & Noh, J. Time-dependent phase transition of self-assembled monolayers formed by thioacetyl-terminated tolanes on Au (111). *Jpn. J. Appl. Phys.* **45**, 5906–5910 (2006).

115. Zhang, Z. S. *et al.* Heat capacity measurements of two-dimensional self-assembled hexadecanethiol monolayers on polycrystalline gold. *Appl. Phys. Lett.* **84**, 5198–5200 (2004).

116. Lewis, M., Tarlov, M. & Carron, K. Study of the photooxidation process of self-assembled alkanethiol monolayers. *J. Am. Chem. Soc.* **117**, 9574–9575 (1995).

117. Schoenfisch, M. H. & Pemberton, J. E. Air stability of alkanethiol self-assembled monolayers on silver and gold surfaces. *J. Am. Chem. Soc.* **120**, 4502–4513 (1998).

118. Thompson, W. R. & Pemberton, J. E. Characterization of octadecylsilane and stearic acid layers on Al₂O₃ surfaces by Raman spectroscopy. *Langmuir* **11**, 1720–1725 (1995).

119. Arkles, B. in *Silicon Compounds: Register and Review* 5th edn (eds Anderson, R., Larson, G. L. & Smith, C.) 59–60 (Huls America, Piscataway, New Jersey, 1991).

120. Endo, T. *et al.* Multiple label-free detection of antigen-antibody reaction using localized surface plasmon resonance-based core-shell structured nanoparticle layer nanochip. *Anal. Chem.* **78**, 6465–6475 (2006).

121. Endo, T., Kerman, K., Nagatani, N., Takamura, Y. & Tamiya, E. Label-free detection of peptide nucleic acid-DNA hybridization using localized surface plasmon resonance based optical biosensor. *Anal. Chem.* **77**, 6976–6984 (2005).

122. Lyandres, O. *et al.* Real-time glucose sensing by surface-enhanced Raman spectroscopy in bovine plasma facilitated by a mixed decanethiol/mercaptohexanol partition layer. *Anal. Chem.* **77**, 6134–6139 (2005).

123. Stuart, D. A. *et al.* In vivo glucose measurement by surface-enhanced Raman spectroscopy. *Anal. Chem.* **78**, 7211–7215 (2006).

124. Shah, N. C., Lyandres, O., Walsh, J. T., Glucksberg, M. R. & Van Duyne, R. P. Lactate and sequential lactate-glucose sensing using surface-enhanced Raman spectroscopy. *Anal. Chem.* **79**, 6927–6932 (2007).

125. Liu, G., Doll, J. & Lee, L. High-speed multispectral imaging of nanoplasmonic array. *Opt. Express* **13**, 8520–8525 (2005).

126. Liu, G. L., Long, Y. T., Choi, Y., Kang, T. & Lee, L. P. Quantized plasmon quenching dips nanospectroscopy via plasmon resonance energy transfer. *Nature Methods* **4**, 1015–1017 (2007).

127. Mock, J. J. Shape effects in plasmon resonance of individual colloidal silver nanoparticles. *J. Chem. Phys.* **116**, 6755–6759 (2002).

128. Novotny, L. in *Progress in Optics* (ed. Wolf, E.) Vol. 50, 137–184 (Elsevier, Amsterdam, 2007).

129. Taubner, T., Hillenbrand, R. & Keilmann, F. Performance of visible and mid-infrared scattering-type near-field optical microscopes. *J. Microsc.* **210**, 311–314 (2003).

130. Stöckle, R. M., Suh, Y. D., Deckert, V. & Zenobi, R. Nanoscale chemical analysis by tip-enhanced Raman spectroscopy. *Chem. Rev. Lett.* **318**, 131–136 (2000).

131. Pettinger, B., Ren, B., Picardi, G., Schuster, R. & Ertl, G. Nanoscale probing of adsorbed species by tip-enhanced Raman spectroscopy. *Phys. Rev. Lett.* **92**, 96101 (2004).

132. Haes, J., Haynes, C. L. & Van Duyne, R. P. Nanosphere lithography: Self-assembled photonic and magnetic materials. *Mater. Res. Soc. Symp.* **636**, D4.8 (2001).

Acknowledgements

This research was supported by the National Science Foundation (grants EEC-0647560, CHE-0414554, DMR-0520513 and BES-0507036), the National Cancer Institute (1 U54 CA119341-01), a Ruth L. Kirschstein National Research Service Award (5 F32 GM077020) to J.N.A., and a Ryan Fellowship to W.P.H.

Materials for electrochemical capacitors

Electrochemical capacitors, also called supercapacitors, store energy using either ion adsorption (electrochemical double layer capacitors) or fast surface redox reactions (pseudo-capacitors). They can complement or replace batteries in electrical energy storage and harvesting applications, when high power delivery or uptake is needed. A notable improvement in performance has been achieved through recent advances in understanding charge storage mechanisms and the development of advanced nanostructured materials. The discovery that ion desolvation occurs in pores smaller than the solvated ions has led to higher capacitance for electrochemical double layer capacitors using carbon electrodes with subnanometre pores, and opened the door to designing high-energy density devices using a variety of electrolytes. Combination of pseudo-capacitive nanomaterials, including oxides, nitrides and polymers, with the latest generation of nanostructured lithium electrodes has brought the energy density of electrochemical capacitors closer to that of batteries. The use of carbon nanotubes has further advanced micro-electrochemical capacitors, enabling flexible and adaptable devices to be made. Mathematical modelling and simulation will be the key to success in designing tomorrow's high-energy and high-power devices.

PATRICE SIMON[1,2] AND YURY GOGOTSI[3]

[1]Université Paul Sabatier, CIRIMAT, UMR-CNRS 5085, 31062 Toulouse Cedex 4, France
[2]Institut Universitaire de France, 103 Boulevard Saint Michel, 75005 Paris, France
[3]Department of Materials Science & Engineering, Drexel University, 3141 Chestnut Street, Philadelphia 19104, USA
e-mail: simon@chimie.ups-tlse.fr; gogotsi@drexel.edu

Climate change and the decreasing availability of fossil fuels require society to move towards sustainable and renewable resources. As a result, we are observing an increase in renewable energy production from sun and wind, as well as the development of electric vehicles or hybrid electric vehicles with low CO_2 emissions. Because the sun does not shine during the night, wind does not blow on demand and we all expect to drive our car with at least a few hours of autonomy, energy storage systems are starting to play a larger part in our lives. At the forefront of these are electrical energy storage systems, such as batteries and electrochemical capacitors (ECs)[1]. However, we need to improve their performance substantially to meet the higher requirements of future systems, ranging from portable electronics to hybrid electric vehicles and large industrial equipment, by developing new materials and advancing our understanding of the electrochemical interfaces at the nanoscale. Figure 1 shows the plot of power against energy density, also called a Ragone plot[2], for the most important energy storage systems.

Lithium-ion batteries were introduced in 1990 by Sony, following pioneering work by Whittingham, Scrosati and Armand (see ref. 3 for a review). These batteries, although costly, are the best in terms of performance, with energy densities that can reach 180 watt hours per kilogram. Although great efforts have gone into developing high-performance Li-ion and other advanced secondary batteries that use nanomaterials or organic redox couples[4-6], ECs have attracted less attention until very recently. Because Li-ion batteries suffer from a somewhat slow power delivery or uptake, faster and higher-power energy storage systems are needed in a number of applications, and this role has been given to the ECs[7]. Also known as supercapacitors or ultracapacitors, ECs are power devices that can be fully charged or discharged in seconds; as a consequence, their energy density (about 5 Wh kg^{-1}) is lower than in batteries, but a much higher power delivery or uptake (10 kW kg^{-1}) can be achieved for shorter times (a few seconds)[1]. They have had an important role in complementing or replacing batteries in the energy storage field, such as for uninterruptible power supplies (back-up supplies used to protect against power disruption) and load-levelling. A more recent example is the use of electrochemical double layer capacitors (EDLCs) in emergency doors (16 per plane) on an Airbus A380, thus proving that in terms of performance, safety and reliability ECs are definitely ready for large-scale implementation. A recent report by the US Department of Energy[8] assigns equal importance to supercapacitors and batteries for future energy storage systems, and articles on supercapacitors appearing in business and popular magazines show increasing interest by the general public in this topic.

Several types of ECs can be distinguished, depending on the charge storage mechanism as well as the active materials used. EDLCs, the most common devices at present, use carbon-based active materials with high surface area (Fig. 2). A second group of ECs, known as pseudo-capacitors or redox supercapacitors, uses fast and reversible surface or near-surface reactions for charge storage. Transition metal oxides as well as electrically conducting polymers are examples of

When citing this review, please cite the original version as shown on the contents page of this chapter.

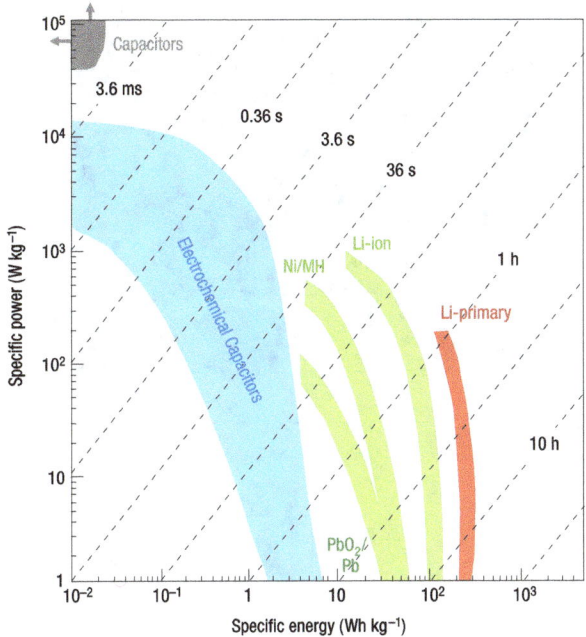

Figure 1 Specific power against specific energy, also called a Ragone plot, for various electrical energy storage devices. If a supercapacitor is used in an electric vehicle, the specific power shows how fast one can go, and the specific energy shows how far one can go on a single charge. Times shown are the time constants of the devices, obtained by dividing the energy density by the power.

pseudo-capacitive active materials. Hybrid capacitors, combining a capacitive or pseudo-capacitive electrode with a battery electrode, are the latest kind of EC, which benefit from both the capacitor and the battery properties.

Electrochemical capacitors currently fill the gap between batteries and conventional solid state and electrolytic capacitors (Fig. 1). They store hundreds or thousands of times more charge (tens to hundreds of farads per gram) than the latter, because of a much larger surface area (1,000–2,000 m² g⁻¹) available for charge storage in EDLC. However, they have a lower energy density than batteries, and this limits the optimal discharge time to less than a minute, whereas many applications clearly need more[9]. Since the early days of EC development in the late 1950s, there has not been a good strategy for increasing the energy density; only incremental performance improvements were achieved from the 1960s to 1990s. The impressive increase in performance that has been demonstrated in the past couple of years is due to the discovery of new electrode materials and improved understanding of ion behaviour in small pores, as well as the design of new hybrid systems combining faradic and capacitive electrodes. Here we give an overview of past and recent findings as well as an analysis of what the future holds for ECs.

ELECTROCHEMICAL DOUBLE-LAYER CAPACITORS

The first patent describing the concept of an electrochemical capacitor was filed in 1957 by Becker[9], who used carbon with a high specific surface area (SSA) coated on a metallic current collector in a sulphuric acid solution. In 1971, NEC (Japan) developed aqueous-electrolyte capacitors under the energy company SOHIO's licence for power-saving units in electronics, and this application can be considered as the starting point for electrochemical capacitor use in commercial devices[9]. New applications in mobile electronics, transportation

(cars, trucks, trams, trains and buses), renewable energy production and aerospace systems[10] bolstered further research.

MECHANISM OF DOUBLE-LAYER CAPACITANCE

EDLCs are electrochemical capacitors that store the charge electrostatically using reversible adsorption of ions of the electrolyte onto active materials that are electrochemically stable and have high accessible SSA. Charge separation occurs on polarization at the electrode–electrolyte interface, producing what Helmholtz described in 1853 as the double layer capacitance C:

$$C = \frac{\varepsilon_r \varepsilon_0 A}{d} \quad \text{or} \quad C/A = \frac{\varepsilon_r \varepsilon_0}{d} \qquad (1)$$

where ε_r is the electrolyte dielectric constant, ε_0 is the dielectric constant of the vacuum, d is the effective thickness of the double layer (charge separation distance) and A is the electrode surface area.

This capacitance model was later refined by Gouy and Chapman, and Stern and Geary, who suggested the presence of a diffuse layer in the electrolyte due to the accumulation of ions close to the electrode surface. The double layer capacitance is between 5 and 20 μF cm⁻² depending on the electrolyte used[11]. Specific capacitance achieved with aqueous alkaline or acid solutions is generally higher than in organic electrolytes[11], but organic electrolytes are more widely used as they can sustain a higher operation voltage (up to 2.7 V in symmetric systems). Because the energy stored is proportional to voltage squared according to

$$E = \tfrac{1}{2} CV^2 \qquad (2)$$

a three-fold increase in voltage, V, results in about an order of magnitude increase in energy, E, stored at the same capacitance.

As a result of the electrostatic charge storage, there is no faradic (redox) reaction at EDLC electrodes. A supercapacitor electrode must be considered as a blocking electrode from an electrochemical point of view. This major difference from batteries means that there is no limitation by the electrochemical kinetics through a polarization resistance. In addition, this surface storage mechanism allows very fast energy uptake and delivery, and better power performance. The absence of faradic reactions also eliminates the swelling in the active material that batteries show during charge/discharge cycles. EDLCs can sustain millions of cycles whereas batteries survive a few thousand at best. Finally, the solvent of the electrolyte is not involved in the charge storage mechanism, unlike in Li-ion batteries where it contributes to the solid–electrolyte interphase when graphite anodes or high-potential cathodes are used. This does not limit the choice of solvents, and electrolytes with high power performances at low temperatures (down to −40 °C) can be designed for EDLCs. However, as a consequence of the electrostatic surface charging mechanism, these devices suffer from a limited energy density. This explains why today's EDLC research is largely focused on increasing their energy performance and widening the temperature limits into the range where batteries cannot operate[9].

HIGH SURFACE AREA ACTIVE MATERIALS

The key to reaching high capacitance by charging the double layer is in using high SSA blocking and electronically conducting electrodes. Graphitic carbon satisfies all the requirements for this application, including high conductivity, electrochemical stability and open porosity[12]. Activated, templated and carbide-derived carbons[13], carbon fabrics, fibres, nanotubes[14], onions[15] and nanohorns[16] have been tested for EDLC applications[11], and some of these carbons are shown in Fig. 2a–d. Activated carbons are the most widely used materials today, because of their high SSA and moderate cost.

Activated carbons are derived from carbon-rich organic precursors by carbonization (heat treatment) in inert atmosphere

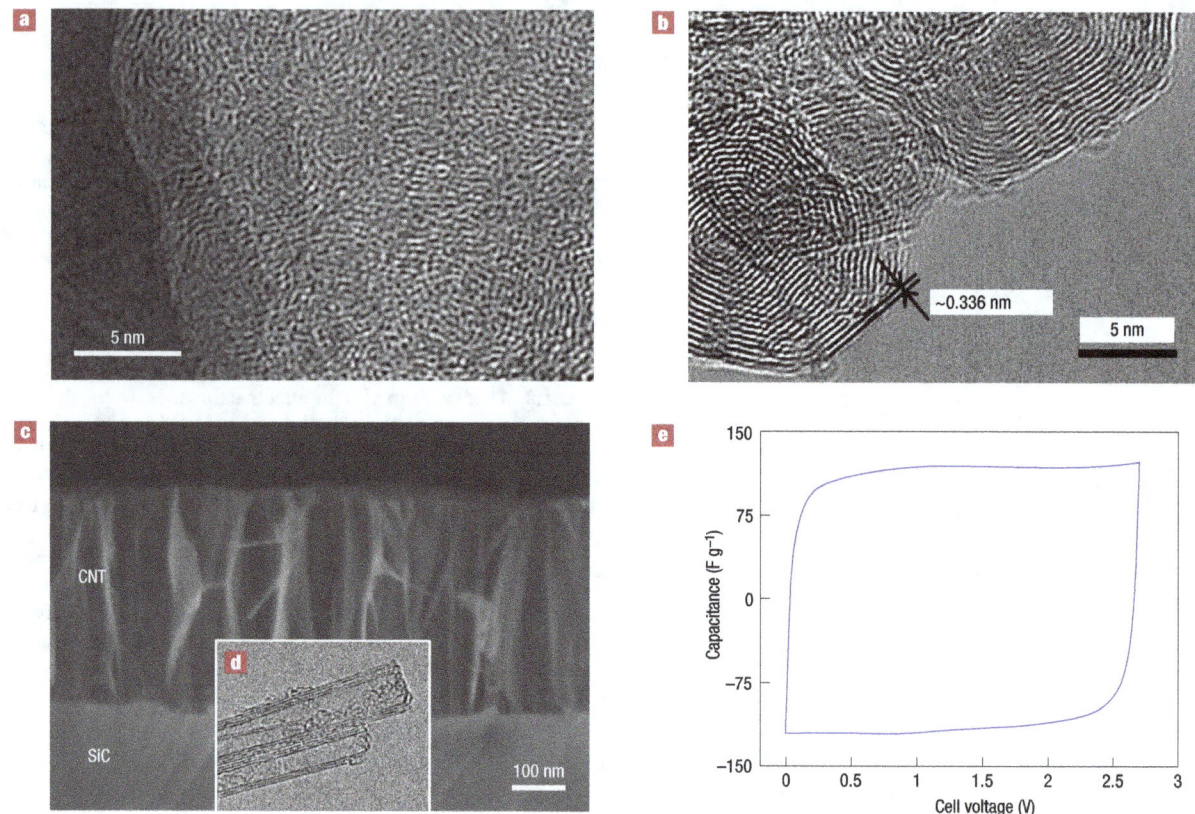

Figure 2 Carbon structures used as active materials for double layer capacitors. **a**, Typical transmission electronic microscopy (TEM) image of a disordered microporous carbon (SiC-derived carbon, 3 hours chlorination at 1,000 °C). **b**, TEM image of onion-like carbon. Reproduced with permission from ref. 80. © 2007 Elsevier. **c**, Scanning electron microscopy image of an array of carbon nanotubes (labelled CNT) on SiC produced by annealing for 6 h at 1,700 °C; inset, **d**, shows a TEM image of the same nanotubes[72]. **e**, Cyclic voltammetry of a two-electrode laboratory EDLC cell in 1.5 M tetraethylammonium tetrafluoroborate NEt$_4^+$,BF$_4^-$ in acetonitrile-based electrolyte, containing activated carbon powders coated on aluminium current collectors. Cyclic voltammetry was recorded at room temperature and potential scan rate of 20 mV s^{-1}.

with subsequent selective oxidation in CO_2, water vapour or KOH to increase the SSA and pore volume. Natural materials, such as coconut shells, wood, pitch or coal, or synthetic materials, such as polymers, can be used as precursors. A porous network in the bulk of the carbon particles is produced after activation; micropores (<2 nm in size), mesopores (2–50 nm) and macropores (>50 nm) can be created in carbon grains. Accordingly, the porous structure of carbon is characterized by a broad distribution of pore size. Longer activation time or higher temperature leads to larger mean pore size. The double layer capacitance of activated carbon reaches 100–120 F g^{-1} in organic electrolytes; this value can exceed 150–300 F g^{-1} in aqueous electrolytes, but at a lower cell voltage because the electrolyte voltage window is limited by water decomposition. A typical cyclic voltammogram of a two-electrode EDLC laboratory cell is presented in Fig. 2e. Its rectangular shape is characteristic of a pure double layer capacitance mechanism for charge storage according to:

$$I = C \times \frac{dV}{dt} \tag{3}$$

where I is the current, (dV/dt) is the potential scan rate and C is the double layer capacitance. Assuming a constant value for C, for a given scan rate the current I is constant, as can be seen from Fig. 2e, where the cyclic voltammogram has a rectangular shape.

As previously mentioned, many carbons have been tested for EDLC applications and a recent paper[11] provides an overview of

what has been achieved. Untreated carbon nanotubes[17] or nanofibres have a lower capacitance (around 50–80 F g^{-1}) than activated carbon in organic electrolytes. It can be increased up to 100 F g^{-1} or greater by grafting oxygen-rich groups, but these are often detrimental to cyclability. Activated carbon fabrics can reach the same capacitance as activated carbon powders, as they have similar SSA, but the high price limits their use to speciality applications. The carbons used in EDL capacitors are generally pre-treated to remove moisture and most of the surface functional groups present on the carbon surface to improve stability during cycling, both of which can be responsible for capacitance fading during capacitor ageing as demonstrated by Azais *et al.*[18] using NMR and X-ray photoelectron spectroscopy techniques. Pandolfo *et al.*[11], in their review article, concluded that the presence of oxygenated groups also contributes to capacitor instability, resulting in an increased series resistance and deterioration of capacitance. Figure 3 presents a schematic of a commercial EDLC, showing the positive and the negative electrodes as well as the separator in rolled design (Fig. 3a,b) and flat design (button cell in Fig. 3c).

CAPACITANCE AND PORE SIZE

Initial research on activated carbon was directed towards increasing the pore volume by developing high SSA and refining the activation process. However, the capacitance increase was limited even for the most porous samples. From a series of activated carbons with different pore sizes in various electrolytes, it was shown that there was no linear relationship between the SSA and the capacitance[19–21]. Some

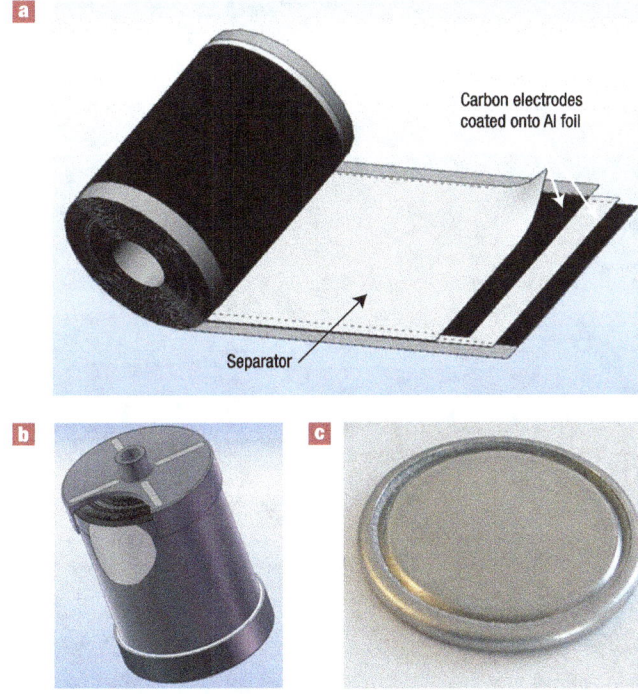

Figure 3 Electrochemical capacitors. **a**, Schematic of a commercial spirally wound double layer capacitor. **b**, Assembled device weighing 500 g and rated for 2,600 F. (Photo courtesy of Batscap, Groupe Bolloré, France.) **c**, A small button cell, which is just 1.6 mm in height and stores 5 F. (Photo courtesy of Y-Carbon, US.) Both devices operate at 2.7 V.

studies suggested that pores smaller than 0.5 nm were not accessible to hydrated ions[20,22] and that even pores under 1 nm might be too small, especially in the case of organic electrolytes, where the size of the solvated ions is larger than 1 nm (ref. 23). These results were consistent with previous work showing that ions carry a dynamic sheath of solvent molecules, the solvation shell[24], and that some hundreds of kilojoules per mole are required to remove it[25] in the case of water molecules. A pore size distribution in the range 2–5 nm, which is larger than the size of two solvated ions, was then identified as a way to improve the energy density and the power capability. Despite all efforts, only a moderate improvement has been made. Gravimetric capacitance in the range of 100–120 F g^{-1} in organic and 150–200 F g^{-1} in aqueous electrolytes has been achieved[26,27] and ascribed to improved ionic mass transport inside mesopores. It was assumed that a well balanced micro- or mesoporosity (according to IUPAC classification, micropores are smaller than 2 nm, whereas mesopores are 2–50 nm in diameter) was needed to maximize capacitance[28].

Although fine-tuned mesoporous carbons failed to achieve high capacitance performance, several studies reported an important capacitive contribution from micropores. From experiments using activated carbon cloth, Salitra et al.[29] suggested that a partial desolvation of ions could occur, allowing access to small pores (<2 nm). High capacitance was observed for a mesoporous carbon containing large numbers of small micropores[30–32], suggesting that partial ion desolvation could lead to an improved capacitance. High capacitances (120 F g^{-1} and 80 F cm^{-3}) were found in organic electrolytes for microporous carbons (<1.5 nm)[33,34], contradicting the solvated ion adsorption theory. Using microporous activated coal-based carbon materials, Raymundo-Piñero et al.[35] observed the same effect and found a maximum capacitance for pore size at 0.7 and 0.8 nm for aqueous and organic electrolytes, respectively. However, the most convincing evidence of capacitance increase in pores smaller than the solvated ion size was provided by experiments using carbide-derived carbons (CDCs)[36–38] as the active material. These are porous carbons obtained by extraction of metals from carbides (TiC, SiC and other) by etching in halogens at elevated temperatures[39]:

$$TiC + 2Cl_2 \rightarrow TiCl_4 + C \qquad (4)$$

In this reaction, Ti is leached out from TiC, and carbon atoms self-organize into an amorphous or disordered, mainly sp^2-bonded[40], structure with a pore size that can be fine-tuned by controlling the chlorination temperature and other process parameters. Accordingly, a narrow uni-modal pore size distribution can be achieved in the range 0.6–1.1 nm, and the mean pore size can be controlled with sub-ångström accuracy[41]. These materials were used to understand the charge storage in micropores using 1 M solution of NEt$_4$BF$_4$ in acetonitrile-based electrolyte[42]. The normalized capacitance (μF cm^{-2}) decreased with decreasing pore size until a critical value close to 1 nm was reached (Fig. 4), and then sharply increased when the pore size approached the ion size. As the CDC samples were exclusively microporous, the capacitance increase for subnanometre pores clearly shows the role of micropores. Moreover, the gravimetric and volumetric capacitances achieved by CDC were, respectively, 50% and 80% higher than for conventional activated carbon[19–21]. The capacitance change with the current density was also found to be stable, demonstrating the high power capabilities these materials can achieve[42]. As the solvated ion sizes in this electrolyte were 1.3 and 1.16 nm for the cation and anion[16], respectively, it was proposed that partial or complete removal of their solvation shell was allowing the ions to access the micropores. As a result, the change of capacitance was a linear function of $1/b$ (where b is the pore radius), confirming that the distance between the ion and the carbon surface, d, was shorter for the smaller pores. This dependence published by Chmiola et al.[42] has since been confirmed by other studies, and analysis of literature data is provided in refs 43 and 44.

CHARGE-STORAGE MECHANISM IN SUBNANOMETRE PORES

From a fundamental point of view, there is a clear lack of understanding of the double layer charging in the confined space of micropores, where there is no room for the formation of the Helmholtz layer and diffuse layer expected at a solid–electrolyte interface. To address this issue, a three-electrode cell configuration, which discriminates between anion and cation adsorption, was used[45]. The double layer capacitance in 1.5 M NEt$_4$BF$_4$-acetonitrile electrolyte caused by the anion and cation at the positive and negative electrodes, respectively, had maxima at different pore sizes[45]. The peak in capacitance shifted to smaller pores for the smaller ion (anion). This behaviour cannot be explained by purely electrostatic reasons, because all pores in this study were the same size as or smaller than a single ion with a single associated solvent molecule. It thus confirmed that ions must be at least partially stripped of solvent molecules in order to occupy the carbon pores. These results point to a charge storage mechanism whereby partial or complete removal of the solvation shell and increased confinement of ions lead to increased capacitance.

A theoretical analysis published by Huang et al.[43] proposed splitting the capacitive behaviour in two different parts depending on the pore size. For mesoporous carbons (pores larger than 2 nm), the traditional model describing the charge of the double layer was used[43]:

$$C/A = \frac{\varepsilon_r \varepsilon_0}{b \ln\left(\frac{b}{b-d}\right)} \qquad (5)$$

where b is the pore radius and d is the distance of approach of the ion to the carbon surface. Data from Fig. 4 in the mesoporous range

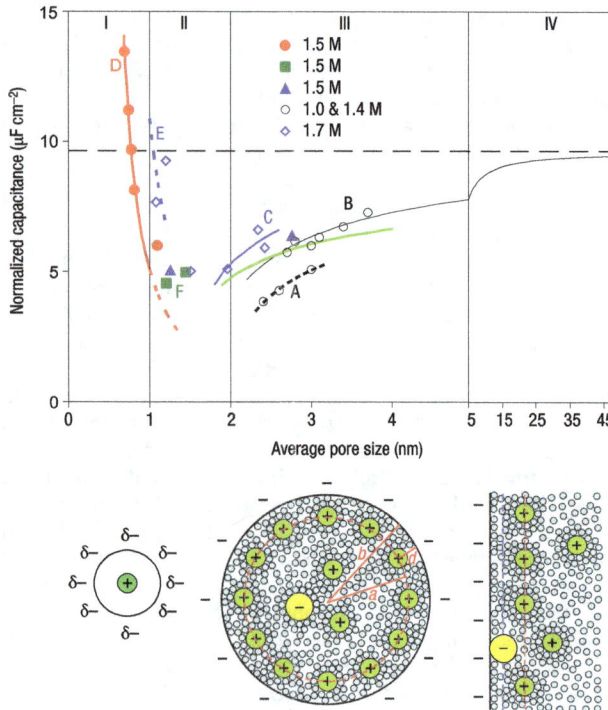

Figure 4 Specific capacitance normalized by SSA as a function of pore size for different carbon samples. All samples were tested in the same electrolyte (NEt$_4$⁺,BF$_4$⁻ in acetonitrile; concentrations are shown in the key). Symbols show experimental data for CDCs, templated mesoporous carbons and activated carbons, and lines show model fits[43]. A huge normalized capacitance increase is observed for microporous carbons with the smallest pore size in zone I, which would not be expected in the traditional view. The partial or complete loss of the solvation shell explains this anomalous behaviour[42]. As schematics show, zones I and II can be modelled as an electric wire-in-cylinder capacitor, an electric double-cylinder capacitor should be considered for zone III, and the commonly used planar electric double layer capacitor can be considered for larger pores, when the curvature/size effect becomes negligible (zone IV). A mathematical fit in the mesoporous range (zone III) is obtained using equation (5). Equation (6) was used to model the capacitive behaviour in zone I, where confined micropores force ions to desolvate partially or completely[44]. A, B: templated mesoporous carbons; C: activated mesoporous carbon; D, F: microporous CDC; E: microporous activated carbon. Reproduced with permission from ref. 44. © 2008 Wiley.

(zone III) were fitted with equation (5). For micropores (<1 nm), it was assumed that ions enter a cylindrical pore and line up, thus forming the 'electric wire in cylinder' model of a capacitor. Capacitance was calculated from[43]

$$C/A = \frac{\varepsilon_r \varepsilon_0}{b \ln\left(\frac{b}{a_0}\right)} \tag{6}$$

where a_0 is the effective size of the ion (desolvated). This model perfectly matches with the normalized capacitance change versus pore size (zone I in Fig. 4). Calculations using density functional theory gave consistent values for the size, a_0, for unsolvated NEt$_4$⁺ and BF$_4$⁻ ions[43].

This work suggests that removal of the solvation shell is required for ions to enter the micropores. Moreover, the ionic radius a_0 found by using equation (6) was close to the bare ion size, suggesting that ions could be fully desolvated. A study carried out with CDCs in a

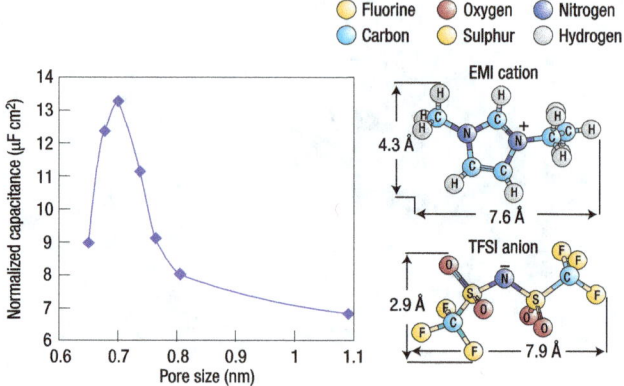

Figure 5 Normalized capacitance change as a function of the pore size of carbon-derived-carbide samples. Samples were prepared at different temperatures in ethyl-methylimidazolium/trifluoro-methane-sulphonylimide (EMI,TFSI) ionic liquid at 60 °C. Inset shows the structure and size of the EMI and TFSI ions. The maximum capacitance is obtained when the pore size is in the same range as the maximum ion dimension. Reproduced with permission from ref. 46. © 2008 ACS.

solvent-free electrolyte ([EMI⁺,TFSI⁻] ionic liquid at 60 °C), in which both ions have a maximum size of about 0.7 nm, showed the maximum capacitance for samples with the 0.7-nm pore size[46], demonstrating that a single ion per pore produces the maximum capacitance (Fig. 5). This suggests that ions cannot be adsorbed on both pore surfaces, in contrast with traditional supercapacitor models.

MATERIALS BY DESIGN

The recent findings of the micropore contribution to the capacitive storage highlight the lack of fundamental understanding of the electrochemical interfaces at the nanoscale and the behaviour of ions confined in nanopores. In particular, the results presented above rule out the generally accepted description of the double layer with solvated ions adsorbed on both sides of the pore walls, consistent with the absence of a diffuse layer in subnanometre pores. Although recent studies[45,46] provide some guidance for developing materials with improved capacitance, such as elimination of macro- and mesopores and matching the pore size with the ion size, further material optimization by Edisonian or combinatorial electrochemistry methods may take a very long time. The effects of many parameters, such as carbon bonding (sp versus sp^2 or sp^3), pore shape, defects or adatoms, are difficult to determine experimentally. Clearly, computational tools and atomistic simulation will be needed to help us to understand the charge storage mechanism in subnanometre pores and to propose strategies to design the next generation of high-capacitance materials and material–electrolyte systems[47]. Recasting the theory of double layers in electrochemistry to take into account solvation and desolvation effects could lead to a better understanding of charge storage as well as ion transport in ECs and even open up new opportunities in areas such as biological ion channels and water desalination.

REDOX-BASED ELECTROCHEMICAL CAPACITORS

MECHANISM OF PSEUDO-CAPACITIVE CHARGE STORAGE

Some ECs use fast, reversible redox reactions at the surface of active materials, thus defining what is called the pseudo-capacitive behaviour. Metal oxides such as RuO$_2$, Fe$_3$O$_4$ or MnO$_2$ (refs 48, 49), as well as electronically conducting polymers[50], have been extensively studied in the past decades. The specific pseudo-capacitance exceeds

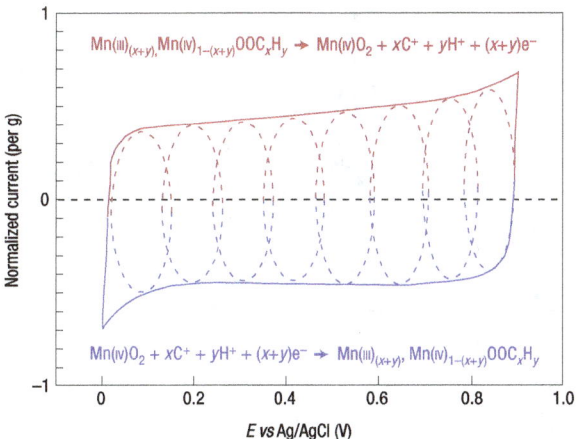

$$Mn(III)_{(x+y)}, Mn(IV)_{1-(x+y)}OOC_xH_y \rightarrow Mn(IV)O_2 + xC^+ + yH^+ + (x+y)e^-$$

$$Mn(IV)O_2 + xC^+ + yH^+ + (x+y)e^- \rightarrow Mn(III)_{(x+y)}, Mn(IV)_{1-(x+y)}OOC_xH_y$$

Figure 6 Cyclic voltammetry. This schematic of cyclic voltammetry for a MnO$_2$-electrode cell in mild aqueous electrolyte (0.1 M K$_2$SO$_4$) shows the successive multiple surface redox reactions leading to the pseudo-capacitive charge storage mechanism. The red (upper) part is related to the oxidation from Mn(III) to Mn(IV) and the blue (lower) part refers to the reduction from Mn(IV) to Mn(III).

that of carbon materials using double layer charge storage, justifying interest in these systems. But because redox reactions are used, pseudo-capacitors, like batteries, often suffer from a lack of stability during cycling.

Ruthenium oxide, RuO$_2$, is widely studied because it is conductive and has three distinct oxidation states accessible within 1.2 V. The pseudo-capacitive behaviour of RuO$_2$ in acidic solutions has been the focus of research in the past 30 years[1]. It can be described as a fast, reversible electron transfer together with an electro-adsorption of protons on the surface of RuO$_2$ particles, according to equation (7), where Ru oxidation states can change from (II) up to (IV):

$$RuO_2 + xH^+ + xe^- \leftrightarrow RuO_{2-x}(OH)_x \qquad (7)$$

where $0 \leq x \leq 2$. The continuous change of x during proton insertion or de-insertion occurs over a window of about 1.2 V and leads to a capacitive behaviour with ion adsorption following a Frumkin-type isotherm[1]. Specific capacitance of more than 600 F g^{-1} has been reported[51], but Ru-based aqueous electrochemical capacitors are expensive, and the 1-V voltage window limits their applications to small electronic devices. Organic electrolytes with proton surrogates (for example Li$^+$) must be used to go past 1 V. Less expensive oxides of iron, vanadium, nickel and cobalt have been tested in aqueous electrolytes, but none has been investigated as much as manganese oxide[52]. The charge storage mechanism is based on surface adsorption of electrolyte cations C$^+$ (K$^+$, Na$^+$...) as well as proton incorporation according to the reaction:

$$MnO_2 + xC^+ + yH^+ + (x+y)e^- \leftrightarrow MnOOC_xH_y \qquad (8)$$

Figure 6 shows a cyclic voltammogram of a single MnO$_2$ electrode in mild aqueous electrolyte; the fast, reversible successive surface redox reactions define the behaviour of the voltammogram, whose shape is close to that of the EDLC. MnO$_2$ micro-powders or micrometre-thick films show a specific capacitance of about 150 F g^{-1} in neutral aqueous electrolytes within a voltage window of <1 V. Accordingly, there is limited interest in MnO$_2$ electrodes for symmetric devices, because there are no oxidation states available at less than 0 V. However, it is suitable for a pseudo-capacitive positive electrode in hybrid systems,

which we will describe below. Other transition metal oxides with various oxidation degrees, such as molybdenum oxides, should also be explored as active materials for pseudo-capacitors.

Many kinds of conducting polymers (polyaniline, polypyrrole, polythiophene and their derivatives) have been tested in EC applications as pseudo-capacitive materials[50,53,54] and have shown high gravimetric and volumetric pseudo-capacitance in various non-aqueous electrolytes at operating voltages of about 3 V. When used as bulk materials, conducting polymers suffer from a limited stability during cycling that reduces the initial performance[9]. Research efforts with conducting polymers for supercapacitor applications are nowadays directed towards hybrid systems.

NANOSTRUCTURING REDOX-ACTIVE MATERIALS TO INCREASE CAPACITANCE

Given that nanomaterials have helped to improve Li-ion batteries[55], it is not surprising that nanostructuring has also affected ECs. Because pseudo-capacitors store charge in the first few nanometres from the surface, decreasing the particle size increases active material usage. Thanks to a thin electrically conducting surface layer of oxide and oxynitride, the charging mechanism of nanocrystalline vanadium nitride (VN) includes a combination of an electric double layer and a faradic reaction (II/IV) at the surface of the nanoparticles, leading to specific capacitance up to 1,200 F g^{-1} at a scan rate of 2 mV s^{-1} (ref. 56). A similar approach can be applied to other nano-sized transition metal nitrides or oxides. In another example, the cycling stability and the specific capacitance of RuO$_2$ nanoparticles were increased by depositing a thin conducting polymer coating that enhanced proton exchange at the surface[57]. The design of specific surface functionalization to improve interfacial exchange could be suggested as a generic approach to other pseudo-redox materials.

MnO$_2$ and RuO$_2$ films have been synthesized at the nanometre scale. Thin MnO$_2$ deposits of tens to hundreds of nanometres have been produced on various substrates such as metal collectors, carbon nanotubes or activated carbons. Specific capacitances as high as 1,300 F g^{-1} have been reported[58], as reaction kinetics were no longer limited by the electrical conductivity of MnO$_2$. In the same way, Sugimoto's group have prepared hydrated RuO$_2$ nano-sheets with capacitance exceeding 1,300 F g^{-1} (ref. 59). The RuO$_2$ specific capacitance also increased sharply when the film thickness was decreased. The deposition of RuO$_2$ thin film onto carbon supports[60,61] both increased the capacitance and decreased the RuO$_2$ consumption. Thin film synthesis or high SSA capacitive material decoration with nano-sized pseudo-capacitive active material, like the examples presented in Fig. 7a and b, offers an opportunity to increase energy density and compete with carbon-based EDLCs. Particular attention must be paid to further processing of nano-sized powders into active films because they tend to re-agglomerate into large-size grains. An alternative way to produce porous films from powders is by growing nanotubes, as has been shown for V$_2$O$_5$ (ref. 62), or nanorods. These allow easy access to the active material, but can only be produced in thin films so far, and the manufacturing cost will probably limit the use of these sophisticated nanostructures to small electronic devices.

HYBRID SYSTEMS TO ACHIEVE HIGH ENERGY DENSITY

Hybrid systems offer an attractive alternative to conventional pseudo-capacitors or EDLCs by combining a battery-like electrode (energy source) with a capacitor-like electrode (power source) in the same cell. An appropriate electrode combination can even increase the cell voltage, further contributing to improvement in energy and power densities. Currently, two different approaches to hybrid systems have emerged: (i) pseudo-capacitive metal oxides with a capacitive carbon electrode, and (ii) lithium-insertion electrodes with a capacitive carbon electrode.

Numerous combinations of positive and negative electrodes have been tested in the past in aqueous or inorganic electrolytes. In most

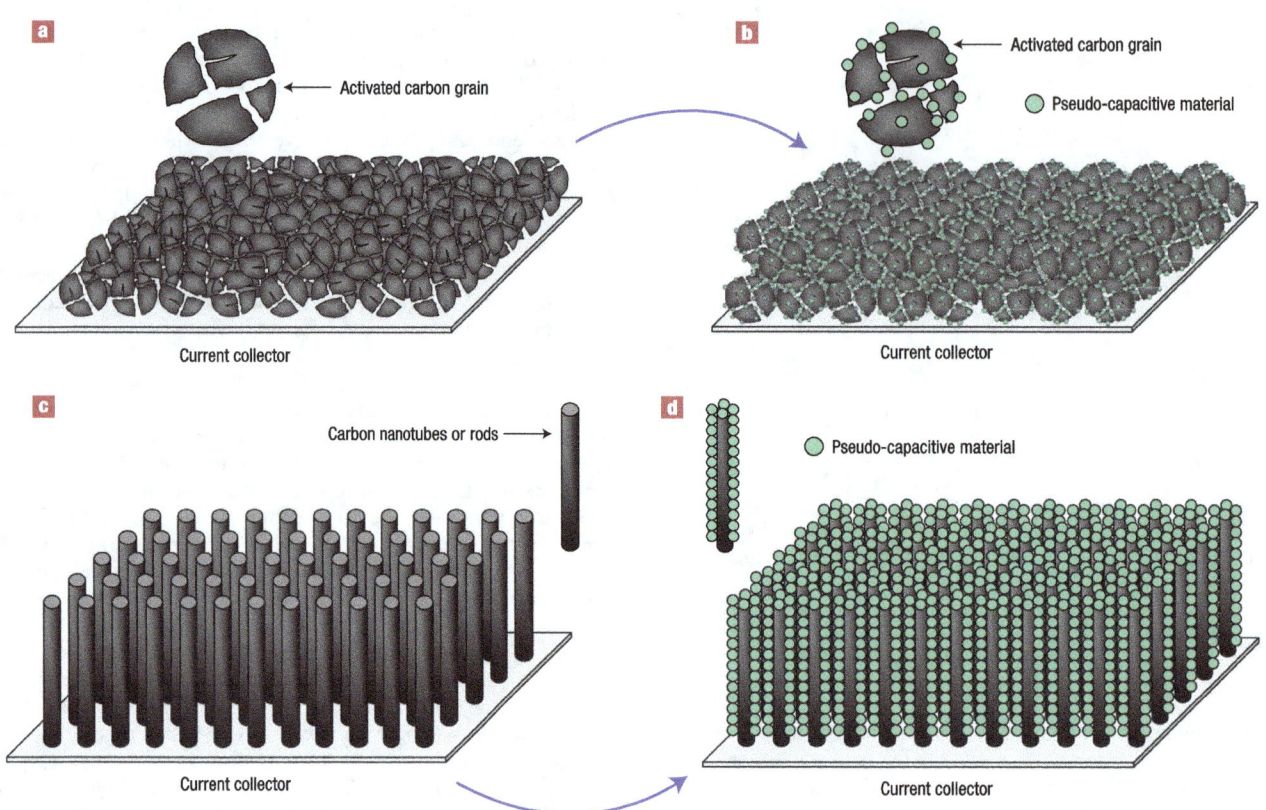

Figure 7 Possible strategies to improve both energy and power densities for electrochemical capacitors. **a**, **b**, Decorating activated carbon grains (**a**) with pseudo-capacitive materials (**b**). **c**, **d**, Achieving conformal deposit of pseudo-capacitive materials (**d**) onto highly ordered high-surface-area carbon nanotubes (**c**).

cases, the faradic electrode led to an increase in the energy density at the cost of cyclability (for balanced positive and negative electrode capacities). This is certainly the main drawback of hybrid devices, as compared with EDLCs, and it is important to avoid transforming a good supercapacitor into a mediocre battery[63].

MnO$_2$ is one of the most studied materials as a low-cost alternative to RuO$_2$. Its pseudo-capacitance arises from the III/IV oxidation state change at the surface of MnO$_2$ particles[58]. The association of a negative EDLC-type electrode with a positive MnO$_2$ electrode leads to a 2-V cell in aqueous electrolytes thanks to the apparent water decomposition overvoltage on MnO$_2$ and high-surface-area carbon. The low-cost carbon–MnO$_2$ hybrid system combines high capacitance in neutral aqueous electrolytes with high cell voltages, making it a green alternative to EDLCs using acetonitrile-based solvents and fluorinated salts. Moreover, the use of MnO$_2$ nano-powders and nanostructures offers the potential for further improvement in capacitance[64]. Another challenge for this system is to use organic electrolytes to reach higher cell voltage, thus improving the energy density.

A combination of a carbon electrode with a PbO$_2$ battery-like electrode using H$_2$SO$_4$ solution can work at 2.1 V (ref. 65), offering a low-cost EC device for cost-sensitive applications, in which weight of the device is of minor concern.

The hybrid concept originated from the Li-ion batteries field. In 1999, Amatucci's group combined a nanostructured lithium titanate anode Li$_4$Ti$_5$O$_{12}$ with an activated carbon positive electrode, designing a 2.8-V system that for the first time exceeded 10 Wh kg^{-1} (ref. 66). The titanate electrode ensured high power capacity and no solid-electrolyte interphase formation, as well as long-life cyclability thanks to low volume change during cycling. Following this pioneering

work, many studies have been conducted on various combinations of a lithium-insertion electrode with a capacitive carbon electrode. The Li-ion capacitor developed by Fuji Heavy Industry is an example of this concept, using a pre-lithiated high SSA carbon anode together with an activated carbon cathode[63,67]. It achieved an energy density of more than 15 Wh kg^{-1} at 3.8 V. Capacity retention was increased by unbalancing the electrode capacities, allowing a low depth of charge/discharge at the anode. Systems with an activated carbon anode and anion intercalation cathode are also under development. The advent of nanomaterials[55] as well as fast advances in the area of Li-ion batteries should lead to the design of high-performance ECs. Combining newly developed high-rate conversion reaction anodes or Li-alloying anodes with a positive supercapacitor electrode could fill the gap between Li-ion batteries and EDLCs. These systems could be of particular interest in applications where high power and medium cycle life are needed.

CURRENT COLLECTORS

Because ECs are power devices, their internal resistance must be kept low. Particular attention must be paid to the contact impedance between the active film and the current collector. ECs designed for organic electrolytes use treated aluminium foil or grid current collectors. Surface treatments have already been shown to decrease ohmic drops at this interface[68], and coatings on aluminium that improve electrochemical stability at high potentials and interface conductivity are of great interest.

The design of nanostructured current collectors with an increased contact area is another way to control the interface between current collector and active material. For example, carbon can be produced in a variety of morphologies[12], including porous films and nanotube brushes that can be grown on various current collectors[69] and that can serve as substrates for further conformal deposition (Fig. 7c and d) of active material. These nano-architectured electrodes could outperform the existing systems by confining a highly pseudo-capacitive material to a thin film with a high SSA, as has been done for Li-ion batteries[70] where, by growing Cu nano-pillars on a planar Cu foil, a six-fold improvement in the energy density over planar electrodes has been achieved[70]. Long's group[64] successfully applied a similar approach to supercapacitors by coating a porous carbon nano-foam with a 20-nm pseudo-capacitive layer of MnO_2. As a result, the area-normalized capacitance doubled to reach 1.5 F cm^{-2}, together with an outstanding volumetric capacitance of 90 F cm^{-3}. Electrophoretic deposition from stable colloidal suspensions of RuO_2 (ref. 71) or other active material can be used for filling the inter-tube space to design high-energy-density devices which are just a few micrometres thick. The nano-architectured electrodes also find applications in micro-systems where micro-ECs can complement micro-batteries for energy harvesting or energy generation. In this specific field, it is often advantageous to grow self-supported, binder-less nano-electrodes directly on semiconductor wafers, such as Si or SiC (ref. 72; Fig. 2c).

An attractive material for current collectors is carbon in the form of a highly conductive nanotube or graphene paper. It does not corrode in aqueous electrolytes and is very flexible. The use of nanotube paper for manufacturing flexible supercapacitors is expected to grow as the cost of small-diameter nanotubes required for making paper decreases. The same thin sheet of nanotubes[14] could potentially act as an active material and current collector at the same time. Thin-film, printable and wearable ECs could find numerous applications.

FROM ORGANIC TO IONIC LIQUID ELECTROLYTES

EC cell voltage is limited by the electrolyte decomposition at high potentials. Accordingly, the larger the electrolyte stability voltage window, the higher the supercapacitor cell voltage. Moving from aqueous to organic electrolytes increased the cell voltage from 0.9 V to 2.5–2.7 V for EDLCs. Because the energy density is proportional to the voltage squared (equation (2)), numerous research efforts have been directed at the design of highly conducting, stable electrolytes with a wider voltage window. Today, the state of the art is the use of organic electrolyte solutions in acetonitrile or propylene carbonate, the latter becoming more popular because of the low flash point and lower toxicity compared with acetonitrile.

Ionic liquids are room-temperature liquid solvent-free electrolytes; their voltage window stability is thus only driven by the electrochemical stability of the ions. A careful choice of both the anion and the cation allows the design of high-voltage supercapacitors, and 3-V, 1,000-F commercial devices are already available[73]. However, the ionic conductivity of these liquids at room temperature is just a few milliSiemens per centimetre, so they are mainly used at higher temperatures. For example, CDC with an EMI/TFSI ionic liquid electrolyte has been shown[46] to have capacitance of 160 F g^{-1} and ~90 F cm^{-3} at 60 °C. In this area, hybrid activated carbon/conducting polymer devices also show an improved performance with cell voltages higher than 3 V (refs 74–76).

For applications in the temperature range –30 °C to +60 °C, where batteries and supercapacitors are mainly used, ionic liquids still fail to satisfy the requirements because of their low ionic conductivity. However, the choice of a huge variety of combinations of anions and cations offers the potential for designing an ionic liquid electrolyte with an ionic conductivity of 40 mS cm^{-1} and a voltage window of >4 V at room temperature[77]. A challenge is, for instance, to find an alternative to the imidazolium cation that, despite high conductivity,

undergoes a reduction reaction at potential <1.5 V versus Li^+/Li. Replacing the heavy bis(trifluoromethanesulphonyl)imide (TFSI) anion by a lighter (fluoromethanesulphonyl)imide (FSI) and preparing ionic liquid eutectic mixtures would improve both the cell voltage (because a protecting layer of AlF_3 can be formed on the Al surface, shifting the de-passivation potential of Al above 4 V) and the ionic conductivity[77]. However, FSI shows poor cyclability at elevated temperatures. Supported by the efforts of the Li-ion community to design safer systems using ionic liquids, the research on ionic liquids for ECs is expected to have an important role in the improvement of capacitor performance in the coming years.

APPLICATIONS OF ELECTROCHEMICAL CAPACITORS

ECs are electrochemical energy sources with high power delivery and uptake, with an exceptional cycle life. They are used when high power demands are needed, such as for power buffer and power saving units, but are also of great interest for energy recovery. Recent articles from Miller et al.[7,10] present an overview of the opportunities for ECs in a variety of applications, complementing an earlier review by Kötz et al.[9]. Small devices (a few farads) are widely used for power buffer applications or for memory back-up in toys, cameras, video recorders, mobile phones and so forth. Cordless tools such as screwdrivers and electric cutters using EDLCs are already available on the market. Such systems, using devices of a few tens of farads, can be fully charged or discharged in less than 2 minutes, which is particularly suited to these applications, with the cycle life of EDLC exceeding that of the tool. As mentioned before, the Airbus A380 jumbo jets use banks of EDLCs for emergency door opening. The modules consist of an in series/parallel assembly of 100-F, 2.7-V cells that are directly integrated into the doors to limit the use of heavy copper cables. This application is obviously a niche market, but it is a demonstration that the EDLC technology is mature in terms of performance, reliability and safety.

The main market targeted by EDLC manufacturers for the next years is the transportation market, including hybrid electric vehicles, as well as metro trains and tramways. There continues to be debate about the advantage of using high power Li-ion batteries instead of ECs (or vice versa) for these applications. Most of these discussions have been initiated by Li-ion battery manufacturers who would like their products to cover the whole range of applications. However, ECs and Li-ion batteries should not necessarily be seen as competitors, because their charge storage mechanisms and thus their characteristics are different. The availability of the stored charge will always be faster for a supercapacitor (surface storage) than for a Li-ion battery (bulk storage), with a larger stored energy for the latter. Both devices must be used in their respective time-constant domains (see Fig. 1). Using a Li-ion battery for repeated high power delivery/uptake applications for a short duration (10 s or less) will quickly degrade the cycle life of the system[10]. The only way to avoid this is to oversize the battery, increasing the cost and volume. In the same way, using ECs for power delivery longer than 10 s requires oversizing. However, some applications use ECs as the main power and energy source, benefiting from the fast charge/discharge capability of these systems as well as their outstanding cycle life. Several train manufacturers have clearly identified the tramway/metro market segment as extremely relevant for EC use, to power trains over short distances in big cities, where electric cables are clearly undesirable for aesthetic and other reasons, but also to recover the braking energy of another train on the same line, thanks to the ECs' symmetric high power delivery/uptake characteristics.

For automotive applications, manufacturers are already proposing solutions for electrical power steering, where ECs are used for load-levelling in stop-and-go traffic[78]. The general trend is to increase the hybridization degree of the engines in hybrid electric vehicles, to allow fast acceleration (boost) and braking energy recovery. The on-board energy storage systems will be in higher demand, and a combination of batteries and EDLCs will increase the battery cycle life, explaining why EDLCs are viewed as a partner to Li-ion batteries for this market[78]. Currently, high price limits the use of both Li-ion batteries and EDLC in large-scale applications (for example for load levelling). But the surprisingly high cost of materials used for EDLC is due to a limited number of suppliers rather than intrinsically high cost of porous carbon. Decreasing the price of carbon materials for ECs, including CDC and AC, would remove the main obstacle to their wider use[79].

SUMMARY AND OUTLOOK

The most recent advances in supercapacitor materials include nanoporous carbons with the pore size tuned to fit the size of ions of the electrolyte with ångström accuracy, carbon nanotubes for flexible and printable devices with a short response time, and transition metal oxide and nitride nanoparticles for pseudo-capacitors with a high energy density. An improved understanding of charge storage and ion desolvation in subnanometre pores has helped to overcome a barrier that has been hampering progress in the field for decades. It has also shown how important it is to match the active materials with specific electrolytes and to use a cathode and anode with different pore sizes that match the anion or cation size. Nano-architecture of electrodes has led to further improvements in power delivery. The very large number of possible active materials and electrolytes means that better theoretical guidance is needed for the design of future ECs.

Future generations of ECs are expected to come close to current Li-ion batteries in energy density, maintaining their high power density. This may be achieved by using ionic liquids with a voltage window of more than 4 V, by discovering new materials that combine double-layer capacitance and pseudo-capacitance, and by developing hybrid devices. ECs will have a key role in energy storage and harvesting, decreasing the total energy consumption and minimizing the use of hydrocarbon fuels. Capacitive energy storage leads to a lower energy loss (higher cycle efficiency), than for batteries, compressed air, flywheel or other devices, helping to improve storage economy further. Flexible, printable and wearable ECs are likely to be integrated into smart clothing, sensors, wearable electronics and drug delivery systems. In some instances they will replace batteries, but in many cases they will either complement batteries, increasing their efficiency and lifetime, or serve as energy solutions where an extremely large number of cycles, long lifetime and fast power delivery are required.

doi:10.1038/nmat2297

References

1. Conway, B. E. *Electrochemical Supercapacitors: Scientific Fundamentals and Technological Applications* (Kluwer, 1999).
2. Service, R. F. New 'supercapacitor' promises to pack more electrical punch. *Science* **313**, 902–905 (2006).
3. Tarascon, J.-M. & Armand, M. Issues and challenges facing rechargeable lithium batteries. *Nature* **414**, 359–367 (2001).
4. Brodd, R. J. *et al.* Batteries, 1977 to 2002. *J. Electrochem. Soc.* **151**, K1–K11 (2004).
5. Armand, M. & Tarascon, J.-M. Building better batteries. *Nature* **451**, 652–657 (2008).
6. Armand, M. & Johansson, P. Novel weakly coordinating heterocyclic anions for use in lithium batteries. *J. Power Sources* **178**, 821–825 (2008).
7. Miller, J. R. & Simon, P. Electrochemical capacitors for energy management. *Science* **321**, 651–652 (2008).
8. US Department of Energy. *Basic Research Needs for Electrical Energy Storage* <www.sc.doe.gov/bes/reports/abstracts.html#EES2007> (2007).
9. Kötz, R. & Carlen, M. Principles and applications of electrochemical capacitors. *Electrochim. Acta* **45**, 2483–2498 (2000).
10. Miller, J. R. & Burke, A. F. Electrochemical capacitors: Challenges and opportunities for real-world applications. *Electrochem. Soc. Interf.* **17**, 53–57 (2008).
11. Pandolfo, A. G. & Hollenkamp, A. F. Carbon properties and their role in supercapacitors. *J. Power Sources* **157**, 11–27 (2006).
12. Gogotsi, Y. (ed.) *Carbon Nanomaterials* (CRC, 2006).
13. Kyotani, T., Chmiola, J. & Gogotsi, Y. in *Carbon Materials for Electrochemical Energy Storage Systems* (eds Beguin, F. & Frackowiak, E.) Ch. 13 (CRC/Taylor and Francis, in the press).
14. Futaba, D. N. *et al.* Shape-engineerable and highly densely packed single-walled carbon nanotubes and their application as super-capacitor electrodes. *Nature Mater.* **5**, 987–994 (2006).
15. Portet, C., Chmiola, J., Gogotsi, Y., Park, S. & Lian, K. Electrochemical characterizations of carbon nanomaterials by the cavity microelectrode technique. *Electrochim. Acta*, **53**, 7675–7680 (2008).
16. Yang, C.-M. *et al.* Nanowindow-regulated specific capacitance of supercapacitor electrodes of single-wall carbon nanohorns. *J. Am. Chem. Soc.* **129**, 20–21 (2007).
17. Niu, C., Sichel, E. K., Hoch, R., Moy, D. & Tennent, H. High power electrochemical capacitors based on carbon nanotube electrodes. *Appl. Phys. Lett.* **70**, 1480 (1997).
18. Azaïs, P. *et al.* Causes of supercapacitors ageing in organic electrolyte. *J. Power Sources* **171**, 1046–1053 (2007).
19. Gamby, J., Taberna, P. L., Simon, P., Fauvarque, J. F. & Chesneau, M. Studies and characterization of various activated carbons used for carbon/carbon supercapacitors. *J. Power Sources* **101**, 109–116 (2001).
20. Shi, H. Activated carbons and double layer capacitance. *Electrochim. Acta* **41**, 1633–1639 (1995).
21. Qu, D. & Shi, H. Studies of activated carbons used in double-layer capacitors. *J. Power Sources* **74**, 99–107 (1998).
22. Qu, D. Studies of the activated carbons used in double-layer supercapacitors. *J. Power Sources* **109**, 403–411 (2002).
23. Kim, Y. J. *et al.* Correlation between the pore and solvated ion size on capacitance uptake of PVDC-based carbons. *Carbon* **42**, 1491 (2004).
24. Izutsu, K. *Electrochemistry in Nonaqueous Solution* (Wiley, 2002).
25. Marcus, Y. *Ion Solvation* (Wiley, 1985).
26. Jurewicz, K. *et al.* Capacitance properties of ordered porous carbon materials prepared by a templating procedure. *J. Phys. Chem. Solids* **65**, 287 (2004).
27. Fernández, J. A. *et al.* Performance of mesoporous carbons derived from poly(vinyl alcohol) in electrochemical capacitors. *J. Power Sources* **175**, 675 (2008).
28. Fuertes, A. B., Lota, G., Centeno, T. A. & Frackowiak, E. Templated mesoporous carbons for supercapacitor application. *Electrochim. Acta* **50**, 2799 (2005).
29. Salitra, G., Soffer, A., Eliad, L., Cohen, Y. & Aurbach, D. Carbon electrodes for double-layer capacitors. I. Relations between ion and pore dimensions. *J. Electrochem. Soc.* **147**, 2486–2493 (2000).
30. Vix-Guterl, C. *et al.* Electrochemical energy storage in ordered porous carbon materials. *Carbon* **43**, 1293–1302 (2005).
31. Eliad, L., Salitra, G., Soffer, A. & Aurbach, D. On the mechanism of selective electroadsorption of protons in the pores of carbon molecular sieves. *Langmuir* **21**, 3198–3202 (2005).
32. Eliad, L. *et al.* Assessing optimal pore-to-ion size relations in the design of porous poly(vinylidene chloride) carbons for EDL capacitors. *Appl. Phys. A* **82**, 607–613 (2006).
33. Arulepp, M. *et al.* The advanced carbide-derived carbon based supercapacitor. *J. Power Sources* **162**, 1460–1466 (2006).
34. Arulepp, M. *et al.* Influence of the solvent properties on the characteristics of a double layer capacitor. *J. Power Sources* **133**, 320–328 (2004).
35. Raymundo-Pinero, E., Kierzek, K., Machnikowski, J. & Beguin, F. Relationship between the nanoporous texture of activated carbons and their capacitance properties in different electrolytes. *Carbon* **44**, 2498–2507 (2006).
36. Janes, A. & Lust, E. Electrochemical characteristics of nanoporous carbide-derived carbon materials in various nonaqueous electrolyte solutions. *J. Electrochem. Soc.* **153**, A113–A116 (2006).
37. Shanina, B. D. *et al.* A study of nanoporous carbon obtained from ZC powders (Z = Si, Ti, and B). *Carbon* **41**, 3027–3036 (2003).
38. Chmiola, J., Dash, R., Yushin, G. & Gogotsi, Y. Effect of pore size and surface area of carbide derived carbon on specific capacitance. *J. Power Sources* **158**, 765–772 (2006).
39. Dash, R. *et al.* Titanium carbide derived nanoporous carbon for energy-related applications. *Carbon* **44**, 2489–2497 (2006).
40. Urbonaite, S. *et al.* EELS studies of carbide derived carbons. *Carbon* **45**, 2047–2053 (2007).
41. Gogotsi, Y. *et al.* Nanoporous carbide-derived carbon with tunable pore size. *Nature Mater.* **2**, 591–594 (2003).
42. Chmiola, J. *et al.* Anomalous increase in carbon capacitance at pore size below 1 nm. *Science* **313**, 1760–1763 (2006).
43. Huang, J. S., Sumpter, B. G. & Meunier, V. Theoretical model for nanoporous carbon supercapacitors. *Angew. Chem. Int. Ed.* **47**, 520–524 (2008).
44. Huang, J., Sumpter, B. G. & Meunier, V. A universal model for nanoporous carbon supercapacitors applicable to diverse pore regimes, carbons, and electrolytes. *Chem. Eur. J.* **14**, 6614–6626 (2008).
45. Chmiola, J., Largeot, C., Taberna, P.-L., Simon, P. & Gogotsi, Y. Desolvation of ions in subnanometer pores, its effect on capacitance and double-layer theory. *Angew. Chem. Int. Ed.* **47**, 3392–3395 (2008).
46. Largeot, C. *et al.* Relation between the ion size and pore size for an electric double-layer capacitor. *J. Am. Chem. Soc.* **130**, 2730–2731 (2008).
47. Weigand, G., Davenport, J. W., Gogotsi, Y. & Roberto, J. in *Scientific Impacts and Opportunities for Computing* Ch. 5, 29–35 (DOE Office of Science, 2008).
48. Wu, N.-L. Nanocrystalline oxide supercapacitors. *Mater. Chem. Phys.* **75**, 6–11 (2002).
49. Brousse, T. *et al.* Crystalline MnO₂ as possible alternatives to amorphous compounds in electrochemical supercapacitors. *J. Electrochem. Soc.* **153**, A2171–A2180 (2006).
50. Rudge, A., Raistrick, I., Gottesfeld, S. & Ferraris, J. P. Conducting polymers as active materials in electrochemical capacitors. *J. Power Sources* **47**, 89–107 (1994).
51. Zheng, J. P. & Jow, T. R. High energy and high power density electrochemical capacitors. *J. Power Sources* **62**, 155–159 (1996).
52. Lee, H. Y. & Goodenough, J. B. Supercapacitor behavior with KCl electrolyte. *J. Solid State Chem.* **144**, 220–223 (1999).
53. Laforgue, A., Simon, P. & Fauvarque, J.-F. Chemical synthesis and characterization of fluorinated polyphenylthiophenes: application to energy storage. *Synth. Met.* **123**, 311–319 (2001).
54. Naoi, K., Suematsu, S. & Manago, A. Electrochemistry of poly(1,5-diaminoanthraquinone) and its application in electrochemical capacitor materials. *J. Electrochem. Soc.* **147**, 420–426 (2000).
55. Arico, A. S., Bruce, P., Scrosati, B., Tarascon, J.-M. & Schalkwijk, W. V. Nanostructured materials for advanced energy conversion and storage devices. *Nature Mater.* **4**, 366–377 (2005).
56. Choi, D., Blomgren, G. E. & Kumta, P. N. Fast and reversible surface redox reaction in nanocrystalline vanadium nitride supercapacitors. *Adv. Mater.* **18**, 1178–1182 (2006).

57. Machida, K., Furuuchi, K., Min, M. & Naoi, K. Mixed proton–electron conducting nanocomposite based on hydrous RuO_2 and polyaniline derivatives for supercapacitors. *Electrochemistry* **72**, 402–404 (2004).

58. Toupin, M., Brousse, T. & Belanger, D. Charge storage mechanism of MnO_2 electrode used in aqueous electrochemical capacitor. *Chem. Mater.* **16**, 3184–3190 (2004).

59. Sugimoto, W., Iwata, H., Yasunaga, Y., Murakami, Y. & Takasu, Y. Preparation of ruthenic acid nanosheets and utilization of its interlayer surface for electrochemical energy storage. *Angew. Chem. Int. Ed.* **42**, 4092–4096 (2003).

60. Miller, J. M., Dunn, B., Tran, T. D. & Pekala, R. W. Deposition of ruthenium nanoparticles on carbon aerogels for high energy density supercapacitor electrodes. *J. Electrochem. Soc.* **144**, L309–L311 (1997).

61. Min, M., Machida, K., Jang, J. H. & Naoi, K. Hydrous RuO_2/carbon black nanocomposites with 3D porous structure by novel incipient wetness method for supercapacitors. *J. Electrochem. Soc.* **153**, A334–A338 (2006).

62. Wang, Y., Takahashi, K., Lee, K. H. & Cao, G. Z. Nanostructured vanadium oxide electrodes for enhanced lithium-ion intercalation. *Adv. Funct. Mater.* **16**, 1133–1144 (2006).

63. Naoi, K. & Simon, P. New materials and new configurations for advanced electrochemical capacitors. *Electrochem. Soc. Interf.* **17**, 34–37 (2008).

64. Fischer, A. E., Pettigrew, K. A., Rolison, D. R., Stroud, R. M. & Long, J. W. Incorporation of homogeneous, nanoscale MnO_2 within ultraporous carbon structures via self-limiting electroless deposition: Implications for electrochemical capacitors. *Nano Lett.* **7**, 281–286 (2007).

65. Kazaryan, S. A., Razumov, S. N., Litvinenko, S. V., Kharisov, G. G. & Kogan, V. I. Mathematical model of heterogeneous electrochemical capacitors and calculation of their parameters. *J. Electrochem. Soc.* **153**, A1655–A1671 (2006).

66. Amatucci, G. G., Badway, F. & DuPasquier, A. in *Intercalation Compounds for Battery Materials* (ECS Proc. Vol. 99) 344–359 (Electrochemical Society, 2000).

67. Burke, A. R&D considerations for the performance and application of electrochemical capacitors. *Electrochim. Acta* **53**, 1083–1091 (2007).

68. Portet, C., Taberna, P. L., Simon, P. & Laberty-Robert, C. Modification of Al current collector surface by sol-gel deposit for carbon-carbon supercapacitor applications. *Electrochim. Acta* **49**, 905–912 (2004).

69. Talapatra, S. *et al.* Direct growth of aligned carbon nanotubes on bulk metals. *Nature Nanotech.* **1**, 112–116 (2006).

70. Taberna, L., Mitra, S., Poizot, P., Simon, P. & Tarascon, J. M. High rate capabilities Fe_3O_4-based Cu nano-architectured electrodes for lithium-ion battery applications. *Nature Mater.* **5**, 567–573 (2006).

71. Jang, J. H., Machida, K., Kim, Y. & Naoi, K. Electrophoretic deposition (EPD) of hydrous ruthenium oxides with PTFE and their supercapacitor performances. *Electrochim. Acta.* **52**, 1733 (2006).

72. Cambaz, Z. G., Yushin, G., Osswald, S., Mochalin, V. & Gogotsi, Y. Noncatalytic synthesis of carbon nanotubes, graphene and graphite on SiC. *Carbon* **46**, 841–849 (2008).

73. Tsuda, T. & Hussey, C. L. Electrochemical applications of room-temperature ionic liquids. *Electrochem. Soc. Interf.* **16**, 42–49 (2007).

74. Balducci, A. *et al.* High temperature carbon–carbon supercapacitor using ionic liquid as electrolyte. *J. Power Sources* **165**, 922–927 (2007).

75. Balducci, A. *et al.* Cycling stability of a hybrid activated carbon//poly(3-methylthiophene) supercapacitor with *N*-butyl-*N*-methylpyrrolidinium bis(trifluoromethanesulfonyl)imide ionic liquid as electrolyte. *Electrochim. Acta* **50**, 2233–2237 (2005).

76. Balducci, A., Soavi, F. & Mastragostino, M. The use of ionic liquids as solvent-free green electrolytes for hybrid supercapacitors. *Appl. Phys. A* **82**, 627–632 (2006).

77. Endres, F., MacFarlane, D. & Abbott, A. (eds) *Electrodeposition from Ionic Liquids* (Wiley-VCH, 2008).

78. Faggioli, E. *et al.* Supercapacitors for the energy management of electric vehicles. *J. Power Sources* **84**, 261–269 (1999).

79. Chmiola, J. & Gogotsi, Y. Supercapacitors as advanced energy storage devices. *Nanotechnol. Law Bus.* **4**, 577–584 (2007).

80. Portet, C., Yushin, G. & Gogotsi, Y. Electrochemical performance of carbon onions, nanodiamonds, carbon black and multiwalled nanotubes in electrical double layer capacitors. *Carbon* **45**, 2511–2518 (2007).

Acknowledgements

We thank our students and collaborators, including J. Chmiola, C. Portet, R. Dash and G. Yushin (Drexel University), P. L. Taberna and C. Largeot (Université Paul Sabatier), and J. E. Fischer (University of Pennsylvania) for experimental help and discussions, H. Burnside (Drexel University) for editing the manuscript and S. Cassou (Toulouse) for help with illustrations. This work was partially funded through the Department of Energy, Office of Basic Energy Science, grant DE-FG01-05ER05-01, and through the Délégation Générale pour l'Armement.

Future lab-on-a-chip technologies for interrogating individual molecules

Harold Craighead[1]

Advances in technology have allowed chemical sampling with high spatial resolution and the manipulation and measurement of individual molecules. Adaptation of these approaches to lab-on-a-chip formats is providing a new class of research tools for the investigation of biochemistry and life processes.

A range of powerful technologies exists for observing and identifying individual molecules. Well-established forms of microscopy such as electron microscopy and optical fluorescence microscopy have allowed us to image and, with the appropriate labels, identify the chemical nature of individual molecules. More recently, scanning probe microscopes have expanded the possibilities for observing and interacting directly with individual molecules. The fine probe of a scanning system can localize electrical measurements to a selected molecule, or measure the mechanical properties of a single biopolymer. Modern material-processing technologies, analogous to those that have been so successful in converting electronics to a 'chip-based' technology, are being explored for possible chemical or biological research lab-on-a-chip approaches[1].

Interest in scaled-down analytical processes, combined with advances in microfluidics, is motivating various chip-based methods in which analyses can be carried out more rapidly and at lower cost via small-scale systems than with current laboratory bench-scale methods. The new research approaches discussed below are motivated by the possibility of observing new phenomena or obtaining more detailed information from biologically active systems. For this, a class of research systems is evolving that uses a range of new physical and chemical approaches to biomolecular analysis. The complexity of life processes and the richness of molecular biology provide fertile ground for research with these new lab-on-a-chip approaches. The development of these new chip-based technologies is changing the nature of the questions that we can ask and for which we can seek experimental answers at the molecular level.

The developing chips are formed using technologies for inorganic-device processing combined with synthetic chemistry and biochemistry. The chips are beginning to integrate electrical, optical and physical measurements with fluid handling to create a new class of functional chip-based systems. The systems provide spatial localization that is relevant to observing, for example, the function of a single active enzyme, the activity of a single receptor on a cell surface, or the release of molecules from a single vesicle exocytotic event in an immune-system cell. A long-term result of this research could be a new class of fluid-handling chip systems engineered to use or analyse individual molecules. This could form the basis for ultra-sensitive sensors and medical diagnostic systems.

Molecular imaging and probes

The scanning probe instruments, such as the atomic force microscope and the scanning tunnelling microscope, have allowed the execution of a class of experiments in which individual molecules can be manipulated and probed. They are providing insights into biomolecular systems previously addressed only by indirect investigations. Scanning tunnelling microscopy can image molecules on surfaces and directly observe their conformation and structure[2,3]. Atomic force microscopy (AFM) can image molecules in a natural hydrated condition and capture images while the molecules are functioning[4-6]. The AFM has also been used to observe the folding and unfolding of individual protein molecules[7-9]. Observing the function of active molecules in a condition as close as possible to *in vivo* circumstances can provide new insights into how living systems function at the molecular level.

Ashkin was the first to use gradient forces from optical beams to trap particles and exert controlled forces on small beads, viruses and bacteria[10,11]. The technique, which came to be known as 'optical tweezers', proved to be a powerful tool for applying controlled forces to individual molecules[12-15]. Typical experiments involve chemically binding a molecule of interest to a bead and chemically linking the other end of the molecule to a surface or another bead. The optical tweezers can be used in fluids and, as a result, can be used for biomolecules in a natural hydrated and functioning condition. This permits force–distance measurements as a biopolymer is unfolded, or the investigation of the breaking of bonds in specific reactions. This technique is now widely used in biophysical studies of single molecules. It has done a great deal to move researchers towards thinking of single molecules as accessible objects that can be selected and studied for their individual properties. It has also helped motivate chip-based approaches for individual-molecule analysis.

Patch-clamp technology, another form of localized electrical probe, revolutionized our ability to study active ion channels in cell membranes[16]. The technology uses a glass pipette drawn down to form a micrometre-scale aperture that can make a tight, electrically insulating seal to a cell membrane. With the electrical current conducted and measured through the pipette, electrical charge is constrained to conducting paths inside the sealed region of the pipette tip. This was a revolutionary technique that allowed the observation of single-molecule ion channels in living cell membranes.

Although powerful and productively used in experimental research, the single probe methods use tedious manipulation of the individual probes and do not take advantage of the integration, parallelism and automation available from a chip-based system. They do, however, set the stage for migration of these approaches to chip-based methods, and possibly further integration into research systems. Researchers are beginning to develop and use engineered devices that are often fabricated in large arrays for accessing individual molecules.

Developing technologies for lab-on-a-chip integration

Technologies exist that unite lithographic approaches for processing hard materials with soft material processing, fluidics and biochemical patterning. These technologies, which are related to those used in commercial device manufacturing, reliably produce systems with feature

[1]Applied and Engineering Physics, 205 Clark Hall, Cornell University, Ithaca, New York 14853, USA.

When citing this review, please cite the original version as shown on the contents page of this chapter.

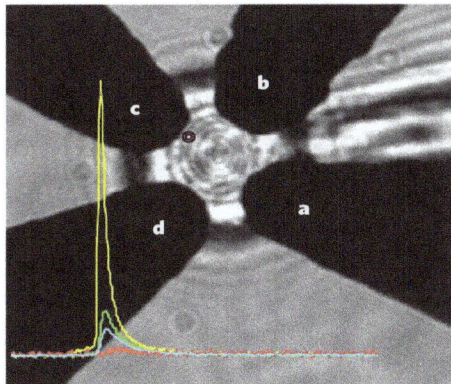

Figure 1 | Optical micrograph of a 4-electrode electrochemical detector array on which a chromaffin cell has been placed. The superimposed trace is the measured current as a function of time at electrodes **a–d** (**a**, red; **b**, green; **c**, yellow; **d**, blue). The small circle drawn near electrode **c** indicates the location of the fusion pore opening, calculated from the relative current collected from the four electrodes[41].

sizes of less than 100 nm[17–22]. They are also used to create complex and massively parallel electronic devices, creating billions of essentially identical devices in a chip format. Optical sources and detectors have also been miniaturized and made parallel, and are available in consumer products such as video cameras and displays. This is a powerful and well-developed technology that can be exploited in the research arena today, and possibly in the future in a new class of highly-functional chip-like devices for biochemical analysis.

At this stage of research, many different materials are being explored and used. Polymers are attractive for many microfluidic uses because they are easily manufactured by embossing or moulding[23]. They can also be bonded to other surfaces and are therefore often used to form a fluid channel on another surface, such as silicon, on which an electronic device has been formed. Single-molecule approaches using electrical or optical detection may create other material requirements with regard to electrical or optical properties. Semiconductors and metals are obviously necessary components of electrical detection schemes. Semiconductor nanowires and carbon nanotubes, for example, are being studied as sensor components[24–26]. Mechanical devices are also being explored and integrated in fluid systems[27–30]. In addition, porous media for sample concentration and filtration can be formed in fluid channels[31]. The situation is nothing like that of silicon microelectronics technology, in which the technology is mature and material technologies are monolithic. However, some common materials and processing techniques are evolving, often on the basis of silicon compounds and polymers, that suggest the possibility for greater integration of fluid handling with optical, electrical and mechanical devices.

Electrical chips with high spatial resolution

The development of the planar patch clamp is an example of the conversion of a very successful probe technique to a chip-type device[32–39]. In this approach, rather than using the narrow aperture of a drawn glass pipette, an electrical probe is formed by an etched opening, of the order of a micrometre, that can be readily fabricated in a thin membrane formed on a planar surface. The macroscopic electrodes are then placed on opposite sides of the aperture, with all current forced to traverse the narrow aperture. The entity of interest, typically an ion channel in a lipid layer spanning the aperture, is thus electrically isolated. Schmidt, Mayer and Vogel[32] demonstrated such a device in a silicon nitride membrane supported on an etched silicon wafer, used to observe the activity of a single voltage-gated ion channel of the peptide alamethicin integrated in a lipid bilayer. The use of a fabrication approach with a lithographically defined aperture formed a reliable aperture size, and opened up the possibility of making arrays of many apertures for localized electrochemical measurement. Although they do not replace the traditional

moveable probe technique, microfabricated planar patch-clamp-like devices such as this are being used. Their planar geometry allows them to be engineered into sensor configurations.

Electronic-chip-fabrication technology can be used to create arrays of electrodes for probing discrete electrical events in living systems. Lindau's group, for example, described an electrode array for spatio-temporal resolution of single exocytotic events in living cells[40,41]. Using a patterned metallic electrode array, on which the cells were placed, electrochemical signals were recorded from the oxidation of catecholamines released by the cell during events in which a vesicle contained in the cell fused with the cell membrane, releasing its contents. By measuring the relative strengths of the current signals from each electrode, the researchers could calculate the position of each individual exocytotic event as well as the total quantity of ions released and the time taken by the event. This level of quantitative detail is difficult to obtain by other methods and, again, the format allows for scaling the number of groups of identical electrodes for measurements, in principal, on large arrays of cells. This is of interest for a number of basic cell-biology studies, and is also of potential utility in obtaining more detailed information about the effects of potential drugs on cellular function.

Control of the structure and design of the arrays also enables integration with complementary measurement approaches. With the same types of electrode array (Fig. 1), Lindau and colleagues were also able to record simultaneously fluorescence images of fusion-pore events in the cells. This brought to bear the full capability of chemically selective fluorescent dyes and optical imaging to verify the interpretation of the electrochemical imaging. This also indicates the value of this type of lab-on-a-chip in terms of engineering the planar devices for integration for multiple analytical processes.

Of course, the capabilities of integrated electronics extend well beyond those of passive electrodes; researchers can incorporate active electronic devices to allow more complex functions, such as signal amplification, to be carried out. Many groups have used microfabricated electrode arrays for extracellular recording of action potentials from

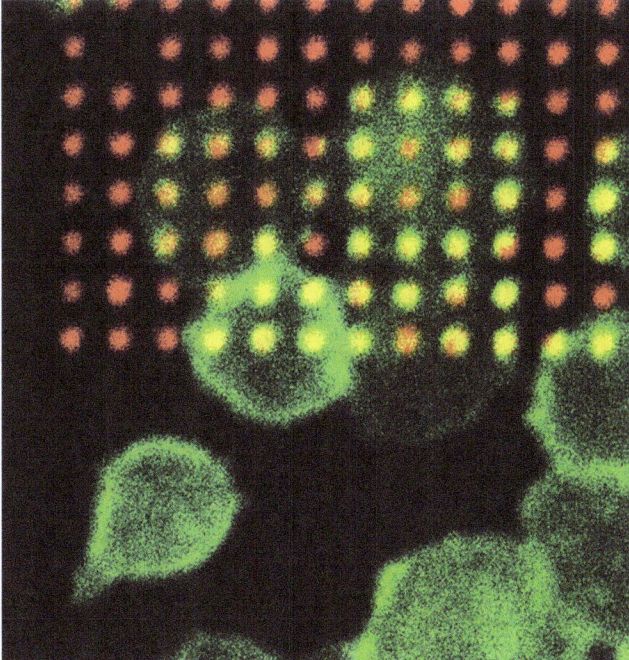

Figure 2 | Fluorescence micrograph of micrometre-size patterned lipid bilayers containing specific ligands used to cluster mast-cell receptors. Clustered receptors (green) stimulate transmembrane signalling events, and the patterns (red) allow visualization of spatial regulation of reorganizing cellular components. Co-concentrated red and green fluorescence appears yellow[45,46]. (Image courtesy of B. Baird, Cornell University, USA.)

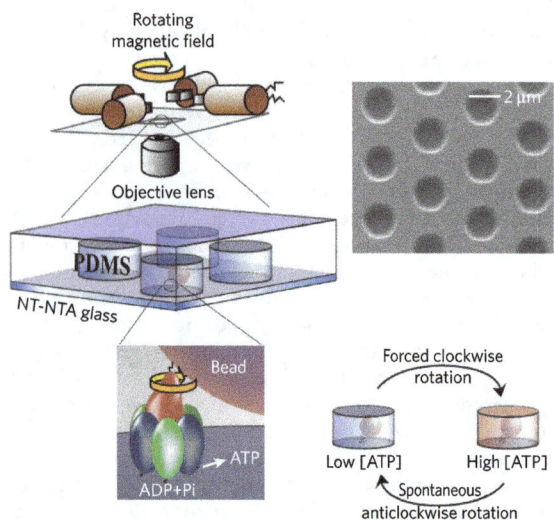

Figure 3 | Schematic of F₁ATPase rotary motor enzyme to which a magnetic bead is attached. The bead can be used to drive or observe the motion of the motor. Each enzyme is isolated in a reaction chamber of radius ~1 μm (shown schematically and in an electron micrograph) that contains the reaction products[47]. (Image courtesy of H. Noji, University of Tokyo, Japan.)

neurons cultured on the electrode arrays. The Fromhertz group has used arrays of electrodes connected to on-chip arrays of transistors to amplify voltage signals near their source in order to improve the signal-to-noise ratio[42]. With the provision of active electronics to the chip, additional signal processing and logic functions could be incorporated into the device.

In addition to the integration of electrical function with optical analysis, the chip-oriented approach can be coupled with biochemical patterning to interface more effectively to biological systems. For example, James *et al.* used methods for patterning proteins in registry with electrode arrays[43,44]. The patterned proteins guided the development of neuronal-cell growth on the electrode surface that was used to record signals from selected points in an organized cell culture. The ability to use patterned cell-growth factors, chemotactic agents or selective binding factors, such as antibodies, provides a method for interfacing the powerful analytical capabilities engineered into the inorganic devices with organized cellular systems.

As noted in the Review on cell biology in this issue (page 403), patterning of active biomaterials on surfaces can be used for cell-based lab-on-a chip assays. However, patterning at subcellular dimensions makes it possible to orient cells with respect to chip elements. The patterns themselves can be used to investigate cell responses to localized stimuli and to observe molecular-scale responses. For example, Orth *et al.* used a high-resolution method of patterning to create patterned lipid bilayers containing antigens to mammalian immune-system cells[45]. The cell-surface receptors clustered in the same pattern and resulted in cell activation. Wu *et al.* used this approach to investigate targeting in transmembrane signalling events[46]. Using fluorescence microscopy with exogeneous or genetically encoded fluorescent probes, the stimulated redistribution membrane, cytoskeletal and cytoplasmic components were tracked (Fig. 2). This allowed the testing of hypotheses about the formation of so-called lipid rafts, and demonstrated decoupling of the inner and outer leaflets of the cell membrane. Similar to the previous examples, the ability to create large numbers of essentially identical chemical stimuli allowed the system to be replicated and presented to numerous cells. The use of the patterned biochemical chip in conjunction with fluorescence microscopy provided an unambiguous visualization of the cellular response to defined stimuli and the capability to quantify it.

Planar devices for single-molecule observation

Simple chips with nanoscale features can be used as windows onto the activity of single molecules, extending the capability for observation of life processes to the molecular scale. The Noji group created an array of individual femtolitre chemical chambers on a chip for studies of the activity of individual F₁ATPase enzyme molecules bound to a surface[47,48]. In this work, the group formed small, optically transparent polymeric chambers to isolate the chemistry associated with an individual enzyme (Fig. 3). The ATPase enzyme is a rotary motor, and by attaching a magnetic bead to the rotating part of the molecule, the rotation of the molecule could be observed optically. The torque on the magnetic bead could be controlled by application of a rotating magnetic field. The researchers were able to count the rotations of the motor and calculate the efficiency as the enzyme motor consumed its ATP fuel. Because the volume of solution containing ATP was very small, the individual enzyme's activity in depleting the available energy source could be observed in reasonable experimental timescales. In a demonstration of incredible control, the individual motors could be driven in the opposite direction by the rotating magnetic field to synthesize countable ATP molecules.

This experiment provides unique insights into the activity of functioning biomolecules, directly addressing questions of kinetics, rates and efficiencies. This is an example of a potentially revolutionary class of chip-based devices in which the observation and control of large numbers of individual molecules are used as detectors of identifiable individual chemical events or components of a chemical synthesis system.

Active enzymes can be optically isolated for measurement by techniques that allow the observation of individual reaction events at high concentrations. The previously discussed approach used physical confinement of reactants to limit the reaction to a small volume. With optical confinement to a small volume surrounding an active enzyme, the observation of optically stimulated or interrogated processes can be localized to observe chemical activity specific to the isolated enzyme. Levene *et al.* described an array of subwavelength-diameter metallic apertures (Fig. 4) that confine light to dimensions well below the diffraction limit in three dimensions[49]. Light incident on this metallic structure, which can be considered as a metallic waveguide operated at a frequency below the cutoff frequency, is rapidly attenuated in the direction of propagation. Because it has no propagating waveguide modes, it is termed a zero-mode waveguide. With a diameter of a few tens of nanometres, this attenuation length becomes less than the film thickness. The transverse dimensions of the light are, of course, limited by the dimensions of the aperture. Other means of optical confinement include the use of metallic tips to concentrate optical excitation[50].

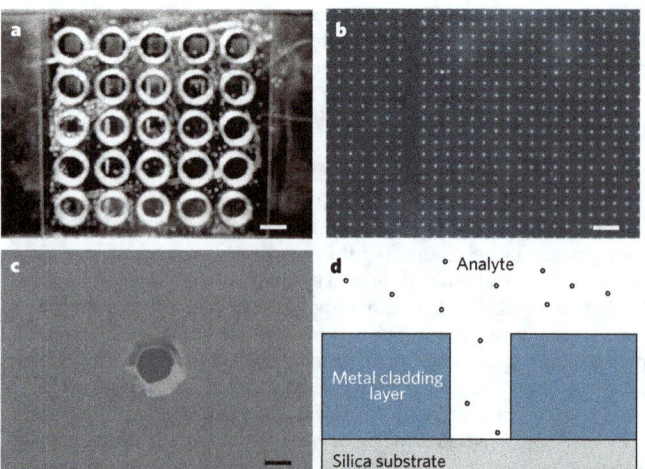

Figure 4 | Arrays of metallic apertures used for optical observation of individual molecules. Array of polymer wells containing arrays of zero-mode waveguides (**a**) shown in electron micrograph (**b**) and in higher magnification (**c**). Also shown (**d**) is a schematic cross-section of the metallic aperture with analyte solution on the metal side[49].

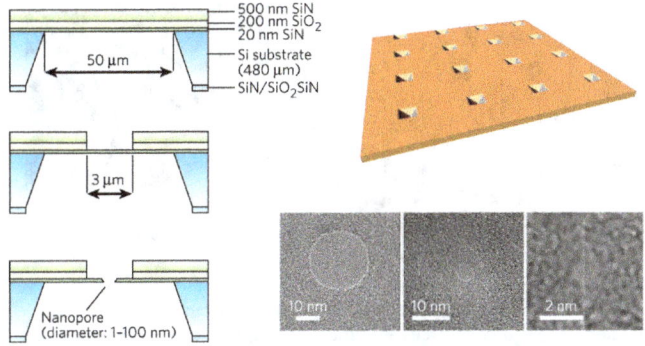

Figure 5 | Nanopores used for DNA translocation studies. Left: cross-sectional schematic of an engineered nanopore in silicon nitride showing three phases of the pore fabrication. Top right: AFM image of an array of engineered pores formed on a chip. Bottom right: transmission electron micrograph of three different pores. (Images courtesy of C. Dekker[83], University of Delft, The Netherlands, and the American Chemical Society.)

The net effect is an optical excitation volume of the order of zeptolitres (10^{-21}), significantly smaller than could, for example, be obtained by total internal reflection illumination, which can confine light to the evanescent field in one direction only. If the light is used to stimulate fluorescence in molecules in a solution covering the zero-mode waveguide at a concentration, C, the average number of molecules, N, in the optical excitation volume, V, is $N=CV$. With an excitation volume of 10^{-21} litres, $N=1$ at a concentration of 10^{21} litres^{-1}, or 1.7 mM. The behaviour of a single optically labelled entity can therefore be observed at high concentrations. If the fluorescent molecule is freely diffusing in the liquid, then the residence time in the optical excitation volume is determined by the diffusion coefficient of the molecule. By observing this time for individual molecules, the concentration of entities with different diffusion coefficients can be determined. This can be used, for example, to measure the degree of oligomerization of a biopolymer[51]. The differences in temporal fluorescent behaviour are even more pronounced in cases in which a species is permanently bound, as has been observed for the enzymatic synthesis of double-stranded DNA by DNA polymerase[49]. In this case, non-bound fluorescent species are only transiently fluorescent during their diffusive motion through the excitation region. Bound species are fluorescent for times limited by fluorescent bleaching or other processes for termination of the fluorescence. With an array, a large number of independent observation volumes can be used for the observation of enzymatic or other optically distinguishable chemical events[49–53].

Devices consisting of nanoscale pores in an electrically insulating layer are being pursued as probes of molecular structure[54–59]. In these devices, ion current passing through the narrow pore is modulated, as it passes through the pore, by the presence or nature of a large biopolymer molecule such as DNA. Naturally occurring membrane proteins such as α-haemolysin have well-defined pore dimensions, of the order of a nanometre. Engineered pores, fabricated by various forms of etching, can form similar apertures in an insulating membrane. Figure 5 shows a schematic and electron micrograph, from the Dekker group, showing apertures a few nanometres in diameter formed in a silicon nitride membrane[56]. When DNA in solution is electrically driven through such a pore, the presence of the DNA influences the passage of other mobile ions in solution through the narrow pore, and the transit of the DNA can be detected by changes in the current measured through the pore. In the simplest case, if the electrically driven speed of the molecule is known, the length of an extended DNA molecule can be inferred from the transit time. Confirmation of the molecule is observed as discrete differences in the current levels corresponding to the condition of a single strand — or a double, or triple, and so on — of a folded DNA molecule passing through the pore. Because of the strong motivation for rapid genetic sequencing or analysis, efforts are under way to extract the identity of DNA bases by resolvable electrical differences as they pass through the pore, but the resolution requirements, signal-to-noise ratio in the current signal, and the subtle differences in electrical signatures of different bases make this a difficult task. The nanopores, however, are simple but powerful biophysical probes of molecular confirmation, and a means of observing and controlling the forces on biopolymers in fluid.

Single-molecule optical analysis in flowing systems

Optical techniques are being developed for observing, detecting, analysing and quantifying single molecules. Labelling a molecule of interest with a bright fluorescent dye molecule or luminescent semiconductor particle allows selected molecules to be located and tracked by optical microscopy. Combining chemically-specific binding with the label makes it possible to identify a specific molecular species. The general approach of selective fluorescent labelling is widely used in imaging and immuno-assays, and, as a result, labelling chemistry is well developed. Adapting the fluorescent approaches to analysis requires integration with fluidics and dealing with the noise limits inherent in single-molecule detection. In addition to the chemical identification imparted by specific binding chemistry, single-molecule methods can also be used to identify other features of a molecule, such as the diffusion rate, electrophoretic mobility, or rate of passage through a mechanical constriction or pore. In appropriate systems, the single-molecule approach can provide unique possibilities in quantification and dealing with limited samples, such as one from a single cell. Further to identification of chemical species or character, a significant application of single-molecule analysis is geared towards extracting genetic information from nucleic acids. A set of techniques is being developed for analysing individual molecules, with the goal of more rapid genetic or RNA expression analysis.

Optical approaches allow molecules of interest to be highlighted by labelling, but to make single-molecule approaches practical this must be coupled to microfluidic systems to efficiently deliver molecules to the optical analytical system[60–70]. Confocal optical approaches provide a direct method with which to excite and detect fluorescence from a microfluidic system. A basic system is shown in the schematic of Fig. 6. Current research approaches tend to use large-scale microscopes, lasers

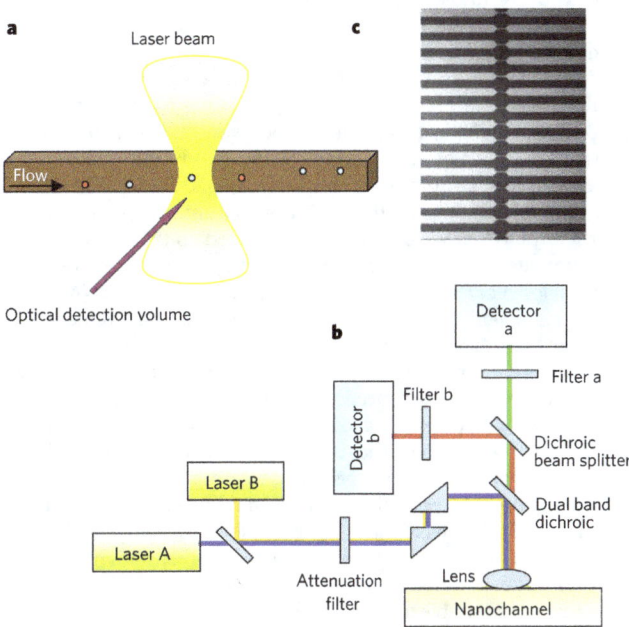

Figure 6 | Fluidic channel system for single-molecule optical measurements. a, Schematic of a fluid channel with optical excitation and detection volume. **b,** Schematic of optical set up. **c,** Image of 15 parallel fluid channels (pale grey) with constricted analysis region (centre)[61].

and detectors because of the availability of these systems. It should be clear, however, that the optical systems could also be miniaturized into more integrated and functional chips.

One unique single-molecule approach is to count individual molecules as they pass an interrogation region. With spectrally identifiable labels, the exact concentrations of different species could be obtained by directly counting the numbers of each molecular species. As signal-to-noise ratio is important, it is desirable to limit the size of the excitation volume to reduce background from scattering or intrinsic fluorescence of unlabelled species in the excitation volume. This motivates the use of confinement of the excitation by limiting either the optical illumination size or the physical size of the channel. Decreasing the optical excitation below the channel width reduces the detection efficiency and makes it possibile that not all molecules in the solution will be detected. Reducing the physical size of the channel improves the optical signal-to-noise ratio, and increases the interaction with the channel surfaces. The optimum in the situation is dictated by the nature of the analysis to be performed, including the concentration range of the molecules, the volume of sample and the strength of surface interactions. It seems likely that devices will be designed for specific applications.

Single-molecule approaches make it possible to detect single binding events, in addition to simply counting differentially labelled species. For example, individual binding events can be detected by simultaneously detecting the presence of fluorescence from two differently labelled molecules. Labels can be formed by pairs of molecules in which excitation energy is transferred from the donor member of the pair to the acceptor of fluorescent excitation in a fluorescence-resonance-energy-transfer or Förster-resonance-energy-transfer (FRET) process. In this case, the emission will be observed only when the binding takes place, bringing donor and acceptor into close proximity. In a microfluidic system in which the molecule of interest has a charge and an electric field is used to drive the molecule, a measure of the electrophoretic mobility can be determined from the transit time through a channel of known length[64]. Changes in mobility could indicate the difference between an unbound labelled capture molecule and a labelled capture molecule bound to a protein. The combination of microfluidics and electrically or pressure driven fluid systems creates integrated design possibilities. Parallelism from many fluid and optical channels could be designed with modern optoelectronic technology. Planar waveguide excitation, out of plane diffractive optics, laser arrays and detector arrays are all readily available to be adapted for miniaturized systems.

In addition to counting and determining the chemical identity of molecules from specific binding chemistry, the conformation and physical properties of individual molecules can be investigated in nanofluidic systems[71–80]. This can be considered to be the chip-based analogue of the probe and optical-tweezer techniques discussed above. A number of studies have been done on the sorting of DNA molecules by size in nanostructures. When associated with restriction digests, the sizes of the DNA fragments contain information on the sequence information in the original longer DNA molecules. Rapid measurement of DNA fragment size was demonstrated in a series of nanostructures with dimension smaller than the radius of gyration of the DNA molecule in free solution. In the so-called entropic traps[75,78], the rate at which a molecule could enter the constriction was length-dependent, and this dependence was used to separate bands of DNA in electrophoretically driven microfluidic channels with nanoscale constrictions. The separation was much more rapid than that achievable with conventional gel-based techniques.

Similar entropic forces in nanochannels have been used to control forces on individual molecules and to elongate them for study[71–80]. When DNA is driven into a fluid channel with transverse dimensions much less than the radius of gyration, or, even more critically, when the dimensions become comparable to the persistence length of the polymer, the confinement exerts forces on the molecule that push it from the constrained region[73] (Fig. 7). The effect of the entropic forces can be used to measure effectively the length of the molecule by the rate of passage into the constrained region[76]. Control of the molecular confirmation also allows analysis of DNA in the context of the overall chromosome.

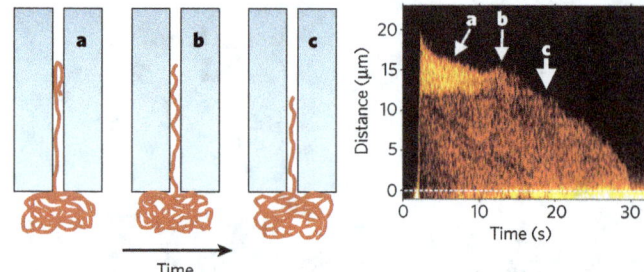

Figure 7 | Schematic and images of DNA retraction from a nanochannel. Diagram (left) showing three different time stages (**a**–**c**) of the entropically driven motion of a single DNA molecule at the interface between a microchannel and a nanochannel. After being electrophoretically driven into the channel as a 'hair pin' loop, the molecule unfolds as it is entropically pulled from the nanochannel by forces acting only on the long end. The image on the right shows stacked fluorescence images of a molecule as a function of time, demonstrating the recoiling and straightening process. The looped (**a**) and then the straightened (**b**, **c**) condition of the molecule are indicated. Once straightened, the molecule can be manipulated and positioned in the channel as an unfolded linear molecule[73].

Towards this end, Austin's group demonstrated the activity of restriction enzymes, cutting DNA at prescribed sequence locations, in an extended single molecule in a nanochannel[77].

Because of interest in their underlying sequence, DNA and RNA represent important targets for single-molecule studies, but the same concepts and physical methods of characterizing biomolecules are being extended to proteins and other compounds. In cases in which chemical differences accompany the physical differences, the situation is more complex, complicating the interpretation of phenomena for analysis. The approach for analysis, however, combining microfluidics, nanoscale structures and optical detection, creates a range of opportunities for molecular analysis.

Outlook for highly functional integrated systems

The current types of research-chip device comprise fairly simple structures, as we have seen above, such as two-dimensional arrays of apertures or reaction chambers to isolate individual molecules for study. In these simple initial devices the advantages of massively parallel replication of identical units with 10^5, 10^6 or more individual reaction or observation areas can be seen. In fluid channel systems that permit one-dimensional arrays, the degree of parallelism could easily be ~10^3. It is clear that the scale and levels of integration are compatible with active optical and electronic devices in today's integrated electronic and optoelectronic systems, and carrying out millions of parallel optical or electronic measurements is a realistic possibility. Because optical beams can traverse free space and fluid systems can be made transparent, the integration with external optical systems is possibly the most straightforward initial direction of integration. Electrical techniques require direct contact with the analyte and generally also require wires for interconnects, which creates more challenges for integration. The advances in this area are likely to be important for determining the breadth of utility of such lab-chip devices.

Opportunities for single-molecule analysis include the ability to characterize small volumes of complex mixtures such as one would obtain from a single cell. The ability to characterize the differences in populations of cells could be greatly advanced by lab-on-a-chip approaches that could rapidly quantify the levels of a set of proteins of interest in a selected cell at a particular time. As previously noted, the ability to handle cells and present controlled stimuli to cells is developing along with techniques that could help to analyse the function of these cells. In a similar way, rapid detection of the expression of RNA or the genetic make-up of individual cells could be accessed by such analytical systems. The capacity for sensitive and rapid analysis of the complex chemical content of body fluids could enable new medical diagnostic approaches, identifying the markers for disease at a stage at which treatment can be most effective.

The goal of substantially more rapid and inexpensive sequencing of entire genomes is now well established. The '$1,000 genome' is now a stated target of the US National Institutes of Health[81]. A range of approaches to this goal is being considered, including single-molecule approaches. The combination of greater understanding of biopolymer physics and the ability to integrate fluidics is presenting new opportunities and research directions.

A valuable device would allow rapid, less invasive analysis of complex biological fluids for medical diagnosis and monitoring of therapies. A range of technologies is being directed towards this general goal. The tool box of lab-on-a-chip methods, and particularly single-molecule approaches, may, in the long term, enable the engineering of a number of highly functional devices for specific diagnostic targets.

Biological and medical applications are viewed as areas of high potential impact of single-molecule approaches. Perhaps a biological analogy could provide insights into how engineered systems with controllable enzymes could be used to 'manufacture' small amounts of a large number of compounds in combinatorial approaches. Large libraries of such synthesized compounds could then be studied by comparably miniaturized systems. In any case, the existence of natural biological systems that have the ability to sense remarkably small numbers of molecules in chemically noisy environments provides guidance on how to best engineer artificial systems to operate most effectively in environments in which the statistics of signalling with small numbers of molecules[82] and time limits associated with molecular motion are the fundamental limits of operation. Living cells have evolved clearly viable approaches to sensing and reacting to their environment on the basis of the analysis of single-molecule events.

1. Manz, A. et al. Miniaturized total chemical analysis systems — a novel concept for chemical sensing. Sens. Actuators B 1, 244–248 (1990).
2. Muller, D. J., Amrein, M. & Engle, A. Adsorption of biological molecules to a solid support for scanning probe microscopy. J. Struct. Biol. 119, 172–188 (1997).
3. Stipe, B. C. A variable temperature scanning tunneling microscope capable of single-molecule vibrational spectroscopy. Rev. Sci. Inst. 70, 133–136 (1999).
4. Hansma, H. & Hoch, J. Biomolecular imaging with the atomic force microscope. Annu. Rev. Biophys. Biomol. Struct. 23, 115–134 (1994).
5. Bustamante, C. & Keller, D. Scanning force microscopy in biology. Phys. Today 48, 32–38 (1995).
6. Reif, M. et al. Single molecule force spectroscopy on polysaccharides by atomic force microscopy. Science 275, 1295–1297 (1997).
7. Reif, M. et al. Reversible unfolding of individual titin immunoglobulin domains by AFM. Science 276, 1109–1112 (1997).
8. Carrion-Vazquez, M. et al. Mechanical and chemical unfolding of a single protein: a comparison. Proc. Natl Acad Sci. USA 96, 3694–3699 (1999).
9. Fisher, T. E. Stretching single molecules into novel confirmations using the atomic force microscope. Nature Struct. Biol. 7, 719–724 (2000).
10. Ashkin, A. et al. Observation of a single-beam gradient force optical trap for dielectric particles. Opt. Lett. 11, 288–290 (1986).
11. Ashkin, A. et al. Optical trapping and manipulation of single cells using infrared laser beams. Nature 330, 769–771 (1987).
12. Wang, M. D. et al. Stretching DNA with optical tweezers. Biophys. J. 72, 1335–1346 (1997).
13. Grier, D. G. A revolution in optical manipulation. Nature 424, 810–816 (2003).
14. Sischka, A. et al. Compact microscope-based optical tweezers system for molecular manipulations. Rev. Sci. Inst. 74, 4827–4831 (2003).
15. Mehta, A. et al. Single-molecule biomechanics with optical methods. Science 283, 1689–1695 (1999).
16. Neher E. & Sakmann, B. Single-channel currents recorded for membrane of denervated frog muscle fibres. Nature 260, 799–802 (1976).
17. Mijatovic, D., Eijkel, J. C. & van den Berg, A. Technologies for nanofluidic systems: top-down vs. bottom-up — a review. Lab Chip 5, 492–500 (2005).
18. Craighead, H. G. Nanoelectromechanical systems. Science 290, 1532 (2000).
19. Beebe, D. J., Mensing, G. A. & Walker, G. M. Physics and applications of microfluidics in biology. Annu. Rev. Biomed. Eng. 4, 261–286 (2002).
20. Quake, S. R. & Scherer, A. From micro- to nanofabrication with soft materials. Science 290, 1536–1540 (2000).
21. Schwarz, M. A. & Hauser, P. C. Recent developments in detection methods for microfabricated analytical devices. Lab Chip 1, 1–6 (2001).
22. Liu, J., Hansen, C. & Quake, S. R. Solving the "world to chip" interface problem with a microfluidic matrix. Anal. Chem. 75, 4718–4723 (2003).
23. Ng, J. M. K., Gitlin, I., Stroock, A. D. & Whitesides, G. M. Components for integrated poly(dimethylsiloxane) microfluidic systems. Electrophoresis 23, 3461–3473 (2002).
24. Zheng, G. et al. Multiplexed electrical detection of cancer markers with nanowire sensor arrays. Nature Biotechnol. 23, 1294–1301 (2005).
25. Patolsky, F. & Lieber, C. M. Nanowire nanosensors. Mater. Today 8, 20–28 (2005).
26. Rosenblatt, S. High performance electrolyte-gated carbon nanotube transistors. Nano Lett. 2, 869–972 (2002).
27. Sazonova, V. et al. A tunable carbon nanotube electromechanical oscillator. Nature 431, 284–287 (2005).
28. Burg, T. & Manalis, S. Suspended microchannel resonators for biomolecular detection. Appl. Phys. Lett. 83, 2698–2701 (2003).
29. Verbridge, S. et al. Suspended glass nanochannels coupled with microstructures for single molecule detection. J. Appl. Phys. 97, 124317–124320 (2005).
30. Ilic, B. et al. Enumeration of DNA molecules bound to a nanomechanical oscillator. Nano Lett. 5, 925–929 (2005).
31. Song, S., Singh, A. K. & Kirby, B. J. Electrophoretic concentration of proteins at laser-patterned porous membranes. Anal. Chem. 76, 4589–4592 (2004).
32. Schmidt, C., Mayer, M. & Vogel, H. A chip-based biosensor for the functional analysis of single ion channels. Angew. Chem. Int. Ed. 39, 3137–3140 (2000).
33. Suzuki, H. et al. Highly reproducible method of planar lipid bilayer reconstitution in polymethyl methacrylate microfluidic chip. Langmuir 22, 1937–1942 (2006).
34. Fertig, N. et al. Microstructured glass chip for ion-channel electrophysiology. Phys. Rev. E. Stat. Nonlin. Soft Matter Phys. 64, 040901-1–040901-4 (2001).
35. Fertig, N., Blick, R. H. & Behrends, J. C. Whole cell patch clamp recording performed on a planar glass chip: electrophysiology on a chip. Biophys. J. 82, 3056–3062 (2002).
36. Klemic, K. B. et al. Micromolded PDMS planar electrode allows patch-clamp electrical recordings from cells. Biosens. Bioelectron. 17, 597–604 (2002).
37. Groves, J. T., Ulman, N. & Boxer, S. G. Micropatterning fluid bilayers on solid supports. Science 275, 651–653 (1997).
38. Pantoja, R. et al. Bilayer reconsistution of voltage-dependent ion channels using a microfabricated silicon chip. Biophys. J. 81, 2389–2394 (2001).
39. Xu, L. et al. Ion–channel assay technologies: quo vadis? Drug Discov. Today 6, 1278–1287 (2001).
40. Dias, A. et al. An electrochemical detector array to study cell biology on the nanoscale. Nanotechnology 13, 285–289 (2002).
41. Hafez, I. et al. Electrochemical imaging of fusion pore openings by electrochemical detector arrays. Proc. Natl Acad. Sci. USA 102, 13879–13884 (2005).
42. Zech, G. & Fromhertz, P. Noninvasive neuroelectronic interfacing wth synaptically connected snail neurons immobilized on a semiconductor chip. Proc. Natl Acad. Sci. USA 98, 10457–10462 (2001).
43. James, C. D. et al. Extracellular recordings from patterned neuronal networks using planar microelectrode arrays. IEEE Trans. Biomed. Eng. 51, 1640–1648 (2004).
44. Oliva, A. A. et al. Patterning axonal guidance molecules using a novel strategy for microcontact printing. Neurochem. Res. 28, 1639–1648 (2003).
45. Orth, R. N. et al. Mast cell activation on patterned lipid bilayers of subcellular dimensions. Langmuir 19, 1599–1605 (2003).
46. Wu, M. et al. Visualization of plasma membrane compartmentalization with patterned lipid bilayers. Proc. Natl Acad. Sci. USA 101, 13798–13803 (2004).
47. Rondelez, Y. et al. Highly coupled ATP synthesis by F_1-ATPase single molecules. Nature 433, 773–777 (2005).
48. Rondelez, Y. et al. Microfabricated arrays of femtoliter chambers allow single molecule enzymology. Nature Biotechnol. 23, 361–365 (2005).
49. Levene, M. J. et al. Zero-mode waveguides for single-molecule analysis at high concentrations. Science 299, 682–686 (2003).
50. Gerton, J. M., Wade, L. A., Lessard, G. A., Ma, Z. & Quake, S. R. Tip-enhanced fluorescence microscopy at 10 nanometer resolution. Phys. Rev. Lett. 93, 180801 (2004).
51. Samiee, K. T. et al. λ-Repressor oligomerization kinetics at high concentrations using fluorescence correlation spectroscopy in zero-mode waveguides. Biophys. J. 88, 2145–2153 (2005).
52. Samiee, K. et al. Zero mode waveguides for single molecule spectroscopy on lipid membranes. Biophys. J. 90, 3288–3299 (2005).
53. Rigneault, H. J. et al. Enhancement of single molecule fluorescence detection in subwavelength apertures. Phys. Rev. Lett. 95, 117401–117404 (2005).
54. Kasianowicz, J. et al. Characterization of individual polynucleotide molecules using a membrane channel. Proc. Natl Acad. Sci. USA 93, 13770–13773 (1996).
55. Deamer, D. W. & Branton, D. Characterization of nucleic acids by nanopore analysis. Acc. Chem. Res. 35, 817–825 (2002).
56. Storm, A. J. et al. Fast DNA translocation through a solid-state nanopore. Nano Lett. 5, 1193–1197 (2005).
57. Heng, J. B. et al. Sizing DNA using a nanometer-diameter pore. Biophys. J. 87, 2905–2911 (2004).
58. Li, J. et al. DNA molecules and configurations in a solid state nanopore microscope. Nature Mater. 2, 611–615 (2003).
59. Mathe, J. et al. Nanopore unzipping of individual DNA hairpin molecules. Biophys. J. 87, 3205–3212 (2004).
60. Goodwin, P. M., Nolan, R. L. & Cai, H. Single-molecule spectroscopy for nucleic acid analysis: a new approach for disease detection and genomic analysis. Curr. Pharm. Biotechnol. 5, 271–278 (2004).
61. Stavis, S. M. Single molecule studies of quantum dot conjugates in a submicrometer fluidic channel. Lab Chip 5, 337–343 (2005).
62. Stavis, S. M. Detection and identification of nucleic acid engineered fluorescent labels in submicrometre fluidic channels. Nanotechnology 16, S314–S323 (2005).
63. Foquet, M. DNA fragment sizing by single-molecule detection in submicrometer-sized closed fluidic channels. Anal. Chem. 74, 1415–1422 (2002).
64. Stavis, S. M. Single-molecule mobility and spectral measurements in submicrometer fluidic channels. J. Appl. Phys. 98, 044903-1–044903-5 (2005).
65. Campbell, L. C. et al. Electrophoretic manipulation of single DNA molecules in nanofabricated capillaries. Lab Chip 4, 225–229 (2004).
66. Mogensen, K. B., El-Ali, J., Wolff, A. & Kutter, J. P. Integration of polymer waveguides for optical detection in microfabricated chemical analysis systems. Appl. Opt. 42, 4072–4079 (2003).
67. Foquet, M. E. et al. Fabrication of microcapillaries and waveguides for single molecule detection. Proc. SPIE 3258, 141–147 (1998).
68. Dittrich, P. S. & Manz, A. Single-molecule fluorescence detection in microfluidic channels — the Holy Grail in μTAS? Anal. Bioanal. Chem. 382, 1771–1782 (2005).

69. Tegenfeldt, J. *et al.* Micro- and nanofluidics for DNA analysis. *Anal. Bioanal. Chem.* **378,** 1678–1692 (2004).
70. Kricka, L. Revolution on a square centimeter. *Nature Biotechnol.* **16,** 513–514 (1998).
71. Tegenfeldt, J. O. *et al.* Stretching DNA in nanochannels. *Biophys. J.* **86,** 596A (2004).
72. Tegenfeldt, J. O. *et al.* The dynamics of genomic-length DNA molecules in 100-nm channels. *Proc. Natl Acad. Sci. USA* **101,** 10979–10983 (2004).
73. Mannion, J. *et al.* Conformational analysis of single DNA molecules undergoing entropically induced motion in nanochannels. *Biophys. J.* **90,** 4538–4545 (2006).
74. Kung, C.-Y. *et al.* Confinement and manipulation of individual molecules in attoliter volumes. *Anal. Chem.* **70,** 658–661 (1998).
75. Han, J. *et al.* Entropic trapping and escape of long DNA molecules at submicron constrictions. *Phys. Rev. Lett.* **83,** 1688–1691 (1999).
76. Cabodi, M. *et al.* Entropic recoil separation of long DNA molecules. *Anal. Chem.* **74,** 5169–5174 (2002).
77. Riehn, R. *et al.* Restriction mapping in nanofluidic devices. *Proc. Natl Acad. Sci. USA* **102,** 10012–10016 (2005).
78. Han, J. & Craighead, H. G. Separation of long DNA molecules in a microfabricated entropic trap array. *Science* **288,** 1026–1029 (2000).
79. Perkins, T. T., Smith, D. E., Larson, R. G. & Chu, S. Stretching of a single tethered polymer in a uniform flow. *Science* **268,** 83–87 (1994).
80. Bakajin, O. B. *et al.* Electrodynamic stretching of DNA in confined environments. *Phys. Rev. Lett.* **80,** 2737–2740 (1998).
81. Service, R. The race for $1000 genome. *Science* **311,** 1544–1546 (2006).
82. Bialek, W. & Setayeshgar, S. Physical limits to biochemical signaling. *Proc. Natl Acad. Sci. USA* **102,** 10040–10045 (2005).
83. Krapf, D. *et al.* Fabrication and characterization of nanopore-based electrodes with radii down to 2 nm. *Nano Lett.* **6,** 105–109 (2006).

Author Information Reprints and permissions information is available at npg.nature.com/reprintsandpermissions. The author declares competing financial interests: details accompany the paper on www.nature.com/nature. Correspondence should be addressed to the author (hgc1@cornell.edu).

Science and technology for water purification in the coming decades

Mark A. Shannon[1,4], Paul W. Bohn[1,2], Menachem Elimelech[1,3], John G. Georgiadis[1,4], Benito J. Mariñas[1,5] & Anne M. Mayes[1,6]

One of the most pervasive problems afflicting people throughout the world is inadequate access to clean water and sanitation. Problems with water are expected to grow worse in the coming decades, with water scarcity occurring globally, even in regions currently considered water-rich. Addressing these problems calls out for a tremendous amount of research to be conducted to identify robust new methods of purifying water at lower cost and with less energy, while at the same time minimizing the use of chemicals and impact on the environment. Here we highlight some of the science and technology being developed to improve the disinfection and decontamination of water, as well as efforts to increase water supplies through the safe re-use of wastewater and efficient desalination of sea and brackish water.

The many problems worldwide associated with the lack of clean, fresh water are well known: 1.2 billion people lack access to safe drinking water, 2.6 billion have little or no sanitation, millions of people die annually—3,900 children a day—from diseases transmitted through unsafe water or human excreta[1]. Countless more are sickened from disease and contamination. Intestinal parasitic infections and diarrheal diseases caused by waterborne bacteria and enteric viruses have become a leading cause of malnutrition owing to poor digestion of the food eaten by people sickened by water[2,3]. In both developing and industrialized nations, a growing number of contaminants are entering water supplies from human activity: from traditional compounds such as heavy metals and distillates to emerging micropollutants such as endocrine disrupters and nitrosoamines. Increasingly, public health and environmental concerns drive efforts to decontaminate waters previously considered clean. More effective, lower-cost, robust methods to disinfect and decontaminate waters from source to point-of-use are needed, without further stressing the environment or endangering human health by the treatment itself.

Water also strongly affects energy and food production, industrial output, and the quality of our environment, affecting the economies of both developing and industrialized nations. Many freshwater aquifers are being contaminated and overdrawn in populous regions—some irreversibly—or suffer saltwater intrusion along coastal regions. With agriculture, livestock and energy consuming more than 80% of all water for human use, demand for fresh water is expected to increase with population growth, further stressing traditional sources. The shift to biofuels for energy may add further demands for irrigation and refining. Alarmingly, within 30 years receding glaciers may cause major rivers such as the Brahmaputra, Ganges, Yellow (which already at times no longer runs to the sea) and Mekong rivers, which serve China, India and Southeast Asia, to become intermittent, imperilling over 1.5 billion people during the dry months[4,5]. Even industrialized nations in North America and Europe, and those in Andean countries in South America, could see major disruptions to agriculture, hydroelectric and thermoelectric generation, and municipal water supplies from reductions in snowmelt and/or loss of glaciers[6,7]. In the coming decades, water scarcity may be a watchword that prompts action ranging from wholesale population migration to war, unless new ways to supply clean water are found.

Fortunately, a recent flurry of activity in water treatment research offers hope in mitigating the impact of impaired waters around the world. Conventional methods of water disinfection, decontamination and desalination can address many of these problems with quality and supply. However, these treatment methods are often chemically, energetically and operationally intensive, focused on large systems, and thus require considerable infusion of capital, engineering expertise and infrastructure, all of which precludes their use in much of the world. Even in highly industrialized countries, the costs and time needed to develop state-of-the-art conventional water and wastewater treatment facilities make it arduous to address all the problems. Furthermore, intensive chemical treatments (such as those involving ammonia, chlorine compounds, hydrochloric acid, sodium hydroxide, ozone, permanganate, alum and ferric salts, coagulation and filtration aids, anti-scalants, corrosion control chemicals, and ion exchange resins and regenerants) and residuals resulting from treatment (sludge, brines, toxic waste) can add to the problems of contamination and salting of freshwater sources. Moreover, chemically intensive treatment methods in many regions of the world cannot be used because of the lack of appropriate infrastructure.

However, even within central Europe there has been a movement towards reducing chemical treatment via engineered 'natural' systems for drinking-water production in order to reduce residual chemicals in the distribution systems[8]. Fortunately there is much more that science and technology can do to mitigate environmental impact and increase efficiency because current treatment methods are still far from natural-law limits in their ability to separate compounds, deactivate or remove deleterious pathogens and chemical agents, transport water molecules, and move ions against concentration gradients. Our expectation is that by focusing on the science of the aqueous interface between constituents in water and the materials used for treatment, new, sustainable, affordable, safe and robust

[1]NSF STC WaterCAMPWS, University of Illinois at Urbana-Champaign, Urbana, Illinois 61801, USA. [2]Department of Chemical and Biomolecular Engineering and Department of Chemistry, University of Notre Dame, Notre Dame, Indiana 46556, USA. [3]Department of Environmental and Chemical Engineering, Yale University, New Haven, Connecticut 06520, USA. [4]Department of Mechanical Science and Engineering, University of Illinois at Urbana-Champaign, Urbana, Illinois 61801, USA. [5]Department of Civil and Environmental Engineering, University of Illinois at Urbana-Champaign, Urbana, Illinois 61801, USA. [6]Department of Materials Science and Engineering, Massachusetts Institute of Technology, Cambridge, Massachusetts 02139, USA.

When citing this review, please cite the original version as shown on the contents page of this chapter.

methods to increase supplies and purify water can be developed and implemented to serve people throughout the world.

Here, we highlight some of the science and next-generation systems being pursued: to disinfect water, removing current and emerging pathogens without intensive use of chemicals or production of toxic byproducts; to sense, transform, and remove low-concentration contaminants in high backgrounds of potable constituents at lower cost; and to re-use wastewater and desalinate water from sea and inland saline aquifers, all of which hold great promise for effectively increasing water supplies. To realize these challenging goals, many open research questions need to be addressed. Our thesis is that research will enable improved disinfection, decontamination, re-use and desalination methods to work in concert to improve health, safeguard the environment, and reduce water scarcity, not just in the industrialized world, but in the developing world, where less chemical- and energy-intensive technologies are greatly needed.

Disinfection

An overarching goal for providing safe water is affordably and robustly to disinfect water from traditional and emerging pathogens, without creating more problems due to the disinfection process itself. Waterborne pathogens have a devastating effect on public health, especially in the developing countries of sub-Saharan Africa and southeast Asia[9]. Waterborne infectious agents responsible for these diseases include a variety of helminthes, protozoa, fungi, bacteria, rickettsiae, viruses and prions[10]. While some infectious agents have been eradicated or diminished, new ones continue to emerge and so disinfecting water has become increasingly more challenging. Viruses are of particular concern, accounting, together with prions, for nearly half of all emerging pathogens in the last two to three decades[9]. Enteric viruses received less attention in the past compared with bacterial pathogens (for example, *Vibrio cholerae*) and protozoan parasites (for example, *Cryptosporidium parvum*), partly because they were difficult to detect, and partly because free chlorine (the main disinfectant used worldwide because of its potency and low cost) was very effective in inactivating them. However, free chlorine is ineffective in controlling waterborne pathogens such as *C. parvum* and *Mycobacterium avium*. *M. avium* in particular is ubiquitous in biofilms within water distribution systems around the world, with remarkable resistance to chlorine at the high pH and low temperature of natural water. Indeed, ageing and deterioration of drinking water distribution systems and the associated growing of biofilms within them has emerged as a key infrastructure rehabilitation challenge: significant resources are needed to maintain and upgrade distribution systems. In the USA, where large numbers of such old systems exist, disinfectants are required to suppress pathogens within the system. Halogenated disinfection strategies for treatment and distribution systems produce toxic disinfection by-products (DBPs) such as trihalomethanes and haloacetic acids. Recent US disinfection regulations[11,12] require the control of *C. parvum* oocysts while minimizing the formation of certain DBPs, which might force some drinking-water utilities to discard free chlorine disinfection and implement alternative technologies.

Therefore, the effective control of waterborne pathogens in drinking water calls for the development of new disinfection strategies, including multiple-barrier approaches that provide reliable physicochemical removal (for example, coagulation, flocculation, sedimentation, and media or membrane filtration) along with effective photon-based and/or chemical inactivation. The 1993 outbreak of cryptosporidiosis in Milwaukee, Wisconsin, USA, in which approximately 400,000 people were infected and more than 100 died, was a wake-up call for the US drinking-water industry. They were reminded that relying exclusively on physicochemical removal, which can suffer from malfunctions arising from defects in manufacturing or operation, can have a devastating effect on public health.

The use of light from visible to ultraviolet (UV) to photochemically inactivate pathogens has recently seen a resurgence in interest,

notwithstanding the historical use of sunlight to disinfect water. Sequential disinfection schemes such as UV/combined chlorine and ozone/combined chlorine are being considered by many drinking-water utilities as the inactivation component of their multiple-barrier treatment plants because, compared with free chlorine, both UV and ozone are very effective in controlling *C. parvum* oocysts. In addition, combined chlorine can provide a residual in distribution systems without forming high levels of regulated DBPs. However, changing disinfection technologies has raised new concerns because viruses, although effectively controlled by ozone, are resistant to both UV and combined chlorine disinfection. Moreover, ozone can form the DBP carcinogen bromate ion in water containing bromide ions, and combined chlorine can form other unregulated DBPs, for example, haloacetonitriles and iodoacetic acid[13,14], that may be more toxic and carcinogenic than those associated with free chlorine.

The situation in developing countries is similar. International agencies and non-governmental organizations have introduced the use of sunlight irradiation of water within PET (polyethylene terephthalate) bottles to kill pathogens, and are promoting the use of sodium hypochlorite for point-of-use disinfection of drinking water in rural areas (for example, the CDC SafeWater System)[15]. Although these initiatives have lowered the incidence of gastrointestinal disease, owing to the lack of adequate sanitation, the source waters in these areas contain ammonia and organic nitrogen that react with the sodium hypochlorite to form combined chlorine species that are ineffective in inactivating viruses. Furthermore, relatively high levels of toxic DBPs can form in the presence of high concentrations of organic matter associated with inadequate sanitation.

To develop alternatives to chlorine (free and combined) and UV disinfection for the control of waterborne viruses requires significant advances in understanding how viruses are inactivated by the benchmark (chlorine and UV) methods and by any new technologies. The goal is to match or improve on the positive aspects of chlorine and UV disinfection while avoiding the negative effects. To do so requires several questions to be answered. It is well established that both UV light and the free chlorine species hypochlorous acid (HOCl) react with various amino acids in the virus capsid proteins as well as with the nucleic acid protected by the capsid[16,17]. However, the actual limiting step (that is, the molecular target and its level of damage) responsible for inactivation is not yet known. Developing a process that targets that inactivation mechanism may create a new, safe, and robust disinfection method.

For example, many species of adenovirus—the waterborne pathogens with highest resistance to UV inactivation—use their fibre head (Fig. 1a) to attach to the amino-terminal D1 domain of the coxsackievirus-and-adenovirus receptor (CAR) of host cells[18]. Amino acid sequence alignments have shown that the hydrophobic side group of tyrosine and ionizable basic side groups of histidine and lysine in the fibre head associate with CAR amino acids (Fig. 1b) and thus play a role in the attachment of adenovirus to the host cell[19]. Consequently, oxidization of the phenolic group of tyrosine, and formation of reactive chloramines with the amino groups of histidine and lysine[20–23] could contribute to changing the conformation of the adenovirus head protein and inhibiting binding to receptors, thus effectively inactivating the virion. HOCl also reacts with nucleic acid and amino acid residues involved in many steps of the infection cycle of viruses, such as cell entry (endocytosis and endosomal lysis), intracellular trafficking and nuclear delivery, in the case of adenovirus[24]. Thus, even if the virus penetrates the cell, the infection cycle could be inhibited at some subsequent step. A potential problem with this strategy is that once inside the cell, the virion might manipulate the host cell to repair the damage and subsequently complete the infection cycle[25]. Consequently, disinfection processes that target the proteins responsible for attachment and penetration would avoid the unwanted possibility of genome repair.

A new generation of disinfection processes to control viruses should be capable of selective reactions with the key residues in proteins responsible for binding to host cell receptor molecules. Heterogeneous processes are envisioned that would use complementary nanostructured and functionalized surfaces that mimic the structure and functionality of the receptors of target protein residues. These structures should have both high affinity and specificity and be relatively inert to the large amounts of organic matter ubiquitous in natural water. The surfaces of new materials could be designed with arrays of sites that serve to trap all waterborne viral pathogens via binding to host receptors.

A futuristic disinfection method involves the combined use of photons and engineered nanostructures. Although UV is effective for inactivating waterborne bacteria and protozoa cysts and oocysts, it is not very effective for viral pathogens. However, UV light is capable of activating photocatalytic materials such as titania (TiO_2), which are capable of inactivating viruses. Furthermore, new photocatalysts such as TiO_2 doped with nitrogen (TiON), or

co-doped with nitrogen and a metal such as palladium, can be activated with visible light[26] (which could potentially inactivate viruses and other waterborne pathogens with much lower energy use than UV), or even with sunlight (for deployment anywhere with bright sunlight). Of particular interest are materials and systems that use low-cost visible lamp light and sunlight to achieve sufficiently high throughput. Low throughput rates have thus far limited adoption of photoinactivation. Throughput rates depend on factors such as incident light flux and wavelength, absorption length through water, geometry, reactor hydrodynamics, contact efficiency of species in water on the photocatalysts and, critically, the inactivation kinetics. Moreover, we need to improve our understanding of the mechanisms for the interactions of pathogens, in particular virions, with excited photocatalyst surfaces and adherent active moieties, such as hydroxyl radicals and superoxides. The physicochemical structure of such surfaces would need to be optimized for maximum selective affinity of target viral capsid molecular motifs.

Once these new materials are developed, they can be engineered into flow-through reactors for high-throughput systems. The configuration and associated cost of such systems could make them economically viable for applications ranging from large water-treatment plants supplying potable and non-potable water to point-of-use systems with segregated lines dedicated to human consumption and hygiene. Antiviral photocatalysts could be immobilized on fibres and foams of various materials[27–29], or incorporated into membranes[30]. Optical fibres could be used to bring photons into compact configurations such as monolithic reactors[31]. Reactors incorporating visible-light photocatalysts could be designed using sunlight as the source of photons[32,33], a configuration that would be particularly beneficial in developing countries. The resulting systems would provide a barrier against all pathogens by inactivating viruses and trapping any larger bacteria and protozoa cysts and oocysts with relatively high resistance to light and photocatalytic inactivation, all without producing DBPs or extensive use of chemicals.

Decontamination

The overarching goal for the future of decontamination is to detect and remove toxic substances from water affordably and robustly. Widely distributed substances, such as arsenic, heavy metals, halogenated aromatics, nitrosoamines, nitrates, phosphates, and so on are known to cause harm to humans and the environment. Two key problems are that the amount of suspected harmful agents is growing rapidly, and that many of these compounds are toxic in trace quantities. To detect their presence and remove them in the presence of safe and natural constituents that are 3 to 9 orders of magnitude more concentrated is challenging, expensive, and unreliable at present. Chemically treating the total volume of water to transform or remove a specific trace compound is also expensive and potentially itself harmful. Moreover, the treatment does not necessarily remove other harmful compounds, and safe constituents may interfere with the remediation. Thus, new methods to detect toxic compounds and decontaminate water are urgently needed.

The problems of detecting and accurately measuring toxic compounds in water and of selectively removing only these compounds are tightly linked. Both are affected by the particular combination of micropollutant classes (heavy metals, As(III/V), BTEX, pharmaceutical derivatives, agricultural chemicals, endocrine disrupters, and so on)[34] relevant to a specific water source. Furthermore, viable avenues for both detection and treatment are tied to the resource base available. Approaches to speciation of As(III/V) or elemental profiling relevant to western Europe are simply not an option for Bangladesh[35,36] or Benin[37]. Powerful methods of monitoring low concentrations of contaminants are invariably built around sophisticated laboratory instrumentation. It is extremely challenging to develop robust, low-cost, effective means of chemical sensing relevant to the water contamination problems of developing nations. Similarly, affordably treating toxic compounds in water, such as by

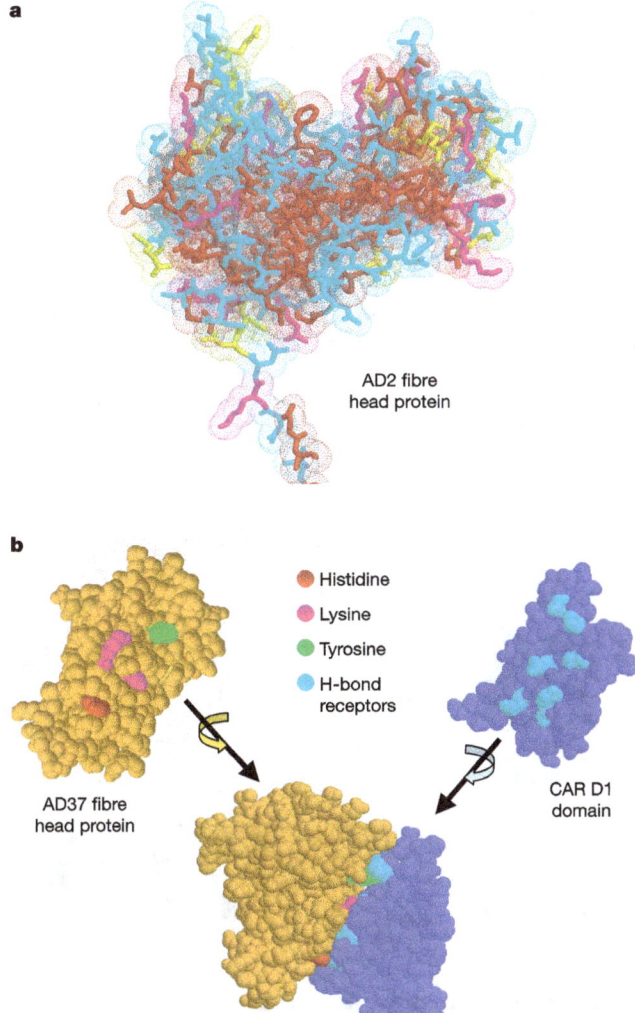

a

AD2 fibre
head protein

b

- ● Histidine
- ● Lysine
- ● Tyrosine
- ● H-bond receptors

AD37 fibre
head protein

CAR D1
domain

Figure 1 | Waterborne virus attachment head and receptor on host cell.
a, Schematic of adenovirus-2 attachment fibre head showing amino acids with basic (magenta, orange, purple), acidic (yellow), and hydrophobic (red) side groups. **b,** Schematic of adenovirus-37 fibre head protein attaching to the D1 domain of the coxsackievirus-and-adenovirus receptor (CAR) on the host cell. Basic ionizable (histidine, lysine) and hydrophobic (tyrosine) side groups of AD37 amino acids and those in CAR D1 domain involved in hydrogen bonding are highlighted. The figure was developed with Protein Explorer[98] based on a structure described in ref. 99 (courtesy of Martin A. Page, University of Illinois).

reducing As(III/V) concentrations to levels currently thought of as safe (<10 parts per billion), without producing toxic waste disposal issues has proved to be a major challenge. But although these goals are beset by severe technical difficulties, they also present exciting opportunities for the research community.

Speciation remains a challenging detection problem. For example, As(III) is estimated to be ~50 times more toxic than As(V), so both As(III) and total As must be measured. Anodic stripping voltammetry[38] has sufficiently low limits of detection (LOD = $1.2 \, \mu g \, l^{-1}$) to be practical and is capable of measuring As(III) in the presence of large excesses of As(V). Alternatively, ion exchange separations may be combined with hydride generation atomic spectroscopy to measure As(III) and As(V) separately[39]. But neither method is suitable for untrained workers. These methods also demonstrate the related generic problem of the LOD dynamic reserve. The temptation, given that detailed dose–response data frequently do not exist (especially at low concentrations of toxic species), is to regulate to the existing analytical capabilities, which can create new problems. For example, if the total As concentration is regulated at a maximum contaminant level of $10 \, \mu g \, l^{-1}$, then the $1.2 \, \mu g \, l^{-1}$ LOD of As(V) represents only an eightfold dynamic reserve. It might not be possible to achieve a tenfold or greater dynamic reserve between the LOD and the maximum contaminant level using detection methods suitable for use by untrained workers to enhance human health.

Beyond these quantitative issues lies the dichotomy between the capabilities for detecting target compounds and for identifying potentially troublesome non-target species. Even powerful multidimensional analytical methods, such as liquid chromatography-mass spectrometry (LC-MS), struggle to characterize waters containing significant amounts of non-target species. These compounds must often be pre-concentrated by factors of 10^2 to 10^3 and can only be assayed accurately in the presence of a small number of potential compounds whose liquid chromatography retention behaviour is known[40]. Such problems point to the critical need to develop molecular recognition motifs (sensor reagents) that can be combined with micro-nanofluidic manipulation[41] and data telemetry to accomplish single-platform chemical sensing having the requisite figures of merit to be competitive with bench-scale instrumentation. In this regard the recent combination of catalytic DNA (DNAzyme) in a micro-nanofluidic platform is of considerable interest. Functional DNA, obtained through *in vitro* selection, can be used to bind metal ions with high affinity (yielding parts-per-trillion LODs) and specificity (>10^6-fold over other cations)[42]. When synthetically elaborated with proximal fluorophore and quencher, the resulting molecular beacon construct (see Fig. 2) may be placed in microfluidic formats to achieve the double selection of a chemical separation followed by a highly specific molecular recognition event[43]. Significant opportunities exist to exploit the *in vitro* selection process to achieve similar performance characteristics for a wide range of micropollutants.

Biosensing strategies are also beginning to be applied to waterborne pathogens. For example, capillary waveguide integrating biosensors have been applied to detect waterborne *Escherichia coli* O157:H7, an enterohaemorrhagic bacterium[44]. However, given the large fraction of the contaminated-water death toll that is due to waterborne pathogens, there is enormous potential for future development of bio-based measurement schemes.

Detection and remediation of toxic compounds are inextricably linked, as treatment of anionic micropollutants demonstrates. Determination of the anionic constituents of aqueous systems remains among the most challenging analytical problems. Typically, anions are determined by ion chromatography coupled with conductance detection, which is universal but does not have the sensitivity required in all instances. Sensitivity can be grafted on through the use of LC-MS[45] at considerable added expense, although lowering the limits of detection from the ~$5 \, \mu g \, l^{-1}$ level to ~$0.05 \, \mu g \, l^{-1}$ may well justify the cost. On the remediation side, compelling research opportunities surround is the development of high-specificity

synthetic anion transporters although these have thus far been focused on biomedical applications[46]—for particularly refractory micropollutant species, such as ClO_4^- and NO_3^-. Anions also illustrate the complexity of designing effective treatment/remediation strategies. For example, disinfection of water sources with ozone (O_3) is highly effective, but if the water contains significant amounts of Br^-, oxidation to the problematic BrO_3^- takes place[47], effectively substituting one water contamination problem for another.

Similar strategic considerations affect the treatment versus removal decision. Whether to treat water via a chemical or biochemical conversion of a micropollutant to an innocuous form, or to remove the toxic contaminant via adsorption, chelation and filtration, or another method is a decision that rests largely on matching the problem to the sophistication of the available technology and the resource base to support the use of the technology, as well as how far the target concentration is below the maximum contaminant level. The use of Sono filter technology at local wells in Bangladesh and reverse osmosis (RO) systems in central plants in the USA may both represent optimized solutions for As removal within the context of the local problem[48]. However, here, too, opportunities exist for research to make an impact.

The treatment protocols used widely and envisioned for the future all encompass a complex interplay of elementary steps such as transport, partitioning, reaction and conversion, and release. To this end, fundamental advances in understanding these processes will necessarily involve sophisticated modelling to assess the way in which the basic steps are coupled most effectively[49], and modelling-based predictions of potential removal activity[50]. Modelling is essential to optimize multi-step strategies—for example, the capture of As by monodisperse Fe_3O_4 nanocrystals followed by magnetic separation of the waste stream[51]—which are often the most effective, or perhaps the only, possible approaches.

Another critical problem involves unintended transformations of non-targeted pollutants. For example, treatment of wastewater with

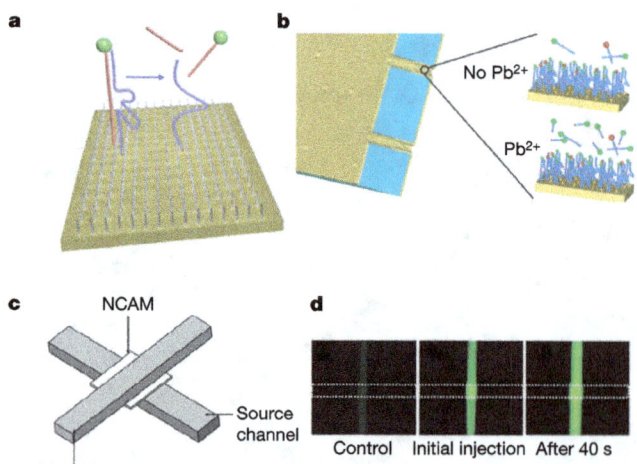

Figure 2 | Lead DNA sensor with a micro-nanofluidic device. Immobilized DNAzyme sensor and micro-nanofluidic devices for detection of Pb^{2+} by fluorescently labelled 17E DNAzyme. **a**, Schematic of immobilized DNAzyme showing catalytic beacon signalling of reaction on the surface, releasing fluorophore into solution for detection. **b**, Schematic of DNAzyme immobilized within the pores of a nanocapillary array membrane (NCAM) with inset showing the mode of ratiometric fluorescence signalling in the absence or presence of Pb(II). **c**, Schematic representation of orthogonal microfluidic channels separated by a NCAM flow gate. **d**, Fluorescence micrographs of the receiving channel before injection of Pb^{2+} sample into a receiving channel containing ~1 μM 17E DNAzyme, after initial Pb^{2+} injection and after 40 s of total injection time for a DNAzyme–NCAM microfluidic device.

Cl$_2$ or monochloramine can oxidize dimethylhydrazine to the suspected carcinogen N-nitrosodimethylamine (NDMA). These unintended secondary effects would seem to argue for separation over transformation strategies, but the relative cost and effectiveness of each approach needs to be considered on a case-by-case basis. A promising workaround focuses on exploiting biology to effect either transformation, such as the biodegradation of NDMA by monooxygenase-expressing microorganisms[52], or removal, as exemplified by the Fe-specific siderophile desferrioxamine-B produced by *Streptomyces pilosus*. Desferrioxamine-B exhibits stability constants in excess of 10^{26} for Th(IV) and Pu(IV), and so may be useful in actinide remediation strategies[53]. Of course, organism-oriented strategies must also be vetted to ensure that they do not introduce other undesirable secondary effects.

Finally, an ubiquitous problem in remediation strategies is the cost or use of critical components that are consumed in stoichiometric reaction, which spurs interest in catalytic treatment approaches to convert organic compounds to innocuous N$_2$, CO$_2$ and H$_2$O. Major anion pollutants such as nitrates and perchlorates are now removed via ion exchange resins or RO, leaving a deleterious brine to be disposed of. Next-generation remediation may use bi-metallic active catalysts to mineralize the brine, such as Pd-Cu/γ-alumina catalysed reduction of NO$_3^-$ (ref. 54). Future efforts may include incorporating active nanocatalysts in a membrane barrier to transform anions at low concentrations in a hybrid process. The combination of modelling and experiments can reveal the mechanisms of these reduction reactions, helping to identify potentially transformative catalytic remediation strategies.

Re-use and reclamation

The overarching goal for the future of reclamation and re-use of water is to capture water directly from non-traditional sources such as industrial or municipal wastewaters and restore it to potable quality. Of all the water withdrawn from rivers, lakes and aquifers, the majority is returned to the environment. Agricultural and livestock users return the least at ~30–40%, whereas industrial users return ~80–90%, power generation returns considerably more at ~95–98%, and public and municipal users return ~75–85%. The rest is lost to the atmosphere or is consumed in biological or chemical processes. A large part of the cost of water for human use is pumping, transport and storage (particularly in developing countries whose citizens often spend substantial time acquiring water). Thus recovering water at or close to the point of use should be very efficient. However, unlike the decontamination of trace compounds just discussed, wastewater contains a wide variety of contaminants and pathogens, and has a very high loading of organic matter, all of which must be removed or transformed to harmless compounds.

Municipal wastewaters are commonly treated by activated sludge systems that use suspended microbes to remove organics and nutrients, and large sedimentation tanks to separate the solid and liquid fractions. This level of treatment produces wastewater effluent suitable for discharge to surface waters or for restricted irrigation and some industrial applications. Similarly, biological treatment via

traditional trickling filters and aquacultures have been used extensively to reduce solids and remove ammonia and nitrites from water. Typically, these biological treatment systems are large with long water residence times. A technology now actively being pursued is membrane bioreactors (MBRs)[55–57]. This technology combines suspended biomass, similar to the conventional activated sludge process, with immersed microfiltration or ultrafiltration membranes that replace gravity sedimentation and clarify the wastewater effluent. MBRs can produce high-quality effluent that is suitable for unrestricted irrigation and other industrial applications.

MBRs have also the potential for use in developing countries to address the pressing need for improved sanitation[55]. Possible applications in developing countries include the direct treatment of raw sewage, particularly in rapidly growing megacities, and the extraction of valuable resources from sewage, namely clean water, nutrients (mostly N and P), and energy. The small footprint, flexible design, and automated operation of MBRs make them ideal for localized, decentralized sewage treatment in the developing world.

One of the growing applications of MBRs is as pretreatment for RO, which, when followed by UV disinfection (or, potentially, visible-light-activated photocatalysts), can produce water for direct or indirect potable use (Fig. 3). Current wastewater re-use systems use a conventional activated sludge process, followed by a microfiltration MBR pretreatment of the secondary effluent, which has high quantities of suspended and dissolved solids. The effluent water from the MBR still partially contains dissolved species and colloidial substances that act to foul the membranes of the subsequent RO system used as a final barrier to contaminants in the product water. Employing a 'tight' ultrafiltration membrane in the MBRs lets through fewer dissolved solids than does microfiltration, allowing the RO system to operate with significantly less fouling. Futuristic direct re-use systems envisioned involve only two steps: a single-stage MBR with an immersed nanofiltration membrane (obviating the need for an RO stage), followed by a photocatalytic reactor to provide an absolute barrier to pathogens and to destroy low-molecular-weight organic contaminants that may pass the nanofiltration barrier.

A major obstacle to the efficient application of MBRs in current or next-generation re-use systems is membrane fouling, particularly when it leads to flux losses that cleaning cannot restore[56,58]. Fouling in MBRs is primarily caused by microbe-generated extracellular polymeric substances, most notably polysaccharides, proteins and natural organic matter. The development of economical, high-flux, non-fouling membranes is therefore needed before viable MBR processes, as well as other membrane-based approaches for wastewater reclamation, can be achieved.

Fouling of polymer membranes is influenced by membrane chemistry and morphology. Polymers used in porous membrane manufacture have chemical and mechanical stability, but are generally hydrophobic in nature, and as a consequence are highly susceptible to adsorption of organic foulants. Commercial methods to reduce fouling largely involve graft polymerization of hydrophilic monomers on the membrane surface[59]. The resulting 'brush' of hydrated

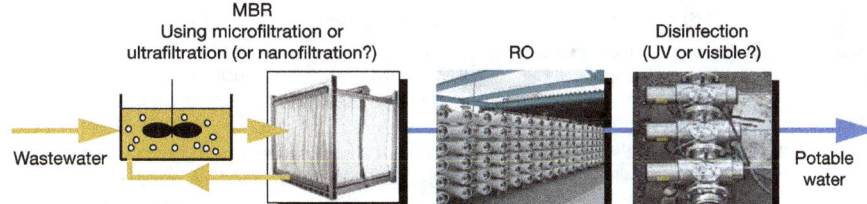

Figure 3 | Membrane bioreactor treatment system for direct conversion to potable water. Depiction of a next generation MBR-based treatment method that can potentially take wastewater from municipal, agricultural, livestock or other high-organic-content sources and convert it to potable

water. Future methods may be able to omit the RO step with a nanofiltration membrane, and follow with a visible light disinfection step to ensure that all pathogens, including viruses, are inactivated.

chains serves as a steric-osmotic barrier to foulant adsorption, but reduces intrinsic permeability owing to partial blocking of surface pores, while internal pores may go unmodified and remain prone to fouling[60]. The extra manufacturing steps also add to membrane cost.

Alternative *in situ* approaches to membrane surface modification under development may generate more efficacious brush layers without the drawbacks of surface graft polymerization[61–64]. Comb copolymers, having hydrophobic backbones and hydrophilic side chains, function as macromolecular surfactants when added to membrane casting solutions[62], lining membrane surfaces and internal pores during the conventional immersion precipitation process used in membrane manufacture. Order-of-magnitude flux enhancements[63,65] and complete resistance to irreversible fouling by the three classes of extracellular polymeric substance foulants[65,66], recently demonstrated for such ultrafiltration membranes (see Fig. 4), offer substantial promise for decreasing operational costs of wastewater treatment through reduced membrane cleaning and replacement and increased process efficiency.

Next-generation membranes offer further opportunities for improved contaminant retention or recovery of valuable constituents from wastewaters, without intensive chemical treatment and while reducing the need for subsequent decontamination. These advanced filtration processes require membranes with much narrower pore size distributions than those derived from immersion precipitation, in addition to fouling-resistant surface/pore chemistries[61,67]. Approaches under investigation include block copolymers, graft/comb copolymers, or lyotropic liquid crystals that self-assemble to form nanodomains that are highly permeable to water[68–71], or can be selectively removed to create nanopores for water passage[72,73]. Such nanostructured materials may be implemented as thin-film coatings on conventional ultrafiltration or microfiltration membrane supports[70,72,74], or on novel high-flux base membrane structures, such as electrospun nanofibres[75]. Recently, for example, rigid star amphiphiles with 1–2 nm hydrophobic cores and hydrophilic side chains were coated onto polyethersulphone ultrafiltration membranes to obtain nanofiltration membranes with comparable or better rejection of As(III) and water permeability several times greater than commercial nanofiltration membranes[76]. The commercial viability of this new class of thin-film composite

membranes for water re-use hinges on the development of inexpensive coatings, chemistries and scalable processing methods that can reproducibly achieve the desired membrane structure and yield fluxes comparable to today's ultrafiltration membranes.

Desalination

The overarching goal for the future of desalination is to increase the fresh water supply via desalination of seawater and saline aquifers. These sources account for 97.5% of all water on the Earth, so capturing even a tiny fraction could have a huge impact on water scarcity. Through continual improvements, particularly in the last decade, desalination technologies can be used reliably to desalinate sea water as well as brackish waters from saline aquifers and rivers.

Desalination of all types, though, is often considered a capital-[77] and energy-intensive[78] process, and typically requires the conveyance of the water to the desalination plant, pretreatment of the intake water, disposal of the concentrate (brine), and process maintenance. Estimated costs of pumping from sea intake to the desalination plant vary widely with geographical location, height and distance from the source water[77]. But so far, the total cost and increased environmental concerns have limited the widespread adoption of desalination technologies. Nevertheless, for a state-of-the-art RO system that uses as little as ~2.2 kW h of electrical energy to produce a thousand litres of drinking water inside the desalination plant, the total energy usage to desalinate water will be ~0.005 kW h l^{-1}, which includes the electrical plus modest conveyance energy needs. Putting this estimated energy use into perspective reveals that supplying even 50 litres a day per capita of drinking water at 0.25 kW h can be a small fraction of the daily energy required per capita (ranging from 3.2 kW h in China to 30 kW h in the USA) for living in a world with strained environmental resources (see http://telstar.ote.cmu.edu/environ/m3/s3/02needs.shtml).

The major desalination technologies currently in use are based on membrane separation via RO and thermal distillation (multistage flash and effect distillation), with RO accounting for over 50% of the installed capacity[77,78]. Conventional thermal desalination processes are inefficient in their use of energy and suffer particularly from corrosion, as well as scaling that also affects RO. Even where fuel is readily available and low-cost, high capital and operational costs limit adoption. Therefore, the market share of large conventional thermal desalination plants will probably decline. However, for family and very small community systems in remote locations, especially in the developing world, solar thermal distillation and humid air desalination technologies may find an increasing role, particularly in inland semi-arid areas with access to saline lakes and aquifers[79,80]. These thermal technologies may also find small-scale applications in locations without ready sources of energy, other than solar.

Although RO systems have a relatively low rate of energy consumption, they use high-cost electrical energy. RO desalination, however, can take advantage of low-grade heat energy to increase flux through the membrane for a given pressure drop. This thermally enhanced RO finds applications in locations where such heat energy is available, typically using waste heat from a co-located electrical power generation and RO desalination plant. We also envision hybrid desalination plants that would combine thermally enhanced RO and thermal desalination to lower electrical energy consumption per unit of product water further, while achieving higher water recoveries than can RO alone. Desalinating inland saline waters, which are present on most continents in quantities similar to fresh water, can also be used to increase water supplies, but disposal of the residual concentrate is a major problem. Hybrid desalination technologies that concentrate precipitates and salts while extracting the water with membranes can potentially process the brine. Two alternative desalination technologies currently under investigation, forward osmosis[81] and membrane distillation[82], can also use low-grade heat

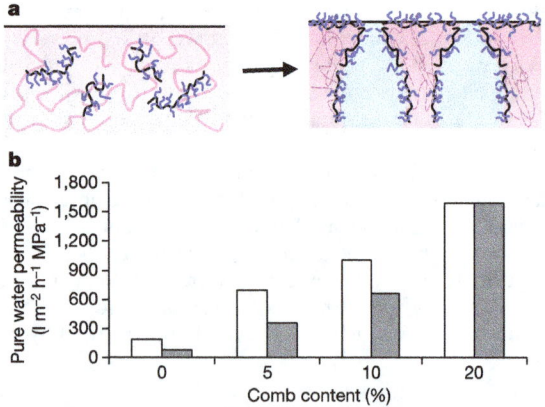

Figure 4 | Comb copolymer amphiphiles for fouling-resistant membranes. **a**, Schematic illustration of *in situ* approach using comb copolymer amphiphiles to modify ultrafiltration membrane surfaces and internal pores during membrane casting. **b**, Pure water permeability of polyacrylonitrile ultrafiltration membranes incorporating 0–20% comb copolymer additive having a polyacrylonitrile backbone and polyethylene oxide side chains. White bars show the initial pure water permeability, and grey bars show the pure water permeability after 24 h of dead-end filtration of 1,000 mg per litre of bovine serum albumin in phosphate buffered saline, followed by a deionized water rinse. Initial flux and flux recovery increase with comb additive content. Membranes exhibit complete resistance to irreversible fouling at 20% comb content (data from ref. 65).

energy and may be used alone, or as hybrid systems with RO, to achieve high water recoveries.

For large-scale desalination, RO has advanced significantly in the past decade, particularly owing to the development of more robust membranes and very efficient energy recovery systems. As a result, the reduction in energy consumption of RO desalination has been remarkable[77,78,83,84]. The specific (per unit of produced potable water) energy of desalination has been reduced from over $10\,kW\,h\,m^{-3}$ in the 1980s to below $4\,kW\,h\,m^{-3}$ (refs 78, 83). Any desalination system will be most energy efficient if it involves a reversible thermodynamic process, which is independent of the system and mechanisms used.

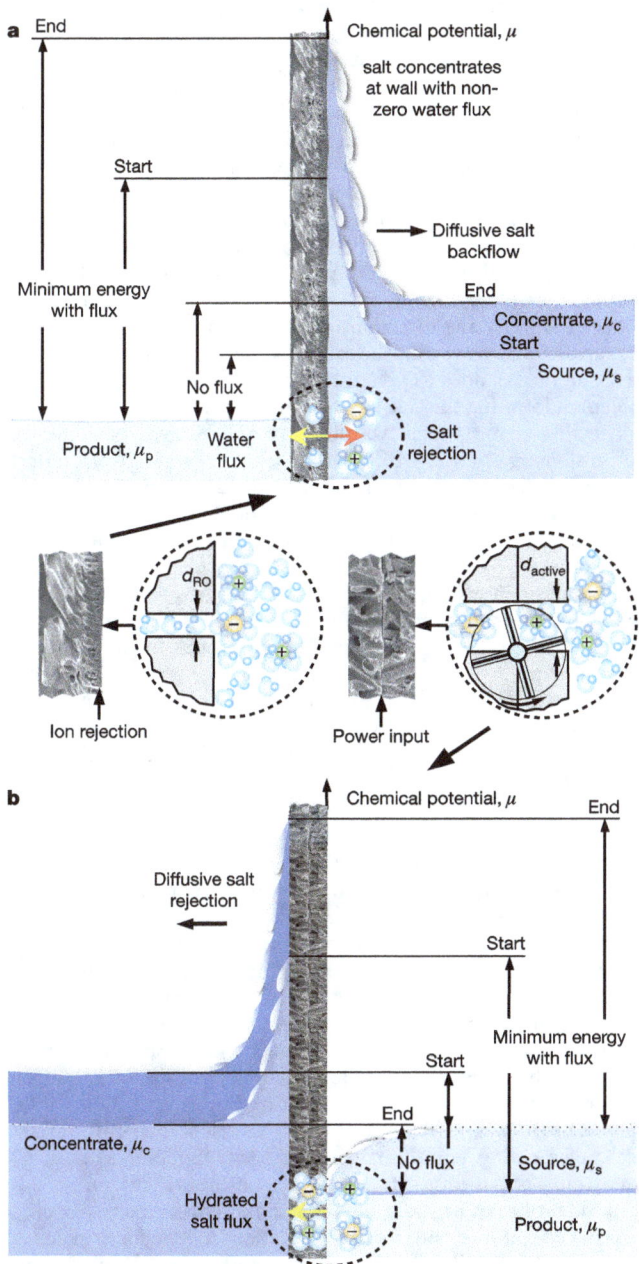

Figure 5 | Reverse osmosis and active desalination membrane processes. Concentration gradients in RO (**a**) and active (**b**) desalination membranes. The energy levels marked with 'start' and 'end' correspond to the evolution of each process. The darker blue colour denotes higher concentration. Insets depict different mechanisms of salt ion separation. The active process with energy input shows a conceptual strategy for overcoming the Born barrier with fixed charges.

From the free-energy change on removing a small amount of pure water from a mixture of water and salt, the theoretical lower bound of the energy needed for desalination can be estimated[85]. For zero per cent recovery, that is, the removal of a relatively small amount of water from a very large amount of sea water, the calculated theoretical minimum energy for desalination is $0.70\,kW\,h\,m^{-3}$ of fresh water produced. This theoretical minimum increases to 0.81, 0.97 and $1.29\,kW\,h\,m^{-3}$ for recoveries of 25, 50 and 75%, respectively, suggesting that further improvements in the energy efficiency of RO desalination are still possible.

Although RO is currently the state-of-the-art desalination technology, there are several challenges and opportunities that could result in additional reductions in the total cost per unit of product water. Among the major challenges of RO desalination are membrane fouling, relatively low recovery for sea water desalination (less than ~55%), which results in large volumes of concentrated brine, and relatively low removal of low-molecular-weight contaminants, most notably boron in sea water. Future RO desalination membranes will ideally have high water flux per unit of pressure applied, near-complete rejection of dissolved species, low fouling propensity, and tolerance to oxidants used in pretreatment for biofouling control. The total cost for RO involves three strongly interrelated components that depend upon region, source water, and energy sources: capital (infrastructure, equipment, membrane replacement), energy (thermal and electrical), and operation (pretreatment, post-treatment, concentrate disposal and cleaning). Lowering the flux to save energy may be offset by the consequent increase in capital costs. Increasing the permeability of the RO membranes decreases both capital and energy costs, but may increase the cost for pretreatment and cleaning. Improvements must be made in all three components to lower the total cost of the product water.

Recent work by the Affordable Desalination Coalition[78,84] has demonstrated a remarkably low specific energy of seawater desalination, at $1.58\,kW\,h\,m^{-3}$, under ideal conditions (that is, new membranes, no fouling, and low water flux) at 42% recovery. This value is relatively close to the theoretical minimum energy for seawater desalination at that recovery, suggesting that next-generation fouling-resistant RO membranes will be able to desalinate sea water with lower energy consumption. Asymmetric membranes depicted in Fig. 5 (a) currently used for RO have relatively large random pore size distributions. Therefore, the separation layer is thicker than ideal to ensure adequate salt rejection, reducing the flux rate. Desalination typically involves ions with small hydration diameter d_H, which require pores with a hydraulic diameter of $d_{RO} < d_H$ to exclude them, increasing mainly enthalpic energy requirements. The energy of desalination depends critically on pore diameters, and the chemical affinity of water and ions with the pore wall.

Approaching the theoretical minimum energy is impractical for desalination plants, because it would require huge facilities with high capital costs. Moreover, in real desalination processes, energy is lost because of inherent thermodynamic irreversibilities that arise from diffusion, viscous dissipation and flux-rate-dependent losses. To reduce the energy needed for desalination, the rate of entropy generation Φ must be minimized because the energy consumed by irreversible processes is $\sim T\Phi$, where T is the absolute temperature. Φ can be expressed as $J_v\Delta p + J_d\Delta\pi + J_a\Delta a$, where J_v is the total volumetric flux of water plus solute, J_d is the flux of solute relative to the water, Δp is the pressure difference from frictional losses ($\Delta p \propto J_v$), $\Delta\pi$ is the osmotic pressure difference across the membrane, and $J_a\Delta a$ is the entropy generation from any active ion pumping. The higher the flux, the higher the salt concentration will be at an RO barrier, increasing $\Delta\pi$ and Δp. Interestingly, for the same separation performance (that is, the same $J_d\Delta\pi$), if $J_a\Delta a$ is sufficiently small, the entropy generation using active membranes could be less than for RO for a water recovery of much more than 50%, because the fluid flux in RO will be higher than in active transport ($J_{v|RO} \gg J_{v|active}$), and the diffusion of salts driven by concentration gradients will act in favour of the separation.

Membranes with a uniform pore distribution and a more permeable separation layer can potentially maintain or improve salt rejection while increasing the flux in RO. Recent research on the transport of water through hydrophobic double-walled carbon nanotubes is promising, demonstrating water fluxes that are over three orders of magnitude higher than those predicted from continuum hydrodynamic models (refs 86–89; also O. Bakajin, personal communication, 23 October 2007). The high flux may be due to the carbon nanotubes' atomically smooth, hydrophobic walls allowing considerable slip of water through the pores. The preliminary work of ref. 89 reported unusually high water flux through microfabricated membranes comprised of aligned carbon nanotubes ~3 μm long with an inner diameter of ~1.6 nm. Further measurements with these membranes reveal[90] salt rejection coefficients that match or exceed those of commercially available nanofiltration membranes, while exceeding their flux by up to four times. But such membranes may be difficult and costly to manufacture, prone to defect formation, and might have a high propensity for fouling given their hydrophobic nature.

The high performance of membranes based on carbon nanotubes[86,87,89], however, reveals an important pore characteristic shared by biological ion channels: hydrophobic pores ~1 nm in diameter. These cores allow the ion hydration shell to remain intact, thereby reducing the enthalpic translocation energy to be closer to the entropic loss for confining an ion in a pore. Decreasing the pore diameter much below 1 nm creates a large free-energy barrier, which arises from stripping the hydration shell off the ion and water molecules that need to overcome a Born energy barrier. Modifying the surfaces of the membrane, as discussed for nanofiltration membranes, can alter the surface properties, and thus potentially decrease the energy barrier.

Technological challenges to incorporating carbon nanotube materials include the functionalization of the mouth of the pores to increase selectivity and potentially reduce hydrophobicity at the surface, integration of the active layer with robust support substrates, scaling up the fabrication of the ion channel and carbon-nanotube-based membranes and increasing the pore density per area of the active layer, and decreasing the cost of membrane fabrication. Still, the costs of such membranes could eventually be affordable with future improvements in carbon nanotube synthesis and membrane processing.

Aquaporins (water channels) and ion channels of biological cells have also motivated the search for alternative approaches to engineering membranes with high water flux and selectivity[91,92]. De novo synthesis of ion channels[93] and the development of low-molecular-weight anion transporters is an emerging topic in supramolecular chemistry[94]. Based on an array of aligned carbon nanotubes with hollow graphitic cores embedded within a solid polymer film, the first biomimetic protein channel controlled by the same mechanism of phosphorylation/dephosphorylation that occurs in nature has also been recently reported[95]. However, much work remains to incorporate these futuristic materials into large-area membranes at competitive costs.

Even if a perfect membrane could be created, with no pressure drop required for complete salt rejection, the increase of flux rates for RO is ultimately limited by the concentration polarization layer at the membrane (see Fig. 5a), which constitutes an additional impedance to fluid flow. The higher the flux of water, the higher the gradient in solute concentration on the rejection side. The polarization impedance can be reduced via tangential fluid flow, but can never be eliminated. What is worse, the transmembrane chemical potential difference increases along the direction of the tangential flow (from 'start' to 'end' in Fig. 5a), while the transmembrane pressure difference decreases because of pressure losses, resulting in additional irreversible losses with higher fluxes.

To see how active systems might compare to RO systems, we can extrapolate from the energetics of existing ion channels. Biological

channels transfer 10^7 ions per pore per second, and measurements corroborated by systematic computer simulations reveal that the free-energy barrier for biological potassium channels is 2–5 kcal mol^{-1} (depending on the type of K$^+$ channel), which corresponds to a specific energetic requirement of 2.55–6.4 kW h m^{-3} of water produced from sea water with a salt concentration of 32,000 parts per million. For potassium ions, the lower bound is near the current energetic costs for RO, but is still much higher than the theoretical minimum. However, these channels are ion- and charge-specific, and there is a significant energetic cost for the exclusion of the other ions. If active nanopores of dimensions greater than 1 nm are created that pass a multiplicity of anions and cations, as depicted in Fig. 5b, the energetics can potentially drop by more than a factor of two below that of biological channels.

Bio-inspired systems for active transport provide another route towards improving the energetics of desalination. In contrast to conventional desalination whereby water is 'pushed' through an RO membrane by a pressure gradient, or in electrodialysis whereby hydrated anions and cations are forced through their respective ion-selective membranes by electrokinetic action, active ion separation involves pumping of both hydrated anions and cations through the same membrane via modulation of pore potentials, against a chemical potential, leaving desalinated product water. As Fig. 5b illustrates, membranes that actively 'pull' hydrated ions through the barrier reverse the direction of the concentration polarization layer, and should not suffer the same decrease of performance with increasing flux as does RO. Nature also provides a solution to the problem of lowering the cost of overcoming electrostatic barriers in engineered systems, which typically involve dielectric membranes. The Born energy barrier[96] to move an ion of charge q from water to the low-dielectric-constant membrane ($\varepsilon \approx 2$ for typical biological membranes) is $\Delta G \approx q^2/d_H(1/\varepsilon - 1/80)$. This barrier can be offset by the energy liberated if the penetrating ion meets a counter-ion buried inside the membrane ($\Delta G' = -q^2/\varepsilon d_H$), or by the pump, as depicted in the insets of Fig. 5. It has been suggested[97] that for biological ion pumps the energy cost of burying the counter ion is paid for by actively manipulating charged groups in the proteins within the pumps. However, whether passive or active, high-permeability membranes with high resistance to fouling are needed, as well as new strategies for synthesizing membranes with multiple functions to screen small molecules, and to resist stresses and chemical degradation.

Conclusion

The work highlighted here, plus the tremendous amount of additional research being conducted on every continent that could not be mentioned, is sowing the seeds of a revolution in water purification and treatment. We believe that advancing the science of water purification can aid in the development of new technologies that are appropriate for different regions of the world. That said, the sheer enormity of the problems facing the world from the lack of adequate clean water and sanitation means that much more work is needed to address the challenges particular to developing nations, which suffer a diversity of socio-economical-political-traditional constraints, and require a broader approach incorporating sustainable energy sources and implementing educational and capacity building strategies. Consortiums of governments at all levels, businesses and industries, financial and health organizations, water and environment associations, and educational and research institutions need to focus increasing attention towards solving these water problems. While better water resource management, improved efficiencies, and conservation are vital for moderating demand and improving availability, it is our belief that improving the science and technology of water purification can help provide cost-effective and robust solutions.

1. Montgomery, M. A. & Elimelech, M. Water and sanitation in developing countries: including health in the equation. Environ. Sci. Technol. 41, 17–24 (2007).

2. Lima, A. A. M. *et al.* Persistent diarrhea signals a critical period of increased diarrhea burdens and nutritional shortfalls: a prospective cohort study among children in northeastern brazil. *J. Infect. Dis.* **181**, 1643–1651 (2000).

3. Behrman, J. R., Alderman, H. & Hoddinott, J. Hunger and malnutrition. in *Copenhagen Consensus—Challenges and Opportunities (London, 2004)* OCLC 57489365 (London School of Hygiene and Tropical Medicine, 2004); ⟨http://www.copenhagenconsensus.com/Files/Filer/CC/Papers/Hunger%5Fand%5FMalnutrition%5F070504.pdf⟩.

4. Singh, P. & Bengtson, L. The impact of warmer climate on melt and evaporation for the rainfed, snowfed and glacierfed basins in the Himalayan region. *J. Hydrol.* **300**, 140–154 (2005).

5. Shiyin, L., Wenxin, S., Shen, Y. & Li, G. Glacier changes since the Little Ice Age maximum in the western Qilian Shan, northwest China, and consequences of glacier runoff for water supply. *J. Glaciol.* **49**, 117–124 (2003).

6. Barnett, T. P., Adam, J. C. & Lettenmaier, D. P. Potential impacts of a warming climate on water availability in snow-dominated regions. *Nature* **438**, 303–309 (2005).

7. Bradley, R. S., Vuille, M., Diaz, H. F. & Vergara, W. Threats to water supplies in the tropical Andes. *Science* **312**, 1755–1756 (2006).

8. van der Kooij, D. in *Heterotrophic Plate Counts and Drinking-water Safety: The Significance of HPCs for Water Quality and Human Health* (eds Bartram, J., Cotruvo, J., Exner, M., Fricker, C. & Glasmacher, A.) 199–232 (IWA Publishing, World Health Organization, Geneva, 2003).

9. Pitman, G. K. *Bridging Troubled Waters—Assessing The World Bank Water Resources Strategy* (World Bank Publications, Washington DC, 2002).

10. World Health Organization. *Emerging Issues in Water and Infectious Disease* 1–22 (World Health Organization, Geneva, 2003).

11. United States Environmental Protection Agency. 40 CFR parts 9, 141 & 142 National Primary Drinking Water Regulations: Long term 2 enhanced surface water treatment rule; final rule. *Federal Register* **71**, 653–702 (2006).

12. United States Environmental Protection Agency. 40 CFR parts 9, 141, & 142 National Primary Drinking Water Regulations: Stage 2 disinfectants and disinfection byproducts rule; final rule. *Federal Register* **71**, 388–493 (2006).

13. Krasner, S. W. *et al.* Occurrence of a new generation of disinfection byproducts. *Environ. Sci. Technol.* **40**, 7175–7185 (2006).

14. Muellner, M. G. *et al.* Haloacetonitriles vs. regulated haloacetic acids: are nitrogen-containing DBPs more toxic? *Environ. Sci. Technol.* **41**, 645–651 (2007).

15. Centers for Disease Control and Prevention. *Safe Water Systems for the Developing World: A Handbook for Implementing Household-Based Water Treatment and Safe Storage Projects* (CDC, Atlanta, 2000).

16. Simonet, J. & Gantzer, C. Inactivation of poliovirus 1 and f-specific RNA phages and degradation of their genomes by UV irradiation at 254 nanometers. *Appl. Environ. Microbiol.* **72**, 7671–7677 (2006).

17. Nuanualsuwan, S. & Cliver, D. O. Capsid functions of inactivated human picornaviruses and feline calicivirus. *Appl. Environ. Microbiol.* **69**, 350–357 (2003).

18. Coyne, C. B. & Bergelson, J. M. CAR: A virus receptor within the tight junction. *Adv. Drug Deliv. Rev.* **57**, 869–882 (2005).

19. Seiradake, E., Lortat-Jacob, H., Billet, O., Kremer, E. J. & Cusack, S. Structural and mutational analysis of human Ad37 and canine adenovirus 2 fiber heads in complex with the D1 domain of coxsackie and adenovirus receptor. *J. Biol. Chem.* **281**, 33704–33716 (2006).

20. Hawkins, C. L., Pattison, D. I. & Davies, M. J. Hypochlorite-induced oxidation of amino acids, peptides and proteins. *Amino Acids* **25**, 259–274 (2003).

21. Nightingdale, Z. D. *et al.* Relative reactivity of lysine and other peptides-bound amino acids to oxidation by hypochlorite. *Free Radic. Biol. Med.* **29**, 425–433 (2000).

22. Bergt, C., Fu, X., Huq, N. P., Kao, J. & Heinecke, J. W. Lysine residues direct the chlorination of tyrosines in YXXK motifs of apolipoprotein A-I when hypochlorous acid oxidizes high density lipoprotein. *J. Biol. Chem.* **279**, 7856–7866 (2004).

23. Pattison, D. I. & Davies, M. J. Kinetic analysis of the role of histidine chloramines in hypochlorous acid mediated protein oxidation. *Biochemistry* **44**, 7378–7387 (2005).

24. Medina-Kauwe, L. K. Endocytosis of adenovirus and adenovirus capsid proteins. *Adv. Drug Deliv. Rev.* **55**, 1485–1496 (2003).

25. Yates, M. V., Malley, J., Rochelle, P. & Hoffman, R. Effect of adenovirus resistance on UV disinfection requirements: A report on the state of adenovirus science. *J. Am. Water Works Assoc.* **98**, 93–106 (2006).

26. Li, Q., Liang, W. & Shang, J. K. Enhanced visible-light absorption from PdO nanoparticles in nitrogen-doped titanium oxide thin films. *Appl. Phys. Lett.* **90**, 063109 (2007).

27. Fu, P., Luan, Y. & Dai, X. Preparation of activated carbon fibers supported TiO_2 photocatalyst and evaluation of its photocatalytic reactivity. *J. Mol. Catal. Chem.* **221**, 81–88 (2004).

28. Medina-Valtierra, J., Garcia-Servin, J., Frausto-Reyes, C. & Calixto, S. The photocatalytic application and regeneration of anatase thin films with embedded commercial TiO_2 particles deposited on glass microrods. *Appl. Surf. Sci.* **252**, 3600–3608 (2006).

29. Changrani, R. G. & Raupp, G. B. Two-dimensional heterogeneous model for a reticulated-foam photocatalytic reactor. *Am. Inst. Chem. Eng. J.* **46**, 829–842 (2000).

30. Molinari, R., Palmisano, L., Drioli, E. & Schiavello, M. Studies on various reactor configurations for coupling photocatalysis and membrane processes in water purification. *J. Membr. Sci.* **206**, 399–415 (2002).

31. Lin, H. & Valsaraj, K. T. Development of an optical fiber monolith reactor for photocatalytic wastewater treatment. *J. Appl. Electrochem.* **35**, 699–708 (2005).

32. Blanco-Galvez, J., Fernandez-Ibanez, P. & Malato-Rodriguez, S. Solar photocatalytic detoxification and disinfection of water: Recent overview. *J. Solar Energy Eng.* **129**, 4–15 (2007).

33. Gill, L. W. & McLoughlin, O. A. Solar disinfection kinetic design parameters for continuous flow reactors. *J. Solar Energy Eng.* **129**, 111–118 (2007).

34. Schwarzenbach, R. P. *et al.* The challenge of micropollutants in aquatic systems. *Science* **313**, 1072–1077 (2006).

35. Sarkar, S. *et al.* Well-head arsenic removal units in remote villages of Indian subcontinent: Field results and performance evaluation. *Water Res.* **39**, 2196–2206 (2005).

36. Khan, A. H. *et al.* Appraisal of a simple arsenic removal method for groundwater of Bangladesh. *J. Environ. Sci. Health Part A* **35**, 1021–1041 (2000).

37. Silliman, S. E., Boukari, M., Crane, P., Azonsi, F. & Neal, C. R. Observations on elemental concentrations of groundwater in central Benin. *J. Hydrol.* **335**, 374–388 (2007).

38. Rasul, S. B. *et al.* Electrochemical measurement and speciation of inorganic arsenic in groundwater of Bangladesh. *Talanta* **58**, 33–43 (2002).

39. Chen, Z. L., Akter, K. F., Rahman, M. M. & Naidu, R. Speciation of arsenic by ion chromatography inductively coupled plasma mass spectrometry using ammonium eluents. *J. Sep. Sci.* **29**, 2671–2676 (2006).

40. Sultan, J. & Gabryelski, W. Structural identification of highly polar nontarget contaminants in drinking water by ESI-FAIMS-Q-TOF-MS. *Anal. Chem.* **78**, 2905–2917 (2006).

41. Kuo, T.-C. *et al.* Gateable nanofluidic interconnects for multilayered microfluidic separational systems. *Anal. Chem.* **75**, 1861–1867 (2003).

42. Liu, J. W. *et al.* A catalytic beacon sensor for uranium with parts-per-trillion sensitivity and millionfold selectivity. *Proc. Natl Acad. Sci. USA* **104**, 2056–2061 (2007).

43. Chang, I. H. *et al.* Miniaturized lead sensor based on lead-specific DNAzyme in a nanocapillary interconnected microfluidic device. *Environ. Sci. Technol.* **39**, 3756–3761 (2005).

44. Zhu, P. X. *et al.* Detection of water-borne *E. coli* O157 using the integrating waveguide biosensor. *Biosens. Bioelectron.* **21**, 678–683 (2005).

45. Snyder, S. A. *et al.* Role of membranes and activated carbon in the removal of endocrine disruptors and pharmaceuticals. *Desalination* **202**, 156–181 (2007).

46. Davis, A. P., Sheppard, D. N. & Smith, B. D. Development of synthetic membrane transporters for anions. *Chem. Soc. Rev.* **36**, 348–357 (2007).

47. Snyder, S. A., Vanderford, B. J. & Rexing, D. J. Trace analysis of bromate, chlorate, iodate, and perchlorate in natural and bottled waters. *Environ. Sci. Technol.* **39**, 4586–4593 (2005).

48. Garelick, H., Dybowska, A., Valsami-Jones, E. & Priest, N. D. Remediation technologies for arsenic contaminated drinking waters. *J. Soils Sediments* **5**, 182–190 (2005).

49. Schideman, L. C., Marinas, B. J., Snoeyink, V. L. & Campos, C. Three-component competitive adsorption model for fixed-bed and moving-bed granular activated carbon adsorbers. Part I. Model development. *Environ. Sci. Technol.* **40**, 6805–6811 (2006).

50. Magnuson, M. L. & Speth, T. F. Quantitative structure—Property relationships for enhancing predictions of synthetic organic chemical removal from drinking water by granular activated carbon. *Environ. Sci. Technol.* **39**, 7706–7711 (2005).

51. Yavuz, C. T. *et al.* Low-field magnetic separation of monodisperse Fe_3O_4 nanocrystals. *Science* **314**, 964–967 (2006).

52. Fournier, D., Hawari, J., Streger, S. H., McClay, K. & Hatzinger, P. B. Biotransformation of N-nitrosodimethylamine by *Pseudomonas mendocina* KR1. *Appl. Environ. Microbiol.* **72**, 6693–6698 (2006).

53. Kraemer, S. M., Xu, J. D., Raymond, K. N. & Sposito, G. Adsorption of Pb(II) and Eu(III) by oxide minerals in the presence of natural and synthetic hydroxamate siderophores. *Environ. Sci. Technol.* **36**, 1287–1291 (2002).

54. Chaplin, B. P., Roundy, E., Guy, K. A., Shapley, J. R. & Werth, C. J. Effects of natural water ions and humic acid on catalytic nitrate reduction kinetics using an alumina supported Pd-Cu catalyst. *Environ. Sci. Technol.* **40**, 3075–3081 (2006).

55. Daiger, G. T., Rittmann, B. E., Adham, S. & Andreottola, G. Are membrane bioreactors ready for widespread application? *Environ. Sci. Technol.* **39**, 399A–406A (2005).

56. Yang, W. B., Cicek, N. & Ilg, J. State-of-the-art of membrane bioreactors: worldwide research and commercial applications in North America. *J. Membr. Sci.* **270**, 201–211 (2006).

57. Bixio, D. *et al.* Wastewater reuse in Europe. *Desalination* **189**, 89–101 (2006).

58. Kimura, K., Yamato, N., Yamamura, H. & Watanabe, Y. Membrane fouling in pilot-scale membrane bioreactors (MBRs) treating municipal wastewater. *Environ. Sci. Technol.* **39**, 6293–6299 (2005).

59. Ulbricht, M. & Belfort, G. Surface modification of ultrafiltration membranes by low temperature plasma. 2. Graft polymerization onto polyacrylonitrile and polysulfone. *J. Membr. Sci.* **111**, 193–215 (1996).

60. Carroll, T., Booker, N. A. & Meier-Haack, J. Polyelectrolyte-grafted microfiltration membranes to control fouling by natural organic matter in drinking water. *J. Membr. Sci.* **203**, 3–13 (2002).

61. Deratani, A., Li, C. L., Wang, D. M. & Lai, J. Y. New trends in the preparation of polymeric membranes for liquid filtration. *Ann. Chim.-Sci. Mater.* **32**, 107–118 (2007).

62. Hester, J. F., Banerjee, P. & Mayes, A. M. Preparation of protein-resistant surfaces on poly(vinylidene fluoride) membranes via surface segregation. *Macromolecules* **32**, 1643–1650 (1999).

63. Hester, J. F. & Mayes, A. M. Design and performance of foul-resistant poly(vinylidene fluoride) membranes prepared in a single step by surface segregation. *J. Membr. Sci.* **202**, 119–135 (2002).

64. Wang, Y. Q. *et al.* Remarkable reduction of irreversible fouling and improvement of the permeation properties of poly(ether sulfone) ultrafiltration membranes by blending with pluronic F127. *Langmuir* **21**, 11856–11862 (2005).

65. Asatekin, A., Kang, S., Elimelech, M. & Mayes, A. M. Anti-fouling ultrafiltration membranes containing polyacrylonitrile-*graft*-poly(ethylene oxide) comb copolymer additives. *J. Membr. Sci.* **298**, 136–146 (2007).

66. Kang, S., Asatekin, A., Mayes, A. M. & Elimelech, M. Protein antifouling mechanisms of PAN UF membranes incorporating PAN-g-PEO additive. *J. Membr. Sci.* **298**, 42–50 (2007).

67. Ulbricht, M. Advanced functional polymer membranes. *Polymer* **47**, 2217–2262 (2006).

68. Akthakul, A., Salinaro, R. F. & Mayes, A. M. Antifouling polymer membranes with sub-nanometer size selectivity. *Macromolecules* **37**, 7663–7668 (2004).

69. Zhou, M., Kidd, T. J., Noble, R. D. & Gin, D. L. Supported lyotropic liquid crystal polymer membranes: promising materials for molecular-size-selective aqueous nanofiltration. *Adv. Mater.* **17**, 1850–1853 (2005).

70. Asatekin, A. *et al.* Antifouling nanofiltration membranes for membrane bioreactors from self-assembling graft copolymers. *J. Membr. Sci.* **285**, 81–89 (2006).

71. Revanur, R., McCloskey, B., Breitenkamp, K., Freeman, B. D. & Emrick, T. Reactive amphiphilic graft copolymer coatings applied to polyvinylidene fluoride ultrafiltration membranes. *Macromolecules* **40**, 3624–3630 (2007).

72. Yang, S. Y. *et al.* Nanoporous membranes with ultrahigh selectivity and flux for the filtration of viruses. *Adv. Mater.* **18**, 709–712 (2006).

73. Phillip, W. A., Rzayev, J., Hillmyer, M. A. & Cussler, E. L. Gas and water liquid transport through nanoporous block copolymer membranes. *J. Membr. Sci.* **286**, 144–152 (2006).

74. Nunes, S. P., Sforca, M. L. & Peinemann, K.-V. Dense hydrophilic composite membranes for ultrafiltration. *J. Membr. Sci.* **106**, 49–56 (1995).

75. Yoon, K. *et al.* High flux ultrafiltration membranes based on electrospun nanofibrous PAN scaffolds and chitosan coating. *Polym.* **47**, 2434–2441 (2006).

76. Lu, Y., Suzuki, T. & Zhang, W. Moore, J. S. &Mariñas, B. J. Nanofiltration membranes based on rigid star amphiphiles. *Chem. Mater.* **19**, 3194–3204 (2007).

77. Zhou, Y. & Tol, R. S. J. Evaluating the costs of desalination and water transport. *Wat. Resour. Res.* **41**, W03003,–1–10 (2005).

78. Veerapaneni, S., Long, B., Freeman, S. & Bond, R. Reducing energy consumption for seawater desalination. *J. Am. Water Works Assoc.* **99**, 95–106 (2007).

79. Morgan, L. A. *et al.* Solar distillation: a promising alternative for water provision with free energy, simple technology and a clean environment. *Desalination* **116**, 45–56 (1998).

80. Bourounia, K., Chaibib, M. T. & Tadrist, L. Water desalination by humidification and dehumidification of air: state of the art. *Desalination* **137**, 167–176 (2001).

81. McCutcheon, J. R., McGinnis, R. L. & Elimelech, M. A novel ammonia-carbon dioxide forward (direct) osmosis desalination process. *Desalination* **174**, 1–11 (2005).

82. Mathioulakis, E., Belessiotis, V. & Delyannis, E. Desalination by using alternative energy: review and state-of-the-art. *Desalination* **203**, 346–365 (2007).

83. Alonitis, S. A., Kouroumbas, K. & Vlachakis, N. Energy consumption and membrane replacement cost for seawater RO desalination plants. *Desalination* **157**, 151–158 (2003).

84. Seacord, T. F., Coker, S. D. & MacHarg, J. Affordable desalination collaboration 2005 results. In *International Desalination And Water Reuse Quarterly* (Green Global Publications, Anaheim, California, 2006).

85. Spiegler, K. S. & El-Sayed, Y. M. The energetics of desalination processes. *Desalination* **134**, 109–128 (2001).

86. Hummer, G., Rasaiah, J. C. & Nowotyta, J. P. Water conduction through the hydrophobic channel of a carbon nanotube. *Nature* **414**, 188–190 (2001).

87. Kalra, A., Garde, S. & Hummer, G. Osmotic water transport through carbon nanotube membranes. *Proc. Natl Acad. Sci. USA* **100**, 10175–10180 (2003).

88. Hinds, B. J. *et al.* Aligned multiwalled carbon nanotube membranes. *Science* **303**, 62–65 (2003).

89. Holt, J. K. *et al.* Fast mass transport through sub-2-nanometer carbon nanotubes. *Science* **312**, 1034–1037 (2006).

90. Fornasiero, F. *et al.* Ion exclusion by sub 2-nm carbon nanotube pores. *Proc. Natl. Acad. Sci. USA.* (in the press).

91. Walz, T., Smith, B. L., Zeidel, M. L., Engel, A. & Agre, P. Biologically-active 2-dimensional crystals of aquaporin chip. *J. Biol. Chem.* **269**, 1583–1586 (1994).

92. Qiao, R., Georgiadis, J. G. & Aluru, N. R. Differential ion transport induced electroosmosis and internal recirculation in heterogeneous osmosis membranes. *Nano Lett.* **6**, 995–999 (2006).

93. Ishida, H., Donowaki, K., Inoue, Y., Qi, Z. & Sokabe, M. Synthesis and ion channel formation of novel cyclic peptides containing a non-natural amino acid. *Chem. Lett. Jpn* **26**, 935–954 (1997).

94. Davis, A. P., Sheppard, D. N. & Smith, B. D. Development of synthetic membrane transporters for anions. *Chem. Soc. Rev.* **36**, 348–357 (2007).

95. Nednoor, P., Gavalas, V. G., Chopra, N., Hinds, B. J. & Bachas, L. G. Carbon nanotube based biomimetic membranes: mimicking protein channels regulated by phosphorylation. *J. Mater. Chem.* **17**, 1755–1757 (2007).

96. Parsegian, A. Energy of an ion crossing a low dielectric membrane: solutions to four relevant electrostatic problems. *Nature* **221**, 844–846 (1969).

97. Facciotti, M. T., Rouhani-Manshadi, S. & Glaeser, R. M. Energy transduction in transmembrane ion pumps. *Trends Biochem. Sci.* **29**, 445–451 (2004).

98. Martz, E. Protein explorer: easy yet powerful macromolecular visualization. *Trends Biochem. Sci.* **27**, 107–109 (2002).

99. van Raaij, M. J., Louis, N., Chroboczek, J. & Cusack, S. Structure of the human adenovirus serotype 2 fiber head domain at 1.5 Å resolution. *Virology* **262**, 333–343 (1999).

Acknowledgements We acknowledge the US National Science Foundation Science and Technology Center, *WaterCAMPWS*, Center for Advanced Materials for the Purification of Water with Systems.

Author Information Reprints and permissions information is available at www.nature.com/reprints. Correspondence and requests for materials should be addressed to M.A.S. (mshannon@uiuc.edu).